Atlas of Tropical Carbonates

José A. Méndez Dot • José Méndez Baamonde

Atlas of Tropical Carbonates

As Exemplified by Facies of Venezuela: Past and Present

Springer

José A. Méndez Dot
Center for Oceanology and Antartic Studies
Venezuelan Institute for Scientific Research
Caracas, Venezuela

José Méndez Baamonde
Institute of Earth Sciences, Faculty of Sciences
Central University of Venezuela
Caracas, Venezuela

ISBN 978-3-031-25493-2 ISBN 978-3-031-25494-9 (eBook)
https://doi.org/10.1007/978-3-031-25494-9

Short description of the Image above: Aerial view of the transition between the eastern and southern barrier reefs (top and bottom, respectively). Locality: Los Roques Archipelago, southeast Caribbean Sea, offshore central mainland coasts of Venezuela. Short description of the Image below: Larger benthic foraminifer *Miogypsinita* sp. showing intraparticle primary porosity (Image Width = 2.4 mm, Plane polarized light). Rock unit: Perla Limestone, Gulf of Venezuela Basin, Late Oligocene-Early Miocene. Copyright: José A. Méndez Dot and José Méndez Baamonde

Editorial Contact: Qiao Shu
This Springer imprint is published by the registered company Springer Nature Switzerland AG
The registered company address is: Gewerbestrasse 11, 6330 Cham, Switzerland

To Liliane, our beloved mother and wife
In memory of Dayana Reyes Tristancho

Foreword

When we had the opportunity to see, for the first time, a bunch of photographs of Venezuelan carbonates from the collection by Professor Dr. José Méndez Baamonde and his son Dr. José Alejandro Méndez Dot, we got very impressed by its quality and variety, and we were also aware of the didactic value of the archive that we had before our eyes and the importance of disseminating it in the best possible way to reach a broad and diverse audience. That material was just a small sample, the tip of the iceberg, of a vast and systematic compilation—an exceptional photographic record of present-day, recent, and ancient carbonates from a Caribbean region, Venezuela, for which there was very little previous documentation in international publications despite its huge interest from different geological, prospective, and conservationist perspectives.

We encouraged José Alejandro, who at that time was starting a long-term postdoctoral stay at the Complutense University of Madrid, to persevere in their idea of publishing that material as an illustrated guidebook. Now, several years later, we congratulate him for this profuse photographic material to light as the *Atlas of Tropical Carbonates: As Exemplified by Facies of Venezuela—Past and Present.*

But this book is much more than an excellent collection of photographs of the under-advertised sedimentary carbonates from Venezuela. It comprises a detailed and up-to-date revision of the main carbonate successions in that region, framed in their geodynamic and paleogeographic context. The reported materials come from outcrops and subsurface cores and include sediments and rocks belonging to Permian, Cretaceous, Paleogene, and Neogene times. Additionally, the book considers the Quaternary and present-day environments of the southern margin of the Caribbean Sea, which are excellent examples of active carbonate systems, including growing coral reefs.

This atlas is the result of an entire dedicated life, more than 40 years of work by Prof. José Méndez Baamonde at the Central University of Venezuela in Caracas, and his continuous involvement with the oil industry of Venezuela. Dr. José Alejandro Méndez Dot, doctorated in Geological Sciences (2017) by the same university, joined his father's work in the last decade. On him has fallen most of the tasks of bringing structure to all that material collected over the years and framing it in the current context of carbonate sedimentology and the complex regional geology of Venezuela.

This atlas comes at a time when tropical carbonates are once again in the economic and scientific spotlight. Besides their fundamental role in the exploration and exploitation of conventional hydrocarbon reservoirs (Venezuela is well known for its large hydrocarbon reserves and very prolific oil-producing basins), it is the enormous potential of tight carbonates as unconventional resources, as well as the use of carbonate bodies as geological reservoirs for storage of CO_2 and other gases, or their potential in projects seeking energy alternatives to hydrocarbons such as geothermal resources.

Besides these energy-related fields, there is the renewed interest in other scientific disciplines such as the reconstruction of past climatic and environmental changes, mainly for the most recent periods but also for the deep-time, including the intriguing hyperthermal episodes that punctuate the history of the Earth. The episodes, well exemplified in this atlas by the record of the mid-Cretaceous Oceanic Anoxic Events in Venezuela, are currently regarded as potential analogues and extreme references for the current anthropogenic global warming and its possible future evolution.

Furthermore, in the framework of global change, monitoring of present-day tropical carbonate formations has become a key task for early warning in processes such as ocean acidification, rising temperatures, pollution, or unsustainable coastal development. The present-day carbonate systems in Venezuela are among the best preserved on the planet, and understanding them is key for their management, protection, and survival. This book is well set within the current scientific framework of its scientific discipline. Its format and layout make it manageable and accessible, completing the repertoire of reference books on sedimentary rocks.

The book you have in your hands is one of those that, unfortunately, are rarely made today. The calm and care with which it has been written distance it from current trends, where the immediacy of scientific results and short-term research projects prevail. The book has the appeal and charm of classic reference works, those that we sedimentologists continue to consult daily after many years.

Dear readers, we present to you a unique beautifully illustrated book that is the result from decades of work in the now-resurgent field of sedimentary tropical carbonates, from a region that is unjustly underreported.

Center for Oceanology and Antartic Studies Javier Martín-Chivelet
Venezuelan Institute for Scientific Research
Caracas, Venezuela

Institute of Earth Sciences (ICT), Faculty of Sciences Idoia Rosales
Central University of Venezuela (UCV)
Caracas, Venezuela

Preface

The understanding of marine carbonates is a scientific field in constant evolution, as evidenced by the countless publications that deal with ancient and modern carbonate facies. The current state-of-the-art carbonate conceptual models increasingly offer researchers more tools for framing multidisciplinary studies on specific carbonate deposits into a global context.

Atlas of Tropical Carbonates: As Exemplified by Facies of Venezuela—Past and Present is a comprehensive book that provides both an up-to-date bibliographic review and a broad photographic set of carbonate rocks (limestones and dolostones) and sediments from Venezuela, a country in northernmost South America. The carbonate facies in this atlas span from the late Paleozoic to Modern times. During this time interval of approximately 275 million years, carbonate production and deposition in what is now Northern Venezuela (the Venezuelan territory north of the Guayana Shield and the Orinoco River, including offshore settings) has always occurred in tropical latitudes.

This book is intended for professors, researchers, and graduate students in the area of Geosciences, especially Geology and Geochemistry, whose investigations are focused on sedimentology and stratigraphy. It is also a valuable resource for oil industry professionals, as well as biologists and ecologists who have an interest in the study of modern coral reefs. It is the outcome of more than 40 years of research on ancient and modern carbonate facies of Venezuela. It comprises 23 chapters that cover the most significant aspects of this topic. Through our investigations, many of which are unpublished until now, and with the support of numerous previous publications, we have gathered our thoughts (data, results, and conclusions) on our decades of work in both ancient and modern carbonate facies.

With this publication comprising carbonate facies of Venezuela from different ages of the Phanerozoic, our goal is to contribute to the understanding of the paleo-evolution (tectonics, geography, oceanography, climate, biota, bathymetry, trophic and redox conditions, and ecology) and present-day characteristics and controlling factors of marine carbonate systems (both coastal and offshore) from a tropical Caribbean perspective. Furthermore, carbonate sedimentation in Venezuela, which occurred during late Paleozoic, Mesozoic, and Cenozoic times, has always been linked to global factors to varying degrees. Therefore, carbonate facies of Venezuela can be correlated with many time-equivalent, ancient carbonate units and modern carbonate systems worldwide.

Ancient carbonate facies of Venezuela exhibit a broad spectrum of lithologic, stratigraphic, biotic, geochemical, and diagenetic characteristics, forming a very diverse and complex array of facies. These varied features make these facies an interesting mixture of typical "tropical carbonates" and facies often associated with "extra-tropical carbonates." They reflect the diverse marine environmental conditions (e.g., nutrient and temperature gradients, redox state, relative sea level, CO_2 concentrations, Mg^{2+}/Ca^{2+} ratio, bathymetry, turbidity, siliciclastic input) as well as the tectonic and depositional settings in which these facies were deposited during the three different eras of the Phanerozoic. Moreover, Venezuela is renowned by its vast petroleum (oil and gas) reserves. In this context, Cretaceous source rocks (i.e., black shales deposited during Oceanic Anoxic Events 1a, 1b, 1d, 2, and 3), along with Cretaceous (Aptian, Albian, and Turonian) and Cenozoic (Oligocene and Miocene) hydrocarbon reservoirs, represent key topics of interest for this book.

Regarding modern facies, Venezuela has some of the most well-developed and biotically diverse reef complexes in the South Caribbean. These shallow offshore and coastal carbonate systems are natural laboratories to carry out multidisciplinary studies, mainly in the fields of geology, geochemistry, geophysics, and biology, with modern anthropogenic global warming and associated ocean acidification and the accelerated current sea-level rise being the topics that probably attracts the most attention. These Caribbean reef archipelagos and islands, many of them well studied, contrast with modern carbonate facies currently depositing in the Venezuelan Atlantic continental shelf and slope, in front of to the pro-deltaic facies of the Orinoco delta. This is a vast, offshore, poorly studied region in which carbonate deposition, from neritic to bathyal and abyssal environments, is occurring under a different tectonic and depositional setting, and environmental conditions compared to the Venezuelan Caribbean domain. The little available information on these deposits describes very different facies and skeletal grain associations consisting of a varied carbonate biota, which points to a whole new world to discover in the field of modern carbonates in the Venezuelan Atlantic domain.

We apply conceptual models (marine carbonate factories and grain associations) to all the carbonate facies considered in this book. These concepts link carbonate precipitation modes, with the main ones being biologically-controlled mineralization by carbonate skeleton-producing organisms, biologically-induced mineralization by microbes, and abiotic precipitation, to the environmental controlling parameters of marine carbonates. These models, some of which are new approaches to classical ones, have been created to understand and distinguish the wide spectrum between tropical and nontropical carbonates. In this context, our view of carbonates of Venezuela under these concepts adds a new perspective on the comprehension of these facies in a regional and global context. Secular seawater chemistry fluctuations and associated alternations of Calcite-Aragonite seas throughout the Phanerozoic and their relationship with the dominant mineralogy and diversification of calcifying marine organisms considered in this book are also discussed.

Caracas, Venezuela José A. Méndez Dot
Caracas, Venezuela José Méndez Baamonde

Synopsis and Structure of the Book

This book focuses on ancient and modern carbonate facies of Venezuela. The ancient facies, either exposed in mountain ranges or buried in the subsurface of petroliferous basins, belong to the Maracaibo, Eastern, Barinas-Apure, Falcón, and Gulf of Venezuela basins. The modern facies include several offshore and coastal reef complexes in the Venezuelan Caribbean, as well as offshore facies in the Venezuelan Atlantic continental shelf and slope.

In general, ancient carbonate units of Venezuela from Middle Permian (Cisularian to Guadalupian), Cretaceous (from Barremian to Campanian), Paleogene (Paleocene and Eocene), and Neogene (Oligocene, Miocene, and Quaternary) ages are shallow in nature. These were deposited in a variety of platforms and ramps. Depositional environments in these marine settings remained mainly in the neritic zone, ranging from the euphotic (upper photic, good light) zone to the oligophotic (lower photic, poor light) zone. A few facies within those units contain features (planktonic biota and non-carbonate authigenic minerals) that suggest deposition occurred in environments below the oligophotic zone. During marine transgressions associated with the occurrence of the Cretaceous Oceanic Anoxic Events, "pelagic realm-like" conditions were created on epeiric platforms. Organic-rich rocks corresponding to these globally studied events, the so-called "black shales," were deposited at paleodepths ranging from the mesophotic (or intermediate photic) zone down to upper bathyal depths in the aphotic zone.

The variety of limestone facies considered in this book includes photozoan- and heterozoan-dominated, mixed photozoan-heterozoan carbonates, facies dominated by benthic or planktonic biota or a combination of both, mud-rich to mud-lean, and organic-lean to organic-rich facies. Some limestones are rich in nonskeletal grains, while others are rich in non-carbonate diagenetic minerals. Dolostones are common, mainly in specific intervals in the Cretaceous. Some limestones and a few dolostones have preserved a combination of primary and secondary porosities. Mixed carbonate-siliciclastic facies (e.g., sandy limestones, dolomitic shales) are also considered. All of the above indicates a broad spectrum of lithologic, environmental, diagenetic, stratigraphic, biotic, and geochemical characteristics. These characteristics are expected for a variety of ancient carbonate facies having very different geological ages and deposited under different tectonic and depositional settings, as well as these differences also lead to differences in paleogeographic, paleoclimatic, trophic, and redox conditions, and bathymetry (depth of the seafloor).

Regarding modern facies, Caribbean reef complexes (photozoan-dominated carbonates but with a significant contribution of heterozoan biota) developed in oligotrophic (crystalline and nutrient-poor) waters within the euphotic zone contrast with offshore Equatorial Atlantic facies. The latter are unconsolidated carbonate deposits that differ in composition from skeletal facies (heterozoan-dominated carbonates) depositing in eutrophic (turbid and nutrient-rich) waters within the mesophotic (intermediate photic) zone to *Globigerina* ooze and deep benthic foraminifera facies depositing from the slope break down to abyssal depths.

Chapter 1: *Introduction to Tropical Carbonates of Venezuela* introduces the reader to carbonate sedimentation in Venezuela (ancient and modern facies), its present-day local and regional geography, the distribution of its main petroliferous basins, mountain ranges, and coastal and offshore modern carbonate complexes. We discuss processes associated with the deposition of Venezuelan carbonate facies in a regional and global context.

Chapter 2: *Carbonate Facies of Venezuela: From the Late Paleozoic to Modern Times* (by far the longest one) focuses on the tectonic and sedimentary history of this country (from the Late Paleozoic to Modern times), including the processes involved in the formation of its petroleum basins and their sedimentary infilling; present-day tectonic, geomorphology, and sedimentology of modern carbonate systems; and detailed descriptions of ancient and modern carbonate facies considered in the following photographic chapters. This chapter serves as a reference so that, during the consultation of the following photographic chapters, the reader can easily access information about the basins, ancient rock units, and modern systems considered in this book.

It is noteworthy that:

- Some of the rock units described as part of the bibliographic review are not included in the following photographic chapters (mainly non-carbonate sedimentary facies).
- Emphasis has been placed on the Cretaceous carbonate units in the Maracaibo Basin, to which the larger number of photographic examples shown belong (followed by examples of the Oligo-Miocene Perla Limestone in the Eastern Gulf of Venezuela sub-Basin).
- We apply concepts of carbonate factories and grain associations only to ancient rock units and modern settings considered in the following photographic chapters.

In Chaps. 3, 4, 5, 6, 7, 8, 9, 10, 11, 12, 13, 14, 15, 16, 17, 18, 19, 20, 21, 22, and 23, this book focuses on photographic examples of ancient and modern carbonate facies. Images are accompanied by descriptive captions.

Chapter 3: *Modern Reef Complexes of Venezuela* is the sole chapter that exclusively focuses on modern Caribbean reef islands and archipelagos of this country. It consists of panoramic and close-up photographs (kilometer, meter, and centimeter scales; a total of 110 images). We show a combination of aerial, land, and underwater views from a variety of environments (e.g., barrier reefs, reef flats, fringing reefs, sabkha flats, hypersaline lagoons), as well as some of the most significant and striking carbonate and non-carbonate biota and biochemical occurrences (e.g., scleractinian corals, octocorals, calcareous green algae, echinoderms, microbialites, mangrove communities, non-calcareous macroalgae).

In Chaps. 4, 5, 6, 7, 8, 9, 10, 11, 12, 13, 14, 15, 16, 17, 18, 19, 20, 21, 22, and 23, this book focuses on carbonate petrography, i.e., the description of photomicrographs of thin sections (a total of 526 photomicrographs). These have horizontal axes ranging from 0.20 to 10 mm in length and were taken using a digital photographic system adapted to a transmitted light polarizing microscope. Photomicrographs were taken using plane-polarized light and cross-polarized light. A small number of photomicrographs (3) were taken using a binocular magnifying glass. These images consist of dry samples of modern organisms (foraminifera and bryozoans). In addition, 23 images consist of cm-sized close views of dry samples of modern organisms (18 scleractinian corals, 4 calcareous algae, and 1 gastropod).

As previously mentioned, images are accompanied by descriptive captions. We consider the available geochemical data for some of the carbonate rocks and sediments (e.g., carbonate and total organic carbon content; and elemental/isotope/biomarker composition, concentrations, and ratios). However, we emphasize the use of petrography through the visible features of the samples (e.g., type of grains, color of the matrix, calcareous biota, diagenetic precipitates) as a tool for interpreting environmental conditions, diagenetic processes, and other topics of interest. Furthermore, each petrographic set is preceded by an introduction consisting of several sections. These documents discuss generalities and specifications about the main subject matter and have been written with the intention of being general references for carbonate rocks and sediments (not only for carbonate facies of Venezuela). Accordingly, we use photographic examples of carbonates of Venezuela to put all the ideas discussed in the introductions into real context.

Chapter 4: *Carbonate Rock Components and Microfacies Classification* introduces the reader to the identification of the variety of carbonate components (or constituents), as well as the textures and fabrics of carbonate rocks, with the main goal of classifying (naming) the samples selected using traditional classification schemes. In Chaps. 5, 6, 7, 8, 9, 10, 11, 12, 13, and 14, we focus on fossilized, carbonate skeletal organisms whose $CaCO_3$ skeletons are secreted by biologically-controlled mineralization processes: Chap. 5: *Calcareous Algae* (green and red), Chap. 6: *Foraminifera* (larger and smaller benthics, and planktonics), Chap. 7: *Scleractinian Corals* (zooxanthellate), Chap. 8: *Molluscs* (bivalves and gastropods), Chap. 9: *Echinoderms* (crinoids and echinoids), Chap. 10: *Bryozoans* (stenolaemates and cheilostomes), Chap. 11: *Brachiopods* (rhynchonelliformeans), Chap. 12: *Ostracods* (podocopans), Chap. 13: *Barnacles* (balanomorphans), and Chap. 14: *Serpulids* (Serpulidae). In the introductions for each of these carbonate-producing organisms, we discuss major groups and distribution throughout the Phanerozoic, skeletal mineralogy, ecology and environment, morphology and diagnostic elements, marine carbonate factories, and skeletal grain associations (we mention only those represented or notably absent in Venezuelan carbonate facies to which examples shown in the photographic sets correspond), among other topics of interest. Chapter 15: *Non-skeletal Components* groups a variety of carbonate components (e.g., micrite, pellets and peloids, ooids, oncoids and cortoids, aggregates and clasts, and microbial carbonates) whose formation involves varied processes and modes of carbonate precipitation. As in Chaps. 5, 6, 7, 8, 9, 10, 11, 12, 13, and 14, we discuss their origins and varieties, mineralogy, and abundance throughout the Phanerozoic. Chapters 16: *Diagenesis* focuses on marine, meteoric, and shallow-burial diagenetic processes. Chapter 17: *Burial Diagenesis* focuses on diagenetic processes in the intermediate- and deep-burial realms. These two chapters, focused on diagenetic processes, are key to understanding ancient carbonate facies and their paragenetic relationships (the relative timing of features in the rocks). Chapter 18: *Dolomite and Dolomitization* deals with dolomite (the only abundant double carbonate mineral) and dolostones (the second most abundant carbonate rock in the sedimentary record), and dolomitization (the replacement of limestones by dolostones). We discuss the "Dolomite Problem," dolomitization models, dolomite crystal chemistry and textures, among other topics. Chapter 19: *Porosity* focuses on types and origins of porosity, as well as pore types and systems, which are of prime interest to hydrocarbon explorationists. Finally, in Chaps. 20, 21, 22, and 23, we deal with non-carbonate diagenetic minerals (authigenic or detrital) that are common in carbonate rocks: Chap. 20: *Ca-Phosphate*, Chap. 21: *Pyrite*, Chap. 22: *Glauconite*, and Chap. 23: *Silica*. These minerals are key for microfacies analysis as they are excellent indicators of paleoenvironmental conditions and of paragenetic relationships.

The supplementary material included in the *Appendices* is divided into two main sections, in which we develop key concepts and conceptual models that are applied to all carbonate facies considered in this book. In *Appendix A, Marine Carbonate Factories*, we place emphasis on how we apply the concepts of carbonate factories and modes of carbonate precipitation to the tropical carbonate facies of Venezuela. In this section, we also deal with climate-independent grain associations, photozoan and heterozoan, for light-dependent and light-independent benthic carbonate-producing organisms, respectively. We also discuss climate-dependent skeletal grain associations (e.g., chlorozoan, coralgal, rhodalgal, bimol, molechfor), among other topics of interest. In *Appendix B, Calcite and Aragonite Seas*, we discuss these two distinct chemical states of the global ocean and their effects on the skeletal mineralogy (low-Mg calcite, high-Mg calcite, and aragonite) of carbonate-producing organisms throughout the Phanerozoic. This conceptual model (from its origin to its amended views) is key to understanding the mineralogy of carbonate biota since the eve of controlled mineralization, as well as its relation to the dominant carbonate biota throughout the Phanerozoic, at a global scale and in carbonate facies of Venezuela (24 photographic examples of carbonate biota from Venezuela are included).

Acknowledgments

The publication of this book has been accomplished primarily thanks to: Professor Dr. Javier Martín-Chivelet, who originally acknowledged the potential in our photographic material and provided guidance and mentorship as a specialist in carbonate rocks and sediments.

Dr. Alexis Vizcaino, our original editor, for his expertise and patience and for kindly accepting this book in the format we devised.

Professor Dr. Maurice E. Tucker, who reviewed the entire manuscript and agreed with all of our ideas, which is a great honor for us.

Gerardo J. Méndez Dot, who provided key technical support in graphing and designing figures included in Chaps. 1, 2, 4, 16, 18, and 19.

The Springer Nature Production Team: Qiao Shu (Responsible Editor) Shirin Farhana B. (Production Editor), Ilakya Raghuraj (Product Manager), and Coral Zhou (Project Coordinator)

We express sincere thanks to many professionals (mainly geologists, geochemists, geophysicists, biologists, and photographers) that contributed to the publication of this work:

Marvin Baquero, Gilberto Soto, Rommel Whilchy, Dayana Reyes Tristancho[†], Franck A. Audemard, Patricia Lugo, María de Salazar, Javier Pérez, Ramón S. Sifontes[†], Franco Urbani, Estrella Villamizar, Carolina Bastidas, Ana Yranzo, Jeannette Pérez, Robert Schaper, Wayne Wright, Fernando L. Valencia, Liliana López, Grony Garbán, Olga Rey, Bruno Granier, Pablo Ricaurte, Andreina Angulo, Ismael Coronado, Ricardo Mateos, Irene Truskowski, Eduard Albert-Villanueva, Alejandro Escalona, Juan Carlos Laya, Idoia Rosales, Sebastián Grande, Guillermo Centeno Bordones, María Eugenia Glem, Sergio Cobarrubia, Adolphe O. Debrot, Jaap van Eenennaam, Kevin Newton, Eliaira Rodríguez, Víctor J. Herrera Rivero, Leyjer A. Hernández Avilez, Bladimir Rodriguez, Humberto Ramírez, Daniel A. Petrash, Florentin Maurrasse, Lenin A. González, Arturo Espinoza, Katya Reategui, Sonia Campos Soto, Maialen Lopez Elorza, Roselis Salazar Ramírez, Samuel Narciso, José Román Villalobos, Aníbal Pimentel, Dalisbett Núñez, Eduard Linares, Michael Smith, Andrea Rojas, Jesús Pérez, Gabriel Gotopo, and Iván Rodríguez.

The following universities, scientific institutes, and Oil & Gas companies provided key support for reaching our goal: Central University of Venezuela (UCV), Complutense University of Madrid (UCM), Venezuelan Institute for Scientific Research, Petróleos de Venezuela S.A., PDVSA-INTEVEP, Petroregional del Lago S.A., and Cardon IV S.A.

We thank the Instituto Nacional de Parques (INPARQUES) de Venezuela, especially Toribio Mata, and the Instituto Nacional de los Espacios Acuáticos (INEA), especially Ana Suárez Pacheco, María Arratia, Sugerlys Palacios, Felipe Verdú, Kenny Herrera, Ronnis Escalona, Luis E. López, Petra Salazar, Steven Castillo, and Yusbelys Gutiérrez. The logistic support of these institutions was priceless during the field campaigns to Los Roques Archipelago, Morrocoy National Park, and Aves Island. We also thank to Captain Klianto Párraga, commander of the vessel GC-24 of the Venezuelan Army, and all the crew of that vessel, as well as to lieutenants José Belisario and Luis Reyes, chiefs of the Naval Scientific Base Simón Bolívar in Aves Island, and all the crew of that base.

Family and friends also played a vital role during the process of researching and writing of this work. We are truly grateful to Liliane Dot, Gerardo J. Méndez Dot, Noralba Contreras, Víctor R. Moratinos, Antonio J. del Nogal, Fabrice Foucher, Bernardo Gasch Arocha, Jean Carlos Dot, María Alejandra Dot, Ana Gabriela Peña, and Pierre Barbeyron.

Contents

Abbreviations

ACC	Amorphous Calcium Carbonate
APSL	Above Present Sea Level
BAB	Barinas-Apure Basin
BPSL	Below Present Sea Level
BSB	Barinas Sub-Basin
BSR	Bacterial Sulfate Reduction
CCOP	Caribbean-Colombian Oceanic Plateau
COP	Caribbean Oceanic Plateau
CP	Caribbean Plate
CPL	Cross-Polarized Light
EDS	Energy Dispersive Spectometry
EGVSB	Eastern Gulf of Venezuela Sub-Basin
EIR	Eastern Interior Range
EPS	Extracellular Polymeric Substances
ESL	Eustatic sea level
EVB	Eastern Venezuela Basin
FB	Falcón Basin
HI	Hydrogen Index
HMC	High-Mg Calcite
ICC	International Chronostratigraphic Chart
IMC	Intermediate-Mg Calcite
IW	Image Width
ka	Thousand Years Ago
kyr	Thousand Years
LBF	Larger Benthic Foraminifera
LIG	Last Interglacial
LIPs	Large Igneous Provinces
LMC	Low-Mg Calcite
LME	Large Marine Ecosystem
Ma	Million Years Ago
MB	Maracaibo Basin
MCEs	Mesophotic Coral Ecosystems
MFS	Maximum Flooding Surface
MFTB	Monagas Fold and Thrust Belt
MIS	Marine Isotope Stage
Myr	Million Years
OAEs	Oceanic Anoxic Events
OMZ	Oxygen Minimum Zone
OP	Oceanic Plateau
PBZ	Plate Boundary Zone
PF	Planktonic Foraminifera
PPL	Plane-Polarized Light

RSL	Relative sea level
SBF	Smaller Benthic Foraminifera
SST	Sea Surface Temperature
TOC	Total Organic Carbon
VHMC	Very High-Mg Calcite
WGVSB	Western Gulf of Venezuela Sub-Basin

1 Introduction to Tropical Carbonates of Venezuela

Through the Phanerozoic, the most extensive shallow (neritic) carbonate systems have developed in seas and oceans located between 30° N and 30° S. Since at least the Late Carboniferous, approximately 320 million years ago (Ma), the Paleozoic terranes that today conform the territory of Venezuela have remained in tropical latitudes between approximately 15° S and 15° N. Furthermore, plate tectonic dynamics have maintained some regions of the northern territory of that country in transitional (continental to marine) settings. First in the northern edge of Gondwana during the late Paleozoic (~ 280–270 Ma, Middle Permian), prior to the final assembly of the Pangea supercontinent. Later, after the breakup of Pangea, in the northern margin of the South American plate from the beginning of the Early Cretaceous (~ 140 Ma, Berriasian-Valanginian boundary) until Modern times. Therefore, the combination of tropical paleo-latitudes and the privileged position within continental margins allowed the generation of a variety of carbonate depositional settings (platforms, ramps and coastal environments). In those settings, carbonate systems were developed during Permian, Cretaceous, Paleogene, Neogene, and Quaternary times.

Nowadays, the Greater Caribbean is part of the Tropical Atlantic Realm and the Tropical Northwestern Atlantic Province. The latter comprises nine ecoregions (Spalding et al. 2007): Western Caribbean, South Western Caribbean, Eastern Caribbean, South Caribbean, Greater Antilles, Bermuda, Bahamian, Southern Gulf of Mexico, and Floridian. Together, those regions represent the second largest coral reef bioprovince on Earth, after the Indo-Pacific (Veron 1995). The Caribbean Sea Large Marine Ecosystem (LME) (Fig. 1.1) is a semi-enclosed tropical sea of about 3.3 million km^2. It is bordered to the north by the Bahamas and the Florida Keys; to the northwest by the Yucatán Peninsula in Mexico; to the west by Central America; to the south by South America; and to the east by the Lesser Antilles Island chain, being the Barbados island the easternmost locality. The Caribbean Sea LME is also bordered by the Gulf of Mexico LME to the northwest and by the North Brazil Shelf LME to the southeast (Agard et al. 2007).

Carbonate production in tropical and subtropical latitudes have maintained from the Late Jurassic to Modern times in some areas of the Gulf of Mexico and the Western Caribbean, such as in Yucatán and Veracruz (Mexico), south Texas and south Florida (U.S.A.), as well as in the Bahama Platform (e.g., Budd and Loucks 1981; Jansa 1981; Wilson 1990; Galloway 2008). Venezuela is one of the countries hosting some of the best-developed Quaternary reef complexes (or systems) in the South Caribbean. These tropical carbonate systems, grouped under the name of the Venezuelan Antilles, consist of archipelagos and islands which have grown over isolated oceanic guyots (underwater mountains with a flat, platform top) located offshore mainland Venezuela (Figs. 1.1 and 1.2). A few modern reef complexes have developed along coastal environments. Together, these complexes host one of the richest diversities of calcareous biota and coral species in the whole Caribbean.

From late Paleozoic times, carbonate sedimentation in Venezuela has been linked to global-scale factors (e.g., tectonics and geodynamics, eustatic sea level, climate, seawater chemistry, geography, and ocean circulation). Therefore, the most significant ancient carbonate units in this country can be correlated to globally widespread, time-equivalent carbonate facies. For example, using current fossil knowledge of larger benthic foraminifera, Middle Permian carbonates cropping out in several localities in the Venezuelan Andes (Fig. 1.2) are correlated to carbonate strata in South, Central, and North America, as well as in Europe, the Middle East, and Asia (e.g., Ross and Ross 1985; Ross 1986; Silva-Pineda et al. 2003; Vachard et al. 2010; Ota and Isozaki 2006; Gaillot and Vachard 2007; Miall and Blakey 2008, Miall and Blakey 2019; Laya and Tucker 2012, 2015; Avendaño-Pazos et al. 2024).

Venezuela is also a country well known for its large hydrocarbon reserves. The main effective source rocks (commonly termed "black shales") are part of thick, regional car-

J. A. Méndez Dot, J. Méndez Baamonde, *Atlas of Tropical Carbonates*, https://doi.org/10.1007/978-3-031-25494-9_1

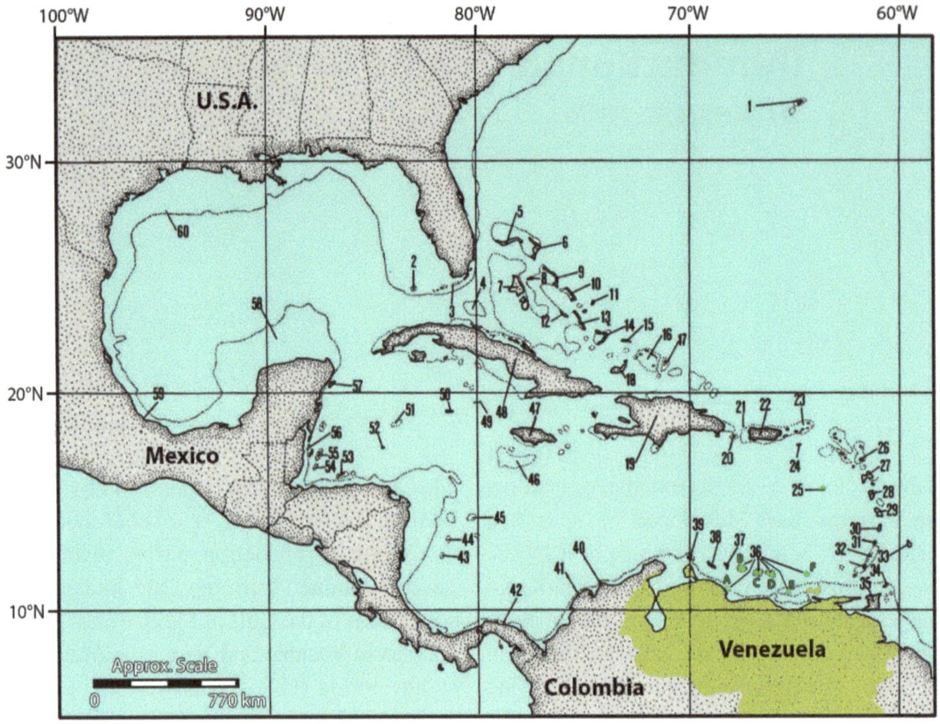

Fig. 1.1 Location of the main Quaternary carbonate complexes in the Greater Caribbean and the Gulf of Mexico. The Venezuelan offshore islands and archipelagos are colored in green. **1.** Bermuda Islands; **2.** Dry Tortugas (Florida); **3.** Florida Keys; **4.** Cay Sal Bank (Bahamas); **5.** Grand Bahama; **6.** Great Abaco (Bahamas); **7.** Andros Island (Bahamas); **8.** New Providence Island (Nassau); **9.** Eleuthera Island (Bahamas); **10.** Cat Island (Bahamas); **11.** San Salvador (Bahamas); **12.** Great Exuma Island; **13.** Long Island (Bahamas); **14.** Acklins Island (Bahamas); **15.** Mayaguana Island (Bahamas); **16.** Caicos Bank; **17.** Grand Turk Island; **18.** Great Inagua Island; **19.** Haiti and Dominican Republic; **20.** Mona Island (Puerto Rico); **21.** Desecheo Island (Puerto Rico); **22.** Puerto Rico; **23.** Virgin Islands; **24.** Saint Croix, Virgin Islands; **25.** Aves Island (Venezuela); **26.** Antigua; **27.** Guadeloupe; **28.** Dominica; **29.** Martinique; **30.** Saint Lucia; **31.** Saint Vincent; **32.** Grenadines; **33.** Barbados; **34.** Tobago; **35.** Trinidad; **36.** Venezuelan carbonate complexes: **A.** Morrocoy; **B.** Las Aves Archipelagos; **C.** Los Roques Archipelago; **D.** La Orchila Island; **E.** La Tortuga Island; **F.** La Blanquilla Island; **37.** Bonaire; **38.** Curaçao; **39.** Aruba; **40.** Santa Marta (Colombia); **41.** Cartagena (Colombia); **42.** San Blas Islands (Panama); **43.** San Andrés Island (Colombia); **44.** Old Providence Island (Colombia); **45.** Serrana Bank; **46.** Pedro Bank; **47.** Jamaica; **48.** Cuba; **49.** Cayman Brac and Little Cayman; **50.** Grand Cayman Island; **51.** Misteriosa Bank; **52.** Swan Island (Honduras); **53.** Roatán (Honduras); **54.** Glovers Reef (Belize); **55.** Lighthouse Reef (Belize); **56.** Belize; **57.** Cozumel (Mexico); **58.** Campeche Bank (Mexico); **59.** Veracruz (Mexico); **60.** Flower Garden Banks (Texas). It highlights the location of The Bermuda Islands at 32° N (~ 1450 km northeast of the Bahamas), whose reefs have grown completely isolated in the western North Atlantic Ocean due to the influence of the warm Gulf Stream current. The latter flows north from the Caribbean and passes just to the west of Bermuda on its path to the North Atlantic. Data modified from Colin 1978. Figure modified from Méndez-Baamonde 2009

bonate and mixed carbonate-siliciclastic successions whose deposition encompassed the so-called mid-Cretaceous (from Barremian to Santonian, about 125 to 85 Ma; Larson and Erba 1999). During much of the Early and Late Cretaceous (the Barremian-Campanian interval), deposition of these successions occurred in a passive margin, where epeiric (or epicontinental) seas and associated basins were generated along the continental margins of the northern edge of the South American plate during sustained high eustatic sea level. In Northern Venezuela (i.e., the Venezuelan territory north of the Guayana Shield and the Orinoco River; Figs. 1.2 and 1.3), passive margin units encompassing the Barremian-Campanian interval are found both in the subsurface and cropping out in the mountain ranges delimiting the Maracaibo Basin (the Venezuelan Andes, and the Perijá and Trujillo ranges) and the Eastern Venezuela Basin (the Eastern Interior Range) (Fig. 1.2). Located northwest and northeast Venezuela, respectively, these basins are two of the most prolific oil-producing regions in the world. In general, the lithology of these passive margin units consists of a variety of shallow, organic-lean carbonates and mixed carbonate-siliciclastic facies, and organic-rich facies (carbonate- or shale-dominated black shales); the latter characterized by high total organic carbon (TOC) content (wt.%). Shallow carbonates and black shales were also deposited during Late Cretaceous times (from Turonian to Campanian) in the Barinas-Apure Basin (mainly in the Barinas sub-Basin), in southwest Venezuela. However, the thickness and extension of both carbonates and black shale units in the Barinas area are smaller compared to those in the previously mentioned

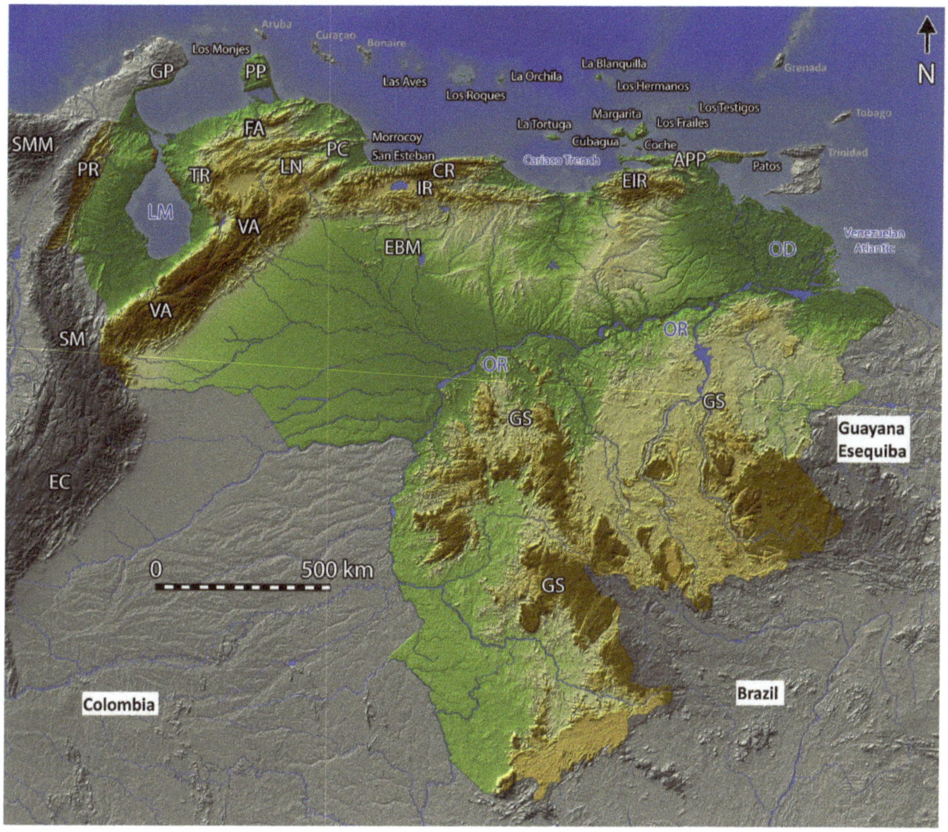

Fig. 1.2 Physical and relief map of Venezuela (mainland and offshore) highlighting mountain ranges, offshore archipelagos and islands (the Venezuelan Antilles, which are indicated with white letters with black borders; the ones considered in this book are: Las Aves, Los Roques, and La Blanquilla), coastal reef complexes (Morrocoy and San Esteban), major hydrographic features (blue letters), major urban areas (red spots), as well as mountain ranges in northeast Colombia. Other islands in the image not belonging to Venezuela are indicated with light gray letters. Abbreviations: *APP* Araya-Paria Peninsula, *CR* coastal range, *EBM* El Baúl Massif, *EC* Eastern Cordillera, *EIR* Eastern Interior Range, *FA* Falcón Anticlinorium, *GP* Guajira Peninsula, *GS* Guayana Shield, *IR* interior range, *LN* Lara Nappes, *PC* Precambrian Complexes (Yumare, El Guayabo), *PP* Paraguaná Peninsula, *PR* Perijá Range, *TR* Trujillo Range, *SM* Santander Massif, *SMM* Santa Marta Massif, *VA* Venezuelan Andes, *LM* Lake Maracaibo, *OD* Orinoco Delta, *OR* Orinoco River. (Original map by Arid Ocean Maps. © 2007–2022 Michael Schmeling)

basins. Carbonate sedimentation also occurred in Venezuela during the active margin stage of the Cenozoic, both in the Falcón Basin (units of Eocene, Oligocene and Miocene ages) in northwest Venezuela, and in the offshore Gulf of Venezuela Basin (Oligo-Miocene limestones) to the north of Falcón Basin. The Cenozoic siliciclastic reservoirs in the Orinoco Heavy Oil Belt complete the Venezuelan main petroliferous basins (Fig. 1.3). See Lugo and Audemard (2021) for a full review on the petroleum systems in Northern Venezuela.

Limestones and dolostones contain about half of the world's oil and gas reserves and also store much of the total volume of groundwaters (e.g., Lucia 2007; Ahr 2008). About 85% of the worldwide effective source rocks (consisting mainly of limestones and shales as lithologic end members) were deposited during the prolonged Warm Greenhouse conditions of the Late Jurassic and most of the Cretaceous (see Takashima et al. 2006; Trabucho-Alexandre et al. 2012; Nozaki et al. 2013; Gernon et al. 2024). Widespread black shale deposition was essentially linked to the flourishing of carbonate systems as new marine environments were generated worldwide (epeiric seas) following major rifting and drifting processes during the Late Jurassic and Early Cretaceous (see Trabucho-Alexandre et al. 2012; Bartok et al. 2015; Martín-Chivelet et al. 2019a). From the opening of the North Atlantic Ocean and the Gulf of Mexico during the Late Jurassic, followed by the opening of the proto-Caribbean during the Early Cretaceous, carbonate systems expanded widely along southern U.S.A., eastern Mexico and northern South America, as well as in the tropical Atlantic along eastern South America and northwest Africa. Shallow carbonate systems were also widely developed in the peri-Tethyan margins and on isolated highs in the Tethys domain during the Jurassic and the Cretaceous. These deposits are nowadays in the Middle East and south Asia, along northern

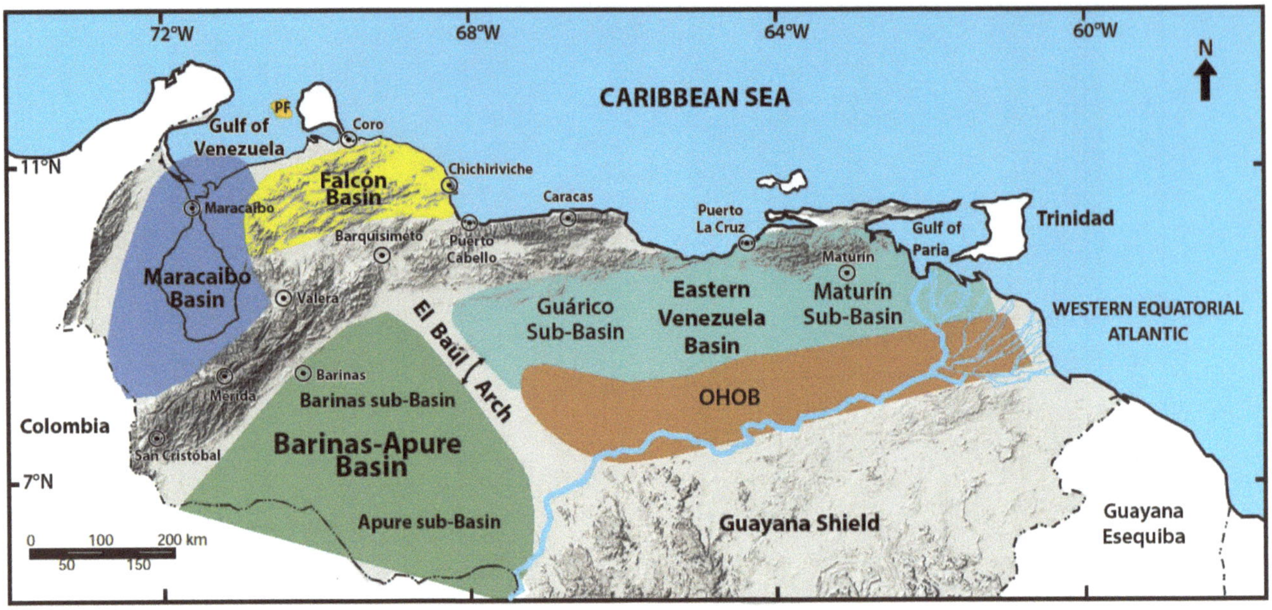

Fig. 1.3 Main petroliferous basins of Venezuela based on the distribution of their sedimentary provinces. The location of some of the most important cities of the country is indicated. Barinas (state of Barinas); Barquisimeto (state of Lara); Caracas (Capital District); Coro (state of Falcón); Maracaibo (state of Zulia); Maturín (state of Monagas); Mérida (state of Mérida); Puerto La Cruz (state of Anzoátegui); San Cristóbal (state of Táchira); Valera (state of Trujillo). Abbreviations: *OHOB* Orinoco Heavy Oil Belt, *PF* Perla Field. Figure modified from Pérez de Mejía et al. 1980

Africa, and in most Europe (see Philip 2003). In many of all above mentioned carbonate domains, the neritic and benthic-dominated carbonate production in platforms and ramps was also coeval to carbonate accumulation in hemipelagic and pelagic environments.

In Venezuela, most of the hydrocarbon reservoirs are hosted in Cenozoic sandstones. However, there are significant hydrocarbon volumes stored in carbonate reservoirs. These porous rock intervals consist of limestones and dolostones of Cretaceous age in the Maracaibo Basin and the Barinas sub-Basin, as well as Oligo-Miocene limestones in the Gulf of Venezuela Basin. Carbonate facies store much of the Cretaceous hydrocarbons in the Middle East (e.g., Ehrenberg et al. 2008), northern Africa (e.g., Radwan et al. 2024), Mexico (e.g., Bourdet et al. 2010), and south U.S.A. (e.g., Loucks and Peng 2021). Although petroleum reservoirs in the North Sea are dominated by sandstones of Permian, Jurassic and Paleogene ages, Cretaceous carbonate reservoirs (chalks) are also significant (e.g., Hardman 1982; D'Heur 1984; Mortensen et al. 1998; Megson and Tygesen 2005; Bergfors et al. 2020).

From the beginning of the second half of the Late Cretaceous (~ 86–80 Ma; Santonian to middle Campanian) and throughout the Cenozoic, an increase in the global area covered by collisional plate boundaries led to global-scale orogenic episodes (e.g., the Himalayan, Andean, Laramide, Alpine and Pyrenean orogenesis), causing the eustatic fall of the sea level and global paleogeographic and paleo-oceanographic readjustments (see Rona and Richardson 1978; Seton et al. 2012; Spikings et al. 2014; Martín-Chivelet et al. 2019b). As a consequence, major changes in the configuration of oceanic basins in the Atlantic realm ended the passive margin setting in what is now the Gulf of Mexico and northern South America. At global scale, orogenically driven CO_2 drawdown via silicate weathering (Kidder and Worsley, 2012) was accompanied by a decrease in marine carbonate production as well as in deposition of petroleum source rocks compared to Cretaceous times. However, relatively high sea level and minimum exposure of emergent areas during the Late Oligocene and the Miocene (Iturralde-Vinent 2006) allowed the development of carbonates systems in the Mediterranean region, southeast Asia, offshore China and India, Iran, Colombia and Venezuela (see Pomar et al. 2015 and references therein).

Tropical carbonate ancient rocks and modern sediments considered in this book are geographically located in Northern Venezuela. Such region lies today in a complex, active tectonic province where several mountain ranges are bounding the main petroliferous basins of this country. The Venezuelan Andes (also known as the Mérida Andes) is the most prominent mountain range in that region. The northern extension of the Andes, or Northern Andes (e.g., Taboada et al. 2000) in Ecuador, Colombia (encompassing the Western, Central and Eastern cordilleras) and Venezuela (Fig. 1.4), has an origin derived from the interaction between the continental South American plate with the oceanic

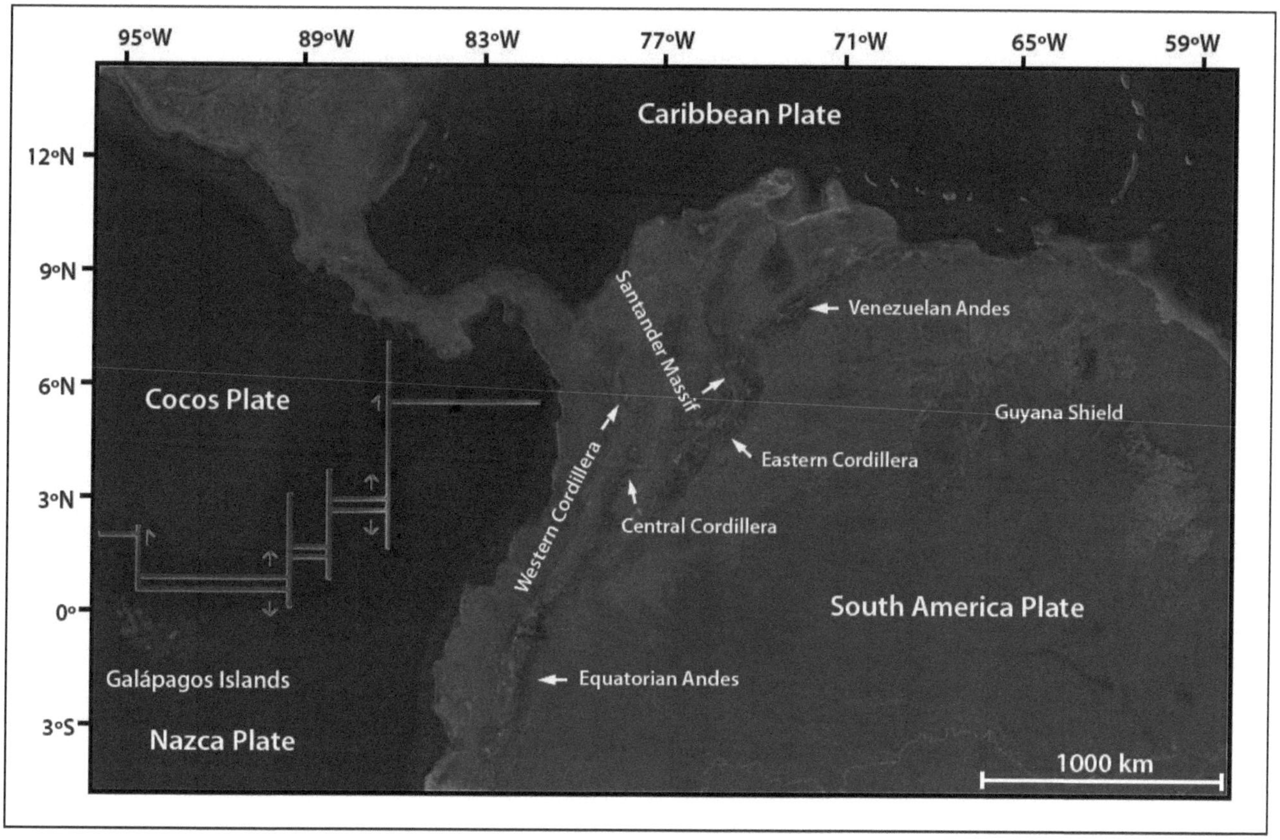

Fig. 1.4 Satellite image of northwest South America highlighting the mountain ranges (or "cordilleras") that conform the northern extension of the Andes Mountains. The Central, Eastern and Western cordilleras of Colombia and the Santander Massif make up the Colombian Andes. Location of the plate boundary between Cocos and Nazca plates according to Audemard (2014). Original image taken from Google Earth. Imagery date: 12/2020 by Landsat/Copernicus

Caribbean and Nazca plates. In the eastern Pacific, the Nazca and Cocos plates formed from the brake-up of the Farallon plate at ~23 Ma (Lonsdale 2005; Spikings et al. 2010). The origin and evolution of the Caribbean plate and its interaction with the northern margin of South America from the Late Cretaceous throughout the Cenozoic have played a key role in the geomorphologic evolution of that region and its Mesozoic and Cenozoic petroleum basins. These topics will be discussed in detail in Chap. 2.

In norther Colombia and Venezuela, plates interaction during the Neogene and Quaternary has also involved other minor tectonic blocks (e.g., the Maracaibo and North Andean blocks; see Villagómez and Spikings 2013; Audemard 2014; Audemard and Castilla 2016). In the case of Northern Venezuela, the collision and long-lasting interaction between the Caribbean plate and the northern margin of South America, along with the related diachronous exhumation of mountain ranges, has played, as previously mentioned, a key regional-scale role in the tectonic evolution of petroliferous basins. Those processes caused the burial of Cretaceous source rocks, which reached the petroleum (oil and gas) window maturation state during various episodes in the Cenozoic (see Escalona and Mann 2006, 2011; Lugo and Audemard 2021).

In addition to the occurrence of major global orogenic events and the eustatic fall of the sea level from the end of the Cretaceous Period, the Cenozoic Era (from 66 Ma to Present) has been, as noted previously, characterized by a substantial reduction in the widespread deposition of marine petroleum source rocks. This decrease in the deposition of organic-rich facies was directly linked to the non-occurrence of events of global- or regional-scale anoxia (or deoxygenation) in the oceans. Those events, known as Oceanic Anoxic Events (OAEs; see Schlanger and Jenkyns 1976; Jenkyns 1980, 2010; Arthur et al. 1990; Arthur and Sageman 1994; Leckie et al. 2002; see also Nozaki et al. 2013), promoted worldwide deposition of black shales during the mid-Cretaceous, including the most prolific effective source rocks in Venezuela. Their importance as petroleum resources and paleoclimate archives makes black shales and corresponding OAEs a relevant subject in this book. From the Santonian to early Campanian, plate tectonic readjustments resulted in an increased oceanic global circulation, highlighting the progressive aperture of the Atlantic Equatorial Seaway between eastern South America and western Africa. The aperture of that seaway started at ~113 Ma (Aptian-Albian boundary; Eagles 2007; Moulin et al. 2010) and its complete opening and deepening occurred at ~82 Ma, resulting in a complete water exchange between North Atlantic and South Atlantic oceans (Friedrich and Erbacher 2006; Wagreich 2012).

During the Cenozoic, the primary global factor that led to the cessation of the occurrence of OAEs and consequently to a substantial reduction in deposition of black shales, compared to mid-Cretaceous times, was a drastic decrease in the production rate of oceanic crust at mid-ocean ridges and from mantle-plume activity and the emplacement of Large Igneous Provinces (LIPs). During the Warm Greenhouse of the mid-Cretaceous (see Hay and Floegel 2012; Wendler and Wendler 2016), long-lasting submarine and/or subaerial volcanism, with some eruptive episodes lasting more than 5 million years (Myr), resulted in a switch to extreme, volcanogenic global warming conditions, or Hothouse climate (Kidder and Worsley 2010; Scotese et al. 2021). This extreme and short-lived climate mode (˂ 1 to c. 3 Myr; Hay and Floegel 2012) was episodically achieved by injecting large quantities of carbon dioxide (CO_2) to the ocean-atmosphere system (e.g., Larson 1991; Coffin and Eldholm 1994; Larson and Erba 1999; Kerr 1998; Stanley 1999; Jones and Jenkyns 2001; Jahren 2002; Erba 2004; Erba and Tremolada 2004; Tejada et al. 2009; Erba et al. 2015; Foster et al. 2018; Scotese et al. 2021; Papadomanolaki et al. 2022; Percival et al. 2024).

The Hothouse climate (or Hyperthermal; e.g. Bottini and Erba 2018; Foster et al. 2018), as trigger of global carbon perturbations that led to the occurrence of the main Cretaceous OAEs, were also characterized by: increased seawater acidification (Erba et al. 2010; Bauer et al. 2017; He et al. 2023; Jones et al. 2023; Percival et al. 2024); increased methane (CH_4) concentrations in the ocean/atmosphere (from dissociation of methane hydrates, thermo-and/or biogenic methane from sill intrusions, or degassed directly from the mantle; e.g., Beerling et al. 2002; Jahren 2002; Jenkyns 2003; Kuroda et al. 2007; Adloff et al. 2020); thermal expansion-driven transgressions (see Kidder and Worsley 2010); intensified continental weathering, accelerated hydrological cycle and enhanced nutrients discharge to the oceans (Jenkyns 2010, 2018; van Helmond et al. 2013; Charbonnier et al. 2018; see also Gernon et al. 2024); increased thermohaline stratification and ocean stagnation (Erbacher et al. 2001; Wilson and Norris 2001; Meyers 2006; Westermann et al. 2013); high CO_2-driven oceanic primary productivity (the biological pump; e.g., Erba 2004; Kuypers et al. 2004) coupled with high amounts of available biolimiting metals (e.g., Fe and Zn; see Larson and Erba 1999; Erba 2004; Erba et al. 2015); bottom marine water anoxia/anoxia-euxinia and the expansion of the oxygen minimum zone (OMZ) from deep basins towards shallow marine environments; and high rates of organic carbon preservation and burial (e.g., Kuypers et al. 2002; Jenkyns 2010; Erba et al. 2015; Tessin et al. 2015; Raven et al. 2018). In some marine settings, the shallowing of anoxic or anoxic-euxinic conditions reached the photic zone (e.g., Sinninghe Damsté and Köster 1998; Pancost et al. 2004; Wagner et al. 2004; van Bentum et al. 2009; Monteiro et al. 2012; Tzortzaki et al. 2013; Petrash et al. 2015).

Despite everything mentioned above, Cenozoic black shales, lithologically dominated by clastic shales, and characterized by lower TOC content compared to their Cretaceous counterparts, are less abundant but globally distributed, mainly in Oligocene and Miocene strata (Klemme and Ulmishek 1991; Trabucho-Alexandre et al. 2012; Sorkhabi 2016). In Venezuela, Cenozoic organic-rich rocks deposited during the ongoing Icehouse climate mode, which has prevailed since the Eocene-Oligocene boundary (~ 34 Ma), are represented by effective source rocks deeply buried in the offshore Gulf of Venezuela Basin (Castillo et al. 2017; Lugo and Audemard 2021).

1.1 Carbonate Rocks and Sediments of Venezuela

Ancient carbonate rocks in Venezuela are nowadays both cropping out in some of the main mountain ranges and in the subsurface of the main petroleum basins of this country. Those units were deposited in a variety of platform and ramp systems of Permian, Cretaceous, Paleogene and Neogene ages. Many of those calcareous facies are interdigitated, in different proportions, with siliciclastic sediments.

During the Precambrian, carbonate sedimentation took place locally in Southern Venezuela, in what is today the Guayana Shield (Figs. 1.2 and 1.3), located south of the Orinoco River. Guayana is a vast, tectonically inactive shield that corresponds to the northern extension of the Amazonian Craton, also encompassing northwestern Brazil, southeastern Colombia and northeastern Perú (see Sarmiento-Rojas 2019). However, in this igneous-metamorphic province, whose rocks ages range between 3700 Ma and 1400 Ma (Montgomery and Hurley 1978; González de Juana et al. 1980), carbonate deposits were metamorphosed to marbles (e.g., dolomitic marbles of the Guaicuripia Formation, Imataca Complex of Archean to Early Paleoproterozoic age). Therefore, the absence of non-metamorphosed carbonate deposits in this ancient shield is total (Edmond et al. 1995).

The most ancient carbonate unit considered in this book is the Palmarito Formation of Middle Permian. This thick carbonate succession crops out in several localities along the Venezuelan Andes and locally in the Perijá Range. The Palmarito Formation represents the largest and most significant carbonate Paleozoic unit in Venezuela. It was deposited in a ramp developed in the northern margin of Gondwana (Laya and Tucker 2012).

A wide variety of carbonate and mixed carbonate-siliciclastic facies with regional extension were deposited in the Maracaibo and Eastern Venezuela basins (Fig. 1.3) during much of the Early and Late Cretaceous (~ 125–80 Ma). In the former basin, events of global- and/or regional-scale anoxia are well represented by a series of organic-rich rock intervals that correspond to the Early and Late Cretaceous OAEs (Aptian and Albian, and Cenomanian to Santonian, respectively), from the OAE 1a to the OAE 3. The OAEs 1a (~ 120 Ma; global in nature) and 1b (~ 114–109 Ma; regional-scale significance) are well recorded in the Cogollo Group (Méndez-Baamonde et al. 2008, 2009; Méndez-Dot et al. 2015; Méndez-Dot 2017; this book). The latter is the most significant shallow-water carbonate succession in the Cretaceous of Venezuela. It consists, from base to top, of the Apón, Lisure and Maraca formations, and encompasses approximately 25 Myr of sedimentation, from the early Barremian to the latest Albian. In the Eastern Venezuela Basin, the El Cantil, Borracha and Chimana formations are the time and partly lithologic equivalents of the Cogollo Group.

The OAE 1d (regional-scale significance), spanning the Albian-Cenomanian boundary (~ 100.5 Ma), is found in the lower part of the Capacho Formation, which crops out in the Venezuelan Andes (Rodríguez-Cuicas et al. 2019, 2020). This anoxic event is also found in the lower part of the Querecual Formation, which crops out in the Eastern Interior Range (Fig. 1.2). This formation and its time and lithologic equivalent La Luna Formation in the Maracaibo Basin (the latter unit crops out in the Venezuelan Andes, and the Perija and Trujillo ranges) are organic-rich expanded sections of anomalous thicknesses, which encompass the OAE 2 (global in nature) spanning the Cenomanian-Turonian boundary (~ 94 Ma) (Tribovillard et al. 1991; Pérez-Infante et al. 1996; Erlich et al. 1999; De Romero et al. 2003; Truskowski 2006; Caroprese 2009), and the OAE 3 (regional-scale significance) spanning the Coniacian-Santonian boundary (~ 86.3 Ma) (Rey et al. 2004; Caroprese 2009; Machado et al. 2016). The La Luna and Querecual formations have been historically the main source rocks in Venezuela, representing two of the so-called "world-class source rocks".

Carbonate sedimentation in the Barinas sub-Basin (Fig. 1.3) is minor and did not begin until the Late early Cretaceous (Turonian). Among Late Cretaceous siliciclastic-dominated units in this sub-basin, the most significant carbonate and mixed carbonate-siliciclastic units are the dolomite reservoirs of the O Member (Turonian), Escandalosa Formation; and the Navay Formation (early Coniacian to early Campanian). The lower member of this unit, the La Morita Member (early Coniacian to early Santonian), is a post-OAE 2 black shale interval that is considered the main effective source rock in the Barinas sub-basin.

During the Cenozoic, carbonate sedimentation in Venezuela was geographically restricted, as it was conditioned by the active margin generated due to the oblique collision of the Caribbean plate against the northern margin of South America. This collision, initiated in the Campanian, led to the uplift of mountain ranges in Northern Venezuela; first the Venezuelan Caribbean Ranges (from west to east: the Perijá, Coastal, Interior, and Eastern Interior ranges), and later the Venezuelan Andes (Fig. 1.2; see Ostos et al. 2005). The exhumation of these ranges promoted a long-term siliciclastic-dominated sedimentation throughout the Cenozoic. The most significant carbonate units are located in the Falcón Basin (Fig. 1.3). This units, of Oligo-Mioceneand and/or Miocene ages, are the San Luis, Churuguara and Capadare formations, and the Cauderalito Member of the Agua Clara Formation (Macellari 1995; Albert-Villanueva et al. 2017; Blanco and Mann 2021; Soto 2025; Soto et al. 2023). The lateral extension of these units is lesser compared to the regional extension of Cretaceous carbonate units in the Maracaibo and Eastern Venezuela basins. The other major carbonate rock unit of Cenozoic age in Venezuela is the Late Oligocene-Early Miocene Perla Limestone, found in the offshore Perla Field (Pomar et al. 2015; Soto et al. 2023). This field is located in the eastern province of the Gulf of Venezuela Basin (the Eastern Gulf of Venezuela sub-Basin), to the west of the Paraguaná Peninsula (Fig. 1.2). In terms of porosity and permeability, the Perla Limestone contains the best carbonate hydrocarbon reservoirs in Venezuela (Valencia and Laya 2020), storing large proven reserves of gas and condensate (Castillo et al. 2017).

During the Quaternary, several oceanic reef complexes have developed over offshore submarine rises (or bathymetric highs) in the Venezuelan Caribbean (Fig. 1.2). Modern facies in these carbonate reef islands and archipelagos host some of the most biodiverse and pristine ecosystems of the Greater Caribbean coral reef bioprovince. Notably, the Los Roques and Las Aves archipelagos are highlighted due to their rich diversity of calcareous biota and variety of carbonate depositional environments; La Blanquilla Island due to its well-exposed, large limestone terrace formed during the last interglacial period; and Aves Island for its remoteness respect to mainland Venezuela (see locality 25 in Fig. 1.1).

Although, not strictly related to the Holocene carbonate complexes considered in this book, the Cariaco Trench (Fig. 1.2), also known as the Cariaco Basin (see Escalona et al. 2011), is one of the most relevant hydrologic features in the Venezuelan Caribbean. Located right south of La Tortuga Island, this trench is a semi-enclosed basin with maximum depths of nearly 1400 m. It is one of the few localities in the world where ocean anoxia occurs at relatively shallow depths (low-oxygen conditions occur at depths >250 m, and fully anoxic conditions at depths >450 m). Therefore, the Cariaco Trench is regarded as a modern analog for understanding oceanic anoxia in the past (see Lin et al. 2006; Taylor et al. 2006; Scranton et al. 2006; Quijada et al. 2016; Muller-Karger et al. 2019).

Venezuela has approximately 4000 km of coastlines, with the longest Caribbean shoreline (~ 2718 km) of all the countries in the region (Weil 2003; Méndez-Baamonde 2007). Coastal and shoreline environments are highly influenced by siliciclastic input from the Venezuelan Caribbean Ranges. However, significant Holocene coral reef complexes and associated reef communities have been able to thrive in specific areas along those Caribbean coastlines. Those complexes are the Morrocoy National Park in the eastern shorelines of the state of Falcón (just in front of the town of Chichiriviche), and the San Esteban National Park in the shorelines of the state of Carabobo (in front of the city of Puerto Cabello) (see Figs. 1.2 and 1.3).

In the Venezuelan Atlantic (Fig. 1.2), modern facies of *Globigerina* ooze and deep-water smaller benthic foraminifera (plus molluscs and bryozoans) are currently accumulating from the middle neritic zone down to bathyal depths on the offshore Venezuelan Atlantic continental shelf and slope. This vast, offshore carbonate setting is located in front of the prodeltaic facies of the Orinoco Delta. It highlights the occurrence of both benthic and planktonic carbonate biota in this continental platform, as coastal and inner shelf environments are largely influenced by a gigantic siliciclastic sediment discharge from the Orinoco Delta; as well as by the influx of a significant volume of siliciclastic sediments carried by the Amazon River in northeast Brazil and distributed along the Orinoco Delta platform by the littoral Guayana current (Méndez-Baamonde 2000; Warne et al. 2002a, b; Weber et al. 2016).

The sedimentary basins, rock outcrops, coastal localities, offshore islands and archipelagos, and shelf environments in Northern Venezuela, which correspond to all the photographic examples shown in this book, are indicated in Fig. 1.5. Examples of ancient carbonate facies correspond to rock units in the subsurface of some of the petroliferous basins and to rock units cropping out in several mountain ranges. Figure 1.6 compiles some of the most significant mapped rock units from Fig. 1.5. The selection criteria for the units included in the former figure are based on the following: the carbonate units to

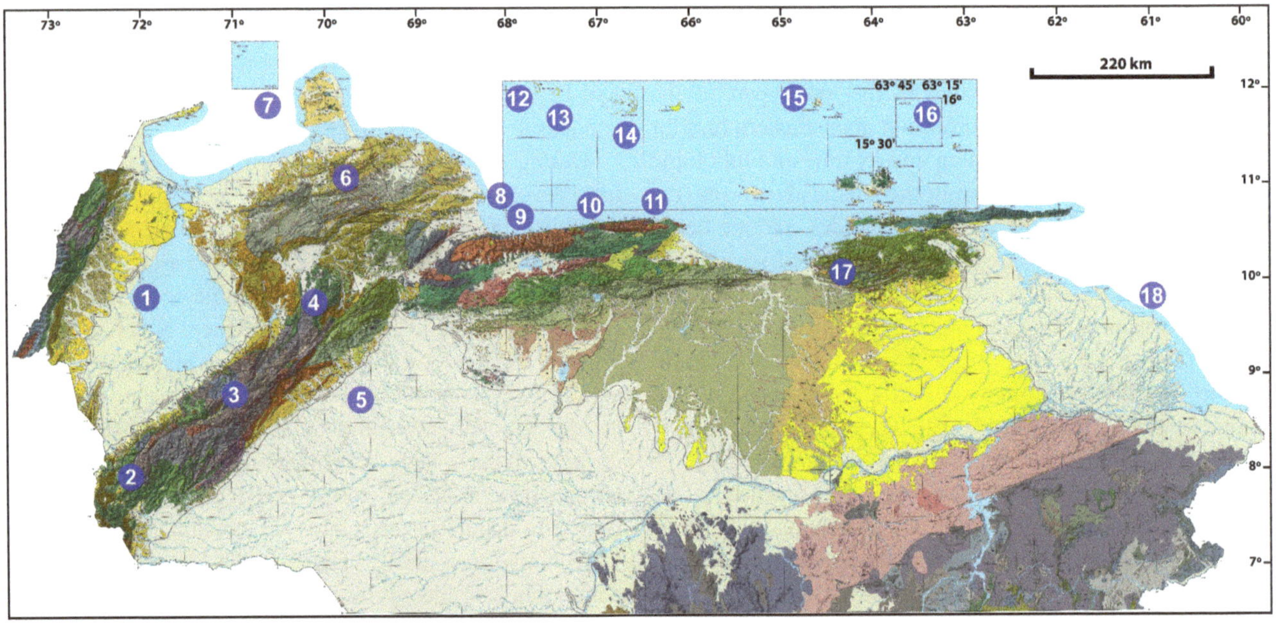

Fig. 1.5 Geologic-shaded relief map of Venezuela showing the geographic distribution of carbonate rocks and sediments considered in this book. The numbered blue circles indicate the approximate locations of cores (subsurface), rock outcrops, shoreline localities, offshore archipelagos and islands, and shelf and slope environments corresponding to the photographic examples shown from Chaps. 3–23. The ages of the considered ancient rock units and lithified or unconsolidated/loose Late Pleistocene and Holocene sediments are indicated. (1) Several cores in the Maracaibo Basin, state of Zulia. Core sections located in eastern, central, and western Lake Maracaibo, and western Maracaibo Basin (the subsurface area between the western margin of Lake Maracaibo and the Perijá Range). Units considered: Cogollo Group (Apón, Lisure and Maraca formations; Early Cretaceous), La Luna Formation *sensu stricto* (Late Cretaceous), Tres Esquinas Member of the La Luna Formation (Late Cretaceous), and Guasare Formation (Paleocene); (2) One outcrop in the Venezuelan Andes, state of Táchira. Units considered: Táchira Ftanita and Tres Esquinas members of the La Luna Formation (Late Cretaceous); (3) Two outcrops in the Venezuelan Andes, state of Mérida. Units considered: Palmarito Formation (Middle Permian) and La Luna Formation *sensu stricto*; (4) One outcrop in the Venezuelan Andes, state of Trujillo. Unit considered: Palmarito Formation; (5) Three cores in the Barinas sub-Basin, state of Barinas (core sections located in the surrounding areas of the city of Barinas). Units considered: O Member of the Escandalosa Formation and La Morita Member of the Navay Formation (Late Cretaceous); (6) Several outcrops in the Falcón Anticlinorium (Falcón Basin), state of Falcón. Units considered: Santa Rita and Jarillal formations (Middle Eocene), Churuguara and San Luis formations (Late Oligocene-Early Miocene), and Capadare Formation (Middle Miocene); (7) Two cores of the Perla Field, Eastern Gulf of Venezuela sub-Basin. Units considered: Perla Limestone (Late Oligocene-Early Miocene); (8) Morrocoy National Park (Holocene); (9) Larga Island, Carabobo State shoreline (Holocene); (10) Henri Pittier National Park, Aragua State shoreline (Holocene); (11) La Guaira State shoreline (Holocene); (12) Aves de Sotavento Archipelago (Holocene); (13) Aves de Barlovento Archipelago (Holocene); (14) Los Roques Archipelago (Late Pleistocene and Holocene); (15) La Blanquilla Island (Late Pleistocene and Holocene); (16) Aves Island (Late Pleistocene and Holocene); (17) Three outcrops in the Eastern Interior Range (Eastern Venezuela Basin), states of Anzoátegui and Monagas. Units considered: El Cantil Formation (Early Cretaceous), and Querecual and San Antonio formations (Late Cretaceous); (18) Venezuelan Atlantic Continental Shelf and Slope (Holocene). Figure modified from Hackley et al. (2006, Sheet 1)

which the photographic examples shown in this book correspond; units that occupy a large surface area; and units of significance, either because of their age (e.g., ancestral Precambrian units) or because they crop out in some of the main mountain ranges of Venezuela and have been well studied. Figure 1.7 shows the nomenclature for the geographic division of the Venezuelan territory used by Hackley et al. (2006) and applied in Fig. 1.6 to refer to the selected rock units.

Cenozoic

- Alluvium. Llanos, Amacuro, Islands (Pleistocene to Holocene)
- Unconsolidated sediments. Andes-Lara, North-Center, Zulia (Pleistocene to Holocene)
- Mesa Fm. Guayana, Eastern (Pleistocene)
- El Milagro Fm and equivalents. Zulia (Pliocene to Pleistocene)
- Chaguaramas, Capaya fms. Eastern (Oligocene to Miocene)
- Carapita, Naricual fms. Eastern (Oligocene to Miocene)
- Churuguara, San Luis fms. Falcón (Miocene)
- Agua Clara, Capadare, Urumaco, La Vela fms. Falcón (Miocene)
- Agua Negra Grp. Falcón, Andes-Lara (Eocene)
- Mirador Fm. Zulia, Andes-Lara (Eocene)
- Misoa Fm. Zulia (Eocene)
- Matatere, Trujillo fms. Andes-Lara (Paleocene to Eocene)

Mesozoic

- La Quinta Fm. Zulia, Andes-Lara (Jurassic)
- Las Mercedes. North-Center (Jurassic to Cretaceous). Metamorphic
- Las Brisas Fm. North-Center (Jurassic to Cretaceous). Metamorphic
- Villa de Cura Association. North-Center (Jurassic to Cretaceous). Meta-volcanosedimentary
- Cogollo Grp, Río Negro Fm. Zulia, Andes-Lara (Early Cretaceous)
- La Luna, Colón, Mito Juan fms. Zulia, Andes-Lara (Late Cretaceous)
- Querecual, San Antonio, San Juan fms. Eastern (Late Cretaceous)
- Siquisique Ophiolitic Complex. Andes-Lara, North-Center (Cretaceous)
- Loma de Hierro Ophiolitic Complex. North-Center (Cretaceous)
- Extrusive and intrusive mafic rocks. North-Center, Andes-Lara (Mesozoic)

Paleozoic

- Caparo, El Horno fms. Andes-Lara (Ordovician to Silurian)
- Siliceous intrusive rocks. Zulia, Andes-Lara (lower Paleozoic)
- San Julián Complex, Ávila Metamorphic Association. North-Center (Ordovician to Permian)
- Ávila Metamorphic Association. North-Center (Paleozoic)
- Mucuchachí Association. Andes-Lara (Carboniferous)
- El Baúl Granitic Association. North-Center (Carboniferous to Permian)
- Palmarito Fm. Andes-Lara (Carboniferous to Permian)
- Sabaneta Fm. Andes-Lara (Carboniferous to Permian)
- Río Palmar Fm. Zulia (Carboniferous to Permian)
- Siliceous intrusive rocks. Andes-Lara (middle Paleozoic)

Precambrian

- Gneis, Imataca Complex. Guayana (Archean to Early Proterozoic)
- Siliceous intrusive rocks, Imataca Complex. Guayana (Archean to Early Proterozoic)
- Supamo Complex. Guayana (Early Proterozoic). Intrusive
- Cuchivero Grp. Guayana (Early Proterozoic). Siliceous intrusive
- Metasedimentary rocks. Guayana (Early Proterozoic)
- Middle Roraima Grp. Guayana (early to middle Proterozoic)
- Upper Roraima Grp. Guayana (early to middle Proterozoic)
- Parguaza Granite. Guayana (Middle Proterozoic)
- Peña de Mora Augengneis, Ávila Metamorphic Association. North-Center (Middle Proterozoic)
- Yumare Complex. North-Center (Proterozoic). Metamorphic and intrusive

Fig. 1.6 List of some of the most significant mapped units shown in Fig. 1.5 (the Geologic-shaded relief map of Venezuela). The units are listed according to their names (formation, group, complex, association), the age, and the geographic area/areas that these occupy within the Venezuelan territory. The geographic division of the Venezuelan territory is numbered from 1 to 10 in Fig. 1.7. Lithology of rock units (e.g., igneous, metamorphic) is indicated (when necessary) only for non-sedimentary units. Abbreviations: *Fm* Formation, *fms* formations, *Grp* Group. (List compiled from Hackley et al. 2006, Sheet 2)

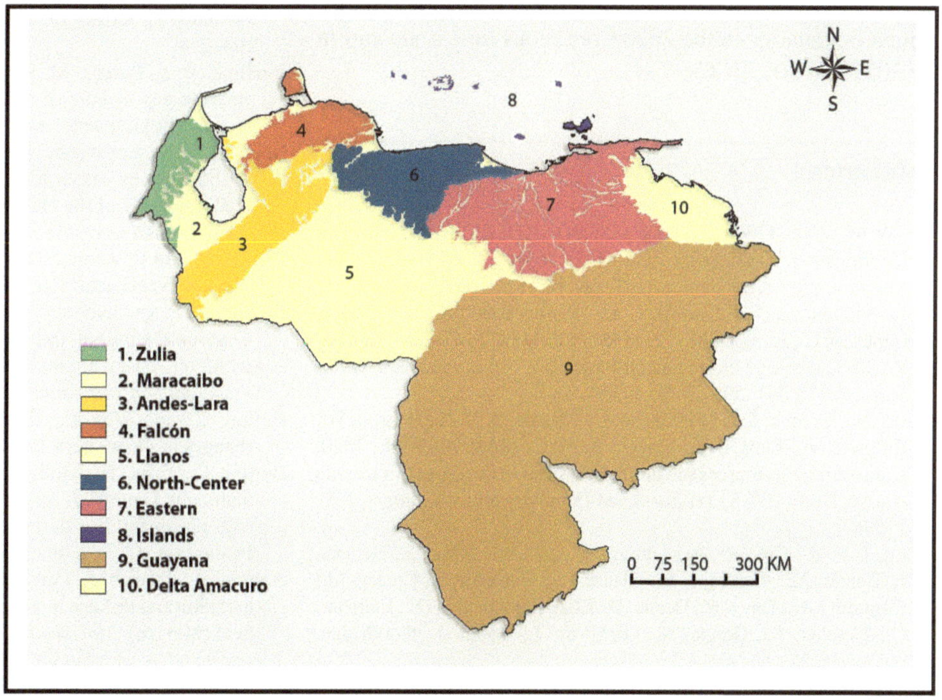

Fig. 1.7 Geographic division of the Venezuelan territory as used in Fig. 1.6. (Figure modified from Hackley et al. 2006, Sheet 1)

1.2 On the Geologic Time Scale

Regarding the Geologic Time Scale used in this book, it is noteworthy that in the International Chronostratigraphic Chart (ICC) published by the International Commission on Stratigraphy in February 2022 (Cohen et al. 2013a, b, c; updated v2022/02), the Barremian-Aptian boundary was taken at 121.4 Ma. This age is 3.6 Myr apart (i.e., 125 Ma) from the Barremian-Aptian boundary in the ICC of year 2021 (Cohen et al. 2013a, b, c; updated v2021/10); 4.9 Myr apart (i.e., 126.3 Ma) from that boundary in the Geologic Time Scales of Gradstein et al. (2012) and Ogg et al. (2016); in full agreement with the Geomagnetic Polarity Time Scale of Ogg (2020), as well as with the Cretaceous Time Scale of Gale et al. (2020) (Chaps. 5 and 27 in The Geologic Time Scale of Gradstein et al. 2020; respectively); and in almost full agreement with the Geomagnetic Polarity Time Scale for the Late Jurassic-Early Cretaceous (~ 160–120 Ma) of Malinverno et al. (2012), who placed the Barremian-Aptian boundary at 121 Ma. The latter time scale has been used by several authors during the past decade (e.g., Bottini et al. 2015; Erba et al. 2015; Jenkyns 2018) in publications involving the study of the early Aptian OAE 1a (~ 120 Ma).

Furthermore, in the ICC published in April 2023 (Cohen et al. 2013a, b, c; updated v2023/04), the Barremian-Aptian boundary is also placed at 121.4 Ma, but the Hauterivian-Barremian boundary was taken from 129.4 Ma to 125.8 Ma. In the Cretaceous time scale of Gale et al. (2020), this boundary is placed at 126.5 Ma. For all above mentioned, we have corrected the previously accepted ages of the Hauterivian, Barremian, and Aptian sedimentary units in the Maracaibo and Eastern Venezuela basins (which are defined by the occurrences of planktonic and benthic foraminifera) considering the ages of the Hauterivian-Barremian and Barremian-Aptian boundaries in the 2020 Cretaceous time scale and in the 2022 and 2023 ICCs.

References

Ahr, W. M., 2008. GEOLOGY OF CARBONATE RESERVOIRS: The Identification, Description, and Characterization of Hydrocarbon Reservoirs in Carbonate Rocks. Texas A&M University, 277 pp.

Albert-Villanueva, E., González, L., Bover-Arnal, T., Ferràndez-Cañadell, C., Esteban, M., Fernández-Carmona, J., Calvo, R., Salas, R., 2017. Geology of the Falcón Basin (NW Venezuela). *Journal of Maps*, *13*(2), 491–501.

Adloff, M., Greene, S. E., Parkinson, I. J., Naafs, B. D. A., Preston, W., Ridgwell, A., Lunt, D. J., Jiménez, J. M. C., Monteiro, F. M., 2020. Unravelling the sources of carbon emissions at the onset of Oceanic Anoxic Event (OAE) 1a. *Earth and Planetary Science Letters*, *530*, 115947.

Agard, J. B. R., Cropper, A., Aquing, P., Attzs, M., Arias, F., Beltrán, J., Bennett, E., Carnegie, R., Clauzel, S., Corredor, J., Creary, M., Cumming, G., Davy, B., Deane, D., Elias-Samlalsingh, N., Fletcher, G., Fletcher, K., Garcia, K., Garraway, J., Gobin, J., Goodridge, A., Gray, A., Hart, S., Haughton, M., Heileman, S., Insanally, R., Jordon, L. A., Kumar, P., Laurent, S., Lumkong, A., Mahon, R., McDonald, F., Mendoza, J., Mohammed, A., Mohammed, E., McShine, H., Mitchell, A., Oderson, D., Oxenford, H., Pantin, D., Parsram, K., Phillips, T., Pichs, R., Potter, B., Rios, M., Rivera-Arriaga, E., Singh, A., Singh, J., Singh-Renton, S., Robertson, L., Schill, S., Toro, C., Trotman, A., Villasol, A., Vina-Davila, N., Walling, L., Warner, G., Zahedi, K., Zurek, M., 2007. Caribbean Sea Ecosystem Assessment (CARSEA). Caribbean Marine Studies, Special Edition, 85 pp.

Arthur, M. A., Jenkyns, H. C., Brumsack, H. J., and Schlanger, S. O., 1990. Stratigraphy, geochemistry, and paleoceanography of organic-carbon rich Cretaceous sequences, *in* Ginsburg, R. N., and Beaudoin, B., editors, Cretaceous Resources, Events and Rhythms: Dordrecht, Kluwer Academic, NATO AS1 Series, 304, 75–119.

Arthur, M. A., and Sageman, B. B., 1994. Marine Black Shales: Depositional Mechanisms and Environments of Ancient Deposits. *Annual Review of Earth and Planetary Sciences*, 22, 499–551.

Audemard, F. A., 2014. Active block tectonics in and around the Caribbean: a review. In: Schmitz, M., Audemard, F.A., Urbani, F. (Eds.), El Límite Noreste de la Placa Suramericana - Estructuras Litosféricas de la Superficie al Manto (The Northeastern Limit of the South American Plate - Lithospheric Structures from Surface to the Mantle). Editorial Innovación Tecnológica. Facultad de Ingeniería-Universidad Central de Venezuela/FUNVISIS, pp. 29–77.

Audemard, F, A., and Castilla, R., 2016. Present-day stress tensors along the southern Caribbean plate boundary zone from inversion of focal mechanism solutions: A successful trial. *Journal of South American Earth Sciences*, *71*, 309–319.

Avendaño-Pazos, J. J., Torres-Martínez, M. A., Lara-Peña, R. A., Navas-Parejo, P., 2024. Biostratigraphy of fusulinids from the lower Permian of central Sonora, Mexico. Paleoenvironmental implications. *Journal of South American Earth Sciences*, *148*, 105133.

Bartok, P., Mejia-Hernandez, M. C., Ismael, M., 2015. Paleogeographic constraints on Middle- to Late-Jurassic tectonic reconstruction of the Maya Block of southern Mexico and equivalent strata of northwestern South America, *in* C. Bartolini and P. Mann, eds., Petroleum Geology and potential of the Colombian Caribbean Margin. *American Association of Petroleum* Geologists, *Memoir 108*, 201–216.

Bauer, K. W., Zeebe, R. E., & Wortmann, U. G., 2017. Quantifying the volcanic emissions which triggered Oceanic Anoxic Event 1a and their effect on ocean acidification. *Sedimentology*, *64*, 204–214.

Beerling, D. J., Lomas, M. R., Gröcke, D. R., 2002. On the nature of methane gas-hydrate dissociation during the Toarcian and Aptian oceanic anoxic events. *American Journal of Science*, *302*, 28–49.

Bergfors, S. N., Schovsbo, N. H., Feilberg, K. L., 2020. Classification of and changes within formation water types in Danish North Sea Chalk: A study of the Halfdan, Dan, Kraka and Valdemar oil reservoirs. *Applied Geochemistry*, *122*, 104702.

Blanco, J., and P. Mann, 2021, Subsurface Geology of La Vela Basin, Offshore Venezuela: Examples of basement and carbonate-hosted, liquid and gas hydrocarbon reservoirs, *in* C. Bartolini, ed., South America–Caribbean–Central Atlantic plate boundary: Tectonic evolution, basin architecture, and petroleum systems. *American Association of Petroleum Geologists Memoir 123*, 411–440.

Bottini, C., and Erba, E., 2018. Mid-Cretaceous paleoenvironmental changes in the western Tethys. *Climate of the Past*, *14*, 1147–1163.

Bottini, C., Erba, E., Tiraboschi, D., Jenkyns, H. C., Schouten, S., Sinninghe Damsté, J. S., 2015. Climate variability and relationship with ocean fertility during the Aptian Stage. *Climate of the Past Discussion*, *11*, 383–402.

Bourdet, J., Pironon, J., Levresse, G., Tritlla, J., 2010. Petroleum accumulation and leakage in a deeply buried carbonate reservoir, Níspero field (Mexico). *Marine and Petroleum Geology*, *27*, 126–142.

Budd, D. A., and Loucks, R. G., 1981. Smackover and Lower Buckner formations, South Texas: depositional systems on a Jurassic carbonate ramp. University of Texas at Austin. *Bureau of Economic Geology Report of Investigations No. 112*. Austin, TX.

Caroprese, C. G., 2009. Bioestratigrafía y paleoecología de la Formación Querecual en el Oriente de Venezuela. Central University of Venezuela. Magister Dissertation. 95 pp.

Castillo, V., Benkovics, L., Cobos, C., Demuro, D., Franco, A., 2017. Perla field: The largest discovery ever in Latin America, In: R. K. Merrill and C. A. Sternbach, eds., Giant fields of the decade 2000–2010: *American Association of Petroleum Geologists, Memoir 113*, 141–152.

Charbonnier, G., Boulila, S., Spangenberg, J. E., Adatte, T., Föllmi, K. B., Laskar, J., 2018. Obliquity pacing of the hydrological cycle during the Oceanic Anoxic Event 2. *Earth and Planetary Science Letters*, *499*, 266–277.

Coffin, M. F., and Eldholm, O., 1994. Large igneous provinces: crustal structure, dimensions, and external consequences. *Reviews of Geophysics*, *32*(1), 1–36.

Cohen, K. M., Finney, S. C., Gibbard, P. L. Fan, J. X., 2013a; updated v2021/10. The ICS International Chronostratigraphic Chart. *Episodes 36*, 199–204.

Cohen, K. M., Finney, S. C., Gibbard, P. L. Fan, J. X., 2013b, updated v2022/02. The ICS International Chronostratigraphic Chart. *Episodes 36*, 199–204.

Cohen, K. M., Finney, S. C., Gibbard, P. L. Fan, J. X., 2013c, updated v2023/04. The ICS International Chronostratigraphic Chart. *Episodes 36*, 199–204.

Colin, P. L., 1978. Caribbean reef invertebrates and plants. A field guide to the invertebrates and plants occurring on coral reefs of the Caribbean, the Bahamas, and Florida. T. F. H. Publications Inc, Neptune City, NJ, 512 pp.

De Romero, L. M., Truskowski, I., Bralower, T. J., Bergen, J. A., Odreman, O., Zachos, J. C., Galea-Álvarez, F. A., 2003. An integrated calcareous microfósil biostratigraphic and carbon-isotope stratigraphic Framework for the La Luna Formation, Western Venezuela. *Palaios, 18*, 349–366.

D'Heur, M., 1984. Porosity and hydrocarbon distribution in the North Sea chalk reservoirs. *Marine and Petroleum Geology*, *1*(3), 211–238.

Eagles, G., 2007. New angles on South Atlantic opening. *Geophysical Journal International*, *168*, 353–361.

Edmond, J. M., Palmer, M. R., Measures, C. I., Grant, B., Stallard, R. F., 1995. The fluvial geochemistry and denudation rate of the Guayana Shield in Venezuela, Colombia, and Brazil. *Geochimica et Cosmochimica Acta*, *59*, 3301–3325.

Ehrenberg, S. N., Aqrawi, A. A., Nadeau, P. H., 2008. An overview of reservoir quality in producing Cretaceous strata of the Middle East. *Petroleum Geoscience*, *14*, 307–318.

Erba, E., 2004. Calcareous nannofossils and Mesozoic oceanic anoxic events. *Marine Micropaleontology*, *52*, 85–106.

Erba, E., and Tremolada, F., 2004. Nannofossil carbonate fluxes during the Early Cretaceous: phytoplankton response to nutrification episodes, atmospheric CO_2, and anoxia. *Paleoceanography*, 19, PA1008.

Erba, E., Bottini, C., Weissert, H. J., Keller, C. E., 2010. Calcareous nannoplankton response to surface-water acidification around Oceanic Anoxic Event 1a. *Science*, *329*, 428–432.

Erba, E., Duncan, R. A., Bottini, C., Tiraboschi, D., Weissert, H., Jenkyns, H. C., Malinverno, A., 2015. Environmental consequences of Ontong Java Plateau and Kerguelen Plateau volcanism. In: Neal, C.R., Sager, W.W, Sano, T., and Erba, E., eds., The Origin, Evolution, and Environmental Impact of Oceanic Large Igneous Provinces. *Geological Society of America, Special Paper 511*, 271–303.

Erbacher, J., Huber, B. T., Norris, R. D., Markey, M., 2001. Increased thermohaline stratification as a possible cause for an ocean anoxic event in the Cretaceous period. *Nature*, *409*, 325–327.

Erlich, R., Macsotay, O., Nederbragt, A., Lorente, M.A., 1999. Palaeoecology, palaeogeography and depositional environments of upper cretaceous rocks ofwestern Venezuela. *Palaeogeography, Palaeoclimatology, Palaeoecology*, *153*, 203–238.

Escalona, A., and Mann, P., 2006. An overview of the petroleum system of the Maracaibo basin. *American Association of Petroleum Geologists Bulletin, 90, 4*, 657–678.

Escalona, A., and Mann, P., 2011. Tectonics, basin subsidence mechanisms, and paleogeography of the Caribbean-South American plate boundary zone. *Marine and Petroleum Geology*, *28*, 8–39.

Escalona, A., Mann, P., Jaimes, M., 2011. Miocene to recent Cariaco basin, offshore Venezuela: Structure, tectonosequences, and basin-forming mechanisms. *Marine and Petroleum Geology*, *28*, 177–199.

Foster, G. L., Hull, P., Lunt, D. J., Zachos, J. C., 2018. Placing our current 'hyperthermal' in the context of rapid climate change in our geological past. *Philosophical Transactions of the Royal Society A*, *376*, 20170086.

Friedrich, O., and Erbacher, J., 2006. Benthic foraminifera assemblages from Demerara Rise (ODP Leg 207, western tropical Atlantic): Possible evidence for a progressive opening of the Equatorial Atlantic Gateway. *Cretaceous Research*, *27*, 377–397.

Gaillot, J., and Vachard, D., 2007. The Khuff Formation (Middle East) and time-equivalents in Turkey and South China: biostratigraphy from Capitanian to Changhsingian times (Permian), new foraminiferal taxa, and palaeogeographical implications/La Formación Khuff (Oriente Medio) y sus equivalentes cronológicos de Turquía y China del Sur: bioestratigrafía, nuevos taxones de foraminíferos e implicaciones paleogeográficas del Capitaniense al Changhsingiense (Pérmico Medio-Superior). *Coloquios de Paleontología*, *57*, 37–223.

Gale, A. S., Mutterlose, J., Batenburg, S., 2020. The Cretaceous Period. In: Gradstein, F. M., Ogg, J. G., Schmitz, M. D., Ogg, G. M. (Eds.), Geologic Time Scale 2020, pp. 1023–1086, Amsterdam, The Netherlands, Elsevier B.V.

Galloway, W. E., 2008. Depositional evolution of the Gulf of Mexico Sedimentary Basin. In: Miall, A. D. (ed.), The Sedimentary Basins of the United States and Canada: Sedimentary Basins of the World, v. 5, K. J. Hsu, Series Editor, Elsevier Science Amsterdam, 505–550.

Gernon, T. M., Mills, B. J. W., Hincks, T. K., Merdith, A. S., Alcott, L. J., Rohling, E. J., Palmer, M. R., 2024. Solid Earth forcing of Mesozoic oceanic anoxic events. *Nature Geoscience*, 1–25.

González de Juana, C., Iturralde, J., Picard, X., 1980. Geología de Venezuela y de sus cuencas petrolíferas. Foninves Editions. Caracas, Venezuela. Volumes I and II, 1031 pp.

Gradstein, F. M., Ogg, G., Schmitz, M., 2012. The Geologic Time Scale 2012 2-Volume Set. Elsevier, 1176 pp.

Gradstein, F. M., Ogg, J. G., Schmitz, M. D., Ogg, G. M., 2020. The Geologic Time Scale 2020. Elsevier, Amsterdam.

Hackley, P., Urbani, F., Karlsen, A. W., Garrity, C. P., 2006. Mapa geológico de Venezuela a escala 1:750.000. *U.S. Geological Survey, Open File Report 2006-1109*.

Hardman, R. F. P., 1982. Chalk reservoirs of the North Sea. *Bulletin of the geological Society of Denmark*, *30*, 119–137.

Hay, W. W., and Floegel, S., 2012. New thoughts about the Cretaceous climate and oceans. *Earth-Science Reviews*, *115*, 262–272.

He, T., Kemp, D. B., Li, J., Ruhl, M., 2023. Paleoenvironmental changes across the Mesozoic–Paleogene hyperthermal events. *Global and Planetary Change*, *222*, 104058.

Iturralde-Vinent, M. A., 2006. Meso-Cenozoic Caribbean paleogeography: implications for the historical biogeography of the region. *International Geology Review*, *48*, 791–827.

Jahren, A. H., 2002. The biochemical consequences of the mid-Cretaceous superplume. *Journal of Geodynamic*, *34*, 177–191.

Jansa, L. F., 1981. Mesozoic carbonate platforms and banks of the eastern North American margin. *Marine Geology*, *44*, 97–117.

Jenkyns, H. C., 1980. Cretaceous anoxic events: from continents to oceans. *Journal of the Geological Society*, *137*(2), 171–188.

Jenkyns, H. C., 2010. Geochemistry of oceanic anoxic event. *Geochemistry, Geophysics, Geosystems, 118*(3), Q03004.

Jenkyns, H. C., 2003. Evidence for rapid climate change in the Mesozoic–Palaeogene greenhouse world. *Philosophical Transactions of the Royal Society of London, Series A, 361*, 1885–1916.

Jenkyns, H. C., 2018. Transient cooling episodes during Cretaceous Oceanic Anoxic Events with special reference to OAE 1a (Early Aptian). *Philosophical Transactions of the Royal Society A*, 376, 20170073.

Jones, C. E., and Jenkyns, H. C., 2001. Seawater strontium isotopes, oceanic anoxic events, and sea floor hydrothermal, activity in the Jurassic and Cretaceous. *American Journal of Science, 301, 2*, 112–140.

Jones, M. M., Sageman, B. B., Selby, D., Jacobson, A. D., Batenburg, S. J., Riquier, L., MacLeod, K. G., Huber, B. T., Bogus, K. A., Tejada, M. L. G., Kuroda, J., Hobbs, R. W., 2023. Abrupt episode of mid-Cretaceous ocean acidification triggered by massive volcanism. *Nature Geoscience, 16*(2), 169–174.

Kerr, A. C., 1998. Oceanic plateau formation: A cause of mass extinction and black shale deposition around the Cenomanian-Turonian boundary? *Journal of Geological Society of London, 155, 4*, 619–626.

Kidder, D. L., and Worsley, T. R., 2010. Phanerozoic large igneous provinces (LIPs), HEATT (Haline Euxinic Acidic Thermal Transgression) episodes and mass extinctions. *Palaeogeography, Palaeoclimatology, Palaeoecology, 295*, 162–191.

Kidder, D. L., and Worsley, T. R., 2012. A human-induced hothouse climate. *GSA Today, 22*(2), 4–11.

Klemme, H. D., and Ulmishek, G. F., 1991. Effective petroleum source rocks of the world: Stratigraphic distribution and controlling depositional factors. *American Association of Petroleum Geologists Bulletin, 75*, 1809–1,851.

Kuroda, J., Ogawa, N. O., Tanimizu, M., Coffin, M. F., Tokuyama, H., Kitazato, H., Ohkouchi, N., 2007. Contemporaneous massive subaerial volcanism and late cretaceous Oceanic Anoxic Event 2. *Earth and Planetary Science Letters, 256*, 211–223.

Kuypers, M. M. M., Pancost, R. D., Nijenhuis, I. A., Sinninghe Damsté, J. S., 2002. Enhanced productivity led to increased organic carbon burial in the euxinic North Atlantic basin during the late Cenomanian oceanic anoxic event. Paleoceanography, 17(4), 1051.

Kuypers, M. M. M., van Breugel, Y., Schouten, S., Erba, E., Sinninghe Damsté, J.S., 2004. N_2-fixing cyanobacteria supplied nutrient N for Cretaceous oceanic anoxic events. *Geology, 32*, 853–856.

Larson, R. L., 1991. Geological consequences of superplumes. *Geology, 19*, 963–966.

Larson, R., and Erba, E., 1999. Onset of the mid-Cretaceous greenhouse in the Barremian-Aptian: Igneous events and the biological, sedimentary and geochemical responses. *Paleoceanography, 14*, 663–678.

Laya, J. C., and Tucker, M. E., 2012. Facies analysis and depositional environments of Permian carbonates of the Venezuelan Andes: Palaeogeographic implications for Northern Gondwana. *Palaeogeography, Palaeoclimatology, Palaeoecology, 331*, 1–26.

Laya, J. C., and Tucker, M., 2015. Petroleum potential of Permian carbonates in northern South America, In: C. Bartolini and P. Mann, eds., Petroleum geology and potential of the Colombian Caribbean Margin. *American Association of Petroleum Geologists, Memoir 108*, 471–498.

Leckie, M., Bralower, T., Cashman, R., 2002. Oceanic anoxic events and plankton evolution: biotic response to tectonic forcing during the mid-Cretaceous. *Paleoceanography, 17*, 1–29.

Lin, X., Wakeham, S. G., Putnam, I. F., Astor, Y. M., Scranton, M. I., Chistoserdov, A. Y., Taylor, G. T., 2006. Comparison of vertical distributions of prokaryotic assemblages in the anoxic Cariaco Basin and Black Sea by use of fluorescence in situ hybridization. *Applied and Environmental Microbiology, 72*(4), 2679–2690.

Lonsdale, P., 2005. Creation of the Cocos and Nazca plates by fission of the Farallon plate. *Tectonophysics, 404*, 237–264.

Loucks, R. G., and Peng, S., 2021. Matrix reservoir quality of the Upper Cretaceous Austin Chalk Group and evaluation of reservoir-quality analysis methods; northern onshore Gulf of Mexico, USA. *Marine and Petroleum Geology, 134*, 105323.

Lucia, F. J., 2007. Carbonate Reservoir Characterization. 2nd edition Springer-Verlag, Berlin, 342 pp.

Lugo, J., and Audemard, F., 2021. Cretaceous to Neogene arc–continent collision, orogenic float development, and implications on the petroleum systems of Northern Venezuela, *in* C. Bartolini, ed., South America–Caribbean– Central Atlantic plate boundary: Tectonic evolution, basin architecture, and petroleum systems. *American Association of Petroleum Geologists Memoir 123*, 1–84.

Macellari, C. E., 1995. Cenozoic sedimentation and tectonics of the southwestern Caribbean pull-apart basin, Venezuela and Colombia, *in* A. J. Tankard, R. Suárez S., and H. J. Welsink, Petroleum basins of South America. *American Association of Petroleum Geologists, Memoir 62*, 757–780.

Machado, M. C., Chemale Jr, F., Kawashita, K., Rey, O., Moura, C. A. V., 2016. Isotope studies of carbonate rocks of La Luna Formation (Venezuela) to constrain the oceanic anoxic event 3 (OAE3). *Journal of South American Earth Sciences, 72*, 38–48.

Malinverno, A., Hildebrandt, J., Tominaga, M., Channell, J. E. T., 2012. M-sequence geomagnetic polarity time scale (MHTC12) that steadies global spreading rates and incorporates astrochronology constraints. *Journal of Geophysical Research, 117*, B06104.

Martín-Chivelet, J., López-Gómez, J., Aguado, R., Arias, C., Arribas, J., Arribas, M. E., Aurell, M., Bádenas, B., Benito, M. I., Bover-Arnal, T., Casas-Sainz, A., Castro, J. M., Coruña, F., de Gea, G. A., Fornós, J. J., Fregenal-Martínez, M., García-Senz, J., Garófano, D., Gelabert, B., Giménez, J., González-Acebrón, L., Guimerà, J., Liesa, C. L., Mas, R., Meléndez, N., Molina, J. M., Muñoz, J. A., Navarrete, R., Nebot, M., Nieto, L. M., Omodeo-Salé, S., Pedrera, A., Peropadre, C., Quijada, I. E., Quijano, M. L., Reolid, M., Robador, A., Rodríguez-López, J. P., Rodríguez-Perea, A., Rosales, I., Ruiz-Ortiz, P. A., Sàbat, F., Salas, R., Soria, A. R., Suarez-Gonzalez, P., Vilas, L., 2019a. The Late Jurassic–Early Cretaceous Rifting. In: Quesada, C., Oliveira, J. T. (Eds.), *The Geology of Iberia: A Geodynamic Approach, Regional Geology Reviews*. Springer International Publishing, Cham, 169–249.

Martín-Chivelet, J., Floquet, M., García-Senz, J., Callapez, P. M., López-Mir, B., Muñoz, J. A., Barroso-Barcenilla, F., Segura, M., Ferreira Soares, A., Morgado Dinis, P., Fonseca Marques, J., Arbués, P., 2019b. Late Cretaceous Post-Rift to Convergence, in: The Geology of Iberia: A Geodynamic Approach, edited by: Quesada, C. and Oliveira, J. T., Regional Geological Reviews, Springer Nature, 285–376.

Megson, J. B., and Tygesen, T., 2005. The North Sea Chalk: an under-explored and underdeveloped play. In: Petroleum Geology: North-West Europe and Global Perspectives—Proceedings of the 6th Petroleum Geology Conference (eds. A. G. Doré & B. A. Vining), pp. 159–68. London: Geological Society.

Méndez-Baamonde, J., 2000. Geología del Delta del Orinoco. INGEOMIN. Caracas, Venezuela. *Boletín de Geología. Publicación Especial N° 13*, 126 pp.

Méndez-Baamonde, J., 2007. Costas litorales del Caribe y del Atlántico, islas y archipiélagos. Las profundidades marinas. In: *Geo Venezuela*, Tomo II: Medios Físicos y Recursos Ambientales. Caracas: Fundación Empresas Polar, 184–237.

Méndez-Baamonde, J., 2009. Carbonatos—Origen y Sedimentación. University Printing of the Central University of Venezuela. Caracas, Venezuela. 270 pp.

Méndez-Baamonde, J., Baquero, M., Méndez-Dot, J. A., 2008. Calizas pelágicas derivadas de eventos oceánicos anóxicos en la Formación Apón del Grupo Cogollo. Cuenca de Maracaibo, Venezuela. In: II ALAGO. Memorias del Congreso Latinoamericano de Geoquímica Orgánica, Caracas, 6–18.

Méndez-Baamonde, J., Baquero, M., Méndez-Dot, J. A., 2009. Eventos Oceánicos Anóxicos OAE 1a y OAE 1b en la Formación Apón del Grupo Cogollo durante el Cretácico Temprano. Cuenca de Maracaibo, Venezuela. *Revista Latinoamericana de Geoquímica Orgánica, 1*, 6–18.

Méndez-Dot, J. A., 2017. Sedimentación de los Eventos Oceánicos Anóxicos 1a y 1b del Cretácico en el Grupo Cogollo de la Cuenca de Maracaibo: determinación del potencial como rocas madre. Doctoral Dissertation. Central University of Venezuela, 229 pp.

Méndez-Dot, J. A., Méndez-Baamonde, J., Reyes, D., Whilchy, R., 2015. The Cogollo Group and the Oceanic Anoxic Events 1a and 1b, Maracaibo Basin, Venezuela, *Brazilian Journal of Geology, 45* (Suppl. 1), 41–61.

Meyers, P. A., 2006. Paleoceanographic and paleoclimatic similarities between Mediterranean sapropels and Cretaceous black shales. *Palaeogeography, Palaeoclimatology, Palaeoecology, 235*, 305–320.

Miall, A. D., and Blakey, R. C., 2008. The Phanerozoic tectonic and sedimentary evolution of North America. *Sedimentary basins of the world, 5*, 1–29.

Miall, A. D., and Blakey, R. C., 2019. The Phanerozoic tectonic and sedimentary evolution of North America. In: Miall AD, editor. The Sedimentary Basins of the United States and Canada. 2nd ed. Elsevier. pp. 1–38.

Monteiro, F. M., Pancost, R. D., Ridgwell, A., Donnadieu, Y. 2012. Nutrients as the dominant control on the spread of anoxia and euxinia across the Cenomanian-Turonian oceanic anoxic event (OAE2): model-data comparison. *Paleoceanography, 27*, PA4209.

Montgomery, C. W., and Hurley, P. M., 1978. Total-rock U-Pb and Rb-Sr systematics in the Imataca Series, Guayana Shield, Venezuela. *Earth and Planetary Science Letters, 39*, 281–290.

Mortensen, J., Engstrom, F., Lind, I., 1998. The relation among porosity, permeability, and specific surface of chalk from the Gorm field, Danish North Sea. *SPE Reservoir Evaluation & Engineering, 1*, 245–251.

Moulin, M., Aslanian, D., Unternehr, P., 2010. A new starting point for the South and Equatorial Atlantic Ocean. *Earth-Science Reviews, 98*, 1–37.

Muller-Karger, F. E., Astor, Y. M., Benitez-Nelson, C. R., Buck, K. N., Fanning, K. A., Lorenzoni, L., Montes, E., Rueda-Roa, D. T., Scranton, M. I., Tappa, E., Taylor, G. T., Thunell, R. C., Troccoli, L., Varela, R., 2019. The Scientific Legacy of the Cariaco Ocean Time-Series Program. *Annual Review of Marine Science, 11*, 413–437.

Nozaki, T., Kato, Y., Suzuki, K., 2013. Late Jurassic ocean anoxic event: evidence from voluminous sulphide deposition and preservation in the Panthalassa. *Scientific Reports, 3*, 1889.

Ogg, J. G. 2020. Geomagnetic Polarity Time Scale. In: Gradstein, F. M., Ogg, J. G., Schmitz, M. D., Ogg, G. M. (Eds.), Geologic Time Scale 2020, pp. 159–192, Amsterdam, The Netherlands, Elsevier B.V.

Ogg, J. G., Ogg, M. G., Gradstein, F. M., 2016. A concise geologic time scale: 2016. Elsevier. 226 pp.

Ostos, M., Yoris, F., Avé-Lallemant, H. G., 2005. Overview of the southeast Caribbean–South American plate boundary zone, *in* Avé Lallemant, H.G., and Sisson, V.B., eds., Caribbean–South American plate interactions, Venezuela: *Geological Society of America Special Paper, 394*, 53–89.

Ota, A., and Isozaki, Y., 2006. Fusuline biotic turnover across the Guadalupian–Lopingian (Middle–Upper Permian) boundary in mid-oceanic carbonate buildups: biostratigraphy of accreted limestone in Japan. *Journal of Asian Earth Sciences, 26*, 353–368.

Pancost, R. D., Crawford, N., Magness, S., Turner, A., Jenkyns, H. C., Maxwell, J. R., 2004. Further evidence for the development of photic-zone euxinic conditions during Mesozoic oceanic anoxic events. *Journal of Geological Society, London, 161*, 353–364.

Papadomanolaki, N. M., van Helmond, N. A., Pälike, H., Sluijs, A., Slomp, C. P., 2022. Quantifying volcanism and organic carbon burial across Oceanic Anoxic Event 2. *Geology, 50*(4), 511–515.

Percival, L. M., Matsumoto, H., Callegaro, S., Erba, E., Kerr, A. C., Mutterlose, J., Suzuki, K., 2024. Cretaceous large igneous provinces: from volcanic formation to environmental catastrophes and biological crises. *Geological Society, London, Special Publications, 544*, SP544-2023-88.

Pérez de Mejía, D., Kiser, G. D., Maximowitsch, B., Young, G., 1980. Geología de Venezuela. In: Felder, B. (Coord.), Brie, A., Gartner, J., Hepp, V., Hrabie, M., Kervella, M., Mons, F., Mowat, G., Neville, N., Plomb, J., Sadras, W., Tejada, A., Trassard, J., Vidal, J. Y., Zinat, D.: Evaluación de Formaciones en Venezuela. Schlumberger Surenco S.A.; 1ra. Ed., 287 pp.

Pérez-Infante, J., Farrimond, P., Furrer, M., 1996. Global and local controls influencing the deposition of the La Luna Formation (Cenomanian-Campanian), western Venezuela. *Chemical Geology, 130*, 271–288.

Petrash, D. A., Lalonde, S. V., González-Arismendi, G., Gordon, R. A., Méndez, J. A., Gingras, M. K., Konhauser, K. O., 2015. Can Mn–S redox cycling drive sedimentary dolomite formation? A hypothesis. *Chemical Geology, 404*, 27–40.

Philip, J., 2003. Peri-Tethyan neritic carbonate areas: distribution through time and driving factors. *Palaeogeography, Palaeoclimatology, Palaeoecology, 196*, 19–37.

Pomar, L., Esteban, M., Martinez, W., Espino, D., Castillo de Ott, V., Benkovics, L., Castro-Leyva, T., 2015. Oligocene-Miocene carbonates of the Perla field, offshore Venezuela: Depositional model and facies architecture, In: C. Bartolini and P. Mann, eds., Petroleum geology and potential of the Colombian Caribbean Margin. *American Association of Petroleum Geologists, Memoir 108*, 647–674.

Quijada, M., Riboulleau, A., Faure, P., Michels, R., Tribovillard, N., 2016. Organic matter sulfurization on protracted diagenetic timescales: the possible role of anaerobic oxidation of methane. *Marine Geology, 381*, 54–66.

Radwan, M. S., Gobashy, M., Dahrough, S., Raslan, S., 2024. Exploring carbonate reservoirs potential, north Egypt. *Contributions to Geophysics and Geodesy, 54*(1), 49–66.

Raven, M. R., Fike, D. A., Gomes, M. L., Webb, S. M., Bradley, A. S. McClelland, H. -L. O., 2018. Organic carbon burial during OAE2 driven by changes in the locus of organic matter sulfurization. *Nature Communications, 9*, 3409.

Rey, O., Simo, J., Lorente, M. A., 2004. A record of long- and short-term environmental and climatic change during OAE3: La Luna Formation, Late Cretaceous (Santonian-early Campanian), Venezuela. *Sedimentary Geology, 170*, 85–105.

Rodríguez-Cuicas, M. E., Montero-Serrano, J. C., Garbán, G., 2019. Paleoenvironmental changes during the late Albian oceanic anoxic event 1d: An example from the Capacho Formation, southwestern Venezuela. *Palaeogeography, Paleoclimatology, Paleoecology, 521*, 10–29.

Rodríguez-Cuicas, M. E., Montero-Serrano, J. C., Garbán, G., 2020. Geochemical and mineralogical records of late Albian oceanic anoxic event 1d (OAE-1d) in the La Grita Member (southwestern Venezuela): Implications for weathering and provenance. *Journal of South American Earth Sciences, 97*, 102408.

Rona, P. A., and Richardson, E. S., 1978. Early Cenozoic global plate reorganization. *Earth and Planetary Science Letters, 40*, 1–11.

Ross, C. A., 1986. Paleozoic evolution of southern margin of Permian basin. *Geological Society of America Bulletin, 97*, 536–554.

Ross, C. A., and Ross, J. R., 1985. Late Paleozoic depositional sequences are synchronous and worldwide. *Geology, 13*, 194–197.

Sarmiento-Rojas, L. F., 2019. Cretaceous stratigraphy and paleo-facies maps of northwestern South America. In F. Cediel & R. P. Shaw (Eds.), Geology and tectonics of Northwestern South America. *Frontiers in Earth Sciences* (pp. 673–747). Cham: Springer.

Schlanger, S. O., and Jenkyns, H. C., 1976. Cretaceous oceanic anoxic events: causes and consequences. *Geologie en Mijnbouw, 55*, 179–184.

Scotese, C. R., Song, H., Mills, B. J. W., van der Meer, D. G., 2021. Phanerozoic paleotemperatures: the earth's changing climate during the last 540 million years. *Earth-Science Reviews, 215*, 103503.

Scranton, M. I., McIntyre, M., Astor, Y., Taylor, G. T., Muller-Karger, F. E., Fanning, K. A., 2006. Temporal variability in the nutrient chemistry of the Cariaco Basin. In Past and Present Water Column Anoxia, ed. LN Neretin, pp. 139–60. Dordrecht.: Springer.

Seton, M., Müller, R. D., Zahirovic, S., Gaina, C., Torsvik, T., Shephard, G., Talsma, A., Gurnis, M., Turner, M., Maus, S., Chandler, M., 2012. Global continental and ocean basin reconstructions since 200 Ma. *Earth-Science Reviews, 113*, 212–270.

Silva-Pineda, A., Buitrón-Sánchez, B. E., Arellano-Gil, J., Vachard, D., Ramírez, J., 2003. Continental and marine Permian biota of south-central Mexico: a synthesis. In: C. Bartolini, R. T. Buffler, and J. Blickwede, eds. The circum-Gulf of Mexico and the Caribbean: hydrocarbon habitats, basin formation, and plate tectonics. *American Association of Petroleum Geologists, Memoir 79*, 462–475.

Sinninghe Damsté, J. S., and J. Köster., 1998. A euxinic southern North Atlantic Ocean during the Cenomanian/Turonian oceanic anoxic event. *Earth and Planetary Science Letters, 158*, 165–173.

Sorkhabi, R., 2016. Rich petroleum source rocks *GeoExPro, 6*(6), 16–22.

Soto, G., 2025, in press. Utilidad de los Macroforaminíferos en la Industria Petrolera. Caso de Estudio: Cuenca de Falcón. *Visión Tecnológica*. 16 pp.

Soto, G., Ortega, S., Marín, Z., Farías, A., Medina, C., 2025, in press. Estudio Bioestratigráfico de la Unidades Oligo-Miocénicas de la Región Falconiana. *Memorias del II Congreso Venezolano de Geociencias*. Caracas, 20–24 de Marzo 2023. 31 pp.

Spalding, M. D., Fox, H. E., Allen, G. A., Davidson, N., Ferdaña, Z. A., Finlayson, M., Halpern, B. S., Jorge, M. A., Lombana, A., Lourie, S. A., Martin, K. D., McManus, E., Molnar, J., Recchia, C. A. Robertson, J., 2007. Marine ecoregions of the world: a bioregionalization of coastal and shelf areas. *BioScience, 57*, 573–583.

Spikings, R. A., Crowhurst, P. V., Winkler, W., Villagomez, D., 2010. Syn-and post-accretionary cooling history of the Ecuadorian Andes constrained by their *in-situ* and detrital thermochronometric record. *Journal of South American Earth Sciences, 30*, 121–133.

Spikings, R., Cochrane, R., Villagomez, D., Van der Lelij, R., Vallejo, C., Winkler, W., Beate, B., 2014. The geological history of northwestern South America: from Pangaea to the early collision of the Caribbean Large Igneous Province (290–75 Ma). *Gondwana Research, 27*, 95–139.

Stanley, S. M. 1999. Earth System History. W. H. Freeman and Company, New York, 615 pp.

Taboada, A., Rivera, L. A., Fuenzalida, A., Cisternas, A., Philip, H., Bijwaard, H., Olaya, J., Rivera, C., 2000. Geodynamics of the northern Andes: Subductions and intracontinental deformation (Colombia). *Tectonics, 19*, 787–813.

Takashima, R, Nishi, H., Huber, B. T., Leckie, R. M., 2006. Greenhouse World and the Mesozoic Ocean. *Oceanography, 19*, 64–75.

Tessin, A., Hendy, I., Sheldon, N., Sageman, B., 2015. Redox-controlled preservation of organic matter during "OAE 3" within the Western Interior Seaway. *Paleoceanography, 30*(6), 702–717.

Taylor, G. T., Iabichella-Armas, M., Varela, R., Muller-Karger, F., Lin, X., Scranton, M.I., 2006. Microbial ecology of the Cariaco Basin's oxic/anoxic interface: the U.S.-Venezuelan CARIACO times series program. In: Neretin, L.N. (Ed.), Past and Present Water Column Anoxia. NATO Science Series. Springer, The Netherlands, pp. 473–499.

Tejada, M. L. G., Suzuki, K., Kuroda, J., Coccioni, R., Mahoney, J. J., Ohkouchi, N., Sakamoto, T., Tatsumi, Y., 2009. Ontong Java Plateau eruption as a trigger for the early Aptian oceanic anoxic event. *Geology, 37*(9), 855–858.

Trabucho-Alexandre, J., Hay, W. W., de Boer, P. L., 2012. Phanerozoic environment of black shale deposition and the Wilson Cycle. *Solid Earth, 3*, 29–42.

Tribovillard, N. P., Stephan, J. F., Manivit, H., Reyre, Y., Cotillon, P., Jautée, E., 1991. Cretaceous black shales of Venezuelan Andes: preliminary results on stratigraphy and paleoenvironmental interpretations. *Palaeogeography, Palaeoclimatology, Palaeoecology, 81*, 313–321.

Truskowski, I., 2006. Calibración bioestratigráfica y correlación con datos geoquímicos de la Formación Querecual en su localidad tipo. Specialization Dissertation, Central University of Venezuela. 106 pp.

Tzortzaki, E., Karakitsios, V., Tsikos, H., 2013. Biomarker evidence for intermittent photic zone euxinia in the Aptian–Albian organic sedimentary record from the Ionian Zone (Epirus, Greece). *Organic Geochemistry*, 64, 84–93.

Vachard, D., Pille, L., Gaillot, J., 2010. Palaeozoic Foraminifera: systematics, palaeoecology and responses to global changes. *Revue de Micropaléontologie, 53*, 209–254.

Valencia, F. L. and Laya, J. C., 2020. Deep-burial dissolution in an Oligocene-Miocene giant carbonate reservoir (Perla Limestone), Gulf of Venezuela Basin: Implications on microporosity development. *Marine and Petroleum Geology, 113*, 104144.

van Bentum, E. C., Hetzel, A., Brumsack, H. J., Forster, A., Reichart, G. J., Damsté, J. S. S., 2009. Reconstruction of water column anoxia in the equatorial Atlantic during the Cenomanian–Turonian oceanic anoxic event using biomarker and trace metal proxies. *Palaeogeography, Palaeoclimatology, Palaeoecology*, 280, 489–498.

van Helmond, N. A., Sluijs, A., Reichart, G. J., Sinninghe Damsté, J. S., Slomp, C. P., Brinkhuis, H., 2013. A perturbed hydrological cycle during Oceanic Anoxic Event 2. *Geology, 42*, 123–126.

Veron, J. E. N., 1995. Corals in Space and Time. Sydney: Southwood Press. 321 pp.

Villagómez, D., and Spikings, R., 2013. Thermochronology and tectonics of the Central and Western Cordilleras of Colombia: Early Cretaceous–Tertiary evolution of the northern Andes. *Lithos*, *160*, 228–249.

Wagner, T., Sinninghe Damsté, J. S., Hofmann, P., Beckmann, B., 2004. Euxinia and primary production in Late Cretaceous eastern equatorial Atlantic surface waters fostered orbitally driven formation of marine black shales. *Paleoceanography*, *19*, PA3009.

Wagreich, M., 2012. "OAE 3" - regional Atlantic organic carbon burial during the Coniacian Santonian. *Climate of the Past*, *8*, 1447.

Warne, A. G., Guevara, E. H., Aslan, A., 2002a. Late quaternary evolution of the Orinoco Delta, Venezuela. *Journal of Coastal Research*, 225–253.

Warne, A. G., Meade, R. H., White, W. A., Guevara, E. H., Gibeaut, J., Smyth, R. C., Aslan, A., Tremblay, T., 2002b. Regional controls on geomorphology, hydrology, and ecosystem integrity in the Orinoco Delta, Venezuela. *Geomorphology*, *44*, 273–307.

Weber, B., Mota, A., Helenes, J., Ramírez, R., Valencia, Y., 2016. Age and provenance of Late Miocene-Early Pliocene sedimentary rocks from the Patao high hydrocarbon reservoir offshore NE Venezuela–U-Pb detrital zircon age, Sm-Nd isotope, and biostratigraphic data. *Journal of Natural Gas Science and Engineering*, *31*, 459–473.

Weil, E. 2003. The corals and coral reefs of Venezuela. In: Cortés J. (ed.), *Latin American Coral Reefs*. Elsevier, pp. 303–330.

Wendler, J. E., and Wendler, I., 2016. What drove sea-level fluctuations during the mid-Cretaceous greenhouse climate? *Palaeogeography, Palaeoclimatology, Palaeoecology*, *441*, 412–419.

Westermann, S., Stein, M., Matera, V., Fiet, N., Fleitmann, D., Adatte, T., Föllmi, K. B., 2013. Rapid changes in the redox conditions of the western Tethys Ocean during the early Aptian oceanic anoxic event. *Geochimica et Cosmochimica Acta*, *121*, 467–486.

Wilson, J. L., 1990. Basement structural controls on Mexican carbonate facies in northeastern Mexico—A review, in Tucker, M.E., et al., eds., Carbonate platforms, facies, sequences, and evolution. *International Association of Sedimentologists*, *Special Publication 9*, 235–255.

Wilson, P. A., and Norris, R. D., 2001. Warm tropical ocean surface and global anoxia during the mid-Cretaceous period. *Nature*, *412*, 425–429.

Carbonate Facies of Venezuela: From the Late Paleozoic to Modern Times

2

This chapter is by far the longest of the book. It provides a synthesis of the tectonic and sedimentary history of Northern Venezuela, from the Late Paleozoic (Middle Permian) to modern times. The chapter includes an updated review of the major tectonic processes that led to the creation of the most prolific petroleum basins in Venezuela (excluding the siliciclastic reservoirs of the Orinoco Heavy Oil Belt). It also covers the sedimentary infilling of those basins, the exhumation of the main mountain ranges, and the present-day tectonic setting, geomorphology, and sedimentary environments that frame the modern Caribbean and Atlantic carbonate systems of Venezuela. Particular emphasis is placed on the Maracaibo Basin and its Cretaceous sedimentary units (the most represented in the following photographic chapters). Rock units of great scientific and/or economic importance include the hydrocarbon source rocks of the Machiques Member, the La Luna Formation, and the Querecual Formation, deposited during the Cretaceous anoxic events. Also included are the Perla Limestone (contained in the largest gas field in Latin America) and the carbonate complexes of the Venezuelan Caribbean (among the largest and most biodiverse of the Eastern and Southern Caribbean). This chapter serves as a reference for readers consulting the following photographic chapters.

2.1 Late Paleozoic Carbonates in Northwest Gondwana

Prior to the final assembly of the Pangea Supercontinent, a vast seaway existed between Laurentia and Gondwana. During the Early to Middle Permian and after the finishing of the late Paleozoic Ice Age, the combined effect of rapid glacioeustatic sea-level rise and basin subsidence (López-Gamundí 2010) led to the creation of extensive epeiric (or epicontinental) seas and peri-cratonic shallow marine environments along that seaway. At the southern margin of that vast shelve environment in northwest Gondwana (Fig. 2.1), carbonate sedimentation in the aragonite seas of the late Paleozoic (Aragonite II) extended from what is now western Venezuela, to the west and southwest, nowadays Colombia, Ecuador, Perú and Bolivia (Laya and Tucker 2012, 2015).

2.1.1 Ancient Mucuchachí Basin

Permian carbonate sedimentation in Venezuela is represented by the Palmarito Formation, which was deposited in the Mucuchachí Basin (Fig. 2.1). The latter was an ancient foreland tropical basin which occupied approximately the present area of the Venezuelan Andes. It is believed to have existed since the Middle Ordovician (Arnold 1966). The Mucuchachí Basin was developed within a compressional tectonic setting related to the formation of Pangea, perhaps also to the Alleghenian orogenesis (the final collision between Gondwana and Laurentia during Pennsylvanian-Permian; Miall and Blakey 2008, 2019). The basin has been interpreted as a narrow, tectonic-foreshortened depression, oriented northeast–southwest, where the thick supraterane succession of Upper Paleozoic rocks was deposited (Arnold 1966; Laya and Tucker 2012). The crystalline basement of the Venezuelan Andes consists of early Paleozoic amphibolite-facies metamorphic rocks of the Iglesias Complex (rocks mainly of Ordovician and Silurian ages) (Tazzo-Rangel et al. 2019). Fine-grained metasediments (slate and fine-grained quartzite) of the Mucuchachí Formation of Late Devonian to Early Carboniferous age unconformably overlie the high-grade regional metamorphic basement and represent the rock basement of the Mucuchachí Basin (Arnold 1966; Laya and Tucker 2012) (Fig. 2.2).

Unconformably overlying the Mucuchachí Formation, the sedimentary infilling of the Mucuchachí Basin consists of continental-fluvial deposits of the Sabaneta Formation (Middle Pennsylvanian sandstones and conglomerates) and carbonates of the Palmarito Formation of Middle Permian age (Fig. 2.2). These continental to marine-ramp deposits are in unconformable contact and separated by a long hiatus (~ 60 Myr) with overlying Jurassic red beds of the La Quinta Formation (van der Lelij et al. 2016a).

J. A. Méndez Dot, J. Méndez Baamonde, *Atlas of Tropical Carbonates*, https://doi.org/10.1007/978-3-031-25494-9_2

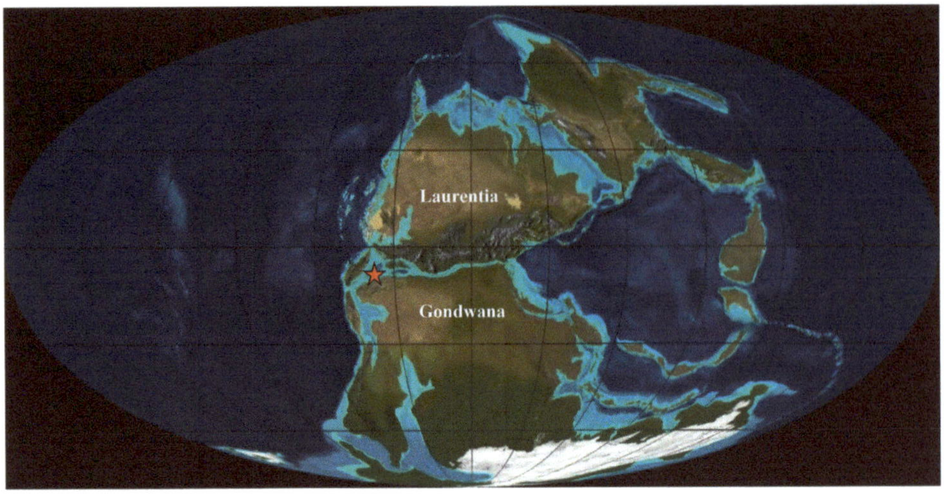

Fig. 2.1 Mollweide projection for the Early Permian Artinskian stage (280 Ma) modified from Blakey (2007). The red star represents the paleogeographical location of the Mucuchachí Basin, following reinterpretations of Laya and Tucker (2012). (Figure slightly modified from Laya et al. 2013a)

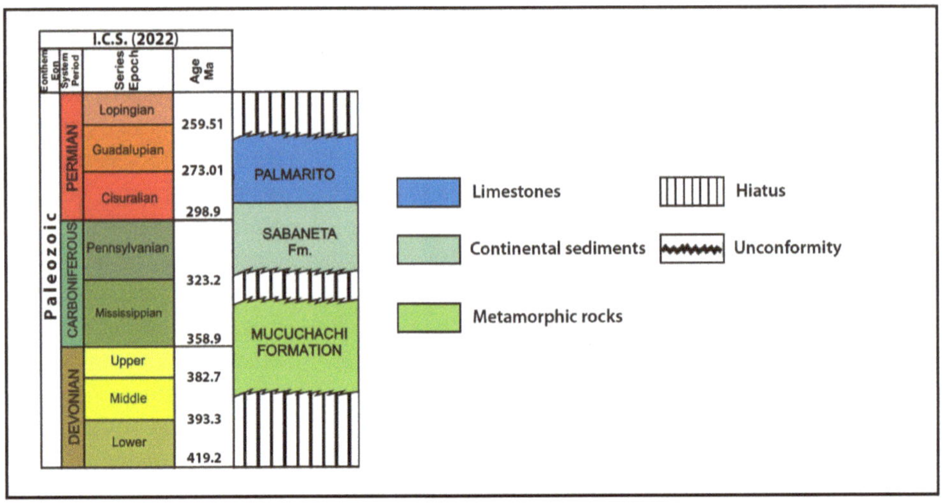

Fig. 2.2 Paleozoic strata in the southern Venezuelan Andes. Abbreviations: *Fm* formation. Figure modified from Laya and Tucker (2012)

2.1.2 Palmarito Formation (Middle Permian: Cisularian to Guadalupian)

Middle Permian carbonates of the Palmarito Formation (Cisularian to Guadalupian) crop out along the Venezuelan Andes (states of Táchira, Mérida, Trujillo and Barinas), and to a lesser extent in the Perijá Range in northwesternmost Venezuela (see Fig. 2.2). Thicknesses of this late Paleozoic carbonate unit are between ~400–550 m (Arnold 1966; Ministerio de Energía y Minas 1997; Laya and Tucker 2012). This carbonate succession consists of a variety of mud- and grain-supported facies, from mudstones to packstones and boundstones, as well as minor proportions of mixed carbonate-siliciclastic facies. The main bioclastic components (or constituents) are crinoids, bryozoans, brachiopods, benthic foraminifera (mainly fusulinids and secondary biseriamminids), sponge spicules, molluscs (bivalves and gastropods), ostracods, trilobites, calcareous algae and rugose corals. Dolomitized facies are moderately common (Laya and Tucker 2012). In this book, photographic examples of the Palmarito Formation correspond to outcrops in the surrounding areas of the cities of Mérida and Valera (see Fig. 2.3).

Transition from the continental Sabaneta Formation to marine-calcareous Palmarito sediments occurred in a carbonate ramp that dipped basinwards towards the north, facing the open ocean. The formation evolved from the underlying fluvial deposits (Sabaneta Formation) through to ostracod-rich tidal-flat deposits, to mid-outer ramp deposits (rich in fusulinids, calcareous algae, crinoids, bryozoans and brachiopods) and to outer ramp deposits (rich in sponge spicules, radiolarians and brachiopods). Much of the Palmarito succession shows a general transgressive trend, with a few thinner, more-regressive intervals (Laya and Tucker 2012), as long-term eustatic sea level (ESL) was rising, probably as a result of the deglaciation taking place at the southern polar region during Middle Permian times (Fielding et al. 2008).

Martin (1996) proposed a global increase in nutrient supply to the oceans, reaching a mesotrophic state during the Permian, which allowed the dominance of heterotrophic calcifiers in the carbonate factory (Wright and Burchette 1996). Related to this, the nature of the carbonate benthic factory of the Palmarito Formation suggests that trophic conditions during deposition of this shallow tropical carbonate succession varied from oligotrophic to meso-eutrophic. The photozoan

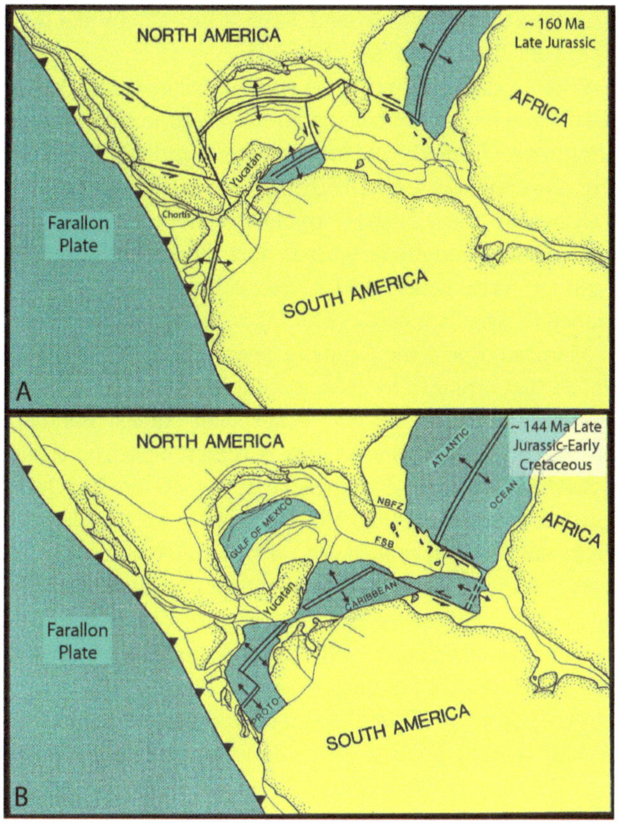

Fig. 2.3 Paleogeographic sketches for the Middle and Late Jurassic. (**a**) Initial opening stage of the proto-North Atlantic Ocean during Late Jurassic at ~160 Ma. A minor opening process also started in what would later be the proto-Caribbean Sea. (**b**) Initial opening stage of the proto-Caribbean and the Gulf of Mexico at ~144 Ma. Figure modified from Ross and Scotese (1988)

(light-dependent) grain association between larger benthic foraminifera (LBF) and calcareous green algae (Photozoan Association *sensu* James 1997) is common and representative of shallow, tropical waters, i.e., the Photozoan T-factory of Michel et al. (2019) [this factory includes the T factory of Schlager 2003, 2005, in which the letter T stands for "tropical" and "top-of-the-water-column"]. Nevertheless, the Heterozoan Association (*sensu* James 1997) dominated by light-independent calcifiers (crinoids, brachiopods, bryozoans, ostracods, molluscs and sponges), i.e., the Heterozoan factory (Michel et al. 2019), prevails in the overall sedimentation. Therefore, the Palmarito Formation is an example of mixed heterozoan-photozoan carbonates whose deposition in shallow-tropical waters was not only controlled by water temperature, but also by other factors such as trophic conditions.

Furthermore, the neritic lime-mud production (i.e., allomicrite) in the Palmarito ramp was also significant, as mudstones, wackestones, and muddy packstones dominate the succession. Micrite matrices in many microfacies of the Palmarito Formation are well preserved, showing little or no traces of neomorphism/recrystallization. The latter suggest an original calcitic composition. We estimate the biologically-induced precipitation of fine-grained (micro- to cryptocrystalline) $CaCO_3$ as the main mode of precipitation of these micritic sediments. In this context, light-colored matrices precipitated via photosynthesis (i.e., the neritic lime-mud factory of Pomar and Hallock 2008). In some dark-colored micritic microfacies with fair to good TOC content (~ 0.5 to 1.5%; e.g., ostracods- and sponge spicules-rich microfacies) deposited under low-oxygen conditions, microbially-induced benthic precipitation of micrite (i.e., the benthic automicrite factory of Pomar and Hallock 2008; which includes the M factory of Schlager 2003, 2005; the letter M alludes to mud-mound, micrite and microbes) was the likely dominating mode of carbonate precipitation (see Appendix A for a detailed treatment of the concepts of Marine Carbonate Factories, and of the Photozoan and Heterozoan Grain Associations).

High primary productivity resulting from an increased influx of nutrients to the carbonate ramp and the development of a warmer period (based on $\delta^{13}C$ and $\delta^{18}O$ on brachiopod shells, respectively) are proposed by Laya et al. (2013a) for the middle part of the Palmarito Formation. This rock interval was dated by the same authors using $^{87}Sr/^{86}Sr$ ratios on well-preserved brachiopod shells, placing the middle part of the succession at ~273.5–270.5 Ma (early Guadalupian). Climate proxies ($\delta^{13}C$ and $\delta^{18}O$) suggest a relation between the deglaciation of the southern polar region and the increase in the primary productivity at tropical paleolatitudes (~ 10° S), where the sedimentation of the Palmarito Formation took place (Laya et al. 2013b).

In order to evaluate the source rock potential of pre-Cretaceous rocks in western Venezuela, Lugo (2008) carried out an organic geochemistry study of the Palmarito Formation. The samples studied have TOC average content of 0.73%, with a maximum value of 1.49%. The Tmax (thermal maturity parameter for organic carbon-bearing rocks) average value determined by Rock-Eval Pyrolysis is 490 ° C, indicating that kerogen in the samples is overmature. Nevertheless, the sampling methods applied by Lugo (2008) were restricted to the lower part of the succession and only from outcrops in the state of Mérida. Therefore, the thermal history of the Palmarito Formation may have been quite different from the rest of the basin, leaving the possibility of potential petroleum source rocks elsewhere (Laya and Tucker 2015).

2.2 Jurassic Rifting and the Origin of the Proto-Caribbean

By the Late Permian (~ 260 Ma), the closure of the Late Carboniferous-Permian seaway was completed. Orogenic processes linked to the collision between Gondwana and Laurentia started during the Middle Mississippian (~ 340 Ma), representing the beginning of a period of ~150 Myr during which the Pangea Supercontinent dominated all aspects of Earth history (Blakey 2003).

The Pangea breakup started during the latest Triassic to Early Jurassic. In western Pangea, this global process involved the drifting between the Americas. Continental margin rifting and plates rotation linked to the Pangaea breakup also involved a series of rigid blocks (e.g., Maya and Chortís blocks) (Pindell and Kennan 2009; Escalona et al. 2021a). By the Middle to Late Jurassic, the opening processes began to form new oceanic crust and oceanic basins between North America and Northwest Africa plates, and to a lesser extent, between the Americas (Fig. 2.3a) (Bartok 1993; Meschede and Frish 1998; Moulin et al. 2010; Bartok et al. 2015). The existence of localized, sabkha settings and nascent marine environments between the Americas during the Middle Jurassic (~169–166 Ma, Bathonian) was confirmed by $^{87}Sr/^{86}Sr$ dating of several evaporite deposits in the Gulf of Mexico, and in the Gulf of Paria between Venezuela and Trinidad (Pindell et al. 2020). By Late early Jurassic (~ 155 Ma, late Oxfordian), the Gulf of México was an open marine environment (see Bartok 1993; Pindell 1994; Meschede and Frisch 2002; Bird et al. 2007; Pindell et al. 2020). Subsequently, this extensional process spread to the southeast during the Kimmeridgian (~ 152 Ma), creating the proto-Caribbean through rifting between Yucatán and northern South America (Fig. 2.3b) (Bartok 1993; Villamil and Pindell 1998; Giunta et al. 2002a, b; Dilek 2003; Escalona et al. 2021a).

From the Late Jurassic, and prior to the beginning of carbonate sedimentation in Northern Venezuela during the Early Cretaceous, some areas of the Gulf of Mexico and the eastern proto-Caribbean, and most of the eastern proto-Caribbean had already become major carbonate provinces (see Bartok et al. 2015). Among widespread carbonate deposits in Yucatán, eastern Mexico and southern U.S.A., it highlights the Bahama Platform (see localities 4 to 15 in Fig. 1.1). The latter consists of shallow-water carbonate sedimentation since the Jurassic, and around 5 km of limestones and dolomites (see Tucker and Wright 1990).

Prior to the opening of the proto-Caribbean, a series of extensional basins (grabens) with southwest-northeast orientation were formed in what is today Northern Venezuela (Fig. 2.4) (van der Lelij 2013; Baquero 2015; van der Lelij et al. 2016a). These Jurassic structures include the Machiques and Uribante troughs in what is now the Perijá Range and southern Venezuelan Andes, respectively, the Urdaneta Graben in what is now western Lake.

Maracaibo, and the Espino graben. The latter is a large graben that crosses North Venezuela from the Apure to the Guárico sub-basins. In the Venezuelan Andes, the basal infilling of the Uribante graben is represented by volcanic rocks (biotite bearing rhyodacite tuffs) of the La Quinta Formation, dated at 202.0 Ma (U/Pb in zircon; van der Lelij 2013; van der Lelij et al. 2016a). This latest Triassic age suggests that the beginning of the extensional regime that led to the breakup between the Americas began ~50 Myr before the early stages of creation of the oceanic floor in what would later be the proto-Caribbean (Baquero 2015).

The oldest Mesozoic sedimentary rocks in Northern Venezuela correspond to the tectono-sedimentary infilling of these Jurassic grabens (Fig. 2.4). In the Perijá Range, this infilling consists of red beds and minor limestones, and associated volcanic rocks (rhyolitic tuffs, volcanic ashes and basalts) of the La Ge Group (from base to top, the Tinacoa, Macoita, and La Quinta formations). In the literature, thick Jurassic volcano-sedimentary strata across North Venezuela are mostly referred to as the La Quinta Formation (Parnaud et al. 1995a; Ministerio de Energía y Minas 1997; Yoris and Ostos 1997; Villamil and Pindell 1998; James 2000; Ostos et al. 2005; Ramos 2009; Escalona and Mann 2011). This intracontinental syn-rift succession also crops out in the

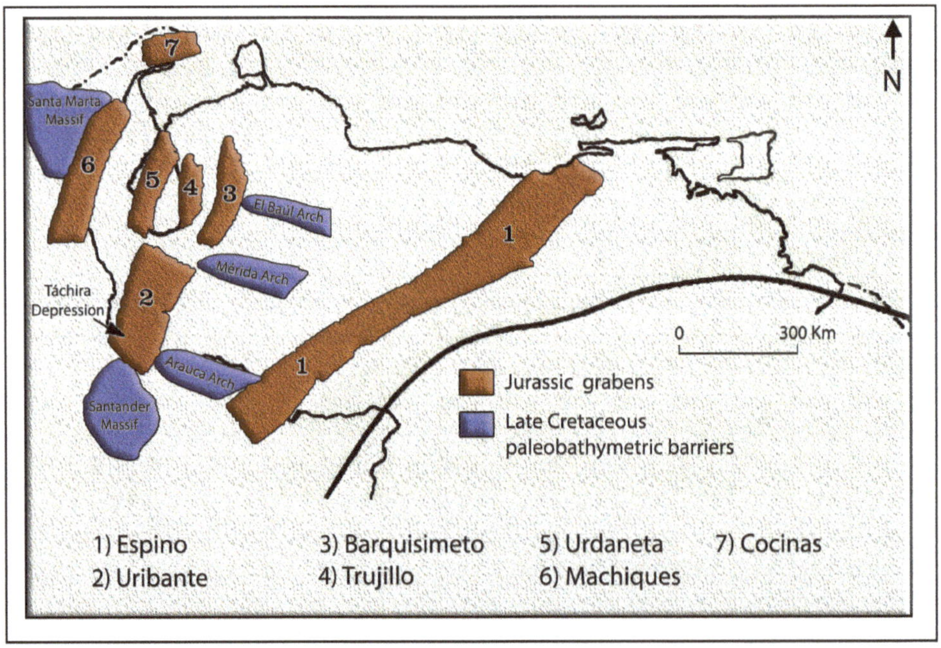

Fig. 2.4 Tectonic reconstruction of major Mesozoic tectonic features in Northern Venezuela. Location of Late Cretaceous paleobathymetric barriers according to Erlich et al. 2000. These barriers became positive areas in the Paleogene. Figure modified from Ministerio de Energía y Minas (1997)

Venezuelan Andes, and has been reported in deep cores in Lake Maracaibo and the Espino Graben (Baquero 2015). La Quinta Formation has also been well studied on the Colombian flank of the Perijá Range. Its heterogenic lithology consists of siltstones, conglomerates, sandstones, shales, lenses of fossiliferous limestones, and interbedded volcanoclastic and basaltic-andesite volcanic rocks (Maze 1984; Schubert 1986; Nova et al. 2019; Bayona et al. 2020).

The thickness of this volcano-sedimentary succession varies widely, from a few hundreds of meters in both the Venezuelan Andes and the Perijá Range to up to 3400 m in the Venezuelan Andes (Schubert 1986), and more than 6 km in Perijá (see Nova et al. 2019). In the Espino Graben (Fig. 2.4), it is ~1650 m thick (Feo-Codecido et al. 1984). The environment of deposition has been interpreted as alluvial and fluvial with episodic volcanic events (Maze 1984). Limestone and marl beds found in the sedimentary succession of the La Quinta Formation in the Venezuelan Andes have been interpreted as deposited in ephemeral lakes developed in an extensional basin (Hernández 2003; Arminio et al. 2004). The age of this unit has been a matter of discussion for decades. As previously mentioned, the oldest age reported for its base is latest Triassic (van der Lelij 2013). An earliest Jurassic age (200.72 Ma) is reported in the Venezuelan Andes by Langer et al. (2014), who found a theropod (*Tachiraptor admirabilis*) within a tuffaceous siltstone. Most authors consider an Early to Middle Jurassic age for the sedimentary facies of this unit (see Nova et al. 2019 and references therein). In the Venezuelan Andes, Hernández (2003) consider a Late Jurassic age for the top. The La Quinta Formation is separated from the overlying Cretaceous passive margin sedimentation by a regional unconformity.

2.3 Cretaceous Passive Margin Stage in Venezuela

Despite the arrival of Pacific-Caribbean terranes in NW Colombia during the Late Cretaceous generated some degree of tectonic loading and flexure in the western region of the Maracaibo Basin, since the Jurassic and during the Early Cretaceous, much of northern South America behaved as a passive continental margin. This is evidenced by subsidence curves, the age of synorogenic turbidite units, stratigraphic and structural studies, and plate motion histories (Villamil and Pindell 1998; Sarmiento-Rojas 2019).

During the Early Cretaceous, and once created the proto-Caribbean, the separation between South American and Africa plates during the late Valanginian (~ 135 Ma) represented the initial phase in the formation of the South Atlantic Ocean (see Dilek 2003; Müller et al. 2008; Moulin et al. 2010) and the onset of the Cretaceous passive margin stage along Northern Venezuela (Baquero 2015). This stage was a long period of subsidence and thermal relaxation that lasted until middle Campanian in western Venezuela, and until Oligocene times in eastern Venezuela (see Parnaud et al. 1995a, b). During most of the Cretaceous (from Hauterivian to Campanian), vast epicontinental basins were created north of the Guayana Shield as the progressive separation of the two Americas created new marine basins.

Regional sedimentation along Northern Venezuela encompassing most of the Early and Late Cretaceous (from ~132 to 80 Ma) was promoted by the remoteness of plate boundaries (Bartok 1993; Meschede and Frish 1998; Moulin et al. 2010; Bartok et al. 2015), by sustained regional subsidence, and by a succession of high-order marine transgressions during an eustatic highstand that created an expanding epeiric sea. Passive margin sedimentation is well represented in the three Cretaceous oil-bearing basins in Venezuela. These are the Maracaibo, Eastern Venezuela and Barinas-Apure basins, whose Cretaceous sedimentary infillings are correlated with the worldwide eustatic changes of the sea level (Macellari 1988; Tribovillard et al. 1991; Martínez and Hernández 1992; Vahrenkamp et al. 1993; De Romero and Galea-Álvarez 1995; Carrillo et al. 1995; Parnaud et al. 1995a, b; Pérez-Infante et al. 1996; James 2000; Rey et al. 2004; Helenes et al. 1998; Villamil and Pindell 1998; Crespo de Cabrera et al. 1999; Helenes and Somoza 1999; Erlich et al. 1999a, 2000; Ruiz et al. 2000; Bralower and Lorente 2003; Erlich et al. 2003; De Romero et al. 2003; Macsotay et al. 2003; Parra et al. 2003; Truskowski 2006; Méndez-Baamonde et al. 2009; Garbán 2010; Méndez-Dot et al. 2015; Machado et al. 2016; Rodríguez-Cuicas et al. 2019, 2020; this book).

Passive margin sedimentation in Northern Venezuela was initiated with deposition of basal siliciclastic facies of the Río Negro (Hauterivian to early Barremian) and Barranquín (Hauterivian to late Barremian) formations in the Maracaibo and Eastern Venezuela basins, respectively (González de Juana et al. 1980; Macellari 1988; Erlich and Barrett 1992; Parnaud et al. 1995a, b). Those units also represent the final infilling of the Jurassic grabens and are the oldest Cretaceous sedimentary rocks in those basins.

The Guayana Shield in Southern Venezuela (consisting of Archean, Paleoproterozoic and Mesoproterozoic rocks) was the main source of siliciclastic sediments during the Cretaceous passive margin stage (e.g., Noguera et al. 2011, 2017), notably the Roraima Formation (Rodríguez-Cuicas et al. 2020). Detrital zircons in Hauterivian to Barremian sandstones of western and eastern Venezuela (Río Negro and Barranquín formations, respectively) yielded U/Pb ages from the late Paleozoic (~ 541 to 400 Ma), the Proterozoic (from ~2000 to 600 Ma), and as old as late Archean (~ 2731 Ma). A secondary source of siliciclastic sediments during the Early Cretaceous, as well as during the Paleocene and Early Eocene, were some positive areas located in what is today western and central Venezuela and eastern Colombia (the Mérida and El Baúl arches, and the Santander Massif, respectively; Fig. 2.4) (Noguera et al. 2011, 2017). The siliciclastic sediments were carried out from the Guayana Shield and distributed across all Northern Venezuela by the paleo-Orinoco River. The latter was a vast fluvio-deltaic system with north-flowing orientation that existed during Jurassic, Cretaceous and Cenozoic times, and whose many branches drained the Guayana Shield, eastern Colombia and Northern Venezuela (see Kasper and Larue 1986; Hoorn et al. 1995; Escalona and Mann 2011; Noguera et al. 2011; Weber et al. 2016).

By the end of deposition of the Río Negro and Barranquín formations, the influence of the Barremian ESL rise (see Haq 2014) is evidenced by the transition from basal siliciclastic units to mixed siliciclastic-carbonate sedimentation of early Barremian age. Such lithologic transition marks the early stage of a juvenile, shallow epeiric sea. Figure 2.5a exemplifies the paleogeography of the proto-Caribbean and the Gulf of Mexico at ~119 Ma (early Aptian). High and stable sea level during Aptian and Albian stages (Haq 2014) were linked to high rates of oceanic crust production (e.g., Larson 1991; Caldeira and Rampino 1991; Kaiho and Saito 1994; Leckie

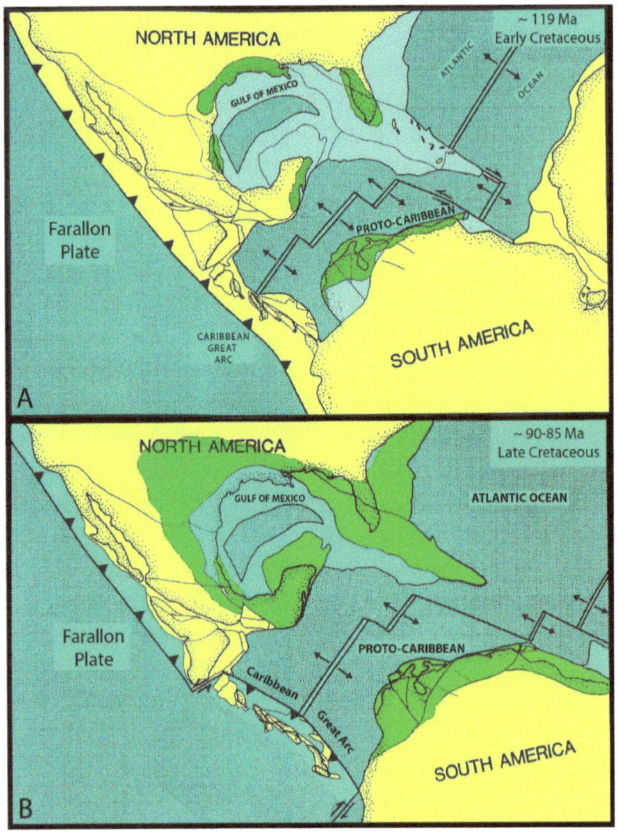

Fig. 2.5 Paleogeographic sketches for the Early and Late Cretaceous. (**a**) Extension of the passive margin sedimentation (areas colored in green) along the proto-Caribbean and the Gulf of Mexico during the early Aptian. (**b**) Extension of the passive margin sedimentation (areas colored in green) during Turonian to Santonian. Figure modified from Ross and Scotese (1988)

et al. 2002), which included the emplacement of two prominent LIPs. The Ontong Java, Manihiki and Hikurangi plateaus, inferred to have formed a single plateau called Ontong Java Nui in the southwestern Pacific Ocean between ~125–118 Ma (Larson and Erba 1999; Erba and Tremolada 2004; Tejada et al. 2009; Bottini et al. 2015; Bodin et al. 2015; Erba et al. 2015; Kuroda et al. 2015; Matsumoto et al. 2021, 2024; Percival et al. 2024); the bulk of it formed at around 122 Ma (Neal et al. 2019). The Kerguelen Plateau in the south Indian Ocean, having a continuous volcanic activity between ca. 122 to 90 Ma (Jiang et al. 2021); the bulk of it formed during ~119–109 Ma (Coffin et al. 2002; Frey et al. 2003; Matsumoto et al. 2020). Sustained submarine and/or subaerial volcanic activity, combined with high ESL, led to the onset of a long period characterized by a Warm Greenhouse climate, punctuated by switches to Hothouse climate, and the creation of extensive epeiric and highly fertile carbonate platforms in tropical and subtropical latitudes worldwide.

During the Early Cretaceous (Calcite II), extensive carbonate platforms were developed not only in northern South America, but in eastern Central America, and southern North America (Fig. 2.5a) (e.g., Palacas et al. 1984; Wilson and Ward 1993; Lehmann et al. 1998, 1999; Villamil and Pindell 1998; Rosenfeld 2003; Schafhauser et al. 2003; Mancini and Puckett 2005; Phelps et al. 2014; Silva-Tamayo et al. 2016). In northwestern Venezuela, the episodic switches from a long-lasting Warm Greenhouse to a Hothouse climate (or Hyperthemal conditions) triggered by LIP eruptions were combined with ideal local factors, which promoted deposition of black shales linked to the OAEs 1a (~ 120 Ma) and 1b (~ 114–109 Ma) in the Maracaibo Basin.

In Northern Venezuela, Early Cretaceous carbonate sedimentation (from early Barremian to latest Albian) is represented only in the Maracaibo and Eastern Venezuela basins, as the Cretaceous marine transgression did not reach the Barinas-Apure Basin until late Albian times. In that basin, fully marine conditions were not reached until early Turonian (Helenes et al. 1998). Sustained high ESL during the first half of the Late Cretaceous reached up to ~240–250 m above present sea level (APSL) (Haq 2014). This high-amplitude sea-level rise led to the largest expansion of the epeiric seas into the flooded landmasses in northern South America and southern North America, and hence, to the largest extension of the passive margin sedimentation (Fig. 2.5b).

Regional sedimentation of black shales occurred during the time interval between late Albian to late Santonian (~ 99 to 83 Ma; the peak of the Calcite II sea in terms of paleoclimatic disturbances), a time interval during which large amounts of organic matter production, preservation and burial led to the deposition of many of the most effective petroleum source rocks worldwide (e.g., Klemme and Ulmishek 1991; Kuroda and Ohkouchi 2006; Takashima et al. 2006; Trabucho-Alexandre et al. 2012). In the American continent, those organic-rich rocks are spread along southern U.S.A. (e.g., Joo and Sageman 2014), eastern Mexico (e.g., Duque-Botero and Maurrasse 2004; Núñez-Useche et al. 2014, 2016, 2020), and along a large area in northern South America spanning Colombia, Venezuela, and Suriname (Erlich et al. 2003).

In this context, the most extreme of the Cretaceous Hothouse climates (or hyperthermals) is well recorded in the sedimentary strata of Northern Venezuela. The beginning of widespread, long-lasting deposition of organic-rich limestones and shales (La Luna and Querecual formations, in the west and east, respectively) was roughly coeval to submarine volcanic emplacement of the Caribbean plate (also commonly known in the literature as the Caribbean Oceanic Plateau; e.g., Révillon et al. 2000) and the Madagascar flood basalts, as well as to the second pulse of the Ontong Java Plateau. The bulk of oceanic crust production correlates closely with the Cenomanian-Turonian boundary (e.g., Sinton and Duncan 1997; Kerr 1998; Snow et al. 2005; Kuroda et al. 2007; Turgeon and Creaser 2008; Adams et al. 2010; Kidder and Worsley 2010; Jones et al. 2021; Papadomanolaki et al. 2022; Percival et al. 2024); although in the case of the Caribbean plate, an older period of plateau magmatism occurred during the late Albian to early Cenomamian (Wright and Wyld 2011; see also Riel et al. 2023). Carbon cycle disturbances linked to the extrusion of enormous volume of ultramafic and mafic lava and related hydrothermal activity during the formation of the Caribbean plate are considered the main trigger for the massive deposition of black shales across the Cenomanian-Turonian boundary (OAE 2), not only in the proto-Caribbean and the Gulf of Mexico, but worldwide (e.g., Leckie et al. 2002; Snow et al. 2005; Turgeon and Creaser 2008; Kidder and Worsley 2010; Jenkyns 2010; Blättler et al. 2011; Flögel et al. 2011; Jenkyns et al. 2017; Beil et al. 2018; Joo et al. 2020; Laugié et al. 2021).

Later in the middle Campanian, the motion of the newly-formed Caribbean plate caused the end of the Cretaceous passive margin stage in northern South America, as its collision with the northwestern margin of the South American plate began at ~80–75 Ma. That collision of plates marked the end of the long period of undisturbed sedimentation promoted by the absence of nearby plate boundaries (e.g., Parnaud et al. 1995a; Lugo and Audemard 2021).

2.4 Origin of the Caribbean Plate

For a better understanding of the evolution and present-day tectonic setting of Northern Venezuela, a review of the origin of the Caribbean plate (CP) is necessary. This plate is an oceanic plateau (OP), i.e., a large mafic igneous province comprising oceanic crust that is much thicker than 5–7 km formed by normal seafloor spreading (e.g., Condie and Abbott 1999). OPs form when a mantle plume head ruptures the lithosphere and erupts voluminous basalt. These outpourings are often referred to as LIPs (e.g., Whattam and Stern 2015; Riel et al. 2023). The CP, also called by many authors as the Caribbean OP (COP; e.g., Sinton and Duncan 1997; Hastie and Kerr 2010) or Caribbean-Colombian OP (e.g., Kerr et al. 2004; Wright and Wyld 2011), conforms the oceanic floor of the Caribbean Sea. It originated from intensive plume-derived magmatism that led to its anomalous thickness, which varies between 8 to 20 Km (e.g., Mauffret and Leroy 1997; Driscoll and Diebold 1998; van der Lelij et al. 2010), reaching a maximum thickness of ~26 km in the Aves Ridge (Christeson et al. 2008). Overall, magmatic activity (plume- and subduction-arc-derived) associated with the formation of this thick OP occurred from the Early Cretaceous (~ 140 Ma, Berriasian) to the early Paleogene (e.g., Whattam and Stern 2015).

Figure 2.6 shows the eastern region of the CP and its main features. In this region, which also corresponds to the Venezuelan Basin, the CP is rimmed on the north by the Greater Antilles (Cuba, Haiti, Dominican Republic, Jamaica, and Puerto Rico), on the east by the Lesser Antilles (a group of around 19 islands and archipelagos that includes Grenada, Martinique, Dominica, Guadalupe, Montserrat, and Barbados), and on the south by the Leeward Antilles (the Venezuelan Antilles, Aruba, Curaçao, and Bonaire; see Fig. 1.2). The Aves and Beata ridges, to the east and west, respectively, are major bathymetric highs. Most of the Greater Antilles contain magmatic arc assemblages of Early Cretaceous through Eocene ages (Pindell et al. 2006). The Lesser Antilles island arc contains the modern Caribbean Great Arc produced by type-B subduction of the Atlantic Ocean floor beneath the eastern margin of the CP (see Ughi 2021). The Tobago Basin is a forearc basin (Aitken et al. 2011). Many authors state that the Grenada Basin is a back-arc basin formed after rifting of the Lesser Antilles island arc (e.g., Bouysse 1988; Pindell and Kennan 2009; Padron et al. 2021), and whose opening was completed at the beginning of the Late Eocene (Garrocq et al. 2021). An alternative model suggests that this basin and the Tobago Basin were originally a single forearc basin that was subsequently divided by an uplifted and intruded Lesse Antilles arc during the Early to Middle Miocene (Aitken et al. 2011). The Aves Ridge is an extinct volcanic arc segment of the Caribbean Great Arc, composed of arc magmatic rocks ranging in age from ~80 to 59 Ma (Bouysse 1988; Neill et al. 2011; Wright and Wyld 2011). The Beata Ridge (up to 20 km thick) divides the Caribbean into the Colombian Basin to the west (not shown in Fig. 2.6) and the Venezuelan Basin to the east and has a Late Cretaceous COP origin (Révillon et al. 2000). The southern margin of the Eastern CP comprises a complex tectonic area delimited to the north by the Leeward Antilles subduction and to the south by northern mainland Venezuela and

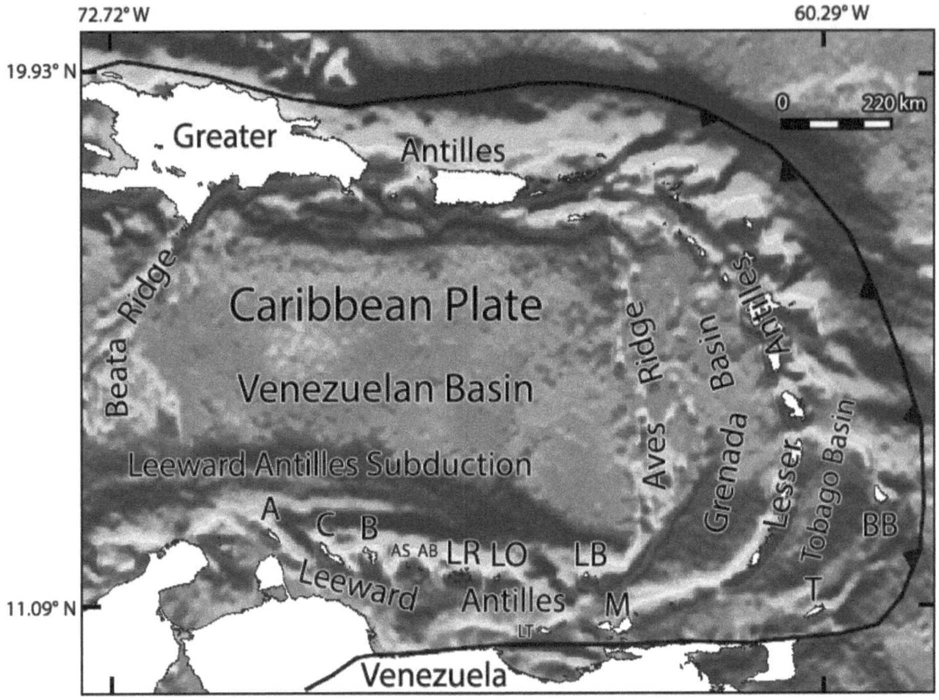

Fig. 2.6 Shaded relief physiography of the eastern Caribbean region, showing basic features, locations, and plate boundaries. Abbreviations: *A* Aruba, *AB* Aves de Barlovento, *AS* Aves de Sotavento, *B* Bonaire, *BB* Barbados, *C* Curaçao, *LB* La Blanquilla, *LO* La Orchila, *LR* Los Roques, *LT* La Tortuga, *M* Margarita, *T* Tobago. (Original figure from Smith and Sandwell 1997. Figure modified from Wright and Wyld (2011)

northern Trinidad (Fig. 2.6). This tectonic setting hosts many of the carbonate units considered in this book and will be further discussed in detail.

The Pacific-derived hypothesis is the most accepted for the origin of the CP. This hypothesis proposes that this OP was formed in the eastern Pacific region, on the Farallon plate, and then drifted into its present position between the two Americas (e.g., Pindell and Dewey 1982; Ross and Scotese 1988; Pindell 1994; Pindell and Kennan 2009; Whattam and Stern 2015; Escalona et al. 2021a, 2021b). Several authors state that the main pulse of plume magmatism (mafic lavas) that led to the formation of the COP occurred from 94 to 88 Ma in the eastern Pacific oceanic floor (e.g., Sinton et al. 1998; Kerr et al. 2004; Snow et al. 2005; Nerlich et al. 2014), when it erupted from the Galápagos hotspot (e.g., Duncan and Hargraves 1984; Hauff et al. 1997; Sinton et al. 1998; Kerr et al. 2004; Pindell and Kennan 2009; Hastie and Kerr 2010; Nerlich et al. 2014; Whattam and Stern 2015). A longer phase of plume magmatism for OP construction (95 to 85 Ma, middle Cenomanian to middle Santonian) is proposed by Whattam and Stern (2015) based on the ages of mafic units in several localities in Central America (NW Costa Rica through Panamá), Ecuador and Colombia, and the Leeward Antilles (Aruba and Curaçao).

Prior to the formation of the COP as a consequence of long-lived plume- and arc-derived magmatic activity during the first half of the Late Cretaceous, a younger magmatic episode and OP construction in the eastern Pacific occurred between 140 and 110 Ma (Hoernle et al. 2004; Whattam and Stern 2015; Riel et al. 2023). This early Cretaceous plume magmatism accounted for the formation of the "old Caribbean plateau" (Riel et al. 2023), which is thought to have been significantly less voluminous (only 8%) compared to the OP magmatism during the Late Cretaceous (Whattam and Stern 2015).

Figure 2.7 shows two prominent models proposed for the evolution of the "Caribbean Great Arc" (or the "Great Arc of the Caribbean") right after the COP was formed. Prior to the main phase of formation of this OP, the eastern margin of the Pacific-Farallon plate was separated from the proto-Caribbean (the region of oceanic crust created by seafloor spreading between North and South America) by the Caribbean Great Arc (see Fig. 2.5). This volcanic arc was initiated during Early Cretaceous times (~ 135 Ma, Valanginian), shortly after the beginning of the formation of the old Caribbean plateau (Riel et al. 2023; see also Pindell et al. 2012; Boschman et al. 2014), by the development of an east-dipping subduction plate boundary, i.e., subduction of the Farallon plate beneath the proto-Caribbean seafloor (e.g., Burke 1988; Boschman et al. 2014; Escalona et al. 2021a; Riel et al. 2023). Thus, the first model (Fig. 2.7a) proposes that the subduction of the Farallon plate beneath the proto-Caribbean suffered a polarity reversal once the COP was formed and then collided against the Great Arc (model based on Burke 1988; Kerr et al. 2003; Thompson et al. 2004). Therefore, the Late Cretaceous Caribbean Great Arc was built on top of the older arc as a result of polarity reversal (Escalona et al. 2021a, 2021b). In the second model

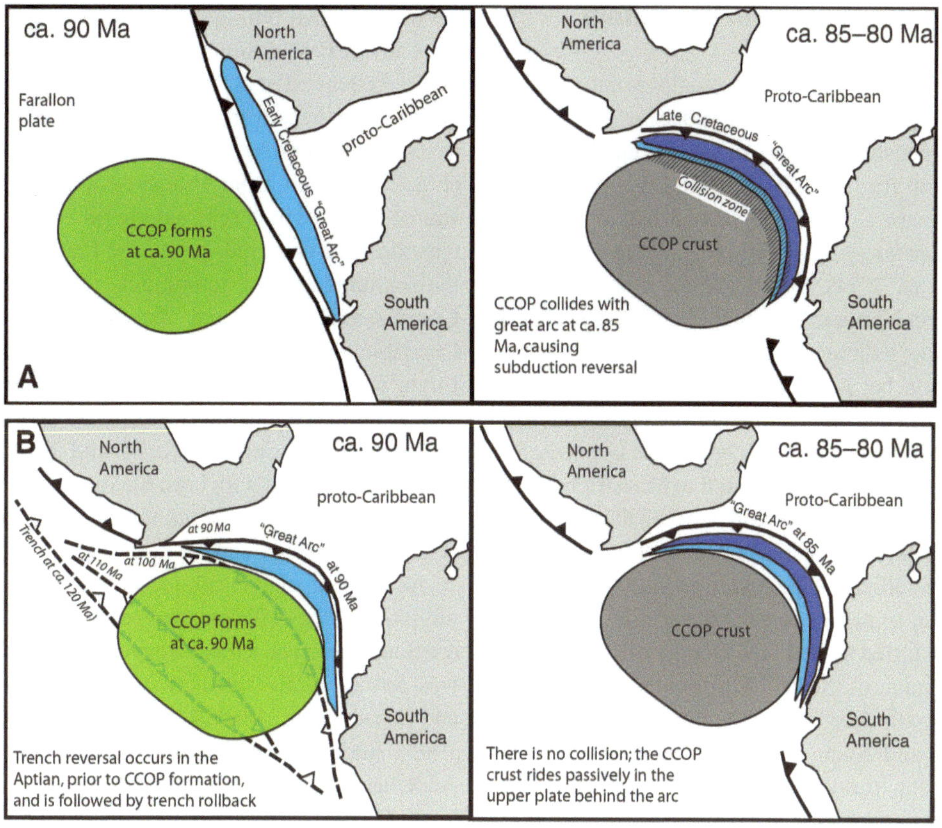

Fig. 2.7 Two prominent models for evolution of the Caribbean Plate (CP), named in this figure as the Caribbean-Colombian Oceanic Plateau (CCOP), and the Caribbean "Great Arc" during the Late Cretaceous. Both models show two time slices, ca. 90 Ma on the left and ca. 85 Ma on the right. (**a**) The Caribbean Great Arc of the Antilles develops above an east dipping subduction zone in the Early Cretaceous. The CCOP mostly forms at ca. 90 Ma on the Farallon Plate which is subducting to the east beneath the Great Arc. When the thick plateau crust collided with the Great Arc in the Late Cretaceous (ca. 85–80 Ma), subduction reversal occurred, and subsequent Great Arc magmatism occurred above a west-dipping subduction zone. (Model based on Burke 1988, Kerr et al. 2003, and Thompson et al. 2004). (**b**) Subduction reversal occurs in the Aptian, at ca. 120 Ma, prior to the development of the CCOP. Post-120 Ma magmatism in the Great Arc occurs above a west-dipping subduction zone. The CCOP forms in the upper plate of this subduction zone, behind the arc, at ca. 90 Ma. There is no collision. (Model based on Pindell et al. 2006. Figure taken from Wright and Wyld 2011)

(Fig. 2.7b), subduction reversal occurred in the early Aptian (~ 120 Ma), prior to the formation of the COP, which did not collide against the Great Arc (model based on Pindell et al. 2006). Although we lean towards the first model, in both scenarios, the COP and its overlying intra-oceanic arc collided and accreted to South America at ~80–75 Ma (Wright and Wyld 2011; Spikings et al. 2014; Escalona et al. 2021b).

Once the bulk of the COP was formed, plume-induced subduction initiation (i.e., westward subduction of the proto-Caribbean seafloor beneath the COP) and arc inception led to a new arc-magmatic phase. Along northwestern South America (Ecuador and Colombia) and the Leeward Antilles, arc-related intrusions are dominated by intermediate and felsic lithologies (Vallejo et al. 2009; Neill et al. 2011; van der Lelij et al. 2010; Wright and Wyld 2011; Whattam and Stern 2015; Baquero et al. 2017; Urbani 2018), although arc-related basalts are present in Ecuador (Vallejo et al. 2009). The age of the transition from OP to arc magmatism varies among different localities. In western Ecuador it occurred by 87–85 Ma (Vallejo et al. 2009; Whattam and Stern 2015), while in western Colombia it started by 92 Ma (Villagómez et al. 2011a). In Aruba and Curaçao, it was initiated by 89 Ma (Wright and Wyld 2011). In Gran Roque Island (northeast Los Roques Archipelago), a locality whose outcropping crystalline rocks are far less studied, the age of the oldest reported arc-related igneous intrusion is 65 Ma (van der Lelij et al. 2010; see also Baquero et al. 2017; Urbani 2018). Lithologies and ages of igneous and meta-igneous rock units cropping out in the Leeward Antilles will be discussed in Sect. 2.11.6.

An alternative hypothesis proposes the *in situ* formation of the Caribbean crust by the generation of multiple oceanic spreading centers between the two Americas since mid-Cretaceous times (e.g., Meschede and Frish 1998; James 2009a, 2009b). We consider the prevailing understanding that the CP formed in the eastern Pacific Ocean and migrated between the Americas.

2.5 Active Margin Stage in Venezuela

The CP has played a key role in the tectonic evolution of the Mesozoic and Cenozoic oil-producing basins in northern South America. Since the beginning of the oblique collision of the Caribbean Great Arc and the CP behind it against northwestern South America during the second half of the Late Cretaceous (Campanian, ~ 80–75 Ma) (e.g., Wright and Wyld 2011; Spikings et al. 2014), the still ongoing active margin stage commenced in Northern Venezuela. A marked change in the lithology of the sedimentary facies in the oil-producing basins in Venezuela began by the Maastrichtian (the latest stage of the Cretaceous), predominating organic-lean, siliciclastic-dominated sedimentation throughout the Cenozoic.

During Late Cretaceous (Campanian to Maastrichtian) to early Paleogene times, the interaction between the Caribbean and South American plates evolved from an oblique subduction to an oblique collision (see Audemard 2009). In northwestern Venezuela, transpressional deformations within the Maracaibo Basin started in the Late Paleogene-Early Eocene (e.g., the development of the NW-SE oriented Burro Negro right-lateral strike-slip tear fault in the northeast area of that basin) (Escalona and Mann 2006a, 211). From the early Neogene to Present, transpression (NW-SE trending) has dominated across northernmost Venezuela. Although transpression has been the dominant process at plate boundary scale during most of the Neogene, transtension has also occurred in two pull-apart basins, the Cariaco Trough and the Gulf of Paria (Audemard 2009; see also Escalona et al. 2011; Audemard et al. 2025). The oblique collision and diachronic northeastward (Late Cretaceous to Eocene) and eastward (Eocene to Present) migration of the collision zone (Fig. 2.8) caused, from west to east, the burial of Cretaceous source rocks (which reached the oil window during various episodes that occurred during the Paleogene and Neogene); the exhumation of mountain ranges (the Venezuelan Caribbean Ranges); and the associated development of Cenozoic foreland basins (Escalona and Mann 2006b, 2011; Lugo and Audemard 2021). The progressive eastward migration of the paleo-Orinoco River System to the current position of the Orinoco River and delta plain was also directly associated with both the diachronic migration of the collision zone and the exhumation of mountain ranges (Fig. 2.8; Escalona and Mann 2011).

As a result of the oblique convergence and diachronous migration of the collision zone between the CP and northwestern South America, a broad convergent plate deformation zone was formed along Northern Venezuela, Western Colombia, and Ecuador (e.g., Pindell and Barrett 1990; Lugo and Mann 1995; Audemard and Serrano 2001; Ostos et al. 2005; Audemard 2009; Escalona and Mann 2011). This margin is known as the Caribbean-South American plate boundary zone (PBZ), or the Southern Caribbean PBZ. This zone is a complex tectonically-faulted region consisting of several amal-

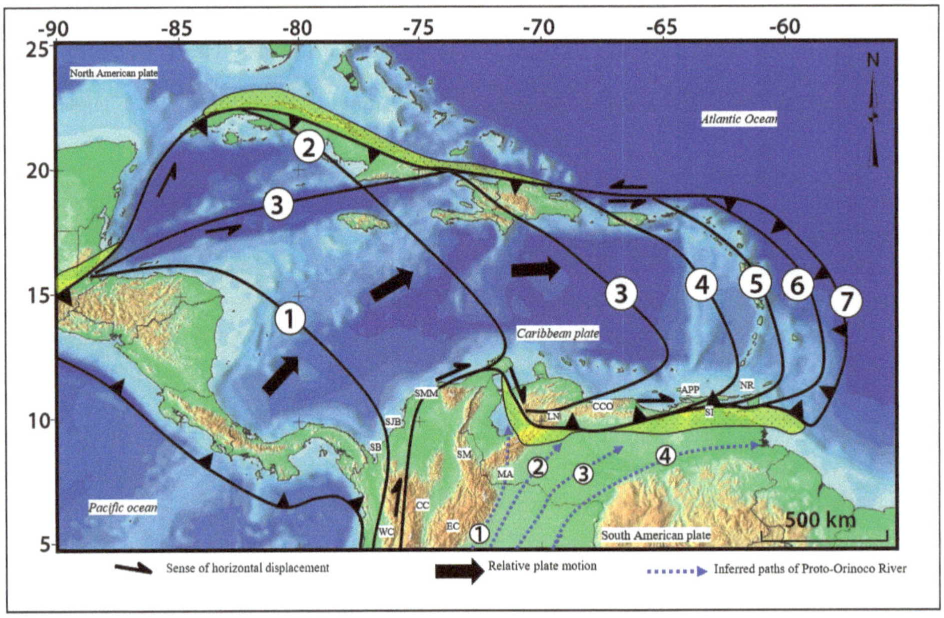

Fig. 2.8 Diachronous northeastward (~ 80–60 Ma) and eastward (~ 60 Ma to Present) displacement of the CP relative to the North and South American plates with numbered, solid black lines representing the inferred locations of the leading edge of the CP at these times: 1 = Late Cretaceous (~ 80 Ma); 2 = Middle Paleocene (~ 60 Ma); 3 = Middle Eocene (~ 44 Ma); 4 = Middle Oligocene (~ 30 Ma); 5 = Middle Miocene (~ 14 Ma); 6 = Pliocene (~ 5 Ma); 7 = Modern. Yellow, stippled areas represent foreland basins produced by diachronous oblique convergence between the Caribbean, North American and South American plates. Abbreviations: *APP* Araya-Paria Peninsula, *CC* Central Cordillera, *CCO* Coastal Range, *EC* Eastern Cordillera, *LN* Lara Nappes, *MA* Mérida Andes, *NR* Northern Range of Trinidad, *SB* Sinu Belt, *SI* Interior Range, *SJB* San Jacinto Belt, *SM* Santander Massif, *SMM* Santa Marta Massif, *WC* Western Cordillera. Blue, dashed lines represent inferred paths of the proto-Orinoco River at the following times: 1 = Paleocene (~ 60 Ma); 2 = Middle Eocene (~ 44 Ma); 3 = Middle Miocene (~ 14 Ma); 4 = Modern. Original figure modified from Lugo and Mann (1995). Figure slightly modified from Escalona and Mann (2011)

gamated tectonic blocks, different in origin, composition, size, and geometry, whose tectonic histories are linked to the eastward displacement of the CP relative to South America (see van der Lelij et al. 2010; Escalona and Mann 2011; Audemard 2014; Audemard and Castilla 2016).

Present-day Southern Caribbean PBZ hosts several structural blocks (Fig. 2.9) separated by major strike-slip faults. In Northern Venezuela, this PBZ (up to 600 km wide in the west) comprises the Maracaibo Triangular Block, Bonaire Block, and the Venezuelan Caribbean Ranges Block (Audemard, 2002, 2014; Audemard et al. 2004; Audemard and Castilla, 2016). The Maracaibo Triangular Block is an autochthonous fragment of continental crust that includes the Maracaibo Basin (van der Lelij et al. 2010). This block is bounded by the left-lateral strike-slip Santa Marta-Bucaramanga fault in Colombia and right-lateral strike-slip Boconó fault in Venezuela and separated on the north from the Bonaire Block by the right-lateral strike-slip Oca-Ancón fault. The Venezuelan Caribbean Ranges Block is bounded to the north and west by a large right-lateral strike-slip fault system and the Caribbean ranges and Venezuelan Andes (the Interior, Eastern Interior, and Coastal ranges in the north, and the Venezuelan Andes in the west; Fig. 2.10) and by a large thrust boundary in the south. The Bonaire Block is bounded to the north by the Southern Caribbean Deformed Belt. This block consists of different basement provinces, which include igneous and metaigneous rocks of CP-affinity in the Venezuelan Antilles; late Paleozoic allochthonous terranes accreted to the northwestern edge of South America during the Late Cretaceous to Paleogene (e.g., the rock basement of the Perla Field); and autochthonous Mesozoic and Cenozoic sedimentary rocks in the Western Gulf of Venezuela sub-Basin (WGVSB), north of the Oca-Ancón fault (Baquero, 2015; Baquero et al. 2015; Navas-Ramírez, 2018; Lugo and Audemard, 2021).

All the above-mentioned indicates that a large area of Northern Venezuela and the whole Leeward Antilles are tectonically located between two prominent tectonic features within the Southern Caribbean PBZ (Fig. 2.10): (**a**) to the north, the Southern Caribbean Deformed Belt, which is an accretionary prism formed due to the subduction of the CP beneath the northern margin of the Bonaire Block, causing the Leeward Antilles subduction; this south-dipping subduction has been active since at least the Late Oligocene (~ 28-26 Ma; Baquero, 2015); (**b**) to the south, a large right lateral strike-slip fault system, which is active since the Early late Miocene (~ 17-15 Ma in the west and becomes progressively younger towards the east); it consists, from west to east, of the Oca-Ancón, San Sebastián, and El Pilar fault system, with the posterior incorporation (~ 5-3 Ma) of the NE-SW Boconó fault (Audemard et al. 2004; Audemard, 2009; Fournier et al. 2017).

During Cenozoic times, the migrating oblique collision between the CP and northern Soth America have caused the

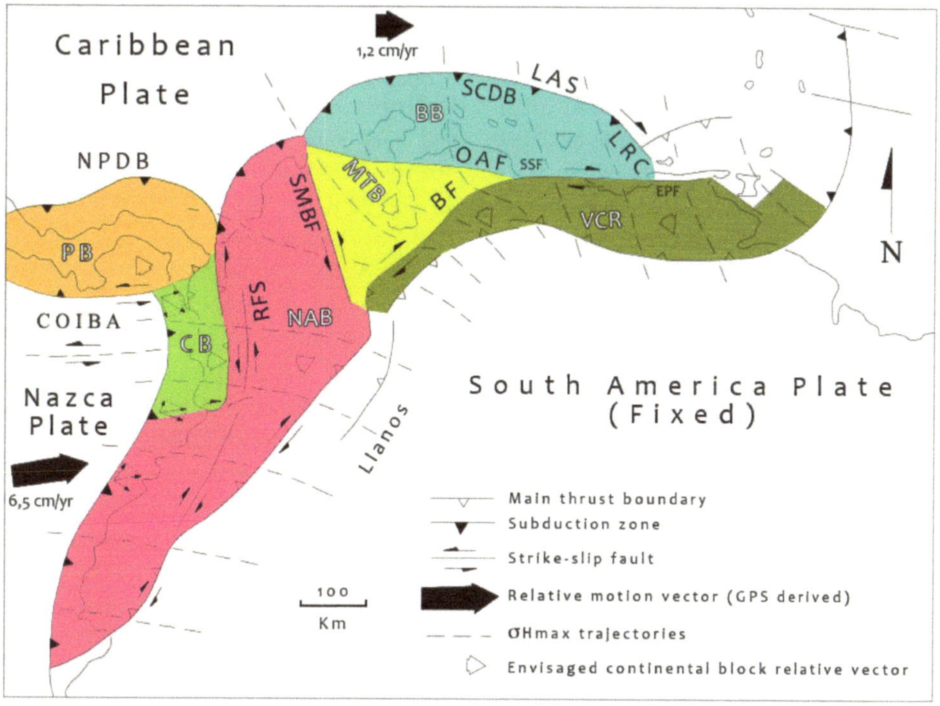

Fig. 2.9 Tectonic blocks in the Caribbean-South America plate boundary zone (PBZ) defined by Audemard (2002) from geologic data and geodynamic criteria. Large solid black arrows are plate vectors and hollow arrows are envisaged continental block vectors, both relative to South America. Major tectonic blocks within the PBZ are: Bonaire Block (BB), Chocó Block (CB), Maracaibo Triangular Block (MTB), North Andean Block (NAB), Panamá Block (PB), Venezuelan Caribbean Ranges (VCR). Major faults reported are: Boconó (BF), El Pilar (EPF), Oca-Ancón (OAF), Romeral (RFS), Santa Marta-Bucaramanga (SMBF), San Sebastián (SSF). Other features are: Leeward Antilles subduction (LAS), Los Roques Canyon (LRC), North Panamá Deformed Belt (NPDB), Southern Caribbean Deformed Belt (SCDB). Figure modified from Audemard (2014). After Franck A. Audemard, personal communication (2020)

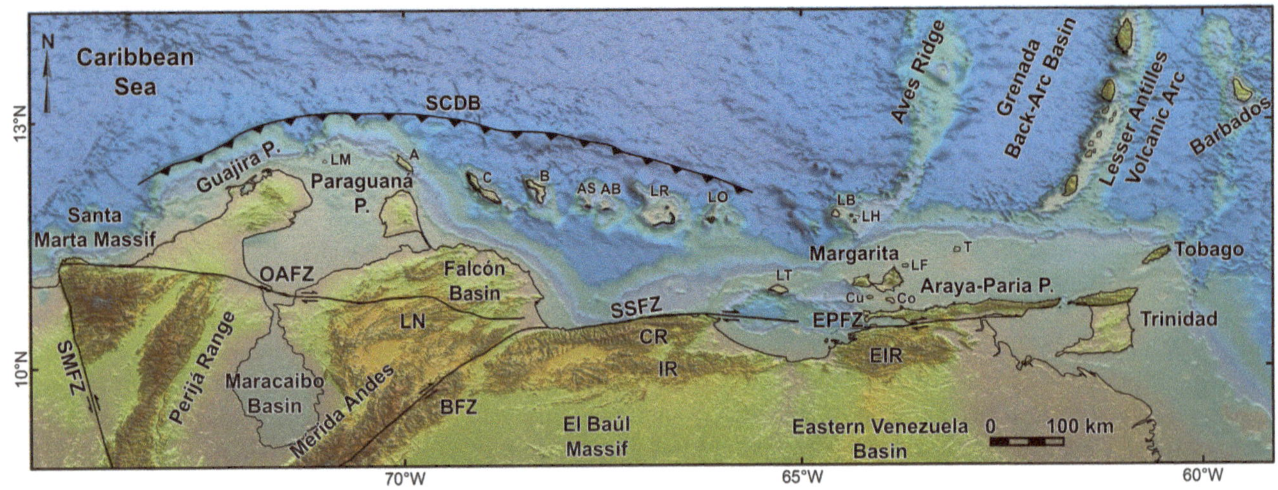

Fig. 2.10 Digital elevation map of northernmost South America (modified from the Shuttle Radar Topography Mission obtained by NASA—see Farr et al. 2007; for details) showing the main tectonic features and fault systems within the Southern Caribbean PBZ. Mountain ranges in northern Venezuela and main offshore islands in the Southern Caribbean PBZ are indicated. Abbreviations: *A* Aruba, *AB* Aves de Barlovento, *AS* Aves de Sotavento, *B* Bonaire, *BFZ* Boconó Fault Zone, *C* Curaçao, *Co* Coche, *CR* Coastal Range, *Cu* Cubagua, *EIR* Eastern Interior Range, *EPFZ* El Pilar Fault Zone, *IR* Interior Range, *LB* La Blanquilla, *LF* Los Frailes, *LH* Los Hermanos, *LM* Los Monjes, *LN* Lara Nappes, *LO* La Orchila, *LR* Los Roques, *LT* La Tortuga, *SCDB* Southern Caribbean Deformed Belt, *SMFZ* Santa Marta Fault Zone, *SSFZ* San Sebastián Fault Zone, *OAFZ* Oca-Ancón Fault Zone, *P* Peninsula, *T* Los Testigos. (Figure modified from Founier et al. 2017)

diachronous exhumation of mountain ranges from west to east in Northern Colombia (the Santa Marta Massif) and Northern Venezuela (the Perijá, Coastal, Interior, and Eastern Interior ranges) (Fig. 2.10). Deformational episodes started in Colombia with the first exhumation pulse of the Santa Marta Massif during the Paleocene (~ 65–58 Ma; Villagómez et al. 2011b), or during the Middle Paleocene to Middle Eocene (~ 60–45 Ma; Patiño et al. 2019), followed by pulses at 50–40 Ma, 40–20 Ma and 25–16 Ma (Villagómez et al. 2011b; Patiño et al. 2019). These orogenic pulses were roughly coeval to the emplacement of the Lara Nappes (an imbricated thrust system of Caribbean-affinity with SSE vergence overthrusting Cretaceous passive margin units) south of the Falcón Basin from the Early Paleocene to the Early-Middle Eocene (Audemard 2009; Baquero et al. 2009; Baquero 2015). The first exhumation pulse of the Perijá Range (an orogen consisting of autochthonous rocks with maximum elevations of 3630 m) occurred during the Early to Middle Eocene (~ 53–45 Ma), followed by pulses in the Late Oligocene-Early Miocene (~ 27–22 Ma) and the Pliocene-Early Pleistocene (~ 5–3 Ma) (see Kellogg 1984; Shagam et al. 1984; Bermúdez et al. 2019). Later in the Early late Miocene (~ 17–15 Ma), and coeval to the inversion of the Falcón Basin and the exhumation of the Falcón Anticlinorium, dextral wrenching along the Southern Caribbean PBZ started in western Venezuela with the activation of the Oca-Ancón fault system and progressed eastward to the El Pilar fault by ~12–10 Ma (Audemard 2009; see also Audemard et al. 2025). The main exhumation pulse of the Coastal Range (allochthonous rocks, maximum elevations of 2765 m) and the Interior Range (para-autochthonous rocks) in north-central Venezuela occurred during the Early to Middle Miocene (~ 19.7–13.6 Ma) (Sisson et al. 2005). To the east, the exhumation of the Eastern Interior Range (para-autochthonous rocks, maximum elevations of 2660 m) and the Araya-Paria Peninsula (allochthonous rocks) occurred in the Middle Miocene to the Early Pliocene (~ 13–5 Ma) (see Locke and Garver 2005; Ostos et al. 2005; Cruz et al. 2007; Arkle et al. 2021).

Towards south, the exhumation the Santander Massif, located in the Colombian-Venezuelan border (see Figs. 1.2 and 1.4), started at slow rates in the Middle Eocene (~ 40 Ma), and then accelerated from the Early Miocene (~ 18 Ma) until the Pliocene (van der Lelij et al. 2016b; Amaya et al. 2017). Regarding the main orogenesis during Cenozoic times in Venezuela, the exhumation of the Venezuelan Andes started in the Late Miocene (~ 10 Ma). Uplift of this NE-SW-trending orogen, with maximum elevations of 4980 m, has been related to a complex geodynamic interaction between the CP, the Panamá Arc, the South American plate, and the continental Maracaibo Triangular Block. A second exhumation pulse of the Venezuelan Andes in the Pliocene (~ 5–3 Ma) (see Audemard and Audemard 2002; Audemard 2003; Bermúdez et al. 2010, 2011) was coeval to the last stages of the Panamá arc collision and suturing against northwestern South America (Audemard 2009), accounting for the final closure of the Panamá Isthmus (Lunt et al. 2008). Furthermore, the Andean orogenesis led to the partitioning of western Venezuela into the modern Maracaibo and Barinas-Apure foreland basins.

Present-day tectonics of northern South America is complex, since involves the interaction between the oceanic Caribbean and Nazca plates and the continental South America plate with several minor blocks (Panamá, Chocó,

North Andean, Bonaire, Maracaibo and Venezuelan Caribbean Ranges). Furthermore, northeast mainland Venezuela lies tectonically near the southern margin of the subduction zone between the oceanic Caribbean and Atlantic plates. In addition, a wide consensus establishes that the CP moves east at about 20 mm/yr relative to South America (see Christeson et al. 2008 and references therein).

2.6 Maracaibo Basin: Tectonic Setting and Cretaceous and Early Paleogene Stratigraphy

The Maracaibo Basin (MB), situated in western Venezuela, is a supergiant basin, in terms of oil reserves and production (Mann et al. 2006). This intermountain lowland region, which contains a thick sedimentary infilling in the subsurface, is bounded by four mountain chains (Fig. 2.11). On the northwest, it is limited by the Perijá Range, on the southwest by the Santander Massif in Colombia, on the southeast by the Venezuelan Andes, and on the northeast by the Trujillo Range. The latter separates the MB from the Falcón Basin. The MB is bounded to the north by the Oca-Ancón fault (Fig. 2.11), which separates that basin from the Maracaibo Province in the Gulf of Venezuela (Navas-Ramírez 2018). Relatively small in size (50,000 km^2) compared to the more extensive Gulf of Mexico giant hydrocarbon province, the MB is one of most prolific hydrocarbon producing regions in the world (Talukdar and Marcano 1994; Goddard and Talukdar 2002; Mann et al. 2006), hosting a large number of giant oil fields (Fig. 2.12). Its sedimentary infilling crops out in the Venezuelan Andes, and the Perijá and Trujillo ranges.

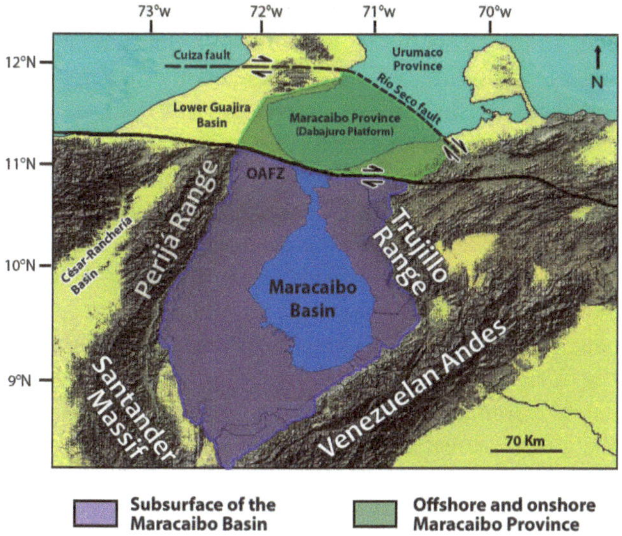

Fig. 2.11 Location map of the Maracaibo Basin (MB), northwest Venezuela. Abbreviations: *OAFZ* Oca-Ancón fault zone. Location of the Cuiza and Río Seco fault according to Rincón et al. (2007); Castillo et al. (2017); and Navas-Ramírez (2018). Figure modified from Méndez-Dot et al. (2015)

The crystalline basement of the MB consists of late Paleozoic autochthonous rocks with continental affinity (Feo-Codecido et al. 1984; Baquero 2015). Although this basin is traditionally considered as being bounded to the north by the Oca-Ancón fault, its autochthonous basement extends north of this strike-slip dextral fault (which began to form in the Early late Miocene) into the Maracaibo Province (Fig. 2.12; Baquero 2015; Navas-Ramírez 2018). The latter encompasses the offshore Dabajuro Platform, as well as the onshore Venezuelan Guajira (northern Zulia State) and onshore northwestern Falcón State, which together comprise the WGVSB. Several exploratory wells drilled during the past three decades in onshore Maracaibo Province (Fig. 2.12) have revealed the continuation of the MB Cretaceous passive margin units in that area (the Cogollo Group and the La Luna Formation; see Baquero 2015; Castillo et al. 2017; Lugo and Audemard 2021) at depths ranging from 2900 to 4200 m (9,600 to 13,800 ft). These Cretaceous units also extend across the Dabajuro Platform. The latter is a shallow marine, poorly-studied embayment with water depths between 30 to 50 m thick (Méndez-Baamonde 2007a; Valencia 2016), where the world class source rock, La Luna Formation, is at

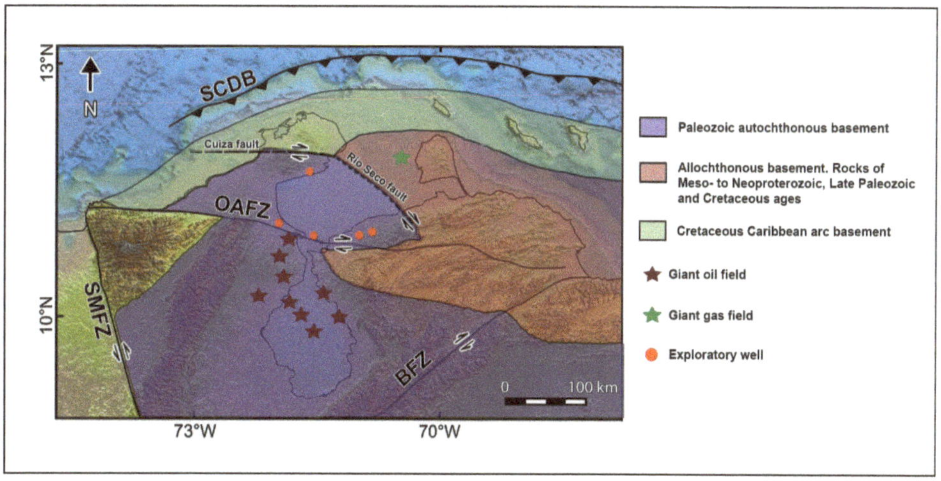

Fig. 2.12 Map showing different basement provinces proposed in western Venezuela and the Leeward Antilles. Abbreviations: *BFZ* Boconó fault zone, *OAFZ* Oca-Ancón fault zone, *SCDB* Southern Caribbean Deformed Belt; *SMFZ* Santa Marta fault zone. (Basemap modified from Fournier et al. 2017. Distribution of basement provinces slightly modified from Navas-Ramírez 2018)

~4500 m deep based on seismic interpretation (Navas-Ramírez 2018). This scenario suggests the presence of effective Late Cretaceous source rocks along the subsurface of the Maracaibo Province, representing great exploratory opportunities for the petroleum industry in Venezuela in that area.

In the Gulf of Venezuela Basin, the Río Seco fault (also known as the Algodones-Río Seco fault) is oriented northwest–southeast and defines the tectonic boundary between the Maracaibo Province of autochthonous basement in the west, and the Urumaco Province of allochthonous basement in the east. This main transfer fault zone (~ 200 km long) connects to the northwest with the Cuiza fault in the Guajira Peninsula (Figs. 2.11 and 2.12; Castillo et al. 2017; Navas-Ramírez 2018). In that area, the Cogollo Group and La Luna Formation extend into the subsurface of Colombia (the Lower Guajira Basin) and also crop out in the Colombian Guajira (e.g., Silva-Tamayo et al. 2016; Avendaño-Sánchez et al. 2021).

Present-day topographic configuration of the Maracaibo foreland basin is derived from successive tectonic episodes that began with the emplacement of the Lara Nappes from the early Paleogene and have continued until Modern times. Progressive tectonic inversion of pre-Cretaceous, rift-related normal faults led to the uplift of the mountain chains that surround the basin (see Kellogg 1984; Lugo and Mann 1995; Audemard and Audemard 2002; Audemard 2003; Escalona and Mann 2006b; Chacín et al. 2005; Escalona and Mann 2011; Amaya et al. 2017).

The Early Cretaceous Maracaibo platform had an estimated length of ~450 km for southwest to northeast and ~ 180 km from southeast to northwest (Vahrenkamp et al. 1993). These dimensions, combined with shallow paleodepths, which increased gradually from east to west and from south to north, led to the deposition of rock units having lateral-regional extension. Cretaceous rock units and lithologies in the subsurface of western Lake Maracaibo and western MB (the latter is the subsurface area between the western margin of Lake Maracaibo and the Perijá Range) are summarized in Fig. 2.13. In the westernmost areas of the MB, the sedimentary infilling of the Machiques Trough (see Fig. 2.4; also known as the Machiques Basin; Nova et al. 2019) crops out in the Perijá Range (Renz 1982; Sarmiento-Rojas 2019). The greater average thicknesses of all Cretaceous passive margin units in Perijá compared to thicknesses of those units in the subsurface are associated with a larger subsidence maintained in the Machiques Trough area (Martínez and Hernández 1992; Vahrenkamp et al. 1993).

Cretaceous passive margin stratigraphy and real thicknesses of rock units ranging in age from Barremian to Santonian are shown in a series of Gamma Ray logs of oil wells across the MB (Figs. 2.14 and 2.15). Total thickness of the Cretaceous strata in well Alpuf-6 (Fig. 2.14; well ALP-6 in Fig. 2.15) is ~440 m (1,450 ft). Such thickness decreases to ~275 m (900 ft) towards the east and southeast and increases towards the southwest, reaching ~650 m (2150 ft) (Fig. 2.15).

2.6.1 Río Negro Formation (Hauterivian to early Barremian)

Siliciclastic facies of the Río Negro Formation mark the final infilling of the Jurassic grabens in western Venezuela and represent the basal shallow marine unit of the Cretaceous passive margin basin (Macellari 1988; Parnaud et al. 1995a). This fluvial and alluvial sedimentation consists of sandstones and conglomerates interfingered with siltstones and shale facies. Río Negro has thicknesses of several hundred meters in the Jurassic grabens, where it unconformably overlies the La Quinta Formation. In the Perijá Range (Machiques Trough), this unit is about 1500 m thick (González de Juana et al. 1980; Ministerio de Energía y Minas 1997; Yoris and Ostos 1997). On the contrary, its thickness is drastically reduced (down to a few tens of meters) in the peneplain topographic depression of the MB (Fig. 2.15). In the Venezuelan Andes, the age of this formation is estimated to range from late Valanginian-early Hauterivian to early Barremian (Macellari 1988). In the subsurface of the MB, this unit is considered to be Hauterivian to early Barremian in age (Murat and Azpiritixaga 1995). The age of the base is diachronic, from older ages in thicker sections in the Perijá Range (probably reaching ages as old as Valanginian), to younger ages in less thicker sections in Lake Maracaibo (Hauterivian to early Barremian).

Porous sandstones in the Río Negro Formation are the deepest Cretaceous hydrocarbon reservoirs in the MB (e.g., Yoris and Ostos 1997; James 2000), reaching depths down to 5500 m (18,200 ft). These facies host porosity values up to 18% and store light oil in the Urdaneta West field, western Lake Maracaibo (Méndez-Dot 2017). Geochemistry of oil samples from the Río Negro reservoir in that oil field suggests that those light oils were generated by La Luna Formation (Escobar et al. 2022). This scenario implies long hydrocarbon migration distances towards greater depths.

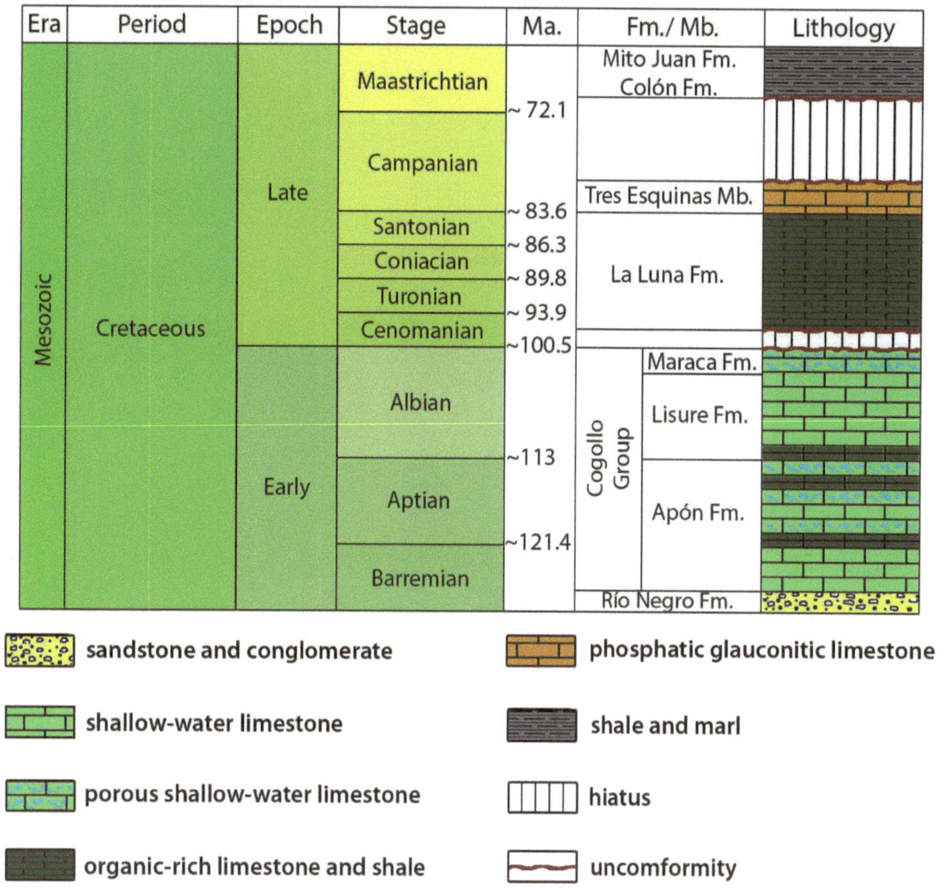

Fig. 2.13 Generalized stratigraphic scheme showing Cretaceous sedimentary units in western Lake Maracaibo and western MB (subsurface and Perijá Range). Abbreviations: *Fm* formation, *Mb* member

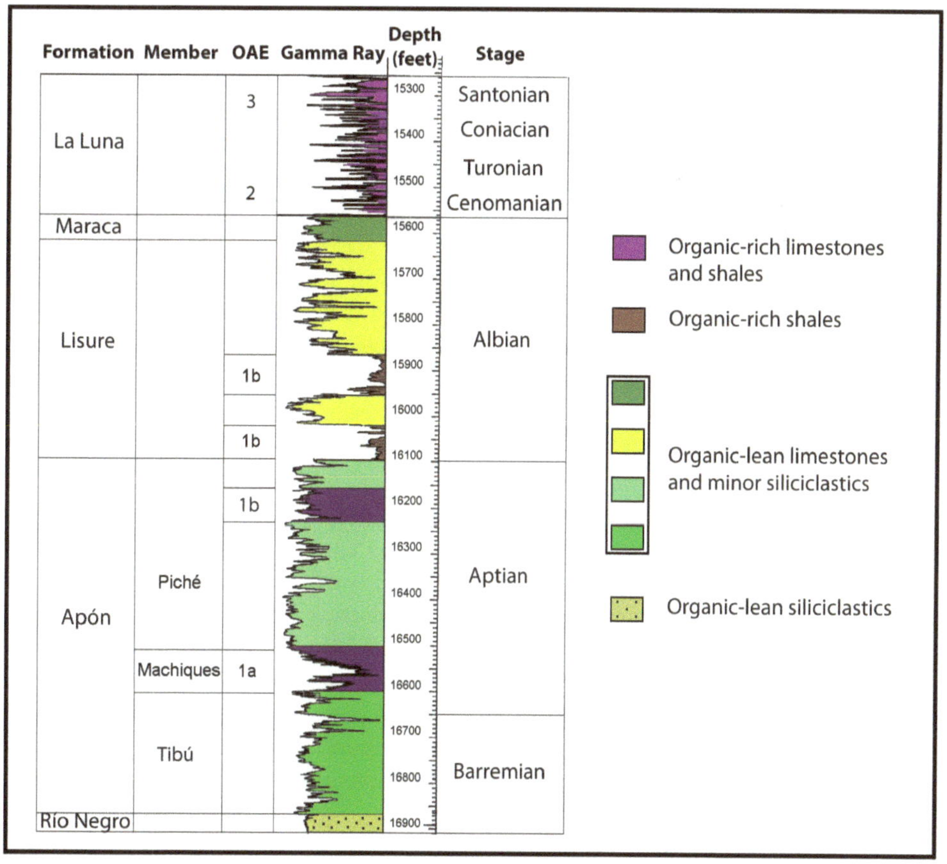

Fig. 2.14 Gamma Ray log register of well Alpuf 6 (western MB) showing differentiated Cretaceous rock units ranging in age from Barremian to Santonian. Stratigraphic location of the Oceanic Anoxic Events (OAEs) 2 and 3 within the La Luna Formation *sensu stricto* according to Davis et al. (1999) and Liborius-Parada (2019)

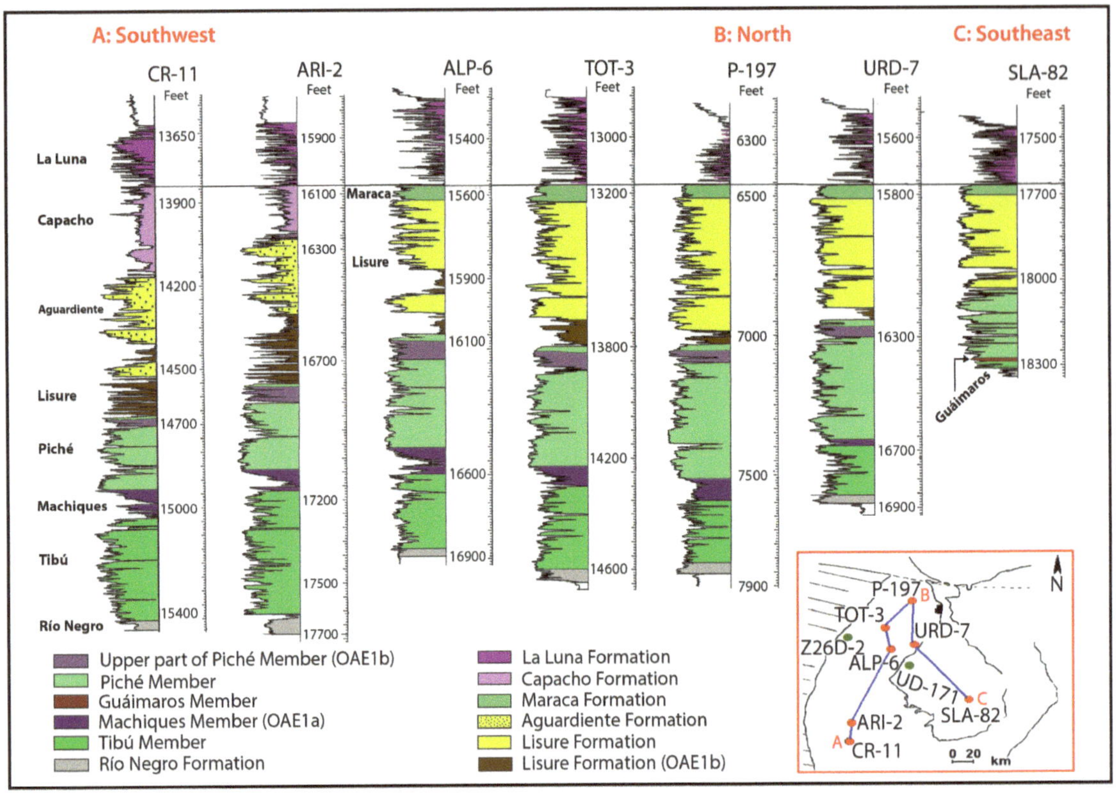

Fig. 2.15 Stratigraphic correlation of Cretaceous rock units within the subsurface of the MB from gamma ray log registers

2.6.2 The Cogollo Group (early Barremian to latest Albian)

Shallow tropical carbonates of the Cogollo Group represent approximately 25 Myr of passive margin sedimentation. This succession was deposited in tropical paleo-latitudes (~ 8–11° N) during the Warm Greenhouse and episodic Hothouse climates that prevailed during the second half of the Early Cretaceous (Barremian, Aptian and Albian stages). These carbonates encompass the subsurface of the MB and also crop out in several localities in the Venezuelan Andes and the Perijá and Trujillo ranges (e.g., Rutsch and Salvador 1954; Renz 1958; González de Juana et al. 1980; Bartok et al. 1981; Murat and Azpiritixaga 1995; Méndez-Dot et al. 2015).

The Cogollo Group consists of three formations: the Apón Formation (consisting of the Tibú, Machiques and Piché members) of early Barremian to late Aptian age, the Lisure Formation of early Albian to late Albian age, and the Maraca Formation of latest Albian age (Figs. 2.13 and 2.14). Deposition of this carbonate succession began in a shallow, distally steepened ramp in the early Barremian. This depositional setting evolved to an epeiric platform in the early-late Aptian, and became drowned during Cenomanian times. This rich variety of carbonate lithofacies, deposited during high ESL, consist mainly of transgressive and highstand deposits. Low-amplitude sea-level fall episodes are recorded in a few grain-supported intervals that underwent subaerial exposure, leading to the creation of secondary porosity and/or the precipitation of void-filling cements. In some oil fields, deep-burial (mesogenetic) dissolution played a significant role in the creation of secondary porosity (Pöppelreiter et al. 2005). Interfingering of siliciclastic sediments are common.

Black shale intervals within this shallow carbonate succession were deposited during maximum flooding events linked to episodic Hothouse climate. These organic-rich intervals are the sedimentary expression of the carbon cycle perturbations linked to the OAEs 1a and 1b. In this book, the photographic examples shown belonging to formations and members of the Cogollo Group correspond to several cores located in the subsurface of western MB (e.g. cores Z26D-2, TOT-3 and ALP-6), and Lake Maracaibo (e.g., core UD-171; Fig. 2.15).

In the subsurface of the MB, the Cogollo Group consists of inner-middle ramp facies of Barremian age and inner-middle platform facies of Aptian and Albian age. Time-equivalent, outermost facies of this carbonate setting were deposited in what is nowadays northeast and northernmost

Colombia. Those facies extend into the Lower Guajira and César-Ranchería basins, north and west of the Perijá Range, respectively (see Fig. 2.11) (see Martínez and Hernández 1992; Silva-Tamayo et al. 2016; Patarroyo and Götz 2020; Vargas-Escudero et al. 2021).

Carbonate production in this shallow, epeiric sea occurred almost exclusively in the upper photic (or euphotic) zone. In the overall, benthic carbonate production was dominated by heterozoan biota, mainly molluscs (aragonitic bivalves, gastropods and oysters) and echinoids, with the secondary contribution of green algae (photoautotrophs) and LBF (mixotrophs, i.e., heterotrophs containing photosymbionts), as well as other heterotrophic calcifiers such as smaller benthic foraminifera (SBF), bryozoans, ostracods, serpulids, and sponge spicules. Furthermore, the neritic lime-mud production was significant, as mudstones, wackestones and muddy-packstones dominate the lithology throughout the succession. Such predominance of lime-mud as component of the tropical, shallow-water carbonate factory also includes automicrites and other biologically-induced micritic precipitates (e.g., microbial crusts) and non-skeletal grains (micritic and glauconitic peloids, ooids, oncoids, cortoids, aggregates and clasts). These components are typical of many Early Cretaceous, warm (CO_2-rich waters) tropical platforms. Abiotic carbonate precipitation is represented by low-Mg calcite in the form of cements (void-filling porosities) or as sparry calcite of neomorphic origin.

In the overall, this long-lasting (~ 25 Myr) shallow carbonate succession is characterized by the scarcity of corals, which by pre-Albian times were still common-to-dominant reef builders in euphotic carbonates deposited under oligotrophic conditions; and the abundance of both heterotrophic calcifiers and micrite-dominated fabrics (in many cases organic-rich). The latter suggests that trophic conditions varied widely between mesotrophic and eutrophic. Mesotrophic conditions favored the flourishing of calcareous green algae (see Hallock 2001; Mutti and Hallock 2003), while variations between meso-eutrophic to fully eutrophic conditions favored the dominance of heterotrophic biota and microbially-induced carbonate production. Furthermore, hypertrophic conditions (i.e., significant phytoplankton densities in the water column; see Mutti and Hallock 2003) during deposition of organic-rich rock intervals favored the occurrence of both opportunistic planktonic foraminifera that thrived in the upper photic zone and low-oxygen (symbiont-barren) SBF. For all above mentioned, the three main modes of carbonate precipitation, i.e., biologically-controlled, biologically-induced and abiotic (see Appendix A for a detailed treatment of carbonate precipitation modes) contributed to sedimentation of shallow facies of the Cogollo Group. The latter is indicative of the wide spectrum of environmental factors that controlled deposition of this carbonate succession.

2.6.2.1 Apón Formation, Tibú Member (early Barremian to earliest Aptian)

The Apón Formation represents the beginning of carbonate sedimentation deposited during the Cretaceous passive margin stage in the MB (Figs. 2.13 and 2.14). Its sedimentation encompassed from the early Barremian to the entire Aptian Stage and it began with deposition of the basal Tibú Member. The latter represents the transition from siliciclastic facies of the Río Negro Formation to carbonate-dominated sedimentation. Arenaceous limestones that lack of calcareous biota characterize the gradual contact between Río Negro and Tibú. The lithology quickly grades to a wackestone characterized by the appearance of the larger benthic foraminifer *Choffatella decipiens* at the lower and middle parts of the Tibú Member across the entire basin. The latter according to the identification of this species in a large number of cores (Bartok et al. 1981; Azpiritxaga 1991; Vahrenkamp et al. 1993; Murat and Azpiritixaga 1995; this book).

The species *C. decipiens* represents the only biostratigraphic marker for the beginning of carbonate sedimentation in the MB. The stratigraphic position of the "Choffatella level", which is up to 32 m thick (or 105 feet) in the subsurface of western MB, coincides with the occurrence of this species in Barremian facies of the Barranquín Formation (Carvajal-Chitty and Rey 2020). The latter is the Cretaceous basal unit in the Eastern Venezuela Basin, as well as the time and lithologic equivalent unit of the Río Negro and the lower part of the Tibú Member. We estimate an early to middle Barremian age (~ 126–123 Ma) for the Choffatella level in the subsurface of the MB. Prior to this book, the precise stratigraphic position of the former Barremian-Aptian boundary (formerly placed at 125 Ma in the ICC of year 2021; Cohen et al. 2013a, b, c; updated v2021/10) in the uppermost part of the Río Negro Formation or the lowermost part of the Tibú Member remained undetermined. The latter is due to the absence of both calcareous biota in the former siliciclastic unit and planktonic markers in the latter carbonate unit (Renz 1982; Murat and Azpiritixaga 1995). In 2022, the Barremian-Aptian boundary was taken at ~121.4 Ma by Cohen et al. (2013a, b, c; updated v2022/02), which is the same age for that boundary in the Cretaceous Time Scale of Gale et al. (2020). The latter places the Barremian-Aptian boundary in the MB at the upper part of the Tibú Member (see Fig. 2.20).

The lithology of the Tibú Member is dominated by wackestone, packstones and mudstones containing molluscs (mainly aragonitic bivalves and oysters, plus gastropods), echinoids and SBF, as well as secondary ostracods, coralline algae and serpulid tubes. Scarce green algae are present towards the west of the basin (e.g., cores Alpuf-6 and Z26D-2). The upper part contains partially dolomi patches of dolomitic mudstones and wackestones and moderately organic-rich matrices. The sedimentation shows a clear transgressive trend, as mud-supported fabrics dominate the lithology, although packstones and grainstones can be locally common (these proportions vary across the basin). Thicknesses of the Tibú Member are very variable, from more than 215 m (~ 700 ft) in the west of the basin, to approximately 10 m (~ 30 ft) in the east, towards ancient shorelines (Fig. 2.15). The geometry and regional subsidence of the distal steepened ramp led to greater thicknesses in western MB, an area with higher accommodation space respect to the central and eastern areas of the basin (Méndez-Baamonde 1989a, 2007b; Méndez-Dot et al. 2015).

The Tibú Member is characterized by the dominance of both heterotrophic calcifiers (the Heterozan factory) and mud-supported fabrics; by the abundance of SBF in the east and their significant decrease towards the west; by the absence of green algae in facies of the east and central areas of the MB and their appearance as scarce components towards west; and by the absence of corals. Meso-eutrophic to fully eutrophic (nutrient-rich) conditions prevailed during deposition of this unit. Runoff-sourced nutrients (accompanied by input of siliciclastic sediments) was a primary controlling factor during deposition of this heterozoan-muddy ramp. Micritic matrices vary from light-colored in the lower part to moderately dark-colored, plus abundant pyrite framboids in the upper part. The latter suggests micrite calcification via photosynthesis during times of lower-nutrient concentrations (the neritic lime-mud factory), and the contribution microbially-induced micrite precipitation during times of low-oxygen and eutrophic conditions (the benthic automicrite factory).

Furthermore, in the upper part of this member in cores Alpuf-6 and Z26D-2 (Fig. 2.15), there are between two and three thin organic-rich intervals. In both cores, the first interval has TOC content of 6.77% and 3.36%, respectively. The second interval (separated from the beginning of the overlying Machiques Member by 10 m in core Alpuf-6, and by 22 m in core Z26D-2) has TOC content between 1.1% and 1.3%. It highlights such organic enrichments prior to the beginning of the OAE 1a, evidencing the influence of local forcing factors (mainly high runoff-sourced nutrients) during deposition of this unit. Pre-OAE 1a black shales are also well recorded in the western Tethys (northern Italy; Giorgioni et al. 2015).

2.6.2.2 Apón Formation, Machiques-Guáimaros Member (early Aptian)

Conformably overlying the Tibú Member, the Machiques Member of early Aptian age is an organic-rich rock interval having average thickness between ~25 and 40 m in the subsurface of the MB (Méndez-Dot et al. 2015; Méndez-Dot 2017), and more than 50 m in the Perijá Range (Audemard and Serrano 2001). This unit is the sedimentary expression of the OAE 1a (Selli Event; ~ 120 Ma) (Figs. 2.14 and 2.15). In four of the most studied sections of the Selli Event, which are located in the western Tethys (norther Italy and southeastern Spain), the North Atlantic (west of Portugal), and the Gulf of Mexico (northeastern Mexico), this anoxic event had a duration of 1.11 to 1.47 Myr (Li et al. 2008; Malinverno et al. 2010; Martínez-Rodríguez et al. 2024). Towards east and southeast limits of the MB, the Machiques Member grades into less organic-rich shales and dolomitic shales of the Guáimaros Member (Renz 1958; Ford and Houbolt 1963; Murat and Azpiritixaga 1995) (Fig. 2.15). This review is focused on the Machiques Member, since all photographic examples shown belonging to this organic-rich interval correspond to cores located in western MB and western Lake Maracaibo.

The Machiques Member consists of an alternation of organic-rich mudstones, wackestones, shales and dolostones. Shales, calcareous shales and dolomitic shales dominate at the base and lower part of the unit, which contains much higher average TOC content. Less organic-rich mudstones, clayey mudstones, dolomitic mudstones and massive dolostones dominate the upper part. Calcareous biota is dominated by planktonic foraminifera of the genera *Hedbergella*, *Globigerinelloides* (e.g., *G. blowi* and *G. barri*) and *Leupoldina*, which are common in the lower and middle parts of the unit. In early Aptian black shales, the occurrence of *Globigerinelloides blowi* as well as several species of the genus *Leupoldina* (*L. cabri, L. pustulans, L. reicheli*) is globally associated with the OAE 1a (e.g., Premoli Silva and Sliter 1999; Premoli Silva et al. 1999; Verga and Premoli Silva 2002; Premoli Silva and Verga 2004; Coccioni et al. 2006; Sarfi et al. 2014; Castro et al. 2019; Socorro and Maurrasse 2020, 2022). SBF (buliminids) are present but less common. Low-oxygen ammonite assemblages (the majority belonging to the genus *Deshayesites*) in the Perijá Range and the Venezuelan Andes support the early Aptian age of the Machiques Member (Renz 1982). Radiolarians, phosphatized fish remains, sponge spicules, ostracods, molluscs, as well as terrigenous quartz and muscovite are secondary or accessory components. Despite all mentioned above, many mudstones, shales and dolostones are barren in skeletal components.

Deposition of black shales of the Machiques Member was promoted by climate and ocean perturbations related to massive Ontong Java Nui submarine volcanism in the southwestern Pacific during the late Barremian and early Aptian (see Bottini et al. 2015; Erba et al. 2015; Matsumoto et al. 2021; Percival et al. 2024); by a major transgression and formation of a maximum flooding surface (MFS) recognized across the entire MB (Murat and Azpiritxaga 1995; Parnaud et al. 1995a); and by the development of bottom marine water anoxia within an expanded OMZ towards shallow depths into the photic zone (Fig. 2.16) (Méndez-Baamonde 1989a, 2009;

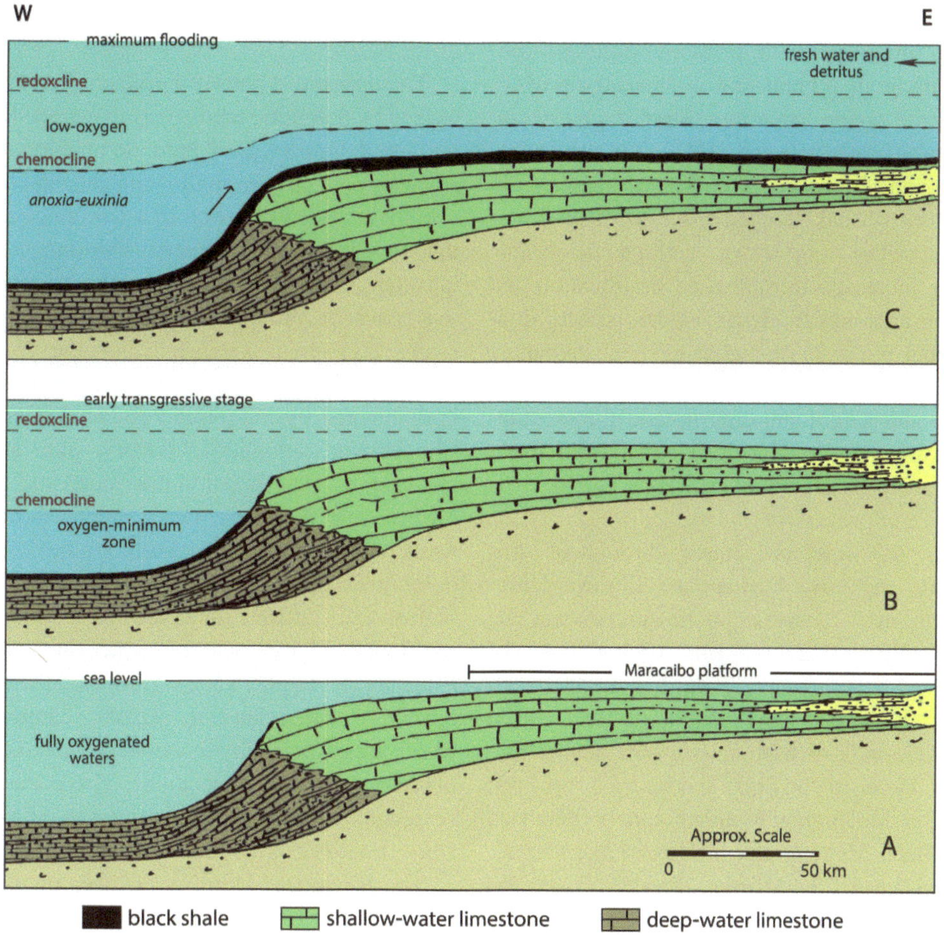

Fig. 2.16 A redox stratified model for depositional environment of black shales corresponding to the OAE 1a (Machiques Member) in the MB. (**a**) A pre-anoxic condition and normal oxygenation of the ocean waters. (**b**) Initiation of the marine transgression followed by development of anoxia in deep-basin waters to the west (W) and low-oxygen conditions towards east (E) on the shallow epeiric platform (**c**) Maximum flooding and expansion of the oxygen minimum zone (OMZ) over the epeiric platform; black shale deposition under anoxic-euxinic conditions promoted by accelerated hydrologic cycle (high influx of freshwater carrying siliciclastic sediments and nutrients) and deceleration of thermohaline circulation. Expansion of the OMZ represented in this model also applies for deposition of all black shales within the Cogollo Group, as well as to the base of the La Luna Formation. *Redoxcline*: water depth interval in which chemical redox potential undergoes its largest change (see Taylor et al. 2006), from upper oxic to lower anoxic waters; lower waters characterized by high concentrations of Fe^{2+} (or ferruginous conditions). *Chemocline*: boundary zone below which a body of water is characterized by the presence of free H_2S, or anoxia-euxinia (see Meyer and Kump 2008). It is noteworthy that, in this model, deposition of black shales to the west of the Maracaibo platform, in what is now northeast Colombia, is hypothetical, as it was not a widespread phenomenon. Black shales corresponding to the OAE 1a are found in the Paja Formation, Eastern Cordillera of Colombia (Gaona-Narvaez et al. 2013a; see also Rivera et al. 2018). Original figure modified from Méndez-Dot et al. (2015). Figure modified from Petrash et al. (2016)

Méndez-Dot et al. 2015; Méndez-Dot 2017). Paleodepths within the mesophotic zone probably never exceeded 80 m. Opportunistic planktonic foraminifera (adapted to stressful conditions; i.e., seawaters charged with excess of volcanogenic CO_2) were able to thrive only in the uppermost part of the water column, as bottom and intermediate waters were depleted in oxygen. In the Cogollo Group, the partial development of "pelagic realm-like" conditions (i.e., flourishing of planktonic foraminifera on shelves or ramps) only occurred during deposition of black shales of the OAEs 1a and 1b.

The main lithologic characteristics of the Machiques Member are the highly variable content (wt.%) of carbonate (between 5–90%), TOC (between 0,1 and 10% in the subsurface, Méndez-Dot 2017; and up to 5,5% in the Perijá Range (Alberdi-Genolet and Tocco 1999), and clay minerals. Kerogen quality (expressed as Hydrogen Index; HI) is also very variable between Type I (microbially-derived; HI > 600), Type II (mixture of microbial- and phytoplankton-derived; HI between 300 and 600), Type II-III (mixture of marine- and terrestrial-derived; HI between 200–300), and Type III (terrestrial-derived; HI between 50 and 200), but with a predominance of Type II kerogen. Such lithologic gradation between organic-rich limestones and shales suggests short-term paleoclimatic events leading to marked changes in the acceleration of the hydrologic cycle. The latter caused variations in the influx of both meteoric (nutrient-rich) waters and clay minerals (continental weathering-derived), and in the carbonate precipitation rate. Stratigraphic variations in the content of carbonate and TOC and in the quality of the organic matter, combined with some organic geochemical indicators (visual kerogen assessment and biomarkers), suggest alternating anoxic and anoxic-euxinic conditions during increased thermohaline stratification and hypertrophication; and suboxic-dysoxic (low-oxygen) conditions during time intervals of higher calcium carbonate precipitation rate. Abundant early diagenetic (authigenic) mineralization products (dolomite, pyrite and fluorapatite) also characterizes this unit. Authigenic calcitic microspheres of microbial origin are also common, as well as silicification processes (selective or pervasive).

Variability in dolomite abundance throughout this organic-rich unit suggests fluctuations in evaporitic-hypersaline conditions. Widespread photic zone anoxia and transient shallow marine euxinia during deposition of the Cogollo black shales was proven by Petrash et al. (2016) on a set of organic-rich samples corresponding to the Cogollo Group black shales (bulk-rock biomarker parameters, rare earth element geochemistry, and iron speciation determined on samples of the Machiques Member and organic-rich intervals in the Piché Member and the Lisure Formation). Biomarker analysis carried by Méndez-Dot (2017) on a set of samples of the Cogollo Group black shales (TOC > 2.5%) supports those interpretations, including the expansion of the OMZ and the shallowing of the chemocline (or O_2-H_2S interface) into the photic zone, and the occurrence of anoxia-euxinia during increased redox stratification (Fig. 2.16). Carbon isotope excursion ($^{13}C_{carb}$) obtained by Espín-Gil (2009) in a set organic rich samples corresponding to Machiques Member may be correlated with high-resolution curves obtained in several localities worldwide which represent the overall global carbon isotopic signature of the OAE 1a (e.g., Menegatti et al. 1998; Bralower et al. 1999; Leckie et al. 2002; Gaona-Narvaez et al. 2013b; Naafs et al. 2016; Castro et al. 2019; Socorro and Maurrasse 2020).

Shallow depths within the photic zone, combined with tropical paleolatitudes, promoted high rates of sedimentation and organic-matter production, leading to deposition of this expanded organic-rich black shale succession. The thickness of the black shale succession of the Machiques Member as a record of the OAE 1a is comparable to several expanded sections corresponding to that anoxic event in Spain (de Gea et al. 2008; Najarro et al. 2011; Quijano et al. 2012; Gaona-Narvaez et al. 2013b; Aguado et al. 2014; Naafs et al. 2016; Ruiz-Ortiz et al. 2016; Sánchez-Hernández and Maurrasse 2016; Castro et al. 2019, 2021; Socorro and Maurrasse 2020, 2022; Herdocia and Maurrasse 2023) and northeast Mexico (Bralower et al. 1999; Li et al. 2008; Núñez-Useche et al. 2015, 2020).

The uppermost part of the Machiques Member is characterized by dolomitic mudstones that grade towards the top to massive dolomites. Such increase in dolomite crystal content suggests both upward shallowing paleodepths and increasing salinities, most likely due to a low-amplitude sea-level fall that reduced the bathymetry. This late transgressive-early highstand deposits (according to the definition of "Deposits" in a Sequence Stratigraphy context by Catuneanu et al. 2011) represents a regional sequence boundary.

As for carbonate benthic production, this black shale unit is dominated by microbially-induced micritic boundstones (the benthic automicrite factory, plus associated microspheres), as well as dolomitic boundstones and dolostones formed due to microbially-influenced dolomitization of micritic-dominated facies. Furthermore, dolomite crystals in dolomitic shales are also microbially-influenced in origin (se Petrash et al. 2016). Calcareous biota is mainly represented by planktonic foraminifera, which are components of the P-factory of Reijmer (2021) (in which the letter P stands for "planktic" and relates to planktic carbonate production). These are occasionally associated with silica-secreting radiolarians (planktic association). The Heterozoan factory is scarcely to moderately represented by SBF, ammonites, and some low-oxygen molluscs.

The Machiques Member has been previously reported as a possible source rock in the western MB by Alberdi-Genolet and Tocco (1999), James (2000), and Goddard and Talukdar (2002). Organic geochemistry maturation parameters indicate that this rock interval is an effective source rock in western MB and western Lake Maracaibo. Its stratigraphic location suggests that, at basin scale, this bituminous unit has contributed to the generation of light oil stored in underlying sandstones of the Río Negro Formation and in overlying packstones of the Piché Member (Méndez-Dot et al. 2015; Méndez-Dot 2017).

2.6.2.3 Apón Formation, Piché Member (early-late Aptian to latest Aptian)

Overlying the Machiques Member, the Piché Member consists of a classic example of alternations of transgressive (mud-supported fabrics) and highstand (grain-supported fabrics) deposits. The variety of skeletal components includes green algae (Bryopsidales and Dasycladales), aragonitic bivalves, gastropods, oysters (calcitic bivalves) and less abundant rudists, SBF (mainly smaller miliolids), echinoids, bryozoans, ostracods, serpulids and sponge spicules. Micritic peloidal grains are abundant, while glauconite peolids are present but less common. Dolomitic mudstones and massive dolomites are locally common at the lower part of the succession (Méndez-Baamonde 1989a, 2007b).

The regional extension of the shallow epeiric platform setting is expressed in the homogeneous thickness of the Piché Member across the subsurface of the MB, with an average of 100–120 m thick (~ 330–400 ft) (Fig. 2.15). This unit was deposited during relatively high and stable sea level prevailing during the Aptian (Haq 2014) that promoted deposition of a carbonate succession consisting of transgressive-regressive ("T-R") sequences (also referred to as "T-R cycles"; see Catuneanu et al. 2011). Mud-supported fabrics (wackestones and less common mudstones) and dolomites at the lower parts of the sequences grade to grain-supported fabrics (mainly packstones) at the upper parts. This general transgressive trend is alternated by thinner intervals affected by forced regressions, as subaerial exposure of the upper part of highstand deposits during low-amplitude sea-level fall episodes led to the creation of secondary porosity and cementation (porosity-destructive) processes. These diagenetically altered grain-supported fabrics represent maximum regressive surfaces (or sequence boundaries) that bound the sequences. In some oil fields in Lake Maracaibo, such late highstand rock intervals with secondary porosities (which depending on the core in the subsurface vary from 3 to 6 intervals across the basin) store light oil generated by both underlying Machiques Member and overlying late Aptian and early Albian black shales.

In some cores belonging to oil wells located in western Lake Maracaibo (e.g., core UD-171) and western MB (e.g., core Z26D-2), there are several thin organic-rich horizons of early-late Aptian age with TOC content ranging from 1% to 3%. These black shale horizons, consisting mainly of mudstones and dolomitic mudstones, are stratigraphically located between the Machiques Member (early Aptian) and the organic-rich interval located at the upper part of the Piché Member (late Aptian). Consistently, a series of black shale horizons stratigraphically located between the organic-rich intervals that correspond to the OAEs 1a and 1b are reported worldwide (e.g., Föllmi 2012; Núñez-Useche et al. 2020). In the western Tethys, early-late Aptian black shale horizons have been grouped under the name of Fallot Episode (see Föllmi 2012).

In the middle and middle-upper parts of the Piché Member, distinctive conical tests of the genus *Mesorbitolina* (probably *M. texana*) become abundant in a rock interval of about 5–10 m thick (~ 15–30 ft.; this thickness may vary depending on the well and the location within the basin). This larger benthic foraminifer represents the only a biostratigraphic marker for the middle Aptian in the MB, as planktonic foraminifera are absent in this interval.

A brief review of the stratigraphy of the late Aptian to early Albian interval in the subsurface of the MB is necessary before continuing with the description of the Piché Member. In the MB, a series of prominent organic-rich rock intervals spanning the Aptian-Albian boundary, and linked to the OAE 1b (Méndez-Dot et al. 2015; Méndez-Dot 2017; this book) hyperthermal spikes, are well recorded. The occurrence of the OAE 1b is widely associated with having been triggered by the Kerguelen Plateau submarine and subaerial volcanism (see Bottini et al. 2015; Erba et al. 2015; Matsumoto et al. 2020). The late Aptian-early Albian interval in the MB is characterized by an alternation of organic-rich and organic-lean rock intervals (Figs. 2.13 and 2.14). Like many other basins worldwide, placing the Aptian-Albian boundary (~ 113 Ma) in that basin has been a matter of debate. In the absence of both high-resolution biostratigraphy and C isotope records, the stratigraphic position of the black shale intervals are the key to placing the boundary between the Aptian and Albian stages.

In agreement with previous studies carried in the Cogollo Group strata (Azpiritxaga 1991; Vahrenkamp et al. 1993; Murat and Azpiritixaga 1995; Parnaud et al. 1995a; Erlich et al. 1999a; Espín-Gil 2009), we have placed the Aptian-Albian boundary at the contact between the organic-lean interval at the top of the Piché Member and the organic-rich shale interval at the base of the Lisure Formation. Furthermore, the occurrence of the OAE 1b (~ 114–109 Ma) spans the Aptian-Albian boundary and includes multiple prominent black shale horizons which are the sedimentary expression of this OAE (Coccioni et al. 2015). In concordance with recent studies focused on the stratigraphy of the OAE 1b sub-events and on the Aptian-Albian boundary in the Vocontian Basin (southeast France) (Petrizzo et al. 2012; Kennedy et al. 2014, 2017), and the Umbria-Marche Basin (central Italy) (Coccioni et al. 2015; Sabatino et al. 2015; Ferraro 2017), we consider the following: the organic-rich interval in the upper part of the Piché Member is the equivalent of the OAE 1b first sub-event of late Aptian age (Jacob Event; ~ 114 Ma); the organic-rich interval at the lowermost part of the Lisure Formation is the equivalent of the OAE 1b second-sub-event of early Albian age (Kilian Event; ~ 113 Ma) (see also Gale et al. 2020; Gutiérrez-Puente et al. 2021, 2022).

The "*Orbitolina* level" is separated by the OAE 1b first sub-event by ~20 to 30 m in some cores (e.g., UD-171 and SOL-6) of the Urdaneta West field (western Lake Maracaibo). The TOC content of this black shale interval of late Aptian age is up to 13% (average of 3.16% in core Z26D-2 in western MB). The lithology is dominated by limestones, dolomitic limestones, clayey limestones, calcareous shales and dolomitic shales. The calcareous biota consists of planktonic foraminifera (mainly the genera *Paraticinella*, *Hedbergella* and *Globigerinelloides*) and SBF (buliminids), and less common inoceramids, oysters and echinoid spines. Microbial microspheres, quartz grains, and authigenic minerals (pyrite, fluorapatite and amorphous silica) are common. This organic-rich interval, which overlies and underlies shallow organic-lean facies, was deposited in shallow paleodepths within the photic zone which probably never exceeded 60 m. The expansion of the OMZ and redox stratified model represented in Fig. 2.16 is also applicable to shallow black shales corresponding to the OAE 1b in the Cogollo Group. The thickness of this black shale interval is approximately 26 m in the western area of the MB. The latter is, to our knowledge, the greatest thickness reported for the OAE 1b first sub-event (Méndez-Dot et al. 2015; Méndez-Dot 2017). Marine carbonate factories and components represented in this organic-rich interval are the same as previously discussed for the Machiques Member (see Sect. 2.6.2.2), although ammonites, sponge spicules, ostracods, and silica-secreting radiolarians are absent.

In the Piché Member *sensu stricto* (i.e., facies not corresponding to the OAE 1b first sub-event), mudstones (some of them partially dolomitized), wackestones and muddy packstones dominate the lithology, while grainstones are less common. Abundant heterotrophic calcifiers and non-skeletal grains (e.g., peloids, pellets, and subordinate ooids) in those muddy textures place this unit in the Heterozoan and the Biochemical-T factories of Michel et al. (2019) (the Biochemical T-factory includes the neritic lime-mud factory). Furthermore, common organic-rich (dark colored) micritic matrices and dark micritic peloids derived from pervasive marine micritization suggests a significant contribution of microbially-induced precipitation of micrite (the benthic automicrite factory). Thin, muddy dolomitic intervals fall into the benthic automicrite factory. Although LBF and green algae are common constituents of some wackestones and packstones, the prevailing meso-eutrophic conditions did not allow the predominance of the Photozoan T-factory. Mixed photozoan-heterozoan association between LBF/green algae (photozoan) and echinoids/molluscs (heterozoan) are common. Furthermore, abiotic precipitation of carbonate is represented by LMC cements partially or completely occluding moldic pores.

2.6.2.4 Lisure Formation (early to late Aptian)

Sedimentation in an epeiric platform continued with deposition of the Lisure Formation during the early and middle Albian. The late Albian age of the upper part of this unit is indicated by the ammonite *Mortoniceras* (Renz 1982). This shallow carbonate and mixed carbonate-siliciclastic succession has similar thickness throughout the subsurface of the MB, with an average between 100–140 m (~ 350–450 ft.; Fig. 2.15). This trend indicates a balance between subsidence and sedimentation rate across the MB. Deposition of the Lisure Formation began with the early Albian transgression in western Venezuela (Parnaud et al. 1995a), representing a prominent MFS across the subsurface of the MB. As noted earlier, organic-rich shales at the base of this formation correspond to the second sub-event of the OAE 1b (the early Albian Kilian Event). The lithology of this early Albian organic-rich rock interval is dominated by shales, calcareous shales, dolomitic shales, and clayey limestones. Carbonate content is represented by dolomite crystals, planktonic foraminifera (mainly the genus *Microhedbergella*), low-oxygen SBF (buliminids), varied amounts of automicrites, and authigenic microspheres. In some intervals of the lower part of the unit, quartz and muscovite are secondary constituents. It highlights the abundance of pyrite framboids severely darkening the variety of microfacies. In core Z26D-2, this shale interval is approximately 23.5 thick (77 ft) and has TOC content between 1.9% and 7.2% (Méndez-Dot 2017); while in core Alpuf-6 (Fig. 2.14) it is about 23 m thick (75 ft), with TOC content between 1.2% and 2.1% (Davis et al. 1994). Shallow paleodepths within the upper photic zone probably never exceeded 60 m during deposition of this black shale interval. Towards southwest in the MB, the expanded nature of the shallow black shales of the Cogollo Group is clearly observed in well logs of sections CR-11 and ARI-2 (Fig. 2.15), in which this bituminous interval has thicknesses of about 37 m (120 ft) and 67 m (220 ft).

Furthermore, in some sections located in western MB, petrophysical log analysis (gamma ray, density, resistivity and sonic signals) indicate that a second and slightly younger black shale interval is present in the lower part of Lisure Formation (Fig. 2.15). In core Alpuf-6, this shale interval of about 24 m thick (80 ft) has "fair" TOC content between 0.58 and 0.64% (Davis et al. 1994). This interval most likely records the OAE 1b third sub-event (Paquier Event; ~ 111 Ma). Although the available geochemical data for this interval is, till date, poor, the pretrophysical signals are clear and indicate the presence of an organic-rich interval at the same stratigraphic level throughout western MB. As it is seen in Fig. 2.15, the presence of one or two organic-rich intervals at the base and lower part of the Lisure Formation across the subsurface of western MB records the complexity that the multiphase OAE 1b also exhibits in many sections studied in the ancient Tethys and the Atlantic Ocean (see Trabucho-Alexandre et al. 2011; Coccioni et al. 2015). In the case of section ARI-2, prolonged anoxic conditions and deposition of one expanded, organic-rich interval encompassing the occurrence of Kilian and Paquier events might also be a plausible hypothesis. The thicknesses here reported for the early Albian black shales intervals are, to our knowledge, only surpassed by those reported by Talbi et al. (2019) in the southern Tethys margin (Tunisia).

From the finishing of the early Albian anoxia, the middle and upper parts of the Lisure Formation consist of a succession of T-R sequences. The lithologies of both transgressive deposits (mainly muddy wackestones and less common

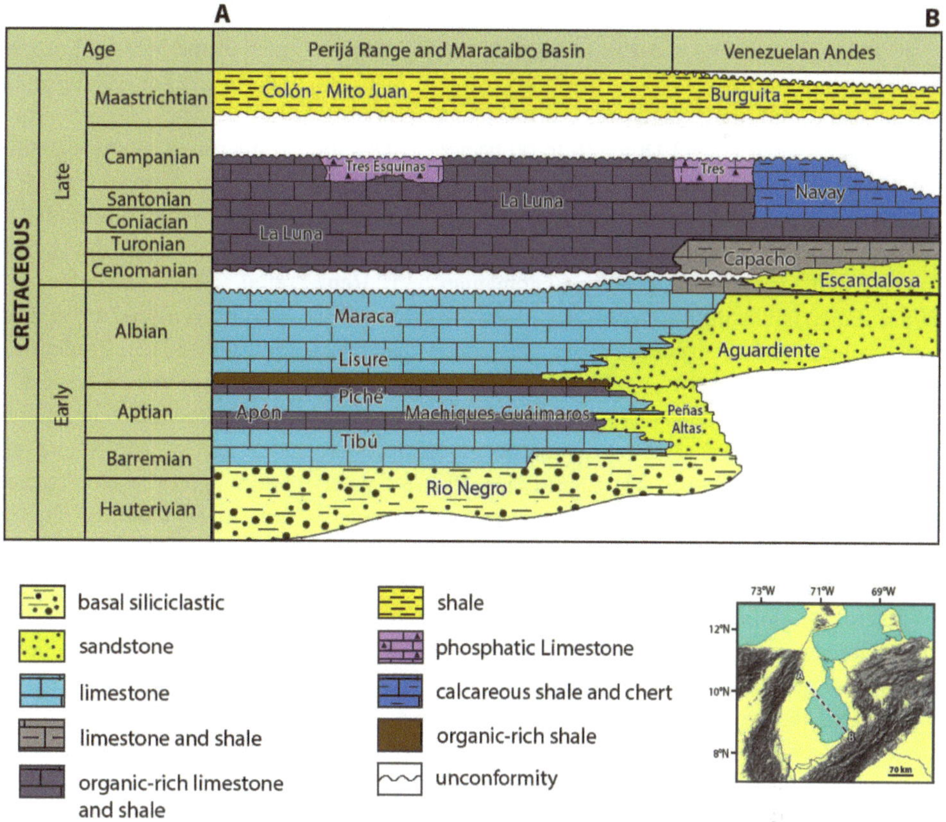

Fig. 2.17 Stratigraphic correlation chart for the Cretaceous in the MB. (Figure compiled and modified from Parnaud et al. 1995a and Rey et al. 2004))

mudstones) and highstand deposits (packstones and grainstones) have regional lateral extension in the subsurface of the MB and in the Perijá Range. Packstones and cemented (non-porous) grainstone textures (e.g., oolites) represent shallow banks that prograded towards the basin. Calcareous biota is dominated by molluscs (aragonitic bivalves, gastropods and oysters), echinoids and green algae (dasycladalean algae are more common than bryopsidalean), while bryozoans, ostracods, red algae, smaller miliolids and serpulids are secondary components. Scarce planktonic foraminifera are locally found in mud-supported fabrics. The Lisure Formation is also characterized by the abundance of non-skeletal grains, which include ooids, micritic and glauconitic peloids, oncoids, cortoids, aggregates and clasts (intraclasts and extraclasts). Such abundance of those components, many of them microbially-mediated in origin, suggests very shallow and agitated depositional settings where reworking and redeposition of grains were common processes. Some wackestones and packstones have considerable amounts of glauconitic grains. These are commonly accompanied by hardground intraclasts, which points to episodic, short-termed decrease in the overall sedimentation rate and the development of hardground surfaces. The Lisure Formation has higher siliciclastic content compared to the rest of the Cogollo Group, as thin sandstone and shale intervals interfinger with muddy-bioclastic limestones. Towards southwest in the MB, a marked lateral facies variation results in the appearance of the Aguardiente Formation of late Albian age, which overlies the early Albian black shales in wells ARI-2 and CR-1 (Fig. 2.15). Such marked transition from carbonate- to siliciclastic-dominated sedimentation also extends to more littoral environments towards south and southeast (Fig. 2.17).

As in facies of the Piché Member *sensu stricto*, carbonate production during deposition of the Lisure Formation was dominated by: the Heterozoan factory (abundant heterotrophic calcifiers, mainly molluscs and echinoids); the Biochemical T-factory (abundant mud- and grain-supported fabrics dominated by non-skeletal grains, mainly ooids); and the benthic automicrite factory (dark-colored micritic matrices and associated pyrite framboids, as well as dark micritic peloids, and micritic oncoids and cortoids). Both green and red algae (the Photozoan T-factory) were able to thrive locally under mesotrophic conditions. The former formed locally packstone textures. Mixed photozoan-heterozoan associations with echinoids, molluscs, bryozoans and ostracods are common. In some grain-supported fabrics, abiotic carbonate precipitation in the form LMC cements led to occlusion of substantial volumes of interparticle primary porosities.

2.6.2.5 Maraca Formation (latest Albian)

Conformably overlying the Lisure Formation, the Maraca Formation marks the end of the shallow carbonate platform sedimentation of the Cogollo Group, having an average thickness between 12 and 18 m (40 to 60 ft) across the subsurface of the MB (Fig. 2.15). While a late Albian age is widely accepted for base of this unit, the age of the top remains a controversy, varying from latest Albian (e.g., Castillo and Mann 2006) to early Cenomanian (e.g., Renz 1982). Paleodepths during deposition of this shallow-water unit probably never exceeded 10 meters deep (Méndez-Baamonde 2007b, 2009). The lithology consists of calcareous and glauconitic sandstones at the base, which grade to dominating packstones and less common grainstones and wackestones towards the top. Light-colored oyster packstones are the dominant lithofacies in this unit; while rudists, aragonitic bivalves, echinoids, green algae, gastropods, bryozoans and SBF are secondary components. The variety of heterozoan-dominated grain associations is accompanied by minor amounts of ooids, carbonate clast, and varying amounts of quartz grains.

Towards the top, the appearance of planktonic foraminifera in oyster-dominated limestones in some cores across the basin confirms the transgressive nature of this bioclastic succession (Davis et al. 1994; this book). A regional unconformity associated with subaerial exposure of the middle-upper part of the Maraca Formation is recognized across the MB, hence bounding shallow oyster biostrome facies from overlying organic-rich strata (Fig. 2.17) (Parnaud et al. 1995a). This regional erosional event affected the uppermost part of the shallow succession, thereby erasing or leaving little evidence of the early Cenomanian sedimentary strata (Canache et al. 1994; Rodríguez-Cuicas et al. 2019). In some oil fields in Lake Maracaibo, secondary (moldic and vuggy) and primary (interparticle) porosities characterize grain-supported fabrics in the middle and upper parts of the Maraca Formation. These porous rock intervals store light oil generated by the immediately overlying La Luna Formation (Méndez-Dot et al. 2015; Méndez-Dot 2017).

The regional unconformity at the contact between the Maraca and La Luna formations also includes the presence of a hiatus (or gap) of ~4 Myr (Canache et al. 1994) in the stratigraphic succession covering the early Cenomanian interval. In this context, different data sets have indicated tectonic uplift in western Venezuela during the early Cenomanian. This hiatus has been interpreted to be associated with the early stages of the interaction between the Caribbean Great Arc with the northwestern border of the South American plate (Martínez and Hernández 1992; Parnaud et al. 1995a). The associated flexural deformation in front of the applied load caused the development of a foreland basin to the west of the Perijá Range. Consequently, a forebulge uplift occurred in the Perija and Lake Maracaibo areas, resulting in the emergence and erosion of the upper part of the Maraca Formation during the early Cenomanian (Parnaud et al. 1995a; Rodríguez-Cuicas et al. 2019).

The Maraca Formation was deposited in a very shallow (uppermost photic zone) heterozoan-dominated, tropical platform. A peculiarity of the Cogollo Group, and especially of the Maraca Formation, is the abundance of low-Mg calcitic oysters to the detriment of rudists (see Pöppelreiter et al. 2005). The latter molluscs were flourishing in shallow platforms elsewhere at that time (e.g., Skelton 2003; Skelton and Gili 2012). Production of allomicrite (the neritic lime-mud factory) was less significant in this heterozoan-dominated unit compared to the rest of the Cogollo Group. In this unit of late Albian age, green algae are not common (although occasionally present), and scleractinian corals are absent. By that time, the Caribbean LIP eruption was already commenced; partial pressure of atmospheric CO_2 (pCO_2) were near reaching the highest peak of the Mesozoic (which occurred in the early Turonian); and seawater Mg^{2+}/Ca^{2+} ratio rose low values of c. 1, shortly after reaching the lowest peak of the Calcite II sea interval during the Aptian (Stanley and Hardie 1998; see Appendix B for a detailed treatment of Calcite and Aragonite Seas and their influences on the fates and the skeletal mineralogy of calcareous organisms throughout the Phanerozoic). In addition to the dominating Heterozoan factory, the Biochemical T-factory is represented in the Maraca Formation by moderately common non-skeletal grains (ooids, aggregates and clasts). LMC cements (abiotic carbonates) are partially occluding primary and secondary pores in many grainstone textures.

2.6.3 Capacho Formation (late Albian to late Cenomanian)

Although photographic examples of the Capacho Formation are not included in this book, we consider this unit of great importance. This formation of late Albian to late Cenomanian age crops out in the Venezuelan Andes (states of Táchira, Mérida and Barinas), where it overlies limestones of the Maraca Formation or sandstones of the Aguardiente Formation; and underlies the La Luna Formation (Figs. 2.17 and 2.18) (Renz 1958; González de Juana et al. 1980; Ministerio de Energía y Minas 1997; De Romero et al. 2003; Rodríguez-Cuicas et al. 2019, 2020). In addition, towards southwest in the subsurface of the MB, the stratigraphic configuration changes due to the appearance of the Capacho Formation overlying sandstones of the Aguardiente Formation (Fig. 2.15).

The Capacho Formation is divided into the lower La Grita Member (5–10 m thick) and the upper Seboruco Member (up to 120 m thick). The lithology of the former member is very similar to the La Luna Formation *sensu stricto*, and consists

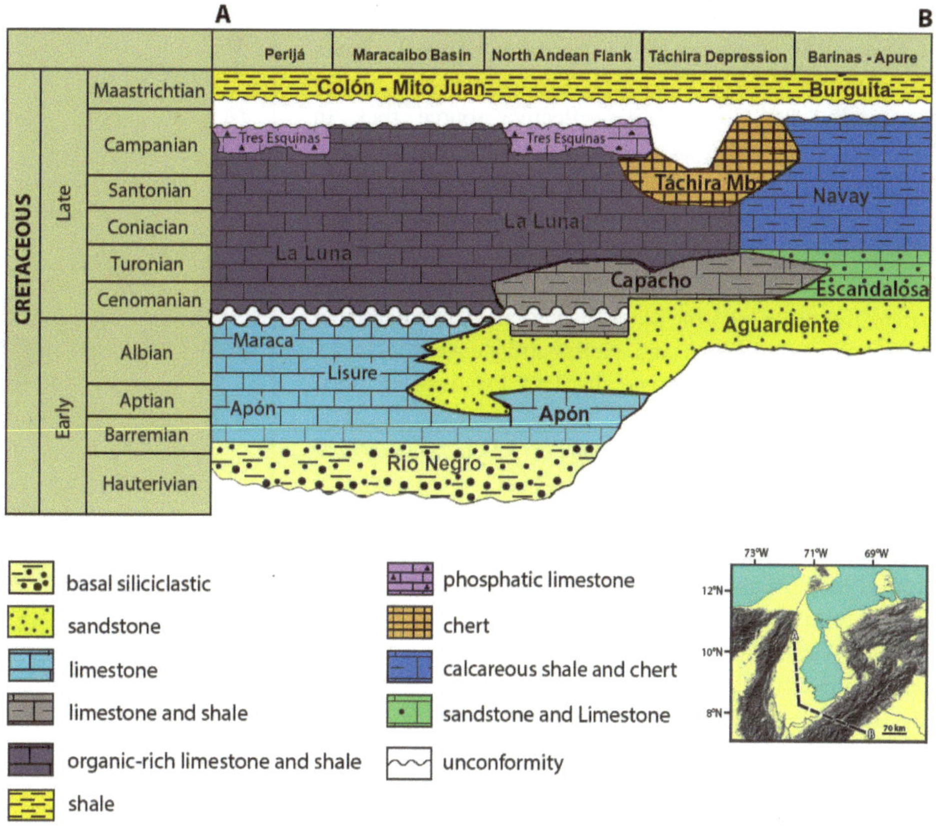

Fig. 2.18 Stratigraphic correlation chart for the Cretaceous in the MB and western Barinas-Apure Basin. (Figure compiled and modified from Garbán et al. 2017 and Rodríguez-Cuicas et al. 2019)

of an alternation of black limestones and black to gray calcareous shales, while the overlying Seboruco Member is characterized by organic-lean shales with low carbonate content (Renz 1958; González de Juana et al. 1980). Deposition of this unit encompassing the Venezuelan Andes occurred in a semi-restricted sub-basin bounded by paleobathymetric barriers, with a poor-water circulation and low-oxygen conditions especially prevailing during deposition of the lower member (Macellari 1988; Erlich et al. 1999a, b; Rodríguez-Cuicas et al. 2019).

In that context, the OAE 1d (Breistroffer Event; ~ 103–99.5 Ma) is well recorded in an outcrop located southwest of the city of Mérida (see Fig. 1.3), where the La Grita Member is 7 m thick and has TOC content up to 10%.

Planktonic foraminiferal assemblages are dominated by the genera *Muricohedbergella*, *Planoheterohelix*, *Globigerinelloides*, *Ticinella* and, and less common, *Planomalina*, *Whiteinella* and *Rotalipora*. Redox sensitivity acceleration of the hydrologic cycle and an increase in continental chemical weathering rates during black shale deposition (Rodríguez-Cuicas et al. 2019, 2020). Petrophysical signals also indicate a substantial organic enrichment in sections of the Capacho Formation in the subsurface of western MB (wells CR-11 and Ari-2 in Fig. 2.15). Such sections underlie La Luna Formation and most probably record the OAE 1d.

Regarding the early Cenomanian stratigraphic gap in the MB strata, Rodríguez-Cuicas et al. (2019) confirms the absence of marker foraminifera species of early Cenomanian age (*Rotalipora brotzeni-Rotalipora reicheli Zone*) in the Capacho Formation. The organic-rich interval corresponding to La Grita Member consists of rocks of late Albian and middle Cenomanian ages (*Rotalipora appenninica* zone, and lower part of the *Rotalipora cushani* Zone, respectively; the zonal schemes used by those authors were that of the Sliter 1989; and Premoli-Silva and Sliter 1995). Early Cenomanian beds from this member were likely impacted by submarine erosion and/or non-deposition, thereby erasing the sedimentary record of this time interval.

2.6.4 La Luna Formation *sensu stricto* (middle-late Cenomanian to late Santonian)

From latest Albian to middle Cenomanian (~ 99.5 to 96 Ma), a rapid ESL rise (between ~205 to 215 m APSL Haq 2014) caused the drowning of late Aptian, shallow facies of the Maraca Formation. Sustained sea-level rise, reaching up to ~240–250 m APSL during late Cenomanian and early Turonian (Haq 2014) led to the largest extension of the Cretaceous epeiric sea in Northern Venezuela. The combination of several global and local factors during the Phanerozoic highest sea level promoted the deposition of the La Luna Formation in western Venezuela. This organic-rich unit is historically the most prolific effective source rock in Venezuela (e.g., Talukdar and Marcano 1994; James 2000; Goddard and Talukdar 2002; López and Lo Mónaco 2017). Owed to this bituminous unit, the MB is considered to be one of the richest petroleum habitats in the world (Brooks 1990). In this book we use the term "La Luna Formation *sensu stricto*" for differentiating black shale facies of this unit from its uppermost siliceous and phosphatic-glauconitic members, the Táchira Ftanita and the Tres Esquinas member, respectively.

The extension of the La Luna Formation *sensu stricto* spans the entire subsurface in the MB (Fig. 2.15), and also crops out in the Perijá Range and the Venezuelan Andes (Figs. 2.17 and 2.18) (e.g., Renz 1982; González de Juana et al. 1980; Erlich et al. 1999a, 2000; this book). It also extends into the subsurface of northwestern Falcón State (Baquero et al. 2009), as well as in north Lara State, to the west of the city of Barquisimeto (see Fig. 1.3). In that area (the Barquisimeto Trough) (see Fig. 2.4), La Luna Formation crops out as a para-autochthonous slab of the one stable eastern margin of the Late Cretaceous MB passive margin. The detached slab was emplaced towards east under the influence of the beginning of the Caribbean orogeny during the early Paleogene (Renz 1982; Tribovillard et al. 1991). Photographic examples of the La Luna Formation *sensu stricto* correspond to several cores located in the subsurface of western Lake Maracaibo (e.g., UD-171, Sol-6). We also show one example corresponding to an outcrop in the state of Mérida, west of the city of Mérida (see Fig. 1.3).

The wide extension of the "La Luna Sea" (Erlich et al. 2000) along the northern border of South America (Fig. 2.19) is not only recorded in La Luna Formation time-equivalents in Venezuela (the Querecual and Navay formations in the Eastern Venezuela and the Barinas-Apure basin, respectively). Widespread deposition of La Luna Formation time and lithologic equivalents also occurred in north and east Colombia (outcrops in the Eastern Cordillera and the subsurface of the Lower Guajira, César-Ranchería, Llanos, Lower Magdalena and Catatumbo basins) (see Martínez and Hernández 1992; Cooper et al. 1995; Villamil and Arango 1998; Villamil and Pindell 1998; Villamil et al. 1999; Rangel et al. 2000, 2003; Rangel and Hernández 2007; Patarroyo et al. 2017; Rivera et al. 2018; Avendaño-Sánchez et al. 2021; Paez-Reyes et al. 2021); as well as in Ecuador, Trinidad, Suriname and northern Brazil (see Erlich et al. 2003; Liborius-Parada 2019).

In the subsurface of the MB, La Luna Formation overlies shallow-water, oyster-dominated limestones of the Maraca Formation (Figs. 2.14 and 2.15). As previously mentioned, this stratigraphic configuration changes towards southwest

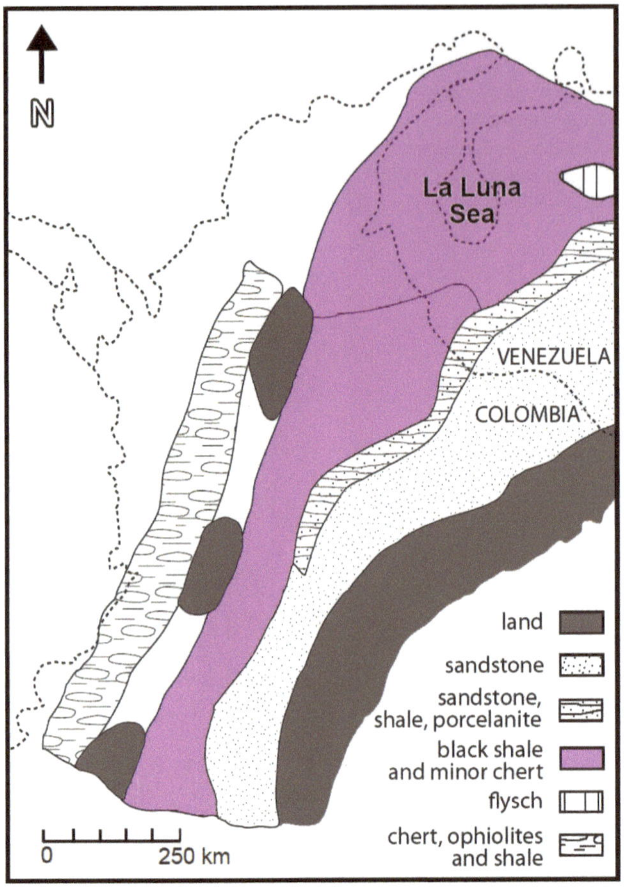

Fig. 2.19 Depositional facies in northwestern South America for the Turonian-Santonian period, highlighting in purple the extension of the "La Luna Sea". Dashed line indicates present-day shoreline and political boundary between Venezuela and Colombia. Original figure modified from Macellari and De Vries (1987). (Figure modified from Davis et al. 1999)

in the subsurface, as La Luna Formation is underlain by the Capacho Formation. This stratigraphy is also found towards southeast in the Venezuelan Andes (Figs. 2.17 and 2.18) (Macellari 1988; De Romero et al. 2003; Rodríguez-Cuicas et al. 2019, 2020).

Since its formal description by Hedberg and Sass (1937), the lithology of La Luna Formation has been generally described as an alternation of dark, organic-rich and highly laminated beds of limestones, marls and shales (Renz 1958; Tribovillard et al. 1991; Pérez-Infante et al. 1996; Erlich et al. 1999a; Bralower and Lorente 2003; Rey 2004; Rey et al. 2004). Carbonate content (wt.%) is higher at the base and lower parts of the unit (Pérez-Infante et al. 1996; Davis et al. 1994, 1999), although this tendency varies widely, depending on the section studied (see De Romero et al. 2003; Liborius-Parada 2019). In the central part of the basin, the La Luna Formation contains generally more than 50% shale, while in the rest of the basin, this unit is generally more than 50% limestone. The reason for the localized dominance of shale over limestone is unclear, but could be related to the lack of preservation of limestone below an OMZ (Erlich et al. 1999a). Furthermore, TOC content (wt.%) decreases towards east and southeast areas of the MB (Pérez-Infante et al. 1996), where both paleodepths and reducing/anoxic conditions were lower.

The thickness of the La Luna Formation *sensu stricto* ranges between 100 and 200 m in the MB, with an increasing tendency from south to north (Ministerio de Energía y Minas 1997). In the subsurface of the MB, this organic-rich unit has average thickness between ~80–120 m (~ 250–400 ft) (Fig. 2.15). Average TOC content is ~5.6%, with most values ranging between 2.5 and 10.8%, and up to 13% (Talukdar and Marcano 1994; Liborius-Parada 2019). Thicknesses of about 180 m are found in the eastern front of the Perijá Range (Ministerio de Energía y Minas 1997). In the Perijá foothills, this bituminous unit is about 160 m thick (Pérez-Infante et al. 1996), with TOC content between 0,5% and 8,5% (Pérez-Infante et al. 1996; Alberdi-Genolet and Tocco 1999). The La Luna para-authochthonous slab in the Barquisimeto Trough is about 200 m thick (Tribovillard et al. 1991). Above-mentioned thicknesses indicate high sedimentation rates and extraordinary conditions for sustaining organic matter productivity and anoxia during the long-lasting pelagic realm-like conditions.

The age of the base of the La Luna Formation has been a matter of debate. Combined biostratigraphy (planktonic foraminifera and calcareous nannoplankton) and carbon-isotope stratigraphy indicate that black shale deposition began in the middle to late-middle Cenomanian (~ 96 Ma) in western MB and probably no earlier than this in the eastern part of the basin (De Romero et al. 2003). This age confirmed those estimated by previous studies (see Bralower and Lorente 2003 and references therein), and coincides with the regional unconformity followed by a stratigraphic gap of ~4 Myr during the early to middle Cenomanian (Canache et al. 1994). According to De Romero et al. (2003), from the contact with the Maraca Formation to 11.2 m upward in section, the La Luna Formation in core Totumo-3 (well TOT-3 in Fig. 2.15) is assigned to the *Rotalipora cushmani* zone. Within this interval, black shales from this core have TOC content up to 6.65% (Méndez-Dot 2017). This organic-rich interval is the expression of the middle Cenomanian Event, which is considered a prelude to the OAE 2 (see Coccioni and Galeotti 2003; Gertsch et al. 2010; Joo et al. 2020). In addition, volcanic ash horizons (~ 1 m thick) reported in the lower part of the La Luna Formation from different sections in the MB (Parnaud et al. 1995a; Davis et al. 1999; Liborius-Parada, 2019) and in Colombia (Richards, 1968; Orth et al. 1993) are believed to have originated from the activity of the Caribbean Great Arc, which during Cenomanian times was located near the western margin of northern South America (Parnaud et al. 1995a). Liborius-Parada (2019) places this thin interval underlying the beginning of the black shale interval that represents the sedimentary expression of the OAE 2.

The transgressive surface at the base of the La Luna Formation is followed upward in section by the interval encompassing the Cenomanian-Turonian boundary (93.9 Ma) and the early Turonian, which corresponds to the MFS associated to the occurrence of the OAE2. This surface represents the maximum flooding during the entire Cretaceous and perhaps including the entire Phanerozoic (Sarmiento-Rojas 2019). Black shales corresponding to the OAE 2 are located in the lower part of the unit [e.g., core Alpuf-6, Davis et al. 1994, 1999, see Fig. 2.14; and core Totumo-3, see De Romero et al. 2003]. Prolonged anoxia and black shale deposition, alternated with minor suboxic intervals, continued during the entire Turonian and Coniacian stages until the late Santonian, encompassing the occurrence of the OAE 3 in the Coniacian-Santonian boundary interval. This anoxic event is located in the uppermost part of this organic-rich unit, both in the Venezuelan Andes (an outcrop west of the city of Mérida) (Rey 2004; Rey et al. 2004; Machado et al. 2016) and in the subsurface of the MB (Liborius-Parada 2019; see Fig. 2.14). The uppermost black shale facies in this unit are considered to be late Santonian in age, about 2.5 to 2.7 Myr posterior to the Coniacian-Santonian boundary (86.3 Ma).

For all mentioned in the two previous paragraphs, regional-scale sedimentation of the La Luna formation *sensu stricto* began during the middle-late Cenomanian and lasted until the late Santonian. During this time interval of about 12 Myr, deposition of this organic-rich expanded unit encompassed the OAEs 2 and 3. This long-lasting black shale sedimentation began prior to the onset of the OAE 2 in the Cenomanian-Turonian boundary (Bonarelli Event; ~ 94 Ma). This OAE (global in nature) had an estimated duration of 430 to 930 thousand years (kyr) (Sageman et al. 2006; Voigt et al. 2008; Gangl et al. 2019). The age and duration of the OAE 3 varies by locality (Locklair et al. 2011). With a total duration of ~2.3 Myr (87.3–84.6 Ma), the OAE 3 is not a short-lived event like the early Aptian OAE 1a or the OAE 2 (Wagner et al. 2004; Lowery et al. 2017). The geographical extension of the last of the Cretaceous anoxic events was restricted to the equatorial Atlantic and adjacent basins, shelves and epeiric seas (Wagreich 2012). In the Western Interior Basin (mid-west U.S.A.), the duration of the OAE 3 encompassing the Coniacian-Santonian boundary interval ranges from ~5.5 to 6 Myr, spanning ~3.4 Myr in the Coniacian and ~ 2.3 Myr in the Santonian (Locklair and Sageman 2008). This scenario is similar to that of the OAE 3 in the subsurface of the MB.

Probably the most highlighting textural feature of the La Luna Formation *sensu stricto* is the typical microlaminated planktonic foraminiferal ooze fabric throughout western

Venezuela (Erlich et al. 1999a). Such mm-scale lamination, which alternates with intervals with lower content of planktonic foraminifera and less apparent lamination, is observed from the lowermost pre-OAE 2 interval to the uppermost OAE 3 interval. Variations in environmental conditions during regional deposition of this organic-rich expanded unit are evidenced by variations in foraminiferal assemblages and abundance. Planktonic foraminifera are dominated by the genera *Globigerinelloides*, *Muricohedbergella*, *Planoheterohelix*, *Whiteinella*, *Archaeoglobigerina*, and *Marginotruncana* (Parra et al. 2003); while SBF of the genera *Gavelinella*, *Lenticulina*, *Orthokarstenia*, *Siphogenerinoides*, *Bolivinoides*, *Bolivina*, and *Gabonita* become more abundant in the upper part (Rey et al. 2004). Localized chert layers and phosphatic lenses, and the sporadic apparition of radiolarians, fish remains and low-oxygen inoceramid molluscs (and even less abundant gastropods and brachiopods) are secondary lithologic characteristics of the La Luna Formation (e.g., Zapata et al. 2000; Parra et al. 2003; Macsotay et al. 2003; Rey et al. 2004; Liborius-Parada 2019). In addition, ammonites are common in Cenomanian to Santonian intervals in the Perijá Range and in northeastern Venezuelan Andes (states of Trujillo and Lara) (Renz 1982; Erlich et al. 2000). Prolonged anoxic to anoxic-euxinic conditions were alternated by shorter suboxic intervals during deposition of this thick black shale unit, allowing preservation of high-quality organic matter (Type I, Type II, and Type II-S kerogens) (Trivobillard et al. 1991; Mongenot et al. 1996; Alberdi-Genolet and Tocco 1999; Erlich et al. 1999b; Gallango et al. 2002; Lo Mónaco et al. 2002; Machado et al. 2016; López and Lo Mónaco 2017; Liborius-Parada 2019).

A rapid response to the Hothouse climate in the ocean-atmosphere system (i.e., excess of volcanogenic CO_2, elevated paleotemperatures, and high concentration of nutrients and biolimiting metals in the oceans) linked to the geographic proximity of the COP submarine volcanism probably had more impact in northern South America than in any other marine continental shelf in the world during the Late Cretaceous. According to Bice et al. (2006), pCO_2 estimations during deposition of Cenomanian and Turonian black shales in Suriname, just southeast Venezuela, are up to 2400 ppmv. Possible errors in temperature and CO_2 estimations, and possible deficiencies in the model used by those authors, might level up that value up to 3500 ppmv or more. Furthermore, those authors estimate sea-surface paleotemperatures between 33° and 42 °C [see also Hay 2017 and Scotesse et al. (2021) for excellent reviews on the Cenomanian-Turonian thermal maximum]. As indicated for the black shale intervals in the Cogollo Group, regional black shale deposition along the MB from the middle-late Cenomanian indicates the expansion of the OMZ over the drowned late Albian shallow platform facies (Méndez-Baamonde 1989b). The occurrence of regional anoxia and widespread black shale deposition was promoted by the following combined factors: sustained increasing sea level; episodic nutrient-loaded continental runoff and upwelling events as sources of nutrients for the La Luna Sea (Tribovillard et al. 1991; Pérez-Infante et al. 1996; Erlich et al. 2000); high rates of marine organic matter productivity; and the development of paleobathymetric barriers (the Santa Marta and Santander massifs, and the El Baúl and Mérida archs; see Fig. 2.4) which caused poor-water circulation and limited-vertical ventilation (e.g., Macellari 1988; Lugo and Mann 1995; Erlich et al. 1999a, 2000; Rey et al. 2004).

In the Perijá foothills, Pérez-Infante et al. (1996) estimate paleowater depths of about 200–300 m for the La Luna Formation. In the Venezuelan Andes (state of Mérida), Rey et al. (2004) consider that sedimentation of La Luna during the Santonian (a 22 m thick Santonian section corresponding to the OAE 3) occurred in upper neritic to upper bathyal depths. In the Barquisimeto Trough, Tribovillard et al. (1991) described La Luna Formation as deposited in a pelagic environment which was relatively shallow (near storm wave base) and stratified. In the subsurface of the MB, by the beginning of the organic-rich sedimentation, paleodepths associated to the late Albian platform drowning were most likely within the photic zone and these kept increasing during the Turonian. We estimate persistent photic zone anoxia in the lowermost carbonate-dominated part of this organic-rich sequence. From biomarker analysis (aryl isoprenoids ratio) determined in core samples from wells located in western MB, Liborius-Parada (2019) suggests the occurrence of intermittent photic zone anoxia, which decreases towards the top of the formation. Short-term paleoclimatic variations linked to Milankovitch cycles are thought to be responsible for the highly variable content (%) of both carbonate and TOC. Alternated warm-humid and warm-dry climate turned in a clear relation between an increase in TOC content in shales and calcareous shales compared to the decrease in TOC content in limestones and clayey limestones (Pérez-Infante et al. 1996; Rey et al. 2004).

Thicknesses of the La Luna Formation in the subsurface of the MB are comparable (although higher in average) to organic-rich expanded sections deposited in Demerara Rise, offshore Suriname. These organic-rich successions deposited during the late Albian to the early Santonian have been studied in several Ocean Drilling Program sites, having thickness of ~90 m and TOC content up to 28% (e.g., Erbacher et al. 2005; Brumsack 2006; Meyers et al. 2006, 2009; Nederbragt et al. 2007; van Bentum et al. 2009; Friedrich et al. 2008; Hetzel et al. 2009; Kraal et al. 2010). Thicknesses of the La Luna Formation *sensu stricto* and Demerara Rise organic-rich expanded sections are exceeded by black shales of the Querecual Formation in the Eastern Venezuela Basin, and by black shales in the Tarfaya Basin, southwest Morocco (see Sect. 2.8.5).

Carbonate production during deposition of organic-rich facies of the La Luna Formation was dominated by the P-factory and the benthic automicrite factory. The first factory is represented by abundant planktonic foraminifera and calcareous nannoplankton as main carbonate components. Prolonged pelagic realm-like, stressed conditions (CO_2-rich waters, hypertrophy and anoxia/anoxia-euxinia) led to massive flourishing or opportunistic (*r*-selected) and intermediate (*r/k*) species of planktonic foraminifera. Planktonic carbonate production was accompanied by benthic precipitation of microbially-induced micrite. Such micritic sediments were precipitated under low-oxygen (suboxic) and anoxic conditions. In many microfacies, the high organic matter content masks the mineral matrix, thus preventing, partially or totally, its visualization with the polarizing microscope. The Heterozoan factory is represented mainly by moderately common symbiont-barren benthic foraminifera, and the sporadic occurrence of low-oxygen molluscs and ammonites and even less abundant gastropods and brachiopods.

In addition, specific intervals within the La Luna Formation are proposed as an unconventional hydrocarbon reservoir by Liborius-Parada (2019). Based on a variety of rock parameters (TOC, HI, water saturation, thickness, Brittle Index, and various inorganic and organic proxies), the author states that the upper part, and to a lesser extent, the lower part of this unit in the cores studied, host the organofacies with the best qualities for unconventional reservoir prospection (see also Liborius-Parada et al. 2023).

2.6.5 La Luna Formation, Táchira Ftanita Member (middle Santonian to early Campanian)

Towards southwest in the Venezuelan Andes, in the state of Táchira, the uppermost part of the La Luna Formation *sensu stricto* (e.g., laminated black calcareous shales corresponding to the OAE 3) grades transitionally to siliceous facies of the Táchira Ftanita Member (Fig. 2.18). In different outcropping sections, this unit is from 15 m thick (De Romero and Galea-Álvarez 1995) to up to 80–100 thick (Garbán 2010), and predominantly consists of radiolarian-rich, rhythmically bedded black cherts (5 to 20 cm), with interbedded black calcareous shales, phosphatic beds, marls, and dark gray limestones (Renz 1958; Garbán 2010). Chert-dominated facies are characterized by two matrices, one formed by fine-grained cryptocrystalline quartz and another by megagranular quartz that is lighter in color (Garbán 2010). Based on biostratigraphy of planktonic foraminifera carried on different outcropping sections, the accepted age for this unit is early-mid Santonian for the base and late Santonian to early Campanian for the top (see Romero and Galea-Álvarez,

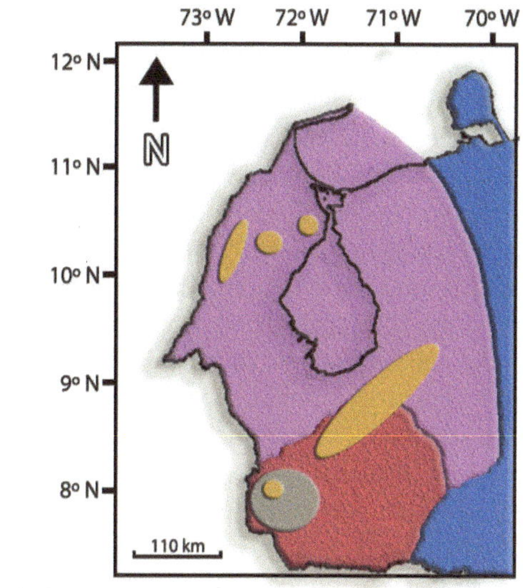

Fig. 2.20 Depositional provinces proposed for western Venezuela during Turonian to Campanian times, highlighting the area of deposition of the Táchira Ftanita Member (middle Santonian to early Campanian) and the Tres Esquinas Member (early to middle Campanian). (Original figure modified from Macellari and De Vries 1987. Distribution of depositional provinces slightly modified from Garbán 2010)

1995; De Romero et al. 2003; Garbán 2010). Photographic examples of the Táchira Ftanita Member correspond to one outcrop in the state of Táchira, northwest of the city of San Cristobal (see Fig. 1.3).

The Táchira Ftanita Member ("ftanita" meaning "phthanite", which is a fine-grained, siliceous sedimentary rocks by definition), is one of the most interesting sedimentary rock intervals in western Venezuela. Its siliceous-dominated lithology represents a unique group of lithofacies deposited in a marine environment restricted exclusively to the Táchira Depression (Fig. 2.20). The latter was, indeed, a bathymetric depression created by the Mérida Arch (see Fig. 2.4), which formed a restricted platform setting where ideal conditions for deposition of siliceous sediments were created during the middle-late Santonian (Erlich et al. 2000).

Although this member represents a notorious lithological variation respect to the underlying La Luna Formation *sensu stricto*, this siliceous unit also shows some textural and compositional features typical of a black shale condensed section. The main of which are the microlaminated, organic-rich textures and abundant globular planktonic foraminifera and low-oxygen SBF (mainly buliminids). Recrystallized and poorly preserved radiolarian tests are the main bioclastic components (Garbán 2010). Calcareous biota is in most cases recrystallized and/or coated by diagenetic silica precipitates (amorphous silica, megaquartz and chalcedony) (Marcucci 1973; Garbán 2010).

The origin of this bedded chert unit is directly linked to biological uptake of dissolved silica by radiolarians, in combination with changes in the intensity of upwelling currents which influenced the influx of dissolved silica and hence the rhythmicity of the succession. The silicious matrix has been interpreted to be directly originated from diagenetic alteration of opaline radiolarian tests (Garbán 2010; Garbán et al. 2017). The pH of the bottom stagnant waters was the main factor controlling the dissolution of calcitic foraminifera tests. Local pH values were primarily controlled by the concentration of CO_2 derived from organic matter decay and by the concentration of bicarbonate ions derived from dissolution of the calcareous tests. Periodic or occasional flushing of the basin by oxygenated upwelling currents removed CO_2 from bottom waters and permitted deposition of calcareous tests (Marcucci 1973).

Paleoenvironmental variations in the water column, from anoxic-euxinic to dysoxic conditions, derived in deposition of nearly pure cherts (> 80% SiO_2), calcareous shales and marls, being the last two lithofacies those with the highest TOC content, with values up to 4.80% (Garbán 2010). Upper bathyal paleodepths (300–400 m deep) are estimated for the deposition of this bedded chert (Grony Garbán, personal communication 2015).

In this siliceous unit, the P-factory was a major contributor to both carbonate and silica production, as planktonic foraminifera and radiolarians (planktic association) are common-dominant skeletal components. The Heterozoan factory is also well represented by common-abundant symbiont-barren benthic foraminifera. It highlights the capacity of these low-oxygen heterotrophs to thrive under such stressed, locally acidic bottom-water conditions. In addition, although far less common than siliceous matrices, microbial automicrite precipitated in the aphotic zone make up the micritic matrix of some clayey limestones.

2.6.6 La Luna Formation, Tres Esquinas Member (early to middle Campanian)

The Tres Esquinas Member is the uppermost member of the La Luna Formation (e.g., Galea-Álvarez 1989). This thin unit (less than 5 m thick) consists of phosphatic foraminiferal wackestones and packstones and less common cherts with abundant glauconitic peloids. This member represents a marked change in the long-lasting anoxic-suboxic depositional environment of the La Luna Formation *sensu stricto* (see Parra et al. 2003), as its deposition has been associated with a variation in paleoclimate and paleoceanography in western Venezuela during late Santonian to early Campanian. An abrupt decrease in TOC content respect to facies of the La Luna formation *sensu stricto* (as well as to some organic-rich facies of the Táchira Ftanita Member) suggests that dysaerobic conditions prevailed during deposition of this unit (De Romero and Galea-Álvarez 1995; Erlich et al. 1999a; Erlich et al. 2000; Parra et al. 2003).

Contrary to regional extension of black shales of the La Luna Formation *sensu stricto* in western Venezuela (i.e., Venezuelan Andes, subsurface of the MB and Perijá Range), deposition of the Tres Esquinas Member was mainly restricted to (Fig. 2.20): (**a**) the southwest of the Venezuelan Andes, where the most studied sections of this unit are found (e.g., De Romero and Galea-Álvarez 1995; De Romero et al. 2003; Parra et al. 2003); although González de Juana et al. (1980) and De Romero et al. (2003) report this member towards northeast in the state of Trujillo; (**b**) the eastern flank of the Perijá Range, as well as a few cores in the subsurface of western MB (see González de Juana et al. 1980; Davis et al. 1999; De Romero et al. 2003; Parra et al. 2003; Liborius-Parada 2019). Photographic examples of the Tres Esquinas Member correspond to one outcrop in the state of Táchira, northwest of the city of San Cristobal (see Fig. 1.3); and to one core in western MB.

Thicknesses of the Tres Esquinas Member vary between 1 and 4 m in various outcropping sections in the Venezuelan Andes (De Romero and Galea-Álvarez 1995; Parra et al. 2003; Rey et al. 2004; Garbán 2010); while in the subsurface of western MB it is 3 m thick (Parra et al. 2003). In outcrops in the state of Mérida, this unit [of 3 m thick, reported by De Romero and Galea-Álvarez 1995 and Parra et al. 2003; and of 1 m thick, reported by Rey et al. 2004] overlies the La Luna Formation *sensu stricto*; while in the state of Táchira (4 m thick) it overlies the Táchira Ftanita Member (Fig. 2.18; Garbán 2010; Garbán et al. 2017). Based on planktonic foraminifera biostratigraphy, De Romero and Galea-Álvarez (1995) suggest a late Santonian-early Campanian age for this condensed section in the state of Mérida, while towards southwest in the state of Táchira those authors suggest and older age (middle to late Campanian), which implies that deposition of this phosphatic-glauconitic unit was diachronous.

Lithofacies of the Tres Esquinas Member are characterized by prominent mineralization processes (fluorapatite, glauconite, pyrite and silica). The biota is dominated by planktonic foraminifera (mainly *Planoheterohelix, Muricohedbergella, Globotruncana*) and SBF (*Bolivinoides, Orthokarstenia, Gavelinella, Praebulimina, Lenticulina,* among others). Abundant reworked peloidal grains (including phosphatized ooids, peloids and fish remains) and sand-

and silt-sized siliciclastic sediments, as well as high degree of bioturbation, are common lithologic features of this Campanian unit (De Romero and Galea-Álvarez 1995; Parra et al. 2003).

Deposition of this phosphatic horizon is inferred to have occurred during a major transgression that affected western Venezuela in late Santonian-early Campanian times; however, interpretations of its depositional environment vary (Erlich et al. 2000). In the Venezuelan Andes, this thin horizon was deposited in a dysaerobic environment at a suggested paleobathymetry of 300–500 m (upper bathyal) (De Romero and Galea-Álvarez 1995). From the end of deposition of organic-rich facies of La Luna Formation (*sensu stricto* and Táchira Ftanita Member), the finishing of anoxia has been linked to the progressive uplift of the Central Cordillera of Colombia; as well as to a more frequent or intense seasonal upwelling that led to an enhanced bottom water oxygenation and vertical mixing of the water column (Erlich et al. 2000; Bralower and Lorente 2003; Parra et al. 2003).

According to Erlich et al. (2000), and following the description of condensed and allochthonous phosphorites by Föllmi et al. (1992) and Föllmi (1996), the origin of this phosphorite unit appear to be allochthonous for many of the microfacies, due to the abundance of reworked and redeposited allochems. However, textural variations (mainly in lamination) typical of condensed phosphorites in some microfacies belonging to different sections in the Venezuelan Andes suggest a condensed (or autochthonous) origin. The latter points to a complex origin for this unit.

2.6.7 Maracaibo Basin Active Margin Stage

The tectonic setting of the northwestern border of South America (Colombia and Venezuela) changed during the middle Campanian (~ 80 to 75 Ma). The oblique collision of the migrating Caribbean Great Arc against the northwest corner of the South American plate and the beginning of the exhumation of the Central Cordillera of Colombia marked the end of passive margin sedimentation and the beginning of an active margin stage. These regional tectonic episodes were of critical importance for the thermal maturation of Cretaceous source rocks and the deposition of Cenozoic sandstone reservoirs (Parnaud et al. 1995a; Lugo and Mann 1995; Macsotay et al. 2003; Escalona and Mann 2006b). As shown in Figs. 2.17 and 2.18, a regional unconformity and a long hiatus (~ 8 Myr) characterizes the contact between passive margin and active margin strata in western Venezuela, which reflect the tectonic disturbances associated with the newly-derived Caribbean-South American plate boundary.

2.6.8 Colón and Mito Juan Formations (Maastrichtian)

After a long period of a transgressive trend sedimentation, a long-lasting regressive cycle started in the early Maastrichtian. Regional-scale sedimentation of the Colón Formation (early Maastrichtian; up to 900 m thick) and overlying sandstones and shales of the Mito Juan Formation (late Maastritchtian, up to 300 m thick) (Ministerio de Energía y Minas 1997) encompasses the subsurface of the MB, the Venezuelan Andes and the Perijá Range (see Figs. 2.17 and 2.18). Dark, fossiliferous shales of the Colón Formation (rich in foraminifera) were deposited in a foredeep basin (Lugo and Audemard 2021). This shale unit, which is considered a regional seal for hydrocarbons across the subsurface of the MB (Parnaud et al. 1995a), grades to shallower sandstones and shales of the Mito Juan Formation, being the sandstone facies hydrocarbon reservoirs in some oil fields in the MB. The abundance of sandstones and a marked change in the foraminiferal assemblages are the main criteria for differentiating these two siliciclastic units (Sutton 1946).

2.6.9 Guasare Formation (Early Paleocene)

The Guasare Formation of Early Paleocene age is a mixed carbonate-siliciclastic succession which hosts the most significant Cenozoic carbonate facies in the MB. It overlies conformably Maastrichtian thick shales and sandstones of the Colón and Mito Juan formations, whose deposition posterior to the long-lasting Late Cretaceous La Luna Sea led to a progressive shallowing of depositional environments. By the early Paleogene, deposition of the Guasare succession took place in the Maracaibo platform province (nowadays north Perijá Range and most of the subsurface of the MB). The latter was bounded to the south, southwest and southeast by a wide deltaic province (nowadays the southern limits of the MB, the Venezuelan Andes and the Santander Massif in Colombia); and by a deep turbiditic province to the east and northeast (nowadays the easternmost MB and the Trujillo Range), where deep turbidite facies (the Trujillo flysch) were deposited. In general, this Paleocene unit consists of shallow calcareous banks interdigitated with progradational siliciclastic facies (González de Juana et al. 1980). We only show two photographic examples of this unit. Those examples belong to one core in the subsurface of western MB.

Mixed carbonate and siliciclastic facies of the Guasare Formation are carbonate-dominated at the base, while the middle and upper parts are siliciclastic-dominated (González de Juana et al. 1980; Hernández-Pérez 1997). The thickness of this unit ranges between ~190 m in the subsurface of the

MB, to up to 400 m in the northern Perijá Range (González de Juana et al. 1980; Ministerio de Energía y Minas 1997). Calcareous lithofacies consist of shallow-marine, bivalve-dominated floatstones-rudstones also containing gastropods, oysters, echinoderms, foraminifera, ostracods, calcareous algae and corals (the Heterozoan factory prevails over both the Photozoan T-factory and the neritic lime-mud factory), with variable amounts of glauconite and sand-sized quartz grains; and tidal flat black mudstones, differentiated in black laminated and black bioturbated, with siderite and pyrite nodules, poorly preserved foraminifera tests and intervening thick laminae of fine-grained sandstones. Environmental reconstruction indicates that sedimentation of the Guasare Formation occurred in a dynamic delta setting with tidal influence. In such setting, delta lobe progradation, abandonment, transgression and renewed progradation were the dominant controlling factors on sedimentation (Hernández-Pérez 1997).

2.6.10 Maracaibo Basin: Source Rocks and Hydrocarbon Reservoirs

The main effective (thermally mature) source rock in the MB is the La Luna Formation *sensu stricto* (Talukdar and Marcano 1994; Gallango et al. 2002; Goddard and Talukdar 2002; Escalona and Mann 2006b; Mann et al. 2006; López and Lo Mónaco 2017; among others). This Late Cretaceous unit (middle-late Cenomanian to late Santonian) encompasses the OAE 2 and the OAE 3. Secondary Cretaceous source rocks that have contributed to the generation and accumulation of hydrocarbons are represented by the Machiques Member (OAE 1a), the upper part of Piche Member (OAE 1b first sub-event), and the base of the Lisure Formation (OAE 1b second sub-event) (Méndez-Dot et al. 2015; Méndez-Dot 2017; this book). Above-mentioned organic-rich units reached the thermal maturation window as a consequence of two main subsidence episodes. The first occurred during the Paleogene (60–40 Ma). Tectonically-induced subsidence was triggered by the oblique collision between the South American and Caribbean plates, leading to burial of source rocks to depths of 4–5 km in the central and northeastern areas of the MB. The second episodes occurred in the Neogene (from 20 Ma to Present). Source rocks were buried to depths of 2–6 Km in southwestern area of the MB during the Neogene uplift the of the Perijá Range-Venezuelan Andes (Escalona and Mann 2006b; see also Lugo and Audemard 2021).

Most of the hydrocarbon volumes generated by Cretaceous source rocks migrated up-section to thick and highly porous siliciclastic sandstone reservoirs of Eocene (Misoa and Mirador Formations) and Miocene (Lagunillas and La Rosa formations) ages (Lugo and Mann 1995; Parnaud et al. 1995a; Yoris and Ostos 1997; Goddard 2006; Escalona and Mann 2006b; Mann et al. 2006; Lugo and Audemard 2021; among others). Cretaceous reservoirs include late Aptian (Piché Member) and latest Albian (Maraca Formation) limestones, characterized by secondary-dissolution porosities, subordinate primary porosity, and structurally-controlled fractures (see Sánchez et al. 1999; Nelson et al. 2000; Pöppelreiter et al. 2005; Castillo and Mann 2006; Méndez-Dot et al. 2015; Méndez-Dot 2017). Barremian sandstones of the Río Negro Formation represent the deepest Cretaceous reservoirs in the basin (Escobar et al. 2022; Méndez-Dot 2017), reaching depths down to 5650 m (18,500 ft) in western Lake Maracaibo. In addition, migration of hydrocarbons towards greater depths reached the fractured Paleozoic basement in a few oil fields of the MB (Nelson et al. 2000; Koning 2003).

The depths of the reservoirs are directly associated with the quality of the stored hydrocarbons. Both API gravity and total sulfur content (wt.%) of the produced crude oils from Cretaceous reservoirs have an approximate range between 25°-40° API and 0.3–2.3%, respectively (Gallango and Tocco 1994; Escobar et al. 2022; Petroregional del Lago, S.A., internal data); while oils produced from Paleogene and Neogene reservoirs have an approximate range of 10°-40° API and total sulfur content of 1.5–6.5% (Gallango and Tocco 1994; López et al. 1995; López and Lo Mónaco 2004; Escobar et al. 2011; Lugo and Audemard 2021). Bacterial biodegradation and/or water washing acted on many Cenozoic, less deep reservoirs, increasing the relative concentration of heavier and sulfur-rich hydrocarbons at the expense of lighter degraded/washed hydrocarbons.

2.7 Barinas-Apure Basin: Tectonic Setting and Cretaceous Stratigraphy

From the late Albian to the early Campanian, the area occupied by the Barinas-Apure Basin (BAB), located in southwest Venezuela (Fig. 2.21), was part of the same epeiric sea that encompassed the Maracaibo and Eastern Venezuela basins. The exhumation of the Venezuelan Andes partitioned western Venezuela into the post-Middle Miocene Maracaibo and Barinas-Apure foreland basins (Parnaud et al. 1995a). The latter basin occupies an area of approximately 95,000 km^2, being the second largest oil-producing basin in Venezuela (see Chacín et al. 2005).

The BAB is located northwest of the Guayana Shield, spanning a vast area commonly named "Los Llanos" (Los Llanos meaning The Lowlands; see Fig. 1.7), being characterized by a lowland topography and the absence of major mountain ranges. The basin is bounded by the Venezuelan Andes to the west and northwest. El Baúl Massif (state of Cojedes) (Fig. 2.21), which is an igneous-metamorphic Paleozoic belt (rock ages ranging between 494 to 283 Ma; Viscarret et al. 2009, 2012), marks the northeastern limits of the basin. Towards southwest in Colombia, Los Llanos Basin is also a post-Middle Miocene foreland basin (Bayona et al. 2008) that represents not only the geographic continuation of the flat land topography, but the extension of both the Cretaceous passive margin strata and overlying Cenozoic siliciclastic units (Cooper et al. 1995; Villamil and Pindell 1998; Villamil et al. 1999; Rangel et al. 2000). For reasons related to oil production, the BAB is operationally divided in the Barinas sub-Basin (BSB) to the north and northwest, and the Apure sub-Basin to the south and southeast (Fig. 2.21).

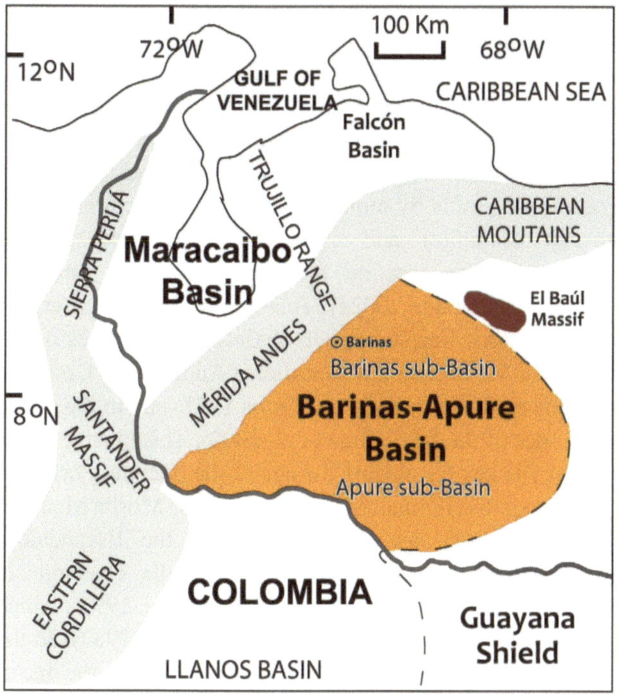

Fig. 2.21 Location of the Barinas-Apure Basin (BAB) and modern adjacent tectonic provinces (Mérida Andes, Perijá Range, Trujillo Range, Caribbean Mountains, Santander Massif and Eastern Cordillera). (Original figure modified from Talukdar and Marcano 1994 and Macellari 1988. Figure modified from Rey et al. 2004)

In this book we show photographic examples belonging to rock units in the subsurface of the BSB. The units considered are the O Member of the Escandalosa Formation, and the La Morita Member of the Navay Formation). Examples shown correspond to three cores in the Borburata Field, northeast of the city of Barinas (Fig. 2.21). Therefore, the following discussion is based on the stratigraphy of the subsurface of both the northwest BSB (i.e., oil fields located in the surrounding areas of the city of Barinas; e.g., Helenes et al. 1998), and the northern border of that sub-basin, northeast of the city of Barinas (e.g., Ruiz et al. 2000).

Late Cretaceous passive margin strata in the BSB crop out along the eastern flank of the Venezuelan Andes (e.g., González de Juana et al. 1980; Ministerio de Energía y Minas 1997) and are buried in the subsurface of that sub-basin (e.g., Parnaud et al. 1995a). Those sedimentary rocks are unconformably overlying Jurassic continental sediments and associated volcanic rocks of the La Quinta Formation. During Barremian, Aptian and early Albian, the inland location of the BSB was still out of reach of the expansive epeiric sea. Therefore, Cogollo Group equivalent carbonate sedimentation is absent, being the sandstones and siltstones of the Aguardiente Formation of late Albian to early-middle Cenomanian age the basal sedimentary unit (Helenes et al. 1998) (Fig. 2.22). The Cretaceous strata in western BSB

Fig. 2.22 Stratigraphic scheme for Cretaceous sedimentary units deposited in the western Barinas sub-Basin (BSB). Only major unconformities are indicated. Abbreviations: *Fm* formation, *Mb* member

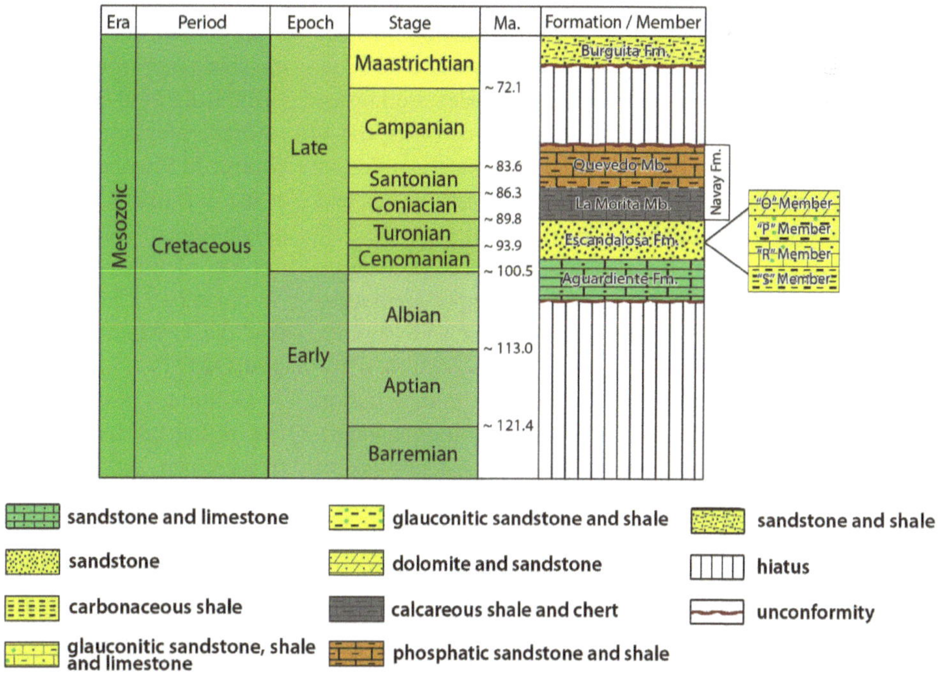

have an approximate thickness of ~395 m (1,300 ft) (Helenes et al. 1998). Burial depths in the main oil fields in both Barinas and Apure sub-basins are between 2100 and 4000 m (7,000 to 11,000 ft) (Chigne and Hernández 1990; Helenes et al. 1998; Pérez-Pérez et al. 2011). Thicknesses decrease continuously eastwards towards the Guayana Shield (Chigne and Hernández 1990). In general, carbonate sedimentation in the BAB is minor compared to Cretaceous carbonate units in the MB. Once fully marine conditions were reached, shallow-nearshore environments in the former basin were highly influenced by the proximity of the Guayana Shield as source of terrigenous sediments.

2.7.1 Escandalosa Formation (middle Cenomanian to late Turonian)

In the subsurface of the BSB, the Escandalosa Formation consists of a siliciclastic-dominated succession at the lower and middle parts, and limestones, massive dolomites, dolomitic limestones and dolomitic shales at the upper part. This unit overlies basal siliciclastics of the Aguardiente Formation (Fig. 2.22), representing the time-equivalent facies of both the lowermost part of the La Luna Formation in the subsurface of the MB and of the upper part of the Capacho Formation in the Venezuelan Andes (see Fig. 2.18). Its thickness is approximately 120 m in the subsurface of the surrounding areas of the city of Barinas (Helenes et al. 1998; Méndez-Baamonde 2002a; Pérez-Pérez et al. 2011; Reverón and Roomer 2014).

The depositional environment of the Escandalosa Formation in western BSB has been interpreted as a siliciclastic-dominated tidal flat (Reverón and Roomer 2014) which evolved towards the top to a carbonate-dominated, hypersaline peritidal setting (Méndez-Baamonde 2002a). The basal dark, micaceous-carbonaceous shales are known as the "S" Member. In ascending stratigraphic order, the S Member is followed by the "R" Member, which consists of an interdigitation of glauconitic-rich sandstones, shales and limestones (González de Juana et al. 1980); and the "P" Member, which consists of an interdigitation of glauconitic sandstones and shales. This member is an important hydrocarbon reservoir in the BSB (e.g., Reverón and Roomer 2014).

The uppermost "O" Member consists of an interdigitation of dolomites, sandstones and shales with poor fossil content. Marked variations in the lithology, from pure sandstones to dolomitic shales and pure dolostones, combined with local but prominent authigenic mineralization processes (pyrite, fluorapatite and glauconite), indicate variations in paleodepths, siliciclastic influx, carbonate precipitation and redox conditions. Within this member, porous facies consist of a combination of pure dolostones, sandy dolostones, dolomitic sandstones and dolomitic shales. The highest porosity and permeability values are present in massive dolostones hosting intercrystalline porosity. Other secondary porosities such as vug and intracrystalline porosities and microfractures are common. In the Borburata field (northeast of the city of Barinas), porosity values in the O Member range between 8% and 15%, having average thickness of ~20–25 m (~ 70–80 ft). Massive dolomitization occurred as the tidal flat environment (mainly shallow subtidal) evolved to a hypersaline-sabkha environment, where dolomitization occurred in an intertidal setting. Combined massive dolomitization (i.e., mole by mole replacement of calcium carbonate by dolomite) and short-termed subaerial-to-aerial exposure led to creation of those secondary porosities (Méndez-Baamonde 2002a). This dolomitic interval grades to fossiliferous limestones and shales and dolomites of the Guayacán Member in the Venezuelan Andes (e.g., González de Juana et al. 1980; Yoris and Ostos 1997) and in the northern border of the BSB (Ruiz et al. 2000). This change in the lithology is key for placing the age of the boundary between the Escandalosa Formation and overlying La Morita Member.

The accepted age for the base of the Escandalosa Formation (the S Member) is early-middle Cenomanian (Fig. 2.22) (Parnaud et al. 1995a; Helenes et al. 1998), which corresponds to the 97-Ma MFS of Haq et al. (1987). In the MB, this MFS is absent due to erosion. Regarding the age of the top of the unit, there is a short geochronologic discrepancy (~ 2 Myr) among different authors. According to Parnaud et al. (1995a, b), the MFS at the base of the overlying La Morita Member is well defined in the BSB, being equivalent to the 88-Ma datum of Haq et al. (1987) (which corresponds to the Coniacian-Santonian boundary by the time of that publication). Based on analysis of planktonic foraminifera, calcareous nannoplankton and palynomorphs on seven subcropping sections in the northern border of BSB, Ruiz et al. (2000) place the boundary between light-colored, planktonic-rich limestones and dolomites of the Guayacán Member and dark shales of the La Morita Member at the Turonian-Coniacian boundary (~ 89.8 Ma). Based on palynological assemblages, Helenes et al. (1998) consider the top of this unit in western BSB to be middle to late Turonian (~ 91.5 Ma) in age. All the above scenarios indicate that the O Member was deposited during Turonian times post-OAE 2.

Dolomites of the O Member were formed due to evaporative reflux, as hypersaline seawater ascended through the fine-grained, upper peritidal sediments (the sabkha-style dolomite; see Warren 2000; Machel 2004). The variety of authigenic minerals formed during early marine diagenesis suggest changing environmental conditions. Glauconite is commonly coating dolomite crystals and also filling intercrystalline pores; fluorapatite is replacing both fine-grained matrices and fish remains; pyrite framboids are coating dolomite crystals, glauconite and fluorapatite. Such mineralization processes, added to anoxic-suboxic conditions prevailing in adjacent basins to the west and east (the Maracaibo and Eastern Venezuela basins, respectively) suggest fluctuating conditions from dysoxia (formation of glauconite) to suboxia and even transient anoxia. All the above suggests that dolomites of the O Member were originated by microbially-mediated processes. We include massive dolostones of the O Member, which are the product of the replacement of micritic sediments under low-oxygen and hypersaline conditions, in the benthic automicrite factory. In the case of dolomitic shale facies, the nucleation and growth of euhedral dolomite crystals involved microbial mediation under conditions very similar to those proposed by Petrash et al. (2016) for the origin of Aptian and Albian dolomitic shales and dolomitic marlstones in the Cogollo Group.

2.7.2 Navay Formation (early Coniacian to early Campanian)

The Navay Formation is the thickest Cretaceous passive margin unit in the subsurface of the BSB (González de Juana et al. 1980; Parnaud et al. 1995a), representing about 10 Myr of sedimentation (Fig. 2.22). This formation is also the main source rock in that sub-basin (e.g., Tocco et al. 1997). In the subsurface of the surrounding areas of the city of Barinas, this unit is 152 m (~ 500 ft) thick (Helenes et al. 1998). Towards northeast, in the northern border of the BSB, its thickness is reduced to ~80 m (260 ft) (Ruiz et al. 2000). This succession of dark shales, cherts and sandstones is the time-equivalent of the middle and upper part of the La Luna Formation (see Figs. 2.17 and 2.18).

The maximum flooding in the BSB occurred posterior to the beginning of the anoxia in the MB. As previously mentioned, Helenes et al. (1998) dated a first MFS at 91.5 Ma (middle to late Turonian); while according to Ruiz et al. (2000), that MFS occurred approximately 1.7 Myr later, in the Turonian-Coniacian boundary. Both studied areas are geographically near, and thicknesses reported for the organic-rich La Morita basal Member are similar (~ 30 m). In both scenarios, the uppermost facies of this member span the Coniacian-Santonian boundary interval (Fig. 2.22) (i.e., the OEA3).

The basal La Morita Member is the most important oil-prone unit in the BSB (Chigne and Hernández 1990; Parnaud et al. 1995a, b; Tocco et al. 1997; Macsotay et al. 2003; Rojas 2012). This black shale succession consists of organic-rich and laminated shales, siliceous shales and phosphatic limestones. Microfossils consist of planktonic foraminifera, SBF and radiolarians, as well as diatoms and dinoflagellates. Foraminiferal assemblages are dominated by planktonics of the genera *Planoheterohelix*, *Muricohedbergella*, *Globigerinelloides*, *Whiteinella*, *Dicarinella*, *Hastigerinoides*, and *Archaeoglobigerina*; while SBF are dominated by the genera *Bulimina*, *Bolivina*, and *Gavelinella*. Phosphatic peloids and phosphatized fish remains are common (Ruiz et al. 2000).

The organic enrichment of the La Morita Member is lower (TOC content between 0.20% and 4.7%) (Tocco et al. 1997; Ruiz et al. 2000) compared to the La Luna and Querecual formations. Even so, its TOC content ranges between "moderate" and "excellent", according to the generative potential parameters of a source rock (Peters et al. 2005). High quality organic matter predominates (Type I and Type II kerogen) (Tocco et al. 1997; Bralower and Lorente 2003). Redox conditions during deposition of these black shales varied from anoxic to dysoxic (Ruiz et al. 2000).

The Late Cretaceous transgression flooded the shallow epeiric platform towards the Guayana Shield and created pelagic realm-like conditions for the flourishing of opportunistic planktonic foraminifera and low-oxygen SBF. In addition, the abundance of radiolarians, diatoms and dinoflagellates suggests the influence of episodic, silicious-rich upwelling currents (Ruiz et al. 2000). Cherts of the La Morita Member, which are composed of a microcrystalline siliceous matrix, are very similar to those with fine-grained matrices in the Táchira Ftanita Member, but slightly older in age.

Carbonate production in the La Morita Member is similar to that of both the La Luna *sensu stricto* and the Táchira Ftanita Member, but paleodepths were shallower (probably within the middle to outer neritic zone). The planktic association between planktonic foraminifera and radiolarians is common (the P-factory). The Heterozoan factory is represented by common low-oxygen SBF. Dark-colored matrices (traduced in organic enrichment) and common fluorapatite mineralization processes suggest eutrophic to hypertrophic conditions.

The increasing influence of upwelling currents and related partial oxygenation of the water column during late Santonian to early Campanian is clearly registered in the upper Quevedo Member. In the northern border of the BSB, this unit is ~50 m (160 ft), where this succession consists of dark shales, phosphatic shales and limestones, sandy phosphorites, sandstones and cherts (Ruiz et al. 2000). Towards southwest in the western area of the same sub-basin, the thickness goes up to 120 m (400 ft) (Helenes et al. 1998). Benthic foraminifera, phosphate nodules, and peloidal grains are the main components, as well as broken and abraded mollusc fragments (Erlich et al. 1999a; Ruiz et al. 2000). A recent study carried by Linares (2020) indicates substantial phosphatization in a succession of sandstones (~ 40 m thick) of the Quevedo Member cropping out in the eastern flank of the southern Venezuelan Andes (state of Táchira). The phosphatized interval is ~28 m thick and the samples studied have P_2O_5 content up to 23%, being classified as phosphorites, and phosphatic sandstones and siltstones. The author states that such P enrichment allows consider this interval as an ore phosphate deposit of economic importance. Results obtained by Linares (2020) are concordant with lithofacies descriptions of the Quevedo Member made by Erlich et al. (1999a, 2000), Ruiz et al. (2000), and Macsotay et al. (2003).

2.7.3 Barinas-Apure Basin Active Margin Stage

The top of the Quevedo Member throughout the BAB is bounded by a regional unconformity (Fig. 2.22). Deformation episodes related to the oblique collision of the CP against northwestern South America during the Campanian led to the migration of the foreland depocenter in the MB to a less deep forebulge in the BAB (Parnaud et al. 1995a). The strata overlying the Campanian unconformity is the Burguita Formation of middle to late Maastrichtian age (Helenes et al. 1998). The latter unit represents the beginning of the active margin sedimentation. In the subsurface of the BAB, this succession of sandstones and siltstones has thicknesses between 0 and 180 m, as the top was unevenly eroded throughout this large basin [Feo-Codecido (1972) in Ministerio de Energía y Minas 1997]. Such erosive event was followed by an ~11 Myr hiatus and major unconformity between these Maastrichtian facies and Middle Eocene sandstones of the Gobernador Formation (Helenes et al. 1998).

2.7.4 Barinas-Apure Basin: Source Rocks and Hydrocarbon Reservoirs

La Morita Member is the main oil-prone unit in the BSB (e.g., González de Juana et al. 1980; Tocco et al. 1997; Yoris and Ostos 1997; Erlich et al. 1999a, 2000). The lower part of the Quevedo Member is also considered as an effective source rock by some authors (Goddard 2006; Pérez-Pérez et al. 2011; Rojas 2012). Turonian porous facies of the P and O members of the Escandalosa Formation are the most significant Cretaceous reservoirs, followed by basal Albian sandstones of the Aguardiente Formation. Eocene sandstones of the Gobernador Formation are the most productive oil reservoir in the BSB, while Oligocene sandstones of the Guafita Formation are the main reservoirs in the Apure sub-Basin (see Parnaud et al. 1995a; Yoris and Ostos 1997; López et al. 1998; Pérez-Pérez et al. 2011; Goddard 2006; Rojas 2012).

The API gravity of the oils produced by BAB reservoirs varies in a wide range between 12° and 32° API (Yoris and Ostos 1997; López et al. 1998; Goddard 2006). Biomarkers analysis indicates that those oils have been sourced from marine, algal, and bacterial organic matter with variations in terrigenous organic matter input. Additionally, geochemical parameters such as the V/Ni ratio, $\delta^{13}C$ of the saturate and aromatic fractions suggest a greater contribution of terrigenous organic matter in the more siliciclastic dysoxic lithofacies for the crude oils belonging to the La Victoria and Guafita fields (located towards southwest, near the Colombian border in the Apure sub-Basin). In contrast, the crude oils from the BSB fields were generated from more carbonate-rich, anoxic lithofacies with a low contribution of terrigenous organic matter (López et al. 1998; López and Lo Mónaco 2017).

The source rocks of crude oils from the BSB still need to be identified in more detail. This suggests that, although the main source rocks in Venezuelan oil-producing basins are well established, there are still doubts concerning the lateral and vertical variations in organic and inorganic facies and paleo-oxygenation conditions for that sub-basin. Therefore, it is also possible that other source rocks might have contributed to the generation and accumulation of hydrocarbons (López and Lo Mónaco 2017).

2.8 Eastern Venezuela Basin: Tectonic Setting and Cretaceous Stratigraphy

The onshore Eastern Venezuela Basin (EVB) is the second richest hydrocarbon province in South America after the MB, with a total area (onshore and offshore) of about 200,000 Km^2. This includes the state of Delta Amacuro (i.e., the Orinoco Delta plain) and the Venezuelan Atlantic continental shelf (Fig. 2.23; see Di Croce et al. 1999). The EVB is bounded by the Guayana Shield and the course of the Orinoco River to the south, by the Atlantic oceanic crust to the east, by metamorphic rocks of the Araya-Paria Peninsula to the northeast, and by the Interior Range to the northwest. In the northeast area of the basin, sedimentary units of Cretaceous, Paleogene, and early Neogene ages crop out in the Eastern Interior Range (EIR). This range includes Cretaceous rock sections outcropping on the islands of Chimana Grande and La Borracha (Fig. 2.23), with some of them among the most studied sections in the basin. The El Baúl Arch separates the westernmost area of the EVB from the BAB (see Fig. 1.3).

The tectonic complexity of the EVB is inherited from its geographic location, just south of the southeastern margin of the Southern Caribbean PBZ (i.e., the eastern margin of the Venezuelan Caribbean Ranges block; see Figs. 2.9 and 2.10), and southwest of the west-dipping subduction of the Atlantic plate beneath the eastern margin of the CP (Fig. 2.23; see Ughi 2021). This subduction zone has led to the creation of the Lesser Antilles arc (the present-day active volcanic arc) and the Barbados accretionary prism. All the above indicates that the tectonic region in which the EVB lies comprises a zone of interaction between the northeastern margin of the

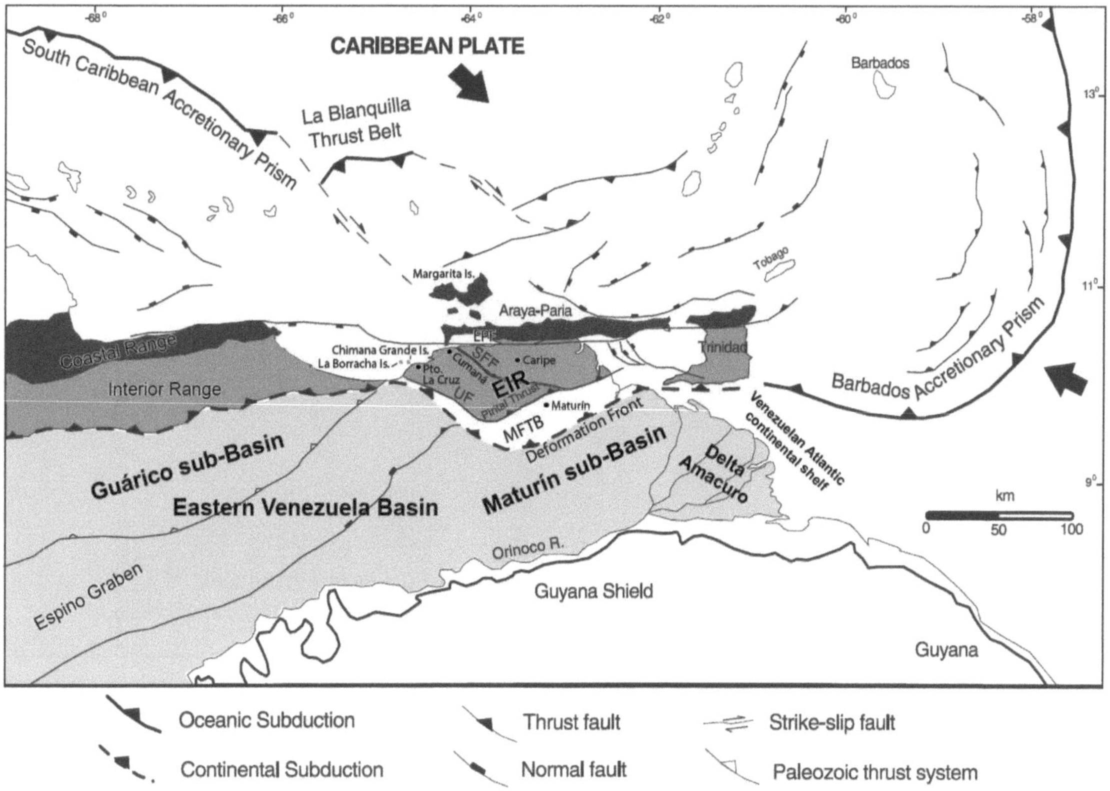

Fig. 2.23 Tectonic map of northeastern Venezuela, showing the petroleum provinces and uplifted terranes of the Eastern Venezuela Basin (EVB); as well as the interaction between the oceanic CP and the continental South America plate. Abbreviations: *EIR* Eastern interior range, *EPF* El Pilar fault, *Is* Island, *R.* River, *UF* Urica fault, *SFF* San Francisco fault. (Original figure simplified from Ysaccis and Audemard 2000. Figure modified from Jácome et al. 2003a)

continental South American plate, the oceanic CP, and the western margin of the oceanic Atlantic plate at tropical-Caribbean latitudes.

The late Paleogene and Neogene evolution of the EVB have been controlled by the oblique collision of the CP against northeast South America, which occurred during Oligocene times. This zone of convergence involves the subduction of South America under the CP in northeastern Venezuela (see Summa et al. 2003). This subduction-collision has led to the exhumation of both allochthonous metamorphic units of the Araya-Paria Peninsula and para-autochthonous sedimentary units comprising the EIR. The latter units, although deposited locally on the northern margin of the South American plate, were transported southward short distances within the basin (Lugo and Audemard 2021).

The western and eastern domains of the EVB are separated by the Espino Graben (Fig. 2.23) (James 2000; Jácome et al. 2003a; Parra et al. 2011; González et al. 2017). This graben is a Jurassic intracontinental basin with SW-NE orientation and approximately 80–120 km wide and over 500 km in length. It contains a thick sedimentary succession (between 3000 and 5000 m) of Jurassic, Cretaceous and Cenozoic rocks (Audemard and Serrano 2001).

Based on hydrocarbon production, tectonic and stratigraphic differences, and subsidence history, the EVB is divided into three petroliferous provinces (Fig. 2.23): the Guárico sub-Basin to the west; in the east, the Maturín sub-Basin (or the Maturín Foreland Basin) to the south and the Monagas Fold and Thrust Belt (MFTB) to the north, which are the two most prolific petroleum provinces in the basin (see Erlich and Barrett 1992; Parnaud et al. 1995b; James 2000; Jácome et al. 2003a, b; Bartok 2003; Summa et al. 2003; Parra et al. 2011).

In this section and following sub-sections, the discussion is focused on Cretaceous passive margin units located in several outcropping sections in the EIR (i.e., the northeastern area of the EVB), and the subsurface of the MFTB (Fig. 2.23). We discuss the stratigraphy and major lateral facies variations from four closely spaced areas within this complex foreland fold and thrust belt system (Fig. 2.24): (**a**) outcrops in the islands of Chimana Grande and La Borracha, Mochima National Park, northern Anzoátegui State (e.g., Crespo de Cabrera et al. 1999; Carvajal-Chitty and Rey 2020); (**b**) one outcrop in northeast Anzoátegui State, which contains the type section of the Querecual Formation at the Querecual River (e.g., Truskowski 2006); (**c**) outcrops in northern

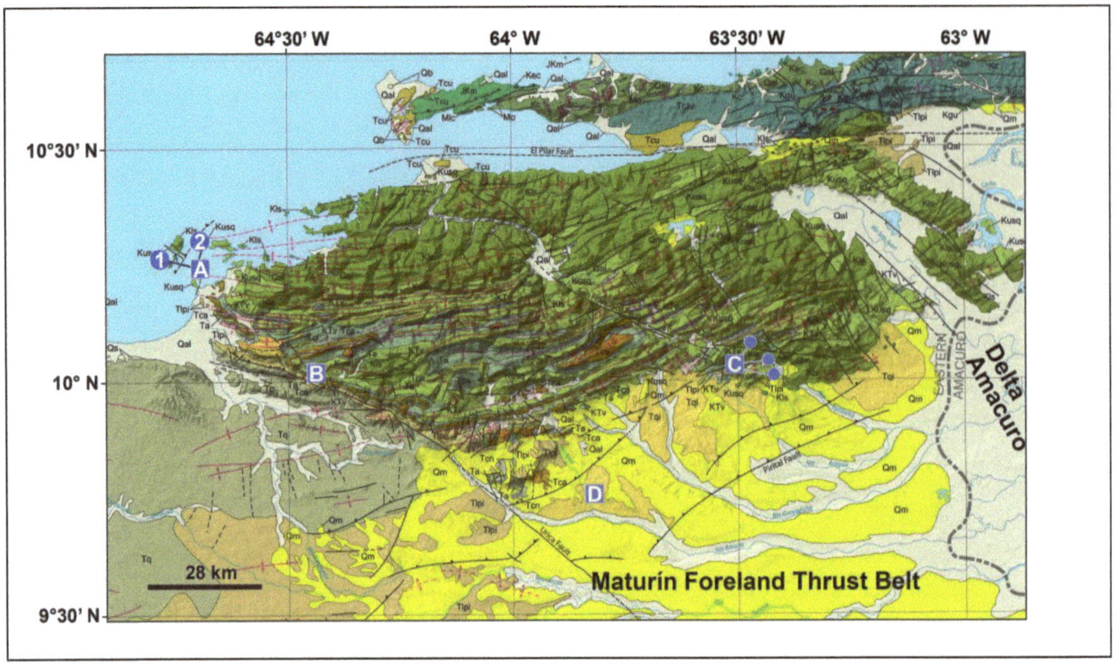

Fig. 2.24 Close view of the geologic-shaded relief map of Venezuela in its northeastern region. The locations of Cretaceous rock successions, on which we have based the stratigraphy of the northeastern area of the EVB shown in Fig. 2.25, are indicated. (**a**) Outcrops in two islands of the Mochima National Park, northern Anzoátegui State: 1) La Borracha Island; 2) Chimana Grande Island. (**b**) Outcrop in northeast Anzoátegui State; the type section of the Querecual Formation at the Querecual River; (**c**) Three rock outcropping sections at García Hill, northern Monagas State; (**d**) Subsurface of western Monagas foothills, northwest Monagas State. We list the most significant mapped sedimentary units in the image. Kls = Early Cretaceous: Barranquín, Borracha, El Cantil, and Chimana formations; KTv = Cretaceous to Eocene: Vidoño Formation; Kusq = Late Cretaceous: San Juán, San Antonio, and Querecual formations; Qm = Pleistocene: Mesa Formation; Ta = Oligocene: Areo Formation; Tca = Eocene: Caratas Formation; Tcl = Eocene to Oligocene: Caratas and Los Jabillos formations; Tlpi = Miocene to Pliocene: Las Piedras Formation; Tq = Oligocene and Miocene: Capaya, La Pica, and Chaguaramas formations; Tqi = Pliocene: Quiriquire Formation; Qal = Pleistocene and Holocene: alluvium. (Figure modified from Hackley et al. 2006, Sheet 1)

Monagas State at García Hill (south of the town of Caripe; see Fig. 2.23) (e.g., Carvajal-Chitty and Rey 2020); (**D**) the subsurface in western Monagas foothills, northwest Monagas State (west of the city of Maturín) (e.g., Helenes and Somoza 1999).

The tectono-sedimentary history of the EVB is similar to that of the MB. The sedimentary infilling of these basins consists of the same megasequences (i.e., rift, passive margin, and oblique collision megasequences), but with a diachronism from west to east in the finishing of the passive margin stage. This is due to the eastward migration of the collision zone between the Caribbean and South American plates. The thrust belt load applied to the northern border of South America began earlier in western Venezuela. First in the MB during the Late Cretaceous (middle Campanian), and later in the Falcón-northern Lara area during the Paleocene. This involved the progressive migration of the foredeep towards the east (see Parnaud et al. 1995a, b). As mentioned in Sect. 2.5, such diachronism also involved the exhumation of mountain ranges in Northern Venezuela, first in the west and later in the east; as well as the beginning of dextral wrenching along the Southern Caribbean PBZ, first in the west with the activation of the Oca-Ancón fault by ~17–15 Ma, and later in the east with the activation of the El Pilar fault by ~12–10 Ma (see Audemard 2009; Fajardo et al. 2023).

In the EVB, the passive margin stage is considered to have finished during Oligocene times, although this age becomes younger towards the easternmost areas of the basin (Parnaud et al. 1995b; Lugo and Audemard 2021). The sedimentary strata cropping out in the EIR consist mostly of passive margin units deposited from the Early Cretaceous to the Oligocene (see Fig. 2.24 and the mapped units listed in the footnote of that figure). The latter, combined with the age of the beginning of the exhumation of this range in the Middle Miocene (~ 13 Ma; see Locke and Garver 2005; Cruz et al. 2007; Arkle et al. 2021), suggests an Early to Middle Miocene age for the beginning of the active margin stage in the northeastern area of the EVB. The east-west right-lateral strike-slip El Pilar master fault separates, at the surface, para-autochthonous sedimentary units of the EIR to the south from allochthonous metamorphic units of the Araya-Paria Peninsula to the north. The NW-SE Urica and San Francisco faults are secondary synthetic Riedle shears (Figs. 2.23 and 2.24; Summa et al. 2003; Fajardo et al. 2020, 2023).

The origin of the Neogene foreland Maturín sub-Basin is directly linked to the uplift of the EIR and the MFTB (also known as the Monagas foothills, or the Maturín foredeep), with these two uplifted terranes separated by the Pirital Thrust (Fig. 2.23). From the Early Miocene until Modern times, tectonic overloading and oblique subduction of the South American plate underneath the CP in northeast Venezuela has caused compressive deformation and shortening, accounting for the uplift of the EIR. To the south of the "Deformation Front" (Fig. 2.23), the tectonic loading associated with the EIR thrust-sheet forced northern South America to flex downward between the Guayana Shield and the El Pilar Fault, accounting for the creation of the Maturín sub-Basin. This topographic depression is regarded as a good example of a peripheral foreland basin (Bartok 2003; Jácome et al. 2003a, 2003b).

The areas now occupied by the Espino Graben, the EIR and the MFTB were main depocenters for sediment accumulation during the Cretaceous to early Paleogene passive margin stage, while towards the Guayana Shield (to the south), the accommodation space reduced progressively. In the EIR, the most complete and studied sections of the Cretaceous passive margin strata in eastern Venezuela crop out (e.g., Hedberg 1950; González de Juana et al. 1980; Di Croce et al. 1999; Crespo de Cabrera et al. 1999; Truskowski 2006; Caroprese 2009; Carvajal-Chitty and Rey 2020). Depositional marine environments represented in such Cretaceous strata range from continental and nearshore in the south to outer neritic and upper bathyal in the north.

In the subsurface of the MFTB, a complex structural regime is dominated by the Late Miocene Pirital Thrust deepening to the north and several right lateral strike-slip faults (Helenes and Cabrera 2003; Fajardo et al. 2020). Most of the discovered hydrocarbons in that area are located adjacent to present-day structural deeps, south of the Pirital Thrust (Fig. 2.23; see Summa et al. 2003).

2.8.1 Barranquín Formation (Hauterivian to late Barremian)

Siliciclastic-dominated facies of the Barranquín Formation of Hauterivian to late Barremian age represents the basal Cretaceous passive margin unit in northern EVB (Fig. 2.25). This unit is the time-equivalent of the Río Negro Formation and of the base of the Tibú Member in the MB. The

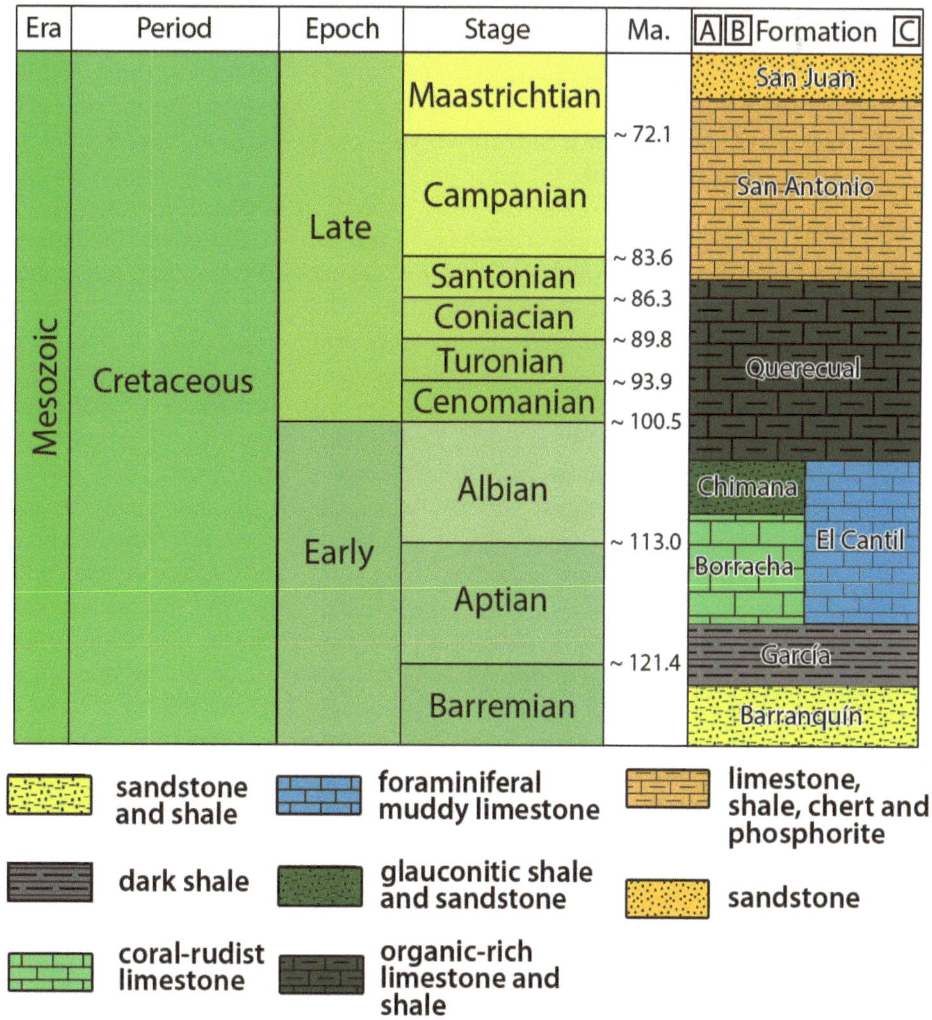

Fig. 2.25 Stratigraphic scheme for the Cretaceous strata cropping out in the Eastern Interior Range (rocks ranging in age from Barremian to Maastrichtian; the Hauterivian interval of the Barranquín Formation is not included). Letters A and B (localities in the northeast Anzoátegui State) indicate the stratigraphy at the islands of La Borracha and Chimana Grande, and the type section of the Querecual Formation, respectively. The letter C indicates the stratigraphy at García Hill (north Monagas State). See the specific geographic locations of those localities in Fig. 2.24

Barranquín Formation crops out across the entire EIR. Although a Hauterivian to late Barremian age is generally accepted for this unit (e.g., Parnaud et al. 1995b; Helenes and Somoza 1999; taking into account that in the ICC of Cohen et al. 2013a, b, c; updated v2023/04, the age of the Hauterivian-Barremian boundary was taken at 125.8 Ma), an older age for the base is likely understood (probably Valanginian), since the lower contact is unknown. The lithology mainly consists of sandstones and shales and occasional limestones towards the top. The thickness varies widely, from 350 m to up to 1800 in the EIR (see González de Juana et al. 1980; Minsiterio de Energía y Minas 1997); being ~700 m thick in the subsurface of north Monagas State (Helenes and Somoza 1999). The combined occurrence of the planktonic foraminifer *Globigerinelloides blowi* and the larger benthic foraminifer *Choffatella decipiens* in a siliciclastic rock interval at the lower part of the studied section by Carvajal-Chitty (2015) and Carvajal-Chitty and Rey (2020) at García Hill suggest an early to early-late Barremian age. This interval is the correlative to the Choffatella level at the base of the Tibú Member in the MB. Nevertheless, the long stratigraphic time range of both *C. decipiens* and *G. Blowi*, combined with the absence of other biostratigraphic markers, prevents the estimation of a more precise age for those rock intervals.

2.8.2 García Formation (late Barremian to early-middle Aptian)

In the EIR, the Barranquín Formation is overlying by marine dark shales, calcareous shales and marls of the García Formation (Fig. 2.25). This unit was deposited during the peak of the early Aptian marine transgression in eastern Venezuela. At García Hill (Fig. 2.24), this succession is approximately 90 m thick. Paleodepths are estimated between 60 to 150 m (Falcón 1989a, b). In La Borracha Island, the occurrence of *Choffatella decipiens* at the lower part of the García Formation (Carvajal-Chitty 2015; Carvajal-Chitty and Rey 2020) indicates a wider range of appearance for this species in the EVB compared to the MB, encompassing the late Barremian (and probably the early Aptian?). The same stratigraphic time range for this species was previously considered by Bolli et al. (1994) in the EIR and Trinidad. Furthermore, at García Hill, the occurrence of the planktonic formainifer *Leupoldina cabri* in a rock interval of approximately 3 m at the lower part of the studied section by Carvajal-Chitty (2015) and Carvajal-Chitty and Rey (2020) allows correlate that interval with the lower-middle part of the Machiques Member. *L. Cabri* has a short stratigraphic time range in the early Aptian (see Verga and Premoli Silva 2002; Premoli Silva and Verga 2004). According to Gale et al. (2020), the *Leupoldina cabri* zone encompasses a time interval of ~2 Myr (from ~121.8 to 119.8 Ma).

2.8.3 El Cantil (early-middle Aptian to late Albian) and Borracha (early-middle Aptian to early Albian) Formations

Carbonate-dominated sedimentation in north EVB began with deposition of the El Cantil and Borracha formations. The former unit is both partly time and lithologic equivalent of the middle and middle-upper parts of the Cogollo Group (i.e., the Piché Member and the Lisure Formation); while the second unit is dominated by coral-rudist limestones, representing lithofacies and grain associations absent in western Venezuela during Early Cretaceous times.

The El Cantil Formation is a mixed carbonate-siliciclastic succession which crops out in eastern and northeastern EIR. Its thickness varies between 300 and 850 m (Ministerio de Energía y Minas 1997); towards south in the subsurface of north Monagas State, this unit is ~800 m thick. Deposition of the El Cantil Formation was diachronic from south to north, as it overlies late Barremian siliciclastics of the Barranquín Formation in the south (the subsurface of western Monagas foothills; Helenes and Somoza 1999); while towards northeast at García Hill, it overlies early-middle Aptian siliciclastics of the García Formation (Fig. 2.25) (Carvajal-Chitty and Rey 2020). This shallow-marine unit consists of mud- and grain-supported limestones interfingered with shales and sandstones (Erlich and Barrett 1992). Carbonate biota consists of LBF and smaller SBF (e.g., orbitolinids and smaller miliolids, respectively), oysters, aragonitic bivalves, gastropods, echinoderms and green algae (Carvajal-Chitty and Rey 2020).

Towards the coastlines to the north and northwest in the EIR, the lower part of the El Cantil Formation grades to reef and biostromic limestones of the Borracha Formation (Fig. 2.25). In the islands of Chimana Grande and La Borracha, this unit of 190 m maximum thickness (González de Juana et al. 1980) overlies the García Formation, and it consists of limestones and sandy limestones containing rudists, oysters, corals, and aragonitic bivalves, as well as minor glauconitic sandstones (Carvajal-Chitty and Rey 2020). The Borracha Formation encompasses the early-middle Aptian to early Albian interval, with a possible middle Albian age for the top. Nevertheless, and like for the El Cantil Formation, the lack of planktonic markers in the late Aptian to middle Albian interval prevents the consideration of a more precise age for the top of those units. The geographic extension of the García Formation is restricted to the northernmost part of the EIR, from the islands of the Mochima National Park, to the surrounding areas of the city of Cumaná, state of Sucre (Fig. 2.23).

Although we only show one photographic example of the El Cantil Formation, we consider this unit and the Borracha Formation to be of great paleoecological importance within the Cretaceous passive margin stage in Venezuela. In the case of the El Cantil Formation, similarities in lithofacies and grain associations suggest similar paleodepths and trophic gradients respect to the Piché Member (Apón Formation) and the Lisure Formation in the MB. Nevertheless, organic-rich intervals corresponding to the OAE 1b are missing in the El Cantil Formation, indicating that local factors did not allow the development of anoxic conditions that prevailed in the MB during late Aptian and early Albian times. Regarding the Borracha Formation, abundant limestones consisting of corals and rudists represent a photozoan-heterozoan association that is absent in the entire Cogollo Group. In the latter, the ecological niche of rudists was occupied by oysters, while corals are very scarce or absent. Cretaceous rudist-rich platforms are included by Michel et al. (2019) in the Photozoan T-factory, as these molluscs were restricted to shallow-tropical, oligotrophic waters.

2.8.4 Chimana Formation (middle Albian to late Albian)

Towards the coastlines in the north area of the EIR, glauconitic shales, sandstones and limestones of the Chimana Formation overlie the Borracha Formation (Fig. 2.25). In the Chimana Grande Island, this unit is ~120 m thick (Hedberg 1950). The Chimana Formation is the shale-dominated equivalent of the upper part of the El Cantil Formation in the marine environments of the north EIR (González de Juana et al. 1980). These glauconitic shales conformably overlie shallow-water platform sedimentation. Deepening of depositional environments was accompanied by a change in paleoenvironmental conditions, from oxygenated waters in the euphotic zone, to dysoxic conditions in the mesophotic zone (~ 40 to 100 m deep). Ammonites and planktonic foraminiferal assemblages in this dark-glauconitic shale unit suggest low-oxygen conditions and a middle Albian age (Renz 1982). In the islands of Chimana Grande and La Borracha, planktonic foraminiferal assemblages suggest a middle to late Albian age for this glauconitic unit (Carvajal-Chitty and Rey 2020).

2.8.5 Querecual Formation (late Albian to middle Santonian)

In the north area of the EIR, a sharp contact between the Chimana and the Querecual formations, characterized by marked darkening of the facies upward in section formations marks the beginning of a long period of organic-rich sedimentation (~ 17 Myr). By the late Albian, a progressive and sustained increase in paleodepths as consequence of the ESL rise and associated marine transgression were combined with the perturbations linked to the early stages of the Caribbean LIP magmatism. These factors triggered deposition of black shales of the Querecual Formation. This unit is the main source rock in Eastern Venezuela (e.g., López and Lo Mónaco 2017), and together with La Luna Formation, represents one of the most prolific source rocks deposited during the mid-Cretaceous worldwide. Photographic examples belonging to the Querecual Formation correspond to the type section of this unit, located in northeast Anzoátegui State (Fig. 2.24).

The Querecual Formation crops out along the northern area of the EIR. Towards northwest, this organic-rich unit overlies the Chimana Formation (localities A and B in Fig. 2.24), while towards east and south it overlies the El Cantil Formation (localities C and D in Fig. 2.24, respectively; see also Fig. 2.25). Although far less studied in the subsurface compared to the La Luna Formation due to the absence of cores, the Querecual Formation is understood to encompass the subsurface of the MFTB (see Schneider et al. 2004). Hedberg (1937) first described this bituminous unit as an alternation of organic-rich limestones, marls and shales with abundant planktonic foraminifera, very similar to the La Luna Formation *sensu stricto* in terms of lithology, but average carbonate content (wt.%) in the Querecual Formation is higher. Therefore, limestones (mudstones, wacketones and packstones) and clayey limestones predominate over shales and calcareous shales (Crespo de Cabrera et al. 1999; Truskowski 2006; Caroprese 2009).

Average TOC content is ~2.4%, ranging between 0.5% and 7% (Alberdi-Genolet and Lafargue 1993; Lugo 2002; Truskowski 2006; Caroprese 2009). As in organic-rich microfacies of the La Luna Formation *sensu stricto*, the high content of preserved organic matter masks the mineral matrix in many microfacies of the Querecual Formation, thus preventing, partially or totally, its visualization with the polarizing microscope. Variations in trace metals enrichment indicate that the increase in the content of both TOC and total sulfur is related to anoxic-euxinic conditions (Lugo 2002). High quality marine organic matter (predominance of Type II over Type III kerogen), and upwelling indicators such as phosphatic sediments, cherts and radiolarians, characterize this organic-rich succession (Crespo de Cabrera et al. 1999).

La Luna and Querecual formations are chronologic and lithologic equivalents, however the beginning of the organic-rich sedimentation was diachronic from east to west. Detailed planktonic and benthic foraminifera biostratigraphy carried by Crespo de Cabrera et al. (1999) and Caroprese (2009) in the same section of the Querecual Formation in the Chimana Grande Island; as well as by Truskowski (2006) in the type section of that formation (see Figs. 2.23 and 2.24), indicates that deposition of that organic-rich unit began during the late Albian and lasted until middle Santonian (Fig. 2.25); while the accepted age for the base of the La Luna Formation is middle-late Cenomanian. Albian OP magmatism reported by Villagómez et al. (2011a), Wright and Wyld (2011), and Whattam and Stern (2015), associated with the formation of the CP, might explain the deposition of black shales in the EVB during pre-Cenomanian times, although such diachronism in the onset of anoxia with respect to that of the La Luna Formation has not yet been studied in detail.

According to Truskowski (2006), the base of the Querecual Formation is late Albian in age (*Rotalipora appenninica* Zone; the zonal schemes used were that of Sliter 1989; and Premoli Silva and Sliter 1999). A slightly younger late-early Albian age for the base of this bituminous unit is proposed by Crespo de Cabrera et al. (1999) (*Rotalipora ticinensis* Zone; the zonal schemes used were that of Caron et al. 1985; and Sliter (1989), as well as by Caroprese (2009) (*Ticinella primula* Zone, *Rotalipora subticinensis* Subzone; the zonal scheme used was that of Premoli Silva and Sliter 1999). A stratigraphic gap (or hiatus) was identified by Crespo de Cabrera et al. (1999) and Truskowski (2006) for the early Cenomanian interval, coinciding with the same missing interval in western Venezuela at the base of the La Luna Formation and at the lower part of the Capacho Formation. More recently, the presence of the calcareous nannofossil *Eiffelithus turriseiffelli*, and the planktonic foraminifer *Ticinella primula*, at the base of Querecual Formation highlight a 103 Ma age for the section in La Borracha Island. Towards southeast at García Hill, the presence of the planktonic foraminifera *Rotalipora appenninica* and *Ticinella primula* at the base of the same formation gives a 102 Ma age (Carvajal-Chitty and Rey 2020).

In any of the previous scenarios, the OAE 1d (~ 103 to 99.5 Ma) encompassing the Albian-Cenomanian boundary interval is well recorded at the base or the lower part of the Querecual Formation. Several organic-rich expanded sections of the Querecual Formation have been studied in the north area of the EIR, with thicknesses between 420 and 800 m (e.g., Yoris 1989; Alberdi-Genolet and Lafargue 1993; Crespo de Cabrera et al. 1999; Lugo 2002; Truskowski 2006; Caroprese 2009). Those sections also record the OAEs 2 and 3, encompassing the Cenomanian-Turonian and the Coniacian-Santonian boundary intervals, respectively. Such anomalous thicknesses indicate extraordinary basin subsidence, uncommonly high sedimentation rates, and sustained organic matter production rates for prolonging anoxia. Thicknesses of the Querecual Formation are comparable with expanded sections in Tarfaya Basin, southwest Morocco. In that basin, Cenomanian and Turonian black shale sections in several cores of exploration wells are up to 800 m thick, with TOC content up to 20% (e.g., Kuhnt et al. 2001; Kolonic et al. 2002, 2005; Lüning et al. 2004; Tsikos et al. 2004; Keller et al. 2008; Kraal et al. 2010; Poulton et al. 2015; Beil et al. 2018).

The long-lasting pelagic realm-like, stressed conditions during deposition of the Querecual Formation are well recorded in the foraminiferal assemblages as the dominating biota. Planktonic assemblages are dominated by the genera *Muricohedbergella*, *Planoheterohelix*, *Globigerinelloides*, *Ticinella*, *Globotruncana*, *Rotalipora*, *Whiteinella*, *Marginotruncana*, and *Dicarinella*. Low-oxygen SBF are common in rock intervals deposited under suboxic-dysoxic conditions, especially in the upper part, being these intervals more common than in La Luna Formation *sensu stricto*. The most abundant genera are *Gavelinella*, *Gabonita*, *Nodosaria*, *Dentalina*, *Praebulimina*, *Lenticulina*, *Orthokarstenia*, *Bolivina*, *Neobulimina*, *Fissurina*, *Pyramidina* and *Siphogenerinoides*. Radiolarians are moderately abundant, especially at the lower part of the succession. Phosphatized fish remains and fragments of *Inoceramus* are present in minor amounts (Crespo de Cabrera et al. 1999; Truskowski 2006; Caroprese 2009).

Paleobathymetric reconstructions for the deposition of the Querecual Formation have been a matter of debate. Detailed descriptions of foraminifera assemblages have led to different interpretations. Carrillo et al. (1995) estimate paleodepths between the outer-neritic to upper bathyal zones. Truskowski (2006) estimates greater paleodepths, between the upper and middle-lower bathyal zone. In contrast, Caroprese (2009) estimates shallower depositional environments, between the middle and outer neritic zones (100–200 m). Considering the similarities between the time-equivalent La Luna and Querecual formations, in terms of depositional settings (within the same mid-Cretaceous epeiric sea), lithology, TOC content, foraminiferal assemblages, and other components (e.g., radiolarian and *Inoceramus*), we agree more with the first scenario.

As for carbonate production, the carbonate factories that contributed to the sedimentation of the Querecual formation are, by definition, the same that dominated sedimentation of the La Luna Formation *sensu stricto*, i.e., the P-factory, the benthic automicrite factory, and the Heterozoan factory. In the Querecual Formation, a greater contribution of the last two of these factories is understood due to the overall predominance of limestones over shales and more abundant low-oxygen SBF.

2.8.6 San Antonio Formation (middle Santonian to middle Maastritchtian)

Along the entire EIR, the Querecual Formation grades conformably upward-section into the San Antonio Formation of middle Santonian to late Campanian age (Fig. 2.25) (e.g., Carrillo et al. 1995; Crespo de Cabrera et al. 1999). Such stratigraphy extends into the subsurface of north Monagas State, being understood to encompass the subsurface of the MFTB. These two units have been historically grouped under the name of the Guayuta Group (e.g., Hedberg 1950; González de Juana et al. 1980; Alberdi-Genolet and Lafargue

1993; López and Lo Mónaco 2017) in reference to the Late Cretaceous bituminous and siliceous-rich strata in eastern Venezuela.

The San Antonio Formation is about 350 m thick and its heterogenic lithology consists of an intercalation of sandstones, dark shales and silty shales, limestones, calcareous sandstones and cherts (Hedberg 1950; Marcucci 1973; González de Juana et al. 1980; Carrillo et al. 1995; Helenes and Somoza 1999). The paleoenvironments represented are marine, mainly middle neritic, with occasional transgressions to outer neritic (Helenes and Somoza 1999).

The abundance of cherts indicates that the San Antonio Formation is the time and partly lithologic equivalent of the Táchira Ftanita Member (see Marcucci 1973). In the same way, abundant phosphatized intervals, mainly consisting of peloidal packstones and wackestones, indicate that this unit is the time and partly lithologic equivalent of the Tres Esquinas Member (uppermost part of the La Luna Formation) and the Quevedo Member (middle and upper part of the Navay Formation) in western Venezuela. Such similarities in chert and phosphatic facies between western and eastern Venezuela suggest similar paleoenvironmental conditions during Santonian and Campanian times along Northern Venezuela. A strong influence of a more active upwelling system after the finishing of the anoxia is widely accepted as the main factor that triggered deposition of these granular phosphate limestones. Such regional phosphate mineralization is associated with the Campanian phosphorite giant episode, during which extensive phosphorites were deposited in the Middle Eastern Tethys (Egypt, Jordan, Saudi Arabia, Israel, and Syria), Morocco, and northern South America (Shields et al. 2000).

The photographic examples shown belonging to the San Antonio Formation correspond to grain-supported, phosphatized microfacies. Petrographic characteristics of these examples suggest an allochthonous origin for these poorly-studied granular phosphorites. Phosphatization processes most likely occurred under low-oxygen (probably dysoxic) and eutrophic conditions.

2.8.7 Latest Cretaceous to Early Cenozoic Siliciclastic Units

Contrary to the Maracaibo and the Barinas-Apure basins, the uppermost Cretaceous and the Paleogene strata in the EVB were deposited in a passive margin setting. In that context, conformably overlying the San Antonio Formation, siliciclastic facies of the San Juan Formation of middle Maastrichtian to early Paleogene age represent the uppermost Cretaceous strata in the EIR (Fig. 2.25), as well as in the subsurface of the MFTB. This unit is overlying in places by alternating siltstones and dark-gray shales (Vidoño Formation) of Paleocene age, and/or by Eocene glauconitic sandstones and siltstones (Caratas Formation) (Parnaud et al. 1995b; Di Croce et al. 1999; Helenes and Somoza 1999; Helenes and Cabrera 2003).

2.8.8 Eastern Venezuela Basin Active Margin Stage

As mentioned earlier, diachronism from west to east of the thrust belt load applied to the northern border of the South American continent led to an eastward migration of the foredeep during the late Paleogene and early Neogene. In eastern Venezuela, the passive margin sedimentation ended during the Late Oligocene, when it was terminated by the collision of the CP against the South American plate and the basin changed into a foreland basin (Parnaud et al. 1995b; Di Croce et al. 1999; Lugo and Audemard 2021). An unconformity separates passive margin Paleogene units from active margin Neogene strata (Helenes and Somoza 1999; Helenes and Cabrera 2003).

2.8.9 Eastern Venezuela Basin: Source Rocks and Hydrocarbon Reservoirs in its Northern Area

The Guayuta Group (Querecual and San Antonio formations) are thought to have generated over 90% of discovered hydrocarbons in the EVB, not including the Orinoco Heavy Oil Belt (Summa et al. 2003). Of those two units, the Querecual Formation is, by far, the main Cretaceous effective source rock in this basin. In addition, Late Oligocene-Early Miocene shales of the Naricual Formation are proposed by López and Lo Mónaco (2017) as secondary source rock contributing to the accumulation of hydrocarbons in the EVB. Rapid tectonically driven burial of Cretaceous passive margin rocks during Late Oligocene-Middle Miocene caused maximum burial depths (Erlich and Barrett 1992; Parnaud et al. 1995a, b; Di Croce et al. 1999). Hydrocarbon generation started in the Early Miocene (Parnaud et al. 1995b).

In the northern area of the EVB (northern Maturín sub-Basin and the MFTB), several thick and highly porous Cenozoic sandstone units are the main reservoirs. It highlights the Oligocene Los Jabillos, Merecure and Naricual formations (which have produced crude oils ranging from 36° to 38° API), and the La Pica Formation of Miocene age (which has produced crude oils ranging from 17.5° to 31° API). Late Cretaceous sandstones of the San Juan Formation (which has produced crude oils from 37° to 40° API) are the oldest sedimentary reservoirs in that area of the basin (Goddard 2006).

2.9 Falcón Basin: Tectonic Setting and Stratigraphy

The Falcón Basin (FB), also referred to as the "Falcón Anticlinorium" (e.g., González de Juana et al. 1980; Boesi and Goddard 1991), is an inverted graben located in northwest Venezuela. This WSW-ENE trending fold and thrust belt consists of Middle Eocene to Middle Miocene rock units, which are separated from the prolific oil-producing MB by the NNW-SSE trending of the Trujillo Range (Fig. 2.26) (Baquero et al. 2009). The anticlinorium covers approximately ~36,000 km^2 (Wheeler 1963; Audemard 2001), with approximate dimensions of 300 km long and 120 km wide (Wheeler 1963). It is bounded to the south by Cretaceous and early Paleogene igneous and metamorphic rocks (the Lara Nappes), and to the north by Neogene siliciclastic units in the Coro Plain (Fig. 2.26). The latter is a coastal plain flanked to the north by the Urumaco Trough (in the Gulf of Venezuela), the Coro High (in the Paraguaná Peninsula), and the La Vela Bay (in the Caribbean Sea).

2.9.1 Origin and Infilling of the Falcón Basin

The FB lies within the Southern Caribbean PBZ (see Figs. 2.9 and 2.10), a tectonic-faulted region. Therefore, this basin has inherited a complex tectono-stratigraphic history which has been a matter of debate by several researchers during the last decades (e.g., Audemard 1993, 1998; Macellari 1995; Bezada et al. 2008; Baquero et al. 2009; Baquero 2015; Albert-Villanueva et al. 2017a; among others). The FB is a polyphase basin whose creation and evolution have been controlled by multi tectonic and magmatic episodes associated to the eastward migration of the Caribbean Great Arc and the CP during the Cenozoic (Baquero et al. 2009; Baquero 2015). The latter implies the creation and migration of several depocenters during Paleogene and Neogene times.

The deepwater proto-FB, south of the Falcón Anticlinorium, was a foreland basin that existed during early Paleogene times. This basin was created as a consequence of the emplacement of the Lara Nappes above passive margin strata in what is now northern Lara state (see Fig. 2.10). The imbricated thrusts of the Lara Nappes consist of intensively deformed, sedimentary and low-grade metamorphic rocks (slates, phyllites, and meta-sandstones) of Early and Late Cretaceous and early Paleocene ages; as well as plutonic and volcanic mafic rocks of Early to mid-Cretaceous ages (the Santa Ana and the Siquisique ophiolitic complexes). These terranes crop out across a large boundary zone south of the Falcón Anticlinorium (Fig. 2.26) (Audemard 2001; Ostos et al. 2005; Baquero et al. 2009; Escalona and Mann 2011; Urbani 2018).

The emplacement of the Lara Nappes, which display a southeastward tectonic transport, created a deep foreland basin that was filled with thick turbidite facies of the Matatere Formation from the Early Paleocene to the Early

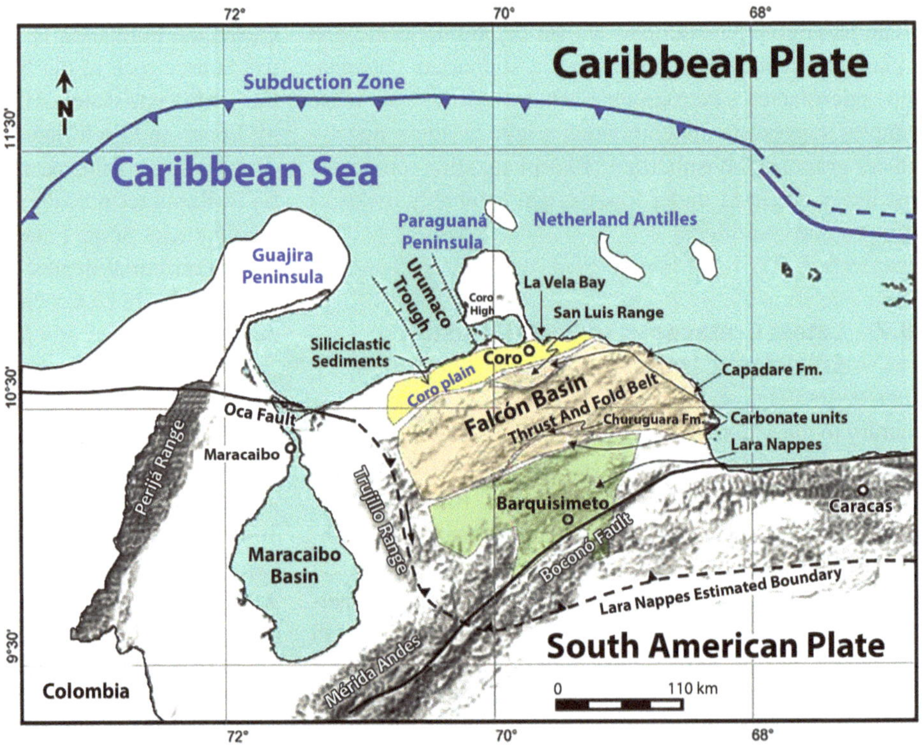

Fig. 2.26 Tectonic framework of the Falcón Basin (FB). The Falcón Thrust Fold and Belt, or Falcón Anticlinorium, is bounded by the Lara Nappes to the south (area colored in green); and by Late Miocene to Quaternary siliciclastic and minor carbonate units to the north (area colored in yellow). (Figure modified from Baquero et al. 2009)

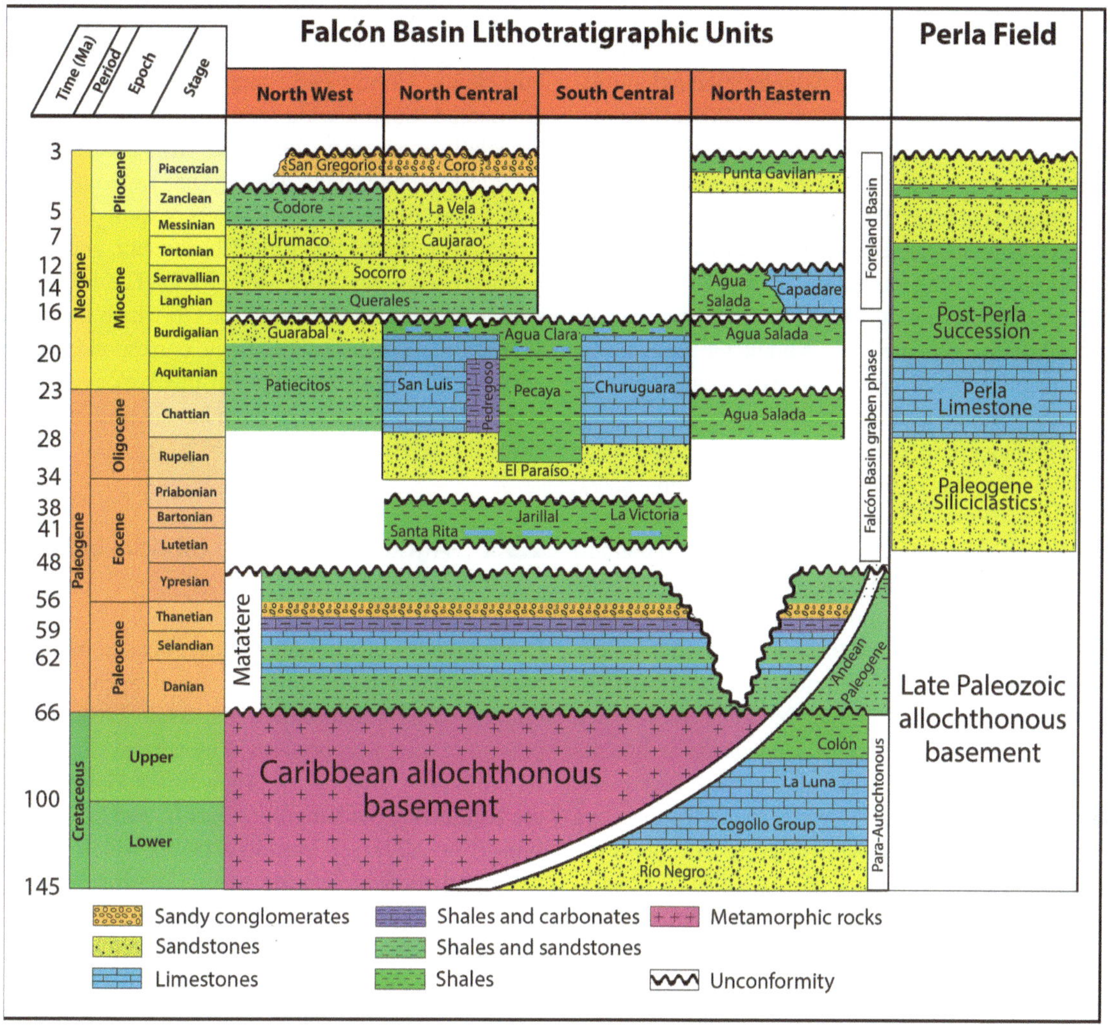

Fig. 2.27 Chronostratigraphic chart showing the correlation between the FB and the Eastern Gulf of Venezuela sub-Basin in the Perla Field area. Stratigraphy of the FB compiled and modified from Baquero et al. (2015), Albert-Villanueva et al. (2017a), and Valencia and Laya (2020). Stratigraphy of the Perla Field according to Valencia and Laya (2020). Late Paleozoic (Permian) allochthonous basement of the Perla Field according to Baquero et al. (2015). Time scale slightly modified from Cohen et al. (2013a, b, c; updated v2022/2). It is noteworthy that the stratigraphy and geographic distribution of lithostratigraphic units of the FB varies substantially among different authors

Eocene (Fig. 2.27) (Baquero et al. 2009; Noguera et al. 2017). The thrusted and folded Matatere Formation overlies allochthonous Cretaceous metamorphic Caribbean terranes (Fig. 2.27) (e.g., Audemard 1991; Bezada et al. 2008; Albert-Villanueva et al. 2017a). The Matatere Formation is a turbidite succession of heterogeneous lithology (see Urbani et al. 2016) which extensively crops out south of the Falcón Anticlinorium. This syn-orogenic unit is considered as part of the Lara Nappes (Baquero et al. 2009) and locally contains Cretaceous passive margin limestones (Albert-Villanueva et al. 2017a). In that area, those carbonate units (the La Luna Formation and the Cogollo Group) are considered to be para-autochthonous in origin and exhibit low-grade metamorphism (Marvin Baquero, personal communication 2022).

The majority of sedimentary units considered in this book belonging to the FB were deposited during its graben phase, which occurred from the Middle Eocene to the Early Miocene. For that time interval, early theories of the FB's formation interpreted it as a pull-apart basin formed in a dextral strike-slip fault system (Muessig 1984). This interpretation was accepted for various authors (e.g., Boesi and Goddard 1991; Macellari 1995). However, the geometry and dimensions of the FB necessary to produce local transtension during Middle Eocene to Early Miocene times has not been convincingly demonstrated, as there is no evidence of an active strike-slip fault in the Falcón area during those times (Audemard 2009). The geometry of the basin and its facies distribution suggests that this basin was formed by north-south trending extension rather than east-west. Thus, as an alternative hypothesis for the opening of the FB, Audemard (1993, 1998) proposed a back-arc origin associated with the migrating Caribbean Great Arc and Atlantic subduction roll-back (see also Bezada et al. 2008; Audemard 2009). Orogenic collapse of the Caribbean nappe pile was involved in the graben formation. The back-arc origin of the FB is demonstrated in detail by Baquero et al. (2015) through the determination of two main extensional episodes, and four tectono-magmatic episodes associated to the migration of the CP. The first extensional episode was initiated in the Middle Eocene. The second episode, initiated in the Early-Middle Oligocene, was of a greater magnitude and opened an extensional basin (a graben) that underwent a rapid subsidence.

During its graben phase, the FB was an active depocenter from the Middle Eocene to the Early Miocene (Fig. 2.27). During that time interval, the basin had an elongate shape, with a WSW-ENE trend (Bezada et al. 2008; Audemard 2009; Baquero et al. 2009). The greater subsidence along the basin's axis promoted deposition of marine dark shales, while shallower environments in the northern and southern limits of the basin allowed the sedimentation of limestones and other shallow-water facies (Díaz de Gamero 1977; see Figs. 2.26 and 2.27). The source of siliciclastic sediments during deposition of siliciclastic and mixed carbonate-siliciclastic units infilling the FB were primarily the Lara Nappes, as well as the ancestral Mérida Arc and the proto-Santander Massif (Noguera et al. 2011, 2017). In addition, sources of siliciclastic sediments to the northwest are proposed by Xie et al. (2010) during Eocene times, such as the Santa Marta Massif and some positive areas related to the eastward movement of the Caribbean Great Arc.

In the Middle Miocene, a compressive stage as a result of the convergence between the CP and northern South America led to the inversion of the FB. The compression generated by the collision-subduction reactivated pre-existing faults and turned the basin into a positive zone (the Falcón Anticlinorium), leading to the exhumation of the sedimentary units deposited during the graben phase (Audemard 2009; Baquero et al. 2009; Baquero 2015). Tectonic inversion of the FB resulted in a northward migration of the depocenter and the formation of the Falcón Foreland Basin (Fig. 2.27). In this newly-formed basin, sedimentation of siliciclastic-dominated units (e.g., La Vela, Urumaco, Querales, Socorro, Caujarao, Coro, Paraguaná, and San Gregorio formations) took place from the Middle Miocene to Late Pleistocene (Macellari 1995; Smith et al. 2010; Albert-Villanueva et al. 2017a).

Photographic examples of carbonate units in the FB considered in this book correspond to rock units of Middle Eocene, Early Miocene and Middle Miocene ages. These units crop out in several localities in the Falcón Anticlinorium. Unlike the regional extension of Cretaceous passive margin sedimentation in the MB, the geometry of the FB controlled facies distribution, leading to a great lateral variability of lithofacies, from carbonate-dominated to siliciclastic-dominated. Such a wide range of lithologic composition is especially marked in rock units deposited during Early to Middle Miocene times. During that time interval, the most significant carbonate rock units exposed in the Falcón Anticlinorium (the San Luis, Churuguara, and Capadare formations) were deposited (Fig. 2.27). The following discussion is focused on the formations to which the photographic examples included in this book correspond, as the highly variable lithology along the FB involves a large number of rock units (the majority of them having a siliciclastic compositional predominance).

2.9.2 Agua Negra Group (Middle to Late Eocene)

During the Middle to Late Eocene, the syn-extensional sedimentation associated with the initial opening of the FB is dominated by siliciclastic units, while carbonate facies are present in minor proportions. These basal siliciclastic deposits are represented by the Agua Negra Group. These rocks are included in the Santa Rita, Jarillal and La Victoria formations (Fig. 2.27) (see Macellari 1995), which were deposited in littoral to outer marine neritic environments created as consequence of a marine transgression during the Early to Middle Eocene (Baquero et al. 2009).

The Santa Rita Formation of Middle Miocene age crops out in south-central and western FB. It unconformably overlies the early Paleogene turbidite succession of the Matatere Formation (Baquero et al. 2009; Albert-Villanueva et al. 2017a). The lithology of this unit is heterogeneous, as it consists of conglomerates, sandstone, marls and interbedded limestones (Macellari 1995), having maximum thickness of ~450 m. Calcareous biota is dominated by LBF (genera *Nummulites*, *Discocyclina*, *Lepidocyclina*) and red algae, as well as secondary agglutinated SBF, bryozoans and echinoderms. Skeletal assemblages suggest that the deposition of limestones facies occurred in a middle-outer shelf setting at depths between 60 and 80 m (Pitelli and Molina 1989).

Photographic examples belonging to limestones of the Santa Rita Formation correspond to grain-supported fabrics containing red algae and LBF, which are the main components of the Photo C-factory of Michel et al. (2019). On a global scale, from Eocene to Middle Miocene times, this photozoan association was common of ramps and platforms in which carbonate production occurred in the mesophotic and oligophotic zones (intermediate depths within the photic zone, and low-light zone, respectively) (see following Sects. 2.9.3.1, 2.9.3.2, and 2.10.1 for a detailed treatment of the development of the Photo C-factory in Cenozoic carbonate facies of Venezuela).

The Santa Rita Formation grades transitionally to the Jarillal Formation of late Middle Eocene age. This unit crops out in south-central FB and it consists of dark shales with sandstones intercalations and minor carbonate horizons, with thickness up to 375 m (Macellari 1995). Dark shales of the Jarillal Formation are rich in foraminifera (Wheeler 1963), mainly SBF and planktonic foraminifera. This unit was deposited at depths between 100 and 200 m, representing a MFS (Pitelli and Molina 1989). Based on palynomorph assemblages, those authors established the age of the Jarillal Formation as Middle late Eocene. Photographic examples belonging to shales of the Jarillal Formation correspond to components of the P-factory (planktonic foraminifera) and the Heterozoan factory (SBF such as *Lenticulina* and *Uvigerina*). Foraminiferal assemblages and prominent authigenic precipitation of fluorapatite suggest low-oxygen conditions during deposition.

The upper part of the Agua Negra Group is represented by the La Victoria Formation of Late Eocene age. This unit crops out in western FB and it consists of dark shales and interbedded sandstones and minor carbonate units, having thicknesses up to 500 and 820 m (Sutton 1946; Pitelli and Molina 1989). This unit is characterized by a poor fossil content, as only a few agglutinated benthic foraminifera have been reported (Pitelli and Molina 1989).

2.9.3 Major Carbonate Units in the Falcón Basin Graben Phase

During the Oligocene and Early Miocene, the thicker sedimentary units in the FB were deposited in the deep central axis, while thicknesses decrease towards the margins of the basin. The lithofacies likewise change in the same direction (Wheeler 1963). The most significant carbonate units in the FB were deposited in shallower paleoenvironments away from the deep central axis. These are the Churuguara and San Luis formations, which are well-developed, tropical examples of globally widespread carbonate sedimentation in the Oligo-Miocene aragonite seas (Aragonite III).

The San Luis and Churuguara formations are time-equivalent, mixed carbonate-siliciclastic facies deposited in the northern and southern areas of the FB graben phase, respectively (e.g., Wheeler 1963; Macellari 1995). Detailed biostratigraphy of LBF carried by Albert-Villanueva (2016) (species of the genera *Miogypsina*, *Heterostegina*, *Operculinoides*, *Lepidocyclina*, *Miosorites*, and *Annulosorites*) suggests that these two units were deposited strictly during the Early Miocene (see also Albert-Villanueva et al. 2017b). From Aquitanian to Burdigalian (23–16 Ma), deposition at both south and north margins of the basin encompassed six transgressive-regressive third order sequences. Carbonate production occurred mainly in the meso-oligophotic zone (~ 40–100 m deep). Prior to such study, the San Luis and Churuguara formations were interpreted as have been deposited during Oligo-Miocene times (Fig. 2.27).

From a geodynamic and tectono-sedimentary point of view, deposition of these formations is considered to have begun during Oligocene times (Marvin Baquero, personal communication 2022). The polyphase tectonic development of the FB proposed by Baquero et al. (2009) and Baquero et al. (2015) suggests that the extensional phase initiated in the Eocene that gave origin to the initial graben intensified in the Early Oligocene, creating an extended crestal collapse graben in the central part of the basin, where accommodation of thick, bathyal shales of the Pecaya Formation occurred (Fig. 2.27). As an isostatic response, local uplift towards north and south created ideal conditions for deposition of the San Luis and Churuguara formations, respectively.

Regarding discrepancies on the age of the base of those carbonate units among various authors, we do not have a firm opinion, since our previous works on ancient carbonate facies of Venezuela have been focused on the MB and the BSB. Therefore, as Sect. 1.10 (FB tectonic setting and stratigraphy) is a bibliographic review, in Fig. 2.27 we consider a Late Oligocene to Early Miocene age for the San Luis Formation, as stated by Macellari (1995) and Johnson et al. (2009); and an Early Oligocene to Early Miocene age for the Churuguara Formation, as stated by Wheeler (1963) and Díaz de Gamero (1977). However, in the text of this chapter and for the rest of this book (including the appendices), we follow the interpretations of Albert-Villanueva (2016), and consider those carbonate-siliciclastic units as strictly Early Miocene in age. Such a study is a multidisciplinary approach (e.g., biostratigraphy, sequence stratigraphy, facies analysis) completely focused on the San Luis and Churuguara formations.

2.9.3.1 San Luis Formation (Early Miocene)

The San Luis Formation was deposited in northern FB and it crops out in the north-central part of the Falcón Anticlinorium (Figs. 2.26 and 2.27), comprising the San Luis Range (maximum elevations of 1500 m) of ~60 km long and ~ 13 km wide. This carbonate-siliciclastic succession is ~800 m thick. Its deposition occurred over a structural high known as the Coro High, which extends towards the north into the Paraguaná Peninsula (Fig. 2.26) (Wheeler 1963; Boesi and Goddard 1991; Albert-Villanueva et al. 2017a). In general, the lithology of this unit consists of massive limestones interbedded with shales, sandstones and conglomeratic sandstones (Wheeler 1963; Díaz de Gamero 1977).

The sedimentation of the San Luis Formation spanned the peak of a major (second-order) transgressive event occurred during the Early Miocene (~ 23 Ma), a time during which carbonates of the Churuguara Formation were also growing in southern FB. This thick carbonate-siliciclastic succession encompasses both distal and proximal facies of a transitional deltaic system towards the north, to a carbonate ramp setting towards the south and southeast. The variety of carbonate textures includes fine-grained wackestones and packstones, as well as floatstones, rudstones and grainstones, with a predominance of grain-supported depositional textures. The most abundant bioclasts are LBF, coralline red algae (or coralline algae), molluscs, gastropods, corals, and planktonic foraminifera. Siliciclastic facies (shales, sandstones and conglomerates), which predominate in the northern part of the San Luis Range, correspond to delta front and delta plain depositional environments.

Facies variations described by Albert-Villanueva (2016) and Albert-Villanueva et al. (2017b) indicate marked lateral changes in lithofacies and grain associations along the distally-steepened ramp. These changes allow differentiate major depositional environments as follow: *inner ramp* facies dominated by corals and accompanied by green and red algae, LBF (photozoan association) and molluscs (mixed photozoan-heterozoan associations); *middle ramp* facies dominated by LBF (lepidocyclinids, amphisteginids, heterosteginids and operculinids) and red algae (photozoan association; rhodalgal lithofacies *sensu* Carannante et al. (1988); see also Pomar et al. (2004) and Flügel (2010); see Appendix A for a detailed treatment of the variety of skeletal grain associations represented in carbonate facies of Venezuela); this association indicates depths in the mesophotic zone; *outer ramp* deposits dominated by fine-grained packstones of planktonic foraminifera accompanied by glauconitic grains and LBF reworked from the middle ramp.

According to the above grain associations, there is a clear relation between carbonate production and depositional environments and related paleodepths in the San Luis ramp. Components of the Photozoan T-factory (corals and secondary green algae) dominated the inner ramp deposits. These sparse, isolated coral mounds developed in a foredelta environment. It is noteworthy that siliciclastic input did not inhibit coral development. Nevertheless, these coral mounds grew at depths between the lower euphotic and upper mesophotic zones, which differentiates these deposits from the massive coral growth in modern barrier reefs at depths in the uppermost euphotic zone (Eduard Albert-Villanueva, personal communication, 2021). The Photo C-factory dominated carbonate production in the middle ramp. In that setting, paleodepths within the mesophotic (intermediate light) zone (~ 40 to 80 m deep), and water temperature and energy level controlled by bathymetry (or depth of the seafloor), exerted a strong control in the flourishing of coralline algae and LBF to the detriment of scleractinian corals and green algae.

It is noteworthy that Michel et al. (2019) created the concept of the Photo C-factory for facies dominated by red algae and/or LBF, which are particularly abundant in the warm-temperate or subtropical province of the Cenozoic, including Modern times. Components of the P-factory dominate San Luis distal, outer ramp facies. In this setting, deposition of planktonic foraminifera, accompanied by the precipitation of glauconite, occurred in the oligophotic zone, or possibly in the uppermost aphotic zone. The micritic sediments in wackestones and packstones textures (including floatstones and rudstones) are dominated by allomicrite precipitated by the neritic lime-mud factory.

In addition to the importance of the San Luis Formation as a carbonate sedimentary unit, the San Luis Range limestones are well known in Venezuela for hosting large vertical caves, locally termed "Haitones". These caves are the product of massive karst dissolution, highlighting the El Guarataro Haitón, down to 305 m, as a part of a large and interconnected karst-aquifer system (see Galán and Herrera 2017).

2.9.3.2 Churuguara Formation (Early Miocene)

The Churuguara Formation was deposited in southern FB and it crops out in the south-central areas of the Falcón Anticlinorium (Figs. 2.26 and 2.27), having thicknesses from a few tens of meters to up to 1270 m (Wheeler 1963; Díaz de Gamero 1977; Albert-Villanueva 2016; Albert-Villanueva et al. 2017a). In general, the lithology consists of variable series of interbedded sandy limestones grading to calcareous sandstones, limestones, argillaceous-fossiliferous limestones, sandstones and glauconitic sandstones, siltstones, and silty shales (Wheeler 1963; Díaz de Gamero 1977). In the overall, siliciclastic and mixed carbonate-siliciclastic facies dominate over pure carbonate facies.

Albert-Villanueva (2016) describes outcropping sections of the Churuguara Formation as consisting of two carbonate levels intercalated within a shale-dominated marine succession. These sections are part of a mountain system that represents the transition zone from the Venezuelan Andes to the Coastal Range, and constitute a faulted and folded belt. Tectonic deformations in the Churuguara Range began in the latest Burdigalian in response to the lateral strike-slip movement of the Oca-Ancón fault and to the beginning of the tectonic inversion of the FB (Boesi and Goddard 1991; Macellari 1995; Baquero et al. 2009; Escalona and Mann 2011).

Carbonate depositional textures in the Churuguara Formation are dominated by floatsotnes and rudstones, although fine-grained packstones are also present. The most abundant bioclasts are LBF, coralline algae (both branching and rhodoliths), echinoids, molluscs and secondary smaller miliolids, corals and green algae. The lowermost part of the succession of early Aquitanian age is dominated by siliciclastics. The overlying late Aquitanian facies consist of a mixed carbonate-siliciclastic heterozoan ramp dominated by echinoids. Lateral changes in lithofacies and grain associations in this Aquitanian ramp allow differentiate major depositional environments (Albert-Villanueva 2016; Albert-Villanueva et al. 2017b): *lagoon* facies dominated by siliciclastic sedimentation; *littoral barrier* facies consisting of fine-grained argillaceous packstones and grainstones of molluscs (bivalves and gastropods) and echinoids (heterozoan association) accompanied by soritids (larger miliolids) and smaller miliolids (both types of miliolids are common of the euphotic zone); *inner-proximal ramp* facies consisting of packstones-grainstones of echinoids accompanied by molluscs (heterozoan association) and reworked LBF (heterozoan-photozoan association); *inner-distal ramp* facies consisting of fine-grained muddy packstones of echinoids accompanied by red algae and LBF (heterozoan-photozoan association); *middle ramp* facies consisting of muddy packstones of LBF (amphisteginids, heterosteginids, lepidocyclinids, operculinids, and myogypsinids) and red algae (photozoan association) accompanied by echinoids (mixed photozoan-heterozoan association); these grain associations indicate paleodepths in the meso-oligophotic zone; *outer ramp* facies (below the slope) consisting of gravity flow deposits dominated by floatstones-rudstones of LBF (mainly lepidocyclinids and operculinids) embedded in a micritic matrix and accompanied by red algae, planktonic foraminifera and glauconite grains.

The above-described facies lateral changes indicate that lithofacies and grain associations during deposition of the Aquitanian Churuguara ramp were controlled by bathymetry and trophic levels. The Heterozoan factory (echinoids and secondary molluscs) dominated carbonate production in the littoral barrier and in the inner-proximal and inner-distal ramp. The latter suggests prevailing eutrophic conditions associated with the high influx of siliciclastic sediments. Towards greater depths into the photic zone, from the inner-distal ramp to the middle ramp, the diminish in heterozoan biota is accompanied by the appearance, towards fully dominance, of components of the Photo C-factory (red algae and LBF). This change in the carbonate biota indicates reduced nutrient levels and optimum conditions of bathimetry and water temperature for the flourishing of light-dependent carbonate producers typical of the meso-oligophotic zone. In distal facies of the outer ramp, carbonate production by photozoan biota was discouraged by low light conditions. Therefore, lithofacies in this environment are a combination of components of the Photo C-factory displaced down the slope, and components of the P-factory precipitated in the upper part of the water column. The neritic lime-mud factory dominated the precipitation of micrite.

Separated from the late Aquitanian carbonate-dominated ramp by siliciclastic sedimentation of early Aquitanian age, the younger middle Burdigalian part of the succession is also a carbonate-dominated ramp. Lateral changes in lithofacies and grain associations and major depositional environments are (Albert-Villanueva 2016; Albert-Villanueva et al. 2017b): *lagoon* facies consisting of shales and minor sandstones; *littoral barrier* facies consisting of packstones and grainstones of reworked corals and bivalves (photozoan-heterozoan association); *inner ramp* facies consisting of packstones and grainstones of green algae, corals, red algae, soritids (photozoan association) and smaller miliolids (photozoan-heterozoan association); *middle-proximal ramp* facies consisting of floatstones and rudstones of rhodoliths and reworked bivalves, LBF, corals and serpulids (mixed photozoan-heterozoan association); *middle-distal ramp* facies consisting of packstones and grainstones of LBF (amphisteginids, heterosteginids, lepidocylinids, and miogypsinids) and red algae (photozoan association/rhodalgal lithofacies); this grain association indicates depths in the meso-oligophotic zone; *outer ramp* facies consisting of floatstone-rudstones of LBF (mainly *Lepidocyclina*).

Contrary to the younger Aquitanian ramp, lateral changes of lithofacies in the Burdigalian Churuguara ramp indicate an almost total dominance of photozoan biota. Components of the Photozoan T-factory (corals and green algae) in association with shallow-water components of the Heterozoan factory (bivalves and smaller miliolids) dominate shallower facies corresponding to the littoral barrier and the inner ramp. Towards greater depths in the photic zone, from the middle-proximal and middle-distal ramp to the outer ramp, lithofacies are dominated by components of the Photo C-factory. Again, this indicates optimal conditions for the flourishing of photozoan biota typical of the meso-oligophotic zone. The neritic lime-mud factory dominated the precipitation of micrite.

2.9.4 Falcón Basin Foreland Phase

By the Early late Miocene, the pattern of sedimentation in the FB changed significantly as a result of rapid uplift of the Middle Eocene and Oligocene-Early Miocene units. Tectonic compression started at ~16 Ma (Langhian) (Fig. 2.27), causing the exhumation of east-west trending Anticlinorium fold belt and the generation of the Falcón Foreland Basin to the north. This northward migration of the depocenter, combined with the proximity of an uplifting siliciclastic sediments source, led to the deposition of siliciclastic-dominated units from Middle Miocene to the latest Pliocene (~ 16–2.6 Ma) (Macellari 1995; Audemard 2001; Baquero et al. 2009; Escalona and Mann 2011; Albert-Villanueva et al. 2017a).

2.9.4.1 Capadare Formation (Middle Miocene)

The youngest Neogene carbonate examples shown in this book correspond to shallow-water facies of the Capadare Formation of Middle Miocene age. Deposition of this succession of 150–300 m thick (Macellari 1995), also known in the literature as the "Capadare Limestone", was restricted to the northeastern areas of the FB (Figs. 2.26 and 2.27). Near the coastlines in the northeastern area of the Falcón Anticlinorium, this unit crops out in several localities (Macellari 1995; Albert-Villanueva et al. 2017a). The Capadare Limestone also crops out towards southeast along the same coastline, in the surrounding areas of the town of Chichiriviche (see Fig. 1.3) (Goddard 1974; Méndez-Baamonde 2007a). Albert-Villanueva et al. (2017a) consider this carbonate unit as a syn-orogenic succession, as it was deposited during the beginning of tectonic inversion of the FB, therefore belonging to the Falcón Foreland Basin phase (Fig. 2.27). Based on planktonic foraminiferal assemblages, the age of the Capadare Limestone was established by Wheeler (1963; in Albert-Villanueva et al. 2017a) and Díaz de Gamero (Díaz de Gamero 1977; in Ministerio de Energía y Minas 1997) as Middle Miocene.

The Capadare Limestone consists mainly of wackestones and packstones, with grainstones in minor proportions. The siliciclastic content of this fossiliferous unit is low, indicating fully marine conditions (Lorente 1978). The most abundant bioclasts are molluscs (aragonitic bivalves, gastropods and oysters), coralline algae, LBF, and echinoids, with secondary planktonic foraminifera, barnacles (balanids), ostracods, serpulids, and briozoans (Lorente 1978; Soto 2025; Soto et al. 2025). Patchy dolomite facies as well as meteoric cements and secondary porosities are moderately common diagenetic features. Authigenic Ca-phosphate (fluorapatite) replacing grains (e.g., LBF and molluscs) and matrix is common in muddy facies.

Carbonate production in the Capadare Limestone was dominated by three factories: the Heterozoan, the Photo-C, and the neritic lime-mud factories. Moderate micrite production, combined with abundant LBF and red algae (photosymbiotic and photoautotrophic calcifiers, respectively), suggest water depths in the mesophotic zone. We do not estimate a predominance of light-dependent over light-independent calcifiers, and vice versa, as further investigations must be carried in order to determine that ratio. Nevertheless, the lithologic descriptions in Lorente (1978) and in Ministerio de Energía y Minas (1997) suggest an important contribution of both heterozoan and photozoan carbonate biota. In addition, moderately common authigenic fluorapatite in some muddy microfacies suggest low-energy and nutrient-rich conditions during early diagenesis.

In the upper part of the Capadare Formation there is a phosphatized limestone interval of about 35–40 m thick (González de Juana et al. 1980). Phosphatization of biolcastic packstones and wackestones occurred in the telogenetic environment, as "guano-derived", phosphate-rich brines percolated through highly porous and permeable sandstones confining the phosphatized limestone intervals (Albert-Villanueva 2016). Analog insular (guano-derived) phosphate facies having very similar petrographic features are found in Mio-Pliocene carbonates of the Seroe Domi Formation in Curaçao (Ten Have et al. 1982; Stienstra 1985). These secondary phosphate rocks and overlain P-rich soil are the most significant phosphate deposits of economic ore potential in Venezuela (Rodríguez 1981). Photographic examples shown belonging to this phosphatized interval exhibit distinctive vacuolar porosity.

2.9.5 Falcón Basin Source Rocks and Reservoirs

The potential of the FB as an oil-producing basin is minor compared to the Maracaibo, Eastern Venezuela and Barinas-Apure basins. This is mainly due to the absence of an effective source rock with the required thermal maturity and substantial TOC content. Furthermore, no petroleum system has been yet established for the FB, primarily because of tectonic and stratigraphic complexity. Rock units in the Falcón Anticlinorium that have been considered as possible source rocks correspond to shale units of Late Eocene to Early Miocene age containing type II-III kerogen (Santa Rita, Jarillal, Cerro Misión, El Paraiso, Pecaya, Castillo, Agua Clara, Pedregoso and Cerro Pelado formations; Ghosh et al. 1997; Audemard and Serrano 2001; Baquero et al. 2009; Montero-Serrano et al. 2010). Another source rock that has been considered is the Late Cretaceous La Luna Formation, which locally crops out in southeast Falcón State, south of the Falcón Anticlinorium (Baquero et al. 2009).

Oil wells in the FB are located north of the Falcón Anticlinorium, in the Coro Plain (Fig. 2.26; Baquero et al. 2009; Albert-Villanueva et al. 2017a). Historically, the most productive reservoirs in the Falcón Foreland Basin are the Middle Miocene calcareous sandstones of the Socorro Formation and the Late Miocene sandy limestone facies of the Caujarao Formations (Goddard 2006; see Fig. 2.27). Oil-source rock correlation of seeps in the Coro Plain indicates a clear difference in lithology and depositional environments respect to La Luna source rock in the MB. Thermal calibration of six wells (five drilled and one pseudowell) was modelled using vitrinite reflectance (%Ro) values and Tmax values (°C) from Rock Eval pyrolysis. Those thermal maturity parameters indicate that Late Eocene to Early Miocene shales deposited in the tectonic graben generated hydrocarbons during the Oligocene-Early Miocene (Baquero et al. 2009). The same parameters determined in the easternmost extension of the La Luna Formation in western Falcón State indicate that this Cretaceous unit is overmature. Diachronism between oil generation and the development of structural traps during Late Miocene and Pliocene resulted in the loss of much of the hydrocarbon generated volumes (see Ghosh et al. 1997; Baquero et al. 2009).

An example of an overmature organic-rich rock unit in the FB is the Pedregoso Formation. It crops out in a short and narrow area in northern Falcón Anticlinorium, bordering the southern margin of the San Luis Range (Fig. 2.27). This succession of shales and limestones (~ 125 m thick) of Early Miocene age consists of moderately organic-rich pelagic shales with rhythmically interbedded limestones and rare sandstones (Wheeler 1963; Díaz de Gamero 1977). A complete organic and inorganic study of the source-rock potential of this unit was carried by Montero-Serrano et al. (2010). Results obtained by these authors indicate that: fully anoxic conditions were not reached during deposition; TOC content range between 0.11 and 1.25 (calculated original TOC value is ~1.8%); the Pedregoso Formation is a "spent" oil source rock (dry gas window stage of hydrocarbon generation), as sedimentary burial caused organic matter overmaturity.

2.10 Perla Field: Tectonic Setting and Stratigraphy

The Perla field is an offshore giant gas/condensate carbonate reservoir in the Eastern Gulf of Venezuela sub-Basin (EGVSB). This reservoir is part of a Late Oligocene to Early Miocene carbonate succession named the Perla Limestone. These offshore carbonates are geographically located approximately 28 km from the western coast of the Paraguaná Peninsula (Valencia 2016), and tectonically within the Urumaco Province (Fig. 2.28). The Perla world-class giant field was discovered in 2009, being considered the largest gas field in Latin America and whose discovery triggered a Play opener for oil and gas exploration in the Southern Caribbean domain (Castillo et al. 2017).

Carbonates of the Perla Limestone grew over a basement high located between two geological highs, the Coro High to the southeast and the Los Monjes High to the northwest (Fig. 2.28). The Urumaco Trough to the south is the main tectonic element of this geologic province. This highly subsiding, elongate area with northwest-southeast orientation was formed due to extensional and transtensional stress during the Oligocene-Late Miocene interval. The Urumaco Trough contains over 6100 m (20,000 ft) of sedimentary rocks and is an important kitchen for hydrocarbon generation (Pomar et al. 2015; Castillo et al. 2017).

In the Urumaco Province, the structural style dominated by northwest-southeast trending faulting orientation involves a set of right-lateral faults (Algodones-Río Seco, Sabaneta, Urumaco, Los Médanos, La Soledad and Los Monjes-Sur faults, stand among those). These synthetic Riedel shear faults resulted from right-lateral movement along the Oca-Ancón fault (Boesi and Goddard 1991; Audemard et al. 1999) and formed a series of geological highs and troughs. Among these are the Urumaco Trough, the Coro High and the La Vela Bay structure (the latter is a trough with a local southern horst) (Fig. 2.28) (Boesi and Goddard 1991; Johnson et al. 2009). The basement high over which the Perla Limestone was deposited is also defined by a NW-SE faulted anticline-like structure (Castillo et al. 2017).

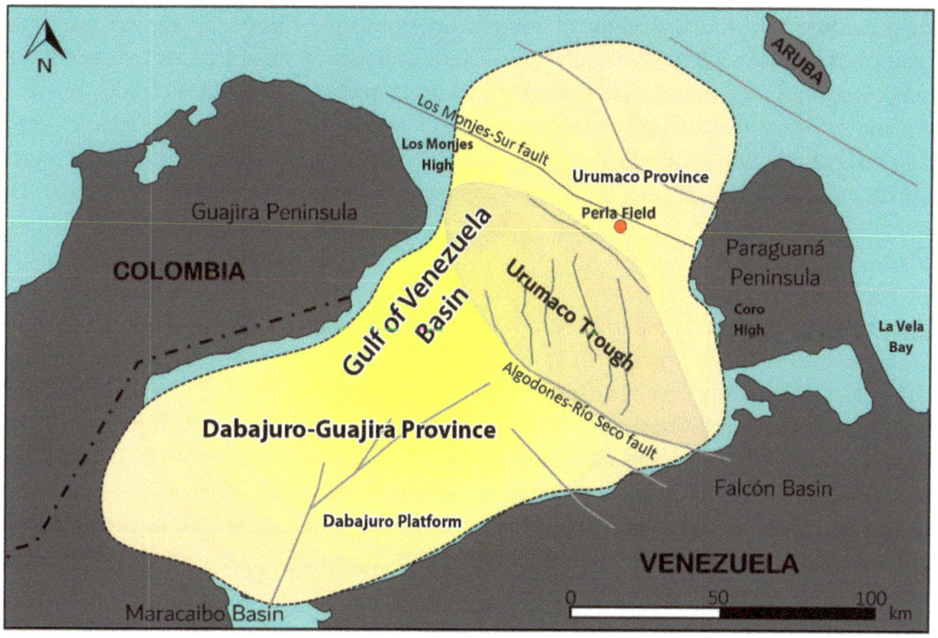

Fig. 2.28 Tectonic framework of the Perla Field, which lies over a tectonic high located in the Urumaco Province. The latter comprises the Eastern Gulf of Venezuela sub-Basin (EGVSB), while the Dabajuro-Guajira Province (Maracaibo Province in Fig. 2.11) includes the Western Gulf of Venezuela sub-Basin (WGVSB). (Original figure modified from Malavé and Contreras 2013. Figure slightly modified from Valencia and Laya 2020)

The Perla Limestone succession is immediately underlain by a thick Eocene-Oligocene siliciclastic succession (mostly sandstones) that filled the basement topography (Fig. 2.27) (Pomar et al. 2015; Valencia and Laya 2020). These sandstones are in turn underlain by an allochthonous basement consisting of late Paleozoic (Permian) igneous rocks, with Early Eocene igneous intrusions (Baquero 2015).

2.10.1 Perla Limestone (Late Oligocene to Early Miocene)

The Perla Limestone is a carbonate succession of about 250 m thick which is laying in ~30–50 m of water column. These middle-outer neritic facies were deposited from the Late Oligocene to the Early Miocene (Chattian to Burdigalian) (Fig. 2.27). Grain-supported fabrics dominate the overall sedimentation. Coralline algae, both branching (mainly Corallinaceae and secondary Sporolithaceae) and rhodoliths, and LBF (mainly of the genera *Amphistegina*, *Miogypsina*, *Heterostegina* and *Lepidocyclina*) are the outstanding components in terms of volume (commonly up to 80%; rhodalgal lithofacies), with secondary contribution of echinoderms, bryozoans, balanids, molluscs, and encrusting benthic foraminifera, as well as minor contribution of green algae, corals, planktonic foraminifera, serpulid tubes, ostracods, and braquiopods (Pinto et al. 2014; Pomar et al. 2015; Valencia 2016; Valencia and Laya 2020; Soto et al. 2025; this book). SBF, mainly of the genera *Quinqueloculina* and *Textularia*, are scarce but present (Soto et al. 2025). Photographic examples of the Perla Limestone included in this book correspond to two cores displaying a wide variety of carbonate biota and diagenetic processes.

Based on seismic interpretation, electric logs, and detailed stratigraphic facies analysis (macroscopic description of cores and petrographic description of microfacies), the Perla Limestone has been interpreted as a gentle distally steepened ramp, bulging in the middle-outer ramp. Some of the petrographic evidence that support this ramp model is the abundance of both rhodolith-rich deposits and reworked grains, as well as evidence of sediment remobilization in the mid-ramp and gravity flows accumulated in the outer ramp (Pinto et al. 2011, 2014; Pomar et al. 2015).

The main lithofacies that have been defined through the study of well cores are summarized, from the base to the top of the succession (Pomar et al. 2015; Valencia 2016): siliciclastic sandstones, mixed sandstone-carbonates, packstones-wackestones with siliciclastics, coral floatstone-rudstone, skeletal packstones and grainstones, coarse and fine branching red algae floatstone-rudstone, rhodolithic rudstones and floatstones, LBF rudstones and grainstones. The lower part of the succession consists mainly of lithofacies occurring in transgressive-regressive sequences bounded by erosional surfaces. Distal facies in the upper part, volumetrically less significant, are dominated by lithofacies rich in planktonic foraminifera and calcareous nannofossils and have gray-black coloration due to the precipitation of authigenic minerals (pyrite, phosphate and glauconite). These lithofacies are interpreted as gravity flows deposited below the chemocline (Pomar et al. 2015).

As for bathymetric light zones, the Perla Limestone is interpreted as an oligophotic ramp. Only a small proportion of carbonates (2–3% of the rock volume) were produced in the euphotic (shallow-water) zone. The red algae (photoautotrophs) and LBF (mixotrophs) photozoan association was globally dominant during the Late Oligocene to Middle Miocene times, especially in the 40 to ≥100 m depth range (the oligophotic zone) in clear waters (see Pomar 2001). Although the oligophotic zone is regarded as a low-energy environment (below the base of wave action), repetitive high-energy events affecting the Perla gentle steepened ramp caused turbulence and sediment erosion and redeposition farther downslope. Furthermore, the muddy matrix is generally absent or scarcely distributed in most of the coarse-grained Perla carbonates (Pomar et al. 2015).

For all the above, the Perla Limestone is the best developed ancient expression in Venezuela of a tropical ramp in which carbonate production was dominated by the Photo C-factory, i.e., oligophotic biota that can thrive in environments with low light levels. The dominance of this factory implies the scarce contribution of corals and green algae (the Photozoan T-factory). The Heterozoan factory is well represented by echinoids, bryozoans and balanids, and less common molluscs and smaller miliolids, and although their abundance is secondary, these components are locally abundant in some microfacies. Water temperature, light availability, and energy level, controlled by bathymetry, were the main controlling factors that promoted deposition of this carbonate succession. Lithofacies and grain associations represented in the Perla Limestone are common of Oligo-Miocene carbonates deposited in warm-temperate, aragonite sea (Aragonite III) of the Mediterranean region (e.g., Bassi et al. 2006; Bassi and Nebelsick 2010; Pomar et al. 2012, 2017). Abiotic carbonate precipitation is volumetrically significant and represented in this unit by a variety of low-Mg calcite cements (void-filling primary and secondary pores) precipitated mainly in the shallow-burial diagenetic environment (see Valencia and Laya 2020). The content of micritic sediments (fine-grained carbonate matrix) in this ramp is low. In many microfacies, the matrix is altered by neomorphism/recrystallization. Nevertheless, micritic matrices and micritic peloidal grains in some packstones and less common wackestones are clearly differentiated.

2.10.2 Diagenesis and Secondary Porosities in the Perla Reservoir

The porous lithofacies in the Perla Limestone consist of a variety of depositional and diagenetic textures, hosting a combination of primary (interparticle and intraparticle) and secondary (e.g., microporosity, vugs, dissolution-enhanced primary pores) porosities (Pinto et al. 2014; Valencia and Laya 2020). Diagenesis played a key role in both the creation and reduction of porosity. Overall, the Perla Limestone suffered minor meteoric alterations in the upper photic zone due to the bathymetry of the depositional setting in the oligophotic zone, away from meteoric (near-surface) alterations. Therefore, major diagenetic processes occurred both in the shallow-burial environment and in the intermediate- and deep-burial realms. The best reservoir lithofacies are characterized by the development of pervasive secondary porosities (mainly microporosity and vugs) created in the mesogenetic environment due to the ascent of diagenetic (CO_2-rich) fluids from the basement. Selective dissolution of metastable grains during earlier diagenesis was responsible for creating moldic pores as the result of mineral stabilizations. A significant volume of porosity was destroyed by cementation (mainly blocky calcite) in the shallow-burial environment. This, combined with physical and chemical compaction, were the main porosity destructive processes. However, pressure-dissolution features, such as non-sutured dissolution seams and stylolites, could have also served as conduits for later mesogenetic dissolution and the creation of microporosity in the deep-burial realm. Additionally, intercrystalline porosity associated with the formation of high-temperature, microcrystalline dolomite crystals is also moderately common (see Valencia and Laya 2020).

2.10.3 Source Rock and Related Hydrocarbon Generation Episode for the Perla Reservoir

The Perla Field reservoir stores enormous reserves of light hydrocarbons, both gas (mainly wet gas composed of ethane and propane) and condensate (light oil with an API gravity greater than 45° API). Geochemical characterization of hydrocarbons (bitumen in rock samples, as well as gas and condensate samples) performed in core samples of the Perla succession suggests the kerogen that generated the light hydrocarbons stored in the Perla reservoirs consists mainly of marine organic matter, with a contribution of terrestrial organic matter. However, the source rock has not been found in any of the rock successions underlying and overlying the Perla Limestone, being the deep Urumaco Trough the most likely area for the presence of the effective source rock (or rocks) (Acosta 2016; Pirela 2017; Lugo and Audemard 2021).

Subsidence in the Urumaco Trough area began around Middle to Late Oligocene times (Baquero 2015; Acosta 2016; Valencia 2016). Geochemical data indicate that the gas stored in the Perla field was generated from the Late Miocene to Present (Castillo et al. 2017; see also Lugo and Audemard 2021). The sedimentary units in the Urumaco Trough consist of a basal unit of Early Miocene age, which is the time-equivalent of the Agua Clara Formation deposited in the FB graben phase (see Fig. 2.27); as well as units deposited during the FB Foreland phase, from Middle Miocene to Early Pliocene (e.g., Cerro Pelado, Socorro, Caujarao, and Urumaco formations; Macellari 1995). Time-equivalent shale facies of the Agua Clara Formation located in the deepest part of the Urumaco Trough is the most likely effective source rock that generated the hydrocarbons stored in the Perla Field (Pirela 2017). This giant gas field occupies a unique structural position over the Perla High and in direct connection to the main hydrocarbon kitchen to the southwest (Pomar et al. 2015).

2.11 Quaternary Carbonate Complexes of the Venezuelan Caribbean

Venezuela has one of the most diverse and best-developed group of Quaternary reef complexes in the South Caribbean. The distribution of these carbonate systems has been controlled by a combination of geographic, geomorphologic, tectonic, climatic and biogeochemical factors, leading to their development in both coastal and offshore environments. The offshore complexes are the largest and present the richest variety of carbonate biota. They have grown away from continental influences, mainly represented by the riverine siliciclastic and organic matter influxes, as well as anthropogenic impact. Furthermore, these complexes are barely affected by coastal upwelling along the Caribbean coastlines of Venezuela. This seasonal system is very active along the Southern Caribbean Sea LME subarea, encompassing Venezuela, as well as Trinidad and Tobago (see Villamizar and Cervigón 2017).

Offshore Venezuelan reef complexes (or systems) comprise a group of archipelagos and islands in the southeastern Caribbean Sea, north of the Caribbean shorelines of mainland Venezuela (Fig. 2.29). These complexes have grown on a series of aligned guyots (underwater mountains with a flat, platform top) and are tectonically located in the northern area of the Southern Caribbean PBZ (see Figs. 2.9 and 2.10; see also Fig. 1.2). From east to west, these complexes are named: Los Hermanos Archipelago; the islands of La Blanquilla, La Tortuga and La Orchila; the archipelagos Los Roques, Aves de Barlovento and Aves de Sotavento; and Los Monjes Islands. Collectively, these archipelagos and islands are known as the Venezuelan Antilles.

Some of the main factors inherited from offshore marine waters of the Caribbean that have contributed to the growth of these modern reef complexes over seamounts are: tropical latitudes, shallow-water depths within the upper photic zone, crystalline oligotrophic waters, elevated seawater total alkalinity, pH values commonly ranging between 8.1 and 8.4, average salinity values between 33% and 37‰ (pH and salinity reach values of up to 9.5 and 190‰ in some hypersaline lagoons, respectively), and optimum average marine water temperatures between 24 and 28 °C.

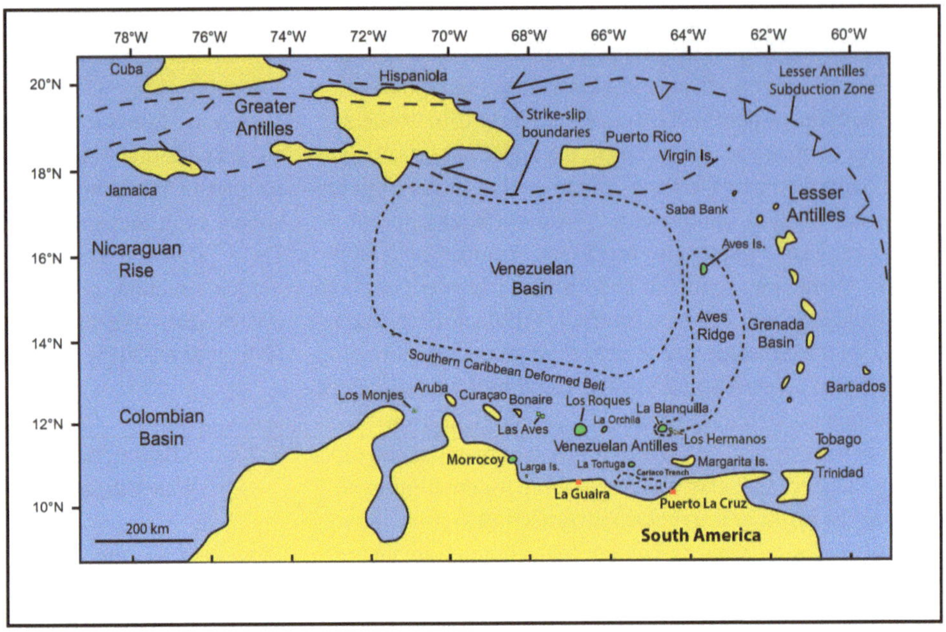

Fig. 2.29 Simplified map of the eastern Caribbean based on GEOMapApp (Haxby et al. 2010), showing the location of the Venezuelan Antilles and Aves Island, as well as major features such as the Aves Ridge, the Venezuelan and Grenada basins, the Lesser Antilles subduction zone and the Greater Antilles. Abbreviation: *Is* Island. (Figure modified from Neill et al. 2011)

Among these complexes, the archipelagos Los Roques and Las Aves (Aves de Barlovento and Aves de Sotavento) stand out for their variety of environments and biodiversity of calcareous organisms. These atoll-like archipelagos lie in a biogeographically unique area of the Caribbean. Their privileged location safeguards them from most Western Atlantic hurricane damage, as well as from the most severe Caribbean coral bleaching events (Miloslavich et al. 2010; Debrot et al. 2019). The only Venezuelan offshore reef island that does not belong to the Venezuelan Antilles is Aves Island. This small, isolated island is located much further north, outside the Southern Caribbean PBZ, closer to the northern end of the Lesser Antilles arc (Fig. 2.29). On the other hand, the Morrocoy National Park and Larga Island (San Esteban National Park) are the best-developed coastal reef systems in Venezuela (Fig. 2.29; see also Fig. 1.2). In spite of their coastal nature, they exhibit many morphological similarities with respect to Los Roques and Las Aves archipelagos, as well as a rich and diverse carbonate biota. Regarding the islands of Aruba, Curaçao and Bonaire, known as the ABC Islands or the Netherland Antilles, they also have well-developed Neogene and Quaternary reefs, and are located north of the coasts of the state of Falcón, within the same PBZ.

During the Quaternary Period, variations in the global (eustatic) ESL associated to glaciations and interglaciations (or glacials and interglacials, respectively) have been reconstructed in detail for the last 800 kyr using a variety of proxies and models (Lisiecki and Raymo 2005; Spratt and Lisiecki 2016; Past Interglacials Working Group of PAGES 2016). In that context, glacial cycles (a glacial cycle consists of a glaciation and an interglaciation) during that interval produced sea-level changes of approximately 130 m (Fig. 2.30), primarily associated with the growth and retreat of continental ice sheets in 100 kyr cycles (Spratt and Lisiecki 2016). Carbonate facies (e.g., coral reef systems) formed during sea-level highstands over the last 130 kyr have been intensively studied (see Murray-Wallace and Woodroffe 2014). Numerous exposed limestone terraces, notches and benches distributed in tropical latitudes worldwide record the last interglacial (LIG) period (between ~130–116 thousand years ago, ka). During this highstand, also known as the Marine Isotope Stage (MIS) 5e (Fig. 2.30), the global sea level was a few meters (up to + 9 m) APSL (Lisiecki and Raymo 2005; Hearty et al. 2007; Woodroffe and Webster 2014; Past Interglacials Working Group of PAGES 2016; Rovere et al. 2016a; Spratt and Lisiecki 2016; Chauveau et al. 2024).

Carbonate terraces were also formed during following interstadials within the last glaciation, i.e., MISs 5c, 5a and 3 (Fig. 2.30). During these interstadials, the global sea level was below present sea level (BPSL). Interstadials are defined as brief intervals of warmer climate and lower-amplitude sea-level highstands during a general trend of glacial cooling (see Murray-Wallace and Woodroffe 2014). During the following last maximum glacial (MIS 2; Fig. 2.30), which occurred between ~21–18 ka, the sea level fell down to ~120–134 m BPSL (see Siddall et al. 2008; Clark et al. 2009; Scotese 2014; Lambeck et al. 2014). This ESL lowstand exposed numerous calcareous terraces and ancient shorelines in the whole Caribbean. Subaerial exposure of Late Pleistocene carbonates led to erosional processes and development of karst morphologies.

From approximately 18 ka until Modern times, the ESL rise (Fig. 2.30) has facilitated the growth of younger carbonate facies. In the Caribbean, the modern reef complexes began their growth at ~7 ka. At that time, the rate of the relative sea

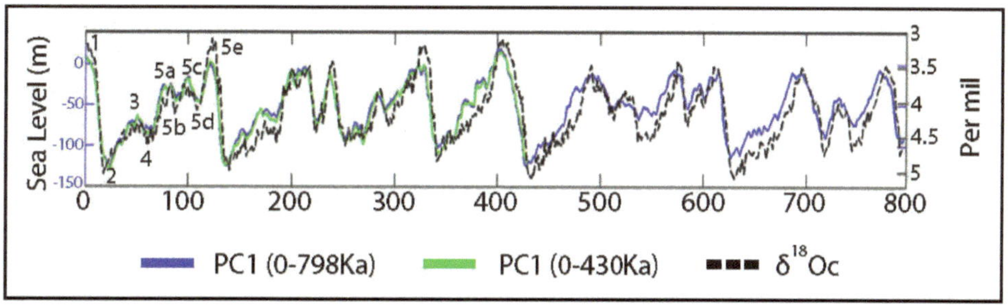

Fig. 2.30 Long (blue) and short (green) Late Pleistocene sea-level stacks compared to the LR04 (black) benthic $\delta^{18}O$ stack of Lisiecki and Raymo (2005). LR04 is a high-resolution record composed of $\delta^{18}O$ measurements mostly from SBF *Uvigerina peregrina* or *Cibicidoides wuellerstorfi*. This Late Pleistocene sea-level stack (average) has been constructed from 57 globally distributed ocean sediment cores. For the construction of the blue and green curves, Spratt and Lisiecki (2016) used a wide variety of proxies and models. These authors performed principal component analysis on seven records from 0 to 430 ka and five records from 0 to 798 ka. Marine Isotope Stages (MISs) 1-5e are indicated. Figure slightly modified from Spratt and Lisiecki (2016). Figure available via license: Creative Commons Attribution 4.0 International

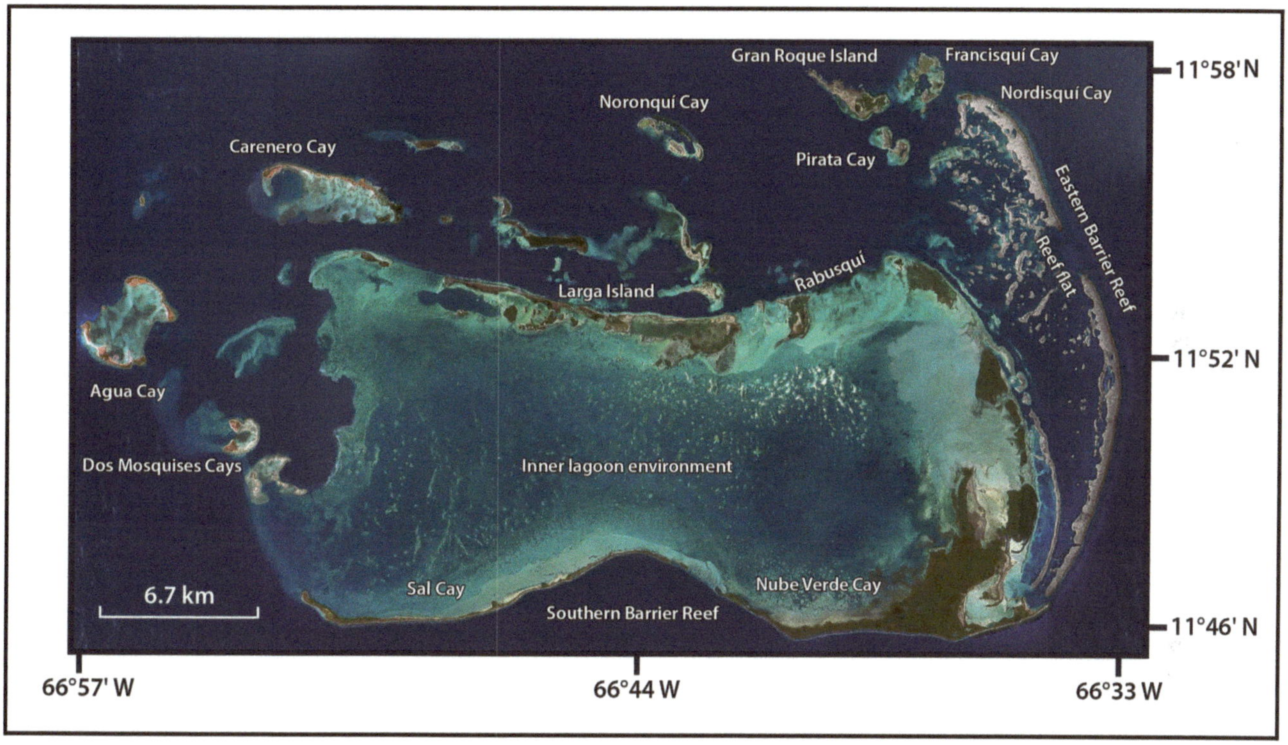

Fig. 2.31 Satellite image of Los Roques Archipelago indicating the major environments (barrier reefs, reef flat, inner lagoon), as well as some of the most important cays or supratidal islands. The meta-igneous basement only crops out on Gran Roque Island (in the northeast). Image taken from Google Earth. Imagery date: 3 April 2015. Data: SIO, NOAA, U.S. Navy, NGA, GEBCO. Image © 2022 Maxar Technologies

level (RSL) rise slowed from a maximum of 10.9 ± 0.6 m/kyr in Suriname and a minimum of 7.4 ± 0.7 m/kyr in south Florida to a rate of less than 2.4 ± 0.4 m/kyr (Khan et al. 2017). This decrease in the rate of the RSL rise promoted aggradation of carbonate facies in the upper photic zone and led to the growth of barrier reefs and fringing reefs, thus initiating the formation of modern reef islands and archipelagos.

The following sections (2.11.1, 2.11.2, 2.11.3, 2.11.4, and 2.11.5) are focused on the Venezuelan Caribbean reef complexes to which photographic examples shown in the following chapters correspond (the islands of La Orchila, La Tortuga, and Los Monjes, and Los Hermanos Archipelago are not included).

2.11.1 Los Roques Archipelago

Los Roques Archipelago National Park is Venezuela's largest coral reef complex. It consists of two large barrier reefs and about 50 cays or supratidal islands (Fig. 2.31), hosting one of the highest reef species diversities in the Southern Caribbean. The archipelago is located approximately 150 km north of

La Guaira, which is a coastal city in north-central Venezuela (Fig. 2.29). This reef complex has grown over an oceanic guyot that is bounded to the south by the Bonaire Trench (down to 1800 m), to the north by the Los Roques Trench (down to 2700 m), to the west by a canyon that separates it from Aves de Barlovento Archipelago (down to 800 m), and to the east by a canyon that separates it from La Orchila Island (down to 1300 m) (Méndez-Baamonde 1984).

Like any modern coral reef system whose main morphologic features are located slightly above or below the sea level, Los Roques is the result of complex symbiosis between carbonate-producing organisms (photoautotrophs, mixotrophs, and heterotrophs), photosynthetic microorganisms (e.g., eukaryotic phytoplankton, zooxanthellae, cyanobacteria) and the non-calcareous flora and fauna. The biologically-controlled carbonate production under predominating oligotrophic conditions (see Pérez-Castresana et al. 2014), and the growth of non-calcareous organisms (mangrove communities) that retain the loose sediments, are in natural balance with the subsidence to which the reef complex is being subject. The igneous and meta-igneous basement of the archipelago only crops out in Gran Roque Island, located in the northeast area of the archipelago (Fig. 2.31). The average temperature is ~28 °C, with a relative humidity of 80%, having a dry climate with an annual average rainfall of approximately 250 mm/yr. Vertical water transparency is relatively high down to depths of 25 m (Villamizar et al. 2003)

The Holocene carbonate reef facies has grown over Pleistocene carbonate reliefs and bioherms. The cays and associated fringing reefs and many other secondary environments are part of a carbonate platform that is limited by two main barrier reefs. These large, massive bioconstructions have formed an atoll-like complex which covers an estimated area of approximately 894 Km^2 (Weil 2003). The eastern barrier reef, which extends from north to south, has a length of about 25 km (Fig. 2.31) and it protects the inner areas from swells and currents. The waves action is derived from the trade winds that come from the east. This barrier has prograded eastward, from the intertidal zone to approximately 50 meters deep. Such depth is the approximate limit of the geologic domain of the barrier. The southern (meridional) barrier reef extends from east to west from approximately 37 km (Fig. 2.31), and it has grown parallel to the direction of the winds. This submerged barrier is protected from the direct action of waves and winds, having a very different morphology respect to the eastern barrier reef. The lower energy of the setting has allowed the growth of large cays with well-developed supratidal environments and extensive mangrove communities along the leeward sides.

The largest environment in the archipelago is the vast central lagoon, where depths range between 1 and 9 m (4–5 m average), having the shape of an irregular oval. Another main environment is the reef-flat (Fig. 2.31), which is a transitional setting that extends from the submerged crest of the eastern barrier reef to the inner lagoon environment (or back-reef lagoon). The reef flat is a medium-energy environment, with very shallow depths, especially in the immediately back-reef environment, where the average depth is less than 2 m. Very shallow depths characterize the shoreface zones along the leeward side of the cays. On the contrary, in the windward side of some of the supratidal islands (where fringing reefs actively grow), depths go down to 25–30 m. In Dos Mosquises Sur, a cay located in the southwestern border of the archipelago, such depth reaches 40 m (Jeannette Pérez, personal communication, 2022). Maximum depths in the inner areas of the archipelago (~ 45–50 m) are reached in a deep channel west of Gran Roque Island (Méndez-Baamonde 1984).

In general, the inner zones of Los Roques consist of a variety of subtidal, intertidal and supratidal environments. The subtidal settings encompass the majority of the area covered by the archipelago, including the inner lagoon and its numerous sand banks and patch reefs (or biohermal buildups that have not formed supratidal islands); the offshore areas surrounding the cays; as well as perennial hypersaline lagoons in the inner areas of some cays (where the growth of clotted microbialites is common; see Petrash et al. 2012). The intertidal settings include sandy beaches in the foreshore zone, shallow environments affected by tides (such as some areas of the reef flat and some partially exposed fringing reefs), extensive seagrass meadows (e.g., *Thalassia* sp.), numerous beachrock deposits, as well as ephemeral hypersaline lagoons (characterized by microbial mat deposits, the growth of laminated microbialites and the precipitation of gypsum, halite and protodolomite; Méndez-Dot 2011; Petrash et al. 2015). The supratidal settings include the backshore (where dune sands form); storm terraces formed during sporadic torrential rains at the windward side of the cays; as well large areas in the interior of the supratidal islands (either covered by xerophilous vegetation, or in the form tidal flats surrounding both perennial and ephemeral hypersaline lagoons).

Los Roques Archipelago is, by far, the most studied and better documented of all offshore Venezuelan reef complexes. It is home to about 69 species of scleractinian corals (Weil 2003). Among the great diversity of reef-building scleractinians (also commonly termed "hexacorals", "stony" corals, or "true" corals) stand out *Acropora palmata*, *Acropora cervicornis*, *Diploria labyrinthiformis*, *Pseudodiploria strigosa*, *Pseudodiploria clivosa*, *Siderastrea siderea*, *Montastraea cavernosa*, *Orbicella annularis* (formerly *Montastraea annularis*), *Porites astreoides*, *Porites porites*, *Agaricia agaricites*, *Agaricia fragilis*, *Favia fragum*, *Meandrina meandrites*, *Madracis formosa*, *Madracis decactis*, *Oculina diffusa*, *Colpophyllia natans*, *Mussa angulosa*, *Manicina areolata*, *Eusmilia fastigiata*, *Tubastrea aurea*, *Mycetophyllia lamarckiana*, as well as hydrozoans ("not true" corals) such as *Millepora complanata* (Méndez-Baamonde 1978, 1984; Villamizar et al. 2003, 2014; Weil 2003).

Massive *in situ* hypercalcification by scleractinian zooxanthellate corals forming the barrier reefs and fringing reefs has provided protection for the flourishing of a diverse calcareous biota. Among calcifiers stand out octocorals (e.g., *Gorgonia flabellum*, *Pseudopterogorgia* sp.); green and red algae (e.g., *Halimeda opuntia*, *Halimeda incrassata* and *Penicillus* sp., and *Neogoniolithon* sp., respectively); molluscs, mainly bivalves and gastropods (e.g., *Arca imbricata* and the "Queen Conch" *Aliger gigas*, respectively), and chitons (see Work 1969); echinoderms, mainly echinoids (e.g., *Diadema antillarum* and *Equinometra lucunter*) and ophiuroids (brittle stars and serpent stars) (see Noriega and Fuentes-Carrero 2014), less abundant sea stars (e.g., *Oreaster reticulatus*; Villamizar and Zamarro 2002) and holothuroids (sea cucumbers), and rare crinoids (feather stars); benthic foraminifera (e.g., *Homotrema rubrum*, *Amphistegina* sp. and *Peneroplis* sp.); and serpulid worm tubes.

At platform scale, the Photozoan T-factory widely dominates carbonate production in all Venezuelan reef complexes (which is intrinsic to all tropical reef complexes in modern aragonite sea, Aragonite III). In addition to dominant photosymbiotic scleractinian corals (reef builders) and photoautotrophic green algae (major sediment producers), we include red algae as common components of this factory in those modern carbonate complexes. These photoautotrophs are volumetrically important reef builders in moderate- to high-energy settings, where they thrive in association with scleractinian corals (photozoan association), forming "coralgal" facies (*sensu* Pomar et al. 2012). Symbiont-bearing LBF are important sediment producers. Although volumetrically less significant, the contribution of the Heterozoan factory (mainly gastropods, bivalves, and echinoderms) is significant. As for the origin of lime-mud, we estimate bryopsidalean green algae (mainly *Halimeda* and *Penicillus*) as a major source of carbonate muds. Moreover, we have visualized the calcium carbonate "rain" in several locations in the archipelago, indicating the contribution of $CaCO_3$ precipitated in the water column (the neritic lime-mud factory). Such rain or clouds of "milky white" (the so-called "whitings", or floating microcrystalline carbonate mud) may result from either biologically-induced precipitation by cyanobacterial activity, or abiotic (inorganic) precipitation directly from seawater.

In respect of modes of carbonate precipitation in peritidal settings, the Biochemical T-factory prevails in supratidal sabkha flats and ephemeral hypersaline lagoons, where high salinities and evaporation rates promote the formation of carbonate-evaporitic crusts containing important amounts of gypsum and halite. On the other hand, in turbid and commonly low-oxygen waters of perennial hypersaline lagoons, the microbially-induced precipitation of carbonate (the benthic automicrite factory) predominates, promoting the growth of microbialites (stromatolites and thrombolites).

Biodiversity in Los Roques Archipelago includes diverse non-calcifying marine biota as important components of reef communities. Among these stand out benthic macroalgae of the phyla Chlorophyta (green algae), Rhodophyta (red algae) and Ochrophyta (which includes brown algae) (see Gómez et al. 2014); sponges (about 60 species), mainly demosponges (e.g., *Aplysina fistularis* and *Xestospongia muta*) (see Díaz et al. 1987; Villamizar 2002); polychaetes (see Fernández et al. 2012); and decapod crustaceans (see Grajal and Laughlin 1984). The bulk of marine fauna consist of a wide variety of fishes (ichthyological fauna), with more than 300 identified species only in shallow depths (< 20 m; see Cervigón and Alcalá 2002); lobsters (see Posada 2002); as well as fauna of great ecological importance, such as marine turtles (e.g., *Chelonia mydas* and *Eretmochelys imbricata*; see Guada 2002). The vegetation consists of xerophilous communities that grow in very dry conditions in the supratidal areas of the cays (e.g., *Melocactus caesius* and *Sesuvium portulacatrum*); halophilous communities (mangroves) that grow in sandy soils flooded with high salinity seawater (e.g., red and black mangroves; *Rhizophora mangle* and *Avicennia germinans*, respectively); and seagrasses (mainly *Thalassia testudinum*) that grow in shallow meadows, hence comprising the so-called "seagrass meadows" (see Villamizar and Zamarro 2002). The avifauna is also very diverse, with approximately 92 identified bird species belonging to 30 families (see Lentino and Rodner 2002).

The rich biodiversity of Los Roques Archipelago makes it a unique ecosystem within the Venezuelan territory. It was declared a national park in 1972, in tribute to the one hundred years of the creation of the first national park in the world, the Yellowstone National Park (Zamarro 2002). The preservation of the coral reef complex and its rich fauna is not only due to its remoteness respect to the Venezuelan coastline, but to its location outside the geographic area routinely influenced by hurricanes (Villamizar et al. 2003), being the uncontrolled tourism activities, and the consequences of global warming and related seawater acidification, the major threats for the coral reef communities.

In order to protect the pristine state of the ecosystems, the archipelago has been divided into different zones. Among them stand out the integral protection zones (e.g., Nube Verde cay and Larga island), where only scientific investigation is allowed, so that coral reefs, mangrove communities and turtles nesting areas are totally protected from any anthropic disturbance; the primitive marine zones (e.g., the Eastern Barrier Reef, Sal Cay), which are ecologically sensitive areas; and the recreational zones (e.g., Gran Roque Island, Fransisquí and Agua cays), where ecotourism is allowed (see Zamarro 2002; Martinez et al. 2021).

During the last two decades, several studies have focused on the status of coral reef communities in Los Roques

Archipelago. Monitoring of both barrier reefs and numerous fringing reefs and patch reefs in a large number of sites have included the study of coral reproduction, settlement rates and survival rates (Humanes and Bastidas 2015); the documentation of coral bleaching events in 2005 (Villamizar et al. 2008) and 2010 (Bastidas et al. 2012) and coral diseases (e.g., the white plague and white band diseases), as well as the potential recovery of some scleractinian species (Cróquer et al. 2003, 2005, 2016; García et al. 2003; Cavada-Blanco et al. 2020; Martinez et al. 2021). Until a little more than a decade ago, some of those obtained results suggested that Los Roques was arguably one of the most well-preserved and pristine sites in the Caribbean (see Jackson et al. 2014). On the other hand, for almost the entire last decade, systematic monitoring of coral reefs in that archipelago has not been possible due to lack of funding.

Like in all major coral reef bioprovinces, coral reef ecosystems in the Greater Caribbean have been hit-hard by modern anthropogenic global warming and associated coral bleaching events (e.g., Eakin et al. 2010; Heron et al. 2016). Coral mortality has globally intensified during the last decade, which includes the long-lasting 2014–2017 global-scale coral bleaching event (Eakin et al. 2019, 2022). Consequently, widespread loss of coral cover in Los Roques Archipelago has increased in the last few years. We do not know the extent of the damage caused by this latest global bleaching event in that coral reef complex. However, coral reefs of Los Roques are without doubt the most diverse and healthy in Venezuela (Estrella Villamizar and Ana Yranzo Duque, personal communication 2022).

2.11.2 Aves Island

Aves Island is a small reef island (Fig. 2.32) located in the northeast Caribbean. Its remote location with respect to mainland Venezuela, approximately 650 km NNW of La Guaira, makes it the northernmost territory of this country (see Fig. 2.29; see also Fig. 1.1). This island, together with La Blanquilla Island and Los Hermanos Archipelago, is the only emergent feature of the long submarine Aves Ridge. This oceanic ridge is an ancient volcanic arc with a broadly north-south trend, approximately 550 Km long and 150 km wide on its thickest section (Arnaiz-Rodríguez and Audemard 2018). Aves Ridge is bounded to the west by the Venezuelan Basin (maximum depths down to 5500 m) and to the east by the Grenada Basin (maximum depths down to 3500 m) (see Fig. 2.29) (Neill et al. 2011; Ughi 2021). These surrounding great depths, and the fact that Aves Island is the only positive topographic feature in northern Aves Ridge, make it a key location in the Caribbean for geological, geochemical, geophysical, and biological research.

The dimensional measurements of the supratidal island can vary depending on the author and the method used to take the

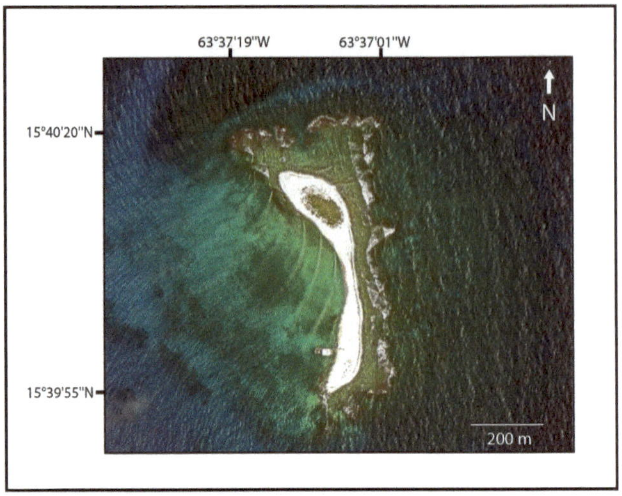

Fig. 2.32 Satellite image of Aves Island. This small reef island (~ 600 m in length) is located approximately 650 km north of continental Venezuela. Image taken from Google Earth. (Imagery Date: 22 August 2017. Image © 2022 Maxar Technologies)

measurements. According to Arteaga (2002), the approximate length of the island is 580 m, and its width is 130 m at its widest point and 30 m at its narrowest. From an image captured in August 2021, Heatwole et al. (2022) determined that the length and width vary from 555 m (north-south extend) and 221 m (east-west extend), respectively, to 569 m and 158 m (the latter measurements were taken with the longest edge aligning with the longest axis of the island). The width of the island varies even more, towards smaller dimensions (when the line traced runs strictly over the island and not over water), from 121 m to 111 m (see Heatwole et al. 2022; for details on dimensional measurements performed by those authors on Aves Island; see also Casanova 2016). Taking into account the partially emerged crest of the fringing reef at both northernmost and southernmost margins of Aves Island (Fig. 2.32), the maximum length is approximately 650–700 m.

Accumulation and remobilization of loose, sand-sized sediments is a constant, dynamic sedimentary process on the small platform along the western-leeward side of the island. The lithified Holocene carbonate facies (fringing reef, back-reef limestones, and beachrocks) grew on the top of an eroded limestone unit most likely formed during the LIG (Fig. 2.33). These Late Pleistocene and Holocene limestone units overlie Early-Middle Pleistocene and Pliocene limestones. Furthermore, dredged carbonate samples from several sites on the Aves Ridge indicate that parts of the ridge were in shallow water during the Eocene and Early Miocene (see Garrocq et al. 2021; Padron et al. 2021). At Aves Island, the succession of pre-Holocene limestones was formed over the uppermost portion of the underwater igneous guyot. This submarine ellipsoidal structure constitutes the "Aves High" (Fig. 2.34).

Studies of combined seismic reflection and refraction and bathymetry carried on the Aves Island platform (Aguilera and Gallovich 1970; Laredo and Picard 1977) suggest that the suc-

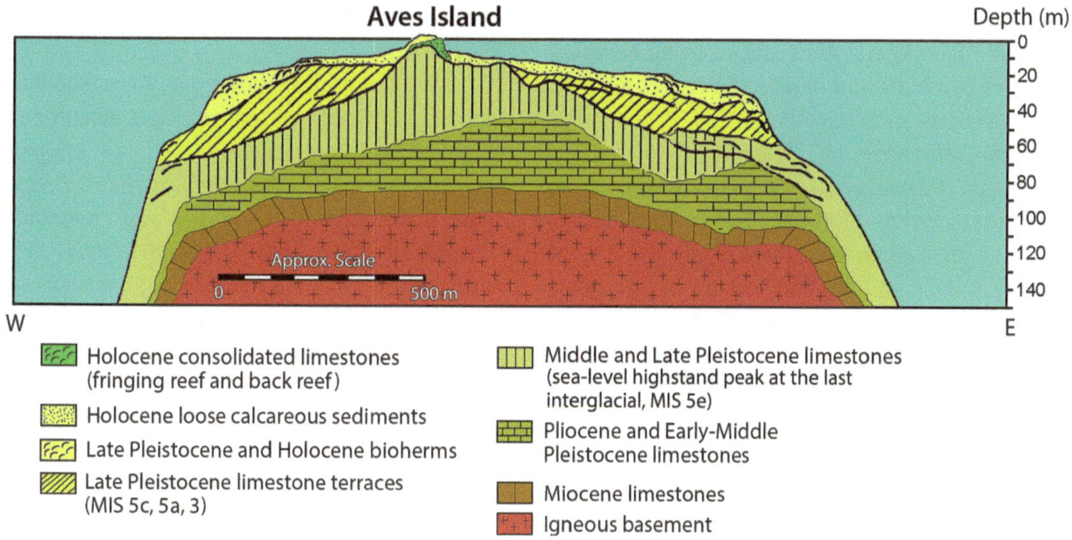

Fig. 2.33 Stratigraphic sketch illustrating the rock basement of Aves Island, which makes up the uppermost portion of the Aves High. The Holocene fringing reef (highlighted in green) lies directly on the eroded limestones formed during the last interglacial (LIG) sea-level highstand peak (MIS 5e). In this sketch, the contact between Miocene limestones and the igneous basement is hypothetical. Deposition of shallow-water limestones over the oceanic guyot could have begun prior to Miocene times. Abbreviations. *W* west, *E* east. (Figure modified from Méndez-Baamonde 2006)

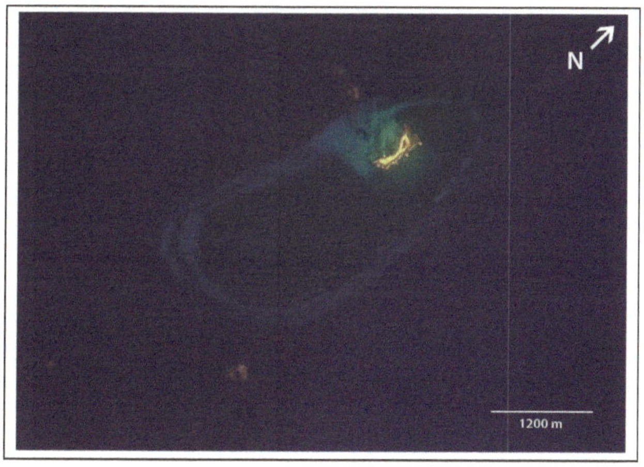

Fig. 2.34 NASA astronaut image of Aves Island and the Aves High. Note both the fringing reef bordering the island and the shallow platform along the leeward (western) side. The Aves High has more than 5 km in length along its north-south orientation. Also note the light-colored, ring-like morphologies (carbonate terraces) outlining the uppermost peripheries of the high. Image courtesy of the Earth Science and Remote Sensing Unit, NASA Johnson Space Center. Imagery date: 27 May 2006

cession of pre-Holocene limestone units overlying the igneous guyot is estimated to be about 80 to 120 m thick (Fig. 2.33). We estimate that the real limestone thickness could be higher, but this must be corroborated by new seismic studies. Changes in the slope marked by the presence of the uppermost section of the youngest limestone terrace are located approximately between −15 to −26 m deep, which is the approximate depth interval at which the slope break increases markedly towards verticality (see Yranzo et al. 2014).

The Holocene fringing reef has grown with a nearly north-south orientation along the eastern limits of the island (Figs. 2.32 and 2.34). The reef is not entirely continuous but presents several openings (see Yranzo et al. 2014). It consists mainly of scleractinian corals and encrusting coralline algae (the genus *Lithothamnion* is abundant). Such coralgal buildup is characteristic of tropical reef complexes that have grown in strong hydrodynamic energy settings. Accumulation and accretion of loose sediments in the supratidal island reach ~3.0–3.5 m above the RSL during off-season hurricanes. In February 2024, Villalobos et al. (2024) determined that the highest elevation of the island is located in its northern area, with an orthometric height of 3.472 m (the orthometric height is the vertical distance from a specific point on the Earth's surface to the geoid, which is a reference surface approximating mean sea level). This height was obtained using the Earth Gravitational Model 2008 (EGM2008; see Pavlis et al. 2012). Supratidal loose sediments (mainly sands and granules) consist mainly of fragments of calcareous green and red algae, molluscs, corals, echinoderms, gastropods, and tests of larger benthic foraminifera. Coral gravel (pebbles and cobbles) removed from the shallow subtidal zone are frequently found along the margins of the island. The abundance of *Homotrema rubrum* encrusting coral gravel is easily noticed due to characteristic pink-reddish color of this encrusting benthic foraminifer. Several beachrock deposits, exhibiting a variety of morphological characteristics, are situated in the backreef setting in the north, northeast and east areas of the island. Several photographic examples of these Holocene deposits are shown in Chap. 3.

Recorded coral species diversity in Aves Island consists of 36 and 13 species of scleractinian corals (hermatypic corals) and octocorals (soft corals), respectively, at depths between 1.5 and 21 m around Aves Island, being necessary further studies of benthic communities in deeper areas. The most abundant species of scleractinians are *Porites astreoides*, *Pseudodiploria strigosa*, *Siderastrea siderea* and *Orbicella faveolata* (formerly *Montastraea faveolata*); while *Antillogorgia americana* (formerly *Pseudopterogorgia americana*) is the most abundant octocoral species. Another important reef builder in this reef island is the genus *Millepora* (mainly *M. Squarrosa* and *M. Alcicornis*), the "fire coral" (Yranzo et al. 2014). The diversity of sponges includes 77 identified species, being *Amphimedon compressa* and *Agelas sventres* (Class Demospongiae) the most abundant (Villamizar et al. 2019).

Aves Island is the sole offshore Venezuelan island situated in the Caribbean hurricane belt (although in the belt's eastern corridor and not in its core). Consequently, this island has been affected by numerous hurricanes and tropical storms during the last decades (e.g., hurricanes David, Allen, and Dean, which occurred in 1979, 1980 and 2007, respectively). These disruptive events resulted in the mobilization and dispersion of the supratidal loose sediments, altering the island's immediate visual perception during short-term periods. The island returns to its normal sedimentary dynamic once the disturbances finish. See Heatwole et al. (2022) and Scherer (2024) for a historical review of cartography and the impact of hurricanes on Aves Island.

During these ideal conditions, Aves Island is well known for being a wildlife refuge, serving as host for migratory bird species, including seagulls (e.g., *Onychoprion fuscatus*, *Onychoprion anaethetus*, *Anous stolidus*, *Leucophaeus atricilla*), turnstone birds (*Arenaria interpres*) and frigate birds (*Fregata magnificens*) (Gremone and Gómez 1983). Furthermore, the island is one of the main nesting localities for the green turtle (*Chelonia mydas*) in the Eastern Caribbean Sea (Weil 2003; Vera and Buitrago 2012; Heatwole et al. 2022).

2.11.3 Las Aves Archipelagos

The Las Aves Archipelagos (also known as the Las Aves Islands) are two atoll-like, reef complexes named Aves de Barlovento (to the east) and Aves de Sotavento (to the west) (Fig. 2.35) which lie over oceanic guyots between the Los Roques Archipelago and the Bonaire Island (see Fig. 2.29). These archipelagos are located approximately 170 km north of the central main coasts of Venezuela and 75 km east of the island of Bonaire. Both complexes are separated by a channel of average depths of 400 m and maximum depths down to 800 m, and by deep canyons (> 1000 m) from Los Roques

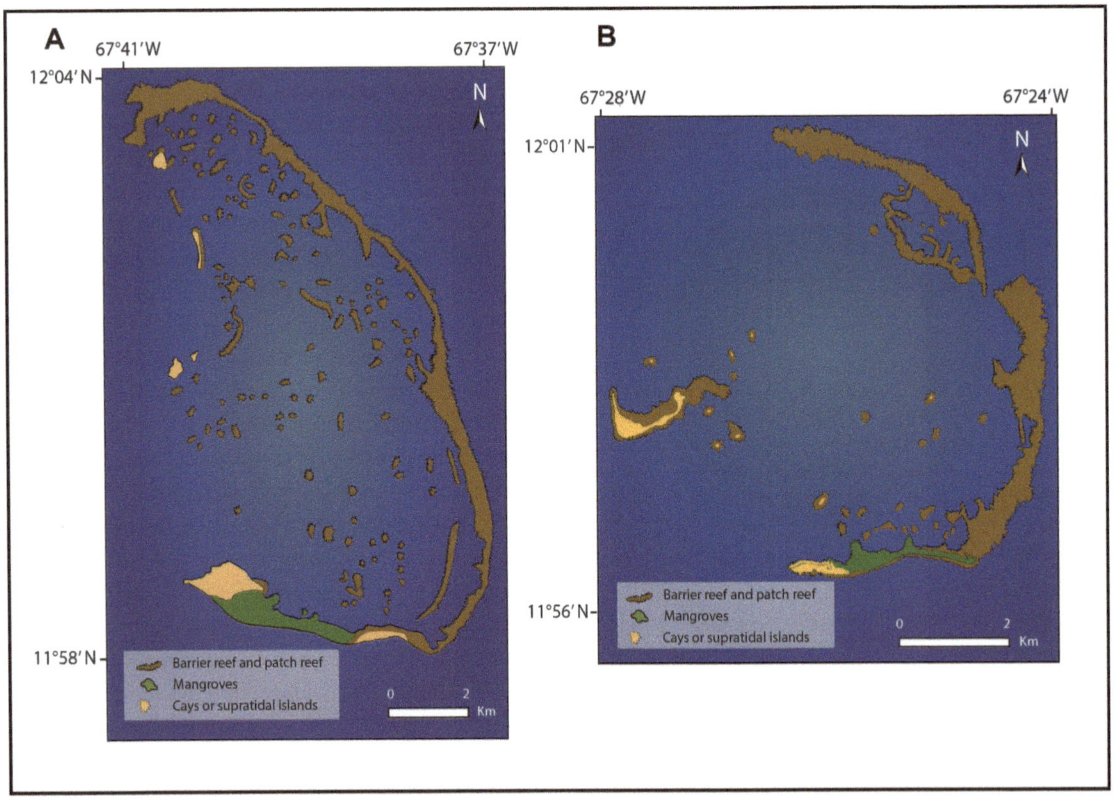

Fig. 2.35 Maps of the Las Aves Archipelagos. (**a**) Aves de Sotavento Archipelago; (**b**) Aves de Barlovento Archipelago. Both archipelagos have developed a semi-atoll morphology. The number of cays or supratidal islands in both complexes is minor, while subtidal patch reefs are more numerous. (Source map: U.S. Navy Survey 1939. Figure modified from Méndez-Baamonde et al. 2009)

Archipelago to the east and Bonaire to the west. They are very similar to Los Roques in terms of morphology and major sedimentary environments, consisting of well-developed eastern barrier reefs and associated reef flats, southern barrier reefs, and central lagoon environments, but their dimensions are smaller.

Aves de Sotavento ("sotavento" meaning leeward) covers an area of ~80 km^2, with a ~ 15 km long eastern barrier reef that has grown continuously, without the development of mouths or openings (Fig. 2.35a). The latter feature, combined with the energy protection granted by Aves de Barlovento ("barlovento" meaning windward) just to the east (Fig. 2.35b), led to the development of a more enclosed inner lagoon environment with shallower depths, averaging 3–4 m deep and a maximum of ~18 m deep. These factors promoted the development of a greater number of patch reefs in the inner lagoon compared to Aves de Barlovento. The latter archipelago covers an area of ~72 km^2, and consists of a barrier reef of ~12 km long and an inner lagoon with an average depth of 18–20 m and maximum depths down to ~60 m (Debrot et al. 2019). Behind the barrier reefs in both atoll-like complexes, a low number of cays or supratidal island have developed.

In both archipelagos, common Caribbean reef builders (*Acropora cervicornis*, *Orbicella annularis*, *Siderastrea* sp., *Diploria* sp., *Pseudodiploria* sp., *Millepora alcicornis* and *Millepora complanata*) make up the bulk of barrier reefs (Méndez-Baamonde 1978, 1984). These uninhabited and poorly-studied reef complexes should be considered as candidates for natural sanctuary areas or a National Coral Reef Reserve (Weil 2003).

2.11.4 La Blanquilla Island

La Blanquilla Island is located approximately 170 km north of the city of Puerto La Cruz (northeast mainland Venezuela) and 90 km northwest of Margarita Island, where it emerges at the southwestern extreme of the Aves Ridge (see Fig. 2.29), as it has grown over a guyot named La Blanquilla High (Bezada et al. 2010). This island has a diameter of ~10 km, covering an area of ~53 km^2. Its main morphologic features are three well-exposed coral reef terraces of Pleistocene age (Fig. 2.36).

Those terraces make up the La Blanquilla Formation and consist of fringing reef facies (boundstone textures consisting mainly of hexacorals) towards the edges of the island and back-reef facies towards the center, where a great variety of fossiliferous and muddy facies are found (Méndez-Baamonde 1984). The lower terrace is named Falucho Member (Maloney 1971). It crops out almost across the whole island, being well exposed in the eastern and northern areas (Fig. 2.36), where it has formed a steep-cliff morphology, as it is constantly subjected to the direct action of the waves.

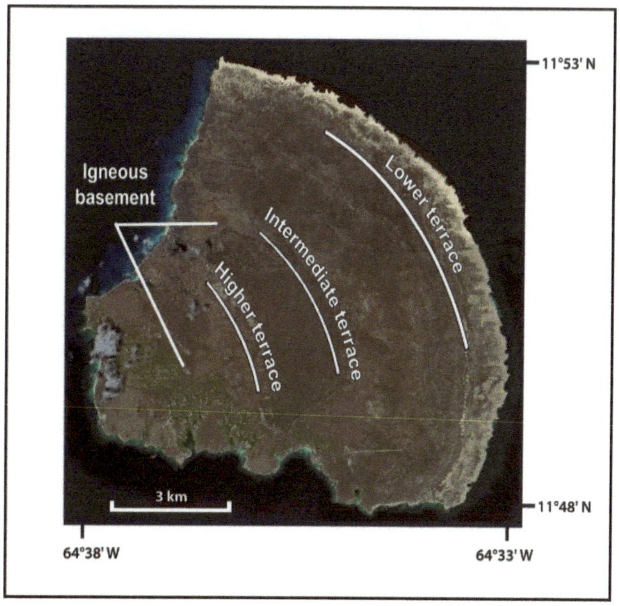

Fig. 2.36 Satellite image of La Blanquilla Island highlighting the three exposed calcareous terraces. Location of the igneous basement according to Schubert and Moticska (1973). (Image taken from Google Earth. Imagery date: 15 May 2015. Data: SIO, NOAA, U.S. Navy, NGA, GEBCO. Image © 2022 CNES/Airbus, Maxar Technologies, TerraMetrics)

This terrace consists of a variety of reef-building scleractinian corals, mainly *Acropora palmata*, *Montastraea cavernosa*, *Orbicella annularis*, *Diploria labyrinthiformis*, *Acropora cervicornis*, and *Siderastrea siderea* (Méndez-Baamonde 1997).

The shape of the island along the eastern coast is, from southeast to north, a semi-arch whose periphery is outlined by the lower limestone terrace. The arched shape is an erosional feature inherited from the strong, direct action of both trade winds and waves. Steep cliffs along the high-energy coastline are about 7–9 m high, which is the average height of the lower, younger terrace. Towards the center of the island, first the intermediate and then the higher terraces crop out, also showing a semi-arch morphology (Fig. 2.36). Along the southern coast, the morphology of the island consists of a series of embayments. The low energy of these environments has allowed the development of sandy beaches, as well as the local growth of mangrove communities.

Modern colonies of scleractinian corals are well developed along shallow waters of the leeward-south coast of the island (47 identified scleractinian species), where reefs consist of common reef-building species. Among most abundant species stand out *Orbicella faveolata*, *Orbicella annularis*, *Orbicella franksi* (formerly *Montastraea franksi*), *Siderastrea siderea*, *Pseudodiploria strigosa*, *Porites astreoides*, *Porites porites*, *Acropora palmata*, *Acropora cervicornis*, and *Dendrogyra cylindrus*, as well as *Millepora complanata* (Weil 2003).

2.11.5 Morrocoy

Sheltered from significant input of riverine terrigenous detritus to the coastlines, the Morrocoy National Park contains the best-developed, coastal carbonate reef system in Venezuela (see Weiss et al. 1978). This park is located in eastern Falcón State (see Fig. 2.29). Its southern area is on the western margin of the Triste Gulf (in front of the town of Tucacas), which is a large inlet of the west-central Caribbean coasts of Venezuela (Bone et al. 1998); while it extends to the north along the shoreface and shallow offshore zones in front of the town of Chichiriviche (Fig. 2.37).

Morrocoy is a transitional ecosystem that encompasses continental, insular, and marine environments. The most significant positive morphologic feature in the area is the Chiririviche Hill. With a maximum elevation of approximately 284 meters, this hill consists mainly of Middle Miocene limestones of the Capadare Formation (Goddard 1974). The inshore zone of Morrocoy (the Golfete de Cuare and mangrove swamp areas; Fig. 2.37) is characterized by low-wave activity, high turbidity, shallow lagoon waters, abundant mangrove communities (e.g., *Rhizophora mangle*), and heterotrophic carbonate biota (molluscs and echinoderms).

The seaward zone is connected to the open ocean and is characterized by coral communities, moderate swell, low turbidity, and depths down to 20 meters (Bone et al. 1998). In Morrocoy, there are approximately 12 reef islands with well-developed fringing reefs (some of the most significant are indicated in Fig. 2.37). These are the main coral reef formations (Weil 2003). Sand banks are also numerous. Coral diversity includes 58 scleractinian corals. The most abundant species are *Orbicella faveolata*, *Orbicella annularis*, *Montastraea cavernosa*, *Orbicella franksi*, *Colpophyllia natans*, *Pseudodiploria strigosa*, *Pseudodiploria clivosa*, *Diploria labyrinthiformis*, *Porites astreoides*, *Porites porites*, *Acropora palmata*, *Acropora cervicornis*, *Stephanocoenia intersepta*, *Madracis decactis*, *Madracis mirabilis*, *Siderastrea siderea*, and *Solenastrea bournoni* (Bone et al. 1998; Weil 2003; Yranzo-Duque et al. 2020). The genus *Orbicella*, notably *O. faveolata*, is the main reef-builder in this coastal coral reef system (Yranzo-Duque et al. 2025). Like all tropical reef systems, carbonate-producing biota in Morrocoy is also well represented by abundant calcareous (green and red) algae, echinoderms, mollusks, and benthic foraminifera. Seagrass meadows in the upper shoreface zones, and beachrock deposits in the intertidal zones are also common.

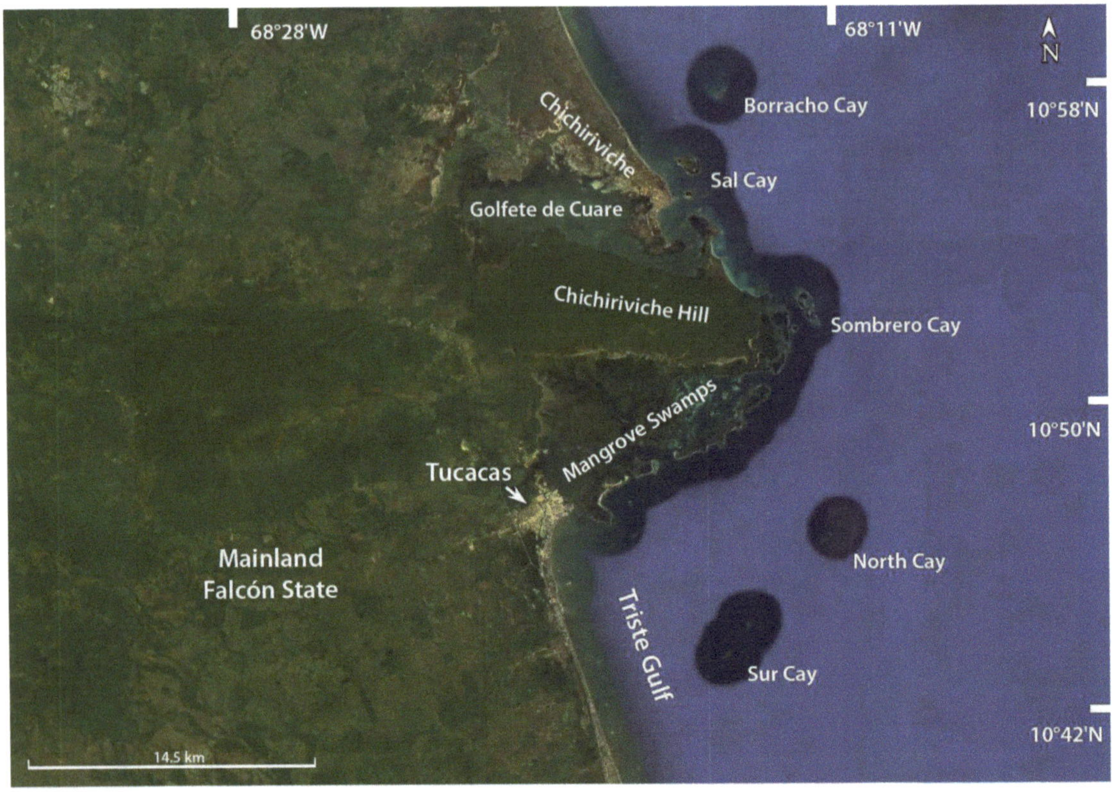

Fig. 2.37 Satellite image of eastern Falcón State, which includes the Morrocoy National Park. Some of the main reef islands (cays) and morphological features in the area are indicated. (Image taken from Google Earth. Imagery date: 13 December 2015. © 2023 CNES/Airbus, Maxar Technologies)

Like in Los Roques Archipelago, coral reefs in Morrocoy have been affected by bleaching events and diseases, mainly yellow band disease. In addition, the coastal nature of this national park makes it more vulnerable to anthropogenic disturbances. The increase in the riverine discharge of sediments (mainly from the Tocuyo and Aroa rivers, located north and south to the park, respectively) linked to deforestation activities; the accumulation of different types of waste (from plastic to oil and boat fuel); and the discharge of wastewater directly into the sea are some of the most harmful disturbances that affect Morrocoy's varied ecosystems (see Yranzo-Duque et al. 2020). Furthermore, recent oil spills from a refinery located ~60 southeast Morrocoy represent the greatest danger to this modern carbonate system.

2.11.6 Rock Basements of the Leeward Antilles

Although most of the CP is submerged beneath the Caribbean Sea, many Late Cretaceous igneous complexes cropping out in Central America, northwest South America (Ecuador and Colombia), and the southeastern Caribbean (the Leeward Antilles) are interpreted as CP fragments (Whattam and Stern 2015). A few mafic units represent rare remnants of the proto-Caribbean seafloor [i.e., not plume- but mid-ocean ridge-derived), which is the case of the Siquisique Ophiolites cropping out in northern Lara state (northwest Venezuela). This unit mainly consists of gabbros and basalts with overlying dark cherts. According to studies of several outcropping sections by different authors, this fragment of oceanic crust has a wide range of ages, from the Middle Jurassic (Bajocian-Bathonian) to mid-Cretaceous (Albian-Coniacian) (Kerr et al. 2009, 2021; Sandoval et al. 2023 and references therein).

The Leeward Antilles consist of the Venezuelan Antilles and the ABC Islands (Aruba, Bonaire, and Curaçao), the latter also known as the Netherland Antilles or the Dutch Antilles (see Figs. 2.10 and 2.29). Quaternary reefs complexes in these islands have developed over a set of chained guyots. The tips of these bathymetric highs in the southeastern Caribbean crop out in the ABC islands (only Aruba and Curaçao are considered in this section) and in some of the Venezuelan Antilles. The ones considered in this section are the islands of Gran Roque, La Blanquilla, and La Orchila. Crystalline rocks cropping out in these islands consist of a variety of igneous (mafic, intermediate, and felsic lithologies) and meta-igneous rocks, many of which are considered fragments of the CP. Roughly, magmatic processes involved in the formation of these units are divided into plume activity (mafic lavas), which formed the COP (mainly basalts and gabbros ranging in age from Albian to Coniancian), and subsequent subduction-arc younger magmatic activity that generated intrusive units (intermediate and felsic lithologies) ranging in age from the Coniacian to Late Paleocene (~ 89 to 59 Ma). In addition to the subduction of the proto-Caribbean beneath the COP that led to the inception of arc magmatic activity, the magmas from which intermediate and felsic intrusions in the Leeward Antilles crystallized also resulted from the interaction between the Caribbean Great Arc and its COP basement and the northern margin of South America (see Neill et al. 2011; Wright and Wyld 2011). Furthermore, the presence of continental blocks of Meso- to Neoproterozoic, Late Paleozoic, and Jurassic ages underlying the proto-Caribbean seafloor and to the east of the Caribbean Great Arc might have also served as a source for the generation of intermediate and felsic magmas (Marvin Baquero, personal communication 2025; see also Baquero 2015; Baquero et al. 2015).

The submarine ascent of these seamounts occurred as consequence of the tectonic processes related to the initial oblique collision and subsequent eastward movement of the CP respect to the South American plate from the Late Cretaceous. Regional uplifting of crystalline basement units outcropping today intensified since the Late Oligocene, due to the development of the Southern Caribbean Deformed Belt and the South Caribbean Accretionary Prism (see Figs. 2.10 and 2.23); and then during the Pleistocene, due to tectonic events related to the closure of the Panamá Isthmus (see Reijmer et al. 2002; Schneider and Schmittner 2006).

In Aruba and Curaçao, the lithology, geochemistry, and ages of outcropping igneous rocks provide clear insights about their origins. Ages of COP rocks of the Aruba Lava Formation range between 103–89 Ma (see Whattam and Stern 2015), the bulk of it formed at 94–89 Ma (White et al. 1999; Wright and Wyld 2011). Arc intrusions (the Aruba Batholith) encompass an age range of 90–82 Ma (White et al. 1999; Wright and Wyld 2011). In Curaçao, the Curaçao Lava Formation consists of OP rocks derived from two different plume magmatism episodes. The older rocks, containing Albian ammonites and dated at 112.7 ± 7.3 Ma (U/Pb; Humphrey, 2010), represent a fragment of the old Caribbean plateau; while rocks dated at 89.5 and 88.9 Ma (Turonian-Coniancian, Ar/Ar; Sinton et al. 1998) represent a fragment of the Late Cretaceous COP. On the other hand, arc magmatism is represented by diorite intrusions dated at 87–85 Ma (U/Pb; Wright and Wyld 2011).

In Gran Roque Island, located in northeast Los Roques, the crystalline basement of this archipelago crops out. It consists of metamafic rocks and intermediate and felsic intrusions. The oldest rocks on Gran Roque are amphibolites (van der Lelij et al. 2010), considered a metadiabase unit by Santamaría and Schubert (1974) and Ostos and Sisson (2005). This unit forms the central and eastern hills and is derived either from mid-ocean ridge basalts (Santamaría and Schubert 1974; Ostos and Sisson 2005), i.e., the proto-Caribbean seafloor, or OP rocks (Giunta et al. 2002a), i.e., the old CP or the COP. It was dated at 130 ± 14 Ma (K/Ar; Santamaría and Schubert 1974), although this age has not been confirmed by new dating. In the eastern hill, this unit is intruded by a metagabbro (lower greenschist facies) dated at 87 Ma [Coniacian, U/Pb in zircon; Humphrey (2012) in Urbani (2018)]. The same metagabbro forms the western hill of the island (Ostos and Sisson 2005; Urbani 2018). According to Ostos and Sisson (2005), the composition of the metagabbro from Gran Roque indicates that this unit may have formed in either a mid-ocean ridge or island arc setting. On the contrary, geochemistry of rare earth elements carried by Humphrey (2012; in Urbani 2018) on the metadiabase and metagabbro units suggests that those units have an OP origin. According to van der Lelij et al. (2010), the age and nature of the exposed amphibolites (metadiabase) is still unclear and may either represent COP material or mid-ocean ridge basalts which were trapped between west dipping subduction zone and the COP. In the eastern hill, a pophyritic andesite dike (inland arc-affinity) was dated at 63.6 Ma (U/Pb in zircon; Baquero et al. 2017). The eastern and central hills are also intruded by quartz-diorite dikes (island arc-affinity) dated at 65.6 Ma [U/Pb in zircon; Wright and Wyld (2011); see also Urbani 2018] and at 64.8 Ma (Ar/Ar in biotite; van der Lelij et al. 2010). The western hill is intruded by pegmatite, aplite (Ostos and Sisson 2005), and quartz-diorite (Urbani 2018) dikes of island arc-affinity.

In La Blanquilla, which is the southernmost exposure of the Aves Ridge (see Fig. 2.29), no COP mafic rocks are found. This, along with the fact that outcropping igneous units are of intermediate and felsic composition, proves that this island was formed by arc magmatism. The crystalline basement of La Blanquilla consists of two plutonic units. The older unit is a granitoid dated at 76 Ma (Campanian, U/Pb in zircon). The younger unit is a tonalite dated at 59 Ma (Late Paleocene, U/Pb in zircon) (see Neill et al. 2011; Wright and Wyld 2011; Baquero et al. 2017). Wright and Wyld (2011) hypothesize that the two dated plutons may approximate the initiation and termination of arc magmatism associated to the formation of the Aves Ridge.

In La Orchila Island, located 56 Km east of Los Roques Archipelago (see Figs. 2.10 and 2.29), at least three geologic terrains of distinct origins and ages are juxtaposed. The older one is a unit of continental affinity (a biotitic quartz-plagioclase schist) dated at 271 Ma (Middle Permian, U/Pb in Zircon; see Urbani 2018). The second terrain, also of continental affinity, is a metasedimentary unit (a garnet-bearing muscovite-quartz schist) intruded by mafic and intermediate igneous rocks, now metamorphosed, which include a tonalitic gneiss dated at 177 Ma (Early Jurassic, Toarcian, U/Pb in Zircon; Baquero et al. 2017). The third terrain is a metadiabase-metatonalite, which Urbani (2018) suggests is probably the time-equivalent of COP rocks in Aruba. This metaigneous unit is intruded by intermediate igneous rocks of arc affinity (porphyritic andesite, spessartite lamprophyre, and kersantite lamprophyre) dated at 57 Ma (Baquero et al. 2017).

2.11.7 Late Pleistocene Carbonate Terraces in the Leeward Antilles, Aves Island, and Barbados

The sea-level highstand of the LIG (MIS 5e; 130–116 ka; e.g., Chauveau et al. 2024) is one of the most enduring and significant coastal geomorphological expressions of global sea-level changes in the more recent geological record (Murray-Wallace and Woodroffe 2014). The LIG is among the most studied periods in Earth's history (e.g., Vézina et al. 1999; Hearty et al. 2007; Rohling et al. 2008; Woodroffe and Webster 2014; Rovere et al. 2016a; Polyak et al. 2018; Chutcharavan and Dutton 2021; Hallmann et al. 2021; Dutton et al. 2022; Rovere et al. 2023; Chauveau et al. 2024). During this highstand, the global sea-level was between +2 and +9 m APSL (Dutton and Lambeck 2012; Dyer et al. 2021; Dumitru et al. 2023).

Limestone terraces formed during the LIG period are prominent on many tropical shorelines. The elevation of these terraces, often composed of fringing reef or skeletal sands, has been used to assess the rate of vertical displacement and determine local rates of uplift or subsidence. In turn, such elevations vary geographically not only due to regional-local tectonic processes, but also due to glacio-isostatic adjustments that result in RSL depart from ESL (see Rovere et al. 2016b). In sites considered tectonically stable, terraces formed during the LIG are typically found at 0–10 m of elevation. Terrace elevations above 10 m are considered to be the result of uplift. Submerged terraces can either reflect subsidence, topographic complexity of the surface during the LIG, or locally variable erosion caused by subsequent dissolution (see Murray-Wallace and Woodroffe 2014; Woodroffe and Webster 2014). Furthermore, the best ESL indicators are from tectonically stable localities that are geographically distant, or "far field", from former ice-sheet margins (see Dutton and Lambeck 2012).

The uranium-thorium dating method (or U-series) is the most widely applied for Late Pleistocene marine carbonates. U-series measurements are made in coral skeletal aragonite (see Edwards et al. 2003; Hibbert et al. 2016; Chutcharavan and Dutton 2021; for complete reviews on this dating method and its diagenetic considerations). We highlight two recent papers. The first, by Chutcharavan and Dutton (2021), presents a comprehensive, global compilation of published U-series-dated fossil coral sea-level indicators and associated elevation measurements and sample data for the LIG period. The second, by Rovere et al. (2023, The World Atlas of Last Interglacial Shorelines), provides a global database of sea-level proxies and samples dated to MIS 5e.

On La Blanquilla Island, there are three exposed carbonate terraces located at elevations between 0–10 m, 11–15 m and 15–30 above RSL. The lower terrace was dated at 131 ± 7 ka. This age indicates that the terrace was formed during the LIG (Schubert and Szabo 1978). Photographic examples of this terrace are shown in Chap. 3. The intermediate and higher terraces are estimated to be Pleistocene in age but older than the lower one. On La Orchila Island, the most extensive morphologic feature is a limestone terrace that belongs to the La Orchila Formation. This terrace, located at elevations between 1–3 m above RSL, and dated at 130 ± 7 ka, was also formed during the LIG (Schubert and Valastro Jr 1976).

On the southwest margin of Gran Roque Island, there is a poorly exposed limestone terrace consisting of three well-differentiated calcarenite beds exhibiting a retrogradational (backstepping) stacking pattern. These beds are located between 0 and 2 m above RSL, and have been estimated to have formed contemporaneously with the terraces formed during the LIG on the islands of La Blanquilla and La Orchila (Schubert and Valastro Jr 1976; Méndez-Baamonde 1984, 1997). Two photographic examples of this terrace are shown in Chap. 3. In the southern barrier reef of Los Roques, a series of submerged limestone terraces are located at approximate depths of – 12 m, – 24 m, and – 55 to – 65 m. These terraces were most likely formed during MISs 5c, 5a and 3 (Méndez-Baamonde 1984, 1985a, b, 1997, 2002b). These interstadials within the last glaciation occurred at ~105 ka, ~ 84 ka, and ~ 60 ka, respectively. The ESL was located at ~10 m BPSL during MIS 5c, at ~15 to 20 m BPSL during MIS 5a (Murray-Wallace and Woodroffe 2014), and at 55 to 90 m BPSL during MIS 3 (Chappell 2002; Siddall et al. 2008; Murray-Wallace and Woodroffe 2014). Analogous set of terraces is also found at similar depths in the southern barrier reefs of the Las Aves Archipelagos. Furthermore, a series of karst morphologies located at depths of approximately – 110 to – 120 m in the southern barrier reef of Los Roques Archipelago were most likely formed during the last maximum glacial (MIS 2), which occurred between 21 and 18 ka. These submerged caves were discovered during the early 1980s by then PhD candidate José Méndez Baamonde. The explorational dives, down to maximum depths of approximately – 300 to – 320 m, were made with the submarine of the ship Calypso, which was commanded by the explorer and researcher Jacques Cousteau.

In Aves Island, the uppermost section of the lithified limestone unit that underlies the loose sandy sediments comprising the supratidal island is exposed along the south and southwestern margins. This unit, known as the Aves Island calcarenite, is currently situated in the uppermost shoreface. It consists of a series of thin calcarenite beds (a bedset) having approximately 8–12 cm thick, and displaying a retrogradational (backstepping) stacking pattern and onlap terminations. This stratal pattern extends for several meters deep into the lower shoreface, where beds are covered by Holocene carbonate facies. The retrogradational geometry, driven by RSL rise, indicates that the bedset is part of an aggradational parasequence (see Catuneanu et al. 2011; Catuneanu 2019; for complete reviews on stratal stacking patterns and terminations and associated RSL changes). It is worth mentioning that the degree of lithification of individual beds comprising this calcarenite unit is considerably higher than that of the Holocene beachrock deposits on the island. Several photographic examples of this unit are shown in Chap. 3. We recently carried a petrographic study of this calcarenite. It consists of reworked grains of coralline algae, molluscs (bivalves and gastropods), green algae (*Halimeda*), LBF (mainly *Amphistegina* and *Peneroplis*), scleractinian corals, echinoid spines, and minor SBF.

The Aves Island calcarenite has been interpreted to have formed during the LIG (Méndez-Baamonde 2006; see Fig. 2.33). Assuming that deposition of this unit occurred between 130 and 116 ka (MIS 5e), we consider the following scenarios: (a) the uppermost beds were originally situated a few meters APSL, probably between 2 and 9 m of elevation; later, during the last glaciation, subaerial exposure of the calcarenite unit due to the eustatic fall of the sea level led to erosion and subsequent dissolution of the uppermost beds; (b) the current position of the bedset in the uppermost shoreface (just below high tide) may suggest mild subsidence of the Late Pleistocene carbonate facies since the end of the LIG; (c) a combination of erosion and dissolution of the uppermost beds, along with mild subsidence of the calcarenite unit from the Late Pleistocene to Modern times. Small subsidence rates proposed for Aves Island by Schubert and Laredo (1984) suggest that the modern carbonate complex may have subsided approximately between 4.5 to 7 m during the last 116 kyr. Arnaiz-Rodríguez and Audemard (2018) consider that Aves Ridge (a remnant island arc) has been inactive for ~ 40 Myr, and during the first 20 Myr after its cessation, the structure experienced thermal subsidence and contraction. These authors further state that subsequent subsidence is likely due to gradual accumulation of carbonate and clastic sediments over the ridge. Garrocq et al. (2021) consider that the subsidence history of the igneous basement dates back to the Early Eocene (~56 Ma), with an accelerated subsidence from the Late Miocene to Present. The precise age of deposition of the Aves Island calcarenite, currently under investigation, is expected to be determined through forthcoming U/Th series dating of small, isolated coral colonies found in growth position within the uppermost beds of the unit.

Caribbean tectonic activity is well represented in the Bonaire Island, where carbonate terraces from Late Miocene to Pliocene ages are well exposed, reaching elevations up to 140 m over RSL. In this island, the lower terrace, formed during the LIG, is located at 8 m above RSL (Laya et al. 2018). In Curaçao, the lower limestone terrace that corresponds to the LIG, and located between 6 and 12 m of elevation, was originally dated by Schubert and Szavo (1978). The age of the terrace was later confirmed by Hamelin et al. (1991), as well as by Muhs et al. (2012). The highest limestone terrace in Curaçao is located between 90 and 150 m of elevation and can be correlated to Miocene highest terraces in Bonaire and Aruba (Laya et al. 2018).

In Barbados, which lies in front of the eastern margin of the Lesser Antilles Island arc (Fig. 2.29; see also Figs. 2.6 and 2.10), over the Barbados Ridge Accretionary Complex, the elevations of the LIG limestone terraces vary between 6 and 61 m (Muhs and Simmons 2017). This island has been one of the most important locations in the world for providing geologic records of Quaternary sea-level history. Its numerous exposed limestone terraces at different elevations are related to a late Quaternary differential uplift rate ranging from 0.07 to 0.52 m/kyr (Matthews 1973; Westbrook 1982; Taylor and Mann 1991; Radtke and Schellmann 2005; Peltier et al. 2015; Tawil-Morsink et al. 2022).

2.11.8 Holocene Carbonate Facies in the North-Central Coasts of Venezuela

The growth of corals reefs and associated reef communities, as well as the formation of beachrocks along the north-central Caribbean coasts of Venezuela, is limited to a few localities. This is because this long coastal area is highly influenced by factors (both naturals and anthropogenic) that prevents the development of carbonate deposits of volumetric relevance. The main of those factors are high riverine influx of both siliciclastic sediments and nutrients, and an active coastal upwelling system. Such natural factors increase seawater turbidity and nutrient levels. The "human" factor is the anthropogenic disturbances linked to coastal activities and nearby cities (La Guaira, Caracas, Maracay, Valencia, Puerto Cabello, Puerto La Cruz, Cumaná, among others).

North-Central Venezuela is characterized by an irregular topography defined by the presence of the Coastal Range. This mountain range has parallel orientation (W-E) respect to the coastlines (see Fig. 2.10), and it consists of two parallel-oriented to each other main ranges; the Interior Range to the south and the Coastal Range to the north. The latter is 350 m long and 80 km wide (Urbani 2018), with elevations up to 2750 m. Steep cliff morphologies prevail along the mainland coasts, as the distance between the northern flanks of the mountain chain and the shorelines is often only tens to a few hundreds of meters. The nearby source of terrigenous sediments to the shorelines has allowed the formation of more than 200 km of quartz sandy beaches. Furthermore, the humid tropical weather of the area promotes an active hydrologic cycle, with a high discharge of nutrient-rich freshwaters from steep streams and alluvial fans that flow downhill from the range. However, a point in favor of the local growth of reef communities along the coastlines of north-central Venezuela is that the intensity of the upwelling system in that area is less strong than in eastern Venezuela (Villamizar and Cervigón 2017).

Well-developed coral reefs are represented by three small reef islands located to the northeast of the city of Puerto Cabello (see Figs. 1.2 and 1.3), at ~1.5 km away from the shorelines. These are the islands of Alcatraz, Santo Domingo and Larga (see Fig. 2.29), which belong to the San Esteban National Park. The latter island is the most studied of the three islands. Like Los Roques, Aves de Barlovento and Aves de Sotavento archipelagos, Larga Island consists of an arch-shaped reef along the windward-east side of the island, a central lagoon, and a low-energy beach along the leeward-west side. Supratidal storm terraces and associated back-central plains, ephemeral hypersaline lagoons, and mangrove forests are the most significant sub-environments (Picard and Goddard 1972).

Scleractinian coral colonies in Larga Island consist of at least 19 species, including *Acropora palmata*, *Agaricia agaricites*, *Orbicella annularis*, *Montastrea cavernosa*, *Orbicella faveolata*, *Pseudodiploria strigosa*, *Porites asteroides*, *Tubastrea aurea*, *Eusmilia fastigiata* and *Scolymia lacera* (Weil 2003; Rodríguez-Quintal et al. 2011). A varied coral reef-associated biota has been documented in this island, including a rich diversity of molluscs (bivalves and gastropods), sponges and polychaetes (Núñez-Flores et al. 2010; Rivolta et al. 2015; Álvarez-Barco et al. 2019). A small contribution of siliciclastic sediments comes from the Coastal Range (less than 1% of the total volume of sediments; Picard and Goddard 1972), with pollution that comes from the nearby coastline and uncontrolled tourism the major threat to the reef communities.

To the east of the city of Puerto Cabello and along the same coastline, moderate to high-energy beach environments characterize the north-central coastlines of Venezuela. Scattered but isolated coral bioherms have been able developed locally at shallow depths (< 7 m deep) in the states of Aragua (Henri Pittier National Park) and La Guaira (see Fig. 1.5). Furthermore, beachrock deposits are also locally moderately common. In such deposits, quartz grains sourced from adjacent sandy-quartz beaches are cemented together with calcareous grains (e.g., molluscs, red algae, echinoids) by aragonite and high-Mg calcite cements.

2.11.9 Holocene Carbonate Facies in the Venezuelan Atlantic Continental Shelf and Slope

The Venezuelan Atlantic coastlines and continental shelf represent the southeasternmost part of the EVB. This large transitional, continental to marine setting faces the Orinoco

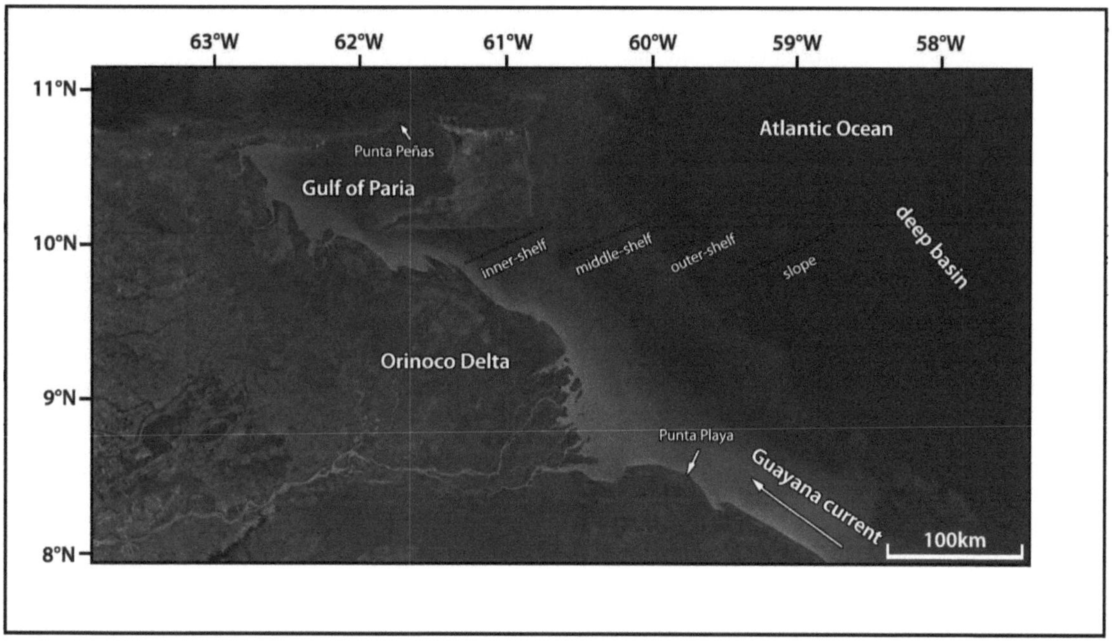

Fig. 2.38 Satellite image of northeast Venezuela, which includes the Orinoco Delta plain and the Venezuelan Atlantic continental shelf. Image taken from Google Earth. (Imagery date: 30 December 2015 by Landsat/Copernicus)

Delta and belongs to the Western Equatorial Atlantic, being located at tropical latitudes, approximately between 8° N and 11° N (Fig. 2.38). During the past two decades, offshore hydrocarbon exploration activities carried in this poorly-studied continental shelf have allowed the scientific community to conduct multidisciplinary baseline studies. Some of these studies have led to the discovery of modern ecosystems currently flourishing from the middle to the outer shelf and slope, including an interesting fauna of polychaetes (Bone et al. 2007, 2011; Márquez-Rojas et al. 2013), and benthic and planktonic calcifiers (Morales 2004; Kawka et al. 2008; Molina et al. 2008; this book).

The Venezuelan Atlantic coastlines extend for approximately 1008 km (Méndez-Baamonde 2007a), from the easternmost part of the Paria Peninsula at Punta de Peñas towards southeast along the Gulf of Paria and the Orinoco Delta shoreline (Fig. 2.38). Contrary to the tectonically active margin setting prevailing along Northern Venezuela, the southeasternmost part of the EVB (the Orinoco Delta floodplain and delta plain) is a subsiding passive margin setting. This area has maintained away from the tectonic activity linked to the Southern Caribbean PBZ (to the northwest) and the Barbados accretionary prism (to the northeast) during Neogene and Quaternary times. During that time interval, the area nowadays comprising the Orinoco Delta plain (also known as the Delta Amacuro; see Figs. 1.7 and 2.23) has been an important sediment depocenter (Warne et al. 2002a) that has become overfilled by clastic sedimentation since the Pliocene (Castillo and Mann 2021). Sedimentary processes along these Atlantic coastlines and inner shelf are dominated by the gigantic siliciclastic sediment discharge and associated freshwater and organic matter input from the Orinoco Delta, as well as by a significant sediment contribution from the SE-NW flow of the Guayana coastal current. The latter carries large amounts of sediments from the Amazon delta northwestward (Fig. 2.38) (see Méndez-Baamonde 2000, 2007a; Warne et al. 2002a, b; Aslan et al. 2003; Medina et al. 2005; Weber et al. 2016).

As for the Orinoco River-Delta system, the watercourse of the river covers an approximate area of 10^6 Km^2, representing the fourth largest river of the world in terms of discharge [Alongi (1998) in Bone et al. 2011].

The floodplain and delta plain comprise one of the world's largest tropical-wetland complexes, covering an approximate area of 22,000 km^2 (Hamilton and Lewis Jr. 1990). Holocene siliciclastic facies of the Orinoco Delta have prograded eastward over Pleistocene siliciclastic deposits (the Mesa Formation) (Méndez-Baamonde 2000), forming a triangular to trapezoidal-shaped depocenter (Warne et al. 2002b). The pro-delta deposits extend offshore as much as 60 km from the coast (McClelland Engineers, 1979; in Warne et al. 2002a).

In the early 2000s, the Venezuelan Navy carried out a series of dredging campaigns on the seafloor in front of the Orinoco delta-front facies and towards east. The dredges spanned water depths from the middle-neritic (60 m deep) to the outer-neritic zones and the slope break (from ~130 m deep), as well as the continental slope (down to 300 m depth) and the deep basin (lower bathyal depths down to ~3000 m deep). Measured variables such as salinity, organic matter content (wt.%) of the sediments, as well as granulometry of unconsolidated sediments on the seafloor (sand, silt and clay fractions) (see Medina et al. 2005; Bone et al. 2011) indicate that shallow neritic environments (0–50 m deep) are directly influenced by the high siliciclastic sediment input (and associated freshwater and particulate organic matter) of both the Orinoco River and the Guayana current. Optimal environmental conditions for the massive flourishing of marine biota beging occurring from the middle shelf at depths of approximately 60 m, where fully marine conditions are reached (Kawka et al. 2008; Molina et al. 2008; Bone et al. 2011).

Among the variety of benthic calcifiers, which include bivalves, gastropods, echinoderms, bryozoans, ostracods and coral fragments, benthic foraminifera have been the most studied group so far. Results obtained by Kawka et al. (2008) and Molina et al. (2008) from the study of a set of dry samples consisting of loose carbonate bioclasts indicate that benthic foraminifera are more abundant than planktonic foraminifera at depths between 60 and 130 m. The larger benthic foraminifer (symbiont-bearing) *Amphistegina lessonii* and the smaller miliolid (symbiont-barren) *Quinqueloculina lamarckiana* are among the dominant species. Along the continental slope and towards greater depths below the photic zone, the planktonic/benthic ratio in foraminifera increases. From depths of approximately 200 m and down to 1800 to 3000 m, assemblages of deep-sea SBF are dominated by species of *Bulimina*, *Cibicidoides*, *Uvigerina*, *Sigmoilopsis* and *Hoeglundina* (Morales 2004). Regarding planktonic foraminifera, *Globigerina* oozes are dominated by the genera *Globigerinoides*, *Globigerina*, *Orbulina* and *Globorotalia*, among others. Photographic examples shown belonging to these Holocene facies correspond to loose bioclasts of molluscs, bryozoans, and benthic and planktonic foraminifera. So far, the available data indicate the contribution of the Heterozoan, Photozoan, Photo-C, and P-factories to the carbonate production in this poorly-studied shelf.

References

Acosta, N., 2016. Potencial de generación de hidrocarburos en secuencias terciarias a través de la caracterización y el modelado geoquímico 2D en el Golfo de Venezuela. Magister Dissertation. Central University of Venezuela, 144 pp.

Adams, D. D., Hurtgen, M. T., Sageman, B. B., 2010. Volcanic triggering of a biogeochemical cascade during Oceanic Anoxic Event 2. *Nature Geoscience*, *3*, 201–204.

Aguado, R., de Gea, G. A., Castro, J. M., O'Dogherty, L., Quijano, M. L., Naafs, B. D. A., Pancost, R. D., 2014. Late Barremian–early Aptian dark facies of the Subbetic (Betic Cordillera, southern Spain): Calcareous nannofossil quantitative analyses, chemostratigraphy and palaeoceanographic reconstructions. *Palaeogeography, Palaeoclimatology, Palaeoecology*, *395*, 198–221.

Aguilera, L., and Gallovich, E., 1970. Ensayos Sísmicos de Refracción efectuados en la Isla de Aves. *Memorias de la Sociedad de Ciencias Naturales de La Salle*, *30*(87), 206–212.

Aitken, T., Mann, P., Escalona, A., Christeson, G. L., 2011. Evolution of the Grenada and Tobago basins and implications for arc migration. *Marine and Petroleum Geology*, *28*, 235–258.

Alberdi-Genolet, M., and Lafargue, E., 1993. Vertical variations of organic matter content in Guayuta Group (upper Cretaceous) interior mountain belt eastern Venezuela. *Organic Geochemistry*, *20*, 425–436.

Alberdi-Genolet, M., Tocco, R., 1999. Trace metals and organic geochemistry of the Machiques Member (Aptian-Albian) and La Luna Formation (Cenomanian-Campanian), Venezuela. *Chemical Geology*, *160*, 19–38.

Albert-Villanueva, E., 2016. Facies y secuencias deposicionales mixtas carbonático-siliciclásticas del Mioceno inferior de la Cuenca de Falcón (noroeste de Venezuela) como modelo exploratorio en el Caribe. Doctoral Dissertation. University of Barcelona. Volumes 1 and 2.

Albert-Villanueva, E., González, L., Bover-Arnal, T., Ferràndez-Cañadell, C., Esteban, M., Fernández-Carmona, J., Calvo, R., Salas, R., 2017a. Geology of the Falcón Basin (NW Venezuela). *Journal of Maps*, *13*(2), 491–501.

Albert-Villanueva, E., Bover-Arnal, T., Ferràndez-Cañadell, C. and Salas, R., 2017b. Comment on "Tectonic and environmental factors controlling on the evolution of Oligo-Miocene shallow marine carbonate factories along a tropical SE Circum-Caribbean" by Silva-Tamayo et al. (2017). *Journal of South American Earth Sciences*, *91*, 394–396.

Álvarez-Barco, J. A., Rodríguez-Quintal, J. G., Rodríguez, C. T., Narciso-Fejure, S. E., 2019. Moluscos asociados al arrecife coralino de Isla Larga, Parque Nacional San Esteban, Estado Carabobo, Venezuela. *Memoria de la Fundación La Salle de Ciencias Naturales*, *77*(185), 5–30.

Amaya, S., Zuluaga, C. A., Bernet, M., 2017. New fission-track age constraints on the exhumation of the central Santander Massif: Implications for the tectonic evolution of the Northern Andes, Colombia. *Lithos*, *282*, 388–402.

Arkle, J. C., Weber, J., Enkelmann, E., Owen, L. A., Govers, R., Jess, S., Deniso, C., O'Sullivan, P. B., Donelick, R. A., 2021. Exhumation of the Coastal Metamorphic Belt Above the Subduction-to-Transform Transition, in the Southeast Caribbean Plate Corner. *Tectonics*, *40*(8), e2020TC006414.

Arminio, J. F., Hernandez, M., Pilloud, A., Audemard, F., 2004. New insights on the Jurassic rift succession of the Mérida Andes, Venezuela: implications for new petroleum systems in northern South America. In: AAPG International Conference. Cancun, Mexico.

Arnaiz-Rodríguez, M. S., and Audemard, F., 2018. Isostasy of the Aves Ridge and neighbouring basins. *Geophysical Journal International*, *215*, 2183–2197.

Arnold, H. C., 1966. Upper Paleozoic Sabaneta–Palmarito sequence of Mérida Andes, Venezuela. *American Association of Petroleum Geologist Bulletin*, *50*, 2366–2387.

Arteaga ACJ, 2002. Isla de Aves. Bastión Venezolano en el Mar Caribe. *Dirección de Geografía y Cartografía de las Fuerzas Armadas (DIGECAFA) de Venezuela*. Caracas, Venezuela: Fuerte Tiuna.

Aslan, A., White, W. A., Warne, A. G., Guevara, E. H., 2003. Holocene evolution of the western Orinoco Delta, Venezuela. *Geological Society of America Bulletin*, *115*(4), 479–498.

Audemard, F. E., 1991. Tectonics of Western Venezuela. Doctoral Dissertation. Rice University, Houston, TX, 245 pp.

Audemard, F.A., 1993. Néotectonique, sismotectonique et aléa sismique du nord-ouest du Vénézuela (systéme de failles d'Oca–

Ancón). Ph.D. Thesis. Université Montpellier II, Montpellier, France, 369 pp.

Audemard, F.A., 1998. Evolution géodynamique de la façade nord sud-Américaine: Nouveaux apports de l'histoire géologique du bassin de Falcón, Vénézúela. In: Presented at the 14th Caribb. Geol. Conference, Port of Spain, Trinidad, 1995.

Audemard, F. A., 2001. Quaternary tectonics and present stress tensor of the inverted northern Falcón Basin, northwestern Venezuela. *Journal of Structural Geology*, *23*, 431–453.

Audemard, F. A., 2002. Ruptura de los grandes sismos históricos venezolanos de los siglos XIX y XX, revelados por la sismicidad instrumental contemporánea. XI Congreso Venezolano de Geofísica, Caracas, Venezuela, Nov. 17-20, 2002. 8 pp.

Audemard, F. A., 2003. Geomorphic and geologic evidence of ongoing uplift and deformation in the Mérida Andes, Venezuela. *Quaternary International*, 101–102C, 43–65.

Audemard, F. A., 2009. Key issues on the post-Mesozoic southern caribbean plate boundary. In: James, K. H., Lorente, M. A., Pindell, J. (Eds.), Origin and Evolution of the Caribbean Plate. *Geological Society, London, Special Publications*, *328*, 567–584.

Audemard, F. A., 2014. Active block tectonics in and around the Caribbean: a review. In: Schmitz, M., Audemard, F.A., Urbani, F. (Eds.), El Límite Noreste de la Placa Suramericana - Estructuras Litosféricas de la Superficie al Manto (The Northeastern Limit of the South American Plate - Lithospheric Structures from Surface to the Mantle). Editorial Innovación Tecnológica. Facultad de Ingeniería-Universidad Central de Venezuela/FUNVISIS, pp. 29–77.

Audemard, F. E., and Serrano, I. C., 2001. Future petroliferous provinces of Venezuela, In: M. W. Downey, J. C. Threet, and W. A. Morgan, eds., Petroleum provinces of the twenty-first century. *American Association of Petroleum Geologists*, *Memoir 74*, 353–372.

Audemard, F. E., and Audemard, F. A., 2002. Structure of the Mérida Andes, Venezuela: relations with the South America-Caribbean geodynamic interaction. *Tectonophysics*, *345*, 299–327.

Audemard, FA and Castilla, R., 2016. Present-day stress tensors along the southern Caribbean plate boundary zone from inversion of focal mechanism solutions: A successful trial. *Journal of South American Earth Sciences*, *71*, 309–319.

Audemard, F. A., Bousquet, J. C., Rodŕiguez JA, 1999. Neotectonic and paleoseismicity studies on the Urumaco Fault, northern Falcón Basin, northwestern Venezuela. *Tectonophysics*, *308*, 23–35.

Audemard, F. A., Romero, G., Rendón, H., Cano, V., 2004. Quaternary fault kinematics and stress tensors along the southern Caribbean from fault-slip data and focal mechanism solutions. *Earth Science Review*, *69*, 181–233.

Audemard, F. A., Alvarado, L., Arnaiz, M., Romero, G., Reinoza-Gómez, C. E., Martínez-González, A. I., Sánchez-Gamboa, A. K., Lebrun, J. F., Rodríguez, L. M., Rodríguez, J. A., Philippon, M., Singer, A., González, J., Lotuffo, M., 2025. Sismicidad de profundidad intermedia de la losa de subducción de las Antillas Menores, próxima a la Península de Paria, Venezuela, *Boletín de Geología*, *47*(1).

Avendaño-Sánchez, G. M., García-González, M., Cruz-Guevara, L. E., Cruz-Ceballos, L. F., 2021. Thermal evolution of Los Cuervos formation in the southern area of the Cesar sub-basin (Colombia), based on geochemical and petrophysical data. *Earth Sciences Research Journal*, *25*(2), 179–192.

Azpiritxaga, I., 1991. Carbonate depositional styles controlled by siliciclastic influx and relative sea level changes, Lower Cretaceous, Central Lake Maracaibo, Venezuela. Magister Dissertation. University of Texas at Austin, 151 pp.

Baquero, M., 2015. Evolución geodinámica del noroccidente de Venezuela, basado en nuevos datos de geocronología, geoquímica e isotópicos. Doctoral Dissertation. Central University of Venezuela, 355 pp.

Baquero, M., Acosta, J., Kassabji, E., Zamora, J., Sousa, J. C., Rodríguez, J., Grobas, J., Melo, L., Schneider, F., 2009. Polyphase development of the Falcón Basin in northwestern Venezuela: implications for oil generation. In: James, K.H., Lorente, M. A., Pindell, J. (Eds.), Origin and Evolution of the Caribbean Plate. *Geological Society, London, Special Publications*, *328*, 587–612.

Baquero, M., Grande, S., Urbani, F., Cordani, U., Hall, C., Armstrong, R., 2015. New Evidence for Putumayo Crust in the Basement of the Falcon Basin and Guajira Peninsula, Northwestern Venezuela, In: C. Bartolini and P. Mann, Petroleum Geology and Potential of the Colombian Caribbean Margin. *American Association of Petroleum Geologists, Memoir 108*, 105–138.

Baquero, M., Mann, P., Grande, S., Urbani, F., Audermard, F., Mendi. D., Melo, L., 2017. New geochronology and radiometric age dates improve the definition and continuity of accreted tectonic terranes of northern Venezuela and the Lesser Antilles. Accessed at https://www.researchgate.net/publication/319252889_New_geochronology_and_radiometric_age_dates_improve_the_definition_and_continuity_of_accreted_tectonic_terranes_of_northern_Venezuela_and_the_Lesser_Antilles

Bartok, P., 1993. Prebreakup geology of the Gulf of Mexico-Caribbean: its relation to Triassic and Jurassic rift systems of the region. *Tectonics, 12*, 441–459.

Bartok, P., 2003. The peripheral bulge of the Interior Range of the Eastern Venezuela Basin and its impact on oil accumulations, in C. Bartolini, R. T. Buffler, and J. Blickwede, eds., The Circum-Gulf of Mexico and the Caribbean: Hydrocarbon habitats, basin formation, and plate tectonics. *American Association of Petroleum Geologists*, *Memoir 79*, 925–936.

Bartok, P., Reijers, T. J. A., Juhasz, I., 1981. Lower Cretaceous Cogollo Group, Maracaibo Basin, Venezuela: sedimentology, diagenesis, and petrophysics. *American Association of Petroleum Geologists Bulletin*, *65*, 1110–1134.

Bartok, P., Mejia-Hernandez, M. C., Ismael, M., 2015. Paleogeographic constraints on Middle- to Late-Jurassic tectonic reconstruction of the Maya Block of southern Mexico and equivalent strata of northwestern South America, *in* C. Bartolini and P. Mann, eds., Petroleum Geology and potential of the Colombian Caribbean Margin. *American Association of Petroleum* Geologists, *Memoir 108*, 201–216.

Bassi, D., and Nebelsick, J. H., 2010. Components, facies and ramps: redefining Upper Oligocene shallow water carbonates using coralline red algae and larger foraminifera (Venetian area, northeast Italy). *Palaeogeography, Palaeoclimatology, Palaeoecology*, *295*, 258–280.

Bassi, D., Carannante, G., Murru, M., Simone, L., Toscano, F., 2006. Rhodalgal/bryomol assemblages in temperate-type carbonate, channelized depositional systems: the Early Miocene of the Sarcidano area (Sardinia, Italy). *Geological Society of London, Special Publications*, *255*, 35–52.

Bastidas, C., Bone, D., Croquer, A., Debrot, D., Garcia, E., Humanes, A., Ramos, R., Rodríguez, S., 2012. Massive hard coral loss after a severe bleaching event in 2010 at Los Roques, Venezuela. *Revista de Biología Tropical*, *60*, 29–37.

Bayona, C., Cortes, M., Jaramillo, C., Ojeda, G., Aristizabal, J. J., Reyes, A., 2008. An integrated analysis of an orogen–sedimentary basin pair: Latest CretaceousCenozoic evolution of the linked Eastern Cordillera orogen and the Llanos foreland basin of Colombia. *Geological Society of America Bulletin*, *120* (9/10), 1171–1197.

Bayona, G., Bustamante, C., Nova, G., Salazar–Franco AM, 2020. Jurassic evolution of the northwestern corner of Gondwana: Present knowledge and future challenges in studying Colombian Jurassic rocks. In: Gómez, J. & Pinilla–Pachon, A.O. (editors), The Geology of Colombia, Volume 2 Mesozoic. *Servicio Geológico Colombiano, Publicaciones Geológicas Especiales 36*, 171–207.

Beil, S., Kuhnt, W., Holbourn, A. E., Aquit, M., Flögel, S., Chellai, E. H., Jabour, H., 2018. New insights into Cenomanian paleoceanography and climate evolution from the Tarfaya Basin, southern Morocco. *Cretaceous Research*, *84*, 451–473.

Bermúdez, M. A., Kohn, B. P., Van Der Beek, P. A., Bernet, M., O'Sullivan, P. B., Shagam, R., 2010. Spatial and temporal patterns of exhumation across the Venezuelan Andes: Implications for Cenozoic Caribbean geodynamics. *Tectonics*, *29*, TC5009.

Bermúdez, M. A., Van Der Beek, P., Bernet, M., 2011. Asynchronous Miocene–Pliocene exhumation of the central Venezuelan Andes. *Geology*, *39*, 139–142.

Bermúdez, M. A., Bernet, M., Kohn, B., Brichau, S., 2019. Exhumation-Denudation History of the Maracaibo Block, Northwestern South America: Insights from Thermochronology. In: Cediel F., Shaw R.P. (Eds.), Geology and Tectonics of Northwestern South America. Frontiers in Earth Sciences. Springer, Cham, 879–898.

Bezada, M. J., Schmitz, M., Jácome, M. I., Rodríguez, J., Audemard, F., Izarra, C., BOLIVAR Active Seismic Working Group, 2008. Crustal structure in the Falcón Basin area, northwestern Venezuela, from seismic and gravimetric evidence. *Journal of Geodynamics*, *45*, 191–200.

Bezada, M., Magnani, M., Zelt, C., Schmitz, M., Levander, A., 2010. The Caribbean–South American plate boundary at 65°W: Results from wide-angle seismic data. *Journal of Geophysical Research*, *115*, B08402.

Bice, K. L., Birgel, D., Meyers, P. A., Dahl, K. A., Hinrichs, K. U., Norris, R. D., 2006. A multiple proxy and model study of Cretaceous upper ocean temperatures and atmospheric CO_2 concentrations. *Paleoceanography*, *21*, PA2002.

Bird, D. E., Hall, S. A., Burke, K., Casey, J. F., Sawyer, D. S., 2007. Early Central Atlantic Ocean seafloor spreading history. *Geosphere*, *3*,282–298.

Blakey, R. C. 2003. Carboniferous–Permian palaeogeography of the assembly of Pangea. In: Wong, T. E. (ed.) XVth International Congress on Carboniferous and Permian Stratigraphy, Utrecht, 10–16 August. Royal Netherlands Academy of Arts and Sciences, Amsterdam, 443–456.

Blakey, R. C. 2007. Carboniferous–Permian palaeogeography of the assembly of Pangea. In: Wong, T. E. (ed.) XVth International Congress on Carboniferous and Permian Stratigraphy, Utrecht, 10–16 August. Royal Netherlands Academy of Arts and Sciences, Amsterdam, 443–456.

Blättler, C. L., Jenkyns, H. C., Reynard, L. M., Herderson, G. M., 2011. Significant increases in global weathering during Oceanic Anoxic Events 1a and 2 indicated by calcium isotopes. *Earth and Planetary Science Letters*, *309*, 77–88.

Bodin, S., Meissner, P., Janssen, N. M., Steuber, T., Mutterlose, J., 2015. Large igneous provinces and organic carbon burial: Controls on global temperature and continental weathering during the Early Cretaceous. *Global and Planetary Change*, *133*, 238–253.

Boesi, T., and Goddard, D., 1991. A new geologic model related to the distribution of hydrocarbon source rocks in the Falcón basin, northwestern Venezuela, In: K. T. Biddle, ed., Active margin basins. *American Association of Petroleum Geologists*, *Memoir 52*, 303–319.

Bolli, H. M., Beckmann, J. P., Saunders, J. B., 1994. Benthic Foraminiferal Biostratigraphy of the South Caribbean Region. Cambridge University Press, Cambridge, UK, 408 pp.

Bone, D., Perez, D., Villamizar, A., Penchaszadeh, P., Klein, E., 1998. Parque Nacional Morrocoy, Venezuela. *CARICOMP—Caribbean Coral Reef, Seagrass and Mangrove Sites. Paris: United Nations Educational, Scientific and Cultural Organization*, 151–159.

Bone, D., Chollett, I. Rodríguez, C., 2007. Macrobentos de Aguas Profundas en la costa Atlántica venezolana. *Interciencia*, *32*(7), 477–481.

Bone, D., Rodriguez, C. T. Chollett, I., 2011. Polychaeta Diversity in the Continental Shelf Off the Orinoco River Delta, Venezuela. Pp 87–97. In: Grillo, O. & Venora, G. Changing Diversity in Changing Environments. In: Tech, Rijeka, Croatia, 392 pp.

Boschman, L. M., van Hinsbergen, D. J., Torsvik, T. H., Spakman, W., Pindell, J. L., 2014. Kinematic reconstruction of the Caribbean region since the Early Jurassic. *Earth-Science Reviews*, *138*, 102–136.

Bottini, C., Erba, E., Tiraboschi, D., Jenkyns, H. C., Schouten, S., Sinninghe Damsté, J. S., 2015. Climate variability and relationship with ocean fertility during the Aptian Stage. *Climate of the Past Discussion*, *11*, 383–402.

Bouysse, P., 1988. Opening of the Grenada back-arc Basin and evolution of the Caribbean Plate during the Mesozoic and early Paleogene. *Tectonophysics*, *149*, 121–143.

Bralower, T. J., and Lorente, M. A., 2003. Paleography and stratigraphy of La Luna Formation and related Cretaceous anoxic depositional systems, *Palaios*, *18*, 301–304.

Bralower, T. J., Cobabe, E., Clement, B., Sliter, W. V., Osburn, CL Longoria, J., 1999. The record of global change in mid-Cretacous (Barremian-Albian) sections from the Sierra Madre, northeastern Mexico. *Journal of Foraminiferal Research*, *29*, 418–437.

Brooks, J., 1990. Classic petroleum provinces. *Geological Society, London, Special Publications*, *50*, 1–8.

Brumsack, H. J., 2006. The trace metal content of recent organic carbon-rich sediments: implications for Cretaceous black shale formation. *Palaeogeography*, *Palaeoclimatology*, *Palaeoecology*, *232*, 344–361.

Burke, K., 1988. Tectonic evolution of the Caribbean. *Annual Review of Earth and Planetary Sciences*, *16*, 201–230.

Caldeira, K., and Rampino, M. R., 1991. The mid-Cretaceous super plume, carbon dioxide, and global warming. *Geophysical Research Letters*, *18*, 987–990.

Canache, M., Pilloud, A., Truskowski, I., Crux, J., Gamarra, S., 1994. Revisión estratigráfica de la sección Cretácica del Río Maraca, Sierra de Perijá, Venezuela. Memorias del V Simposio Bolivariano. Exploración Petrolera en las Cuencas Sub-Andinas. Puerto La Cruz, Venezuela, 240–241.

Carannante, G., Esteban, M., Milliman, J. D., Simone, L., 1988. Carbonate lithofacies as paleolatitude indicators: problems and limitations. *Sedimentary Geology*, *60*, 333–346.

Caron, M., Bolli, H. M., Saunders, J. B., Perch-Nielsen, K., 1985. Cretaceous planktic foraminifera *Plankton Stratigraphy*, *1*, 17–86.

Caroprese, C. G., 2009. Bioestratigrafía y paleoecología de la Formación Querecual en el Oriente de Venezuela. Central University of Venezuela. Magister Dissertation. 95 pp.

Carrillo, M., Paredes, I., Cruz, J. A., Crespo de Cabrera, S., 1995. Aptian to Maastrichtian paleobathymetric reconstruction of the Eastern Venezuelan Basin. *Marine Micropaleontology*, *26*, 405–418.

Carvajal-Chitty, H. I., 2015. Paleoecología del Aptiense-Albiense del Oriente de Venezuela. Doctoral Dissertation. Central University of Venezuela, 163 pp.

Carvajal-Chitty, H., and Rey, O., 2020. Aportes a la bioestratigrafía del Aptiense–Albiense en la cuenca oriental de Venezuela con base en foraminíferos y nanopláncton calcáreo. *Caribbean Studies*, *48*(1), 135–163.

Casanova, E. E., 2016. Dinámica de la comunidad planctónica en relación a las perturbaciones naturales en el refugio de fauna silvestre de Isla de Aves, Venezuela. Doctoral Dissertation. Central University of Venezuela. 185 pp.

Castillo, M. V., and Mann, P., 2006. Deeply buried, early cretaceous paleokarst terrane, Southern Maracaibo basin, Venezuela. *American Association of Petroleum Geologists Bulletin*, *90*, 567–579.

Castillo, K., and Mann, P., 2021. Structure, stratigraphy, and hydrocarbon potential of the easternmost part of the Eastern Venezuelan foreland basin, *in* C. Bartolini, ed., South America–Caribbean– Central

Atlantic plate boundary: Tectonic evolution, basin architecture, and petroleum systems. *American Association of Petroleum Geologists Memoir 123*, 591–620.

Castillo, V., Benkovics, L., Cobos, C., Demuro, D., Franco, A., 2017. Perla field: The largest discovery ever in Latin America, In: R. K. Merrill and C. A. Sternbach, eds., Giant fields of the decade 2000–2010: *American Association of Petroleum Geologists, Memoir 113*, 141–152.

Castro, J. M., de Gea, G. A., Quijano, M. L., Aguado, R., Froehner, S., Naafs, B. D. A., Pancost, R. D., 2019. Complex and protracted environmental and ecological perturbations during OAE 1a-Evidence from an expanded pelagic section from south Spain (Western Tethys). *Global and Planetary Change, 183*, 103030.

Castro, J. M., Ruiz-Ortiz, P. A., de Gea, G. A., Aguado, R., Jarvis, I., Weissert, H., Molina, J. M., Nieto, L. M., Pancost, R. D., Quijano, M. L., Reolid, M., Skelton, P. W., López-Rodríguez, C., Martínez-Rodríguez, R., 2021. High-resolution C-isotope, TOC and biostratigraphic records of OAE1a (Aptian) from an expanded hemipelagic cored succession, Western Tethys: a new stratigraphic reference for global correlation and palaeoenvironmental reconstruction. *Paleoceanography and Paleoclimatology, 36*(3), e2020PA004004.

Catuneanu, O., 2019. Model-independent sequence stratigraphy. *Earth-Science Reviews, 188*, 312–388.

Catuneanu, O., Galloway, W. E., Kendall, C. G. S. C., Miall, A. D., Posamentier, H. W., Strasser, A., Tucker, M. E., 2011. Sequence stratigraphy: methodology and nomenclature. *Newsletters on Stratigraphy, 44*, 173–245.

Cavada-Blanco, F., Cappelletto, J., Agudo-Adriani, E., Martinez, S. J., Rodríguez, J, Cróquer A., 2020. Status of the pillar coral *Dendrogyra cylindrus* in Los Roques National Park, Southern Caribbean. bioRxiv. https://doi.org/10.1101/2020.09.15.297770.

Cervigón, F., and Alcalá, A., 2002. La fauna icnológica. In: Guía del Parque Nacional Archipiélago Los Roques. Ed. Javier Zamarro. Agencia Española de Cooperación Internacional, Caracas, Venezuela, 125–136.

Chacín, L., Jácome, M. I., Izarra, C., 2005. Flexural and gravity modelling of the Mérida Andes and Barinas–Apure Basin, Western Venezuela. *Tectonophysics, 405*, 155–167.

Chappell, J., 2002. Sea level changes forced ice breakouts in the Last Glacial cycle: new results from coral terraces. *Quaternary Science Reviews, 21*, 1229–1240.

Chauveau, D., Georgiou, N., Cerrone, C., Dean, S., Rovere, A., 2024. Sea-level oscillations within the Last Interglacial: insights from coral reef stratigraphic forward modelling. *Quaternary Science Reviews, 336*, 108759.

Chigne, N., and Hernández, L., 1990. Main aspects of petroleum exploration in the Apure area of southwestern Venezuela, 1985–1987. *Geological Society, London, Special Publications, 50*, 55–75.

Christeson, G.L., Mann, P., Escalona, A. Aitken, T. J., 2008. Crustal structure of the Caribbean–northeastern South America arc–continent collision zone. *Journal of Geophysical Research, 113*, B08104.

Chutcharavan, P. M., and Dutton, A., 2021. A global compilation of U-series-dated fossil coral sea-level indicators for the Last Interglacial period (Marine Isotope Stage 5e). *Earth System Science Data, 13*, 3155–3178

Clark, P. U., Dyke, A. S., Shakun, J. D., Carlson, A. E., Clark, J., Wohlfarth, B., Mitrovica, J. X., Hostetler, S. W., and McCabe, A. M., 2009. The Last Glacial Maximum. *Science*, 325, 710–714.

Coccioni, R., and Galeotti, S., 2003. The mid-Cenomanian Event: prelude to OAE 2. *Palaeogeography, Palaeoclimatology, Palaeoecology, 190*, 427–440.

Coccioni, R., Luciani, V., Marsili, A., 2006. Cretaceous oceanic anoxic events and radially elongated chambered planktonic foraminifera: Paleoecological and paleoceanographic implications. *Palaeogeography, Palaeoclimatology, Paleoecology, 235*, 66–92.

Coccioni, R., Sabatino, R., Frontalini, F., Gardin, S., Sideri, M., Sprovieri, M., 2015. The neglected history of the Oceanic Anoxic Event 1b: insights and new data from the Poggio le Guaine section (Umbria-Marche basin). *Stratigraphy,* 11, 245–282.

Coffin, M. F., Pringle, M. S., Duncan, R. A., Gladczenko, T. P., Storey, M., Müller, R. D., Gahagan, L. A., 2002. Kerguelen hotspot magma output since 130 Ma. *Journal of Petrology, 43*, 1121–1137.

Cohen, K. M., Finney, S. C., Gibbard, P. L. Fan, J. X., 2013a; updated v2021/10. The ICS International Chronostratigraphic Chart. *Episodes 36*, 199–204.

Cohen, K. M., Finney, S. C., Gibbard, P. L. Fan, J. X., 2013b, updated v2022/02. The ICS International Chronostratigraphic Chart. *Episodes 36*, 199–204.

Cohen, K. M., Finney, S. C., Gibbard, P. L. Fan, J. X., 2013c, updated v2023/04. The ICS International Chronostratigraphic Chart. *Episodes 36*, 199–204.

Condie, K. C., and Abbott, D. H., 1999. Oceanic plateaus and hotspot islands: identification and role in continental growth. *Lithos, 46*, 1–4.

Cooper, M. A., Addison, F. T., Alvarez, R., Coral, M., Graham, R. H., Hayward, A. Howe, S., Martinez, J., Naar, J., Peñas, R., Pulham, A. J., Taborda, A.B, Pulham, A. J., Taborda, A., 1995. Basin development and tectonic history of the Llanos Basin, Eastern Cordillera, and middle Magdalena Valley, Colombia. *American Association of Petroleum Geologists, Bulletin 79*, 1421–1442.

Crespo de Cabrera, S., Sliter, W., Jarvis, I., 1999. Integrated foraminiferal biostratigraphy and Chemostratigraphy of the Querecual Formation (Cretaceous) Eastern Venezuela. *Journal of Foraminiferal Research, 29*, 487–499.

Cróquer, A., Pauls, S. M., Zubillaga, A. L., 2003. White plague disease outbreak in a coral reef at Los Roques National Park, Venezuela. *Revista de Biología Tropical, 51*, 39–45.

Cróquer, A., Weil, E., Zubillaga A. L., Pauls, S. M., 2005. Impact of an outbreak of white plague II on a coral reef in the Archipelago Los Roques National Park, Venezuela. *Caribbean Journal of Science, 41*, 815–823.

Cróquer, A., Cavada-Blanco, F., Zubillaga, A. L., Agudo-Adriani, E. A., Sweet, M., 2016. Is *Acropora palmata* recovering? a case study in Los Roques National Park, Venezuela. *PeerJ, 4*, e1539.

Cruz, L., Fayon, A., Teyssier, C., Weber, J., 2007. Exhumation and deformation processes in transpressional orogens: The Paria Peninsula, SE Caribbean–South American plate boundary. In: Till, A.B., Roeske, S.M., Sample, J.C., Foster, D. (eds.). Exhumation associated with strikeslip systems. *Geological Society of America, Special Paper 434*, 149–165.

Davis, C., Pratt, L., Mompart, L. Murat, B. 1994. Sedimentary geology and carbon isotope stratigraphy of Cretaceous marine strata in western Venezuela. In: *Proceedings of the 5th Simposio Bolivariano: Exploración Petrolera en las Cuencas Subandinas, Puerto La Cruz, Venezuela*, March 13–16, pp. 432–446.

Davis, C., Pratt, L., Sliter, W., Mompart, L. Murat, B. 1999. Factors influencing organic carbon and trace metal accumulation in the Upper Cretaceous La Luna Formation of the western Maracaibo Basin, Venezuela. In: Barrera, E. & Johnson, C. C. (eds.) Evolution of the Cretaceous Ocean– Climate System. *Geological Society of America, Special Papers, 332*, 203– 230.

Debrot, A. O., Yranzo, A., Arocha, D., 2019. Los Roques and Las Aves archipelagos, Venezuela: A marine ecological and conservation reconnaissance of two little-known southeastern caribbean oceanic archipelagos. *Atoll Research Bulletin*, 622, 1–27.

de Gea, G. A., Aguado, R., Castro, J. M., Molina, J. M., O'Dogherty, L., Ruiz-Ortiz, P. A., 2008. Lower Aptian Subbetic organic-rich facies, radiolarites, and associated deposits: the local expression of Oceanic Anoxic Event 1a (Carbonero Formation, southern Spain). *Cretaceous Research, 29*, 861–870.

De Romero, L. M., and Galea-Álvarez, F. A., 1995. Campanian *Bolivinoides* and microfacies from the La Luna Formation, western Venezuela. *Marine Micropaleontology*, *26*, 385–404.

De Romero, L. M., Truskowski, I., Bralower, T. J., Bergen, J. A., Odreman, O., Zachos, J. C., Galea-Álvarez, F. A., 2003. An integrated calcareous microfósil biostratigraphic and carbon-isotope stratigraphic Framework for the La Luna Formation, Western Venezuela. *Palaios, 18*, 349–366.

Díaz de Gamero, M. L., 1977. Estratigrafía y micropaleontología del Oligoceno y Mioceno Inferior del centro de la cuenca de Falcón, Venezuela. GEOS-Universidad Central de Venezuela 22, 3–60.

Díaz, M. C., Alvarez, B., Van Soest, R. W. M., 1987. New species of Demospongiae (Porifera) from the national park "Archipiélago de los Roques", Venezuela. *Bijdragen tot de Dierkunde*, *57*(1), 31–41.

Di Croce, J., Bally, A., Vail, P., 1999. Sequence stratigraphy of the Eastern Venezuelan Basin, In: P. Mann, ed., Caribbean Basins, Sedimentary Basins of the World (K. Hsu, series editor): Amsterdam, Elsevier, 419–476.

Dilek, Y., 2003. Ophiolite pulses, mantle plumes and orogeny. *Geological Society, London, Special Publications*, *218*, 9–19.

Dumitru, O. A., Dyer, B., Austermann, J., Sandstrom, M. R., Goldstein, S. L., D'Andrea, W. J., Cashman, M., Creel, R., Bolge, L., Raymo, M. E, 2023. Last interglacial global mean sea level from high-precision U-series ages of Bahamian fossil coral reefs. Quat. Sci. Rev. 318, 108287.

Driscoll, N. W., and Diebold, J. B., 1998. Deformation of the Caribbean region: One plate or two? *Geology*, *26*, 1043–1046.

Duncan, R. A., and Hargraves, R. B., 1984. Plate tectonic evolution of the Caribbean region in the mantle reference frame. In: Bonini, W. E., Hargraves, R. B. & Shagam, R. (eds.) The Caribbean–South American Plate Boundary and Regional Tectonics. *Geological Society of America*, Memoirs, *162*, 81–93.

Duque-Botero, F., Maurrasse, F. J. -M. R., 2004. Cyanobacterial productivity, variations in the organic carbon, and facies of the Indidura Formation (Cenomanian-Turonian), northeastern Mexico, *Journal of Iberian Geology, 31*, 85–98.

Dutton, A., and Lambeck, K., 2012. Ice Volume and Sea Level During the Last Interglacial. *Science*, *337*, 216–219.

Dutton, A., Villa, A., and Chutcharavan, P. M., 2022. Compilation of Last Interglacial (Marine Isotope Stage 5e) sea level indicators in The Bahamas, Turks and Caicos, and the east coast of Florida, USA. *Earth System Science Data*, 14, 2385–2399.

Dyer, B., Austermann, J., D'Andrea, W. J., Creel, R. C., Sandstrom, M. R., Cashman, M., Rovere, A., Raymo, M. E., 2021. Sea-level trends across The Bahamas constrain peak last interglacial ice melt. *Proceedings of the National Academy of Sciences*, *118*(33), e2026839118.

Eakin, C. M., Morgan, J. A., Heron, S. F., Smith, T. B., Liu, G., Alvarez-Filip, L., et al. 2010. Caribbean corals in crisis: record thermal stress, bleaching, and mortality in 2005. *PloS one*, *5*(11), e13969.

Eakin, C. M., Sweatman, H., Brainard, R. E., 2019. The 2014–2017 global-scale coral bleaching event: insights and impacts. *Coral Reefs*, *38*, 539–545.

Eakin, C. M., Devotta, D., Heron, S., Connolly, S., Liu, G., Geiger, E., et al. 2022. The 2014-17 global coral bleaching event: The most severe and widespread coral reef destruction. Research Square. https://doi.org/10.21203/rs.3.rs-1555992/v1.

Edwards, R. L., Gallup, C. D., Cheng, H., 2003. Uranium-series Dating of Marine and Lacustrine Carbonates. *Reviews in Mineralogy and Geochemistry*, *52*, 363–405.

Erba, E., and Tremolada, F., 2004. Nannofossil carbonate fluxes during the Early Cretaceous: phytoplankton response to nutrification episodes, atmospheric CO_2, and anoxia. *Paleoceanography*, 19, PA1008.

Erba, E., Duncan, R. A., Bottini, C., Tiraboschi, D., Weissert, H., Jenkyns, H. C., Malinverno, A., 2015. Environmental consequences of Ontong Java Plateau and Kerguelen Plateau volcanism. In: Neal, C.R., Sager, W.W, Sano, T., and Erba, E., eds., The Origin, Evolution, and Environmental Impact of Oceanic Large Igneous Provinces. *Geological Society of America, Special Paper 511*, 271–303.

Erbacher, J., Friedrich, O., Wilson, P. A., Birch, H., Mutterlose, J., 2005. Stable organic carbon isotope stratigraphy across Oceanic Anoxic Event 2 of Demerara Rise, western tropical Atlantic. *Geochemistry, Geophysics, Geosystems*, *6*. Q06010.

Erlich, R., and Barrett, S., 1992. Petroleum geology of the Eastern Venezuela Basin, In: R. Macqueen and D. Leckie, eds., Foreland basins and fold belts. *American Association of Petroleum Geologists, Memoir 55*, 341–362.

Erlich, R., Macsotay, O., Nederbragt, A., Lorente, M.A., 1999a. Palaeoecology, palaeogeography and depositional environments of upper cretaceous rocks ofwestern Venezuela. *Palaeogeography, Palaeoclimatology, Palaeoecology*, *153*, 203–238.

Erlich, R., Palmer-Koleman, S., Lorente, M.A., 1999b. Geochemical characterization of oceanographic and climatic changes recorded in Upper Albian to Lower Maastrichtian strata, western Venezuela. *Cretaceous Research*, *20*, 547–581.

Erlich, R. N., Nederbragt, A. J., Lorente, M. A., 2000. Birth and death of the Late Cretaceous "La Luna Sea", and origin of the Tres Esquinas phosphorites. *Journal of South American Earth Sciences*, *13*, 21–45.

Erlich, R. N., Villamil, T., Keens-Dumas, J., 2003. Controls on the deposition of Upper Cretaceous organic carbon–rich rocks from Costa Rica to Suriname. In: C. Bartolini, R. T. Buffler, and J. Blickwede, eds., The Circum-Gulf of Mexico and the Caribbean: Hydrocarbon habitats, basin formation, and plate tectonics. *American Association of Petroleum Geologists*, *Memoir 79*, 1–45.

Escalona, A., and Mann, P., 2006a. Tectonic controls of the right-lateral Burro Negro tear fault on Paleogene structure and stratigraphy, northeastern Maracaibo Basin. *AAPG bulletin*, *90*(4), 479–504.

Escalona, A., and Mann, P., 2006b. An overview of the petroleum system of the Maracaibo basin. *American Association of Petroleum Geologists Bulletin, 90*(4), 657–678.

Escalona, A., and Mann, P., 2011. Tectonics, basin subsidence mechanisms, and paleogeography of the Caribbean-South American plate boundary zone. *Marine and Petroleum Geology*, *28*, 8–39.

Escalona, A., Mann, P., Jaimes, M., 2011. Miocene to recent Cariaco basin, offshore Venezuela: Structure, tectonosequences, and basin-forming mechanisms. *Marine and Petroleum Geology*, *28*, 177–199.

Escalona, A., Norton, I. O., Lawver, L. A., Gahagan, L., 2021a. Quantitative Plate tectonic reconstructions of the Caribbean region from Jurassic to present. In C. Bartolini (Ed.), South America-Caribbean-Central Atlantic Plate Boundary: Tectonic Evolution, Basin Architecture, and Petroleum Systems. *American Association of Petroleum Geologists Memoir 123*, 239–263.

Escalona, A., Ahmad, S. S., Watson, L., 2021b, Late Cretaceous–Pliocene Paleogeography of the Circum-Caribbean Region Based on Quantitative Plate Reconstruction and Georeferenced Databases. In C. Bartolini (Ed.), South America–Caribbean–Central Atlantic plate boundary: Tectonic evolution, basin architecture, and petroleum systems. *American Association of Petroleum Geologists Memoir 123*, 513–538.

Escobar, M., Márquez, G., Inciarte, S., Rojas, J., Esteves, I., Malandrino, G., 2011. The organic geochemistry of oil seeps from the Sierra de Perijá eastern foothills, Lake Maracaibo Basin, Venezuela. *Organic Geochemistry*, *42*, 727–738.

Escobar, M., Díaz, J. G., Márquez, G., Boente, C., Tocco, R., 2022. The La Luna/Río Negro(.) Petroleum System at the Urdaneta West

Field, Lake Maracaibo Basin, NW Venezuela: 1d Basin Modelling and Secondary Oil Migration. *Journal of Petroleum Geology*, *45*(3), 283–302.

Espín-Gil, M. G., 2009. A sedimentological and stratigraphic study of the Cogollo Group limestones (Lower Cretaceous), Maracaibo Basin, Western Venezuela. Magister Dissertation, Kingston University. Chapter 5: Geochemistry, 82–116.

Fajardo, A., Aubourg, C., Nivière, B., Regard, V., Uzcátegui, R., 2020. Neotectonic evolution of Northeastern Venezuela. Geomorphological and seismic analysis in the Monagas Fold and Thrust Belt. *Journal of South American Earth Sciences*, *104*, 102867.

Fajardo, A., Aubourg, C., Niviere, B., Uzcátegui, R., Demory, F., 2023. Neotectonic Evolution of Northeastern Venezuela. Paleomagnetic Evidence of Block Rotation of the Serranía del Interior Range. *Boletín de la Academia de Ciencias Físicas, Matemáticas y Naturales*, *Vol. LXXXIII*, n.°1, 1–15.

Falcón, R. A., 1989a. Estudio bioestratigráico preliminar mediante foraminíferos en el Miembro García de la Formación El Cantil. Cretácico Inferior de Venezuela Oriental. Central University of Venezuela *Geos*, *29*, 36–47.

Falcón, R. A., 1989b. Revisión y redefinición del término "Miembro García" coma "Formación García", Cretácico Inferior de Venezuela Oriental. Central University of Venezuela. *GEOS*, *29*, 48–58.

Farr, T. G., Rosen, P. A., Caro, E., Crippen, R., Duren, R., Hensley, S., Kobrick, M., Paller, M., Rodriguez, E., Roth, L., Seal, D., Shaffer, S., Shimada, J., Umland, J., Werner, M., Oskin, M., Burbank, D., Alsdorf, D., 2007. The Shuttle radar Topography mission. Rev. Review of *Geophysics*, *45*, 1–33.

Feo-Codecido, G., Smith, F., Aboud, N., Di Giacomo, E., 1984. Basement and Paleozoic rocks of the Venezuelan Llanos Basin. *Geological Society of America Memoir*, *162*, 175–187.

Fernández, P. V., Bone, D., Rodríguez, C., 2012. Cryptofaunal polychaetes from oceanic and continental reefs in Venezuela. *Bulletin of Marine Science*, *88*, 339–352.

Ferraro, S., 2017. Oceanic Anoxic Event (OAE) 1b (late Aptian-early Albian): evolutionary, palaeoecological, palaeoceanographic and palaeoclimatic implication. Università degli Studi di Urbino Carlo Bo. Doctoral Dissertation. 130 pp.

Fielding, C. R., Frank, T. D., Birgenheier, L. P., Rygel, M. C., Jones, A. T., Roberts, J., 2008. Stratigraphic imprint of the Late Palaeozoic Ice Age in eastern Australia: a record of alternating glacial and nonglacial climate regime. *Journal of the Geological Society*, *165*, 129–140.

Flögel, S., Wallmann, K., Poulsen, C.J., Zhou, J., Oschlies, A., Voigt, S., Kuhnt, W., 2011. Simulating the biogeochemical effects of volcanic CO_2 degassing on the oxygen-state of the deep ocean during the Cenomanian/Turonian Anoxic Event (OAE2). *Earth and Planetary Science Letters*, *305*, 371–384.

Flügel, E, 2010. Microfacies of Carbonate Rocks—Analysis, Interpretation and Application. Springer, Heidelberg-Dordrecht-London-New York, 984 pp.

Föllmi, K. B., 1996. The phosphorus cycle, phosphogenesis and marine phosphate-rich deposits. *Earth-Science Reviews*, *40*, 55–124.

Föllmi, K. B., 2012. Early Cretaceous life, climate and anoxia. *Cretaceous Research*, *35*, 230–257.

Föllmi, K. B., Garrison, R. E., Ramírez, P. C., Zambrano-Ortiz, F., Kennedy, W. J., Lehner, B. L., 1992. Cyclic phosphate-rich successions in the upper Cretaceous of Colombia. *Palaeogeography, Palaeoclimatology, Palaeoecology*, *93*, 151–182.

Ford, A. B., and Houbolt, J. J. H. C., 1963. The microfacies of the Cretaceous of western Venezuela: Leiden, The Netherlands, E. J. Brill, 55 pp.

Fournier, H. W., Lee, J. K., Urbani, F., Grande, S., 2017. The tectonothermal evolution of the Venezuelan Caribbean Mountain System: $^{40}Ar/^{39}Ar$ age insights from a Rodinian-related rock, the Cordillera de la Costa and Margarita Island. *Journal of South American Earth Sciences*, *80*, 149–173.

Frey, F. A., Coffin, M. F., Wallace, P. J., Weis, D., 2003. Leg 183 synthesis: Kerguelen Plateau–Broken Ridge — a large igneous province, In: Frey, F. A., Coffin, M. F., Wallace, P. J., Quilty, P. G., (Eds.), Proceedings of the Ocean Drilling Program, Scientific Results 183, 1–48.

Friedrich, O., Erbacher, J., Moriya, K., Wilson, P. A., Kuhnert, H., 2008. Warm saline intermediate waters in the Cretaceous tropical Atlantic Ocean. *Nature Geoscience*, *1*, 453.

Galán, C., and Herrera, F., 2017. Ríos subterráneos y acuíferos kársticos de Venezuela: inventario, situación y conservación. In: Rodríguez-Olarte, D. (Ed.). Ríos en Riesgo de Venezuela. Volumen 1. Colección Recursos Hidrobiológicos de Venezuela. Barquisimeto, Lara, Venezuela: Universidad Centroccidental Lisandro Alvarado. 28 pp.

Gale, A. S., Mutterlose, J., Batenburg, S., 2020. The Cretaceous Period. In: Gradstein, F. M., Ogg, J. G., Schmitz, M. D., Ogg, G. M. (Eds.), Geologic Time Scale 2020, pp. 1023–1086, Amsterdam, The Netherlands, Elsevier B.V.

Galea-Álvarez, F. A., 1989, Microfacies, edad y ambiente de sedimentación de la Formación La Luna, Flanco Norandino, Venezuela: Contribuciones de Los Simposios Sobre el Cretácico de América Latina, Parte A, Eventos y Registro Sedimentario, Buenos Aires, pp. A57–A73.

Gallango, O., and Tocco, R., 1994. Análisis Geológico Integrado de las Cuencas Barinas-Maracaibo. Síntesis Geoquímica de Crudos. Informe Técnico INTEVEP, S.A. Filial de Petróleos de Venezuela, S.A. 76 pp.

Gallango, O., Novoa, E., Bernal, A., 2002. The petroleum system of the central Perijá belt, western Venezuela. *American Association of Petroleum Geologists*, *Bulletin 86*, 1263–1284.

Gangl, S. K., Moy, C. M., Stirling, C. H., Jenkyns, H. C., Crampton, J. S., Clarkson, M. O., Ohneiser, C., Porcelli, D., 2019. High-resolution records of Oceanic Anoxic Event 2: Insights into the timing, duration and extent of environmental perturbations from the palaeo-South Pacific Ocean. *Earth and Planetary Science Letters*, *518*, 172–182.

Gaona-Narvaez, T., Maurrasse, F. J. -M. R., Etayo-Serna, F., 2013a. Geochemistry, palaeoenvironments and timing of Aptian organic-rich beds of the Paja Formation (Curití, Eastern Cordillera, Colombia). Geological Society, London, Special Publication, 382, 31-48.

Gaona-Narvaez, T., Maurrasse, F. J. -M. R., Moreno-Bedmar, J. A., 2013b. Stable carbon-isotope stratigraphy and ammonite biochronology at Madotz, Navarra, northern Spain: implication for the timing and duration of oxygen depletion during OAE-1a. *Cretaceous Research*, *40*, 143-157.

Garbán, G. J., 2010. Geoquímica y estratigrafía del Miembro Ftanita de Táchira (Cretácico tardío), Formación La Luna, Venezuela Occidental: Implicaciones Paleooceanográficas y paleoambientales. Doctoral Dissertation, Central University of Venezuela, 402 pp.

Garbán, G., Martínez, M., Márquez, G., Rey, O., Escobar, M., Esquinas, N., 2017. Geochemical signatures of bedded cherts of the upper La Luna Formation in Táchira State, western Venezuela: Assessing material provenance and paleodepositional setting. *Sedimentary Geology*, *347*, 130–147.

García, A., Cróquer, A., Pauls, S. M., 2003. Estado actual de las enfermedades y otros signos de deterioro coralino en siete arrecifes del Parque Nacional Archipiélago de Los Roques, Venezuela. *Revista de Biología Tropical*, *51* (Supl. 4), 173–180.

Garrocq, C., Lallemand, S., Marcaillou, B., Lebrun, J. -F., Padron, C., Klingelhoefer, F., Laigle, M., Münch, P., Gay, A., Schenini, L., Beslier, M-O., Cornée, J-J., Mercier de Lépaine, B., Quillévéré, F.,

BouDagher-Fadel, M. 2021. Genetic relations between the Aves Ridge and the Grenada back-arc Basin, East Caribbean Sea. *Journal of Geophysical Research: Solid Earth, 126*, e2020JB020472.

Gertsch, B., Adatte, T., Keller, G., Tantawy, A. A. A., Berner, Z., Mort, H. P., Fleitmann, D., 2010. Middle and late Cenomanian oceanic anoxic events in shallow and deeper shelf environments of western Morocco. *Sedimentology, 57*, 1430–1462.

Ghosh, S., Pestman, P., Meléndez, L., Truskowski, I., Zambrano, E., 1997. Evolución tectonoestratigráfica y sistemas petrolíferos de la Cuenca de Falcón, Venezuela Noroccidental. Proceedings VIII Congreso Geológico Venezolano, vol. 1, pp. 317–329.

Giorgioni, M., Keller, C. E., Weissert, H., Hochuli, P. A., Bernasconi, S. M., 2015. Black shales–from coolhouse to greenhouse (early Aptian). *Cretaceous Research, 56*, 716–731.

Giunta G., Beccaluva, L., Coltorti, M., Siena, F., Vaccaro, C., 2002a. The sourthern margin of the Caribbean plate in Venezuela: tectono-magmatic setting of the ophiolitic units and kinematic evolution. *Lithos, 63*, 19–40.

Giunta, G., Beccaluva, L., Coltoroti, M., Siena, F., 2002b. The peri-Caribbean ophiolites: structure, tectono-magmatic significance and geodynamic implications. *Caribbean Journal of Earth Science, 36*, 1–20.

Goddard, D., 1974. CONSIDERACIONES GEOLOGICAS DEL AREA TUCACAS-CHICHIRIVICHE Y NOTAS SOBRE EL HOLOCENO. *Boletín de la Asociación Venezolana de Geología, Minería y Petrolera, 17*(1,2,3), 10p. Accessed at https://www.researchgate.net/publication/323858323_Consideraciones_Geologicas_Del_Area_Tucacas_Chichiriviche_y_Notas_Sobre_el_Holoceno

Goddard, D. A., 2006. Venezuela Sedimentary Basins: Principal Reservoirs & Completion Practices. Accessed at researchgate.net/publication/276832554_Venezuela_Sedimentary_Basins_Principal_Reservoirs_Completion_Practices

Goddard, D. A., and Talukdar, S. C., 2002. Cretaceous fine-grained Mudstones of the Maracaibo Basin, Venezuela. *Gulf Coast Association of Geological Societies Transactions, 52*, 1093–1101.

Gómez, S., García, M., Villamizar, E., Narváez, M., 2014. Macroalgas bénticas asociadas a arrecifes coralinos submareales en el Parque Nacional Archipiélago Los Roques, Venezuela, *Acta Biológica Venezuelica, 34*(2), 245–255.

González de Juana, C., Iturralde, J., Picard, X., 1980. Geología de Venezuela y de sus cuencas petrolíferas. Foninves Editions. Caracas, Venezuela. Volumes I and II, 1031 pp.

González, W., Sigismondi, M., Jácome, M., Izarra, C., Graterol, V., 2017. Magnetic characterization and signature of the basement of eastern Venezuela: Espino Graben. In: *SEG Technical Program Expanded Abstracts 2017*. Society of Exploration Geophysicists, 1859–1864.

Grajal, P. A., and Laughlin, G. R., 1984. Decapod crustaceans inhabiting live and dead colonies of three species of Acropora in the Roques Archipelago, Venezuela. *Bijdragen tot de Dierkunde, 54*(2), 220–230.

Gremone, C., and Gómez, J., 1983. Isla de Aves como área de desove de la tortuga verde *Chelonia midas*. Fundación para la Defensa de la Naturaleza, FUDENA, Caracas. 25 pp.

Guada, H. J., 2002. Tortugas marinas del parque nacional Archipiélago Los Roques. In: Guía del Parque Nacional Archipiélago Los Roques. Ed. Javier Zamarro. Agencia Española de Cooperación Internacional, Caracas, Venezuela, 137–139.

Gutiérrez-Puente, N. A., Barragán, R., Núñez-Useche, F., 2021. Paleoenvironmental changes and biotic response to Aptian–Albian episodes of accelerated global change: evidence from the western margin of the proto-North Atlantic (central-eastern Mexico). *Cretaceous Research, 126*, 104883.

Gutiérrez-Puente, N. A., Barragán, R., Núñez-Useche, F., Enciso-Cárdenas, J. J., Martínez-Yáñez, M., Camacho-Ortegón, L. F., 2022. Accumulation of organic-rich sediments associated with Aptian–Albian oceanic anoxic events in central-eastern Mexico. *Journal of South American Earth Sciences, 118*, 103907.

Hackley, P., Urbani, F., Karlsen, A. W., Garrity, C. P., 2006. Mapa geológico de Venezuela a escala 1:750.000. *U.S. Geological Survey, Open File Report 2006-1109*.

Hallmann, N., Camoin, G., Webster, J. M., Humblet, M., 2021. A standardized database of Marine Isotopic Stage 5e sea-level proxies on tropical Pacific islands. *Earth System Science Data, 13*, 2651–2699.

Hallock, P., 2001. Coral reefs, carbonate sediments, nutrients, and global change. In: The history and sedimentology of ancient reef systems. Springer, Boston, MA, 387–427.

Hamelin, B., Bard, E., Zindler, A., Fairbanks, R.G., 1991. $^{234}U/^{238}U$ mass spectrometry of corals: how accurate is the U-Th age of the last interglacial period? *Earth and Planetary Science Letters 106*, 169–180.

Hamilton, S. K., and Lewis, Jr., W. M., 1990. Physical characteristics of the fringing floodplain of the Orinoco River, Venezuela. *Interciencia, 15*, 491–500.

Haq, B. U., 2014. Cretaceous eustacy revisited. *Global and Planetary Change, 113*, 44–58.

Haq, B. U., Hardenbol, J., Vail, P. R., 1987, Chronology of fluctuating sea levels since the Triassic. *Science, 235*, 1156–1167.

Hastie, A. R., and Kerr, A. C., 2010. Mantle plume or slab window?: Physical and geochemical constraints on the origin of the Caribbean oceanic plateau. *Earth-Science Reviews, 98*(3–4), 283–293.

Hauff, F., Hoernle, K., Schmincke, H. U, Werner, R., 1997. A mid-Cretaceous origin for the Galapagos hotspot: volcanological, petrological and geochemical evidence from Costa Rican oceanic crustal segments. *Geologische Rundschau, 86*, 141–155.

Haxby, W. F., Melkonian, A. K., Coplan, J., Chan, S. -M., Ryan, W. B. F., 2010. GeoMapApp freeware software, v.2.3. Lamont–Doherty Earth Observatory, Palisades, NY.

Hay, W., 2017. Toward understanding Cretaceous climate—an updated review. *Science China Earth Sciences, 60*, 5–19.

Hearty, P. J., Hollin, J. T., Neumann, A. C., O'Leary, M. J., McCulloch, M., 2007. Global sea-level fluctuations during the Last Interglaciation (MIS 5e). *Quaternary Science Reviews, 26*, 2090–2112.

Heatwole, H., Torres F., King, S. K., Debrot, A. O., Torres, J., 2022. The history of a remote Caribbean cay, Isla Aves, and its terrestrial biota: a review and new data. *Atoll Research Bulletin, 630*, 38.

Hedberg, H. D., 1937, Stratigraphy of the río Querecual section of northeastern Venezuela. *Geological Society of America Bulletin, 48(12)*, 1971–2024.

Hedberg, H. D., 1950. Geology of the eastern Venezuela basin (Anzoátegui-Monagas-Sucre-eastern Guárico portion). *Geological Society of America Bulletin, 61*(11), 1173–1216.

Hedberg, H.D., and Sass, L. C., 1937. Synopsis de las formaciones geológicas de la parte occidental de la Cuenca de Maracaibo, Venezuela: Boletín de Geología y Mineralogía [Venezuela], Servicio Técnico de Geología y Minería, Caracas, v. 2–4, 83–84.

Helenes, J., and Somoza, D., 1999. Palynology and sequence stratigraphy of the Cretaceous of eastern Venezuela. *Cretaceous Research, 20*, 447–463.

Helenes, J., and Cabrera, D., 2003. Oligocene-Miocene palynomorph assemblages from eastern Venezuela. *Palynology, 27*, 5–25.

Helenes, J., De Guerra, C., Vasquez, J., 1998. Palynology and chronostratigraphy of the Upper Cretaceous in the subsurface of the Barinas area, western Venezuela. *American Association of Petroleum Geologists Bulletin, 82*, 1308–1328.

Herdocia, C., and Maurrasse, F. J. -M. R, 2023. Geochemical factors associated with deposition of lower Aptian organic-rich sediments during OAE1a in the Basque-Cantabrian Basin, northern Spain. Frontiers in Geochemistry, 1:1080169.

Hernández, M., 2003. Análisis Geológico Integrado en la Facies No-Roja de la Formación La Quinta (Sección de Carretera Jají-San Juan), Edo Mérida. BSC Thesis. Universidad Central de Venezuela, pp. 280.

Hernández-Pérez, E., 1997. Sedimentology of a mixed carbonate-siliciclastic succession: the Guasare Formation, Maracaibo Basin, Venezuela. Magister Dissertation, University of Calgary, 206 pp.

Heron, S. F., Maynard, J. A., Van Hooidonk, R., Eakin, C. M., 2016. Warming trends and bleaching stress of the world's coral reefs 1985–2012. *Scientific Reports*, *6*, 1–14.

Hetzel, A., Böttcher, M. E., Wortmann, U. G., Brumsack, H. J. 2009. Paleo-redox conditions during OAE 2 reflected in Demerara Rise sediment geochemistry (ODP Leg 207). *Palaeogeography, Palaeoclimatology, Palaeoecology*, *273*, 302–328.

Hibbert, F. D., Rohling, E. J., Dutton, A., Williams, F. H., Chutcharavan, P. M., Zhao, C., Tamisiea, M. E., 2016. Coral indicators of past sea-level change: A global repository of U-series dated benchmarks. *Quaternary Science Reviews*, *145*, 1–56.

Hoernle, K., Hauff, F., van den Bogaard, P., 2004. 70 m.y. history (139–69 Ma) for the Caribbean large igneous province. *Geology*, *32*(8), 697–700.

Hoorn, C., Guerrero, J., Sarmiento, G. A., Lorente, M. A., 1995. Andean tectonics as a cause for changing drainage patterns in Miocene northern South America. *Geology, 23*, 237–240.

Humanes, A., and Bastidas, C., 2015. *In situ* settlement rates and early survivorship of hard corals: a good year for a Caribbean reef. *Marine Ecology Progress Series*, *539*, 139–151.

Humphrey, C. G., 2010. In-situ U-Pb secondary ion mass spectrometry (IN-SIMS) geochronology from the Leeward Antilles islands of Aruba, Curaçao, Bonaire, and Gran Roque: Implications for the temporal evolution of the Caribbean large igneous province (CLIP) and early arc magmatism. Master's thesis, University of Georgia, Athens, 83 pp.

Jackson, J. B. C., Donovan, M. K., Cramer, K. L., Lam, V. V., 2014. Status and trends of Caribbean coral reefs. *Global Coral Reef Monitoring Network, IUCN, Gland, Switzerland*, 1970–2012.

Jácome, M. I., Kusznir, N., Audemard, F., Flint, S., 2003a. Formation of the Maturín Foreland Basin, eastern Venezuela: Thrust sheet loading or subduction dynamic topography. *Tectonics*, *22*(5), 1046.

Jácome, M. I. N., Kusznir, N. J., Audemard, F., Flint, S., 2003b. Tectonostratigraphic evolution of the Maturín Foreland Basin: Eastern Venezuela, in C. Bartolini, R. T. Buffler and J. F. Blickwede, eds., The circum-Gulf of Mexico and the Caribbean: hydrocarbon habitats, basin formation, and plate tectonics. *American Association of Petroleum Geologists, Memoir 79*, 735–749.

James, N. P., 1997. The cool-water carbonate depositional realm. In: James, N. P., Clarke, J., (eds.), Cool-water carbonates, vol 56. SEPM Society for Sedimentary Geology, Tulsa, 1–20.

James, K. H., 2000. The Venezuelan Hydrocarbon habitat, part 1: Tectonics, Structure, Palaeoceanography and Source Rocks. *Journal of Petroleum Geology,* 23(1), 5–53.

James, K. H., 2009a. *In situ* origin of the Caribbean: discussion of data. In: James, K.H., Lorente, M.A., Pindell, J. (Eds.), Origin and Evolution of the Caribbean Plate. *Geological Society*, *London*, *Special Publications*, *328*, 77–125.

James, K. H., 2009b. Evolution of Middle America and the *in situ* Caribbean plate model. In: James, K.H., Lorente, M.A., Pindell, J. (Eds.), Origin and Evolution of the Caribbean Plate. *Geological Society*, *London*, *Special Publications*, *328*, 127–138.

Jenkyns, H. C., 2010. Geochemistry of oceanic anoxic event. *Geochemistry, Geophysics, Geosystems*, *118*(3), Q03004.

Jenkyns, H. C., Dickson, A. J., Ruhl, M., Van den Boorn, S. H., 2017. Basalt-seawater interaction, the Plenus Cold Event, enhanced weathering and geochemical change: deconstructing Oceanic Anoxic Event 2 (Cenomanian–Turonian, Late Cretaceous). *Sedimentology*, *64*, 16–43.

Jiang, Q., Jourdan, F., Olierook, H. K., Merle, R. E., Whittaker, J. M., 2021. Longest continuously erupting large igneous province driven by plume-ridge interaction. *Geology*, *49*, 206–210.

Johnson, K. G., Sánchez-Villagra, M. R., Aguilera, O. A., 2009. The Oligocene–Miocene transition on coral reefs in the Falcón Basin (NW Venezuela). *Palaios*, *24*, 59–69.

Jones, M. M., Sageman, B. B., Selby, D., Jicha, B. R., Singer, B. S., Titus, A. L., 2021. Regional chronostratigraphic synthesis of the Cenomanian-Turonian Oceanic Anoxic Event 2 (OAE2) interval, Western Interior Basin (USA): New Re-Os chemostratigraphy and $^{40}Ar/^{39}Ar$ geochronology. *Geological Society of America Bulletin*, *133*(5–6), 1090–1104.

Joo, Y. J., and Sageman, B. B., 2014. Cenomanian to Campanian carbon isotope chemostratigraphy from the Western Interior Basin, USA. *Journal of Sedimentary Research*, *84*, 529–542.

Joo, Y. J., Sageman, B. B., Hurtgen, M. T., 2020. Data-model comparison reveals key environmental changes leading to Cenomanian-Turonian Oceanic Anoxic Event 2. *Earth-Science Reviews*, *203*, 103123.

Kaiho, K., and Saito, S., 1994. Oceanic crust production and climate during the last 100 Myr. *Terra Nova*, *6*, 376–384.

Kasper, D. A., and Larue, D. K., 1986. Paleogeographic and tectonic implications of quartzose sandstones of Barbados. *Tectonics*, *5*, 837–854.

Kawka, V., Molina, A., Gómez, M. G., Méndez, J., Yanes, C., 2008. ESTUDIO GEOQUÍMICO AMBIENTAL DE LA ESPECIE DE FORAMINÍFERO BENTÓNICO RECIENTE *Amphistegina lessonii*, EN SEDIMENTOS SUPERFICIALES DE LA PLATAFORMA CONTINENTAL ATLÁNTICA VENEZOLANA. *Isla*, *2*(6).

Keller, G., Adatte, T., Berner, Z., Chellai, E. H., Stueben, D., 2008. Oceanic events and biotic effects of the Cenomanian-Turonian anoxic event, Tarfaya Basin, Morocco. *Cretaceous Research*, *29*, 976–994.

Kellogg, J. N., 1984. Cenozoic tectonic history of the Sierra de Perijá, Venezuela-Colombia, and adjacent basins. In: Bonini, W.E., Hargraves, R.B., Shagam, R. (Eds.), The Caribbean-South American Plate Boundary and Regional Tectonics. *Geological Society of America*, *Memoir 162*, 239–261.

Kennedy, W. J., Gale, A. S., Huber, B. T., Petrizzo, M. R., Bown, P., Barchetta, A., Jenkyns, H. C., 2014. Integrated stratigraphy across the Aptian/Albian boundary at Col de Pré- Guittard (south-east France): A candidate Global Boundary Stratotype Section. Cretaceous *Research*, *51*, 248–259.

Kennedy, J. W., Gale, A. S., Huber, B. T., Petrizzo, M. R., Bown, P., Jenkyns, H. C., 2017. The global boundary stratotype section and point (GSSP) for the base of the Albian stage, of the Cretaceous, the Col de Pré-Guittard section, Arnayon, Drôme, France. *Episodes*, *40*, 177–188.

Kerr, A. C., 1998. Oceanic plateau formation: A cause of mass extinction and black shale deposition around the Cenomanian-Turonian boundary? *Journal of Geological Society of London, 155, 4*, 619–626.

Kerr, A. C., White, R. V., Thompson, P. M. E., Tarney, J., Saunders, A.D., 2003. No oceanic plateau—No Caribbean Plate? The seminal role of an oceanic plateau in Caribbean Plate evolution, *in* Bartolini, C., Buffl er, R.T., and Blickwede, J.F., eds., The circum-Gulf of Mexico and Caribbean region; Hydrocarbon habitats, basin formation, and plate tectonics: Tulsa, *American Association of Petroleum Geologists*, *Memoir 79*, 126–168.

Kerr, A. C., Tarney, J., Kempton, P. D., Pringle, M., Nivia, A., 2004. Mafic pegmatites intruding oceanic plateau gabbros and ultramafic cumulates from Bolivar, Colombia: evidence for a 'wet' mantle plume? *Journal of Petrology*, *45*, 1877–1906.

Kerr, A. C., Neill, I., Urbani, F., Spikings, R., Barry, T., Tarney, J., 2009. The Siquisique basalts and gabbros, Los Algodones, Venezuela: Late Cretaceous oceanic plateau formed within the proto-Caribbean Plate. In *AGU Fall Meeting Abstracts (V41C-2193)*.

Kerr, A., Kevin R. Chamberlain, & Barquero, M. 2021. Edades U-Pb del Gabro de la Ofiolita de Siquisique. In: Urbani, F. Editor. Geología del Norte de los estados Lara y Yaracuy. Caracas: *Academia Nacional de la Ingeniería y el Hábitat y Fundación Geos*, 402–404.

Khan, N. S., Ashe, E., Horton, B. P., Dutton, A., Kopp, R. E., Brocard, G., Engelhart, S. E., Hill, D. F., Peltier, W. R., Vane, C. H., Scatena,

F. N., 2017. Drivers of Holocene sea-level change in the Caribbean. *Quaternary Science Reviews*, *155*, 13–36.

Kidder, D. L., and Worsley, T. R., 2010. Phanerozoic large igneous provinces (LIPs), HEATT (Haline Euxinic Acidic Thermal Transgression) episodes and mass extinctions. *Palaeogeography, Palaeoclimatology, Palaeoecology, 295*, 162–191.

Klemme, H. D., and Ulmishek, G. F., 1991. Effective petroleum source rocks of the world: Stratigraphic distribution and controlling depositional factors. *American Association of Petroleum Geologists Bulletin*, *75*, 1809–1,851.

Kolonic, S., Sinninghe Damsté, J. S., Böttcher, M., Kuypers, M. M. M., Kuhnt, W., Beckmann, B., G. Scheeder, Wagner, T., 2002. Geochemical characterization of Cenomanian/Turonian black shales from the Tarfaya Basin (SW Morocco). *Journal of Petroleum Geology,* 25, 325–350.

Kolonic, S., Wagnet, T., Forster, A., Sinninghe Damsté, J. S., Walsworth-Bell, B., Erba, E., Turgeon, S., Brumsack, H., Chellai, E., Tsikos, H., Kuhnt, W., Kuypers, M. M. M., 2005. Black shale deposition on the northwest African shelf during the Cenomanian/Turonian oceanic anoxic event: Climate coupling and global organic carbon burial, *Paleoceanography*, *20*, PA1006.

Koning, T., 2003. Oil and gas production from basement reservoirs: examples from Indonesia, USA and Venezuela. *Geological Society, London, Special Publications*, *214*, 83–92.

Kraal, P., Slomp, C. P., Forster, A., Kuypers, M. M. M., 2010. Phosphorus cycling from the margin to abysal depths in the proto-Atlantic during oceanic anoxic event 2. *Palaeogeography, Palaeoclimatology, Palaeoecology, 295, 1*, 42–54.

Kuhnt, W., Chellai, E. H., Holbourn, A., Luderer, F., Thurow, J., Wagner, T., El Albani, A., Beckmann, B., Herbin, J. -P., Kawamura, H., Kolonic, S., Nederbragt, S., Street, C., Ravilious, K., 2001. Morocco Basin's sedimentary record may provide correlations for Cretaceous paleoceanographic events worldwide. *Eos, 82*, 361–364.

Kuroda, J., and Ohkouchi, N., 2006. Implication of spatiotemporal distribution of black shales deposited during the Cretaceous Oceanic Anoxic Event-2. *Paleontological Research*, *10*, 345–358.

Kuroda, J., Ogawa, N. O., Tanimizu, M., Coffin, M. F., Tokuyama, H., Kitazato, H., Ohkouchi, N., 2007. Contemporaneous massive subaerial volcanism and late cretaceous Oceanic Anoxic Event 2. *Earth and Planetary Science Letters, 256*, 211–223.

Kuroda, J., Ihoriya, N., Hori, R.S., Ogawa, N.O., Ikehara, M., Tanimizu, M., Ohkouchi, N., 2015. Geochemistry of Aptian bedded chert succession from the deep Pacific basin: New insights into Cretaceous oceanic anoxic event (OAE) 1a, In: Neal, C. R., Sager, W. W., Sano, T., and Erba, E., eds., The Origin, Evolution, and Environmental Impact of Oceanic Large Igneous Provinces. *Geological Society of America, Special Paper 511*, 305–328.

Lambeck, K., Rouby, H., Purcell, A., Sun, Y., Sambridge, M., 2014. Sea level and global ice volumes from the Last Glacial Maximum to the Holocene. *Proceedings of the National Academy of Sciences*, *111*, 15296–15303.

Langer, M.C., Rincón, A.D., Ramezani, J., Solórzano, A., Rauhut, O.W., 2014. New dinosaur (Theropoda, stem-Averostra) from the earliest Jurassic of the La Quinta Formation, Venezuelan Andes, *Open Science*, *1*(2), 140184.

Laredo, M. and Picard, X., 1977. Aplicación a la ingeniería de métodos geofísicos marinos (método sparker). Memorias del Congreso Geológico Latinoamericano II, Caracas, Noviembre 1973. Boletín de Geología. Publicación Especial, 7(IV), 3161–3163.

Larson, R. L., 1991. Latest pulse of Earth: Evidence for a mid-Cretaceous superplume. *Geology*, *19*, 547–550.

Larson, R., and Erba, E., 1999. Onset of the mid-Cretaceous greenhouse in the Barremian-Aptian: Igneous events and the biological, sedimentary and geochemical responses. *Paleoceanography, 14*, 663–678.

Laugié, M., Donnadieu, Y., Ladant, J. -B., Bopp, L., Ethé, C., Raisson, F., 2021. Exploring the impact of Cenomanian paleogeography and marine gateways on oceanic oxygen. *Paleoceanography and Paleoclimatology*, *36*(7), e2020PA004202

Laya, J. C., and Tucker, M. E., 2012. Facies analysis and depositional environments of Permian carbonates of the Venezuelan Andes: Palaeogeographic implications for Northern Gondwana. *Palaeogeography, Palaeoclimatology, Palaeoecology*, *331*, 1–26.

Laya, J. C., and Tucker, M., 2015. Petroleum potential of Permian carbonates in northern South America, In: C. Bartolini and P. Mann, eds., Petroleum geology and potential of the Colombian Caribbean Margin. *American Association of Petroleum Geologists, Memoir 108*, 471–498.

Laya, J. C., Tucker, M. E., Gröcke, D. R., Perez-Huerta, A., 2013a. Carbon, oxygen and strontium isotopic composition of low-latitude Permian carbonates (Venezuelan Andes): climate proxies of tropical Pangea. *Geological Society, London, Special Publications*, *376*, 367–385.

Laya, J. C., Tucker, M. E., Perez-Huerta, A., 2013b. Metre-scale cyclicity in Permian ramp carbonates of equatorial Pangea (Venezuelan Andes): Implications for sedimentation under tropical Pangea conditions. *Sedimentary Geology*, *292*, 15–35.

Laya, J. C., Sulaica, J., Teoh, C. P., Whitaker, F., Gabellone, T., Tucker, M. E., Tesch, P., Miller, B., Prince, K., Izaguirre, I., 2018. Controls on Neogene carbonate facies and stratigraphic architecture of an isolated carbonate platform–the Caribbean Island of Bonaire. *Marine and Petroleum Geology*, *94*, 1–18.

Leckie, M., Bralower, T., Cashman, R., 2002. Oceanic anoxic events and plankton evolution: biotic response to tectonic forcing during the mid-Cretaceous. *Paleoceanography*, *17*, 1–29.

Lehmann, C., Osleger, D. A., Montanez, I. P., 1998. Controls on cyclostratigraphy of Lower Cretaceous carbonates and evaporites, Cupido and Coahuila platforms, northeastern Mexico. *Journal of Sedimentary Research*, *68*, 1109–1130.

Lehmann, C., Osleger, D. A., Montañez, I. P., Sliter, W., Vanneau, A. A., Banner, J., 1999. Evolution of Cupido and Coahuila carbonate platforms, early Cretaceous, northeastern Mexico. *Geological Society of America Bulletin*, *111*, 1010–1029.

Lentino, M., and Rodner, C., 2002. Los Roques: una muestra de la riqueza de nuestra avifauna insular. In: Guía del Parque Nacional Archipiélago Los Roques. Ed. Javier Zamarro. Agencia Española de Cooperación Internacional, Caracas, Venezuela, 143–165.

Liborius-Parada, A., 2019. Applied petroleum source rock evaluation and high-resolution sequence stratigraphy for unconventional reservoirs in La Luna Formation (Cenomanian-Santonian) Northwest Maracaibo Basin, Venezuela. Doctoral Dissertation. University of Oklahoma, 345 pp.

Liborius-Parada, A., Philp, R. P., Slatt, R., 2023. La Luna Formation, Venezuela. A Prospective Unconventional Reservoir. *Interpretation*, *11*(4), 23 pp.

Li, Y. -X., Bralower, T. J., Montañez, I. P., Osleger, D. A., Arthur, M. A., Bice, D. M., Herbert, T. D., Erba, E., Premoli Silva, I., 2008. Toward an orbital chronology for the early Aptian oceanic anoxic event (OAE1a, ~120 Ma), Earth Planetary Science Letters, 271, 88–100.

Linares, E., 2020. Caracterización mineralógica y geoquímica de rocas fosfáticas de la Formación Navay, Estado Táchira, Venezuela. Magister Dissertation. Central University of Venezuela. 127 pp.

Lisiecki, L. E., and Raymo M. E., 2005. A Pliocene-Pleistocene stack of 57 globally distributed benthic $\delta^{18}O$ records. *Paleoceanography*, *20*, PA1003.

Locke, B., and Garver, J., 2005. Thermal evolution of the eastern Serranía del Interior foreland fold and thrust belt, northeastern Venezeula, based on apatite fission-track analyses. In: Avé Lallemant, H., Sisson, V. (Eds.), Caribbean-South America Plate Interactions, Venezuela. *Geological Society of America Special Paper 394*, 315–328.

Locklair, R. E., and Sageman, B. B., 2008. Cyclostratigraphy of the Upper Cretaceous Niobrara Formation, western interior, USA: a Coniacian–Santonian orbital timescale. *Earth and Planetary Science Letters*, *269*, 540–553.

Locklair, R., Sageman, B., Lerman, A., 2011. Marine carbon burial flux and the carbon isotope record of Late Cretaceous (Coniacian–Santonian) Oceanic Anoxic Event III. *Sedimentary Geology*, *235*, 38–49.

Lo Mónaco, S. L., López, L., Rojas, H., Garcia, D., Premovic, P., Briceño, H., 2002. Distribution of major and trace elements in La Luna Formation, southwestern Venezuelan basin. *Organic Geochemistry*, *33*, 1593–1608.

López, L., and Lo Mónaco, S. L., 2004. Geochemical implications of trace elements and sulfur in the saturate, aromatic and resin fractions of crude oil from the Mara and Mara Oeste fields, Venezuela. *Fuel*, *83*, 365–374.

López, L., and Lo Mónaco, S. L., 2017. Vanadium, nickel and sulfur in crude oils and source rocks and their relationship with biomarkers: Implications for the origin of crude oils in Venezuelan basins. *Organic Geochemistry*, *104*, 53–68.

López, L., Mónaco, S. L., Galarraga, F., Lira, A., Cruz, C., 1995. V/Ni ratio in maltene and asphaltene fractions of crude oils from the west Venezuelan basin: correlation studies. *Chemical Geology*, *119*, 255–262.

López, L., Lo Mónaco, S., Richardson, M., 1998. Use of molecular parameters and trace elements in oil-oil correlation studies, Barinas sub-basin, Venezuela. *Organic Geochemistry*, *29*, 613–629.

López-Gamundí, O. R., 2010. Transgressions related to the demise of the Late Paleozoic Ice Age: Their sequence stratigraphic context. *In*: Late Paleozoic glacial events and postglacial transgressions in Gondwana. *The Geological Society of America, Special Paper 468*, 1–35.

Lorente, M. A., 1978. La caliza de Capadare y su relación con la Cuenca de Agua Salada, Edo. Falcón. Bachelor Degree Dissertation. Central University of Venezuela. 161 pp.

Lowery, C. M., Leckie, R. M., Sageman, B. B., 2017. Micropaleontological evidence for redox changes in the OAE3 interval of the US Western Interior: Global vs. local processes. *Cretaceous Research*, *69*, 34–48.

Lugo, P., 2002. Distribución de elementos mayoritarios y traza en rocas del Grupo Guayuta (formaciones Querecual y San Antonio): implicaciones paleoambientales. Magister Dissertation. Central University of Venezuela, 108 pp.

Lugo, P., 2008. Cuantificación y caracterización de la materia orgánica en rocas pre-Cretácicas de los Andes Venezolanos. Doctoral Dissertation. Central University of Venezuela, 136 pp.

Lugo, J., and Mann, P., 1995. Jurassic-Eocene tectonic evolution of Maracaibo Basin, Venezuela. In: Tankard, A., Suarez, S., Welsink, H. (Eds.), Petroleum Basins of South America. *American Association of Petroleum Geologists, Memoir*, *62*, 699–725.

Lugo, J., and Audemard, F., 2021. Cretaceous to Neogene arc–continent collision, orogenic float development, and implications on the petroleum systems of Northern Venezuela, *in* C. Bartolini, ed., South America–Caribbean– Central Atlantic plate boundary: Tectonic evolution, basin architecture, and petroleum systems. *American Association of Petroleum Geologists Memoir 123*, 1–84.

Lüning, S., Kolonic, S., Belhadj, E. M., Belhadj, Z., Cota, L., Baric, G., Wagner, T., 2004. Integrated depositional model for the Cenomanian-Turonian organic-rich strata in North Africa, *Earth Science Reviews, 64*, 51–117.

Lunt, D. J., Valdes, P. J., Haywood, A., Rutt, I. C., 2008. Closure of the Panama Seaway during the Pliocene: implications for climate and Northern Hemisphere glaciation. *Climate Dynamics*, *30*, 1–18.

Macellari, C. E., 1988. Cretaceous paleogeography and depositional cycles of western South America. *Journal of South American earth Sciences*, *1*(4), 373–418.

Macellari, C. E., 1995. Cenozoic sedimentation and tectonics of the southwestern Caribbean pull-apart basin, Venezuela and Colombia, *in* A. J. Tankard, R. Suárez S., and H. J. Welsink, Petroleum basins of South America. *American Association of Petroleum Geologists, Memoir 62*, 757–780.

Macellari, C. E., and De Vries, T. J., 1987. Late Cretaceous upwelling and anoxic sedimentation in northwestern South America: *Palaeogeography, Palaeoclimatology, Palaeoecology*, 59, 279–292.

Machado, M. C., Chemale Jr, F., Kawashita, K., Rey, O., Moura, C. A. V., 2016. Isotope studies of carbonate rocks of La Luna Formation (Venezuela) to constrain the oceanic anoxic event 3 (OAE3). *Journal of South American Earth Sciences*, *72*, 38–48.

Machel, H. G., 2004. Concepts and models of dolomitization: a critical reappraisal. *Geological Society, London, Special Publications*, *235*, 7–63.

Macsotay, O., Erlich, R., Peraza, T., 2003. Sedimentary structures of the La Luna, Navay and Querecual Formations, Upper Cretaceous of Venezuela. *Palaios*, *18*, 334–348.

Malavé, B. J., and Contreras, O. J., 2013. New contributions to the geodynamics of the caribbean through structural transects in the Gulf of Venezuela and its implications in the definition of petroleum systems. In: American Association of Petroleum Geologists International Conference and Exhibition, Cartagena, Colombia, Article #30286.

Malinverno, A., Erba, E., Herbert, T. D., 2010. Orbital tuning as an inverse problem: chronology of the early Aptian oceanic anoxic event 1a (Selli Level) in the Cismon APTICORE. *Paleoceanography*, *25*, PA2203.

Maloney, N. J., 1971. Geología de la isla de la Blanquilla y notas sobre el Archipiélago de los Hermanos. *Acta Científica Venezolana*, *22*, 6–10.

Mancini, E. A., and Puckett, T. M., 2005. Jurassic and Cretaceous transgressive-regressive (TR) cycles, northern Gulf of Mexico, USA. *Stratigraphy*, *2*, 31–48.

Mann, P., Escalona, A., Castillo, M. V., 2006. Regional geologic and tectonic setting of the Maracaibo supergiant basin, western Venezuela. *American Association of Petroleum Geologists Bulletin*, *90*, 445–477.

Marcucci, E., 1973. Origin and silica-carbonate relations of cherts of the Upper Cretaceous of Venezuela. Magister Dissertation. Rice University, Houston Texas. 81 pp.

Márquez-Rojas, B., Díaz-Díaz, O., Balza, M. A., 2013. Holoplanctonic polychaetes (Annelida: Polychaeta) from Venezuela. *Pan-American Journal of Aquatic Science*, *8*, 160–165.

Martin, R. E., 1996. Secular increase in nutrient levels through the Phanerozoic: implications for productivity, biomass, and diversity of the marine biosphere. *Palaios*, *11*, 209–219.

Martínez, J. I., and Hernández, R., 1992. Evolution and drowning of the Late Cretaceous Venezuelan carbonate platform. *Journal of South American Earth Sciences*, *5*, 197–210.

Martinez, S. J., Cavada-Blanco, F., Cappelletto, J., Agudo-Adriani, E., Aldo, C., 2021. Distribution, abundance, and health indicators of the critically endangered coral species *Acropora cervicornis* in Los Roques National Park, 2014. *Advances in Oceanography and Limnology*, *12*, 10005.

Martínez-Rodríguez, R., Batenburg, S. J., Castro, J. M., de Gea, G. A., Nieto, L. M., Ruiz-Ortiz, P. A., Robinson, S., 2024. Integrated cyclostratigraphy of the Cau core (SE Spain)-A timescale for climate change during the early Aptian Anoxic Event (OAE 1a) and the late Aptian. *Global and Planetary Change*, *233*, 104361.

Matsumoto, H., Kuroda, J., Coccioni, R., Frontalini, F., Sakai, S., Ogawa, N. O., Ohkouchi, N., 2020. Marine Os isotopic evidence for multiple volcanic episodes during Cretaceous Oceanic Anoxic Event 1b. *Scientific Reports*, *10*, 12601.

Matsumoto, H., Coccioni, R., Frontalini, F., Shirai, K., Jovane, L., Trindade, R., Savian, J. F., Tejada, M. L. G., Gardin, S., Kuroda, J.,

2021. Long-term Aptian marine osmium isotopic record of Ontong Java Nui activity. Geology, 49(9), 1148–1152

Matsumoto, H., Goto, K. T., Shimoda, G., Watanabe, Y., Shirai, K., Tejada, M. L. G., Ishikawa, A., Ando, A., Sano, T., Kuroda, J., Suzuki, K., 2024. Brief and intensive volcanic emissions from Ontong Java Nui heralded Oceanic Anoxic Event 1a. *Communications Earth & Environment*, *5*, 155.

Matthews, R. K., 1973. Relative elevation of late Pleistocene high sea level stands: Barbados uplift rates and their implications. *Quaternary Research*, *3*, 147–153.

Mauffret, A., and Leroy, S., 1997. Seismic stratigraphy and structure of the Caribbean igneous province: *Tectonophysics*, *283*, 61–104.

Maze, W., 1984. Jurassic La Quinta Formation in the La Sierra de Perijá, northwestern Venezuela; geology and tectonic environment of red beds volcanic rocks. In: W. Bonini, R. Hargraves, and R. Shagam, eds., The Caribbean –South American plate boundary and regional tectonics: *Geological Society of America, Memoir 162*, 263–282.

Medina, E., Francisco, M., Sternberg, L., Anderson, W. T., 2005. Isotopic signatures of organic matter in sediments of the continental shelf facing the Orinoco Delta: Possible contribution of organic carbon from savannas. *Estuarine, Coastal and Shelf Science*, *63*, 527–536.

Méndez-Baamonde, J., 1978. Archipiélago Los Roques/Islas de Aves. Cuadernos Lagoven. Caracas, Venezuela. 48 pp.

Méndez-Baamonde, J., 1984. Los Roques: Un ejemplo de Sedimentación de Carbonatos. Morfología, Sedimentología, Organismos, Mineralogía y Geoquímica. Relación con Las Aves y La Blanquilla. Doctoral Dissertation. Central University of Venezuela. Volume I, 205 pp.

Méndez-Baamonde, J., 1985a. Terrazas submarinas del frente arrecifal en el Archipiélago Los Roques, Aves de Barlovento, Aves de Sotavento y La Blanquilla. Memorias del VI Congreso Geológico de Venezuela, 5549–5569.

Méndez-Baamonde, J., 1985b. Transgresiones y regresiones marinas en el Pleistoceno tardío. Subsidencia y levantamiento en los Roques Las Aves y La Blanquilla. Memorias del VI Congreso Geológico de Venezuela, 5570–5597.

Méndez-Baamonde, J., 1989a. Modelo depositacional del Grupo Cogollo. Talud externo, márgenes y plataforma. Memorias del VII Congreso Geológico Venezolano. Tomo II. Barquisimeto. Venezuela, 828–850.

Méndez-Baamonde, J., 1989b. La Formación La Luna. Características de una Cuenca Anóxica en una Plataforma de Aguas Someras. Memorias del VII Congreso Geológico Venezolano. Tomo II. Barquisimeto. Venezuela, 851–866.

Méndez-Baamonde, J., 1997. El Cuaternario. Ministerio de Energía y Minas. Dirección General Sectorial de Servigeomin. Caracas. Venezuela. *Boletín de Geología*, *31*, 74–115.

Méndez-Baamonde, J., 2000. Geología del Delta del Orinoco. INGEOMIN. Caracas, Venezuela. *Boletín de Geología. Publicación Especial N° 13*, 126 pp.

Méndez-Baamonde, J., 2002a. Caracterización sedimentológica del Miembro "O" en el área de Borburata. Descripción detallada de los núcleos BOR-12 y BOR-14. PDVSA, Distrito Sur-Barinas. Internal Report. 87 pp.

Méndez-Baamonde, J., 2002b. Origen y Formación del Archipiélago Los Roques. In: Guía del Parque Nacional Archipiélago Los Roques. Ed. Javier Zamarro. Agencia Española de Cooperación Internacional, Caracas, Venezuela, 85–99.

Méndez-Baamonde, J., 2006. Isla de Aves: equilibrio entre la subsidencia, aumento del nivel del mar, acreción y progradación. Research study for Promotion to Full Professor, Faculty of Sciences, Central University of Venezuela. 55 pp.

Méndez-Baamonde, J., 2007a. Costas litorales del Caribe y del Atlántico, islas y archipiélagos. Las profundidades marinas. In: *Geo Venezuela*, Tomo II: Medios Físicos y Recursos Ambientales. Caracas: Fundación Empresas Polar, 184–237.

Méndez-Baamonde, J., 2007b. Aspectos generales de la sedimentación del Grupo Cogollo durante el Aptiense y Albiense en la Cuenca del Lago de Maracaibo y Perijá. Venezuela. Memorias del IX Congreso Geológico de Venezuela, Caracas, 20–21.

Méndez-Baamonde, J., 2009. Carbonatos—Origen y Sedimentación. University Printing of the Central University of Venezuela. Caracas, Venezuela. 270 pp.

Méndez-Baamonde, J., Baquero, M., Méndez-Dot, J. A., 2009. Eventos Oceánicos Anóxicos OAE 1a y OAE 1b en la Formación Apón del Grupo Cogollo durante el Cretácico Temprano. Cuenca de Maracaibo, Venezuela, *Revista Latinoamericana de Geoquímica Orgánica, 1*, 6–18.

Méndez-Dot, J. A., 2011. Modelos del Holoceno en Venezuela para la formación de dolomita singenética en ambientes de sabkha y lagunas hipersalinas del Archipiélago Los Roques y su utilidad para interpretar facies de dolomita del pasado geológico. Magister Dissertation. Central University of Venezuela. 194 pp.

Méndez-Dot, J. A., 2017. Sedimentación de los Eventos Oceánicos Anóxicos 1a y 1b del Cretácico en el Grupo Cogollo de la Cuenca de Maracaibo: determinación del potencial como rocas madre. Doctoral Dissertation. Central University of Venezuela, 229 pp.

Méndez-Dot, J. A., Méndez-Baamonde, J., Reyes, D., Whilchy, R., 2015. The Cogollo Group and the Oceanic Anoxic Events 1a and 1b, Maracaibo Basin, Venezuela, *Brazilian Journal of Geology, 45* (Suppl. 1), 41–61.

Menegatti, A. P., Weissert, H., Brown, R. S., Tyson, R. V., Farrimond, P., Strasser, A., Caron, M., 1998. High resolution $\delta^{13}C$ stratigraphy through the early Aptian "Livello Selli" of the Alpine Tethys, *Paleoceanography, 13*, 530–545.

Meschede, M., and Frish, W., 1998. A plate-tectonic model for the Mesozoic and early Cenozoic history of the Caribbean plate. *Tectonophysic, 296*, 269–291.

Meschede, M., and Frisch, W., 2002. The evolution of the Caribbean Plate and its relation to global plate motion vectors: Geometric constraints for an Inter-American origin, in Jackson, T. A., ed., Caribbean geology into the third millennium: Transactions of the 15th Caribbean Geological Conference, University of West Indies Press, Kingston, Jamaica, 1–14.

Meyer, K. M., and Kump, L. R., 2008. Oceanic euxinia in Earth history: causes and consequences. *Annual Review of Earth and Planetary Sciences*, *36*, 251–288.

Meyers, P., Bernasconi, S., Forster, A., 2006. Origins and accumulation of organic matter in expanded Albian to Santonian black shale sequences on the Demerara Rise, South American Margin. *Organic Geochemistry*, *37*, 1816–1830.

Meyers, P. A., Bernasconi, S. M., Yum, J. -G., 2009. 20 My of nitrogen fixation during deposition of mid-Cretaceous black shales on the Demerara Rise, ecuatorial Atlantinc Ocean. *Organic Geochemistry, 40*, 158–166.

Miall, A. D., and Blakey, R. C., 2008. The Phanerozoic tectonic and sedimentary evolution of North America. *Sedimentary basins of the world*, *5*, 1–29.

Miall, A. D., and Blakey, R. C., 2019. The Phanerozoic tectonic and sedimentary evolution of North America. In: Miall AD, editor. The Sedimentary Basins of the United States and Canada. 2nd ed. Elsevier. pp. 1–38.

Michel, J., Laugié, M., Pohl, A., Lanteaume, C., Masse, J. P., Donnadieu, Y., Borgomano, J., 2019. Marine carbonate factories: a global model of carbonate platform distribution. *International Journal of Earth Sciences*, *108*, 1773–1792.

Miloslavich, P., Díaz, J. M., Klein, E., Alvarado, J. J., Díaz, C., Gobin, J., Escobar-Briones, E., Cruz-Motta, J. J., Weil, E., Cortés, J., Bastidas, A. C., Robertson, R., Zapata, F., Martín, A., Castillo, J., Kazandjian,

A., Ortiz, M., 2010. Marine biodiversity in the Caribbean: regional estimates and distribution patterns. *PloS One*, *5*, e11916.

Molina, A., Kawka, V., Gómez, M. G., Méndez, J., Yanes, C., 2008. ESTUDIO GEOQUÍMICO AMBIENTAL DE LA ESPECIE DE FORAMINÍFERO BENTÓNICO *Quinqueloculina lamarckiana* PRESENTE EN SEDIMENTOS RECIENTES DE LA PLATAFORMA CONTINENTAL ATLÁNTICA VENEZOLANA. *Isla*, *2*(6).

Ministerio de Energía y Minas, 1997. Léxico Estratigráfico de Venezuela. República de Venezuela. Dirección General Sectorial de ServiGeomin. Dirección de Geología. *Boletín de Geología, Publicación Especial N°. 12, Tomos 1 y 2*. 828 pp.

Mongenot, T., Tribovillard, N. P., Desprairies A., Lallier-Vergés, E., Laggoun-Defarge, F., 1996. Trace elements and paleoenvironmental markers in strongly mature hydrocarbon source rock: the Cretaceous La Luna Formation of Venezuela. *Sedimentary Geology*, *103*, 23–37.

Montero-Serrano, J. C., Martínez, M., Riboulleau, A., Tribovillard, N., Marquez, G., Gutiérrez-Martín, J. V., 2010. Assessment of the oil source-rock potential of the Pedregoso Formation (Early Miocene) in the Falcón Basin of northwestern Venezuela. *Marine and petroleum geology*, *27*, 1107–1118.

Morales, F. J., 2004. Caracterización de conjuntos de foraminíferos bentónicos en sedimentos recientes de ambientes de talud continental del area de la Fachada Atlántica de Venezuela: implicaciones geoambientales. Magister Dissertation, Central University of Venezuela. 251 pp.

Moulin, M., Aslanian, D., Unternehr, P., 2010. A new starting point for the South and Equatorial Atlantic Ocean. *Earth-Science Reviews*, *98*, 1–37.

Muessig, K. W., 1984. Structure and Cenozoic tectonics of the Falcón Basin, Venezuela, and adjacent areas. *Geological Society of America Memoirs*, *162*, 217–230.

Muhs, D. R., Pandolfi, J. M., Simmons, K. R., Schumann, R. R., 2012. Sea-level history of past interglacial periods from uranium-series dating of corals, Curaçao, Leeward Antilles islands. *Quaternary Research*, *78*, 157–169.

Muhs, D. R., and Simmons, K. R., 2017. Taphonomic problems in reconstructing sea-level history from the late Quaternary marine terraces of Barbados. *Quaternary Research*, *88*, 409–429.

Müller, R. D., Sdrolias, M., Gaina, C., Roest, W. R., 2008. Age, spreading rates, and spreading asymmetry of the world's ocean crust. *Geochemistry, Geophysics, Geosystems*, *9*, Q04006.

Murat, B., and Azpiritixaga, I., 1995. Aplicación del análisis secuencial en los carbonatos del Cretácico en la Cuenca de Maracaibo. *Boletín de la Sociedad Venezolana de Geólogos*, *20*, 7–29.

Murray-Wallace, C. V., and Woodroffe, C. D., 2014. Quaternary Sea-Level Changes — A Global Perspective. Cambridge University Press, Cambridge, 484 pp.

Mutti, M., and Hallock, P., 2003. Carbonate systems along nutrient and temperature gradients: some sedimentological and geochemical constraints. *International Journal of Earth Sciences*, *92*, 465–475.

Naafs, B. D. A., Castro, J. M., De Gea. G. A., Quijano, M. L., Schmidt. D. N., Pancost, R. D., 2016. Gradual and sustained carbon dioxide release during Aptian Oceanic Anoxic Event 1a. *Nature Geoscience*, 9, 135–139.

Najarro, M., Rosales, I., Moreno-Bedmar, J. A., de Gea, G. A., Barrón, E., Company, M., Delanoy, G., 2011. High resolution chemo- and biostratigraphic records of the Early Aptian oceanic anoxic event in Cantabria (N Spain): Paleoceanographic and paleoclimatic implications. *Palaeogeography, Palaeoclimatology, Palaeoecology*, *299*, 137–158.

Navas-Ramírez, R., 2018. Tectonostratigraphic evolution of the Gulf of Venezuela. Magister Dissertation. University of Stavanger. 93 pp.

Nederbragt, A. J., Thurow, J., Pearce, R., 2007. Sediment composition and cyclicity in the mid-Cretaceous at Demerara Rise, ODP Leg 207. *In* Mosher, D.C., Erbacher, J., and Malone, M.J. (Eds.), *Proceedings of the Ocean Drilling Program, Scientific Results*, 207: College Station, TX (Ocean Drilling Program), 1–31.

Neal, C. R., Coffin, M. F., Sager, W. W., 2019. Contributions of scientific ocean drilling to understanding the emplacement of submarine large igneous provinces and their effects on the environment. Oceanography, *32* (1), 176–192.

Neill, I., Kerr, A. C., Hastie, A. R., Stanek, K. P., Millar, I. L., 2011. Origin of the Aves Ridge and Dutch–Venezuelan Antilles: interaction of the Cretaceous 'Great Arc' and Caribbean–Colombian Oceanic Plateau? *Journal of the Geological Society*, *168*, 333–348.

Nelson, R. A., Moldovanyi, E. P., Matcek, C. C., Azpiritxaga, I., Bueno, E., 2000. Production characteristics of the fractured reservoirs of the La Paz field, Maracaibo Basin, Venezuela. *American Association of Petroleum Geologists Bulletin*, *84*, 1791–1809.

Nerlich, R., Clark, S. R., Bunge, H.-P., 2014. Reconstructing the link between the Galapagos hotspot and the Caribbean Plateau. *GeoResJ*, *1*, 1–7.

Noguera, M., Wright, J. E., Urbani, F., Pindell, J. L., 2011. U-Pb Geochronology of Detrital Zircons from the Venezuelan Passive Margin: Implications for an Early Cretaceous Proto-Orinoco River System and Proto-Caribbean Oceanic Basin Paleogeography. *Geologica Acta*, 9, 1–22.

Noguera, M., Wright, J., Fournier, H., Urbani, F., Baquero, M., 2017. Geocronología U-Pb de Cristales de Zircón Detríticos de la Formación Matatere, Estados Lara y Yaracuy. *Academia Nacional de la Ingeniería y el Hábitat, Boletín 37*, 950–983.

Noriega, N., and Fuentes-Carrero, Y., 2014. Contribución al conocimiento de la diversidad de los equinodermos del Parque Nacional Archipiélagos de los Roques, Venezuela, *Acta Biológica Venezuelica*, *34*(2), 285–292.

Nova, G., Chaparro, G. A. B., Silva-Tamayo, J. C., Cardona, A., Rapalini, A., Cortes, P. C. M., Eisenhauer, A., Dussan, K. T., Valencia, V. A., Ramírez, V., Montes, C., 2019. Jurassic break-up of the Peri-Gondwanan margin in northern Colombia: Basin formation and implications for terrane transfer. *Journal of South American Earth Sciences*, *89*, 92–117.

Núñez-Flores, M., Rodríguez-Quintal, J. G., Cristina Díaz, M., 2010. Distribución de esponjas (Porifera) a lo largo de un gradiente de profundidad en un arrecife coralino, Parque Nacional San Esteban, Carabobo, Venezuela. *Revista de Biología Tropical*, *58*, 175–187.

Núñez-Useche, F., Barragán R., Moreno-Bedmar J. A., Canet, C., 2014. Mexican archives for the major Cretaceous Oceanic Anoxic Events. *Boletín de la Sociedad Geológica Mexicana*, *66*, 491–505.

Núñez-Useche, F., Barragán, R., Moreno-Bedmar, J. A., Canet, C., 2015. Geochemical and paleoenvironmental record of the early to early late Aptian major episodes of accelerated change: evidence from Sierra del Rosario, Northeast Mexico. Sedimentary Geology 324, 47–66.

Núñez-Useche, F., Canet, C., Barragán, R., Alfonso, P., 2016. Bioevents and redox conditions arround the Cenomanian-Turonian anoxic event in Central Mexico. *Palaeogeography, Palaeoclimatology, Palaeoecology*, *449*, 205–226.

Núñez-Useche, F., Barragán, R., Torres-Martínez, M. A., López-Zúñiga, P. A., Moreno-Bedmar, J. A., Chávez-Cabello, G., Canet, C., Chacon-Baca, E., 2020. Response of the western proto-North Atlantic margin to the early Aptian Oceanic Anoxic Event (OAE) 1a: an example from the Cupido platform margin-Gulf of Mexico, NE Mexico, *Cretaceous Research*, *113*, 104488.

Orth, C. J., Attrep Jr, M., Quintana, L. R., Elder, W. P., Kauffman, E. G., Diner, R., Villamil, T., 1993. Elemental abundance anomalies in the late Cenomanian extinction interval: a search for the source (s). *Earth and Planetary Science Letters*, *117*, 189–204.

Ostos, M., and Sisson, V. B., 2005. Geochemistry and tectonic setting of igneous and metaigneous rocks of northern Venezuela. *Geological Society of America Special Paper*, *394*, 119–156.

Ostos, M., Yoris, F., Avé-Lallemant, H. G., 2005. Overview of the southeast Caribbean–South American plate boundary zone, *in* Avé Lallemant, H.G., and Sisson, V.B., eds., Caribbean–South American plate interactions, Venezuela: *Geological Society of America Special Paper*, *394*, 53–89.

Padron, C., Klingelhoefer, F., Marcaillou, B., Lebrun, J. F., Lallemand, S., Garrocq, C., Laigle, M., Roest, W 2021. Deep structure of the Grenada Basin from wide-angle seismic, bathymetric and gravity data *Journal of Geophysical Research: Solid Earth*, *126*, e2020JB020472.

Paez-Reyes, M., Carvajal-Ortiz, H., Sahoo, S. K., Varol, O., Miller, B. V., Hughes, G. W., Gaona-Narvaez, T., Patarroyo, G. D., Curtis, J. H., Lerma, I. Copeland, P., 2021. Assessing the contribution of the La Luna Sea to the global sink of organic carbon during the Cenomanian-Turonian Oceanic Anoxic Event 2 (OAE2). *Global and Planetary Change*, *199*, 103424.

Palacas, J. G., Anders, D. R., King, J. D., 1984. Florida Basin—A prime example of carbonate source rocks of petroleum, in J. G. Palacas, ed., Geochemistry and source rock potential of carbonate rocks. *American Association of Petroleum Geologists, Studies in Geology 18*, 71 -96.

Papadomanolaki, N. M., van Helmond, N. A., Pälike, H., Sluijs, A., Slomp, C. P., 2022. Quantifying volcanism and organic carbon burial across Oceanic Anoxic Event 2. *Geology*, *50*(4), 511–515.

Parnaud, F., Gou, Y., Pascual, J. C., Capello, M. A., Truskowski, I., Passalacqua, H., 1995a. Stratigraphic synthesis of western Venezuela. In: Tankard, A.J., Suárez, S., Welsink, H.J. (Eds.), Petroleum Basins of South America. *American Association of Petroleum Geologists, Memoir 62*, 667–679.

Parnaud, F., Gou, Y., Pascual, J. -C., Truskowski, I., Gallango, O., Passalacqua, H., Roure, F., 1995b. Petroleum geology of the central part of the Eastern Venezuela basin, *in* A. J. Tankard, R. Suárez S., and H. J. Welsink, Petroleum basins of South America. *American Association of Petroleum Geologists*, *Memoir 62*, 741–756.

Parra, M., Moscardelli, L., Lorente, M. A., 2003. Late Cretaceous anoxia and lateral microfacies changes in the Tres Esquinas Member, La Luna Formation, Western Venezuela. *Palaios, 18*, 321–333.

Parra, M., Sánchez, G. J., Montilla, L., Guzmán, O. J., Namson, J., Jácome, M. I., 2011. The Monagas fold–thrust belt of eastern Venezuela. part I: structural and thermal modeling. *Marine and Petroleum Geology*, *28*, 40–69.

Past Interglacials Working Group of PAGES, 2016. Interglacials of the last 800,000 years. *Review of Geophysics*, *54*, 162–219.

Patarroyo, P., and Götz, S., 2020. Depósitos del Aptiano inferior con amonitas y rudistas, Punta Espada, Alta Guajira (Uribia–Colombia). Litoestratigrafía y apreciaciones regionales. *Boletín de Geología*, *42*, 227–241.

Patarroyo, G. D., Torres, G. A., Rincón, D. A., Cárdenas, C. P., Márquez, R. E., 2017. Bioestratigrafía e inferencias paleoambientales de las asociaciones de foraminíferos en las formaciones cretácicas La Luna-Colón (Cuenca del Catatumbo, Colombia). *Boletín de Geología*, *39*, 25–40.

Patiño, A. M., Parra, M., Ramírez, J. C., Sobel, E. R., Glodny, J., Almendral, A., Echeverri, S., 2019. Thermochronological constraints on Cenozoic exhumation along the southern Caribbean: The Santa Marta range, northern Colombia. In B. K. Horton & A. Folguera (Eds.), Andean Tectonics (pp. 103–132). Elsevier.

Pavlis, N. K., Holmes, S. A., Kenyon, S. C., Factor, J. K., 2012. The development and evaluation of the Earth Gravitational Model 2008 (EGM2008). *Journal of Geophysical Research*: *Solid Earth*, *117*, B04406.

Peltier, W. R., Argus, D.F., Drummond, R., 2015. Space geodesy constrains ice age terminal deglaciation: the global ICE-6G_C (VM5a) model. *Journal of Geophysical Research*: *Solid Earth*, *120*, 450–487.

Percival, L. M., Matsumoto, H., Callegaro, S., Erba, E., Kerr, A. C., Mutterlose, J., Suzuki, K., 2024. Cretaceous large igneous provinces: from volcanic formation to environmental catastrophes and biological crises. *Geological Society, London, Special Publications*, *544*, SP544-2023-88.

Pérez-Castresana, G., Villamizar, E., Varela, R., Fuentes, Y., 2014. Descripción preliminar del fitoplancton en seis arrecifes coralinos del Parque Nacional Archipiélago de Los Roques. *Acta Biológica Venezuelica*, *34*(2), 293–309.

Pérez-Infante, J., Farrimond, P., Furrer, M., 1996. Global and local controls influencing the deposition of the La Luna Formation (Cenomanian-Campanian), western Venezuela. *Chemical Geology*, *130*, 271–288.

Pérez-Pérez, A., D'onofrio, L., Bosch, M., Zapata, E., 2011. Association between magnetic susceptibilities and hydrocarbon deposits in the Barinas-Apure Basin, Venezuela. *Geophysics*, *76*, 35–41.

Peters, K. E., Walters C., Moldowan J. M., 2005. The Biomarker Guide. Biomarkers and isotopes in the environmental and human history. Volume I. Second Edition. Cambridge University Press, 471 pp.

Petrash, D. A., Gingras, M. K., Lalonde, S. V., Orange, F., Pecoits, E., Konhauser, K. O., 2012. Dynamic controls on accretion and lithification of modern gypsum-dominated thrombolites, Los Roques, Venezuela. *Sedimentary Geology*, *245*, 29–47.

Petrash, D. A., Lalonde, S. V., González-Arismendi, G., Gordon, R. A., Méndez, J. A., Gingras, M. K., Konhauser, K. O., 2015. Can Mn–S redox cycling drive sedimentary dolomite formation? A hypothesis. *Chemical Geology*, *404*, 27–40.

Petrash, D. A., Gueneli, N., Brocks, J. J., Méndez-Dot, J. A., González-Arismendi, G., Poulton, S. W., Konhauser, K., 2016. Black shale deposition and early diagenetic dolomite cementation during Oceanic Anoxic Event 1: The mid-Cretaceous Maracaibo Platform, north-western South America. *American Journal of Science, 316*, 669–711.

Petrizzo, M.R., Huber, B.T., Gale, A.S., Barchetta, A., Jenkyns, H. C., 2012. Abrupt planktonic foraminiferal turnover across the Niveau Kilian at Col de Pré-Guittard (Vocontian Basin, southeast France): new criteria for defining the Aptian/Albian boundary. *Newsletters on Stratigraphy*, *45*, 55–74.

Phelps, R. M., Kerans, C., Loucks, R. G., Da Gama, R. O., Jeremiah, J., Hull, D., 2014. Oceanographic and eustatic control of carbonate platform evolution and sequence stratigraphy on the Cretaceous (Valanginian–Campanian) passive margin, northern Gulf of Mexico. *Sedimentology*, *61*, 461–496.

Picard, X., and Goddard, D., 1972. Sedimentación de carbonatos en las islas Larga, Alcatraz y Santo Domingo, Edo. Carabobo, *Asociación Venezolana de Geología, Minería y Petróleo. Boletín Informativo*, *15*, 5–6.

Pindell, J. L., 1994. Evolution of the Gulf of Mexico and the Caribbean. In: Donovan S.K. and Jackson, T. A. (eds.) Caribbean Geology: an introduction, University of the West Indies Publishers Association/ University of the West Indies Press, Kingston, Jamaica, 13–39.

Pindell, J., and Dewey, J. F., 1982. Permo-Triassic reconstruction of western Pangea and the evolution of the Gulf of Mexico/Caribbean region. *Tectonics*, *1*, 179–211.

Pindell, J. L., and Barrett, S. F., 1990. Geological evolution of the Caribbean region: a Plate tectonic perspective. In: DENGO, G. & CASE, J. E. (eds.) The Caribbean Region. The Geology of North America, H. Geological Society of America, 405–432.

Pindell, J. L., and Kennan, L., 2009. Tectonic evolution of the Gulf of Mexico, Caribbean and northern South America in the mantle reference frame: an update. In: James, K.H., Lorente, M.A., Pindell, J. (Eds.), Origin and Evolution of the Caribbean Plate. *Geological Society*, *London*, *Special Publications*, *328*, 1–55.

Pindell, J., Kennan, L., Stanek, K. P., Maresch, W. V., Draper, G., 2006. Foundations of Gulf of Mexico and Caribbean evolution: eight controversies resolved. *Geologica Acta: an international earth science journal*, *4*, 303–341.

Pindell, J., Maresch, W. V., Martens, U., & Stanek, K. (2012). The Greater Antillean Arc: Early Cretaceous origin and proposed relationship to Central American subduction mélanges: implications for models of Caribbean evolution. *International Geology Review, 54*, 131–143.

Pindell, J., Weber, B., Elrich, W. H., Cossey, S., Bitter, M., Molina-Garza, R., Graham, R., Erlich, R. 2020. Strontium Isotope Dating of Evaporites and the Breakup of the Gulf of Mexico and Proto-Caribbean Seaway. *Geological Society of America. Special Paper 546*, 21.

Pinto, J., Ortega, S., Marin, Z., Berrios, I., Perez, A., Pirela, M., 2011. Controls on the newly-discovered gas accumulations in the Miocene "Perla" carbonate bank, Gulf of Venezuela: A preliminary assessment: I South American Oil and Gas Congress, SPE Western Venezuela Section, Maracaibo, 36, 1–4.

Pinto, D., Díaz, N., Tang, G., Arends, A., Ramírez, R., Pomar, L., Padrón V., 2014. INTERPRETACION DE RAMPAS CARBONATICAS ASOCIADAS A PALEOISLAS DURANTE EL OLIGOCENO-MIOCENO EN EL GOLFO DE VENEZUELA. CASO DE ESTUDIO: CAMPO PERLA. *Memorias del I Congreso Venezolano de Gas Natural (ICVGAS), Porlamar, Venezuela, 24-28 Noviembre.*

Pirela, M., 2017. Caracterización de fluidos y estimación de la posible dirección de llenado de yacimiento. Caso de estudio: Campo Perla, Cuenca del Golfo de Venezuela. Magister Dissertation. Central University of Venezuela, 134 pp.

Pitelli, R., and Molina, A., 1989. El Eoceno Medio Tardío y Eoceno Tardío de la parte occidental de la cuenca de Falcón. *Boletín de la Sociedad Venezolana de Geología, 36*, 5–12.

Polyak, V. J., Onac, B. P., Fornos, J. J., Hay, C., Asmerom, Y., Dorale, J. A., Ginés, J., Tuccimei, P., Ginés, A., 2018. A highly resolved record of relative sea level in the western Mediterranean Sea during the last interglacial period. *Nature Geoscience, 11*, 860–864.

Pomar, L., 2001. Types of carbonate platforms: a genetic approach. *Basin Research, 13*, 313–334.

Pomar, L., and Hallock, P., 2008. Carbonate factories: a conundrum in sedimentary geology. *Earth-Science Reviews, 87*, 134–169.

Pomar, L., Brandano, M., Westphal, H., 2004. Environmental factors influencing skeletal grain sediment associations: a critical review of Miocene examples from the western Mediterranean. *Sedimentology, 51*, 627–651.

Pomar, L., Bassant, P., Brandano, M., Ruchonnet, C., Janson, X., 2012. Impact of carbonate producing biota on platform architecture: insights from Miocene examples of the Mediterranean region. *Earth-Science Reviews, 113*, 186–211.

Pomar, L., Esteban, M., Martinez, W., Espino, D., Castillo de Ott, V., Benkovics, L., Castro-Leyva, T., 2015. Oligocene-Miocene carbonates of the Perla field, offshore Venezuela: Depositional model and facies architecture, In: C. Bartolini and P. Mann, eds., Petroleum geology and potential of the Colombian Caribbean Margin. *American Association of Petroleum Geologists, Memoir 108*, 647–674.

Pomar, L., Baceta, J. I., Hallock, P., Mateu-Vicens, G., Basso, D., 2017. Reef building and carbonate production modes in the west-central Tethys during the Cenozoic. *Marine and Petroleum Geology, 83*, 261–304.

Pöppelreiter, M., Balzarini, M. A., De Sousa, P., Engel, S., Galarraga, M., Hansen, B., Márquez, X., Morell, J., Nelson, R., Rodriguez, F., 2005. Structural control on sweet-spot distribution in a carbonate reservoir: Concepts and 3-D models (Cogollo Group, Lower Cretaceous, Venezuela). *American Association of Petroleum Geologists Bulletin, 89*, 1651–1676.

Posada, J. M., 2002. La pesca artesanal, una forma de vida tradicional. In: Guía del Parque Nacional Archipiélago Los Roques. Ed. Javier Zamarro. Agencia Española de Cooperación Internacional, Caracas, Venezuela, 55–66.

Poulton, S.W., Henkel, S., März, C., Urquhart, H., Flögel, S., Kasten, S., Sinninghe Damsté, J. S., Wagner, T., 2015. A continental-weathering control on orbitally driven redox-nutrient cycling during Cretaceous Oceanic Anoxic Event 2. *Geology, 43*, 963–966.

Premoli-Silva, I., and Sliter, W. V., 1995. Cretaceous planktonic foraminiferal biostratigraphy and evolutionary trends from the Bottaccione section, Gubbio, Italy. *Palaeontographia Italica, 82*, 1–89.

Premoli Silva, I., and Sliter, W. V., 1999. Cretaceous paleoceanography: evidence from planktonic foraminiferal evolution. In: Barrera, E., Johnson, C.C. (Eds.), The Evolution of Cretaceous Ocean-Climatic System. *Special Papers-Geological Society of America, 301–328.*

Premoli Silva, I., Erba, E., Salvini, G., Locatelli, C., Verga, D., 1999. Biotic changes in Cretaceous oceanic anoxic events of the Tethys. *The Journal of Foraminiferal Research, 29*, 352–370.

Premoli Silva, I., Verga, D., 2004. Practical Manual of Cretaceous Planktonic Foraminifera. In: Verga, D., Rettori, R. (Eds.), Internationl School on Planktonic Foraminifera, 3 Course, Cretaceous. Universities of Perugia and Milan, Tipografia Pontefelcino, Perugia, Italy.

Quijano, M. L., Castro, J. M., Pancost, R. D., de Gea, G. A., Najarro, M., Aguado, R., Rosales, I., Martín-Chivelet, J., 2012. Organic geochemistry, stable isotopes, and facies analysis of the Early Aptian OAE—New records from Spain (Western Tethys). *Palaeogeography, Palaeoclimatology, Palaeoecology, 365*, 276–293.

Radtke, U., and Schellmann, G., 2005. Timing and magnitude of sea level change during MIS 5 derived from Barbados coral reef terraces: a critical literature review and new data. *Journal of Coastal Research*, 52–62.

Ramos, V. A., 2009. Anatomy and global context of the Andes: Main geologic features and the Andean orogenic cycle. In: Kay, S.M., Ramos, V.A., Dickinson, W.R. (eds.). Backbone of the Americas: Shallow subduction, plateau uplift, and ridge and terrane collision. *Geological Society of America Memoir, 204*, 31–65.

Rangel, A., and Hernández, R., 2007. Thermal maturity history and implications for hydrocarbon exploration in the Catatumbo Basin, Colombia. *CT&F-Ciencia, Tecnología y Futuro, 3*, 7–24.

Rangel, A., Parra, P., Niño, C., 2000. The La Luna formation: chemostratigraphy and organic facies in the Middle Magdalena Basin. *Organic Geochemistry, 31*, 1267–1284.

Rangel, A., Katz, B., Ramirez, V., dos Santos Neto, E. V., 2003. Alternative interpretations as to the origin of the hydrocarbons of the Guajira Basin, Colombia. *Marine and Petroleum Geology, 20*, 129–139.

Reijmer, J. J., 2021. Marine carbonate factories: Review and update. *Sedimentology, 68*, 1729–1796.

Reijmer, J. J., Betzler, C., Kroon, D., Tiedemann, R., Eberli, G. P., 2002. Bahamian carbonate platform development in response to sea-level changes and the closure of the Isthmus of Panama. *International journal of earth sciences, 91*, 482–489.

Renz, O., 1958. Estratigrafía del Cretáceo en Venezuela Occidental. *Boletín del Ministerio de Minas e Hidrocarburos, 5*(10), pp. 3–48.

Renz, O., 1982. The Cretaceous Ammonites of Venezuela. Birkhäuser Verlag, Basel. Editions Maraven, Caracas. 132 pp.

Reverón, J., and Roomer, J., 2014. Probabilistic facies discrimination from simultaneous seismic-inversion results in clastic reservoirs in southwestern Venezuela. *The Leading Edge, 33*, 784–790.

Révillon, S., Hallot, E., Arndt, N. T., Chauvel, C., Duncan, R. A., 2000. A complex history for the Caribbean Plateau: petrology, geochemistry, and geochronology of the Beata Ridge, South Hispaniola, *The Journal of Geology, 108*(6), 641–661.

Rey, O., 2004. Cicloestratigrafía del EAO 3: Controles climáticos y cíclicos sobre la sedimentación del Santoniense, quebrada San Miguel, estado Mérida. Doctoral Dissertation. Central University of Venezuela, 190 pp.

Rey, O., Simo, J., Lorente, M. A., 2004. A record of long- and short-term environmental and climatic change during OAE3: La Luna Formation, Late Cretaceous (Santonian-early Campanian), Venezuela. *Sedimentary Geology, 170*, 85–105.

Richards, H. G., 1968. Cretaceous section in Barco area of northeastern Colombia. *American Association of Petroleum Geologists Bulletin, 52*, 2324–2336.

Riel, N., Duarte, J. C., Almeida, J., Kaus, B. J., Rosas, F., Rojas-Agramonte, Y., Popov, A., 2023. Subduction initiation triggered the Caribbean large igneous province. *Nature Communications*, *14*, 786.

Rincón, D. A., Arenas, J. E., Cuartas, C. H., Cárdenas, A. L., Molinares, C. E., Caicedo, C., Jaramillo, C., 2007. Eocene-Pliocene plantonick foraminifera biostratigraphy from the continental margino of the southwest Caribbean. *Stratigraphy*, *4*, 261–311.

Rivera, H. A., Le Roux, J. P., Sánchez, L. K., Mariño-Martínez, J. E., Salazar, C., Barragán, J. C., 2018. Palaeoredox conditions and sequence stratigraphy of the Cretaceous storm-dominated, mixed siliciclastic-carbonate ramp in the Eastern Cordillera Basin (Colombia): Evidence from sedimentary geochemical proxies and facies analysis. *Sedimentary Geology*, *372*, 1–24.

Rivolta, A., Rodríguez, C., Rodríguez, J. G., Molins, L., 2015. Criptofauna asociada a colonias muertas de *Orbicella annularis* en el Parque Nacional San Esteban, Venezuela. *Boletín del Instituto Oceanográfico de Venezuela*, *54*, 191–200.

Rodríguez, S., 1981. The use of soil geochemistry in outlining phosphate deposits, northern and southern Venezuela. *Journal of Geochemical Exploration*, *15*, 481–488.

Rodríguez-Cuicas, M. E., Montero-Serrano, J. C., Garbán, G., 2019. Paleoenvironmental changes during the late Albian oceanic anoxic event 1d: An example from the Capacho Formation, southwestern Venezuela. *Palaeogeography, Paleoclimatology, Paleoecology*, *521*, 10–29.

Rodríguez-Cuicas, M. E., Montero-Serrano, J. C., Garbán, G., 2020. Geochemical and mineralogical records of late Albian oceanic anoxic event 1d (OAE-1d) in the La Grita Member (southwestern Venezuela): Implications for weathering and provenance. *Journal of South American Earth Sciences*, *97*, 102408.

Rodríguez-Quintal, J. G., Herrera-Reveles, A. N. A., Carmen Rodríguez, M. C., 2011. Estructura de la comunidad íctica arrecifal en el Parque Nacional San Esteban, Venezuela. *Boletín del Instituto Oceanográfico de Venezuela*, *50*, 31–40.

Rohling, E. J., Grant, K., Hemleben, C. H., Siddall, M., Hoogakker, B. A. A., Bolshaw, M., Kucera, M., 2008. High rates of sea-level rise during the last interglacial period. *Nature Geoscience*, *1*, 38–42.

Rojas, E., 2012. Petroleum System of the Barinas Basin: A 2D/3D Basin Modeling Approach. Conference paper. XIII Latin American Congress on Organic Geochemistry. Santa Marta, Colombia. Novienbre 2012.

Rosenfeld, J. H., 2003. Economic potential of the Yucatan Block of Mexico, Guatemala, and Belize, in C. Bartolini, R. T. Buffler, and J. Blickwede, eds., The Circum-Gulf of Mexico and the Caribbean: Hydrocarbon habitats, basin formation, and plate tectonics. *American Association of Petroleum Geologists, Memoir 79*, 340–348.

Ross, M., and Scotese, C., 1988. A Hierarchical Tectonic Model of the Gulf of Mexico and Caribbean Region. *Tectonophysics, 155*, 139–168.

Rovere, A., Raymo, M. E., Vacchi, M., Lorscheid, T., Stocchi, P., Gomez-Pujol, L., Harris, D. L., Casella, E., O'Leary, M. J., Hearty, P. J., 2016a. The analysis of last interglacial (MIS 5e) relative sea-level indicators: reconstructing sea-level in a warmer world. *Earth-Science Review*, *159*, 404e427.

Rovere, A., Stocchi, P., Vacchi, M., 2016b. Eustatic and Relative Sea Level Changes. *Current Climate Change Reports*, *2*, 221–231.

Rovere, A., Ryan, D. D., Vacchi, M., Dutton, A., Simms, A. R., Murray-Wallace, C. V., 2023. The World Atlas of Last Interglacial Shorelines (version 1.0). *Earth System Science Data*, *15*, 1–23.

Ruiz, M., Lorente, M. A., Durán, I., Fasola, A., 2000. Late Cretaceous Upwelling in the Southwest of the Thethys Sea, a Case History from the Barinas Basin, Venezuela. In: *SEPM Research Conference. Paleogeography and Hydrocarbon Potential of La Luna Formation and Related Cretaceous Anoxic Systems*. 6 pp.

Ruiz-Ortiz, P. A., Castro, J. M., de Gea, G. A., Jarvis, I., Molina, J. M., Nieto, L. M., Pancost, R. D., Quijano, M. L., Reolid, M., Skelton, P. W., Weissert, H. J., 2016. New drilling of the early Aptian OAE1a: the Cau core (Prebetic Zone, south-eastern Spain). *Scientific drilling*, *21*, 41–46.

Rutsch, R. F., and Salvador, A., 1954. Mollusks from the Cogollo and La Luna formations (Cretaceous) of the Chejendé area, western Venezuela. *Journal of Paleontology*, 417–426.

Sabatino, N., Coccioni, R., Manta, D. S., Baudin, F., Vallefuoco, M., Traina, A., Sprovieri, M., 2015. High-resolution chemostratigraphy of the late Aptian–early Albian oceanic anoxic event (OAE 1b) from the Poggio le Guaine section (Umbria–Marche Basin, central Italy). *Palaeogeography, Palaeoclimatology, Palaeoecology*, *426*, 319–333.

Sageman, B. B., Meyers, S. R., Arthur, M. A., 2006. Orbital time scale and new C-isotope record for Cenomanian-Turonian boundary stratotype. *Geology*, *34*, 125–128.

Sánchez-Hernández, Y., and Maurrasse, F. J. -M. R., 2016. The influence of regional factors in the expression of oceanic anoxic event 1a (OAE 1a) in the semi-restricted Organyà Basin, south-central Pyrenees, Spain. *Paleogeography, Palaeoclimatology, Palaeoecology, 441*, 582–598.

Sánchez, M., Vásquez, A., Van Alstine, D., Butterworth, J., García, J., Carmona, R., Poquioma, W., Ramones, M., 1999. Application of geomechanics in the development of the naturally fractured carbonates of the Mara Oeste field, Venezuela. 1999 SPE Latin American and Caribbean Petroleum Engineering Conference, Caracas, Venezuela, SPE 54008.

Sandoval, M. I., Baumgartner, P. O., Urbani, F., & Reategui, W. (2023). Cretaceous radiolarians from the Siquisique Ophiolite (northern Venezuela). *Journal of South American Earth Sciences*, *127*, 104364.

Santamaría, F., and Schubert, C., 1974. Geochemistry and geochronology of the southern Caribbean–northern Venezuela plate boundary. *Geological Society of America Bulletin*, *85*, 1085–1098.

Sarfi, M., Ghasemi-Nejad, E., Mahanipour, A., Yazdi-Moghadam, M., Sharifi, M., 2014. Integrated biostratigraphy and geochemistry of the lower Cretaceous Radiolarian Flood Zone of the base of the Garau Formation, northwest of Zagros Mountains, Iran, *Arabian Journal of Geosciences*, *8*, 7245–7255.

Sarmiento-Rojas, L. F., 2019. Cretaceous stratigraphy and paleo-facies maps of northwestern South America. In F. Cediel & R. P. Shaw (Eds.), Geology and tectonics of Northwestern South America. *Frontiers in Earth Sciences* (pp. 673–747). Cham: Springer.

Schafhauser, A., Stinnesbeck, W., Holland, B., Adatte, T., Remane, J., 2003. Lower Cretaceous pelagic limestones in southern Belize: Proto-Caribbean deposits on the southeastern Maya Block, In: C. Bartolini, R. T. Buffler, and J. Blickwede, eds., The Circum-Gulf of Mexico and the Caribbean: Hydrocarbon habitats, basin formation, and plate tectonics. *American Association of Petroleum Geologists, Memoir 79*, 624–637.

Scherer, W., 2024. Isla de Aves. Academia de Ciencias Física, Matemática y Naturales. Caracas, Venezuela. 99 pp.

Schlager, W., 2003. Benthic carbonate factories of the Phanerozoic. *International Journal of Earth Sciences*, *92*, 445–464.

Schlager, W., 2005. Carbonate Sedimentology and Sequence Stratigraphy. SEPM Concepts in Sedimentology and Paleontology #8. Society of Economic Paleontologists and Mineralogists, 200 pp.

Schneider, B., and Schmittner, A., 2006. Simulating the impact of the Panamanian seaway closure on ocean circulation, marine productivity and nutrient cycling. *Earth and Planetary Science Letters*, *246*, 367–380.

Schneider, F., Pagel, M., Hernández, E., 2004. Basin-modeling in complex areas: example from the Eastern Venezuelan foothills. In: Swennen, R., Roure, F. & Granath, J. (Eds.): Deformation, fluid flow and reservoir appraisal in foreland fold-and-thrust belts. *American Association of Petroleum Geologists, Hedberg Series*, *1*, 357–369.

Schubert, C., 1986. Stratigraphy of the Jurassic La Quinta Formation, Mérida Andes, Venezuela: type section. *Zeitschrift der Deutschen Geologischen Gesellschaft*, 391–411.

Schubert, C., and Moticska, P., 1973, Reconocimiento geológico de las islas venezolanas en el Mar Caribe, entre Los Roques y Los Testigos (Dependencias Federales). *Acta Científica Venezolana, 23*, 210–223.

Schubert, C., and Valastro Jr, S., 1976. Quaternary geology of La Orchila Island, central Venezuelan offshore, Caribbean Sea. *Geological Society of America Bulletin, 87*, 1131–1142.

Schubert, C., and Szabo, B. J., 1978. Uranium-series ages of Pleistocene marine deposits on the islands of Curacao and La Blanquilla, Caribbean Sea. *Geologie en Mijnbouw, 57*, 325–332.

Schubert, C., and Laredo, M., 1984. Geology of Aves Island (Venezuela) and subsidence of Aves Ridge, Caribbean Sea. Marine Geology, 59, 305–318.

Scotese, C. R., 2014. *Atlas of Neogene Paleogeographic Maps (Mollweide Projection), Maps 1-7, PALEOMAP Atlas for ArcGIS, Volume 1, the Cenozoic*. PALEOMAP Project, Evanston, IL.

Scotese, C. R., Song, H., Mills, B. J. W., van der Meer, D. G., 2021. Phanerozoic paleotemperatures: the earth's changing climate during the last 540 million years. *Earth-Science Reviews, 215*, 103503.

Shagam, R., Kohn, B. P., Banks, P. O., Dasch, L. E., Vargas, R., Rodriguez, G. I., Pimentel, N., 1984. Tectonic implications of Cretaceous-Pliocene fission-track ages from rocks of the circum-Maracaibo Basin region of western Venezuela and eastern Colombia. *Geological Society of America Memoirs, 162*, 385–412.

Shields, G., Stille, P., Brasier, M. D., 2000. Isotopic records across two phosphorite giant episodes compared: the Precambrian–Cambrian and the Late Cretaceous–Recent. In: Glenn, C.R., Prévôt-Lucas, L., Lucas, J. (Eds.), Marine Authigenesis: from Global to Microbial. *SEPM Special Publications, 66*, 103–115.

Siddall, M., Rohling, E. J., Thompson, W. G., Waelbroeck, C., 2008. Marine isotope stage 3 sea level fluctuations: Data synthesis and new outlook. *Reviews of Geophysics, 46*, 29 pp.

Silva-Tamayo, J. C., Ramirez, C., Lara, M., Sial, A. N., Trujillo, D., Salazar, E., 2016. Sedimentology and chemostratigraphy of a Valanginian carbonate succession from the Baja Guajira Basin, northern Colombia. *Brazilian Journal of Geology, 46*, 163–180.

Sinton, C. W., and Duncan, R. A., 1997. Potential links between ocean plateau volcanism and global ocean anoxia at the Cenomanian-Turonian boundary. *Economic Geology, 92*, 836–842.

Sinton, C. W., Duncan, R., Storey, M., Lewis, J., Estrada, J. J., 1998. An oceanic flood basalt province within the Caribbean plate. *Earth and Planetary Science Letters, 155*, 221–235.

Sisson, V. B., Lallemant, H. G. A., Ostos, M., Blythe, A. E., Snee, L. W., Copeland, P., Wright, J. E., Donelick, R. A., Guth, L. R., 2005. Overview of radiometric ages in three allochthonous belts of northern Venezuela: Old ones, new ones, and their impact on regional geology. *Geological Society of America Special Paper 394*, 91–117.

Skelton, P. W., ed., 2003. The Cretaceous world: Cambridge, Cambridge University Press, 360 pp.

Skelton, P. W., and Gili, E., 2012. Rudists and carbonate platforms in the Aptian: a case study on biotic interactions with ocean chemistry and climate. *Sedimentology, 59*, 81–117.

Sliter, W. V., 1989. Biostratigraphic zonation for Cretaceous planktonic foraminifers examined in thin section. *Journal of Foraminiferal Research, 19*(1), 1–19.

Smith, W., and Sandwell, D., 1997. Measured and Estimated Seafloor Topography, World Data Center for Geophysics & Marine Geology, Boulder Research Publication RP-1, poster, 34′ × 53′.

Smith, C. J., Collins, L. S., Jaramillo, C., Quiroz, L. I., 2010. Marine paleoenvironments of Miocene–Pliocene formations of north-central Falcón State, Venezuela. *The Journal of Foraminiferal Research, 40*, 266–282.

Snow, L. J., Duncan, R. A., Bralower, T. J., 2005. Trace element abundances in the Rock Canyon Anticline, Pueblo, Colorado, marine sedimentary section and their relationship to Caribbean plateau construction and oxygen anoxic event 2, Paleoceanography, 20, PA3005.

Socorro, J., and Maurrasse, F. J. -M. R., 2020. Regional palaeoenvironmental influence on organic matter sequestration and characteristics of carbon isotope segment C5 in a hemipelagic sequence, Organyà Basin, northeast Spain. *The Depositional Record, 6*, 552–580.

Socorro, J., and Maurrasse, F. J. M. R., 2022. Demise of Organic Matter–Rich Facies and Changing Paleoenvironmental Conditions Associated with the End of Carbon Isotope Segment C5 of Oceanic Anoxic Event 1a in the North and Northeastern Iberian Peninsula. *The Journal of Geology, 130*(2), 133–170.

Soto, G., 2025, in press. Utilidad de los Macroforaminíferos en la Industria Petrolera. Caso de Estudio: Cuenca de Falcón. *Visión Tecnológica*. 16 pp.

Soto, G., Ortega, S., Marín, Z., Farías, A., Medina, C., 2025, in press. Estudio Bioestratigráfico de la Unidades Oligo-Miocénicas de la Región Falconiana. *Memorias del II Congreso Venezolano de Geociencias*. Caracas, 20-24 de Marzo 2023. 31 pp.

Spikings, R., Cochrane, R., Villagomez, D., van der Lelij, R., Vallejo, C., Winkler, W., Beate, B., 2014. The geological history of northwestern South America: from Pangaea to the early collision of the Caribbean Large Igneous Province (290–75 Ma). *Gondwana Research, 27*, 95–139.

Spratt, R. M., and Lisiecki, L. E., 2016. A Late Pleistocene sea level stack. *Climate of the Past, 12*, 1079–1092.

Stanley, S. M., and Hardie, L. A., 1998. Secular oscillations in the carbonate mineralogy of reef-building and sediment-producing organisms driven by tectonically forced shifts in seawater chemistry. *Palaeogeography, Palaeoclimatology, Palaeoecology, 144*, 3–19.

Stienstra, P., 1985. Geology of a small rock-phosphate deposit, Ceru Colorado, Aruba, Netherlands Antilles. *Journal of Sedimentary Research, 55*, 96–104.

Summa, L. L., Goodman, E. D., Richardson, M., Norton, I. O., Green, A. R., 2003. Hydrocarbon systems of Northeastern Venezuela: plate through molecular scale-analysis of the genesis and evolution of the Eastern Venezuela Basin. *Marine and Petroleum Geology, 20*(3–4), 323–349.

Sutton, F. A., 1946. Geology of the Maracaibo Basin, Venezuela. *A Bulerican Association of Petroleum Geologists Bulletin, 30*, 1621–1741.

Takashima, R, Nishi, H., Huber, B. T., Leckie, R. M., 2006. Greenhouse World and the Mesozoic Ocean. *Oceanography, 19*, 64–75.

Talbi, R., Lakhdar, R., Smati, A., Spiller, R., Levey, R., 2019. Aptian–Albian shale oil unconventional system as registration of Cretaceous oceanic anoxic sub-events in the southern Tethys (Bir M'Cherga basin, Tunisia). *Journal of Petroleum Exploration and Production Technology, 9*, 1007–1022.

Talukdar, S., and Marcano, F., 1994. Petroleum system of the Maracaibo Basin, Venezuela. *American Association of Petroleum Geologists, Memoir 60*, 463–481.

Tawil-Morsink, K., Austermann, J., Dyer, B., Dumitru, O. A., Precht, W. F., Cashman, M., Goldstein, S. L., Raymo, M. E., 2022. Probabilistic investigation of global mean sea level during MIS 5a based on observations from Cave Hill, Barbados, *Quaternary Science Reviews, 295*, 107783.

Taylor, F. W., and Mann, P., 1991. Late Quaternary folding of coral reef terraces, Barbados. *Geology, 19*, 103–106.

Taylor, G. T., Iabichella-Armas, M., Varela, R., Muller-Karger, F., Lin, X., Scranton, M.I., 2006. Microbial ecology of the Cariaco Basin's oxic/anoxic interface: the U.S.-Venezuelan CARIACO times series program. In: Neretin, L.N. (Ed.), Past and Present Water Column Anoxia. NATO Science Series. Springer, The Netherlands, pp. 473–499.

Tazzo-Rangel, M. D., Weber, B., González-Guzmán, R., Valencia, V. A., Frei, D., Schaaf, P., Solari, L. A., 2019. Multiple metamorphic events in the Palaeozoic Mérida Andes basement, Venezuela: insights from U–Pb geochronology and Hf–Nd isotope systematics. *International Geology Review, 61*, 1557–1593.

Tejada, M. L. G., Suzuki, K., Kuroda, J., Coccioni, R., Mahoney, J. J., Ohkouchi, N., Sakamoto, T., Tatsumi, Y., 2009. Ontong Java Plateau eruption as a trigger for the early Aptian oceanic anoxic event. *Geology*, *37*(9), 855–858.

Ten Have, T., Heijnen, W., Nickel, E., 1982. Alterations in guano phosphates and Mio-Pliocene carbonates of table mountain Santa Barbara, Curacao. *Sedimentary Geology*, *31*, 141–165.

Thompson, P. M. E., Kempton, P. D., White, R. V., Saunders, A. D., Kerr, A. C., Tarney, J., Pringle, M. S., 2004. Elemental, Hf-Nd isotopic and geochronological constraints on an Island arc sequence associated with the Cretaceous Caribbean plateau: Bonaire, Dutch Antilles. *Lithos*, *74*, 91–116.

Tocco, R., Parnaud, F., Gallango, O., Alberdi, M., Passalacqua, H., 1997. Geochemical modelling of the principal source rocks of the Barinas and Maracaibo basins, western Venezuela. *Boletín de la Sociedad Venezolana de Geólogos*, *22*, 17–28.

Trabucho-Alexandre, J., van Gilst, R, Rodríguez-López, J. P., de Boer, P., 2011. The sedimentary expression of the oceanic anoxic event 1b in the North Atlantic. *Sedimentology, 58*, 1217–1246.

Trabucho-Alexandre, J., Hay, W. W., de Boer, P. L., 2012. Phanerozoic environment of black shale deposition and the Wilson Cycle. *Solid Earth*, *3*, 29–42.

Tribovillard, N. P., Stephan, J. F., Manivit, H., Reyre, Y., Cotillon, P., Jautée, E., 1991. Cretaceous black shales of Venezuelan Andes: preliminary results on stratigraphy and paleoenvironmental interpretations. *Palaeogeography*, *Palaeoclimatology*, *Palaeoecology*, *81*, 313–321.

Truskowski, I., 2006. Calibración bioestratigráfica y correlación con datos geoquímicos de la Formación Querecual en su localidad tipo. Specialization Dissertation, Central University of Venezuela. 106 pp.

Tsikos, H., Jenkyns, H. C., Walswort-Bell, B., Petrizzo, M. R., Forster, A., Kolonic, S., Erba, E., Premoli Silva, I., Baas, M., Wagner, T., Sinninghe Damsté, J. S., 2004. Carbon-isotope stratigraphy recorded by the Cenomanian–Turonian oceanic anoxic event: correlation and implications based on three key-localities. *Geological Society of London*, 161, 711–720.

Tucker, M. E., and Wright, V. P., 1990. Carbonate Sedimentology. Blackwell Publishing. 482 pp.

Turgeon, S. C., and Creaser, R. A., 2008. Cretaceous oceanic anoxic event 2 triggered by a massive magmatic episode. *Nature, 454*, 323–327.

Ughi, A., 2021. Gravity and Heat Flow Analysis of the Oceanic Subduction Zone in the Eastern Caribbean. In C. Bartolini (Ed.), South America-Caribbean-Central Atlantic Plate Boundary: Tectonic Evolution, Basin Architecture, and Petroleum Systems. *American Association of Petroleum Geologists Memoir 123*, 387–409.

Urbani, F., 2018. Una Revisión de los Terrenos Geológicos del Sistema Montañoso del Caribe, Norte de Venezuela. Boletín de Geología Volumen XXIII, Número 36, 118–216.

Urbani, F., Mendi, D., Gómez, A., Valletta, G., Martínez, G., Reátegui, W., 2016. New studies of Matatere Formation, Lara and Yaracuy states, Venezuela - Nuevos estudios sobre la Formación Matatere, estados Lara y Yaracuy. Accessed at https://www.researchgate.net/publication/301748697_New_studies_of_Matatere_Formation_Lara_and_Yaracuy_states_Venezuela_-_Nuevos_estudios_sobre_la_Formacion_Matatere_estados_Lara_y_Yaracuy

Vahrenkamp, V. C., Franssen, R. C. W. M., Grötsch, J., Munoz, P. J., 1993. Maracaibo Platform (Aptian-Albian), northwestern Venezuel. I: Simo, J. A. T., Scott, R.W., Masse, J. P., editors, Cretaceous Carbonate Platforms. *American Association of Petroleum Geologist, Memoir 25*, 25–33.

Valencia, F., 2016. Diagénesis de los carbonatos del campo PERLA, Golfo de Venezuela: Implicaciones en las propiedades petrofísicas. Magister Dissertation. Simón Bolívar University. Caracas, Venezuela, 167 pp.

Valencia, F. L. and Laya, J. C., 2020. Deep-burial dissolution in an Oligocene-Miocene giant carbonate reservoir (Perla Limestone), Gulf of Venezuela Basin: Implications on microporosity development. *Marine and Petroleum Geology*, *113*, 104144.

Vallejo, C., Winkler, W., Spikings, R.A., Luziuex, L., Heller, F., Bussy, F., 2009. Mode and timing terrane accretion in the forearc of the Andes in Ecuador. In: Kay, S.M., Ramos, V.A., Dickinson, W.R. (Eds.), Backbone of the Americas: shallow subduction, plateau uplift, and ridge and terrane collision. Geological Society of America Memoir 204, 197–216.

van Bentum, Hetzel, Brumsack, H. J., Forster, A., Reichart, G. J., Sinninghe Damsté, J. S., 2009. Reconstruction of water column anoxia in the equatorial Atlantic during the Cenomanian-Turonian oceanic anoxic event using biomarker and trace metal proxies. *Paleogeography, Palaeoclimatology, Palaeoecology, 280*, 489–498.

van der Lelij, R., 2013. Reconstructing north-western Gondwana with implications for the evolution of the Iapetus and Rheic oceans: a geochronological, thermochronological and geochemistry study. Doctoral Dissertation, University of Geneva. 221 pp.

van der Lelij, R., Spikings, R. A., Kerr, A. C., Kounov, A., Cosca, M., Chew, D., Villagomez, D., 2010. Thermochronology and tectonics of the Leeward Antilles: Evolution of the southern Caribbean Plate boundary zone. *Tectonics*, *29*, TC6003.

van der Lelij, R., Spikings, R., Ulianov, A., Chiaradia, M., Mora, A., 2016a. Palaeozoic to Early Jurassic history of the northwestern corner of Gondwana, and implications for the evolution of the Iapetus, Rheic and Pacific Oceans. *Gondwana Research*, *31*, 271–294.

van der Lelij, R., Spikings, R., Mora, A., 2016b Thermochronology and tectonics of the Mérida Andes and the Santander Massif, NW South America, Lithos, 248, 220–239.

Vargas-Escudero, M. A., Ríos-Reyes, C. A., García-González, M., Ortiz-Orduz, A., 2021. Diagenesis and thermal maturity of the Cogollo Group sedimentary rocks in the ANH-CR-Montecarlo-1X well, Cesar-Ranchería Basin, Colombia. *Andean Geology*, *48*, 472–495.

Vera, V., and Buitrago, J., 2012. Actividad reproductiva de *Chelonia mydas* (Testudines: Cheloniidae) en Isla de Aves, Venezuela (2001-2008). *Revista de Biología Tropical*, *60*, 745–758.

Verga, D., and Premoli Silva, I., 2002. Early Cretaceous planktonic foraminifera from the Tethys: the genus Leupoldina. *Cretaceous Research*, *23*, 189–212.

Vézina, J., Jones, B., Ford, D., 1999. Sea-level highstands over the last 500,000 years; evidence from the Ironshore Formation on Grand Cayman, British West Indies. *Journal of Sedimentary Research*, *69*, 317–327.

Villagómez, D., Spikings, R., Magna, T., Kammer, A., Winkler, W., Beltrán, A., 2011a. Geochronology, geochemistry and tectonic evolution of the Western and Central cordilleras of Colombia. *Lithos*, 125, 875–896.

Villagómez, D., Spikings, R., Mora, A., Guzmán, G., Ojeda, G., Cortés, E., van der Lelij, R., 2011b. Vertical tectonics at a continental crust-oceanic plateau plate boundary zone: Fission track thermochronology of the Sierra Nevada de Santa Marta, Colombia. *Tectonics*, *30*. TC4004.

Villalobos, J. R., Pimentel, A., Sánchez, M., 2024. Establecimiento y Actualización de Vértices Geodésicos: Isla de Aves de la República Bolivariana de Venezuela. Instituto Geográfico de Venezuela Simón Bolívar. 5 pp.

Villamil, T., and Arango, C., 1998. Integrated stratigraphy of latest Cenomanian and early Turonian facies of Colombia. In: Pindell, J.L., Drake, C. (Eds.), Paleogeographic Evolution and Non-Glacial Eustacy, Northern South America. *Society of Economic*

Paleontologists and Mineralogists Special Publication, 58, 129–159.

Villamil, T., and Pindell, J. L., 1998. Mesozoic paleogeographic evolution of northern South America: Foundations for sequence stratigraphic studies in passive margin strata deposited during nonglacial times. In: PINDELL, J. L. & DRAKE, C. (eds.) Paleogeographic Evolution and Non-Glacial Eustacy: Northern South America. *SEPM Special Publications*, 58, 283–318.

Villamil, T., Arango, C., Hay, W. W., 1999. Plate tectonic paleoceanographic hypothesis for the Cretaceous source rocks and cherts of northern South America. In: Barrera, E., and Johnson, C.C. (Eds.). Evolution of the Cretaceous Ocean-Climate System. *Geological Society of America, Special Paper 332*, 191–202.

Villamizar, E., 2002. Universo submarino: los arrecifes de coral. In: Guía del Parque Nacional Archipiélago Los Roques. Ed. Javier Zamarro. Agencia Española de Cooperación Internacional, Caracas, Venezuela, 109–124.

Villamizar, A., and Zamarro, J., 2002. La Vegetación. In: Guía del Parque Nacional Archipiélago Los Roques. Ed. Javier Zamarro. Agencia Española de Cooperación Internacional, Caracas, Venezuela, 99–108.

Villamizar, E., and Cervigón, F., 2017. Variability and sustainability of the Southern Subarea of the Caribbean Sea large marine ecosystem. *Environmental Development, 22*, 30–41.

Villamizar, E., Posada, J. M., Gómez, S., 2003. Venezuela. Rapid Assessment of coral reefs in the Archipelago de Los Roques National Park, Venezuela (part 1: stony corals and algae). *Atoll Research Bulletin, 496*, 512–519.

Villamizar, E., Camisotti, H., Rodríguez, B., Pérez, J., Romero, M., 2008. Impacts of the 2005 Caribbean bleaching event at Archipiélago de Los Roques National Park, Venezuela. *Revista de Biología Tropical, 56*, 255–270.

Villamizar, E., Yranzo, A., González, M., Herrera, A. T., Pérez, J., Camisotti, H., 2014. Diversidad y condición de salud de corales pétreos en algunos arrecifes del Parque Nacional Archipiélago Los Roques, Venezuela, *Acta Biológica Venezuelica, 34*(2), 257–279.

Villamizar, E., Yranzo, A., Romero, M., Boadas, H., 2019. Diversity and Structure of Sponge Communities of the Isla de Aves Wildlife Refuge, Caribbean Sea, Venezuela. *Caribbean Journal of Science, 49*, 109–124.

Viscarret, P., Wright, J., Urbani, F., 2009. New U-Pb zircon ages of El Baúl Massif, Cojedes State, Venezuela. *Revista Técnica de la Facultad de Ingeniería. Universidad del Zulia, 32*, 210–221.

Viscarret, P., Urbani, F., Wright, J., 2012. Una nueva geocronología del Macizo El Baúl, estado Cojedes, Venezuela. *GEOS Revista Venezolana de Ciencias de la Tierra, 42*. 1–14.

Voigt, S., Erbacher, J., Mutterlose, J., Weiss, J., Westerhold, T., Wiese, F., Wilmsen, M., Wonik, T., 2008. The Cenomanian—Turonian of the Wunstorf section—(North Germany): global stratigraphic reference section and new orbital time scale for Oceanic Anoxic Event 2. *Newsletters in Stratigraphy, 43*, 65–89.

Wagner, T., Sinninghe Damsté, J. S., Hofmann, P., Beckmann, B., 2004. Euxinia and primary production in Late Cretaceous eastern equatorial Atlantic surface waters fostered orbitally driven formation of marine black shales. *Paleoceanography, 19*, PA3009.

Wagreich, M., 2012. "OAE 3" - regional Atlantic organic carbon burial during the Coniacian Santonian. *Climate of the Past, 8*, 1447.

Warne, A. G., Guevara, E. H., Aslan, A., 2002a. Late quaternary evolution of the Orinoco Delta, Venezuela. *Journal of Coastal Research*, 225-253.

Warne, A. G., Meade, R. H., White, W. A., Guevara, E. H., Gibeaut, J., Smyth, R. C., Aslan, A., Tremblay, T., 2002b. Regional controls on geomorphology, hydrology, and ecosystem integrity in the Orinoco Delta, Venezuela. *Geomorphology, 44*, 273–307.

Warren, J. K., 2000. Dolomite: Occurrence, evolution and economically important associations. *Earth-Science Review, 52*, 1–81.

Weber, B., Mota, A., Helenes, J., Ramírez, R., Valencia, Y., 2016. Age and provenance of Late Miocene-Early Pliocene sedimentary rocks from the Patao high hydrocarbon reservoir offshore NE Venezuela–U-Pb detrital zircon age, Sm-Nd isotope, and biostratigraphic data. *Journal of Natural Gas Science and Engineering, 31*, 459–473.

Weil, E. 2003. The corals and coral reefs of Venezuela. In: Cortés J. (ed.), *Latin American Coral Reefs*. Elsevier, 303–330.

Weiss, M. P., Goddard, D. A., Picard, X., 1978. Marine Geology of Reefs and Inner Shelf, Chichiriviche, Estado Falcón, Venezuela. *Marine Geology, 28*, 211–244.

Westbrook, G. K., 1982. The Barbados Ridge Complex: tectonics of a mature forearc system. *Geological Society, London, Special Publications, 10*, 275–290.

Whattam, S. A., and Stern, R. J., 2015. Late Cretaceous plume-induced subduction initiation along the southern margin of the Caribbean and NW South America: The first documented example with implications for the onset of plate tectonics. *Gondwana Research, 27*, 38–63.

Wheeler, C. B., 1963. Oligocene and lower Miocene stratigraphy of western and northeastern Falcón Basin, Venezuela. *American Association of Petroleum Geologists Bulletin, 47*, 35–68.

White, R. V., Tarney, J., Kerr, A. C., Saunders, A. D., Kempton, P. D., Pringle, M. S., Klaver, G. T., 1999. Modification of an oceanic plateau, Aruba, Dutch Caribbean: implications for the generation of continental crust. *Lithos, 46*, 43–68.

Wilson, J. L., and Ward, W. C., 1993. Early Cretaceous carbonate platforms of Northeastern and East-Central Mexico. In: Simo, T., Scott, R.W., Masse, J.-P. (Eds.), Cretaceous Carbonate Platforms. *American Association of Petroleum Geologists, Memoir 56*, 35–49.

Woodroffe, C. D., and Webster, J. M., 2014. Coral reefs and sea-level change. *Marine Geology, 352*, 248–267.

Work, R. C., 1969. Systematics, ecology, and distribution of the mollusks of Los Roques, Venezuela. *Bulletin of Marine* Science, *19*. 614–711.

Wright, V. P., Burchette, T. P., 1996. Shallow-water carbonates environment. In: Reading, H.G. (Ed.), Sedimentary Environments: Processes, Facies, Stratigraphy, 325–394.

Wright, J. E., and Wyld, S. J., 2011. Late Cretaceous subduction initiation on the eastern margin of the Caribbean-Colombian Oceanic Plateau: One Great Arc of the Caribbean (?). *Geosphere, 7*, 468–493.

Xie, X., Mann, P., Escalona, A., 2010. Regional provenance study of Eocene clastic sedimentary rocks within the South America–Caribbean plate boundary zone using detrital zircon geochronology. *Earth and Planetary Science Letters, 291*, 159–171.

Yoris, F., and Ostos, M., 1997. Petroleum geology of Venezuela. In: Singer, J.M. (Ed.), Well Evaluation Conference. Schlumberger Oilfield Service, Caracas, Venezuela, 140 pp.

Yoris, F., 1989. Consideraciones sobre la Formación Querecual de Venezuela Oriental. *Geos*, N° 29, Caracas, pp. 126–151.

Yranzo, A., Villamizar, E., Romero, M., Boadas, H., 2014. Estructura de las comunidades de corales y octocorales de Isla de Aves, Venezuela, Caribe Nororiental. *Revista de Biología Tropical, 62*, 115–136.

Yranzo, A, Bustillos, F. A., Villamizar, E., Herrera-Reveles, A., Pérez, J., Boadas, H., Pereira, C., Rodríguez, J. G., Narciso, S., Cavada, F., 2020. Tesoros escondidos de Cuare y Morrocoy, Publicación divulgativa de los resultados del Proyecto *Orbicella* Morrocoy. Fundación Segre, EDGE of Existence Programme - Sociedad

Zoológica de Londres, ZSL., Caracas, Venezuela. Publisher: Freddy A. Bustillos R. ISBN: 978-980-18-1257-9, 32 pp.

Ysaccis, R., and Audemard, F. E., 2000. A Neogene Orogenic float in Northern South America: Eastern Venezuela Basin vs. Caribbean Plate. American Association of Petroleum Geologists Annual Meeting, New Orleans, Louisiana.

Yranzo-Duque, A., Herrera-Reveles, A. T., Villamizar, E., Cabada-Blanco, F., Pérez-Benítez, J., Boadas, H., Rodríguez-Quintal, J. G., Pereira, C., Narcizo, S., Bustillos, F. A., 2025. Revealing the status of Orbicella: Main reef-builder of Morrocoy National Park and Cuare Wildlife Refuge, Venezuela, Southern Caribbean. *PLoS ONE*, *20*(2): e0317728, 16 pp.

Zamarro, J., 2002. Prólogo del Editor. In: Guía del Parque Nacional Archipiélago Los Roques. Ed. Javier Zamarro. Agencia Española de Cooperación Internacional, Caracas, Venezuela, 9.

Zapata, E., Lorente, M. A., Rey, O., Padrón, V., 2000. Diagenetic Evaluation of La Luna Formation at Táchira and Mérida States, Western Venezuela. In: *SEPM Research Conference. Paleogeography and Hydrocarbon Potential of La Luna Formation and Related Cretaceous Anoxic Systems.*

Modern Reef Complexes of Venezuela

3

This chapter focuses on modern carbonate facies belonging to some of the main Venezuelan reef complexes in the Caribbean. The following set of photographs corresponds to the Los Roques Archipelago, Aves Island, Las Aves Archipelagos (Aves de Barlovento and Aves de Sotavento), Morrocoy National Park, La Blanquilla Island, and Henri Pittier National Park.

The images shown consist of aerial views (photographs taken from airplanes and drone images) and land views of some of the most significant environments in these complexes. These include barrier reefs, reef flats, fringing reefs, central lagoons, cays or supratidal islands, patch reefs, sandbanks, storm terraces, beachrocks, and mangrove forests. Close views of some of the typical carbonate biota in the Caribbean (e.g., echinoderms and chitons) are shown. Examples of peritidal environments, such as hypersaline lagoons, sabkha flats, and related microbialite structures (stromatolites and thrombolites), as well as some non-calcareous biota (mangroves and macroalgae) are also included. Underwater photographs include some of the most abundant and representative scleractinian corals and octocorals (among other organisms).

Photos included in this chapter were taken during numerous field campaigns carried out from the early 1980s to 2025. Deep marine environments (down to 300 m) were photographed using the submarine of the ship Calypso, which was commanded by the explorer and researcher Jacques Cousteau.

J. A. Méndez Dot, J. Méndez Baamonde, *Atlas of Tropical Carbonates*, https://doi.org/10.1007/978-3-031-25494-9_3

3.1 Modern Reef Complexes of Venezuela: Photographic Set and Descriptive Captions

Los Roques Archipelago

Aerial view of the eastern margin of the Los Roques Archipelago National Park. Major sedimentary environments are numbered from outer to inner settings. (**1**) The eastern outer barrier reef, which outlines the margins of the archipelago along the windward margin; (**2**) The reef flat, a shallow, moderate energy-level environment located in the back-reef; (**3**) The inner barrier reef and associated reef flat; (**4**) A series of aligned mangrove-dominated cays which have formed behind the inner reef flat; (**5**) A shallow subtidal area where clear waters allow visualization of the organic-lean, sandy sedimentation prevailing in that area; (**6**) A second generation of mangrove-dominated cays and associated peritidal environments; (**7**) Coral bioherms (dark colored) have grown in a shallow subtidal environment representing the beginning of the vast inner lagoon in that area.

Los Roques Archipelago

Aerial view of the southeastern margin of the Los Roques Archipelago. This is the transitional area between the eastern barrier reef (convex-shapep at the top) and the southern barrier reef (with mangrove forests in the back reef). At the surface, the barriers are separated by a large opening (or mouth) called "Boca Sebastopol" that allows the connection between the platform and the open ocean. Note the submerged and uninterrupted barrier reef, which extends down the fore-reef slope to depths of 40–50 m.

Los Roques Archipelago

Aerial view of the beginning of the double barrier reef–flat system at Boca de Sebastopol (also seen in the previous image). In this type of double-barrier system, the outer barrier is commonly younger, with a wider reef flat (see Hopley, 2011). The latter is clearly visualized in this image.

Drone image (October 2023) courtesy of Víctor J. Herrera Rivero

Los Roques Archipelago
Aerial view of the southern barrier reef at its easternmost margin and associated back-reef zones. In this area, the back-reef is characterized by dense mangrove communities that have grown along moderate- to low-energy environments, protected from the direct action of winds and waves. The shallow, low-energy shoreface zones contain clear waters. Note the submerged barrier reef and the marked slope inclination (~ 45°), which marks the outer, deeper limits of vigorous scleractinian coral growth.

Drone image (October 2023) courtesy of Víctor J. Herrera Rivero

Los Roques Archipelago
Aerial view of the southern margin of Los Roques, where the southern barrier reef extends for more than 35 km, making it the longest barrier reef in the southeast Caribbean. The mouth in the image is known as "Boca de Cote". This site one of the most outstanding environments in Los Roques, as the low energy of the setting allows divers visualize both the submerged, massive coral buildup that composes the barrier, and the drastic slope down to the lowermost limit of the photic zone.

Photo taken during early 1980s

Los Roques Archipelago
Aerial view of the Gran Roque Island from the leeward side (the Eastern Hill at the bottom). This island is situated in the northeast area of the archipelago and stands as the sole locality within this national park with a population settlement. Notably, Gran Roque is the only island in the entire reef complex where the crystalline rock basement, underlying the entire archipelago, crops out.

Los Roques Archipelago
Gran Roque Island seen from the windward-northern side. The image depicts the western hill, with an approximate elevation of 125 m above RSL. This hill is formed by a metagabbro. This Late Cretaceous (Coniacian) metamafic unit, which represent a fragment of either the COP or the proto-Caribbean seafloor, is intruded by dikes of pegmatite, aplite and quartz-diorite, which represent younger (Early Paleocene) arc magmatic activity.

Los Roques Archipelago

From the top of the eastern hill of Gran Roques Island, the view reveals two small hills (~ 20 m high) cropping out just north of Gran Roque. These hills consist mainly of metadiabase (amphibolite) and are covered by bird guano, giving the dark rocks a peculiar whitish appearance. Phosphatization of these rocks, due to the high phosphorus content of the guano, have formed a conspicuous phosphate green mineral called variscite [AlPO4.2(H2O)] (see Schubert and Moticska 1973; Urbani 2005, 2018).

Los Roques Archipelago

This photo was taken on a walk towards the area on the southwest of Gran Roque Island where lithified limestone terraces crop out slightly above RSL. The view shows the western hill on its southern flank. In this area, outcropping Cretaceous igneous and metaigneous rocks are almost entirely covered by Quaternary alluvium.

Los Roques Archipelago
The image corresponds to the southern flank of the western hill of Gran Roque Island. It shows a limestone terrace that is exposed in this area. Three beds are well-differentiated, exhibiting a backstepping stacking patter. The uppermost bed is located at ~1.8–2 m above RSL. This terrace has been estimated to be time-equivalent to the exposed limestone terraces formed during the LIG in the islands of La Blanquilla and La Orchila (Schubert and Valastro 1976; Méndez-Baamonde 1984, 1997).

Los Roques Archipelago
Another image of the Gran Roque limestone terrace. This photo depicts the uppermost bed (right above the hammer that serves as a scale) and the intermediate bed. In this image, the latter extends as a flat surface deepening seaward and displays irregularly-shaped erosive features resulting from carbonate dissolution. In this and previous image, the displaced blocks immediately overlying the uppermost bed of the terrace consist of igneous blocks cemented by carbonate.

Los Roques Archipelago

Aerial view of the outer reef flat, just behind the reef crest. In this area, typically less than 1.5 m deep, the submerged barrier reef is evident in the image by the breaking waves. Light areas in the image are dominated by loose sediments, ranging from coarse sands to gravels, many of which are coral pebbles and cobbles. The dark, aligned areas are coral constructions termed "spurs", which have locally developed in Los Roques and massively in the Las Aves Archipelagos.

Photo taken during early 1980s

Los Roques Archipelago

This photo was taken on the outermost area of the outer reef flat, where depths do not exceed 60–50 cm. Few scleractinian coral branching colonies (mainly *Acropora* and *Porites*) in this agitated, moderately energy-level setting are slightly exposed during low tide, as seen in the image. Hermatypic corals in this environment are associated with other framework builders such as *Millepora* (the "fire coral") and coralline red algae. See Thornborough and Davis (2011) for a comprehensive review on reef flats.

Los Roques Archipelago

Aerial view of the southern margin of Los Roques Archipelago at Boca de Cote. In this area, extensive mangrove-dominated cays have developed at the back-reef setting along the southern barrier reef. This shallow and low-energy environment transitions rapidly into the vast, deeper inner lagoon, where countless patch reefs are found. The distance between the steep slope at the fore-reef and the inner lagoon in this area is only a few hundred meters.

Los Roques Archipelago

Aerial view depicting an isolated mangroove-dominated cay at Boca de Cote. The dark blue color of the water on the right side of the image indicates drastic deepening along the slope break. This contrasts with very shallow depths, crystalline waters, and noticeable light-colored sedimentation behind the mangrove forest.

Aves de Sotavento Archipelago
Aerial view of the reef-flat of Aves de Sotavento. Dark-colored coral constructions, oriented more or less parallel to the wave direction, are known as "spur" formations (ridges composed of corals and hydrozoans). These aligned reliefs are separated by "grooves" (channels containing skeletal sand and gravel). In general, the spurs are formed by a combination of coral *in situ* growth and the accretion of coral pebble and cobble.

Photo taken during early 1980s

Aves de Barlovento Archipelago
This aerial view depicts the extensive growth of dense mangrove forests in the back-reef environment along the southern barrier reef of Aves de Barlovento. Major similarities in reef zonation (physical distribution of major environments) between Los Roques and Las Aves archipelagos are inherited from the same dominating abiotic controlling factors prevailing in these reef complexes (e.g., depth, light intensity, the action of waves and winds and their energy, water temperature and chemistry).

Photo taken during early 1980s

La Blanquilla Island

Aerial view of the eastern coast of La Blanquilla Island. The exposed limestone unit corresponds to the Falucho Member of the La Orchila Formation. This Late Pleistocene calcareous terrace was formed during the LIG period and crops out along the entire island's coastline. The direct impact of waves has shaped eastern coast into a series of convex embayments. Steep cliffs along this coast reach heights between 4 and 9 m.

Photo taken during early 1980s

La Blanquilla Island

Aerial view of the southern coast of La Blanquilla Island, exhibiting a series of narrow embayments. The low rainfall rate in this area of the southeast Caribbean is noticeable due to the almost total absence of vegetation cover (this is also evident in the previous photo). From this perspective, and towards the western coast of the island (at the top of the image), the low-energy level at the embayments has allowed the development of enclosed beach environments.

Photo taken during early 1980s.

La Blanquilla Island

Aerial view of a narrow embayment developed at the southern coast of La Blanquilla Island. The low-energy level and the local stagnation of the waters within the inlet have promoted the growth of mangrove communities. Accumulation of organic matter at the enclosed embayment is noticeable by the dark color of the water.

Photo taken during early 1980s

La Blanquilla Island

Along the southern coast and toward the western side of La Blanquilla Island, low-energy beach environments (as the one in the image) have developed along successive enclosed embayments. Note the light color of the supratidal sandy beach sediments, which grade towards inland into exposed uppermost sediments of an ephemeral hypersaline lagoon. This mudflat (sabkha) is partially surrounded by mangrove communities.

Photo taken during early 1980s.

La Blanquila Island

Both images depict the lower limestone terrace of La Blanquilla Island (Falucho Member, La Blanquilla Formation) on its southern shoreline. In this area, this terrace (a reef limestone containing abundant, well-preserved corals in growth position) reaches an elevation of approximately 4–5 m above RSL. This terrace, dated at 131 ± 7 ka (U-series) by Shubert and Zsabo (1978), was formed during the LIG (MIS 5e, ~ 130–116 ka). La Blanquilla Island is one of the many localities in the Caribbean (including Bahamas, Florida, Bermuda, Jamaica, Haiti, Yucatán, Curaçao, Bonaire, and Barbados) where this sea-level highstand is well recorded.

Photos courtesy of Franck A. Audemard

Aves Island

Aerial view of Aves Island. Located approximately between 15°39′ and 15°40′ N latitude, this small reef island is the northernmost territory of Venezuela. Oriented roughly N-S (in the image, the bottom and top of the island are the southern and northern areas, respectively), it is the only emergent feature of the northern Aves Ridge. The supratidal island measures approximately 500–550 m in length. Taking into account the fringing reef, which is partially exposed at both the northernmost and southernmost areas, Aves Island's maximum length extends to about 650–700 m.

Drone image (February 2024) courtesy of Sergio Cobarrubia

Aves Island

Aves Island, as observed from its western, leeward side. This view (approximately 400 m away) allows to perceive how little the island rises above RSL, with maximum elevations of 3-3.5 m during off-season hurricanes. The facility built for the personnel who guard the island (also used by researchers as a scientific base), situated in the southwest leeward of the island (up and to the right from this viewpoint), is approximately 20–22 m high.

Aves Island

These two photos, taken from the facility, show the southwestern margin of Aves Island. This is the only area where the lithified limestone unit (the Aves Island calcarenite), which underlies the loose sediments that comprise the supratidal island, is perenially exposed. Note the well-differentiated carbonate beds, displaying a backstepping stacking pattern. This retrogradational geometry, driven by RSL rise, records a transgression as the result of positive and underfilled accommodation. This limestone unit, primarily composed of sand- and fine gravel-sized bioclasts (skeletal grains) of coralline algae, molluscs, green algae, scleractinian corals, echinoids, and tests of LBF, has been interpreted to have been deposited during the LIG (MIS 5e, 130–116 ka), a period when global (eustatic) sea level was +2 to +9 m higher than today (see Dutton and Lambeck 2012). Backstepping pattern is typical of the LIG sea-level highstand (see Blanchon et al. 2009).

Aves Island

Another view of the Aves Island calcarenite from the facility. Note the extension of this limestone unit from the uppermost shoreface to the shallow subtidal zone. In the latter zone, limestone beds exhibit a striking yellowish to dark brown color. Considering that Aves Island is a small reef island located literally in the middle of the ocean, the facility was strategically built in the southwest area of the island for two main reasons: the low energy along the leeward side and the presence of this lithified unit that served as a hard substrate for raising the piles that support the structure.

Aves Island

A land view of the Aves Island calcarenite. Individual carbonate beds in the image are approximately 6–12 cm thick. They become younger from the shallow subtidal to the upper shoreface (from right to left in the image). Considering a LIG age (MIS 5e) for this limestone unit, the uppermost beds were probably between 2 and 9 m APSL at the time of deposition. Later during the last glacial period (peak at MIS 2), the eustatic sea-level fell, resulting in erosion and dissolution of the top of the aggradational parasequence. This interpretation does not take into account for the mild subsidence that this unit may have experienced from the end of the LIG to the present.

Los Roques Archipelago
Aerial view of a mangrove-dominated cay located in the eastern area of Los Roques Archipelago. This morphology, with large dimensions and enclosed and open lagoons, characterizes the cays situated behind the double barrier reef system. The shallow and low-energy nature of the environment has facilitated the rapid expansion of the mangrove forests, which constrain seawater, promoting sediment accumulation and accretion, thereby contributing to the cay's expansion.

Photo taken from Méndez-Baamonde (2009)

Los Roques Archipelago
Aerial view of a small, isolated cay in the central lagoon. The shallow subtidal area surrounding the supratidal island has been colonized by scleractinian corals, forming the fringing reef (visible as dark-colored bodies surrounding the light-colored sandy sediments). The growth of the fringing reef promoted the accumulation and accretion of loose sediments, which constitute the supratidal reef island.

Los Roques Archipelago

Aerial view of a submerged sandbank (at the bottom of the image) located in the inner lagoon. This deposit has formed over a bathymetric high that modifies local currents and promotes sediment accumulation in the inner lagoon environment. At the top of the image, two isolated arch-shaped cays with well-developed supratidal environments are seen.

Los Roques Archipelago

A ground view of a small sandbank, whith its uppermost sediments barely exposed (approximately 15–20 cm above RSL). The accumulated loose sediments mainly consist of fragments (bioclasts) from green and red algae (bryopsidaleans and corallines, respectively), molluscs, echinoids, and scleractinian corals, as well as complete and fragmented tests of benthic foraminifera (predominantly larger symbiont-bearing types and, secondarily, smaller symbiont-barren ones).

Aves de Barlovento Archipelago

An aerial view of a group of cays and patch reefs located in the western area of Aves de Sotavento Archipelago. The subaerial morphologies visible in the image are located in the only area, apart from the barrier reef, where supratidal islands have formed within this archipelago. The marked contrast in colors allows for distinguishing the fringing reefs (dark brown) and associated shallow subtidal environments (light blue) from the deeper surrounding waters (dark blue), which range from 15 to 25 m deep.

Image (November 2016) courtesy of Jaap van Eenennaam and Kevin Newton (from Debrot et al. 2019)

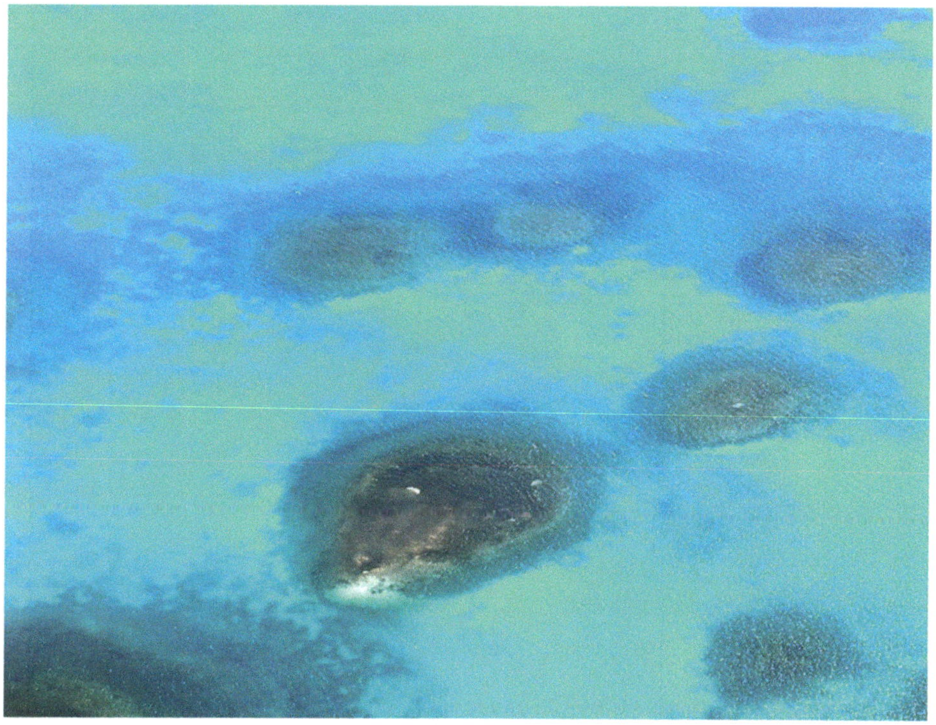

Los Roques Archipelago

Aerial view of several patch reefs (or coral pinnacles) in the inner lagoon environment of Los Roques. These subtidal morphologies exhibit characteristic semicircular to slightly elongate shapes. The image corresponds to an area between Larga Island and the southern barrier reef, where the density of patch reefs is overwhelming, some of them reaching more than 1 km in length.

Los Roques Archipelago

The two photos show aerial views of Rabusquí Cay. The image above corresponds to the southern area of the cay (with Larga Island visible at the top), while the image below corresponds to the northern area. Note the massive submerged fringing reef that borders Rabusquí (visible in the lower part of the image above). In both images, it highlights the presence of perennial hypersaline lagoons in the inner areas. The dark color of the water in these enclosed lagoons indicates high organic matter productivity. Strong bacterial activity in these stagnant, eutrophic (nutrient rich) waters promote the growth of microbialite structures. In contrast, the waters surrounding the cay are crystalline and oligotrophic (nutrient poor).

Los Roques Archipelago

The image shows the Larga Island shoreface-foreshore-backshore, where calcareous dunes along the backshore have formed more or less parallel to the shoreline. These dunes reach heights of up to 3 meters. On Larga Island, these supratidal deposits serve as beach-barrier for a variety of peritidal environments that have developed in the swale immediately behind.

Los Roques Archipelago

A closer view of calcareous dunes formed at the backshore of Larga Island. These deposits consist of well-sorted, fine to medium sand-sized, windblown carbonate grains. These supratidal, wind-derived deposits commonly form a type of grainstone termed "aeolianite", which is cemented by vadose and/or phreatic low-Mg calcite cements.

Morrocoy

These images show the shoreface-foreshore-backshore environments at Borracho Cay. In the image above, note the drastic change in the morphology of the accreted loose sediments. The foreshore (the shore between the low- and high-tide levels) exhibits a noticeable inclination, while the backshore (the zone above the high-tide level) exhibits a nearly flat morphology. The image below shows the foreshore viewed from upper shoreface. The hammer, which serves as a visual scale, is located at a height of about 2 m above RSL. Also note the dense xerophilous vegetation cover on the backshore. See Tucker and Wright (1990) for details on shorelines terminology of modern carbonate environments.

Los Roques Archipelago
This photo corresponds to Agua Cay, situated in the westernmost margin of Los Roques. The image depicts the transition from the upper shoreface to the backshore (the latter environment is covered by herbaceous vegetation), which is marked by a narrow foreshore, where dark algae have accumulated. In the background, calcareous dunes formed in the backshore rise to a height of approximately 3–4 m.

Los Roques Archipelago
This image depicts the transition between the upper shoreface and foreshore in the leeward side of a cay. The photo corresponds to Dos Mosquises Sur Cay (Dos Mosquises Norte Cay is seen in the background), situated in the southwestern border of Los Roques Archipelago. Note the almost total absence of a slope between the surf zone and the exposed sandy sediments in the foreshore, indicating the very low energy of the environment.

Aves Island
Both images show Aves Island shoreface-foreshore transition on the leeward side. The photo above was taken from south to north, while the one below was taken from north to south. Supratidal loose sandy sediments may reach a maximum elevation of approximately 3–3.5 m above high tide. This height is subject to constant variation, as the accretion of loose sediments undergoes a highly dynamic process influenced by sporadic storm events and hurricanes.

Aves Island
The image shows the north and central area of Aves Island from the top of the facility. The presence of the fringing reef on the windward-eastern side of the island is evident from the breaking of the waves. The central area represents the narrowest part of the island. In this portion, the distance between the submerged reef and the shallow waters on the leeward side measures approximately 80–100 m. In the lower part of the image, the carbonate rock basement of the island (the calcarenite unit) is visible as dark, submerged rock bodies.

Los Roques Archipelago
A storm terrace formed at the windward side o kay. The terrace consists of cemented coral rubble (ranging from pebbles to cobbles and boulders in size) and is located in the supratidal zone (above the normal high tide). The blocks were originally part of the adjacent subtidal fringing reef and were removed from their original place of growth during storm episodes. This terrace has reached a height of approximately 2 meters above RSL.

Los Roques Archipelago
The image depicts a large storm terrace formed at the eastern margin of the southern barrier reef of Los Roques. In this area, accumulation of loose coral rubble (mainly pebbles) in the supratidal zone due to constant swelling reaches about 1.5 m high and extends approximately 10 m inland. This is the easternmost termination of probably the largest storm terrace in this archipelago, since the submerged barrier reef extends for more than 30 Km.

Los Roques Archipelago

The image above depicts a large calcareous dune viewed from the inner area of a cay. From the top of the dune, approximately 4 m high, the photo bellow was taken. It depicts a storm terrace (consisting of loose coral pebbles) formed at the foreshore along the windward side of the same cay. The high energy in this environment is evident from the strong swell. Toward the backshore zone, supratidal sediments are covered by herbaceous vegetation that exhibits a striking green color.

Aves Island

The image shows a localized of accumulation of coral pebbles at the foreshore zone on the windward side of Aves Island (northeast area). Loose cobbles were removed from the adjacent subtidal setting during sporadic storm episodes. This narrow zone, with an evident inclination, marks the transition between the subtidal back-reef setting and the backshore. The latter is located at approximately 1–1.2 m of elevation with respect to RSL.

Photograph courtesy of Ana Yranzo Duque

Los Roques Archipelago

The image above depicts coral cobbles and gastropod shells that have accumulated in a storm terrace formed on the windward side of Dos Mosquises Sur Cay. Some of these cobbles are partially covered by the encrusting benthic foraminifera *Homotrema rubrum*. The image below provides a closer view of some of the cobbles, highlighting the pink-reddish color of this symbiont-bearing benthic foraminifera, which is partially covering the surface of a coral. *H. rubrum* is common in many reef complexes throughout the Caribbean. Its striking pink coloration is due to remnants of pigments from a photosynthetic alga (see Strathearn 1986).

Los Roques Archipelago

Beachrock located in the lower intertidal zone of the shoreface of Agustín Cay. Carbonate beds in the image are covered by "fleshy" macroalgae, mainly brown algae (Class Phaeophyceae), and minus green algae. The current position of this beachrock deposit with respect to the foreshore (visible the top of the image) indicates local retrogradation of the shoreline. Beachrocks are excellent indicators of paleo-RSL (i.e., former shore position) for Quaternary deposits (see Kelletat 2006; Vousdoukas et al. 2007; Mauz et al. 2015).

Morrocoy

Both photos show beachrocks belonging to the same location (the leeward side of Borracho Cay). Beachrocks are intertidal deposits defined as beach sands that are cemented to produce hard sheets of limestone, which generally dip seaward (James and Jones 2015). The current position of these lithified carbonate beds, right at the intertidal zone (or the shoreface-foreshore transition), suggests a late Holocene age for these deposits.

Morrocoy

Beachrock formed on the northern side of Sal Cay shows well-differentiated beds deepening seaward. The older, exposed beds at the swash zone (the wave-washed part of the foreshore; lower part of the image), are subject to meteoric diagenesis but are still affected by surf spray. Younger prograding-aggrading bands, which have formed seaward, indicate a local shoreline regression.

Morrocoy

Semicircular to elliptical holes, partially filled with sandy sediments in this beackrock, are termed potholes. These features are formed by the dissolution of lithified carbonate due to the action of slightly acidic meteoric waters. In the case of the example depicted, dissolution was also facilitated by the presence of organic matter in the form of macroalgae. Their interaction with marine or meteoric waters produces CO_2 and organic acids, which promotes the dissolution of $CaCO_3$.

Aves Island

View of the northeast margin of Aves Island from the sandy supratidal island. Seaward, first note the presence of a beachrock deposit in the intertidal zone (shown in the following photo), and later the partially exposed fringing reef of the island, whose presence is also evident due to the visible breaking of the waves at the top of the image.

Aves Island

The image shows a close view of the beachrock deposit depicted in the previous photo. Note that only the tip of the uppermost bands is located above RSL at the time this photo was taken. The characteristic deepening seaward of carbonate bands is clearly noticeable. Individual bands are approximately 7–12 cm thick.

Aves Island

The images depict a beachrock, situated in the northeast margin of the island, and currently located the intertidal zone. Like examples shown in the previous page, this beachrock was also deposited in the back reef. In the image above, the uppermost section of this Holocene *in situ* deposit is slightly above RSL (~ 10–15 cm) at the time this photo was taken. It is noteworthy that tides around Aves Island typically range from 0.2 to 0.3 m at most (see Schubert and Laredo 1984). In the image below, note the distinctive holes with sharp outlines, formed due to localized dissolution of the exposed beachrock bed. Sporadic rainfalls serve as the sole source of meteoric water on the island.

Aves Island
Just south the beachrock deposit showed in the previous page, this beachrock in the northeast of the island occupies a large area, extending from the intertidal to the supratidal zone. The effects of erosion and meteoric diagenesis due to exposure are noticeable, as seen in both images by distinct features, such as rounded margins, pits, potholes and slots.

Aves Island
Another example of the beachrock deposit depicted in the two preceding images. The effects of diagenesis due to exposure have given this deposit a flat upper surface. This beachrock was described by Schubert and Laredo (1984) as a "flat platform". We estimate that this deposit is probably the older lithified Holocene beachrock on Aves Island. Within this image, loose, sandy sediments are covering the inlandmost section of the lithified flat surface.

Los Roques Archipelago

Chitons of the genus *Acanthopleura granulata* (Phylum Mollusca, Class Polyplacophora) inhabiting a large and deep bore-hole in a beachrock located in the surf zone. Shells in the image, of approximately 8 cm-long, consist of eight plates or valves which articulate with one another. The shell plates are encircled by a skirt known as a girdle, which is a strong but flexible muscular structure that holds the chiton valves in place. See Maggenti et al. (2008) and Agudo-Padrón (2021) for details on chiton shell elements.

Morrocoy

Another chiton from the same species (*Acanthopleura granulata*). This specimen is about 5.5 cm-long (the length of the razor is 5.7 cm). As in the previous image, note the small borehole ("homing scar") within which the chiton is hiding. Homing scars are pits in which these small-scale bioeroders accommodate, (see Kázmér and Taboroši 2012). In Morrocoy and Los Roques, these mollusks display a rich diversity (see Work 1969; Agudo-Padrón 2021, 2023).

Morrocoy

This photo was taken at a beachrock on the surf zone of the leeward side of Borracho Cay. It depicts brown algae of the genus *Padina* (Phylum Ochrophyta, Class Phaeophyceae, Family Dictyotaceae) showing a thin and delicate spiral shape of the thalli. These are light-brown to whitish in color, exhibit apparent zonation, and are thinly calcified. *Padina* is an important bioindicator for seawater acidification, as it commonly darkens when pH values drop below normal values (María Eugenia Glem, personal communication 2023).

Morrocoy

Both images show a hard substrate in the intertidal zone (at the leeward side of Sombrero Cay) colonized by non-calcified green and red macroalgae (phyla Chlorophyta and Rhodophyta, respectively). The latter (genus *Laurencia*, Family Rhodomelaceae) contain the red pigments phycoerythrin and phycocyanin. Pigments of chlorophyll a and b are responsible for the striking coloration of the green algae in the image (some of them of the genus Phyllodictyon, Order Cladophorales) (María Eugenia Glem, personal communication 2023).

Aves Island

Green algae, probably of the Family Ulvaceae, covering a large coral rubble. Note the intense green color exhibited by these non-calcifying photosynthetic algae. The green algae depicted in this image are remarkably thin and flexible, a characteristic that becomes evident when waves brake over them. In addition to pigments Chlorophyll a and b, the primary photosynthetic pigments found in green algae, carotenes also contribute to the to the overall green coloration while providing shades of yellow and orange.

Aves Island

This *Halimeda opuntia* shrub (Order Bryopsidales) measures approximately 40 cm wide. These living green algae are attached to a hard substrate in the back reef environment at depths of less than 1 m. The shrub consists of small aragonitic flakes, ranging in size from approximately 3–7 mm. The genus *Halimeda* (mainly *H. opuntia* and *H. incrassata*) is a major sediment-producer in modern tropical carbonate settings. After death, flakes undergo reduction in size, resulting in sediments ranging from clay to sand and gravel.

Photo courtesy of Eliaira Rodríguez

Los Roques Archipelago
Echinoids (Class Echinoidea) in the image belong to the species *Echinometra lucunter* (commonly known as the "rock boring urchin"). They are inhabiting small boreholes in a partially exposed fringing reef. Such holes serve as shelter for the articulated skeletons, which consist of high-density, moveable spines covering the coronal test plates. Calcified elements of echinoids are composed of high-Mg calcite. The main function of the spines is to protect the animal from predators.

Los Roques Archipelago
An underwater image of *Diadema antillarum* inhabiting a borehole in a massive hexacoral buildup. These and previous image exemplify the bioerosive nature of modern epifaunal echinoids. *D. antillarum* is commonly black in color, but occasionally blue, as seen in the specimen in the image. Together with *Echinometra lucunter*, *D. antillarum* is the most abundant species of the Class Echinoidea in Los Roques Archipelago (Noriega and Fuentes–Carrero 2014).

Los Roques Archipelago
View of a mangrove forest that borders a cay. The trees in the image reach elevations of approximately 12–15 m. This halophilic vegetation is common in low-energy environments, where the mangrove takes advantage of the muddy subtidal sediments to grow rapidly. This colonization process promotes stabilization of unconsolidated sediments and accumulation of organic matter. Coral reefs and mangroves rarely adjoin one another. When these communities are directly adjacent, one or both are usually highly stressed (see Kjerfve et al. 1998).

Los Roques Archipelago
A view of a mangrove swamp, mainly *Rhizophora mangle*, from a small boat. The strong roots of the mangrove trees have developed adaptation mechanisms to grow in stressful conditions. In this particular shallow, low-water circulation environment, the strong H_2S “rotten” smell increases when the muddy, organic-rich sediments of the bottom are partially removed. This indicates the local development of anoxic-euxinic conditions at very shallow depths (less than 2 m deep) within the uppermost, seafloor sediments.

Morrocoy

A close view of a community of *Rhizophora mangle*, commonly known as the "red mangle". Note the numerous thick, arched and branched roots that grow directly from the moderately organic-rich muddy substrate in a very low-energy and partially flooded environment. In Morrocoy, the red mangle is common in swamp areas, where this species forms dense and impenetrable masses at tidal level. See Steyermark (1994) for a full review on the flora of this national park.

Los Roques Archipelago

A close view of the species *Avicennia germinans*, commonly known as the black mangle, exhibiting characteristic lower black roots (or prop roots). Its aerial roots (pneumatophores) capture oxygen, allowing the plant to breathe even when partly submerged (see Villamizar and Zamarro 2002; for a full review on mangroves in Los Roques). The tree in the image reaches approximately 4 m in height (not entirely visible in the image).

Los Roques Archipelago
Hypersaline lagoon during the rainy season. In Los Roques, Rainfall is scarce and rarely exceeds 270 mm/yr. This lagoon is part of a restricted sabkha flat located in the inner area of a cay. The water layer in the central part of the lagoon is approximately 30–50 cm thick and is composed of a mixture of marine hypersaline marine waters and rain fall. The sabkha flat is surrounded by halophilic vegetation that reaches a height of 3–4 m.

Los Roques Archipelago
The ephemeral nature of enclosed hypersaline lagoons in Los Roques become evident during the dry season. The image shows the exposed top sabkha sediments, as the thin water layer has completely evaporated. The uppermost layer of sediments in the central area of the lagoon consists of a thin crust of halite directly overlying thick microbial mat deposits. In the outer sabkha, the uppermost sediments consist of a partially hardened crust composed of a mixture of aragonite, high-Mg calcite, gypsum and halite. See Warren (2016) for a full review on sabkhas and evaporitic environments.

Los Roques Archipelago

Evaporation of hypersaline waters of the lagoons during the dry season leads to the precipitation of halite (NaCl). This whitish/vitreous thin layer, approximately ~1 cm thick, directly overlays the muddy and organic-rich unconsolidated uppermost sediments of the central sabkha flats. Those sediments consist of a complex mixture of microbial mats, microbially-mediated carbonates, and gypsum. The halite layer is ephemeral, as it is completely washed during the rainy season.

Los Roques Archipelago

The pink color that characterizes the central areas of ephemeral hypersaline lagoons in Los Roques is caused by the presence of haloarchaea (also designated halobacteria), which colonize the lagoon surface. This surface mainly consists of a thin crust of halite less than 1 cm thick. Haloarchaea, the predominant microbial biota in these extreme habitats, thrive at high NaCl concentrations, as well as elevated temperatures and pH levels, and produce C50 carotenoids (bacterioruberins) responsible for the distinctive red-to-pink pigmentation (see Ventosa 2006).

Los Roques Archipelago

Shrinkage cracks formed in the sabkha flat perimeter that surrounds a hypersaline lagoon in Gran Roque Island. These sedimentary structures, also named "desiccation mudcracks" (see Plummer and Gostin 1981; Harazim et al. 2013), develop in muddy sediments due to intense evaporation rates that promote desiccation of the muddy sediments. In the case of the structures visible in the image, their formation is controlled by a combination of microbial activity, high salinity of porewaters, the shallow depth of the water table, and their periodic subaerial exposure.

Los Roques Archipelago

The two images depict a perennial hypersaline lagoon in Pirata Cay, which is located southeast of Gran Roque Island. These restricted waters reach maximum depths of 1.6 m. In February 2009, measured values for total alkalinity, pH and salinity in the surface waters were 169 mM/L, between 8.7 and 9.3, and an average of approximately 90‰, reaching up to 140‰ in the surrounding sabkha flat, respectively. Hypersaline superficial waters with increased pore-water alkalinity and pH are oversaturated with respect to aragonite, calcite, and dolomite (see Petrash et al. 2012).

Photo below courtesy of Daniel A. Petrash

Los Roques Archipelago
Domal-shaped subtidal thrombolites within the hypersaline lagoon of Pirata Cay. The organosedimentary structures visible in the image, up to 4 m in size, occur in the deepest areas of the lagoon. The size of the thrombolites decreases from less than 10 cm in diameter toward shallower depths. Microbes (including sulfide-oxidizing bacteria, sulfate-reducing bacteria, cyanobacteria) play a key role in the formation and preservation of these thrombolites (see Petrash et al. 2012).

Photo courtesy of Daniel A. Petrash (from Petrash et al. 2012).

Los Roques Archipelago
A subtidal thrombolite collected by hand in shallow waters of the hypersaline lagoon in Pirata Cay. This coarse-grained microbialite is composed of authigenic gypsum (constituting 40–55% of bulk mineralogy), aragonite, and high-Mg calcite. The lithification of this thrombolite is thought to have proceeded through a series of complex dissolution/precipitation and replacement reactions. These reactions were mediated by endolithic bacterial communities as well as by extracellular polymeric substances comprising the microbial mat (see Petrash et al. 2012).

Los Roques Archipelago

The image above shows subtidal to intertidal stromatolite structures surrounding the shallowest areas of the Pirata lagoon. These carbonate-dominated microbialites are lithified. Partially exposed beds are approximately 3 cm thick. Note the variation of the fabrics in the image below (scale bar is 2.7 cm-size), from micritic-laminated in the middle-upper part to clotted-thrombolitic in the middle-lower part. The main components include micritized *Halimeda* flakes, mollusc fragments and benthic foraminifera tests, as well microbially-induced automicrite bounding grains together and lithifying the structure.

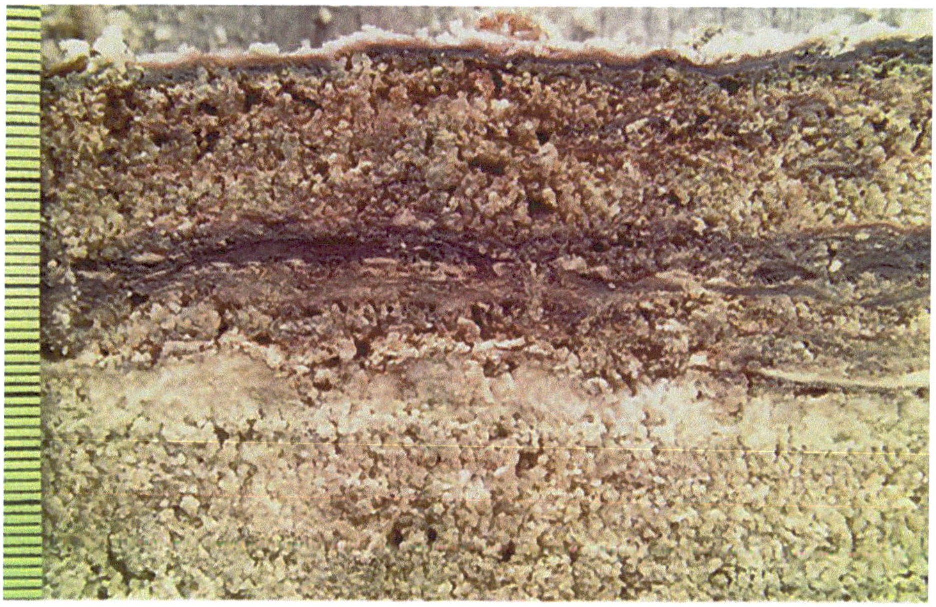

Los Roques Archipelago

Close view of an intertidal laminated stromatolite. The structure (~ 5 cm thick) was collected by hand in a supratidal sabkha flat. The photo was taken *in situ* and then the structure was put back in place, as it was poorly lithified. The coarse-grained grains (or granules) that constitute the laminated-clotted fabric are most likely composed of a combination of Ca-carbonate, gypsum and halite. Note the wavy-laminated microbial mats of purple color, which are probably the product of autotrophic oxygenic bacteria. See Konhauser (2007) for details on the stages and biogeochemical reactions involved in modern marine intertidal stromatolite formation.

Photograph taken during early 1980s

Los Roques Archipelago

Pseudodiploria strigosa showing characteristic domal and robust morphology. Species within this genus (*P. Strigosa* and *P. Clivosa*) are part of the so-called "brain corals", a colloquial term used to group corals with meandrine-sinuous, "cerebral" morphology of the calcified valleys and ridges. *Pseudodiploria* (especially *P. strigosa*) is one of the most abundant reef-building scleractinians in Los Roques. It is mainly found at shallow depths (< 6 mm deep), although colonies also grow locally at intermediate depths (down to 25 m) in the fore-reef.

Aves Island

Pseudodiploria strigosa, exhibiting striking yellow color, is accompanied by an octocoral of the genus *Antillogorgia*. These coral genera are among the most abundant in Aves Island. At shallow depths (< 10 m) on this island, octocorals are particularly abundant in areas characterized by flat seafloor, as seen in this image. Additionally, the growth of adjoined colonies of brain corals and octocorals occurs in environments protected from direct wave action. See Yranzo et al. (2014) for a review on communities of hermatypic corals and octocorals in Aves Island.

Photograph courtesy of Ana Yranzo Duque

Los Roques Archipelago

Acropora palmata exhibiting characteristic branching, ramose three-like morphology. Individual branches flatten towards their ends and they grow inclined at low angles from single, thick stalks. Commonly known as the "elkhorn" coral, *A. palmata* is one major reef-building coral in the entire Caribbean. It grows massively in shallow and high-energy environments, including the reef crest. In the barrier reefs of the Los Roques Archipelago, this species stands out as the most dominant scleractinian coral, especially at depths between 1 and 3 meters.

Photograph taken during early 1980s

Los Roques Archipelago

Dendrogyra cylindrus, commonly known as the "pillar" coral, exhibiting characteristic robust morphology consisting of vertical cylindrical pillars with blunt ends, reaching heights of up to 2.5 m. This species is among the most common hexacorals in deeper sites (depths greater than 6.5 m) of the vas inner lagoon environment of the Los Roques (see Villamizar et al. 2003). *D. cylindrus* is widely distributed in tropical waters throughout the entire Caribbean (see Montaggioni and Braithwaite 2009).

Photograph taken during early 1980s

Los Roques Archipelago
Numerous adjoined colonies of *Orbicella annularis* (formerly *Montastraea annularis*) at depths of approximately 5–7 m. This species is characterized by a robust domal morphology and striking spherical, closely-spaced corallites. In the Los Roques, *O. annularis* is common in settings below the fair-weather wave base. It often found in association with other domal scleractinian corals such as *Orbicella faveolata* and *Montastraea cavernosa.*

Photograph taken during the early 1980s

Los Roques Archipelago
Close view of *Montastraea cavernosa.* This scleractinian coral is characterized by protuberant skeletal "cup-like" corallites. These are covered by organic living polyp tissue, which host zooxanthellae microalgae. These photosynthetic symbionts are responsible for the striking purple coloration (in this species, the color may also appear as green, gray, brown or orange). *M. cavernosa* is among the most common reef-building corals in the Los Roques Archipelago, especially in deeper sites.

Photograph taken during the early 1980s

Henri Pittier National Park

Alcyonarian octocoral colonies. These soft corals are collectively known as gorgonians (Subclass Alcyonaria, Order Gorgonacea). Note the two different morphologies of branching corals in the photo. The most abundant form in the image is the arborescent shrubs of delicate appearance (*Antillogorgia*, probably *A. Americana* or *A. acerosa*). More robust forms with a horn-like shape and erect habit (probably *Pseudoplexaura* sp.) stand out due to their yellow coloration.

Los Roques Archipelago

These colonies of *Antillogorgia* (probably *A. bipinnata* or *A. Americana*) exhibit delicate morphology with weakly calcified, fragile skeletons. These skeletons are composed of minute high-Mg calcite sclerites (0.10 to 1 mm in length), which are the only skeletal elements in gorgonians that fossilize. The organic red tissues partially covering the octocorals in the image result from the activity of cyanobacteria (Ana Yranzo Duque, personal communication 2022).

Los Roques Archipelago

Octocorals of the genus *Pseudoplexaura*, probably *P. flagellosa*. The two colonies are attached to a massive coral bioherm. These soft corals are characterized by a relatively robust stalk that branches into multiple "horn-like" thinner ramifications. In Los Roques Archipelago, gorgonians (e.g., *Pseudoplexaura* and *Pseudopterogorgia*) are common in low-energy, deeper sites, where they grow from hard substrates (such as in this example) or from soft muddy flat bottoms.

Los Roques Archipelago

Octocoral *Eunicea* (probably *E. tournefortii*), commonly known as the "purple tree" gorgonian, exhibiting characteristic bushy morphology. The thin, cylindrical branches dichotomously ramify from the basal part of the structure. The latter is not seen in the image, as the colony's stalk has grown into a small cavity between hexacoral bodies. Similar to the genera *Antillogorgia* and *Pseudoplexaura*, *Eunicea* is widely distributed in the Caribbean (see Sánchez 2009).

Los Roques Archipelago

Gorgonia flabellum in the center of the image surrounded by numerous gorgonian forms. Along with *Gorgonia ventalina*, this species is commonly known as the "Venus Sea Fan". These delicate soft corals consist of a lattice of several main flattened branches oriented in one plane, forming a network of small branchlets. Their size can reach 1.5–2 m in length. Species of the Venus Sea Fan are widely distributed throughout the Caribbean, inhabiting shallow marine settings, commonly at depths not exceeding 15–20 m.

Photograph taken during the early 1980s

Aves Island

The Venus Sea Fan in the image is surrounded by numerous colonies of gorgonians. These community of octocorals covers massive hexacoral bioherms, which serve as substrate for the growth of such delicate morphologies. The photograph was taken at a depth of 15 m in the northwest area of the coral reef of Aves Island. This site is characterized by a marked slope that descends to approximately 70 m, then drops drastically to 100 m (see Yranzo et al. 2014).

Photograph courtesy of Ana Yranzo Duque

Los Roques Archipelago

A sea star (Class Asteroidea) of the species *Oreaster reticulatus* (commonly known as the "red cushion" sea star), approximately 30–35 cm long, is resting on a sandy substrate at a depth of about 1.6 m deep. The specimen exhibits characteristic orange-brownish color. This epifaunal echinoderm has a thick body with five broad, flat arms covered with small, pointed spines arranged in rows. Common from intertidal environments down to depths of about 10 m, this species is the most emblematic echinoderm in Los Roques.

Photograph courtesy of Leyjer A. Hernández Avilez

Los Roques Archipelago

Close view (approximately 15–17 cm in width) of a brittle star (Class Ophiuroidea, Order Ophiurida) exhibiting striking red-white color and resting on a scleractinian coral. Note the delicate morphology of this ophiuroid, consisting of a central disk and five thin arms. In Los Roques, the class Ophiuroida is well represented by at least 20 species (see Noriega and Fuentes–Carrero 2014).

Photograph courtesy of Humberto Ramírez

Henri Pittier National Park

Colony of serpulids (or polychaete worms living within calcareous tubes). Individuals in the image, approximately 4–7 cm in diameter, are encrusting a hard coral substrate. The visible organic tissues exhibiting varied pigmentations and coiling forms are termed radiole. These feather-like head structures are joined to the operculum, which is the system that provides the serpulid worm with protection from the external environment and acts as a filter for food and water exchange.

Los Roques Archipelago

Demosponges (Phylum Porifera; Class Demospongiae) in the image belong to the species *Aplysina fistularis*, which is indigenous to the Greater Caribbean. The cylindrical, hollow bodies of these spicule-less sponges (Order Verongida) are composed of a "spongin" skeleton (a tough organic compound that readily decomposes after death; James and Jones 2015). The skeleton of this species is chitinous in composition (Machałowski et al. 2021).

Photograph taken during the early 1980s

Aves Island

Aves Island reef at depth of 40 m. The image portrays crystalline waters with optimal light penetration at upper mesophotic depths, showcasing a variety of sponges (green, blue and red in color) accompanied by octocorals. See Villamizar et al. (2019) for a complete review on the diversity and structure of the sponge community of Aves Island.

Photograph courtesy of Ana Yranzo Duque (from Villamizar, 2024)

Los Roques Archipelago

The image depicts the deepest facies of a former fore-reef and the beginning of the slope, extending down to approximately 110 m. This boundary is characterized by a drastic transition between the robust bodies of scleractinian corals and the fine-grained sediments that compose the slope. The massive coral buildup visible in the image was deposited during the last glaciation, the sea-level lowstand peak of which occurred between 21 and 18 ka (MIS 2). During that time interval, ESL was approximately 120–130 m BPSL.

Photograph taken during the early 1980s

Los Roques Archipelago
These peculiar echinoids, commonly known as "sand dollars" (Order Clypeasteroida), were photographed at 280–300 m depth. The circular tests or coronal plates of the specimens in the image (probably belonging to the genera *Clypeaster*, *Leodia*, or *Mellita*; see Francisco and Pauls 2008) measure approximately 8–10 cm in diameter. These flat tests, exhibiting pentameral symmetry, were initially formed at shallow depths and were subsequently displaced down the slope of the southern barrier reef of Los Roques to depths below the photic zone.

Photograph taken during the early 1980s

Los Roques Archipelago
Crinoid at depth of approximately 140–150 m. As in the two previous images, the light in this photo emanates from the flash of the camera. This image shows the arborescent and delicate morphology of a feather star. This deep-water specimen remains unattached (or unstalked) to the soft substrate. Its crown comprises multiple arms (undivided branches extending from the calyx), each adorned with small pinnules.

Photograph taken during the early 1980s

References

Agudo-Padrón, A. I., 2021. Chitons (Mollusca: Polyplacophora) from the Venezuelan coasts, Southern Caribbean: A checklist and brief review in advance of their knowledge. *Aquaculture and Fisheries Studies*, *3*(1), 3 pp.

Agudo-Padrón, A. I., 2023. Chitons (Mollusca, Polyplacophora) present in the territorial waters of Venezuela, Southeast Caribbean: a general review and update of their knowledge: Quítons (Mollusca, Polyplacophora) ocorrentes nas águas territoriais da Venezuela, Sudeste do Caribe: uma revisão geral e atualização do seu conhecimento. *Brazilian Journal of Animal and Environmental Research*, *6*(2), 1018–1039.

Blanchon, P., Eisenhauer, A., Fietzke, J., Liebetrau, V., 2009. Rapid sea-level rise and reef back-stepping at the close of the last interglacial highstand. Nature, 458, 881–884.

Debrot, A. O., Yranzo, A., Arocha, D., 2019. Los Roques and Las Aves archipelagos, Venezuela: A marine ecological and conservation reconnaissance of two little-known southeastern caribbean oceanic archipelagos. *Atoll Research Bulletin*, 622, 1–27.

Dutton, A., and Lambeck, K., 2012. Ice volume and sea-level during the last interglacial. *Science*, *337*, 216–219.

Francisco, V., and Pauls, S. M., 2008. Especies del Orden Clypeasteroida (Echinodermata: Echinoidea) de las costas de Venezuela. *Revista de Biología Tropical*, *56* (Suppl. 3), 215–228.

Harazim, D., Callow, R. H., Mcilroy, D., 2013. Microbial mats implicated in the generation of intrastratal shrinkage ('synaeresis') cracks. *Sedimentology*, *60*, 1621–1638.

Hopley, D., 2011. Double and triple reef fronts. In: Encyclopedia of Modern Coral Reefs. Hopley D (Ed.). Springer, pp. 325–326.

James, N. P., and Jones, B., 2015. Origin of Carbonate Sedimentary Rocks. John Wiley & Sons, Ltd, Chinchester, 464 pp.

Kázmér, M., and Taboroši, D., 2012. Bioerosion on the small scale–examples from the tropical and subtropical littoral. *Monostori Jubilee Volume Hantkeniana*,7,37–94.

Kelletat, D., 2006. Beachrock as sea-level indicator? Remarks from a geomorphological point of view. *Journal of Coastal Research*, *22*, 1558–1564.

Kjerfve, B., Ogden, J. C., Garzón-Ferreira, J., Jordán-Dahlgren, E., De Meyer, K., Penchaszadeh, P., Wiebe, W. J., Woodley, J. D., Zieman, J. C., 1998. CARICOMP—A Caribbean Network of Marine Laboratories, Parks, and Reserves for Coastal Monitoring and Scientific Collaboration. In: Caribbean coral reef, seagrass and mangrove sites (edited by Björn Kjerfve), 1–16.

Konhauser, K. O., 2007. Introduction to Geomicrobiology. Blackwell Publishing. 433 pp.

Machałowski, T., Rusak, A., Wiatrak, B., Haczkiewicz-Leśniak, K., Popiel, A., Jaroszewicz, J., Żak, A., Podhorska-Okołów, M., Jesionowski, T., 2021. Naturally formed chitinous skeleton isolated from the marine demosponge *Aplysina fistularis* as a 3D scaffold for tissue engineering. *Materials*, 14, 2992.

Maggenti, M. A. B., Maggenti, A., Scott, G., 2008. Dictionary of Invertebrate Zoology. *Zea E-Books*. *61*. 379 pp.

Mauz, B., Vacchi, M., Green, A., Hoffmann, G., Cooper, A., 2015. Beachrock: a tool for reconstructing relative sea level in the far-field. *Marine Geology*, *362*, 1–16.

Méndez-Baamonde, J., 1984. Los Roques: un ejemplo de sedimentación de Carbonatos. Morfología, sedimentología, organismos, mineralogía y geoquímica. Relación con Las Aves y La Blanquilla. Doctoral Dissertation. Central University of Venezuela. Volume I, 205 pp.

Méndez-Baamonde, J., 1997. El Cuaternario. Ministerio de Energía y Minas. Dirección General Sectorial de Servigeomin. Caracas. Venezuela. *Boletín de Geología*, *31*, 74–115.

Méndez-Baamonde, J., 2009. Carbonatos – Origen y Sedimentación. University Printing of the Central University of Venezuela. Caracas, Venezuela. 270 pp.

Montaggioni, L., and Braithwaite, C. J. R., 2009. Quaternary Coral Reef Systems: History, Development Processes and Controlling Factors, 5. Elsevier, Amsterdam. 520 pp.

Noriega, N., and Fuentes–Carrero, Y., 2014. Contribución al conocimiento de la diversidad de los equinodermos del Parque Nacional Archipiélagos de los Roques, Venezuela. *Acta Biológica Venezuelica*, *34*(2), 285–292.

Petrash, D. A., Gingras, M. K., Lalonde, S. V., Orange, F., Pecoits, E., Konhauser, K. O., 2012. Dynamic controls on accretion and lithification of modern gypsum-dominated thrombolites, Los Roques, Venezuela. *Sedimentary Geology*, *245*, 29–47.

Plummer, P. S., and Gostin, V. A., 1981. Shrinkage cracks; desiccation or synaeresis? *Journal of Sedimentary Research*, *51*, 1147–1156.

Sánchez, J. A., 2009. Systematics of the candelabrum gorgonian corals (Eunicea Lamouroux; Plexauridae; Octocorallia; nidaria). *Zoological Journal of the Linnean Society*, *157*, 237–263.

Schubert, C., and Moticska, P., 1973, Reconocimiento geologic de las islas venezolanas en el Mar Caribe, entre Los Roques y Los Testigos (Dependencias Federales). *Acta Científica Venezolana*, *23*, 210–223.

Schubert, C., and Valastro Jr, S., 1976. Quaternary geology of La Orchila Island, central Venezuelan offshore, Caribbean Sea. *Geological Society of America Bulletin*, *87*, 1131–1142.

Schubert, C., and Szabo, B. J., 1978. Uranium-series ages of Pleistocene marine deposits on the islands of Curacao and La Blanquilla, Caribbean Sea. *Geologie en Mijnbouw*, *57*, 325–332.

Schubert, C., and Laredo, M., 1984. Geology of Aves Island (Venezuela) and subsidence of Aves Ridge, Caribbean Sea. *Marine Geology*, *59*, 305–318.

Strathearn, G.E., 1986. *Homotrema rubrum*: Symbiosis Identified by Chemical and Isotopic Analyses. *Palaios*, *1*, 48– 54.

Steyermark, J. A. 1994. Flora del Parque Nacional Morrocoy. Fundación Instituto Botánico de Venezuela y Agencia Española de Cooperación Internacional. Caracas, Venezuela. 415 p.

Thornborough, K. J., and Davies, P. J., 2011. Reef flats. In: Hopley, D. (Ed.), Encyclopedia of Modern Coral Reefs: Structure. Form and Process. Springer, Dordrecht, New York, pp. 869–876.

Tucker, M. E., and Wright, V. P., 1990. Carbonate Sedimentology. Blackwell Publishing. 482 pp.

Urbani, F., 2005. Geología de la isla del Gran Roque, Parque Nacional Los Roques, Venezuela: Guía de Excursión. *Revista Venezolana de Ciencias de la Tierra*, *29*, 1–35.

Urbani, F., 2018. Una Revisión de los Terrenos Geológicos del Sistema Montañoso del Caribe, Norte de Venezuela. Boletín de Geología Volumen XXIII, Número 36, 118–216.

Ventosa, A., 2006. Unusual micro-organisms from unusual habitats: hypersaline environments. Prokaryotic Diversity: Mechanisms and Significance (Logan NA, Lappin-Scott HM & Oyston PCF, eds), pp. 223–253. Cambridge University Press, Cambridge.

Villamizar, A., and Zamarro, J., 2002. La Vegetación. In: Guía del Parque Nacional Archipiélago Los Roques. Ed. Javier Zamarro.

Embajada de España en Venezuela y Ministerio del Ambiente y de los Recursos Naturales. Caracas. 2002, 99–108.

Villamizar, E., Posada, J. M., Gómez, S., 2003. Venezuela. Rapid Assessment of coral reefs in the Archipelago de Los Roques National Park, Venezuela (part 1: stony corals and algae). *Atoll Research Bulletin*, *496*, 512–519.

Villamizar, E., Yranzo, A., Romero, M., Boadas, H., 2019. Diversity and Structure of Sponge Communities of the Isla de Aves Wildlife Refuge, Caribbean Sea, Venezuela. *Caribbean Journal of Science*, *49*, 109–124.

Villamizar, E., 2024. Arrecifes Coralinos en Venezuela: Frágil Patrimonio Natural. Fundación Empresas Polar. Editorial Arte. ISBN: 978-980-379-431-6. 129 pp.

Vousdoukas, M. I., Velegrakis, A. F., Plomaritis, T. A., 2007. Beachrock occurrence, characteristics, formation mechanisms and impacts. *Earth-Science Reviews*, *85*, 23–46.

Warren, J. K., 2016. Evaporites: A Geological Compendium. Second Edition. Springer International Publishing Switzerland. 1813 pp.

Work, R. C., 1969. Systematics, ecology, and distribution of the mollusks of Los Roques, Venezuela. *Bulletin of Marine* Science, *19*. 614–711.

Yranzo, A., Villamizar, E., Romero, M., Boadas, H., 2014. Estructura de las comunidades de corales y octocorales de Isla de Aves, Venezuela, Caribe Nororiental. *Revista de Biología Tropical*, *62*, 115–136.

Carbonate Rock Components and Microfacies Classification

4

This is a key chapter in this book, as it is the first one that is entirely focused on carbonate petrography and mcrofacies analysis, which will be the theme for the rest of the book. Its purpose is to introduce the reader to the identification of the variety of carbonate components (or constituents), as well as the textures and fabrics of carbonate rocks (limestones and dolostones), with the main goal of classifying (naming) the samples selected using traditional, widely used classification schemes. We also introduce some terms for classifying mixed carbonate-siliciclastic and sedimentary siliceous rocks, as well as organic-rich rocks of varying lithologies. From this chapter and throughout the rest of the book, the vast majority of photographic examples consist of photomicrographs (or mm-scale images) originated from views of thin sections taken using a polarizing microscope. A small number of images taken using a binocular magnifying glass (mm-scale), as well as close views of dry samples of modern organisms (cm-scale), are included. Each image is accompanied by a headline indicating the geological age and basin/locality corresponding to each example, and a caption describing the most significant visible features in the images. For each image, the full length of the horizontal axis, or the image width (IW), is indicated, as well as the type of light used to take the photomicrograph, i.e., plane-polarized light (PPL) or cross-polarized light (CPL).

In this chapter, we place emphasis on the identification of components (carbonate and non-carbonate), as well as on the classification of the samples. Naming carbonate microfacies is essential to carrying out microfacies analysis. Furthermore, from this chapter and through the rest of the book, in the descriptive captions we frequently comment on depositional environments (e.g., bathymetry, energy level); paleoenvironmental conditions (e.g., redox and trophic conditions); the abundance of skeletal and non-skeletal grains and authigenic minerals in a given carbonate rock unit considered in this book; major grain associations (photozoan and heterozoan); diagenetic processes; among other factors and features.

4.1 Major Marine Carbonate Minerals

Before delving into the subject matter of this chapter, a brief review of the major marine carbonate minerals is necessary. The most important carbonate minerals in seawater reactions, and therefore in modern marine sediments (biogenic and abiogenic), are the $CaCO_3$ polymorphs aragonite and calcite. Aragonite is both denser and more soluble (metastable, i.e., thermodynamically less stable) than calcite. These two phases also differ in their tendency to accept divalent cations other than Ca^{2+} in solid solution. Cations such as Sr^{2+} and Ba^{2+} whose ionic radii exceed that of Ca^{2+} are more frequently accommodated in the orthorhombic aragonite structure. Conversely, smaller cations such as Mg^{2+} are found more frequently in the hexagonal calcite structure. Marine calcite may contain variable amounts of $MgCO_3$ in solid solution (Morse et al. 2007).

Marked differences in the magnesium content of calcite result in the mineralogical distinction between low-Mg calcite (LMC; stable phase) and high-Mg calcite (HMC; metastable phase), which are usually separated by a value of 4 mol% $MgCO_3$ (Tucker and Wright 1990; Flügel 2010). Some authors distinguish calcites with more than 4 mol% $MgCO_3$ in intermediate-Mg calcite (IMC; 4–12 mol% $MgCO_3$), and HMC (12–18 mol% $MgCO_3$) (see James and Jones 2015). Other authors (e.g., Smith et al. 2006) consider different boundaries to define IMC (4–8 mol% $MgCO_3$) and HMC (greater than 8 mol% $MgCO_3$). HMC *sensu lato* (i.e., calcite having ≥4 mol% $MgCO_3$) is the other major calcium carbonate polymorph in modern marine sediments. Biogenic HMC may be composed of up to 20 mol% $MgCO_3$ (Morse et al. 2007); although Scholle and Ulmer-Scholle (2003) and Flügel (2010) state that this value may reach up to 30% in some tropical coralline (red) algae.

Dolomite [$CaMg(CO_3)_2$] is the most important double carbonate (Morse et al. 2007). It has a hexagonal crystal structure and, when stochiometric, contains equal proportions of calcium and magnesium. In an ideal dolomite, the

J. A. Méndez Dot, J. Méndez Baamonde, *Atlas of Tropical Carbonates*, https://doi.org/10.1007/978-3-031-25494-9_4

crystal structure is highly organized (ordered) with repeated stacked layers of Ca, CO_3, and Mg. In these ideal structures, the Ca and Mg are in discrete layers (James and Jones 2015). Dolomite is essentially an authigenic mineral (not biogenic in origin, although microbes can influence its precipitation under certain conditions). It can be formed syngenetically (synsedimentary) or during many different diagenetic stages (from early-marine to late-burial diagenesis).

4.2 Carbonate Components

The study of marine carbonate rocks and sediments through the use of the petrographic (polarizing) microscope begings with differentiating the carbonate components: grains (or allochems), matrix, authochthonous (*in situ*) organisms and bioconstructions, authigenic precipitates (cements, dolomite, recrystallized carbonate, and microbial carbonates), and open pores (when present). Non-carbonate components (terrigenous, authigenic, and organic) may be present in different proportions relative to their carbonate counterparts.

4.2.1 Grains

Carbonate grains, or mechanically deposited "allochems" (LMC, HMC, and aragonite in composition) are roughly separated in: (**a**) skeletal grains (or bioclasts; e.g., calcareous algae, foraminifera, molluscs, corals, bryozoans, among others); (**b**) non-skeletal grains, such as ooids (concentric grains formed around a nucleus), peloids (carbonate grains of uncertain origin), oncoids (microbially coated grains), aggregates (several grains cemented together; also termed compound grains), and clasts (single or compound reworked-redeposited grains).

The size of the grains roughly varies from: micrometer-scale grains (not visible/recognizable with the polarizing microscope), such as coccoliths, some planktonic foraminifera, as well as undifferentiated grains resulting from the breakdown of skeletal material; millimeter-scale skeletal grains (these may be preserved as complete fossils or as fragments of an individual fossil) and non-skeletal grains, commonly ranging from 0.1–0.5 mm to 10 mm (and up to a few centimeters in size), and representing the vast majority of carbonate fossils in shallow platform/ramp settings.

4.2.2 Matrix

The matrix is the fine-grained, synsedimentary, mechanically or autochthonously deposited material between grains. It is mainly composed of micrite. Micrite (carbonate mud or lime-mud) is the finely crystalline $CaCO_3$ (see Scholle and Ulmer-Scholle 2003; Flügel 2010). Three categories can be distinguished according to differences in crystal (grain) size and composition: *Micrite*, with a crystal size limit of <4 μm (Folk 1959), and characterized by microcrystalline (barely perceptible crystal size) or cryptocrystalline (imperceptible crystal size) crystal textures; *Microspar*, generally 5- to 20-μm-sized (Folk 1965) (it can be as coarse as 30 μm), produced by neomorphism and/or recrystallization of micrite (not primary precipitates); *Pseudospar*, neomorphic or recrystallized calcite fabric with average crystal size larger than 30–50 μm (Folk 1965) (neomorphism and recrystallization processes will be discussed in detail in Chap. 16). Furthermore, some "coarse-grained" matrices consist of small bioclasts or peloids forming a coalescent structure, resulting in "bioclastic" or "peloidal" matrices.

The three main marine carbonate precipitation modes can be involved in the formation of micritic sediments: (**a**) abiotic, or inorganic precipitation; (**b**) biologically-induced precipitation carried by eukaryotic phytoplankton and cyanobacteria and other bacteria; (**c**) biologically-controlled precipitation, through disintegration of skeletal material (see Tucker and Wright 1990; Schlager 2005; Flügel 2010). Two major modes of micrite genesis are widely considered: *automicrite*, which refers to autochthonous micrite precipitated *in situ* on the seafloor or within the sediments (mainly biologically-induced in origin); *allomicrite*, or micrite formed by precipitation in the water column (biologically-induced or abiotic); disintegration of algae and invertebrate skeletons (biologically-controlled in origin); accumulation of calcareous plankton (biologically-controlled); byproduct of bioerosion; secretions by fish (James and Jones 2015). In this chapter we do not emphasize on the origin of micrites (or carbonate muds), as this topic will be discussed in detail in Chap. 15.

In addition, the composition of a given matrix may vary widely according to the content (wt.%) of carbonate and clay minerals; as well as to the content of TOC. In general, the higher the TOC content, the darker the matrix will look.

4.2.3 Authocthonous Organisms and Bioconstructions

Differentiating between marine carbonate grains (detrital) and autochthonous organisms (found in growth position) is key for classifying some limestones. Carbonate-producing organisms such as scleractinian (reef-building) corals, and encrusters such as encrusting coralline algae, serpulids, and some benthic foraminifera and bryozoans, may be found in situ (not mechanically transported). Moreover, these organisms, together with processes linked to microbially-induced $CaCO_3$ precipitation, may form bioconstructions at different areal and volumetric scales and under different environmental conditions (e.g., trophic and redox conditions, bathymetry). Coral reefs and microbialites represent two end-members (opposite ends of the spectrum) among bioconstructions. Accurate diagnosis of the origin of the variety of carbonate components in a given microfacies results in the differentiation between allochthonous and autochthonous limestones.

4.2.4 Cements and other Authigenic Carbonates

A cement is defined as a passively precipitated mineral between grains (as distinct from matrix) that serves to bind them together. Carbonate cements (composed of crystals) are authigenic in origin, i.e., they precipitate *in situ* within rocks or sediments, rather than having been transported and deposited (see James and Jones, 2015). Cementation mainly originates from abiotic (or inorganic) precipitation of carbonate in empty spaces or pores within the rock due to supersaturation of pore fluids with respect to the mineral. They also form, on a smaller volumetric scale, due to biologically-induced precipitation of carbonate (microbial cementation). The type of cement (texture, crystal size) and its mineralogical composition (LMC, HMC, aragonite, or dolomite) will depend mainly on the diagenetic environment in which the precipitation process occurs (marine, meteoric, or burial environments), as well as on the pore type and size, and on the abiotic or microbial nature of the process. In addition to cements and automicrite (the latter in the form of matrix, cement, replacement, or coating on grains), authigenic precipitation of carbonates is predominantly represented by dolomite and its varied crystal textures and origins, as well as by recrystallized carbonates. The latter are diagenetic products (via dissolution-reprecipitation) of previously formed carbonate sediments, and like dolomite, they can be formed in a wide range of diagenetic stages. Authigenic carbonates are regarded as excellent indicators of the hydrogeochemical and redox conditions during diagenesis, from early-marine to late-burial. Regarding the use of some terms commonly associated with carbonate cementation and/or recrystallization, it is worth mentioning that the term "sparry calcite" or "sparite" is commonly used in a loose manner to refer to diagenetic LMC crystals, either formed as cement or formed due to neomorphism and/or recrystallization of previously formed $CaCO_3$ sediments. Furthermore, microspar (finely crystalline LMC), more commonly associated with recrystallization of micritic sediments, can also form as a primary precipitate, filling very small pores (e.g., micropores and small intraparticle pores). Carbonate diagenesis and cementation will be discussed in detail in Chaps. 16, 17 and 19, while dolomite and dolomitization is the focus of Chap. 18.

4.2.5 Pores

In rocks, open pores are small to minute openings or passageways, while porosity is the percentage of bulk volume of a rock that is occupied by interstices, whether isolated or connected (Bates and Jackson 1980). Primary porosity can be formed during both pre-depositional (e.g., intraparticle pores in foraminifera, corals, and other bioclasts) and depositional stages (mainly interparticle pores) and it is genetically related to the original depositional texture of the rock. Secondary porosity is formed at any diagenetic stage after deposition and it is strictly related to diagenetic processes (see Choquette and Pray 1970). Porosity types and porosity creation and destruction processes will be discussed in detail in Chap. 19.

4.2.6 Non-Carbonate Components

Carbonate rocks may contain non-carbonate mineral components in widely varying proportions. In many cases, the most abundant group of these components is siliciclastic (terrigenous) minerals. The most common are quartz, feldspar, muscovite, and clays (illite, kaolinite, smectite, montmorillonite, and mixed-layer clays); some clays may also be formed authigenically during burial diagenesis. Silica may be represented by biogenic silica (radiolarians, sponge spicules, and diatoms, originally as opal-A) and diagenetic silica (finely crystalline sedimentary quartz, or chert). Other minerals (authigenic, detrital, or evaporitic) include Ca-phosphate (fluorapatite, found in replaced carbonate and non-carbonate components, fish remains, and some brachiopod shells), pyrite, glauconite, gypsum, anhydrite, halite, and hematite. In general, non-carboante authigenic minerals are regarded as good paleoenvironmental markers (e.g., redox, trophic, and evaporitic conditions, sedimentation rate). The non-mineral component of some carbonate facies is organic matter, which can be in the form of kerogen or bitumen.

4.3 The Microfacies Concept

As previously mentioned, from this chapter this book is focused on the description of mm-scale images and their components. Therefore, defining the concept of "microfacies" is necessary. As originally defined by Brown (1943) and again independently by Cuvillier (1952), the term microfacies referred only to petrographic and paleontological criteria studied in thin-sections. However, later Flügel (2010) expanded the concept of microfacies as the total of all sedimentological and paleontological data which can be described and classified from thin sections, peels, polished slabs or rock samples.

4.4 Texture and Fabric

To classify carbonate microfacies is essential to make a proper use of the concepts of texture and fabric. The *texture* collectively refers to size, shape, and arrangement of components in a carbonate rock. The *fabric* describes orientation in space of particles, matrix, and cement (definitions taken from James and Jones 2015). For avoiding ambiguities, in this chapter and through the rest of the book, regarding classification of microfacies we use term *texture* for describing depositional and diagenetic textures; although we also use this term for describing carbonate grains, precipitates, recrystallization products, and dolomite crystals. On the other hand, we use the term *fabric* mainly for describing packing densities and patterns (e.g., mud- or grain-supported, fitted/condensed, clotted, agglutinated, fenestral, microbial, among others).

4.5 Classification Schemes for Carbonate Microfacies

An ideal classification scheme for carbonate microfacies combines objective, quantifiable description of readily observable features that are grouped into named categories. At the same time, it is desirable to have groupings that incorporate a maximum level of genetic or interpretive significance (groupings that reflect mechanisms of formation, environments of deposition, and the like; Scholle and Ulmer-Scholle 2003).

Most petrographic analyses for limestones are for the purpose of environmental interpretations. The color of the rock is probably the most easily distinguishable property that allows infer some of the main paleoenvironmental conditions of the depositional and/or diagenetic environment. The two most important features of a limestone seen in thin section are the actual grain properties, including composition, granulometry and morphometry (shape and size); and the rock fabric, which is the arrangement of grains and/or crystals (Tucker and Wright 1990).

For classifying carbonate microfacies we combine descriptive, genetic and diagenetic approaches. The textural classification falls within the descriptive branch, as it is based on differentiating the main components (grains, matrix, *in situ* organisms, and cements), relative abundance (%) between them and the packing of those components (mud-supported or grain-supported fabrics). The genetic classification requires an understanding of the processes responsible for the origin of a given rock type [see Flügel (2010) for a complete review on classification of carbonate microfacies). In this book we use the classifications of Dunham (1962) and Embry and Klovan (1971), with the addition of some modifier terms applied to describe diagenetically altered rocks.

4.5.1 Dunham Classification Scheme

In this book we mostly use the classification scheme of Dunham (1962), since it is, in our opinion, the most practical and easy to apply; and it is used worldwide, especially in the Oil and Gas industry (e.g., Lucia et al. 2003). This classification is based on the depositional textures of the rock (Fig. 4.1), but also carries genetic implications, since it uses the relative abundance (%) of the rock's main components, their packing and their autochthonous or allochthonous origin for environmental implications.

The scheme consists of two main groups: rocks with recognizable depositional textures and rocks with no recognizable depositional textures. The classification also distinguishes between: rocks whose components were not bound together

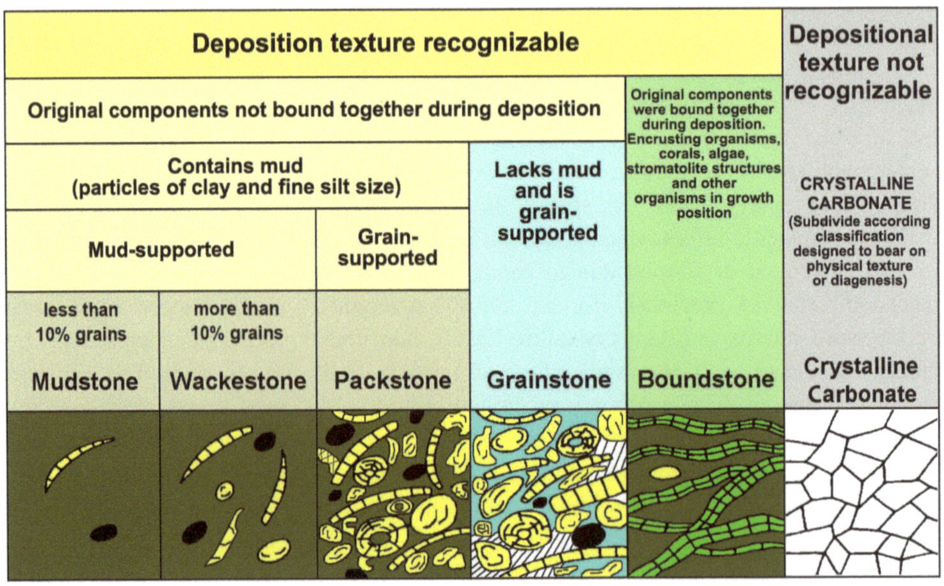

Fig. 4.1 Diagram showing the six stages spectrum of Dunham classification of carbonate rocks according to their depositional textures (Dunham 1962). Figure reprinted by permission of the AAPG whose permission is required for further use

during deposition (i.e., allochthonous rocks), and rocks supported by a primary framework and whose original components were bound together during deposition (i.e., autochthonous rocks). The four main rock types are named according to the relative percentage of grains and micrite (or lime-mud): *Mudstone*, *Wackestone*, *Packstone* and *Grainstone*. The micrite content decreases from left to right, which in most cases is accompanied by an increase of the energy level. It is noteworthy that grainstones can be partially or totally cemented, or completely lack of porosity-filling cement.

Autochthonous limestones showing signs of being lithified or organically bound during deposition, such as coral reefs and microbialites (stromatolites and thrombolites), are named *Boundstone*. We consider beachrocks to be boundstones, as these deposits result from rapid *in situ* lithification in the intertidal zone, and in many cases occur under microbial activity via induced cementation of $CaCO_3$ (see Mauz et al. 2015; Haredy et al. 2019). Recrystallized rocks that have gone through diagenetic processes leading to changes in crystal sizes and textures (i.e., dolomitization and recrystallization) are named *Crystalline Carbonate*. As previously mentioned, further modifiers can be applied, not only for diagenetic processes, but also to describe grain types and dominant organisms, although these are not all explicitly codified by Dunham (Scholle and Ulmer-Scholle 2003). The classification of dolomites and dolostones (mineral and rock, respectively) is covered in Chap. 18.

The name of the rock (e.g., mudstone, packstone, boundstone) is preceded by the dominant grains (e.g., green algae wackestone, peloidal packstone); by the main characteristics of the fabric (e.g., laminated mudstone, sparse/packed wackestone, condensed/fitted packstone); by the main diagenetic processes or products (e.g., cemented grainstone, recrystallized packstone, pyritic mudstone); thus implying both depositional texture of the rock, as well as subsequent diagenetic alterations. We use the adjectives "skeletal" or "bioclastic" for grouping three or more groups of calcifiers. Compositional features of the matrix (e.g., organic-rich, phosphatized) are also indicated. In addition, gradation between rock types (e.g., wackestone-packstone) may also be considered. For naming mixed carbonate-siliciclastic rocks we use adjectives (sandy, silty, clayey, argillaceous) to refer to the siliciclastic content (e.g., sandy wackestone, clayey mudstone).

4.5.2 Embry and Klovan Classification Scheme

Embry and Klovan (1971) extended and modified the Dunham classification scheme, by introducing five new genetic-descriptive terms (Fig. 4.2). First, Embry and Klovan's classification differentiates mud- and grain-supported, allochthonous limestones according to the size of the grains: less than 2 mm (mudstone, wackestone, packstone and grainstone) and greater than 2 mm, i.e., mud-supported *floatstone* (corresponding to Dunham's wackestone) and grain-supported *rudstone* (corresponding to Dunham's packstone and grainstone; it is noteworthy that *rudstones* can be either muddy or mud-lean, as well as totally/partially cemented or not). Secondly, this classification divides boundstone rocks in three categories based on the genesis of the autochthonous limestones: *bafflestone*,

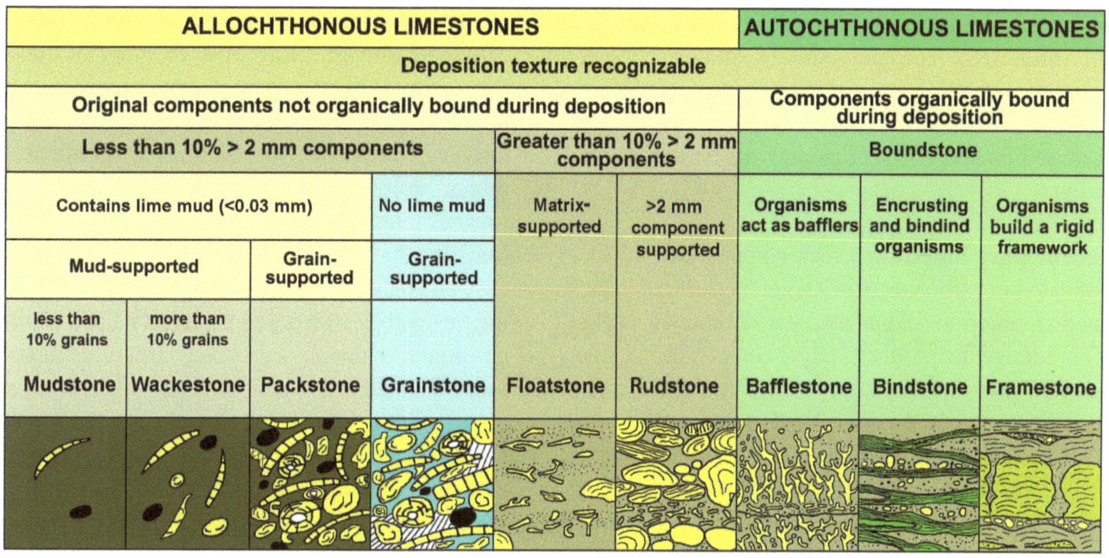

Fig. 4.2 Diagram showing the nine stages spectrum of Embry and Klovan expanded classification of carbonate rocks (Embry and Klovan 1971). Figure reprinted by permission of CSPG whose permission is required for further use. © 1971 - Canadian Society of Petroleum Geologists

bindstone, and *framestone* (see Flügel 2010, for details on boundstone types). *Bafflestones* are carbonate rocks containing *in situ* stalk-shaped and branching fossils that trapped sediment during deposition by acting as bafflers, slowing down water movement and allowing sediment to settle (e.g., encrusting red algae and serpulid tubes). *Bindstones* are carbonate rocks containing *in situ* lamellar, tabular, or encrusting fossils and/or allochems (skeletal or non-skeletal) that encrusted and bound sediment during deposition. In bindstones the matrix, not the *in situ* fossils, forms the supporting framework of the rock (e.g., stromatolites). *Framestones* are carbonate rocks containing *in situ*, massive sedentary fossils (e.g., scleractinian corals) that built a rigid, three-dimensional supporting framework at the time of deposition.

4.6 Terminology for Classifying Carbonate, Mixed Carbonate-Siliciclastic and Siliceous Rocks

We use a group of terms to classify carbonates and mixed carbonate-siliciclastic and siliceous sedimentary rocks based on the content (wt.%) of carbonate minerals ($CaCO_3$ polymorphs and dolomite), measured geochemically or estimated optically. Carbonate content is not only easy to measure but also among the most standard and useful geochemical parameters in the study of marine sedimentary rocks. Depending on the lithology, the remaining proportion is assumed to consist of siliciclastic-terrigenous sediments (mainly quartz and clay minerals and subordinate feldspar and muscovite), sedimentary silica (biogenic or authigenic), authigenic minerals (e.g.,fluorapatite, pyrite, glauconite, and gypsum), and organic matter. Some of these terms follow Bates and Jackson (1980), but we have defined the carbonate cut-off values (%). The grain size of siliciclastic sediments is also considered.

For mixed carbonate-siliciclastic rocks, when the content of carbonate sediments is 50% or greater, the terms used are the following:

Limestone (or pure limestone): rock composed of 80% or more of calcium carbonate ($CaCO_3$) in the form of grains (allochems), micritic sediments, autochthonous organisms, or authigenic products (e.g., cement, microspar, pseudospar); plus or minus dolomite crystals and siliciclastic sediments (mainly quartz and clays and subordinate muscovite and feldspar).

Dolomite or Dolostone: rock composed of 50% or more of dolomite [$CaMg(CO_3)_2$] crystals.; plus or minus calcium carbonate (mainly in the form of micrite and subordinate allochems) and siliciclastic sediments (mainly clays and subordinate quartz).

Dolomitic Mudstone: rock composed of 50% or more of calcium carbonate in the form of micritic sediments comprising the matrix, with dolomite crystals (10% or more); plus or minus siliciclastic sediments (mainly clays and subordinate quartz).

Clayey Limestone: rock composed of 50% or more of calcium carbonate (micrite, allochems, cement, pseudospar), as well as considerable amounts of clay minerals (i.e., clay-sized sediments) comprising the matrix (the adjective “clayey” can be replaced by “argillaceous”); plus or minus dolomite crystals and siliciclastic sediments (quartz, feldspar, and muscovite).

Sandy Limestone: rock composed of 50% or more of calcium carbonate, as well as considerable amounts of sand-sized quartz grains; plus or minus dolomite crystals and siliciclastic sediments (feldspar, muscovite, and clays).

Silty Limestone: rock composed of 50% or more of calcium carbonate, as well as considerable amounts of silt-sized quartz grains; plus or minus dolomite crystals and siliciclastic sediments (feldspar, muscovite, and clays).

Sandy Dolomite: rock composed of 50% or more of dolomite crystals, as well as considerable amounts of sand-sized quartz grains; plus or minus calcium carbonate (micritic sediments and grains) and silicilastic sediments (feldspar, muscovite, and clays).

Clayey Dolomite: rock composed of 50% or more of dolomite crystals, as well as considerable amounts of clay minerals comprising the matrix; plus or minus calcium carbonate (mainly micritic sediments and subordinate grains) and siliciclastic sediments (quartz, feldspar, and muscovite).

When the content of carbonate sediments is lower than 50%, the terms used are the following:

Calcareous Shale: rock composed of less than 50% calcium carbonate and therefore 50% or more of clay minerals (e.g., illite, kaolinite, smectite, montmorillonite) comprising the matrix, with $CaCO_3$ (10% or more) in the form of micrite, allochems, or authigenic products; plus or minus dolomite crystals and siliciclastic sediments (quartz, feldspar, and muscovite).

Dolomitic Shale: rock composed of less than 50% dolomite and therefore 50% or more of clay minerals comprising the matrix, with dolomite crystals (10% or more); plus or minus calcium carbonate (mainly micritic sediments and subordinate grains) and siliciclastic sediments (quartz, feldspar, and muscovite).

Calcareous Sandstone: rock composed of less than 50% calcium carbonate and therefore 50% or more of sand-sized quartz grains, with $CaCO_3$ (10% or more) in the form of micrite, allochems, or authigenic products; plus or minus silicilastic sediments (feldspar, muscovite, and clays).

Dolomitic Sandstone: rock composed of less than 50% dolomite and therefore 50% or more of sand-sized quartz grains, with dolomite crystals (10% or more); plus or minor silicilastic sediments (feldspar, muscovite, and clays) and calcium carbonate (micritic sediments and allochems).

For mixed carbonate-sedimentary siliceous rocks, when the content of carbonate sediments is lower than 50%, the terms used are the following:

Chert: rock composed of less than 50% calcium carbonate and therefore 50% or more of sedimentary quartz (microcrystalline or cryptocrystalline quartz and/or radiolarian tests and other silica-secreting organisms); plus or minus siliciclastic-terrigenous sediments (clays and quartz) and calcium carbonate (mainly micritic sediments and tests of planktonic and smaller benthic foraminifera).

Calcareous Chert: rock composed of less than 50% calcium carbonate and therefore 50% or more of sedimentary quartz, with $CaCO_3$ (10% or more) mainly in the form of tests of foraminifera and subordinate micrite; plus or minus siliclastic sediments (clays and quartz).

4.7 Black Shales

We classify some microfacies as "black shales". Black shale is a descriptive (non-genetic) term used in reference to fine-grained, laminated, organic-rich rocks (generally TOC > 1%) deposited in a broad swathe of sedimentary environments that does not take into account the proportion (%) between carbonate, siliciclastic minerals, and quartz of sedimentary origin (see Arthur and Sageman 1994; Wignall 1994; Trabucho-Alexandre et al. 2012). In this book we use this term in reference to organic-rich rocks (plus considerable or minor amounts of planktonic foraminifera) that were deposited during the Cretaceous OAEs.

4.8 Carbonate Rock Components and Microfacies Classification: Photographic Set and Descriptive Captions

In all photos, IW = image width (the true length of the horizontal axis of the view)

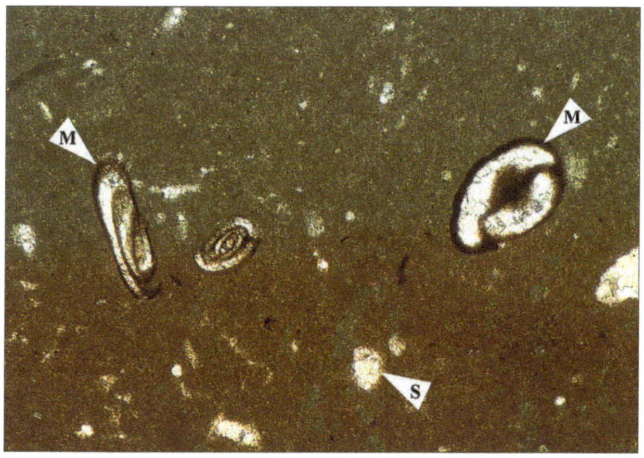

Cretaceous (middle Aptian). Piché Member, Apón Formation. Maracaibo Basin

Smaller miliolids (M) and sponge spicules (S) mudstone. This mud-supported fabric (less than 10% grains), as all mudstones (which are characterized by scarce grains floating in the micrite matrix), was deposited in a low-energy depositional environment. Dark-colored, calcitic walls of the miliolids are clearly differentiated from the slightly organic-rich microcrystalline matrix.

IW = 2.4 mm. PPL

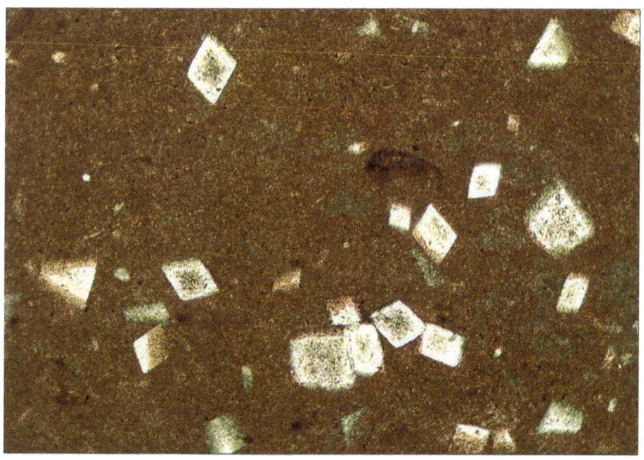

Cretaceous (early Aptian). Machiques Member, Apón Formation. Maracaibo Basin

Dolomitic, organic-rich mudstone. This micritic limestone is barren of carbonate grains, a common feature of some microfacies of the Machiques Member (OAE 1a) that were deposited under anoxic and hypertrophic conditions. The formation of dolomite (rhombic) crystals replacing the lime-mud matrix is indicative of slightly hypersaline conditions during early diagenesis. The light-red coloration of the matrix is due to Alizarin Red S staining technique, which is used to distinguish calcite from dolomite crystals, as the latter preserve their original color.

IW = 2.4 mm. PPL

Cretaceous (early Aptian). Machiques Member, Apón Formation. Maracaibo Basin

Organic-rich, pyritic mudstone. The abundance of pyrite framboids (black spots) scattered through the micritic matrix is the most prominent feature of this sample, which is also barren of carbonate skeletal components. These authigenic precipitates indicate anoxic-euxinic conditions (free H_2S/HS^- in the water column) and sufficient availability of reduced iron (Fe^{2+}). A very low-energy depositional environment and stressful conditions for the development of calcareous biota are associated with deposition of this microfacies.

IW = 1.3 mm. PPL

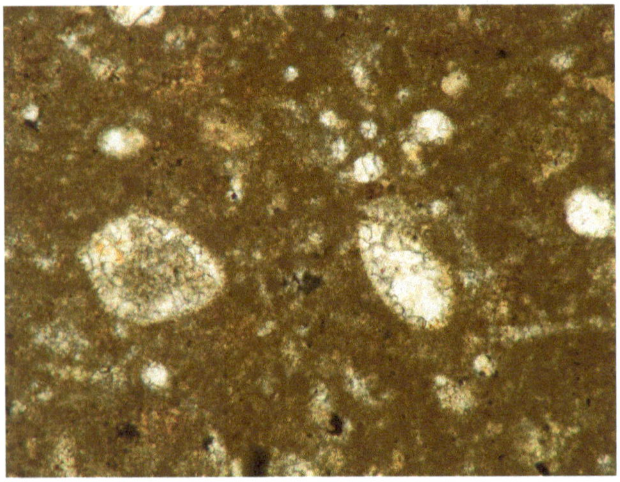

Cretaceous (early Aptian). Machiques Member, Apón Formation. Maracaibo Basin

This mud-supported fabric would, in most cases, be classified as a wackestone, though it may also be interpreted as a mudstone-wackestone gradation. The micritic matrix is moderatelly enriched in organic matter (1.1% TOC). Larger bioclasts consist of egg-shaped ostracods, whereas the smaller spherical grains are sponge spicules (heterozoan association). This grain association, together with the TOC content of the sample and its stratigraphic position in the upper part of the OAE 1a, suggest deposition under low-oxygen conditions (disoxic to suboxic).

IW = 1.1 mm. PPL

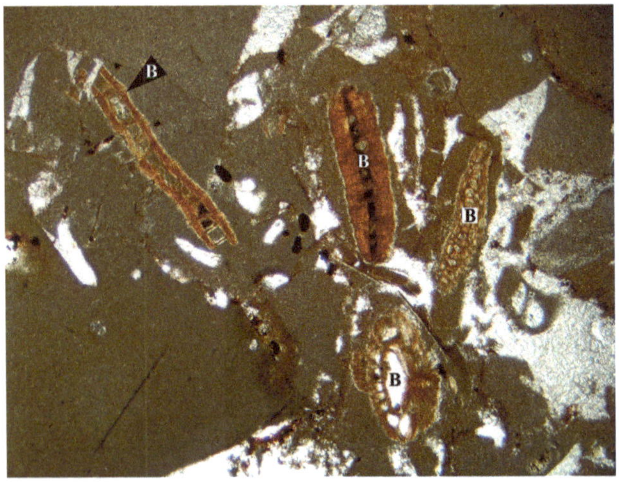

Early Miocene. Churuguara Formation. Falcón Basin

LBF (B), patchy sparry, sparse wackestone. This mud-supported fabric shows scattered patches of microspar (white precipitate). Tests of LBF (symbiont-bearing, light-dependent) represent more than 10% of the components. These are replaced by fluorapatite (Ca-phosphate). This mineral exhibits a characteristic brownish-orange color and its precipitation suggests a very low-energy setting during early marine diagenesis. Intermediate depths within the photic zone (60–80 m) are associated with the deposition of this microfacies.

IW = 7.5 mm. PPL

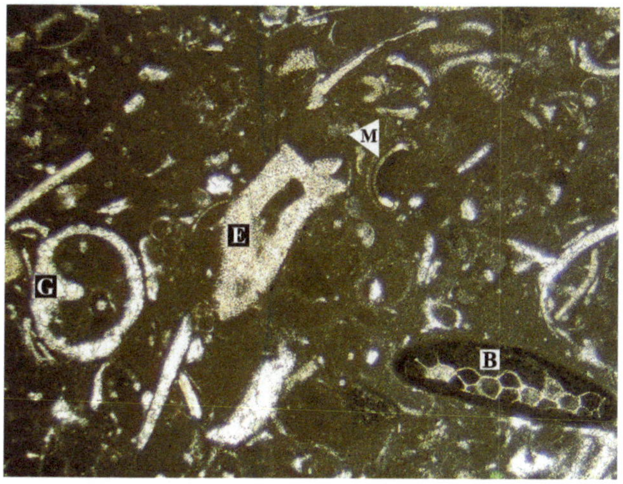

Cretaceous (middle Albian). Lisure Formation. Maracaibo Basin

Skeletal wackestone containing different-sized fragments of bivalve molluscs (M), echinoderms (E), gastropod (G), and a bryozoan (B) (heterozoan association) coated by microbially-induced micrite. The grain content is greater than 10%, but these are clearly supported by microcrystalline micrite. Mud-supported, skeletal microfacies dominated by heterotrophic (light-independent) calcifiers, such as this example, are very common in shallow carbonates of the Lisure Formation.

IW = 4.2 mm. PPL

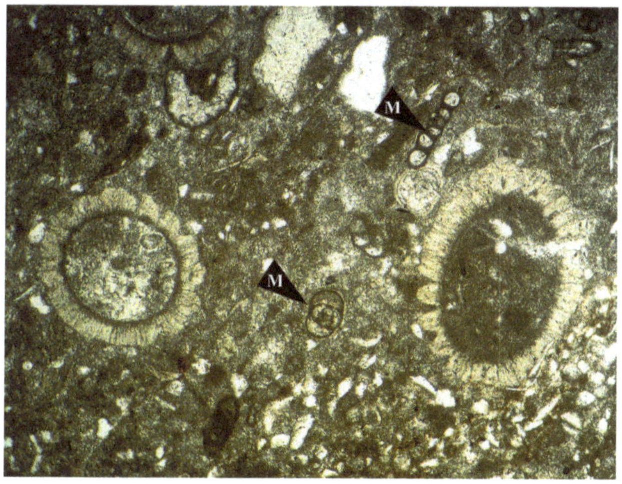

Cretaceous (middle Aptian). Piché Member, Apón Formation. Maracaibo Basin

Dasycladalean green algae (ring-like grains) and smaller miliolids (M) (seen through different cuts), sandy wackestone. Abundant quartz grains (small whitte grains) are scattered through the matrix. In general, green algae (photoautotrophs) thrive almost exclusively in the upper photic (or euphotic) zone; while smaller miliolids (symbiont-barren benthic foraminifera; heterotrophs) are common in shallow-lagoonal muddy facies. This photozoan-heterozoan association is quite common in the Piché Member.

IW = 2.4 mm. PPL

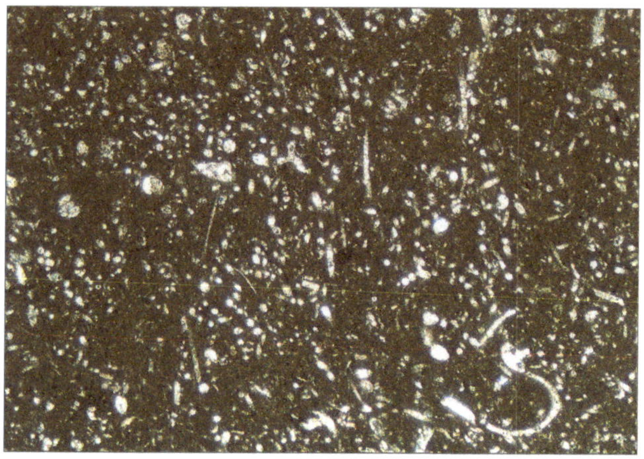

Cretaceous (middle Aptian). Piché Member, Apón Formation. Maracaibo Basin

Sponge spicules wackestone. Although small-sized, spherical to elongate spicules are abundant in the image, these are floating in the matrix, hence comprising a mud-supported fabric. Spicule-rich (or spiculite) microfacies in the Cogollo Group were deposited in low-energy environments. They are characterized by a moderate dark organic-rich matrix, which is clearly seen in the image. The latter suggests mesotrophic to eutrophic conditions and most likely strong microbial activity for the flourishing of these heterotrophs.

IW = 4.2 mm. PPL

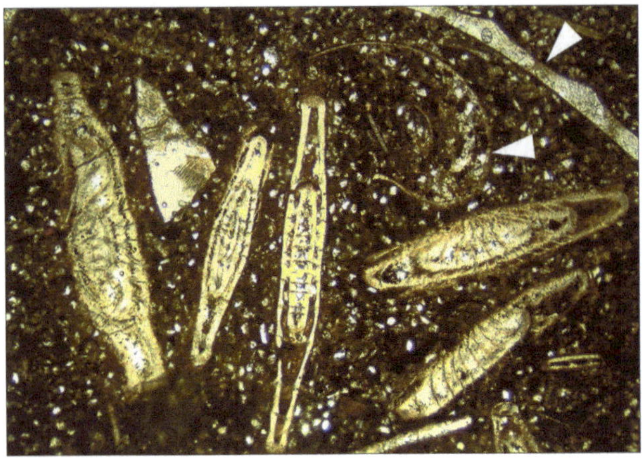

Early Miocene. San Luis Formation. Falcón Basin

LBF and molluscs, sandy wackestone. Tests of LBF (large and flat) are the main bioclastic components. Some of these tests are larger than 2 mm; hence, this sample is also classified as a floatstone. Large and thin fragments of molluscs (heterotrophs), indicated with white arrows, are secondary components. The moderate organic richness of the matrix is noted by its dark color. Quartz grains are scattered in the matrix, representing an important influx of siliciclastic sediments from the shoreline. This photozoan-heterozoan association is common in Miocene carbonates of the FB.

IW = 7.5 mm. PPL

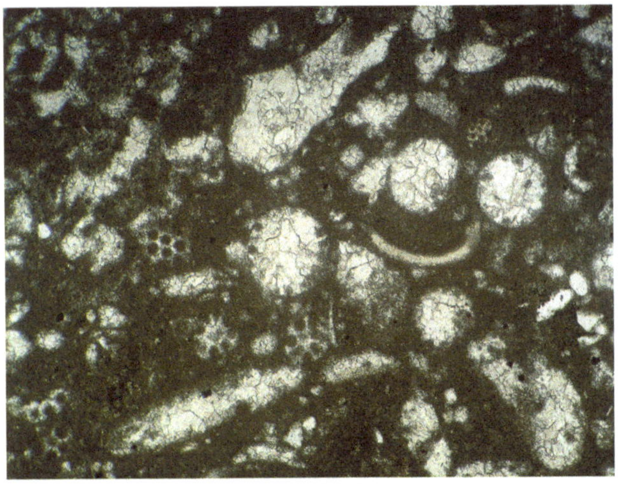

Cretaceous (middle Aptian). Piché Member, Apón Formation. Maracaibo Basin

Green algae wackestone-packstone. Some green algal grains preserved characteristic tubules (small circular elements filled with lime-mud). In most grains, the aragonite dissolution and subsequent precipitation of LMC led to their preservation as structureless, spherical to elongate grains. The grain content relative to the micritic matrix is clearly greater than 10% and probably ranges between 30–40% (visual estimation). The subjective use of Dunham's classification allows the classification of this sample as a wackestone-packstone gradation, or as a packed wackestone.

IW = 2.4 mm. PPL

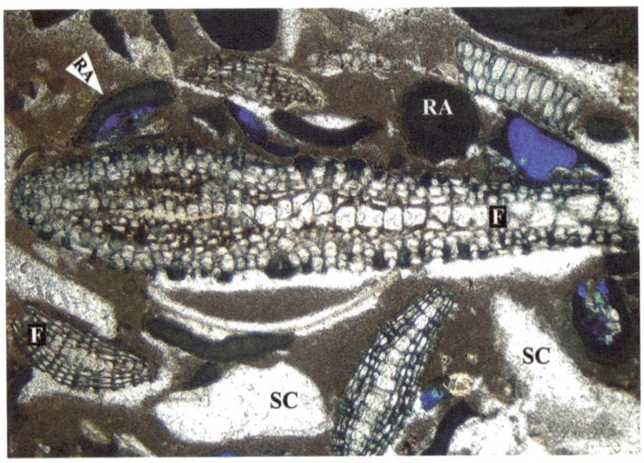

Early Miocene. Churuguara Formation. Falcón Basin

Skeletal wackestone-packstone. Skeletal grains in the image (some of them larger than 2 mm) consist of LBF (F) and red algae (RA) (photozoan association). Patches of sparry calcite (SC) were formed from selective dissolution of aragonitic bioclasts (most probably molluscs) and further dissolution of micritic sediments surrounding them, followed by subsequent precipitation of LMC. Some elongate grains are in tangential contact with each other. The relative abundance between matrix (brown) and grains (including leached grains) in this sample appears visually quite similar.

IW = 4.2 mm. PPL

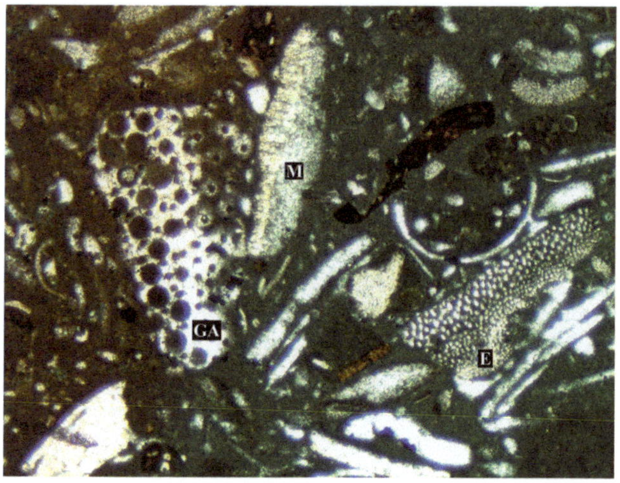

Cretaceous (middle Aptian). Piché Member, Apón Formation. Maracaibo Basin

This skeletal wackestone-packstone gradation consists of fragments of molluscs (M), echinoids (E), and dasycladalean green algae (GA) (heterozoan-photozoan association, but heterozoan dominated). Contacts between poorly sorted grains are abundant, but the fabric is not entirely grain-supported, as small grains floating in the micritic matrix are also abundant.

IW = 2.4 mm. PPL

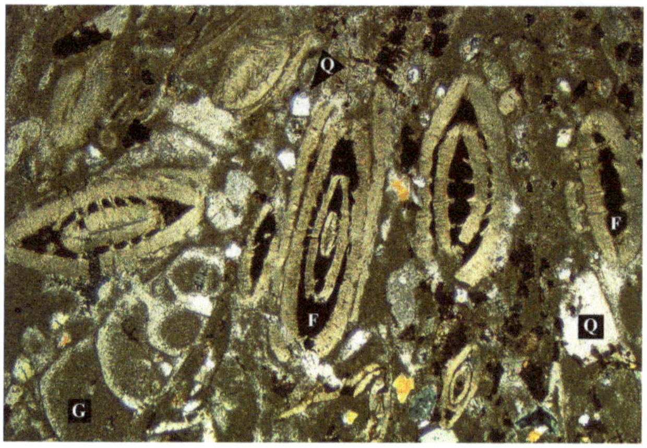

Early Miocene. Churuguara Formation. Falcón Basin

LBF (F) and gastropod (G), sandy packstone. Some of the skeletal grains reach 2 mm in length, therefore this microfacies can also be named a rudstone. This is an example of a grain-supported texture in which the abundance of benthic skeletal grains and lime-mud indicates different energy levels during deposition, making packstone an ambiguous term. Grain-supported fabrics are commonly associated with moderate to high-energy environments. However, for this sample the energy level was not enough to wash away all the micritic sediments. Quartz (Q) grains exhibit characteristic gray to white and occasionally yellow color under CPL.

IW = 4.2 mm. CPL

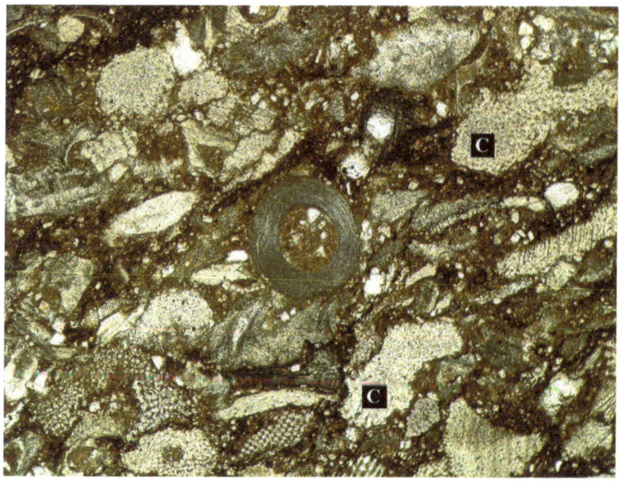

Middle Permian (Cisularian to Guadalupian). Palmarito Formation. Venezuelan Andes

Crinoidal, condensed packstone. Crinoid (C) grains (heterotrophs) exhibit irregular shapes. These are accompanied by a well-preserved brachiopod spine (heterozoan association) in the center of the image (hollow spherical grain). Physical (mechanicaL) and chemical compaction processes formed this packed fabric. The former is evident due to the sutured contacts along grains, whilte the latter is noticed due to abundantl solution seams (or microstylolites) between grains. In spite of the bioclastic content, the presence of lime-mud between grains is clearly evidenced in this grain-supported fabric.

IW = 2.4 mm. PPL

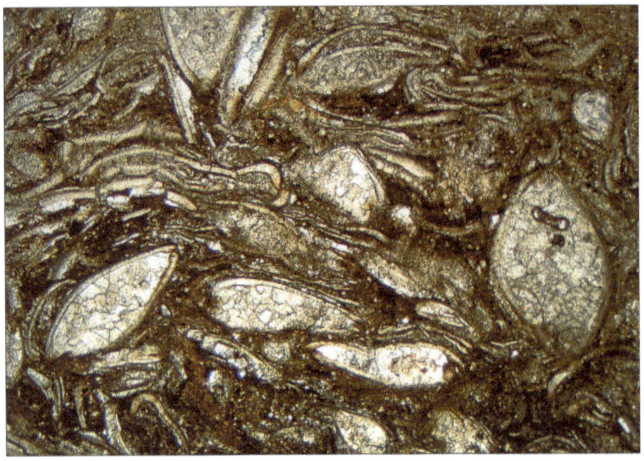

Middle Permian (Cisularian to Guadalupian). Palmarito Formation. Venezuelan Andes

Ostracods, condensed packstone. This grain-supported fabric consists of a high density of ostracod carapaces (heterotrophs). Most individual valves are preserved as disarticulated elements, as most ostracod carapaces were fractured due to mechanical (burial) compaction. Some carapaces remained articulated, as coarse-grained LMC infillings (cloudy white in color) prevented their fracturing. The dark-brown color of the micritic matrix suggests moderate organic richness and high nutrient levels during deposition.

IW = 4.2 mm. PPL

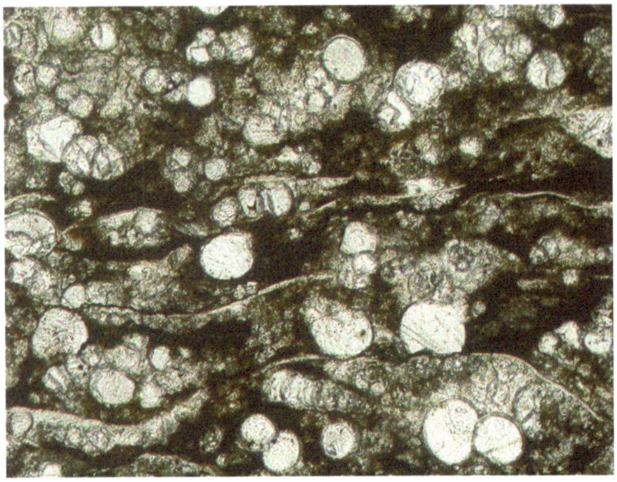

Cretaceous (middle to late Cenomanian). La Luna Formation sensu stricto. Maracaibo Basin

Planktonic foraminifera, organic-rich packstone. Although the high organic matter content masks the optical visualization of the mineral matrix, the abundance of globular planktonic foraminifera tests indicates a carbonate-dominated composition for this sample. Contrary to moderate- to high-energy level associated with many grain-supported fabrics, microfacies rich in planktonic foraminifera indicate a very low-energy level. Hypertrophic conditions, high organic matter productivity, and seafloor anoxia are associated with deposition of this black shale.

IW = 1.3 mm. PPL

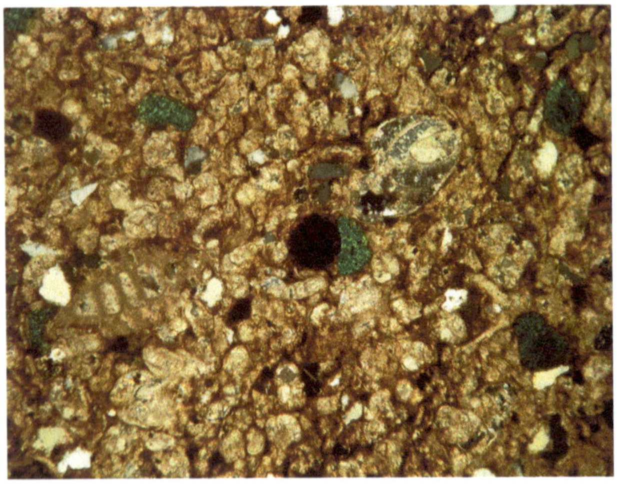

Cretaceous (early Campanian). Tres Esquinas Member, La Luna Formation. Venezuelan Andes

Foraminiferal, glauconitic, sandy packstone. Bioclasts in the image are a combination of SBF and planktonic foraminifera. The abundance of micritic sediments is low but noticeable due to their dark brown color between grains. Glauconitic (green) grains are scattered through the sample. Angular, fine sand-sized quartz grains are differentiated by their white color. In general, the presence of glauconitic grains as indicators of dysoxic conditions and the decrease in organic richness compared to underlying La Luna Formation *sensu stricto* (organic-rich facies) characterizes the lithology of the Tres Esquinas Member.

IW = 2 mm. CPL

Photomicrograph courtesy of Grony Garbán

Cretaceous (middle Albian). Lisure Formation. Maracaibo Basin

Non-skeletal, patchy sparry, muddy packstone. Different-sized grains in the image consist of peloids (micritic and glauconitic), aggregate grains cemented together by muddy sediments, and elongate to rounded bioclastic fragments (echinoids and molluscs) coated by micritic and oolitic cortices. Although the sample is classified as a packstone, the micritic content is significant. Patches of sparry calcite are locally infilling small pores and/or replacing the micritic matrix. Glauconite (green) grains were formed from the replacement of micritic peloids and/or ooids.

IW = 4.2 mm. PPL

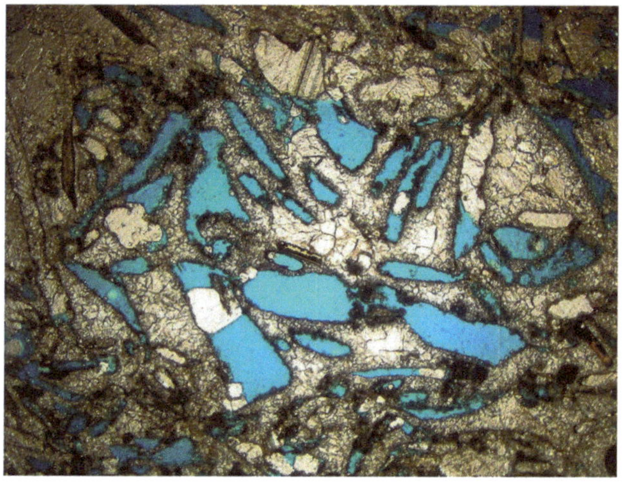

Cretaceous (middle Aptian). Piché Member, Apón Formation. Maracaibo Basin

Recrystallized, porous wackestone-packstone. The original depositional texture was altered in the meteoric environment. Selective dissolution of aragonitic bioclasts (mainly molluscs) created moldic pores. These represent substantial secondary porosity. Coarse-grained pseudospar crystals are replacing the matrix. Few pores are partially filled by calcite cement. This sample corresponds to a highstand deposits in the Cogollo carbonate succession which underwent subaerial exposure. This stratigraphic interval has produced light oil generated by the overlying mature source rock (the OAE 1b first sub-event).

IW = 4.2 mm. PPL

Cretaceous (Campanian). San Antonio Formation. Eastern Interior Range

Phosphatized peloids, sandy condensed packstone. Peloidal grains exhibit inequigranular sizes and are rounded to angular and elongate in shape. The original texture of these allochems was obliterated due to abrasion, reworking and phosphatization. Some of the peloidal grains were blackened due to the precipitation of pyrite postdating phosphatization. Deposition of this granular phosphate microfacies involved the action of P-rich upwelling currents, hydraulic reworking and concentration of the allochthonous grains in a final deposit. In this sample, the content of micritic sediments is very low.

IW = 4.2 mm. PPL

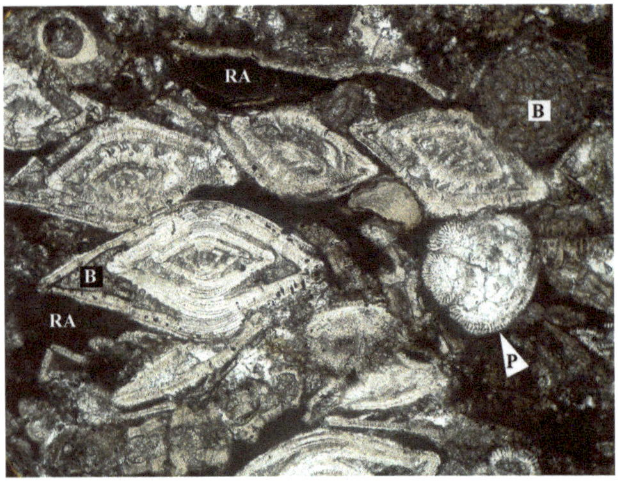

Late Oligocene-Early Miocene. Perla Limestone. Eastern Gulf of Venezuela Sub-Basin

Skeletal, condensed grainstone. Bioclasts consist of LBF (B) (mainly tests of *Amphistegina* sp.), red algae (RA) (squeezed dark grains) and a planktonic foraminifer (P). Note the grain-supported, condensed fabric, with grains in sutured contact and partly interpenetrated, evidencing physical and chemical compaction, respectively. This sample lacks of micritic sediments between grains, a lithologic compositional feature not so easy to recognize, especially in microfacies that have undergone mechanical compaction.

IW = 2.4 mm. PPL

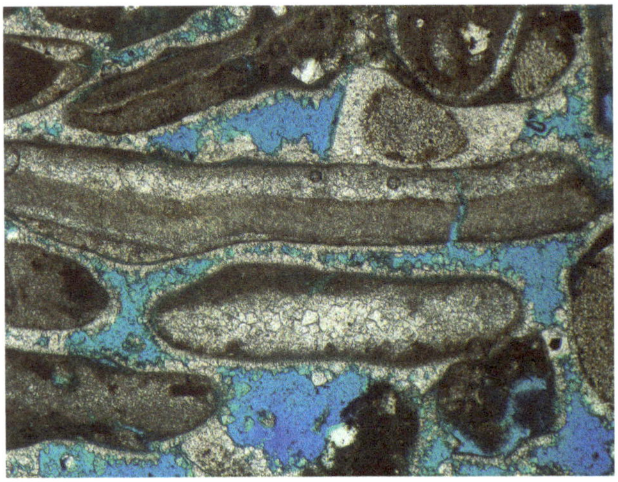

Cretaceous (latest Albian). Maraca Formation. Maracaibo Basin

Porous grainstone. Elongate grains in the image (probably green algae and/or molluscs) were micritized prior to final deposition and later partially meteorically altered to sparry calcite. The absence of micrite sediments between grains led to the preservation of primary interparticle porosity, in combination with partial (or incomplete) cementation. Note that bladed crystals are rimming the grains and meniscus cement is precipitated at grain contacts, but most porosity remains preserved. This grain-supported, mud-lean fabric is an example of some of the best reservoir lithofacies in the Maraca Formation.

IW = 2.4 mm. PPL

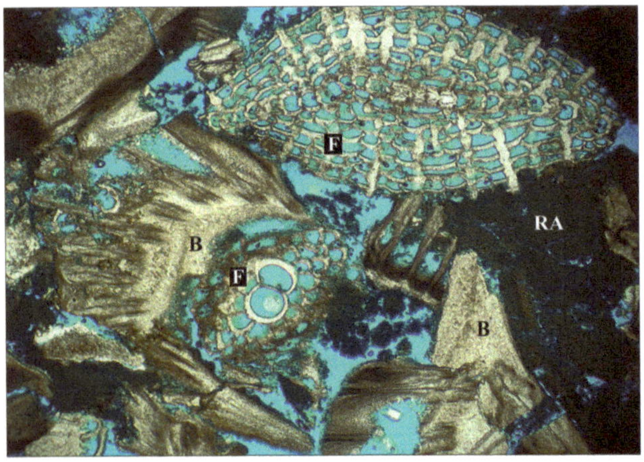

Late Oligocene-Early Miocene. Perla Limestone. Eastern Gulf of Venezuela Sub-Basin

Skeletal porous grainstone. Bioclasts in the image consist of LBF (F), balanid barnacles (B) (heterotrophs), and red algae (RA) (photozoan-heterozoan association). As in the previous example, the lack of micritic sediments during deposition of the skeletal components, combined with the lack of precipitation of pore-filling cement led to preservation of both interparticle and intraparticle primary porosity. This grain-supported, mud-lean fabric is an excellent example of some of the best carbonate reservoirs in Venezuela.

IW = 4.2 mm. PPL

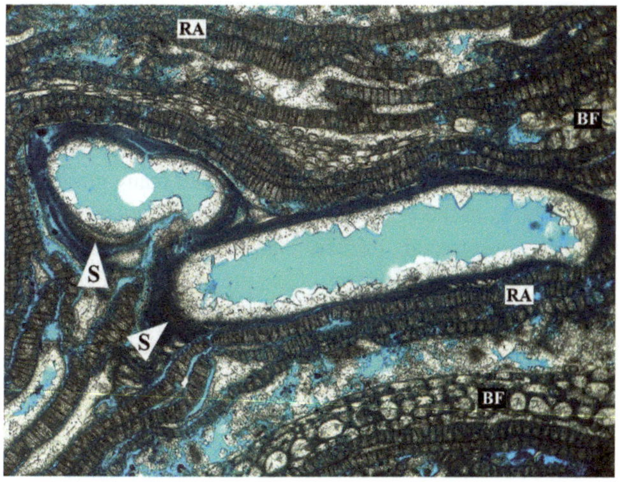

Late Oligocene-Early Miocene. Perla Limestone. Eastern Gulf of Venezuela Sub-Basin

This rhodolith grainstone consists mainly of non-geniculate red algae (RA), with secondary encrusting benthic foraminifera (BF) (photozoan association) and serpulid (S) tubes. This mud-lean fabric can also be classified as a rudstone, as the organims in the image are larger than 2 mm. Rhodoliths in the Perla Limestone can reach up to 10 cm in diameter, far exceeding the field of view of the microscope. In this unit, free-living nodular rhodoliths are associated with agitated, basinward depositional environments. Large intraparticle primary pores are internally rimmed by dogtooth cement.

IW = 2.4 mm. PPL

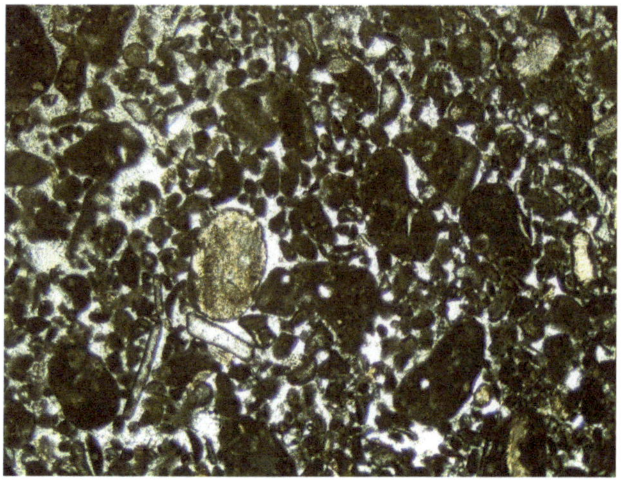

Cretaceous (middle Albian). Lisure Formation. Maracaibo Basin

Non-skeletal and echinoderm cemented grainstone. Precipitation of meteoric calcite cement (white precipitate between grains) completely occluded primary interparticle porosity. Poorly sorted, rounded to irregularly-shaped non-skeletal grains in the image are a combination of micritized skeletal grains (smaller grains) and aggregate grains (larger grains composed of several grains cemented toguether by micrite sediments). Abrasion, reworking and redeposition of these muddy grains in a mud-lean environment led to the formation of this grain-supported fabric. Some echinoderms (light-colored grains) are differentiated because they escaped micritization.

IW = 7.5 mm. PPL

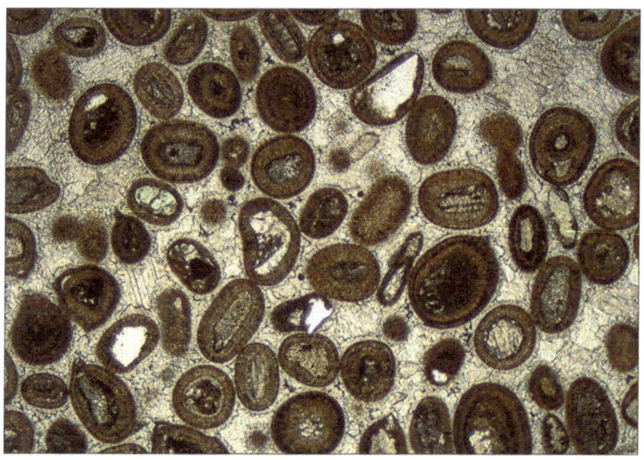

Cretaceous (middle Albian). Lisure Formation. Maracaibo Basin

Oolitic cemented grainstone. Well-sorted ooid grains are composed of a variety of nuclei (mainly carbonate grains and subordinate quartz grains). The original primary interparticle porosity (up to 40% volume) was completely occluded by the precipitation of calcite cement during meteoric diagenesis. Good sorting of the spehrical to elliptical grains and the total absence of lime-mud indicate a strong hydrodynamic control on the deposition of this grain-supported fabric.

IW = 7.5 mm. PPL

Late Oligocene-Early Miocene. Perla Limestone. Eastern Gulf of Venezuela Sub-Basin

Skeletal, cemented grainstone. Grains in the image consist of LBF (F), coralline red algae (RA), and echinoderms (E) (photozoan-heterozoan association, but photozoan dominated). Poor sorting of grains is a common feature of many grain-supported fabrics in the Perla Limestone. As in the two previous examples, substantial primary interparticle porosity was occluded by the precipitation of calcite cement. Microfacies dominated by LBF and red algae (plus secondary heterotrophs) characterize the overall Perla Limestone sedimentation.

IW = 7.5 mm. PPL

Late Pleistocene. Aves Island

This porous calcarenite (grainstone) corresponds to the carbonate basement that immediately underlies the unconsolidated, supratidal sand sediments of Aves Island. It consists mainly of grains of red algae (RA) and molluscs (M), while grains of bryopsidalean green algae (GA) and scleractinian corals (SC) are secondary components, thus making up a photozoan-heterozoan association. Note that the interpaticle pores (P) (black colored spots) are internally rimmed by isopachous rims of marine cement (C). The precipitation of these abiotic rims of aragonitic composition lithified this shallow, subtidal microfacies.

IW = 4.3 mm. CPL

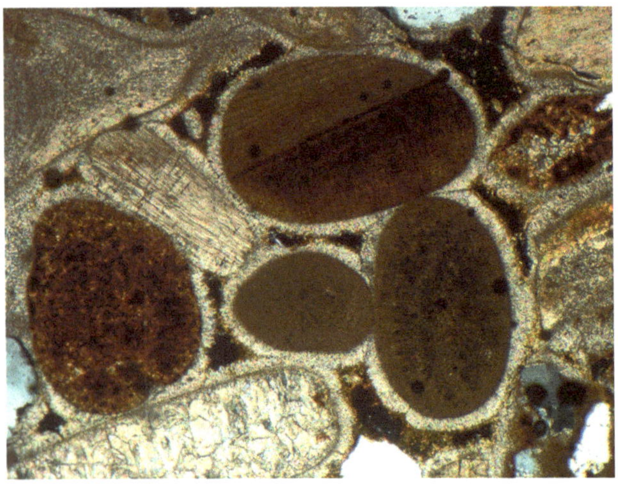

Holocene. Larga Island. Henri Pittier National Park. Aragua State Shoreline

We consider beachrocks under a dual vision. In this beachrock with interparticle porosity (black-colored spots), allochems (grains) in the image consist of red algae (dark and rounded) and molluscs (light brown). Precipitation of isopachous crusts of HMC cement rimming the grains binds them together. This process occurred in the intertidal zone and promoted the *in situ* lithification, which is why we consider beachrocks as boundstone. If one considers that the grains are not organically bound, but cemented by abiotic precipitates, this microfacies can be classified as a grainstone.

IW = 2.4 mm. CPL

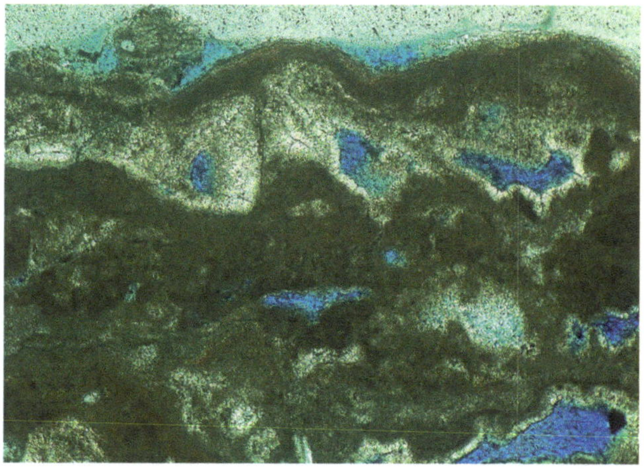

Holocene. Los Roques Archipelago

Micritic boundstone. The photomicrograph shows the uppermost, micritic part of a laminated stromatolite. The structure has grown *in situ* in the shallower areas of a perennial hypersaline lagoon. This microbial fabric can also be classified as a bindstone, since the microbial-induced precipitation of micrite promoted the stabilization of the structure. Note the undulated micritic layer at the top, which is a characteristic textural feature of cyanobacterial-induced organomicrites (automicrites). Marine syngenetic cement is filling fenestral pores created by combined localized carbonate dissolution and organic matter decay.

IW = 1.3 mm. PPL

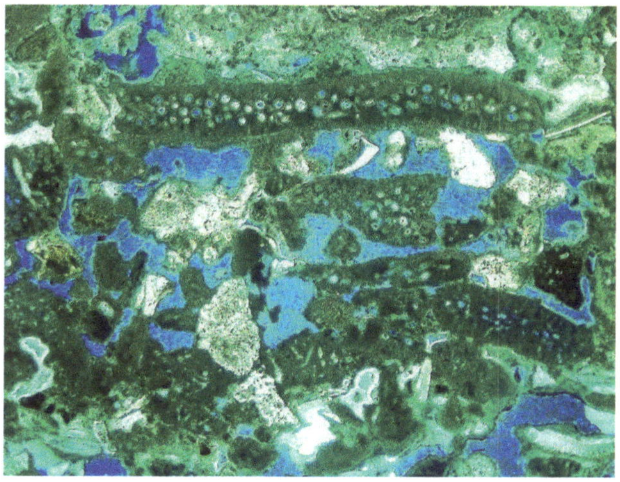

Holocene. Los Roques Archipelago

Green algae-dominated boundstone/bindstone. This photomicrograph corresponds to the lower part of the same laminated stromatolite deposit shown in the previous example. A lower magnification was used for this image to capture the fenestral fabric, which is characteristic of intertidal stromatolites. This bindstone was formed due to trapping/binding and microbial cementation of allochems, mainly *Halimeda* grains. Automicrite, precipitated between grains, is the lithifying cement. Empty pores within the structure are a combination of intraparticle, interparticle, and fenestral pores.

IW = 2.4 mm. PPL

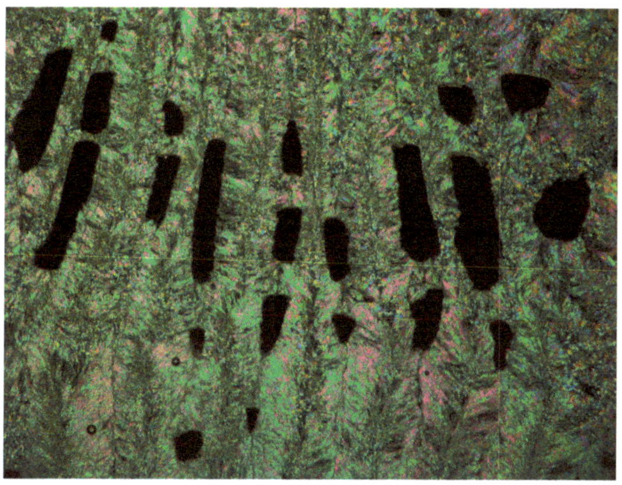

Holocene. Aves Island

This sample corresponds to a scleractinian coral colony found in growth position. It is therefore classified as a boundstone/framestone. The image shows a small portion of the three-dimensional skeletal arrangement of the species *Montastraea cavernosa*. The radiating, fibrous aragonitic bundles that compose the skeleton exhibit high birrefringence under CPL, in contrast to the black spots, which represent intraskeletal pores.

IW = 4.2 mm. CPL

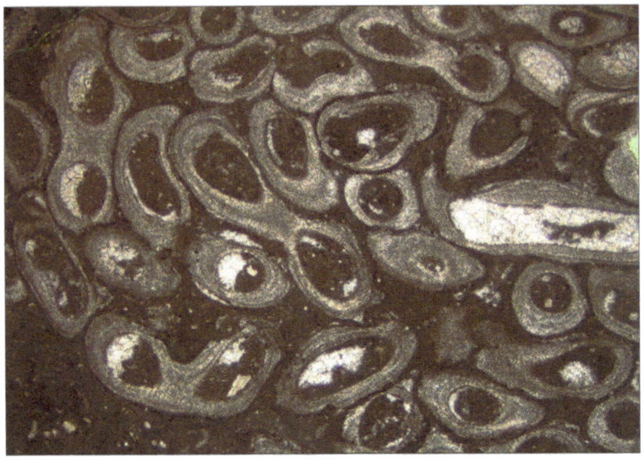

Cretaceous (middle Aptian). Piché Member, Apón Formation. Maracaibo Basin

Serpulids tubes, muddy boundstone. Well-preserved tubes retained their autochthonous arrangement. They grew adjoined to each other, forming an *in situ* growth arrangement that served as a baffle for the accumulation of micritic sediments. Therefore, this microfacies can be also named a bafflestone. Some of the tubes exhibit geopetal infillings (micritic sediments at the bottom and LMC cement at the top), indicating that such infillings postdate the attachment of the tubes originally hosting the filter-feeding (heterotrophic) worms. This sample can be easily misinterpreted as a packstone.

IW = 4.2 mm. PPL

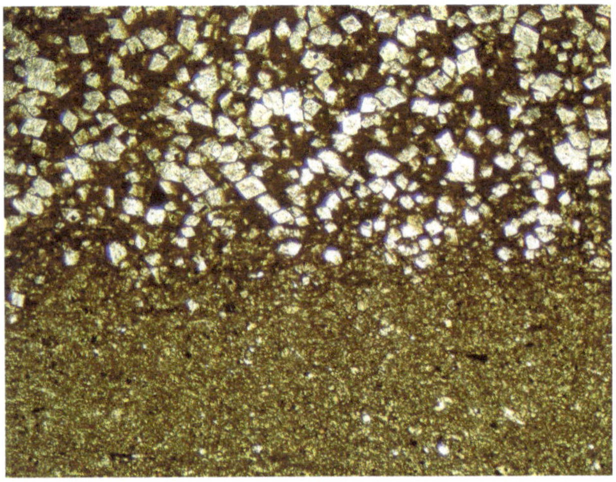

Cretaceous (early Aptian). Machiques Member, Apón Formation. Maracaibo Basin

Transition from an organic-rich micropeloidal mudstone (bottom) to a dolomitic mudstone (top). Well-developed dolomite crystals become abruptly abundant in the upper part of the image. These crystals are the product of the replacement of the organic-rich micritic matrix. Such lithologic change at the top of the Machiques Member has a lateral regional extension along the subsurface of the MB.

IW = 7.5 mm. PPL

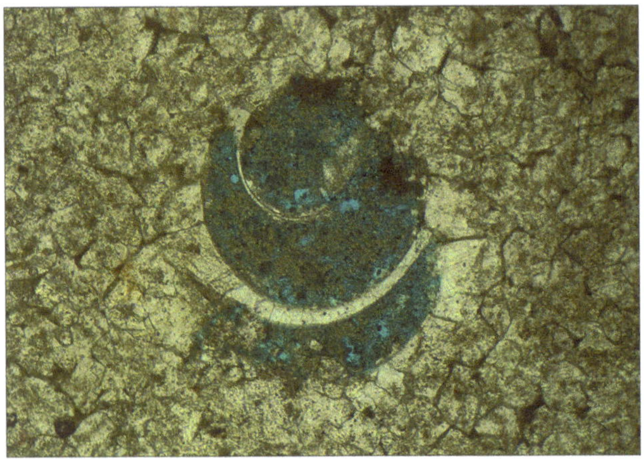

Cretaceous (early-late Aptian). Piché Member, Apón Formation. Maracaibo Basin

Crystalline limestone consisting of a sutured mosaic of tightly packed dolomite crystals. The original depositional texture is not recognizable due to complete replacement of the original calcium carbonate substrate by dolomite. In the center of the image, the shape of a gastropod coiled shell is distinguished and hosts intraparticle biomoldic porosity. Massive dolostone microfacies in the Cogollo Group are associated with transient hypersaline-evaporitic conditions developed in peritidal settings.

IW = 2.4 mm. PPL

Cretaceous (Turonian). O Member, Escandalosa Formation. Barinas Sub-Basin

Oolitic, crystalline limestone. Massive dolomitization (shallow hypersaline and high evaporative conditions) created a packed fabric consisting of a mixture of subhedral and anhedral crystals. Ooids in the image are totally replaced by hematite. The precipitation of this ferruginous oxide mineral during meteoric diagenesis, which exhibits characteristic opaque optical behavior, postdates dolomitization and also most likely selective pyritization of these non-skeletal grains during marine diagenesis.

IW = 4.2 mm. PPL

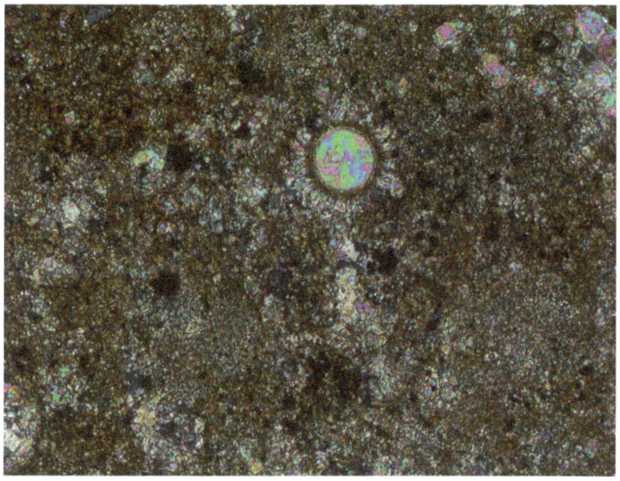

Cretaceous (early Aptian). Machiques Member, Apón Formation. Maracaibo Basin

Recrystallized mudstone. Recrystallization of the micritic matrix led to an increase in crystal size, thus forming a mosaic of small-sized calcite crystals that are visually perceptible. A small microsphere of microbial, authigenic origin exhibiting high birefringence stands out in the absence of allochems. Scattered pyrite framboids, as well as carbonate crystals replaced by silica (low birefringence crystals of gray color), have given this recrystallized texture a heterogeneous and very striking visual appearance.

IW = 0.5 mm. CPL

Early Paleocene. Guasare Formation. Maracaibo Basin

Crystalline limestone. This diagenetic texture consists of coarse-grained LMC crystals of irregular shapes and different sizes derived from recrystallization (crystal enlargement), hence obliterating the original texture of the microfacies. Crystals exhibit twin lamellae (closely-spaced parallel lines exhibiting conspicuous colors), which are features commonly associated to recrystallization during burial.

IW = 1.3 mm. CPL

Cretaceous (Turonian). O Member, Escandalosa Formation. Barinas Sub-Basin

This recrystallized limestone consists of different-sized dolomite crystals (which are the product of $CaCO_3$ replacement) surrounding a large LMC crystal in the center (which is the product of recrystallization) with no visible crystal boundaries and exhibiting twin lamellae. As in the previous example, the original depositional texture of the rock was completely obliterated.

IW = 2.4 mm. PPL

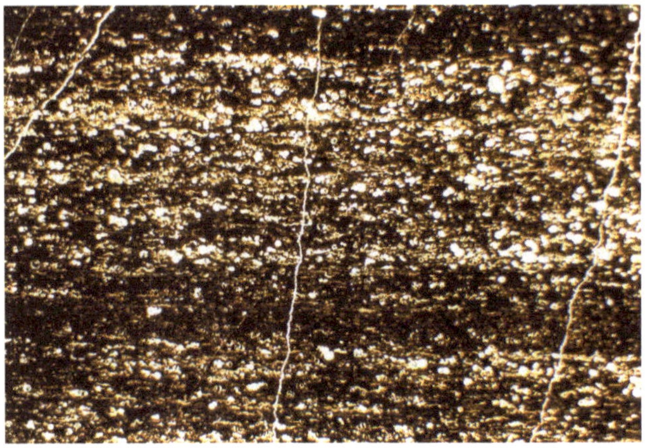

Cretaceous (early Santonian). La Luna Formation *sensu stricto*. Venezuelan Andes

Microlaminated, organic-rich calcareous shale of the upper part of the La Luna Formation *sensu stricto* (OAE 3). The high organic matter content of this black shale (5.5% TOC) masks the mineral matrix, thus preventing its optical visualization. Measured carbonate content is 38%. Carbonate allochems consist mainly of planktonic foraminifera. The mineral matrix is composed of clay minerals and quartz grains (Rey 2004; Rey et al. 2004). Note the high degree of lamination, which indicates a very low energy during deposition.

IW = 5 mm. PPL
Photomicrograph courtesy of Olga Rey

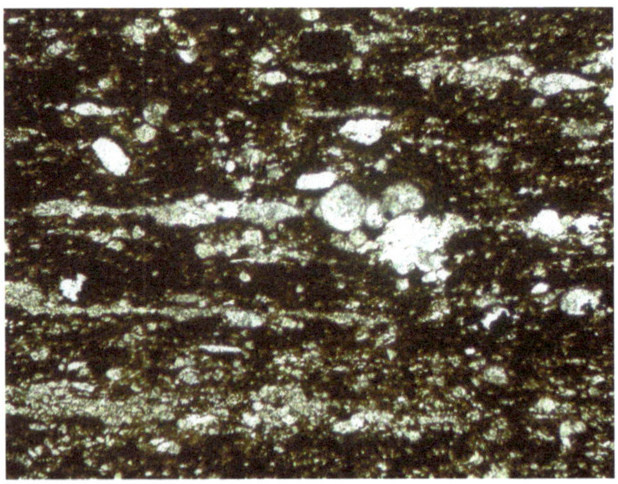

Cretaceous (early Aptian). Machiques Member, Apón Formation. Maracaibo Basin

Laminated, dolomitic black shale having 4.20% TOC and 22.73% carbonate. The carbonate content of this thermally mature calcareous shale (T_{Max} = 444 °C) corresponding to OAE 1a is mainly represented by small-sized, microbially-influenced dolomite crystals and a few identifiable planktonic foraminifera. As in the previous example, the high organic matter content of this black shale masks the optical visualization of the mineral matrix.

IW = 1.3 mm. PPL

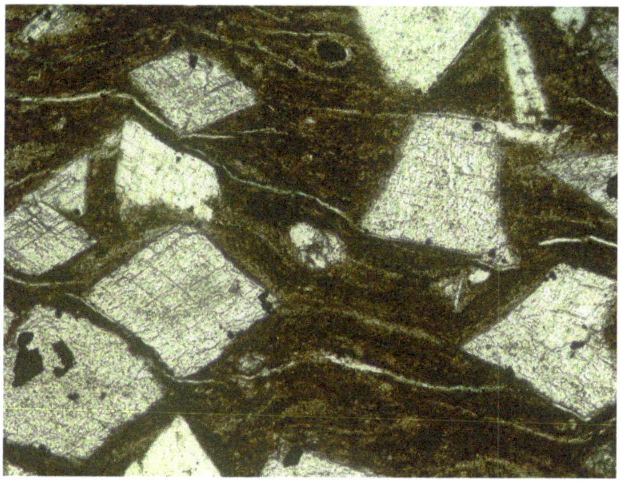

Cretaceous (early Albian). Lisure Formation. Maracaibo Basin

Pyritic, organic-rich dolomitic shale of the OAE 1b second sub-event (base of the Lisure Formation). Euhedral (rhombic) dolomite crystals are embedded in a highly laminated, wavy organic-rich matrix composed of clay minerals. The abundance of pyrite framboids has substantially darkened the sample, indicating anoxic-euxinic conditions. This microfacies is an excellent example of clay-dominated matrices deposited during short-term increased freshwater influx carrying clay minerals and nutrients, which is a common lithologic feature of shallow black shales in the Cogollo Group.

IW = 1.3 mm. PPL

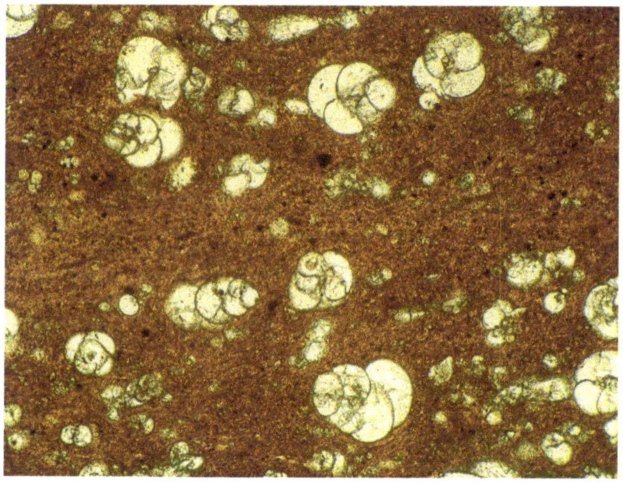

Cretaceous (late Santonian). Táchira Ftanita Member, La Luna Formation. Venezuelan Andes

Calcareous chert. Foraminifera tests represent the carbonate component of this sample. Tests are a combination of SBF (larger globular tests) and planktonic foraminifera (smaller globular tests). The moderate organic-rich matrix is composed mainly of cryptocrystalline quartz (chert) of sedimentary origin. Partially reducing conditions, combined with paleodepths within the upper bathyal zone, and the influence of upwelling currents, are associated with the deposition of this microfacies.

IW = 2.4 mm. PPL

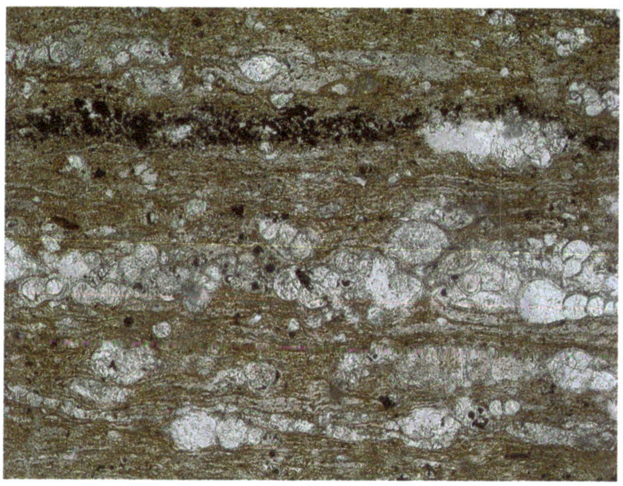

Cretaceous (Coniacian). La Morita Member, Navay Formation. Barinas Sub-Basin

Pyritic, organic-rich calcareous shale. Microlamination (mm-scale) in microfacies deposited during maximum flooding episodes is, in general, more apparent when the matrix is dominated by clay minerals (such as this example) compared to chert- or micrite-dominated matrices. Planktonic foraminifera represent the carbonate component of this sample. The globular tests are oriented horizontally to the bedding plane of the rock. Although pyrite framboids indicate anoxic-euxinic conditions, the organic richness of this sample is lower compared to previous examples shown of the La Luna formation *sensu stricto*.

IW = 1.3 mm. PPL

Cretaceous (Turonian). O Member, Escandalosa Formation. Barinas Sub-Basin

Sandy, glauconitic dolostone. The cloudy appearance of rhombic dolomite (D) crystals is easily distinguished from poorly sorted quartz (Q) grains of white limpid color. Early diagenetic dolomitization of the originally micritic fabric postdates deposition of quartz grains. Precipitation of glauconite (green areas) and pyrite framboids (black spots) postdates dolomitization, indicating a very low energy environment and partially reducing conditions during early diagenesis.

IW = 1.3 mm. PPL

Holocene. La Guaira State Shoreline

Sandy limestone. This mixed carbonate-siliciclastic microfacies is a beachrock formed in the north-central, Caribbean shoreline of Venezuela. A combination of HMC and aragonite marine cements makes up the matrix of this sample, cementing sand-sized quartz grains (white to gray) and carbonate skeletal grains (brown). Sandy, quartz-rich beaches along the north-central coasts of Venezuela represent an immediate source of siliciclastic sediments for the beachrocks and many other mixed, silicilastic-dominated deposits along those coasts.

IW = 2.4 mm. CPL

Cretaceous (middle Albian). Lisure Formation. Maracaibo Basin

Sandy, oolitic muddy limestone. This mixed siliciclastic-carbonate microfacies is composed of medium-sorted quartz grains (white to cream color grains having angular to subangular shapes) and ooids, all embedded in a micritic matrix. In this sample, the quartz content relative to carbonate is near 50%; therefore this sample can also be classified as an oolitic muddy sandstone. Ooid grains are micritized, resembling peloidal grains.

IW = 4.2 mm. CPL

Cretaceous (early Albian). Lisure Formation. Maracaibo Basin

Calcareous, glauconitic sandstone. Short-termed high siliciclastic influx episodes during early Albian times led to deposition of this mixed microfacies at the lower part of the Lisure Formation in western Lake Maracaibo. Pink-stained grains are fragments of LMC oysters. This is an excellent example of the use of Alizarin Red S as a tool for differentiating calcium carbonate substrates from quartz grains, as siliciclastic substrates do not stain pink. Glauconite (green) grains were formed by the from precipitation of this mineral coating quartz grains.

IW = 4.2 mm. PPL

References

Arthur, M. A., and Sageman, B. B., 1994. Marine Black Shales: Depositional Mechanisms and Environments of Ancient Deposits. *Annual Review of Earth and Planetary Sciences*, 22, 499–551.

Bates, R. L., and Jackson, J. A., 1980. Glossary of Geology. Second Edition. American Geological Institute. Falls Church, Virginia. 751 pp.

Brown, J. S., 1943. Suggested use of the word microfacies. *Economic Geology*, *38*, 325.

Choquette, P. W., and Pray, L. C., 1970. Geologic nomenclature and classification of porosity in sedimentary carbonates. *American Association of Petroleum Geologists Bulletin*, *54*, 207–250.

Cuvillier, J., 1952. La notion de "microfaciès" et ses applications. *VII Convegno Nationale del Metano et del Petrolio, Prestampa, Sezione I*, 3–7.

Dunham, R. J., 1962. Classification of carbonate rocks according to depositional texture. In: Ham, W.E. (Ed.), Classification of Carbonate Rocks. *American Association of Petroleum Geologists, Memoir 1*, 108–121.

Embry, A. F., Klovan, J. E., 1971. A late Devonian reef tract on northeastern Banks Island, N.W.T. *Bulletin of Canadian Petroleum Geology*, *19*, 730–781.

Flügel, E, 2010. Microfacies of Carbonate Rocks—Analysis, Interpretation and Application. Springer, Heidelberg-Dordrecht-London-New York, 984 pp.

Folk, R. L., 1959, Practical petrographic classification of limestones. *American Association of Petroleum Geologists Bulletin*, *43*, 1–38.

Folk, R. L., 1965. Some aspects of recrystallization in ancient limestones, In: L. C. Pray, and R. S. Murray, eds., Dolomitization and Limestone Diagenesis: Tulsa, OK, *SEPM Special Publication*, *13*, 14–48.

Haredy, R. A., Ghandour, I. M., Erginal, A. E., Bozcu, M., 2019. Beachrock cementation patterns along the Gulf of Aqaba coast, Saudi Arabia. *Arabian Journal for Science and Engineering*, *44*, 479–487.

James, N. P., and Jones, B., 2015. Origin of Carbonate Sedimentary Rocks. John Wiley & Sons, Ltd, Chinchester, 464 pp.

Lucia, F. J., Kerans, C., Jennings, J. W., 2003. Carbonate reservoir characterization. *Journal of Petroleum Technology*, *55*, 70–72.

Mauz, B., Vacchi, M., Green, A., Hoffmann, G., Cooper, A., 2015. Beachrock: a tool for reconstructing relative sea level in the far-field. *Marine Geology*, *362*, 1–16.

Morse, J. W., Arvidson, R. S., Lüttge, A., 2007. Calcium Carbonate Formation and Dissolution. *Chemical Reviews*, *107*, 342–381.

Rey, O., 2004. Cicloestratigrafía del EAO 3: Controles climáticos y cíclicos sobre la sedimentación del Santoniense, quebrada San Miguel, estado Mérida. Doctoral Dissertation. Central University of Venezuela, 190 pp.

Rey, O., Simo, J., Lorente, M. A., 2004. A record of long- and short-term environmental and climatic change during OAE3: La Luna Formation, Late Cretaceous (Santonian-early Campanian), Venezuela. *Sedimentary Geology*, *170*, 85–105.

Schlager, W., 2005. Carbonate Sedimentology and Sequence Stratigraphy. SEPM Concepts in Sedimentology and Paleontology #8. Society of Economic Paleontologists and Mineralogists, 200 pp.

Smith, A. M., Key Jr, M. M., Gordon, D. P., 2006. Skeletal mineralogy of bryozoans: taxonomic and temporal patterns. *Earth-Science Reviews*, *78*, 287–306.

Scholle, P. A., and Ulmer-Scholle, D., 2003. A Color Guide to the Petrography of Carbonate Rocks: Grains, Textures, Porosity, Diagenesis. *American Association of Petroleum Geologists, Memoir 77*, 474 pp.

Trabucho-Alexandre, J., Hay, W. W., De Boer, P. L., 2012. Phanerozoic environments of black shale deposition and the Wilson Cycle. *Solid Earth*, *3*(1), 29–42.

Tucker, M. E., and Wright, V. P., 1990. Carbonate Sedimentology. Blackwell Publishing. 482 pp.

Wignall, P. B., 1994. Black Shales. Clarendon Press, Oxford. 127 pp.

5 Calcareous Algae

Marine calcareous algae constitute a varied group of aquatic benthic plants that couple photosynthesis and calcification to produce organic matter (e.g., CH_2O) and O_2 and secrete their $CaCO_3$ skeletons. Therefore, they require light in order to carry the controlled mineralization of calcium carbonate, and hence inhabit environments within the photic zone. In modern tropical and subtropical marine shelf environments, skeletal remains of calcium-carbonate-depositing benthic algae make up a great proportion of the sediments. These photoautotrophs have been significant producers of carbonate sediments since the Ordovician (e.g., Wray 1977a; Bucur 1999).

Precipitation of carbonate minerals is a long-term sink of CO_2, but in the short term, it can serve as a source (Hallock 2015; Pomar et al. 2022). Marine calcification is a sink of C that induces an increase in the dissolved CO_2 level. On the other hand, carbonate dissolution acts as a sink of CO_2 (e.g., Frankignoulle et al. 1994; Killops and Killops 2005; Ries 2010; Granier 2012). In the case of marine calcareous algae (green and red), calcification acts as both a sink and a source for CO_2, playing an important ecological role in the global carbon budget. In a very simplified view, daylight calcification is coupled to photosynthesis, which involves CO_2 uptake and O_2 production; while at night, algae respiration involving aerobic oxidation of organic carbon releases CO_2 and can promote partial dissolution of diurnally precipitated carbonates (see Granier 2012; Pomar et al. 2022). Considering the full marine spectrum where calcareous algae calcification is favored (encompassing from warm to temperate-cool waters, and from the euphotic zone to the oligophotic zone), the net balance (sink or source of CO_2) between calcification/photosynthesis and dissolution/respiration processes on a global scale is still largely unknown (see Rees et al. 2007; Basso 2012).

Living algae are classified according to their pigments, type of flagellation, and chloroplast ultrastructure. In contrast, fossil calcareous algae, green and red, are only classified according to the morphologies of thalli and diagnostic features, which include specialized cavities that contain the reproductive organs during life. As a consequence, many fossil algae cannot be related to a precise group of living algae, and the *incertae sedis* are numerous among them (Vachard 2021).

5.1 Calcareous Green Algae

Among the calcareous green algae (Phylum Chlorophyta), marine forms make up less than 15% of the total flora, as freshwater forms dominate. Due to the predominance of green pigments (chlorophyll a and b), most Chlorophyta have a distinctive grass-green color, except where it is masked by calcification (Wray 1977a).

5.1.1 Major Groups and Stratigraphic Distribution

Marine calcareous green algae (Phylum Chlorophyta) are subdivided in two orders (Granier 2012): Dasycladales and Bryopsidales. True Dasycladales appeared in the Middle Ordoviacian (Vachard 2021); the Cambrian "dasycladales" are very disputable (e.g., Jiang and Zhang 2021). Members of the Order Bryopsidales appeared in the Ordovician (Bucur 1999; Vachard 2021). Members of both orders remain extant in modern oceans.

Dasycladales major peaks of abundance occurred in the Devonian (Calcite I) (e.g., Bucur 1999; Tosolini et al. 2012),

J. A. Méndez Dot, J. Méndez Baamonde, *Atlas of Tropical Carbonates*, https://doi.org/10.1007/978-3-031-25494-9_5

the Carboniferous (e.g., Mamet 1991; Vachard et al. 2012) and Permian (transition from Calcite I to Aragonite II) (Sartorio and Venturini 1988; Aguirre and Riding 2005; Flügel 2010; Abdelghany et al. 2019; Vachar 2021); the Middle and Late Triassic (Aragonite II), and the Late Jurassic to Early Cretaceous (Calcite II) (Sartorio and Venturini 1988; Elliott 1984, 1991; Flügel 1991, 2010; Okla 1991; Yilmaz 2000; Aguirre and Riding 2005; Carras et al. 2006; Stanley and Hardie 1998; Marino and Santantonio 2010; Bucur et al. 2012; Granier and Lethiers 2017; Steuber et al. 2023); and in the Paleocene and Eocene (Calcite II) (e.g., Aguirre and Riding 2005; Ries 2010; Afzal et al. 2011). Although far less abundant than modern Bryopsidales, dasycladalean algae (e.g., genus *Acetabularia*) are common in some modern marine settings of the Caribbean (see Granier 2012).

Bryopsidales were common in the Devonian (Bucur 1999; Vachard 2021), the Late Triassic (Aragonite II), the Middle-Late Jurassic (Flügel 1988, 1991, 2010), and the Early Cretaceous (Calcite II) (e.g., Bucur 1999; Flügel 1988; Bucur et al. 2012; Ioan et al. 2018). They became abundant from the Oligocene to the present (Aragonite III), being among the main carbonate sediment-producing organisms in modern tropical oceans (e.g., Wray 1977a; Martín et al. 1997; Stanley and Hardie 1998; Kiessling et al. 2003; Stanley 2006; Ries 2010; Granier 2012; Pomar et al. 2017) (see Appendix B for a detailed treatment of the concepts of Calcite and Aragonite Seas).

5.1.2 Skeletal Mineralogy

Marine calcareous green algae have the capability of secreting a calcareous “skeleton”, or coating. The $CaCO_3$ precipitated to form the coating is generally aragonite (the orthorhombic form), but there are short periods in the geologic record during which its calcite variant (the rhombohedral form) existed contemporaneously in discrete species (Granier 2012).

5.1.3 Ecology and Environment

Marine calcareous green algae are benthic photoautotrophs that attach to soft or firm substrates and flourish mainly in the upper photic (or euphotic) zone (from 0 to ~40 m deep) of tropical and subtropical waters (e.g., Pomar 2001). Calcification in green algae occurs predominantly in normal saline waters, although they thrive locally in hypersaline settings (see Westphal et al. 2010 and references therein). Photosynthetic enhancement of aragonite precipitation by green algae and zooxanthellate corals is most advantageous in shallow, tropical waters that are supersaturated with respect to $CaCO_3$ (Hallock 2001; see also Steuber et al. 2023). On the other hand, cooler waters (at higher latitudes and/or at greater depths) require greater energy expenditure to precipitate and maintain aragonite and therefore favor calcite-secreting organisms, i.e., coralline red algae, LBF, and light-independent heterotrophs (Hallock 1997).

Modern bryopsidalean algae (the most common are the genera *Halimeda*, *Penicillus*, and *Udotea*) are dominant in tropical, low-energy, muddy lagoons (behind topographic highs), where they can constitute up to 77% of the sediments of Pacific and Caribbean coral systems (Hillis-Colinvaux 1980). These calcified plants represent a major source of calcareous mud, as their fragile skeletons disaggregate after death (Wray 1977a; Martín et al. 1997; Scholle and Ulmer-Scholle 2003; Flügel 2010; Granier 2012).

Halimeda bioherms are abundant in the outer shelf (20 to 40 m deep) of the Northern Great Barrier Reef (Australia), being the largest actively accumulating *Halimeda* deposits worldwide (Davies and Marshall 1985; Marshall and Davies 1988; Rees et al. 2007; McNeil et al. 2020). *In situ Halimeda* deposits are also found in deeper, clear waters (40–70 m) within the mesophotic zone of the Nicaraguan Rise (Nicaragua, Southwest Caribbean) (Hine et al. 1987) and the Kalukalukuang Bank, eastern Java Sea (Indonesia) (Phipps and Roberts 1988; Roberts et al. 1988).

Photozoan carbonate facies (*sensu* James 1997) consisting of predominant light-dependent calcareous green algae and scleractinian corals (chlorozoan association *sensu* Lees and Buller 1972) dominate modern-tropical, coral-reef complexes (Hallock 2001; Schlager 2003; Mutti and Hallock 2003; Pomar et al. 2004; Michel et al. 2019). The shift from oligotrophic, chlorozoan association, to chloralgal association (*sensu* Lees 1975), dominated by green algae (plus foraminifera, calcareous red algae and molluscs), is thought to be the result of community response to increasing nutrient supplies (Hallock 2001). Therefore, calcareous green algae also thrive/thrived under oligo-mesotrophic to mesotrophic conditions (Mutti and Hallock 2003), which is the case of the muddy green algae-rich facies of the Cogollo Group (Aptian and Albian in the MB, northwest Venezuela).

5.1.4 Carbonate Factories and Skeletal Grain Associations

Calcareous green algae are components of the Photozoan T-factory of Michel et al. (2019) [this factory includes the Tropical factory (T factory) of Schlager (2003, 2005); see also the Photozoan factory of James and Jones 2015)], as well as occasional minor components of the Heterozoan factory of those authors. The Photozoan T-factory is dominated by the Photozoan Association of James (1997), which is characterized by an association of skeletons of corals, calcareous algae, symbiont-bearing benthic foraminifera, some calcareous sponges and stromatoporoids. Secondary components belonging to the Heterozoan Association *sensu* James (1997) can also be moderately common to common.

This section discusses only the skeletal grain associations involving calcareous green algae that are either represented or notably absent in Venezuelan carbonate facies/microfacies, as illustrated by the examples in the photographic set. In Aptian and Albian carbonates of the Piché Member and the Lisure Formation (MB), respectively, green algae-dominated microfacies (or chloralgal lithofacies, plus abundant micritic peloids) are locally rich in LBF (photozoan association). The photozoan-heterozoan association with smaller miliolids, molluscs and echinoderms (plus bryozoans) is also locally common. Chlorozoan lithofacies are absent due to the scarcity/absence of scleractinian corals.

Carbonate production in modern reef complexes (e.g., Los Roques Archipelago and Aves and Larga islands) is largely dominated by the Photozoan T-factory. Therefore, the photozoan association between green algae and scleractinian corals (chlorozoan sediments; plus coralline algae and LBF) predominates at platform scale. Nevertheless, the photozoan-heterozoan association between green algae and heterotrophic calcifiers (bivalves, gastropods and echinoids) is also common.

5.1.5 Morphology and Diagnostic Elements

Dasycladalean algae exhibit varied morphologies. Shrubs are several centimeters tall. Thalli (or stems) are segmented or unsegmented. Unbranched thalli (segmented or not) exhibit various shapes, including rod-shaped forms, clavate pear-shaped forms, stalked thalli formed by whorls of rays. Branched thalli may exhibit dichotomous-tube-like forms and articulate forms appearing like beads on a string. Individual segments of segmented thalli may separate on death of the organism and form spherical, hollow grains with primary and secondary laterals arranged perpendicularly to the main axis, and radially-oriented spherical elements (utricles, or wall perforations). Irregularly-shaped grains exhibiting randomly arranged utricles are also common (see Wray 1977a; Scholle and Ulmer-Scholle 2003; Radoičić et al. 2005; Carras et al. 2006; Radoičić and Schlagintweit 2007; Flügel 2010; Granier et al. 2014, 2017).

Bryopsidalean algae are plants consisting of cm-sized (about 5 to 15 cm high) segmented thalli attached to the substrate by rhizoids. They are texturally characterized by well-differentiated peripheral (outer) zone (cortex) and an inner region (medulla). The peripheral cortex is heavily calcified. The inner medulla region is less calcified than the outer cortex and hosts closely spaced, thick filaments or oriented parallel to the axis of the thallus, interwoven or branched. In the cortex, filaments are shorter in length and are oriented perpendicular to the margins. The filaments are poorly-calcified elements, originally composed of organic tissue and intervening calcified areas, and commonly form tubules (utricles) (see Wray 1977a; Flügel 1988, 2010; Scholle and Ulmer-Scholle 2003; Schlagintweit 2010; Ioan et al. 2018).

5.1.6 Preservation as Fossils

Fragments (grains) of fossilized green algae are commonly reduced in size due to disaggregation, fragmentation, and mechanical abrasion. They may also experience reworking prior to their final deposition. They are mainly preserved as molds (either neomorphized/recrystallized or filled by LMC cements), or as micritized products. Tubules are filled with sparry calcite or lime-mud. The size of dasycladalean and bryopsidalean microfossils commonly range between 0.5 and 7–8 mm (but occasionally up to 20 mm long). Shapes vary between rounded and elongate.

5.2 Calcareous Red Algae

The red algae are a large group of dominantly benthic, marine plants; only a relatively few species live in freshwater (Wray 1977a). The red color of these calcareous algae is the result of pigments phycoerythrin and phycocyanin that can mask other pigments such as chlorophyll (a green pigment) and beta-carotene (a yellow pigment) (Hamid et al. 2015).

5.2.1 Major Groups and Stratigraphic Distribution

In this section, we deal with calcareous red algae of five groups belonging to the Phylum Rhodophyta. The orders Corallinales, Sporolithales and Hapalidiales, which range in age from the Early Cretaceous to the Holocene, are collectively referred to as "coralline algae", or "corallines" (see Rösler et al. 2017; Peña et al. 2020). The Family Elianellaceae (age range: Cambrian to Middle Eocene; Granier and Dias-Brito 2016a, 2016b; Vachard et al. 2016; Aguirre et al. 2020), which has been tentatively assigned to the Order Rhodogorgonalles (Granier and Dias-Brito 2016a); the latter is the fourth order of calcareous red algae (Peña et al. 2020). The Order Archaeolithophyllales (Family Archaeolithophyllaceae; Corrochano et al. 2013), which embraces a late Paleozoic, non-systematic group commonly referred to in the literature as "Ancestral Corallines".

We include the extinct family 'Solenoporaceae' in this review, as studies focused on 'calcareous red algae' belonging to this family are widely referred to in the literature. 'Solenoporaceans' (age range: mid-Ordovician to Miocene) are a heterogeneous group that includes a great variety of organisms, such as algae, bryozoans, chaetetid sponges, benthic foraminifera, hydrozoans, and cyanobacteria-like fossils (Aguirre et al. 2007). Since Riding (2004) demonstrated that the type genus *Solenopora* is a chaetetid sponge, not an alga, Granier and Dias-Brito (2016a) introduced the Family Elianellinceae to embrace the algal fossils of the 'Solenoporaceae'. Since the true nature of the family is uncertain, we refer to it (the family, and especially the calcareous red algae within it) within inverted commas following Aguirre et al. (2007).

Members of the Corallinales (families Corallinaceae, Lithophyllaceae and Mastophoraceae; Peña et al. 2020) first appeared during the latest Aptian (~ 112.86 Ma) (Yang et al. 2016; Rösler et al. 2017; Peña et al. 2020), however they have a poor fossil record in Cretaceous, Paleocene and Eocene strata (Calcite II) (Aguirre et al. 2010). By the Late Oligocene to the Early Miocene, they were abundant in the Mediterranean (e.g., Brandano et al. 2009, 2010) and in the Caribbean (e.g., Pomar et al. 2015; Albert-Villanueva 2016). The bloom of Corallinales occurred during the Late Oligocene and Miocene (Aragonite III), as they became globally dominant carbonate producers, even locally replacing corals as dominant reef builders (Halfar and Mutti 2005; Pomar and Hallock 2008). Members of the family Corallinaceae (e.g., genera *Corallina*, *Spongites*, *Lithoporella*, *Lithophyllum*, *Neogoniolithon*) are the main corallines in both modern and ancient low- to mid-latitude shallow-water algal assemblages at least since the Miocene (Rösler et al. 2017).

Members of the Order Sporolithales (Family Sporolithacea, genera *Sporolithon* and *Heydrichia*; Le Gall et al. 2010) are known as far back as the Valanginian (~ 137 Ma), about 24 Myr before the first appearance of members of the Order Corallinales (Peña et al. 2020; see also Tomás et al. 2007; Woelkerling et al. 2014; Silva-Tamayo et al. 2016). They were moderately common in the Cretaceous (Calcite II), especially from the Albian to the Coniacian (Aguirre et al. 2000; Carannante et al. 1995; Simone et al. 2012). Species diversity and abundance of sporolithaceans decreased dramatically in the Early Paleocene and continued to decline throughout most of the Cenozoic to a low in the Late Pliocene (Aguirre et al. 2000). Their present-day distribution is restricted to tropical latitudes (Braga and Bassi 2007; Darrenougue et al. 2013).

Members of the Order Hapalidiales (Family Hapalidiaceae; Nelson et al. 2015) appeared during the Aptian (~ 117 Ma; Peña et al. 2020). Like the Corallinales, the Hapalidiales has a poor fossil record in Cretaceous strata. After the Cretaceous-Tertiary mass extinction, this family underwent the highest increase in abundance during the Paleocene (Aguirre et al. 2007), remaining common throughout the Cenozoic. Present-day distribution of hapalidiaceans (e.g., genera *Lithothamnion*, *Mesophyllum*) includes tropical and extra-tropical marine settings.

Extinct 'solenoporaceans' were moderately common during the early Paleozoic (Ordovician and Silurian; Calcite I) (e.g., Brooke and Riding 1998) and late Paleozoic (Devonian to Permian; from Calcite I to Aragonite II) (e.g., Wray 1977b; Chen et al. 2020). Their abundance increased in the Late Triassic (Aragonite II) (Stanton and Flügel 1987; Flügel 1991, 2010; Senowbari-Daryan et al. 2008) and Late Jurassic (Calcite II) (Flügel 1991, 2010; Bucur et al. 2005). 'Solenoporaceans' were locally common in the Early Cretaceous (Misra and Kumar 1988; Masse and Arnaud Vanneau 1999; Misra et al. 2009; Vachard 2021), and in the early Cenozoic (Calcite II) (e.g., Aguirre et al. 2007; Baceta et al. 2011).

The Family Elianellaceae was introduced by Granier and Dias-Brito (2016a) in order to resolve taxonomical problems linked to *incertae sedis* red algae formerly belonging to the Family 'Solenoporaceae' (e.g., *Elianella elegans*, *Cordilites cretosus*, *Parachaetetes* asvapatii, and *Marinella lugeoni*). Calcareous red algae so far described as Elianellaceae are particularly common during the Late Carboniferous (Aragonite II), the Late Jurassic, the Early and mid-Cretaceous, the Paleocene, and the Middle Eocene (Calcite II) (Granier and Dias-Brito 2016a, b; Vachard et al. 2016; Aguirre et al. 2020). Recently, Pliocene branching calcareous red algae described by Dorsey et al. (2023) display internal structure characteristic of the Family Elianellaceae, which might extent the occurrence of members of this family to early Neogene times.

Ancestral corallines (a non-systematic group) were common during the late Paleozoic (Carboniferous and Permian). This group includes calcareous red algae that resemble members of the family Corallinaceae (e.g., Wray 1977a, b; Vachard and Aretz 2004; Cózar et al. 2005; Huck et al. 2006; Flügel 2010; Corrochano et al. 2013). Furthermore, several ancestral corallines formerly included within the informal group "phylloid" algae (the term phylloid was introduced by Pray and Wray 1963; for those Carboniferous to Triassic algae of dubious systematic affinities) are now included in the Order Archaeolithophyllales (see Vachard and Kabanov 2007). The main representative of this order is the genus *Archaeolithophyllum* (see Corrochano et al. 2013; Vachard 2021). The latter is reported mainly in strata of the Early and Late Carboniferous (transition from Calcite I to Aragonite II) and Permian (Aragonite II) (see Wray 1977b; Cózar and Vachard 2003; Cózar et al. 2005).

5.2.2 Skeletal Mineralogy

Coralline algae skeletons are composed of HMC. The $MgCO_3$ content (%) varies strongly, commonly over 8%, and up to 30% in some tropical forms (Scholle and Ulmer-Scholle 2003; Flügel 2010). Some temperate water, non-geniculate species incorporate low amounts of aragonite (Smith et al. 2012). Granier and Dias-Brito (2016a, b) consider HMC as the probably original mineralogy of the Elianellaceae. Based on the morphological similarities between *Archaeolithophyllum* and recent calcitic corallinaceans, and the similar cathodoluminescence pattern and trace element composition of the algal thalli and the high-Mg radiaxial-fibrous calcite, it is suggested that *Archaeolithophyllum* was probably composed originally of HMC (Corrochano et al. 2013).

5.2.3 Ecology and Environment

In modern oceans ($m\mathrm{Mg}^{2+}/\mathrm{Ca}^{2+}$ ratio of ~5.2; e.g., Ries 2010; Hasiuk and Lohmann 2010), coralline algae are common from tropical to polar oceans at all depths within the photic zone, being major framework builders in temperate and cold-water benthic ecosystems (see Basso 2012 and references therein). Under similar temperatures, corals exist preferably in oligotrophic conditions, while coralline algae are more abundant in mesotrophic and even slightly eutrophic conditions (see Halfar et al. 2004; Halfar and Mutti 2005; Bassi and Nebelsick 2010; Klicpera et al. 2015). Such wider range of temperature (controlled by latitude and/or bathymetry) and trophic conditions under which red algae thrive allow these light-dependent calcifiers associate to heterotrophic calcifiers in heterozoan carbonates *sensu* James (1997). The latter have differentiated markedly, throughout the Mesozoic and Cenozoic, these HMC algae from other photoautotrophic calcifiers (aragonitic green algae and scleractinian corals) (see Westphal et al. 2010; Michel et al. 2018).

In modern Caribbean reef complexes, encrusting corallines are secondary but important reef framework builders. They are common in the reef flat and in many other moderate- to high-energy settings (e.g., cryptic habitats between massive hexacoral buildups) in Los Roques Archipelago, where they form bioherms at depths less than one meter. Some of these *in situ* bioconstructions reach several tens of meters in length. Modern coralline algal buildups are locally found at depths between 50 and 120 m in a subtropical transitional area in East Australia (off Fraser Island, between the tropical Great Barrier Reef and temperate, cooler waters to the south) (Lund et al. 2000; Davies et al. 2004). As for modern rhodolith facies, they are abundant in slope settings (between 60 and 90 m deep) located in tropical-to-transitional waters in the Bahamas (San Salvador Island) (Littler et al. 1991); in shallow-temperate waters in the southwestern Gulf of California (south Baja California) (Steller and Foster 1995; Marrack 1999); in lower euphotic to mesophotic depths (30–60 m) of shelves in most of the islands of the Lesser Antilles (Reid and Macintyre 1988); among other localities.

During the Cenozoic, but especially during the Oligocene and the Miocene, red algae were dominating skeletal components of carbonate shelves and ramps. In high latitudes they were more common in shallow waters (Aguirre et al. 2000). In tropical and subtropical ramps, shelves and slopes, rhodalgal lithofacies (*sensu* Carannante et al. 1988), characterized by red algae (encrusting and rhodoliths) as major components, dominated the carbonate benthic factory in the meso-oligophotic zone (40 to ≥100 m) (Aguirre et al. 2000; Pomar 2001; Mutti and Hallock 2003; Halfar and Mutti 2005; Pomar et al. 2004, 2015; Braga 2017). This distribution pattern, also common in some modern carbonate shelves, is seemingly related primarily to latitude and bathymetry (or depth of the seafloor) that control water temperature (see Carannante et al. 1988), although other factors (e.g., trophic conditions, salinity, CO_2 concentrations, water energy, substrate requirements, among others) also play important roles (Carannante et al. 1988; Pomar et al. 2004; Westphal et al. 2010).

5.2.4 Carbonate Factories and Skeletal Grain Associations

Calcareous red algae are components of the Photo C-, and Photozoan T-factories of Michel et al. (2019), as well as occasional to common secondary components of the Heterozoan factory of these authors (see Appendix A for a detailed treatment of these carbonate benthic factories).

The Photo C-factory is dominated by calcareous red algae and symbiont-bearing benthic foraminifera and typically occurs in the warm-temperate or subtropical province. Nevertheless, this carbonate factory is well-represented in tropical, Cenozoic carbonates of Venezuela. In this country, the most representative rock units dominated by components of this factory and deposited at paleolatitudes between approximately 8° to 10° N are: (a) the Perla Limestone, a distally steepened oligophotic ramp of Late Oligocene to Early Miocene age in the EGVSB (Pomar et al. 2015); (b) middle- to outer-ramp facies of the San Luis and Churuguara formations of Early Miocene age in the FB, and deposited in the meso-oligophotic zone (Albert-Villanueva 2016). Water temperature, light availability, and energy level, controlled by bathymetry, were primary local factors controlling deposition of those units in which heterotrophic calcifiers are secondary but important components. Prior to the carbonate factories new approach by Michel et al. (2019), many Cenozoic red algae-dominated facies were described as Heterozoan carbonates *sensu* James (1997), e.g., Brandano et al. (2009), Silva-Tamayo et al. (2017), Albert-Villanueva et al. (2017).

We consider calcareous red algae as common/dominant components of the Photozoan T-factory (see Appendix A for a detailed discussion). In modern shallow-water tropical settings, carbonate production by coralline algae is, in the overall, volumetrically and quantitatively (GT $CaCO_3$ yr.$^{-1}$) less compared to carbonate production by green algae and scleractinian corals. Nevertheless, these HMC calcifiers are prominent $CaCO_3$-producers and bioconstructors in coral reefs worldwide (see Narayan et al. 2022). It highlights their abundance in Venezuelan reef complexes of the Southern Caribbean and in Aves Island.

This section discusses only the skeletal grain associations involving calcareous red algae represented in Venezuelan carbonate facies/microfacies, as illustrated by the examples in the photographic set. In Middle Permian carbonates of the Palmarito Formation (Venezuelan Andes), red algae (ancestral and 'solenoporaceans') are secondary components of photozoan-dominated microfacies (in association with LBF, green algae and rugose corals); as well as secondary components of heterozoan-dominated carbonates (in association with gastropods, bryozoans, crinoids and brachiopods) (see Laya and Tucker 2012). In Albian carbonates of the Lisure Formation (MB), red algae are scarce to moderately common components of heterozoan and mixed heterozoan-photozoan microfacies, in association with molluscs, echinoids, bryozoans, and green algae (plus non-skeletal grains such as ooids, oncoids, and peloids).

As previously mentioned, in Cenozoic facies of the FB and the EGVSB (e.g., San Luis, Churuguara, and Capadare formations, and PL, respectively), red algae are dominant components of photozoan carbonates in association with LBF (the Photo C-factory). They are also secondary components of photozoan carbonates dominated by scleractinian corals and green algae (Photozoan T-factory, plus LBF such as soritids), as well as secondary components of mixed photozoan-heterozoan carbonates in association with LBF, bryozoans, echinoids, barnacles (balanids), and molluscs. Depending on the content of red algae versus other bioclastic components (both photozoan and heterozoan biota), many of these facies can be described as rhodalgal lithofacies (dominated by red algae, plus LBF and bryozoans) or as rhodechfor lithofacies *sensu* Hayton et al. (1995) (dominated by red algae, echinoids, LBF and bryozoans).

In modern reef complexes (e.g., Los Roques Archipelago and Aves Island), carbonate production, at platform scale, is largely dominated by the photozoan association between corals and green algae (chlorozoan association). Nevertheless, the photozoan association between scleractinian corals and coralline algae, or coralgal lithofacies *sensu* Pomar et al. (2012), predominates locally in moderate- to high-energy environments (e.g., the reef flat of Los Roques and the fringing reef of Aves Island). In some Quaternary lithified facies (carbonate basement and beachrocks) of Aves Island, red algae grains are the most abundant components of mixed photozoan-heterozoan microfacies. They are accompanied by photozoan components such as calcareous green algae (*Halimeda*) and scleractinian corals (plus LBF), as well as molluscs (bivalves and gastropods), which represent the most common heterozoan components (with minor echinoids).

5.2.5 Morphology and Diagnostic Elements

The heavy calcified HMC plant body of coralline algae is known as thallus (plural, thalli). Their morphology is indicative of certain energy conditions as corallines growing in high-energy conditions have robust fused framework with thick crusts, branches and columns, while the corallines growing in moderate-energy conditions have delicate framework with thin branches and crusts. Corallines architecturally have two groups, the *geniculate* (articulated) and the *non-geniculate* (non-articulated or encrusting) forms. Geniculate corallines (e.g., *Corallina*, *Amphiroa*) have erect and branched (tree-like) growth forms, varying in shape from cylindrical to flattened, and can reach lengths of several centimeters to up to 20 cm in length. They consist of rigid calcified segments (*intergenicula*) joined by shorter and more flexible non-calcified segments (*genicula*). The non-calcified cells of the genicula decay quickly after death, and the calcified segments then fall apart and disperse in the sediment. This structural collapse makes the recognition and correct taxonomic identification of geniculate corallines rather difficult. In non-geniculate corallines (e.g., *Neogoniolithon*, *Lithothamnion*), the thalli are completely calcified and consequently have a high preservation potential (see Aguirre et al. 2010; Kundal 2011). Woelkerling et al. (1993) have suggested 10 types of growth forms for non-geniculate corallines. These are: encrusting (plants crustose and flattened or applanate or sleeve-like, up to 10 cm long), unconsolidated (free) filaments, layered, warty, lumpy, fruticose, discoid, foliose, ribbon-like, and arborescent (see also Nebelsick and Bassi 2000; Kundal 2011). In addition, rhodoliths consist of free-living, nodular aggregates of non-geniculate corallines which can be produced whenever crustose or branching corallines either settle on non-cohesive particulate substrate or are detached from hard substrate (see Nebelsick and Bassi 2000). Rhodolith buildups (ranging from less than 3 cm to up to 30 cm in diameter) can consist of different coralline algal species, as well as intergrowths of corallines with other encrusting organisms such as benthic foraminifera, bryozoans, serpulids, and, occasionally barnacles (see Aguirre and Braga 2022). The Corallinales includes both geniculate and non-geniculate, whereas Sporolithales and Hapalidiales are strictly non-geniculate (Nelson et al. 2015).

At mm-scale (under the polarizing microscope), coralline algae (and most other red algae) are easy to identify due to their characteristic very fine-scale reticulate (boxwork) cellular internal structure. The thallus is composed of filaments making up a plant body with a characteristic internal (latticework) structure. The filaments, in turn, are composed of small HMC cells. In non-geniculate corallines, the thalli may be organized in in a dorsiventral manner, an isobilateral manner, or in a combination of dorsiventral and radial. In dorsiventral organization (the most widespread), two types of anatomincal construction, *monomerous* and *dimerous* are evident. In monomerous construction, the thallus consists of a core region (formerly called hypothallus) and a peripheral region (formerly called perithallus), which is commonly thicker. The core can be *coaxial*, i.e., the filaments are arrangement in tiers, and *non-coaxial*, i.e., the filaments are curved or plumose (the filaments curve outwards or towards the dorsal and ventral surface of a crustose part of a thallus). The dimerous organization involves two distinct groups of cell filaments oriented more or less at right angles to one another. The basal (dorsal) filaments are called *primegenous* (formerly called hypothallus), from which the *postigenous* filaments (formerly called perithallus) arise dorsally (ventrally) at right angles. In geniculate corallines, calcified segments (intergenicula) have a central core region (formerly called medulla), which is surrounded on both sides by the peripheral region (formerly called cortex and commonly thinner). The intergenicula appear to have monomerous construction, i.e., there is a single system of filaments in which some filaments contribute to a core region that runs more or less parallel to the surface of the thallus, while other filaments curve outwards and contribute collectively to a more peripheral region. The other main morphologic feature in corallines is the conceptacle. These cavities (or structures) are commonly, spherical, sub-spherical, ovate, or elliptical in shape and house the reproductive organs. In non-geniculate forms, conceptacles are located in the peripheral region, while in geniculate forms, they may be present in the core region or in the peripheral region (see Kundal 2011). Some other morphologic features in geniculate and non-geniculate corallines will be mentioned and described in the photographic set of this chapter (see Johnson 1964; Wray 1977a; Townsend and Adey 1990; Misra et al. 2001; Scholle and Ulmer-Scholle 2003; Saunders and Hommersand 2004; Brandano et al. 2005; Flügel 2010; Woelkerling et al. 2014; Chelaru and Bucur 2016; Hamad 2020; for reviews on morphology of coralline algae).

In regard to the classification of members of the Coranillales, Sporolithales and Hapalidiales at the species level, systematic paleontology should be reserved for specialists (Flügel 2010). In general, to classify a coralline red alga, the following morphologic features must be taken into consideration: conceptacles, sporangia (structures that produce spores which are found within the conceptacles and contain tetraspores or carpospores), tetrasporangial conceptacles (specialized structures/cavities that house tetrasporangia), and the type of connection between adjacent cells (cell fusions or secondary pit-connections) (see Johansen 1981; Woelkerling 1988; Irvine and Chamberlain 1994; Misra et al. 2001; Harvey et al. 2003; Kato et al. 2011; Kundal 2011; Rösler et al. 2017; Caragnano et al. 2018; Peña et al. 2020).

Sporolithales (having cell fusions and secondary pit-connections) differs from other orders in producing tetrasporangia within calcified sporangial compartments and in having tetrasporocytes that undergo cruciate cleavage (Le Gall et al. 2010). In contrast, in Hapalidiales (having cell fusions or no connections), zonate sporangia are developed in conceptacles with multiporate roofs, while in Corallinales (having cell fusions/secondary pit-connections) the zonate sporangia are developed in uniporate conceptacles (Harvey et al. 2003; Nelson et al. 2015; Peña et al. 2020).

Extinct members of the Family Elianellaceae share some morphological features (Granier and Dias-Brito 2016b): they are either encrusting or form free structures; they have a framework made of erected filaments, more or less densely juxtaposed, with common to rare horizontal partitions, defining a more or less preserved lattice-network with columns and rows; they are deprived of fossilized reproductive organs (which were possibly external).

Members of the order Archaeolithophyllales share some morphologic features that resemble coralline algae. Among the main feature of the thallus of the genus *Archaeolithophyllum*, the following stand out: it is composed of chains of cells that may fuse laterally or also form secondary pit-connections; it is composed of an upper layer with polygonal cells, and a lower layer with elongated cells; it has conceptacles that are roofed by a layer of cells and contain tetraspores or carpospores (see Vachard and Kabanov 2007; Corrochano et al. 2013).

5.2.6 Preservation as Fossils

The more Mg present in calcite, the more soluble and vulnerable to dissolution or recrystallisation it is (Smith et al. 2012). Furthermore, HMC with Mg substitution greater than 12% is more soluble than calcite; (Andersson et al. 2008). Despite their mineralogic metastability, fossil calcareous red algae commonly preserve their original textural features. The loss of Mg during dissolution of HMC red algae grains and subsequent precipitation of LMC calcite is a mineralogical stabilization process that is still poorly understood (see Tucker and Wright 1990).

Calcareous red algae microfossils commonly consist of cylindrical, flattened, branched, nodular, columnar, domed, layered, protuberant, and lumpy forms. Like calcareous green algae, red algal grains undergo disaggregation, fragmentation, and mechanical abrasion, and may also experience reworking prior to their final deposition. The grains range in size from sand-sized, commonly medium sand (0.25 mm), to gravel-sized (up to 10 mm or more). Rhodoliths, which commonly exhibit multilayered structures and nodular to spheroidal/discoidal shapes, may reach up to 30 centimeters in length. HMC cells making up the filaments are commonly well preserved and visually distinguishable with the polarizing microscope, especially in Cenozoic specimens, although their degree of preservation depends on the intensity of diagenesis undergone by the microfacies in question. Conceptacles in corallines are not always visible or preserved, but when they are, they are typically filled with sparry calcite.

5.3 Calcareous Algae: Photographic Set and Descriptive Captions

In all photos, IW = image width (the true length of the horizontal axis of the view)

Holocene. Los Roques Archipelago
Dried sample of the bryopsidalean alga *Halimeda opuntia*. The fragile structure of this heavy calcified alga is composed of individual, flattened and cylindrical to lobate aragonitic plate sheds. These form a rhizoid and dense structure consisting of branched clumps, ribbed-leaf like segments arranged in different planes with respect to a central thallus axis (structure not seen in the image). *Halimeda* thallus attaches to both soft and hard substrates in a wide variety of environments, commonly of moderate- to low-energy level.

IW = 7 cm.

Holocene. Los Roques Archipelago
Dried sample of the bryopsidalean alga *Halimeda incrassata*. Note the rhizoid bulb at the top that branches from the thallus at the bottom. The branches are made up of thin, calcified aragonitic segments linked together by small, noncalcified joints. Species of *Halimeda* can grow to a height of about 20 cm and are major sources of carbonate gravel- and sand-sized sediments in modern tropical carbonate reef complexes. Postmortem disaggregation of their fragile skeletons forms small individual plate sheds (or "flakes"), which are prone to abrasion and size diminishing.

IW = 7 cm.

Holocene. Los Roques Archipelago
Dried sample of the bryopsidalean alga *Penicillus* sp. This erect alga is composed of interwoven aragonitic filaments forming a rhizoidal base, a thin rigid stem, and a brush-like crown of free-branched filaments. The bushy crown, which can grow between 10 and 15 cm tall, frequently disaggregates into micron-sized aragonite needles. Thus, *Penicillus* is considered to be the source of much of the carbonate clay-sized sediments in modern tropical carbonate reef complexes.

IW = 5 cm.

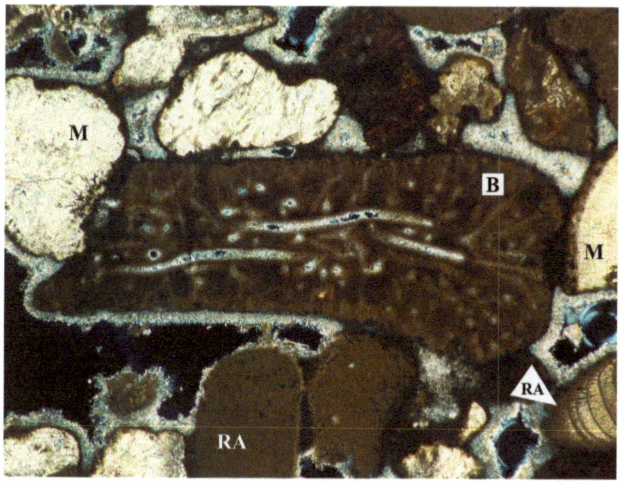

Late Pleistocene. Aves Island

A bryopsidalean (B) algal grain (*Halimeda*), accompanied by red algae (RA) (photozoan association) and mollusc (M) grains (photozoan-heterozoan association) in a porous calcarenite microfacies in which allochems are cemented together by isopachous rims of marine fibrous aragonite. The *Halimeda* grain preserves its brown-reddish color, which is characteristic of young (Late Pleistocene and Holocene) fossils that have undergone minimal meteoric diagenetic alteration. Note the long tubules in the medulla region, oriented parallel to long axis of the grain.

IW = 2.3 mm. CPL

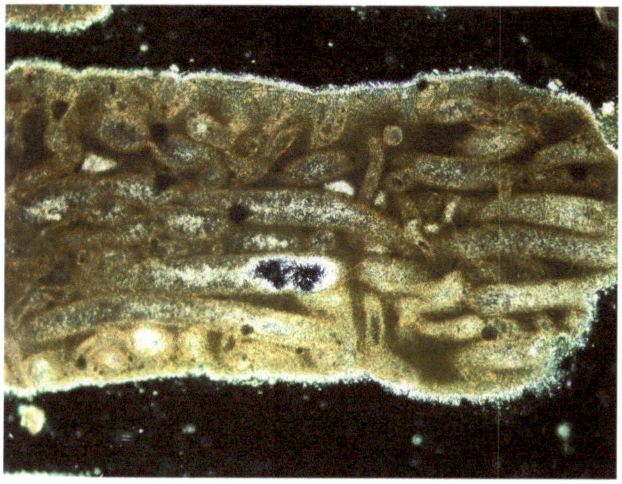

Holocene. Larga Island. Carabobo State Shoreline

A close-up view of a loose grain of *Halimeda*. The grain was collected in the foreshore environment. Thick medullary tubules exhibit a characteristic parallel to interwoven orientation with respect to the long axis. As in the previous example, note that these originally poorly calcified elements are partially filled by minute needles of marine aragonite acicular cement, which are also coating the grain.

IW = 1.3 mm. CPL

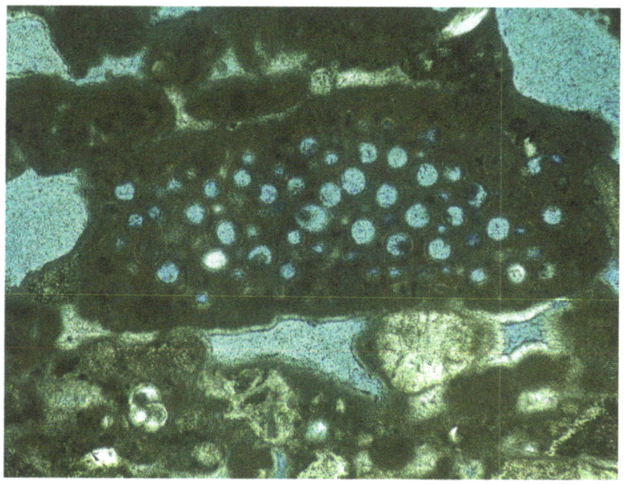

Holocene. Los Roques Archipelago

This *Halimeda* grain was carried by the wind from the surrounding areas of a perennial hypersaline lagoon and subsequently trapped within a laminar stromatolite structure. Abrasion and reworking into an adjacent environment diminished the size of the grain. The characteristic brown to reddish color of this aragonitic alga was obliterated by micritization carried out by endolithic cyanobacteria living in the benthos of the hypersaline lagoon. Note the spherical utricles, which are not filled by marine cements and are instead dyed by blue epoxy filling the intraparticle pores.

IW = 1.3 mm. PPL

Cretaceous (middle Aptian). Piché Member, Apón Formation. Maracaibo Basin

Fairly-preserved single plate shed of a bryopsidalean alga in a green algae grainstone. The grain is seen in longitudinal section. The tubules (utricles) in the central medulla region are oriented parallel to the long axis of the grain. These small inner structures were filled by a combination of sparry calcite (white tubules) and micrite (dark tubules). In the peripheral cortex, diagenetic stabilization (dissolution-reprecipitation) is prominent, and most tubules, which are oriented perpendicular to the margins of the grain, are filled by micritic sediments.

IW = 2.4 mm. PPL

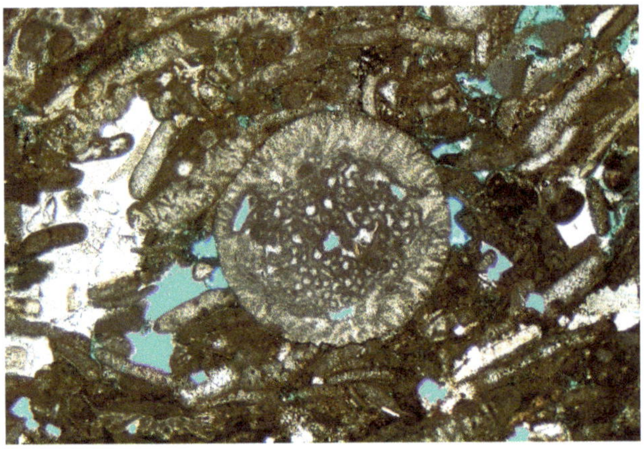

Cretaceous (middle Aptian). Piché Member, Apón Formation. Maracaibo Basin

Transverse section of a bryopsidalean algal grain in the center of the image (probably *Boueina* sp.). The cross-section through the single plate determines the shape in which the tubules within the central medulla region appear. Thus, unlike the elongate shapes in the previous example, the tubules are spherical. Note the well differentiation between the less calcified inner medulla, where localized small intraparticle pores have formed, and the heavily calcified peripheral cortex.

IW = 2.4 mm. PPL

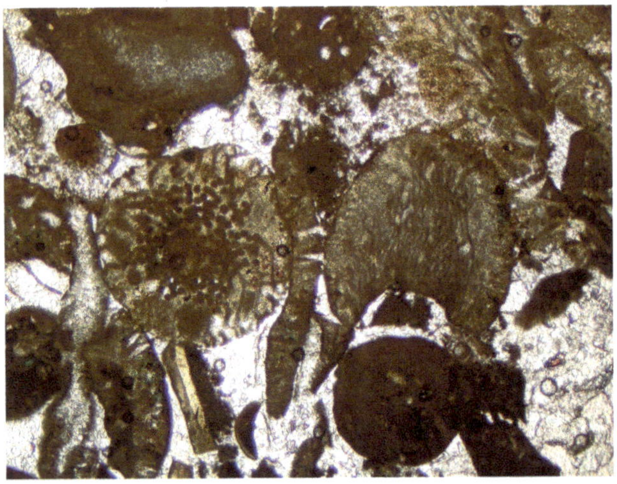

Cretaceous (middle Albian). Lisure Formation. Maracaibo Basin

Transverse (left) and longitudinal (right) sections of bryopsidalean algae grains in a cemented grainstone. The preservation of textural features (medullary filaments) and skeletal morphology (well-differentiated cortex and medulla) is not always observed in fossil bryopsidalean grains due to metastability of aragonite and the fragility of the skeletons. In the case of these two grains, the infilling of the tubules with synsedimentary lime-mud promoted preservation of such diagnostic elements.

IW = 2.4 mm. PPL

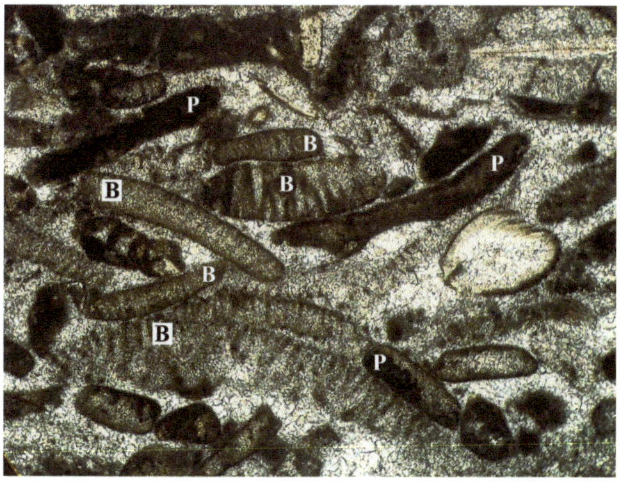

Cretaceous (middle Albian). Lisure Formation. Maracaibo Basin

Bryopsidalean (B) algae grains in the image exhibit identifiable medullary filaments (tubules). Note that these grains are visibly altered by meteoric diagenesis, which is indicated by sparry calcite partially replacing the aragonite coatings. Such process postdates the infilling of their tubules with dark micritic sediments. Other peloidal (P) or micritized-textureless grains in this grain-supported fabric could also be of green algae origin.

IW = 4.2 mm. PPL

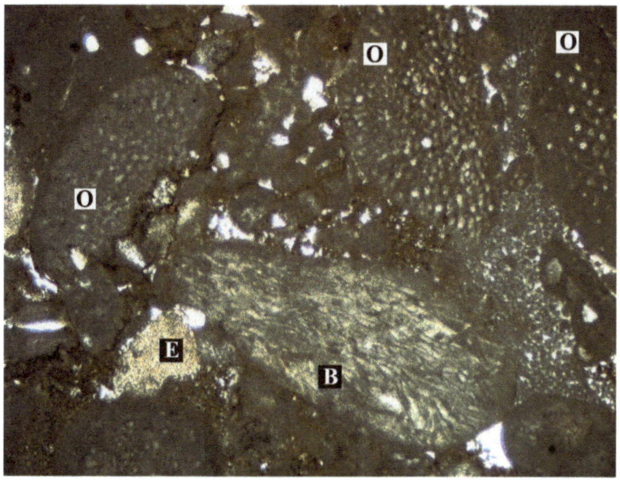

Cretaceous (middle Aptian). Piché Member, Apón Formation. Maracaibo Basin

A bryopsidalean (B) algal grain, accompanied by several tests of orbitolinids (O) (photozoan/chloroforam association) and an echinoid (E) grain (photozoan/-heterozoan association), in a packed packstone rich peloidal grains (small micritic grains of uncertain origin). Deposition of this microfacies is associated with mesotrophic or slightly eutrophic conditions prevailing during middle Aptian times in the Maracaibo platform.

IW = 4.2 mm. PPL

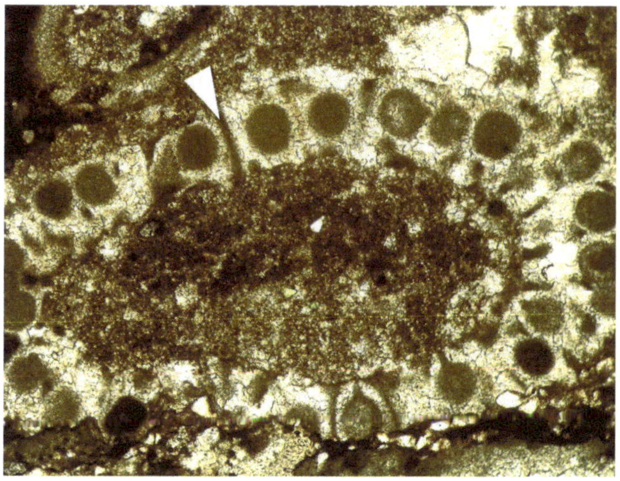

Cretaceous (middle Aptian). Piché Member, Apón Formation. Maracaibo Basin

Transverse section of a segment of a dasycladalean alga (*Neomeris* sp.) showing its cylindrical cavity (in lieu of its main axis and its primary sterile laterals). The circular elements filled with micritic sediments are called gametophores (or sporangia). These are the reproductive organs of the alga. The original aragonitic coating of the alga was dissolved and the mold was later cemented by drusy calcite. The secondary sterile laterals (white arrow) are filled with lime-mud and arranged perpendicularly to the main axis.

IW = 2.4 mm. PPL

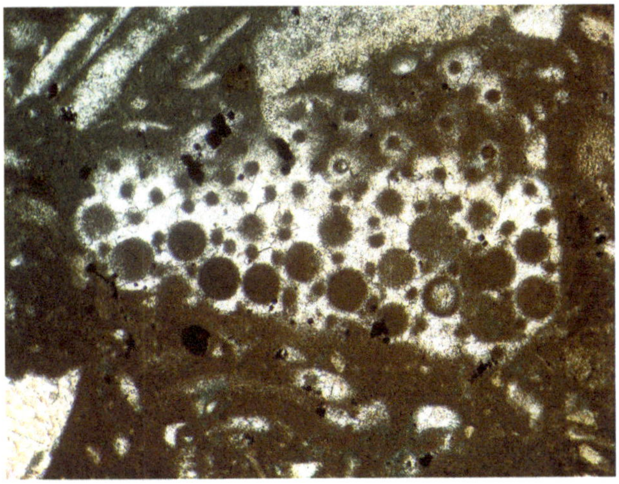

Cretaceous (middle Aptian). Piché Member, Apón Formation. Maracaibo Basin

Fragment of a dasycladalean alga (*Neomeris* sp.). Note the different-sized, circular sporangia. In life, these can be fertile or sterile, being the latter commonly smaller in diameter (see Granier et al. 2017). As in the previous example, the tubules were filled by synsedimentary micrite. Postdating dissolution of the aragonitic coating of the alga was followed by precipitation of LMC.

IW = 2.4 mm. PPL

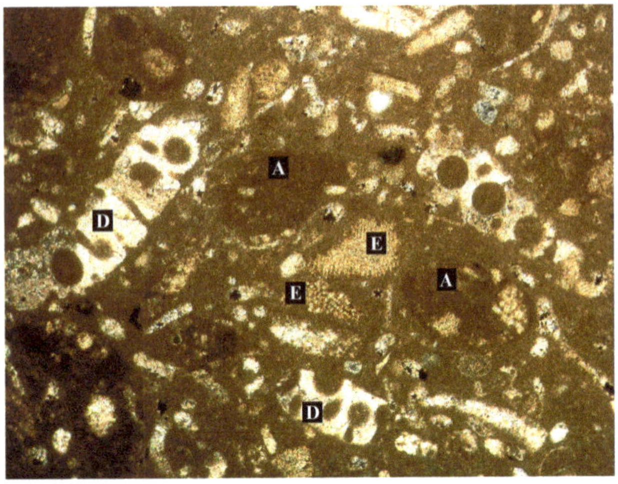

Cretaceous (middle Aptian). Piché Member, Apón Formation. Maracaibo Basin

Bioclastic wackestone-packstone consisting of fragments of dasycladalean (D) algae, echinoderms (E), as well as micritic aggregates (A). Spherical elements and laterals perpendicularly oriented to the margins are diagnostic elements for identifying the algae fragments. In the Piché Member, dasycladaceans are common of muddy (grain-supported or not) microfacies (such as this and previous examples) corresponding to transgressive and catch-up deposits. In addition, the photozoan-heterozoan association between green algae and echinoderms is common in shallow facies of this member.

IW = 4.2 mm. PPL

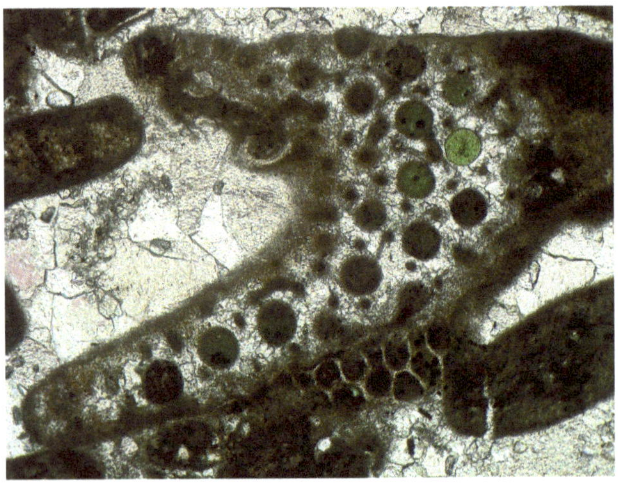

Cretaceous (middle Albian). Lisure Formation. Maracaibo Basin

This dasycladalean algal grain exhibits branching-dichotomous morphology. Note the micritic infillings of the spherical elements, indicating that this grain was originally deposited in a muddy environment and later redeposited in a mud-lean environment. The latter is indicated by blocky calcite void-filling interparticle pores in the image. Localized precipitation of glauconite (green) within the tubules was promoted by the organic richness of the micritic sediments infilling the tubules.

IW = 2.4 mm. PPL

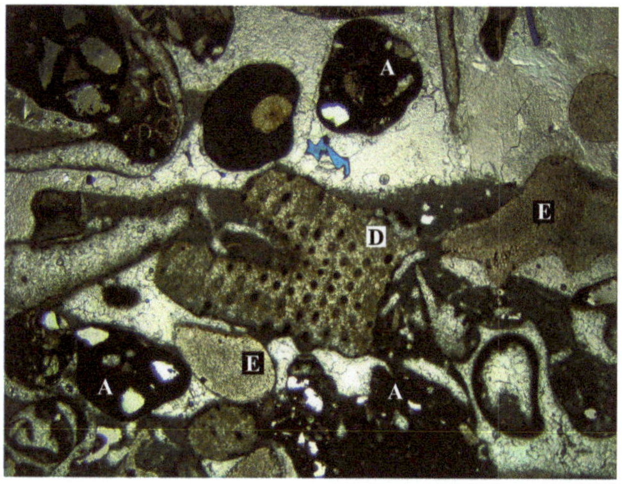

Cretaceous (late Albian). Lisure Formation. Maracaibo Basin

In this grainstone, the central dasycladalean (D) algal grain exhibits well-preserved dichotomous morphology, which is a characteristic feature of some genera of this group of calcifying photoautotrophs. In the Lisure formation, such distinctive morphologic feature is, in most cases, not preserved due to abrasion and fragmentation. The green algal grain is accompanied by echinoid (E) grains (photozoan-heterozoan association) and rounded aggregates (A).

IW = 7.5 mm. PPL

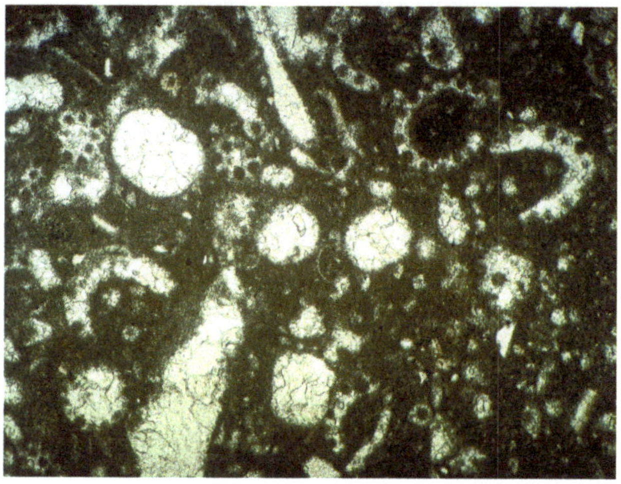

Cretaceous (middle Aptian). Piché Member, Apón Formation. Maracaibo Basin

Wackestone-packstone of fragments of dasycladalean algae (*Terquemella* or *Edgellia* sp.; Bruno Granier, personal communication 2023). Some textural features (spherical elements) are distinguished in some of the grains in the image. The appearance of these diagenetically stabilized grains contrast with the well-preserved state (non-recrystallized) of the LMC micritic matrix.

IW = 2.4 mm. PPL

Cretaceous (middle Aptian). Piché Member, Apón Formation. Maracaibo Basin

Transverse section through a dasycladalean alga of the genus *Salpingoporella* in a sandy micritic limestone. Note the well-preserved, radiating fibrous calcite crystals that make up the ring-like thallus. Primary laterals (L) are arranged perpendicularly to the main axis. *Salpingoporella* is one of the very few Dasycladales that has a calcitic coating instead of an aragonitic one. It is known from the Berriasian to the Aptian (Bruno Granier, personal communication 2023)

IW = 1.3 mm. PPL

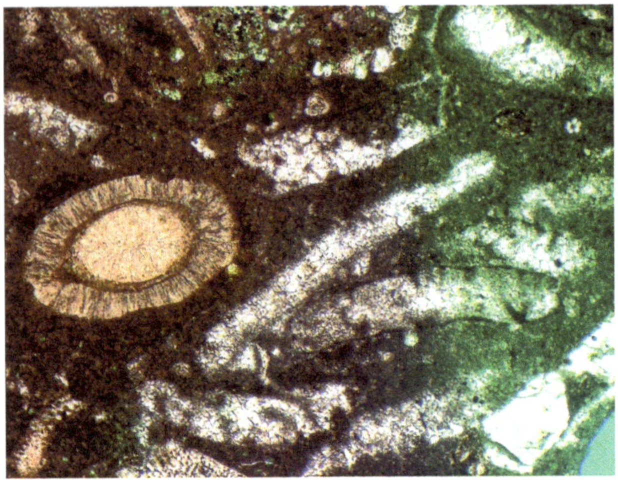

Cretaceous (middle Aptian). Piché Member, Apón Formation. Maracaibo Basin

Green algae wackestone-packstone. The well-preserved state of the fibrous thallus of the single *Salpingoporella* grain (spherical grain to the left) contrasts with the diagenetically-stabilized, elongate green algae grains. Note the dark-colored micritic matrix, which shows no traces of recrystallization. This feature highlights the metastability of aragonitic green algae grains, which are prone to losing most of their morphologic features via dissolution-reprecipitation, even in microfacies that have not undergone through significant meteoric alteration.

IW = 2.4 mm. PPL

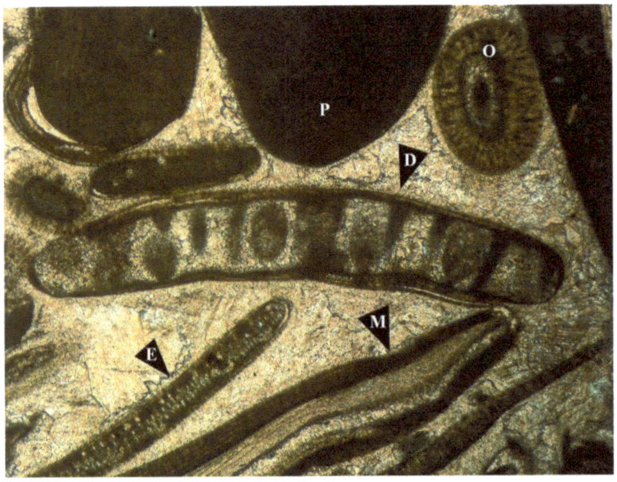

Cretaceous (middle Albian). Lisure Formation. Maracaibo Basin

The elongate, central dasycladalean (D) grain is coated by a thin oolitic cortex, forming a superficial ooid. The green algal grain is easily identified by its characteristic morphologic elements (sporangia and laterals), which are filled with micritic sediments. The other superficial ooids in the image were formed by coating an echinoderm (E) and a mollusc (M). These peculiar ooids, along with peloidal (P) grains and radial ooids (O), originally belonged to a mixed photozoan-heterozoan muddy deposit and were later redeposited in a mud-lean environment.

IW = 2.4 mm. PPL

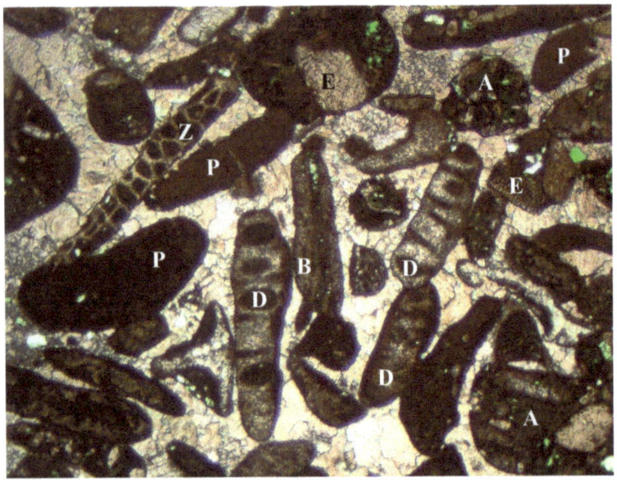

Cretaceous (middle Albian). Lisure Formation. Maracaibo Basin

This grainstone contains fragments of dasycladalean (D) and bryopsidalean (B) algae. In the Lisure Formation, the former are more abundant than the latter, reflecting a global trend in Cretaceous strata (see Steuber et al. 2023). Bryozoans (Z) and echinoids (E) are also visible in the image, along with dark peloidal (P) grains and aggregates (A). This photomicrograph illustrates a common Lisure microfacies, comprising a mix of redeposited skeletal and non-skeletal grains.

IW = 4.2 mm. PPL

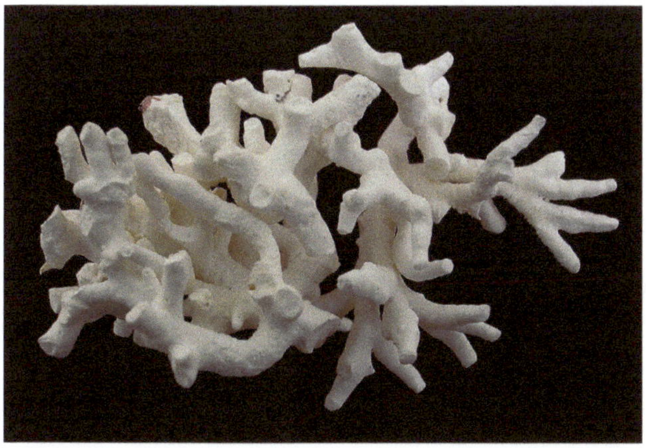

Holocene. La Blanquilla Island

Dried sample of a cm-sized fragment of a non-geniculate coralline alga (probably of the genus *Neogoniolithon* sp). The fragment exhibits characteristic branching habit of modern encrusting coralline algae. Robust, and fruticose branches are composed of HMC (18 mol% $MgCO_3$; measured value). Forms exhibiting this habit are abundant in a wide variety of depositional environments in Los Roques Archipelago, but especially in shallow, agitated settings with moderate-energy level.

IW = 8 cm.

Holocene. Henri Pittier National Park. Aragua State Shoreline

Longitudinal section of a segment of a modern geniculate coralline (genus *Amphiroa*), cemented together with other grains by isopachous rims and bridges of marine cements. The red algal grain exhibits monomerous thallus with a coaxial core and peripheral filaments. Note the well-differentiated dark, thin peripheral region (or cortex) and inner core region (medulla). The cortex filaments are composed of small cells, barely visible under this magnification. The core is composed of alternating rows of long and short cells.

IW = 1.3 mm. CPL

Late Pleistocene. Aves Island

Close view of the core region (medulla) of a modern geniculate coralline (*Amphiroa* sp.) seen in longitudinal section. As mentioned for the previous image, note the arrangement of cells, composed of alternating long and short cells more or less parallel-oriented to the long axis, while a series of arched long-celled filaments cross transversely to that axis. These are morphological diagnostic features of the genus Amphiroa. In Aves Island, coralline algae (geniculate and non-geniculate) are the most abundant skeletal components of both the carbonate basement and beachrock deposits.

IW = 0.6 mm. CPL

Middle Eocene. Santa Rita Formation. Falcón Basin

Calcified segments of geniculate corallines in the image (genus *Corallina*) exhibit distinctive cylindrical shapes. Cell rows in the core regions join regularly (i.e., cells of different rows are more or less equally large). The peripheral region in the genus *Corallina* is very thin, and commonly not preserved or visible, as is seen in the two segments in the image.

IW = 4.2 mm. PPL

Middle Miocene. Capadare Formation. Falcón Basin

Transverse section through a grain of a coralline alga. Filaments in the peripheral region are arranged in a radiating pattern and appear as roughly equant-sized, rectangular cells. In the core region, individual cells appear as small spherical sections. Phosphatization of the grain, predating meteoric diagenesis, is evident from its orange-brownish color. This specimen, like most ancient red algae shown in this photographic set, exemplifies the common well-preserved state (distinguishable filaments and cells) of ancient (pre-Pleistocene) fossils of this group.

IW = 1.3 mm. PPL

Early Miocene. San Luis Formation. Falcón Basin

The non-geniculate, encrusting red alga depicted in the image (*Lithophyllum* or *Mesophyllum*) exhibits a well-differentiated core region (CR) or hypothallus, and peripheral region (PR) or perithallus. The thallus has a monomerous organization with a coaxial core region. Note the differences in arrangement, shape and length between calcified cells in the core and peripheral regions (larger and more or less rectangular, and smaller and more or less square, respectively). In this example, conceptacles in the peripheral region are not visible. The pink coloration of the entire image results from the thin section having been stained with Alizarin Red S.

IW = 1.1 mm. PPL

Photomicrograph courtesy of Gilberto Soto

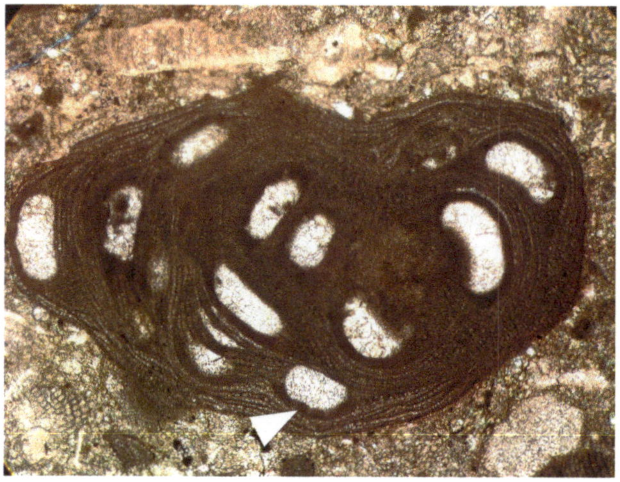

Early Miocene. Churuguara Formation. Falcón Basin

A non-geniculate coralline algal nodular grain (a small rhodolith) embedded in a bioclastic matrix. The individual filaments forming the peripheral region are well differentiated. The small-sized, calcified cells comprising the filaments are barely distinguishable at this magnification. The conceptacles (ellipsoidal to "bean-like" shapes) are multiporate, as indicated by identifiable multiple pores on the roof of some of these sparry-filled cavities (white arrow).

IW = 2.4 mm. PPL

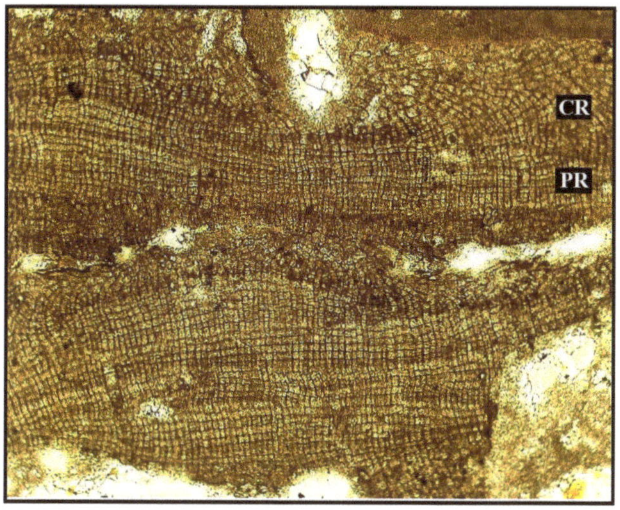

Holocene. La Blanquilla Island

A close view of the monomerous thallus of a modern non-geniculate coralline alga exhibiting a well-differentiated core region (CR) and peripheral region (PR). Note the non-coaxial core filaments (i.e., filaments are not aligned in tiers but are curved). The latter is a diagnostic morphological feature of the genus *Lithothamnion*, which is common component of reef facies in La Blaquilla and Aves islands. However, in this image, conceptables are not visibly distinguishable.

IW = 1.3 mm. PPL

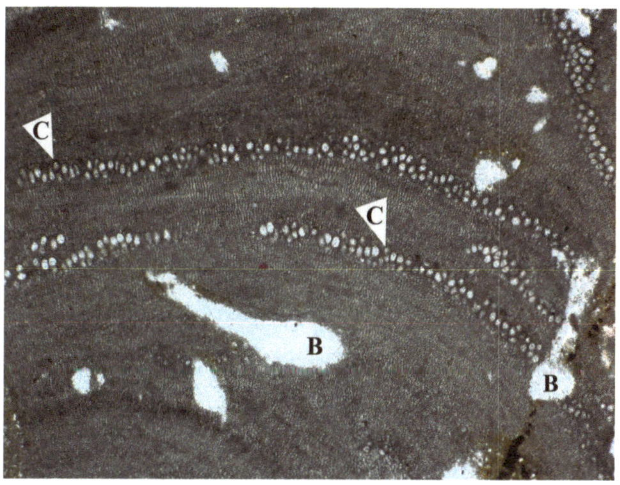

Early Miocene. San Luis Formation. Falcón Basin

The image depicts the thick peripheral region of a lumpy growth form of the non-geniculate genus *Sporolithon*. Note the barely noticeable, curved arrangement of filaments, composed of tiny, elongate cells. Numerous conceptacles (C), cylindrical to ovoid in shape, are arranged in rows (sori). Larger cavities filled with sparry calcite are borings (B). In coralline algae, these bioerosive features are commonly the product of the activity of specific groups of macroborers, which include sponges, bivalves, and polychaete worms.

IW = 2.3 mm. PPL
Photomicrograph courtesy of Gilberto Soto

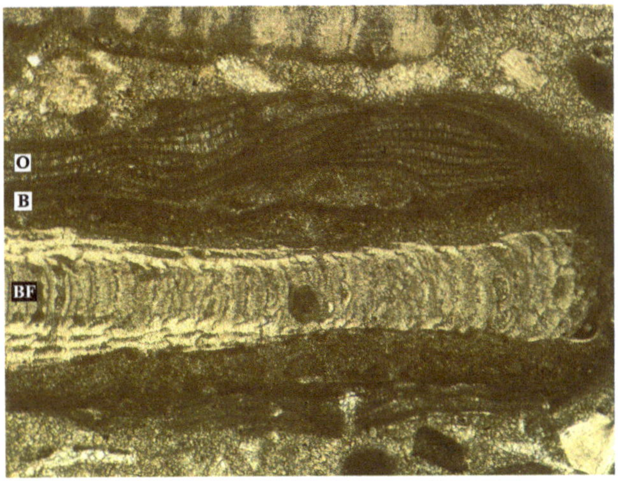

Middle Eocene. Santa Rita Formation. Falcón Basin

In this image, a non-geniculate coralline with a dimerous organization is encrusting a larger benthic foraminifer (BF). Individual outer (O) (postigenous) filaments and their calcified cells (square-like) are visibly well-differentiated. In contrast, the basal (B) (primigineous) filaments do not exhibit well differentiation. This coralline alga-LBF photozoan association is a common example of carbonate production in the meso-oligophotic zone during the Eocene-Miocene interval in the FB. The Santa Rita Formation was deposited in an outer ramp at depths between 60 and 80 m (Pitelli and Molina 1989).

IW = 2.4 mm. PPL

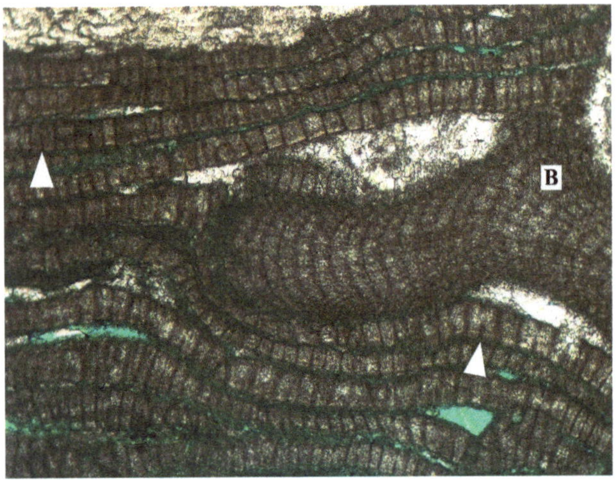

Late Oligocene-Early Miocene. Perla Limestone. Eastern Gulf of Venezuela Sub-Basin

Filaments of an encrusting (non-geniculate) coralline with a dimerous organization are surrounding a branching (B) red algal grain. Calcified cells comprising the filaments or the encrusting form (probably of the genus *Lithoporella*) are visually well differentiated under this magnification. These exhibit square to rectangular shapes, with some adjacent cells connected by cell fusions (white arrows). Empty gaps between layers are either filled with blue-dyed epoxy or LMC cement.

IW = 1.3 mm. PPL

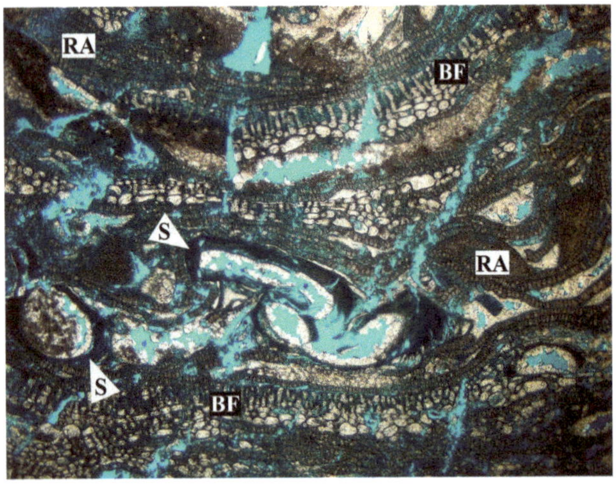

Late Oligocene-Early Miocene. Perla Limestone. Eastern Gulf of Venezuela Sub-Basin

A rhodolith consisting of filaments of red algae (RA) intergrown with encrusting benthic foraminifera (BF) and secondary serpulid (S) tubes. Note the abundant microfractures (grain breakages) cross-cutting the large components, formed during burial diagenesis. Mobilization and redeposition along the gently-steeped ramp in the oligophotic zone are associated with the deposition of highly porous rhodolith rudstones in the Perla Limestone (see Pomar et al. 2015).

IW = 4.2 mm. PPL

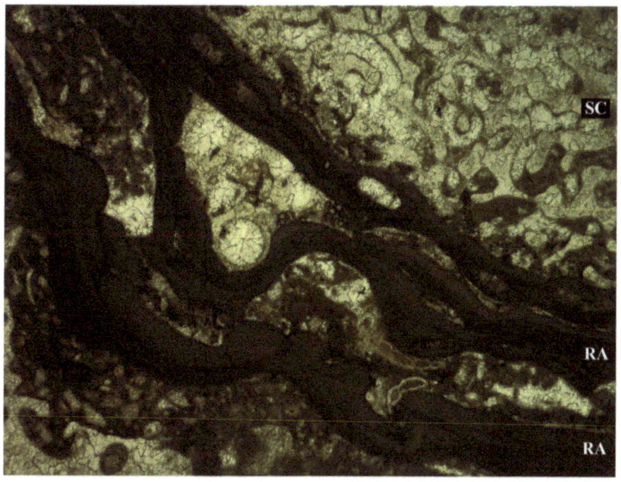

Early Miocene. San Luis Formation. Falcón Basin

A coralline red alga (RA) (probably of the genus *Lithothamnion*) displaying a layered to foliose growth form, is encrusting a scleractinian coral (SC) (photozoan association; coralgal lithofacies sensu Pomar et al. 2012). Note the growth for of the alga, consisting of several lamellate branches arranged horizontally or at various angles to one another. In the San Luis Formation, the photozoan association between corals and coralline red algae is often accompanied by green algae and LBF (as well as molluscs) in facies of the inner ramp (see Albert-Villanueva 2016).

IW = 4.3 mm. PPL
Photomicrograph courtesy of Gilberto Soto

Early Miocene. Churuguara Formation. Falcón Basin

Some growth forms in corallines can be visually striking. This is the case with these two red algae grains, which present warty protuberance forms. Note that the grains consist of lamellae decreasing in diameter from the basal part to the upper part. Their original reticulate cell arrangements underwent dissolution-reprecipitation (HMC to LMC), forming a pseudospar fabric that is visible in the uppermost lamellae

IW = 4.2 mm. PPL

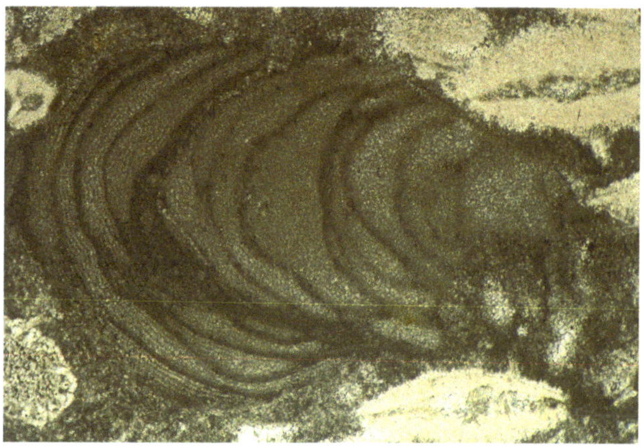

Middle Eocene. Santa Rita Formation. Falcón Basin

Another example of a striking form of coralline algae preservation. In this case, note the protuberant form of the grain, with individual lamellae decreasing in diameter from left to right in this view. As with many fossilized corallines, the absence of visible diagnostic features (e.g., thallus organization and conceptacles) prevents the classification of this alga. The genera *Mesophyllum* (order Hapalidiales) and *Lithophyllum* (order Corallinales) have been reported in microfacies of the Santa Rita Formation by Pitelli and Molina (1989).

IW = 4.2 mm. PPL

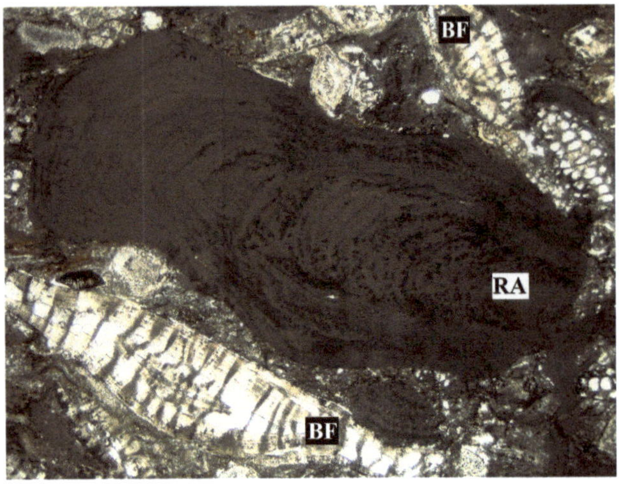

Late Oligocene-Early Miocene. Perla Limestone. Eastern Gulf of Venezuela Sub-Basin

A gravel-sized red algal (RA) grain (non-geniculate), exhibiting a lumpy protuberance growth form, accompanied by tests of LBF (BF) (photozoan association), forming a packed fabric. This image exemplifies both the dominance of these oligophotic carbonate producers (corallines are photoautotrophs and LBF are mixotrophs) and the Photo C-factory of Michel et al. (2019) in the overall sedimentation of the Perla Limestone.

IW = 7.5 mm. PPL

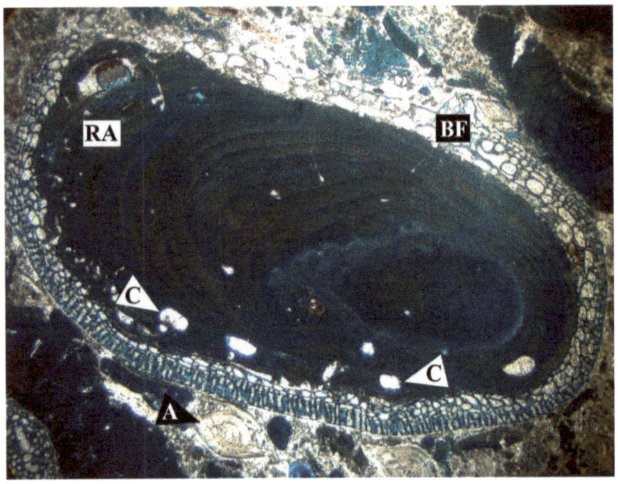

Late Oligocene-Early Miocene. Perla Limestone. Eastern Gulf of Venezuela Sub-Basin

Another gravel-sized red alga (RA) grain (non-geniculate). It is coated by encrusting benthic foraminifera (BF), probably of the genus *Acervulinidae*, and embedded in a bioclastic matrix in which tests of the large benthic foraminifer *Amphistegina* (A) are abundant. The dark-blue color of this large alga grain is due to pervasive microporosity developed in the deep-burial environment (see Valencia and Laya, 2020). Also, note the conceptacles (C) in the peripheral region of the grain.

IW = 7.5 mm. PPL

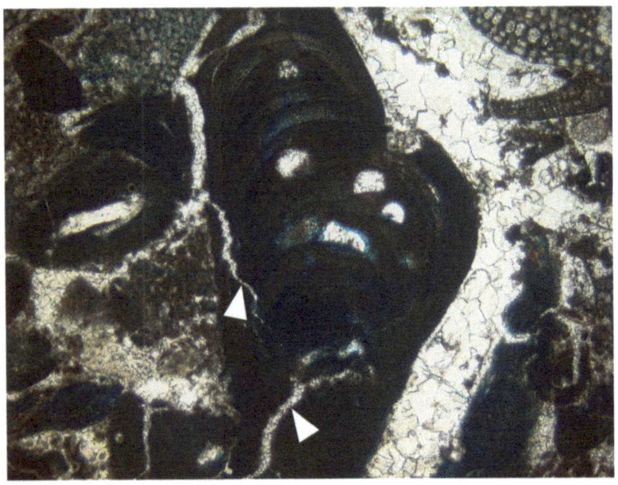

Late Oligocene-Early Miocene. Perla Limestone. Eastern Gulf of Venezuela Sub-Basin

The coralline algal grain in the center of the image exhibits lumpy/fruticose growth form and conceptacles (spherical to ovate white spots). In addition to microporosity, indicated by the dark blue color, the grain is cross-cut by breakages (white arrows), or microfractures, which represent another diagenetic feature developed in the deep-burial environment. The variety of diagenetic processes present in coralline algae-rich facies of the Perla Limestone will be discussed in more detail in Chaps. 16, 17, and 19.

IW = 7.5 mm. PPL

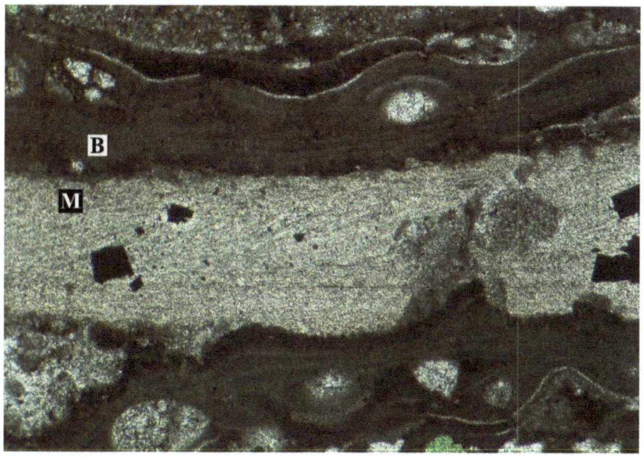

Cretaceous (middle Albian). Lisure Formation. Maracaibo Basin

In this image, an encrusting Cretaceous coralline alga is coating an elongated calcitic mollusc (M) shell. Tiny cells of the basal (B) filaments are distinctly visible under this magnification and are oriented perpendicular to the mollusc shell. Filaments in the perithallus are arranged in an undulating pattern. Conceptacles, filled with sparry calcite, exhibit distinctive ovate shapes. Coralline red algae are, in fact, not abundant in Cretaceous carbonate facies of Venezuela.

IW = 2.4 mm. PPL

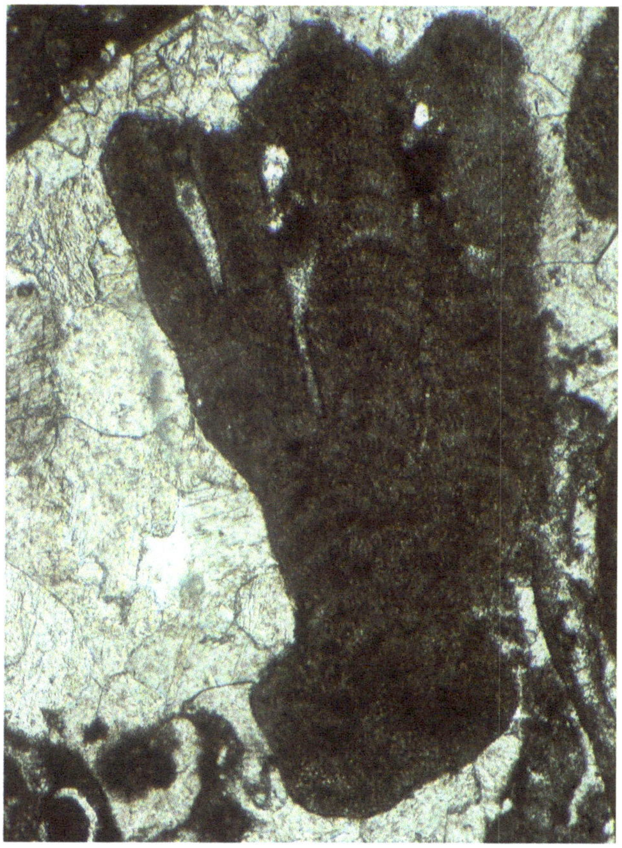

Cretaceous (middle Albian). Lisure Formation. Maracaibo Basin

Red alga *Marinella lugeoni* (Family Elianellaceae) showing characteristic dark appearance, and erect, digitiform and branching thallus. Note the subspherical shape of the basal thallus (bottom) and the thick, closely-arranged branches in the upper part. The thallus is composed of juxtaposed filaments, with a fabric that looks "hairy" rather than forming a lattice-like. This species, like all members of the family Elianellaceae, is deprived of fossilized reproductive organs, which were possibly external (see Granier and Dias-Brito 2016b).

IW = 2.2 mm. PPL

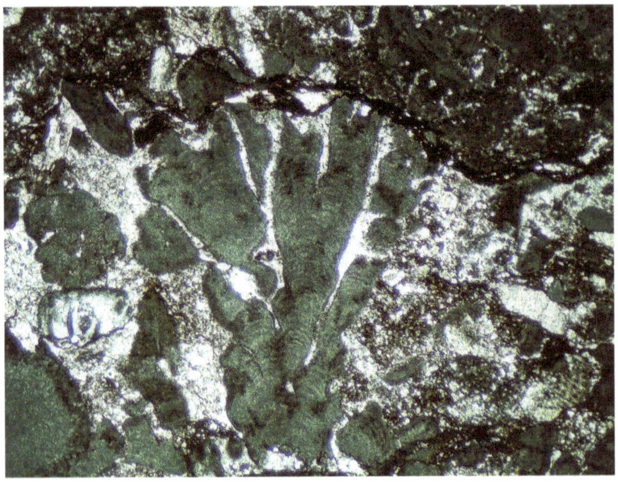

Cretaceous (middle Albian). Lisure Formation. Maracaibo Basin

Another example of *Marinella lugeoni* exhibiting a digitiform branching habit. Note the upward thickening of the branches, a morphologic feature that characterizes this species, also evident in the previous example. The precipitation of LMC cement of meteoric origin (white precipitate) held the branching structure together, despite visible signs of fragmentation. This species of red alga is particularly common in Late Jurassic and Early Cretaceous strata. See Leinfelder and Werner (1993), for additional photographic examples and a detailed morphologic description of this red alga.

IW = 7.5 mm. PPL

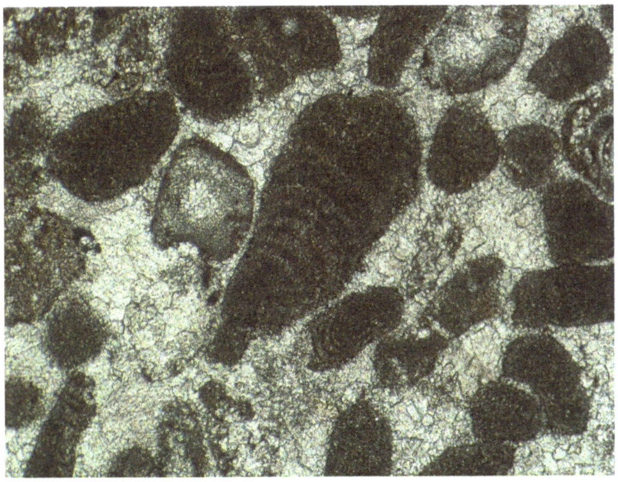

Cretaceous (middle Albian). Lisure Formation. Maracaibo Basin

A red algae grainstone. The rounded to elongate grains exhibit an almost complete loss of textural diagnostic features due to fragmentation, abrasion, and meteoric diagenesis before their final deposition. Nevertheless, diagnostic cell rows are visible in some grains, particularly in the elongate central one, which is likely derived from the breakage of a *Marinella lugeoni* branching thallus. Red algae are not volumetrically significant in Cretaceous carbonate facies in Venezuela, but they are locally moderately common in the Lisure Formation.

IW = 2.4 mm. PPL

Middle Permian (Cisularian to Guadalupian). Palmarito Formation. Venezuelan Andes

Rounded red alga grain. The original cellular structure of this gravel-sized, "ancentral" alga grain is relatively well-preserved, suggesting an original HMC composition. This, along with the sample's age (Middle Permian) and the presence of conceptacles (ovoid cavities filled by cloudy sparry calcite), serves as diagnostic criteria for classifying this grain as *Archaeolithophyllum* sp. This genus has been reported in facies of the Palmarito Formation by Laya and Tucker (2012).

IW = 2.4 mm. PPL

References

Abdelghany, O., Fowler, A. R., Hashem, W., 2019. Oman Exotic limestones in the UAE-Oman border area with implications for Middle-Upper Permian Neo-Tethyan rifting. *Journal of Asian Earth Sciences*, *181*, 103879.

Afzal, J., Williams, M., Leng, M. J., Aldridge, R. J., 2011. Dynamic response of the shallow marine benthic ecosystem to regional and pan-Tethyan environmental change at the Paleocene–Eocene boundary. *Palaeogeography, Palaeoclimatology, Palaeoecology*, *309*, 141–160.

Aguirre, J., and Riding, R., 2005. Dasycladalean algal biodiversity compared with global variations in temperature and sea level over the past 350 Myr. *Palaios*, *20*, 581–588.

Aguirre, J., and Braga, J. C., 2022. Middle Miocene (Serravallian) rhodoliths and coralline algal debris in carbonate ramps (Betic Cordillera, S Spain). *Frontiers in Earth Science*, *10*, 958148.

Aguirre, J., Riding, R., Braga, J. C., 2000. Diversity of coralline red algae: origination and extinction patterns from the Early Cretaceous to the Pleistocene. *Paleobiology*, *26*, 651–667.

Aguirre, J., Baceta, J. I., Braga, J. C., 2007. Recovery of marine primary producers after the Cretaceous–Tertiary mass extinction: Paleocene calcareous red algae from the Iberian Peninsula. *Palaeogeography, Palaeoclimatology, Palaeoecology*, *249*, 393–411.

Aguirre, J., Perfectti, F., Braga, J. C., 2010. Integrating phylogeny, molecular clocks, and the fossil record in the evolution of coralline algae (Corallinales and Sporolithales, Rhodophyta). *Paleobiology*, *36*, 519–533.

Aguirre, J., Braga, J. C., Pujalte, V., Orue-Etxebarria, X., Salazar-Ortiz, E., Rincón-Martínez, D., Abad, M., Pérez-Valera, F., 2020. Middle Eocene Rhodoliths from Tropical and mid-latitude Regions. *Diversity*, *12*, 117.

Albert-Villanueva, E., 2016. Facies y secuencias deposicionales mixtas carbonático-siliciclásticas del Mioceno inferior de la Cuenca de Falcón (noroeste de Venezuela) como modelo exploratorio en el Caribe. Doctoral Dissertation. University of Barcelona. Volumes 1 and 2.

Albert-Villanueva, E., Bover-Arnal, T., Ferràndez-Cañadell, C. and Salas, R., 2017. Comment on "Tectonic and environmental factors controlling on the evolution of Oligo-Miocene shallow marine carbonate factories along a tropical SE Circum-Caribbean" by Silva-Tamayo et al. (2017). *Journal of South American Earth Sciences*, *91*, 394–396.

Andersson, A. J., Mackenzie, F. T., Bates, N. R., 2008. Life on the margin: implications of ocean acidification on Mg-calcite, high latitude and cold-water marine calcifiers. *Marine Ecology Progress Series*, *373*, 265–273.

Baceta, J. I., Pujalte, V., Wright, V. P., Schmitz, B., Arenas, C., Pomar, L., Colombo, F., 2011. Carbonate platform models, sea-level changes and extreme climatic events during the Paleocene-early Eocene greenhouse interval: a basin-platform-coastal plain transect across the southern Pyrenean basin. *Geo-Guías*, *7*, 101–150.

Bassi, D., and Nebelsick, J. H., 2010. Components, facies and ramps: redefining Upper Oligocene shallow water carbonates using coralline red algae and larger foraminifera (Venetian area, northeast Italy). *Palaeogeography, Palaeoclimatology, Palaeoecology*, *295*, 258–280.

Basso, D., 2012. Carbonate production by calcareous red algae and global change. *Geodiversitas*, *34*, 13–33.

Braga, J. C., 2017. Neogene rhodoliths in the Mediterranean basins, In: R. Riosmena-Rodríguez, W. Nelson, and J. Aguirre (eds.), Rhodolith/Maërl Beds: A Global Perspective: Springer, Berlin Heidelberg, *Coastal Research Library*, *15*, 169–193.

Braga, J. C., and Bassi, D., 2007. Neogene history of *Sporolithon* Heydrich (Corallinales, Rhodophyta) in the Mediterranean region. *Palaeogeography, Palaeoclimatology, Palaeoecology*, *243*, 189–203.

Brandano, M., Vannucci, G., Pomar, L., Obrador, A., 2005. Rhodolith assemblages from the lower Tortonian carbonate ramp of Menorca (Spain): environmental and paleoclimatic implications. *Palaeogeography, Palaeoclimatology, Palaeoecology*, *226*, 307–323.

Brandano, M., Frezza, V., Tomassetti, L., Cuffaro, M., 2009. Heterozoan carbonates in oligotrophic tropical waters: the Attard member of the lower coralline limestone formation (Upper Oligocene, Malta). *Palaeogeography, Palaeoclimatology, Palaeoecology*, *274*, 54–63.

Brandano, M., Corda, L., Castorina, F., 2010. Facies and sequence architecture of a tropical foramol-rhodalgal carbonate ramp: Miocene of the central Apennines (Italy). In: Mutti, M., Piller, W. E., Betzler, C. (Eds.), Carbonate Systems During the Olicocene-Miocene Climatic Transition. *International Association of Sedimentologists, Special Publication 42*, 107–128.

Brooke, C., and Riding, R., 1998. Ordovician and Silurian coralline red algae. *Lethaia*, *31*, 185–195.

Bucur, I. I., 1999. Stratigraphic significance of some skeletal algae (Dasycladales, Caulerpales) of the Phanerozoic. In: Farinacci, A., Lord, A. R. (Eds.), Depositional Episodesand Bioevents. Università La Sapienza, Rome, pp. 53–104.

Bucur, I. I., Hoffmann, M., Kolodziej, B., 2005. Upper Jurassic-Lowermost Cretaceous benthic algae from Tethys and the European platform: a case study from Poland. *Revista Española de Micropaleontología*, *37*, 105–129.

Bucur, I. I., Rashidi, K., Senowbari-Daryan, B., 2012. Early Cretaceous calcareous algae from central Iran (Taft Formation, south of Aliabad, near Yazd). *Facies*, *58*, 605–636.

Caragnano, A., Foetisch, A., Maneveldt, G. W., Millet, L., Liu, L. C., Lin, S. M., Rodondi, G., Payri, C. E., 2018. Revision of Corallinaceae (Corallinales, Rhodophyta): recognizing *Dawsoniolithon* gen. nov., *Parvicellularium* gen. nov. and Chamberlainoideae subfam. nov. containing *Chamberlainium* gen. nov. and *Pneophyllum*. Journal of Phycology 54, 391–409.

Carannante, G., Esteban, M., Milliman, J. D., Simone, L., 1988. Carbonate lithofacies as paleolatitude indicators: problems and limitations. *Sedimentary Geology*, *60*, 333–346.

Carannante, G., Cherchi, A., Simone, L., 1995. Chlorozoan versus foramol lithofacies in Upper Cretaceous rudist limestones. *Palaeogeography, Palaeoclimatology, Palaeoecology*, *119*, 137–154.

Carras, N., Conrad, M. A., Radoicic, R., 2006. *Salpingoporella*, a common genus of Mesozoic Dasycladales (calcareous green algae). *Revue de Paléobiologie*, *25*, 457.

Chelaru, R., and Bucur, I. I., 2016. The taxonomy of middle Miocene red algae from the Gârbova de Sus Formation (Transylvanian Basin, Romania). *Carnets Geol.*, *16*(11), 307.

Chen, W., Wang, Y., Huang, Y., Wang, T., Yi, Z., Kiessling, W., 2020. Reef-building red algae from an uppermost Permian reef complex as a fossil analogue of modern coralline algal ridges. *Facies*, *66*, 1–14.

Corrochano, D., Vachard, D., Armenteros, I. 2013. New insights on the red alga *Archaeolithophyllum* and its preservation from the Pennsylvanian of the Cantabrian Zone (NW Spain). *Facies*, *59*, 949–967.

Cózar, P., and Vachard, D., 2003. *Neoprincipia* nov. gen., a new Mississippian red alga, and remarks on the Archaeolithophyllaceae (Rhodophyta). *Geobios*, *36*, 505–517.

Cózar, P., Somerville, I. D., Medina-Varea, P., 2005. Note on the earliest occurrences of the calcareous algae *Paraepimastopora* and *Archaeolithophyllum* in Mississippian rocks. *Coloquios de Paleontología*, *55*, 7–20.

Darrenougue, N., De Deckker, P., Payri, C., Eggins, S., Fallon, S., 2013. Growth and chronology of the rhodolith-forming, coralline red alga Sporolithon durum. *Marine Ecology Progress Series*, *474*, 105–119.

Davies, P. J., and Marshall, J. F., 1985. Halimeda bioherms—low energy reefs, northern Great Barrier Reef. Proceedings of the 5th International Coral Reef Symposium, 5, 1–7.

Davies, P. J., Braga, J. C., Lund, M., Webster, J. M., 2004. Holocene deep water algal buildups on the eastern Australian shelf. *Palaios, 19*, 598–609.

Dorsey, R. J., Braga, J. C., Gardner, K., McDougall, K., O'Connell, B., 2023. Early Pliocene marine transgression into the lower Colorado River valley, southwestern USA, by re-flooding of a former tidal strait. *Geological Society, London, Special Publications, 523*, 369–397.

Elliott, G. F., 1984. Climatic tolerance in some aragonitic green algae of the post-Palaeozoic. *Palaeogeography, Palaeoclimatology, Palaeoecology, 48*, 163–169.

Elliott, G. F., 1991. Dasycladalean algae of the Palaeozoic and Mesozoic. In: Riding, R. (ed.): *Calcareous Algae and Stromatolites*, 125–130. Springer-Verlag, Berlin.

Flügel, E., 1988. *Halimeda*: paleontological record and palaeoenvironmental significance. *Coral Reefs, 6*, 123–130.

Flügel, E., 1991. Triassic and Jurassic marine calcareous algae: a critical review. *Calcareous algae and stromatolites*, 481–503.

Flügel, E, 2010. Microfacies of Carbonate Rocks—Analysis, Interpretation and Application. Springer, Heidelberg-Dordrecht-London-New York, 984 pp.

Frankignoulle, M., Canon, C., Gattuso, J. P., 1994. Marine calcification as a source of carbon dioxide: Positive feedback of increasing atmospheric CO_2. *Limnology and Oceanography, 39*, 458–462.

Granier, B., 2012. The contribution of calcareous green algae to the production of limestones: a review. *Geodiversitas, 34*, 35–60.

Granier, B., and Dias-Brito, D., 2016a. On the fossil alga *Elianella elegans* PFENDER & BASSE, 1948, and its so-called lookalikes, with description of *Elianella brasiliana* n. sp Revision of the Juliette PFENDER Collection. Part 1. *Carnets de Géologie*, 213–229.

Granier, B., and Dias-Brito, D., 2016b. On the fosil alga *Marinella lugeoni* Pfender, 1939, nom. cons., and its seven unfortunate avatars. Revision of the Juliette Pfender collection. Part 2. Revision of the Jesse Harlan Johnson collection. Part 2. *Carnets de Géologie, 16*, 231–45.

Granier, B. R. C., and Lethiers, A., 2017. *Draconisella mortoni* sp. nov., a *Mizzia*-like Dasycladalean alga from the Lower Cretaceous of Oman. *Palaeontologia Electronica, 20*(1), 1–14.

Granier, B., Dias-Brito, D., Bucur, I. I., 2014. *Clypeina tibanai*, sp. nov. (Polyphysacea, Dasycladales, Chlorophyta), mid-Cretaceous green alga from the Potiguar Basin, Brazilian margin of the young South Atlantic Ocean. *Geologica Acta: an international earth science journal, 12*, 227–237.

Granier, B., Azerêdo, A., Ramalho, R., 2017. Taxonomic revision of *Cylindroporella*? lusitanica Ramalho, 1970: In search for the origins of the Family Dasycladaceae. Island Arc, *26*, e12176.

Halfar, J., and Mutti, M., 2005. Global dominance of coralline red-algal facies: a response to Miocene oceanographic events. *Geology, 33*, 481–484.

Halfar, J., Godinez-Orta, L., Mutti, M., Valdez-Holguín, J. E., Borges, J. M., 2004. Nutrient and temperature controls on modern carbonate production: an example from the Gulf of California, Mexico. *Geology, 32*, 213–216.

Hallock, P., 1997. Reefs and reef limestones in Earth history. In: Birkeland, C. (Ed.), Life and Death of Coral Reefs. Chapman and Hall, New York, 13–42.

Hallock, P., 2001. Coral reefs, carbonate sedimentation, nutrients, and global change. In: Stanley, G.D. (Ed.), The History and Sedimentology of Ancient Reef Ecosystems. Kluwer Academic/Plenum Publishers, pp. 387–427.

Hallock, P., 2015. Changing influences between life and limestones in earth history. In: Birkeland, C. (Ed.), Coral Reefs in the Anthropocene. Springer, Dordrecht, pp. 17–42.

Hamad, M., 2020. Coralline red algae from the Lower Pliocene Shagra Formation of Wadi Wizer, Red Sea coast, Egypt: Biofacies analysis, systematics and palaeoenvironmental implications. *Geopersia, 10*, 1–21.

Hamid, N., Ma, Q., Boulom, S., Liu, T., Zheng, Z., Balbas, J., Robertson, J., 2015. Seaweed minor constituents. In: Tiwari BK, Troy DJ (eds.) Seaweed sustainability: food and non-food applications. Elsevier, London, pp. 193–241.

Harvey, A., Broadwater, S., Woelkerling, W. J., Mitrovski, P. J., 2003. *Choreonema* (Corallinales, Rhodophyta): 18s rDNA phylogeny and resurrection of the Hapalidiaceae for the subfamilies Choreonematoideae, Austrolithoideae, and Melobesioideae. *Journal of Phycology, 39*, 988–998.

Hasiuk, F. J., and Lohmann, K. C., 2010. Application of calcite Mg partitioning functions to the reconstruction of paleocean Mg/Ca. *Geochimica et Cosmochimica Acta, 74*, 6751–6763.

Hayton, S., Nelson C., Hood S., 1995. A skeletal assemblage classification system for non-tropical carbonate deposits based on New Zealand Cenozoic limestones. *Sedimentary Geology, 100*. 123–141.

Hillis-Colinvaux, L., 1980. Ecology and taxonomy of *Halimeda*: primary producer of coral reefs. *Advances in Marine Biology, 17*, 1–327.

Hine, A. C., Hallock, P., Harris, M. W., Mullins, H. T., Belknap, D. F., Jaap, W. C., 1987. *Halimeda* bioherms along an open seaway: Miskito channel, Nicaraguan rise, SW Caribbean Sea. *Coral Reefs, 6*, 173–178.

Huck, S., Aretz, M., Herbig, H. G., Aurag, A., Milhi, A., 2006. Facies and depositional setting of Mississippian (Late Viséan) microbial mounds in the Jerada massif, NE Morocco. *Kölner Forum für Geologie und Paläontologie, 15*, 43–44.

Ioan IB, Milan N S., Emanoil, S., Divna, J., George, P., Svetlana, P., 2018. Rediscovery of the type locality of the Udoteacean alga *Boueina hochstetteri* Toula, 1884, in the Lower Cretaceous of Serbia. *Carnets de géologie, 18*, 123–137.

Irvine, L. M., Chamberlain, Y. M., 1994. Seaweeds of the British Isles. Volume 1. Rhodophyta Part 2B Corallinales, Hildenbrandiales. HMSO, London.

James, N. P., 1997. The cool-water carbonate depositional realm. In: James, N. P., Clarke, J., (eds.), Cool-water carbonates, vol 56. SEPM Society for Sedimentary Geology, Tulsa, 1–20.

James, N. P., and Jones, B., 2015. Origin of Carbonate Sedimentary Rocks. John Wiley & Sons, Ltd, Chinchester, 464 pp.

Johansen, H. W., 1981. Coralline Algae, a First Synthesis. CRC Press, Boca Raton, Florida.

Jiang, H.X., and Zhang, Y. Y, 2021. Earliest calcified green algae from the 520 Ma old Cambrian dolostones in Xinjiang, China. *Biopetrology, 1*(3), 9–18.

Johnson, J. H., 1964. Fossil and Recent calcareous algae from Guam. US Geological Survey Professional Paper, 403-G, 1–40.

Kato, A., Baba, M., Suda, S., 2011. Revision of the Mastophoroideae (Corallinales, Rhodophyta) and polyphyly in nongeniculate species widely distributed on Pacific coral reefs. Journal of Phycology, 47, 662–672.

Kiessling, W., Flügel, E., Golonka, J. A. N., 2003. Patterns of Phanerozoic carbonate platform sedimentation. *Lethaia, 36*, 195–225.

Killops, S. D., and Killops, V. J., 2005. An Introduction to Organic Geochemistry. Wiley Blackwell. Second Edition. 393 pp.

Klicpera, A., Michel, J., Westphal, H., 2015. Facies patterns of a tropical heterozoan carbonate platform under eutrophic conditions: the Banc d'Arguin, Mauritania, *Facies, 61*, 421.

Kundal, P., 2011. Generic distinguishing characteristics and stratigraphic ranges of fossil corallines: an update. *Journal Geological Society of India*, 78, 571–586.

Laya, J. C., and Tucker, M. E., 2012. Facies analysis and depositional environments of Permian carbonates of the Venezuelan Andes: Palaeogeographic implications for Northern Gondwana. *Palaeogeography, Palaeoclimatology, Palaeoecology*, *331*, 1–26.

Lees, A., 1975. Possible influence of salinity and temperature on modern shelf carbonate sedimentation. *Marine Geology*, *19*, 159–198.

Lees, A., and Buller, A. T., 1972. Modern temperate-water and warm-water shelf carbonate sediments contrasted. *Marine Geology*, *13*, M67–M73.

Le Gall, L., Payri, C. E., Bittner, L., Saunders, G. W., 2010. Multigene phylogenetic analyses support recognition of the Sporolithales ord. nov. *Molecular Phylogenetics and Evolution*, *54*, 302–305.

Leinfelder, R. R., and Werner, W., 1993. Systematic position and palaeoecology of the Upper Jurassic to Tertiary alga Marinella lugeoni Pfender: Zitteliana, v. 20 (Hagn/Herm-Festschrift), pp. 105–122.

Littler, M. M., Littler, D. S., Hanisak, M. D., 1991. Deep-water rhodolith distribution, productivity, and growth history at sites of formation and subsequent degradation. *Journal of Experimental Marine Biology and Ecology*, *150*, 163–182.

Lund, M., Davies, P. J., Braga, J. C., 2000. Coralline algal nodules off Fraser Island, eastern Australia. *Facies*, *42*, 25–34.

Mamet, B. L. 1991. Carboniferous calcareous algae. In: Riding, R. (ed.) Calcareous algae and stromatolites. Springer-Verlag, Berlin, 370–451.

Marino, M., and Santantonio, M., 2010. Understanding the geological record of carbonate platform drowning across rifted Tethyan margins: Examples from the Lower Jurassic of the Apennines and Sicily (Italy). *Sedimentary Geology*, *225*, 116–137.

Marrack, E. C., 1999. The relationship between water motion and living rhodolith beds in the southwestern Gulf of California, Mexico, *Palaios*, 159–171.

Marshall, J. F., and Davies, P. J., 1988. *Halimeda* bioherms of the northern Great Barrier Reef. *Coral Reefs*, *6*, 139–148.

Masse, J. P., and Arnaud Vanneau, A., 1999. Early Cretaceous calcareous algae of the mid-Pacific Mountains. *Revue de Micropaléontologie*, *42*, 57–69.

Martín, J. M., Braga, J. C., Riding, R., 1997. Late Miocene *Halimeda* alga-microbial segment reefs in the marginal Mediterranean Sorbas Basin, Spain. *Sedimentology*, *44*, 441–456.

McNeil, M. A., Nothdurft, L. D., Dyriw, N. J., Webster, J. M., Beaman, R. J., 2020. Morphotype differentiation in the Great Barrier Reef Halimeda bioherm carbonate factory: internal architecture and surface geomorphometrics. *The Depositional Record*, *7*, 176–199.

Michel, J., Borgomano, J., Reijmer, J. J., 2018. Heterozoan carbonates: When, where and why? A synthesis on parameters controlling carbonate production and occurrences. *Earth-Science Reviews*, *182*, 50–67.

Michel, J., Laugié, M., Pohl, A., Lanteaume, C., Masse, J. P., Donnadieu, Y., Borgomano, J., 2019. Marine carbonate factories: a global model of carbonate platform distribution. *International Journal of Earth Sciences*, *108*, 1773–1792.

Misra, P. K., and Kumar, P, 1988. Fossil algae from the Cretaceous of Yaragur, Tiruchirapalli District, Tamil Nadu. *Palaeobotanist*, *37*, 36–51.

Misra, P. K., Jauhri, A. K., Singh, S. K., Kishore, S., Chowdhury, A., 2001. Coralline algae from the Oligocene of Kachchh, Gujarat, India. *Journal of the Palaeontological Society of India*, *46*, 59–76.

Misra, P. K., Kishore, S., Singh, S. K., Jauhri, A. K., 2009. Rhodophycean algae from the lower Cretaceous of the Cauvery Basin, South India. *Journal of the Geological Society of India*, *73*, 325–334.

Mutti, M., and Hallock, P., 2003. Carbonate systems along nutrient and temperature gradients: some sedimentological and geochemical constraints. *International Journal of Earth Sciences*, *92*, 465–475.

Narayan, G. R., Reymond, C. E., Stuhr, M., Doo, S., Schmidt, C., Mann, T., Westphal, H., 2022. Response of large benthic foraminifera to climate and local changes: Implications for future carbonate production. *Sedimentology*, *69*, 121–161.

Nebelsick, J. H., and Bassi, D., 2000. Diversity, growth forms and taphonomy: Key factors controlling the fabric of coralline algae dominated shelf carbonates, in Insalaco, E., Skelton, P. W., and Palmer, T.J., eds., Carbonate Platform Systems: Components and interactions: London, *Geological Society London, Special Publication*, v. *178*, p. 89–107

Nelson, W. A., Sutherland, J. E., Farr, T. J., Hart, D. R., Neill, K. F., Kim, H. J., Yoon, H. S., 2015. Multigene phylogenetic analyses of New Zealand coralline algae: *Corallinapetra novaezelandiae* gen. et sp. nov. and recognition of the Hapalidiales ord. nov. *Journal of Phycology*, 51, 454–468.

Okla, S. M., 1991. Dasycladacean algae from the Jurassic and Cretaceous of central Saudi Arabia. *Micropaleontology*, 183–190.

Peña, V., Vieira, C., Braga, J. C., Aguirre, J., Rösler, A., Baele, G., De Clerck, O., Le Gall, L., 2020. Radiation of the coralline red algae (Corallinophycidae, Rhodophyta) crown group as inferred from a multilocus time-calibrated phylogeny. *Molecular phylogenetics and evolution*, *150*, 106845.

Phipps, C. V. G., and Roberts, H. H., 1988. Seismic characteristics and accretion history of *Halimeda* bioherms on Kalukalukuang Bank, eastern Java Sea (Indonesia). *Coral Reefs*, *6*, 149–159.

Pitelli, R., and Molina, A., 1989. El Eoceno Medio Tardío y Eoceno Tardío de la parte occidental de la cuenca de Falcón. *Sociedad Venezolana de Geología, Boletín 36*, 5–12.

Pomar, L., 2001. Types of carbonate platforms: a genetic approach. *Basin Research*, *13*, 313–334.

Pomar, L., and Hallock, P., 2008. Carbonate factories: a conundrum in sedimentary geology. *Earth-Science Reviews*, *87*(3–4), 134–169.

Pomar, L., Brandano, M., Westphal, H., 2004. Environmental factors influencing skeletal grain sediment associations: a critical review of Miocene examples from the western Mediterranean. *Sedimentology*, *51*, 627–651.

Pomar, L., Bassant, P., Brandano, M., Ruchonnet, C., Janson, X., 2012. Impact of carbonate producing biota on platform architecture: insights from Miocene examples of the Mediterranean region. *Earth-Science Reviews*, *113*, 186–211.

Pomar, L., Esteban, M., Martinez, W., Espino, D., Castillo de Ott, V., Benkovics, L., Castro Leyva, T., 2015. Oligocene-Miocene carbonates of the Perla field, offshore Venezuela: Depositional model and facies architecture, In: C. Bartolini and P. Mann, eds., Petroleum geology and potential of the Colombian Caribbean Margin. *American Association of Petroleum Geologists, Memoir 108*, 647–674.

Pomar, L., Baceta, J. I., Hallock, P., Mateu-Vicens, G., Basso, D., 2017. Reef building and carbonate production modes in the west-central Tethys during the Cenozoic. *Marine and Petroleum Geology*, *83*, 261–304.

Pomar, L., Hallock, P., Mateu-Vicens, G., Baceta, J. I., 2022. Why do bio-carbonates exist?. *Journal of Marine Science and Engineering*, *10*, 1648.

Pray, L. C., and Wray J. L., 1963. Porous algal facies (Pennsylvanian), Honaker Trail, San Juan Canyon, Utah, In: Bass R. O. (ed.), *Shelf Carbonates of the Paradox Basin, a Symposium*. 4th Field Conference, June 12-16, 1963. Guidebook, Four Corners Geological Society, 204–234.

Radoičić, R., Conrad, M. A., Carras, N., 2005. Observations on *Neomeris cretacea* var. *undulata* Raineri followed by *Dissocladella bonardii*, n. sp. (Dasycladales, green algae). *Revue de Paléobiologie*, *24*, 311.

Radoičić, R., and Schlagintweit, F., 2007. *Neomeris mokragorensis* sp. nov.(calcareous alga, dasycladales) from the Cretaceous of Serbia, Montenegro and the Northern Calcareous Alps,(Gosau Group, Austria). *Geoloski anali Balkanskoga Poluostrva*, *68*, 39–51.

Rees, S. A., Opdyke, B. N., Wilson, P. A., Henstock, T. J., 2007. Significance of Halimeda bioherms to the global carbonate budget based on a geological sediment budget for the Northern Great Barrier Reef, Australia. *Coral Reefs*, *26*, 177–188.

Reid, R. P., and Macintyre, I. G., 1988. Foraminiferal-algal nodules from the eastern Caribbean; growth history and implications on the value of nodules as paleoenvironmental indicators. *Palaios*, *3*, 424–435.

Riding, R., 2004. *Solenopora* is a chaetetid sponge, not an alga. *Palaeontology*, *47*, 117–122.

Ries, J. B., 2010. Review: geological and experimental evidence for secular variation in seawater Mg/Ca (calcite-aragonite seas) and its effects on marine biological calcification. *Biogeosciences*, *7*, 2795–2849.

Roberts, H. H., Aharon, P., Phipps, C. V., 1988. Morphology and sedimentology of Halimeda bioherms from the eastern Java Sea (Indonesia). *Coral Reefs*, *6*, 161–172.

Rösler, A., Perfectti, F., Peña, V., Aguirre, J., Braga, J. C., 2017. Timing of the evolutionary history of Corallinaceae (Corallinales, Rhodophyta). *Journal of Phycology*, *53*, 567–576.

Sartorio, D., and Venturini, S. 1988. Southern Tethys biofacies. AGIP, San Donato Milanese, 235 pp.

Saunders, G. W., and Hommersand, M. H., 2004. Assessing red algal supraordinal diversity and taxonomy in the context of contemporary systematic data. *American Journal of Botany*, *91*, 1494–1507.

Schlager, W., 2003. Benthic carbonate factories of the Phanerozoic. *International Journal of Earth Sciences*, *92*, 445–464.

Schlager, W., 2005. Carbonate Sedimentology and Sequence Stratigraphy. SEPM Concepts in Sedimentology and Paleontology #8. Society of Economic Paleontologists and Mineralogists, 200 pp.

Schlagintweit, F., 2010. Gosavisiphon gen. nov. based on Halimeda paucimedullaris SCHLAGINTWEIT & EBLI, 1998: a remarkable macroalga (Udoteaceae?) from the Late Cretaceous of the Northern Calcareous Alps (Austria and Germany) with affinites to Late Paleozoic and Late Triassic phylloids. *Geologia Croatica*, *63*, 27–53.

Scholle, P. A., and Ulmer-Scholle, D., 2003. A Color Guide to the Petrography of Carbonate Rocks: Grains, Textures, Porosity, Diagenesis. *American Association of Petroleum Geologists, Memoir 77*, 474 pp.

Senowbari-Daryan, B., Torabi, H., Rashidi, K., 2008. New solenoporaceans from Upper Triassic (? Norian-Rhaetian) reef limestones in central Iran. *Geologia Croatica*, *61*, 135–157.

Silva-Tamayo, J. C., Ramirez, C., Lara, M., Sial, A. N., Trujillo, D., Salazar, E., 2016. Sedimentology and chemostratigraphy of a Valanginian carbonate succession from the Baja Guajira Basin, northern Colombia. *Brazilian Journal of Geology*, *46*, 163–180.

Silva-Tamayo, J. C., Lara, M. E., Yobo, L. N., Erdal, Y. D., Sanchez, J., Zapata, P. A., 2017. Tectonic and environmental factors controlling on the evolution of Oligo-Miocene shallow marine carbonate factories along a tropical SE Circum-Caribbean. *Journal of South American Earth Sciences*, 78, 213–237.

Simone, L., Carannante, G., Bassi, D., Cherchi, A., 2012. Rudist-bearing rhodalgal facies in the post-Turonian recovery of peri-Tethyan carbonate systems: a case history from the Nurra region (northwestern Sardinia, Italy). *Geodiversitas*, *34*, 167–187.

Smith, A. M., Sutherland, J. E., Kregting, L., Farr, T. J., Winter, D. J., 2012. Phylomineralogy of the coralline red algae: correlation of skeletal mineralogy with molecular phylogeny. *Phytochemistry*, *81*, 97–108.

Stanley, S. M., 2006. Influence of seawater chemistry on biomineralization throughout Phanerozoic time: Paleontological and experimental evidence. *Palaeogeography, Palaeoclimatology, Palaeoecology*, *232*, 214–236.

Stanley, S. M., and Hardie, L. A., 1998. Secular oscillations in the carbonate mineralogy of reef-building and sediment-producing organisms driven by tectonically forced shifts in seawater chemistry. *Palaeogeography, Palaeoclimatology, Palaeoecology*, *144*, 3–19.

Stanton, R. J., and Flügel, E., 1987. Palecology of upper Triassic reefs in the Northern Calcareous Alps: Reef communities. *Facies*, *16*, 157–185.

Steller, D. L., and Foster, M. S., 1995. Environmental factors influencing distribution and morphology of rhodoliths in Bahía Concepción, BCS, México. *Journal of Experimental Marine Biology and Ecology*, *194*, 201–212.

Steuber, T., Löser, H., Mutterlose, J., Parente, M., 2023. Biogeodynamics of Cretaceous marine carbonate production. *Earth-Science Reviews*, 104341.

Tomás, S., Aguirre, A., Braga J C., Martin-Closas, C., 2007. Late Hauterivian coralline algae (Rhodophyta, Corallinales) from the Iberian Chain (E Spain). Taxonomy and the evolution of multisporangial reproductive structures. *Facies*, *53*, 79–95.

Tosolini, A. M., Wallace, M. W., Gallagher, S. J., 2012. Shallow water mud-mounds of the Early Devonian Buchan Group, East Gippsland, Australia. *Sedimentary Geology*, *281*, 208–221.

Townsend, R. A. and Adey, W. H. 1990. Morphology of the Caribbean Alga: *Goniolithon improcerum* Foslie *et* Howe *in* Foslie (Corallinaceae, Rhodophyceae). *Botanica Marina*, *33*, 99–116.

Tucker, M. E., and Wright, V. P., 1990. Carbonate Sedimentology. Blackwell Publishing. 482 pp.

Vachard, D. 2021. Calcareous algae (Rhodophyta and Chlorophyta). In: Alderton, D. and Elias, S.A. (Eds), Encyclopedia of Geology, 389–406. Academic.

Vachard D., and Aretz, M., 2004. Biostratigraphical precisions on the Early Serpukhovian (Late Mississippian), by means of a carbonate algal microflora (cyanobacteria, algae and pseudo-algae) from La Serre (Montagne Noire, France). *Geobios*, *37*, 643–666.

Vachard, D., and Kabanov, P., 2007. *Palaeoaplysinella* gen. nov. and *Likinia* Ivanova and Ilkhovskii, 1973 emend., from the type Moscovian (Russia) and the algal affinities of the ancestral palaeoaplysinaceae n. comb. *Geobios*, *40*(6), 849–860.

Vachard, D., Krainer, K. Lucas, S. 2012. Pennsylvanian (Late Carboniferous) calcareous microfossils from Cedro Peak (New Mexico, USA); Part 1: Algae and Microproblematica. *Annales de Paléontologie*, *98*, 225–252.

Vachard, D., Cózar, P., Aretz, M., Izart, A., 2016. Late Viséan-Serpukhovian foraminifers in the Montagne Noire (France): Biostratigraphic revision and correlation with the Russian substages. *Geobios*, *49*, 469–498.

Valencia, F. L. and Laya, J. C., 2020. Deep-Burial dissolution in an Oligocene-Miocene giant carbonate reservoir (Perla Limestone), Gulf of Venezuela Basin: Implications on microporosity development. *Marine and Petroleum Geology*, *113*, 104144.

Westphal, H., Halfar, J., Freiwald, A., 2010. Heterozoan carbonates in subtropical to tropical settings in the present and past. *International Journal of Earth Sciences*, *99*, 153–169.

Woelkerling, W. J., 1988. The Coralline Red Algae: An Analysis of the Genera and Subfamilies of Nongeniculate Corallinaceae. Oxford University Press, London, Oxford.

Woelkerling, W. J., Irvine, L. M., Harvey, A. S., 1993. Growth-forms in non-geniculate coralline red algae (Coralliinales, Rhodophyta). *Australian Systematic Botany*, *6*, 277-293.

Woelkerling, W. J., Granier, B., Dias-Brito, D., 2014. *Heydrichia* (?) *poignantii*, sp. nov.(Sporolithaceae, Sporolithales, Rhodophyta), a 100 million year old fossil coralline red alga from north-eastern Brazil, and a new Hauterivian record of Sporolithon from Switzerland. *Carnets de Géologie*, *14*, 139–158.

Wray, J. L., 1977a. Calcareous Algae, Developments in Paleontology and Stratigraphy. *Elsevier Scientific Publishing Co., Amsterdam, New York*, *4*, 1–185.

Wray, J. L. 1977b. Late Paleozoic calcareous red algae. In: *Fossil algae: recent results and developments* (ed. E. Flügel), pp. 167–76. Berlin: Springer.

Yang, E. C., Boo, S. M., Bhattacharya, D., Saunders, G. W., Knoll, A. H., Fredericq, S., Graf, L., Yoon, H. S., 2016. Divergence time estimates and the evolution of major lineages in the florideophyte red algae. *Scientific Reports*, *6*, 1–11.

Yilmaz, İ. Ö., 2000. Taxonomic and Paelogeographic Approaches to the Dasyclad Algae in the Upper Jurassic (Kimmeridgian)-Upper Cretaceous (Cenomanian) Peritidal Carbonates of the Fele (Yassıbel) Area (Western Taurides, Turkey). *Turkish Journal of Earth Sciences*, *8*, 81–102.

Foraminifera

6

Foraminifera (Phylum Retaria; e.g., Ruggiero et al. 2015; Burki et al. 2016) are marine protozoa (unicellular eukaryotic organisms) that live either on the marine seafloor (benthics) or in the water column (planktonics) (e.g., Bellier et al. 2010; BouDagher-Fadel 2015, 2018). Carbonate chambered skeletons (= tests) secreted by foraminifera are present in marine sedimentary records since the Ordovician (see Nehrke et al. 2013). Micropaleontologists recognize about 40,000 fossil species preserved in the geological record, with estimates of roughly 10,000 extant species (Erez 2003; Murray 2007; Boudaugher-Fadel 2018).

Biomineralization in foraminifera, which is controlled by soft tissue of the foraminiferid cell called cytoplasm, creates tests of incredible beauty and highly variable structural complexity. Tests can be very simple, with one or only few chambers, or they can be complex, with many chambers arranged in various three-dimensional configurations (Erez 2003). Foraminifera thrive in a wide range of environments, from shallow marine to deep-sea settings and from polar to tropical latitudes. They are important as biostratigraphical indicators in marine rocks of late Paleozoic, Mesozoic and Cenozoic age; and especially important in studies of Mesozoic to Quaternary climate history because isotopes within their $CaCO_3$ tests record changes in temperature and ocean chemistry (e.g., Armstrong and Brasier 2005). Regarding the contribution of carbonate produced by marine organisms to the global carbon budget, foraminifera, together with coccolithophores, has been estimated as the major contributors to the $CaCO_3$ flux (Schiebel 2002; Tambutté et al. 2011).

6.1 Major Groups and Stratigraphic Distribution

Foraminifera (Class Foraminifera; former Order Foraminiferida) are subdivided in orders, superfamilies and families. The two major groups of foraminifera are separated according to their life strategy. The benthic forms are known from the Early Cambrian (Calcite I). The earliest planktonic forms are known from the Late Triassic (Aragonite II) (see Boudaugher-Fadel 2018). Of the roughly 10,000 extant species, benthic forms represent the vast majority (e.g., Loeblich and Tappan 1988; Erez 2003; Bellier et al. 2010). Fewer than 100 extant species of foraminifera are planktonic (Boudaugher-Fadel 2018).

For study purposes, and according to the size of their tests, benthic foraminifera are divided into two informal groups, smaller and larger, being the latter characterized by complex internal morphologies (e.g., Armstrong and Brasier 2005; Hallock and Pomar 2008; Flügel 2010; BouDagher-Fadel 2015). The cut-off value between smaller and larger tests varies depending on the author. Flügel (2010) considers that tests of SBF range from approximately 100 μm to 600 μm (0.1 mm to 0.6 mm); while Flügel (2010) and Renema (2019) consider that tests of LBF are larger than 0.5 mm (and up to 20 cm). Pawlowski et al. (1993) and Lee (1998) consider that tests of LBF are larger than 1 mm (1000 μm), while tests of SBF are smaller than 1 mm but larger than 0.1 mm (100 μm). The latter are characterized by chambers lacking of complex internal constructions or architecture (Bellier et al. 2010). In addition, Pawlowski et al. (1993) also consider tests of benthic foraminifera smaller than 0.1 mm, which are called microforaminifera. SBF make up a substantial proportion of both known fossils and living benthic foraminifera (Armstrong and Brasier 2005). However, due to their small sizes, their study has been always limited to the large fraction of tests (> 150 μm), resulting in a substantial loss of information on species richness, biodiversity and species composition (Schroder-Adams et al. 1987; Lo Giudice-Cappelli and Austin 2019).

It is noteworthy that the division of benthic foraminifera in larger and smaller is merely descriptive and based on the sizes of the tests, as it does not contemplate the ecology (e.g., symbiont-bearing or symbiont-barren) and environment of each species. Nevertheless, due to the positive correlation between the size of the tests and the ecology of the majority

J. A. Méndez Dot, J. Méndez Baamonde, *Atlas of Tropical Carbonates*, https://doi.org/10.1007/978-3-031-25494-9_6

of examples shown in the photographic set, in this chapter we refer to symbiont-bearing benthic foraminifera (belonging to the lineages with morphologically larger forms, and interpreted to have hosted algal symbionts) and symbiont-barren benthic foraminifera as LBF and SBF, respectively. However, this consideration has its exceptions, since some tests of symbiont-bearing LBF have sizes equal to or smaller than 1 mm (e.g., examples shown of the genus *Amphistegina*). The same applies to some tests of symbiont-barren SBF having tests equal to or larger than 1 mm (e.g., an example shown of the genus *Martinottiella*). It should be noted that encrusting benthic foraminifera belonging to the LBF lineages (e.g., species of Neogene *Homotrema*) are not addressed in this chapter. Furthermore, we mention only the foraminifera orders and superfamilies/families to which examples shown in the photographic set belong.

6.1.1 Orders of Larger Benthic Foraminifera

BouDagher-Fadel (2018) deals with 14 orders of LBF. Examples of Paleozoic LBF correspond to **Order Fusulinida** (age range: Early Carboniferous to Late Permian) and **Order Palaeotextulariida** (age range: Late Devonian to Late Permian). Shallow platform/ramp limestones rich in fusulinids (Superfamily Schwagerinoidea) and less common biseriamminids (paleotextulariids of the Family Biseriamminidae) are extensively developed in Late Paleozoic strata (Carboniferous and Permian) (e.g., Sartorio and Venturini 1988; Jenny and Guex 2005; Vachard et al. 2006, 2010; Gaillot and Vachard 2007; Flügel 2010; Laya and Tucker 2012; Boudaugher-Fadel 2018).

Examples of Mesozoic LBF correspond to **Order Textulariida** (age range: Early Cambrian to Holocene). Textulariids (e.g., superfamilies Lituoloidea, Loftusoidea, Orbitolinoidea, Nezzazatoidea, and Ataxophragmioidea) dominate shallow-benthic foraminiferal assemblages in Late Triassic, Jurassic, and Early and mid-Cretaceous strata (e.g., Sartorio and Venturini 1988; Ward et al. 1995; BouDagher-Fadel et al. 2001; Bachmann and Hirsch 2006; Flügel 2010; Boudaugher-Fadel 2018; Boudagher-Fadel and Price 2019; Campos-Soto et al. 2019). They are also moderatelly common in Paleogene strata (superfamilies Textularioidea, Orbitolinoidea, and Coskinolinoidea; e.g., Sartorio and Venturini 1988; Boudaugher-Fadel 2018).

Examples of Cenozoic LBF correspond to **Order Rotaliida** (age range: Triassic to Holocene) and **Order Miliolida** (age range: Carboniferous to Holocene). Rotaliids, which are rare or absent in Triassic and Jurassic strata, became common in the Late Cretaceous; abundant in the Late Paleocene; dominant during the Eocene and the Oligocene (e.g., Superfamily Nummulitoidea); and abundant in the Early Miocene (e.g., Superfamily Asterigerinoidea). Their abundance diminished from the Middle late Miocene to Present, but they are still common in modern settings (e.g., species of *Amphistegina*) (e.g., Sartorio and Venturini 1988; Beavington-Penney 2004; Beavington-Penney and Racey 2004; Renema 2007; Hallock and Pomar 2008; Flügel 2010; Boudagher-Fadel and Price 2013, 2014; Boudaugher-Fadel 2018; Kateb et al. 2018; Narayan et al. 2022; Soto 2025; Soto et al. 2025). Larger miliolids of the Superfamily Alveolinoidea became common in the Late Cretaceous (from Cenomanian), and are abundant in Late Paleocene and Eocene strata (e.g., Sartorio and Venturini 1988; Scheibner and Speijer 2008; Boudagher-Fadel and Price 2021); whereas larger miliolids of the Superfamily Soritoidea became common during the Eocene and Oligocene, and are abundant from the Late Miocene to Present (e.g., *Peneroplis* sp.) (e.g., Sartorio and Venturini 1988; Holzmann et al. 2001; BouDagher-Fadel 2008a, b, 2018).

6.1.2 Orders of Smaller Benthic Foraminifera

Examples of SBF correspond to five different orders: **Miliolida** (age range: Carboniferous to Holocene), **Buliminida** (age range: Triassic to Holocene; e.g., Haynes 1981; Loeblich and Tappan 1992; Sen Gupta 1999; but see Pawlowski et al. 2013; who include buliminids in the Order Rotaliida), **Lagenida** (age range: Late Silurian to Holocene; Boudaugher-Fadel 2018), **Lituolida** (age range: Early Silurian to Holocene; Kaminski and Perdana 2020), **Rotaliida** (age range: Triassic to Holocene), and **Textulariida** (age range: Early Cambrian to Holocene).

Smaller miliolids (e.g., Superfamily Milioloidea; age range: Late Triassic to Holocene) are arguably the most widely-recognized group of SBF in Mesozoic and Cenozoic shallow carbonates. They are abundant in platform-lagoon facies of the Middle and Late Jurassic and the Early Cretaceous (Barremian, Aptian and Albian) (e.g., Sartorio and Venturini 1988; Omaña and Alencáster 2009; Flügel 2010; Filkorn and Scott 2011; Lefort et al. 2011; Schlagintweit et al. 2016; Campos-Soto et al. 2019), as well as common in Paleogene, Neogene and Quaternary strata and modern oceans (e.g., Sartorio and Venturini 1988; Rasser et al. 2005; Vaziri-Moghaddam et al. 2006; Parker and Gischler 2011, 2015). In ancient facies, assemblages of SBF commonly include lagenides (e.g., genera *Lenticulina* and *Nodosaria*), which are common in the Late Triassic and throughout the Jurassic, the Cretaceous and the Pelogene (see Friedrich and Erbacher 2006; Reolid et al. 2008a, b; Friedrich 2010; Farouk et al. 2017, 2020; Azmi 2018; Salazar-Ramírez et al. 2019); lituolides (e.g., *Novalesia*), which are common in the Cretaceous (Velíc 2007; Rami et al. 2012; Omaña et al. 2016); buliminids of the superfamilies Buliminoidea and Bolivinoidea (e.g., *Bulimina* and

Bolivina, respectively), which are common/abundant in the mid-Cretaceous, as well as the Buliminoidea *Uvigerina*, which is common in the Cenozoic (Von Daniels 1986; De Romero and Galea-Álvarez 1995; Koutsoukos and de Klasz 2000; Rey et al. 2004; Roetzel et al. 2006; Truskowski 2006; Friedrich and Erbacher 2006; Friedrich 2010; Giusberti et al. 2016; Farouk et al. 2017, 2020); rotaliids such as *Gavelinella*, which is common in the Late Cretaceous (Koutsoukos and de Klasz 2000; Truskowski 2006; Dubicka and Peryt 2014); rotaliids of the cibicidids group (Family Cibicididae), which are common in the Paleogene (Boersma et al. 1985; Schweizer et al. 2009; Farouk et al. 2017, 2020); and textulariids, which are moderately common in the Cretaceous (e.g., *Marssonella*, *Minouxia*), and common in the Cenozoic (e.g., *Textularia*, *Martinottiella*) (Loeblich and Tappan 1988; Kaminski 2014; Soto 2025; Soto et al. 2025). Modern offshore platforms and deep-sea settings are rich in buliminids, textulariids, cibicidids and lagenides (e.g., Javaux and Scott 2003; Morales 2004; Lisiecki and Raymo 2005; Gooday et al. 2008).

6.1.3 Orders of Planktonic Foraminifera

Planktonic foraminifera (PF) are divided into two orders (BouDagher-Fadel 2015): **Globigerinida** (age range: Late Triassic to Holocene) and **Heterohelicida** (age range: Cretaceous to Holocene). Members of the Order Globigerinida first appeared during the Late Triassic. Earlier planktonics (Superfamily Favuselloidea) were aragonitic, matching the seawater chemistry (Aragonite II) of early Mesozoic times. PF did not become common until the Middle Jurassic. They became abundant during the Early Cretaceous (Berriasian and Valanginian) and dominant during the CO_2-rich Calcite II seas of the mid-Cretaceous (e.g., Leckie 1987; Sartorio and Venturini 1988; Sliter 1989; Hart 1999; Premoli Silva and Sliter 1999; Crespo de Cabrera et al. 1999; Premoli Silva and Verga 2004; Truskowski 2006; Flügel 2010; BouDagher-Fadel 2015; Steuber et al. 2023), a time interval that witnessed the appearance of PF of the Order Heterohelicida (Nederbragt et al. 1998; BouDagher-Fadel 2015). After the massive extinction event in the Cretaceous-Paleocene boundary, PF recovered and became common throughout the Paleocene and the Early to Middle Eocene (e.g., Kelly 2002; Petrizzo 2007; McInerney and Wing 2011). Despite the "descent into the Icehouse", or the global cooling trend that started in the Late Eocene (~ 40–34 Ma) (Zachos et al. 2008; Fenton et al. 2016), PF have remained common during the Aragonite III seas of the second half of the Cenozoic (Bandy 1964; Armstrong and Brasier 2005; Flügel 2010; Fenton et al. 2016; Schiebel et al. 2018). The *Globigerina* ooze, or the pelagic and deep-sea PF assemblages (dominated by the genera *Globigerinoides*, *Globigerina*, *Orbulina* and *Globorotalia*) account for up to 80% of modern carbonate deposition in seas and oceans (Armstrong and Brasier 2005). In addition, Cenozoic PF have arguably the most detailed fossil record of any group (Fenton et al. 2016).

6.2 Skeletal Mineralogy

The vast majority of foraminifera are calcite producers (Erez 2003; de Nooijer et al. 2009). Planktonic species are almost exclusively composed of LMC (e.g., Bender et al. 1975; Scholle and Ulmer-Scholle 2003), although earlier forms from the Late Triassic are aragonitic and then switched to calcite in the Jurassic (Boudaugher-Fadel 2018). The composition of benthic species is more varied, being common both LMC and HMC tests, but dominating the first (de Nooijer et al. 2009). Furthermore, some benthic forms construct their tests of cemented (agglutinated) clastic terrigenous or calcareous sediment grains. Among agglutinators, some are indiscriminate in their selection of building materials, whereas others carefully select calcareous particles (mainly micritic crystals), sponge spicules, quartz grains, mica flakes or other specific components for their tests (Scholle and Ulmer-Scholle 2003). In addition, some Mesozoic LBF (Order Involutinida), SBF (Order Robertinida), and PF (Family Favusellidae) were aragonitic (see Rigaud et al. 2021; Steuber et al. 2023).

6.3 Ecology and Environment

High species diversity of foraminifera is traduced in varied ecological and environmental requirements among the three major groups. SBF (both extinct and extant) are light-independent, heterotrophic calcifiers, and therefore thrive in both the photic and aphotic zones. Most extant, tropical and subtropical LBF and PF harbour algal photosymbionts (or endosymbionts) in a relationship similar to that between scleractinian corals and their zooxanthellae. Therefore, they are light-dependent, symbiont-bearing calcifiers that inhabit the photic zone. By morphologic analogies (among other features, such as isotopic signatures), many planktonic and benthic foraminifera fossils are interpreted to have hosted algal symbionts (e.g., Lee 1998; D'Hondt and Zachos 1998; Hallock 1999; Erez 2003; Armstrong and Brasier 2005; Hallock and Pomar 2008; Pomar and Hallock 2008; BouDagher-Fadel 2015, 2018). The first LBF that likely had photosymbionts were the late Paleozoic fusulinids (see Davydov 2014; Lipps and Stanley Jr. 2016; Shi 2021). Photosymbiosis in PF dates back to the Late Cretaceous (see D'Hondt and Zachos 1998; Abramovich et al. 2003; Bornemann and Norris 2007).

Photosymbiosis is generally regarded as a mutualistic relationship that combines a heterotrophic host and photosynthesizing algae into a mixotrophic holobiont (Takagi et al. 2025). Algal symbiosis is one of the major mechanisms that allow mixotrophic nutrition, the ability to both feed and photosynthesize, in contrast to purely autotrophic and heterotrophic nutrition, which allows one or the other. Therefore, symbiont-bearing (photosymbiotic) foraminifera (LBF and many PF) are mixotrophs. LBF host different types of symbionts, including diatoms, dinoflagellates, chlorophytes, and rhodophytes. PF have a symbiotic relationship only with two types of algal symbionts, either (but not both) dinoflagellates or chrysophytes (the latter commonly referred to as golden-brown algae). The advantages of algal symbiosis and mixotrophy play a major role in foraminiferal reproduction and calcification, not only by the energy acquired through photosynthesis and uptake of CO_2, but to the access to dissolved inorganic nutrients (i.e., NH_4^+, NO_2^-, NO_3^-, PO_4^{3-}) (see Hallock 1999; Pawlowski et al. 2001a, 2001b; Erez 2003; Lee 2006; Pomar and Hallock 2008; BouDagher-Fadel 2015, 2018; Pomar et al. 2017; Renema 2019; Takagi et al. 2025; for reviews on photosymbiosis in foraminifera).

Limiting environmental factors (e.g., latitude, bathymetry, water temperature and transparency, nutrients, salinity, CO_2 concentrations, turbidity, hydrodynamic energy, substrate, food supply, and redox conditions) define the ecological niche of each species of foraminifera. Such factors control species growth zonation patterns and their distribution within the spectrum of marine depositional environments, from shallow, near-shore and inner platform/ramp settings, to mid-outer platform/ramps and edges, slopes and basins (see Haq and Boersma 1998; Armstrong and Brasier 2005; Flügel 2010; Bellier et al. 2010; Schiebel et al. 2018; Renema 2019).

6.3.1 Ecology and Environment: Benthic Foraminifera

Benthic foraminifera live on the seafloor, where they are attached or free-living (sessile). The live at the water-sediment interface or within the first centimeters of the upper unconsolidated sediment layer (infaunal mode of life), or attached on plants or algae some centimeters above the seafloor (epifaunal mode of life) (see Gräfe 2005; Reolid et al. 2008a). They occur within a wide range of depths, from the intertidal zone and shallow neritic platforms to bathyal depths (and even deep-sea abyssal depths). Depth distribution of photosymbiotic benthic foraminifera depends on the type of photosymbionts.

6.3.1.1 Larger Benthic Foraminifera

Symbiont-bearing LBF are largely considered "K-strategists" with limited ability to rapidly increase their population densities, having long life spans, large sizes and delayed reproduction (Hallock 1985; Hottinger 2007). These mixotrophs have inhabited the whole photic zone spectrum throughout the Phanerozoic. Some flourish/flourished in the euphotic zone (e.g., late Paleozoic fusulinids, Cretaceous orbitolinids, and Neogene soritids), some in the mesophotic zone (e.g., some Cenozoic asterigerinids and nummulitids), and some in the oligophotic zone (e.g., some Cenozoic asterigerinids and nummulitids; see Pomar 2001; Hallock and Pomar 2008; Ma et al. 2018; Renema 2019). LBF with green or red symbionts are restricted to shallower depths than those with diatoms that inhabit greater depths (e.g., Leutenegger 1984; Romero et al. 2002; Flügel 2010). Morphologies of LBF are strongly influenced by light availability and water motion, with larger, flatter and more fragile tests characteristic of lower light, low wave-energy environments (Hallock and Pomar, 2008).

The presence of LBF in the fossil record indicates a warm environment, while their absence points to cooler or more nutrient-rich environments (Boudaugher-Fadel 2018). LBF are restricted to a worldwide climatic belt bordered by the 15–20 ° C minimum temperature range, i.e., warm tropical/subtropical waters between 30° N and S of the equator (see Adams et al. 1990; Langer and Hottinger 2000; Schlager 2003, 2005), and are largely associated to oligotrophic (nutrient-poor) waters (e.g., Armstrong and Brasier 2005; Boudaugher-Fadel 2018). Nevertheless, some extant LBF can tolerate water temperatures as low as 10–11 ° C, including *Amphistegina* and *Sorites* in the Mediterranean (Hallock et al. 2011), and *Amphisorus* and *Amphistegina* on the southwest Australian shelf (Li et al. 1999). High rates of nutrient flux tend to discourage photosymbiosis, so LBF which culture symbionts and other oligotrophic species are discouraged (Armstrong and Brasier 2005). Even though, the presence of LBF in some mixed photozoan-heterozoan and heterozoan-dominated carbonates throughout the Phanerozoic suggests that some of these mixotrophs also thrive/thrived in nutrient-intermediate (mesotrophic) and nutrient-rich (eutrophic) conditions (e.g., Bassi and Nebelsick 2010; Núñez-Useche et al. 2020; Narayan et al. 2022; this book).

6.3.1.2 Smaller Benthic Foraminifera

SBF are known to inhabit the most varied and/or extreme marine ecosystems, from shallow intertidal-hypersaline settings to deep basin environments, therefore they are good environmental indicators. The vast majority of SBF are symbiont-barren (not photosymbiotic) heterotrophs, many of them are stress-tolerant opportunistics (see Fajemila et al. 2015). Some live epifaunally (K-strategists) in the photic zone, feeding of diatoms (and other protozoa), bacteria and invertebrates (e.g., *Discorbis*; common from the Eocene to the Holocene). Many prefer to live infaunally ("r-strategists", i.e., well-adapted to an exponential increase in population size; Hallock 1985; Hottinger 2007) feeding on particulate organic matter or grazing upon bacteria. SBF may dominate the biota in eutrophic and low-oxygen (dysoxic to suboxic) conditions (e.g., *Bolivina*, *Bulimina*, *Lenticulina*) (e.g., Erbacher et al. 1999; Gräfe 2005; Armstrong and Brasier 2005; Friedrich and Erbacher 2006; Reolid et al. 2008a; Friedrich 2010; Fajemila et al. 2015), as well as in deep environments below the photic zone (e.g., *Uvigerina*; see Von Daniels 1986). The depth of infaunal mode of life in the sediments is largely controlled by the availability of oxygen and nutrients (upwelling- or continental-sourced) and by the organic flux (food) in the uppermost parts of the sediment column (Van der Zwaan et al. 1999). A few SBF are known to culture photosymbionts (e.g., some species of the genus *Elphidium*, which are common from the Eocene to the Holocene) and are commonly associated to symbiont-barren smaller benthic forms which inhabit the photic zone (e.g., species of *Ammonia*). Smaller miliolids of the Superfamily Milioloidea (e.g., *Quinqueloculina*) represent a common example of symbiont-barren, epifaunal SBF which flourish in the upper photic zone. These shallow-water heterotrophs have been common skeletal componets of near-shore and lagoonal euphotic environments since the Jurassic, including modern oceans (see Renema and Troelstra 2001; Armstrong and Brasier 2005; Vaziri-Moghaddam et al. 2006; Mouanga and Langer 2014; Fajemila et al. 2015).

In general, marine sediments deposited under stressed conditions (i.e., any factor that discourages photosymbiosis) in the neritic/photic zone (from inner-to-outer platform/ramps settings) contain varying amounts of SBF. These become abundant-to-dominant in marine environments characterized by low-oxygen and eutrophic conditions (i.e., high food supply) and/or below the photic zone. However, some genera of SBF thrive as minor components in facies dominated by photozoan grain associations.

6.3.2 Ecology and Environment: Planktonic Foraminifera

PF are free-living floaters that inhabit the water column, being widely distributed from low-tropical to high-polar latitudes of modern oceans, with higher species diversity in the low-mid-latitudes. Biomineralization (calcification) of PF occurs mainly in marine hemipelagic and pelagic environments, away from any continental influence. Therefore, the largest accumulations of tests occur in the outer neritic, slope and deep basin environments (above the calcite compensation depth, which in modern oceans is ~4500 m in the Atlantic and ~ 3000 m in the Pacific), where they become part of the sediments at the seafloor upon death. It is estimated that today 47% of the ocean bottom is covered by PF tests that fall in a constant rain (see Bellier et al. 2010; Schiebel et al. 2018). Most PF thrive in the uppermost tens of meters of the water column, and may, depending on the reproduction cycle, move down to 400 m water depth; certain species are known to occur down to 2000 m water depth (Schiebel and Hemleben 2005; see also Reijmer 2021).

Tests of PF are small, commonly not exceeding 0.6 mm (600 μm) in diameter, and these are made of relatively globular chambers that provide buoyancy. Most PF are restricted to marine salinities of 30–40‰. Water temperature (controlled by latitude and/or bathymetry) and trophic conditions are the main factors controlling species distribution and depth stratification. Assemblages of PF in modern, high-productivity upwelling equatorial regions differ from oligotrophic, subtropical gyres. Many genera of modern PF are adapted to nutrient-poor (oligotrophic) surface waters by hosting photosynthetic symbionts (Edgar et al. 2013). Considering that photosymbiotic PF are mixotrophic, their living depth may be determined by the trade-off between phototrophy and heterotrophy. Generally, it can be assumed that the shallower the waters they inhabit, the more phototrophy-dependent they may be (Takagi et al. 2025). In this context, most modern PF that live oligotrophically in the upper 50 m of the photic zone are photosymbiotic (e.g., many species of *Globigerinoides*, *Globigerina*, *Orbulina*), many of them have globular, spinose and porous chambers. Modern intermediate forms living at 50–100 m are both symbiont-bearing species, with spinose thin walls and adapted to oligotrophic waters (e.g., *Orbulina universa*); and symbiont-barren species with non-spinose, keeled walls that are adapted to more mesotrophic-to-eutrophic waters (e.g., modern *Globigerina bulloides* and species of Cretaceous *Globotruncana*). Deeper forms lacking of photosymbionts and living below 100 m and down to 300 m include species with thicker, non-porous chambers and keeled tests (e.g., modern *Globorotalia* and species of Cretaceous *Marginotruncana* and *Dicarinella*). These species are adapted to low or no light, and cooler, denser and/or more eutrophic water masses (see Leckie 1987; D'Hont and Zachos 1998; Premoli Silva and Sliter 1999; Hart 1999; Armstrong and Brasier 2005; Flügel 2010; Bellier et al. 2010; BouDagher-Fadel 2015; Schiebel et al. 2018).

According to their life-history strategy and reproductive potential, PF have been categorized in opportunistic species ("*r*-selected", e.g., Cretaceous genera *Globigerinelloides*, *Hedbergella* and *Planoheterohelix*) at one extreme, smaller in size and able to rapidly increase their population, commonly associated to unstable (or short-lived) eutrophic and hypertrophic environments; specialist species ("*k*-selected"; e.g., Cretaceous genus *Rotalipora*) at the other extreme, larger in size and associated to stable oligotrophic environments; and intermediate species ("*r*/*k*"; e.g., Cretaceous genera *Dicarinella* and *Whiteinella*), whose strategies vary relative to the two end members (see Hallock 1985; Hallock et al. 1991; Hart 1999; Premoli Silva and Sliter 1999).

In shallow (platform/ramp) carbonate deposits, PF are particularly abundant in rock intervals deposited during episodes of maximum flooding. Widespread deposition of organic-rich lithofacies rich in PF (the so-called black shales) occurred across shelves and basins during several time intervals of the mid-Cretaceous "calcitic sea". Those intervals, known as OAEs, accounted for the deposition of many of the world's most prolific petroleum source rocks (see Takashima et al. 2006).

6.4 Carbonate Factories and Skeletal Grain Associations

Overall, of all the carbonate-secreting organisms considered in this book, foraminifera as a whole group are probably the only ones that are well-represented as common components of all marine carbonate benthic and planktic factories. Members of this group also display the most varied skeletal grain associations. This section discusses only the skeletal grain associations involving foraminifera (the most varied of all carbonate organisms considered in this book) represented in Venezuelan carbonate microfacies, as illustrated by the examples in the photographic set.

6.4.1 Larger Benthic Foraminifera

LBF are components of the Photozoan T- and Photo C-factories of Michel et al. (2019); as well as occasional secondary components of the Heterozoan factory of these authors (see Appendix A for a detailed treatment on the marine carbonate factories considered in this book). These mixotrophs display a wide variety of Photozoan and Heterozoan associations *sensu* James (1997) throughout the Phanerozoic. In Permian carbonates of the Palmarito Formation (Venezuelan Andes), fusulinids and green algae form a common photozoan association (chloroforam lithocaies *sensu* Beauchamp 1994). The photozoan-heterozoan association between fusulinids and heterotrophic calcifiers (crinoids, bryozoans and brachiopods) is also common (see Laya and Tucker 2012).

In Early Cretaceous (Aptian) carbonates of the Piché Member (Apón Formation, MB), the photozoan association between LBF (*Mesorbitolina*) and green algae (plus micritic peloids) is locally common (chloralgal lithofacies *sensu* Lees 1975). Mixed photozoan-heterozoan associations are also locally moderately common. These microfacies consist of LBF (e.g., *Choffatella*, *Mesorbitolina*, *Cuneolina*, *Nezzazata*) and molluscs and/or echinoids (plus smaller miliolids and other SBF). Depending on the content of LBF versus other bioclastic components (both heterozoan and photozoan biota), microfacies containing these mixotrophs can be grouped into foramol lithofacies *sensu* Lees and Buller (1972), molechfor lithofacies *sensu* Carannante et al. (1988), or echinofor lithofacies *sensu* Hayton et al. (1995). It is noteworthy that foramol, molechfor and echinofor are Cenozoic-biased terms in origin, but applicable to Cretaceous facies (e.g., Carannante et al. 1995; Simone et al. 2012; Kaya et al. 2020), and not restricted to non-tropical carbonates (see Pomar et al. 2004; Wilson and Vecsei 2005; Brandano et al. 2009; Klicpera et al. 2015). See Appendix A for definitions of the skeletal grain associations considered in this book.

In Cenozoic carbonates of the FB and EGVSB (e.g., the Churuguara, San Luis and Capadare formations, and the Perla Limestone, respectively), LBF are dominant components of photozoan-dominated carbonates in association with red algae (plus secondary echinoids, bryozoans, and balanid barnacles). This photozoan association is the expression of the Photo C-factory developed at tropical paleolatitutes. Many of these microfacies can be grouped into rhodalgal lithofacies *sensu* Carannante et al. (1988), or into rhodechfor lithofacies *sensu* Hayton et al. (1995). These lithofacies are especially abundant in the Perla Limestone, which is a succession particularly rich in red algae (encrusting and rhodoliths) and LBF (free-living and encrusting). In addition, echinofor lithofacies in which LBF outspace the echinoids are locally common in this unit. On the other hand, in previously mentioned units of the FB, LBF are common components of heterozoan-dominated microfacies in association with echinoids (echinofor lithofacies), molluscs (molechfor lithofacies), serpulids, and plus or minor bryozoans and balanids (see Lorente 1978; Pomar et al. 2015; Albert-Villanueva 2016).

In Los Roques Archipelago, LBF are important sand- and fine gravel-sized sediment producers inhabiting the central lagoon environment in association with calcareous green algae (photozoan association, chloralgal sediments). The photozoan association between these mixotrophs and scleractinian corals is also locally common (chlorozoan sediments), as well as the photozoan-heterozoan association with heterotrophic calcifiers (bivalves, gastropods, and echinoids). At Aves Island, LBF (mainly *Amphistegina* and *Peneroplis* and the encrusting species *Homotrema rubrum*) are secondary components of the lithified Quaternary rocks considered in this book (the carbonate basement and beachrocks). They form both photozoan and heterozoan associations, although mostly photozoan-dominated. Calcareous red algae grains (photozoan) are the most common photozoan components, followed by scleractinian corals (mainly grains and a few small *in situ* colonies) and green algae (*Halimeda*). Molluscs (gastropods and bivalves) are the main heterozoan components, followed by secondary echinoids, and minor SBF (mainly smaller miliolids), brachiopods, and ostracods.

6.4.2 Smaller Benthic Foraminifera

SBF are components of the Heterozoan factory of Michel et al. (2019). The marine productivity as food is the source of energy driving the development of this factory. In Early Cretaceous (Barremian, Aptian and Albian) carbonates of the Cogollo Group (MB), smaller miliolids, occasionally accompanied by other shallow-water SBF (e.g., *Novalesia*), are common of mud-supported microfacies (in many cases moderately- to organic-rich) rich in other heterotrophs, such as sponge spicules, ostracods, molluscs and echinoids (heterozoan association, including foramol and molechfor lithofacies). Smaller miliolids are also occasionally found in association with photozoan biota (green algae and *Mesorbitolina*; chloralgal lithofacies).

In organic-rich, dysoxic-suboxic facies of the mid-Cretaceous (the black shales), stress-tolerant opportunistics (e.g., buliminids) are found in association with low-oxygen molluscs (inoceramids). In those facies, as well as in some facies of late Middle Eocene (Jarillal Formation in the FB), SBF coincided in the final deposit with planktonic (or planktic) biota (PF and radiolarians). In modern facies of the Venezuelan Atlantic continental shelve and slope, SBF (e.g., *Quinqueoloculina lamarckiana*, *Textularia* spp.) start to appear in the middle shelf at depths of ~60 m. At depths within the photic zone, the heterozoan-photozoan association between these symbiont-barren benthic foraminifera and symbiont-bearing *Amphistegina lessoni* is moderately common. From the slope and towards the deep basin, deep-sea SBF (*Bulimina*, *Cibicidoides*, *Uvigerina*, *Sigmoilopsis* and *Hoeglundina*) coincide in the final deposit with PF (Morales 2004).

6.4.3 Planktonic Foraminifera

PF (with and without photosymbionts) are, together with coccolithophores, pteropods (or nektic gastropods), and radiolarians (silica-secreting planktonics), components of the marine Planktic factory (the P-factory) of Reijmer (2021) (see also the Pelagic marine factory of James and Jones 2015). In mid-Cretaceous black shale facies, PF and radiolarians (planktic association) coincided in the final deposit with low-oxygen SBF (see Crespo de Cabrera et al. 1999; Rey et al. 2004; Truskowski 2006; this book). In late Middle Eocene strata (Jarillal Formation), this planktic-heterozoan "association" is also locally common.

In modern reef complexes of the Venezuelan Caribbean, PF are occasionally found in the deeper areas of the central lagoons of Los Roques and Las Aves archipelagos (below the base of wave action), as well as in the deeper, low-energy areas of the fore-reef slope. In the Venezuelan Atlantic continental shelf and slope, PF (*Globigerina* ooze, mainly the genera *Globigerina*, *Globigerinoides*, *Globorotalia*, and *Orbulina*) dominate carbonate sediment composition from outer neritic settings down to lower bathyal-abyssal depths. From depths of approximately 200 m and down to 1800 to 3000 m, globigerinids coincide in the final deposits with far less abundant deep-sea SBF (Morales 2004).

6.5 Morphology and Diagnostic Elements

The taxonomic classification of foraminifera is based on the composition and structure of the test walls and takes the following features into account, in order of importance: wall structure and composition, chamber shape and arrangement, aperture and ornament (Loeblich and Tappan 1988).

Foraminifera main morphologic features are summarized as follows. Tests, from less than 0.1 mm, to up to giant sizes of 20 cm, consist of walls that divide empty chambers. Three types of walls are recognized in living foraminifera with calcitic tests: (**a**) agglutinated walls (e.g., orders Textulariida and Lituolida), which have a composite appearance and are composed of exogenous particles or grains (calcareous and/or clastic terrigenous) cemented together to construct the test. Grain size and composition varies, depending on the genera life's strategy; (**b**) porcelaneous walls, which are calcareous and entirely secreted by the animal, being imperforate and having smooth and homogeneous (uniform) appearance (e.g., Order Miliolida); (**c**) hyaline walls, again calcareous and secreted by the animal, are perforate and composed of carbonate interlocking crystals, which are oriented in prisms (radial hyaline) with their *c* axis normal to the test surface, or randomly oriented (granular hyaline) (e.g., orders Lagenida, Buliminida, Rotaliida, and Globigerinida). This wall type may exhibit vitreous, transparent or translucent appearance, while lamellar microstructures (when present) can be visualized under cross polarized light. These three types of calcitic tests exist as fossils and in addition a fourth type is recognized. Calcareous granular walls, which are common is some late Paleozoic foraminifera which have calcareous tests with a texture called microgranular (e.g., orders Fusulinida and Palaeotextulariida). These walls are composed of microcrystalline calcite crystals without optical alignment. Regarding the chamber spirally or coiled arrangement about an axis, chambers involved in a complete revolution are termed a whorl or a coil. Multichambered tests may have chambers arranged in a single linear chain row (uniserial), or double rows (biserial), or triple rows (triserial). Tests can be planispiral, with chambers arrangement more or less symmetrically in a single plane coil around the growth axis (e.g., *Globigerinelloides*); or trochospiral (e.g., *Hedbergella*), when the spiral does not lie in one plane, but in a snail-like helical spiral. Other morphologic features in foraminifera that can be visualize using a polarizing microscope are: the embryonic chamber (also termed proloculus or protoconch) of megalospheric tests (having a large embryonic chamber) and microspheric tests (having a small embryonic chamber); chamberlets (or small chambers), wall perforations (often visible when using a 40X magnification); apertures and openings to the outside environment; ornamentation (e.g., ridges, keels, spines); involute and evolute forms (chambers overlap towards the coiling axis, thus the majority of previous coils are hidden/not visible; and the majority of previous coils are hidden, respectively); and spiral side and umbilical side of trochospiral tests. Other morphologic features/elements will be indicated in the captions of the photomicrographs (for a full review on morphology of foraminifera see Loeblich and

Tappan 1988; Sliter 1989; Haq and Boersma 1998; Scholle and Ulmer-Scholle 2003; Premoli Silva and Verga 2004; Armstrong and Brasier 2005; Flügel 2010; Bellier et al. 2010; BouDagher-Fadel 2015, 2018).

6.6 Preservation as Fossils

In general, foraminifera tests are well preserved due to their dominating calcitic composition and high resistance to mechanical abrasion, reworking and mechanical compaction. In most cases, adult skeletons (complete or not) preserve identifiable elements/features that allow systematic assignments. Oriented cuts (e.g., longitudinal, transverse, axial, equatorial) through preferably complete foraminifera tests (in most cases the tests size range between ~0.1 mm to up to 10 mm in length) are required to determine the basic morphologic features that allow apply systematic paleontology.

Depending on the section through which the tests are seen, the number of visible chambers as well as the shapes of the tests may vary substantially. The size and preservation state of the tests are key for the classification of individuals. If tests are too small (e.g., globular planktonics), or are diagenetically altered by recrystallization, the diagnostic features might be either difficult to observe with the polarizing microscope or totally obliterated. Chambers may be filled with micrite (occasionally organic-rich), microspar, pseudospar, LMC cements, and occasionally with non-carbonate minerals (e.g., glauconite, silica), or may retain the original intraparticle porosity.

6.7 Geochemical Proxies in Foraminifera

Isotopic and elemental proxies in foraminifera are widely used for reconstructing past seawater temperature. The most traditional proxies are the ratios of stable oxygen and carbon isotopes and Mg/Ca ratio. Variations in the $^{18}O/^{16}O$ ($\delta^{18}O$) ratio have been used in planktonics (e.g., the orbital theory of Pleistocene climate of Imbrie et al. 1984; and the "orbital tunning" of the last 300 ka by Martinson et al. 1987) and benthics. Oxygen isotope stratigraphy (see Grossman 2012) applied to tests of benthic foraminifera (mainly SBF) have been effectively used as signal of seawater temperature (and related interpretations on orbital cycles, glacial-interglacial cycles, and variations in continental ice volume and sea level) for middle- and short-term (Pleistocene and Holocene; Lisiecki and Raymo 2005) and long-term (from 66 Ma to Present; Zachos et al. 2001; Westerhold et al. 2020) time scales.

SBF $\delta^{18}O$ records should produce a better stack than planktonic records (the term "stack" refers to a composite record created by combining multiple individual records) because the deep ocean is more uniform in temperature and salinity than surface water. Less local and regional climatic variability improves the accuracy of alignment and produces a better estimate of average $\delta^{18}O$ change (Lisiecki and Raymo 2005). Furthermore, many SBF secrete their calcitic tests in oxygen isotope equilibrium with ambient seawater (see Grossman 2012; Lea 2014). High-resolution $\delta^{18}O$ records in SBF (species of *Uvigerina* and *Cibicidoides*) from several globally distributed marine cores were used to construct the Late Pleistocene and Holocene $\delta^{18}O$ sea-level stack (from 0 to 798 ka) with defined marine isotope stages (i.e., alternating cool and warm periods, or glacials and intergalacials; Lisiecki and Raymo 2005; Spratt and Lisiecki 2016; see also Past Interglacials Working Group of PAGES 2016). The accurate construction of high-resolution $\delta^{18}O$ sea-level stacks on Quaternary timescales requires the use of paleotemperature equations (see Lea 2014).

The application of oxygen-isotope paleothermometry to pre-Pleistocene times is fraught with difficulties (Haq 2014), such as long-term (secular) changes in the concentration of $\delta^{18}O$ seawater (see Shaviv et al. 2023) and the effects of diagenesis. Therefore, for ancient facies, in addition to the use of mathematical equations and corrections for paleotemperature calculations, the coupling of deep-sea SBF oxygen ($^{18}O/^{16}O$) isotope records with other quantitative and qualitative proxies are used to accurately estimate global average temperatures and to construct paleotemperature curves for the Phanerozoic. The covariance of $^{18}O/^{16}O$ and $^{13}C/^{12}C$ in benthic foraminifera, in combination with values of $^{18}O/^{16}O$ in phosphate fossils (e.g., conodonts), are the most used quantitative proxies. Lithological qualitative-indicators of climate include bauxite and calcrete, which form in warm-wet climate and semi/arid regions, respectively. Multiproxy studies focused on paleoclimate also require mathematical equations and corrections for paleotemperature calculations (see Westerhold et al. 2020; Scotese et al. 2021; Wierzbowski 2021; Shaviv et al. 2023).

Individually, $\delta^{13}C$ in foraminifera is used to interpret the ocean's long-term dissolved inorganic carbon reservoir (longer than 180 ka, which is the residence time of carbon in the oceans), controlled by fluxes of carbon sources and sinks, as well as preferential carbon uptake, paleoproductivity and ocean circulation on shorter time scales (see Curry et al. 1988; Mutti and Hallock 2003; Katz et al. 2010; Hoogakker et al. 2024).

On the other hand, the Mg/Ca ratio in planktonics (e.g., Dekens et al. 2002; Anand et al. 2003; Barker et al. 2005) and benthics (e.g., Lear et al. 2002; Martin et al. 2002) also record the paleotemperature of the sea surface waters and the bottom waters, respectively. The effect of diagenesis places the accurate use of this elemental proxy on LMC tests of foraminifera within an estimated range of 0 to 100 Ma (Lea 2014). The coupling of $\delta^{18}O$ and Mg/Ca ratio in foraminifera tests can be used for reconstructing paleotemperature and related variations in salinity and evaporation-to-precipitation balance (see Elderfield and Ganssen 2000; Henderson 2002; Rosenthal and Linsley 2007; Ferguson et al. 2008; Katz et al. 2010; Lea 2014; Schiebel et al. 2018).

6.8 Foraminifera: Photographic Set and Descriptive Captions

In all photos, IW = image width (the true length of the horizontal axis of the view)

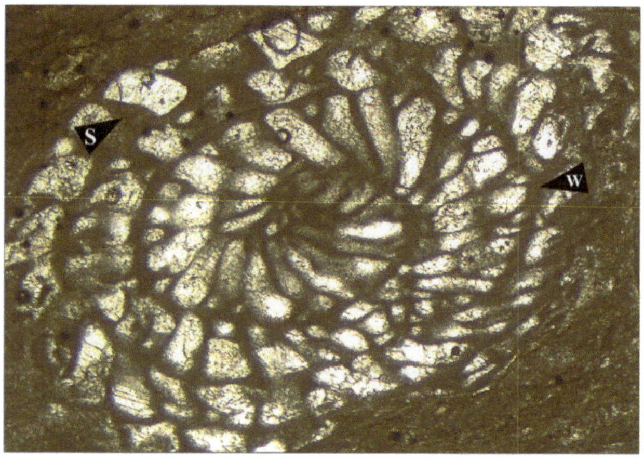

Middle Permian (Cisularian to Guadalupian). Palmarito Formation. Venezuelan Andes

Transverse section of a fusulinid (probably *Parafusulina* or *Triticites*; Order Fusulina, Superfamily Schwagerinoidea). The multichambered, planispiral, and evolute test of this larger benthic foraminifer shows a characteristic spiral form and labyrinthic, dark microgranular walls. The latter are composed of cemented cryptocrystalline lime mud, texturally differentiated from the micritic matrix by its more homogeneous appearance. Wall structure consists of a spiral wall (W) (spirotheca) and radially-oriented septal walls (S) (septotheca) (see Armstrong and Brasier 2005).

IW = 3.1 mm. PPL

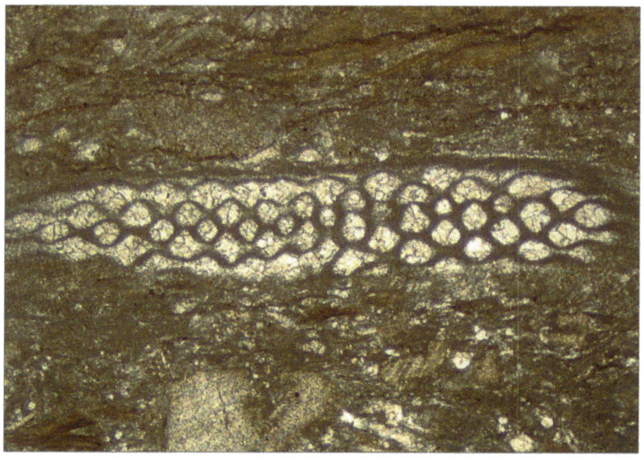

Middle Permian (Cisularian to Guadalupian). Palmarito Formation. Venezuelan Andes

Longitudinal section of a fusulinid. Note the marked differences in element sizes and arrangement compared to the previous transverse section. Small subglobular to ovate chambers (termed septal folds) are divided by inner dark-calcitic microgranular wall structure. The outer wall is very thin and it might be optically confused with the micritic matrix of the sample. Fusulinids are useful biostratigraphic markers for the late Paleozoic, as these LBF were major rock forming elements of carbonate benthic tropical factories from the Late Carboniferous to the latest Permian.

IW = 4.2 mm. PPL

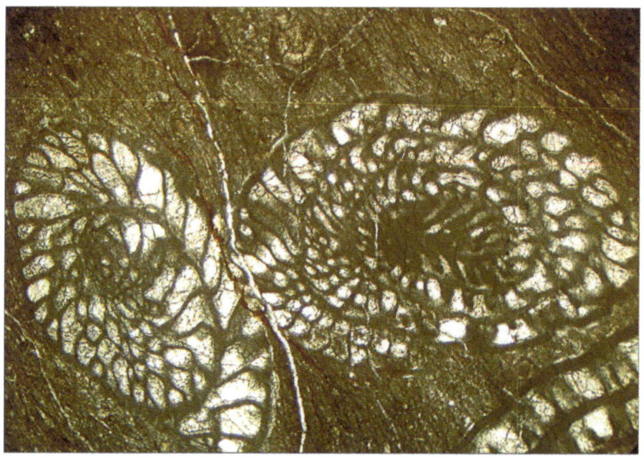

Middle Permian (Cisularian to Guadalupian). Palmarito Formation. Venezuelan Andes

Fusulinid tests in the image are in sutured contact, which combined with abundant solution seams in the micritic matrix, evidences mechanical and chemical compaction, respectively. Despite such burial processes, the enrolled tests did not suffer substantial mechanical deformation. These are ~3.5 mm in length, even though in the Palmarito Formation fusulinid tests reach up to 7 mm. Fusulinid-rich microfacies in this shallow carbonate unit are interpreted to be deposited in open marine conditions in an inner-ramp setting (see Laya and Tucker 2012).

IW = 7.5 mm. PPL

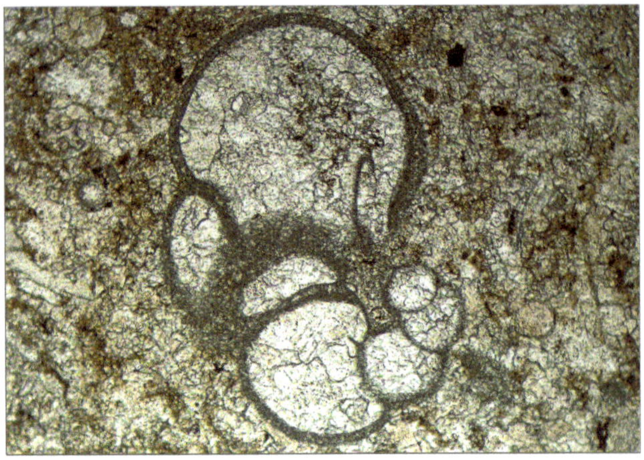

Middle Permian (Cisularian to Guadalupian). Palmarito Formation. Venezuelan Andes

Sagittal section of the genus *Globivalvulina* (Order Palaeotextulariida, Family Biseriamminidae). The test of this larger benthic foraminifer is planispiral and biserial, characterized by different-sized subglobular chambers and aberrant morphology. Dark microgranular walls (gray color) are clearly differentiated from both the recrystallized matrix and chambers' sparry calcite infillings. The Family Biseriamminidae is considered by BouDagher-Fadel (2018) as belonging to the lineages of LBF, but is otherwise considered as SBF by other authors (e.g., Beauchamp 1994).

IW = 1.3 mm. PPL

Middle Permian (Cisularian to Guadalupian). Palmarito Formation. Venezuelan Andes

Axial section of the genus *Paremiratella*. (Order Palaeotextulariida, Family Biseriamminidae). Unlike the previous example, this section allows visualize the biserial chamber arrangement of this test. Note the hemispherical shape of the chambers. Thick microgranular walls (dark gray) divide chambers of the same row, giving the test a widely chamber arrangement. Biseriamminids, including *Globivalvulina* and *Paremiratella*, are biostratigraphic markers for the Middle to Late Permian strata (see Gaillot and Vachard 2007)

IW = 1.3 mm. PPL

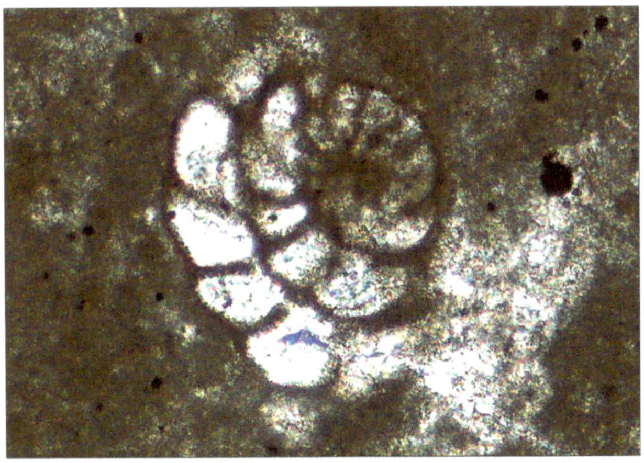

Cretaceous (middle Albian). Lisure Formation. Maracaibo Basin

Transverse section of the genus *Nezzazata* (probably *Nezzazata isabellae*; Order Textulariida, Superfamily Nezzazatoidea). The coiled test (trochospiral) of this larger benthic foraminifer consists of agglutinated, dark-colored walls composed of micritic sediments bounded by organic cements; while chambers, which increase rapidly in height from early to later whorls, are filled with sparry calcite. The genus *Nezzazata* is common of Albian-Turonian strata (e.g., Omidvar et al. 2014). This specimen is an example of a benthic test smaller in size (< 1 mm) but belonging to the LBF lineages.

IW = 1.3 mm. PPL

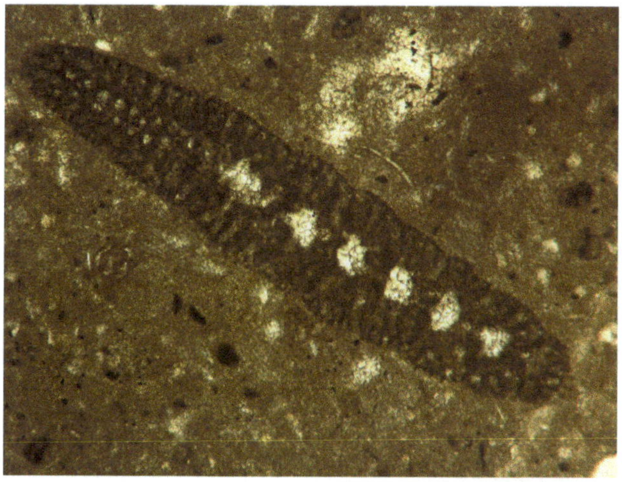

Cretaceous (early to middle Barremian). Tibú Member, Apón Formation. Maracaibo Basin

Longitudinal section of the species *Choffatella decipiens* (Order Textulariida, Superfamily Loftusioidea, Family Choffatellidae; according to BouDagher-Fadel 2018). The elongate, planispiral test of this larger benthic foraminifer consists of an inner endoskeleton, where a series of centrally-aligned chambers are filled with sparry calcite. The exoskeleton consists of numerous small chambers (alveoli), filled by synsedimentary micrite and oriented radially, thus approximatelly perpendicular to the external wall surface. In facies of the Tibú Member, tests of *C. decipiens* may reach up to 4 mm in length.

IW = 1.1 mm. PPL

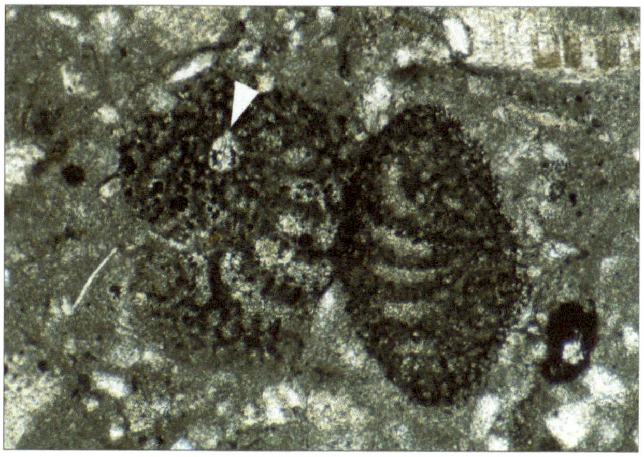

Cretaceous (early to middle Barremian). Tibú Member, Apón Formation. Maracaibo Basin

Choffatella decipiens seen through equatorial and subaxial sections (specimens to the left and right, respectively). The walls of both tests are selectively replaced by pyrite, which highlights the enrolled chamber arrangement of their planispiral forms. Note the protoconch in the equatorial section (white arrow), surrounded by the beginning of the compressed spiral. Visible elongate chambers are filled by synsedimentary micrite. This larger benthic foraminifer is an important biostratigraphic marker for the early to middle Barremian in the MB, since it only appears at the lower and middle parts of the Tibú Member.

IW = 2.4 mm. PPL

Cretaceous (middle Barremian). Tibú Member, Apón Formation. Maracaibo Basin

Longitudinal view of the genus *Cuneolina* (Order Textulariida, Family Cuneolinidae) showing distinctive conical shape. Walls of this larger benthic foraminifer are agglutinated, being constructed by cementation of microcrystalline micritic grains. Septal walls divide each row of chambers (at least 8 from this view). The latter are partitioned by vertical walls called beams, which are perpendicular to the septum, thus dividing chambers into nearly rectangular chamberlets. The genus *Cuneolina* has a wide range of appearance in Cretaceous strata, from Valanginian to Maastritchtian (Boudaugher-Fadel 2018).

IW = 1.3 mm. PPL

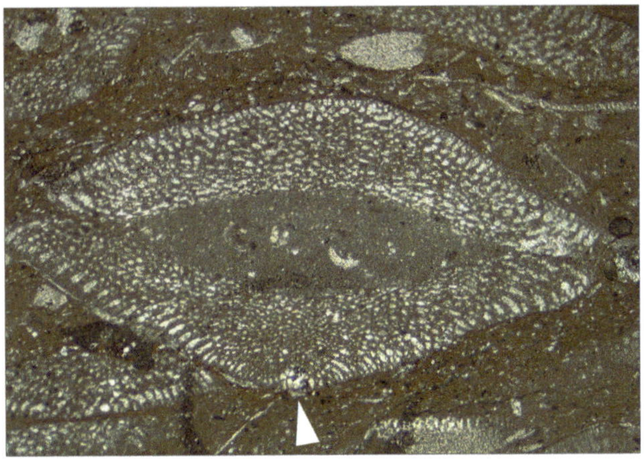

Cretaceous (middle Aptian). Piché Member, Apón Formation. Maracaibo Basin

Longitudinal sections of orbitolinids (*Mesorbitolina*, probably *M. texana*, Order Textulariida, Family Orbitolinidae). Tests of these LBF show distinctive conical, "umbrella" shape. Chambers in the marginal zone (exoskeleton) exhibit elongate shapes and are oriented nearly normal respect to the surface wall. Note that chambers bend slightly toward the apex of the conical test (white arrow), where the embryonic chamber is located (not quite visible). In the MB, this photosymbiotic benthic foraminifer is common in middle Aptian muddy microfacies stratigraphically located between the black shales of the OAEs 1a and 1b first sub-event.

IW = 4.2 mm. PPL

Cretaceous (middle Aptian). Piché Member, Apón Formation. Maracaibo Basin

Transverse section of an orbitolinid embedded in a sandy micritic matrix. The transverse cut allows the visualization of the radial element arrangement. The agglutinated wall structure is composed of microcrystalline micritic sediments. Note the very thin minute cells in the outer marginal zone (exoskeleton), whose thickness was locally reduced due to mechanical abrasion of the test (white arrow). This outer area is followed by a radial arrangement of chamberlets. Towards the center of the grain, the loose pattern of reticulate partitions is characteristic of transverse sections of orbitolinids.

IW = 4.2 mm. CPL

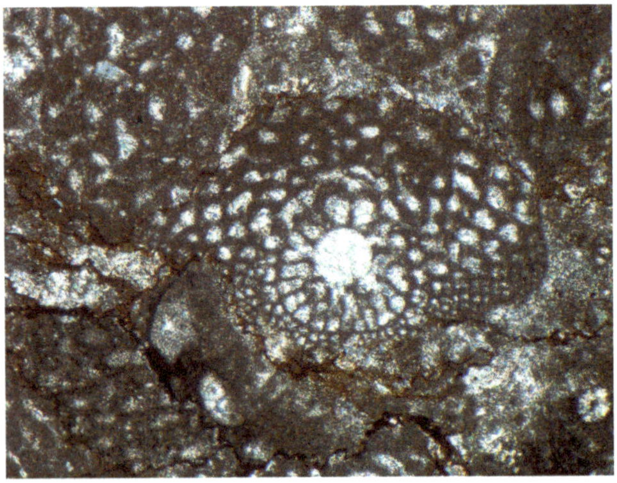

Cretaceous (middle Aptian). Piché Member, Apón Formation. Maracaibo Basin

Mechanically and chemically compacted multichambered tests of the genus *Mesorbitolina* are forming a condensed fabric in which individual tests are in sutured contact with each other along microstylolite surfaces. The central megalospheric test, seen in tangential section, is clearly distinguished by its spherical embryonic chamber (or protoconch), which is immediately surrounded by a series of small chambers (the subembryonic zone) divided by radial partitions.

IW = 2.4 mm. PPL

Early Miocene. Churuguara Formation. Falcón Basin
Longitudinal-axial section of a nummulitid (*Operculinoides*, probably *O. dia*, Order Rotaliida, Superfamily Nummulitoidea, Family Nummulitidae). The test of this larger benthic foraminifer is lenticular (a characteristic morphology of many Cenozoic LBF). The concentric test structure, which originates in the embryonic chamber, consists of thick hyaline walls. The latter are built by thin and tightly enrolled laminae. Note the irregular margins and missing portions of the outer wall of the test. This indicates extensive transport and reworking (see Beavington-Penney, 2004)

IW = 3.1 mm. PPL

Early Miocene. San Luis Formation. Falcón Basin
Longitudinal-axial section of a nummulitid (genus *Operculinoides*, probably *O. panamensis*). This megalospheric test is biconvex (also a characteristic morphology of many Cenozoic LBF), with a less strongly curved periphery compared to the previous nummulitid example. Flatter tests are commonly associated with deeper and lower-energy settings. Also note the robust walls, and V-shaped intraskeletal pores, filled by synsedimentary dark lime mud (partially replaced by glauconite), representing significant primary intraskeletal porosity.

IW = 2.4 mm. PPL

Middle Miocene. Capadare Formation. Falcón Basin
Phosphatized, bioclastic sandy limestone in which a test of a nummulitid (probably an operculinid) in the center of the image is seen through a sub-axial section. The test of this larger benthic foraminifer is planispiral and evolute, consisting of thick, equally biconvex walls and thin, parallel walls oriented perpendicular to the long axis of the test. Significant intraparticle porosity was filled by synsedimentary micrite, which is partially phosphatized.

IW = 4.2 mm. PPL

Early Miocene. Churuguara Formation. Falcón Basin

Transverse section of a nummulitid (*Operculinoide*s, probably *O. dia*) embedded in a bioclastic matrix (consisting mainly of PF). The multichambered test exhibits distinctive spirally-coiled shape. The hyaline wall structure consists of an outer spiral wall, which in nummulites is called marginal cord, and radially-oriented septum walls. In this species, walls are thin related to the size of chambers and test. This relation varies widely in nummulitids, depending mainly on the hydrodynamic conditions (see Beavington-Penney and Racey 2004).

IW = 4.2 mm. PPL

Late Oligocene-Early Miocene. Perla Limestone. Eastern Gulf of Venezuela Sub-Basin

A test of the genus *Amphistegina* (Order Rotaliida, Superfamily Asterigerinoidea, Family Amphisteginidae) showing intraskeletal porosity. Note the large gaps between robust hyaline walls of this lenticular larger benthic foraminifer, originally composed of HMC. Despite pervasive CO_2-driven dissolution that affected this sample during burial diagenesis, the walls of the test are fairly preserved. In general, tests of *Amphistegina* are smaller in average compared to other LBF in the Perla Limestone (e.g. *Lepidocyclina*, *Miogypsina*, *Miogypsinita*, *Heterostegina*).

IW = 1.3 mm. PPL

Holocene. Los Roques Archipelago

Longitudinal-axial section of a modern rotaliid (*Amphistegina* sp.). The test is megalospheric and lenticular, with strongly curved periphery. The latter is a common morphologic feature of extant LBF that inhabit high-energy and high-light environments (see Beavington-Penney and Racey 2004). The view under CPL allows the visualization of the hyaline walls, which are composed of microcrystalline calcite crystals oriented normal to the test walls. The walls are also optically characterized by the alternation of dark and light colors, thus forming pronounced extinction bands.

IW = 1.3 mm. CPL

Late Oligocene-Early Miocene. Perla Limestone. Eastern Gulf of Venezuela Sub-Basin

Longitudinal section of the genus *Lepidocyclina* (probably *L. yurnagunensis*, Order Rotaliida, Superfamily Asterigerinoidea, Family Lepidocyclinidae). The test of this larger benthic foraminifer is megalospheric, discoidal and lenticular, and exhibits distinctive small equatorial chambers along the long axis. A flat wall divides the bilocular, embryonic apparatus into proloculus and deuteroconch. Most rectangular and elongate lateral chambers are filled with sparry calcite (white lean precipitate).

IW = 4.2 mm. PPL

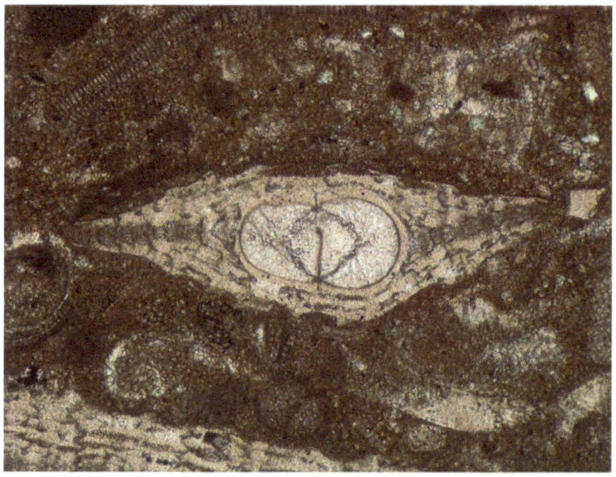

Middle Eocene. Santa Rita Formation. Falcón Basin

Another example of the genus *Lepidocyclina* (probably *L. macdonaldi*). The megalospheric test is lenticular, with equatorial chambers clearly differentiated due to their dark infilling. The central protoconch and deuteroconch are visibly dissimilar in size and separated by a straight wall. Note the markedly larger size of the embryonic apparatus in relation to the length of the test compared with the previous example. This genus appeared from the Middle Eocene to the Early Miocene, being among the most common LBF in Cenozoic microfacies of the FB and the Perla Limestone.

IW = 2.4 mm. PPL

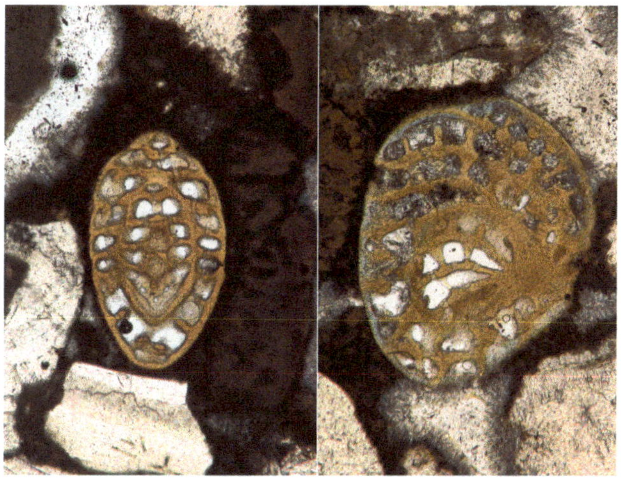

Holocene. Aves Island

Axial (left) and oblique (right) sections of the genus *Peneroplis* (Order Miliolida, Superfamily Soritoidea, Family Peneroplidae) exhibiting a characteristic amber to orangey color. The test of this larger benthic foraminifer is multichambered (i.e., high intraskeletal porosity) and biconvex, with a planispiral coiling pattern. Walls are porcelaneous and imperforate. The color of this genus is characteristic of LBF hosting dinoflagellate photosymbionts. *Peneroplis* has a wide range of appearance (from Eocene to Holocene), being very common in modern, shallow (euphotic zone) inner platforms and lagoons.

IW = 0.55 mm each. PPL

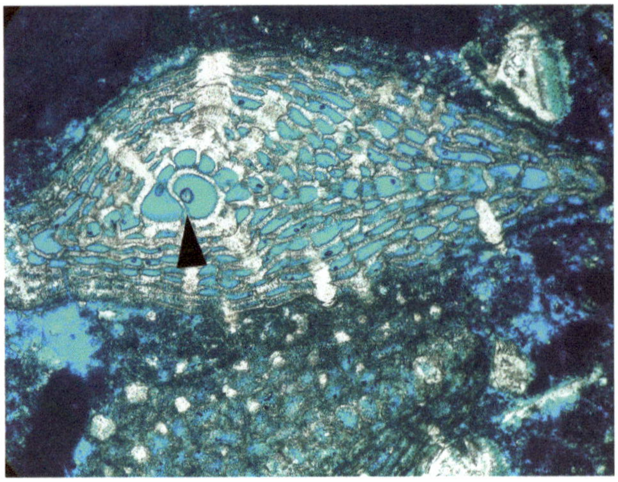

Late Oligocene-Early Miocene. Perla Limestone. Eastern Gulf of Venezuela Sub-Basin

Longitudinal/slightly oblique section of the genus *Miogypsinita* (Order Rotaliida, Superfamily Rotalioidea, Family Miogypsinidae). The megalospheric test of this larger benthic foraminifer is biconvex and planispiral. Well-preserved intraparticle porosity and hyaline calcite walls allow visualize multi-chambered arrangement. The embryonic apparatus is bilocular, and consists of two unequal embryonic chambers (black arrow). The short age range of this genus (Late Oligocene to Early Miocene) makes this genus an excellent biostratigraphic marker.

IW = 2.4 mm. PPL

Late Oligocene-Early Miocene. Perla Limestone. Eastern Gulf of Venezuela sub-Basin

Equatorial section of the genus *Heterostegina* (*H. antillea*; Order Rotaliida, Superfamily Nummulitoidea, Family Nummulitidae) in a bioclastic matrix. This cut allows a clear view of the coiled (planispiral) test of this larger benthic foraminifer and its many chamberlets. These are filled with a combination of microspar and cloudy cements. Note the numerous, thin secondary walls (dark, short partitions termed septulae), more or less perpendicularly oriented with respect to septal walls.

IW = 4.2 mm. CPL

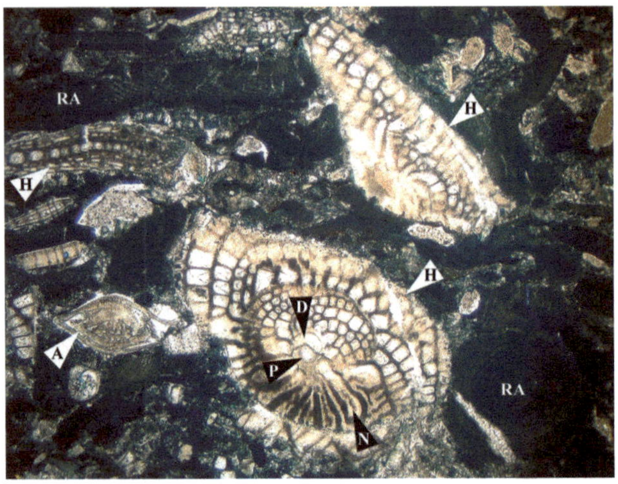

Late Oligocene-Early Miocene. Perla Limestone. Eastern Gulf of Venezuela Sub-Basin

Heterostegina (H) *antillea* tests in this skeletal condensed fabric are seen through different cuts. The equatorial section of the bigger test in the image shows the proloculus (P) and deuteroconch (D) (or second embryonic chamber) surrounded by distinctive elongate nepionic (N) chambers. Note the marked difference in size compared to the *Amphistegina* (A) test. In the Perla Limestone, microfacies dominated by LBF and red algae (RA) exemplify the dominance of carbonate production in the oligophotic zone during deposition of this unit.

IW = 7.5 mm. PPL

Cretaceous (middle Aptian). Piché Member, Apón Formation. Maracaibo Basin

Axial-sagittal section of a smaller miliolid (Order Miliolida, Superfamily Milioloidea). The test of this smaller benthic foraminifer (symbiont barren) consists of porcelaneous (imperforate) walls. These are homogeneous in appearance and visibly darker than the micritic matrix. Elongate and slightly concave-convex chambers increase in size outward and are arranged in varying planes around the central, spherical embryonic chamber. All chambers are filled with coarse-grained calcite spar (white). Note the dark color of the micritic matrix, suggesting eutrophic conditions during deposition.

IW = 0.5 mm. PPL

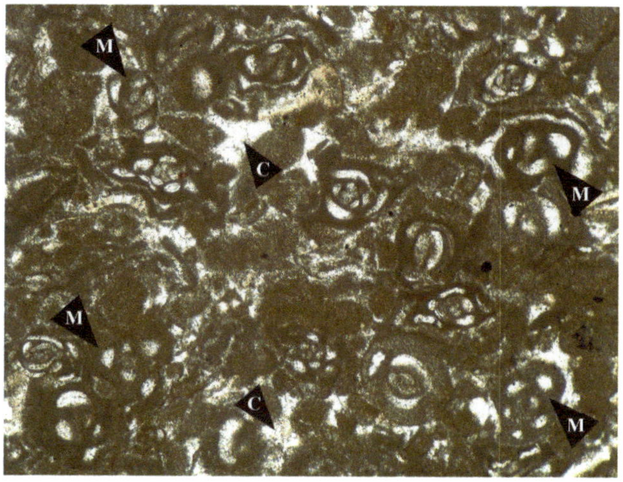

Cretaceous (middle Aptian). Piché Member, Apón Formation. Maracaibo Basin

Smaller miliolids (M) are major components in this wackestone. Characteristic coiling pattern of these SBF are seen through a variety of cuts in this image. Miliolids are common components of muddy-strangressive deposits in the Piché Member. Their occurrence is associated with both light-independent calcifiers (e.g., sponge spicules and other SBF, such as *Sabaudia* sp.) and light-dependent calcifiers (orbitolinids). LMC cement (C) filling interparticle pores must not be confused with chambers of miliolids, which are filled with sparry calcite.

IW = 2.4 mm. PPL

Cretaceous (latest Albian). Maraca Formation. Maracaibo Basin

Axia-sagittal section of the genus *Novalesia* (Order Lituolida, Superfamily Spiroplectamminoidea). The test of this smaller benthic foraminifer is biserial and it thickens towards the final whorl. The dark-colored walls are agglutinated. Septal walls projecting from the outer wall subdivide chambers and are arranged parallel to each other forming a closely arrangement of chambers. This genus has a wide range of appearance in Early Cretaceous strata, from Valanginian to Albian (e.g., Omaña et al. 2016; Yavari et al. 2017).

IW = 0.5 mm. PPL

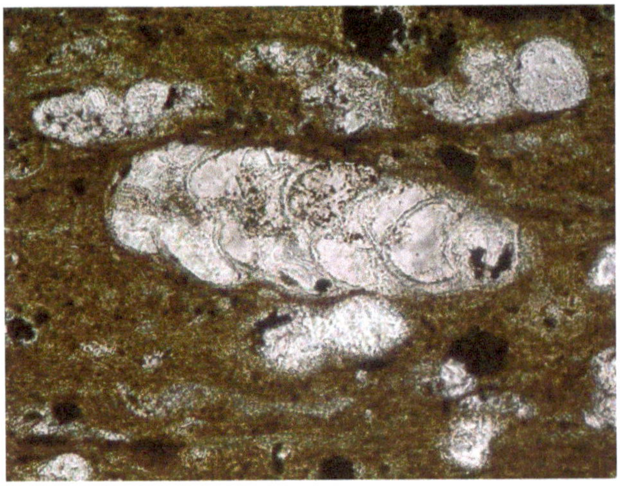

Cretaceous (Coniacian). La Morita Member, Navay Formation. Barinas Sub-Basin

Longitudinal section of the genus *Bolivina* (Order Bulimina, Superfamily Bolivinoidea) embedded in a moderately organic-rich, clayey-pyritic matrix. The elongate test of this smaller benthic foraminifer is biserial. Calcareous hyaline walls are thin and chambers are closely arranged. During deposition of Cretaceous black shales of Venezuela, buliminids (such as this and following examples) were able to thrive in low-oxygen (but not anoxic), eutrophic conditions.

IW = 0.5 mm. PPL

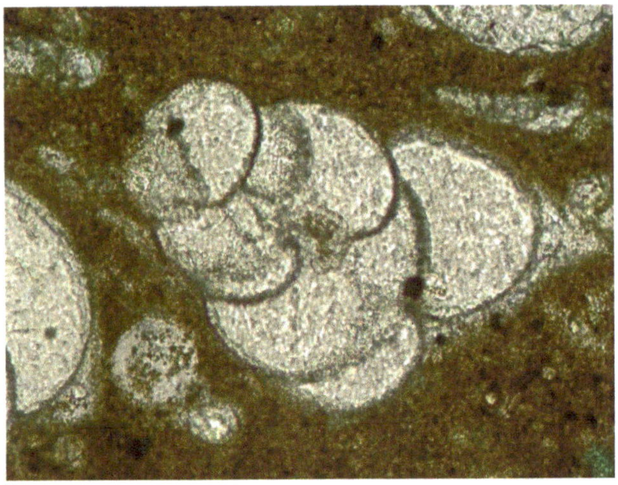

Cretaceous (late Santonian). Táchira Ftanita Member, La Luna Formation. Venezuelan Andes

Longitudinal/oblique section of a buliminid (probably *Praebulimina* sp. or *Bulimina* sp., Order Buliminida, Superfamily Buliminoidea). Note the marked thicken of the test from the initial to the final whorl. The triserial chamber arrangement of this smaller benthic foraminifer differentiates its test from previous biserial *Bolivina*. The interruption of the visible biserial arrangement in the third whorl by a single chamber is the diagnostic criteria for identifying the triserial arrangement of the test in a two-dimensional view (Gilberto Soto, personal communication, 2019).

IW = 0.5 mm. PPL

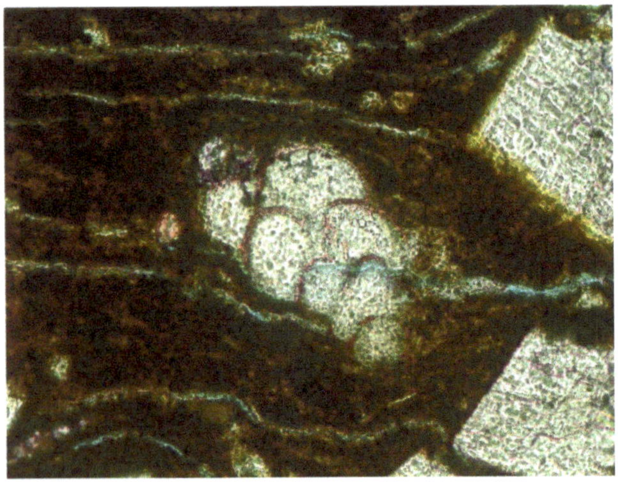

Cretaceous (late Aptian). Piché Member, Apón Formation. Maracaibo Basin

Longitudinal section passing through two chambers of a triserial test of a buliminid. Note that the thin calcareous walls are stained with Alizarin Red S (pink color), hence outlining the globular chambers. The infilling of the latter was replaced by amorphous silica, which is the reason why the whole test is not pink-stained. This pyritic, organic-rich dolomitic shale corresponds to the OAE 1b first sub-event, an interval in which buliminids are common. In the MB, these SBF are also common in black shales of the OAE 1b second sub-event (early Albian), but scarce in black shales of the OAE 1a (early Aptian).

IW = 0.5 mm. PPL

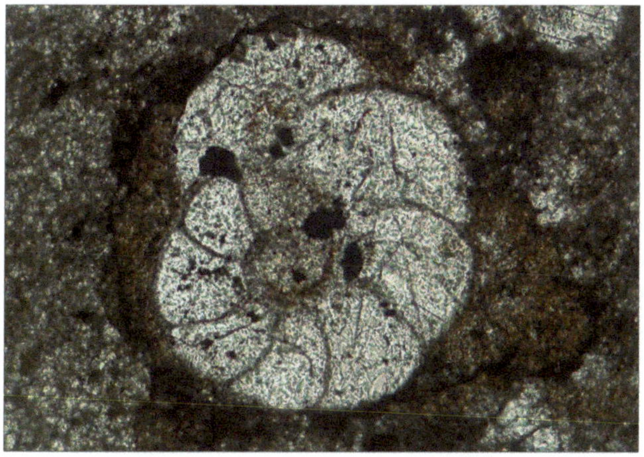

Cretaceous (latest Albian). Querecual Formation. Eastern Interior Range

View from the umbilical side of the genus *Gavelinella* (Order Rotaliid, Family Gavelinellidae) in an organic-rich microfacies that corresponds to the OAE 1d. Note the helical coil of the multichambered, trochospiral test, as well as radially-oriented thin walls and central umbilical boss. Although low-oxygen conditions prevailed since the beginning of deposition of the Querecual Formation, several genera of SBF were able to thrive (e.g., *Dentalina*, *Praebulimina*, *Bolivina*, *Bulimina*, *Neobulimina*, *Orthokarstenia*, *Lenticulina*).

IW = 0.5 mm. PPL

Photomicrograph taken from Truskowski (2006)

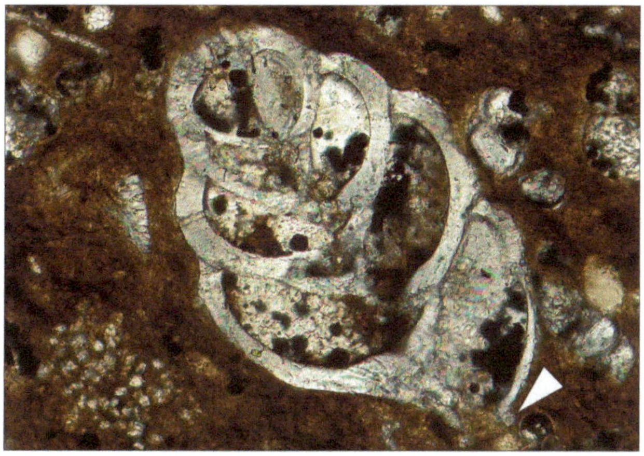

Late Middle Eocene. Jarillal Formation. Falcón Basin

Axial section of the genus *Uvigerina* (Order Buliminida, Superfamily Buliminoidea) embedded in a highly phosphatized matrix. This early Cenozoic buliminid test is differentiated from previous examples of Cretaceous triserial buliminids by its thicker calcareous walls, more elongate chambers, and visible "bottle neck" basal aperture (white arrow). This infaunal genus is commonly associated to depositional environments within the middle-outer neritic to upper bathyal zones, as well as to dysoxic-suboxic conditions and organic-rich sediments (see Van der Zwaan et al. 1986; Abd El-Gaied et al. 2019).

IW = 0.5 mm. PPL

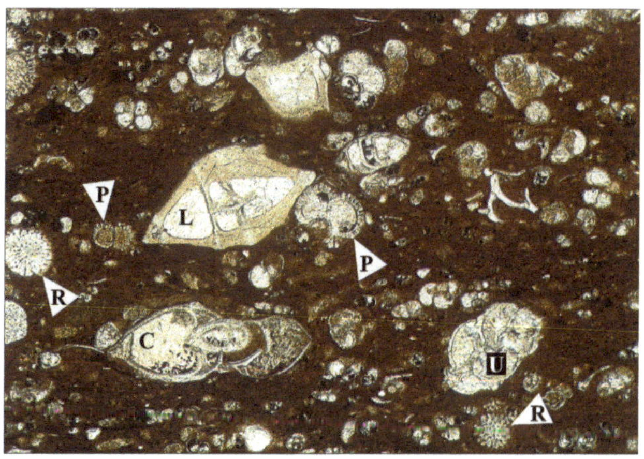

Late Middle Eocene. Jarillal Formation. Falcón Basin

Bioclasts in this organic-rich, phosphatized shale are mainly tests of PF (P) and SBF. The latter include tests of *Lenticulina* sp. (L) (Order Lagenida, Superfamily Nodosarioidea), *Uvigerina* sp. (U), and cibicidids (C) (Order Rotaliida, Family Cibicididae). These carbonate bioclasts are accompanied by radiolarians (R) (silica-secreting organisms). Low-oxygen conditions during a sea-level highstand are associated to deposition of this microfacies.

IW = 2.4 mm. PPL

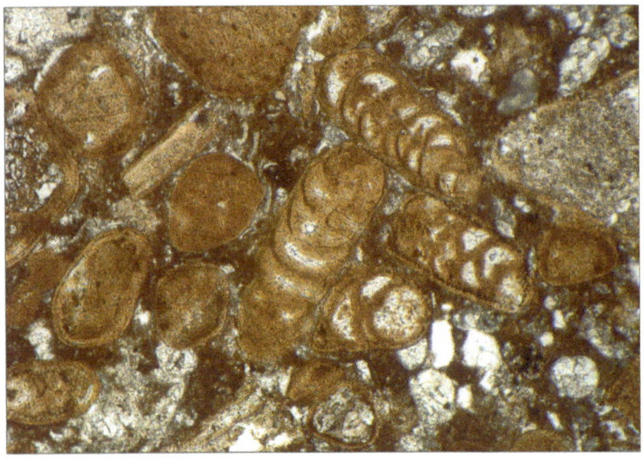

Cretaceous (early Campanian). Tres Esquinas Member, La Luna Formation. Maracaibo Basin

Phosphatized allochems in the image consist of elongate SBF, most likely of the genus *Bolivinoides* (Order Buliminida, Superfamily Bolivinoidea), and peloidal grains (mainly phosphatized ooids). The biserial arrangement of the tests is noticed in the contrasting color of the phosphatized walls (brown-orange color) and silicified chambers infilling (white color). In this unit, the flourishing of SBF (genera *Lenticulina*, *Orthokarstenia*, *Praebulimina* and *Gavelinella*) is associated to dysoxic conditions and upper bathyal depths (see De Romero and Galea-Álvarez 1995; Parra et al. 2003)

IW = 1.3 mm. PPL

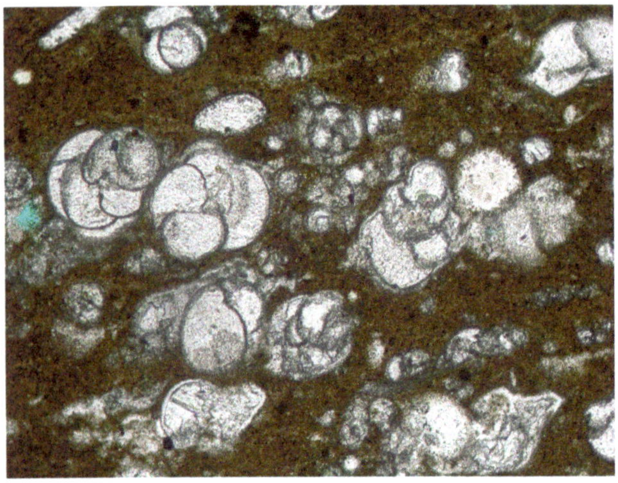

Cretaceous (late Santonian). Táchira Ftanita Member, La Luna Formation. Venezuelan Andes

Allochems in the image are mainly SBF (buliminids) seen through random transverse/oblique sections. Tests are embedded in a moderately organic-rich, pyritic, microcrystalline chert matrix. In addition to upper bathyal paleodepths and dysoxic-suboxic conditions prevailing during deposition of the Táchira Ftanita Member, these SBF were able to thrive infaunally within the siliceous mud. The latter implies highly stressful conditions, including locally acidic pH values, thus making their occurrence an excellent example of the extraordinary capacity of these buliminids for thriving in extreme marine habitats.

IW = 1.3 mm. PPL

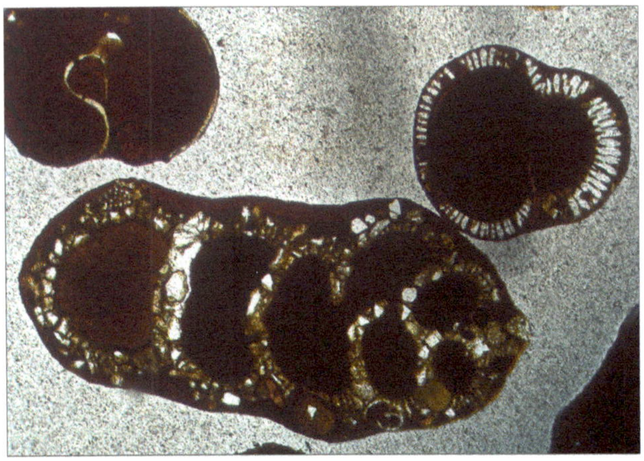

Holocene. Venezuelan Atlantic Continental Shelf

The photomicrograph shows loose grains of a smaller benthic foraminifer (larger grain in the image) and a planktonic foraminifer (up and to the right), accompanied by a gastropod (up and to the left). Note that the walls the smaller benthic test are composed of small quartz grains cemented together. Moderate sorting of the grains indicates a selective process employed by this agglutinated specimen to construct the test. The planktonic specimen with undifferentiated inner chambers shows striking porous radial wall structure.

IW = 1.3 mm. PPL

Holocene. Venezuelan Atlantic Continental Slope

This photomicrograph, which was taken using a binocular magnifying glass, shows a dry mount of tests of foraminifera. The bioclasts were taken at upper bathyal depths. Globular PF (P) in the image include the genera *Globigerina*, *Globigerinoides*, *Globorotalia*, and *Orbulina* (commonly grouped as *Globigerina* oozes; Order Globigerinida) have large sizes (~ 400 to 600 µm). Among visible tests of SBF, it highlights a test the genus *Lenticulina* (L), and an elongate agglutinated (A) tests of the genus *Martinottiella* (Order Textulariida).

IW = 8.2 mm

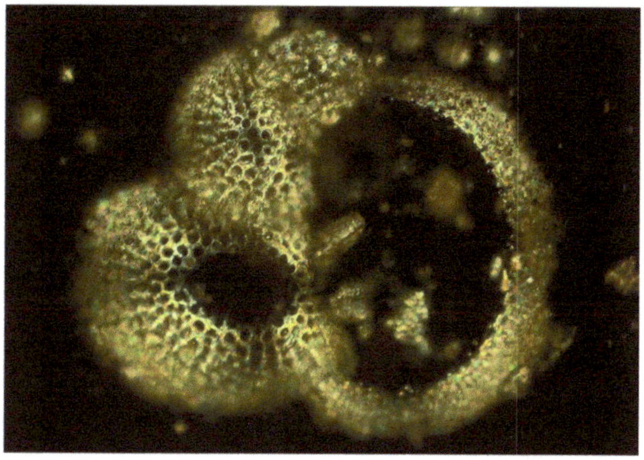

Holocene. Aves de Barlovento Archipelago

Oblique section of a loose test of a globigerinid, probably *Globigerina* sp. or *Globigerinoides* sp. (Order Globigerinida, Family Globigerinidae). The test of this planktonic foraminifer is trochospiral and consists of globular to ovate chambers forming a lobulate peripheral outline. Calcareous walls are coarsely perforated. The latter is a morphologic feature that improves the buoyancy of the test, which is necessary for remaining in the shallower layer of the water column during its life cycle. Wall perforations, as they are seen in this image, are not commonly preserved in ancient foraminifera fossils.

IW = 0.5 mm. CPL

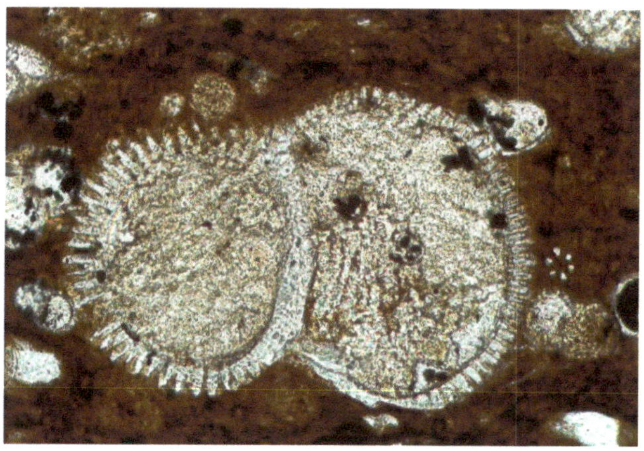

Late Middle Eocene. Jarillal Formation. Falcón Basin

This globigerinid test is embedded in a highly phosphatized matrix. The cut (most likely oblique) is passing through two globular chambers of the multichambered test. Note the thick walls and well-preserved porous radial wall structure. The synsedimentary phosphatized micrite infilling of the chambers is partially replaced to microsparite.

IW = 0.5 mm. PPL

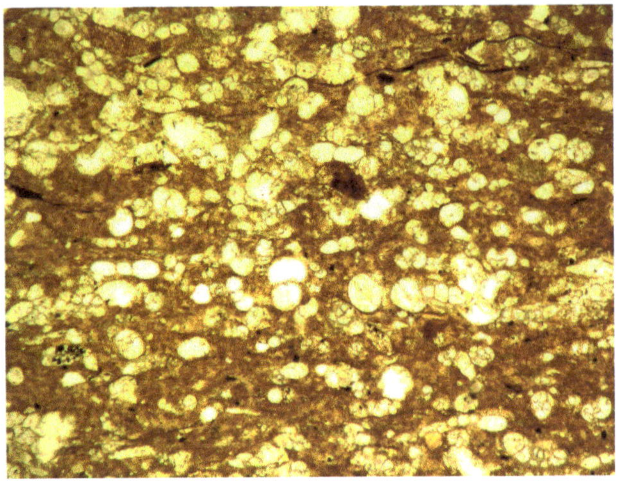

Cretaceous (Coniacian). La Morita Member, Navay Formation. Barinas Sub-Basin

Late Cretaceous planktonic foraminiferal black shale. Globular tests in the image consist of random sections (e.g., transversal, oblique, tangential) through multichambered tests. This and the two following photomicrographs are examples of the massive bloom of small globular opportunistics during major transgressive episodes (i.e., development of pelagic realm-like conditions in the uppermost part of the water column). Those major transgressions were accompanied by the expansion of the OMZ and the development of bottom water anoxia.

IW = 1.7 mm. PPL
Photomicrograph courtesy of PDVSA-INTEVEP

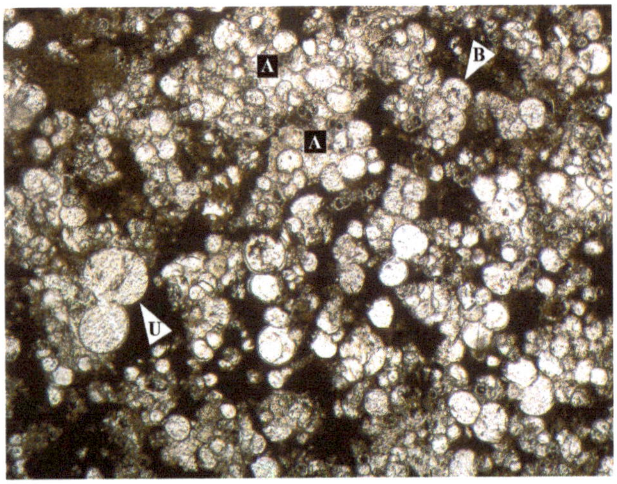

Cretaceous (late Cenomanian). La Luna Formation *sensu stricto*. Venezuelan Andes

Organic-rick, planktonic foraminiferal packstone. High density of tests, combined the effects of compaction led to the formation of aggregates (A) of biogenic carbonate, in which differentiating individual tests is quite difficult. Some biserial (B) tests are distinguished among the crowd of globular specimens. Moreover, in this sample, some of the tests seen as uniserial (U) forms are indeed transversal sections through biserial tests. Note the dark-colored (isotropic) matrix, which differentiates black shales of the La Luna Formation *sensu stricto* from black shales of the La Morita Member. The latter are lighter in color.

IW = 1.3 mm. PPL

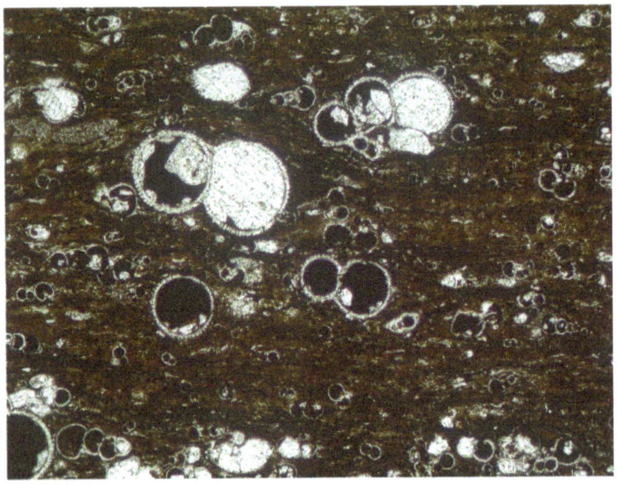

Cretaceous (late Cenomanian-early Turonian). La Luna Formation *sensu stricto*. Maracaibo Basin

Globular PF in this organic-rich calcareous shale belonging to the OAE 2 vary notoriously in size. Well-preserved radial porous wall structure is clearly noticed in larger tests in the image. Most globular chambers are filled by organic matter, while in some chambers this infilling is replaced by pseudospar (white precipitate). Note the wavy lamination of the matrix, which is composed of an alternation of microlamillae of brown-colored clay minerals and dark organic matter. Also note that the tests are oriented parallel to the bedding plane.

IW = 1.3 mm. PPL

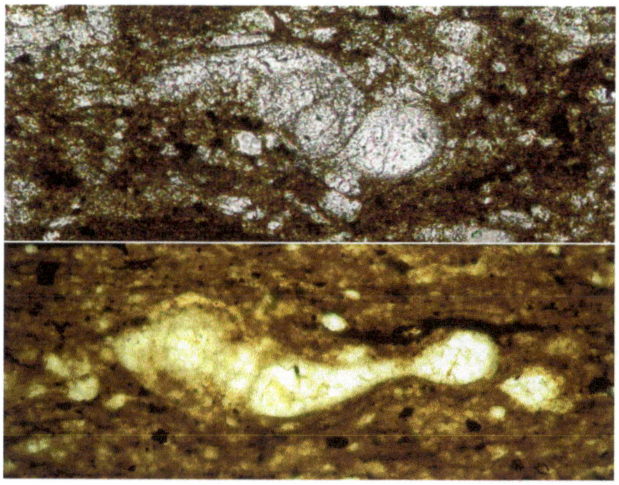

Cretaceous (early Aptian). Machiques Member, Apón Formation. Maracaibo Basin

Both photomicrographs, corresponding to different core sections in the subsurface of western MB, show images of representative *Leupoldina cabri–pustulans* Group (Order Globigerinida, Family Schackoinidae). The test of this planktonic foraminifer is planispiral and characterized by radially and broadly elongate chambers with bulbous or clavate terminations. The elongate chambered morphology, seen in the two images, has been interpreted as an adaptation to low-oxygen concentration in the upper water column during the OAE 1a (see Verga and Premoli Silva 2002; Coccioni et al. 2006).

IW = 0.5 mm (above); 0.6 mm (below). PPL
Photomicrograph below courtesy of Gilberto Soto

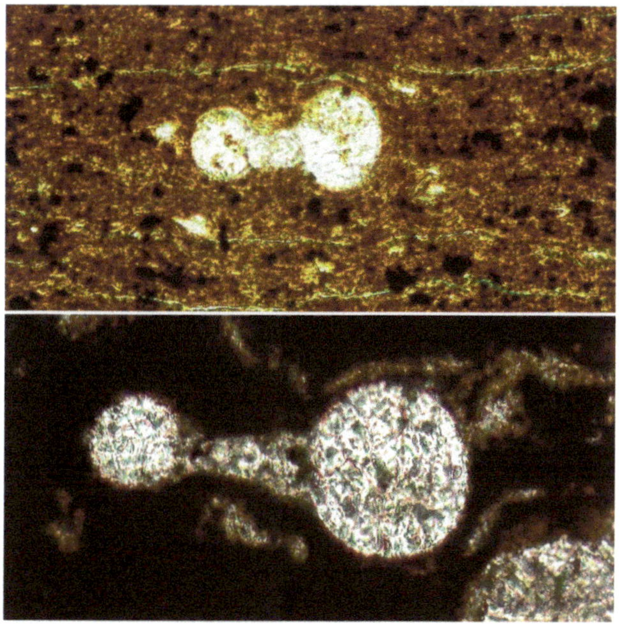

Above – Cretaceous (early Aptian). Machiques Member, Apón Formation. Maracaibo Basin. Below – Cretaceous (late Cenomanian-early Turonian). Querecual Formation. Eastern Interior Range

Both photomicrographs show axial sections of tests of PF of the genus Globigerinelloides (Order Globigerinida, Family Globigerinilloididae). The image above shows the species *Globigerinelloides blowi* (base of the Machiques Member, OAE 1a); while the image below shows the species *Globigerinelloides caseyi* (middle-lower part of the Querecual Formation, OAE 2). Despite differences in age (~ 20 Myr), both tests (planispiral) exhibit characteristic club shape of this superfamily, with an inflated last chamber (to the right), a smaller globular first chamber (to the left), and small middle chambers.

IW = 0.6 mm (above); 0.3 mm (below). PPL
Photomicrograph below taken from Truskowski (2006)

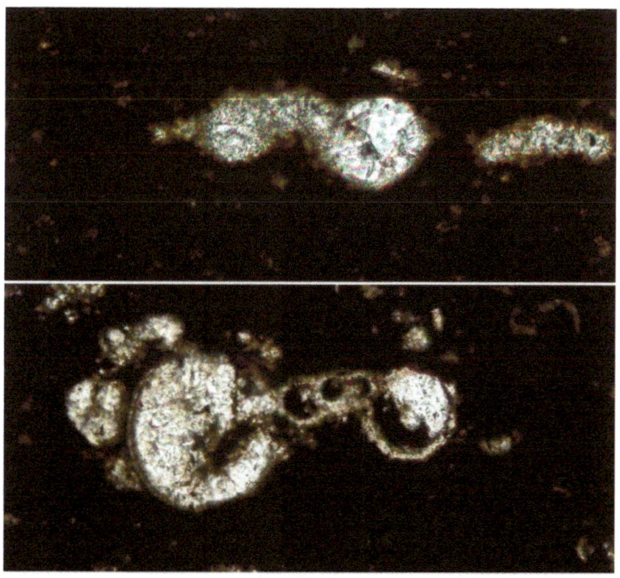

Above – Cretaceous (early Aptian). Machiques Member, Apón Formation. Maracaibo Basin. Below – Cretaceous (middle Turonian). Querecual Formation. Eastern Interior Range

Both images show axial sections of hedbergellids (Order Globigerinida, Family Hedbergellidae). The image above shows the genus *Hedbergella* (probably *H. aptiana*), and corresponds to the lower part of the Machiques Member (OAE 1a). The image below shows the species *Muricohedbergella planispira*, and corresponds to the middle-upper part of the Querecual Formation (post-OAE 2 interval). Tests of these opportunistic PF are low-arched trochospiral. Note the black color of the matrix in both images, indicating high TOC content and hypertrophic-anoxic conditions during deposition.

IW = 0.5 mm (above); 0.3 mm (below). PPL.
Photomicrograph below taken from Truskowski (2006)

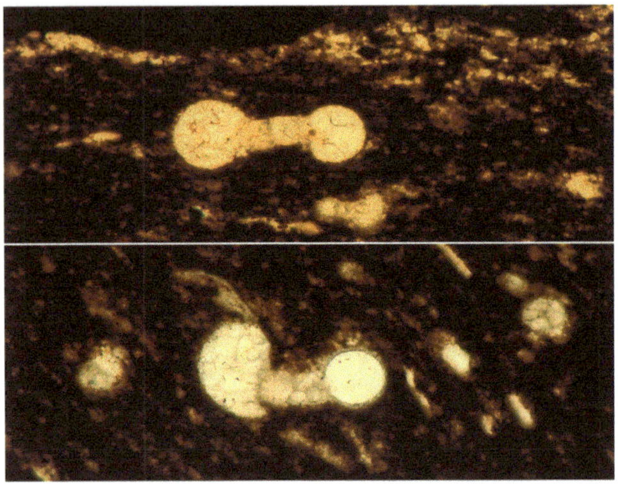

Cretaceous (early Aptian). Machiques Member, Apón Formation. Maracaibo Basin

Both images belong to the same sample (which contains 1.83% TOC and is stratigraphically located at the lower part of the Machiques Member), and show tests of the genus *Globigerinelloides*. Above, *G. blowi*, which is a key biostratigraphic marker for the OAE 1a. Bellow, *G. barri* exhibiting striking inflated last chamber (to the left), which allows easily differentiate it from other species of Aptian *Globigerinelloides*. These opportunistics represent the partial development of pelagic realm-like conditions over the Maracaibo platform during the OAE 1a.

IW = 0.6 mm each. PPL
Photomicrographs courtesy of Gilberto Soto

Cretaceous (late Aptian). Piché Member, Apón Formation. Maracaibo Basin

A test of the genus *Hedbergella* (probably *H. trocoidea*) in the center of the image exhibiting characteristic low trochospiral coil. Like many PF test in shallow black shales of the Cogollo Group, the walls of this test are recrystallized and the chambers are filled with sparry calcite. Therefore, many diagnostic features are not quite visible, such as well-differentiated individual chambers. This sample contains 3.52% TOC and 78% carbonate and belongs to the organic-rich interval in the upper part of the Piché Member (the OAE 1b first sub-event).

IW = 0.5 mm. PPL

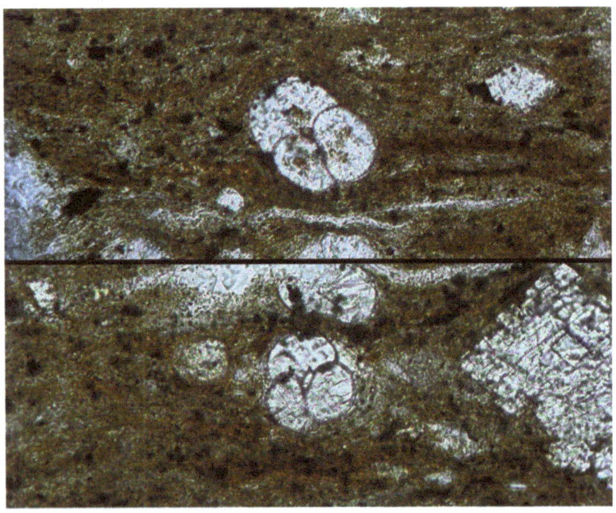

Cretaceous (early Albian). Lisure Formation. Maracaibo Basin

Both images belong to the same sample from the base of the Lisure Formation, and show images of the genus *Microhedbergella* (probably *M. renilaevis*; Order Globigerinida, Family Hedbergellidae). This genus is characterized by a very small test size, thin walls and globular chambers. In this core (Z26D-2), the appearance of this planktonic foraminifer in an organic-rich interval (the OAE 1b second sub-event) having TOC up to 7% and 23 m in thickness marks the beginning of the Albian. See more photographic examples of the genus *Muricohedbergella* in Gutiérrez-Puente et al. (2021).

IW = 0.5 mm each. PPL

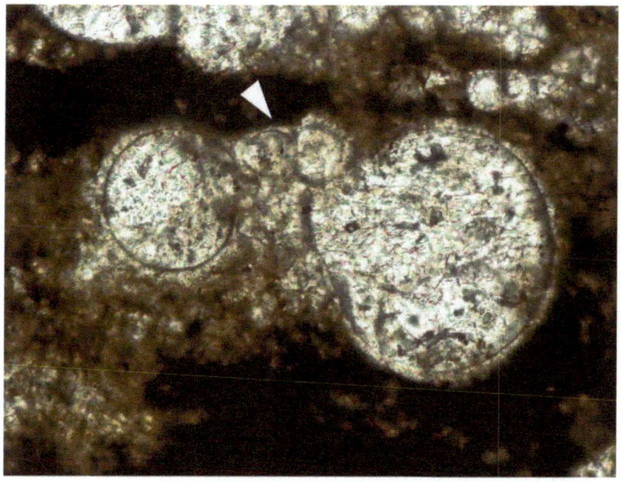

Cretaceous (latest Albian). Querecual Formation. Eastern Interior Range

Axial section of the species *Ticinella roberti* (Order Globigerinida, Family Rotaliporidae). The test of this planktonic foraminifer is trochospirally enrolled and biconvex. Well-preserved thin calcareous walls outline both the inflated chambers at both endings and two small globular chambers in the middle (white arrow). The latter are not aligned respect to the larger chambers, thus giving the test an asymmetrical shape. This sample of late Albian age corresponds to the OAE 1d, an organic-rich interval characterized by the abundance of species of *T. primula* and *T. roberti*.

IW = 0.3 mm. PPL
Photomicrograph taken from Truskowski (2006)

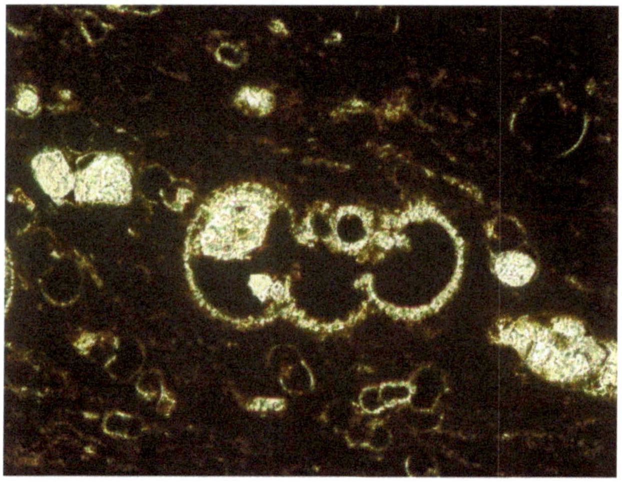

Cretaceous (middle-late Cenomanian). La Luna Formation *sensu stricto*. Maracaibo Basin

Axial section of *Muricohedbergella* sp., with globular chambers filled by synsedimentary, organic-rich matrix. Low trochospiral coiled test of this planktonic foraminifer exhibits moderately thick calcareous walls with barely visible microgranular texture reminiscent of small perforations. This sample corresponds to the base of the La Luna Formation, an organic-rich expanded section whose deposition started during middle to late Cenomanian, highlighting the beginning of fully anoxic conditions in the MB prior to the beginning of the OAE 2 at global scale.

IW = 0.5 mm. PPL

Cretaceous (late Cenomanian). Querecual Formation. Eastern Interior Range

This view from the spiral side of the species *Muricohedbergella planispira* (formerly *Hedbergella planispira*) allows visualize the coiled arrangement of the trochospiral test. Note the small globular early chambers (white arrow) coiling clockwise (from this view), while peripheral later chambers are inflated. *M. planispira* is among the most dominating species of PF in organic-rich microfacies of the Querecual Formation. Its occurrence within this expanded section encompasses the late Albian-Santonian interval.

IW = 0.3 mm. PPL
Photomicrograph taken from Truskowski (2006)

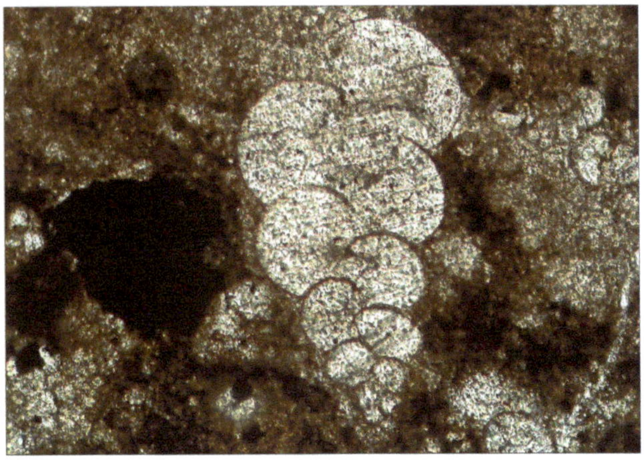

Cretaceous (late Cenomanian). Querecual Formation. Eastern Interior Range

Longitudinal section of the genus *Planoheterohelix reussi*, formerly *Heterohelix reussi* (Order Heterohelicida, Family Heterohelicidae). The elongate planispiral test of this planktonic foraminifer is characterized by biserial chamber arrangement and thin calcareous walls. Globular chambers increase markedly in size towards the final whorl. This opportunistic dweller is an excellent indicator of the presence of a strong OMZ. *Planoheterohelix* is one of the most dominant genera among planktonic assemblages in the Querecual, La Luna and Navay formations.

IW = 0.5 mm. PPL
Photomicrograph taken from Truskowski (2006)

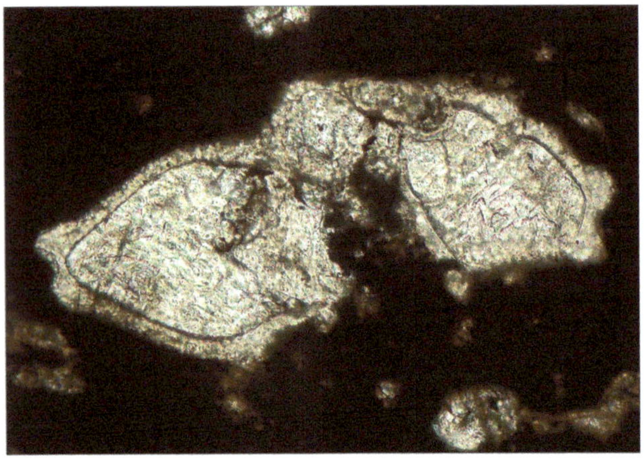

Cretaceous (early Coniacian). Querecual Formation. Eastern Interior Range

Equatorial section of the genus *Marginotruncana*, probably *M. renzi* (Order Globigerinida, Family Globotruncanidae). The test of this k-selected specialist is biconvex (trochospiral) and medium-arched, with a marked chamber curved alignment. Calcareous walls are thick and exhibit prominent keeled outer margins in the larger chambers at both endings. This keeled planktonic foraminifer inhabited depths within the photic zone but below the shallower and dweller-dominated first 50 m of water column.

IW = 0.3 mm. PPL
Photomicrograph taken from Truskowski (2006)

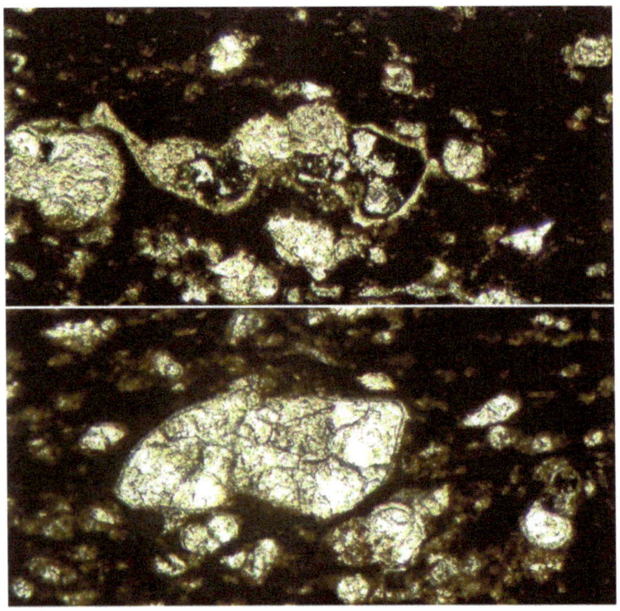

Cretaceous (early Turonian). Querecual Formation. Eastern Interior Range

Both images show equatorial sections of the genus *Dicarinella* (Order Globigerinida, Family Globotruncanellidae). Tests of this r/k intermediate planktonic foraminifer (trochospiral exhibit characteristic concavo-convex, medium-arched shape. Well-preserved walls of ending chambers are keeled, thus giving the tests a pinched peripheral outline. Smaller middle chambers are clearly differentiated in the specimen above. As *Marginotrucana*, *Dicarinella* and other intermediate and specialist genera of PF are far less abundant in the Querecual Formation compared to globular opportunistic genera. However, when present, they are easily differentiated due to their aberrant forms and average larger size.

IW = 0.5 mm each. PPL
Photomicrographs taken from Truskowski (2006)

References

Abd El-Gaied, I. M., Salama, Y. F., Saber, S. G., Sayed, M. M., 2019. Benthic foraminiferal communities of the Eocene platform, north Eastern Desert, Egypt. *Journal of African Earth Sciences*, *151*, 121–135.

Abramovich, S., Keller, G., Stüben, D., Berner, Z., 2003. Characterization of late Campanian and Maastrichtian planktonic foraminiferal depth habitats and vital activities based on stable isotopes. *Palaeogeography, Palaeoclimatology, Palaeoecology*, *202*, 1–29.

Adams, C. G., Lee, D. E., Rosen, B. R., 1990. Conflicting isotopic and biotic evidence for tropical sea-surface temperatures during the Tertiary. *Palaeogeography, Palaeoclimatology, Palaeoecology*, *77*, 289–313.

Albert-Villanueva, E., 2016. Facies y secuencias deposicionales mixtas carbonático-siliciclásticas del Mioceno inferior de la Cuenca de Falcón (noroeste de Venezuela) como modelo exploratorio en el Caribe. Doctoral Dissertation. University of Barcelona. Volumes 1 and 2.

Anand, P., Elderfield, H., Conte, M. H., 2003. Calibration of Mg/Ca thermometry in planktonic foraminifera from a sediment trap time series. *Paleoceanography*, *18*. 1–15.

Armstrong, H. A., and Brasier, M. D., 2005. Microfossils, Second edition. Blackwell Publishing, 296 pp.

Azmi, A., 2018. Late Triassic to early Jurassic microfossils and palaeoenvironments of the Waterloo Mudstone Formation. Doctoral dissertation, University of Birmingham. 355 pp.

Bachmann, M., and Hirsch, F., 2006. Lower Cretaceous carbonate platform of the eastern Levant (Galilee and the Golan Heights): stratigraphy and second-order sea-level change. *Cretaceous Research*, *27*, 487–512.

Bandy, O. L., 1964. Cenozoic planktonic foraminiferal zonation. *Micropaleontology*, 1–17.

Barker, S., Cacho, I., Benway, H., Tachikawa, K., 2005. Planktonic foraminiferal Mg/Ca as a proxy for past oceanic temperatures: a methodological overview and data compilation for the Last Glacial Maximum. *Quaternary Science Reviews*, *24*, 821–834.

Bassi, D., and Nebelsick, J. H., 2010. Components, facies and ramps: redefining Upper Oligocene shallow water carbonates using coralline red algae and larger foraminifera (Venetian area, northeast Italy). *Palaeogeography, Palaeoclimatology, Palaeoecology*, *295*, 258–280.

Beauchamp, B., 1994. Permian climatic cooling in the Canadian Arctic. *Geological Society of America Special Paper*, *288*, 229–246.

Beavington-Penney, S. J., 2004. Analysis of the effects of abrasion on the test of Palaeonummulites venosus: implications for the origin of nummulithoclastic sediments. *Palaios*, *19*, 143–155.

Beavington-Penney, S. J., and Racey, A., 2004. Ecology of extant nummulitids and other larger benthic foraminifera: applications in palaeoenvironmental analysis. *Earth-Science Reviews*, *67*, 219–265.

Bellier, J. -P., Mathieu, P., Granier, B., 2010. Short treatise on foraminiferology (Essential on modern and fossil Foraminifera). Carnets de Géologie - Notebooks on Geology, Brest, Book 2010/02 (CG2010_B02), 104 pp.

Bender, M. L., Lorens, R. B., Williams, D. F., 1975. Sodium, magnesium and strontium in the tests of planktonic foraminifera. *Micropaleontology*, 448–459.

Boersma, A., Bougault, H., Cande, S. C., 1985. Oligocene benthic foraminifers from North Atlantic sites: benthic foraminifers as water-mass indexes in the North and South Atlantic. *Initial Reports of the Deep Sea Drilling Project*, *82*, 611–627.

Bornemann, A., and Norris, R. D., 2007. Size-related stable isotope changes in Late Cretaceous planktic foraminifera: implications for paleoecology and photosymbiosis. *Marine Micropaleontology*, *65*(1–2), 32–42.

BouDagher-Fadel, M. K., 2008a. The Cenozoic larger benthic foraminifera: the Palaeogene. *Developments in Palaeontology and Stratigraphy*, *21*, 297–545.

BouDagher-Fadel, M. K., 2008b. The Cenozoic larger benthic foraminifera: the Neogene. *Developments in Palaeontology and Stratigraphy*, *21*, 419–548.

BouDagher-Fadel, M. K., 2015. Biostratigraphic and Geological Significance of Planktonic Foraminifera (Updated 2nd Edition): UCL Press, London, 298 pp.

Boudaugher-Fadel, M. K., 2018. Evolution and geological significance of larger benthic foraminifera. UCL Press, London, 607 pp.

Boudagher-Fadel, M. K., and Price, G. D., 2013. The phylogenetic and palaeogeographic evolution of the miogypsinid larger benthic foraminifera. *Journal of the Geological Society*, *170*, 185–208.

BouDagher-Fadel, M. K., and Price, G. D., 2014. The phylogenetic and palaeogeographic evolution of the nummulitoid larger benthic foraminifera. *Micropaleontology*, 483–508.

Boudagher-Fadel, M., and Price, G., 2019. Global evolution and paleogeographic distribution of mid-Cretaceous orbitolinids. *UCL Open Environment*. 1–21.

Boudagher-Fadel, M., Price, G., 2021. The geographic, environmental and phylogenetic evolution of the Alveolinoidea from the Cretaceous to the present day. *UCL Open: Environment*, (2):03, 1–34.

Boudagher-Fadel, M. K., Rose, E. P. F., Bosence, D. W. J., Lord, A. R., 2001. Lower Jurassic foraminifera and calcified microflora from Gibraltar, western Mediterranean. *Palaeontology*, *44*, 601–621.

Brandano, M., Frezza, V., Tomassetti, L., Cuffaro, M., 2009. Heterozoan carbonates in oligotrophic tropical waters: the Attard member of the lower coralline limestone formation (Upper Oligocene, Malta). *Palaeogeography, Palaeoclimatology, Palaeoecology*, *274*, 54–63.

Burki, F., Kaplan, M., Tikhonenkov, D. V., Zlatogursky, V., Minh, B. Q., Radaykina, L. V., Smirnov. A., Mylnikov. P., Keeling. P. J., Keeling. P. J., 2016. Untangling the early diversification of eukaryotes: a phylogenomic study of the evolutionary origins of Centrohelida, Haptophyta and Cryptista. *Proceedings of the Royal Society B: Biological Sciences*, *283*. 1–10.

Campos-Soto, S., Benito, M. I., Cobos, A., Caus, E., Quijada, I. E., Suarez-Gonzalez, P., Mas, R., Royo-Torres, R., Alcalá, L., 2019. Revisiting the age and palaeoenvironments of the Upper Jurassic–Lower Cretaceous? Dinosaur-bearing sedimentary record of eastern Spain: implications for Iberian palaeogeography. *Journal of Iberian Geology*, *45*, 471–510.

Carannante, G., Esteban, M., Milliman, J. D., Simone, L., 1988. Carbonate lithofacies as paleolatitude indicators: problems and limitations. *Sedimentary Geology*, *60*, 333–346.

Carannante, G., Cherchi, A., Simone, L., 1995. Chlorozoan versus foramol lithofacies in Upper Cretaceous rudist limestones. *Palaeogeography, Palaeoclimatology, Palaeoecology*, *119*, 137–154.

Coccioni, R., Luciani, V., Marsili, A., 2006. Cretaceous oceanic anoxic events and radially elongated chambered planktonic foraminifera: Paleoecological and paleoceanographic implications. *Palaeogeography, Palaeoclimatology, Palaeoecology*, *235*, 66–92.

Crespo de Cabrera, S., Sliter, W., Jarvis, I., 1999. Integrated foraminiferal biostratigraphy and Chemostratigraphy of the Querecual Formation (Cretaceous) Eastern Venezuela. *Journal of Foraminiferal Research*, *29*, 487–499.

Curry, W. B., Duplessy, J. C., Labeyrie, L. D., Shackleton, N. J., 1988. Changes in the distribution of $\delta^{13}C$ of deep water ΣCO_2 between the last glaciation and the Holocene. *Paleoceanography*, *3*, 317–341.

Davydov, V., 2014. Warm water benthic foraminifera document the Pennsylvanian–Permian warming and cooling events—the record from the Western Pangea tropical shelves. *Palaeogeography, Palaeoclimatology, Palaeoecology*, *414*, 284–295.

Dekens, P. S., Lea, D. W., Pak, D. K., Spero, H. J., 2002. Core top calibration of Mg/Ca in tropical foraminifera: Refining paleotemperature estimation. *Geochemistry, Geophysics, Geosystems*, *3*, 1–29.

de Nooijer, L. J., Toyofuku, T., Kitazato, H., 2009. Foraminifera promote calcification by elevating their intracellular pH. *Proceedings of the National Academy of Sciences*, *106*(36), 15374–15378.

De Romero, L. M., Galea-Álvarez, F. A., 1995. Campanian *Bolivinoides* and microfacies from the La Luna Formation, western Venezuela. *Marine Micropaleontology, 26*, 385–404.

D'Hondt, S., and Zachos, J. C., 1998. Cretaceous foraminifera and the evolutionary history of planktic photosymbiosis. *Paleobiology*, *24*, 512–523.

Dubicka, Z., and Peryt, D., 2014. Classification and evolutionary interpretation of late Turonian–early Campanian *Gavelinella* and *Stensioeina* (Gavelinellidae, benthic foraminifera) from western Ukraine. *The Journal of Foraminiferal Research*, *44*, 151–176.

Edgar, K. M., Bohaty, S. M., Gibbs, S. J., Sexton, P. F., Norris, R. D., Wilson, P. A., 2013. Symbiont 'bleaching' in planktic foraminifera during the Middle Eocene Climatic Optimum. *Geology*, *41*(1), 15–18.

Elderfield, H., and Ganssen, G., 2000. Past temperature and $\delta^{18}O$ of surface ocean waters inferred from foraminiferal Mg/Ca ratios. *Nature*, *405*, 442–445.

Erbacher, J., Hemleben, C., Huber, B. T., Markey, M., 1999. Correlating environmental changes during early Albian oceanic anoxic event 1B using benthic foraminiferal paleoecology. *Marine Micropaleontology*, *38*, 7–28.

Erez, J., 2003. The source of ions for biomineralization in foraminifera and their implications for paleoceanographic proxies. *Reviews in Mineralogy and Geochemistry*, *54*, 115–149.

Fajemila, O. T., Langer, M. R., Lipps, J. H., 2015. Spatial patterns in the distribution, diversity and abundance of benthic foraminifera around Moorea (Society Archipelago, French Polynesia). *PloS One*, *10*, e0145752.

Farouk, S., Ismail, A., Abd El-Rauf, I., Tawfik, M., 2017. Foraminiferal palaeobathymetry and depositional sequences of the subsurface Eocene Apollonia Formation in North Western Desert, Egypt, *Palaeobiodiversity and Palaeoenvironments*, *98*, 403–429.

Farouk, S., Jain, S., Belal, N., Omran, M., Al-Kahtany, K., 2020. Quantitative Middle Eocene benthic foraminiferal biofacies from west-central Sinai, Egypt: Implications to paleobathymetry and sequence stratigraphy. *Marine Micropaleontology*, *155*, 101823.

Fenton, I. S., Pearson, P. N., Dunkley Jones, T., Farnsworth, A., Lunt, D. J., Markwick, P., Purvis, A., 2016. The impact of Cenozoic cooling on assemblage diversity in planktonic foraminifera. *Philosophical Transactions of the Royal Society B: Biological Sciences*, *371*, 20150224.

Ferguson, J. E., Henderson, G. M., Kucera, M., Rickaby, R. E. M., 2008. Systematic change of foraminiferal Mg/Ca ratios across a strong salinity gradient. *Earth and Planetary Science Letters*, *265*, 153–166.

Filkorn, H. F., and Scott, R. W., 2011. Microfossils, paleoenvironments and biostratigraphy of the Mal Paso Formation (Cretaceous, upper Albian), state of Guerrero, Mexico. *Revista Mexicana de Ciencias Geológicas*, *28*, 175–191.

Flügel, E, 2010. Microfacies of Carbonate Rocks—Analysis, Interpretation and Application. Springer, Heidelberg-Dordrecht-London-New York, 984 pp.

Friedrich, O., 2010. Benthic foraminifera and their role to decipher paleoenvironment during mid-Cretaceous Oceanic Anoxic Events–the "anoxic benthic foraminifera" paradox. *Revue de Micropaléontologie*, *53*, 175–192.

Friedrich, O., and Erbacher, J., 2006. Benthic foraminiferal assemblages from Demerara Rise (ODP Leg 207, western tropical Atlantic): possible evidence for a progressive opening of the Equatorial Atlantic Gateway. *Cretaceous Research*, *27*, 377–397.

Gaillot, J. and Vachard, D., 2007. The Khuff Formation (Middle East) and time-equivalents in Turkey and South China: Biostratigraphy from Capitanian to Changhsingian times (Permian), new foraminiferal taxa, and palaeogeographical implications. *Coloquios de Paleontología*, *57*, 37–223.

Giusberti, L., Boscolo-Galazzo, F., Thomas, E., 2016. Variability in climate and productivity during the Paleocene–Eocene Thermal Maximum in the western Tethys (Forada section). *Climate of the Past*, *12*, 213–240.

Gooday, A. J., Nomaki, N. H., Kitazato, H., 2008. Modern deep-sea benthic foraminifera: a brief review of their morphology-based biodiversity and trophic diversity. *Geological Society, London, Special Publications*, *303*, 97–119.

Grossman, E. L., 2012. Oxygen Isotope Stratigraphy, in The Geologic Time Scale 2012, F. M. Gradstein, J. G. Ogg, M. D. Schmitz, and G. M. Ogg (editors). Elsevier, Amsterdam, volume 1, 181–206.

Gräfe, K.-U., 2005. Late Cretaceous benthic foraminifers from the Basque-Cantabrian Basin, Northern Spain. *Journal of Iberian Geology*, *3*, 277–298.

Gutiérrez-Puente, N. A., Barragán, R., Núñez-Useche, F., 2021. Paleoenvironmental changes and biotic response to Aptian–Albian episodes of accelerated global change: evidence from the western margin of the proto-North Atlantic (central-eastern Mexico). *Cretaceous Research*, *126*, 104883.

Hallock, P., 1985. Why are larger foraminifera large? *Paleobiology*, *11*, 195–208.

Hallock, P., 1999. Symbiont-bearing Foraminifera. In: Sen Gupta, B. (Ed.), Modern Foraminifera. Kluwer Press, Amsterdam, pp. 123–139.

Hallock, P., and Pomar, L., 2008. Cenozoic evolution of larger benthic foraminifers: paleoceanographic evidence for changing habitats. In *Proceedings of the 11th International Coral Reef Symposium, Ft. Lauderdale, Florida*, 16–20.

Hallock, P., Premoli-Silva, I., Boersma, A., 1991. Similarities between planktonic and larger foraminiferal evolutionary trends through Paleogene paleoceanographic changes. *Palaeogeography, Palaeoclimatology, Palaeoecology*, *83*, 49–64.

Hallock, P., Triantaphyllou, M., Dimiza, M., Koukousioura O., 2011. An invasive foraminifer in coastal ecosystems of the Eastern Mediterranean: Implications for understanding larger foraminiferal-dominated biofacies in the Cenozoic. In: GSA Annual Meeting in Minneapolis, Paper No. 231-10, *Geological Society of America Abstracts with Programs*, *43*, 556 pp.

Haq, B. U., 2014. Cretaceous eustasy revisited. *Global and Planetary Change*, *113*, 44–58.

Haq, B. U., and Boersma, A., 1998. Introduction to Marine Micropaleontology. Elsevier. 376 pp.

Hart, M. B., 1999. The evolution and biodiversity of Cretaceous planktonic Foraminiferida. *Geobios*, *32*, 247–255.

Haynes J. R., 1981. Foraminifera, Macmillan, London, 433 pp.

Hayton, S., Nelson C., Hood S., 1995. A skeletal assemblage classification system for non-tropical carbonate deposits based on New Zealand Cenozoic limestones. *Sedimentary Geology*, *100*. 123–141.

Henderson, G. M., 2002. New oceanic proxies for paleoclimate. *Earth and Planetary Science Letters*, *203*, 1–13.

Holzmann, M., Hohenegger, J., Hallock, P., Piller, W. E., Pawlowski, J., 2001. Molecular phylogeny of large miliolid foraminifera (Soritacea Ehrenberg 1839). *Marine Micropaleontology*, *43*, 57–74.

Hoogakker, B., Ishimura, T., de Nooijer, L., Rathburn, A., Schmiedl, G., 2024. A review of benthic foraminiferal oxygen and carbon isotopes. *Quaternary Science Reviews*, *342*, 108896.

Hottinger, L., 2007. Revision of the foraminiferal genus *Globoreticulina rahaghi*, 1978, and of its associated fauna of larger foraminifera from the late Middle Eocene of Iran. Notebooks on Geology (Brest), CG2007- A06, 1–51.

Imbrie, J., Hays, J. D., Martinson, D. G., McIntyre, A., Mix, A. C., Morley, J. J., Pisias, N. G., Prell, W. L., Shackleton, N. J., 1984. The orbital theory of Pleistocene climate: support from a revised chronology of the marine $d^{18}O$ record. In: Berger, A., Imbrie, J., Hays,

J., Kukla, G., Saltzman, B. (Eds.), Milankovitch and climate. Reidel Publishing Company, Dordrecht, pp. 269–305.

James, N. P., 1997. The cool-water carbonate depositional realm. In: James, N. P., Clarke, J., (eds.), Cool-water carbonates, vol 56. SEPM Society for Sedimentary Geology, Tulsa, 1–20.

James, N. P., and Jones, B., 2015. Origin of Carbonate Sedimentary Rocks. John Wiley & Sons, Ltd, Chinchester, 464 pp.

Javaux, E. J., and Scott, D. B., 2003. Illustration of modern benthic foraminifera from Bermuda and remarks on distribution in other subtropical/tropical areas. *Palaeontologia Electronica*, *6*(4), 29 pp.

Jenny, C., and Guex, J., 2005. Pathological development in response to environmental stress in some latest Permian foraminifers: preliminary report. *Bulletin de la Societe Vaudoise des Sciences Naturelles*, *89*, 171–184.

Kaminski, M. A., 2014. The year 2010 classification of the agglutinated foraminifera. *Micropaleontology*, *60*(1), 89–108.

Kateb, A. E., Stalder, C., Stainbank, S., Fentimen, R., Spezzaferri, S., 2018. The genus *Amphistegina* (benthic foraminifera): distribution along the southern Tunisian coast. *Bioinvasions Records*, *7*, 391–398.

Katz, M. E., Cramer, B. S., Franzese, A., Hönisch, B., Miller, K. G., Rosenthal, Y., Wright, J. D., 2010. Traditional and emerging geochemical proxies in foraminifera. *The Journal of Foraminiferal Research*, *40*, 165–192.

Kaya, M. Y., Dupont-Nivet, G., Proust, J. N., Roperch, P., Meijer, N., Frieling, J., et al., 2020. Cretaceous evolution of the Central Asian proto-Paratethys Sea: Tectonic, eustatic, and climatic controls. *Tectonics*, *39*, e2019TC005983.

Kelly, D. C., 2002. Response of Antarctic (ODP Site 690) planktonic foraminifera to the Paleocene–Eocene thermal maximum: Faunal evidence for ocean/climate change. *Paleoceanography*, *17*, 23–1.

Kaminski, M. A., and Perdana, P. R., 2020. Early Silurian Benthic Foraminifera from Saudi Arabia–including the oldest known multichambered lituolids. *Stratigraphy*, *17*, 141–185.

Klicpera, A., Michel, J., Westphal, H., 2015. Facies patterns of a tropical heterozoan carbonate platform under eutrophic conditions: the Banc d'Arguin, Mauritania, *Facies*, *61*, 421.

Koutsoukos, E. A., and de Klasz, I., 2000. Late Cretaceous foraminiferal biogeography (Families Bolivinidae, Buliminellidae, Gavelinellidae, Siphogenerinoididae, Turrilinidae) in northeastern Brazilian shelf and central West African basins. *Cretaceous Research*, *21*, 381–405.

Langer, M. R., and Hottinger, L., 2000. Biogeography of selected "larger" foraminifera. *Micropaleontology*, *46*, 105–126.

Laya, J. C., and Tucker, M. E., 2012. Facies analysis and depositional environments of Permian carbonates of the Venezuelan Andes: Palaeogeographic implications for Northern Gondwana. *Palaeogeography, Palaeoclimatology, Palaeoecology*, *331*, 1–26.

Lea, D. W., 2014. Elemental and Isotopic Proxies of Past Ocean Temperatures. *Treatise on Geochemistry: Second Edition. Elsevier Ltd*, 373–397.

Lear, C. H., Rosenthal, Y., Slowey, N., 2002. Benthic foraminiferal Mg/Ca-paleothermometry: A revised core-top calibration. *Geochimica et Cosmochimica Acta*, *66*, 3375–3387.

Leckie, R. M., 1987. Paleoecology of mid-Cretaceous planktonic foraminifera: a comparison of open ocean and epicontinental sea assemblages. *Micropaleontology*, 164–176.

Lee, J. J, 1998. "Living sands" - Larger foraminifera and their endosymbiotic algae. *Symbiosis*, *25*, 71–100.

Lee, J. J. 2006. Algal symbiosis in larger foraminifera. *Symbiosis*, *42*. 63–75.

Lees, A., 1975. Possible influence of salinity and temperature on modern shelf carbonate sedimentation. *Marine Geology*, *19*, 159–198.

Lees, A., and Buller, A. T., 1972. Modern temperate-water and warm-water shelf carbonate sediments contrasted. *Marine Geology*, *13*, M67–M73.

Lefort, A., Lathuiliere, B., Carpentier, C., Huault, V., 2011. Microfossil assemblages and relative sea-level fluctuations in a lagoon at the Oxfordian/Kimmeridgian boundary (Upper Jurassic) in the eastern part of the Paris Basin. *Facies*, *57*, 649–662.

Leutenegger, S., 1984. Symbiosis in benthic foraminifera; specificity and host adaptations. *The Journal of Foraminiferal Research*, *14*, 16–35.

Li, Q., James, N. P., Bone, Y., McGowran, B., 1999. Palaeoceanographic significance of recent foraminiferal biofacies on the southern shelf of Western Australia: a preliminary study. *Palaeogeography, Palaeoclimatology, Palaeoecology*, *147*, 101–120.

Lipps, J. H., Stanley, G. D. Jr., 2016. Photosymbiosis in the Past and Present Reefs. In: Hubbard, D. K., ed., Coral Reefs at the Crossroads. Coral Reefs of the World 6. Springer Science Publishers, Dordrecht. 47–68.

Lisiecki, L. E., and Raymo, M. E., 2005. A Pliocene-Pleistocene stack of 57 globally distributed benthic $\delta^{18}O$ records. *Paleoceanography*, *20*, PA1003.

Loeblich, A. R. J., and Tappan, H., 1988. Foraminiferal genera and their classification. Van Nostrand Reinhold Company, New York, 970 pp.

Loeblich, A. R., and Tappan, H., 1992. Present status of Foraminiferal Classification. In: Takayanagi, Y. & Saito, T (eds.), Studies in Benthic Foraminifera. Tokai University Press, 93–102.

Lo Giudice-Cappelli, E., and Austin, W. E., 2019. Size matters: analyses of benthic foraminiferal assemblages across differing size fractions. *Frontiers in Marine Science*, *6*, 752.

Lorente, 1978. La caliza de Capadare y su relación con la Cuenca de Agua Salada, Edo. Falcón. Bachelor Degree Dissertation. Central University of Venezuela. 161 pp.

Ma, Z. L., Li, Q. Y., Liu, X. Y., Luo, W., Zhang, D. J., Zhu, Y. H., 2018. Palaeoenvironmental significance of Miocene larger benthic foraminifera from the Xisha Islands, South China Sea. *Palaeoworld*, *27*, 145–157.

Martin, P. A., Lea, D. W., Rosenthal, Y., Shackleton, N. J., Sarnthein, M., Papenfuss, T., 2002. Quaternary deep sea temperature histories derived from benthic foraminiferal Mg/Ca. *Earth and Planetary Science Letters*, *198*, 193–209.

Martinson, D. G., Pisias, N. G., Hays, J. D., Imbrie, J., Moore, T. C., Shackleton, N. J., 1987. Age dating and the orbital theory of the ice ages: development of a high-resolution 0 to 300,000-year Chronostratigraphy1. *Quaternary Research*, *27*, 1–29.

McInerney, F. A., and Wing, S. L., 2011. The Paleocene-Eocene Thermal Maximum: A perturbation of carbon cycle, climate, and biosphere with implications for the future. *Annual Review of Earth and Planetary Sciences*, *39*, 489–516.

Michel, J., Laugié, M., Pohl, A., Lanteaume, C., Masse, J. P., Donnadieu, Y., Borgomano, J., 2019. Marine carbonate factories: a global model of carbonate platform distribution. *International Journal of Earth Sciences*, *108*, 1773–1792.

Morales, F. J., 2004. Caracterización de conjuntos de foraminíferos bentónicos en sedimentos recientes de ambientes de talud continental del area de la Fachada Atlántica de Venezuela: implicaciones geoambientales. Magister Dissertation, Central University of Venezuela. 251 pp.

Mouanga, G. H., and Langer, M. R., 2014. At the front of expanding ranges: Shifting community structures at amphisteginid species range margins in the Mediterranean Sea. *Neues Jahrbuch für Geologie und Paläontologie, Abhandlungen*, *271*, 141–150.

Murray, J. W., 2007. Biodiversity of living benthic foraminifera: How many species are there? *Marine Micropaleontology*, *64*, 163–176.

Mutti, M., and Hallock, P., 2003. Carbonate systems along nutrient and temperature gradients: some sedimentological and geochemical constraints. *International Journal of Earth Sciences*, *92*, 465–475.

Narayan, G. R., Reymond, C. E., Stuhr, M., Doo, S., Schmidt, C., Mann, T., Westphal, H., 2022. Response of large benthic foraminifera to climate and local changes: Implications for future carbonate production. *Sedimentology*, *69*, 121–161.

Nederbragt, A. J., Erlich, R. N., Fouke, B. W., Ganssen, G. M., 1998. Palaeoecology of the biserial planktonic foraminifer Heterohelix moremani (Cushman) in the late Albian to middle Turonian Circum-North Atlantic. *Palaeogeography, Palaeoclimatology, Palaeoecology*, *144*, 115–133.
Nehrke, G., Keul, N., Langer, G., De Nooijer, L. J., Bijma, J., Meibom, A., 2013. A new model for biomineralization and trace-element signatures of Foraminifera tests. *Biogeosciences*, *10*, 6759–6767.
Núñez-Useche, F., Barragán, R., Torres-Martínez, M. A., López-Zúñiga, P. A., Moreno-Bedmar, J. A., Chávez-Cabello, G., Canet, C., Chacon-Baca, E., 2020. Response of the western proto-North Atlantic margin to the early Aptian Oceanic Anoxic Event (OAE) 1a: an example from the Cupido platform margin-Gulf of Mexico, NE Mexico. *Cretaceous Research*, *113*, 104488.
Omaña, L., and Alencáster, G., 2009. Lower Aptian shallow-water benthic foraminiferal assemblage from the Chilacachapa range in the Guerrero-Morelos Platform, south Mexico. *Revista Mexicana de Ciencias Geológicas*, *26*, 575–586.
Omaña, L., Alencaster, G., Buitrón, B. E., 2016. Mid-early late Albian foraminiferal assemblage from the El Abra Formation in the El Madroño locality, eastern Valles-San Luis Potosí Platform, Mexico: Paleoenvironmental and paleobiogeographical significance. *Boletín de la Sociedad Geológica Mexicana*, *68*, 477–496.
Omidvar, M., Mehrabi, H., Sajjadi, F., Bahramizadeh-Sajjadi, H., Rahimpour-Bonab, H., Ashrafzadeh, A., 2014. Revision of the foraminiferal biozonation scheme in Upper Cretaceous carbonates of the Dezful Embayment, Zagros, Iran: integrated palaeontological, sedimentological and geochemical investigation. *Revue de Micropaléontologie*, *57*, 97–116.
Parker, J. H., and Gischler, E., 2011. Modern foraminiferal distribution and diversity in two atolls from the Maldives, Indian Ocean. *Marine Micropaleontology*, *78*, 30–49.
Parker, J. H., and Gischler, E., 2015. Modern and relict foraminiferal biofacies from a carbonate ramp, offshore Kuwait, northwest Persian Gulf. *Facies*, *61*(10), 1–22.
Parra, M., Moscardelli, L., Lorente, M. A., 2003. Late Cretaceous anoxia and lateral microfacies changes in the Tres Esquinas Member, La Luna Formation, Western Venezuela. *Palaios*, *18*, 321–333.
Past Interglacials Working Group of PAGES, 2016. Interglacials of the last 800,000 years. *Review of Geophysics*, *54*, 162–219.
Pawlowski, J., Lee, J. J., Gooday, A., 1993. Microforaminifera—perspective on a neglected group of foraminifera. *Archiv für Protistenkunde*, *143*, 271–284.
Pawlowski, J. A. N., Holzmann, M., Fahrni, J. F., Hallock, P., 2001a. Molecular identification of algal endosymbionts in large miliolid foraminifera: 1. Chlorophytes. *Journal of Eukaryotic Microbiology*, *48*, 362–367.
Pawlowski, J. A. N., Holzmannm, M., Fahrni, J. F., Pochon, X., and Lee, J. J., 2001b. Molecular identification of algal endosymbionts in large miliolid foraminifera: 2 Dinoflagellates. *Journal of Eukaryotic Microbiology*, *48*, 368–373.
Pawlowski, J., Holzmann, M., Tyszka, J., 2013. New supraordinal classification of Foraminifera: Molecules meet morphology. *Marine Micropaleontology*, *100*, 1–10.
Pomar, L., 2001. Types of carbonate platforms: a genetic approach. *Basin Research*, *13*, 313–334.
Pomar, L., and Hallock, P., 2008. Carbonate factories: a conundrum in sedimentary geology. *Earth-Science Reviews*, *87*(3–4), 134–169.
Pomar, L., Brandano, M., Westphal, H., 2004. Environmental factors influencing skeletal grain sediment associations: a critical review of Miocene examples from the western Mediterranean. *Sedimentology*, *51*, 627–651.
Pomar, L., Esteban, M., Martinez, W., Espino, D., Castillo de Ott, V., Benkovics, L., Castro-Leyva, T., 2015. Oligocene-Miocene carbonates of the Perla field, offshore Venezuela: Depositional model and facies architecture, In: C. Bartolini and P. Mann, eds., Petroleum geology and potential of the Colombian Caribbean Margin: American *Association of Petroleum Geologists, Memoir 108*, 647–674.
Pomar, L., Baceta, J. I., Hallock, P., Mateu-Vicens, G., Basso, D., 2017. Reef building and carbonate production modes in the west-central Tethys during the Cenozoic. *Marine and Petroleum Geology*, *83*, 261–304.
Petrizzo, M. R., 2007. The onset of the Paleocene–Eocene Thermal Maximum (PETM) at Sites 1209 and 1210 (Shatsky Rise, Pacific Ocean) as recorded by planktonic foraminifera. *Marine Micropaleontology*, *63*, 187–200.
Premoli Silva, I., and Sliter, W. V., 1999. Cretaceous paleoceanography: evidence from planktonic foraminiferal evolution. In: Barrera, E., Johnson, C.C. (Eds.), The Evolution of Cretaceous Ocean-Climatic System. *Special Papers-Geological Society of America, 301–328.*
Premoli Silva, I., Verga, D., 2004. Practical Manual of Cretaceous Planktonic Foraminifera. In: Verga, D., Rettori, R. (Eds.), Internationl School on Planktonic Foraminifera, 3 Course, Cretaceous. Universities of Perugia and Milan, Tipografia Pontefelcino, Perugia, Italy.
Rami, M., Reza Vaziri, M., Taherpour Khalil Abad, Abolfazl Hosseini, S., Carevic, I., Allameh, M., 2012. Microbiostratigraphy of the Lower Cretaceous strata from the Bararig Mountain, SE Iran. *Revista mexicana de ciencias geológicas*, *29*(1), 63–75.
Rasser, M. W., Scheibner, C., Mutti, M., 2005. A paleoenvironmental standard section for Early Ilerdian tropical carbonate factories (Corbieres, France; Pyrenees, Spain). *Facies*, *51*, 218–232.
Reijmer, J. J., 2021. Marine carbonate factories: Review and update. *Sedimentology*, *68*, 1729–1796.
Renema, W., 2007. Fauna development of larger foraminifera in the Cenozoic of Southeast Asia. In: Renema, W. (Ed.), Biogeography, Time and Place: distributions, barriers and islands, vol. 29. Springer, Netherlands, pp. 179–215.
Renema, W., and Troelstra, S. R., 2001. Larger foraminifera distribution on a mesotrophic carbonate shelf in SW Sulawesi (Indonesia). *Palaeogeography, Palaeoclimatology, Palaeoecology*, *175*, 125–146.
Renema, W., 2019. Large benthic foraminifera in low-light environments. In: Loya, Y., Puglise, K. A., Bridge, T. C. L., (eds.). Mesophotic coral ecosystems. Springer, New York, 553–561.
Reolid, M., Nagy, J., Rodríguez-Tovar, F. J., Olóriz, F., 2008a. Foraminiferal assemblages as palaeoenvironmental bioindicators in Late Jurassic epicontinental platforms: relation with trophic conditions. *Acta Palaeontologica Polonica*, *53*, 705–722.
Reolid, M., Rodríguez-Tovar, F. J., Nagy, J., Olóriz, F., 2008b. Benthic foraminiferal morphogroups of mid to outer shelf environments of the Late Jurassic (Prebetic Zone, southern Spain): Characterization of biofacies and environmental significance. *Palaeogeography, Palaeoclimatology, Palaeoecology*, *261*, 280–299.
Rey, O., Simo, J., Lorente, M. A., 2004. A record of long- and short-term environmental and climatic change during OAE3: La Luna Formation, Late Cretaceous (Santonian-early Campanian), Venezuela. *Sedimentary Geology*, *170*, 85–105.
Rigaud, S., Granier, B., Masse, J. P., 2021. Aragonitic foraminifers: an unsuspected wall diversity. *Journal of Systematic Palaeontology*, *19*(6), 461–488.
Roetzel R., Ćorić S., Galović I. & Rögl F. 2006: Early Miocene (Ottnangian) coastal upwelling conditions along the southeastern scarp of the Bohemian Massif (Parisdorf, Lower Austria, Central Paratethys). *Beiträge zur Paläontologie*, *30*, 387–413.
Romero, J., Caus, E., Rosell, J., 2002. A model for the palaeoenvironmental distribution of larger foraminifera based on late Middle

Eocene deposits on the margin of the South Pyrenean basin (NE Spain). *Palaeogeography, Palaeoclimatology, Palaeoecology, 179*, 43–56.

Rosenthal, Y., and Linsley, B., 2007. Mg/Ca and Sr/Ca paleothermometery from calcareous marine fossils. In: Encyclopedia of Quaternary Sciences, edited by S. A. Elias, Elsevier, London. 1723–1731.

Ruggiero, M. A., Gordon, D. P., Orrell, T. M., Bailly, N., Bourgoin, T., Brusca, R. C., Cavalier-Smith, T., Guiry, M. D., Kirk, P. M., 2015. A higher level classification of all living organisms. *PLoS One, 10*, e0119248.

Salazar-Ramírez, R. W., Herrero, C., Goy, A., 2019. Lower Toarcian ammonite and foraminifera assemblages in the San Miguel de Aguayo Section (Basque-Cantabrian Basin, Spain). *Journal of Iberian Geology*, *46*, 39–60.

Sartorio, D., and Venturini, S. 1988. Southern Tethys biofacies. AGIP, San Donato Milanese, 235 pp.

Schiebel, R., 2002. Planktonic foraminiferal sedimentation and the marine calcite budget. *Global Biogeochemical Cycles*, *16*(4):13–21.

Schiebel, R., and Hemleben, C., 2005. Modern planktic foraminifera. *Paläontologische Zeitschrift*, *79*, 135–148.

Schiebel, R., Smart, S. M., Jentzen, A., Jonkers, L., Morard, R., Meilland, J., Michel, E., Coxall. H. K., Hull, P. M., de Gabriel-Thoron, T., Aze, T., Quillévéré, F., Ren, H., Sigman, D. M., Vonhof, H. B., Martínez-García, A., Kucera, M., Bijma, J., Spero, H. J., Haug, G. H., 2018. Advances in planktonic foraminifer research: New perspectives for paleoceanography. *Revue de Micropaleontologie*, *61*, 113–138.

Scheibner, C., and Speijer, R. P., 2008. Late Paleocene–early Eocene Tethyan carbonate platform evolution—A response to long-and short-term paleoclimatic change. *Earth-Science Reviews*, *90*, 71–102.

Schlager, W., 2003. Benthic carbonate factories of the Phanerozoic. *International Journal of Earth Sciences*, *92*, 445–464.

Schlager, W., 2005. Carbonate Sedimentology and Sequence Stratigraphy. SEPM Concepts in Sedimentology and Paleontology #8. Society of Economic Paleontologists and Mineralogists, 200 pp.

Schlagintweit, F., Rosales, I., Najarro, M., 2016. *Glomospirella cantabrica* n. sp., and other benthic foraminifera from Lower Cretaceous Urgonian-type carbonates of Cantabria, Spain: Biostratigraphic implications. *Geologica Acta: an international earth science journal*, *14*, 113–138.

Scholle, P. A., and Ulmer-Scholle, D., 2003. A Color Guide to the Petrography of Carbonate Rocks: Grains, Textures, Porosity, Diagenesis, *American Association of Petroleum Geologists Memoir 77*, 474.

Schroder-Adams, C., Scott, D. B., Medioli, F. S., 1987. Can smaller benthic foraminifera be ignored in paleoenvironmental analyses? *Journal of Foraminiferal Research*, *17*, 101–105.

Schweizer, M., Pawlowski, J., Kouwenhoven, T., Van Der Zwaan, B., 2009. Molecular phylogeny of common cibicidids and related Rotaliida (Foraminifera) based on small subunit rDNA sequences. *The Journal of Foraminiferal Research*, *39*, 300–315.

Scotese, C. R., Song, H., Mills, B. J. W., van der Meer, D. G., 2021. Phanerozoic paleotemperatures: the earth's changing climate during the last 540 million years. *Earth-Science Reviews*, *215*, 103503.

Sen Gupta, B. K., 1999. Systematics of modern foraminifera. In: Sen Gupta, B.K. (Ed.), Modern Foraminifera. Kluwer Academic, London, 7–36.

Shaviv, N. J., Svensmark, H., Veizer, J., 2023. The Phanerozoic Climate. *Annals of the New York Academy of Sciences*, *1519*, 7–19.

Shi, Y., 2021. Heterochronic origin of spherical fusulinid foraminifera in the late Paleozoic. *Paleobiology*, *47*, 115–133.

Simone, L., Carannante, G., Bassi, D., Cherchi, A., 2012. Rudist-bearing rhodalgal facies in the post-Turonian recovery of peri-Tethyan carbonate systems: a case history from the Nurra region (northwestern Sardinia, Italy). *Geodiversitas*, *34*(1), 167–187.

Sliter, W. V., 1989. Biostratigraphic zonation for Cretaceous planktonic foraminifers examined in thin section. *The Journal of Foraminiferal Research*, *19*, 1–19.

Soto, G., 2025, in press. Utilidad de los Macroforaminíferos en la Industria Petrolera. Caso de Estudio: Cuenca de Falcón. *Visión Tecnológica*. 16 pp.

Soto, G., Ortega, S., Marín, Z., Farías, A., Medina, C., 2025, in press. Estudio Bioestratigráfico de la Unidades Oligo-Miocénicas de la Región Falconiana. *Memorias del II Congreso Venezolano de Geociencias*. Caracas, 20–24 de Marzo 2023. 31 pp.

Spratt, R. M., and Lisiecki, L. E., 2016. A Late Pleistocene sea level stack. *Climate of the Past*, *12*, 1079–1092.

Steuber, T., Löser, H., Mutterlose, J., Parente, M., 2023. Biogeodynamics of Cretaceous marine carbonate production. *Earth-Science Reviews*, 104341.

Takagi, H., Nakamura, Y., Schmidt, C., Kucera, M., Saito, H., Moriya, K., 2025. Two waves of photosymbiosis acquisition in extant planktonic foraminifera explained by ecological incumbency. *The ISME Journal*, *19*(1), wrae244.

Takashima, R, Nishi, H., Huber, B. T., Leckie, R. M., 2006. Greenhouse World and the Mesozoic Ocean. *Oceanography, 19*, 64–75.

Tambutté, S., Holcomb, M., Ferrier-Pagès, C., Reynaud, S., Tambutté, É., Zoccola, D., Allemand, D., 2011. Coral biomineralization: From the gene to the environment. *Journal of Experimental Marine Biology and Ecology*, *408*, 58–78.

Truskowski, I., 2006. Calibración bioestratigráfica y correlación con datos geoquímicos de la Formación Querecual en su localidad tipo. Specialization Dissertation. Central University of Venezuela, 106 pp.

Vachard, D., Gaillot, J., Pille, L., Blazejowski, B., 2006. Problems on Biseriamminoidea, Mississippian-Permian biserially coiled foraminifera. A reappraisal with proposals. *Revista Española Micropaleontología*, *38*, 453–492.

Vachard, D., Pille, L., Gaillot, J., 2010. Palaeozoic Foraminifera: systematics, palaeoecology and responses to global changes. *Revue de micropaléontologie*, *53*, 209–254.

Van der Zwaan, G. J., Jorissen, F. J., Verhallen, P. J. J. M., Von Daniels, C. H., 1986. Atlantic-European Oligocene to recent *Uvigerina*: taxonomy, paleoecology and paleobiogeography. *Utrecht Micropaleontological Bulletins*, *35*, 240 pp.

Van der Zwaan, G. J., Duijnstee, A. I., Den Dulk, M., Ernst, S. R., Jannink, N. T., Kouwenhoven, T. J., 1999. Benthic foraminifers: proxies or problems?: a review of paleocological concepts. *Earth-Science Reviews*, *46*, 213–236.

Vaziri-Moghaddam, H., Kimiagari, M., Taheri, A., 2006. Depositional environment and sequence stratigraphy of the Oligo-Miocene Asmari Formation in SW Iran. *Facies*, *52*, 41–51.

Velić, I., 2007. Stratigraphy and palaeobiogeography of Mesozoic benthic foraminifera of the Karst Dinarides (SE Europe)-part 1. *Geologia Croatica*, *60*(1), 1–60.

Verga, D., and Premoli Silva, I., 2002. Early Cretaceous planktonic foraminifera from the Tethys: the genus *Leupoldina*. *Cretaceous Research*, *23*, 189–212.

Von Daniels, C. H., 1986. *Uvigerina* in the NW European Neogene. *Utrecht Micropaleontological Bulletins*, *35*, 67–119.

Ward, W. C., Keller, G., Stinnesbeck, W., Adatte, T., 1995. Yucatán subsurface stratigraphy: Implications and constraints for the Chicxulub impact. *Geology*, *23*, 873–876.

Westerhold. T., Marwan, N., Drury, A.J., Liebrand, D., et al, 2020. An astronomically dated record of Earth's climate and its predictability over the last 66 million years, *Science*, *369*, 1383–1387.

Wierzbowski, H., 2021. Advances and challenges in palaeoenvironmental studies based on oxygen isotope composition of skeletal carbonates and phosphates. *Geosciences*, *11*, 419.

Wilson, M. E. J., and Vecsei, A., 2005. The apparent paradox of abundant foramol facies in low latitudes: their environmental significance and effect on platform development. *Earth-Science Reviews*, *69*, 133–168.

Yavari, M., Yazdi, M., Ghalavand, H., Adabi, M. H., 2017. Urgonian Type Microfossils of the Dariyan Formation, from Southwest of Iran (Northeast of Shiraz). *Journal of Sciences, Islamic Republic of Iran*, *28*, 255–265.

Zachos, J., Pagani, M., Sloan, L., Thomas, E., Billups, K., 2001. Trends, rhythms, and aberrations in global climate 65 Ma to present. *science*, *292*, 686–693.

Zachos, J. C., Dickens, G. R., Zeebe, R. E., 2008. An early Cenozoic perspective on greenhouse warming and carbon-cycle dynamics. *Nature*, *451*, 279–283.

7 Scleractinian Corals

Scleractinian (photosymbiotic) corals are sessile, fully marine metazoans. Metazoans comprise multicellular animals that require oxygen for growth and possess cells that differentiate to form tissues (a characteristic of all animals; Peters et al. 2005). Zooxanthellate corals (due to their symbiotic relationship with zooxanthellae microalgae) are hypercalcifiers that dominate $CaCO_3$ precipitation in modern coral reefs (rigid skeletal structures in which corals are the major framework components; Hallock 2015) as well as in associated reef systems and their diverse depositional environments (e.g., lagoons, reef flats; see Montaggioni and Braithwaite 2009). They also play a crucial role in reef communities, which include all calcifying and non-calcifying biota that thrive in direct or indirect symbiosis with coral reefs (see Grassle et al. 1973; Castro 1988).

The variety of coral reefs (barrier reefs, fringing reefs, atolls, bank reefs, coral pinnacles or patch reefs; Montaggioni and Braithwaite 2009) represent some of the most diverse and fascinating ecosystems in the modern oceans. Tiny coral polyps, living in symbiosis with photosynthetic microalgae, build vast limestone reefs that harbor more plant and animal species than any other ecosystem on Earth (Goreau et al. 1979). These grand and complex bioconstructions, build upon countless generations of *in situ* coral growth, with the subordinate contribution of encrusting calcifiers such as coralline algas and accumulated skeletal debris from other calcareous organisms, are arguably the planet's most spectacular and volumetrically largest biomineralization products (Cohen and McConnaughey 2003).

Scleractinian corals (also commonly known as "stony" corals, "hermatypic" corals, or "hexacorals") have been dominant components of the shallow-water, tropical carbonate benthic factories from the Late Oligocene to Present (Budd 2000). The production and preservation of coral reef limestones is intimately connected to the Earth's biogeochemical cycles, especially of carbon, oxygen, nitrogen and phosphorus (Hallock 1997, 2015). Stony corals are regarded as the major marine contributors to the $CaCO_3$ flux in modern, tropical shallow-marine environments (Tambutté et al. 2011).

7.1 Major Groups and Stratigraphic Distribution

Scleractinian corals (Phylum Cnidaria, Class Anthozoa, Subclass Hexacorallia, Order Scleractinia; see Pratlong et al. 2017) are broadly divisible into two groups: the reef builders (and the focus of this chapter), which grow as large and massive colonies in the euphotic (or upper photic) zone and harbor endosymbiotic photosynthesizing dinoflagellates (zooxanthellae, Family Symbiodiniaceae); and the non-reef builders, or azooxanthellate, which more commonly grow as solitary and thus can extend from warm shallow-waters to cold and deep aphotic waters (see Schuhmacher and Zibrowius 1985; Stanley 2003; Scholle and Ulmer-Scholle 2003; Cohen and McConnaughey 2003; Flügel 2010; Budd et al. 2010; Tambutté et al. 2011).

Photosymbiotic aragonitic corals (age range: Middle Triassic to Holocene; e.g., Stanley and Helmle 2010) began to play a major role in reef-building communities during the Late Triassic (Aragonite II) (e.g., Flügel 1982, 2010; Stanton and Flügel 1987; Stanley 2003, 2006). The end-Triassic Extinction was the most severe extinction crisis ever experienced by scleractinian corals (Flügel and Kiessling 2002). Their recover and persistence as reef builders during the Late Jurassic and Early Cretaceous (Calcite II) (e.g., Scott 1988; Phelps et al. 2014; Silva-Tamayo et al. 2016; Carvajal-Chitty

J. A. Méndez Dot, J. Méndez Baamonde, *Atlas of Tropical Carbonates*, https://doi.org/10.1007/978-3-031-25494-9_7

and Rey 2020) is perhaps the most obvious discrepancy between the fossil record of reef builders and changes in the imputed Mg/Ca ratio (Stanley 2006; Steuber et al. 2023).

A decline of scleractinian corals during the late Early Cretaceous (Aptian and Albian) (e.g., Kiessling et al. 2003) was followed by their displacement by rudists (and subordinated oysters) as the dominant carbonate producers in shallow marine settings by the late Albian. This trend lasted throughout the first half of the Late Cretaceous (from the Cenomanian to the early Santonian); although corals did not entirely disappear (see Scott 1995 and references therein; Carannante et al. 1995; Filkorn and Scott 2011). In fact, corals and rudists did co-occur, although Early Cretaceous corals may have preferred slightly deeper environments than rudists (Scott et al. 1990; see also Steuber et al. 2023). In Venezuela, an Early Cretaceous example of the co-occurrence of corals and rudists is the Aptian-Albian Borracha Formation in the EVB (see Carvajal-Chitty and Rey 2020). Such skeletal association is missing in the Cretaceous of the MB. On a global scale, the scarcity of corals in mid-Cretaceous strata have been attributed to the peak of the Mesozoic Calcite II seas, which was characterized by a global calcite phase, i.e., absolute concentration of Ca (~ 25–30 mM) and Mg/Ca molar ratio (≤ 1) in seawater were at their highest and lowest Mesozoic levels, respectively (see Stanley and Hardie 1998, 1999; Stanley 2006; Ries 2009). In the tropical Atlantic Ocean of Demerara Rise, off Suriname, South America (southeast of Venezuela), the maximum upper ocean temperatures (between 33° and 42 ° C) and the peak of estimate partial pressure of atmospheric CO_2 (pCO_2; up to 2400–3500 ppmv) were reached in the Turonian (Bice et al. 2006).

Paleocene and Eocene were times of relatively low abundance of scleractinian corals (Flügel 2010) compared to more abundant LBF and red algae, although they recovered respect to Late Cretaceous times (Kiessling et al. 2003; Pomar et al. 2017). The Late Eocene and Early Oligocene fossil record of aragonitic corals is poor in the Caribbean, the Western Tethys and the Indo-West Pacific, but their recovery started from the Late Oligocene (e.g., Budd 2000; Renema 2007; Perrin and Kiessling 2010), when decreasing CO_2 concentrations in the ocean/atmosphere system and increasing Mg^{2+}/Ca^{2+} ratios in seawater had reached to Neogene levels (Stanley and Hardie 1998; Pearson and Palmer 2000; Zachos et al. 2001). By the Oligocene-Miocene transition, the biogeography of scleractinian corals was gradually changing from a pan-tropical Tethyan Province to three coral reef provinces of the Western Atlantic–Caribbean, Indo-Pacific and Mediterranean (Perrin and Bosellini 2012). From the Late Miocene to Present (Aragonite III), and after a period of global dominance of coralline algae during much of the Miocene (Burdigalian to early Tortonian; see Halfar and Mutti 2005), corals became the dominant reef builders (e.g., Stanley 2003, 2006; Flügel 2010; Ries 2010). See Appendix B for a detailed treatment of the concepts of Calcite and Aragonite Seas.

The regions in which shallow-water, reef-building scleractinian corals are living in modern oceans are restricted to the intertropical Indo-Pacific and western Atlantic-Caribbean provinces (Montaggioni and Braithwaite 2009). Approximately 757 species of reef-building corals are currently reported for the shallow reefs of the Indo-Pacific and 62 for the Atlantic-Caribbean (Veron et al. 2017; see also Muir and Pichon 2019). The former province, hosting the higher coral species diversity in the world, include the Australian Great Barrier, the largest barrier reef in the world (~ 2500 Km long); as well as the "Coral Triangle", a six million Km^2 area [encompassing Sumatra and Java (southern Indonesia) in the west; Sabah (north Indonesia) and the Philippines in the north, and Papua New Guinea and Solomon Islands in the east]. In the Atlantic-Caribbean Province, the major coral reef complexes are mainly restricted to the Caribbean Sea, highlighting the Mesoamerican Barrier Reef (~ 950 km long, encompassing the Yucatán Peninsula in southeast Mexico, Belize, Guatemala and Honduras), the Florida Keys (~ 320 Km of coral reefs along the Florida coast) and the Bahama Platform (including the Great Bahama Bank and Andros Barrier Reef) (see Tucker and Wright 1990; Spalding 2004; Arrivillaga and Arreola 2016).

Present-day Indo-Pacific and western Atlantic-Caribbean provinces share only a few living genera of scleractinians (8 genera according to Montaggioni and Braithwaite 2009), which include modern prolific Caribbean reef builders (e.g., *Acropora*, *Porites*, *Siderastrea*, *Montastraea*, and *Favia*), all common/abundant in modern coral reefs of Venezuela (see Weil, 2003). This separation of coral fauna resulted from a series of extinction events that occurred from the Eocene (Budd and Johnson 1999; Budd 2000). Some of these events were linked to plate tectonic dynamics and oceanographic changes. Reconfiguration of ocean gateways in both high and low latitudes resulted in a flow reversal across the Panama Seaway in the Early Miocene that would have brought colder waters into the Caribbean, leading to a decline and regional extinction of many scleractinian genera (see von der Heydt and Dijkstra 2005; Johnson et al. 2009). The closure of the Tethyan seaway during the Miocene affected the coral fauna mainly in the Mediterranean (or Eastern Atlantic) (Pomar and Hallock 2008). Finally, the number of coral genera that were shared between the Indo-Pacific and western Atlantic-Caribbean provinces fell from about half to one-third during the Plio-Pleistocene extinction event. The latter was linked to the final closure of the Central America (or Panama) seaway, a gradual process that occurred between 13 and 2.6 Ma (Lunt et al. 2008) which was responsible for the loss of a third of the Caribbean corals (e.g., Budd et al. 1994; Collins et al. 1996; Budd and Johnson 1999).

7.2 Skeletal Mineralogy

Unlike extinct Paleozoic corals of the orders Tabulata and Rugosa, which were calcitic and ranged in age from the Ordovician to the end of the Permian, scleractinians secrete the $CaCO_3$ polymorph that corresponds to the seawater chemistry at the time they acquired skeletons. Therefore, the Triassic-Holocene Scleractinia have aragonitic skeletons, matching the seawater chemistry of the Triassic (Aragonite II) (see Stanley 2003; Taylor 2008).

7.3 Ecology and Environment

Of the three geomorphological settings where extensive coral reef development has occurred geologically, only two remain common today (Potts and Jacobs 2000). *Oceanic*: these include the barrier reefs and fringing reefs common around high volcanic islands (e.g., Tahiti, Jamaica); and atolls, offshore barrier and bank reefs found on shallow platforms (e.g., Los Roques, Las Aves, and Chagos archipelagos, Bermuda). They are usually surrounded by deep (> 1000 m) oceanic waters and isolated by great distances from continental land masses. *Continental shelves*: these settings include the fringing reefs around exposed continental margins and high islands (e.g., Morrocoy and San Esteban national parks, Venezuela); the barrier reef systems common along outer shelf breaks, and the innumerable free-standing patch or platform reefs in intermediate locations (e.g., the Great Barrier Reef system). *Epeiric* (*or epicontinental*): this setting is uncommon (or non-existent) on the modern Earth but it was widespread and probably the dominant reefal environment of the Mesozoic and early Cenozoic. Epeiric seas form during quiet tectonic periods when ocean basins become shallower and ocean waters flood the central cratons of continents. Epeiric platforms are extensive, with widths between 100 and 10,000 km, and are shallow, often less than 50 m deep (see Chap. 2 in Flügel 2010).

Coral-zooxanthellae symbiosis dates as back as Middle Triassic (~ 222 Ma). Therefore, the earliest scleractinians had photosymbionts like their modern equivalents (Stanley and Swart 1995; Stanley and Helmle 2010; Frankowiak et al. 2016; Lipps and Stanley Jr. 2016; Kołodziej et al. 2018). Zooxanthellate corals are sessile (permanently attached) mixotrophs, i.e., they have the ability to both feed and photosynthesize (Hallock 2001). They host algal endosymbionts (zooxanthellae) in a relationship that allows the holobiont (the host-symbiont association) to acquire trophic resources by both feeding and photosynthesizing (Pomar et al. 2017). Zooxanthellae are phototrophic dinoflagellates that occur in many common marine invertebrates (Rowan and Powers 1991, 1992). These unicellular algae photosynthesize within the coral tissue providing corals with most of their energy, while the coral hosts in turn live in shallow, clear waters where zooxanthellae have optimal exposure to sunlight for photosynthesis (e.g., Goreau and Goreau 1959; Stanley and van de Schootbrugge 2009). Nutrients are provided to the coral host in their inorganic form, such as NO_3^-, NO_2^-, NH_4^+ and PO_4^{3-}. These are accumulated in the symbiotic zooxanthellae and transformed into organic molecules before being transferred to the coral (Grover et al. 2002, 2003; Godinot et al. 2009; Pomar et al. 2017). Regarding heterotrophy, and like all marine heterotrophic calcifiers, scleractinian corals feed upon tiny planktonic organisms, from bacteria to mesoplankton (phyto- and zooplankton) (see Houlbrèque and Ferrier-Pages 2009; Pomar et al. 2017).

Regarding calcification, reef-building scleractinian corals are benthic hypercalcifiers that have the ability to extract massive concentrations of Ca^{2+} and CO_3^{2-} from seawater and secrete them as $CaCO_3$ skeletons (Stanley and Hardie 1999). The organism responsible for the skeletal formation is the polyp, which is the organic-collagenous anatomical unit beneath the coral calcareous skeleton. It consists of a sac-shaped structure with a central mouth surrounded by a ring of tentacles and a column as the main body (Tambutté et al. 2011). The polyp is supported by a skeletal cup or "calice" on the upper exoskeleton or "corallum" surface (Budd et al. 2010). In normal marine alkaline waters (salinity of ~35‰ and pH around 8.2–8.4), polyps uptake inorganic carbon from HCO_3^-, a process that releases CO_2 (see Furla et al. 2000; Tansik et al. 2015). Meanwhile, photosynthetic symbionts hosted in the polyp tissue uptake inorganic carbon from CO_2, produce energy and O_2, and synthesize and remove substances that accelerate and inhibit calcification, respectively (see Constantz 1986; Tambutté et al. 2011).

Modern zooxanthellate corals thrive massively mainly in the upper photic zone (i.e., they are components of the Photozoan association *sensu* James 1997), which is commonly from 0 m down to about 20–30 m deep, although its lower limit can be defined as the maximum depth of vigorous hermatypic coral growth. In very clear waters, this is about 40–50 m (Pomar 2001), which indeed is the approximate maximum depth of the fore-reef of the eastern barrier reef of Los Roques Archipelago. Photosynthetic enhancement of aragonite precipitation by zooxanthellate corals and calcareous green algae is most advantageous in warm waters that are supersaturated respect to $CaCO_3$ (Hallock 2001). On the other hand, cooler waters (at greater depths and/or at higher latitudes) favor the precipitation of calcareous (photosynthetic) red algae, symbiont-bearing LBF and heterotrophic calcifiers (Hallock 1997). Therefore, coral reefs grow in shallow tropical and subtropical waters between 25° N and S latitudes, except where extended by warm ocean currents (Scholle and Ulmer-Scholle 2003).

Contrary to calcareous green algae, the growth rate of scleractinian corals is thought to have a positive relationship with water motion, hence they are prone to grow in moderate- to high-energy environments. Coral calcification is also favored by warm, crystalline (low turbidity), oligotrophic (nutrient-poor) waters (as high rates of nutrient flux tend to discourage photosymbiosis; Armstrong and Brasier 2005); little sedimentation or terrigenous input; and little seasonal or inter-annual environmental variations (see Potts and Jacobs 2000 and references therein). Although seawater temperature may range from anomalously low (18 °C) and high (above 30 °C), the range of sea surface temperature (SST) of 23° to 29 °C is considered to be optimum for calcification in scleractinian corals (see Veron 1995; Scholle and Ulmer-Scholle 2003; Tambutté et al. 2011).

In the geological past, scleractinian corals had a broader physiological tolerance and thrived under different environmental conditions. Potts and Jacobs (2000) proposed that the Mesozoic origin of scleractinian-zooxanthellae symbiosis occurred in turbid, inshore, meso- to eutrophic habitats. These authors suggested that: (**a**) the subsequent radiations of corals for thriving in clear, oligotrophic waters were facilitated by adaptations for success in turbid habitats that "pre-adapted" these mixotrophs for life in oceanic settings; and (**b**) turbid-zone coastal/inshore areas have provided ecological and evolutionary continuity and refugia for corals during non-reefal periods (e.g., the Eocene and Early Oligocene) when physicochemical conditions have been too extreme for coral reef growth (see also de Oliveira Soares 2020). In addition to turbid habitats (where light intensity in reduced by substantial amounts of suspended organic matter, siliciclastic sediments, or dissolved materials), the bathymetry (depth of the seafloor) of vigorous formation of scleractinian corals has varied geologically. Thus, the common view that coral bioherms and reefs form massively only in the euphotic zone (in many cases up to sea level, e.g., flat reefs), such as modern coral reefs, does not necessarily apply for coral bioconstructions of the past. From a revision of the specialized literature and their study on Late Eocene coral facies in northern Spain (southern Pyrenees), Morsilli et al. (2012) state that zooxanthellate corals thrived in the mesophotic zone at least during the Late Eocene and until the Late Miocene, and in many cases, in terrigenous-dominated marginal habitats and pro-delta settings. It is noteworthy that, in the geological literature, the mesophotic zone is defined as the zone with sufficient light for coral growth, commonly below wave action, ~ 40 to 80 m deep. This hypothesis was reinforced by later works, such as Pomar et al. (2014), a study focused on a Late Oligocene carbonate ramp in southern Italy), or Albert-Villanueva (2016), a study focused on Early Miocene carbonates of the San Luis Formation (FB), Venezuela. Furthermore, Morsilli et al. (2012) reinforces the hypothesis that zooxanthellate corals thriving in mesophotic conditions may have been the norm rather than the exception in Greenhouse/Calcite seas (from the Middle Jurassic to the latest Eocene), as well as during the transitional state toward Icehouse/Aragonite seas (the Eocene-Oligocene boundary).

Over the last 20 years, a new understanding has emerged regarding the capacity of modern photosymbiotic scleractinian corals to thrive in deeper settings within the photic zone. This shift has been driven by worldwide discoveries and documentation of their remarkable biodiversity within mesophotic coral ecosystems (MCEs). These are light-dependent coral ecosystems found at depths ranging from 30–40 down to 150 m in very clear tropical and subtropical waters (see Hinderstein et al. 2010; Kahng et al. 2010; Pyle and Copus 2019). It is noteworthy that aforementioned depth range is considered the mesophotic zone in specialized coral biologist literature, while in specialized carbonate geologist literature, it spans the mesophotic and oligophotic zones (see Appendix A for details on photic zone bathymetric zonation). Roughly, MCEs are dominated by photosymbiotic (zooxanthellate) corals (particularly platy corals); azooxanthellate scleractinian corals (Kahng et al. 2010); sponges, mainly Demospongia and, to a lesser extent, Calcarea (Pomponi et al. 2019); octocorals, especially gorgonians (Benayahu et al. 2019; Sánchez et al. 2019); coralline algae (mainly non-geniculate and rhodoliths; Kahng et al. 2010; Horta et al. 2016; Pyle and Copus 2019; Spalding et al. 2019; Voerman et al. 2023); calcareous green algae (primarily in upper mesophotic zone; Kahng et al. 2010; Spalding et al. 2019); non-calcifying macroalgae (green, red, and brown; Spalding et al. 2019); LBF (primarily nummulitids and amphisteginids; Renema 2019); and antipatharians (or black corals, primarily in the lower mesophotic zone; Bo et al. 2019). Although MCEs account for approximately 80% of potential coral reef habitat worldwide, very little is known about them in comparison to shallow reefs (Pyle and Copus 2019). In Venezuela, research on MCEs is limited. However, these ecosystems have been well documented in several Caribbean localities, including Bonaire and Curaçao (Frade et al. 2019) and Puerto Rico (Appeldoorn et al. 2019).

7.4 Carbonate Factory and Skeletal Grain Associations

Scleractinian corals are components of the Photozoan T-factory of Michel et al. (2019), which includes the Tropical factory of Schlager (2003, 2005) [see also the Photozoan factory of James and Jones (2015)]. They are also secondary or accessory components of the Photo C-factory of Michel et al. (2019). See Appendix A for a detailed treatment of the marine carbonate benthic factories considered in this book.

This section discusses only the skeletal grain associations involving zooxanthellate corals represented in Venezuelan carbonate facies/microfacies, as illustrated by the examples in the photographic set. In Late Oligocene-Early Miocene carbonates of the Perla Limestone (EGVSB), corals are minor components of photozoan carbonates dominated by red algae and LBF (the Photo C-factory), plus heterotrophic calcifiers (echinoderms, bryozoans, balanid barnacles, molluscs) and minor *Halimeda* (see Pomar et al. 2015). In Early Miocene carbonates of the FB, corals are dominant components of photozoan (inner-ramp) microfacies of the San Luis Formation, in association with green algae (chlorozoan lithofacies *sensu* Lees and Buller 1972), LBF and red algae (plus molluscs) (see Albert-Villanueva 2016).

In Los Roques Archipelago, carbonate production is, at platform scale, largely dominated by the photozoan association between corals and green algae (chlorozoan sediments). Nevertheless, the photozoan association between corals and red algae, or coralgal lithofacies *sensu* Pomar et al. (2012), dominates locally in moderate to high-energy environments (e.g., the reef flat). In some lithified Quaternary facies of Aves Island (carbonate basement and beachrocks), scleractinian corals occur as secondary components in most microfacies. Within these limestones, corals (mainly as bioclasts but also as a few isolated colonies in growth position) form photozoan associations, mainly with coralline algae (coralgal lithofacies) and calcareous green algae (chlorozoan lithofacies), as well as photozoan-heterozoan associations with mollusscs (gastropods and bivalves). LBF and echinoids are secondary heterozoan components of these microfacies.

7.5 Morphology and Diagnostic Elements

Corals can be classified into five basic growth forms: solitary, pseudocolonial, uniserial erect, multiserial encrusting, and multiserial erect (Coates and Jackson 1987). In modern settings, large solitary or colonial coral skeletons form bioherms (or *in situ* bioconstructions), which are commonly from centimeter-to-decimeter-scale to hundreds of kilometers in length. Colonial forms can be robust and domal, flat and encrusting (sheets or mounds), branching or erect (plate- or tree-like). These growth forms reflect specific adaptations to environmental factors (Coates and Jackson 1987; Scholle and Ulmer-Scholle 2003).

The basic building blocks of all parts of all coral skeletons are the sclerodermites, consisting of fine aragonite crystals or elongated fibers symmetrically arranged in three dimensional fans about a calcification center. These aragonite fibers are of ~0.05–4 μm in diameter and are preferentially elongated in the *c*-axis direction. They grow as spherulites radiating outwards in bundles (termed fascicles) from the centers of calcification. A coral colony is essentially a collection of the individual skeletons, or corallites, of its resident polyps (Cohen and McConnaughey 2003). A corallite is a cylindrical extension of an individual calice below the corallum surface. Radially arranged vertical partitions within corallites are called "septa"; the central skeletal structure formed by septal fusion is called "columella; extensions of the septa beyond the wall are termed "costa"; the vertical structure surrounding the corallite is the "wall"; while the skeletal tissue between corallites is termed "coenosteum" (Budd et al. 2010).

The great majority of zooxanthellate corals are multiserial with small, highly integrated corallites. In life, polyps are interconnected beneath the skeletal surface of the colony (the corallum) as reflected in their skeletal structure. Morphology varies widely among colonies, differing mainly in the corallite mean size (diameter) and size-frequency distribution; and the average level of integration of corallites, i.e., the degree of "plumbing" between them (Coates and Jackson 1987).

7.6 Preservation as Fossils

In thin sections, coral skeletons are commonly found as fragmented and abraded grains (mm to cm size) because many scleractinians live in high-energy environments. Nevertheless, in many sedimentary deposits they may also be found in growth position, forming dense rock bodies ranging from centimeters to meters in size and even larger Aragonitic composition means that most pre-Pleistocene fossil remains have poorly preserved skeletal microstructures. Once removed from marine environments, they become susceptible to neomorphism (mineralogic inversion of aragonite to LMC), or dissolution followed by subsequent filling of the molds by LMC under subaerial-meteoric or burial conditions. Skeletal elements (corallites and wall structure between them) are, in many cases, recognizable (e.g., radiating or meandroid septal structure), as these are often preserved as spar-filled molds, while their original highly porous structure in commonly filled with micritic sediments (Scholle and Ulmer-Scholle 2003; Flügel 2010).

7.7 Coral Calcification and Ocean Carbon Cycle

In general, from the equations that represent the biochemical reactions involved in coral calcification and CO_2 fixation via photosynthesis, it can be concluded that for each mole of $CaCO_3$ precipitated, 1 mole of CO_2 is released. This is nearly true in freshwater, but the process is reduced in buffered seawater, with the result that only 0.6 moles of CO_2 is liberated per mole of $CaCO_3$ deposited (Wollast et al. 1980; Ware 1991). The rate of CO_2 released by coral reefs at the ecosystem level is dependent on community configuration (coral-dominated vs algal-dominated reefs) and on the percentage of substratum cover. Regarding the contribution of carbonate (calcite and aragonite) produced by marine organisms, traditionally foraminifera, coccolithophores, and stony corals are regarded as the major marine contributors (see Tambutté et al. 2011 and references therein).

7.8 Ocean Acidification, Marine Heatwaves, and Coral Bleaching

It is well established that aragonite saturation state is heavily dependent upon partial pressure of atmospheric carbon dioxide (pCO_2), and high concentrations of these greenhouse gas in the ocean/atmosphere system produce low saturation states of aragonite in shallow tropical seas (Hallock 2005). CO_2, together with CH_4 and NO_2, is the most significant greenhouse gas. In the past century, human activities, especially the combustion of fossil fuels, combined with widespread changes in land use, have resulted in rapidly increasing pCO_2 that are causing the lowering of the pH of surface seawater (e.g., IPCC 2007; Canadell et al. 2021 and references therein), a process known as "ocean acidification" (e.g., Caldeira and Wickett 2003; de Putron et al. 2011; Westphal et al. 2022). It is worth mentioning that ocean acidification indeed results in a reduced saturation states with respect to the three $CaCO_3$ polymorphs (aragonite, HMC, and LMC) that are involved the biological formation of calcareous skeletons and shells (see Westphal et al. 2022 and references therein).

Burning of fossil fuels has added more CO_2 to the atmosphere in 100 years than reef building did in 15,000 years (Hallock 2015), triggering modern anthropogenic global warming. Such increased pCO_2 has induced a decrease in the concentration of carbonate [CO_3^{2-}] in the oceans, as well as in the aragonite saturation state, and in the buffering capacity of seawater, leading to a subsequent increase in the concentration of bicarbonate [HCO_3^-] (see Egleston et al. 2010; de Putron et al. 2011). All these changes may directly affect coral calcification or indirectly affect calcification by affecting either coral metabolism or coral symbionts, and in turn light-enhanced calcification (Tambutté et al. 2011). Present-day global average pCO_2 has increased from a pre-industrial level of ~280 ppm (e.g., Caldeira and Wickett 2003; Hofmann et al. 2009; Hay 2011; Ganopolski et al. 2016) to 429.35 ppm (February 2026; NOAA Global Monitoring Laboratory at Mauna Loa, Hawaii; https://gml.noaa.gov/ccgg/trends/).

Among phenomena triggered by anthropogenic global warming affecting the biosphere, marine heatwaves are intrinsically related to the health of coral reefs. Marine heatwaves are qualitatively defined as discrete prolonged anomalously warm water events in a particular oceanic location or region, which can be caused by a combination of atmospheric and oceanographic processes (see Hobday et al. 2016; for a full review on qualitative and quantitative definitions and drivers of marine heatwaves). These events can lead to severe and persistent impacts on marine ecosystems, from mass mortality of benthic communities, including coral bleaching, changes in phytoplankton blooms, shifts in species composition and geographical distribution, and toxic algal blooms. Marine heatwaves have become more frequent over the twentieth century. Since the 1980s, they have approximately doubled in frequency and have become more intense and longer. This trend is expected to continue, with marine heatwaves on a global scale becoming more frequent by the end of this century (see Fox-Kemper et al. 2021 and references therein). In recent years, the impact of widespread marine heatwaves has resulted in more frequent coral bleaching events (see Liu et al. 2024; Neely et al. 2024; Spady et al. 2026).

Coral bleaching results from thermal stress, causing breakdown of the symbiosis between corals and their photosymbiotic microalgae (zooxanthellae). This leads to the loss of pigments (i.e., whitening and discoloration) and symbionts, leaving corals physiologically and nutritionally compromised. Bleaching has become a recurrent phenomenon in tropical oceans due to anthropogenic global warming and associated increases in ocean temperatures. Since the occurrence of the first documented coral bleaching events in the first half of the 1980s, they have become significantly widespread and more frequent and severe during the last four decades (e.g., Glynn 1984; Goreau and Hayes 1994; Brown 1997; Loya et al. 2001; Douglas 2003; Baker et al. 2008; Heron et al. 2016; Hughes et al. 2017; Eakin et al. 2019, 2022; van Woesik and Kratochwill 2022; Shelesinger and van Woesik 2023; Goreau and Hayes 2024; Liu et al. 2024; Doherty et al. 2025; Montefalcone et al. 2026). Large-scale coral bleaching is triggered by increases in SST. The term "bleaching hotspot" has been coined to describe SST anomalies that approximate or exceed by 1.0 °C or more the SST expected climatologically during the warmest month of the year at the site as the predictor of coral reef bleaching, as seen from NOAA's satellite-derived observations (see Goreau and Hayes 1994, 2024; Goreau et al. 2000). Corals ultimately die if temperature anomalies become too intense (> 2 °C) or persist for prolonged periods (see Mellin et al. 2024).

Mass coral bleaching events were recorded in the Caribbean in 2005 (e.g., Wilkinson and Souter 2008; Villamizar et al. 2008; Eakin et al. 2010), and 2010 (e.g., Alemu and Clement 2014; Bastidas et al. 2012). In Los Roques Archipelago, reef decline and coral mortality during those bleaching events occurred when seawater temperatures remained above 29 °C for at least 3–4 months (Villamizar et al. 2008, 2014; Bastidas et al. 2012). Although systematic monitoring of coral reefs in that archipelago has not been carried out during the last decade due to a lack of funding, coral cover has notably declined. Such widespread coral mortality has been clearly associated with several global-scale bleaching events that occurred in the past 20 years (see Eakin et al. 2019, 2022; Reimer et al. 2024; Spady et al. 2026).

Current pCO_2 are unprecedented in the last 2 Myr. Atmospheric CO_2 reached 350–450 ppmv during the mid-Pliocene (3.3–3.0 Ma), and possibly 1000 ppmv during the Early Eocene (52–48 Ma). Despite periods over the Earth's history during the past 60 Myr when CO_2 concentrations were significantly higher than at present due to natural processes, multiple line of evidences show that the rate at which CO_2 has increased in the atmosphere since 1900 has been faster (Canadell et al. 2021; see also Scotese et al. 2021). Human burning of fossil fuels can release as much CO_2 in centuries as do LIPs over 10^4-10^5 or longer (Kidder and Worsley 2012). This relentless rise in CO_2 concentrations is a major contributor to the ongoing global temperature increase and the main cause of the strengthening ocean acidification (Canadell et al. 2021; Zheng et al. 2023 and references therein). The scenario for coral reefs in the upcoming decades is uncertain and concerning, as ocean acidification, marine heatwaves, and coral bleaching events are connected processes. Climate model simulations of future global warming for the next several decades predict an unavoidable increase in pCO_2 and global temperatures (beyond the 2 °C limit recommended by Paris Accords by year 2100) due to global population growth and energy use per capita (Scotese et al. 2021).

The answer on how scleractinian corals will response to such strengthening environmental conditions might possibly be linked to their Mesozoic origins. Potts and Jacobs (2000) state that similar environmental conditions (elevated SST and irradiance, possibly low alkalinity seawater) that are invoked as causes for coral bleaching today may have been more widespread and probably much more severe in the epeiric seas of the past. Yet it is most likely that ancestors of modern zooxanthellate corals existed in chronically bleached states during the 150–200 Myr of higher atmospheric CO_2 (and presumably less alkaline oceans) and warmer climate for much of the Mesozoic and Paleogene. Those authors suggest that, in modern oceans the most likely reservoirs for genotypes better adapted to warmer, higher CO_2, more acid oceans lie in turbid inshore habitats, rather than in offshore oceanic settings.

Increasing evidence suggests that nearshore coral reefs exposed to elevated natural turbidity (due to high terrestrial sediment input and the presence of particulate organic matter) may be more resilient to global warming impacts and serve as important conservation hotspot. Two of the most studied turbid coral reefs are the Paluma Shoals Reef Complex (19° S), situated on the inshore central Great Barrier Reef Australia, and the Abrolhos Reefs (18°39' S), situated approximately 70 off the coast of Brazil, on the Brazilian continental shelve (Zweifler et al. 2021). The latter are the largest coral reefs in the South Atlantic. Mies et al. (2020) findings show that corals in the Abrolhos Reefs display critical features that make them less susceptible to coral bleaching, highlighting their higher tolerance to turbidity, as nearly 60% of species are found in turbid conditions; and higher tolerance to nutrient enrichment, as nitrate concentration in the South Atlantic is naturally elevated. There are still many unknowns regarding how turbid reefs will respond to future global and local threats, such as sea-level rise, declining ocean pH, and increased storm severity (Zweifler et al. 2021).

7.9 Coral Reefs and Current Sea-Level Rise

In addition to the anthropogenic global warming-associated phenomena discussed in the previous section, coral reefs also face another hazard: sea-level rise. The Earth is currently in an interglacial period (MIS 1; see Lisiecki and Raymo 2005; Spratt and Lisiecki 2016) within a glacial cycle, which means that sea level is rising due to natural eustatic processes, including orbital parameters, ocean thermal expansion, melting of ice sheets and glaciers, and glacial isostatic adjustment. However, anthropogenic global warming has triggered a rapid acceleration in the rate at which the sea level has been rising since the beginning of the "Age of Oil" in the late nineteenth century. Global mean sea level rose faster in the twentieth century than in any prior century over the last three millennia. Since the late 1960s, the rate of the sea level rise has increased from 2.3 mm per year to a current rate of 3.7 mm per year. As of 2018, global average sea level was about 15–25 cm higher than in 1900, and 7–15 cm higher than in 1971. By 2050, sea level is expected to rise an additional 10–25 cm. Beyond 2050, the amount by which sea level will rise is more uncertain (Fox-Kemper et al. 2021).

The accelerated rate of the sea-level rise is causing an increase in the bathymetry of shallow marine environments, which directly impacts light availability for coral reefs and colonies. Consequently, the higher rates of light attenuation with increasing depths will affect zooxanthellate corals reefs in the euphotic zone and coral colonies in the upper mesophotic zone. Over time, this will result in the displacement of coral colonies from the lower mesophotic zone (i.e., the oligophotic zone) into the aphotic zone. This scenario takes into account that the ESL will continue to rise for several thousand years due to natural processes and astronomical cycles. In the case of turbid coral reefs, the sea-level rise will have an even greater impact. This is because surface light is attenuated more rapidly with increasing turbidity, and as a result, turbid reefs exhibit a shallow photic zone that limits the maximum depth range of coral growth (see Sweifler et al. 2021 and references therein).

According to Storlazzi (2020), based on rates of vertical reef accretion in Hawaii and throughout the Pacific, it is unlikely that reefs there and other locations will keep pace with the rising sea level. Their inability to do so will result in subtle but significant changes in depth-controlled factors (e.g., water energy) which, in turn, control the growth of reef types and their morphology (e.g., colony size and shape, reef growth rates, and species diversity). This raises several important questions, such as: what are the sea-level rise projections for coastal and oceanic tropical settings ("the far field", i.e., distant from the major ice sheets) where coral reefs are located? How will coral reefs respond to rapid sea-level rise at a decadal time-scale? Until Earth scientists improve sea-level projections for tropical settings and quantify rates of net carbonate production and reef accretion potential, the impact of the sea-level rise is difficult to predict (Sweifler et al. 2021).

7.10 Geochemical Proxies in Scleractinian Corals

High temporal resolution measurements of $^{18}O/^{16}O$ ($\delta^{18}O$) and/or Sr/Ca, and the coupling of both determinations in scleractinian corals, are widely used as proxies for SST in tropical ocean regions. These proxies also provide information about the surface-hydrologic balance and can resolve the evaporation-precipitation (rainfall) seasonal balance (e.g., Gagan et al. 2000; Marshall and McCulloch 2002; McGregor and Gagan 2003; Saenger et al. 2008; Lough 2010; Lea 2014). Diagenesis restricts the determination of geochemical proxies on aragonitic corals ranging from 0 to 130 ka, as they are more susceptible to suffer dissolution than skeletons of calcitic calcifiers (e.g., SBF and brachiopods), especially under the influence of meteoric waters (Lea 2014).

Aragonitic corals oxygen isotopic composition is invariably depleted relative to equilibrium by ~1–6‰, presumably because of their different biochemical mechanisms of precipitation (vital effects) as well as the influence of symbiotic zooxanthellae (McConnaughey 1989a, b). Due to this offset, generally, coral $\delta^{18}O$ records are not interpreted directly as temperature records but rather as climate records whose variability reflects some combination of temperature and salinity effects. In this context, the use of oxygen isotope paleotemperature equations (e.g., Gagan et al. 2000) is necessary to use coral $\delta^{18}O$ variations as climate proxies (see Lea 2014).

The use of Sr/Ca in corals as paleothermometer requires the consideration of the following two issues: (**a**) growth rate and symbiont activity have a marked influence on coral Sr/Ca (e.g., Cohen et al. 2001, 2002); (**b**) secular shifts in seawater Sr/Ca, with generally higher values during glacial episodes (e.g., Stoll and Schrag 1998). The relationship between coral Sr/Ca and seawater temperature also requires the use of paleotemperature equations. Calibrations are available for a number of coral species, but mostly for species of *Porites* (see Gagan et al. 1998; Lea 2014).

7.11 Scleractinian Corals: Photographic Set and Descriptive Captions

In all photos, IW = image width (the true length of the horizontal axis of the view)

Holocene. Los Roques Archipelago

Acropora cervicornis, commonly known as the "staghorn" coral, showing characteristic branched morphology. Note the tubular walls that grow normally to slightly inclined from the skeletal surface and enclose the corallites. Branches are fragile and easily broken in high wave activity, and yet *in situ* colonies are abundant in areas of relatively high-water turbulence. In life, colonies of brown/bronze color are arborescent, reaching more than 2 m in diameter. *A. cervicornis* is among the most common species of scleractinians. In Los Roques, *A. cervicornis* is among the most dominant reef builders at depths between 3 to 7 m deep.

IW = 9 cm

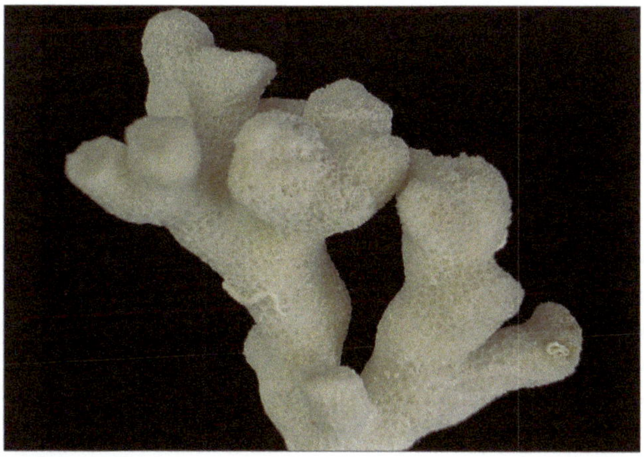

Holocene. Los Roques Archipelago

Porites porites, commonly known as the "hump" coral or "finger" coral, showing characteristic branching growth habit. Individual branches (> 2 cm thick) bifurcate into finger-like lobes. Corallites are very small (less than 2.5 mm in diameter). In life, colonies (of pale bluish-grey, green, purple or pink color) grow into tight clusters and may reach up to 2 m in diameter. *P. porites* dense growths are common in high-energy environments. In Los Roques this species is very common in both shallow sites (< 6.5 m) and deeper sites (7.5–13.5 m) (Villamizar et al. 2003).

IW = 7 cm

Holocene, Los Roques Archipelago

Porites astreoides, commonly known as the "mustard hill" coral (due to its characteristic yellow, yellow-green, or yellow-brown color) showing characteristic shape and closely spaced corallites, thus giving this hexacoral a highly porous appearance. In life, colonies are domal and mounded, and may reach up to 60–80 cm in diameter, being easily differentiated from branching species such as *P. porites, P. furcata* and *P. divaricata* (see Prada et al. 2014). After *Acropora palmata*, *P. astreoides* is the most common scleractinian species in shallow sites in Los Roques.

IW = 12 cm

Holocene. Los Roques Archipelago

Montastraea cavernosa, commonly known as the "great star" coral, showing hemispherical shape derived from its original robust, domal growth habit. Note the cylindrical porous corallites, the central columella and the multiseptate structure. The latter is surrounded by a massive wall. Also note the height of corallite walls (the height of the theca). In life, *M. caversosa* form massive boulders or domes over 1.5 m in diameter exhibiting green, gray, brown or orange colors. This species is among the most common reef building corals in Los Roques, especially in deeper sites.

IW = 14 cm

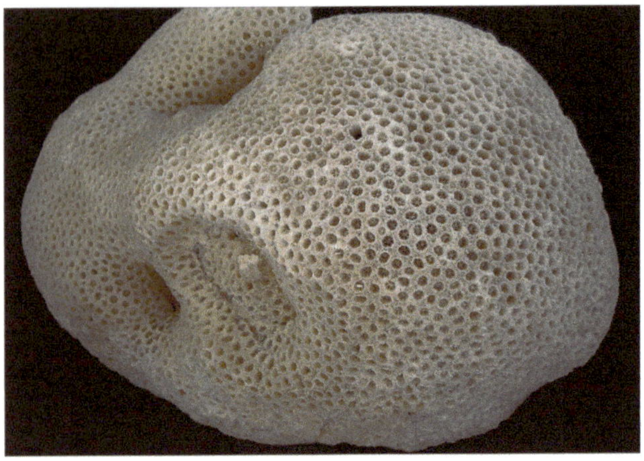

Holocene. Los Roques Archipelago

Orbicella annularis (formerly *Montastraea annularis*; see Budd et al. 2012), commonly known as the "boulder star" coral. In life, colonies (pale green, bronze to gray) are made up of densely grouped nodular bodies, which form massive domal structures up to 2–3 m wide and 1 m high. Corallites are flattened and closely spaced, thus giving colonies a porous appearance. In Los Roques, this species is abundant in both shallow and deeper sites. Together with the genus *Acropora* (*A. palmata* and *A. cervicornis*), *Orcibella* (*O. annularis* and *O. faveolata*) is the dominant reef-building genus in most Caribbean reefs.

IW = 17 cm

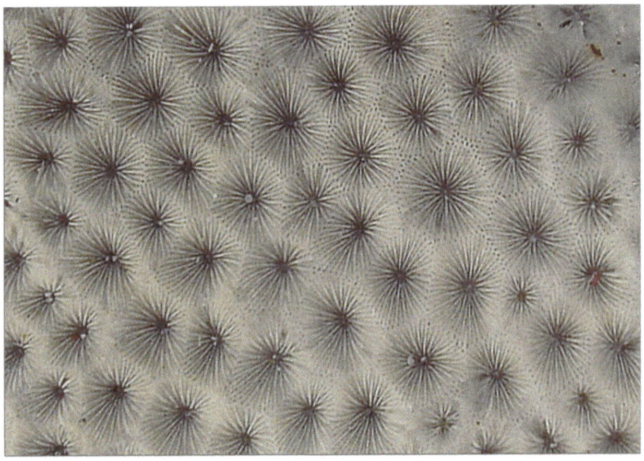

Holocene. Los Roques Archipelago

Close view of *Siderastrea siderea*, commonly known as the "round starlet" coral. Note the highly porous skeletal tissue between the closely spaced corallites and well-developed radiating septa. In life, encrusting colonies of pale brown to pink color grow into large domes or round boulders up to 2 m wide. In the Caribbean Sea, *S. siderea* is associated with other domal colonies, such as *Pseudodiploria strigosa, Favia fragum*, *Porites astreoides* and especially to *Orbicella annularis* (see Foster 1979, 1980).

IW = 4 cm

Holocene. Los Roques Archipelago

Favia fragum, commonly known as the "golfball" coral, due to its small size (commonly <10 cm), hemispherical shape and massive morphology. Porous, tubular corallites are multiseptate, closely spaced and enclosed by thick walls. In life, colonies are light orange-brown to pale green, being commonly found as domes between larger colonies (both domal and branching) at depths from 0 to 25 m. Species of the genus *Favia* are, together with those of the genera *Montastraea* and *Siderastrea*, the oldest Cenozoic coral reef builders in the Caribbean. These genera have existed since the Early Eocene (~ 50 Ma; Montaggioni and Braithwaite 2009).

IW = 9 cm

Holocene. Los Roques Archipelago

Pseudodiploria strigosa (formerly *Diploria strigosa*), commonly known as the "symmetrical brain" coral, exhibiting interlocking massive ridges and cup-shaped, porous, and sinuous wide depressions or valleys, thus giving the colony a meandrine form. The thick columella crosses the depressions longitudinally, while septo-costae extend outside the corallites to the top of the wall. In life, colonies may be of different colors (green, brown, yellowish, blue/gray) and form massive domes up to 2 m in diameter. In Los Roques, *P. strigosa* is among the most common hexacorals in shallow sites.

IW = 12 cm

Holocene. Los Roques Archipelago

Pseudodiploria clivosa (formerly *Diploria clivosa*), commonly known as the "knobby brain" coral. Note the meandrine morphology, similar to *P. strigosa*, but ridges are thinner and sharper and are made up of a tightly woven pattern; while sinuous valleys are less porous and narrower. The absence of trough-like grooves at the top of the ridges helps to differentiate this species from *Diploria labyrinthiformis* (the "grooved brain" coral). In life, *P. clivosa* colonies are green to brown, yellow-brown and, bluish/gray. They form domes up to 1.5 m in diameter, being moderately abundant in both shallow and deeper sites of Los Roques.

IW = 21 cm

Holocene. Los Roques Archipelago

Agaricia agaricites, commonly known as the "lettuce-leaf" coral, due to its rhizoid-lobate, leaf-like morphology. Note the sharp ridges (walls) enclosing the corallites, which are arranged in long meandering valleys. In life, colonies of brown to purple color are encrusting, with thick plates disposed horizontally to the substrate at the base and vertical towards the top, although the disposition of the plates depends on environmental conditions, (e.g., hydrodynamic energy and depth). Plates usually have corallites on both sides. *A. agaricites* is common in both shallow and deeper sites of Los Roques.

IW = 13 cm

Holocene. Los Roques Archipelago

Close view of *Agaricia fragilis*, commonly known as the “fragile saucer” coral, showing meandrine morphology consisting of cup-shaped ridges and corallites aligned concentrically along closely-spaced, continuous valleys. Multiseptate corallites with well-developed costae are clearly visible in the image. Colonies of this species (commonly <20 cm in diameter) are dark brown/bronze to green-yellowish and grow as thin sheets or plates more or less parallel to the substrate. In the Caribbean, this species is mainly found in sheltered cryptic habitats (see Bongaerts et al. 2013; Morais and Santos 2018).

IW = 5 cm

Holocene. Los Roques Archipelago

Meandrina meandrites, commonly known as the “maze” coral, showing conspicuous steep and deep valleys enclosed by thick high walls. These consist of evenly spaced sheet-like septo-costae that divide adjacent valleys and seldom join at the top of the walls. The orientation of the latter varies from parallel to each other and transverse to the valleys, to curved points where straight valleys end and walls converge or bifurcate, thus forming characteristic meandrine morphology. Colonies are typically small (10 to 20 cm) and isolated. *M. meandrites* is moderately common in deeper sites in Los Roques.

IW = 16 cm

Holocene. Los Roques Archipelago

Oculina (probably *O. diffusa*), commonly known as the “ivory bush” coral, exhibiting characteristic branching growth habit. Colonies are short, reaching up to 30 cm in length. Individual cylindrical branches in the image are about 1 to 1.5 cm in diameter. Note the widely spaced cup-like corallites (~ 3–4 mm in diameter), with numerous short radiating septo-costae extending above the corallite walls and giving them a spiky appearance. In life, colonies are yellow-brown. Species of *Oculina* are widely distributed along the Caribbean and the Gulf of Mexico. They inhabit a wide range of depths (~ 6 to 80 m) (see Zarco-Perelló et al. 2013; Reed et al. 2019).

IW = 9 cm

Holocene. Los Roques Archipelago
Madracis decactis, commonly known as the "ten-ray star" coral, showing characteristic branching morphology. Individual branches are short and lobate. Corallites are widely separated, circular in outline, and consisting of ten septa. The decameral arrangement is visible in some of the corallites in the image. In life, colonies are encrusting and form short nodular-lobate masses of pale to dark greenish/brown or yellowish/green color. *M. decactis* commonly grow in sheltered habitats, from shallow to intermediate depths (down to 20 m). This genus is moderately abundant in deeper sites of Los Roques.

IW = 6 cm

Holocene. Los Roques Archipelago
Scolymia (probably *S. lacera*), commonly known as the "scoly" coral, showing characteristic rounded shape. Unlike the majority of scleractinians, species of *Scolymia* are solitary. In life, they exhibit a wide variety of colors (yellow, green, blue, purple, orange). Corallites are large in size, reaching up to 10 cm in diameter. Note the central columella and radial arrangement of numerous, closely spaced septa, some visibly larger. The latter feature, combined with septal teeth (or teeth-like terminations of individual septa), gives these solitary "buttons" a serrate outline.

IW = 8 cm

Holocene. Los Roques Archipelago
Mussa angulosa, commonly known as the "large spiny flower" coral, is the single species of the genus *Mussa*. Note the large size of the three individual corallites in the image (up to 4 cm in diameter). These are tubular and widely spaced, with a concave surface. Septo-costae are radially arranged and evenly spaced, with large prominent teeth and sharp spinose terminations. Colonies, which vary from gray, green, brown or blue color, are flat and form low domes. *M. angulosa* is common in the Caribbean Sea, where inhabits sheltered (cryptic) habitats down to 40–50 m (see Montaggioni and Braithwaite 2009).

IW = 13 cm

Holocene. Los Roques Archipelago

Manicina areolata, commonly known as the "rose" coral, is the only species of the genus *Manicina*. The small elliptical colony exhibits meandrine morphology consisting of broad and deep axial valleys and a continuous columella. Note that septa and costae prolongations from adjacent valleys do not fuse at the top of the walls. Instead, these evenly spaced sheets are separated by a small groove along the crest. Colonies (commonly less than 10 cm wide) of brown-to-pale orange or green-to-gray color, are found either free-living or attached. In the Caribbean, *M. aerolata* is mainly found at intermediate depths (see Johnson 1992; Meesters et al. 2013).

IW = 9 cm

Holocene. Los Roques Archipelago

Eusmilia fastigiata, commonly known as the "smooth flower" coral, is the only species of the genus *Eusmilia*. The base of the colony branches in a series of one, two or three tubular, oval-shaped (or flattened) corallites. These are widely spaced and very large in diameter. Septa are evenly spaced, with septo-costae in two alternating orders, one long and one short. Colonies of yellow-brown to brown or grey color grow into small hemispherical mounds. *E. fastigiata* is found along the Caribbean and the Gulf of Mexico. It is most commonbetween 5 and 30 m deep, especially in back-reefs and reef-lagoon borders (Reyes et al. 2009)

IW = 9 cm

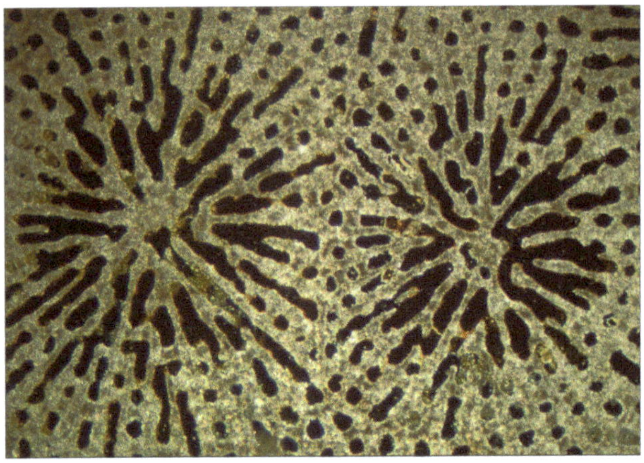

Holocene. Los Roques Archipelago

Closely spaced corallites of the scleractinian coral *Siderastrea siderea* showing partially preserved central columella, radiating multiseptal structure and primary intraparticle porosity. Note the small spherical pores linearly arranged (synapticula) between and surrounding corallites, forming a "synapticulothecate" wall structure, which is characteristic of this genus (see Foster 1979). In general, such diagnostic elements are preserved in modern scleractinians (Holocene in age).

IW = 7.5 mm. CPL

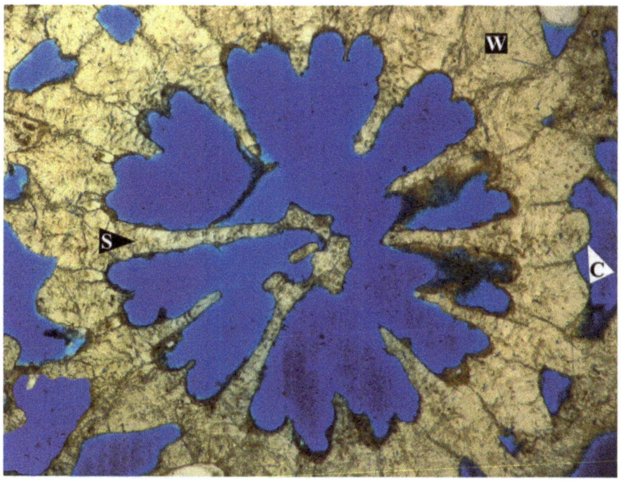

Late Pleistocene. Aves Island

This transverse section of a corallite from the species *Orbicella annularis* exhibits a diameter of 2.6-2.8 mm (the maximum lenght of the intracorallite space), reaching up to 3.5 mm in this species, and shows a well-preserved state of several morphological elements. Among these are the septa (S, radial partitions within the corallite), the wall (W) enclosing the corallite, and the costae (C, conspicuous dentate stuctures extending from the septa across the outer wall).

IW = 2.3 mm. CPL

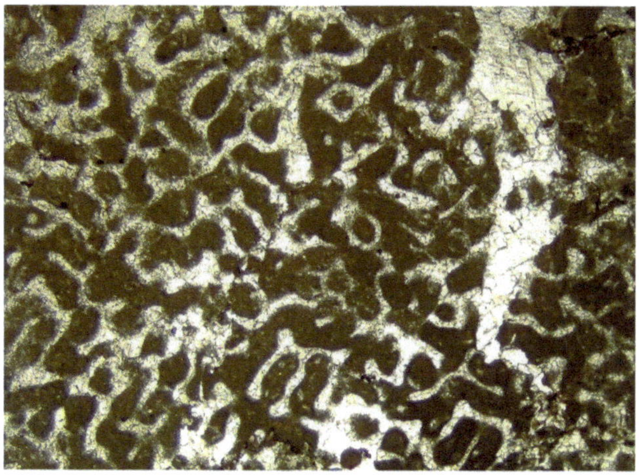

Early Miocene. San Luis Formation. Falcón Basin

This scleractinian coral shows moderatelly well-preserved meandering morphology. The porous structure (probably *Porites* sp.) was filled with micritic sediments, while the originally aragonitic walls underwent neomorphism (mineralogic inversion) and are now preserved as LMC. Note that many walls and meandering passages are floating in a micritic matrix, indicating that the original wall structure experienced localized dissolution. Therefore, the original ratio of calcified skeleton to primary porosity has been modified, resulting in a more porous appearance than the original skeletal framework.

IW = 2.4 mm. PPL

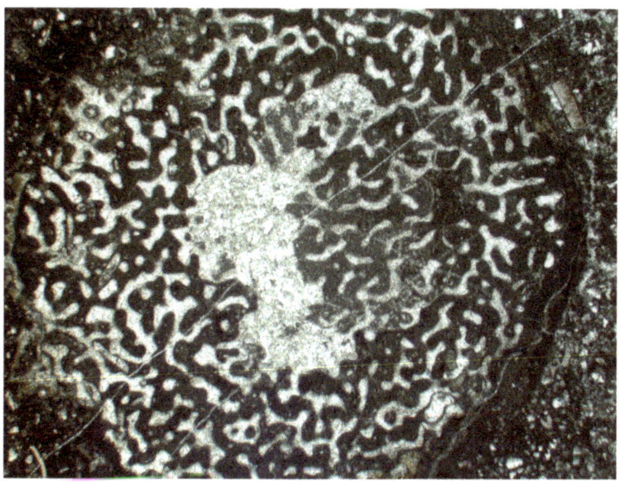

Early Miocene. San Luis Formation. Falcón Basin

Scleractinian coral fragment embedded in a sandy micritic matrix. As in the previous example, the structure is filled with micritic sediments, while the originally aragonitic walls, which exhibit meandering morphology, are neomorphized to LMC. This low magnification allows for the visualization of the grain outline. Note the thick, dark micritic coating giving the grain a rounded shape, suggesting agitation and mechanical abrasion prior to final deposition.

IW = 10 mm. PPL

Photomicrograph courtesy of Eduard Albert-Villanueva

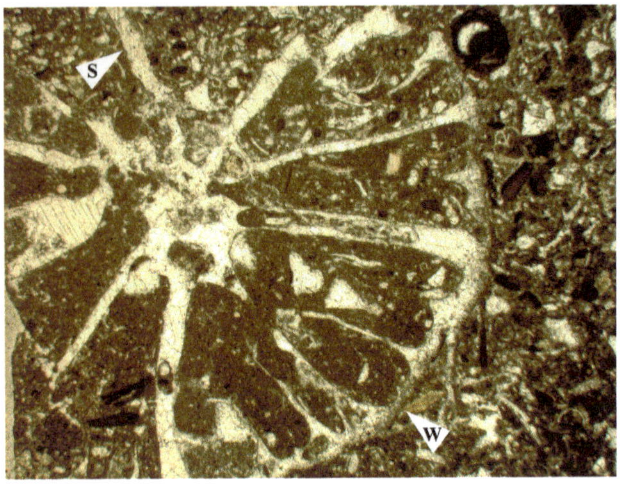

Early Miocene. San Luis Formation. Falcón Basin
Corallite embedded in a bioclastic muddy matrix and exhibiting fair preservation after chemical and mechanical diagenesis. The central columella and thin radiating septa (S) are neomorphized to LMC. Also note the thin outer wall (W) and the spherical shape of the corallite. Contrary to the growth of modern coral reefs in the upper photic zone, the development of coral mounds in the San Luis ramp occurred at depths in the mesophotic zone (Eduard Albert-Villanueva, personal communication 2021).

IW = 10 mm. PPL
Photomicrograph courtesy of Eduard Albert-Villanueva

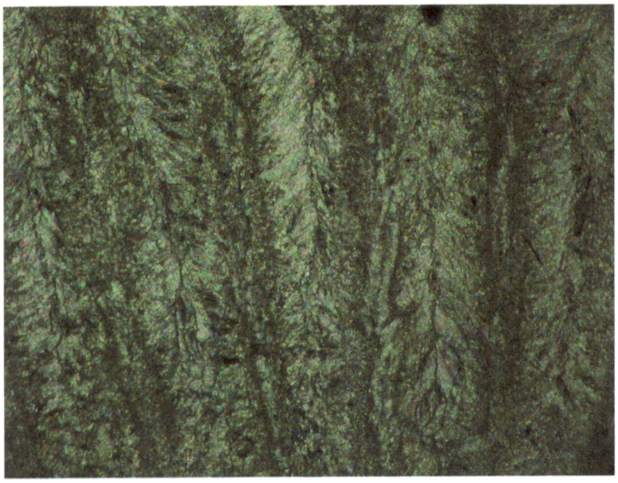

Late Pleistocene. Aves Island
The image shows the intercorallite skeletal tissue of the species *Montastraea cavernosa*. Note the alternation of ridges (costae) and te parallel valleys (intercostal spaces). Individual aragonitic fibers radiate from the centers of calcification, forming clusters or bundles known as "facicles". This well-preserved aragonitic fabric is rarely retained in pre-Pleistocene samples.

IW = 2.3 mm. CPL

References

Albert-Villanueva, E., 2016. Facies y secuencias deposicionales mixtas carbonático-siliciclásticas del Mioceno inferior de la Cuenca de Falcón (noroeste de Venezuela) como modelo exploratorio en el Caribe. Doctoral Dissertation. University of Barcelona. Volumes 1 and 2.

Alemu, J. B., and Clement, Y., 2014. Mass coral bleaching in 2010 in the Southern Caribbean. *PLoS One*, *9*(1), e8382.

Appeldoorn, R. S., Alfaro, M., Ballantine, D. L., Bejarano, I., Ruíz, H. J., Schizas, N. V., Schmidt, W. E., Sherman, C. E., Weil, E., 2019. Puerto Rico. In: Loya, Y., Puglise, K. A., Bridge, T. C. L., (eds.). Mesophotic coral ecosystems. Springer, New York, pp. 111–129.

Armstrong, H. A., and Brasier, M. D., 2005. Microfossils, Second edition. Blackwell Publishing, 296 pp.

Arrivillaga, A., and Arreola, M. E., 2016. The Mesoamerican Barrier Reef System. In: *Marine Transboundary Conservation and Protected Areas*. Routledge, 268–282.

Baker, A. C., Glynn, P. W., Riegl, B., 2008. Climate change and coral reef bleaching: An ecological assessment of long-term impacts, recovery trends and future outlook. *Estuarine, Coastal and Shelf Science*, *80*, 435–471.

Bastidas, C., Bone, D., Cróquer, A., Debrot, D., Garcia, E., Humanes, A., Ramos, R., Rodríguez, S., 2012. Massive hard coral loss after a severe bleaching event in 2010 at Los Roques, Venezuela. *Revista de Biología Tropical*, *60*, 29–37.

Bice, K. L., Birgel, D., Meyers, P. A., Dahl, K. A., Hinrichs, K. U., Norris, R. D., 2006. A multiple proxy and model study of Cretaceous upper ocean temperatures and atmospheric CO_2 concentrations. *Paleoceanography*, *21*. PA2002.

Benayahu, Y., Bridge, T. C., Colin, P. L., Liberman, R., McFadden, C. S., Pizarro, O., Schleyer, M. H., Shoham, E., Reijen, B. T., Weis, M., Tanaka, J., 2019. Octocorals of the Indo-Pacific. In: Loya, Y., Puglise, K. A., Bridge, T. C. L., (eds.). Mesophotic coral ecosystems. Springer, New York, 709–728.

Bo, M., Montgomery, A. D., Opresko, D. M., Wagner, D., Bavestrello, G., 2019. Antipatharians of the Mesophotic Zone: Four Case Studies. In: Loya, Y., Puglise, K. A., Bridge, T. C. L., (eds.). Mesophotic coral ecosystems. Springer, New York, 683–708.

Bongaerts, P., Frade, P. R., Ogier, J. J., Hay, K. B., Van Bleijswijk, J., Englebert, N., Vermeij, M. J. A., Bak, R. P. M, Visser, P. M., Hoegh-Guldberg, O., 2013. Sharing the slope: depth partitioning of agariciid corals and associated Symbiodinium across shallow and mesophotic habitats (2-60 m) on a Caribbean reef. *BMC Evolutionary Biology*, *13*, 1–15.

Brown, B. E., 1997. Coral bleaching: causes and consequences. *Coral Reefs*, *16*, S129–S138.

Budd, A. F., 2000. Diversity and extinction in the Cenozoic history of Caribbean reefs. *Coral Reefs*, *19*, 25–35.

Budd, A. F., and Johnson, K. G., 1999. Origination preceding extinction during late Cenozoic turnover of Caribbean reefs. *Paleobiology*, *25*, 188–200.

Budd, A. F, Stemann, T. A., Johnson K. G., 1994. Stratigraphic distributions of genera and species of Neogene to recent Caribbean reef corals. *Journal of Paleontology*, *68*, 951–977.

Budd, A. F., Romano, S. L., Smith, N. D., Barbeitos, M. S., 2010. Rethinking the phylogeny of scleractinian corals: a review of morphological and molecular data. *Integrative and Comparative Biology*, *50*, 411–427.

Budd, A. F., Fukami, H., Smith, N. D., Knowlton, N., 2012. Taxonomic classification of the reef coral family Mussidae (Cnidaria: Anthozoa: Scleractinia). *Zoological Journal of the Linnean Society*, *166*, 465–529.

Caldeira, K., and Wickett, M. E., 2003). Anthropogenic carbon and ocean pH. *Nature*, *425*, 365.

Canadell, J. G., Monteiro, P. M. S., Costa, M. H., Cotrim da Cunha, L., Cox, P. M., Eliseev, A. V., Henson, S., Ishii, M., Jaccard, S., Koven, C., Lohila, A., Patra, P. K., Piao, S., Rogelj, J., Syampungani, S., Zaehle, S., Zickfeld, K., 2021. Global Carbon and other Biogeochemical Cycles and Feedbacks. In Climate Change 2021: The Physical Science Basis. Contribution of Working Group I to the Sixth Assessment Report of the Intergovernmental Panel on Climate Change [Masson-Delmotte, V., P. Zhai, A. Pirani, S.L. Connors, C. Péan, S. Berger, N. Caud, Y. Chen, L. Goldfarb, M.I. Gomis, M. Huang, K. Leitzell, E. Lonnoy, J.B.R. Matthews, T.K. Maycock, T. Waterfield, O. Yelekçi, R. Yu, and B. Zhou (eds.)]. Cambridge University Press, Cambridge, United Kingdom and New York, NY, USA, pp. 673–816.

Carannante, G., Cherchi, A., Simone, L., 1995. Chlorozoan versus foramol lithofacies in Upper Cretaceous rudist limestones. *Palaeogeography, Palaeoclimatology, Palaeoecology*, *119*, 137–154.

Carvajal-Chitty, H., and Rey, O., 2020. Aportes a la bioestratigrafía del Aptiense–Albiense en la cuenca oriental de Venezuela con base en foraminíferos y nanoplâncton calcáreo. *Caribbean Studies*, *48*, 135–163.

Castro, P., 1988. Animal symbioses in coral reef communities: a review. *Symbiosis*, 5, 161–184.

Cohen, A. L., and McConnaughey, T. A., 2003. Geochemical perspectives on coral mineralization. *Reviews in mineralogy and geochemistry*, *54*, 151–187.

Cohen, A., Layne, G., Hart, S., Lobel, P., 2001. Kinetic control of skeletal Sr/Ca in a symbiotic coral: Implications for the paleotemperature proxy. *Paleoceanography*, *16*, 20–26.

Cohen, A., Owens, K. E., Layne, G. D., Shimizu, N., 2002. The effect of algal symbionts on the accuracy of Sr/Ca paleotemperatures from coral. *Science*, *296*, 331–333.

Collins, L. S., Budd, A. F., Coates, A. G., 1996. Earliest evolution associated with closure of the Tropical American Seaway. *Proceedings of the National Academy of Sciences*, *93*, 6069–6072.

Coates, A. G., and Jackson, J. B., 1987. Clonal growth, algal symbiosis, and reef formation by corals. *Paleobiology*, 363 378.

Constantz, B. R., 1986. Coral skeleton construction; a physiochemically dominated process. *Palaios*, *1*, 152–157.

de Oliveira Soares, M., 2020. Marginal reef paradox: a possible refuge from environmental changes? *Ocean and Coastal Management*, *185*, 105063.

de Putron, S. J., McCorkle, D. C., Cohen, A. L., Dillon, A. B., 2011. The impact of seawater saturation state and bicarbonate ion concentration on calcification by new recruits of two Atlantic corals. *Coral Reefs*, *30*, 321–328.

Doherty, M., L, Johnson, J. V., Goodbody-Gringley, G., 2025. Widespread coral bleaching and mass mortality during the 2023–2024 marine heatwave in Little Cayman. *PLoS One*, *20*(5): e0322636

Douglas, A. E., 2003. Coral bleaching—how and why? *Marine Pollution Bulletin*, *46*, 385–392.

Eakin, C. M., Morgan, J. A., Heron, S. F., Smith, T. B., Liu, G., Alvarez-Filip, L., et al. 2010. Caribbean corals in crisis: record thermal stress, bleaching, and mortality in 2005. *PloS One*, *5*(11), e13969.

Eakin, C. M., Sweatman, H., Brainard, R. E., 2019. The 2014–2017 global-scale coral bleaching event: insights and impacts. *Coral Reefs*, *38*, 539–545.

Eakin, C. M., Devotta, D., Heron, S., Connolly, S., Liu, G., Geiger, E., et al. 2022. The 2014-17 global coral bleaching event: The most severe and widespread coral reef destruction. Research Square. https://doi.org/10.21203/rs.3.rs-1555992/v1

Egleston, E. S., Sabine, C. L., Morel, F. M. M., 2010. Revelle revisited: Buffer factors that quantify the response of ocean chemistry to changes in DIC and alkalinity. *Global Biogeochemical Cycles*, *24*, GB1002.

Filkorn, H. F., and Scott, R. W., 2011. Microfossils, paleoenvironments and biostratigraphy of the Mal Paso Formation (Cretaceous, upper Albian), state of Guerrero, Mexico. *Revista Mexicana de Ciencias Geológicas*, *28*, 175–191.

Flügel, E., 1982. Evolution of Triassic reefs: current concepts and problems. *Facies*, *6*, 297–327.

Flügel, E, 2010. Microfacies of Carbonate Rocks—Analysis, Interpretation and Application. Springer, Heidelberg-Dordrecht-London-New York, 984 pp.

Flügel, E. and Kiessling, W., 2002. Patterns of Phanerozoic reef crises. In Kiessling, W., Flügel, E. & Golonka, J. (eds): Phanerozoic Reef Patterns. SEPM (Society for Sedimentary Geology) Special Publications 72, 691–733.

Foster, A. B., 1979. Phenotypic plasticity in the reef corals *Montastraea annularis* (Ellis & Solander) and *Siderastrea siderea* (Ellis & Solander). *Journal of Experimental Marine Biology and Ecology*, *39*, 25–54.

Foster, A. B., 1980. Environmental variation in skeletal morphology within the Caribbean reef corals *Montastraea annularis* and *Siderastrea siderea*. *Bulletin of Marine Science*, *30*, 678–709.

Fox-Kemper, B., Hewitt, H. T., Xiao, C., Aðalgeirsdóttir, G., Drijfhout, S. S., Edwards, T. L., Golledge, N. R., Hemer, M., Kopp, R. E., Krinner, G., Mix, A., Notz, D., Nowicki, S., Nurhati, I. S., Ruiz, L., Sallée, J. -B., Slangen, A. B. A., Yu, Y., 2021. Ocean, Cryosphere and Sea Level Change. In Climate Change 2021: The Physical Science Basis. Contribution of Working Group I to the Sixth Assessment Report of the Intergovernmental Panel on Climate Change [Masson-Delmotte, V., P. Zhai, A. Pirani, S.L. Connors, C. Péan, S. Berger, N. Caud, Y. Chen, L. Goldfarb, M.I. Gomis, M. Huang, K. Leitzell, E. Lonnoy, J.B.R. Matthews, T.K. Maycock, T. Waterfield, O. Yelekçi, R. Yu, and B. Zhou (eds.)]. Cambridge University Press, Cambridge, United Kingdom and New York, NY, USA, pp. 1211–1362.

Frade, P. R., Bongaerts, P., Baldwin, C. C., Trembanis, A. C., Bak, R. P. M., Vermeij, M. J. A., 2019. Bonaire and Curaçao. In: Loya, Y., Puglise, K. A., Bridge, T. C. L., (eds.). Mesophotic coral ecosystems. Springer, New York, pp. 149–162.

Frankowiak, K., Wang, X.T., Sigman, D.M., Gothmann, A.M., Kitahara, M.V., Mazur, M., Meibom, A., Stolarski, J., 2016. Photosymbiosis and the expansion of shallow-water corals. *Science Advances*, *2* (11), 1–7.

Furla, P., Galgani, I., Durand, I., Allemand, D., 2000. Sources and mechanisms of inorganic carbon transport for coral calcification and photosynthesis. *Journal of Experimental Biology*, *203*, 3445–3457.

Gagan, M. K., Ayliffe, L. K., Hopley, D., Cali, J. A., Mortimer, G. E., Chappell, J., McCulloch, M. T., Head, M. J., 1998. Temperature and surface-ocean water balance of the mid-Holocene tropical western Pacific. *Science 279*, 1014–1018.

Gagan, M. K., Ayliffe, L. K., Beck, J. W., Cole, J. E., Druffel, E. R. M., Dunbar, R. B., Schrag, D. P., 2000. New views of tropical paleoclimates from corals. *Quaternary Science Reviews*, *19*, 45–64.

Ganopolski, A., Winkelmann, R., Schellnhuber, H. J., 2016. Critical insolation–CO_2 relation for diagnosing past and future glacial inception. *Nature*, *529*, 200–203.

Glynn, P. W., 1984. Widespread coral mortality and the 1982–83 El Niño warming event. *Environmental Conservation*, *11*, 133–146.

Godinot, C., Ferrier-Pagès, C., Grover, R., 2009. Kinetics of phosphate uptake by the scleractinian coral *Stylophora pistillata*. *Limnology and Oceanography*, *54*, 1627–1633.

Goreau, T. F., and Goreau, N. I., 1959. The physiology of skeleton formation in corals. II. Calcium deposition by hermatypic corals under various conditions in the reef. *The Biological Bulletin*, *117*, 239–250.

Goreau, T. J., and Hayes, R. L., 1994. Coral bleaching and ocean "hot spots". *Ambio-Journal of Human Environment Research and Management*, *23*(3), 176–180.

Goreau, T. J., and Hayes, R. L., 2024. 2023 Record marine heat waves: coral reef bleaching HotSpot maps reveal global sea surface temperature extremes, coral mortality, and ocean circulation changes. *Oxford Open Climate Change*, *4*(1), kgae005.

Goreau, T. F., Goreau, N. I., Goreau, T. J., 1979. Corals and coral reefs. *Scientific American*, *241*, 124–137.

Goreau, T., McClanahan, T., Hayes, R., Strong, A. L., 2000. Conservation of coral reefs after the 1998 global bleaching event. *Conservation Biology*, *14*(1), 5–15.

Grassle, J. F., Jones, O. A., Endean, R., 1973. Variety in coral reef communities. *Biology and Geology of Coral Reefs*, *2*, 247–270.

Grover, R., Maguer, J. F., Reynaud-Vaganay, S., Ferrier-Pagès, C., 2002. Uptake of ammonium by the scleractinian coral *Stylophora pistillata*: effect of feeding, light, and ammonium concentrations. *Limnology and Oceanography*, *47*, 782–790.

Grover, R., Maguer, J.F., Allemand, D., Ferrier-Pages, C., 2003. Nitrate uptake in the scleractinian coral *Stylophora pistillata*. *Limnology and Oceanography*, *48*, 2266–2274.

Halfar, J., and Mutti, M., 2005. Global dominance of coralline red-algal facies: a response to Miocene oceanographic events. *Geology*, *33*, 481–484.

Hallock, P., 1997. Reefs and reef limestones in Earth history. In: Birkeland, C. (Ed.), Life and Death of Coral Reefs. Chapman and Hall, New York, 13–42.

Hallock, P., 2001. Coral reefs, carbonate sedimentation, nutrients, and global change. In: Stanley, G.D. (Ed.), The History and Sedimentology of Ancient Reef Ecosystems. Kluwer Academic/Plenum Publishers, pp. 387–427.

Hallock, P., 2005. Global change and modern coral reefs: new opportunities to understand shallow-water carbonate depositional processes. *Sedimentary Geology*, *175*, 19–33.

Hallock, P., 2015. Changing influences between life and limestones in earth history. In: Birkeland, C. (Ed.), Coral Reefs in the Anthropocene. Springer, Dordrecht, pp. 17–42.

Hay, W. W., 2011. Can humans force a return to a 'Cretaceous' climate? *Sedimentary Geology*, *235*, 5–26.

Heron, S. F., Maynard, J. A., Van Hooidonk, R., Eakin, C. M., 2016. Warming trends and bleaching stress of the world's coral reefs 1985–2012. *Scientific Reports*, *6*, 1–14.

Hinderstein, L., Marr, J. C. A., Martinez, F. A., Dowgiallo, M. J., Puglise, K. A., Pyle, R. L., Zawada, D. G., Appeldoorn, R., 2010. Theme section on "Mesophotic" coral ecosystems: characterization, ecology, and management. *Coral Reefs*, *29*(2), 247–251.

Hobday, A. J., Alexander, L. V., Perkins, S. E., Smale, D. A., Straub, S. C., Oliver, E. C., Benthuysen, J. A., Burrows, M. T., Donat, M. G., Feng, M., Holbrook, N. J., Moore, P. J., Scannell. H. A., Sen Gupta, A., Wernberg, T., 2016. A hierarchical approach to defining marine heatwaves. *Progress in Oceanography*, *141*, 227–238.

Hofmann, D. J., Butler, J. H., Tans, P. P., 2009. A new look at atmospheric carbon dioxide. *Atmospheric Environment*, *43*, 2084–2086.

Horta, P. A., Riul, P., Amado-Filho, G. M., Gurgel, C. F. D., Berchez, F., Nunes, J. M.C., Scherner, F., Pereira, S., Lotufo, T., Peres, L., 2016. Rhodoliths in Brazil: current knowledge and potential impacts of climate change. *Brazilian Journal of Oceanography*, *64*(SPE2):117–136.

Houlbrèque, F., Ferrier-Pages, C., 2009. Heterotrophy in tropical scleractinian corals. *Biological Reviews*, *84*, 1–17.

Hughes, T. P., Kerry, J. T., Álvarez-Noriega, M., Álvarez-Romero, J. G., Anderson, K. D., et al. 2017. Global warming and recurrent mass bleaching of corals. *Nature*, *543*, 373–77.

IPCC, 2007. Climate change 2007: the physical science. Basis. Contribution of working group I to the fourth assessment report of the Intergovernmental Panel on Climate Change. In: Solomon, S., Qin, D., Manning, M., Chen, Z., Marquis, M., Averyt, K.B., Tignor, M., Miller, H.L. (Eds.), Cambridge University Press, Cambridge, United Kingdom and New-York, NY, USA, 996 pp.

James, N. P., 1997. The cool-water carbonate depositional realm. In: James N. P., Clarke, J., (eds.), Cool-water carbonates, vol 56. SEPM Society for Sedimentary Geology, Tulsa, 1–20.

James, N. P., and Jones, B., 2015. Origin of Carbonate Sedimentary Rocks. John Wiley & Sons, Ltd, Chinchester, 464 pp.

Johnson, K. G., 1992. Population dynamics of a free-living coral: recruitment, growth and survivorship of *Manicina areolata* (Linnaeus) on the Caribbean coast of Panama. *Journal of Experimental Marine Biology and Ecology*, *164*, 171–191.

Johnson, K. G., Sánchez-Villagra, M. R., Aguilera, O. A., 2009. The Oligocene–Miocene transition on coral reefs in the Falcón Basin (NW Venezuela). *Palaios*, *24*, 59–69.

Kahng, S. E., Garcia-Sais, J. R., Spalding, H. L., Brokovich, E., Wagner, D., Weil, E., Hinderstein, L., Toonen, R.J., 2010. Community ecology of mesophotic coral reef ecosystems. Coral Reefs, 29, 255–275.

Kidder, D. L., and Worsley, T. R., 2012. A human-induced hothouse climate. *GSA Today*, 22(2), 4–11.

Kiessling, W., Flügel, E., Golonka, J. A. N., 2003. Patterns of Phanerozoic carbonate platform sedimentation. *Lethaia*, *36*, 195–225.

Kołodziej, B., Salamon, K., Morycowa, E., Szulc, J., Łabaj, M. A., 2018. Platy corals from the Middle Triassic of Upper Silesia, Poland: Implications for photosymbiosis in the first scleractinians. *Palaeogeography, Palaeoclimatology, Palaeoecology*, *490*, 533–545.

Lea, D. W., 2014. Elemental and Isotopic Proxies of Past Ocean Temperatures. *Treatise on Geochemistry: Second Edition. Elsevier Ltd*, 373–397.

Lees, A., and Buller, A. T., 1972. Modern temperate-water and warm-water shelf carbonate sediments contrasted. *Marine Geology*, *13*, M67–M73.

Lisiecki, L. E., and Raymo M. E., 2005. A Pliocene-Pleistocene stack of 57 globally distributed benthic $\delta^{18}O$ records. *Paleoceanography*, *20*, PA1003.

Lipps, J. H., Stanley, G. D. Jr., 2016. Photosymbiosis in the Past and Present Reefs. In: Hubbard, D. K., ed., Coral Reefs at the Crossroads. Coral Reefs of the World 6. Springer Science Publishers, Dordrecht. 47–68.

Liu, B., Foo, S. A., Guan, L., 2024. Optimization of thermal stress thresholds on regional coral bleaching monitoring by satellite measurements of sea surface temperature. *Frontiers in Marine Science*, *11*, 1438087.

Lough, J. M. 2010. Climate records from corals. Wiley Interdisciplinary Reviews: *Climate Change*, *1*, 318–331.

Loya, Y., Sakai, K., Yamazato, K., Nakano, Y., Sambali, H., Van Woesik, R., 2001. Coral bleaching: the winners and the losers. *Ecology letters*, *4*, 122–131.

Lunt, D. J., Valdes, P. J., Haywood, A., Rutt, I. C., 2008. Closure of the Panama Seaway during the Pliocene: implications for climate and Northern Hemisphere glaciation. *Climate Dynamics*, *30*, 1–18.

Marshall, J. F., and McCulloch, M. T., 2002. An assessment of the Sr/Ca ratio in shallow water hermatypic corals as a proxy for sea surface temperature. *Geochimica et Cosmochimica Acta*, *66*, 3263–3280.

McConnaughey, T., 1989a. ^{13}C and ^{18}O isotopic disequilibrium in biological carbonates: I Patterns. *Geochimica et Cosmochimica Acta 53*, 151–162.

McConnaughey, T. A., 1989b. ^{13}C and ^{18}O isotopic disequilibrium in biological carbonates: II. In vitro simulation of kinetic isotopic effects. *Geochimica et Cosmochimica Acta 53*, 163–171.

McGregor, H. V., and Gagan, M. K., 2003. Diagenesis and geochemistry of *Porites* corals from Papua New Guinea: Implications for paleoclimate reconstruction. *Geochimica et Cosmochimica Acta*, *67*, 2147–2156.

Meesters, E. H., Mueller, B., Nugues, M. M., 2013. Caribbean free-living coral species co-occurring deep off the windward coast of Curaçao. *Coral Reefs*, *32*, 109–109.

Mellin, C., Brown, S., Cantin, N., Klein-Salas, E., Mouillot, D., Heron, S. F., Fordham, D. A., 2024. Cumulative risk of future bleaching for the world's coral reefs. *Science Advances*, *10*, eadn9660.

Michel, J., Laugié, M., Pohl, A., Lanteaume, C., Masse, J. P., Donnadieu, Y., Borgomano, J., 2019. Marine carbonate factories: a global model of carbonate platform distribution. *International Journal of Earth Sciences*, *108*, 1773–1792.

Mies, M., Francini-Filho, R. B., Zilberberg, C., Garrido, A. G., Longo, G. O., Laurentino, E., Güth, A. Z., Sumida, P. Y. G., Banha, T. N. S., 2020. South Atlantic coral reefs are major global warming refugia and less susceptible to bleaching. *Frontiers in Marine Science*, *7*, 514.

Montefalcone, M., Bianchi, C. N., Morri, C., Ahmed, H., Pancrazi, I., 2026. Outcomes of the fourth global coral bleaching (2023–2024) in the Maldives. *Coral Reefs*, 1–17.

Montaggioni, L., Braithwaite, C. J. R., 2009. Quaternary Coral Reef Systems: History, Development Processes and Controlling Factors, 5. Elsevier, Amsterdam. 520 pp.

Morais, J., and Santos, B. A., 2018. Limited potential of deep reefs to serve as refuges for tropical Southwestern Atlantic corals. *Ecosphere*, *9*, e02281.

Morsilli, M., Bosellini, F. R., Pomar, L., Hallock, P., Aurell, M., Papazzoni, C. A., 2012. Mesophotic coral buildups in a prodelta setting (late Eocene, southern Pyrenees, Spain): a mixed carbonate–siliciclastic system. *Sedimentology*, *59*, 766–794.

Muir, P. R., Pichon, M., 2019. Biodiversity of reef-building, scleractinian corals. In: Loya, Y., Puglise, K. A., Bridge, T. C. L., (eds.). Mesophotic coral ecosystems. Springer, New York, pp. 589–620.

Neely, K. L., Nowicki, R. J., Dobler, M. A., Chaparro, A. A., Miller, S. M., Toth, K. A. 2024. Too hot to handle? The impact of the 2023 marine heatwave on Florida Keys coral. *Frontiers in Marine Sciences*, *11*, 1489273.

Pearson, P. N., and Palmer, M. R., 2000. Atmospheric carbon dioxide concentrations over the past 60 million years. *Nature*, *406*, 695–699.

Perrin, C., and Kiessling, W., 2010. Latitudinal trends in Cenozoic reef patterns and their relationship to climate. In: Mutti, M., Piller, W., Betzler, C. (Eds.), Carbonate Systems during the Oligocene–Miocene Climatic Transition. *International Association of Sedimentologists*, *Special Publication 42*, 17–34.

Perrin, C., and Bosellini, F. R., 2012. Paleobiogeography of scleractinian reef corals: changing patterns during the Oligocene–Miocene climatic transition in the Mediterranean. *Earth-Science Reviews*, *111*, 1–24.

Peters, K. E., Walters, C., Moldowan, J. M., 2005. The Biomarker Guide. Biomarkers and isotopes in the environmental and human history. Volume I. Second Edition. Cambridge University Press, 471 pp.

Phelps, R. M., Kerans, C., Loucks, R. G., Da Gama, R. O., Jeremiah, J., Hull, D., 2014. Oceanographic and eustatic control of carbonate platform evolution and sequence stratigraphy on the Cretaceous (Valanginian–Campanian) passive margin, northern Gulf of Mexico. *Sedimentology*, *61*, 461–496.

Pomar, L., 2001. Types of carbonate platforms: a genetic approach. *Basin Research*, *13*, 313–334.

Pomar, L., and Hallock, P., 2008. Carbonate factories: a conundrum in sedimentary geology. *Earth-Science Reviews*, *87*(3–4), 134–169.

Pomar, L., Bassant, P., Brandano, M., Ruchonnet, C., Janson, X., 2012. Impact of carbonate producing biota on platform architecture: insights from Miocene examples of the Mediterranean region. *Earth-Science Reviews*, *113*, 186–211.

Pomar, L., Mateu-Vicens, G., Morsilli, M., Brandano, M., 2014. Carbonate ramp evolution during the late Oligocene (Chattian), Salento Peninsula, southern Italy. *Palaeogeography, Palaeoclimatology, Palaeoecology*, *404*, 109–132.

Pomar, L., Esteban, M., Martinez, W., Espino, D., Castillo de Ott, V., Benkovics, L., Castro Leyva, T., 2015. Oligocene-Miocene carbonates of the Perla field, offshore Venezuela: Depositional model and facies architecture, In: C. Bartolini and P. Mann, eds., Petroleum geology and potential of the Colombian Caribbean Margin: *American Association of Petroleum Geologists*, *Memoir 108*, 647–674.

Pomar, L., Baceta, J. I., Hallock, P., Mateu-Vicens, G., Basso, D., 2017. Reef building and carbonate production modes in the west-central Tethys during the Cenozoic. *Marine and Petroleum Geology*, *83*, 261–304.

Pomponi, S. A., Diaz, M. C., Van Soest, R. W. M., Bell, L. J., Busutil, L., Gochfeld, D. J., Kelly, M., Slattery, M., 2019. Sponges. In: Loya, Y., Puglise, K. A., Bridge, T. C. L., (eds.). Mesophotic coral ecosystems. Springer, New York, 563–588.

Potts, D. C. and Jacobs, I. R., 2000. Evolution of reef-building scleractinian corals in turbid environments: a paleo-ecological hypothesis. Proceedings 9th International Coral Reef Symposium, 23–27 October 2000. Vol. 1, 249–254, Bali, Indonesia.

Prada, C., DeBiasse, M. B., Neigel, J. E., Yednock, B., Stake, J. L., Forsman, Z. H., Baums, I. B., Hellberg, M. E., 2014. Genetic species delineation among branching Caribbean Porites corals. *Coral Reefs*, *33*, 1019–1030.

Pratlong, M., Rancurel, C., Pontarotti, P., Aurelle, D., 2017. Monophyly of Anthozoa (Cnidaria): why do nuclear and mitochondrial phylogenies disagree? *Zoologica Scripta*, *46*(3), 363–371.

Pyle, R. L., and Copus, J. M., 2019. Mesophotic coral ecosystems: Introduction and overview. In: Loya, Y., Puglise, K. A., Bridge, T. C. L., (eds.). Mesophotic coral ecosystems. Springer, New York, pp. 3–27.

Reed, J. K., Farrington, S., David, A., Harter, S., Pomponi S., Diaz, M. C., Voss, J. D., Spring, K. D., Hine, A. C., Kourfalou, V., Smith, R. H., Vaz, A. C., Paris, C. B., Hanisak, M. D., 2019. Pulley Ridge, Gulf of Mexico, USA. In: Loya Y, Puglise KA, Bridge TCL (eds.) Mesophotic coral ecosystems. Springer, New York, pp. 57–69.

Renema, W., 2007. Fauna development of larger foraminifera in the Cenozoic of Southeast Asia. In: Renema, W. (Ed.), Biogeography, Time and Place: distributions, barriers and islands, vol. 29. Springer, Netherlands, 179–215.

Renema, W., 2019. Large benthic foraminifera in low-light environments. In: Loya, Y., Puglise, K. A., Bridge, T. C. L., (eds.). Mesophotic coral ecosystems. Springer, New York, 553–561.

Reimer, J. D., Peixoto, R. S., Davies, S. W., Traylor-Knowles, N., Short, M. L., Cabral- Tena, R. A., et al., 2024. The Fourth Global Coral

Bleaching Event: Where do we go from here? *Coral Reef, 43*(4), 1121–1125
Reyes, J., Santodomingo, N., Cairns, S. D., 2009. Caryophylliidae (Scleractinia) from the Colombian Caribbean. *Zootaxa.*
Ries, J. B., 2009. Effects of secular variation in seawater Mg/Ca ratio (calcite–aragonite seas) on $CaCO_3$ sediment production by the calcareous algae *Halimeda, Penicillus* and *Udotea*–evidence from recent experiments and the geological record. *Terra Nova, 21*, 323–339.
Ries, J. B., 2010. Geological and experimental evidence for secular variation in seawater Mg/Ca (calcite-aragonite seas) and its effects on marine biological calcification. *Biogeosciences, 7*, 2795–2849.
Rowan, R., and Powers, D. A., 1991. Molecular genetic identification of symbiotic dinoflagellates(zooxanthellae). *Marine ecology progress series. Oldendorf, 71*, 65–73.
Rowan, R., and Powers, D. A., 1992. Ribosomal RNA sequences and the diversity of symbiotic dinoflagellates (zooxanthellae). *Proceedings of the National Academy of Sciences, 89*, 3639–3643.
Saenger, C., Cohen, A. L., Oppo, D. W., Hubbard, D., 2008. Interpreting sea surface temperature from strontium/calcium ratios in *Montastrea* corals: Link with growth rate and implications for proxy reconstructions. *Paleoceanography, 23*, PA3102.
Sánchez, J. A., Dueñas, L. F., Rowley, S. J., Gonzalez-Zapata, F. L., Vergara, D. C., Montaño-Salazar, S. M., Calixto-Botía, I., Gómez, C. E., Abeytia, R., Colin, P. L., Cordeiro, R. T. S., Pérez, C. D. 2019. In: Loya, Y., Puglise, K. A., Bridge, T. C. L., (eds.). Mesophotic coral ecosystems. Springer, New York, 729–746.
Schlager, W., 2003. Benthic carbonate factories of the Phanerozoic. *International Journal of Earth Sciences, 92*, 445–464.
Schlager, W., 2005. Carbonate Sedimentology and Sequence Stratigraphy. SEPM Concepts in Sedimentology and Paleontology #8. Society of Economic Paleontologists and Mineralogists, 200 pp.
Scholle, P. A., and Ulmer-Scholle, D., 2003. A Color Guide to the Petrography of Carbonate Rocks: Grains, Textures, Porosity, Diagenesis. *American Association of Petroleum Geologists, Memoir 77*, 474 pp.
Schuhmacher, H., Zibrowius, H., 1985. What is hermatypic? A redefinition of ecological groups in corals and other organisms. *Coral Reefs, 4*, 1–9.
Scotese, C. R., Song, H., Mills, B. J. W., van der Meer, D. G., 2021. Phanerozoic paleotemperatures: the earth's changing climate during the last 540 million years. *Earth-Science Reviews, 215*, 103503.
Scott, R. W., 1988. Evolution of Late Jurassic and Early Cretaceous reef biotas. *Palaios, 3*, 184–193.
Scott, R. W., 1995. Global environmental controls on Cretaceous reefal ecosystems. *Palaeogeography, Palaeoclimatology, Palaeoecology, 119*, 187–199.
Scott, R. W., Fernández-Mendiola, P. A., Gili, E., Simo, A., 1990. Persistence of coral-rudist reefs into the late cretaceous. *Palaios, 5*, 98–110.
Silva-Tamayo, J. C., Ramírez, C., Lara, M., Sial, A. N., Trujillo, D., Salazar, E., 2016. Sedimentology and chemostratigraphy of a Valanginian carbonate succession from the Baja Guajira Basin, northern Colombia. *Brazilian Journal of Geology, 46*, 163–180.
Shlesinger, T., and van Woesik, R., 2023. Oceanic differences in coral-bleaching responses to marine heatwaves. *Science of the Total Environment, 871*, 162113.
Spalding, M. D., 2004. A guide to the coral reefs of the Caribbean, 1st edition, University of California Press, Berkeley, 256 pp.
Spalding, H. L., Amado-Filho, G. M., Bahia, R. G., Ballantine, D. L., Fredericq, S., Leichter, J. J., Nelson, W. A., Slattery, M., Tsuda, R. T., 2019. Macroalgae. In: Loya, Y., Puglise, K. A., Bridge, T. C. L., (eds.). Mesophotic coral ecosystems. Springer, New York, 507–536.
Spratt, R. M., and Lisiecki, L. E., 2016. A Late Pleistocene sea level stack. *Climate of the Past, 12*, 1079–1092.
Stanley, G. D., 2003. The evolution of modern corals and their early history. *Earth-Science Reviews, 60*, 195–225.
Stanley, S. M., 2006. Influence of seawater chemistry on biomineralization throughout Phanerozoic time: Paleontological and experimental evidence. *Palaeogeography, Palaeoclimatology, Palaeoecology, 232*, 214–236.
Stanley, G. D., and Swart, P. K., 1995. Evolution of the coral-zooxanthellae symbiosis during the Triassic: a geochemical approach. *Paleobiology, 21*, 179–199.
Stanley, S. M., and Hardie, L. A., 1998. Secular oscillations in the carbonate mineralogy of reef-building and sediment-producing organisms driven by tectonically forced shifts in seawater chemistry. *Palaeogeography, Palaeoclimatology, Palaeoecology, 144*, 3–19.
Stanley, S. M., and Hardie, L.A., 1999. Hypercalcification: paleontology links plate tectonics and geochemistry to sedimentolgy. *GSA Today, 9*, 1 –7.
Stanley, G. D. and van de Schootbrugge, B., 2009. The evolution of the coral-algal symbiosis. In Coral Bleaching: Patterns, Processes, Causes and Consequences. M. van Oppen & J. Lough, Eds., 7–20, Ecological Studies 205. Springer. Berlin-Heidelberg.
Stanley, G. D., and Helmle, K. P., 2010. Middle Triassic coral growth bands and their implication for photosymbiosis. *Palaios, 25*, 754–763.
Stanton, R. J., and Flügel, E., 1987. Palecology of upper Triassic reefs in the Northern Calcareous Alps: Reef communities. *Facies, 16*, 157–185.
Steuber, T., Löser, H., Mutterlose, J., Parente, M., 2023. Biogeodynamics of Cretaceous marine carbonate production. *Earth-Science Reviews*, 104341.
Spady, B. L., Skirving, W. J., De La Cour, J. L., Geiger, E. F., Liu, G., Hoegh-Guldberg, O., Norrie, A., Heron, S. F., Pomeroy, M. W., Kolodziej. G., Brown, K. T., 2026. The 4th global coral bleaching event: ushering in an era of near-annual bleaching. *Coral Reefs, 29*, 1–6.
Stoll, H. M., and Schrag, D. P., 1998. Effects of Quaternary sea level cycles on strontium in seawater. *Geochimica et Cosmochimica Acta 62*, 1107–1118.
Storlazzi, C. D., 2020. Sea-Level Rise and Climate Change Impacts to Reefs. U.S. Geological Survey, Pacific Coastal and Marine Science Center. Available at: https://www.usgs.gov/centers/pcmsc/science/sea-level-rise-and-climate-change-impacts-reefs
Tambutté, S., Holcomb, M., Ferrier-Pagès, C., Reynaud, S., Tambutté, É., Zoccola, D., Allemand, D., 2011. Coral biomineralization: From the gene to the environment. *Journal of Experimental Marine Biology and Ecology, 408*, 58–78.
Tansik, A. L., Fitt, W. K., Hopkinson, B. M., 2015. External carbonic anhydrase in three Caribbean corals: quantification of activity and role in CO_2 uptake. *Coral Reefs, 34*, 703–713.
Taylor, P. D., 2008. Seawater chemistry, biomineralization and the fossil record of calcareous organisms. In Origin and Evolution of Natural Diversity (eds Okada K, Mawatari SF, Suzuki N, Gautam P). University of Hokkaido, Sapporo, Japan, pp. 21–29.
Tucker, M. E., and Wright, V. P., 1990. Carbonate Sedimentology. Blackwell Publishing. 482 pp.
van Woesik, R., and Kratochwill, C., 2022. A global coral-bleaching database, 1980–2020. *Scientific Data, 9*(1), 20
Veron, J. E. N., 1995. Corals in Space and Time. Sydney: Southwood Press. 321 pp.
Veron, J. E. N., Stafford-Smith, M. G., Turak, E., DeVantier, L. M., 2017. Corals of the world. Accessed at http://www.coralsoftheworld.org/coral_geographic/ interactive_map/?version=0.01
Villamizar, E., Posada, J. M., Gómez, S., 2003. Rapid Assessment of Coral Reefs in the Archipiélago de Los Roques National Park,

Venezuela (Part I: Stony Corals and Algae). In Status of Coral Reefs in the western Atlantic: results of initial Surveys, Atlantic and Gulf Rapid Reef Assessment (AGRRA) Program (J.C. Lang, Ed.), *Atoll Research Bulletin*, *496*, 512–529.

Villamizar, E., Camisotti, H., Rodríguez, B., Pérez, J., Romero, M., 2008. Impacts of the 2005 Caribbean bleaching event at Archipiélago de Los Roques National Park, Venezuela. *Revista de Biología Tropical*, *56*, 255–270.

Villamizar, E., Yranzo, A., González, M., Herrera, A. T., Pérez, J., Camisotti, H., 2014. Diversidad y condición de salud de corales pétreos en algunos arrecifes del Parque Nacional Archipiélago Los Roques, Venezuela. *Acta Biológica Venezuelica*, *34*, 257–279.

Voerman, S. E., Marsh, B. C., Bahia, R. G., Pereira-Filho, G. H., Becker, A. C. F., Amado-Filho, G. M.,Ruseckas, A., Turbull, G. A., Samuel, I. D. W., Burdett. H. L., Burdett, H. L., 2023. Dominance of photo over chromatic acclimation strategies by habitat-forming mesophotic red algae. *Proceedings of the Royal Society B*, *290*, 20231329.

von der Heydt, A., and Dijkstra, H. A., 2005. Flow reorganizations in the Panama Seaway: A cause for the demise of Miocene corals? *Geophysical Research Letters*, *32*(2), L02609.

Ware, J. R., 1991. Coral reefs: sources or sinks of atmospheric CO_2? *Coral Reefs*, *11*, 127–130.

Weil, E. 2003. The corals and coral reefs of Venezuela. – In: Cortés J. (ed.), *Latin American Coral Reefs*. Elsevier, 303–330.

Westphal, H., Ries, J. B., Doo, S. S, 2022. The Effect of Ocean Acidification on Skeletal Structures. *Journal of Marine Science and Engineering*, *10*, 786.

Wilkinson, C. R., Souter, D., 2008. Status of Caribbean coral reefs after bleaching and hurricanes in 2005. Global Coral Reef Monitoring Network, 148 pp.

Wollast, R., Garrels, R.M., Mackenzie, F.T., 1980. Calcite-seawater reactions in ocean surface waters. *American Journal of Science*, *280*, 831–848.

Zachos, J., Pagani, M., Sloan, L., Thomas, E., Billups, K., 2001. Trends, rhythms, and aberrations in global climate 65 Ma to present. *Science*, *292*, 686–693.

Zarco-Perelló, S., Mascaró, M., Garza-Pérez, R., Simoes, N., 2013. Topography and coral community of the Sisal Reefs, Campeche Bank, Yucatán, México. *Hidrobiológica*, *23*, 28–41.

Zheng, J., Zhang, H., Zhang, S., 2023. Comparison of atmospheric carbon dioxide concentrations based on GOSAT, OCO-2 observations and ground-based TCCON data. *Remote Sensing*, *15*, 5172.

Zweifler, A., O'leary, M., Morgan, K., Browne, N. K., 2021. Turbid coral reefs: past, present and future—a review. *Diversity*, *13*, 251.

8 Molluscs

Molluscs (Phylum Mollusca) are among the most abundant marine calcareous invertebrates, being also common in freshwater and terrestrial environments. These metazoans first appeared near the Precambian-Cambrian boundary (Parkhaev, 2017), or in the earliest Cambrian (see Maloof et al. 2010; Porter, 2010; Wanninger and Wollesen, 2019; Murdock, 2020; Li et al. 2023), and have been important bioclastic contributors of limestones throughout the Phanerozoic. The phylum includes about 130,000 described extant species and about 70,000 described fossil species. Estimated total species diversity of the modern malacofauna is almost 200,000, thus approximately 70,000 living species are waiting for their formal description (see Parkhaev, 2017).

As a whole group, these heterotrophic calcifiers are very diverse, not only in morphologic and taxonomic terms, but in skeletal mineralogy, life strategies, ecology and environments. The contribution of molluscs to the global carbon budget (historically overlooked) and to the nitrogen cycle is thought to be of substantial influence in shallow marine and coastal ecosystems. This occurs through calcium carbonate production (i.e., calcification) and benthic respiration (i.e., aerobic/dysaerobic oxidation of organic carbon), and CO_2 release by these two processes; and through ammonium regeneration (as N is limiting for primary production) via organic matter remineralization, respectively (see Chauvaud et al. 2003; Martin et al. 2006; Lejart et al. 2012). Although far less significant than foraminifera, coccolithophores and corals in terms of calcium carbonate production in modern marine settings on a global scale, molluscs are considered major contributors to the $CaCO_3$ flux, especially in neritic settings (Tambutté et al. 2011).

8.1 Major Groups and Stratigraphic Distribution

Molluscs are divided in those that bear a conch or shell (bivalves, gastropods, scaphopods, cephalopods, and monoplacophorans), the Conchifera (molluscs with a "true shell"); and those that do not (chitons and aplacophorans), the Aculifera (plate-shelled and shell-less, respectively) (see Vinther, 2015; Parkhaev, 2017; Wanninger and Wollesen, 2019; Murdock, 2020). The subphylum Conchifera is divided in seven classes (see Parkhaev, 2017). In this chapter we consider four groups of the class Bivalvia (bivalve molluscs: aragonitic bivalves, oysters, rudists, and inoceramids) and the class Gastropoda (univalve molluscs).

Bivalves (Class Bivalvia), also called pelecypods (see Bieler and Mikkelsen, 2006), are among the most abundant and widespread of all molluscs in the geologic record, and numerically the second largest class of molluscs after gastropods. They appeared sometime between the Precambrian-Cambrian boundary and the earliest Cambrian (Aragonite I) (Parkhaev, 2008, 2017; Porter, 2010), and were moderately common during the middle and late Paleozoic (see Miller, 1989; Flügel, 2010; Hsieh et al. 2019); were dominant during the Late Triassic (Aragonite II); after the end-Triassic extinction they recovered and became again common in the mid-Late Jurassic and the Cretaceous (Calcite II) (see Hallam, 1977; Bartok et al. 1981; Jablonski, 1997; McRoberts, 2001; Lockwood, 2003; Fraser et al. 2004; Fraiser and Bottjer, 2007; Flügel, 2010; James and Jones, 2015; Méndez-Dot et al. 2015); remaining common during the Cenozoic (Geister, 1992; Crame, 2002; del Carmen-Perrilliat et al. 2010), including modern oceans (James et al. 1994; Flessa et al. 1996; Best and Kidwell, 2000; Bieler and

J. A. Méndez Dot, J. Méndez Baamonde, *Atlas of Tropical Carbonates*, https://doi.org/10.1007/978-3-031-25494-9_8

Mikkelsen, 2006; Flügel, 2010). In general, the use of the terms "bivalve" and "pelecypod" is generic, as each encompasses several orders and a large number of bivalve families (up to 17 orders and 250 families, according to Nevesskaya, 2008). To avoid ambiguities regarding mineralogically "evolved" bivalve molluscs considered in this chapter (those that first appeared in the Mesozoic), we use the term "aragonitic bivalves" to refer to aragonitic bivalve molluscs that first appeared in the eve of controlled biomineralization and were aragonitic (although some evolved to have bimineralic shells). They have been common throughout the Phanerozoic, but especially during post-Paleozoic times.

Oysters (Class Bivalvia, Order Ostreida, Superfamily Ostreoidea) are bivalve molluscs that appeared during the Aragonite II seas of the Early Triassic (Hautmann et al. 2017). They were abundant and locally dominant throughout the rest of the Mesozoic, especially during the Calcite II seas of the Cretaceous (see Ayyasami, 2006; Brigaud et al. 2008; Korte et al. 2009; Abdelhady and Elewa, 2010; Pöppelreiter et al. 2005). Oysters decreased in abundance during the Paleogene and the Neogene, but maintained a similar paleogeographic distribution compared to the Cretaceous peak. The abundance of these calcitic mollusks has increased during the Holocene (Aragonite III), remaining moderately common in modern oceans (see Guo et al. 2018).

Although we do not include photographic examples of rudists in the photographic section of this chapter, given the importance of this bivalve molluscs in the Cretaceous Period, we will consider them in this discussion. Rudists (Class Bivalvia, Order Hippuritida; suborders Requieniidina and Radiolitidina; Skelton, 2013) are an extinct group of bivalve molluscs (age range: Late Jurassic to latest Cretaceous; e.g., Steuber and Löser, 2000). They were dominant during the Calcite II seas of the Cretaceous (see Steuber et al. 2023), and are widely known as having competitively displaced corals from reef frameworks in many tropical and subtropical shallow carbonate settings during the mid-Cretaceous (see Sartorio and Venturini, 1988; Scott, 1995; Carannante et al. 1995; Gili et al. 1995a,b; Skelton et al. 1997; Stanley and Hardie, 1998, 1999, Aqrawi et al. 1998; Johnson et al. 2001; Steuber, 2002; Stanley, 2006; Ries, 2010; Skelton and Gili, 2012). Nevertheless, "reef" is not the proper word for qualify rudist deposits, since rudists are not considered reef-builders because of the absence of growth fabrics that extended significantly above the sediment surface (Gili et al. 1995a; see also Steuber et al. 2023). In this context, the term "rudist bank" is preferable (see Skelton and Gili, 2025).

Inoceramids (Class Bivalvia, Order Myalinida, Family Inoceramidae; Schneider et al. 2013) are a group of extinct bivalve molluscs. They first appeared in the Early Permian (Aragonite II), but flourished during the Cretaceous Period (Calcite II), peaking in the Late Cretaceous. They became extinct in the mid-Maastrichtian, about 2 Myr before the Cretaceous-Paleogene boundary (see Dhondt, 1992; Henderson, 2004; Cobban et al. 2006; Kauffman et al. 2007; Kumagae et al. 2011).

Like the first bivalves, gastropods (Class Gastropoda) appeared sometime between the Precambrian-Cambrian boundary and the earliest Cambrian (Aragonite I) (Parkhaev, 2007, 2008, 2017; Porter, 2010), representing the largest class of both living and fossil molluscs. Although they are known throughout the Phanerozoic fossil record, they are rarely major rock-forming organisms (Scholle and Ulmer-Scholle, 2003), a trend that is clearly recorded in ancient carbonate facies of Venezuela. In general, the distribution of gastropods in the fossil record is similar to that of the aragonitic bivalves. They have a rich fossil record in the middle and late Paleozoic (from Calcite I to Aragonite II) (Frýda et al. 2008; Frýda, 2013); increased gradually in abundance from the early Mesozoic (the "Mesozoic Marine Revolution"; see Vermeij, 1977; Buatois et al. 2016) and throughout the Paleogene (see Sohl, 1987; Sartorio and Venturini, 1988; Nützel, 2005; Allmon. 2007; Flügel, 2010; James and Jones, 2015; Ferrari et al. 2018); peaking in abundance during the Aragonite III seas of the Cenozoic, including modern oceans (Horowitz and Potter, 1971; Scholle and Ulmer-Scholle, 2003; Frýda, 2013; James and Jones, 2015).

8.2 Skeletal Mineralogy

The main mineral components of $CaCO_3$ mollusc shells are calcite and aragonite. Shell layers are composed of monocrystalline units (prisms, tablets, laths or fibers). In addition, it is well known that controlled biomineralization processes in many mollusc shells are associated with organic components, the diversity of which reflects the taxonomic diversity (Dauphin, 2008).

Manybibalve shells (ancient and modern) are purely aragonitic. Some are bimineralic, having interlayered aragonite and calcite; a few are completely calcitic, having commonly less than 1 mol% Mg). Gastropods are/were predominantly aragonitic but, like aragonitic bivalves, evolved bimineralic shells on numerous occasions. Where present, the calcite layer (rarely exceeding 1 mol% Mg) normally is thicker than the aragonite layer (Scholle and Ulmer-Scholle, 2003; Taylor, 2008). Numerous studies from Cambrian mollusks report both aragonitic and calcitic shells (e.g., Kouchinsky, 2000; Kouchinsky et al. 2012; Vendrasco et al. 2010a,b). Nevertheless, it is widely held that the first molluscan shells were aragonitic (Porter, 2010).

Ancient (Jurassic, Cretaceous and Cenozoic) and modern oysters are entirely LMC in composition (Dauphin, 2003; Marin and Luquet, 2005; Hautmann, 2006a,b; Taylor, 2008). Nevertheless, Hautmann (2001, 2006a) proposed that the absence of morphological details of the shell interior in Triassic oysters is due to an originally aragonitic mineralogy of the inner shell layer, which was dissolved during diagenesis. Shell mineralogical data of suggest that there was a shift in the mineralogy of the inner shell layer of oysters from aragonite to calcite across the Triassic–Jurassic boundary (Hautmann, 2004; Hautmann et al. 2008).

All extinct rudists had inner shell layers composed of aragonite and an outer shell layer of calcite with variable thickness and structures (Skelton and Smith, 2000). They have been grouped into taxa that are aragonite-dominated, calcite dominated, or show no distinct dominance of one polymorph (Steuber, 2002; Steuber et al. 2016). Shells consisting of thick, outer LMC layers dominated during the mid-Cretaceous (Steuber, 2000; Steuber et al. 2023).

The extinct Family Inoceramidae, whose members were common in the Calcite II seas of mid-Cretaceous (e.g. the genus *Inoceramus*), had acquired a largely calcitic shell mineralogy already in the Permian (Aragonite II). Therefore, inoceramids may be cited as an example of a family with a mineralogically preadapted shell, which seemed to have profited from the change of seawater chemistry (see Hautmann, 2006b).

Regarding the organic matter content, fossilized mollusc shells retain small amounts of the original organic material, however, from what is known, most mollusc shells are/were originally organomineral in composition. The organic compounds, or shell matrix (mainly proteins, glycoproteins, acidic polysaccharides, chitin and pigments), are concentrated forming an outer thin cortex surrounding the heavy calcified shell. The mineral phase in most mollusc shells usually represents more than 95% of the weight, and the organic fraction less than 5%. However, this small amount of organic material is sufficient to improve both the resistance of shells to fracture and the preservation of original fabrics (Marin and Luquet, 2005 and references therein; see also Gilbert et al. 2022).

8.3 Ecology and Environment

Molluscs are surface-dwelling heterotrophs, i.e., they capture and consume living and dead organic matter as a source for cellular carbon (Armstrong and Brasier, 2005). Therefore, they flourish mainly in eutrophic (nutrients-rich) to mesotrophic (intermediate nutrient levels) conditions, under which elevated nutrient flux (i.e., NH_4^+, NO_3^-, NO_2^-, PO_4^{3-}), from terrestrial runoff or subsurface sources (e.g., upwelling), promotes primary productivity. This is traduced in plankton densities increase and waters transparency decline (even within the euphotic zone), and hence in the flourishing of heterotrophic calcifiers (e.g., molluscs, echinoderms, ostracods) (Hallock, 2001; Mutti and Hallock, 2003). The latter implies that marine molluscs, as all heterozoan carbonate biota, are light-independent organisms, i.e., they may live at all water depths, depending on limiting factors such as dissolved oxygen, substrate requirements, food supply, competitive displacement, salinity and hydraulic energy (see Pomar, 2001).

As a whole group, molluscs are able to thrive in a wide range of environmental conditions (e.g., high- or low-dissolved oxygen concentrations), of depositional environments (from rocky intertidal and coastal areas to open marine and upper bathyal settings), and of depositional systems (from pure carbonate to mixed carbonate-siliciclastic systems) (e.g., Best and Kidwell, 2000; Gutiérrez et al. 2003; Embry et al. 2010; Fortunato, 2015; Schwarz et al. 2018). In bivalves, a rule of thumb is that many epifaunal types are calcitic whereas many infaunal forms are aragonitic (James and Jones, 2015). Many aragonitic bivalves live infaunally (for avoiding predation) as suspension feeders or as borers. Epifaunal forms, those exposed at the water sediment interface and attached to soft/hard substrates; or those which have cast of their attachments to become free-lying, some with the ability to swim, are also moderately common (Harper and Skelton, 1993; Hautmann, 2006b). Contrary to aragonitic bivalves, most evolved fully calcitic and calcitic-dominated molluscs that became dominant in the "calcite" sea of the Jurassic and the Cretaceous (Calcite II) had epifaunal life strategies (Hautmann, 2006b). There is a broad consensus that oysters live attached to the substrate (e.g., Hautmann, 2001, 2006a, 2006b; Márquez-Aliaga et al. 2005; Checa et al. 2006). They are exceptionally well adapted to sessile filter-feeding life in intertidal and shallow waters (sometimes as reef-builders) where environmental conditions fluctuate widely (Guo et al. 2018).

Rudists were essentially filter-feeders that lived epifaunally, completely or partially exposed at the water sediment interface (Gili et al. 1995a). They are usually divided in three morphological types: recumbents, clingers, and elevators. Recumbents were not attached, but the lower valve was stretched out flush with the substrate. Clingers had their lower valve entirely attached to the substrate (stable sediment, other shells, or hard grounds) over a broad zone and the shells were generally horizontally oriented. Elevators were animals whose large lower valve was wholly involved in upward growth, trying to keep ahead of sediment accumulation (James and Jones, 2015). Although all molluscs are/were heterotrophs, Cretaceous rudist-rich platforms occurred in rather oligotrophic oceanic regions (see Michel et al. 2019; Pohl et al. 2019). Rudists are frequently considered to belong to the photozoan group of carbonate producers (Föllmi, 2012). In fact, a few of these Cretaceous bivalves were possibly photosymbiotic (James and Jones, 2015; Steuber et al. 2023). According to Skelton and Gili (2025), the only well-documented evidence for photosymbiosis in rudists is, to date, is limite to Late Cretaceosu genus *Torreites*.

Extinct inoceramid bivalves lived either infaunally (and semi-infaunally) or epifaunally. While most were definitely bottom dwellers that fed on suspended particulate organic matter, a few forms could have been pseudoplanktonic, attaching themselves to various floating objects. In general, they did not thrive in very shallow marine settings (Dhondt, 1992) and were more common in calcareous chalks or offshore mudstones (but see Kumagae et al. 2011 and references therein). They thrived mainly during sea-level highstand episodes. Therefore, inoceramids are common of organic-rich facies deposited under dysoxic-suboxic conditions, especially during the mid-Cretaceous (see Henderson, 2004; Kauffman et al. 2007; Walliser et al. 2018), including black shales of Venezuela (Crespo de Cabrera et al. 1999; Erlich et al. 1999, 2000; Macsotay et al. 2003; Rey et al. 2004; Truskowski, 2006; Caroprese, 2009; Garbán, 2010; this book).

Most gastropods are infaunal, mobile animals which live as grazers and detritus-feeders or burrowers, many of which indeed evolved from epifaunal forms during the Cretaceous and the Cenozoic. Sedentary semi-infaunal or epifaunal life habits are also common. Most gastropods are more commonly associated with shallow marine environments, being able to thrive in a wide range of salinity conditions. Deep-water forms are less abundant but widely distributed in modern and ancient settings (see Sälgeback and Savazzi, 2006; Vermeij, 2017).

In general, the intensification of drilling predation on molluscs is highly correlated with large-scale trends of infaunalization in mollusc prey taxa. Evidence of predation (i.e., drill holes on shells of metazoans) date as back as early Paleozoic. Predation (i.e., carnivorous predators capturing prey capable of defense or resistance) on aragonitic bivalves and gastropods increased during the middle and late Paleozoic, and became intensified in the late Mesozoic (the Mesozoic Marine Revolution; see Buatois et al. 2016) and the Cenozoic. Aragonitic bivalves and gastropods have evolved a variety of strategies to reduce the effects of predation (mainly by crustaceans), including the evolution of thicker shells and various elements of ornamentation (see Vermeij, 1977; Vermeij et al. 1981; Bush et al. 2007; Schweitzer and Feldmann, 2010; Bush and Bambach, 2011; Buatois et al. 2016).

8.4 Carbonate Factories and Skeletal Grain Associations

Molluscs are components of the Heterozoan factory of Michel et al. (2019), as well as occasional to common secondary components of the Photozoan T- and Photo C-factories of these authors (see Appendix A for a detailed treatment on these carbonate benthic factories). The marine productivity as food is the source of energy driving the development of the Heterozoan factory. It is characterized by components of the Heterozoan Association *sensu* James (1997) (echinoderms, bryozoans, ostracods, brachiopods, serpulids, barnacles (balanids), symbiont-barren benthic foraminifera, many calcareous sponges, and plus or minus red algae), and occasional secondary components of the Photozoan Association *sensu* James (1997) (mainly LBF and calcareous green algae; see also Michel et al. (2018)]. Although more commonly associated to both ancient and modern non-tropical settings, heterozoan carbonates occur at all latitudes.

This section discusses only the skeletal grain associations involving molluscs represented in Venezuelan carbonate facies/microfacies, as illustrated by the examples in the photographic set. In Middle Permian carbonates of the Palmarito Formation (Venezuelan Andes), molluscs are secondary components of both heterozoan microfacies (dominated by crinoids, bryozoans, brachiopods and ostracods) and photozoan microfacies (dominated by green algae) (see Laya and Tucker, 2012). In Early Cretaceous (Barremian, Aptian and Albian) carbonates of the Cogollo Group (MB), the heterozoan association between molluscs (mainly aragonitic bivalves and oysters, plus gastropods) and echinoids is the most common grain association. Secondary contribution of other heterotrophic calcifiers (bryozoans, ostracods, serpulids, smaller miliolids, rudists, plus micritic and glauconitic peloids and oncoids/cortoids) is occasionally significant. The mixed heterozoan-photozoan association between molluscs and light-dependent calcifiers (LBF and calcareous algae) is locally occasional/common.

Depending on the content of molluscs versus other bioclastic components (both heterozoan and photozoan biota), heterozoan-dominated microfacies of the Cogollo Group that are rich in molluscs can be grouped into foramol lithofacies *sensu* Lees and Buller (1972), molechfor lithofacies *sensu* Carannante et al. (1988), or bimol (bivalve-dominated) lithofacies *sensu* Hayton et al. (1995). It is noteworthy that these skeletal grain associations are Cenozoic-biased in origin, but applicable to Cretaceous microfacies (e.g., Carannante et al. 1995; Simone et al. 2012), and not restricted to non-tropical carbonates (see Pomar et al. 2004; Wilson and Vecsei, 2005; Brandano et al. 2009; Klicpera et al. 2015). See Appendix A for definitions of the skeletal grain associations considered in this book.

In Aptian and Albian carbonates cropping out in the EIR (north EVB), the heterozoan-photozoan association between rudists and scleractinian corals dominates the lithology of reefal limestones of the Borracha Formation. It should be noted that Cretaceous rudist-rich platforms are included in the Photozoan T-factory, as they were restricted to shallow-tropical, oligotrophic waters (Michel et al. 2019). In time equivalent facies of the El Cantil Formation, molluscs are associated to both photozoan and heterozoan biota (LBF and green algae, and echinoids, respectively) (see Carvajal-Chitty and Rey, 2020). In mid-Cretaceous black shales of the MB, scarce to moderately common molluscs (mainly inoceramids, and minor oysters) are associated to low-oxygen SBF (e.g., buliminids), and coincided in the final deposit with PF.

In Early Miocene carbonates of the FB, molluscs are secondary components of photozoan-dominated, inner-ramp microfacies of the San Luis Formation (in association with corals, plus calcareous algae and LBF; chlorozoan association *sensu* Lees and Buller, 1972). In time equivalent facies of the Churuguara Formation, these heterotrophs are common components of heterozoan-photozoan microfacies of the inner-proximal ramp, in association with echinoids and LBF (foramol lithofacies) (see Albert-Villanueva, 2016). As for Middle Miocene carbonates of the Capadare Formation, molluscs are also common components of heterozoan-photozoan facies, in association with echinoids, LBF and red algae (foramol lithofacies) (see Lorente, 1978). In Late Oligocene-Early Miocene carbonates of the Perla Limestone (EGVSB), molluscs are secondary components of photozoan-dominated microfacies (consisting mainly of coralline algae and LBF) deposited in the oligophotic (poor-light) zone. In these facies, the contribution of heterozoan biota (echinoids, bryozoans, and balanids) is in many cases significant (see Pomar et al. 2015).

In Los Roques Archipelago, carbonate production in largely dominated by the Photozoan T-factory (being scleractinian corals and green algae the major carbonate-producing organisms). Nevertheless, molluscs (bivalves and gastropods, and minor chitons) are secondary but significant carbonate producers (see Work, 1969; Debrot et al. 2019). They associate mainly with echinoids (heterozoan association), as well as with green algae and scleractinian corals (heterozoan-photozoan association). In Late Pleistocene and Holocene lithified facies of Aves Island (carbonate basement and beachrocks, respectivelly), molluscs are abundant components of mixed photozoan-heterozoan carbonates, although the assemblages are photozoan-dominated. Within these facies, coralline algae are the main photozoan components, followed by calcareous green algae (*Halimeda*), scleractinian corals, and LBF. The heterozoan biota is represented by a secondary abundance of echinoids, while SBF, ostracods, and brachiopods are scarce but present. As for Larga Island (at ~1.5 km away from the shorelines of north-central Venezuela), molluscs inhabit both reef-associated and mangrove-associated environments (oligotrophic conditions/photozoan sediments, and eutrophic conditions/heterozoan sediments, respectively) (see Díaz, 1995; Acosta-Balbas et al. 2014 and references therein; Álvarez-Barco et al. 2019).

8.5 Morphology and Diagnostic Elements

Adult bivalve mollusc shells commonly range between 1 mm and 20–25 cm. Some may reach up to 1.5–2 m in length (e.g., giant inoceramids and oysters). Bivalve shells consist of two valves that can be equal or unequal in shape and size. Shells are joined at a hinge by soft muscle tissue. In addition, all bivalve shells possess a thin, outermost proteinaceous layer termed the periostracum. In aragonitic bivalves, individual valves are commonly thin, elongate and curved, and have well-differentiated ventral and dorsal sides. Specialized groups such as oysters and rudists have robust, thicker shells. The latter commonly exhibit more complex morphologies and aberrant forms (e.g., one massive horn-like valve and the other shaped like a cover plate on the top of the horn). Some bivalve molluscs exhibit a plicated morphology, which is similar to that of brachiopods. Shells of bivalves are layered (double or multilayered), although this layered structure is not always visible in thin section, especially in shells consisting of very thin valves. One of the most fundamental structure of shells consists of an inner lamellar layer and an outer prismatic layer (very common in oysters and inoceramids), but groups show enormous variability of structures. Vesicular prismatic fabrics are common in oysters and rudists. Bimineralic shells commonly exhibit mixed fabrics. Shells are composed of densely spaced calcified fibers, which may exhibit varied growth-line patterns. Some microstructure wall types occurring in shells are: crossed-lamellar, foliated, homogeneous, laminar, nacreous, and prismatic (see Majewske, 1969; Scholle and Ulmer-Scholle, 2003; Flügel, 2010; James and Jones, 2015; for details on morphology of bivalves). Some other morphological details on the four distinct bivalve groups considered in this chapter will be given in the captions of the photomicrographs.

Gastropod univalve shells are characterized by conical tubes with a wide variety of coiling patterns, from low- to high-spired and commonly exhibiting external ornamentation. Calcareous unchambered shells (ostracum) consist of whorls arranged around a central axis or pillar termed columella. In the fossil record the shell sizes range from less than 1 mm to several centimeters to decimeters in length. Most shells are composed of an inner, mostly thicker, aragonitic layer. Aragonitic shell microstructures are mainly crossed-lamellar, foliated, nacreous and laminar. Typical appearance of a gastropod section perpendicular to the axis of coiling is characterized by a "projection" of the wall into the central cavity (Flügel, 2010). In life, shells are coated by the periostracum. The color pattern typical of many modern gastropod shells (colloquially known as snails) is formed by different organic pigments which are limited to this proteinaceous outer layer. Shells are also characterized by an aperture (operculum) that connects the soft-bodied animal with the aquatic environment (see Frýda, 2013).

8.6 Preservation as Fossils

Most ancient bivalve molluscs are commonly preserved as mm-sized fragments, especially in shallow, high-energy settings. Occasionally, bivalve molluscs are preserved as articulated shells, which may reach a few centimeters in length. Modern aragonitic bivalves exhibit well-preserved wall structure and fibrous crystals. Ancient aragonitic shells are mostly preserved as molds, either neomorphized/recrystallized or filled by LMC cements, and can be identified by characteristic thin and curved shapes and valves thickening at the dorsal ending. Calcitic and/or bimineralic bivalve shells (oysters, rudists and inoceramids) commonly preserve wall microstructures, which combined with distinctive shell morphology facilitates their identification.

Gastropods (from mm- to cm-sized) may be preserved as complete shells or as fragmented ones. Like aragonitic bivalves, ancient gastropods are preserved as molds that can be recognized by their characteristic shapes. Oriented sections (longitudinal and transversal) are very useful for recognizing these univalve molluscs.

8.7 Molluscs: Photographic Set and Descriptive Captions

In all photos, IW = image width (the true length of the horizontal axis of the view)

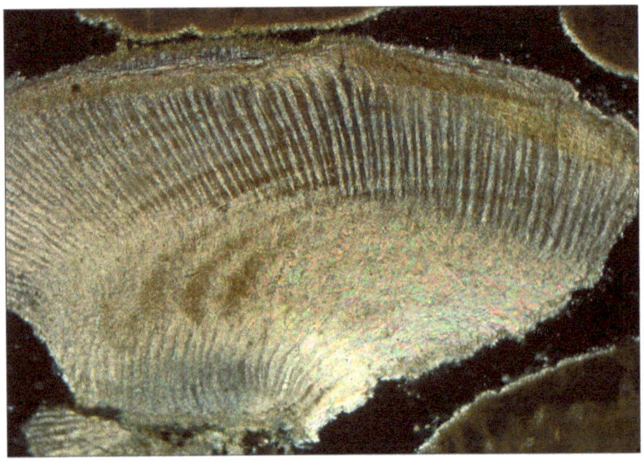

Holocene. Morrocoy National Park
This aragonitic mollusc shell fragment (bivalve or gastropod) shows two types of shell microstructure. The cross-lamellar wall microstructure (above) is composed of parallel prisms oriented perpendicular to, or slightly inclined towards, the shell surface, exhibiting alternation of light and dark extinction bands. This texture grades into a foliated microstructure composed of densely spaced crystals with wavy appearance. Delicate, very thin aragonitic fibrous crystals show a characteristic mixture of colors (brownish-pink-green) when they are viewed under CPL.

IW = 2.4 mm. CPL

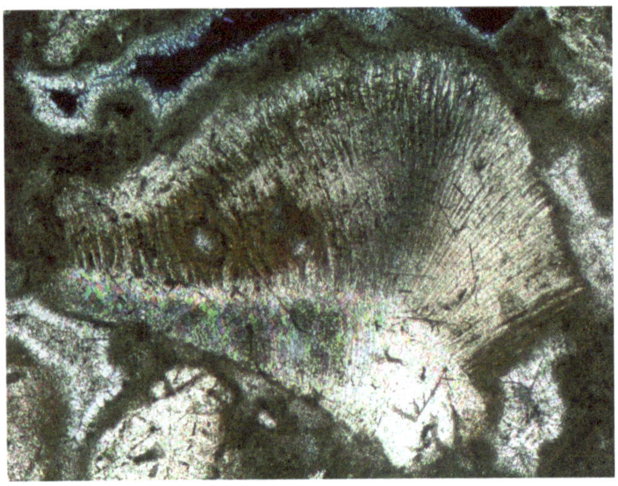

Holocene. Los Roques Archipelago
This is storm-derived mollusc shell fragment became part of a lithified stromatolite structure formed in a hypersaline lagoon. The fragment exhibits a characteristic double-layered wall structure and cross-lamellar microstructure. Calcified fibers are oriented almost perpendicular to the surface of the opposite layer. The thicker layer (above in this view) is composed of larger fibers. The "horn-like" layer (below) is composed of shorter and thinner fibers. The view under CPL enhances the visualization of fibers and their arrangements.

IW = 2.4 mm. CPL

Holocene. Henri Pittier National Park. Aragua State Shoreline

An aragonitic mollusc shell fragment is cemented together with other bioclasts and quartz grains in a beachrock. The shell exhibits a foliated microstructure, with fibers showing wavy extinction and differing in orientation. Like the microstructure shown in the two previous examples (crossed-lamellar), foliated is a common wall microstructure in both bivalves (aragonitic and calcitic) and gastropods. In general, microstructures are not preserved in ancient aragonitic shells due to metastability of this $CaCO_3$ polymorph.

IW = 2.4 mm. CPL

Early Miocene. Churuguara Formation. Falcón Basin

A radial section of a bivalve mollusc shell in a sandy mudstone. Equal valves form a bilaterally symmetric shell that is articulated along a hinge line. In life, the valves are held in position at ventral (V) and dorsal (D) sides by soft and elastic ligament tissue. Both valves show characteristic folding at the dorsal end. The well-preserved state of the valves, without exhibiting features of having undergone neomorphic alterations, suggests an original calcitic composition.

IW = 4.2 mm. CPL

Middle Miocene. Capadare Formation. Falcón Basin

Packstone of aragonitic bivalves. Molluscs grains in the image, originally aragonitic, are preserved as molds filled by coarse-grained, inequigranular LMC. The mineralogical metastability of these grains contrasts with the non-recrystallized micritic matrix, which is composed of LMC microcrystals. Meso-eutrophic (intermediate- to nutrient-rich) conditions are associated with the deposition of this microfacies.

IW = 2.4 mm. PPL

Cretaceous (middle Aptian). Piché Member, Apón Formation. Maracaibo Basin

Another packstone consisting of fragments of aragonitic bivalves. The grains exhibit characteristic elongate and smoothly curved shapes. Thickening at the dorsal (D) side in some of the fragments serves as a diagnostic criterion for identifying this fragmented bivalves. Note the similarities between diagenetically stabilized grains in this example and the previous one, despite a significant age difference (~ 100 Myr), as mm-sized ancient aragonitic bivalves fossilize similarly. Deposition of this microfacies is associated with upper-neritic paleodepths and meso-eutrophic conditions.

IW = 4.2 mm. PPL

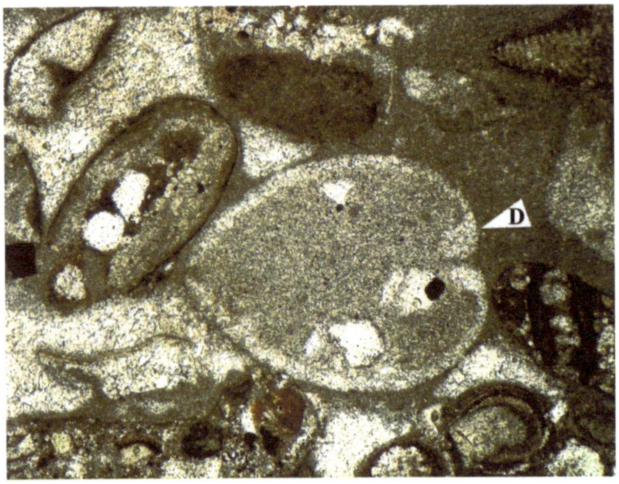

Cretaceous (middle Albian). Lisure Formation. Maracaibo Basin

An originally aragonitic bivalve shell in the center of the image accompanied by non-skeletal and peloidal grains. The valves of the shell held the bilateral-hinge structure, which is uncommon in bivalves with thin and fragile valves, such as this example. Both valves exhibit characteristic thickening at the dorsal (D) side. Synsedimentary micritic infiling and early (near-surface) lithification of the facies prevented mechanical compaction of the shell.

IW = 4.2 mm. PPL

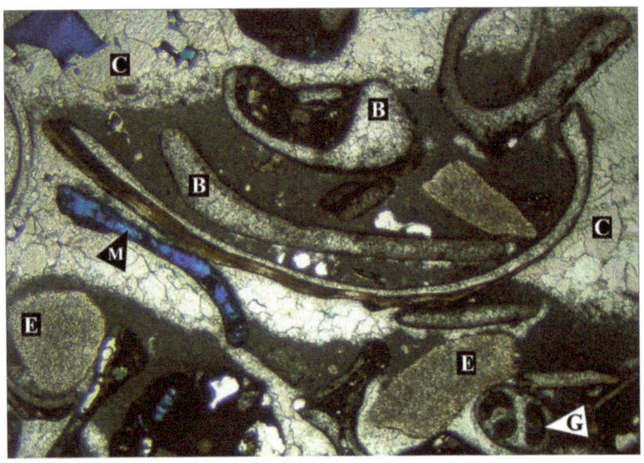

Cretaceous (latest Albian). Maraca Formation. Maracaibo Basin

Large, thin and curved bivalve (B) mollusc grains in this packstone-grainstone are accompanied by echinoid fragments (E) and a gastropod (G) (heterozoan association). This sample exemplifies the dominance of heterotrophic calcifiers in shallow (upper photic zone) facies of the Maraca Formation. A near-surface diagenetic environment is indicated by the dissolution of an aragonitic shell to form a moldic (M) pore and by the precipitation of coarse-grained LMC (C) (mainly blocky calcite) void-filling interparticle pores.

IW = 4.2 mm. PPL

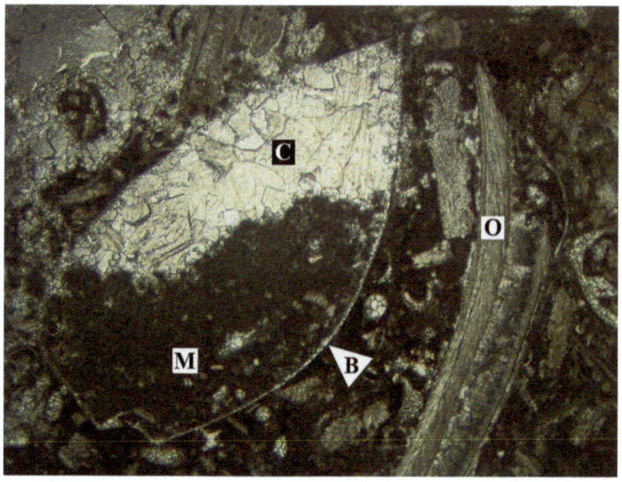

Cretaceous (middle Albian). Lisure Formation. Maracaibo Basin

The bivalve (B) mollusc shell in the image consists of very thin valves, originally aragonitic. Note the geopetal infilling of the shell, consisting of synsedimentary micrite (M) at the bottom and coarse-grained LMC cement (C) at the top, which prevented the shell from crushing during burial. On the right of the image, an oyster (O) shell fragment exhibits well preserved LMC fibers. In general, aragonitic bivalves and calcitic oysters are the most abundant molluscs in shallow Cretaceous carbonate of the MB.

IW = 7.5 mm. PPL

Cretaceous (latest Albian). Maraca Formation. Maracaibo Basin

Side view of an oyster shell fragment in a calcareous sandstone. The LMC composition of the shell is indicated by a well-preserved foliated microstructure, composed of densely-packed calcitic fibers (or lamellae) that show a uniformly oriented, undulated pattern. The view under CPL light highlights the wavy appearance of fibers. Oysters in the Cogollo Group are common in a variety of facies, ranging from carbonate-dominated to less abundant siliciclastic-dominated rock intervals, such as this example.

IW = 4.2 mm. CPL

Cretaceous (middle Albian). Lisure Formation. Maracaibo Basin

A fragment of an oyster shell exhibiting well-preserved LMC fibers. Two different wall fabrics can be differentiated in this example. The prismatic fabric is dark-gray in color, with fibers oriented nearly normal. The apparent darker color of this prismatic fabric is probably due to a higher content of organic components. The foliated fabric is light-beige in color and composed of fibers uniformly oriented at low angle. In Early Cretaceous shallow carbonates of the MB (Barremian to Albian in age), oysters can reach several centimeters in length.

IW = 2.4 mm. PPL

Cretaceous (latest Albian). Maraca Formation. Maracaibo Basin

Detailed view of the crossed-lamellar crystal microstructure of an oyster. Note the calcitic fibers inclined at small angles. A microfracture along the long axis (or *c-axis*) of the shell remained open (black arrow). Elongate and robust morphology of oysters tends to promote the development of pressure dissolution features along sutured contacts. In this case, a microstylolite (or solution seam) was developed along the surface contact between the shell and partially dolomitized carbonate substrate (white arrow).

IW = 1.3 mm. PPL

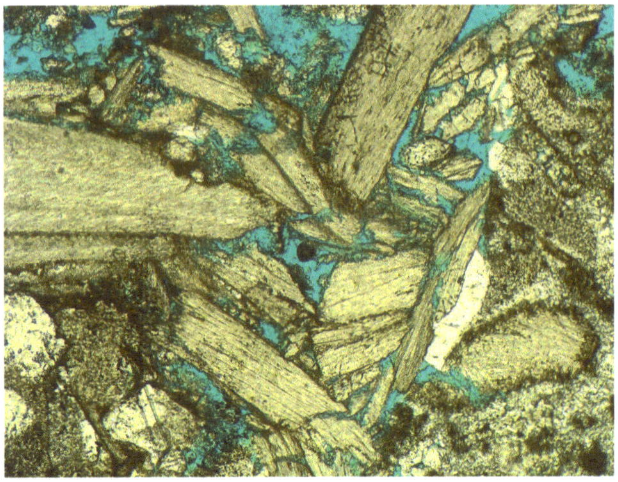

Cretaceous (latest Albian). Maraca Formation. Maracaibo Basin

Fragments of oyster shells are major components in this porous grainstone. Despite fragmentation and mechanical abrasion in a moderate to high-energy environment, the fragments preserved diagnostic textural features. This is an example of characteristic porous highstand deposits in the Maraca Formation, which has produced hydrocarbons generated by the overlying Cenomamian source rock (La Luna Formation).

IW = 2.4 mm. PPL

Cretaceous (late Aptian). Piché Member, Apón Formation. Maracaibo Basin

Oyster shells in this organic-rich limestone, dolomitic were able to thrive under stressed conditions. The dark color of the matrix indicated abundant organic matter and low-oxygen conditions during deposition, while abundant dolomite crystals suggest hypersaline conditions during early marine diagenesis. This sample correspond to the OAE 1b first subevent in western Lake Maracaibo. In this area of the basin, oysters are common in late Aptian black shales, but absent in time-equivalent black shales from cores located in western MB (e.g. cores Z26D-2 and Alpuf-6).

IW = 2.4 mm. PPL

Cretaceous (middle-late Cenomanian). La Luna Formation *sensu stricto*. Maracaibo Basin

A fragment of an oyster shell exhibiting characteristic robust multilayered wall structure. The latter is composed of alternating prismatic (dark grey) and foliated (light colored) layers. The shell is embedded in an organic-rich matrix with abundant small-sized, globular PF. Molluscs (mainly inoceramids and oysters) are occasional components of some low-oxygen microfacies of the La Luna Formation. This and the previous sample exemplify the wide range of environmental conditions under which some molluscs were able to thrive during mid-Cretaceous times.

IW = 7.5 mm. PPL

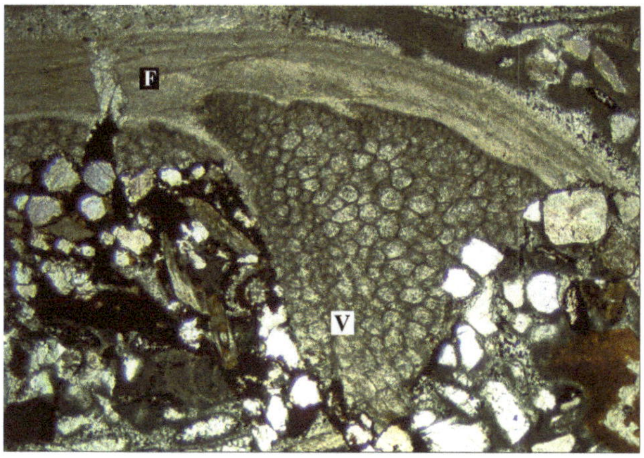

Cretaceous (late Aptian). El Cantil Formation. Eastern Interior Range

A fragment of an oyster shell in a sandy limestone. The shell consists of two well-differentiated layers. The outer, robust foliated (F) layer exemplifies the common appearance of oysters. The inner layer exhibits striking vesicular (V) fabric. The latter, originally highly porous, consists of very thin walls, while the porous tubular structure is filled by sparry calcite.

IW = 4.2 mm. PPL

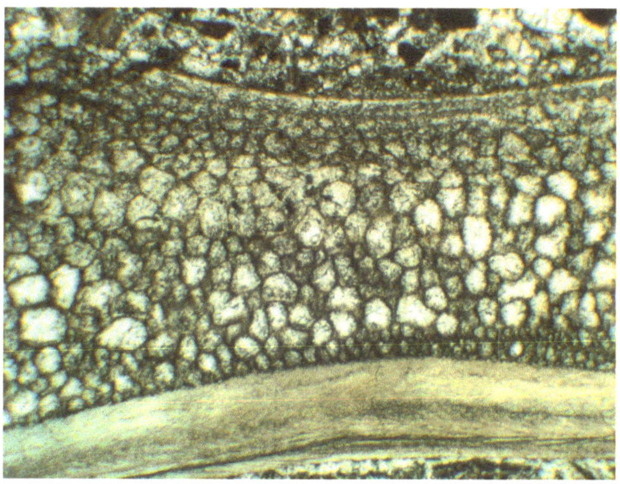

Late Oligocene-Early Miocene. Perla Limestone. Eastern Gulf of Venezuela Sub-Basin

Another oyster exhibiting robust, foliated layer and a vesicular layer. Note the textural similarities between this and previous example, despite both samples being about 100 Myr apart. The vesicular fabric allows the organism to build a thick shell without having to secrete massive quantities of calcite (see Scholle and Ulmer-Scholle, 2003). Oysters are minor components of facies of the Perla Limestone (see Pomar et al. 2015).

IW = 4.2 mm. PPL

Cretaceous (middle Aptian). Piché Member, Apón Formation. Maracaibo Basin

Oyster shell fragments (O) are abundant in this dolostone microfacies. In addition to their resistance to meteoric diagenesis, LMC oyster shells are, in general, not well-suited for dolomite replacement. In the case of this sample, mineralogic stability of the shells (not prone to be replaced by dolomite) during replacement of the original muddy matrix led to preservation of these molluscs.

IW = 2.3 mm. PPL

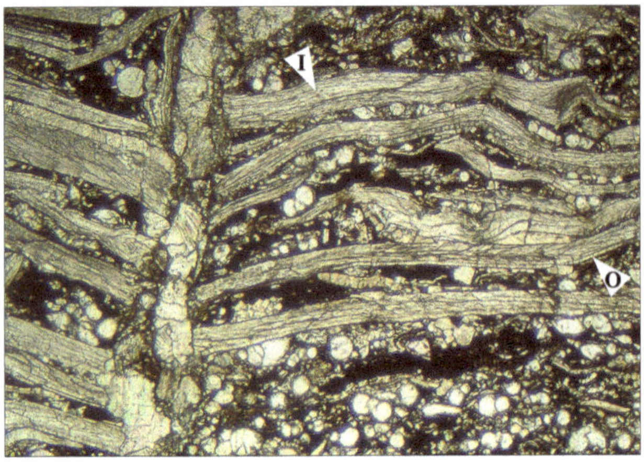

Cretaceous (middle-late Cenomanian). La Luna Formation *sensu stricto*. Maracaibo Basin

Shells of the genus *Inoceramus* in a black shale rich in PF. The horizontally-contorted shells are attached more or less perpendicularly to a recrystallized carbonate substrate. These are very thin relative to their width and consist of an inner (I), foliated-nacreous layer (not commonly preserved) composed of horizontal and low-angle lamellae, and an outer (O), thinner layer composed of perpendicular prisms. Localized contortion of layers is due to burial effects, as this facies is buried down to almost 6000 m.

IW = 2.4 mm. PPL

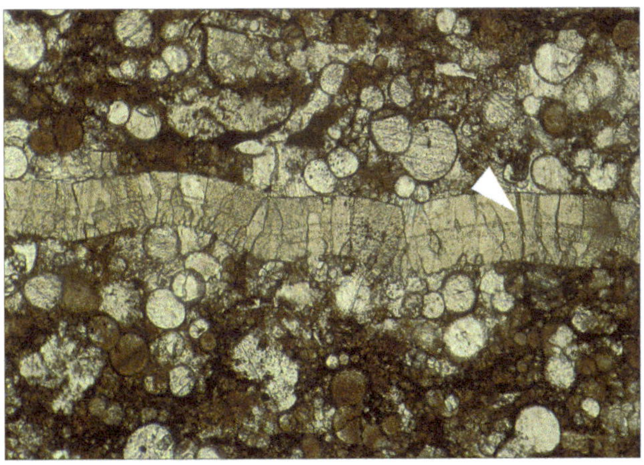

Cretaceous (middle-late Cenomanian). La Luna Formation *sensu stricto*. Maracaibo Basin

The coarsely prismatic outer layer of an inoceramid shell surrounded by globular PF tests. Inclusion-rich calcitic prisms are delimited by characteristic vertical growth lines (white arrow), which are oriented more or less perpendicular to the margins of the shell. Detachment of horizontal and prismatic layers due to taphonomic and diagenetic processes is common in this genus. Additionally, the prismatic layer is commonly better preserved as a fossil. The occurrence of this low-oxygen mollusc in facies of the La Luna Formation *sensu stricto* is associated with thin rock intervals deposited under low-oxygen conditions.

IW = 0.5 mm. PPL

Cretaceous (late Aptian). Piché Member, Apón Formation. Maracaibo Basin

Dual image (PPL and CPL) of an inoceramid shell embedded in a bioclastic and partly dolomitized matrix. The contrasting views allows for the visualization the well-preserved prismatic, calcitic layer. Prisms are rectangular in shape, more or less equant-sized, and exhibit alternating dark and light extinction pattern. As in the previous example, the low energy of the depositional setting allowed the preservation of the elongate and thin structure, since individual prisms can also be found as disaggregated particles. It highlights the capacity of this specialized mollusc to thrive under low-oxygen conditions, as this sample, belonging to the OAE 1b first sub-event, has 2.82% TOC.

IW = 0.5 mm. PPL (above). CPL (below)

Holocene. Los Roques Archipelago

A large shell of the species *Aliger gigas* (formerly *Lobatus gigas*), commonly known in the literature as the "Queen Conch". This heavy calcified aragonitic, adult gastropod shell is 24 cm-long, from the apex (the pointed ending to the left) to the siphonal canal (the sharp and highly-spired ending structure to the right) (see Davis et al. 1993 for more details on this shell morphology). The univalve shell exhibits characteristic coiled and conical shape, with a wide and long aperture, several spinose whorls, translucent color and pink pigmentation. *Aliger gigas* is abundant across the whole Caribbean Sea (see Aldana-Aranda et al. 2014).

IW = 35 cm

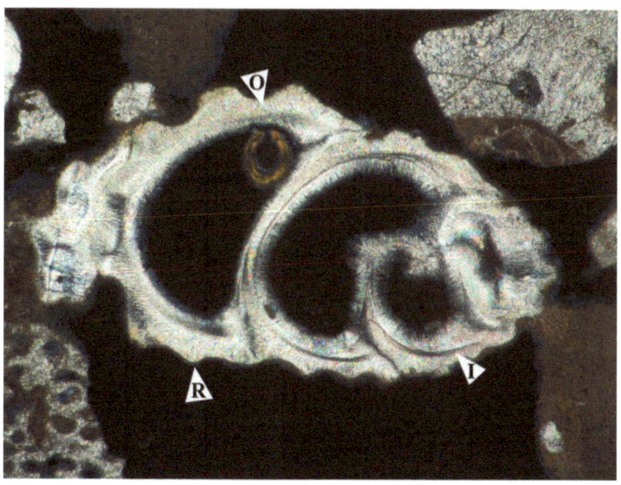

Late Pleistocene. Aves Island

This gastropod shell exhibits a characteristic double-layered wall structure, composed of inner (I) and outer (O) layers, with a crossed-lamellar microstructure. The shell surface ornamentation (R) is a distinctive morphological feature in many gastropods. The unchambered body interior is rimmed by isopachous, acicular aragonite cement (needle-shaped crystals). Gastropods in this lithified unit, the Aves Island Calcarenite, are in some cases preserved as complete shells, such as this specimen. In contrast, bivalve mollusks occur only as fragments.

IW = 2.3 mm. CPL

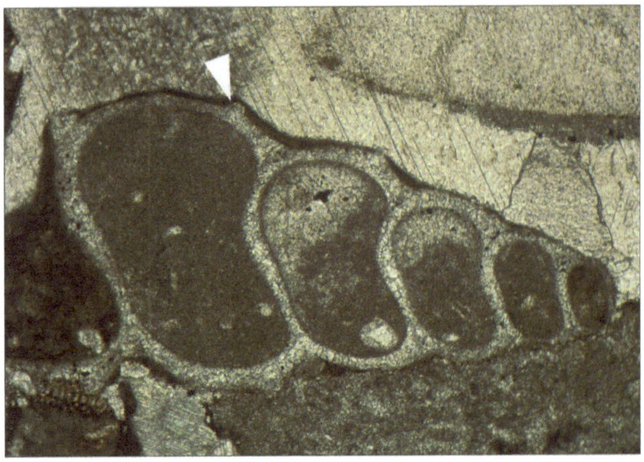

Cretaceous (latest Albian). Maraca Formation. Maracaibo Basin

A longitudinal/oblique section of a gastropod mold, with cavities filled by synsedimentary micrite. The conical and coiled shell exhibits a characteristic gradual enlarging from the initial and to the later whorls. The original texture of the shell was completely lost due to metastability of aragonite. This "mold" form of preservation, resulting from dissolution-reprecipitation, is intrinsic to all ancient gastropods. Traces of the original external ornamentation (white arrow) are also visible.

IW = 2.4 mm. PPL

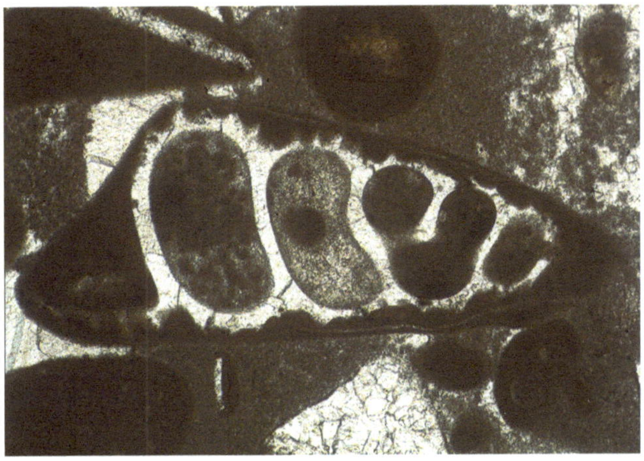

Cretaceous (middle Albian). Lisure Formation. Maracaibo Basin

Another longitudinal/oblique section of a gastropod. Note that the shell is coated with a thick, dark, organic-rich outer micritic envelope. The origin of the latter is probably linked to the original thin organic matter layer that envelops gastropods, termed the periostracum. In this context, the constructive micritization process occurred due to precipitation of calcitic microcrystals around the exposed organic filament surfaces. Both the formation of the micritic envelope and the infilling of the body whorls by micritic sediments occurred prior to the mineralogic stabilization of the aragonitic shell.

IW = 2.4 mm. PPL

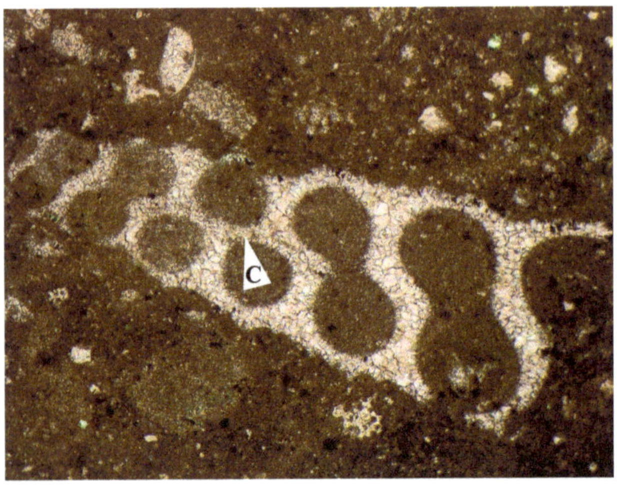

Cretaceous (middle Aptian). Piché Member, Apón Formation. Maracaibo Basin

This conical gastropod shell, seen in longitudinal to slightly oblique section, is composed of coarse-grained LMC derived from dissolution of the original aragonite mineralogy and subsequent precipitation of the stable $CaCO_3$ polymorph. Note the columella (C), which is the central pillar around which the whorls of gastropods are arranged. In thin sections, this structural element is, in many cases, not visible, since the cuts are off the central axis of the grains. When visible, this central column is, in most cases, only seen partially (as in this example).

IW = 2.4 mm. PPL

Cretaceous (latest Albian). Maraca Formation. Maracaibo Basin

Transverse sections of gastropod shells, such as the grain in the center of the image, are of great utility for identifying these univalve molluscs. Characteristic projection of the wall into the central cavity (white arrow) is a useful diagnostic element for identifying gastropods seen in sections perpendicular to the axis of coiling of the shell. The mold of the shell preserved the ornamented exterior consisting of equally spaced nodes.

IW = 2.4 mm. PPL

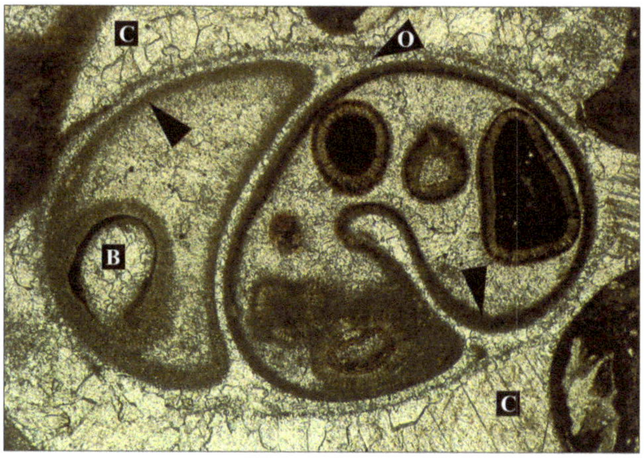

Cretaceous (middle Albian). Lisure Formation. Maracaibo Basin

This oblique section passes through two whorls of a gastropod shell. It shows the diagenetically stabilized ostracum (O) outlining the grain. This fine-grained calcite outlining is clearly distinguishable from the surrounding LMC cement (C) that is filling interparticle porosity. It highlights the constructive micritization process that formed a micritic envelope that outlines the body whorls (black arrows). This process predated meteoric diagenesis and postdated both the infilling of the whorl on the right by ooids and the formation of a small, ovate boring (B) structure in the whorl on the left.

IW = 2.4 mm. PPL

Cretaceous (middle Albian). Lisure Formation. Maracaibo Basin

Gastropod shells in the image are seen in transverse and oblique sections (specimens to the right and left, respectively). These are accompanied by shell fragments (gastropods and/or bivalves) in a wackestone. Note the contrast between the non-recrystallized, organic-lean LMC micritic matrix and diagenetically stabilized bioclasts preserved as molds (whitish). Gastropod-dominated facies are not volumetrically abundant in shallow Cretaceous carbonates facies in the MB.

IW = 2.4 mm. PPL

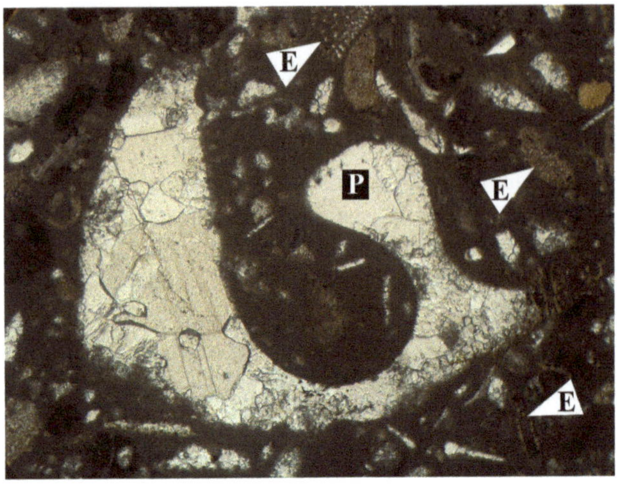

Cretaceous (early Albian). Lisure Formation. Maracaibo Basin

This fragment of a gastropod, derived from the breakage of a large shell, is easily identifiable due to the distinctive shape of the wall projection (P). It is accompanied by numerous smaller echinoid grains (E), all embedded in a slightly dark micritic matrix. The skeletal association between molluscs (gastropods, aragonitic bivalves, and oysters) and echinoderms is the most common in the shallow facies of the Lisure Formation.

IW = 4.2 mm. PPL

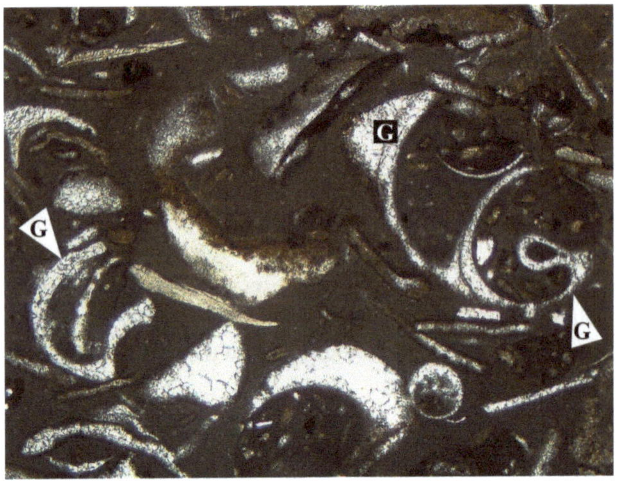

Cretaceous (middle Aptian). Piché Member, Apón Formation. Maracaibo Basin

A mollusc-dominated sparse packstone in which gastropods (G) are major skeletal components. Overall, molluscs are the most abundant heterotrophic calcifiers in shallow Cretaceous carbonates of the MB. As in the previous example, the micritic matrix in this sample is slightly dark in color. This, combined with the absence of photozoan calcifiers (calcareous algae and LBF), suggests that eutrophic conditions prevailed during deposition of this heterozoan microfacies.

IW = 4.2 mm

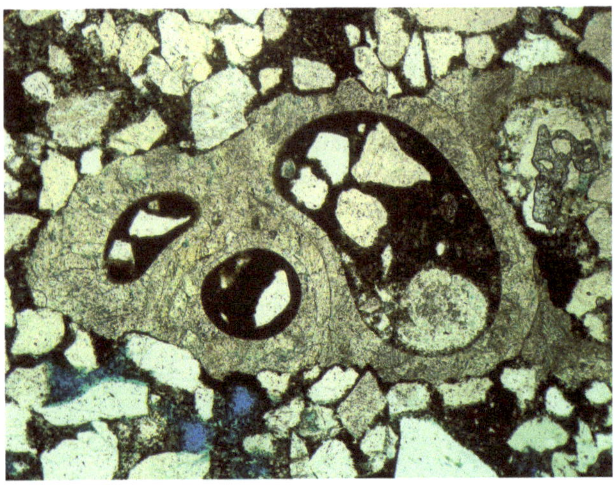

Early Miocene. Churuguara Formation. Falcón Basin

A gastropod shell surrounded and infilled by a sandy-clayey matrix. Note the thick wall structure of the shell. In general, this morphologic feature is common of gastropods that thrive in shallow, agitated environments, which is the case of this sample belonging to littoral siliciclastic-dominated facies of the Churuguara Formation. Such thickness often prevents the shell from fracturing during mechanical abrasion and/or reworking.

IW = 2.4 mm. PPL

Cretaceous (middle Aptian). Piché Member, Apón Formation. Maracaibo Basin

A close view of a thin-walled, small gastropod, seen in transverse section, and embedded in a microscrystalline micritic matrix. A shallow depositional environment, and low-energy and eutrophic conditions are associated to the deposition of this shell in a mudstone-wackestone containing molluscs, SBF, and sponge spicules. Generally, gastropods in Cretaceous carbonate facies of the MB tend to become smaller in size and have thinner wall structures when they correspond to transgressive-muddy deposits, such as this example.

IW = 0.5 mm. PPL

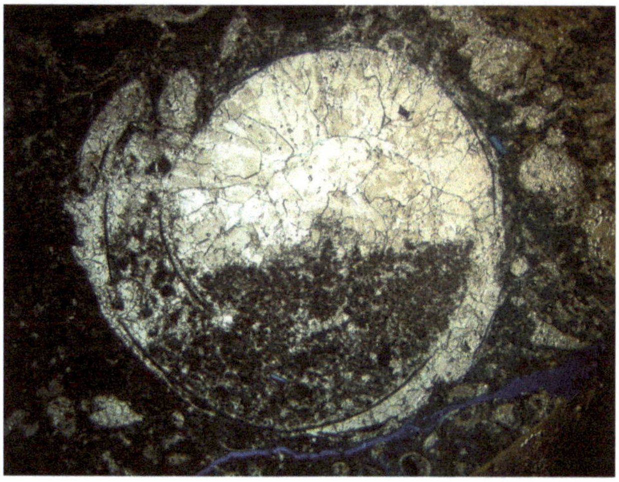

Late Oligocene-Early Miocene. Perla Limestone. Eastern Gulf of Venezuela Sub-Basin

This large, coiled gastropod shell has a geopetal fills that consists of micropeloidal matrix in the lower part and coarse-grained (blocky) calcite cement in the upper portion. In general, gastropod empty calcified shells are ideal for the formation of geopetal structures. In the Perla Limestone, geopetal sediment infills, many containing micropeloids (such as in this example), are associated with interludes of calm within turbulent episodes during sediment transport and deposition of this skeletal-dominated succession (see Pomar et al. 2015).

IW = 7.5 mm. PPL

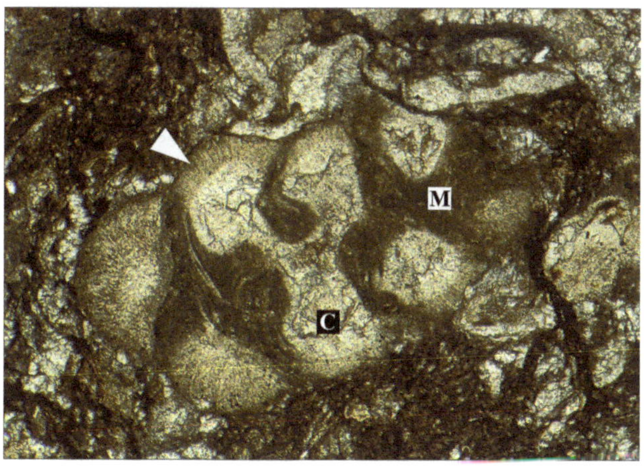

Middle Permian (Cisularian to Guadalupian). Pamarito Formation. Venezuelan Andes

This gastropod shell was filled and replaced by carbonate precipitates of different origins and textures, thus promoting preservation of such distinctive morphology and outline. Most of the longitudinal-central area is filled by dark micritic (M) sediments. Fibrous crystals, probably originally of marine diagenetic origin, are replacing the walls (white arrow). Body whorl apertures are filled by cloudy sparry calcite (C). Gastropods are common secondary bioclasts in wackestones and packstones of the Palmarito Formation (see Laya and Tucker, 2012).

IW = 2.4 mm. PPL

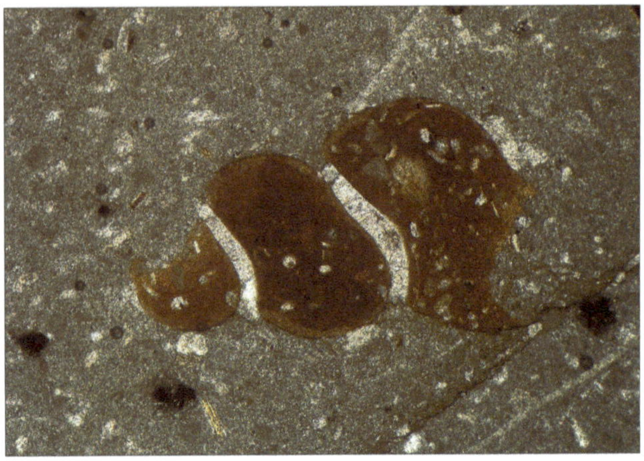

Early Miocene. Churuguara Formation. Falcón Basin

A gastropod exhibiting strongly smooth-curved outlining. Precipitation of fluorapatite (orange-brownish color) during marine diagenesis replaced both the micritic infilling and the walls outlining the margins of the shell. On the other hand, the walls crossing transversally the long axis of the shell (in this view) were not replaced but instead neomorphized during mineralogic stabilization. Gastropods are secondary bioclastic components in Early Miocene facies of the FB, being present in both carbonate-dominated and siliciclastic-dominated microfacies (see Albert-Villanueva, 2016).

IW = 1.3 mm. PPL

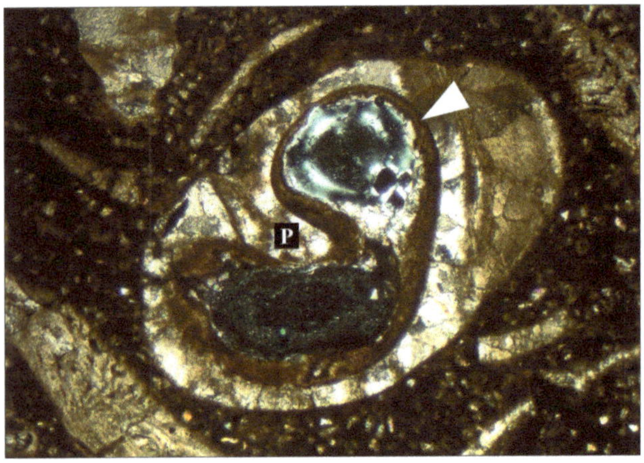

Early Miocene. San Luis Formation. Falcón Basin

The gastropod in the center of the image is embedded in an organic-rich, sandy matrix and accompanied by recrystallized shell fragments. Characteristic projection (P) of the whorl into the central body cavity serves as diagnostic element. Recrystallization of the gastropod shell formed a coarse-grained inequigranular crystalline fabric, which is noticeable under CPL. This view also allows for the recognition of the amorphous silica precipitate that filled the body aperture. The shape of this central cavity is outlined by a dark micritic envelope (white arrow).

IW = 1.3 mm. CPL

References

Acosta-Balbas, V., Betancourt-Tineo, R., Prieto-Arcas, A., 2014. Estructura comunitaria de bivalvos y gasterópodos en raíces del mangle rojo *Rhizophora mangle* (Rhizophoraceae) en isla Larga, bahía de Mochima, Venezuela. *Revista de Biología Tropical*, *62*, 551–565.

Abdelhady, A. A., and Elewa, A. M., 2010. Evolution of the Upper Cretaceous Oysters: Traditional morphometrics approach. In: Elewa, A.M. (ed.), Morphometrics for Nonmorphometricians. Lecture Notes in Earth Sciences, Springer-Verlag, Berlin, pp. 157–176.

Albert-Villanueva, E., 2016. Facies y secuencias deposicionales mixtas carbonático-siliciclásticas del Mioceno inferior de la Cuenca de Falcón (noroeste de Venezuela) como modelo exploratorio en el Caribe. Doctoral Dissertation. University of Barcelona. Volumes 1 and 2.

Aldana-Aranda, D., Oxenford, H. A., Bissada, C., Enriquez-Díaz, M., Brulé, T., Delgado, G. A., Martínez-Morales, I., Frenkiel, L., 2014. Reproductive patterns of queen conch, *Strombus gigas* (Mollusca, Gastropoda), across the Wider Caribbean Region. *Bulletin of Marine Science*, *90*, 813–831.

Allmon, W. D., 2007. Cretaceous marine nutrients, greenhouse carbonates, and the abundance of turritelline gastropods. *The Journal of Geology*, *115*, 509–523.

Álvarez-Barco, J. A., Rodríguez-Quintal, J. G., Rodríguez, C. T., Narciso-Fejure, S. E., 2019. Moluscos asociados al arrecife coralino de Isla Larga, Parque Nacional San Esteban, Estado Carabobo, Venezuela. *Memoria de la Fundación La Salle de Ciencias Naturales*, *77*, 5-30.

Aqrawi, A. A. M., Thehni, G. A., Sherwani, G. H., Kareem, B. M. A., 1998. Mid-Cretaceous rudist-bearing carbonates of the Mishrif Formation: An important reservoir sequence in the Mesopotamian Basin, Iraq. *Journal of petroleum Geology*, *21*, 57–82.

Armstrong, H. A., and Brasier, M. D., 2005. Microfossils, Second edition. Blackwell Publishing, 296 pp.

Ayyasami, K., 2006. Role of oysters in biostratigraphy: A case study from the Cretaceous of the Ariyalur area, southern India. *Geosciences Journal*, *10*, 237–247.

Bartok, P., Reijers, T. J. A., Juhasz, I., 1981. Lower Cretaceous Cogollo Group, Maracaibo Basin, Venezuela: sedimentology, diagenesis, and petrophysics. *American Association of Petroleum Geologists Bulletin*, *65*, 1110–1134.

Best, M. M., and Kidwell, S. M., 2000. Bivalve taphonomy in tropical mixed siliciclastic-carbonate settings. I. Environmental variation in shell condition. *Paleobiology*, *26*, 80–102.

Bieler, R., and Mikkelsen, P. M., 2006. Bivalvia-a look at the branches. *Zoological Journal of the Linnean Society*, *148*, 223–235.

Brandano, M., Frezza, V., Tomassetti, L., Cuffaro, M., 2009. Heterozoan carbonates in oligotrophic tropical waters: the Attard member of the lower coralline limestone formation (Upper Oligocene, Malta). *Palaeogeography, Palaeoclimatology, Palaeoecology*, *274*, 54–63.

Brigaud, B., Pucéat, E., Pellenard, P., Vincent, B., Joachimski, M. M., 2008. Climatic fluctuations and seasonality during the Late Jurassic (Oxfordian–Early Kimmeridgian) inferred from $\delta^{18}O$ of Paris Basin oyster shells. *Earth and Planetary Science Letters*, *273*, 58–67.

Buatois, L. A., Carmona, N. B., Curran, H. A., Netto, R. G., Mángano, N. G., Wetzel, A., 2016. The Mesozoic Marine Revolution, in *The Trace Fossil Record of Major Evolutionary Events, Volume 2*. Editors M. G. Mángano and L. A. Buatois (Dordrecht: Topics in Geobiology), 40, 18–134.

Bush, A. M., and Bambach, R. K., 2011. Paleoecologic megatrends in marine metazoa. *Annual Review of Earth and Planetary Sciences*, *39*, 241–269.

Bush, A. M., Bambach, R. K., Daley, G. M., 2007. Changes in theoretical ecospace utilization in marine fossil assemblages between the mid-Paleozoic and late Cenozoic. *Paleobiology*, *33*, 76–97.

Carannante, G., Esteban, M., Milliman, J. D., Simone, L., 1988. Carbonate lithofacies as paleolatitude indicators: problems and limitations. *Sedimentary Geology*, *60*, 333–346.

Carannante, G., Cherchi, A., Simone, L., 1995. Chlorozoan versus foramol lithofacies in Upper Cretaceous rudist limestones. *Palaeogeography, Palaeoclimatology, Palaeoecology*, *119*, 137–154.

Caroprese, C. G., 2009. Bioestratigrafía y paleoecología de la Formación Querecual en el Oriente de Venezuela. Magister Dissertation. Central University of Venezuela. 95 pp.

Carvajal-Chitty, H., and Rey, O., 2020. Aportes a la bioestratigrafía del Aptiense–Albiense en la cuenca oriental de Venezuela con base en foraminíferos y nanopláncton calcáreo. *Caribbean Studies*, *48*, 135–163.

Chauvaud, L., Thompson, J. K., Cloern, J. E., Thouzeau, G., 2003. Clams as CO_2 generators: the *Potamocorbula amurensis* example in San Francisco Bay. *Limnology and Oceanography*, *48*, 2086–2092.

Checa, A. G., Jiménez-Jiménez, A. P., Márquez-Aliaga, A., Hagdorn, H., 2006. Further comments on the origin of oysters. *Palaeogeography, Palaeoclimatology, Palaeoecology*, *240*, 672–674.

Cobban, W. A., Walaszczyk, I., Obradovich, J. D., McKinney, K. C., 2006. A USGS zonal table for the Upper Cretaceous middle Cenomanian-Maastrichtian of the Western Interior of the United States based on ammonites, inoceramids, and radiometric ages. *U.S. Geological Survey open-file report*, *1250*, 46.

Crame, J. A., 2002. Evolution of taxonomic diversity gradients in the marine realm: a comparison of late Jurassic and recent bivalve faunas. *Paleobiology*, *28*, 184–207.

Crespo de Cabrera, S., Sliter, W. V., Jarvis, I., 1999. Integrated foraminiferal biostratigraphy and chemostratigraphy of the Querecual Formation (Cretaceous), eastern Venezuela. *The Journal of Foraminiferal Research*, *29*, 487–499.

Dauphin, Y., 2003. Soluble organic matrices of the calcitic prismatic shell layers of two pteriomorphid bivalves *Pinna nobilis* and *Pinctada margaritifera*. *Journal of Biological Chemistry*, *278*, 15168–15177.

Dauphin, Y., 2008. The nanostructural unity of mollusc shells. *Mineralogical Magazine*, *72*, 243–246.

Davis, M., Bolton, C. A., Stoner, A. W., 1993. A comparison of larval development, growth, and shell morphology in three Caribbean *Strombus* species. *Veliger*, *36*. 236–244.

Debrot, A. O., Yranzo, A., Arocha, D., 2019. Los Roques and Las Aves archipelagos, Venezuela: A marine ecological and conservation reconnaissance of two little-known southeastern Caribbean oceanic archipelagos. *Atoll Research Bulletin*, *622*, 1–27.

del Carmen-Perrilliat, M., Vega, F. J., Coutiño, M. A., 2010. Miocene mollusks from the Simojovel area in Chiapas, southwestern Mexico. *Journal of South American Earth Sciences*, *30*, 111–119.

Dhondt, A. V., 1992. Cretaceous inoceramid biogeography: a review. *Palaeogeography, Palaeoclimatology, Palaeoecology*, *92*, 217–232.

Díaz, J. M., 1995. Zoogeography of marine gastropods in the southern Caribbean: a new look at provinciality. *Caribbean Journal of Science*, *31*, 104–121.

Embry, J. C., Vennin, E., Van Buchem, F. S. P., Schroeder, R., Pierre, C., Aurell, M., 2010. Sequence stratigraphy and carbon isotope stratigraphy of an Aptian mixed carbonate-siliciclastic platform to basin transition (Galve sub-basin, NE Spain). *Geological Society, London, Special Publications*, *329*, 113–143.

Erlich, R., Macsotay, O., Nederbragt, A., Lorente, M. A., 1999. Palaeoecology, palaeogeography and depositional environments of upper cretaceous rocks of western Venezuela. *Palaeogeography, Palaeoclimatology, Palaeoecology*, *153*, 203–238.

Erlich, R. N., Nederbragt, A. J., Lorente, M. A., 2000. Birth and death of the Late Cretaceous "La Luna Sea", and origin of the Tres Esquinas phosphorites. *Journal of South American Earth Sciences*, *13*, 21–45.

Ferrari, M., El-Hedeny, M., Zakhera, M., El-Sabbagh, A., Al Farraj, S., 2018. Middle–Upper Jurassic marine gastropods from central Saudi Arabia. *Alcheringa: An Australasian Journal of Palaeontology*, *42*, 383–401.

Flessa, K. W., Jablonski, D. A. V. I. D., Erwin, D. H., Lipps, J. H., 1996. The geography of evolutionary turnover: a global analysis of extant bivalves. *Evolutionary Paleobiology*, 376–397.

Föllmi, K. B., 2012. Early Cretaceous life, climate and anoxia. *Cretaceous Research*, *35*, 230–257.

Fraiser, M. L., and Bottjer, D. J., 2007. When bivalves took over the world. *Paleobiology*, *33*, 397–413.

Fraser, N. M., Bottjer, D. J., Fischer, A. G., 2004. Dissecting "Lithiotis" bivalves: implications for the Early Jurassic reef eclipse. *Palaios*, *19*, 51–67.

Flügel, E, 2010. Microfacies of Carbonate Rocks—Analysis, Interpretation and Application. Springer, Heidelberg-Dordrecht-London-New York, 984 pp.

Fortunato, H., 2015. Mollusks: tools in environmental and climate research. *American Malacological Bulletin*, *33*(2), 1–15.

Frýda, J., 2013. Fossil Invertebrates: Gastropods. Reference Module in Earth Systems and Environmental Sciences, Elsevier. 12 pp.

Frýda, J., Nützel, A., Wagner, P. J., 2008. Paleozoic gastropods. In Ponder, W., Lindberg, D.L. (eds.): Phylogeny and Evolution of the Mollusca, 237–268. University of California Press, Berkeley.

Garbán, G.J., 2010. Geoquímica y estratigrafía del Miembro Ftanita de Táchira (Cretácico tardío), Formación La Luna, Venezuela Occidental: Implicaciones Paleooceanográficas y paleoambientales. Doctoral Dissertation, Central University of Venezuela, 402 pp.

Geister, J., 1992. Modern reef development and Cenozoic evolution of an oceanic Island/reef complex: Isla de Providencia (Western Caribbean Sea, Colombia). *Facies*, *27*, 1–69.

Gilbert, P. U. P. A., Bergmann, K. D., Boekelheide, N., Tambutté, S., Mass, T., Marin, F., Adkins, J. F., Erez, J., Gilbert, B., Knutson, V., Cantine M., Hernández, J. O., Knoll, A. H., 2022. Biomineralization: integrating mechanism and evolutionary history. *Science Advances*, *8*(10), eabl9653.

Gili, E., Masse, J. P., Skelton, P. W., 1995a. Rudists as gregarious sediment-dwellers, not reef-builders, on Cretaceous carbonate platforms. *Palaeogeography, Palaeoclimatology, Palaeoecology*, *118*, 245–267.

Gili, E., Skelton, P. W., Vicens, E., Obrador, A., 1995b. Corals to rudists—an environmentally induced assemblage succession. *Palaeogeography, Palaeoclimatology, Palaeoecology*, *119*, 127–136.

Gutiérrez, J. L., Jones, C. G., Strayer, D. L., Iribarne, O. O., 2003. Mollusks as ecosystem engineers: the role of shell production in aquatic habitats. *Oikos*, *101*(1), 79–90.

Guo, X., Li, C., Wang, H., Xu, Z., 2018. Diversity and evolution of living oysters. *Journal of Shellfish Research*, *37*, 755–771.
Hallam, A., 1977. Jurassic bivalve biogeography. *Paleobiology*, *3*, 58–73.
Hallock, P., 2001. Coral reefs, carbonate sediments, nutrients, and global change. In: *The history and sedimentology of ancient reef systems*. Springer, Boston, MA, 387–427.
Harper, E. M., and Skelton, P. W., 1993. The Mesozoic marine revolution and epifaunal bivalves. *Scripta Geologica, Special Issue*, *2*, 127–153.
Hautmann, M., 2001. Taxonomy and phylogeny of cementing Triassic bivalves (families Prospondylidae, Plicatulidae, Dimyidae and Ostreidae). *Palaeontology*, *44*, 339– 373.
Hautmann, M. 2004. Effect of end-Triassic CO_2 maximum on carbonate sedimentation and marine mass extinction. *Facies*, *50*, 257–261.
Hautmann, M., 2006a. Shell morphology and phylogenetic origin of oysters. *Palaeogeography, Palaeoclimatology, Palaeoecology*, *240*, 668–671.
Hautmann, M., 2006b. Shell mineralogical trends in epifaunal Mesozoic bivalves and their relationship to seawater chemistry and atmospheric carbon dioxide concentration. *Facies*, *52*, 417.
Hautmann, M., Benton, M.J., Tomašových, A., 2008. Catastrophic ocean acidification at the Triassic-Jurassic boundary. *Neues Jahrbuch für Geologie und Palaontologie, Abhandlungen*, *249*, 119–127.
Hautmann, M., Ware, D., Bucher, H., 2017. Geologically oldest oysters were epizoans on Early Triassic ammonoids. *Journal of Molluscan Studies*, *83*, 253–260.
Hayton, S., Nelson, C., Hood, S., 1995. A skeletal assemblage classification system for non-tropical carbonate deposits based on New Zealand Cenozoic limestones. *Sedimentary Geology*, *100*. 123–141.
Henderson, R. A., 2004. A mid-Cretaceous association of shell beds and organic-rich shale: bivalve exploitation of a nutrient-rich, anoxic sea-floor environment. *Palaios*, *19*, 156–169.
Hsieh, S., Bush, A. M., Bennington, J. B., 2019. Were bivalves ecologically dominant over brachiopods in the late Paleozoic? A test using exceptionally preserved fossil assemblages. *Paleobiology*, *45*, 265–279.
Horowitz, A. S., and Potter, P. E., 1971. Introductory Petrography of Fossils: New York, Springer, 302 pp.
Jablonski, D., 1997. Body-size evolution in Cretaceous molluscs and the status of Cope's rule. *Nature*, *385*, 250–252.
James, N. P., 1997. The cool-water carbonate depositional realm. In: James, N. P., Clarke, J., (eds.), Cool-water carbonates, vol 56. SEPM Society for Sedimentary Geology, Tulsa, 1–20.
James, N. P., and Jones, B., 2015. Origin of Carbonate Sedimentary Rocks. John Wiley & Sons, Ltd, Chinchester, 464 pp.
James, N. P., Boreen, T. D., Bone, Y., Feary, D. A., 1994. Holocene carbonate sedimentation on the west Eucla Shelf, Great Australian Bight: a shaved shelf. *Sedimentary Geology*, *90*, 161–177.
Johnson, C. C., Sanders, D., Kauffman, E.G., Hay, W.W., 2001. Patterns and processes influencing Upper Cretaceous reefs. In: Kiessling, W., Flügel, E., Golonka, J. (Eds.), Phanerozoic Reef Patterns. *SEPM Special Publication*, 549–585.
Kauffman, E. G., Harries, P. J., Meyer, C., Villamil, T., Arango, C., Jaecks, G., 2007. Paleoecology of giant Inoceramidae (*Platyceramus*) on a Santonian (Cretaceous) seafloor in Colorado. *Journal of Paleontology*, *81*, 64–81.
Klicpera, A., Michel, J., Westphal, H., 2015. Facies patterns of a tropical heterozoan carbonate platform under eutrophic conditions: the Banc d'Arguin, Mauritania. *Facies*, *61*, 421.
Korte, C., Hesselbo, S. P., Jenkyns, H. C., Rickaby, R. E., Spötl, C., 2009. Palaeoenvironmental significance of carbon-and oxygen-isotope stratigraphy of marine Triassic–Jurassic boundary sections in SW Britain. *Journal of the Geological Society*, *166*, 431–445.
Kouchinsky, A., 2000. Shell microstructures in Early Cambrian molluscs. *Acta Palaeontologica Polonica*, *45*(2), 119–150.
Kouchinsky, A., Bengtson, S., Runnegar, B., Skovsted, C., Steiner, M., Vendrasco, M., 2012. Chronology of early Cambrian biomineralization. *Geological Magazine*, *149*(2), 221–251.
Kumagae, T., Maeda, H., Komatsu, T., 2011. Paleoecology of *Inoceramus amakusensis* Nagao et Matsumoto, 1940 (Bivalvia) in a Late Cretaceous shallow clastic sea: The Himenoura Group, Kyushu, Japan. *Cretaceous Research*, *32*, 738–749.
Laya, J. C., and Tucker, M. E., 2012. Facies analysis and depositional environments of Permian carbonates of the Venezuelan Andes: Palaeogeographic implications for Northern Gondwana. *Palaeogeography, Palaeoclimatology, Palaeoecology*, *331*, 1–26.
Lees, A., and Buller, A. T., 1972. Modern temperate-water and warm-water shelf carbonate sediments contrasted. *Marine Geology*, *13*, M67–M73.
Lejart, M., Clavier, J., Chauvaud, L., Hily, C., 2012. Respiration and calcification of *Crassostrea gigas*: contribution of an intertidal invasive species to coastal ecosystem CO_2 fluxes. *Estuaries and Coasts*, *35*, 622–632.
Li, L., Topper., T. P., Betts, M. J., Dorjnamjaa, D., Altanshagai, G., Enkhbaatar, B., Li, G., Skovsted, C. B., 2023. Calcitic shells in the aragonite sea of the earliest Cambrian. *Geology*, *51*, 8–12.
Lorente, 1978. La caliza de Capadare y su relaci6n con la Cuenca de Agua Salada, Edo. Falc6n. Bachelor Degree Dissertation. Central University of Venezuela. 161 pp.
Lockwood, R., 2003. Abundance not linked to survival across the end-Cretaceous mass extinction: patterns in North American bivalves. *Proceedings of the National Academy of Sciences*, *100*, 2478–2482.
Macsotay, O., Erlich, R., Peraza, T., 2003. Sedimentary structures of the La Luna, Navay and Querecual Formations, Upper Cretaceous of Venezuela. *Palaios*, *18*, 334–348.
Majewske, O. P., 1969, Recognition of Invertebrate Fossil Fragments in Rocks and Thin Sections [Internat. Sed. Petrog. Series v. 13]: Leiden, E. J. Brill, 101 pp.
Maloof, A. C., Porter, S. M., Moore, J. L., Dudas, F. O., Bowring, S. A., Higgins, J. A., Fike, D. A., Eddy, M. P., 2010. The earliest Cambrian record of animals and ocean geochemical change. *Geological Society of America Bulletin*, *122*, 1731–1774.
Marin, F., and Luquet, G., 2005. Molluscan biomineralization: The proteinaceous shell constituents of *Pinna nobilis* L. *Materials Science and Engineering, C 25*, 105–111.
Márquez-Aliaga, A., Jiménez-Jiménez, A. P., Checa, A. G., Hagdorn, H., 2005. Early oysters and their supposed Permian ancestors. *Palaeogeography, Palaeoclimatology, Palaeoecology*, *229*, 127–136.
Martin, S., Thouzeau, G., Chauvaud, L., Jean, F., Guérin, L., Clavier, J., 2006. Respiration, calcification, and excretion of the invasive slipper limpet, Crepidula fornicata L.: implications for carbon, carbonate, and nitrogen fluxes in affected areas. *Limnology and Oceanography*, *51*, 1996–2007.
McRoberts, C. A., 2001. Triassic bivalves and the initial marine Mesozoic revolution: A role for predators? *Geology*, *29*, 359–362.
Méndez-Dot, J. A., Méndez-Baamonde, J., Reyes, D., Whilchy, R., 2015. The Cogollo Group and the Oceanic Anoxic Events 1a and 1b, Maracaibo Basin, Venezuela. *Brazilian Journal of Geology, 45* (Suppl. 1), 41–61.
Michel, J., Borgomano, J., Reijmer, J. J., 2018. Heterozoan carbonates: When, where and why? A synthesis on parameters controlling carbonate production and occurrences. *Earth-Science Reviews*, *182*, 50–67.
Michel, J., Laugié, M., Pohl, A., Lanteaume, C., Masse, J. P., Donnadieu, Y., Borgomano, J., 2019. Marine carbonate factories: a global model of carbonate platform distribution. *International Journal of Earth Sciences*, *108*, 1773–1792.
Miller, A. I., 1989. Spatio-temporal transitions in Paleozoic Bivalvia: A field comparison of upper Ordovician and upper Paleozoic bivalve-dominated fossil assemblages. *Historical Biology*, *2*, 227–260.

Murdock, D. J., 2020. The 'biomineralization toolkit' and the origin of animal skeletons. *Biological Reviews*, *95*, 1372–1392.

Mutti, M., and Hallock, P., 2003. Carbonate systems along nutrient and temperature gradients: some sedimentological and geochemical constraints. *International Journal of Earth Sciences*, *92*, 465–475.

Nevesskaya, L. A., 2008. Dynamics of taxonomic diversity of bivalves in the Phanerozoic. *Paleontological Journal*, *42*(4), 335–342.

Nützel, A., 2005. Recovery of gastropods in the Early Triassic. *Comptes Rendus Palevol*, *4*, 501–515.

Parkhaev, P. Y., 2007. The Cambrian "basement" of gastropod evolution. In: The Rise and Fall of the Ediacaran Biota (Vickers-Rich, P. and Komarower, P.L., Eds.). *Geological Society, London, Special Publications*, *286*, 415–421.

Parkhaev, P. Y., 2008. The early Cambrian radiation of Mollusca. In: Phylogeny and Evolution of the Mollusca (eds W. F. Ponder and D. R. Lindberg), pp. 33–70. University of California Press, Berkeley.

Parkhaev, P. Y., 2017. Origin and the early evolution of the phylum Mollusca. *Paleontological Journal*, *51*(6), 663–686.

Pohl, A., Laugié, M., Borgomano, J., Michel, J., Lanteaume, C., Scotese, C. R., Fraud, C., Poli, E., Donnadieu, Y., 2019. Quantifying the paleogeographic driver of Cretaceous carbonate platform development using paleoecological niche modeling. *Palaeogeography, Palaeoclimatology, Palaeoecology*, *514*, 222–232.

Pomar, L., 2001. Types of carbonate platforms: a genetic approach. *Basin Research*, *13*, 313–334.

Pomar, L., Brandano, M., Westphal, H., 2004. Environmental factors influencing skeletal grain sediment associations: a critical review of Miocene examples from the western Mediterranean. *Sedimentology*, *51*, 627-651.

Pomar, L., Esteban, M., Martinez, W., Espino, D., Castillo de Ott, V., Benkovics, L., Castro-Leyva, T., 2015. Oligocene-Miocene carbonates of the Perla field, offshore Venezuela: Depositional model and facies architecture, In: C. Bartolini and P. Mann, eds., Petroleum geology and potential of the Colombian Caribbean Margin: *American Association of Petroleum Geologists, Memoir 108*, 647–674.

Pöppelreiter, M., Balzarini, M. A., De Sousa, P., Engel, S., Galarraga, M., Hansen, B., Márquez, X., Morell, J., Nelson, R., Rodriguez, F., 2005. Structural control on sweet-spot distribution in a carbonate reservoir: Concepts and 3-D models (Cogollo Group, Lower Cretaceous, Venezuela). *American Association of Petroleum Geologists Bulletin*, *89*, 1651–1676.

Porter, S. M., 2010. Calcite and aragonite seas and the de novo acquisition of carbonate skeletons. *Geobiology*, *8*, 256–277.

Rey, O., Simo, J. A., Lorente, M. A., 2004. A record of long-and short-term environmental and climatic change during OAE3: La Luna Formation, Late Cretaceous (Santonian–early Campanian), Venezuela. *Sedimentary Geology*, *170*, 85–105.

Ries, J. B., 2010. Geological and experimental evidence for secular variation in seawater Mg/Ca (calcite-aragonite seas) and its effects on marine biological calcification. *Biogeosciences*, *7*, 2795–2849.

Sälgeback, J., and Savazzi, E., 2006. Constructional morphology of cerithiform gastropods. *Paleontological Research*, *10*, 233–259.

Sartorio, D., and Venturini, S. 1988. Southern Tethys biofacies. AGIP, San Donato Milanese, 235 pp.

Schneider, S., Crampton, J. S., Lukeneder, A., 2013. Propeamussiidae, Inoceramidae, and other Bivalvia from the Lower Cretaceous Puez Formation (Valanginian–Cenomanian; Dolomites, South Tyrol, Italy). *Cretaceous research*, *46*, 216–231.

Scholle, P. A., and Ulmer-Scholle, D., 2003. A Color Guide to the Petrography of Carbonate Rocks: Grains, Textures, Porosity, Diagenesis. *American Association of Petroleum Geologists, Memoir 77*, 474 pp.

Schwarz, E., Veiga, G. D., Álvarez Trentini, G., Isla, M. F., Spalletti, L. A., 2018. Expanding the spectrum of shallow-marine, mixed carbonate–siliciclastic systems: Processes, facies distribution and depositional controls of a siliciclastic-dominated example. *Sedimentology*, *65*, 1558–1589.

Schweitzer, C. E., and Feldmann, R. M., 2010. The Decapoda (Crustacea) as predators on Mollusca through geologic time. *Palaios*, *25*, 167–182.

Scott, R. W., 1995. Global environmental controls on Cretaceous reefal ecosystems. *Palaeogeography, Palaeoclimatology, Palaeoecology*, *119*, 187–199.

Simone, L., Carannante, G., Bassi, D., Cherchi, A., 2012. Rudist-bearing rhodalgal facies in the post-Turonian recovery of peri-Tethyan carbonate systems: a case history from the Nurra region (northwestern Sardinia, Italy). *Geodiversitas*, *34*(1), 167–187.

Skelton, P. W., 2013. Rudist classification for the revised Bivalvia volumes of the 'Treatise on Invertebrate Paleontology'. *Caribbean Journal of Earth Science*, *45*, 9–33.

Skelton, P. W., and Gili, E., 2012. Rudists and carbonate platforms in the Aptian: a case study on biotic interactions with ocean chemistry and climate. *Sedimentology*, *59*, 81–117.

Skelton, P. W., Gili, E., Rosen, B. R., Valldeperas, F. X., 1997. Corals and rudists in the late Cretaceous: A critique of the hypothesis of competitive displacement. *Boletín de la Real Sociedad Española de Historia Natural (Sección Geológica)*, *92*, 225–239.

Stanley, S. M., 2006. Influence of seawater chemistry on biomineralization throughout Phanerozoic time: Paleontological and experimental evidence. *Palaeogeography, Palaeoclimatology, Palaeoecology*, *232*, 214–236.

Stanley, S. M., and Hardie, L. A., 1998. Secular oscillations in the carbonate mineralogy of reef-building and sediment-producing organisms driven by tectonically forced shifts in seawater chemistry. *Palaeogeography, Palaeoclimatology, Palaeoecology*, *144*, 3–19.

Stanley, S. M., and Hardie, L. A., 1999. Hypercalcification: paleontology links plate tectonics and geochemistry to sedimentology. *GSA today*, *9*, 1–7.

Steuber, T., 2000. Skeletal growth rates of Upper Cretaceous rudist bivalves: implications for carbonate production and organism—environment feedbacks. In: Insalaco, E., Skelton, P.W., Palmer, T.J. (Eds.), Carbonate Platform Systems: Components and Interactions. *Geological Society Special Publications*, *178*, pp. 21–32.

Steuber, T., 2002. Plate tectonic control on the evolution of Cretaceous platform-carbonate production. *Geology*, *30*, 259–262.

Steuber, T., Löser, H., 2000. Species richness and abundance patterns of Tethyan Cretaceous rudist bivalves (Mollusca: Hippuritacea) in the Central-Eastern Mediterranean and Middle East, analysed from a palaeontological data base. *Palaeogeography*, *Palaeoclimatology*, *Palaeoecology*, 162, 75–104.

Steuber, T., Scott, R.W., Mitchell, S.F., Skelton, P.W., 2016. In: Stratigraphy and diversity dynamics of Jurassic-Cretaceous Hippuritida (rudist bivalves). Treatise on Invertebrate Paleontology Online, no. 81, Part N, Revised, Volume 1, Chapter 26C. University of Kansas, Lawrence, pp. 1–17.

Steuber, T., Löser, H., Mutterlose, J., Parente, M., 2023. Biogeodynamics of Cretaceous marine carbonate production. *Earth-Science Reviews*, 104341.

Sohl, N. F., 1987. Cretaceous gastropods: contrasts between Tethys and the temperate provinces. *Journal of Paleontology*, *61*, 1085–1111.

Skelton, P. W., and Smith, A. B., 2000. A preliminary phylogeny for rudist bivalves: sifting clades from grades. In: The evolutionary biology of the Bivalvia (Eds. E.M. Harper, J.D. Taylor and J.A. Crame), *Geological Society London Special Publication*, *177*, 97–127

Skelton, P. W., and Gili, E., 2025. Myths and models of rudist ecology. *Geologica Acta*, *23*, 1–22.

Tambutté, S., Holcomb, M., Ferrier-Pagès, C., Reynaud, S., Tambutté, É., Zoccola, D., Allemand, D., 2011. Coral biomineralization: From the gene to the environment. *Journal of Experimental Marine Biology and Ecology*, *408*, 58–78.

Taylor, P. D., 2008. Seawater chemistry, biomineralization and the fossil record of calcareous organisms. In Origin and Evolution of Natural Diversity (eds Okada K, Mawatari SF, Suzuki N, Gautam P). University of Hokkaido, Sapporo, Japan, pp. 21–29.

Truskowski, I., 2006. Calibración bioestratigráfica y correlación con datos geoquímicos de la Formación Querecual en su localidad tipo. Specialization Dissertation, Central University of Venezuela. 106 pp.

Vendrasco, M. J., Porter, S. M., Kouchinsky, A., Li, G., Fernandez, C. Z., 2010a. New data on molluscs and their shell microstructures from the Middle Cambrian Gowers Formation, Australia. *Palaeontology*, *53*(1), 97–135.

Vendrasco M. J., Porter S. M., Kouchinsky A. V., Li, G., Fernandez, C. Z., 2010b. Shell microstructures in early Mollusks. *The Festivus XLII*(*4*), 43–54.

Vermeij, G. J., 1977. The Mesozoic marine revolution: evidence from snails, predators and grazers. *Paleobiology*, *3*(3), 245-258.

Vermeij, G. J., 2017. Life in the arena: Infaunal gastropods and the late Phanerozoic expansion of marine ecosystems into sand. *Palaeontology*, *60*, 649–661.

Vermeij, G. J., Schindel, D. E., Zipser, E., 1981. Predation through geological time: evidence from gastropod shell repair. *Science*, *214*, 1024–1026.

Vinther, J., 2015. The origins of molluscs. *Palaeontology*, *58*, 19–34.

Walliser, E. O., Mertz-Kraus, R., Schöne, B. R., 2018. The giant inoceramid *Platyceramus platinus* as a high-resolution paleoclimate archive for the Late Cretaceous of the Western Interior Seaway. *Cretaceous Research*, *86*, 73–90.

Wanninger, A., and Wollesen, T., 2019. The evolution of molluscs. *Biological Reviews*, *94*, 102–115.

Wilson, M. E. J., and Vecsei, A., 2005. The apparent paradox of abundant foramol facies in low latitudes: their environmental significance and effect on platform development. *Earth- Science Reviews*, *69*, 133–168.

Work, R. C., 1969. Systematics, ecology, and distribution of the mollusks of Los Roques, Venezuela. *Bulletin of Marine* Science, *19*. 614–711.

9 Echinoderms

Echinoderms (Phylum Echinodermata) are calcareous invertebrates restricted to the marine realm. This morphologically diverse group of benthic animals appeared during the Middle Cambrian (e.g., Zhuravlev and Wood, 2008; Porter, 2010) and comprises approximately 7000 living species, and 13,000 fossil species. In modern oceans, echinoderms are epitomized in by familiar sea stars and sea urchins, and exotic sea lilies and feather stars (Pawson, 2007).

The contribution of these carbonate-producing benthic organisms to the global marine carbon budget has been historically overlooked, prevailing the view that calcium carbonate ($CaCO_3$) is predominantly produced and exported by marine photosynthesizers (coccolithophores, scleractinian corals, calcareous algae, and many foraminifera) in the biological pump. Modern echinoderms, which are distributed from tropical to cold waters along shorelines, shelves and deep basin settings, play a major role in carbonate production, with a global $CaCO_3$ contribution amount slightly below to that of the scleractinian corals (see Lebrato et al. 2010; Tambutté et al. 2011).

9.1 Major Groups and Stratigraphic Distribution

Animals within the phylum Echinodermata encompass an enormous morphological diversity (i.e., disparity) inherited from large-scale evolutionary patterns throughout the Phanerozoic (Cary and Hinman, 2017; Deline, 2021). Five extant classes of echinoderms are universally recognized (e.g., Pawson, 2007; Cary and Hinman, 2017): Asteroidea (sea stars, starfish), Ophiuroidea (brittle stars, serpent stars), Holothuroidea (sea cucumber, beche de mer), Crinoidea (sea lilies, feather stars), and Echinoidea (sea urchins, sand dollars, heart urchins). The last two groups are the most important rock builders during the Phanerozoic, and the focus of this chapter.

Systematic paleontology is not often applied to ancient echinoderms, since in most cases the fossilized elements of skeletons have gone through a series of taphonomic and diagenetic processes that prevent their taxonomic classification. Thus, echinoderm fragments are generally listed without any further assessment except as "crinoids" (mainly for Paleozoic, Triassic and Jurassic fossils) or "echinoids" (mainly for Cretaceous and Cenozoic fossils) (Flügel, 2010). It is noteworthy that the five extant classes of echinoderms first appeared in the Ordovician, but the phylum traces its roots back to the Middle Cambrian (see Zamora and Rahman, 2015).

Crinoids appeared during the Early Ordovician (Calcite I) (e.g., Guensburg, 2012) and were major rock forming components of carbonate rocks from the Ordovician to the Permian (Aragonite II) (e.g., Guensburg and Sprinkle, 1992; Ausich, 1998; Buitrón-Sánchez et al. 2007; Laya and Tucker, 2012; Ausich et al. 2015; Wright, 2015; Wright et al. 2017). Following the end-Permian biotic crisis which led to near extinction of crinoids, they recovered during the Early to Middle Triassic. Although they never recovered in terms of abundance compared to Paleozoic times, Mesozoic crinoids peaked in genus diversity in the Middle Jurassic (transition from Aragonite II to Calcite II), but decreased significantly starting in the Cretaceous (Gorzelak et al. 2016). Post-Cretaceous crinoid-dominated deposits are rare (Flügel, 2010). Their abundance has decreased progressively throughout the Cenozoic (James and Jones, 2015). Even so, there are approximately 650 extant species of crinoids (mainly deep-water forms), 100 stalked sea lilies and unstalked 550 feather stars (Pawson, 2007; Summers et al. 2014).

Echinoids appeared during the Late Ordovician (Calcite I), as part of the diversification and rise of other echinoderms during that time (Sprinkle and Guensburg, 2003; Thompson et al. 2021), but were minor rock forming elements throughout the Paleozoic. Few lineages of Permian echinoids crossed the Permian-Triassic boundary (mainly of the clade's stem

J. A. Méndez Dot, J. Méndez Baamonde, *Atlas of Tropical Carbonates*, https://doi.org/10.1007/978-3-031-25494-9_9

group and fewer of the crown group) and started a major diversification during the Aragonite II seas of the Triassic (see Thompson et al. 2017, 2018, 2019). By the Jurassic, the diversity and abundance of echinoids increased, becoming major rock forming elements and reaching their peak during the Calcite II seas of the Cretaceous (e.g., Horowitz and Potter, 1971; Néraudeau et al. 1995; Néraudeau and Mathey, 2000; Smith, 2007; Flügel, 2010; Ali, 2015; Michel et al. 2018). Echinoids abundance and diversity remained high during the entire Cenozoic (late in Calcite II and through Aragonite III) (e.g., McKinney, 1984; McKinney et al. 1992; Néraudeau, 1994; Donovan, 2001; Nebelsick and Kroh, 2002; Kroh and Nebelsick, 2010; Saucede et al. 2013; James and Jones, 2015; Aguilera et al. 2020). A little over 1000 echinoid species inhabit the oceans today (Mongiardino Koch et al. 2018).

9.2 Skeletal Mineralogy

Modern and ancient echinoderm skeletons are/were composed of IMC and HMC. In modern species, the Mg content is lower in crinoids, with values up to 12 wt.% $MgCO_3$; while this value goes up to 18 wt.% in echinoids (Scholle and Ulmer-Scholle, 2003; Dickson, 2004; Gorzelak et al. 2013).

9.3 Ecology and Environment

Echinoderms are filter and detritus-feeding, heterotrophs (light-independent) that thrive/thrived mainly under mesotrophic to eutrophic conditions (see Lukasik et al. 2000; Hallock, 2001; Mutti and Hallock, 2003). These benthic organisms live as burrowers, grazers and bioeroders. More than 80% of the global $CaCO_3$ production from echinoderms in modern oceans occurs between 0 and 800 m, with the highest contribution attributed to the shelf and upper slope (Lebrato et al. 2010).

Crinoids are/were fully marine organisms restricted to waters close to normal salinities (James and Jones, 2015). During the Paleozoic, crinoids were common on shallow platforms and ramps. Their post-Paleozoic distribution is much reduced and shifted from shallow shelf to slope and basinal environments (Scholle and Ulmer-Scholle, 2003). Extant crinoids are mainly found in deep oceans, which makes their detailed functional studies problematic (Cary and Hinman, 2017).

Fossils and extant crinoids (mainly shallow water and deep water, respectively) live/lived stalked or permanently attached (sea lilies), or unstalked (feather stars) (e.g., Cary and Hinman, 2017). Most Paleozoic crinoids were sessile, filter-feeder attached organisms, either to hard substrates, to skeletal fragments, or rooted in soft substrate (Guensburg and Sprinkle, 1992). The apparent low predation pressure on modern crinoids may be a consequence of unique biochemical, morphologic, and behavioral traits. The suggestion that crinoids may be distasteful, or even toxic to predators is frequently repeated. However, observations of deep water modern stalked crinoids suggest that these are possible prey for some fishes that feed on the arms. The crinoid fossil record confirms the impression gained from modern crinoids by revealing possible predation on crinoids. The apparent scarcity of crinoid calyxes in relation to the great volume of columnal material suggests that carnivores were eating crinoids. An alternative explanation for the apparent discrepancy between preserved calyxes and columnals may be that disarticulated calyx plates are so small and rare compared to columnals that they can be easily overlooked by most collectors (see Meyer and Ausich, 1983).

Most echinoids (sea urchins) live in normal marine waters that range from 22 to 38‰ (James and Jones, 2015; but see Russell, 2013). They inhabit/inhabited varied depositional settings and environments, from carbonate-dominated to siliciclastic-dominated settings, and from intertidal to deep environments, although more common of shallow platforms and ramps. Modern echinoids are abundant in tropical, temperate and polar waters (e.g., Nebelsick, 1999; Néraudeau et al. 2001; David et al. 2005; Chiappone et al. 2013; Tyler et al. 2018).

Regularia, or regular echinoids (sea urchins with spherical/globular tests, five-fold radial symmetry, and long spines, among other diagnostic features), are/were epifaunal detritus feeders living especially on hard bottoms. On the other hand, irregularia echinoids (or irregular) are/were infaunal and occur/occurred within or on soft bottoms (Flügel, 2010; Petsios et al. 2021). Regular echinoids are one of the dominant algal grazers in shallow marine communities. They can also be major bioeroders in shallow marine environments. Irregular echinoids, through their deposit-feeding activities, play an important role as bioturbators, modifying the structure of the sediment. Epifaunal prey are generally considered more vulnerable to predation, and epifaunal taxa lacking in other defensive behaviours (e.g., swimming, cementing, spines) often experience higher rates of extinction (see Petsios et al. 2021 and references therein). Large spines also protect epifaunal echinoid tests from failure under impact on hard substrata in high-energy settings (Strathmann, 1981).

Starting in the Early Jurassic, echinoids experienced a relatively rapid radiation of infaunal (irregular) lineages which continued into the early Cenozoic, with infaunal echinoids surpassing the diversity of epifaunal echinoids in the Cretaceous (Kier, 1982; Kroh and Smith, 2010; Petsios et al. 2021). The Mesozoic infaunalization of echinoids may have offered a means to escape increasing predation pressure. The latter was a driving force for epifaunal echinoids to develop antipredatory morphologies (e.g., spine length, width, shape, ornamentation, and density of spines) (Lapic, 2021). The Cenozoic intensification of predation on echinoids by drilling predators (e.g., cassid gastropods) may have contributed to the continued radiation of infaunal echinoid clades in the Cenozoic (Kier, 1982; Petsios et al. 2021).

9.4 Carbonate Factories and Skeletal Grain Associations

Echinoderms (crinoids and echinoids) are components of the Heterozoan factory of Michel et al. (2019), as well as occasional to common secondary components of the Photozoan T- and Photo C-factories of these authors (see Appendix A for a detailed treatment of these carbonate benthic factories). This section discusses only the skeletal grain associations involving echinoderms that are represented in Venezuelan carbonate microfacies, as illustrated by the examples in the photographic set.

In Middle Permian carbonates of the Palmarito Formation (Venezuelan Andes), crinoids are abundant components of Heterozoan carbonates *sensu* James (1997), in association with bryozoans, brachiopods, ostracods and molluscs. These ancient deposits include bryonoderm lithofacies *sensu* Beauchamp (1994) (dominated by bryozoans and crinoids). These heterotrophic calcifiers are also common of photozoan-heterozoan facies (*sensu* James, 1997) dominated by LBF (fusulinids) and calcareous green algae (see Laya and Tucker, 2012).

In Early Cretaceous carbonates of the Cogollo Group (MB), the heterozoan association between echinoids and molluscs (mainly bivalves and oysters, plus gastropods) is the most common grain association. Secondary contribution of other heterotrophic calcifiers (bryozoans, ostracods, serpulids, smaller miliolids, rudists, plus micritic and glauconitic peloids and oncoids/cortoids) is occasionally significant. The mixed heterozoan-photozoan association between echinoids and light-dependent calcifiers (LBF and calcareous algae) is locally common. Depending on the content of echinoids versus other bioclastic components (both heterozoan and photozoan biota), heterozoan-dominated microfacies of the Cogollo Group that are rich in these heterotrophic calcifiers can be grouped into foramol lithofacies *sensu* Lees and Buller (1972), molechfor lithofacies *sensu* Carannante et al. (1988), or echinofor lithofacies *sensu* Hayton et al. (1995). It is noteworthy that foramol, molechfor and echinofor are Cenozoic-biased terms in origin, but applicable to Cretaceous microfacies (e.g., Carannante et al. 1995; Simone et al. 2012; Kaya et al. 2020), and not restricted to non-tropical carbonates (see Pomar et al. 2004; Wilson and Vecsei, 2005; Brandano et al. 2009; Klicpera et al. 2015). See Appendix A for definitions of the skeletal grain associations considered in this book.

In Late Oligocene-Early Miocene carbonates of the Perla Limestone (EGVSB), echinoids are moderately common to common secondary components of photozoan-dominated microfacies in association with red algae and LBF, plus an important contribution of other heterotrophs, mainly bryozoans and barnacles (balanids), and less common molluscs (see Pomar et al. 2015). These microfacies are the expression of the PhotoC-factory developed at tropical paleolatitutes, and can be grouped into rhodalgal lithofacies *sensu* Carannante et al. (1988), or rhodechfor lithofacies *sensu* Hayton et al. (1995). In addition, echinofor lithofacies in which LBF outspace the echinoids are locally common in this unit.

In Early Miocene carbonates of the Churuguara Formation (FB), echinoids are dominant components of heterozoan (littoral barrier and inner-proximal ramp) microfacies in association with molluscs, as well as with smaller and larger miliolids (molechfor lithofacies). In this unit, echinoids are also common components of photozoan-dominated (middle ramp) facies in association with LBF and red algae (rhodechforlithofacies) (see Albert-Villanueva, 2016). In lithified Quaternary facies of Aves Island (carbonate basement and beachrocks), echinoids occur as accessory components within photozoan–heterozoan microfacies dominated by coralline algae and mollusks (bivalves and gastropods), with secondary contributions of green algae (*Halimeda*) and scleractinian corals. In these microfacies, LBF (*Peneroplis* sp., *Amphistegina* sp., and *Homotrema rubrum*) are also present as accessory components.

9.5 Morphology and Diagnostic Elements

Echinoderms as distinctive group of animals may be briefly defined as possessing a skeleton of calcium carbonate in the form of calcite; a unique water-vascular system which mediates feeding, locomotion, and other functions. They are an exceptional group of bilaterians which take many forms of symmetry. Pentameral symmetry, developed from a bilaterally symmetric larva, is the major form and the other forms are derived from it. Many echinoderms show a more or less conspicuous bilateral symmetry superimposed upon a radial pattern. Some taxa of sea stars (Astreoidea) and feather stars (Crinoidea) have "outgrown" bilateral and/or five-part symmetry and may have from more than 10 to up 50 or more arms (see Pawson, 2007; Ji et al. 2012; Li et al. 2020).

Echinoderm skeletons (up to 15 cm in some echinoids and 1 m in some crinoids) are made of multi-plate or multi-element skeletons. Crinoids consist of a root-like attachment, a long stem made of columnals (or stem plates), and a calyx with radiating, feathery arms. Echinoids consist of a globular to discoidal, hollow, endoskeletal tests (coronal plates). In life, echinoid tests are covered with elongate, moveable spines, which normally detach after death

(Scholle and Ulmer-Scholle, 2003). Echinoid spines show a great variety in shape and size (commonly long in regular echinoids, and short in irregular echinoids). In cross section, spines exhibit characteristic radial pattern, and may be hollow (diadematoid type) or solid (cidaroid type). For both crinoids and echinoids, each skeleton element consists of a three-dimensional porous network termed "stereom", which is composed of crystallographically uniformly oriented carbonate crystals forming a highly porous meshwork (Flügel, 2010). Calcitic units comprising the stereom are termed ossicles in crinoids, and plates in echinoids (although often also termed ossicles by some authors). The stereom network architecture and density varies widely. The porosity of the crystalline structure ranges between ~10 and 70% volume, with an average of 50–60% (Kroh and Nebelsick, 2010).

9.6 Preservation as Fossils

Echinoderms are commonly preserved as single fragments of the original articulate multi-plate skeletons. Individual elements exhibit unit extinction under CPL due to crystallographic continuity. Disarticulated elements of crinoids and echinoids are preserved under a wide variety of shapes (symmetric or irregular) and sizes (mm- to cm-sized). Crinoid columnal plates may exhibit spherical to ellipsoidal shapes, pentagonal symmetry, or irregularly-shaped forms. Serrated outlines are common diagnostic features. Plates may preserve or not characteristic hollow center (termed lumen) (see Flügel, 2010). Echinoid plates are perforated and exhibit regularly arranged meshwork of pores and plates. In many cases, mechanical and chemical diagenesis prior to final deposition leads to poor preservation of originally symmetric patterns. Echinoid spines do commonly preserve their radiating "flower-like" structures.

Despite the difference in microstructures related to the stereom architecture (shape size of crystalline units, lattice arrangement and porosity), both crinoids and echinoids individual HMC elements react very similar to meteoric and shallow-burial diagenetic alterations. Chemical reconfiguration of metastable skeletal structures led to the formation of LMC syntaxial products characterized by unit (single crystal) extinction and conspicuous zoning patterns. Echinoderm grains are commonly overgrown by precipitation of syntaxial LMC cement (e.g., Emery and Marshall, 1989; Frank et al. 1995; Nebelsick, 2004; El-Saiy and Jordan, 2007; Mahboubi et al. 2010; Ukar and Laubach, 2016). Diagenetic stabilization of HMC occurs rapidly in geologic terms (10^1 to 10^5 yr). Skeletal HMC loses about half its Mg concentration during such neomorphic processes (Dickson, 2004). Intraparticle pores in echinoderm skeletons that are not filled by syntaxial calcite or micritic sediments are filled by authigenic minerals (e.g., pyrite, glauconite and fluorapatite).

9.7 Geochemical Proxies in Echinoderms

Fossil echinoderm skeletons have long been ignored as potential Mg/Ca seawater proxies. Dickson (2004) showed that preserved echinoderms (i.e., deposited within fine-grained sediments of low porosity and permeability) can be utilized as empirical archives for the accurate monitoring of secular variations (long term phases of 100–200 Ma) of seawater Mg/Ca ratio (between 1.0 and 5.0) throughout the Phanerozoic. Results obtained by this author on the study of unaltered ossicles from the Cambrian to the Eocene suggest that echinoderms with high mol% $MgCO_3$ composition (9–12%) occurred during icehouse cycles (aragonite seas with high Mg/Ca ratio); while echinoderms with low mol% $MgCO_3$ composition (4–6%) occurred during greenhouse cycles (calcite sea systems with low Mg^{2+}/Ca^{2+} ratio) (see also Ries, 2004).

There are, however, serious difficulties with using the skeletal Mg/Ca of echinoderms as paleo-seawater Mg/Ca proxy. Gorzelak et al. (2013) state that physiology (which affects skeletal Mg content) differs significantly among extant echinoderm groups and possibly also differed among fossil echinoderm groups. Environmental changes (temperature, salinity, pH, among other factors) cannot be ascribed solely as factors controlling the distribution of Mg in echinoid skeletons. Therefore, according to those authors, great care should be taken in the use of their skeletons in the past seawater temperature and Mg/Ca ratio reconstructions (see also Gorzelak et al. 2016).

9.8 Echinoderms: Photographic Set and Descriptive Captions

In all photos, IW = image width (the true length of the horizontal axis of the view)

Holocene. Aves Island.

These two images depict the same modern echinoid grain. The grain, part of a beachrock deposit, retains its original, orderly arrangement of interlocking HMC units, indicating that it has not yet undergone significant meteoric diagenesis. In the image on the left, the entire grain exhibits distinct unit extinction (light gray), uniformly darkening upon rotation (image on the right), because the lattice behaves optically as a single crystal, a characteristic feature of both echinoids and crinoids. The intraparticle pores (visible as dark, aligned, elongated stripes) are infilled with micrite sediment.

IW = 0.6 mm and CPL each

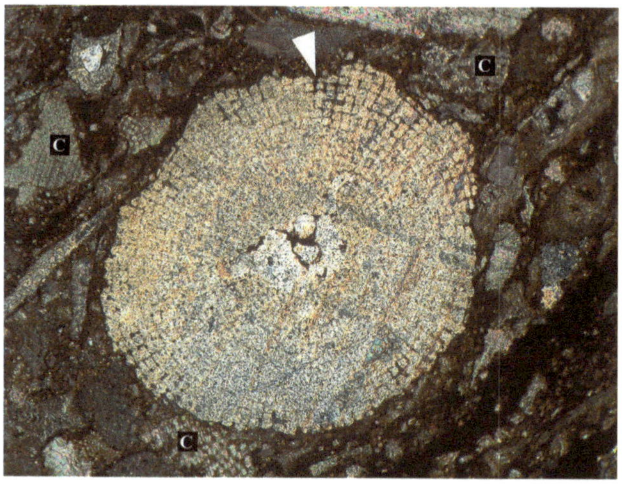

Middle Permian (Cisularian to Guadalupian). Palmarito Formation. Venezuelan Andes

Irregularly rounded crinoid plate showing traces of the central-axial canal (termed lumen) in a crinoidal packstone. The radial arrangement of square-to-rectangular skeletal units (ossicles) is clearly distinguished in the outer margins of the grain (white arrow). Note the syntaxial overgrowth cement, in optical continuity with the host grain, formed by dissolution-reprecipitation of the HMC stereom. Also note the parallel extinction bands crossing the grain, a diagnostic feature of optical continuity. These features are also visible in smaller crinoid (C) grains in the image.

IW = 2.4 mm. CPL

Middle Permian (Cisularian to Guadalupian). Palmarito Formation. Venezuelan Andes

Irregularly-shaped crinoid grain surrounded by a packed grain-supported fabric with abundant pressure solution features. Single crystal continuity of the central grain is apparent due to conspicuous, closely-spaced twinning lamellae crossing the grain. These optical features, are commonly observed on diagenetic-altered echinoderms, both crinoids and echinoids.

IW = 2.4 mm. PPL

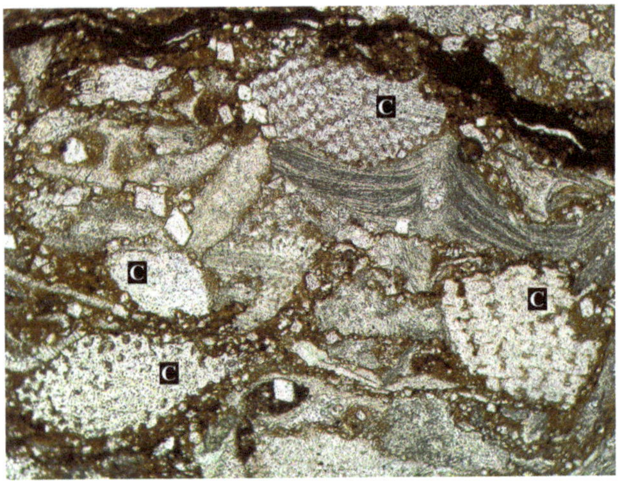

Middle Permian (Cisularian to Guadalupian). Palmarito Formation. Venezuelan Andes

Crinodial packstone in which crinoid (C) fragments exhibit characteristic serrate outlines and dusty appearance. Crinoids are among the most abundant bioclasts in the Palmarito Formation. In this unit, these heterotrophs are common in both carbonate-dominated (from mudstones to packstones and boundstones) and mixed carbonate-siliciclastic microfacies (see Laya and Tucker, 2012).

IW = 2.4 mm. PPL

Middle Permian (Cisularian to Guadalupian). Palmarito Formation. Venezuelan Andes

Close view of a crinoid fragment embedded in a slightly organic-rich muddy matrix. Note the radiating "V-shaped" arrangement of calcitic units (ossicles). Well-preserved microstructure in this fragment indicates the grain escaped from destructive meteoric diagenesis. Pore spaces within the calcified, reticular structure were filled by synsedimentary micrite. Transient low-oxygen conditions promoted the local precipitation of fluorapatite (showing characteristic brown-orange color), which is replacing the micritic infilling.

IW = 1.3 mm. PPL

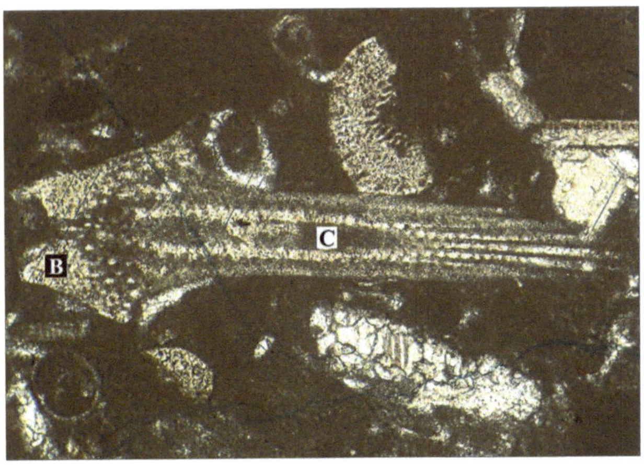

Cretaceous (middle Aptian). Piché Member, Apón Formation. Maracaibo Basin

Longitudinal section of a complete echinoid spine in a mud-supported limestone. The photomicrograph shows the elongate and thin spine (to the right), the hollow center (C), and the characteristic bulbous attachment "ball" (B). The micritic synsedimentary infilling reveals the original highly porous structure of the spine.

IW = 2.4 mm. PPL

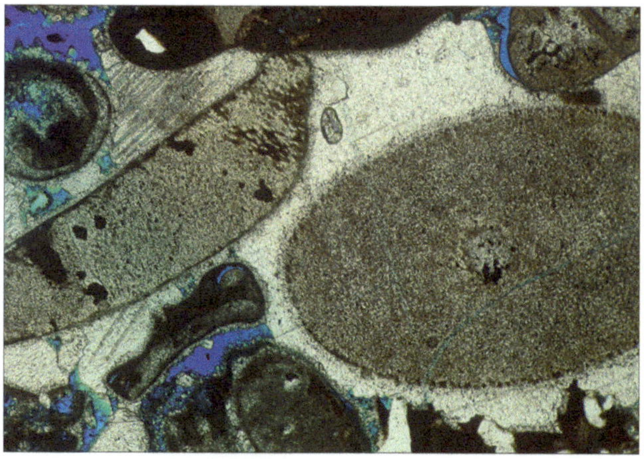

Cretaceous (latest Albian). Maraca Formation. Maracaibo Basin

Echinoid grains showing dusty appearance. The whitish microspar precipitate that is partially covering both grains is predated by micritic infilling of the original highly porous skeletons. Primary interparticle pores surrounding both echinoids are filled by syntaxial overgroth. This LMC cement shows traces of twinning lamellae, indicating optical continuity between the overgrowth and echinoid grains.

IW = 2.4 mm. PPL

Cretaceous (middle Albian). Lisure Formation. Maracaibo Basin

Transverse section of a regular echinoid in a poorly-sorted sandy limestone. Characteristic radial symmetry of the original highly porous meshwork is fairly preserved. Light-colored stereom lattice contrasts with dark micritic infilling. The latter predates final deposition of the grain in a mud-lean environment. It highlights the well-preserved, radial arrangement of thin spines attached to the outer surface of the plate. Syntaxial calcitic overgrowth surrounding the echinoid shows a distinctive plicated outline.

IW = 2.4 mm. PPL

Cretaceous (middle Albian). Lisure Formation. Maracaibo Basin

Wackestone in which different-sized echinoid grains are major skeletal components. Note the area of the large, central echinoid grain indicated with the white arrow, where small calcified units are floating in the micritic matrix. This "micritic way of preservation" could be the result of crystal diminution by diagenetic dissolution (see Neugebauer, 1978). Alternatively, a locally more porous original lattice could explain this distinctive preservation pattern.

IW = 2.4 mm. PPL

Cretaceous (late Barremian). Tibú Member, Apón Formation. Maracaibo Basin

Grains in this dolomitic wackestone are partially dolomitized echinoid fragments. Their original HMC composition most likely served as a source of Mg^{2+} for dolomitization of the micritic matrix. Even so, echinoid grains preserved partially diagnostic serrate morphology. Localized pyritization (black spots), which is very common in wackestones of the Tibú Member, postdates dolomitization and suggests transient low-oxygen conditions during early marine diagenesis.

IW = 2.4 mm. PPL

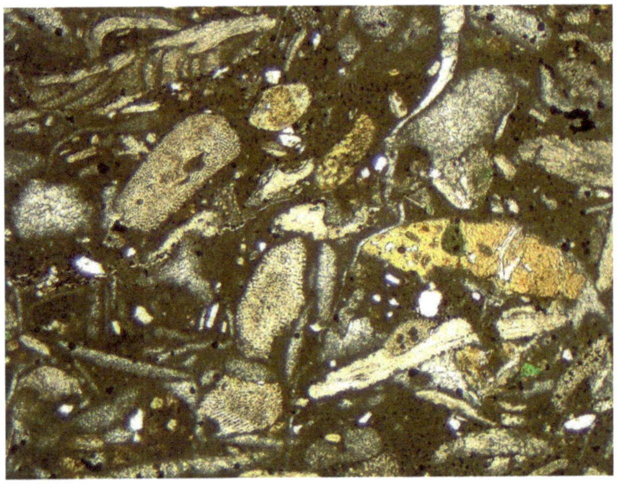

Cretaceous (middle Albian). Lisure Formation. Maracaibo Basin

Packstone of fragments of echinoids and molluscs. This heterozoan grain association is the most common in shallow microfacies of the Cogollo Group. Eutrophic conditions are associated to with the dominance of these heterozoan calcifiers. Abundant pyrite framboids (black spots) locally replacing the micritic matrix suggests transient low-oxygen conditions during early marine diagenesis. Pyrite precipitation was promoted by available organic matter associated to nutrient-rich conditions during deposition.

IW = 4.2 mm. PPL

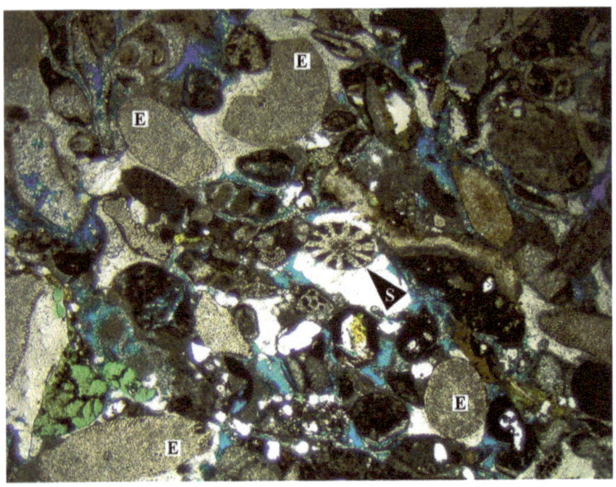

Cretaceous (latest Albian). Maraca Formation. Maracaibo Basin

Echinoid (E) grains are major components in this porous grainstone. Note the echinoid spine (S) in the center of the image exhibiting characteristic flower-like morphology. Gastropods, bryozoans, green algae, aggregate grains and glauconitic peloids are secondary components in this sample. Note the syntaxial calcite overgrowth surrounding the echinoid grain in the image. This meteoric cement is occluding primary interparticle pores.

IW = 7.5 mm. PPL

Cretaceous (middle Aptian). Piché Member, Apón Formation. Maracaibo Basin

An echinoid spine seen in transverse section and exhibiting characteristic circular shape and densely arranged radiating calcified elements (septa). The latter widen outward forming a characteristic flower-like pattern. The ring-like, central canal (termed lumen) serves as a wall for the growth of these wedge-shaped elements. Micritic sediments, partially replaced by glauconite and pyrite, are infilling the spine intraparticle pores. Among individual elements of echinoids, fossil spines are commonly the most well-preserved.

IW = 0.5 mm. PPL

Cretaceous (middle Albian). Lisure Formation. Maracaibo Basin

An equinoid spine served as nucleus for the formation of the ooid in the image. The morphology of this Albian specimen, consisting of 15 radial septa, is similar to that of the previous Aptian one (the latter is about 8 Myr younger). It highlights the well-preserved state of the delicate spine, as the formation of the ooid involved movilization and abrasion in an agitated environment. In the Lisure Formation, echinoids and other heterozoan skeletal components (molluscs and bryozoans) are common nuclei in ooids and other coated grains (oncoids and cortoids).

IW = 0.5 mm. PPL

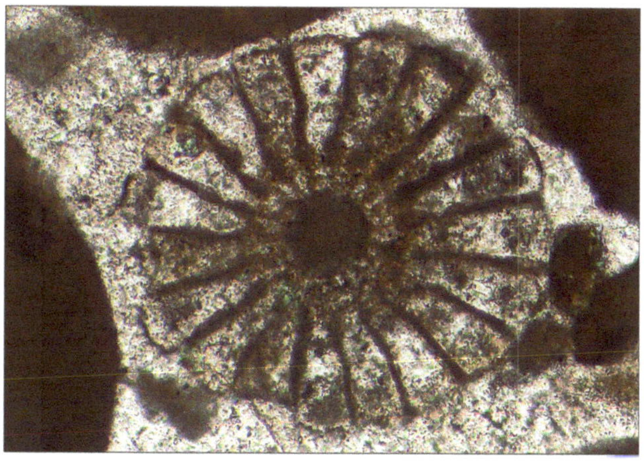

Early Miocene. Churuguara Formation. Falcón Basin

Another transverse section of an echinoid spine. Note the morphologic similarities and differences between this Cenozoic specimen and the two previous Cretaceous examples, despite the age difference of approximately 100 Myr. The increase in septa thickness towards the periphery of the structure is less marked in this early Cenozoic specimen. This spine is also larger in diameter and has a smaller central canal relative to the overall diameter of the structure.

IW = 1.3 mm. PPL

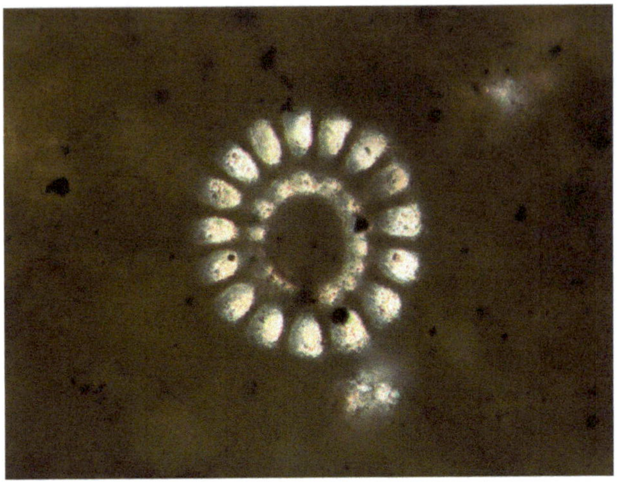

Cretaceous (early-late Aptian). Piché Member, Apón Formation. Maracaibo Basin

This and the following spine are examples of the occurrence of echinoids under nutrient-rich and low-oxygen conditions. In this image, note the dark and homogenous texture of the micritic matrix in which the small and delicate spine is embedded. This sample belongs to a transgressive deposit located at the lower part of the Piché Member, an interval characterized by the heterozoan association between echinoids, SBF and sponge spicules.

IW = 0.5 mm. PPL

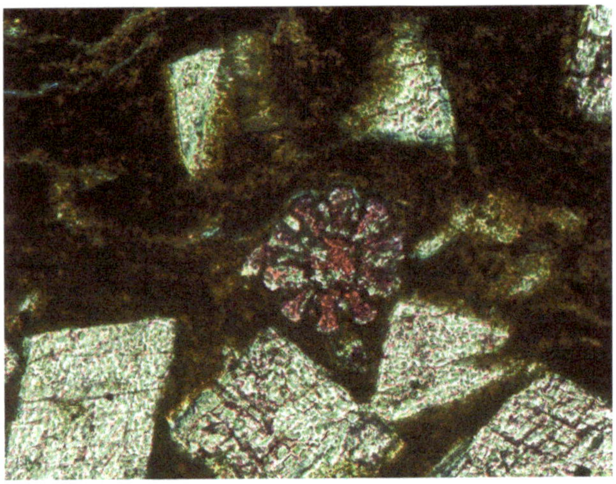

Cretaceous (middle Aptian). Piché Member, Apón Formation. Maracaibo Basin

This sample corresponds to the OAE 1b first sub-event. The pink-stained echinoid spine in the center of the imaged is embedded in a dark, slightly organic-rich, pyritic, clayey matrix. Contrary to previously shown examples of echinoid spines consisting of a hallow central canal, this small spine has a solid center (cidaroid type). In addition to low-oxygen and eutrophic conditions during deposition, euhedral dolomite crystals indicate hypersaline conditions during marine diagenesis, highlighting the echinoid's capacity to thrive under such stressed conditions.

IW = 0.5 mm. PPL

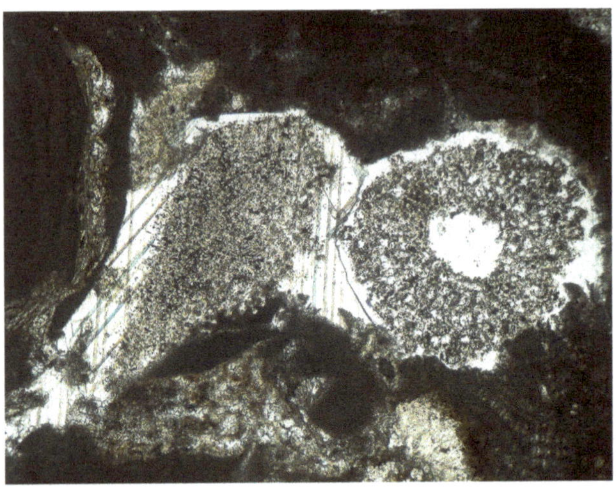

Late Oligocene-Early Miocene. Perla Limestone. Eastern Gulf of Venezuela Sub-Basin

Echinoid fragments in the image are in optical continuity with syntaxial calcite overgrowth and are surrounded by red algae. Note the twin lamellae (parallel lines of different colors), a common feature of syntaxial calcitic overgrowth. Cement precipitation occurred prior to mechanical compaction, as the formation of the overgrowth required empty spaces surrounding the echinoid grains. We selected this example for comparison to previous Cretaceous examples, as diagenetically altered echinoids fossilize similarly, despite differences in age.

IW = 2.4 mm. PPL

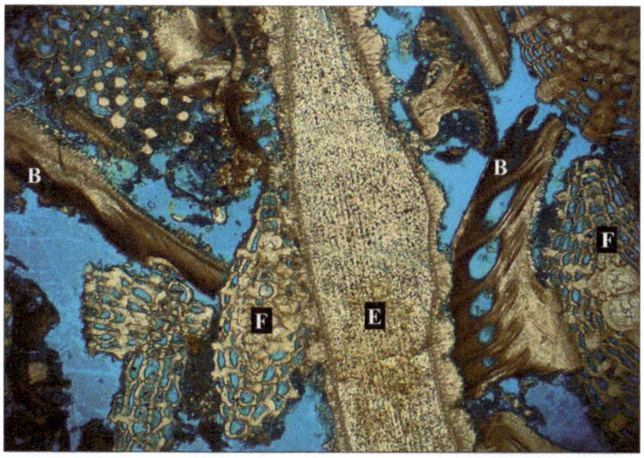

Late Oligocene-Early Miocene. Perla Limestone. Eastern Gulf of Venezuela Sub-Basin

A large echinoid (E) grain in the center of the image, accompanied by highly porous tests of LBF (F) and barnacle (B) fragments (photozoan-heterozoan association) in a porous rudstone. The original highly porous network of the stereom is occluded by pore-filling cement. Traces of the original microstructure of the stereom are visible. Note the syntaxial overgrowth with plicated outline that is rimming the echinoid grain. This rim is contemporary to the precipitation of the cement-filling intraparticle porosity.

IW = 4.2 mm. PPL

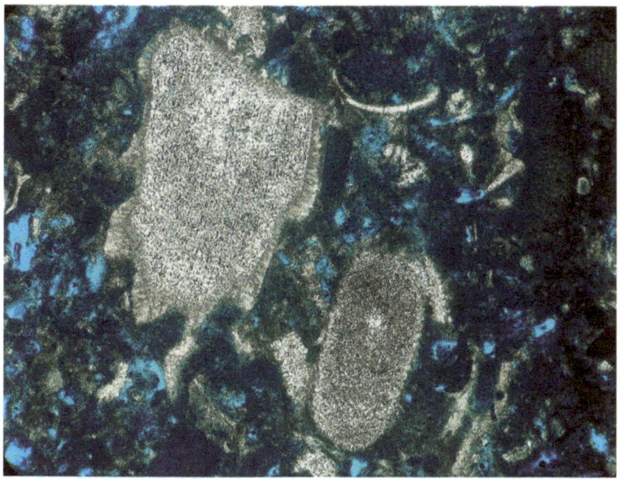

Late Oligocene-Early Miocene. Perla Limestone. Eastern Gulf of Venezuela Sub-Basin

Echinoids in the image (dusty central grains) contrast with surrounding dark-blue stained elements (mainly red algae). Early diagenetic cementation of metastable HMC stereoms created a mineralogically stable, non-porous carbonate substrate, thus preventing their total corrosion during posterior late diagenetic, CO_2-driven dissolution and microporosity development, and hence remaining undyed by blue epoxy filling porosity.

IW = 4.2 mm. PPL

Late Oligocene-Early Miocene. Perla Limestone. Eastern Gulf of Venezuela Sub-Basin

This packed fabric consists of echinoids (E) (light-brown, dusty grains), LBF (F) (dark-brown grains), and red algae (RA) (darker grains). This photozoan-heterozoan association is common in microfacies of the Perla Limestone. Abundant cement (white precipitate) is mainly of the syntaxial type, as is surrounding the echinoderms and adjacent grains. Cementation occurred in the shallow-burial environment and shortly predates mechanical compaction.

IW = 7.5 mm. PPL

References

Aguilera, O., Bencomo, K., de Araújo, O. M. O., Dias, B. B., Coletti, G., Lima, D., da Silva-Caminha, S. A. F., Polck, M., Alves-Martins, M. V., Jaramillo, C., Kuter, V. T., Lopes, R. T., 2020. Miocene heterozoan carbonate systems from the western Atlantic equatorial margin in South America: The Pirabas formation. *Sedimentary Geology*, *407*, 105739.

Albert-Villanueva, E., 2016. Facies y secuencias deposicionales mixtas carbonático-siliciclásticas del Mioceno inferior de la Cuenca de Falcón (noroeste de Venezuela) como modelo exploratorio en el Caribe. Doctoral Dissertation. University of Barcelona. Volumes 1 and 2.

Ali, M. S. M., 2015. Echinoids from the Galala Formation (Upper Cretaceous, Cenomanian) of the Ras Gharib-Sheikh Fadl road, Eastern Desert, Egypt. *Cretaceous Research*, *56*, 569–586.

Ausich, W. I., 1998. Early phylogeny and subclass division of the Crinoidea (Phylum Echinodermata). *Journal of Paleontology*, *72*, 499–510.

Ausich, W. I., Kammer, T. W., Wright, D. F., Cole, S. R., Peter, M. E., Rhenber, E. C., 2015. Toward a phylogenetic classification of the Crinoidea (Echinodermata). Pp. 29–32 in S. Zamora and I. Rábano, eds., Progress in Echinoderm Paleobiology. Cuadernos del museo Geominero, 19, Instituto Geológico y Minero de España, 292 pp.

Beauchamp, B., 1994. Permian climatic cooling in the Canadian Arctic. *Geological Society of America Special Paper*, *288*, 229–246.

Brandano, M., Frezza, V., Tomassetti, L., Cuffaro, M., 2009. Heterozoan carbonates in oligotrophic tropical waters: the Attard member of the lower coralline limestone formation (Upper Oligocene, Malta). *Palaeogeography, Palaeoclimatology, Palaeoecology*, *274*, 54–63.

Buitrón-Sánchez, B. E., Almazán-Vázquez, E., Vachard, D., 2007. Middle Permian crinoids (Echinodermata, Crinoidea) from Cerros Los Monos, Caborca, Sonora, Mexico and paleogeographic considerations. *Revista Mexicana de Ciencias Geológicas*, *24*, 344–353.

Carannante, G., Esteban, M., Milliman, J. D., Simone, L., 1988. Carbonate lithofacies as paleolatitude indicators: problems and limitations. *Sedimentary Geology*, *60*, 333–346.

Carannante, G., Cherchi, A., Simone, L., 1995. Chlorozoan versus foramol lithofacies in Upper Cretaceous rudist limestones. *Palaeogeography, Palaeoclimatology, Palaeoecology*, *119*, 137–154.

Cary, G. A., and Hinman, V. F., 2017. Echinoderm development and evolution in the post-genomic era. *Developmental Biology*, *427*, 203–211.

Chiappone, M., Rutten, L. M., Miller, S. L., Swanson, D. W., 2013. Recent trends (1999–2011) in population density and size of the echinoid *Diadema antillarum* in the Florida Keys. *Florida Scientist*, 23–35.

David, B., Choné, T., Festeau, A., Mooi, R., De Ridder, C., 2005. Biodiversity of Antarctic echinoids: a comprehensive and interactive database. *Scientia Marina*, *69*, 201–203.

Deline, B., 2021. Echinoderm Morphological Disparity: Methods, Patterns, and Possibilities. Elements of Paleontology. 40 pp. Cambridge University Press, Cambridge.

Dickson, J. A. D., 2004. Echinoderm skeletal preservation: calcite-aragonite seas and the Mg/Ca ratio of Phanerozoic oceans. *Journal of Sedimentary Research*, *74*, 355–365.

Donovan, S. K., 2001. Evolution of Caribbean echinoderms during the Cenozoic: moving towards a complete picture using all of the fossils. *Palaeogeography, Palaeoclimatology, Palaeoecology*, *166*, 177–192.

El-Saiy, A. K., and Jordan, B. R., 2007. Diagenetic aspects of tertiary carbonates west of the Northern Oman Mountains, United Arab Emirates. *Journal of Asian Earth Sciences*, *31*, 35–43.

Emery, D., and Marshall, J. D., 1989. Zoned calcite cements: has analysis outpaced interpretation? *Sedimentary Geology*, *65*, 205–210.

Frank, T. D., Lohmann, K. C., Meyers, W. J., 1995. Chronostratigraphic significance of cathodoluminescence zoning in syntaxial cement: Mississippian Lake Valley Formation, New Mexico. *Sedimentary Geology*, *105*, 29–50.

Flügel, E, 2010. Microfacies of Carbonate Rocks – Analysis, Interpretation and Application. Springer, Heidelberg-Dordrecht-London-New York, 984 pp.

Gorzelak, P., Stolarski, J., Mazur, M., Meibom, A., 2013. Micro-to nanostructure and geochemistry of extant crinoidal echinoderm skeletons. *Geobiology*, *11*, 29–43.

Gorzelak, P., Krzykawski, T., Stolarski, J., 2016. Diagenesis of echinoderm skeletons: constraints on paleoseawater Mg/Ca reconstructions. *Global and Planetary Change*, *144*, 142–157.

Guensburg, T. E., 2012. Phylogenetic implications of the oldest crinoids. *Journal of Paleontology*, *86*(3), 455-461.

Guensburg, T. E., and Sprinkle, J., 1992. Rise of echinoderms in the Paleozoic evolutionary fauna: significance of paleoenvironmental controls. *Geology*, *20*, 407–410.

Hallock, P., 2001. Coral reefs, carbonate sedimentation, nutrients, and global change. In: Stanley, G.D. (Ed.), The History and Sedimentology of Ancient Reef Ecosystems. Kluwer Academic/Plenum Publishers, pp. 387–427.

Hayton, S., Nelson C., Hood S., 1995. A skeletal assemblage classification system for non-tropical carbonate deposits based on New Zealand Cenozoic limestones. *Sedimentary Geology*, *100*. 123–141.

Horowitz, A. S., and Potter, P. E., 1971. Introductory Petrography of Fossils: New York, Springer, 302 pp.

James, N. P., 1997. The cool-water carbonate depositional realm. In: James N. P., Clarke, J., (eds.), Cool-water carbonates, vol 56. SEPM Society for Sedimentary Geology, Tulsa, 1–20.

James, N. P., and Jones, B., 2015. Origin of Carbonate Sedimentary Rocks. John Wiley & Sons, Ltd, Chinchester, 464 pp.

Ji, C., Wu, L., Zhao, W., Wang, S., Lv, J., 2012. Echinoderms have bilateral tendencies. *PLoS One*, *7*(1), e28978.

Kaya, M. Y., Dupont-Nivet, G., Proust, J. -N., Roperch, P., Meijer, N., Frieling, J., *et al.*, 2020. Cretaceous evolution of the Central Asian proto-Paratethys Sea: Tectonic, eustatic, and climatic controls. *Tectonics*, *39*, e2019TC005983.

Kier, P. M., 1982. Rapid evolution in echinoids. *Palaeontology*, *25*, 1–9.

Klicpera, A., Michel, J., Westphal, H., 2015. Facies patterns of a tropical heterozoan carbonate platform under eutrophic conditions: the Banc d'Arguin, Mauritania. *Facies*, *61*, 421.

Kroh, A., and Nebelsick, J. H., 2010. Echinoderms and Oligo-Miocene carbonate systems: potential applications in sedimentology and environmental reconstruction. *International Association of Sedimentologists, Special Publication 42*, 201–228.

Kroh, A., and Smith, A. B., 2010. The phylogeny and classification of post-Palaeozoic echinoids. *Journal of systematic Palaeontology*, *8*, 147–212.

Lapic, W. A., 2021. An Asynchronous Mesozoic Marine Revolution: Drilling Versus Durophagy in Post-Paleozoic Echinoids. Magister Dissertation, Miami University.

Laya, J. C., and Tucker, M. E., 2012. Facies analysis and depositional environments of Permian carbonates of the Venezuelan Andes: Palaeogeographic implications for Northern Gondwana. *Palaeogeography, Palaeoclimatology, Palaeoecology*, *331*, 1–26.

Lebrato, M., Iglesias-Rodríguez, D., Feely, R. A., Greeley, D., Jones, D. O., Suarez-Bosche, N., Lampitt, R. S., Cartes, J. E., Green, D. R. H., Alker, B., 2010. Global contribution of echinoderms to the marine carbon cycle: $CaCO_3$ budget and benthic compartments. *Ecological Monographs*, *80*, 441–467.

Lees, A., and Buller, A. T., 1972. Modern temperate-water and warm-water shelf carbonate sediments contrasted. *Marine Geology*, *13*, M67–M73.

Li, Y., Omori, A., Flores, R. L., Satterfield, S., Nguyen, C., Ota, T., *et al.*, 2020. Genomic Insights of Body Plan Transitions from Bilateral to Pentameral Symmetry in Echinoderms. *Communications Biology*, *3*, 371.

Lukasik, J. J, James, N. P, McGowrna, B., Bone Y., 2000. An epeiric ramp: low-energy, cool-water carbonate facies in a Tertiary inland sea, Murray basin, South Australia. *Sedimentology*, *47*, 851–881.

Mahboubi, A., Moussavi-Harami, R., Carpenter, S. J., Aghaei, A., Collins, L. B., 2010. Petrographical and geochemical evidences for paragenetic sequence interpretation of diagenesis in mixed siliciclastic–carbonate sediments: Mozduran Formation (Upper Jurassic), south of Agh-Darband, NE Iran. *Carbonates and Evaporites*, *25*, 231-246.

McKinney, M. L., 1984. Allometry and heterochrony in an Eocene echinoid lineage: morphological change as a by-product of size selection. *Paleobiology*, *10*, 407-419.

McKinney, M. L., McNamara, K. J., Carter, B. D., Donovan, S. K., 1992. Evolution of Paleogene echinoids: a global and regional view. In: Prothero, D.R. & Berggren, W.A. (eds.), Eocene-Oligocene Climatic and Biotic Evolution: 349-367. Princeton University Press, Princeton, New Jersey.

Meyer, D. L., and Ausich, W. I., 1983. Biotic interactions among recent and among fossil crinoids. *Biotic interactions in Recent and Fossil Benthic Communities*, 377–427.

Michel, J., Borgomano, J., Reijmer, J. J., 2018. Heterozoan Carbonates: When, where and why? A synthesis on parameters controlling carbonate production and occurrences. *Earth-Science Reviews*, *182*, 50–67.

Michel, J., Laugié, M., Pohl, A., Lanteaume, C., Masse, J. P., Donnadieu, Y., Borgomano, J., 2019. Marine carbonate factories: a global model of carbonate platform distribution. *International Journal of Earth Sciences*, *108*, 1773–1792.

Mongiardino Koch, N., Coppard, S.E., Lessios, H.A., Briggs, D.E.G., Mooi, R., Rouse, G. W., 2018. A phylogenomic resolution of the sea urchin tree of life. *BMC Evolutionary Biology*, *18* (1), 189.

Mutti, M., and Hallock, P., 2003. Carbonate systems along nutrient and temperature gradients: some sedimentological and geochemical constraints. *International Journal of Earth Sciences*, *92*, 465–475.

Nebelsick, J., 1999. Taphonomic comparison between Recent and fossil sand dollars. *Palaeogeography, Palaeoclimatology, Palaeoecology*, *149*, 349–358.

Nebelsick, J. H., and Kroh, A., 2002. The stormy path from life to death assemblages: the formation and preservation of mass accumulations of fossil sand dollars. *Palaios*, *17*, 378–393.

Nebelsick, J. H., 2004, Taphonomy of Echinoderms: introduction and outlook, in Heinzeller, T., and Nebelsick, J.H., eds., Echinoderms: München: Proceedings of the 11th International Echinoderm Conference: A.A. Balkema Publishers, London, UK, p. 471–477.

Néraudeau, D., 1994. Hemiasterid echinoids (Echinodermata: Spatangoida) from the Cretaceous Tethys to the present-day Mediterranean. *Palaeogeography, Palaeoclimatology, Palaeoecology*, *110*, 319–344.

Néraudeau, D., David, B., Al-Muallem, M. S., 1995. The Cretaceous echinoids from central Saudi Arabia. *Neues Jahrbuch für Geologie und Paläontologie Abhandlungen*, *197*, 399–424.

Néraudeau, D., and Mathey, B., 2000. Biogeography and diversity of South Atlantic Cretaceous echinoids: implications for circulation patterns. *Palaeogeography, Palaeoclimatology, Palaeoecology*, *156*, 71–88.

Néraudeau, D., Goubert, E., Lacour, D., Rouchy, J. M., 2001. Changing biodiversity of Mediterranean irregular echinoids from the Messinian to the present-day. *Palaeogeography, Palaeoclimatology, Palaeoecology*, *175*, 43–60.

Neugebauer, J., 1978. Micritization of crinoids by diagenetic dissolution. *Sedimentology*, *25*(2), 267–283.

Pawson, D. L., 2007. Phylum echinodermata. *Zootaxa*, *1668*, 749–764.

Petsios, E., Portell, R. W., Farrar, L., Tennakoon, S., Grun, T. B., Kowalewski, M., Tyler, C. L., 2021. An asynchronous Mesozoic marine revolution: the Cenozoic intensification of predation on echinoids. *Proceedings of the Royal Society B*, *288*, 1–8.

Pomar, L., Brandano, M., Westphal, H., 2004. Environmental factors influencing skeletal grain sediment associations: a critical review of Miocene examples from the western Mediterranean. *Sedimentology*, *51*, 627–651.

Pomar, L., Esteban, M., Martinez, W., Espino, D., Castillo de Ott, V., Benkovics, L., Castro Leyva, T., 2015. Oligocene-Miocene carbonates of the Perla field, offshore Venezuela: Depositional model and facies architecture, In: C. Bartolini and P. Mann, eds., Petroleum geology and potential of the Colombian Caribbean Margin: *American Association of Petroleum Geologists, Memoir 108*, 647–674.

Porter, S. M., 2010. Calcite and aragonite seas and the de novo acquisition of carbonate skeletons. *Geobiology*, *8*, 256–277.

Ries, J. B., 2004. Effect of ambient Mg/Ca ratio on Mg fractionation in calcareous marine invertebrates: A record of the oceanic Mg/Ca ratio over the Phanerozoic. *Geology*, *32*, 981–984.

Russell, M.P., 2013. Echinoderm responses to variation in salinity. *Advances in Marine Biology*, *66*, 171–212.

Saucede, T., Pierrat, B., Brayard, A., David, B., 2013. Palaeobiogeography of Austral echinoid faunas: a first quantitative approach. *Geological Society, London, Special Publications*, *381*, 117–127.

Scholle, P. A., and Ulmer-Scholle, D., 2003. A Color Guide to the Petrography of Carbonate Rocks: Grains, Textures, Porosity, Diagenesis. *American Association of Petroleum Geologists, Memoir 77*, 474 pp.

Simone, L., Carannante, G., Bassi, D., Cherchi, A., 2012. Rudist-bearing rhodalgal facies in the post-Turonian recovery of peri-Tethyan carbonate systems: a case history from the Nurra region (northwestern Sardinia, Italy). *Geodiversitas*, *34*(1), 167–187.

Smith, A. B., 2007. Intrinsic versus extrinsic biases in the fossil record: contrasting the fossil record of echinoids in the Triassic and early Jurassic using sampling data, phylogenetic analysis, and molecular clocks. *Paleobiology*, *33*, 310–323.

Sprinkle, J., and Guensburg, T. E., 2003. Major expansion of echinoderms in the early Late Ordovician (Mohawkian, middle Caradoc) and its possible causes. pp. 327–332 in G. L. Albanesi, M. S. Beresi & S. H. Peralta (eds) Ordovician from the Andes. Instituto Superior de Correlaci6n en Geol6gica (INSUGEO), Serie Correlaci6n Geol6gica, San Juan, Argentina.

Strathmann, R. R., 1981. The role of spines in preventing structural damage to echinoid tests. *Paleobiology*, *7*, 400–406.

Summers, M. M., Messing, C. G., Rouse, G. W., 2014. Phylogeny of Comatulidae (Echinodermata: Crinoidea: Comatulida): A new classification and an assessment of morphological characters for crinoid taxonomy. *Molecular Phylogenetics and Evolution*, *80*, 319–339.

Tambutté, S., Holcomb, M., Ferrier-Pagès, C., Reynaud, S., Tambutté, É., Zoccola, D., Allemand, D., 2011. Coral biomineralization: From the gene to the environment. *Journal of Experimental Marine Biology and Ecology*, *408*, 58–78.

Thompson, J. R., Petsios, E., Bottjer, D. J., 2017. A diverse assemblage of Permian echinoids (Echinodermata, Echinoidea) and implications for character evolution in early crown group echinoids. *Journal of Paleontology*, *91*, 767–780.

Thompson, J. R., Hu, S. X., Zhang, Q. Y., Petsios, E., Cotton, L. J., Huang, J. Y., Zhou, C. Y., Wen, W., Bottjer, D. J., 2018. A new stem group echinoid from the Triassic of China leads to a revised macroevolutionary history of echinoids during the end-Permian mass extinction. *Royal Society Open Science*, *5*, 171548.

Thompson, J. R., Posenato, R., Bottjer, D. J., Petsios, E., 2019. Echinoids from the Tesero Member (Werfen Formation) of the Dolomites (Italy): implications for extinction and survival of echi-

noids in the aftermath of the end-Permian mass extinction. *PeerJ*, *7*, e7361.
Thompson, J. R., Cotton, L. J., Candela, Y., Kutscher, M., Reich, M., Bottjer, D. J., 2021. The Ordovician diversification of sea urchins: systematics of the Bothriocidaroida (Echinodermata: Echinoidea). *Journal of Systematic Palaeontology*, *19*(20), 1395–1448.
Tyler, C. L., Dexter, T. A., Portell, R. W., Kowalewski, M., 2018. Predation-facilitated preservation of echinoids in a tropical marine environment. *Palaios*, *33*, 478–486.
Ukar, E., and Laubach, S. E., 2016. Syn-and postkinematic cement textures in fractured carbonate rocks: Insights from advanced cathodoluminescence imaging. *Tectonophysics*, *690*, 190–205.
Wilson, M. E. J., and Vecsei, A., 2005. The apparent paradox of abundant foramol facies in low latitudes: their environmental significance and effect on platform development. *Earth-Science Reviews*, *69*, 133–168.
Wright, D. F., 2015. Fossils, homology, and "Phylogenetic Paleo-ontogeny": a reassessment of primary posterior plate homologies among fossil and living crinoids with insights from developmental biology. *Paleobiology*, *41*, 570–591.
Wright, D. F., Ausich, W. I., Cole, S. R., Peter, M. E., Rhenberg, E. C., 2017. Phylogenetic taxonomy and classification of the Crinoidea (Echinodermata). *Journal of Paleontology*, *91*, 829–846.
Zamora, S., and Rahman, I. A., 2015. Deciphering the early evolution of echinoderms with C ambrian fossils. *Palaeontology*, *57*(6), 1105–1119.
Zhuravlev, A. Y., and Wood, R. A., 2008. Eve of biomineralization: controls on skeletal mineralogy. *Geology*, *36*(12), 923–926.

10 Bryozoans

Bryozoans (Phylum Bryozoa; Subphylum Ectoprocta) are calcareous aquatic invertebrate animals, predominantly marine (some inhabit freshwater lakes and rivers), which form colonies of numerous units known as zooids. These colonial animals are locally abundant as fossils in many marine deposits from the Ordovician to the Holocene and contain almost 6000 described living species (Bock and Gordon, 2013). Bryozoans have a rich fossil record due to the presence of a calcareous skeleton in the great majority of species. Compared to the approximately 808 genera of extant bryozoans, an estimated 1289 genera are extinct (Taylor and Waeschenbach, 2015).

In modern oceans, colonies of this highly diverse and taxonomically complex group of metazoans are found from shorelines to bathyal depths and at all latitudes. Nevertheless, major bryozoan deposits (i.e., *in situ* bioconstructions or reef-like structures) are particularly abundant in shelf sediments deposited in temperate and cold-water settings (see Lombardi et al. 2020). The value of bryozoans as paleoenvironmental indicators has been widely acknowledged, but caution is needed particularly in interpreting ancient fauna, mainly because the environmental factors controlling the distribution of modern bryozoans are still poorly understood (see Smith, 1995; Taylor, 2005).

10.1 Major Groups and Stratigraphic Distribution

The Subphylum Ectoprocta appeared during the Early Ordovician (Calcite I). Two classes with calcified skeletons are distinguished: Stenolaemata and Gymnolaemata (see Jiménez-Sánchez et al. 2010; Ma et al. 2014; Taylor and Waeschenbach, 2015; Ernst, 2020). The Class Stenolaemata consists of six orders: Esthonioporata (age range: Early Ordovician to Late Devonian); Fenestrata, Cystoporata, and Cryptostomata (the three of them ranged in age from the Early Ordovician to the end-Permian); Trepostomata (age range: Early Ordovician to the end-Triassic); and Cyclostomata (age range: Early Ordovician to Holocene). The Class Gymnolaemata consists of two orders: Ctenostomata (uncalcified; age range: Early Ordovician to Holocene), and Cheilostomata (age range: Late Jurassic to Holocene). In addition, Bryozoa together with Brachiopoda and Phoronida (a small group of tube-worm marine invertebrates) make up the "lophophorate" phyla. Although phylogenetic relationships between these three groups are unclear or yet to be resolved, they possess a lophophore, which is a ciliated feeding apparatus (a crown of tentacles) (see Taylor et al. 2010).

Stenolaemates dominated bryozoan assemblages during the Paleozoic (e.g., Ross, 1995; Ernst et al. 2006; Hageman and McKinney, 2010; Ma et al. 2014; Ernst, 2020). They were especially abundant in the Ordovician and Silurian (Calcite I), and in the Carboniferous and Permian (transition from calcite I to Aragonite II). All Paleozoic bryozoans were hit hard by the end-Permian mass extinction (Powers and Pachut 2008; Ernst, 2020). Triassic bryozoans are rare and consist predominantly of trepostomes (Boardman 1984; Ernst, 2020; Key et al. 2022). After the end-Triassic mass extinction, a slow evolutionary radiation of bryozoans started in the Middle Jurassic and was only limited to the Order Cyclostomata, which had been less abundant throughout the Paleozoic (Taylor, 2010). Therefore, bryozoan faunas of Jurassic and Early Cretaceous ages are rare and of low diversity on a global scale (see Martha et al. 2019), and consist almost entirely of cyclostomes. Their peak in diversity occurred in the Maastrichtian at the very end of the Cretaceous (McKinney and Taylor 2001; Taylor and Waeschenbach, 2015; Ernst, 2020). Abundance and diversity of cyclostomes decreased drastically during the Cenozoic (Taylor and Di Martino, 2014). Although extant, cyclostomes constitute only a minor component of the modern bryozoan biota, representing less than 10% of species globally (Bock and Gordon 2013).

Cheilostomata is the most successful post-Paleozoic order of bryozoans. Cheilostomes appeared during the Calcite II

J. A. Méndez Dot, J. Méndez Baamonde, *Atlas of Tropical Carbonates*, https://doi.org/10.1007/978-3-031-25494-9_10

seas of the Late Jurassic (e.g., Dzik, 1975; Taylor, 1994). Late Jurassic and Early Cretaceous cheilostomes are rare. They began an explosive diversification in the Late Albian (Martha and Taylor, 2017; Dick et al. 2018). By the Late Cretaceous (Campanian) they had become the dominant order in bryozoan assemblages (e.g., Lidgard et al. 1993; Ernst, 2020). Nevertheless, the two extant bryozoan orders with mineralized skeletons (Cheilostomata and Cyclostomata) were almost equally numerous in the Late Cretaceous (Lidgard et al. 1993). Cheilostomes predominate over cyclostomes in the Cenozoic fossil record (Lidgard et al. 1993; Taylor and Di Martino, 2014; Cook et al. 2018), accounting for about 84% of Holocene bryozoans described species (Bock and Gordon, 2013).

In general, records of Cenozoic tropical bryofaunas are relatively few compared with higher latitudes. Some of the factors that might explain such poor Cenozoic fossil record is that most Cenozoic palaeotropical sediments are still in the tropics today and consequently are often in regions of dense natural vegetation (e.g., tropical rainforests) where fresh exposures are limited. In addition, surface weathering in the humid tropics can extend deeply beneath the land surface, destroying any fossils in the process. Nevertheless, bryozoan assemblage diversity in the modern tropics is not significantly less than in higher latitudes (Taylor and Di Martino, 2014).

10.2 Skeletal Mineralogy

The skeletal mineralogy of bryozoans is summarized by Smith et al. (2006) after an extensive literature review. Most bryozoans are either entirely calcitic (about 65%) or entirely aragonitic (about 15%). The remainder, those with mixed mineralogies, can lie anywhere between the two. Most bryozoans secrete IMC (mean = 5 wt.%$MgCO_3$), though the range lies between 0 and 14 wt.% $MgCO_3$. Stenolaemata bryozoans almost always precipitate LMC. Cheilostomes are dominated by calcitic skeletons (59%), while aragonitic and bimineralic skeletons are present in similar proportions (21% and 20%, respectively). The mean Mg content of the group is 5.6 wt.% $MgCO_3$ and 74% of the specimens contain IMC. In modern aragonite seas, aragonitic and bimineralic skeletons dominate in tropical latitudes, whereas calcitic species are more common at high latitudes (see also Smith, 2014; Taylor et al. 2009, 2015; Taylor and Di Martino, 2014).

10.3 Ecology and Environment

Bryozoans are light-independent, heterotrophic calcifiers commonly found in continental shelf environments, from intertidal to shelf-breaks and upper-slope settings, and to a lesser extent in deep basin environments. Most species prefer normal salinity waters (ranging from 32 to 37‰), although some species can tolerate lower or higher salinities. They are generally found in waters with temperatures with 10 to 30 °C, but Antarctic bryozoans may tolerate temperatures below 0 °C (see Smith, 1995 and references therein). Colonies grow mainly in places where water movement is relatively constant, thus guaranteeing an adequate supply of phytoplankton as a food resource (Taylor 2000; Wood et al. 2012). Conversely, high rates of sedimentation, as well as stagnant, low-oxygen conditions are unfavourable to bryozoans (Lombardi et al. 2020).

Like all marine calcareous heterotrophs, bryozoans feed on particulate organic matter. Therefore, they thrive mainly in settings with high nutrients influx (i.e., mesotrophic to eutrophic conditions) and high plankton productivity (Mutti et al. 1999; Lukasik et al., 2000). Even though, where nutrient supplies are low, bryozoans are most competitive because of their lower maintenance requirements (Lukasik et al. 2000; Harper et al. 2004).

Bryozoans are benthic, colonial suspension-feeders. Most have sessile, epibenthic (or epifaunal) life strategies, living either attached to hard and firm substrates, anchored or as encrusters (on shells, corals, crinoids, red algae, encrusting benthic foraminifera). The free-living growth forms (i.e., infaunal) are characteristic of soft substrates (e.g., sandy bottoms, sea weeds and algae fronds). Sessile attachers/encrusters can be found in all bryozoan orders. Free-living colonies, totally lacking until the Late Cretaceous and abundant in modern oceans, are known only in cheilostomes (Smith, 1995; Taylor, 2005; Taylor and James, 2013).

These heterotrophs are able to develop bioconstructions including biogenic reefs. Frame-building bryozoans have been described elsewhere in the fossil record but in general have become less common through geological time, largely supplanted by corals and sponges. Paleozoic bioconstructions were built mainly by trepostomes and cystoporates. In modern oceans, biogenic bryozoan reefs are well documented in numerous temperate to cold water areas (e.g., Mediterranean Sea, New Zealand, California, Bermuda Islands) (see Lombardi et al. 2020; Ernst, 2020).

Paleozoic bryozoan-rich deposits exhibit a panglobal distribution, including strata corresponding to tropical and subtropical paleolatitudes. From the Middle Triassic and for the rest the Mesozoic and the Cenozoic, bryozoan distribution pattern exhibits a clear latitudinal zonation, as most bryozoan-rich deposits occurred in temperate and cold-water extratropical platforms. Nowadays these heterotrophic calcifiers are minor components in warm, tropical carbonates, but are common in temperate and cold-water heterozoan carbonates (Taylor, 2005; Taylor and Di Martino, 2014).

Regarding their usefulness in paleoenvironmental interpretations, bryozoans have many possible applications. Their presence or absence, diversity and abundance, colonial growth forms and internal morphology (zooid size, branch diameter, skeletal wall thickness) can indicate temperature, salinity, water energy and depth, substrate character, sedimentation rate and marine seasonality (Smith, 1995; Taylor, 2005; Holcová and Zágoršek, 2008; Clark et al. 2010). Modern bryozoans are of great utility in the study of climate change and ocean acidification (Fortunato, 2015).

10.4 Carbonate Factories and Skeletal Grain Associations

Bryozoans are components of the Heterozoan factory of Michel et al. (2019), as well as occasional to common secondary components of the Photo C-factory of these authors (see Appendix A for a detailed treatment of these carbonate benthic factories). This section discusses only the skeletal grain associations involving bryozoans that are either represented or notably absent in Venezuelan carbonate microfacies, as illustrated by the examples in the photographic set.

In Middle Permian carbonates of the Palmarito Formation (Venezuelan Andes), bryozoans are common to abundant components of Heterozoan carbonates *sensu* James (1997), in association with crinoids [bryonoderm lithofacies *sensu* Beauchamp (1994) and Beauchamp and Desrochers (1997)], brachiopods, ostracods, molluscs and sponges; as well as common components of mixed heterozoan-photozoan microfacies in association with LBF (fusulinids) and green algae (see Laya and Tucker, 2012). In Albian carbonates of the Lisure and Maraca formations (MB), bryozoans are moderately common to common components of heterozoan-dominated microfacies in association with molluscs (bivalves, oysters and gastropods), echinoids, ostracods, smaller miliolids and serpulids (plus micritic and glauconitic peloids and oncoids/cortoids). Some of these microfacies can be grouped into bimol or echinofor lithofacies [*sensu* Hayton et al. (1995), dominated by bivalves and echinoids, respectively]. In these units, bryozoans are occasionally associated with photozoan biota (mainly green algae and occasional red algae). See Appendix A for definitions of the skeletal grain associations considered in this book.

In Late Oligocene-Early Miocene carbonates of the Perla Limestone (EGVSB), bryozoans are moderately common to common components of photozoan-dominated microfacies in association with coralline algae and LBF, plus an important contribution of other heterotrophs, mainly echinoids and balanid barnacles, and less common molluscs (see Pomar et al. 2015). These microfacies are the expression of the Photo C-factory developed at tropical paleolatitutes, and can be grouped into rhodalgal lithofacies *sensu* Carannante et al. (1988) or rhodechfor lithofacies *sensu* Hayton et al. (1995). Bryomol lithofacies (*sensu* Nelson, 1988), dominated by bryozoans (> 50%) and molluscs, and very common in Cenozoic non-tropical carbonates worldwide, are absent.

In modern, poorly studied carbonate facies of the Venezuelan Atlantic continental shelf and slope, bryozoans are associated to other heterotrophs, mainly molluscs (bivalves and gastropods) and symbiont-barren benthic foraminifera (e.g., *Quinqueloculina lamarckiana*), as well as to symbiont-bearing benthic foraminifera (e.g., *Amphistegina lessonii*) (see Kawka et al. 2008; Molina et al. 2008).

10.5 Morphology and Diagnostic Elements

Modern bryozoan colonies range from mm-sized to a few cm in size, while modern biogenic calcified structures vary greatly in size, ranging from a few centimeters to many kilometers (Lombardi et al. 2020). There are nine basic bryozoan colony-forms: encrusting, dome-shaped, palmate, foliose, fenestrate, robust branching, delicate branching, articulated and free-living (Taylor and James, 2013; see also Smith, 1995; Hageman et al. 1998). The basic unit of a bryozoan colony (zoarium) is the zooid (solitary animal). Individual zooids live within calcareous tube- or boxlike chambers termed zooecia, which are typically less than 1 mm in diameter and length. The skeleton of a single zooid is termed zooecium, which consists of calcareous layers of zooecial walls. Wall fabrics consist of laminar, foliated or granular crystals. Stenolaemates have wall-parallel skeletal fabrics composed mostly of flattened crystallites. These range from minute granular crystallites (submicrometer in size) to foliated lamellae measuring tens of micrometers long. Cheilostomes may have wall-parallel or wall-perpendicular fabrics, and have greater variety of crystallite forms and arrangements (Taylor and Weedon, 2000; Scholle and Ulmer-Scholle, 2003; Taylor et al. 2010; Batson et al. 2020).

In addition to colony growth-forms (or habits) and wall structure, some of the most relevant significant morphologic features of bryozoans for the identification of the order of a given specimen in thin sections are (this brief review is focused on the photographic examples shown, which correspond to the orders Trepostomata, Cystoporata, Fenestrata, Cyclostomata and Cheilostomata): the shape and size of zooecia (e.g., elongate and/or tubular in cyclostomes, or ovate to spherical "bubble-like" in trepostomes); the arrangement of zooecia (e.g., macroserial, regular "box-like" in cheilostomes, or oligoserial in fenestrates); openings (apertura) developed at one end of the zooecia (e.g., trepostomes); zooecia with internal partitions (e.g., diaphragms in trepostomes), or unpartitioned zooecia (e.g., cheilostomes); presence of other cavities (e.g., fenestrules in fenestrates, cystopores in cystoporates, mesopores in trepostomes, ovicells in cheilostomes); robusts skeletons (e.g., fenestrates), or delicate, thin-walled forms (e.g., cheilostomes); and tubular and spinose walls beyond the exterior of the bryozoan (e.g., acanthorods in trepostomes) (Hageman et al. 1998; Taylor and Weedon, 2000; Scholle and Ulmer-Scholle, 2003; Smith et al. 2006; Wyse-Jackson et al. 2006; Ostrovsky et al. 2008; Taylor and James, 2013; Taylor and Waeschenbach, 2015).

10.6 Preservation as Fossils

Most fossil bryozoans are commonly in the range of less 1 mm to 10 mm in size, although some may reach more than 1 cm. Grains are derived from detachment, breakage (mechanical or biological) and disarticulation of cm-sized structures. Therefore, bryozoan colonies are a common source of sand- and gravel-sized sediments. In most cases zooecia and other cavities are filled by micritic sediments, microspar, pseudospar, calcitic cements, or by authigenic minerals (e.g., glauconite). The preservation of zooecia as intraparticles pores do occur, but is less common.

10.7 Bryozoans: Photographic Set and Descriptive Captions

In all photos, IW = image width (the true length of the horizontal axis of the view)

Holocene. Venezuelan Atlantic Continental Shelf
This and the following photomicrograph were taken using a binocular magnifying glass. The large gravel-sized cheilostome bryozoan grain in the image was derived from the breakage of an originally robust-branching, palmate or articulate colony. Zooecia cavities exhibit oligoserial arrangement. Note that the fragment widens from three to at least five parallel series of ellipsoidal zooecia. Thick zooecial walls (W) confer the colony greater mechanical strength. Thin depressed walls within the zooecial compartments are partially preserved.

IW = 8 mm.

Holocene. Venezuelan Atlantic Continental Shelf
A fragment of a cheilostome bryozoan colony in the center of the image, derived from a free-living colony and surrounded by mollucs. This view shows the bryozoan fragment from the convex upper surface. The wall arrangement consists of a symmetric macroserial arrangement of zooecia. Note that cavities are confined to small, hollow and perforated “tube-like” depressions. In life, free-living bryozoans rest on soft substrate with the concave underside facing the seafloor sediments.

IW = 5.5 mm.

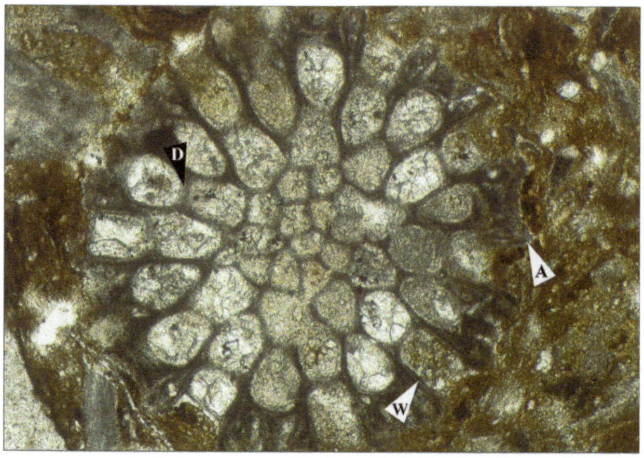

Middle Permian (Cisularian to Guadalupian). Palmarito Formation. Venezuelan Andes

Transverse section through a trepostome bryozoan colony embedded in a slightly organic-rich micritic matrix. The margins of the calcitic exoskeleton (zooecium) exhibiting high optical relief and ridged-spinose outlines are termed acanthorods (A). Thin zooecial walls (W) radiate outward. Transverse partitions crossing zooecia are termed diaphragms (D). These partitions give zooecia oval to ovate shapes. These are smaller (bubble-like) in the center of the colony (endozone). Originally empty cavities are filled with pseudospar.

IW = 2.4 mm. PPL

Middle Permian (Cisularian to Guadalupian). Palmarito Formation. Venezuelan Andes

Trepostome bryozoans in the image are seen in transverse (T) and longitudinal/oblique (larger grain) sections. These are accompanied by crinoids (C) and undifferentiated bioclastic fragments in a packed packstone. The large bryozoan skeleton exhibits well-preserved spherical zooecia in the center (endozone). The outer skeleton (exozone) consists of elongate zooecia, zooecial walls (W) that thicken outward, and wide zooecial openings (O).

IW = 7.5 mm. PPL

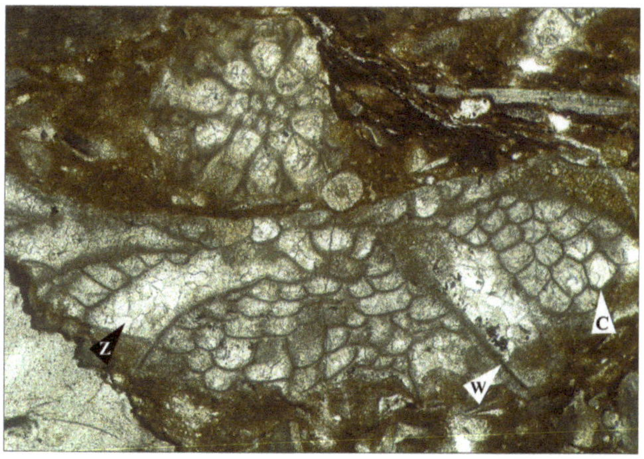

Middle Permian (Cisularian to Guadalupian). Palmarito Formation. Venezuelan Andes

Longitudinal section of a cystoporid bryozoan fragment (Family Fistuliporidae) accompanied by a trepostome seen in transverse section (above). The large bryozoan grain consists of elongate and wide unpartitioned zooecia (Z). These are surrounded by small bubble-like cystopores (C) (extrazooidal skeletal structures). Such distinctive morphology of alternating small and large elements is characteristic of cystoporids. Walls (W) are very thin compared to the volume occupied by micrite or pseudospar infillings.

IW = 4.2 mm. PPL

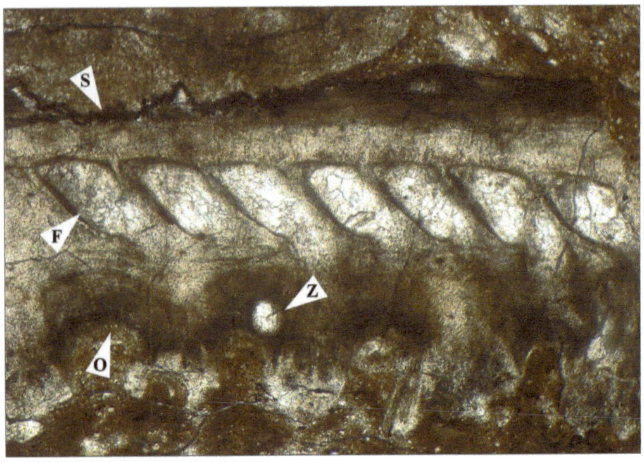

Middle Permian (Cisularian to Guadalupian). Palmarito Formation. Venezuelan Andes

A fragment of a fenestrate bryozoan. Note the small spherical zooecia (Z), contrasting with large, aligned fenestrules (F). These are the openings between branches of a fenestrate bryozoan (see Maggenti et al. 2008). The top margin of the skeleton (from this view) is marked by a sutured contact along a stylotlite (S). The opposite margin of the grain exhibits conspicuous subcircular outlines (O) facing outward from the skeleton and separated by small ridges, a characteristic breakage morphology of fenestrates (see Scholle and Ulmer-Scholle, 2003; for explanatory diagrams and photomicrographic examples of fenestrate bryozoans).

IW = 2.4 mm. PPL

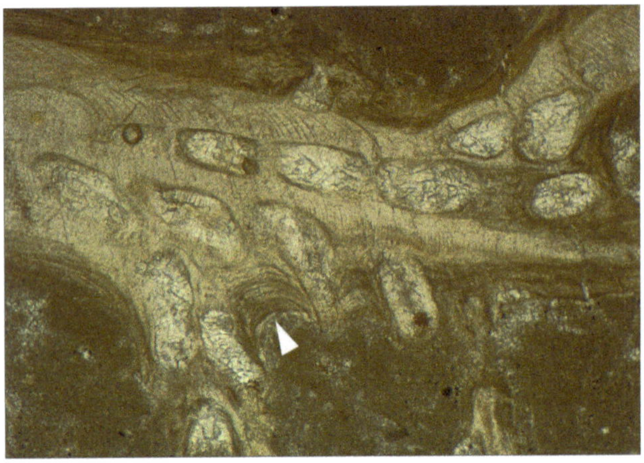

Middle Permian (Cisularian to Guadalupian). Palmarito Formation. Venezuelan Andes

A fenestrate bryozoan fragment embedded in a micritic matrix. The robust skeleton exhibits characteristic branching morphology. Branches are composed of very thin calcitic lamellae. Well-preserved delicate microtexture is visible at the sinuous margins of the ramified structure (white arrow). Fenestrules are aligned in different rows, which are longitudinally oriented along main and lateral branches. The aligned fenestrules serve as diagnostic criteria for the identification this specimen of the Order Fenestrata, as branched morphologies are characteristic in all orders of calcifying bryozoans.

IW = 2.4 mm. PPL

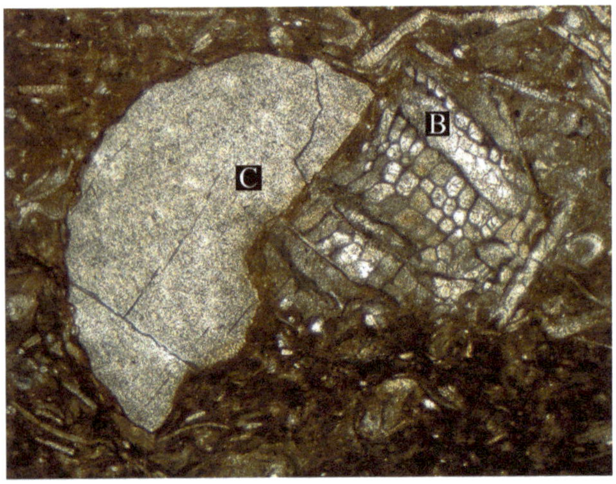

Middle Permian (Cisularian to Guadalupian). Palmarito Formation. Venezuelan Andes

Large fragments of a cystoporid bryozoan (B) and a crinoid (C) are surrounded by numerous, small fragmented bioclasts (disarticulated valves of brachiopods and ostracods). Despite reworking, the cystoporid retained characteristic morphology and diagnostic elements (large zooecia and small cystoporids). The heterozoan association between crinoids and bryozoans (bryonoderm association) is very common in inner ramp facies of the Palmarito Formation (Laya and Tucker, 2012).

IW = 4.2 mm. PPL

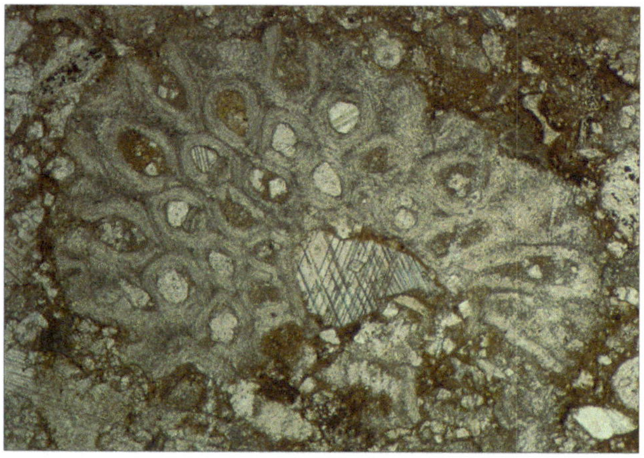

Middle Permian (Cisularian to Guadalupian). Palmarito Formation. Venezuelan Andes

Transverse section of a trepostome bryozoan fragment embedded in a muddy bioclastic matrix. Zooecia are oval to ovate in the center and these become elongate towards the margins of the grain. Well-preserved walls of more or less uniform thickness are composed of very thin, concentric calcitic lamellae. Note that the walls between adjacent zooecium are well differentiated (not fused).

IW = 2.4 mm. PPL

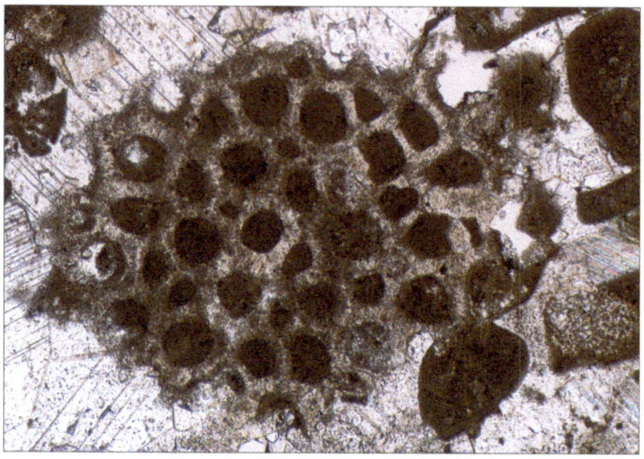

Cretaceous (middle Albian). Lisure Formation. Maracaibo Basin

A reworked cyclostome bryozoan fragment is filled by micritic sediments and surrounded by LMC cement exhibiting twinning lamellae. The grain was reworked and rededosited from a muddy to a mud-lean environment. The zooecia arrangement is macroserial. Zooecial walls are fused (each zooecium is sharing walls with other 6 to 7 analogs). In addition, these are visibly recrystallized, which suggests an original metastable composition (possibly HMC).

IW = 2.4 mm. PPL

Cretaceous (middle Albian). Lisure Formation. Maracaibo Basin

This cyclostome bryozoan exhibits large-shaped zooecia due to the longitudinal orientation of the cut. Thin and delicate zooecial walls bend outwards, almost reaching perpendicularity respect to the margins of the grain. Such delicate walls are associated to low-energy environments. Dark micritic mud infilling the zooecia cavities predates final deposition of the bryozoan grain together with surrounding peloidal grains in a mud-lean environment.

IW = 2.4 mm. PPL

Cretaceous (latest Albian). Maraca Formation. Maracaibo Basin

A cyclostome bryozoan coated by bladed cement in a grainstone with primary porosity. Note the two-sided arrangement, which is termed "bifoliate". It consists of a median calcitic wall that crosses the grain longitudinally, and dichotomous-branching growth of delicate zooecial walls at both sides, thus giving the colony a bilateral (not identical) symmetry. Like the two previous examples, zooecia were filled by micritic sediments prior to redeposition of the grain in a micrite-free depositional environment.

IW = 2.4 mm. PPL

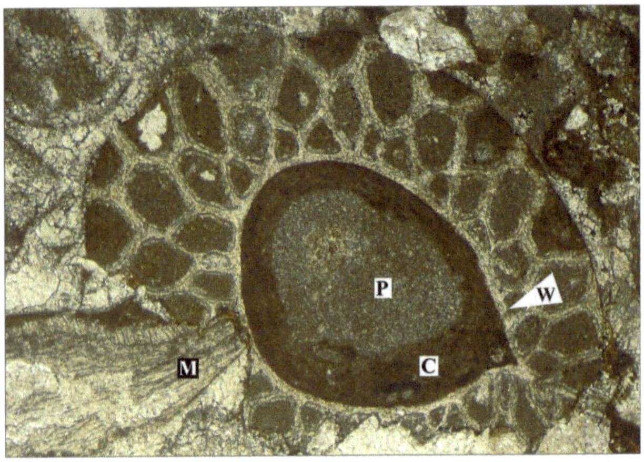

Cretaceous (middle Albian). Lisure Formation. Maracaibo Basin

An encrusting bryozoan accompanied by skeletal and non-skeletal grains. Note that the innermost thin calcitic wall (W) of the bryozoan is coating a cortoid. This coated grain consists of a microbial micritic coating (C) precipitated around a micritic peloid (P). The encrusting growth habit the bryozoan is associated to a moderate- to high-energy level. Later, a decrease in the hydrodynamic energy was necessary for deposition of micritic sediments infilling the zooecia cavities. Down and to the left the bryozoan is truncated by a mollusc (M).

IW = 2.4 mm. PPL

Cretaceous (middle Albian). Lisure Formation. Maracaibo Basin

A bifoliate bryozoan with zooecia cavities filled by moderately organic-rich micritic sediments later replaced by localized precipitation of glauconite (green). Precipitation of pyrite (black spots) postdates glauconitization. Bryozoans in the Lisure Formation are common of muddy facies containing echinoids and molluscs (heterozoan association). For this example, eutrophic conditions are indicated by the authigenic minerals, whose precipitation was promoted by the presence of organic matter in the micritic sediments filling the zooecia in combination with transient low-oxygen conditions during early marine diagenesis.

IW = 1.3 mm. PPL

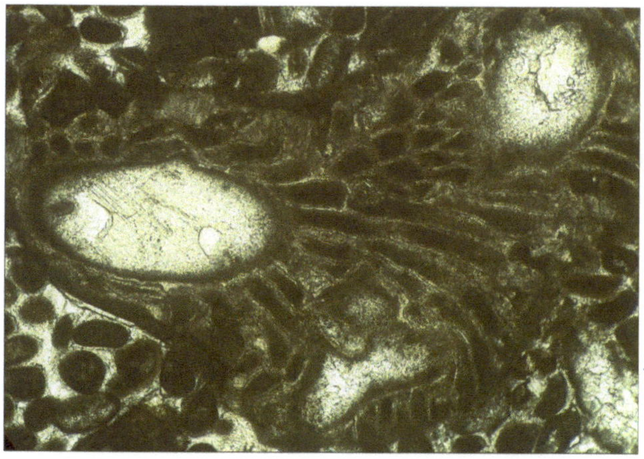

Cretaceous (latest Albian). Maraca Formation. Maracaibo Basin

Oblique section through a bryozoan surrounded by micritic peloidal grains. The larger cavities filled by cloudy sparry calcite are possibly ovicells (also termed brood chambers), which are specialized chambers hosting zooids reproductive specialists. Alternatively, these cavities might also be boring structures used as shelter for small organisms living inside originally smaller zooecia and then enlarged by the host organism. The oblique cut produced a mixture of elongated and small cavities, thus giving the grain a distorted appearance.

IW = 2.4 mm. PPL

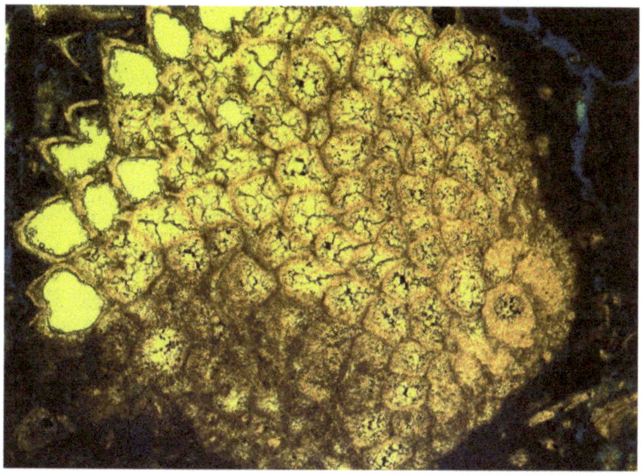

Late Oligocene-Early Miocene. Perla Limestone. Eastern Gulf of Venezuela Sub-Basin

Close view of a small cheilostome bryozoan grain consisting of a macroserial (box-like) arrangement of zooecia forming a closely-spaced nodular colony. Ovate cavities are filled by a combination of two calcite cements. Cloudy blocky calcite is the most abundant, while microspar standout for its white-limpid color (up and to the left). Thick, brown-colored walls are clearly visible at the left and upper left margins of the grain (from this view).

IW = 1.2 mm. PPL
Photomicrograph courtesy of Cardon IV, S.A.

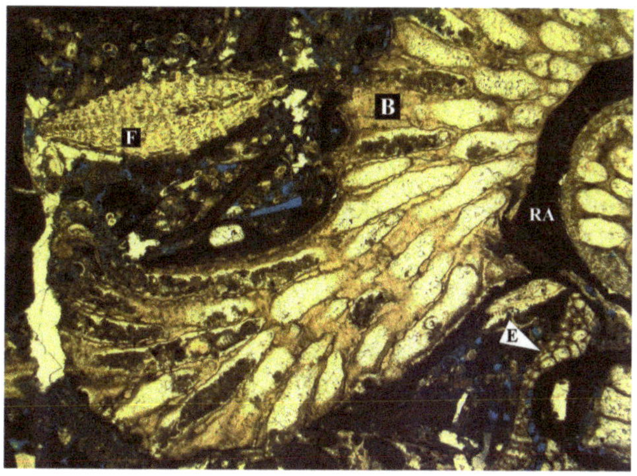

Late Oligocene-Early Miocene. Perla Limestone. Eastern Gulf of Venezuela Sub-Basin

A large bryozoan (B) exhibiting characteristic encrusting growth habit and elongate zooecia. This is accompanied by encrusting red algae (RA), a larger benthic foraminifer (F) and an encrusting (E) benthic foraminifer in a mud-lean rudstone. The photozoan-heterozoan association between encrusting, light-dependent calcifiers (benthic foraminifera and red algae) and bryozoans is common of moderate- to high-energy depositional environments of the Perla distally steepened ramp (see Pomar et al. 2015).

IW = 5 mm. PPL
Photomicrograph courtesy of Cardon IV, S.A.

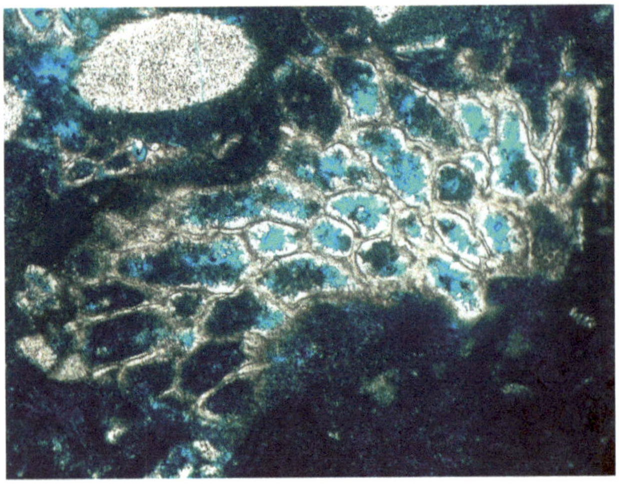

Late Oligocene-Early Miocene. Perla Limestone. Eastern Gulf of Venezuela Sub-Basin

A bryozoan encrusting undifferentiated red algal grains (heterozoan-photozoan association) and accompanied by an echinoid grain (up and to the left). Intraparticle pores within the bryozoan (zooecia cavities) are of secondary origin, as these were derived from burial dissolution of HMC red algae. Such mesogenetic dissolution allows the visualization of the thin, delicate and highly porous structure of the bryozoan.

IW = 2.4 mm. PPL

References

Batson, P. B., Tamberg, Y., Taylor, P. D., Gordon, D. P., Smith, A. M., 2020. Skeletal resorption in bryozoans: occurrence, function and recognition. *Biological Reviews*, *95*, 1341–1371.

Beauchamp, B., 1994. Permian climatic cooling in the Canadian Arctic. *Geological Society of America Special Paper*, *288*, 229–246.

Beauchamp, B., and Desrochers, A., 1997. Permian warm- to very cold-water carbonates and cherts in Northwest Pangea. In: James, N. P, Clarke J. A. D. (eds.), Cool-water carbonates. *SEPM Special Publication*, *56*, 327–347.

Boardman, R. S., 1984. Origin of the post-Triassic Stenolaemata (Bryozoa): a taxonomic oversight. *Journal of Paleontology*, 19–39.

Bock, P. E., and Gordon, D. P., 2013. Phylum Bryozoa Ehrenberg, 1831. In: Zhang, Z.-Q.(Ed.) Animal biodiversity: an outline of higher-level classification and survey of taxonomic richness (Addenda 2013). *Zootaxa*, *3703*, 67–74.

Carannante, G., Esteban, M., Milliman, J. D., Simone, L., 1988. Carbonate lithofacies as paleolatitude indicators: problems and limitations. *Sedimentary Geology*, *60*, 333–346.

Clark, N., Williams, M., Okamura, B., Smellie, J., Nelson, A., Knowles, T., Taylor, P., Leng, M., Zalasiewicz, J., Haywood, A., 2010. Early Pliocene Weddell Sea seasonality determined from bryozoans. *Stratigraphy*, *7*, 199.

Cook, P. L., Bock, P. E., Hayward, P. J., Gordon, D. P., Weaver, H., 2018. Class Gymnolaemata, order Cheilostomata. *Australian Bryozoa*, *2*, 61–279.

Dick, M. H., Sakamoto, C., Komatsu, T., 2018. Cheilostome Bryozoa from the Upper Cretaceous Himenoura Group, Kyushu, Japan. *Paleontological Research*, *22*(3), 239–264.

Dzik, J. 1975. The origin and phylogeny of the cheilostomatous Bryozoa. *Acta Palaeontologica Polonica*, *20*, 395–423.

Ernst, A., 2020. Fossil record and evolution of Bryozoa. In: Schwaha T (ed) Handbook of zoology. Phylum Bryozoa. De Gruyter, Berlin/Boston, pp. 11–56.

Ernst, A., Senowbari-Daryan, B., Rashidi, K., 2006. Lower Permian Bryozoa of the Jamal Formation from Bagh-e Vang (Shotori Mountains, northeast Iran). *Facies*, *52*, 627–635.

Fortunato, H., 2015. Bryozoans in climate and ocean acidification research: a reappraisal of an under-used tool. *Regional Studies in Marine Science*, *2*, 32–44.

Hageman, S. J., Bock, P. E., Bone, Y., McGowran, B., 1998. Bryozoan growth habits: classification and analysis. *Journal of Paleontology*, 418–436.

Hageman, S. J., and McKinney, F. K., 2010. Discrimination of fenestrate bryozoan genera in morphospace. *Palaeontologia Electronica*, *13*, 1–43.

Harper, D. A. T., Cocks, L. R. M., Popov, L. E., Sheehan, P. M., Bassett, M. G., Copper, P., Holmer, L. E., Jin, J., Rong, J. -Y., 2004. Brachiopods. In: Webby, B.D.E., Paris, F., Droser, M.L., Percival, I.G. (Eds.), The Great Ordovician Biodiversification Event. Columbia University Press, New York, pp. 157–178.

Hayton, S., Nelson C., Hood S., 1995. A skeletal assemblage classification system for non-tropical carbonate deposits based on New Zealand Cenozoic limestones. *Sedimentary Geology*, *100*. 123–141.

Holcová, K., and Zágoršek, K., 2008. Bryozoa, foraminifera and calcareous nannoplankton as environmental proxies of the "bryozoan event" in the Middle Miocene of the Central Paratethys (Czech Republic). *Palaeogeography, Palaeoclimatology, Palaeoecology*, *267*, 216–234.

James, N. P., 1997. The cool-water carbonate depositional realm. In: James N. P., Clarke, J., (eds.), Cool-water carbonates, vol 56. SEPM Society for Sedimentary Geology, Tulsa, 1–20.

Jiménez-Sánchez, A., Anstey, R. L., Azanza, B., 2010. Description and phylogenetic analysis of *Iberostomata fombuenensis* new genus and species (Bryozoa, Ptilodictyina). *Journal of Paleontology*, *84*, 695–708.

Kawka, V., Molina, A., Gómez, M. G., Méndez, J., Yanes, C., 2008. ESTUDIO GEOQUÍMICO AMBIENTAL DE LA ESPECIE DE FORAMINÍFERO BENTÓNICO RECIENTE *Amphistegina lessonii*, EN SEDIMENTOS SUPERFICIALES DE LA PLATAFORMA CONTINENTAL ATLÁNTICA VENEZOLANA. *Isla*, *2*(6).

Key, M. M., Wyse Jackson, P. N., Reid, C. M., 2022. Trepostome bryozoans buck the trend and ignore calcite-aragonite seas. *Palaeobiodiversity and Palaeoenvironments*, *102*, 253–263.

Laya, J. C., and Tucker, M. E., 2012. Facies analysis and depositional environments of Permian carbonates of the Venezuelan Andes: Palaeogeographic implications for Northern Gondwana. *Palaeogeography, Palaeoclimatology, Palaeoecology*, *331*, 1–26.

Lidgard, S., McKinney, F. K., Taylor, P. D., 1993. Competition, clade replacement, and a history of cyclostome and cheilostome bryozoan diversity. *Paleobiology*, *19*, 352–371.

Lombardi, C., Taylor, P. D., Cocito, S., 2020. Bryozoans: The 'Forgotten' Bioconstructors. In: *Perspectives on the Marine Animal Forests of the World*. Springer, Cham. 193–217.

Lukasik, J. J., James, N. P., McGowran, B., Bone, Y., 2000. An epeiric ramp: low-energy, cool-water carbonate facies in a Tertiary inland sea, Murray Basin, South Australia. *Sedimentology*, *47*, 851–881.

Ma, J. Y., Buttler, C. J., Taylor, P. D., Rosso, A., 2014. Cladistic analysis of the 'trepostome' Suborder Esthonioporina and the systematics of Palaeozoic bryozoans. *Studi Trentini di Scienze Naturali, 94*, 153–161.

Martha, S. O., and Taylor, P. D., 2017.The oldest erect cheilostome bryozoan: *Jablonskipora* gen. nov. from the upper Albian of southwest England. *Papers in Palaeontology*, 4, 55–66.

Martha, S. O., Taylor, P. D., Rader, W. L., 2019. Early Cretaceous cyclostome bryozoans from the early to middle Albian of the Glen Rose and Walnut formations of Texas, USA. *Journal of Paleontology, 93*, 244–259.

Maggenti, M. A. B., Maggenti, A., Scott, G., 2008. Dictionary of Invertebrate Zoology. *Zea E-Books, 61*. 379 pp.

McKinney, F. K., and Taylor, P. D., 2001. Bryozoan generic extinctions and originations during the last one hundred million years. *Palaeontologia. Electronica, 4*, 26 pp.

Michel, J., Laugié, M., Pohl, A., Lanteaume, C., Masse, J. P., Donnadieu, Y., Borgomano, J., 2019. Marine carbonate factories: a global model of carbonate platform distribution. *International Journal of Earth Sciences, 108*, 1773–1792.

Molina, A., Kawka, V., Gómez, M. G., Méndez, J., Yanes, C., 2008. ESTUDIO GEOQUÍMICO AMBIENTAL DE LA ESPECIE DE FORAMINÍFERO BENTÓNICO *Quinqueloculina lamarckiana* PRESENTE EN SEDIMENTOS RECIENTES DE LA PLATAFORMA CONTINENTAL ATLÁNTICA VENEZOLANA. *Isla, 2*(6).

Mutti, M., Bernoulli, D., Spezzaferri, S. Stille, P., 1999. Lower and Middle Miocene carbonate facies in the central Mediterranean: the impact of paleoceanography on sequence stratigraphy. In: *Advances in Carbonate Sequence Stratigraphy: Application to Reservoirs, Outcrops and Models* (Eds., P. Harris, A. Saller, J. Simo and C.R. Handford), *SEPM Special Publication, 63*, 371–384.

Nelson, C. S., 1988. An introductory perspective on non-tropical shelf carbonates. *Sedimentary Geology, 60*, 3–12.

Ostrovsky, A. N., Taylor, P. D., Dick, M. H., Mawatari, S. F., 2008. Pre-Cenomanian cheilostome Bryozoa: Current state of knowledge, in Okada, H., Mawatari, S.F., Suzuki, N., and Gautam P., eds., Origin and Evolution of Natural Diversity, University of Hokkaido, Sapporo, Japan, pp. 69–74.

Pomar, L., Esteban, M., Martinez, W., Espino, D., Castillo de Ott, V., Benkovics, L., Castro Leyva, T., 2015. Oligocene-Miocene carbonates of the Perla field, offshore Venezuela: Depositional model and facies architecture, In: C. Bartolini and P. Mann, eds., Petroleum geology and potential of the Colombian Caribbean Margin. *American Association of Petroleum Geologists, Memoir 108*, 647–674.

Powers, C. M., and Pachut, J. F., 2008. Diversity and distribution of Triassic bryozoans in the aftermath of the end-Permian mass extinction. *Journal of Paleontology, 82*, 362–371.

Ross, J. R. P., 1995. Permian Bryozoa. In: Scholle, P.A., Peryt, T.M., Ulmer-Scholle, D.S. (eds.). The Permian of Northern Pangea. Springer, New York, 196–209.

Scholle, P. A., and Ulmer-Scholle, D., 2003. A Color Guide to the Petrography of Carbonate Rocks: Grains, Textures, Porosity, Diagenesis. *American Association of Petroleum Geologists, Memoir 77*, 474 pp.

Smith, A. M., 1995. Palaeoenvironmental interpretation using bryozoans: a review. *Geological Society, London, Special Publications, 83*, 231–243.

Smith, A. M., 2014. Growth and calcification of marine bryozoans in a changing ocean. *The Biological Bulletin, 226*, 203–210.

Smith, A. M., Key Jr, M. M., Gordon, D. P., 2006. Skeletal mineralogy of bryozoans: taxonomic and temporal patterns. *Earth-Science Reviews, 78*, 287–306.

Taylor, P. D., 1994. An early cheilostome bryozoan from the Upper Jurassic of Yemen. *Neues Jahrbuch für Geologie und Paläontologie, Abhandlungen, 191*, 331–344.

Taylor, P. D., 2000. Origin of the modern bryozoan fauna. In: Culver, S.J., Rawson, P.F. (Eds.), Biotic Response To Global Change. The Last 145 Million Years. The Natural History Museum & Cambridge University Press, London and Cambridge, pp. 195–206.

Taylor, P. D., 2005. Bryozoans and palaeoenvironmental interpretation. *Journal of the Palaeontological Society of India, 50*, 1–11.

Taylor, P. D., 2010. Barremian bryozoans from Serre de Bleyton (Drôme, SE France). *Annalen des Naturhistorischen Museums in Wien, Serie A, 112*, 673–700.

Taylor, P. D., and Weedon, M. J., 2000. Skeletal ultrastructure and phylogeny of cyclostome bryozoans. *Zoological Journal of the Linnean Society, 128*, 337–399.

Taylor, P. D., and James, N. P., 2013. Secular changes in colony-forms and bryozoan carbonate sediments through geological history. *Sedimentology, 60*, 1184–1212.

Taylor, P. D., and Di Martino, E., 2014. Why is the tropical Cenozoic fossil record so poor for bryozoans. *Studi Trentini di Scienze Naturali, 94*, 249–257.

Taylor, P. D., and Waeschenbach, A., 2015. Phylogeny and diversification of bryozoans. *Palaeontology, 58*, 585–599.

Taylor, P. D., James, N. P., Bone, Y., Kuklinski, P., Kyser, T. K., 2009. Evolving mineralogy of cheilostome bryozoans. *Palaios, 24*, 440–452.

Taylor, P. D., Vinn, O., Wilson, M. A., 2010. Evolution of biomineralisation in 'lophophorates'. *Special Papers in Palaeontology, 84*, 317–333.

Taylor, P. D., Lombardi, C., Cocito, S., 2015. Biomineralization in bryozoans: present, past and future. *Biological Reviews, 90*(4), 1118–1150.

Wood, A. L., Probert, P. K., Rowden, A. A., Smith, A. M., 2012. Complex habitat generated by marine bryozoans: a review of its distribution, structure, diversity, threats and conservation. *Aquatic Conservation: Marine and Freshwater Ecosystems, 22*, 547–563.

Wyse-Jackson, P. N., Mckinney, F. K., Bancroft, A. J., 2006. Fenestrate bryozoan genera based on species from Ireland originally described by Frederick M 'Coy in 1844. *Palaeontology, 49*, 741–767.

11 Brachiopods

Brachiopods, commonly called "lamp-shells", are exclusively marine, sessile invertebrate metazoans with a soft body enclosed in a shell consisting of two unequal valves. They are recorded since the Lower Cambrian and were extremely common fossils throughout the Paleozoic; altogether about 30,000 fossil species have been described. Since the Mesozoic, the diversity and abundance of brachiopods have been dramatically reduced, and today they are considered a minor phylum (Emig et al. 2013). Since some 95% of all brachiopod taxa are extinct, the fossil record is the primary source of data to frame and test models for the evolution of the phylum (Harper et al. 2017).

Although brachiopods are not major rock forming elements in Mesozoic and Cenozoic strata, their occurrence has always remained a key factor for understanding global marine environments in the recent and deep past, as brachiopod shells are used as high-resolution archives to reconstruct the physical and chemical composition of the seawater (Ye et al. 2018a).

11.1 Major Groups and Stratigraphic Distribution

Brachiopods (Phylum Brachiopoda) are divided into three subphyla (Williams et al. 1996): Linguliformea, Craniiformea and Rhynchonelliformea. The first group consists of organophosphatic, inarticulate shells (these have a hinge axis but no hinge line) whose valves are held together only by musculature. The second are calcitic or aragonitic, with a valve-to-valve contact along a hinge line, but valves lack of hinge teeth and socket and are held together by musculature. Rhynchonelliformea brachiopods (former Class Articulata) are the largest subphylum, and the focus of this chapter. They consist of articulated calcitic shells, whose valves are held together by a hinge with a teeth-and-socket mechanism (Scholle and Ulmer-Scholle, 2003; Carlson, 2016). In addition, and as mentioned in the previous chapter, Brachiopoda together with Bryozoa and Phoronida make up the "lophophorate" phyla, which possess a lophophore (i.e., a ciliated feeding apparatus; see Taylor et al., 2010).

The Subphylum Rhynchonelliformea first appeared in the Middle Cambrian (Calcite I), and is divided into five classes (Carlson, 2016). Three of them were short-lived and did not cross the Cambrian-Ordovician boundary. Class Strophomenata (age range: Early Ordovician to latest Permian) is divided in four orders. The orders Orthotetida (age range: Early Ordovician to latest Permian) and Productida (age range: Early Silurian to latest Permian) have the widest stratigraphic range and are the most abundant strophomenates in the Paleozoic record (see also Bretsky, 1969; Shen et al. 2000; Rong and Zhan, 2006; Plotnick et al. 2013; Mottequin et al. 2014).

Class Rhynchonellata consists of several orders, being the extant orders Terebratulida and Rhynchonellida (age range: Late Cambrian to Holocene) the most abundant rhynchonellates in the Paleozoic record (see Copper, 1997; Brock and Yazdi, 2000; Schreiber et al. 2013; Mottequin et al. 2014). Mesozoic abundance and diversity of terebratulids and rhynchonellids decreased compared to Late Paleozoic times (but see Sandy and Campbell 1994; Sandy, 1997; Chen et al. 2005; Matyszkiewicz et al. 2016), especially during the Cretaceous. Nevertheless, they recovered and were locally abundant from Middle Triassic to Middle Jurassic. Rhynchonellids disproportionately declined to become minor components of Cenozoic faunas (see Manojlovic and Clapham, 2021). In modern oceans, terebratulids are by far the most abundant and diverse group of brachiopods (Lee, 2008). Despite the numerous studies focused on recent specimens and their use as geochemical and environmental proxies, articulate brachiopods are considered to be rare in ecosystems of modern oceans, especially in tropical latitudes (Kowalewski et al. 2002).

In general, brachiopods of the classes Strophomenata and Rhynchonellata dominated the Paleozoic benthic communities (see Garbelli, 2017). Their peak of abundance occurred

J. A. Méndez Dot, J. Méndez Baamonde, *Atlas of Tropical Carbonates*, https://doi.org/10.1007/978-3-031-25494-9_11

during the Ordovician, Silurian and Devonian (Calcite I), although they were also common in the Carboniferous and Permian (transition from Calcite I to Aragonite II). Brachiopods were ecologically dominant over molluscs (bivalves and gastropods) throughout the Paleozoic. In terms of abundance, molluscs did not become clearly dominant over brachiopods until after the Permian-Triassic extinction (see Tomasových, 2006; Leighton et al. 2013; Hsieh et al. 2019). Systematic paleontology is not applied to the photographic examples in this chapter.

11.2 Skeletal Mineralogy

Shells of rhynchonelliformean brachiopods are composed of LMC. In life the shells are overlain by a thin organic layer (periostracum), which is never preserved as a fossil. Shells commonly contain a small amount of organic material (proteinaceous membranes), commonly between 2 and 5%, mainly concentrated in the fibrous secondary layer (see Cusack et al. 2008; Garbelli et al. 2012, 2017; Pérez-Huerta et al. 2013; Penman et al. 2013; Immel et al. 2015; Ye et al. 2018a; Roda et al. 2022).

11.3 Ecology and Environment

Brachiopods are heterotrophs (light-independent), sessile benthic filter-feeder organisms that live/lived attached to hard or firm substrates. During the Paleozoic and, to a lesser extent during the Triassic and the Jurassic, articulate brachiopods were common on platforms and ramps, from inner to outer edge/slope environments located in tropical subtropical latitudes. In general, Paleozoic and early Mesozoic rynchonelliformean brachiopods were not associated with pelagic and deep-basin environments. They switched from shallow to deep-water and/or cryptic refugia sometime during the Jurassic. In modern oceans, the occurrence of terebratulids and rhynchonellids is mainly restricted to higher latitudes (temperate to cool waters) (see Kowalewski et al. 2002; Tomasových, 2006; Shen et al. 2009, 2013; Flügel, 2010; Buatois et al. 2016; Harper and Peck, 2004).

Brachiopods feed mainly on phytoplankton. Food resources increased significantly in the Middle Ordovician (although the level of food resources was still far below modern levels), coinciding with the Ordovician brachiopod radiation, which was the most marked interval of diversification in the entire history of the phylum. Brachiopods have among the lowest metabolic rates (i.e., low-food requirement) of any marine organisms. They can survive where others starve from lack of resource supply, making them very resilient in difficult times (see Harper et al. 2004 and references therein). Nevertheless, as functional ecology implies, high marine productivity thus appears to be the steering parameter to produce a large amount of suspension-feeding biota (i.e., brachiopods, crinoids and bryozoans) (Michel et al. 2018).

The hypothesis that increasing predation and bioerosion, from the beginning of the "Mesozoic Marine Revolution" (enhanced predator-prey arms race), as one of the main causes for the failure of brachiopods to recover following the Permo-Triassic mass extinction, and their subsequent purported shift to refugia from predation, has been explored by numerous authors. Before the Mesozoic marine revolution, triggered by the Permo-Triassic mass extinction, the Paleozoic marine fauna had been dominated by brachiopods, rugose corals, tabulate corals, cephalopods, trilobites, and crinoids, and lower actinopterygian fishes and nonneoselachian sharks. The Modern fauna, commencing in the Middle to Late Triassic, was characterized by bivalves, gastropods, malacostracan crustaceans (crabs, lobsters), echinoids, neoselachian sharks, neopterygian bony fishes (including basal holosteans and teleosts), and marine tetrapods (reptiles in the Mesozoic, whales in the Cenozoic) (see Benton and Wu, 2022). According to Harper and Peck (2004), Jurassic large brachiopods occurred in shallow sites, from polar to tropical latitudes, but are absent in later periods from tropical areas. These changes occurred in parallel in both major orders (Rhynchonellida and Terebratulida). Currently, shallow-water temperate and polar brachiopods can be large, but in the tropics, they are small. Despite the historical consideration of brachiopods as unattractive preys (e.g., low-body mass and toxicity of tissues) in comparison to bivalves and gastropods, increasing evidence from the study of predation on Paleozoic, Mesozoic, Cenozoic and modern articulate brachiopods suggests that these benthic heterotrophs are/were also edible and vulnerable preys (e.g., smaller and thin-shelled relative to bivalves). Bioerosion on brachiopod shells is due either to predatory organisms or to epi- and endobionts; the former includes cephalopods, gastropods, arthropods and fishes; the latter includes bivalves, balanid crustaceans, gastropods, endolithic sponges and polychaete annelids (see Leighton, 1999; Harper and Wharton, 2000; Ruggiero and Bitner, 2007; Leighton et al. 2013; Tyler et al. 2013; Harper et al. 2017).

11.4 Carbonate Factory and Skeletal Grain Associations

Brachiopods are components of the Heterozoan factory of Michel et al. (2019), as well as secondary to accessory components of the Photo C-factory of these authors (see Appendix A for a detailed treatment on these carbonate benthic factories). This section discusses only the skeletal grain associations involving brachiopods represented in Venezuelan carbonate microfacies, as illustrated by the examples in the photographic set. In Middle Permian carbonates of the Palmarito Formation (Venezuelan Andes), brachiopods are common to abundant components of Heterozoan carbonates *sensu* James (1997), in association with bryozoans and crinoids (plus ostracods, molluscs and sponges). Some of these microfacies can be grouped into bryonoderm lithofacies [*sensu* Beauchamp (1994) and Beauchamp and Desrochers (1997)] in which brachiopods are secondary components. In this late Paleozoic unit, these heterotrophic calcifiers are moderately common to common components of mixed heterozoan-photozoan microfacies in association with LBF (fusulinids) and green algae (see Laya and Tucker, 2012). In Late Oligocene-Early Miocene carbonates of the Perla Limestone (EGVSB), brachiopods are scarce but previously reported by Pinto et al. (2014) and Soto et al. (2025). They occur in photozoan-dominated microfacies, associated with coralline algae and LBF, with a secondary contribution of heterozoan biota, primarily echinoids. As for brachiopods in Miocene carbonates of the FB (San Luis, Churuguara and Capadare formations), these heterotrophs are very scarce (or absent in most cases), and not previously reported (to our knowledge). Therefore, no skeletal grain association is considered, as further studies need to be carried. In lithified Quaternary facies of Aves Island (carbonate basement and beachrocks), brachiopods occur as minor components within mixed photozoan–heterozoan microfacies dominated by coralline algae and mollusks. In these microfacies, green algae (*Halimeda*) and scleractinian corals are secondary components, while LBF and echinoids are accessory components.

11.5 Morphology and Diagnostic Elements

Shells of rhynchonelliformean brachipods are in the mm- to cm-size range. Shells consist of two equilateral valves but these are unequal in size and shape. The ventral (posterior) and brachial (anterior) valves can be differentiated by their characteristic symmetry. The brachial valve (also termed dorsal) typically has a rounded, elongate central elevation (a fold or ridge). The ventral valve, commonly larger in size, may exhibit flat (or depressed), concavo-convex or planoconvex shapes. The folded and counterpart depressed morphology is common in the Class Strophomenata. The Class Rhynchonellata is characterized by biconvex shells, similar to bivalve molluscs, but the ventral valve is visibly larger than the brachial valve. The pedicle is a stalk-like structure (non-mineralized) emerging between the two valves of most brachiopods, facilitating the attachment to a hard substrate. The position of the pedicle opening is located in the major depression (or sulcus) on the surface of the ventral valve (also termed pedicle valve) (Scholle and Ulmer-Scholle, 2003; Maggenti et al. 2008; Flügel, 2010; Carlson, 2016; Harper et al. 2017).

Microstructures of brachiopod shells are key for differentiating them from bivalve molluscs. The shell fabric consists of layers composed of crystals differing in texture, length and orientation. The shell succession is composed of an outermost, thin prismatic/acicular primary layer, a secondary fibrous or laminar layer and, when present, an innermost columnar/prismatic tertiary layer. The primary layer is not always present because it is only a few micrometers thick, and it easily undergoes missing due to taphonomic/diagenetic effects. The arrangement of LMG fibers (e.g., foliated and/or laminar) are commonly visible in the thick secondary layer. The presence of a columnar tertiary layer underlying the secondary fabric is a morphologic analogous feature in both Strophomenata and Rhynchonellata brachiopods (Williams, 1997; Scholle and Ulmer-Scholle, 2003; Cusack et al. 2008; Garbelli et al. 2012; Immel et al. 2015; Garbelli, 2017; Ye et al. 2018a,b, 2020; Roda et al. 2022). In addition, the secondary layer can be also divided in distinguishable sublayers (posterior, central, anterior) (Penman et al. 2013; Ye et al. 2018a).

Other shell microstructures of relevance are wall perforations. Punctate (endopunctate) shells have small holes (punctae up to 100 μm wide) that perforate the wall and are oriented perpendicular to the shell surface (present mainly in Rhynchonellata). Pseudopunctate shells have stacked, conical plications (pseudopunctae) in the fibers of the secondary layer that look somewhat like punctae (common in Strophomenata). Impunctate shells have primary and secondary layers lacking perforations (common in Rhynchonellida). Spines are common ornamenting features in the Class Strophomenata (Order Productida). These are hollow and have an internal tubular structure (Álvarez and Brunton, 2001; Scholle and Ulmer-Scholle, 2003; Pérez-Huerta et al. 2009; Flügel, 2010; Pérez-Huerta, 2012; Garbelli et al. 2012; Garbelli, 2017; Ye et al. 2020).

11.6 Preservation as Fossils

Brachiopods are commonly preserved as fragments of articulate-bivalve shells, as single valves, and as fragments of the latter. Their sizes commonly range between 0.5-1 mm to a few centimeters in length. Robust shell fragments exhibiting well-preserved LMC fibers are the most common form of preservation. Detached elements are also common. The latter include hollow spines (commonly smaller than 1 mm), prismatic layers, and foliated lamellae.

11.7 Geochemical Proxies in Brachiopods

Articulate brachiopods are, together with belemnites, the only abundant and stratigraphically widespread fossils with LMC skeletons throughout the Phanerozoic. Geochemical determinations in brachiopods (mainly $^{18}O/^{16}O$ and $^{87}Sr/^{86}Sr$) are known for almost the entire Phanerozoic, but mostly for Paleozoic strata (from the Cambrian to Permian). Due to their LMC composition, brachiopod shells are quite resistant to diagenetic alterations. Like many SBF, brachiopods typically secrete their skeletons in oxygen isotope equilibrium with ambient seawater. Therefore, variations in the $\delta^{18}O$ in shells of these heterotrophic calcifiers have been widely used as proxies for marine fossil record changes in seawater temperatures (see Veizer et al. 1999; Brand, 2004; Prokoph et al. 2008; Grossman, 2012; Brand et al. 2019; Letulle et al. 2023).

Although not a direct proxy indicator of past environmental conditions, the Sr (strontium) isotopic composition of brachiopod shells has been utilized as a proxy parameter for plate tectonic processes and their influences on atmospheric CO_2 concentrations (e.g., Vollstaedt et al. 2014). This is because $^{87}Sr/^{86}Sr$ variations reflect principally the waxing and waning of Sr input to the oceans from rivers (continental flux) versus the input from the submarine hydrothermal systems (mantle flux) (Veizer et al. 1999; see also Steuber and Veizer, 2002). Furthermore, the $^{87}Sr/^{86}Sr$ ratio obtained from brachiopod shells can be compared to the $^{87}Sr/^{86}Sr$ seawater calibration curves for the Phanerozoic (see Gradstein and Ogg, 2004; McArthur et al. 2012), thus representing a powerful material for dating ancient carbonate facies. The use of this dating method on brachiopods has been mainly applied for Paleozoic strata (e.g., Van Geldern et al. 2006; Laya et al. 2013; Garbelli et al. 2019).

11.8 Brachiopods: Photographic Set and Descriptive Captions

In all photos, IW = image width (the true length of the horizontal axis of the view)

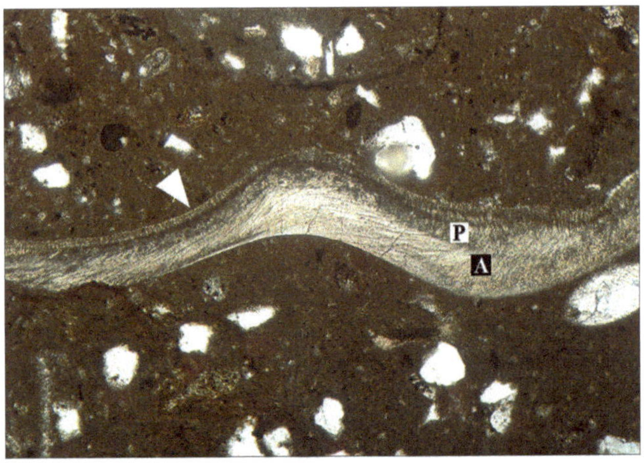

Early Miocene. Churuguara Formation. Falcón Basin
Brachiopod shell exhibiting characteristic central fold and well-preserved layered microstructure. The outer, thinner primary layer (white arrow) is composed of thin, small-sized fibers oriented perpendicular to the shell surface. The thicker, fibrous secondary layer is composed of two well-differentiated sublayers. The secondary posterior (P) and anterior (A) sublayers can be differentiated by calcitic fibers differing in orientation. Such micromorphologic feature indicates different growth stages for this shell, from older fibers in the posterior sublayer to younger fibers in the anterior sublayer (see Pennan et al. 2013; Ye et al. 2018a).

IW = 2.4 mm. PPL.

Early Miocene. San Luis Formation. Falcón Basin
Longitudinal section of a portion of a brachiopod shell exhibiting multilayered wall structure. Unlike the previous example, the thin outer primary layer is not preserved. The inner, thicker secondary layer consists of at least two differentiable sublayers. The secondary posterior (SP) layer exhibits homogeneous wall texture, being composed of fibers oriented nearly horizontal respect to the long axis. The secondary anterior (SA) layer exhibits foliated texture, being composed of calcitic fibers arranged at low angle and showing wavy appearance. The tertiary layer (TL) is composed of more or less equant-sized calcitic prisms oriented perpendicular to the shell margin.

IW = 1.3 mm. CPL

Middle Miocene. Capadare Formation. Falcón Basin

Brachiopod shell fragment showing characteristic perforations or "punctae". These are the small tubules oriented perpendicular to the shell surface. Note the contrast between light brown color of the shell secondary layer and dark brown color of the micritic infilling of perforations. In addition to represent one of the most characteristic morphological shell features in brachiopods, the growth of punctae is involved in vital functions for the organism, such as sensory devices, storage compartments and respiration regulatory mechanism of oxygen (see Pérez-Huerta et al. 2009).

IW = 2.4 mm. PPL

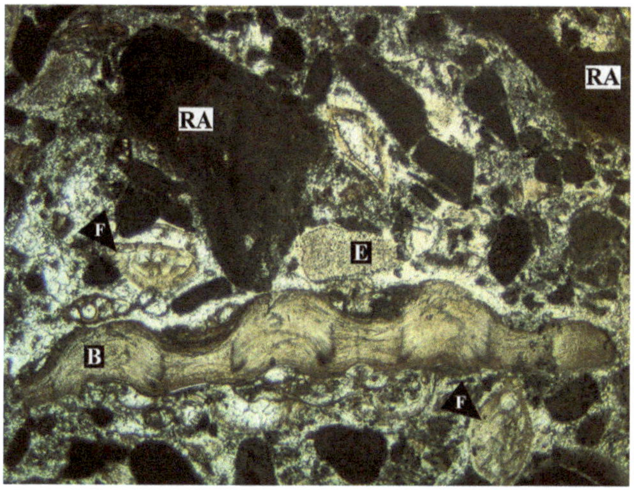

Late Oligocene-Early Miocene. Perla Limestone. Eastern Gulf of Venezuela Sub-Basin

In this poorly-sorted grainstone, the large brachiopod (B) shell stands out due to its plicated morphology, which consists of successive folds on the brachial side. It is accompanied by fragments of red algae (RA) and echinoids (E), as well as tests of LBF (F), forming a photozoan-dominated, mixed photozoan-heterozoan association. In microfacies of the Perla Limestone, brachiopods are scarce, but easily differentiated due to their distinctive morphology and the well-preserved state of their LMC fibers, which exhibit a wavy appearance.

IW = 7.5 mm. PPL

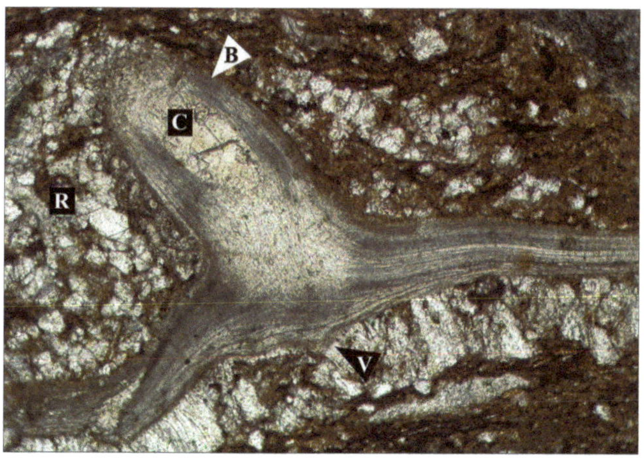

Middle Permian (Cisularian to Guadalupian). Palmarito Formation. Venezuelan Andes

Longitudinal section of a brachiopod shell showing dissimilar bivalve structure. The articulated shell consists of characteristic folded brachial (B) valve and a larger ventral (V) valve. In life, this large depression is attached to the substrate. The foliated microstructure of the shell is composed of long calcitic fibers arranged at low angle. Dark calcitic bands along the shell are alternated by very thin, light-colored recrystallized lamellae. Note the textural differences between the LMC shell and both coarse-grained calcitc infilling the tubular hollow canal (C) and coarse-grained recrystallized (R) carbonate crystals surrounding the shell.

IW = 2.4 mm. PPL

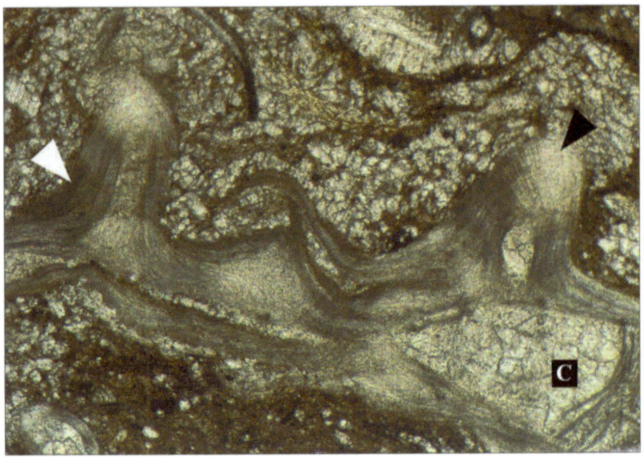

Middle Permian (Cisularian to Guadalupian). Palmarito Formation. Venezuelan Andes

The brachiopod shell in the image exhibits characteristic plicated morphology on the brachial side (as shown in the two previous examples), accompanied a corresponding long depression on the ventral side. Similar to the previous example, well-preserved, dark-gray calcitic lamellae with wavy appearance (white arrow) contrast with light-colored, recrystallized areas of the shell (black arrow). Coarse-grained, cloudy sparry calcite (C) is filling the hollow spaces withing the shell.

IW = 4.2 mm. PPL

Middle Permian (Cisularian to Guadalupian). Palmarito Formation. Venezuelan Andes

In this image, a fragment of a brachiopod shell is seen in longitudinal section and exhibits the secondary (fibrous) thick layer, which consists of two well-differentiated sublayers. The posterior (P) sublayer is composed of fibrous calcite crystals oriented horizontally to the shell margin. The anterior (A) sublayer is composed of dark gray calcitic micro- to crypto-crystals with no distinguishable orientation under this magnification. The darker color of this sublayer is associated with a higher concentration of organic components in the biomineral.

IW = 2.4 mm. PPL

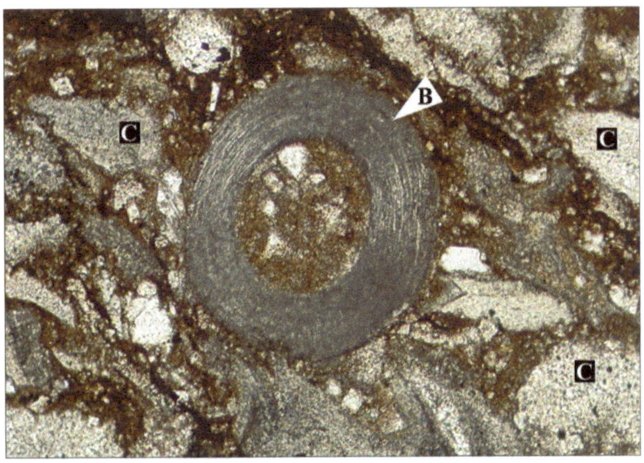

Middle Permian (Cisulrian to Guadalupian). Palmarito Formation. Venezuelan Andes

A brachiopod (B) spine seen in transverse section and accompanied by crinoid (C) grains in a packed packstone. The hollow central canal of the spine is filled by synsedimentary micritic sediments. Radial-concentric and very thin cacitic fibers are well-preserved, contrasting with altered HMC crinoids. The hollow central canal of the spine serves as diagnostic criteria for differentiating brachiopod spines from coated grains (such as ooids and oncoids) that grow around a nucleus. Tubular hollow spines grow on the surface of the ventral valves of producting brachiopods (see Álvarez and Brunton, 2001) and become detached after death of the organism.

IW = 1.3 mm. PPL

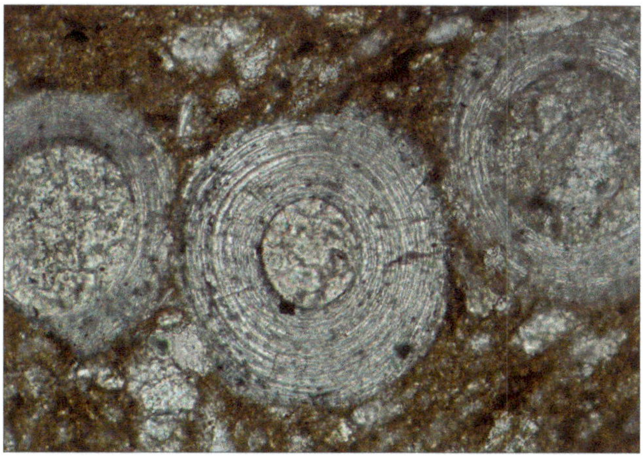

Middle Permian (Cisularian to Guadalupian). Palmarito Formation. Venezuelan Andes

Although the radial-concentric wall structure of the brachiopod spines in the image is visible, these are slightly altered due to recrystallization, giving them a silver coloration (contrasting with the dark-gray color of the spine shown in the previous image). Microtextural preservation of non-recrystallized LMC fibers is a prerequisite for performing isotope analysis in articulate brachiopods. Therefore, scanning electron microscopy, cathodoluminescence petrography, and microsampling techniques should be implemented to target the best-preserved parts of shells (see Grossman, 2012; Laya et al. 2013).

IW = 1.3 mm. PPL;

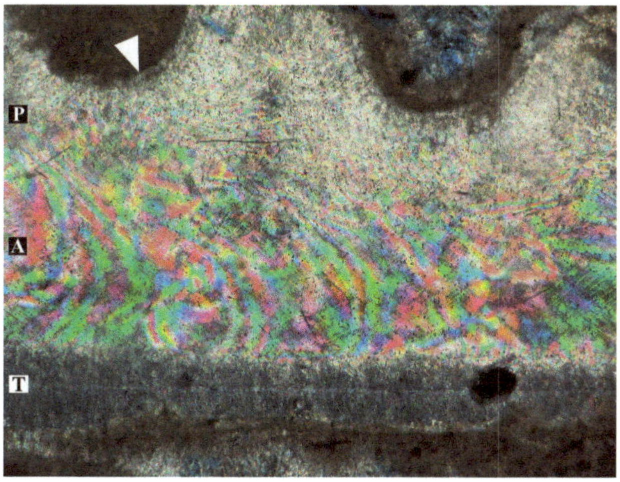

Holocene. Aves Island

Close-up view of a modern brachiopod shell exhibiting well-differentiated shell fabric, namely secondary posterior (P) and anterior (A) sublayers, along with a tertiary (T) layer. These are visually highlighted by differences in shape, length and orientation of the LMC fibers, which result in marked variations in birefringence under CPL. The thin primary layer is present (white arrow) but visually obscured by a micritic envelope riming the shell. Brachiopods are scarce but present in lithified Quaternary facies of Aves Island.

IW = 0.6 mm. CPL

References

Álvarez, F., and Brunton, C. H. C., 2001. Fundamental differences in external spine growth in brachiopods. *Systematics Association Special Volume*, *63*, 108–118.

Beauchamp, B., 1994. Permian climatic cooling in the Canadian Arctic. *Geological Society of America Special Paper*, *288*, 229–246.

Beauchamp, B., and Desrochers, A., 1997. Permian warm- to very cold-water carbonates and cherts in Northwest Pangea. In: James, N. P, Clarke J. A. D. (eds.), Cool-water carbonates. *SEPM Special Publication*, *56*, 327–347.

Benton, M. and Wu, F.-X. 2022. Triassic Revolution. *Frontiers in Earth Science*, *10*, 899541.

Brand, U., 2004. Carbon, oxygen and strontium isotopes in Paleozoic carbonate components: an evaluation of original seawater-chemistry proxies. *Chemical Geology*, *204*, 23–44.

Brand, U., Bitner, M. A., Logan, A., Azmy, K., Crippa, G., Angiolini, L., Colin, P., Griesshaber, E., Harper, E. M., Taddei Ruggiero, E., Häussermann, V., 2019. Brachiopod-based oxygen-isotope thermometer: update and review. *Rivista Italiana di Paleontologia e Stratigrafia*, *125*, 775–78.

Bretsky, P. W., 1969. Central Appalachian Late Ordovician communities. *Geological Society of America Bulletin*, *80*, 193–212.

Brock, G. A., and Yazdi, M., 2000. Palaeobiogeographic affinities of Late Devonian brachiopods from Iran. *Record of the Western Australian Museum, Supplement 58*, 321–334.

Buatois, L. A., Carmona, N. B., Curran, H. A., Netto, R. G., Mángano, N. G., Wetzel, A., 2016. The Mesozoic Marine Revolution, in *The Trace Fossil Record of Major Evolutionary Events, Volume 2.* Editors M. G. Mángano and L. A. Buatois (Dordrecht: Topics in Geobiology), 40, 18–134.

Carlson, S. J., 2016. The evolution of Brachiopoda. *Annual Review of Earth and Planetary Sciences*, *44*, 409–438.

Chen, Z. Q., Kaiho, K., George, A. D., 2005. Early Triassic recovery of the brachiopod faunas from the end-Permian mass extinction: a global review. *Palaeogeography, Palaeoclimatology, Palaeoecology*, *224*, 270–290.

Copper, P., 1997. Articulate brachiopod shellbeds: Silurian examples from Anticosti, eastern Canada. *Geobios*, *30*, 133–148.

Cusack, M., Dauphin, Y., Chung, P., Pérez-Huerta, A., Cuif, J. P., 2008. Multiscale structure of calcite fibres of the shell of the brachiopod Terebratulina retusa. *Journal of Structural Biology*, *164*, 96–100.

Emig, C. C., Bitner, M. A., Alvarez, F., 2013. Phylum Brachiopoda. In: Zhang, Z.-Q. (Ed.) Animal Biodiversity: An Outline of Higher-level Classification and Survey of Taxonomic Richness (Addenda 2013). *Zootaxa*, *3703*, 75–78.

Flügel, E, 2010. Microfacies of Carbonate Rocks—Analysis, Interpretation and Application. Springer, Heidelberg-Dordrecht-London-New York, 984 pp.

Garbelli, C., 2017. Shell microstructures in Lopingian brachiopods: Implications for fabric evolution and calcification. *Rivista Italiana di Paleontologia e Stratigrafia*, *123*, 541–560.

Garbelli, C., Angiolini, L., Jadoul, F., Brand, U., 2012. Micromorphology and differential preservation of Upper Permian brachiopod low-Mg calcite. *Chemical Geology*, *298*, 1–10.

Garbelli, C., Angiolini, L., Shen, S. Z., 2017. Biomineralization and global change: a new perspective for understanding the end-Permian extinction. *Geology*, *45*, 19–22.

Garbelli, C., Shen, S. Z., Immenhauser, A., Brand, U., Buhl, D., Wang, W. Q., Zhang, H., Shi, G. R., 2019. Timing of Early and Middle Permian deglaciation of the southern hemisphere: Brachiopod-based 87Sr/86Sr calibration. *Earth and Planetary Science Letters*, *516*, 122–135.

Gradstein, F., and Ogg, J., 2004. Geologic time scale 2004–why, how, and where next! *Lethaia*, *37*(2), 175–181.

Grossman, E. L., 2012. Oxygen Isotope Stratigraphy, in The Geologic Time Scale 2012, F. M. Gradstein, J. G. Ogg, M. D. Schmitz, and G. M. Ogg (editors). Elsevier, Amsterdam, volume 1, 181–206.

Harper, E. M., and Wharton, D. S., 2000. Boring predation and Mesozoic articulate brachiopods. *Palaeogeography, Palaeoclimatology, Palaeoecology*, *158*, 15–24.

Harper, D. A. T., Cocks, L. R. M., Popov, L. E., Sheehan, P. M., Bassett, M. G., Copper, P., Holmer, L. E., Jin, J., Rong, J.-Y., 2004. Brachiopods. In: Webby, B.D.E., Paris, F., Droser, M.L., Percival, I.G. (Eds.), The Great Ordovician Biodiversification Event. Columbia University Press, New York, pp. 157–178.

Harper, D. A., Popov, L. E., Holmer, L. E., 2017. Brachiopods: origin and early history. *Palaeontology*, *60*, 609–631.

Harper, E. M., and Peck, L. S., 2024. The Demise of Large Tropical Brachiopods and the Mesozoic Marine Revolution. *Royal Society Open Science*, *11*, 231630.

Hsieh, S., Bush, A. M., Bennington, J. B., 2019. Were bivalves ecologically dominant over brachiopods in the late Paleozoic? A test using exceptionally preserved fossil assemblages. *Paleobiology*, *45*, 265–279.

Immel, F., Gaspard, D., Marie, A., Guichard, N., Cusack, M., Marin, F., 2015. Shell proteome of rhynchonelliform brachiopods. *Journal of Structural Biology*, *190*, 360–366.

James, N. P., 1997. The cool-water carbonate depositional realm. In: James N. P., Clarke, J., (eds.), Cool-water carbonates, vol 56. SEPM Society for Sedimentary Geology, Tulsa, 1–20.

Kowalewski, M., Simões, M. G., Carroll, M., Rodland, D. L., 2002. Abundant brachiopods on a tropical, upwelling-influenced shelf (Southeast Brazilian Bight, South Atlantic). *Palaios*, *17*, 277–286.

Laya, J. C., and Tucker, M. E., 2012. Facies analysis and depositional environments of Permian carbonates of the Venezuelan Andes: Palaeogeographic implications for Northern Gondwana. *Palaeogeography, Palaeoclimatology, Palaeoecology*, *331*, 1–26.

Laya, J. C., Tucker, M. E., Perez-Huerta, A., 2013. Metre-scale cyclicity in Permian ramp carbonates of equatorial Pangea (Venezuelan Andes): Implications for sedimentation under tropical Pangea conditions. *Sedimentary Geology*, *292*, 15–35.

Leighton, L. R., 1999. Possible latitudinal predation gradient in middle Paleozoic oceans. *Geology*, *27*, 47–50.

Lee, D. E., 2008. The terebratulides: the supreme brachiopod survivors. *Fossils Strata*, *54*, 241–249.

Leighton, L. R., Webb, A. E., Sawyer, J. A., 2013. Ecological effects of the Paleozoic-Modern faunal transition: Comparing predation on Paleozoic brachiopods and molluscs. *Geology*, *41*, 275–278.

Letulle, T., Gaspard, D., Daëron, M., Arnaud-Godet, F., Vinçon-Laugier, A., Suan, G., Lécuyer, C., 2023. Multi-proxy assessment of brachiopod shell calcite as a potential archive of seawater temperature and oxygen isotope composition. *Biogeosciences*, *20*(7), 1381–1403.

Maggenti, M. A. B., Maggenti, A., Scott, G., 2008. Dictionary of Invertebrate Zoology. *Zea E-Books*, *61*, 379 pp.

Matyszkiewicz, J., Krajewski, M., Kochman, A., Kozłowski, A., Duliński, M., 2016. Oxfordian neptunian dykes with brachiopods from the southern part of the Kraków-Częstochowa Upland (southern Poland) and their links to hydrothermal vents. *Facies*, *62*, 12–40.

Manojlovic, M., and Clapham, M. E., 2021. The role of bioturbation-driven substrate disturbance in the Mesozoic brachiopod decline. *Paleobiology*, *47*(1), 86–100.

McArthur, J. M., Howarth, R. J., Shields, G. A., 2012. Strontium Isotope Stratigraphy, in The Geologic Time Scale 2012, F. M. Gradstein, J. G. Ogg, M. D. Schmitz, and G. M. Ogg (editors). Elsevier, Amsterdam, volume 1, 127–144.

Michel, J., Borgomano, J., Reijmer, J. J., 2018. Heterozoan carbonates: When, where and why? A synthesis on parameters controlling carbonate production and occurrences. *Earth-Science Reviews*, *182*, 50–67.

Michel, J., Laugié, M., Pohl, A., Lanteaume, C., Masse, J. P., Donnadieu, Y., Borgomano, J., 2019. Marine carbonate factories: a global model of carbonate platform distribution. *International Journal of Earth Sciences*, *108*, 1773–1792.

Mottequin, B., Brice, D., Legrand-Blain, M., 2014. Biostratigraphic significance of brachiopods near the Devonian–Carboniferous boundary. *Geological Magazine*, *151*, 216–228.

Pérez-Huerta, A., 2012. Functional morphology and modifications on spine growth in the productid brachiopod *Heteralosia slocomi*. *Acta Palaeontologica Polonica*, *58*, 383–390.

Penman, D. E., Hönisch, B., Rasbury, E. T., Hemming, N. G., Spero, H. J., 2013. Boron, carbon, and oxygen isotopic composition of brachiopod shells: Intra-shell variability, controls, and potential as a paleo-pH recorder. *Chemical Geology*, *340*, 32–39.

Pérez-Huerta, A., Cusack, M., McDonald, S., Marone, F., Stampanoni, M., Mackay, S., 2009. Brachiopod punctae: a complexity in shell biomineralisation. *Journal of Structural Biology*, *167*, 62–67.

Pérez-Huerta, A., Dauphin, Y., Cusack, M. 2013. Biogenic calcite granules—Are brachiopods different? *Micron*, *44*, 395–403.

Pinto, D., Díaz, N., Tang, G., Arends, A., Ramírez, R., Pomar, L., Padrón V., 2014. INTERPRETACION DE RAMPAS CARBONATICAS ASOCIADAS A PALEOISLAS DURANTE EL OLIGOCENO-MIOCENO EN EL GOLFO DE Venezuela. CASO DE ESTUDIO: CAMPO PERLA. *Memorias del I Congreso Venezolano de Gas Natural (ICVGAS), Porlamar, Venezuela, 24–28 Noviembre.*

Plotnick, R. E., Dattilo, B. F., Piquard, D., Bauer, J., Corrie, J., 2013. The orientation of strophomenid brachiopods on soft substrates. *Journal of Paleontology*, *87*, 818–825.

Prokoph, A., Shields, G. A., Veizer, J., 2008. Compilation and time-series analysis of a marine carbonate $\delta^{18}O$, $\delta^{13}C$, $^{87}Sr/^{86}Sr$ and $\delta^{34}S$ database through Earth history. *Earth-Science Reviews*, *87*, 113–133.

Roda, M. S., Griesshaber, E., Angiolini, L., Rollion-Bard, C., Harper, E. M., Bitner, M. A., Milner Garcia, S., Ye, F., Henkel, D., Hässermann, V., Eisenhauer, A., Gnägi, H., Brand, U., Schmahl, W. W., 2022. The architecture of Recent brachiopod shells: diversity of biocrystal and biopolymer assemblages in rhynchonellide, terebratulide, thecideide and craniide shells. *Marine Biology*, *169*(4), 1–52.

Rong, J. Y., and Zhan, R. B., 2006. Surviving the end-Ordovician extinctions: evidence from the earliest Silurian brachiopods of northeastern Jiangxi and western Zhejiang provinces, east China. *Lethaia*, *39*, 39–48.

Ruggiero, E. T., and Bitner, M. A., 2007. Bioerosion on brachiopod shells–a Cenozoic perspective. *Earth and Environmental Science Transactions of the Royal Society of Edinburgh*, *98*, 369–378.

Sandy, M. R., 1997. Mesozoic brachiopods of Mexico-A review and some comments on their paleobiogeographic affinities and paleoecology. *Revista Mexicana de Ciencias Geológicas*, *14*, 11.

Sandy, M. R., and Campbell, K. A., 1994. New rhynchonellid brachiopod genus from Tithonian (Upper Jurassic) cold seep deposits of California and its paleoenvironmental setting. *Journal of Paleontology*, *68*, 1243–1252.

Scholle, P. A., and Ulmer-Scholle, D., 2003. A Color Guide to the Petrography of Carbonate Rocks: Grains, Textures, Porosity, Diagenesis. *American Association of Petroleum Geologists, Memoir 77*, 474 pp.

Schreiber, H. A., Bitner, M. A., Carlson, S. J., 2013. Morphological analysis of phylogenetic relationships among extant rhynchonellide brachiopods. *Journal of Paleontology*, *87*, 550–569.

Shen, S. Z., Archbold, N. W., Shi, G. R., Zhong-Qiang, C., 2000. Permian brachiopods from the SelongXishan section, Xizang (Tibet), China Part 1: Stratigraphy, Strophomenida, Productida and Rhynchonellida. *Geobios*, *33*, 725–752.

Shen, S. Z., Xie, J. F., Zhang, H., Shi, G. R., 2009. Roadian–Wordian (Guadalupian, middle Permian) global palaeobiogeography of brachiopods. *Global and Planetary Change*, *65*, 166–181.

Shen, S. Z., Zhang, H., Shi, G. R., Li, W. Z., Xie, J. F., Mu, L., Fan, J. X., 2013. Early Permian (Cisuralian) global brachiopod palaeobiogeography. *Gondwana Research*, *24*, 104–124.

Soto, G., Ortega, S., Marín, Z., Farías, A., Medina, C., 2025, in press. Estudio Bioestratigráfico de la Unidades Oligo-Miocénicas de la Región Falconiana. *Memorias del II Congreso Venezolano de Geociencias*. Caracas, 20-24 de Marzo 2023. 31 pp.

Steuber, T., and Veizer, J., 2002. Phanerozoic record of plate tectonic control of seawater chemistry and carbonate sedimentation. *Geology*, *30*, 1123–1126.

Taylor, P. D., Vinn, O., Wilson, M. A., 2010. Evolution of biomineralisation in 'lophophorates'. *Special Papers in Palaeontology*, *84*, 317–333.

Tomasových, A., 2006. Differential effects of environmental factors on the ecology of brachiopods and bivalves during the Late Triassic and Jurassic. Doctoral Dissertation, Würzburg University, 430 pp.

Tyler, C. L., Leighton, L. R., Carlson, S. J., Huntley, J. W., Kowalewski, M., 2013. Predation on modern and fossil brachiopods: assessing chemical defenses and palatability. *Palaios*, *28*, 724–735.

Van Geldern, R., Joachimski, M. M., Day, J., Jansen, U., Alvarez, F., Yolkin, E. A., Ma, X. P., 2006. Carbon, oxygen and strontium isotope records of Devonian brachiopod shell calcite. *Palaeogeography, Palaeoclimatology, Palaeoecology*, *240*, 47–67.

Veizer, J., Ala, D., Azmy, K., Bruckschen, P., Buhl, D., Bruhn, F., Carden, G. A. F., Diener, A., Ebneth, S., Godderis, Y., Jasper, T., Korte, C., Pawellek, F., Podlaha, O. G., Strauss, H., 1999. $^{87}Sr/^{86}Sr$, $\delta^{13}C$ and $\delta^{18}O$ evolution of Phanerozoic seawater. *Chemical Geology*, *161*, 59–88.

Vollstaedt, H., Eisenhauer, A., Wallmann, K., Böhm, F., Fietzke, J., Liebetrau, V., Krabbenhöft, A., Farkas, J., Tomasovych, A., Raddatz, J., Veizer, J., 2014. The Phanerozoic $\delta^{88/86}Sr$ record of seawater: New constraints on past changes in oceanic carbonate fluxes. *Geochimica et Cosmochimica Acta*, *128*, 249–265.

Williams, A., 1997. Shell structure. In: R. L. Kaesler (Ed.), Treatise on invertebrate paleontology, Part H, Brachiopods (Revised) (pp. 267–320). Boulder, Colorado, and Lawrence, Kansas: The Geological Society of America and The University of Kansas Press.

Williams, A., Carlson, S. J., Brunton, C. H. C., Holmer, L. E., Popov, L., 1996. A supra-ordinal classification of the Brachiopoda. *Philosophical Transactions of the Royal Society of London. Series B: Biological Sciences*, *351*, 1171–1193.

Ye, F., Crippa, G., Angiolini, L., Brand, U., Capitani, G., Cusack, M., Garbelli, C., Griesshaber, E., Harper, E., Schmahl, W., 2018a. Mapping of recent brachiopod microstructure: A tool for environmental studies. *Journal of Structural Biology*, *201*, 221–236.

Ye, F., Crippa, G., Garbelli, C., Griesshaber, E., 2018b. Microstructural data of six recent brachiopod species: SEM, EBSD, morphometric and statistical analyses. *Data in Brief*, *18*, 300–318.

Ye, F., Garbelli, C., Shen, S., Angiolini, L., 2020. The shell fabric of Palaeozoic brachiopods: patterns and trends. *Lethaia*, doi:https://doi.org/10.1111/let.12412

12 Ostracods

Ostracods (Phylum Arthropoda, Subphylum Crustacea, Class Ostracoda) (e.g., Webber et al. 2010) are arthropods and crustaceans. The former constitutes the largest animal phylum, accounting for about 80% of the total number of species in the animal kingdom (Zhang, 2013), including non-calcifiers such as arachnids and insects, as well as calcareous shell-bearing trilobites (Class Trilobita). On the other hand, crustaceans are the group of extant animals with the widest morphological diversity on the planet (Martin and Davis, 2001), including non-calcifier decapods (e.g., lobsters, crabs, shrimps) and calcareous shell-bearing barnacles. Ostracods are the most diverse living group of crustaceans. They are also the most abundant of fossil arthropods, represented by some 50,000 living and fossil species (Liebau, 2005; Brandão et al. 2019). The Class Ostracoda is one of the best documented groups within the whole animal kingdom, inhabiting marine, freshwater and even semi-terrestrial environments). Their most characteristic feature is a bivalve and well-calcified shell that fossilizes easily (Haq and Boersma, 1998).

12.1 Major Groups and Stratigraphic Distribution

Ostracods have a long and well-documented fossil record from the Ordovician to Present. The classification of the Ostracoda is in a state of flux (i.e., there is more than one classification), as taxonomic standards have changed considerably through time. The affinities of "putative" Cambrian forms are hotly disputed (Armstrong and Brasier, 2005). At the end of the Cambrian and/or in the lowermost Ordovician, the majority of non-calcifying ostracodomorphs were replaced by podocopomorphs with well-calcified carapaces (Liebau, 2005).

In the classification of the higher taxa of the Ostracoda proposed by Liebau (2005), "true" ostracods of the superorder Podocopomorpha include four orders (the "P-orders"): Palaeocopida, Punciocopida (poorly represented in the fossil record), Platycopida, and Podocopida, all ranging in age from the Ordovician to the Holocene (see also Danielopol and Swanson, 2019). In the most widely known ostracods classification (see Horne et al. 2002, 2005; Horne, 2003; Armstrong and Brasier, 2005; Rodriguez-Lazaro and Ruiz-Muñoz, 2012; Danielopol and Swanson, 2019; Jain, 2017), the Class Ostracoda is divided into two subclasses: Myodocopa and Podocopa. Members of the first are weakly calcified, exclusively nekto-planktonic and nekto-benthonic swimming ostracods. Podocopans (the focus of this chapter) are benthic in habit and comprise the orders Palaeocopida, Platycopida, and Podocopida. Members of these three extant orders first appeared in the Early Ordovician (Calcite I) (see Rodriguez-Lazaro and Ruiz-Muñoz, 2012). The Order Palaeocopida comprises essentially the Paleozoic "straight-hinged" ostracods. Palaeocopids are mainly confined to Paleozoic strata (Crasquin and Forel, 2013). They were very diverse and widespread during the Paleozoic but are not not common in modern settings (Horne et al. 2002; Horne, 2003). The Order Platycopida comprises exclusively marine ostracods characterized by a strongly calcified carapace. Platycopids were moderately common in marine settings during the Mesozoic. They are common in modern marine settings. The Order Podocopida comprises the most commonly fossilized group of ostracods and the most abundant in the marine fossil record (although they are also abundant in ancient and modern freshwater and brackish water environments). Well-calcified podocopids were dominant in Paleozoic seas. Podocopida comprises the majority of Mesozoic and Cenozoic fossil ostracods (Armstrong and Brasier, 2005). The vast majority of ostracods encountered in Quaternary sediments are likely to be podocopids (Horne et al. 2002). Criteria for classifying fossil ostracods in thin sections include the shape and size of the carapace, shell overlap, exterior ornamentation and type of hinge. In general, these criteria do not allow systematic assignments (Flügel, 2010). Therefore, systematic paleontology is not applied to the photographic examples in this chapter.

J. A. Méndez Dot, J. Méndez Baamonde, *Atlas of Tropical Carbonates*, https://doi.org/10.1007/978-3-031-25494-9_12

Ostracods reached their peak of abundance during the Calcite II seas of the Silurian (Flügel, 2010), as podocopids radiated in marginal marine environments (including hypersaline and brackish water settings), remaining abundant during the Devonian. Suggestions that the first freshwater ostracods appeared in the Devonian are controversial; undoubted nonmarine/freshwater radiations developed during the Early Carboniferous (Horne, 2003). Ostracods decreased in abundance during the Carboniferous and the Early to Middle Permian. Although they never recovered the abundance reached in the middle Paleozoic, ostracods (mainly podocopids and subordinated platycopids) had few peaks of abundance during the Late Permian-Early Triassic (Aragonite II), and from the Middle Jurassic throughout the Cretaceous (Calcite II) (Horne, 2003; Luger, 2003; Scholle and Ulmer-Scholle, 2003). After a low peak during the Paleogene, ostracods have remained moderately common throughout the Neogene (Aragonite II), including a diversification into the Atlantic and Pacific fauna posterior to the closure of the Panamá isthmus during the Late Pliocene to the Early Pleistocene (Cronin, 1988; Cronin and Ikeya, 1990; Allmon, 2001; Yasuhara et al. 2016). They remain moderately common in modern oceans (although far less abundant than molluscs and echinoderms).

12.2 Skeletal Mineralogy

Podocopan ostracods secrete a bivalve exoskeleton known as carapace. This calcified shell is mainly composed of crystalline LMC embedded in a chitinous and proteinaceous matrix (Dwyer et al. 2002; Scholle and Ulmer-Scholle, 2003; Keyser and Walter, 2004; Armstrong and Brasier, 2005; Yamada and Keyser, 2010). In some taxa, 80–90% of the carapace consists of calcite and 2–15% of chitin and proteins. Apatite precipitated in association with calcite is also common (Dwyer et al. 2002).

12.3 Ecology and Environment

Most ostracods are aquatic benthic heterotrophs that thrive in the neritic zone, and more rarely planktonics in the pelagic realm. These light-independent crustaceans are known for inhabiting freshwater (lakes and rivers; these forms have commonly non- or weakly-calcified carapaces), brackish (estuaries and delta fronts), and marine waters extending to hypersaline settings as well. In modern settings they can be found from tropical to polar latitudes and at different depths, from intertidal settings, to inner-to-outer shelf and bathyal depths (down to 1000–1500 m) (Haq and Boersma, 1998; Scholle and Ulmer-Scholle, 2003; Armstrong and Brasier, 2005; Galindo-Estronza et al. 2016). In general, ostracods are rarely major sediment formers, however, except in stressed (brackish, hypersaline) environments (Scholle and Ulmer-Scholle, 2003), including low-oxygen conditions (Whatley, 1990, 1991).

Whereas the freshwater and less calcified ostracods may spend much time swimming several centimeters above the substrate, marine benthic calcified forms are heavier and tend to be either crawlers, burrowers or interstitial, feeding on detritus or predate diatoms, foraminifera and small polychaete worms. Such ostracods thrive best in muddy sands and silts or on seaweed and seagrasses. Some podocopids live attached to firm substrates, such as larger shells, however they are not common as fossils (Armstrong and Brasier, 2005).

Ostracods present two modes of feeding and respiration: filter-feeding (i.e., the animal feeds on straining suspended organic matter from the water), and deposit-feeding (i.e., the animal feeds on organic matter that has settled on the bottom). Feeding strategies vary/varied depending on the order, as well as on oxygenation on the sea bottom, as marine ostracods are very sensitive to small reductions in oxygen availability. Whatley (1990,1991) showed that during the so-called "kenoxic" events ("kenoxia" is used to denote a dysoxic event when the oxygen level is below 1 mL/L; see Forel et al. 2009), which were much more common throughout the Phanerozoic than "anoxic" events (e.g., Jenkyns, 2010), filter-feeding ostracods dominated over deposit-feeding ostracods, especially during the Late Devonian-Early Carboniferous (Lethiers and Whatley, 1994; Song and Gong, 2019), the Early Triassic (Crasquin-Soleau et al. 2007; Forel et al. 2009), as well as during several low-oxygen episodes in the Jurassic and the Cretaceous (Whatley, 1990, 1991; Whatley et al. 2003). Filter-feeder ostracods in the Paleozoic oceans comprised all palaeocopids and platycopids, as well as some podocopids, such as metacopids (which became extinct in the Toarcian); while deposit-feeder species, abundant in well-oxygenated settings (including modern settings), were mainly podocopids (Lethiers and Whatley, 1994; Forel et al. 2009). Some of the latter evolved specialized adaptations to survive and proliferate under low-oxygen conditions (Lethiers and Whatley, 1994). Platycopids dominated ostracod assemblages during Mesozoic kenoxic events (Whatley, 1991), and remain as the sole extant group of filter-feeding podocopomorphs in modern settings (Horne et al. 2002).

12.4 Carbonate Factory and Skeletal Grain Associations

Ostracods are components of the Heterozoan factory of Michel et al. (2019), and may be accessory components of the Photo C-factory of the same authors (see Appendix A for a detailed treatment on these carbonate benthic factories). This section discusses only the skeletal grain associations involving ostracods represented in Venezuelan carbonate microfacies, as illustrated by the examples in the photographic set. In Middle Permian carbonates of the Palmarito Formation (Venezuelan Andes), ostracods are moderately common to common components of Heterozoan carbonates *sensu* James (1997), in association with brachiopods, bivalves and crinoids (Laya and Tucker, 2012; see also Arnold, 1966). In Early Cretaceous carbonates of the Cogollo Group (early Barremian to late Albian in age), ostracods are also moderately common to common components of heterozoan microfacies in association with molluscs, echinoids and smaller miliolids [molechfor lithofacies *sensu* Carannante et al. (1988), or foramol lithofacies *sensu* Lees and Buller (1972)], plus bryozoans and minor serpulids. See appendix A for definitions of the skeletal grain associations considered in this book.

12.5 Morphology and Diagnostic Elements

Ostracods have carapaces consisting of pairs of bilateral valves, generally equally shaped but different or slightly different in size. The bivalve shells may be ovate, "egg-shaped", or "bean-shaped", and commonly between 0.5 mm and 2 mm large (some modern species may reach up to 8 mm). In the Podocopida one valve (right or left) is usually larger than the other and overlaps the smaller valve at the ventral margin or at all margins. In the Platycopida the right valve overlaps the left valve. In the Palaeocopida valves do not overlap. The valves articulate at the dorsal (or posterior) margin along a hinge structure of interlocking grooves and bars or teeth and sockets. The ventral (or anterior) free margin may show conspicuous calcified morphologic features (i.e., marginal infold) considered as useful taxonomic criteria (e.g., "fish-hook"-like terminations). The valves may show localized thickening and/or strong curving at one or both margins and these are thin and curved, similar to that of some bivalve molluscs but generally smaller in length and thickness. The surface of the valves may be smooth or ornamented (e.g., ridged, spinose). The calcified valves consist of an outer and an inner calcitic lamella. The microstructures are homogeneous prismatic and finely prismatic and form one shell layer that commonly shows extinction bands under cross CPL. Calcitic crystals are oriented perpendicular to the valve margins. In life, the calcareous valves of podocopids are enclosed by a chitinous layer (see Haq and Boersma, 1998; Scholle and Ulmer-Scholle, 2003; Horne et al. 2002; Armstrong and Brasier, 2005; Yamada, 2007; Flügel, 2010; Yamada and Keyser, 2010; Rodriguez-Lazaro and Ruiz-Muñoz, 2012; Naumcheva, 2019).

12.6 Preservation as Fossils

Layers and microstructures in ancient ostracod valves are, in most cases, well preserved due to their LMC composition. Carapaces may be preserved as articulated valves (filled with micrite, microspar, pseudospar, or LMC cement), as disarticulated individual valves, or as fragments. In the last two scenarios, morphological features at the margins of the valves, when well preserved, help to distinguish ostracods from valves of LMC molluscs and brachiopods.

12.7 Geochemical Proxies in Ostracods

Ostracods shell calcification takes place very quickly, typically within a few hours to several days, in geochemical equilibrium with the ambient water. In life, ostracods grow by molting their calcite valves several times (up to eight times) during ontogeny. After molting, the ions used to mineralize the shell are taken directly from the host waters without prior storage. Thus, the chemistry of ostracod shells reflects the chemical conditions of the ambient water (e.g., temperature, salinity, dissolved ion composition, hydrology) at the time of molting and new valve growth (Morishita et al. 2007; Dettman and Dwyer, 2012; Börner et al. 2013 and references therein).

Widespread distribution and abundance of benthic fossil and living calcified ostracods in sediments ranging from the Ordovician to Present make these crustaceans useful geochemical proxies. The isotopic and elemental composition of LMC ostracod valves have been widely used as empirical proxies in paleoenvironmental reconstructions of marine, hypersaline, brackish and freshwater environments. The coupling of $\delta^{18}O$ with Mg/Ca and Sr/Ca ratios is used to interpret changes in bottom water temperature and salinity. The $\delta^{18}O$ and $\delta^{13}C$ compositions may provide information about glacial versus interglacial conditions, and carbon cycle, CO_2 exchange, and productivity, respectively ($\delta^{13}C$ is more useful in freshwater ostracods). The $\delta^{87}Sr$ is a useful proxy for interpreting freshwater influx to marginal marine settings, as well as for recognizing the transition from marine to brackish and freshwater conditions in coastal settings (see Dwyer et al. 2002; Ito et al. 2003; Morishita et al. 2007; Bennett et al. 2011; Dettman and Dwyer, 2012; Börner et al. 2013; Lea, 2014; Devriendt et al. 2017; Jain, 2017).

12.8 Ostracods: Photographic Set and Descriptive Captions

In all photos, IW = image width (the true length of the horizontal axis of the view)

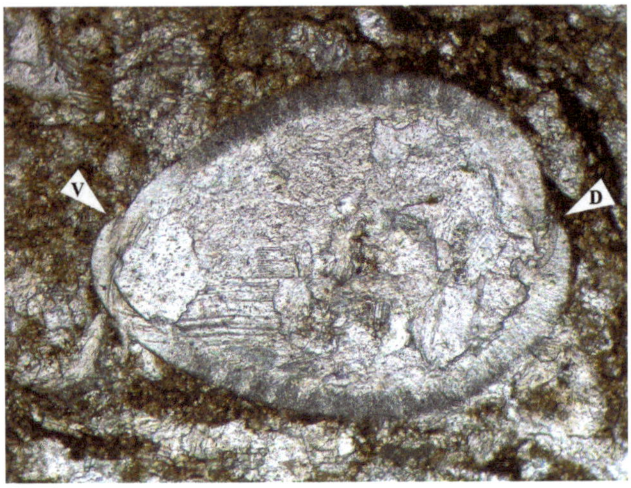

Middle Permian (Cisularian to Guadalupian). Palmarito Formation. Venezuelan Andes

Sagittal view of an ostracod carapace filled by coarse-grained carbonate LMC crystals. The articulated carapace exhibits "egg" shape and bilateral but slightly unequal valves, with a flattened dorsal (D) margin and smoothly pointed ventral (V) margin. The larger valve is overlapping the smaller valve at both margins. LMC valves exhibit well-preserved crystals oriented perpendicular to the margins, as well as characteristic sweeping extinction bands under CPL.

IW = 1.3 mm. CPL

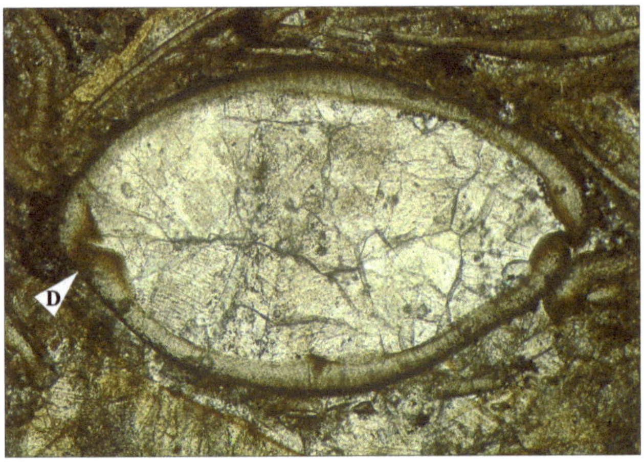

Middle Permian (Cisularian to Guadalupian). Palmarito Formation. Venezuelan Andes

An oval-shaped ostracod carapace surrounded by numerous fragments of ostracod valves. Note the localized thickening of the valves at both margins, especially at the groove articulation of the dorsal (D) one. As in the previous example, the carapace, which is about 1 mm large, is filled with cloudy, inequigranular sparry calcite. Textural features of this LMC infilling suggest it is a void-filling intraparticle meteoric cement, although meteoric cements are not common in the Palmarito Formation.

IW = 1.3 mm. PPL

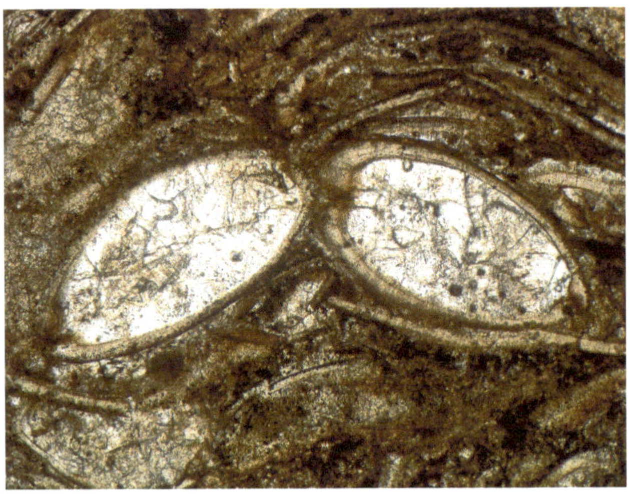

Middle Permian (Cisularian to Guadalupian). Palmarito Formation. Venezuelan Andes

Ostracods in this packstone are preserved as complete shells and as disarticulated or fragmented valves. Note abundant pyrite framboids (black spots) blackening the sample. The latter indicates transient low-oxygen conditions during early marine diagenesis and suggests eutrophic conditions during deposition. Ostracods are common bioclastic components of the base of the Palmarito Formation, a rock interval has been interpreted as deposited in a tidal flat environment (Laya and Tucker, 2012).

IW = 2.4 mm. PPL

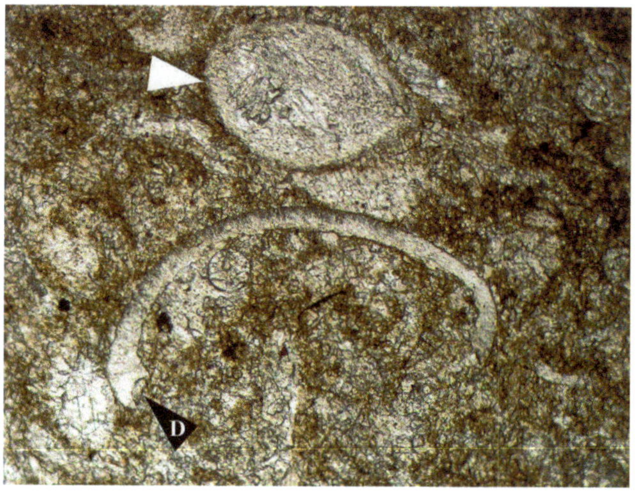

Middle Permian (Cisularian to Guadalupian). Palmarito Formation. Venezuelan Andes

Disarticulated ostracod valve accompanied by a complete ostracod carapace (indicated with the white arrow) in a partially recrystallized wackestone. In addition to the size and curved (or arched) shape, the characteristic groove-shaped termination of the valve at the dorsal (D) margin serves as diagnostic to reliably differentiate this ostracod valve from calcareous valves derived from the breakdown of other heterotrophic calcifiers (e.g., brachiopods and molluscs).

IW = 1.3 mm. PPL

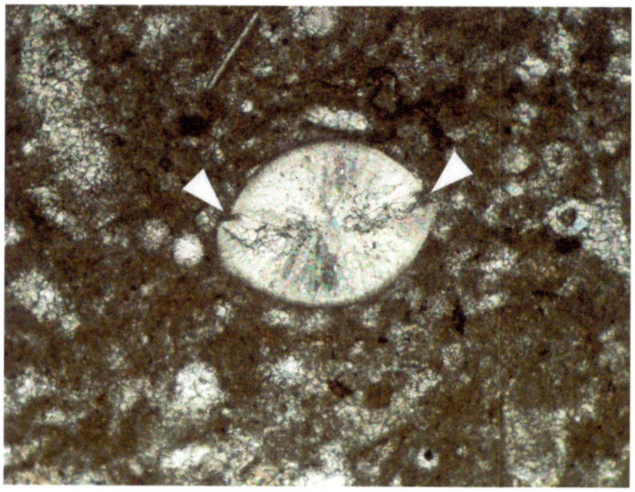

Middle Permian (Cisualrian to Guadalupian). Palmarito Formation. Venezuelan Andes

This ostracod carapace shows valves dissimilar in size, indicated by the location of the joinings of the valves (white arrows). Note that the shell's internal cavity is filled by calcite cement formed in optical continuity with the finely prismatic crystals of the valves. Distinctive extinction bands (light and dark), oriented perpendicular to the margins of the carapace, make this optical continuity especially apparent.

IW = 1.3 mm. CPL

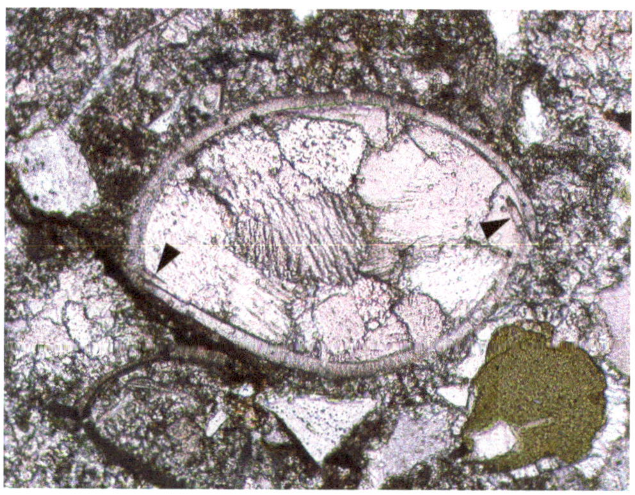

Cretaceous (middle Albian). Lisure Formation. Maracaibo Basin

Ostracod carapace in a dolomitic fabric. Well-preserved thin valves show finely prismatic crystals, highlighting their preservation after dolomitization of the micritic infilling. Note the black arrows pointing to different valves, as the overlapped valve is switched at both margins (the valve that overlaps the other at one margin is being overlapped at the other margin). This switch in the overlapping of valves is also characteristic of some calcified Paleozoic ostracods (see some Carboniferous examples in Bennett et al. 2011).

IW = 1.3 mm. CPL

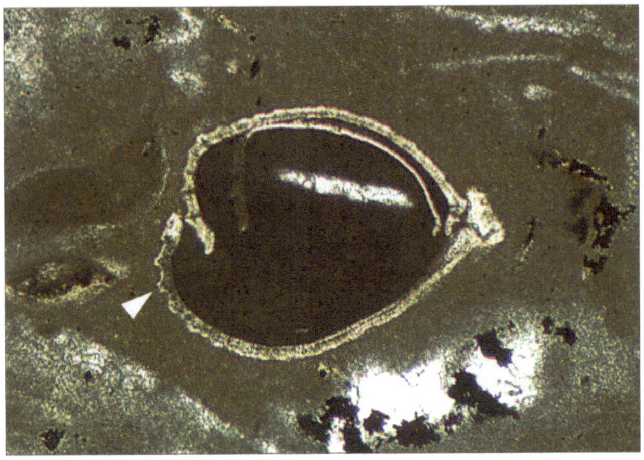

Cretaceous (late Barremian). Tibú Member, Apón Formation. Maracaibo Basin

Ostracod carapace filled with dark micritic sediments. Note distinctive exterior ridged ornamentation of the valves (white arrow). These are clearly dissimilar in length and shape. The larger valve is overlapping the smaller one at both margins and exhibits striking thick, golf-club-like termination at the ventral margin.

IW = 1.3 mm. PPL

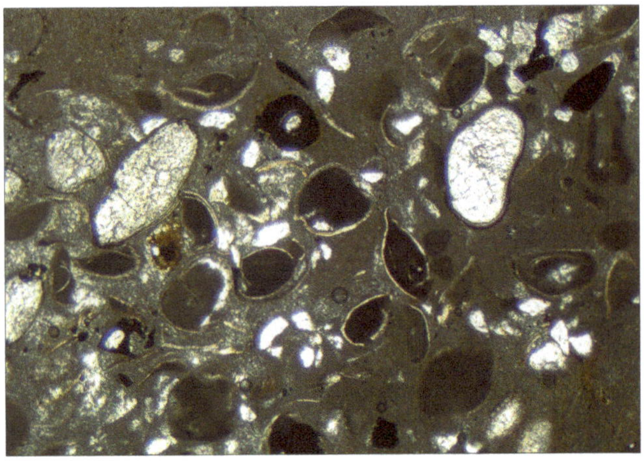

Cretaceous (late Barremian). Tibú Member, Apón Formation. Maracaibo Basin

Ostracods in this sandy wackestone are preserved as articulated carapaces filled either with dark micritic sediments or cloudy sparry calcite, or preserved as disarticulated/fragmented valves. The dark color of the micritic matrix suggests meso-eutrophic conditions during deposition. Ostracods are minor components in shallow facies of the Cogollo Group, but they are moderately common in mudstones and wackestones of the Tibú Member. Ostracod-rich facies in the Cogollo Group have been interpreted as being deposited in a shallow tidal flat environment (see Bartok et al. 1981).

IW = 4.2 mm. PPL

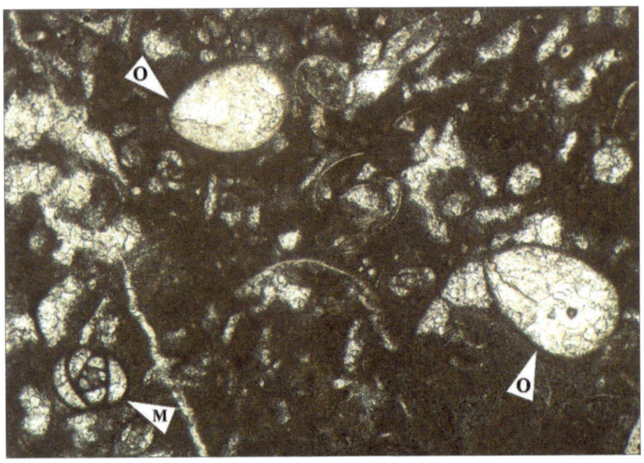

Cretaceous (middle Aptian). Piché Member, Apón Formation. Maracaibo Basin

Ostracods (O) in the image are accompanied by smaller miliolids (M) and undifferentiated, small diagenetically-altered grains in a wackestone. This heterozoan association is moderately common of some transgressive and moderately organic-rich deposits in the Cogollo Group. In that shallow carbonate succession, the flourishing of ostracods is associated to higher nutrient levels compared to other common heterotrophic calcifiers such as many types of molluscs, as well as echinoids, bryozoans and serpulids.

IW = 2.4 mm. PPL

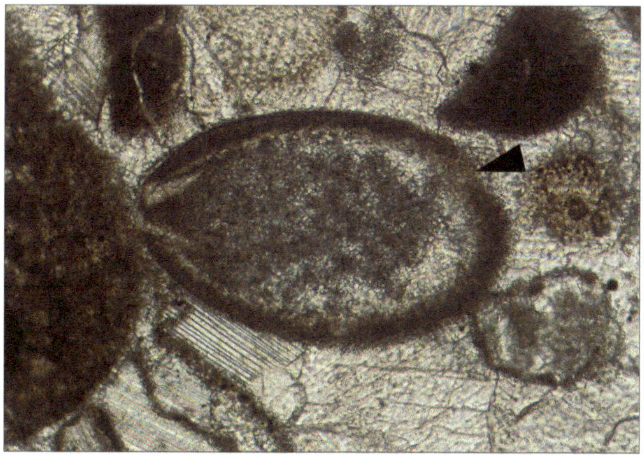

Cretaceous (middle Barremian). Tibú Member, Apón Formation. Maracaibo Basin

Ostracod carapace in a recrystallized matrix. Note the distinctive fish-hook like terminations of both valves (marginal infolds) at the ventral margin (or free margin). The dissimilar size of the valves is indicated by the location of the joining of the valves (or hinge) at the dorsal margin (black arrow), barely visible due to micritization of the carapace. The latter implies that the hinge line, or the distance between both margins (the line along which valves articulate) is shorter than the length of the carapace (see Haq and Boersma 1998; Armstrong and Brasier, 2005; Naumcheva, 2019).

IW = 1.3 mm. PPL

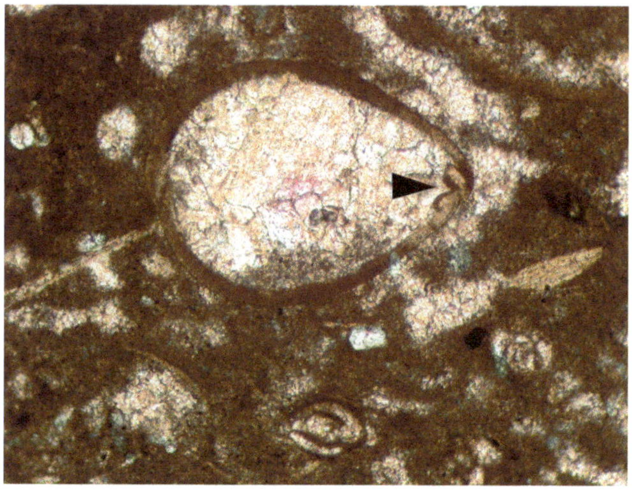

Cretaceous (middle Aptian). Piché Member, Apón Formation. Maracaibo Basin

"Egg-shaped" ostracod carapace with a straight hinge line (or equal to the length of the carapace). Note the distinctive marginal infold (black arrow). These strongly calcified inner lamellae develop along the free margin of both valves and commonly show wide morphological variability among species of calcified ostracods (note the differences between this and previous example). These terminal elements are essential for the animal to close and open its bivalve carapace. Marginal infolds become notable in the adult stage (see Yamada and Keyser, 2010).

IW = 1.3 mm. PPL

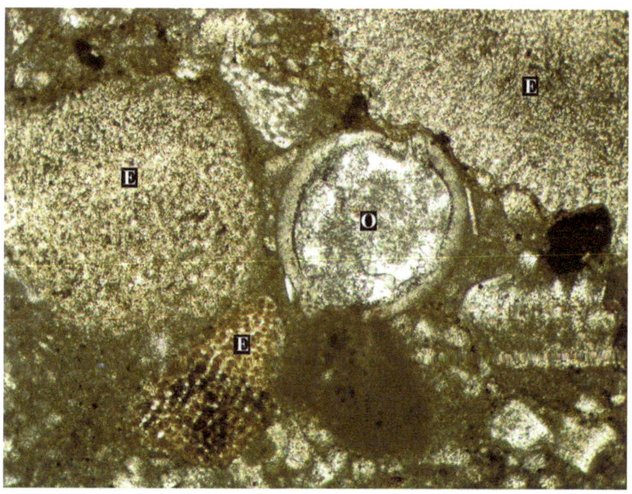

Cretaceous (middle Albian). Lisure Formation. Maracaibo Basin

Ostracod (O) carapace accompanied by echinoids (E) fragments in a muddy dolomitic wackestone. This heterozoan association exemplifies the dominance of heterotrophic calcifiers in facies of the Lisure Formation. In general, among heterotrophs in the Cogollo Group, ostracods (and serpulids) are far less common than echinoids and molluscs.

IW = 2.4 mm. PPL

References

Allmon, W. D., 2001. Nutrients, temperature, disturbance, and evolution: a model for the late Cenozoic marine record of the western Atlantic. *Palaeogeography, Palaeoclimatology, Palaeoecology, 166*, 9–26.

Armstrong, H. A., and Brasier, M. D., 2005. Microfossils, Second edition. Blackwell Publishing, 296 pp.

Arnold, H. C., 1966. Upper Paleozoic Sabaneta–Palmarito sequence of Mérida Andes, Venezuela. *American Association of Petroleum Geologist Bulletin, 50*, 2366–2387.

Bennett, C. E., Williams, M., Leng, M. J., Siveter, D. J., Davies, S. J., Sloane, H. J., Wilkinson, I. P., 2011. Diagenesis of fossil ostracods: Implications for stable isotope based palaeoenvironmental reconstruction. *Palaeogeography, Palaeoclimatology, Palaeoecology, 305*, 150–161.

Bartok, P., Reijers, T. J. A., Juhasz, I., 1981. Lower Cretaceous Cogollo Group, Maracaibo Basin, Venezuela: sedimentology, diagenesis, and petrophysics. *American Association of Petroleum Geologists Bulletin, 65*, 1110–1134.

Börner, N., De Baere, B., Yang, Q., Jochum, K. P., Frenzel, P., Andreae, M. O., Schwalb, A., 2013. Ostracod shell chemistry as proxy for paleoenvironmental change. *Quaternary International, 313*, 17–37.

Brandão, S. N., Angel, M. V., Karanovic, I., Perrier, V., Meidla, T., 2019. World Ostracoda Database. Accessed at http://www.marinespecies.org/ostracoda on 27 March 2019.

Carannante, G., Esteban, M., Milliman, J. D., Simone, L., 1988. Carbonate lithofacies as paleolatitude indicators: problems and limitations. *Sedimentary Geology, 60*, 333–346.

Crasquin, S., and Forel, M. B., 2013. Ostracods (Crustacea) through Permian–Triassic events. *Earth-Science Reviews, 137*, 52–64.

Crasquin-Soleau, S., Shen, S. Z., Li, W. Z., Cao, C. Q., 2007. Ostracods from the Lopingian and Permian–Triassic boundary beds at the Gyanyima section in southwestern Tibet, China. *Palaeoworld, 16*, 222–232.

Cronin, T. M., 1988. Geographical Isolation in Marine Species: Evolution and Speciation in Ostracoda, I. In: Hanai T, Ikeya N, Ishizaki K, eds., Evolutionary biology of ostracoda: its fundamentals and applications. Proceedings of the Ninth International Symposium on Ostracoda, Shizuoka, Japan, 1985. Developments in Palaeontology and Stratigraphy 11, Kodansha, Tokyo and Elsevier, Amsterdam, pp. 871–889.

Cronin, T. M., and Ikeya, N.,1990. Tectonic events and climatic change: Opportunities for speciation in Cenozoic marine Ostracoda. *Causes of Evolution–A Paleontological Perspective. University of Chicago Press, Chicago and London*, 210–248.

Danielopol, D., and Swanson, K. M., 2019. Ostracod phylogeny and evolution—thirty years after "Manawan perspective". Geo-Eco-Marina, *25*. 219–251.

Dettman, D. L. and Dwyer, G. S. 2012. The calibration of environmental controls on elemental ratios in ostracod shell calcite: a critical assessment. In: Horne, D.J., Holmes, J.A., Rodriguez-Lazaro, J. & Viehberg, F. (eds.), *Ostracoda as Proxies for Quaternary Climate Change. Developments in Quaternary Science, 17*. Elsevier, Amsterdam, 147–165.

Devriendt, L. S., McGregor, H. V., Chivas, A. R., 2017. Ostracod calcite records the $^{18}O/^{16}O$ ratio of the bicarbonate and carbonate ions in water. *Geochimica et Cosmochimica Acta, 214*, 30–50.

Dwyer, G. S., Cronin, T. M., Baker, P. A., 2002. Trace elements in marine ostracodes. In: Holmes, J.A., Chivas, A.R. (Eds.), *American Geophysical Union Monograph, 131*, 205–225.

Flügel, E, 2010. Microfacies of Carbonate Rocks—Analysis, Interpretation and Application. Springer, Heidelberg-Dordrecht-London-New York, 984 pp.

Forel, M. B., Crasquin, S., Kershaw, S., Feng, Q. L., Collin, P. Y. 2009. Ostracods (Crustacea) and water oxygenation in the earliest Triassic of South China: implications for oceanic events at the end-Permian mass extinction. *Australian Journal of Earth Sciences, 56*, 815–823.

Galindo-Estronza, A. M. G., Alfaro, M., Schizas, N. V., 2016. Morphological and genetic species diversity in ostracods (Crustacea: Oligostraca) from Caribbean reefs. *Marine Biodiversity, 47*, 37–53.

Haq, B. U., and Boersma, A., 1998. Introduction to Marine Micropaleontology. Elsevier, 376 pp.

Horne, D. J., Cohen, A., Martens, K., 2002. Taxonomy, morphology and biology of Quaternary and living Ostracoda. *The Ostracoda: applications in Quaternary research, 131*, 5–36.

Horne, D. J., 2003. Key events in the ecological radiation of the Ostracoda. *The Paleontological Society Papers, 9*, 181–202.

Horne, D. J., Schön, I., Smith, R., Martens, K., 2005. What are Ostracoda? A cladistic analysis of the extant superfamilies of the subclasses Myodocopa and Podocopa (Crustacea: Ostracoda). In: Koenemann, S. & Jenner, R.A. (Eds), Crustacea and Arthropod Relationships. *Crustacean Issues, 16*, 249–273.

Jain, S., 2017, Fundamentals of Invertebrate Palaeontology: Microfossils. Springer Nature India Private Limited. 323 pp.

James, N. P., 1997. The cool-water carbonate depositional realm. In: James N. P., Clarke, J., (eds.), Cool-water carbonates, vol 56. SEPM Society for Sedimentary Geology, Tulsa, 1–20.

Jenkyns, H. C., 2010. Geochemistry of oceanic anoxic event. *Geochemistry, Geophysics, Geosystems, 11*(3), Q03004.

Ito, E., De Deckker, P., Eggins, S. M., 2003. Ostracodes and their shell chemistry: implications for paleohydrologic and paleoclimatologic applications. *The Paleontological Society Papers, 9*, 119–152.

Keyser, D and Walter, R., 2004. Classification in Ostracods. *Revista Española de Micropaleontología, 36*, 1–11.

Laya, J. C., and Tucker, M. E., 2012. Facies analysis and depositional environments of Permian carbonates of the Venezuelan Andes: Palaeogeographic implications for Northern Gondwana. *Palaeogeography, Palaeoclimatology, Palaeoecology, 331*, 1–26.

Lea, D. W., 2014. Elemental and Isotopic Proxies of Past Ocean Temperatures. *Treatise on Geochemistry: Second Edition. Elsevier Ltd*, 373–397.

Lees, A., and Buller, A. T., 1972. Modern temperate-water and warm-water shelf carbonate sediments contrasted. *Marine Geology, 13*, M67–M73.

Lethiers, F., and Whatley, R., 1994. The use of Ostracoda to reconstruct the oxygen levels of Late Palaeozoic oceans. *Marine Micropaleontology, 24*, 57–69.

Liebau, A., 2005. A revised classification of the higher taxa of the Ostracoda (Crustacea). *Hydrobiologia, 538*, 115–137.

Luger, P., 2003. Paleobiogeography of late Early Cretaceous to early Paleocene marine Ostracoda in Arabia and north to Equatorial Africa. *Palaeogeography, Palaeoclimatology, Palaeoecology, 196*, 319–342.

Martin, J. W., and Davis, G. E., 2001. An updated classification of the Recent crustacea. *Natural History Museum of Los Angeles County, Contributions in Science, 39*, 1–124.

Michel, J., Laugié, M., Pohl, A., Lanteaume, C., Masse, J. P., Donnadieu, Y., Borgomano, J., 2019. Marine carbonate factories: a global model of carbonate platform distribution. *International Journal of Earth Sciences, 108*, 1773–1792.

Morishita, T., Tsurumi, A., Kamiya, T., 2007. Magnesium and strontium distributions within valves of a recent marine ostracode, Neonesidea oligodentata: Implications for paleoenvironmental reconstructions. *Geochemistry, Geophysics, Geosystems, 8*, 11 pp.

Naumcheva, M. A., 2019. A new method for description of smooth ostracod shells, using Permian and Triassic ostracods from Eastern Europe. *Paleontological Journal, 53*, 616–628.

Rodriguez-Lazaro, J., and Ruiz-Muñoz, F., 2012. A general introduction to ostracods: morphology, distribution, fossil record and applications. *Developments in Quaternary Science*, 17, pp. 1–14. Elsevier.

Scholle, P. A., and Ulmer-Scholle, D., 2003. A Color Guide to the Petrography of Carbonate Rocks: Grains, Textures, Porosity, Diagenesis. *American Association of Petroleum Geologists, Memoir 77*, 474 pp.

Song, J. J., and Gong, Y. M., 2019. The filter-feeders signal: response of ostracods to marine environmental changes in the Devonian F–F event. *Palaeoworld*, *29*, 544–551.

Whatley, R. C., 1990. Ostracoda and global events. In: Whatley, R.C., Maybury, C. (Eds.), Ostracoda and global events. Chapman and Hall. 3–24.

Whatley, R., 1991. The platycopid signal: a means of detecting kenoxic events using Ostracoda. *Journal of Micropalaeontology*, *10*, 181–185.

Whatley, R. C., Pyne, R. S., Wilkinson, I. P., 2003. Ostracoda and palaeo-oxygen levels, with particular reference to the Upper Cretaceous of East Anglia. *Palaeogeography, Palaeoclimatology, Palaeoecology*, *194*, 355–386.

Webber, W. R., Fenwick, G. D., Bradford-Grieve, J. M., Eagar, S. H., Buckeridge, J. S., Poore, G. C. B., Dawson, E. W., Watling, L., Jones, J. B., Wells, J. B. J., Bruce, N. L., Ahyong, S. T., Larsen, K., Chapman, M. A., Olesen, J., Ho, J. -S., Green, J. D., Shiel, R. J., Rocha, C. E. F., Lörz, A. -N., Bird, G. J. and Charleston, W. A., 2010. Phylum Arthropoda. Subphylum Crustacea. Shrimps, crabs, lobsters, barnacles, slaters, and kin. In: Gordon D.P. (ed.), New Zealand Inventory of Biodiversity, Vol. 2: Kingdom Animalia, Chapter 8, Canterbury University Press, Christchurch, 98–232.

Yamada, S., 2007. Structure and evolution of podocopan ostracod hinges. *Biological Journal of the Linnean Society*, *92*, 41–62.

Yamada, S., and Keyser, D., 2010. Calcification of the marginal infold in podocopid ostracods. *Hydrobiologia*, *638*, 213–222.

Yasuhara, M., Iwatani, H., Hunt, G., Okahashi, H., Kase, T., Hayashi, H., Irizuki, T., Aguilar, Y. M., Fernando, A. G. S., Renema, W., 2016. Cenozoic dynamics of shallow-marine biodiversity in the Western Pacific. *Journal of Biogeography*, *44*(3), 567–578.

Zhang, Z. Q., 2013. Phylum Arthropoda. In: Zhang, Z. -Q. (Ed.) Animal Biodiversity: An Outline of Higher-level Classification and Survey of Taxonomic Richness (Addenda 2013). *Zootaxa*, *3703*, 17–26.

13 Barnacles

Barnacles (Phylum Arthropoda; Subphylum Crustacea; Class Thecostraca; Subclass Cirripedia; Chan et al. 2021) are sessile, filter-feeding crustaceans (Webber et al. 2010). They are an extremely diverse array of marine invertebrates known since the Silurian (e.g., Newman and Abbott, 1980). These benthic heterotrophs are found as calcified attachers, as soft-bodied burrowers, or as parasitic forms (e.g., hosted by other crustaceans and echinoderms) (see Chan et al. 2021).

Heavily calcifying barnacles (the focus of this chapter) are commonly referred to as balanids, due to distinctive and widespread fossil shell plates and modern specimens of the Order Balanomorpha (e.g., Collareta et al. 2019). The Balanomorpha ("acorn" barnacles) are the largest group of barnacles (by far the most species rich) and rank among the most diverse, commonly encountered and ecologically important marine crustaceans in the world. The large biological and structural diversity exhibited by the acorn barnacles makes them ideal models in studies of ecology and evolution (Pérez-Losada et al. 2014).

13.1 Major Groups and Stratigraphic Distribution

A recent, updated classification of fossil and living barnacles by Chan et al. (2021) divides the Subclass Cirripedia into three infraclasses: Acrothoracica (burrowing barnacles), Rhizocephala (parasitic barnacles), and Thoracica (stalked and acorn barnacles). Heavily calcified species of the Thoracica, Order Balanomorpha, are the most abundant barnacles in both the fossil record and modern settings (Scholle and Ulmer-Scholle, 2003; Collins et al. 2014; Pérez-Losada et al. 2014; Klompmaker et al. 2015; Coletti et al. 2018; Collareta et al. 2019; Chan et al. 2021).

Until two decades ago, the oldest specimens widely accepted as barnacles with mineralized shells were Silurian (Calcite I) in age (Briggs et al., 2005). From their recent phylogenetic analysis and taxonomic review, Chan et al. (2021) consider the oldest certain record of a thoracican cirripede to be from the Early Carboniferous (330 to 320 Ma; see also Buckeridge and Newman, 2006), right at the transition from Calcite I to Aragonite II. Nevertheless, thoracican shell plates were originally phosphatic in composition and only later in the Middle Jurassic (early in Calcite II) switched to calcitic mineralization (see Gale, 2015; Gale and Schweigert, 2015; Chan et al. 2021). The "superorders" Thoracicalcarea (Gale, 2015) and Phosphatothoracica (Gale, 2019) were defined in order to differentiate calcitic and phosphatic forms within the Thoracica.

Balanomorphan (acorn) barnacles (balanids) appeared in the first half of the Late Cretaceous (Calcite II), although they are represented in Cretaceous strata by a single extinct family (Chan et al. 2021). Diversification events in the Eocene, and in the Oligocene to Pliocene led to a marked increase in their abundance (for details see Pérez-Losada et al. 2014; Chan et al. 2021). By Late Oligocene and Miocene (Aragonite III), balanids became locally abundant (e.g., Halfar and Mutti, 2005; Bassi et al. 2006; Pomar et al. 2012, 2015; Buckeridge et al. 2014; Coletti et al. 2018), peaking during the Pliocene and Pleistocene (Aragonite III). In Neogene strata (including modern facies), they are locally major rock forming elements in non-tropical settings (e.g., Kamp et al. 1988; Nelson et al. 1988; Scholle and Ulmer-Scholle, 2003; Aguirre et al. 2008; Massari and D'Alessandro, 2012; Buckeridge et al. 2018).

As for stalked barnacles (e.g., Order Scalpellomorpha), they radiated earlier than balanomorphans, in the Early Cretaceous (Pérez-Losada et al. 2008; Chan et al. 2021). Scalpellomorphs also form calcified multi-plate shells (although generally lighter, less robust shells), but are far less common in both the fossil record (see Webber et al. 2010; Harzhauser and Schlögl, 2012) and modern settings (Ullmann et al. 2018).

J. A. Méndez Dot, J. Méndez Baamonde, *Atlas of Tropical Carbonates*, https://doi.org/10.1007/978-3-031-25494-9_13

13.2 Skeletal Mineralogy

Multi-plate shells of modern balanids are composed of LMC (typically less than 1 mol% $MgCO_3$), but in some cases of HMC is observed. The basal attachment disc may be aragonitic in some warm-water species (Bourget, 1987; Scholle and Ulmer-Scholle, 2003; Ullmann et al. 2018). Micron-scale variations in shell structure (dimension, orientation, and width of calcitic crystallites) result in varied macro-scale features (plate thickness, architecture, and morphology; see Shaw et al. 2024 for details).

13.3 Ecology and Environment

Balanid barnacles are benthic, sessile heterotrophs. These filter-feeders live epifaunally, feeding on phytoplankton and small zooplankton (e.g., microalgae and other suspended organic particles). Their attaching mode, commonly found in moderate- to high-energy depositional environments, is divided into two categories: forms that attach to rocky substrates, from intertidal to subtidal and neritic settings (mainly within the upper neritic zone), and epibionts (i.e., forms that attach to other living organisms), such as those found on shells and some marine invertebrates (see Newman and Abbott, 1980; Kamp et al. 1988; Scholle and Ulmer-Scholle, 2003; Hood et al. 2003; Flügel, 2010; Buckeridge et al. 2014; Pochai et al. 2017; Coletti et al. 2018).

Most modern balanid deposits are generally associated with shallow depths (< 50 m deep), high-energy, hard substrates and plankton-rich waters at middle and high latitudes (e.g., Kamp et al. 1988; Nelson et al. 1988; Hayton et al. 1995; Webber et al. 2010; Coletti et al. 2018). Highly productive oceanic zones, including upwelling-oxygenated zones are ideal for the flourishing of these heterotrophic calcifiers (e.g., Westphal et al. 2010; Klicpera et al. 2015; Reymond et al. 2016; Michel et al. 2018). They also thrive locally at low-latitudes, especially in nutrients-rich (eutrophic) environments (see Carannante et al. 1988; Westphal et al. 2010; Reijmer et al. 2012; Klicpera et al. 2015). In Miocene carbonate ramps in the Mediterranean region (subtropical paleolatitudes), balanids thrived within a wider depth range, as they are common secondary components of carbonate factories developed in euphotic to mesophotic depths (Benisek et al. 2009, 2010; Coletti et al.

2018), as well as in meso-oligophotic depths (Bassi et al. 2006; Pomar et al. 2012).

The extensive fossil record of drill holes of predatory origin on balanid shells suggests that drilling predation on heavily calcified balanomorphans (drill holes attributed mainly to gastropods) was a worldwide phenomenon during the Cenozoic, and continues in modern settings (Klompmaker et al. 2015 and references therein; see also Palmer, 1988, 1990; Klompmaker et al. 2014; Mondal et al. 2021).

13.4 Carbonate Factories and Skeletal Grain Associations

Balanid barnacles are components of the Heterozoan factory of Michel et al. (2019), as well as occasional to common secondary components of the Photo C-factory of these authors (see Appendix A for a detailed treatment of these carbonate benthic factories). This section discusses only the skeletal grain associations involving balanids represented in Venezuelan carbonate microfacies, as illustrated by the examples in the photographic set.

In Late Oligocene-Early Miocene carbonates of the Perla Limestone (EGVSB), balanids are secondary components of photozoan-dominated microfacies in association with red algae and LBF. These photozoan-dominated carbonates are the expression of the Photo C-factory developed at tropical paleolatitutes, having an important contribution of other heterotrophic calcifiers, mainly echinoids and bryozoans, and less common molluscs (see Pomar et al. 2015). Depending on the content of calcareous red algae versus LBF, echinoids and bryozoans, many of these microfacies can be grouped into rhodalgal lithofacies (*sensu* Carannante et al. 1988) or rhodechfor lithofacies (*sensu* Hayton et al. 1995). See appendix A for definitions of the skeletal grain associations considered in this book.

In Early and Middle Miocene carbonates of the FB (San Luis and Capadare formations, respectively), balanids are rare, and not previously reported (to our knowledge). Therefore, no skeletal grain association is considered, as further studies need to be carried. Regarding the flourishing of balanid barnacles in Larga Island (Carabobo State shoreline), these heterotrophic calcifiers are also rare/scarce (or absent in most of the island's environments). We speculate about a biotic association of these crustaceans with molluscs and echinoids [Heterozoan Association *sensu* James (1997)].

13.5 Morphology and Diagnostic Elements

Balanid barnacles are readily distinguishable, particularly if shells remain articulated. Shells range from a few millimeters to up to 80 cm in length (Newman and Abbott, 1980). Multi-plate shells consist of a massive basal plate that attaches to the substrate. This discoidal attachment exhibits distinctive tubular internal structures (parietal tubes) aligned in series. Immovable, compartmentalized, calcitic wall plates (usually 6 to 8 in number) consist of distinctive longitudinal thick walls (septa) and a plicated (molar-like) internal structure. These compartmental plates are attached to the basal plate and articulate laterally by perpendicularly calcitic "bar-like" septa termed radii (singular: radius). Moveable opercular plates (valves) are used to seal off the living chamber during times of stress (e.g., low-tide exposure) (Scholle and Ulmer-Scholle, 2003; Maggenti et al. 2008; Collins et al. 2014; Coletti et al. 2018; Checa et al. 2019; Collareta et al. 2019).

13.6 Preservation as Fossils

After death, balanid barnacle multi-plate shells disarticulate into individual shell plates. These undergo fragmentation and mechanical abrasion, and in some cases reworking and redeposition. Fossil plates range from a few millimeters (commonly >1 mm) to a few centimeters in length and commonly exhibit aberrant morphologies. Most common recognizable elements of fossil barnacles are the massive basal plate; the imbricated and robust wall plates consisting of large compartments separated by thick ridged walls; and foliated opercular valves. These elements are typically well preserved due to their LMC composition.

13.7 Barnacles: Photographic Set and Descriptive Captions

In all photos, IW = image width (the true length of the horizontal axis of the view)

Holocene. Larga Island. Carabobo State Shoreline
Cross section of the basal plate of a balanid barnacle shell. Parietal tubes (porous circular elements) are aligned in a series of six. These tubules are separated by robust ribbed walls (W) exhibiting characteristic plicated morphology. This aligned-porous (or vesicular) structure serves as diagnostic criteria for identifying this crustacean. The shell and other carbonate bioclasts and quartz (Q) grains in the image are coated by early marine cement.

IW = 1.8 mm. CPL

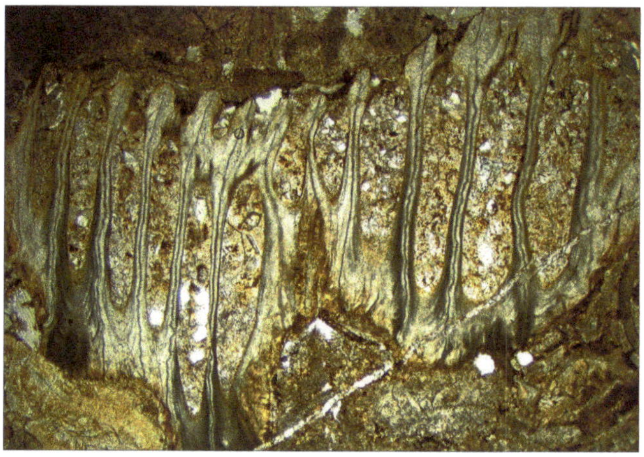

Early Miocene. Capadare Formation. Falcón Basin
Balanid plate exhibiting ribbed and compartmentalized structure. Elongate compartments are separated by thick walls termed longitudinal ribs. These are oriented parallel to each other and showing high optical relief. Compartments are filled by a combination of small bioclastic fragments, micritic sediments and quartz grains. Such infilling is partially phosphatized. Elongate walls and compartments of balanids are easily to differentiate, even after fragmentation and abrasion of the multi-plate shell, due to characteristic imbricated appearance.

IW = 4.2 mm. PPL

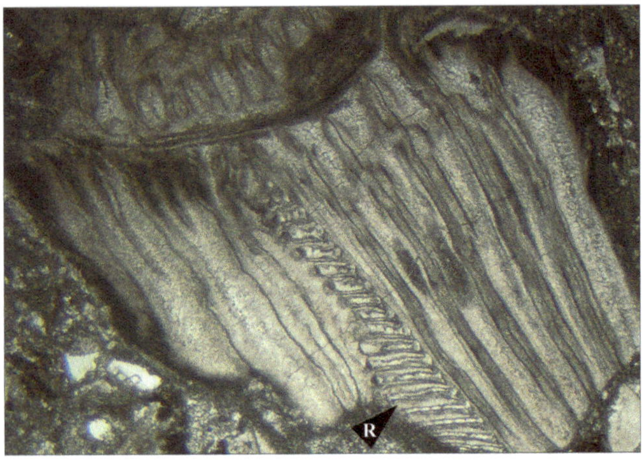

Early Miocene. Capadare Formation. Falcón Basin

Balanid shell fragment exhibiting robust, aberrant morphology. The multi-compartment structure consists of thick, longitudinally ribbed walls oriented parallel to each other. Walls smaller in size and oriented more or less perpendicular to larger ones are termed radii (R). These calcified bar-like elements articulate the compartmentalized plates.

IW = 4.2 mm. PPL

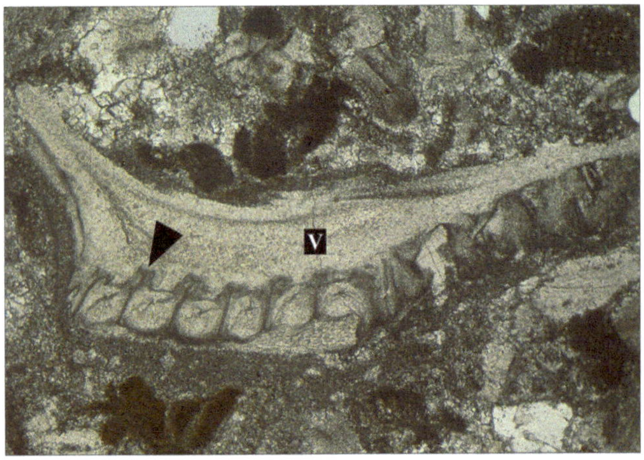

Early Miocene. San Luis Formation. Falcón Basin

Another balanid shell fragment exhibiting aberrant morphology. In this case, the tubular structures in the image (parietal tubes) are joined to a robust, calcitic opercular valve (V) by transverse septa. Note that these walls bifurcate towards the interior of the shell (black arrow), forming sutural margins with the valve. This heavily calcified valve is highly resistant to mechanical abrasion, reworking and destructive chemical diagenesis, and is therefore commonly well-preserved as a fragmented element.

IW = 4.2 mm. PPL

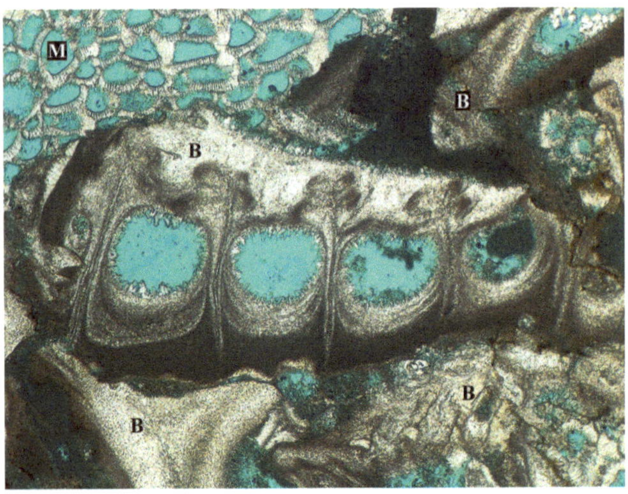

Late Oligocene-Early Miocene. Perla Limestone. Eastern Gulf of Venezuela Sub-Basin

Grainstone in which balanid (B) fragments are the main components. These are accompanied by a large benthic foraminifer of the genus *Miogypsina* (M). Note that the parietal tubes within the central balanid fragment, commonly preserved as empty primary pores, are internally rimmed by bladed cement. Although the balanid fragments in the image are derived from different shell elements of originally articulated shells, they can be easily differentiated from other bioclasticcomponents by their distinctive robustness and well-preserved (not recrystallized).

IW = 2.4 mm. PPL

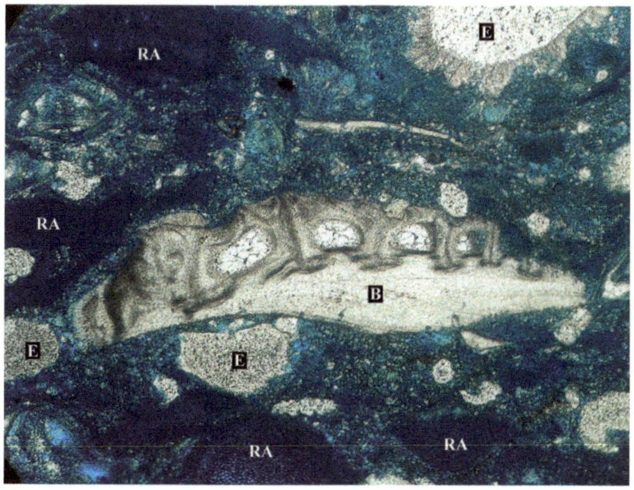

Late Oligocene-Early Miocene. Perla Limestone. Eastern Gulf of Venezuela Sub-Basin

Balanid (B) fragment in the center of the image accompanied by echinoids (E) and red algae (RA) grains in a muddy-dolomitic matrix. This sample shows the highly resistant nature of balanid shells to chemical diagenesis. Note the well-preserved state of the LMC balanid valve and parietal tubes compared to pervasive microporosity development on the matrix and red algal grains (microporosity is indicated by the intense dark-blue dyed).

IW = 4.2 mm. PPL

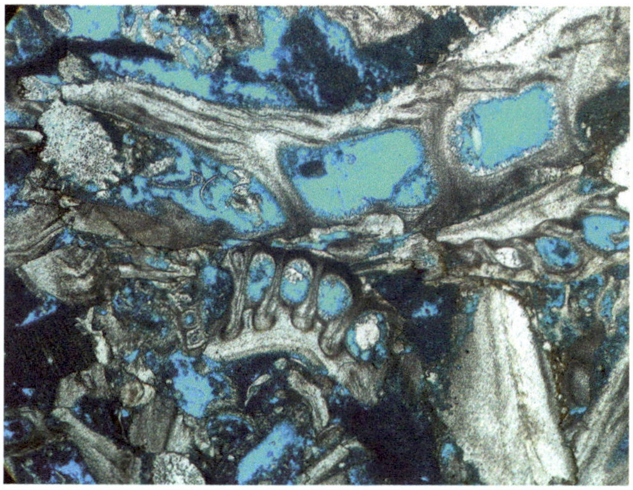

Late Oligocene-Early Miocene. Perla Limestone. Easter Gulf of Venezuela Sub-Basin

Shell fragments of balanids (robust grains) are accompanied by poorly preserved red algal grains (dark-colored, smaller grains) in a grainstone/rudstone. Although these crustaceans are secondary constituents in the Perla Limestone, they are occasionally common components in grain-supported fabrics. In this example, the combined lack of micritic sediments and void-filling calcitic cements led to the preservation of substantial primary intra-shell porosity.

IW = 7.5 mm PPL

Late Oligocene-Early Miocene. Perla Limestone. Eastern Gulf of Venezuela Sub-Basin

Balanid (B) plate fragment in a packed red algae rudstone. This sample shows the highly resistant nature of heavily calcified balanid shell fragments to mechanical compaction. Note the sutured contacts between the robust LMC walls of the shell fragment and squeezed red algae (RA) grains, evidencing the ductile behavior of the latter. Meanwhile, the brittle fragmented shell retained large primary intraparticle pores. Chemical compaction is also noticeable in the image by conspicuous dark stylocumulates.

IW = 2.4 mm. PPL

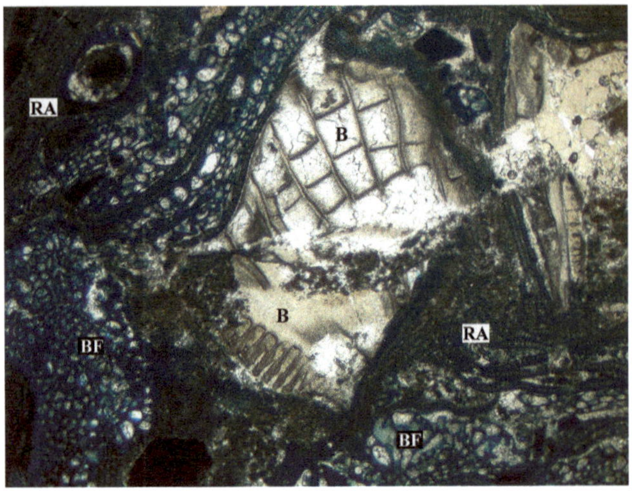

Late Oligocene-Early Miocene. Perla Limestone. Eastern Gulf of Venezuela Sub-Basin

Robusts and highly compartmentalized balanid (B) fragments are surrounded by encrusting red algae (RA) and encrusting benthic foraminifera (BF). This is an example of the common photozoan-heterozoan association between secondary heterotrophs (balanid barnacles) and dominant light-dependent calcifiers (red algae and benthic foraminifera) which characterizes the bulk of Perla oligophotic (low light) ramp.

IW = 7.5 mm. PPL

References

Aguirre J., Martín J. M., Braga J. C., Betzler, C., Berning, B., Buckeridge, J. S., 2008. Densely packed concentration of sessile barnacles (Cirripedia: Sessilia) from the Early Pliocene of SE Spain. *Facies*, *54*, 2, 193–206.

Bassi, D., Carannante, G., Murru, M., Simone, L., Toscano, F., 2006. Rhodalgal/bryomol assemblages in temperate-type carbonate, channelized depositional systems: the Early Miocene of the Sarcidano area (Sardinia, Italy). *Geological Society, London, Special Publications*, *255*, 35–52.

Benisek, M. F., Betzler, C., Marcano, G., Mutti, M., 2009. Coralline-algal assemblages of a Burdigalian platform slope: implications for carbonate platform reconstruction (northern Sardinia, western Mediterranean Sea). *Facies*, *55*, 375–386.

Benisek, M. F., Marcano, G., Betzler, C., 2010. Facies and stratigraphic architecture of a Miocene warm-temperate to tropical fault-block carbonate platform, Sardinia. *International Association of Sedimentologists, Special Publication*, *42*, 129–148.

Bourget, E., 1987. Barnacle shells: composition, structure and growth. In: *Barnacle Biology* (ed. Southward, A. J.), pp. 267–286. Balkema, The Netherlands.

Briggs, D. E. G., Sutton, M. D., Siveter, D. J., Siveter, D. J., 2005. Metamorphosis in a Silurian barnacle. *Proceedings of the Royal Society of London Series B*, *272*, 2365–2369.

Buckeridge, J. S., and Newman, W. A., 2006. A revision of the Iblidae and the stalked barnacles (Crustacea: Cirripedia: Thoracica), including new ordinal, familial and generic taxa, and two new species from New Zealand and Tasmanian waters. *Zootaxa*, *1136*, 1–38.

Buckeridge, J. S., Lee, D. E., Robinson, J. H., 2014. A diverse shallow-water barnacle assemblage (Cirripedia: Sessilia) from the Oligocene of Southland, New Zealand. *New Zealand Journal of Geology and Geophysics*, *57*, 253–263.

Buckeridge, J. S., Beu, A. G., Gordon, D. P., 2018. Depositional environment of the early Pleistocene Castlepoint Formation, New Zealand: a canyon fill in situ. *New Zealand Journal of Geology and Geophysics*, *61*, 524–542.

Carannante, G., Esteban, M., Milliman, J. D., Simone, L., 1988. Carbonate lithofacies as paleolatitude indicators: problems and limitations. *Sedimentary Geology*, *60*, 333–346.

Chan, B. K., Dreyer, N., Gale, A. S., Glenner, H., Ewers-Saucedo, C., Pérez-Losada, M., Kolbasov, G. A., Grandall, K. A., Høeg, J. T., 2021. The evolutionary diversity of barnacles, with an updated classification of fossil and living forms. *Zoological Journal of the Linnean Society*, *193*, 789–846.

Checa, A. G., Salas, C., Rodríguez-Navarro, A. B., Grenier, C., Lagos, N. A., 2019. Articulation and growth of skeletal elements in balanid barnacles (Balanidae, Balanomorpha, Cirripedia). *Royal Society Open Science*, *6*, 190458.

Coletti, G., Bosio, G., Collareta, A., Buckeridge, J., Consani, S., El Kateb, A., 2018. Palaeoenvironmental analysis of the Miocene barnacle facies: case studies from Europe and South America. *Geologica Carpathica, 69*, 573–592.

Collareta, A., Coletti, G., Bosio, G., Buckeridge, J., de Muizon, C., DeVries, T. J., Varas-Malca, R. M., Altamirano-Sierra, A., Urbina-Schmitt, M., Bianucci, G., 2019. A new barnacle (Cirripedia: Neobalanoformes) from the early Miocene of Peru: Palaeoecological and palaeobiogeographical implications. *Neues Jahrbuch für Geologie und Paläontologie-Abhandlungen*, *292*, 321–338.

Collins, J. S., Donovan, S. K., Mellish, C., 2014. An illustrated guide to the fossil barnacles (Cirripedia) from the Crags (Plio-Pleistocene) of East Anglia. *Proceedings of the Geologists' Association*, *125*, 215–226.

Flügel, E, 2010. Microfacies of Carbonate Rocks—Analysis, Interpretation and Application. Springer, Heidelberg-Dordrecht-London-New York, 984 pp.

Gale A. S., 2015. Origin and phylogeny of the thoracican cirripedes family Stramentidae. *Journal of Systematic Paleontology*, *14*, 653–702.

Gale, A. S, 2019. Stalked barnacles (Cirripedia, Thoracica) from the Late Jurassic (Tithonian) Kimmeridge Clay of Dorset, UK; palaeoecology and bearing on the evolution of living forms. *Proceedings of the Geologists' Association*, *130*, 355–365.

Gale, A. S., and Schweigert, G., 2015. A new phosphatic-shelled cirripedes (Crustacea, Thoracica) from the Lower Jurassic (Toarcian) of Germany—the oldest epiplanktonik barnacle. *Palaeontology*, 59, 59–70.

Halfar, J., and Mutti, M., 2005. Global dominance of coralline red-algal facies: a response to Miocene oceanographic events. *Geology*, *33*, 481–484.

Harzhauser, M., and Schlögl, J., 2012. Lepadiform and scalpelliform barnacles from the Oligocene and Miocene of the Paratethys Sea. *Palaeontology*, *55*(5): 923–936.

Hayton, S., Nelson, C., Hood, S., 1995. A skeletal assemblage classification system for non-tropical carbonate deposits based on New Zealand Cenozoic limestones. *Sedimentary Geology*, *100*, 123–141.

Hood, S. D., Nelson, C. S., Kamp, P. J., 2003. Petrogenesis of diachronous mixed siliciclastic-carbonate megafacies in the cool-water Oligocene Tikorangi formation, Taranaki Basin, New Zealand. *New Zealand Journal of Geology and Geophysics*, *46*, 387–405.

James, N. P., 1997. The cool-water carbonate depositional realm. In: James N. P., Clarke, J., (eds.), Cool-water carbonates, vol 56. SEPM Society for Sedimentary Geology, Tulsa, 1–20.

Kamp, P. J., Harmsen, F. J., Nelson, C. S., Boyle, S. F., 1988. Barnacle-dominated limestone with giant cross-beds in a non-tropical, tide-swept, Pliocene forearc seaway, Hawke's Bay, New Zealand. *Sedimentary Geology*, *60*, 173–195.

Klicpera, A., Michel, J., Westphal, H., 2015. Facies patterns of a tropical heterozoan carbonate platform under eutrophic conditions: the Banc d'Arguin, Mauritania. *Facies*, *61*, 421.

Klompmaker, A. A., Portell, R. W., Karasawa, H., 2014. First fossil evidence of a drill hole attributed to an octopod in a barnacle. *Lethaia*, *47*, 309–312.

Klompmaker, A. A., Portell, R. W., Lad, S. E., Kowalewski, M., 2015. The fossil record of drilling predation on barnacles. *Palaeogeography, Palaeoclimatology, Palaeoecology*, *426*, 95–111.

Maggenti, M. A. B., Maggenti, A., Scott, G., 2008. Dictionary of Invertebrate Zoology. *Zea E-Books*. *61*. 379 pp.

Massari, F., and D'Alessandro, A., 2012. Facies partitioning and sequence stratigraphy of a mixed siliciclastic carbonate ramp stack in the Gelasian of Sicily (S Italy): a potential model for ice-house, distally-steepened heterozoan ramps. *Rivista Italiana di Paleontologia e Stratigrafia*, *118*, 503–534.

Michel, J., Borgomano, J., Reijmer, J. J., 2018. Heterozoan carbonates: When, where and why? A synthesis on parameters controlling carbonate production and occurrences. *Earth-Science Reviews*, *182*, 50–67.

Michel, J., Laugié, M., Pohl, A., Lanteaume, C., Masse, J. P., Donnadieu, Y., Borgomano, J., 2019. Marine carbonate factories: a global model of carbonate platform distribution. *International Journal of Earth Sciences*, *108*, 1773–1792.

Mondal, S., Maitra, A., Bose, K., Goswami, P., Bardhan, S., Mallick, S., 2021. Substrate-controlled naticid gastropod drilling predation on recent barnacles from Chandipur, India and its Palaeontological implications. *Historical Biology*, *33*, 1029–1042.

Nelson, C. S., Keane, S. L., Head, P. S., 1988. Non-tropical carbonate deposits on the modern New Zealand shelf. *Sedimentary Geology*, *60*, 71–94.

Newman, W. A., and Abbott, D. P., 1980. Cirripedia: the barnacles. *Intertidal invertebrates of California*, *504*, 535.

Palmer, A. R., 1988. Feeding biology of *Ocenebra lurida* (Prosobranchia: Muricacea): diet, predator-prey size relations, and attack behavior. *Veliger*, *31*, 192–203.

Palmer, A. R., 1990. Predator size, prey size, and the scaling of vulnerability: hatchling gastropods vs. barnacles. *Ecology*, *71*, 759–775.

Pérez-Losada, M., Harp, M., Høeg, J. T., Achituv, Y., Jones, D., Watanabe, H., Crandall, K. A., 2008. The tempo and mode of barnacle evolution. *Molecular Phylogenetics and Evolution*, *46*, 328–346.

Pérez-Losada, M., Høeg, J. T., Simon-Blecher, N., Achituv, Y., Jones, D., Crandall, K. A., 2014. Molecular phylogeny, systematics and morphological evolution of the acorn barnacles (Thoracica: Sessilia: Balanomorpha). *Molecular Phylogenetics and Evolution*, *81*, 147–158.

Pochai, A., Kingtong, S., Sukparangsi, W., Khachonpisitsak, S., 2017. The diversity of acorn barnacles (Cirripedia, Balanomorpha) across Thailand's coasts: The Andaman Sea and the Gulf of Thailand. *Zoosystematics and Evolution*, *93*, 13–34.

Pomar, L., Bassant, P., Brandano, M., Ruchonnet, C., Janson, X., 2012. Impact of carbonate producing biota on platform architecture: insights from Miocene examples of the Mediterranean region. *Earth-Science Reviews*, *113*, 186–211.

Pomar, L., Esteban, M., Martinez, W., Espino, D., Castillo de Ott, V., Benkovics, L., Castro Leyva, T., 2015. Oligocene-Miocene carbonates of the Perla field, offshore Venezuela: Depositional model and facies architecture, In: C. Bartolini and P. Mann, eds., Petroleum geology and potential of the Colombian Caribbean Margin. *American Association of Petroleum Geologists, Memoir 108*, 647–674.

Reijmer, J. J. G., Bauch, T., Schäfer, P., 2012. Carbonate facies patterns in surface sediments of upwelling and non-upwelling shelf environments (Panama, East Pacific). *Sedimentology*, *59*, 1, 32–56.

Reymond, C. E., Zihrul, K. S., Halfar, J., Riegl, B., Humphreys, A., Hildegard, W., 2016. Heterozoan carbonates from the equatorial rocky reefs of the Galapagos Archipelago. *Sedimentology*, *63*, 940–958.

Scholle, P. A., and Ulmer-Scholle, D., 2003. A Color Guide to the Petrography of Carbonate Rocks: Grains, Textures, Porosity, Diagenesis. *American Association of Petroleum Geologists*, *Memoir 77*, 474 pp.

Shaw, J., Kang, Y., Triano, C., Hoppe, C. J., Aldred, N., Metzler, R. A., Dickinson, G. H., 2024. Comparative assessment of shell structural, mechanical, and elemental properties in adult acorn barnacles. *Diversity*, *16*(8), 482.

Ullmann, C. V., Gale, A. S., Huggett, J., Wray, D., Frei, R., Korte, C., Broom-Fendley, S., Littler, K., Hesselbo, S. P., 2018. The geochemistry of modern calcareous barnacle shells and applications for palaeoenvironmental studies. *Geochimica et Cosmochimica Acta*, *243*, 149–168.

Van Syoc, R., 2010. Darwin, barnacles and the Galapagos: a view through a 21st century lens. *Proceedings of the California Academy of Sciences*, *61*, 157.

Webber, W. R., Fenwick, G. D., Bradford-Grieve, J. M., Eagar, S. H., Buckeridge, J. S., Poore, G. C. B., Dawson, E. W., Watling, L., Jones, J. B., Wells, J. B. J., Bruce, N. L., Ahyong, S. T., Larsen, K., Chapman, M. A., Olesen, J., Ho, J. S., Green, J. D., Shiel, R. J., Rocha, C. E. F., Lörz, A. N., Bird, G. J. and Charleston, W. A., 2010. Phylum Arthropoda. Subphylum Crustacea. Shrimps, crabs, lobsters, barnacles, slaters, and kin. In: Gordon D.P. (ed.), New Zealand Inventory of Biodiversity, Vol. 2: Kingdom Animalia, Chapter 8, Canterbury University Press, Christchurch, 98–232.

Westphal, H., Halfar, J., Freiwald, A., 2010. Heterozoan carbonates in subtropical to tropical settings in the present and past. *International Journal of Earth Sciences*, *99*, 153–169.

Serpulids

14

Fossilized worm tubes (composed of mucous, chitin, phosphate, agglutinated particles, or calcium carbonate) associated with the activity of serpulid polychaetes are known since the Cambrian (Vinther et al. 2011). Polychaetes (Class Polychaeta), commonly called bristle worms, make up the largest and most diverse class of the Phylum Annelida (Maggenti et al. 2008). These soft-bodied animals have a very poor paleontological record. Among polychaete annelids, tubes composed exclusively of calcium carbonate and built by serpulid worms have a good chance of being preserved (e.g., Ippolitov et al. 2014; Sanfilippo et al. 2017).

Serpulids, commonly known as "fan worms", are among the more easily recognized annelids in modern settings owing to their colored radiolar crown and their distinctive calcareous tubes (Kupriyanova et al. 2006). Not surprisingly, serpulid tubes have the best fossil record among all annelids (Ippolitov et al. 2014). In addition to contributing to sediment and reef formation, many species of serpulids can build "reefs", or small-scale biocontructions (see Montefalcone et al. 2022). In modern carbonate systems, although not volumetrically as abundant as other heterotrophs (e.g., molluscs and echinoids) or as other encrusters (many species of coralline algae and bryozoans), the occurrence of serpulids should be noticed. Their presence can be used as paleoenvironmental markers of specific conditions (see Ippolitov et al. 2014; Montefalcone et al. 2022; Cristina et al. 2024).

14.1 Major Groups and Stratigraphic Distribution

Among polychaete annelids, calcareous tubes are known in serpulids, sabellids and cirratulids (Vinn et al. 2008a, b; Vinn et al., 2009; Vinn and Mutvei, 2009). The first record of calcareous tubes in both serpulids and sabellids are known from the Permian of Sicily (Sanfilippo et al. 2017). Much younger calcareous cirratulids are known from the Late Oligocene of Mexico (Fischer et al. 2000).

In general, tubes in the family Sabellidae (Order Sabellida) are constructed of mucus and muddy or sandy sediments (Vinn et al. 2008b), while tubes in the family Cirratulidae (Order Terebellida) are constructed of mucus/micritic peloidal lamellae intercalated with lenses of fibrous calcite/aragonite (Fischer et al. 2000). Calcareous tubes in these two families are restricted to a single genus in each family (Fischer et al. 2000; Vinn et al. 2008b; Vinn et al., 2009). In contrast, tubes in the family Serpulidae (Order Sabellida) are entirely built of crystalline calcium carbonate (Vinn et al. 2008a; Ten Hove and Kupriyanova, 2009; Vinn and Mutvei, 2009; Ippolitov et al. 2014; Vinn, 2020).

The earliest serpulids are known from the Permian (Sanfilippo et al. 2017). Prior to this finding, the earliest unequivocal serpulids were known from the Middle Triassic (Vinn et al. 2008a; Ippolitov et al. 2014); although they remained rare during that period. The diversity of these tube-secreting polychaetes rapidly increased from the Early Jurassic up to the Late Jurassic and the Cretaceous, becoming moderately common in marine invertebrate communities, also colonizing reefs and shallow water exposed settings in temperate and tropical areas. They became more common during the Cenozoic, including modern settings, colonizing a wide range of marine environments, from shallow waters to bathyal, as well as abyssal (Sanfilippo et al. 2017 and references therein).

The family Serpulidae is a discrete group of sessile polychaetes that secrete calcareous tubes and belongs to the Order Sabellida. The views on the relationships within Serpulidae have undergone many changes over the years (see Kupriyanova et al. 2006). Serpulidae has been traditionally divided into the subfamilies Serpulinae and Spirorbinae. Although phylogenetic analyses confirmed the position of Spirorbinae as a subfamily of Serpulidae (Kupriyanova, 2003; Kupriyanova et al. 2006; Lehrke et al. 2007), regarding the taxonomy and morphology of calcareous worm tubes, many specialists prefer to separate spirorbin coiled tubes (either dextrally or sinistrally) in a tight flat spiral (the char-

J. A. Méndez Dot, J. Méndez Baamonde, *Atlas of Tropical Carbonates*, https://doi.org/10.1007/978-3-031-25494-9_14

acter that gave the name to the group) from serpulid tubes. The latter have more variable shapes, and coiling, when present, is irregular (see Ten Hove and Kupriyanova, 2009; Ippolitov et al. 2014).

In addition, pre-Cretaceous (Late Ordovician to Middle Jurassic) planispirally coiled calcareous tube-worms routinely identified as *Spirorbis* (Taylor and Vinn, 2006) and traditionally described as serpulids are actually microconchids, an extinct group of lophophorates (Taylor and Vinn, 2006; Ippolitov et al. 2014). Following the guidelines given by Ten Hove and Kupriyanova (2009), in this chapter we refer to calcareous tubes as Serpulidae (not including Spirorbinae).

14.2 Skeletal Mineralogy

Fossil serpulid tubes are either calcitic or aragonitic. Ancient tubes of bimineralic compositions (or a combination of aragonite and calcite) are not known. The first purely aragonitic serpulids are known from the Jurassic. Calcitic serpulids with prismatic structures likely occurred already in the Jurassic based on the similarity of their skeletal structures to modern calcitic prismatic serpulid structures. The mineral composition of the majority of fossil serpulid tubes is not known, but considering the preservation issues it is likely that most Mesozoic serpulid fossils had a calcitic composition. Thus, one can hypothesize that the first serpulids were aragonitic when they appeared in the aragonite seas of the Permian and Triassic (Aragonite II); whereas the appearance and major diversification of calcitic species took place in the calcite seas of the Jurassic and the Cretaceous (Calcite II). In general, Cenozoic serpulid skeletal microstructures are better preserved than older Mesozoic ones (Vinn, 2020), which suggests that the tubes are compositionally calcitic-dominated. Modern serpulid tubes are composed either of HMC (Mg in calcite typically ranging from 6–16 wt.% $MgCO_3$), aragonite or a combination of both polymorphs (Scholle and Ulmer-Scholle, 2003; Flügel, 2010; Smith et al. 2013). Generally, modern tubes have a high amount of organic matter (~7.5 wt.%), consisting of chitin and proteins (Grenier et al. 2024). These compounds form a thin organic layer that internally covers the tubes, but the tubes lack an external organic layer, such as the periostracum of molluscs and brachiopods (see Vinn, 2024).

14.3 Ecology and Environment

Serpulids are a group of sessile, aquatic invertebrates (soft-bodied worms) that inhabit solid calcareous tubes or worm shells that serve as protection against predators and environmental disturbances. The worms are filter-feeding heterotrophs that depend mainly on phytoplankton productivity (e.g., Kupriyanova et al. 2001). They become common in environments with relatively abundant suspended organic matter (Brandano et al. 2009), i.e., eutrophic conditions and heterozoan-dominated carbonate settings (Klicpera et al. 2015). Normally tubes are attached at least partially to a hard substrate. These secondary frame builders (encrusters, bioeroders, bafflers) create microhabitats (i.e., cryptic niches) within and/or in between coral masses, as well as among encrusting coralline algae and other encrusters (e.g., bryozoans, benthic foraminifera, and some molluscs), and in partially lithified surfaces (hardgrounds). Some species are secondarily free-living on soft substrate as adults, while larvae attach to the smallest objects (Ten Hove and Kupriyanova, 2009; Ippolitov et al. 2014).

Since the fossil record of serpulids is, in general, poor (not being among the most common and volumetrically significant group of marine calcifiers), the occurrence of calcareous tube-building Serpulidae is a notable exception (Ippolitov et al. 2014). Tubes associated to carbonate biomineralization of serpulid worms occur in very different environments. They can be common in marine shallow-water environments, from intertidal rocky shores, to inner and outer platform settings, and on isolated seamounts and knolls. They can inhabit carbonate and mixed carbonate-siliciclastic settings. Some species are/were frame-builders (down to 30 m deep) with the ability to grow in groups of tubes tangled together (see Tucker and Wright, 1990; Ten Hove and van den Hurk, 1993; Scholle and Ulmer-Scholle, 2003; Ten Hove and Kupriyanova, 2009; Flügel, 2010; Smith et al. 2012, 2013; Ippolitov et al. 2014; Rossbach et al. 2021; Montefalcone et al. 2022). Serpulid tubes have also been found in subtidal to intertidal environments and in sabkha sediments, where they form bioconstructions associated with microbialites (e.g., Davaud et al. 1994; Dragastan and Richter, 2011); as well as in deep-sea sediments, including hydrothermal vents and hydrocarbon seeps (Vinn et al. 2012, 2013; Kupriyanova et al. 2014). As mentioned above, some serpulids can form bioconstructions, or "reef-like" (*in situ*) structures. These bioconstructions (commonly from several centimeters to up to several meters wide, and up to 1 m thick) form where certain species in dense populations build tubes perpendicular to the substrate in clumps and cement the tubes to each other. This gives serpulids the capability of forming reef-like structures when settling densely. Although more documented in modern temperate and cool marine settings, such bioconstructions also form in tropical waters (see Montefalcone et al. 2022 and references therein). In ancient facies and modern settings, the occurrence of high-density serpulid tubes can be used as paleoenvironmental indicators for interpretations of energy level, sedimentation rate, sea-level changes, water turbidity, trophic and salinity conditions, oxygenation levels, substrate type (hardground availability), and opportunistic colonization (see Ten Hove and Kupriyanova, 2009; Yang et al. 2012; Muthuvairavasamy et al. 2021; Montefalcone et al. 2022; Cristina et al. 2024).

14.4 Carbonate Factories and Skeletal Grain Associations

Serpulids are components of the Heterozoan factory of Michel et al. (2019), as well as accessory components of the Photozoan T- and Photo-C factories of these authors (see Appendix A for a detailed treatment on these carbonate benthic factories). This section discusses only the skeletal grain associations involving serpulids represented in Venezuelan carbonate facies/microfacies, as illustrated by the examples in the photographic set.

In Aptian and Albian carbonates of the Piché Member and the Lisure Formation (MB), respectively, serpulids are secondary components of Heterozoan carbonates *sensu* James (1997), in association with molluscs, echinoids and smaller miliolids (molechfor lithofacies *sensu* Carannante et al. 1988; or foramol lithofacies *sensu* Lees and Buller, 1972), plus bryozoans and ostracods. See Appendix A for definitions of the skeletal grain associations considered in this book.

In Early Miocene carbonates of the Churuguara Formation (FB), serpulids are secondary components of photozoan, red algae-dominated (inner-proximal ramp) facies (rhodalgal lithofacies *sensu* Carannante et al. 1988), plus LBF, corals, and bivalves (see Albert-Villanueva, 2016). In Middle Miocene carbonates of the Capadare Formation (FB), serpulids are common components of heterozoan facies in association with echinoids and molluscs (molechfor lithofacies) (see Lorente, 1978). In Los Roques Archipelago, although not volumetrically significant, the presence of serpulids in photozoan-dominated sediments (chlorozoan facies *sensu* Lees and Buller, 1972) is easily noticed due to their colorful, outer soft-tissue radiole structures encrusting massive scleractinian coral bodies at shallow depths. In Holocene beachrocks of Aves Island, serpulid tubes occur as minor *in situ* components within deposits currently located in the intertidal zone, whose surfaces have been colonized by macroalgae. These grain-supported fabrics consist mainly of grains (allochems) that underwent mechanical abrasion and/or reworking, notably coralline algae, with secondary contributions from molluscs, scleractinian corals, and green algae (a photozoan-heterozoan association, though photozoan-dominated). In these deposits, LBF and echinoids are also minor components.

14.5 Morphology and Diagnostic Elements

Individual fossil tubes are typically from <1 mm to 1 cm in diameter and up to 20 cm long. The resulting shape of unpartitioned tubes depends upon the degree of rotation of the worm within the tube (among other morphologic characters of the worm), being commonly straight, curved or spirally coiled. Tubes may be found isolated or in colonies forming intergrown clusters (Ten Hove and van den Hurk, 1993; Scholle and Ulmer-Scholle, 2003; Ten Hove and Kupriyanova, 2009; Flügel, 2010).

In thin sections, the tubes display distinct shapes depending on the plane of section: circular to oval and hollow (concentric rings) in transverse sections, elliptical in oblique sections, or tubular with chevron-shaped growth lamellae in longitudinal sections. Their exterior surfaces may be smooth or ornamented. In general, tube walls are thin relative to tube diameter, though the ratio varies among species. The most noticeable micromorphological feature of fossil serpulid tubes under a standard polarizing microscope is the concentric wall structure (composed of very thin lamellae, ~5 μm thick) in cross section, and the "chevron-shaped" growth lamellae in longitudinal section. Lenticular gaps between laminae are common diagnostic features. Ultrastructurally, serpulid tubes can be either single-, double-, or multilayered. Tube layers are recognized as separate layers when, in adjacent layers, the orientation, morphology, and size of crystals are different. Crystal sizes are typically less than 1 μm thick and 3 μm long. The frequency of growth lines (abrupt or repetitive changes in the character of an accreting tissue) in the tube wall of Mesozoic and Cenozoic serpulids is in the range of modern species (i.e., 7-37 growth lines per 50 μm). In fossil serpulid tubes, regularly or irregularly arranged prismatic and spherulitic crystals are the most common microstructures (Vinn et al. 2008a; Vinn, 2020, 2024). In modern tubes, the three main microstructures are granular-prismatic and lamello-fibrillar calcite, and fibrous aragonite (Grenier et al. 2024).

In addition, since Cretaceous times several serpulid genera have fully or partially calcified operculum (i.e., the organomineral structure that serves as a tube plug when the worm withdraws into its tube, hence protecting the worm against predators). Although not commonly fossilized and not easily linked with fossil tubes, some opercula (minute cup-shaped structures less than 2 μm thick) composed of regular spherulitic prismatic crystals are characteristic enough for distinguishing genera and species (see Vinn and Ten Hove, 2011; Ippolitov et al. 2014).

14.6 Preservation as Fossils

Depending on their original mineralogy, serpulid tubes may preserve their original textural features. Calcitic tubes are commonly well preserved, while, aragonitic tubes are preserved as molds (neomorphized/recrystallized or filled by LMC cements). Individual tubes are typically less than 10 mm in diameter, with the majority falling in the range of 0.5 to 3 mm. Tubes, when found in clusters, may have retained their autochthonous (*in situ*) growth arrangement. Hollow tubes are commonly filled with micritic sediment, microspar, pseudospar, or LMC cements, and rarely preserved as primary intraparticle pores. Geopetal infilling (micrite at the bottom and sparry calcite at the top) is common and indicates that tubes acted as bafflers in life position.

14.7 Serpulids: Photographic Set and Descriptive Captions

In all photos, IW = image width (the true length of the horizontal axis of the view)

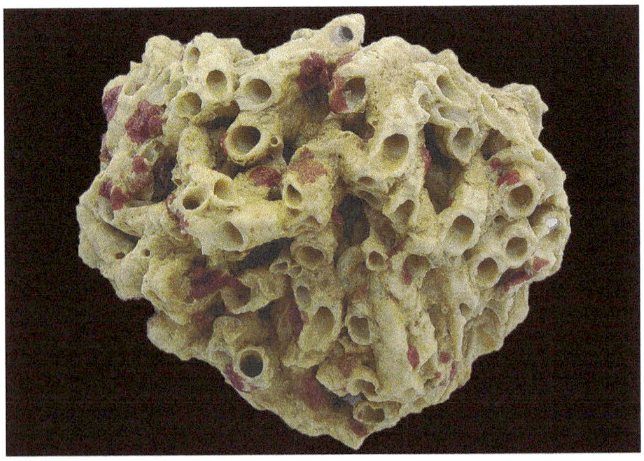

Holocene. Los Roques Archipelago
Cluster of serpulid worm tubes exhibiting characteristic cylindrical forms. The tubes are cemented together by early marine cement, probably biologically induced in origin. The structure, which is an example of a cm-sized serpulid bioconstruction, was collected by hand in a sheltered environment of a fringing reef, where the encrusted structure was found detached from its original growth arrangement. Individual calcitic tubes are approximately 8 mm in diameter, while the tube walls are less than 2 mm thick. Localized encrustations of the photosymbiotic benthic foraminifer *Homotrema rubrum* are noticeable due to their distinctive pink-reddish color.

IW = 19 cm

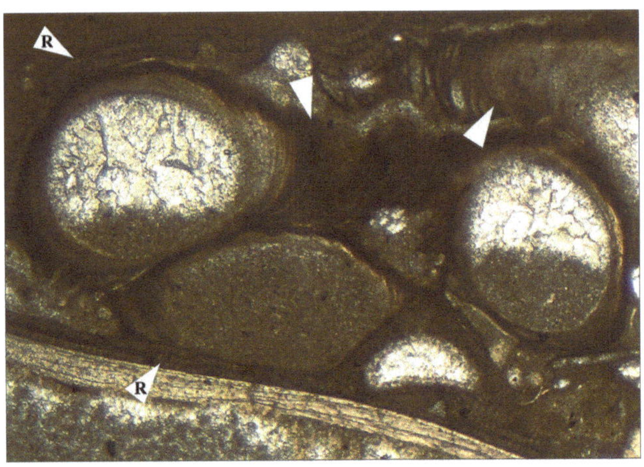

Early Miocene. Churuguara Formation. Falcón Basin
Serpulid tubes in the image are seen through transverse and oblique-to-longitudinal sections. Circular to ovoid shapes represent the transverse component of the cuts through the tubes. The oblique-longitudinal component allow visualization of the tubular structures (white arrows). Note the geopetal infilling of the tubes consisting of micritic sediments at the bottom and LMC cements at the top. Such geopetal fabrics are commonly developed in serpulid tubes and other small cavities (such as gastropod shells). Calcareous tubes are surrounded by rhodoliths (R), representing an encrusting heterozoan-photozoan association.

IW = 2.4 mm. PPL

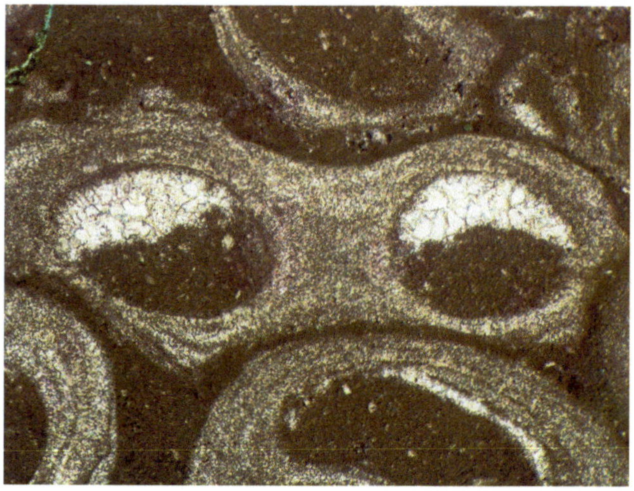

Cretaceous (middle Aptian). Piché Member, Apón Formation. Maracaibo Basin

Serpulid tubes in the image are visibly diagenetically altered, which suggests they underwent neomorphism due to an original metastable composition. Even so, traces of the original concentric wall structure are visible. As in the previous example, note the geopetal infillings (micrite at the bottom and coarse-grained LMC cement at the top), indicating that the tubes acted as bafflers in life position, causing the deposition of the micritic sediments. Autochthonous depositional textures such as this boundstone/bafflestone are actually not common in shallow Cretaceous carbonates of the MB.

IW = 1.3 mm. PPL

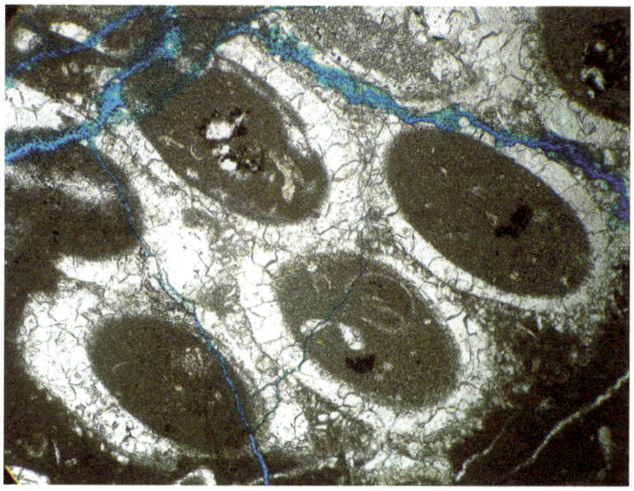

Cretaceous (middle Aptian). Piché Member, Apón Formation. Maracaibo Basin

Cluster of serpulid worm tubes embedded in a micritic matrix. The tubes are adjoined to each other, as these retained the original *in situ* growth arrangement. Their preservation as molds filled with coarse-grained LMC crystals indicates an original metastable mineralogical composition. Note the coarse-grained sizes of the crystals comprising the tubes, as consequence of dissolution of primary carbonate and precipitation of LMC filling the mold.

IW = 2.4 mm. PPL

Cretaceous (middle Albian). Lisure Formation. Maracaibo Basin

Walls (W) of the serpulid tube in the image are well preserved, which suggests an original LMC composition. The geopetal infilling of this large tube consists of muddy bioclastic sediments at the bottom, while the upper cup-shaped part (not entirely visible in the field of the photograph) is filled by coarse-grained calcite crystals exhibiting twinning of the calcite. In shallow Cretaceous carbonates of the MB, serpulids are secondary heterotrophic calcifiers, far less common than echinoids, molluscs and smaller miliolids.

IW = 4.2 mm. PPL

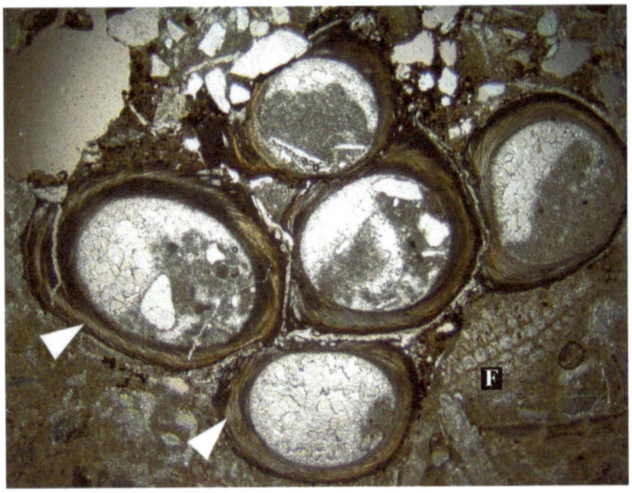

Early Miocene. Churuguara Formation. Falcón Basin

Serpulid tubes are accompanied by a larger benthic foraminifer (F) (heterozoan-photozoan association) in a sandy limestone. Note that the calcitic walls are thin relative to the diameter of the tubes, taking into account that the tubes did not undergo mobilization and abrasion, as they retained their *in situ* growth arrangement (indicated by their geopetal infillings). The layered microstructure of the tube walls, which is made of thin calcitic laminae, is perceptible in this image (white arrows).

IW = 4.2 mm. PPL

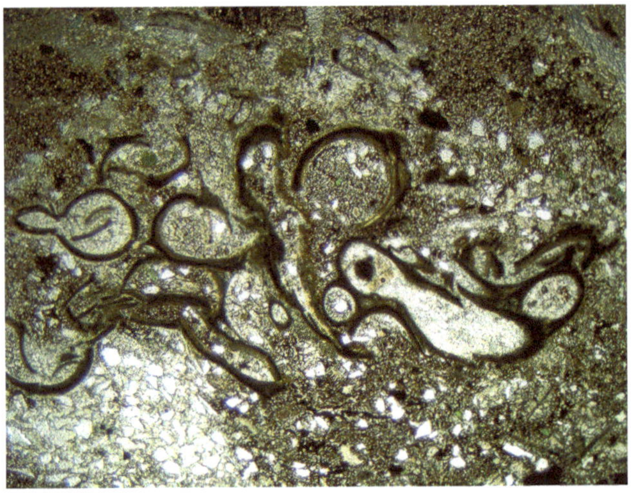

Middle Miocene. Capadare Formation. Falcón Basin

Serpulid tubes in the image are embedded in a sandy, moderately organic-rich, dolomitic matrix. The tubes partially lost the original growth arrangement due to burial compaction. Note that some of the tubes are broken, while others preserve their circular to eliptical shapes, with some even adjoined to each other. The abundance of these heterotrophic calcifiers in facies of the Capadare Formation is substantially greater than in Early Miocene carbonate strata (San Luis and Churuguara formations) in the FB.

IW = 4.2 mm. PPL

Cretaceous (middle Aptian). Piché Member, Apón Formation. Maracaibo Basin

Wall of a serpulid tube surrounded by closely-packed dolomite mosaic. The calcitic wall exhibits characteristic laminar microstructure composed of very thin calcitic lamellae uniformly oriented. Note the lenticular gaps of sparry calcite between successive laminae (resulting from parabolic structure of the precipitates of the outer layer). These elongate, spar-filled lenses paralleling the laminar shell structure are the most uniquely diagnostic characteristic of most serpulid worm tubes (Scholle and Ulmer-Scholle, 2003).

IW = 2.4 mm. PPL

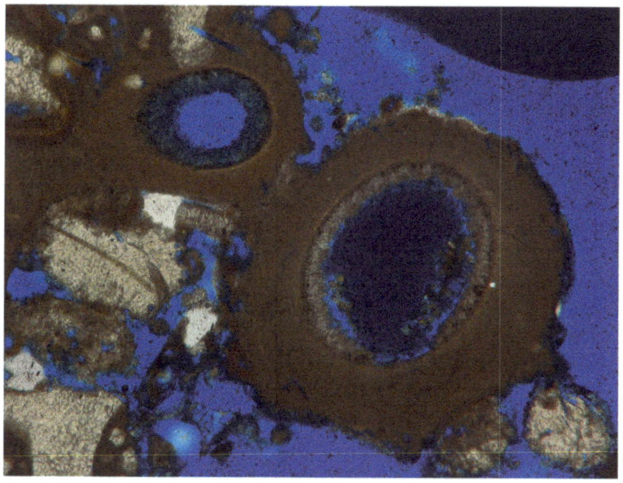

Holocene. Aves Island

Modern serpulid tubes occur in the uppermost beds of intertidal lithified deposits on Aves Island. They are found in association with macroalgae colonizing the beachrock surfaces. The tubes in the image, which preserve their pristine dark brown color, are internally rimmed by isopachous, acicular aragonite cement and are surrounded by interparticle porosity.

IW = 2.4 mm. PPL

References

Albert-Villanueva, E., 2016. Facies y secuencias deposicionales mixtas carbonático-siliciclásticas del Mioceno inferior de la Cuenca de Falcón (noroeste de Venezuela) como modelo exploratorio en el Caribe. Doctoral Dissertation. University of Barcelona. Volumes 1 and 2.

Brandano, M., Frezza, V., Tomassetti, L., Cuffaro, M., 2009. Heterozoan carbonates in oligotrophic tropical waters: the Attard member of the lower coralline limestone formation (Upper Oligocene, Malta). *Palaeogeography, Palaeoclimatology, Palaeoecology*, *274*, 54–63.

Carannante, G., Esteban, M., Milliman, J. D., Simone, L., 1988. Carbonate lithofacies as paleolatitude indicators: problems and limitations. *Sedimentary Geology*, *60*, 333–346.

Cristina, I., Coimbra, R., Teixeira, H., 2024. Serpulid tubes as potential palaeoenvironmental markers. *Zubía*, *42*, 259–264.

Davaud, E., Strasser, A., Jedoui, Y., 1994. Stromatolite and serpulid bioherms in a Holocene restricted lagoon (Sabkha el Melah, southeastern Tunisia), in Bertrand-Sarfati, J., and Monty, C., eds., Phanerozoic Stromatolites II: Dordrecht, The Netherlands, Kluwer Academic Publishers, pp. 131–151.

Dragastan, O. N., and Richter, D. K., 2011. Stromatolites and calcareous algae of Münder Formation (Tithonian-Berriasian) from NW Germany. *Acta Palaeontologica Romaniae*, *7*, 139–168.

Fischer, R., Pernet, B., Reitner, J., 2000. Organomineralization of cirratulid annelid tubes-fossil and recent examples. *Facies*, *42*, 35.

Flügel, E, 2010. Microfacies of Carbonate Rocks—Analysis, Interpretation and Application. Springer, Heidelberg-Dordrecht-London-New York, 984 pp.

Grenier, C., Berent, K., Rodríguez-Navarro, A., Vinn, O., Checa, A. G., 2024. Microstructures, crystallography and growth patterns of serpulid tubeworms (Class Polychaeta). *Marine Biology*, *171*, 23 pp.

Ippolitov, A. P., Vinn, O., Kupriyanova, E. K., and Jäger, M., 2014. Written in stone: history of serpulid polychaetes through time. *Memoirs of Museum Victoria*, *71*, 123–159.

James, N. P., 1997. The cool-water carbonate depositional realm. In: James N. P., Clarke, J., (eds.), Cool-water carbonates, vol 56. SEPM Society for Sedimentary Geology, Tulsa, 1–20.

Klicpera, A., Michel, J., Westphal, H., 2015. Facies patterns of a tropical heterozoan carbonate platform under eutrophic conditions: the Banc d'Arguin, Mauritania. *Facies*, *61*, 421.

Kupriyanova, E. K., 2003. Life history evolution in serpulimorph polychaetes: a phylogenetic analysis. *Hydrobiologia*, *496*, 105–114.

Kupriyanova, E. K., Nishi, E., Ten Hove, H. A., Rzhavsky., 2001. Life-history patterns in serpulimorph polychaetes: ecological and evolutionary perspectives. *Oceanography and Marine Biology, An Annual Review*, *39*, 1–101.

Kupriyanova, E. K., Macdonald, T. A., Rouse, G. W., 2006. Phylogenetic relationships within Serpulidae (Sabellida, Annelida) inferred from molecular and morphological data. *Zoologica Scripta*, *35*, 421–439.

Kupriyanova, E. K., Vinn, O., Taylor, P. D., Schopf, J. W., Kudryavtsev, A. B., Bailey-Brock, J., 2014. Serpulids living deep: calcareous tubeworms beyond the abyss. *Deep-Sea Research Part I: Oceanographic Research Papers*, *90*, 91–104.

Lees, A., and Buller, A. T., 1972. Modern temperate-water and warm-water shelf carbonate sediments contrasted. *Marine Geology*, *13*, M67–M73.

Lehrke, J., Harry, A., Macdonald, T. A., Bartolomaeus, T., Bleidorn, C., 2007. Phylogenetic relationships of Serpulidae (Annelida: Polychaeta) based on [18]S rDNA sequence data, and implications for opercular evolution. *Organisms Diversity & Evolution*, *7*, 195–206.

Lorente, 1978. La caliza de Capadare y su relación con la Cuenca de Agua Salada, Edo. Falcón. Bachelor Dissertation. Central University of Venezuela. 161 pp.

Maggenti, M. A. B., Maggenti, A., Scott, G., 2008. Dictionary of Invertebrate Zoology. *Zea E-Books*, *61*, 379 pp.

Michel, J., Laugié, M., Pohl, A., Lanteaume, C., Masse, J. P., Donnadieu, Y., Borgomano, J., 2019. Marine carbonate factories: a global model of carbonate platform distribution. *International Journal of Earth Sciences*, *108*, 1773–1792.

Montefalcone, M., Oprandi, A., Azzola, A., Morri, C., Bianchi, C. N., 2022. Serpulid reefs and their role in aquatic ecosystems: A global review. *Advances in Marine Biology*, *92*, 1–54.

Muthuvairavasamy, R., Siddiqui, N. A., Rai, J., Desai, B. G., Menier, D., 2021. Opportunistic colonizers in the Kallankurichchi Formation, Cauvery Basin, South India: Implications on Maastrichtian environmental stress. *Geological Journal*, *56*(4), 2060–2071.

Rossbach, F. I., Casoli, E., Beck, M., Wild, C., 2021. Mediterranean Red Macro Algae Mats as Habitat for High Abundances of Serpulid Polychaetes. *Diversity*, *13*, 265.

Sanfilippo, R., Rosso, A., Reitano, A., Insacco, G., 2017. First record of sabellid and serpulid polychaetes from the Permian of Sicily. *Acta Palaeontologica Polonica*, *62*, 25–38.

Scholle, P. A., and Ulmer-Scholle, D., 2003. A Color Guide to the Petrography of Carbonate Rocks: Grains, Textures, Porosity, Diagenesis. *American Association of Petroleum Geologists*, *Memoir 77*, 474 pp.

Smith, A. M., Henderson, Z. E., Kennedy, M., King, T. M., Spencer, H. G., 2012. Reef formation versus solitariness in two New Zealand serpulids does not involve cryptic species. *Aquatic Biology*, *16*, 97–103.

Smith, A. M., Riedi, M. A., Winter, D. J., 2013. Temperate reefs in a changing ocean: skeletal carbonate mineralogy of serpulids. *Marine Biology*, *160*, 2281–2294.

Taylor, P. D., and Vinn, O., 2006. Convergent morphology in small spiral worm tubes ('Spirorbis') and its palaeoenvironmental implications. *Journal of the Geological Society*, *163*, 225–228.

Ten Hove, H. A., van den Hurk, P., 1993. A review of recent and fossil serpulid 'reefs'; actuopalaeontology and the 'Upper Malm' serpulid limestones in NW Germany. *Geologie in Mijnbouw*, *72*. 23–67.

Ten Hove, H. A., and Kupriyanova, E. K., 2009. Taxonomy of Serpulidae (Annelida, Polychaeta): the state of affairs. *Zootaxa*, *2036*, 1–126.

Tucker, M. E., and Wright, V. P., 1990. Carbonate Sedimentology. Blackwell Publishing. 482 pp.

Vinn, O., 2009. The ultrastructure of calcareous cirratulid (Polychaeta, Annelida) tubes. *Estonian Journal of Earth Sciences*, *58*(2), 153–156.

Vinn, O., 2020. Biomineralization of Polychaete Annelids in the Fossil Record. *Minerals*, *10*, 858.

Vinn, O., 2024. Distribution of growth lines in the tube wall of serpulids (Polychaeta, Annelida). *Journal of the Marine Biological Association of the United Kingdom*, *104*, e24, 1–5.

Vinn, O., and Mutvei, H., 2009. Calcareous tubeworms of the Phanerozoic. *Estonian Journal of Earth Sciences*, *58*. 286-296.

Vinn, O., and Ten Hove, H. A., 2011. Microstructure and formation of the calcareous operculum in *Pyrgopolon ctenactis* and *Spirobranchus giganteus* (Annelida, Serpulidae). *Zoomorphology*, *130*, 181-188.

Vinn, O., Ten Hove, H. A., Mutvei, H., Kirsimaee, K., 2008a. Ultrastructure and mineral composition of serpulid tubes (Polychaeta, Annelida). *Zoological Journal of the Linnean Society*, *154*, 633–650.

Vinn, O., Ten Hove, H. A., Mutvei, H., 2008b. On the tube ultrastructure and origin of calcification in sabellids (Annelida, Polychaeta). *Palaeontology*, *51*, 295–301.

Vinn, O., Kupriyanova, E. K., Kiel, S., 2012. Systematics of serpulid tubeworms (Annelida, Polychaeta) from Cretaceous and Cenozoic hydrocarbon-seep deposits in North America and Europe. *Neues Jahrbuch fur Geologie und Palaontologie-Abhandlungen*, *265*, 315.

Vinn, O., Kupriyanova, E. K., Kiel, S., 2013. Serpulids (Annelida, Polychaeta) at Cretaceous to modern hydrocarbon seeps: Ecological and evolutionary patterns. *Palaeogeography, Palaeoclimatology, Palaeoecology*, *390*, 35–41.

Vinther, J., Eibye-Jacobsen, D., Harper, D. A., 2011. An Early Cambrian stem polychaete with pygidial cirri. *Biology Letters*, *7*, 929–932.

Yang, H., Shen, J., Zhang, L., Li, M., Huang, Z., Wang, Y., 2012. Serpulids and their paleoecology of the Paleogene Kalatar Formation in southwest Tarim Basin of China. *Science China Earth Sciences*, *55*(7), 1087–1100.

15 Non-skeletal Components

In this chapter we deal with the following non-skeletal carbonate components (or constituents): **(a)** micrite, or the fine-grained fraction (typically < 4 μm) of the sediments; **(b)** non-skeletal grains, or those that do not strictly consist of complete or fragmented skeletal grains and whose formation may involve varied modes of calcium carbonate ($CaCO_3$) precipitation and the aggregation of one or more grains. These include peloids, pellets, coated grains (ooids, oncoids, and cortoids), aggregates, and clasts (intraclasts and extraclasts). We also consider some examples of microbial carbonates, i.e., the product of microbially-mediated mineralization of $CaCO_3$.

15.1 Carbonate Mud (Micrite)

Carbonate mud is the result of a variety of organic and inorganic precipitation processes together with microbial abrasion. The material is microcrystalline calcite or aragonite, generally shortened to *micrite* (a term that is roughly synonymous with *lime-mud*). The official crystal size range is between 1 and 4 μm (Folk 1959), with material <1 μm called *minimicrite* (e.g., James and Jones 2015). In general, micrite is understood to be the fine-grained matrix of carbonate rocks (i.e., finely crystalline carbonate material) and the fine-grained component of carbonate grains. In standard petrography, the crystal size of micrite ranges from cryptocrystalline (not resolvable with the petrographic microscope) to microcrystalline (individual crystals recognizable with a petrographic microscope) (Flügel 2010).

In modern tropical settings, the mineralogical composition of shallow-marine carbonate muds is dominated by aragonite or HMC, with the proportion of these minerals varying considerably between different localities. For most sedimentological purposes, the term matrix is used to describe the component under 63 μm in size, which corresponds to the usage of mud in descriptions of modern sediments. The split of the <63 μm material into several fractions followed by the study of their mineralogical composition and geochemical signatures commonly reveals different origins related to the size of the particles (see Tucker and Wright 1990; Lasemi and Sandberg 1993; Gischler et al. 2013).

15.1.1 Origins of Carbonate Muds

As mentioned in Chap. 4, the genesis of micritic sediments comprising the matrix of carbonate rocks is either allochthonous (*allomicrite*, "whitings" precipitated in the water column) or autochthonous (*automicrite*, precipitated on the seafloor). In modern tropical and subtropical marine environments, most shallow and deep carbonate muds are allomicritic in origin (i.e., the neritic lime-mud factory of Pomar and Hallock 2008). It is well known that biologically-induced and biologically-controlled $CaCO_3$ precipitation (mainly by cyanobacteria, and by coccolithophores and calcifying dinoflagellates, respectively; the latter two groups appearing in the Mesozoic) are the dominant modes of precipitation for micritic sediments in ancient Phanerozoic as well as in modern marine environments. In some modern settings, the contribution of inorganically-precipitated (abiotic) allomicrite may be locally significant. Furthermore, in shallow tropical marine environments, the skeletal breakdown/disintegration of calcareous green algae (*Halimeda*, *Penicillus*, *Udotea*) represents a significant biogenic source for carbonate muds (see Robbins and Blackwelder 1992; Robbins et al. 1997; Flügel 2010; Gischler et al. 2013). In contrast, the precipitation of micrite on the seafloor (i.e., the benthic automicrite factory of Pomar and Hallock 2008) is not significant in modern oceans (Michel et al. 2019). However, biologically-induced (or microbially-induced) automicrite precipitation is relevant in certain modern tropical deposits (see Hillgärtner et al. 2001; Diaz and Eberli, 2022). These include shallow subtidal grainstones and intertidal beachrocks (nutrient-poor environments), as well as shallow subtidal stromatolites (nutrient-rich and hypersaline environments). In these depos-

J. A. Méndez Dot, J. Méndez Baamonde, *Atlas of Tropical Carbonates*, https://doi.org/10.1007/978-3-031-25494-9_15

its, automicrite precipitation constructs micritic envelopes that coat grains (skeletal and non-skeletal grains) and meniscus bridges that cement and lithify the grains.

"Whitings" (floating needles of lime-mud) that make up the so-called calcium carbonate "rain" are formed where ambient temperatures, salinities, and carbonate supersaturation are naturally high. Among modern environments in which whitings are significant stand out the following: the Persian Gulf, the Turks and Caicos Islands, Florida Bay, and numerous localities on the Bahama Banks (Pomar and Hallock 2008). Although originally considered as precipitated directly from seawater (e.g., Cloud Jr. 1962), whitings in the Bahamas are, at least in part, associated with the activity of cyanobacteria and picoplankton (Robbins and Blackwelder 1992; Robbins et al. 1997; Pomar and Hallock 2008).

Results from a comparative study of modern carbonate mud in six locations (reefs and carbonate platforms) of the Atlantic, Indian, and Pacific oceans, obtained by Gischler et al. (2013), indicate that: **(a)** only mud from the Bahamas (mainly aragonite) and northern Belize (mainly HMC), areas which are characterized by common whitings, is interpreted to be mainly inorganically precipitated; **(b)** in Florida, the Maldives, French Polynesia, and the Great Barrier Reef, carbonate muds are dominated by aragonite mainly derived from the breakdown of skeletal grains and disintegration of bryopsidalean green algae (biogenic origin); **(c)** the coarser grain-size fractions (63–20 and 20–4 μm) are dominated by skeletal fragments with the exception of the Bahaman samples that are composed of peloids (mud grains of uncertain origin); **(d)** the grain-size fraction <4 μm is characterized by the occurrence of small aragonite needles, nanograins, and coccoliths; **(e)** coccoliths (the LMC products of coccolithophores) are common in deeper lagoonal settings of the open ocean (Maldives, French Polynesia); **(f)** the geochemistry of the <4 μm fraction indicates algal and skeletal origins for most of the samples because strontium concentrations range between 2000–8000 ppm and the $\delta^{13}C$ values are not as high as in non-skeletal grains; the Bahaman samples, however, exhibit the highest aragonite contents and strontium concentrations, suggesting abiotic precipitation.

For the origin of many ancient allomicrites, $CaCO_3$ precipitation in the photic zone via photosynthesis by cyanobacteria and eukaryotic phytoplankton was promoted in the past during times characterized by high pCO_2, and concentrations of dissolved O_2, and Ca^{2+} in seawater (Pomar and Hallock 2008). Photosynthetically-precipitated allomicrite mainly occurred in oxygenated and nutrient-poor (oligotrophic) to nutrient-intermediate (mesotrophic) waters.

Regarding the origin of some carbonate micrite facies in ancient tropical and subtropical carbonate platforms, a significant proportion of the lime-mud production is owed to autochthonous precipitation of mirite (mainly biologically-induced and subordinate biologically-influenced). This includes facies deposited during the latest Devonian to Early Carboniferous, Late Permian to Early Triassic, Late Jurassic, and Early Cretaceous. During those time intervals, nutrient-rich (eutrophic) waters low in oxygen, but not anoxic, favored the precipitation of automicrites (Schlager 2005; Flügel 2010). This carbonate production system (or "factory") can be named: **(a)** the mud-mound factory (Schlager 2000, 2003; see also Riding 2002); **(b)** the M factory (Schlager 2005), in which the M alludes to mud-mound, micrite, and microbes; **(c)** the benthic automicrite factory (Pomar and Hallock 2008); this factory is applicable to the entire Phanerozoic (see Kabanov 2009) (see Appendix A for a detailed treatment on micritic carbonate factories). The benthic automicrite factory (photically or aphotically precipitated automicrite) is responsible for the deposition of microbe/sponge automicrite boundstones. These facies represent a particular type of carbonate rock that predominated in many ancient marine environments in which calcification by microbial activity occurred in association with benthic and interstitial degradation of organic matter. These environments were characterized by low pO_2, high pCO_2, and high Ca^{2+} concentration in seawater (Pomar and Hallock 2008). Automicrite boundstones are, by definition, microbialites (see Sect. 15.3). Among the variety of microbes that are known for being able to precipitate $CaCO_3$ under low-oxygen (eutrophic) or anoxic (hypertrophic) conditions, we consider nitrate-reducing bacteria under suboxic conditions, and sulfate-reducing bacteria and some anoxygenic phototrophic bacteria under anoxic conditions as micrite producers (see Konhauser 2007; Dupraz et al. 2009; Bundeleva et al. 2012; Seifan and Berenjian 2019).

In general, it is difficult to distinguish the origin of muds (photically or aphotically, and automicrite or allomicrite) with a polarizing microscope unless associated skeletal components (if present) provide the necessary information (Schlager 2005). The color of the matrix, as a qualitative indicator of the TOC content (wt.%), as well as the presence (or absence) of authigenic minerals (pyrite, fluorapatite, and glauconite), may also provide information about the redox and trophic conditions, and bathymetry (depth of the seafloor). Generally, light-colored micritic matrices with low TOC content are allomicritic in origin, while dark-colored micritic matrices with high TOC content are automicrite-dominated in origin.

15.1.2 Mineralogy of Ancient Carbonate Muds

The micritic matrix of many Paleozoic and Cenozoic limestones consists of microspar, in contrast to Mesozoic limestones with well-preserved micritic microfabrics (not neomorphized/recrystallized). This textural difference has been explained by differences in the original mineralogy of carbonate muds. The dominant mineralogy of mud precursors of micrites has a strong influence on the resulting microfabric. Aragonite-dominated precursor micrites are characterized by microspar fabric (or the product of aggrading neomorphism), aragonite relicts, and high amounts of Sr. Calcite-dominated micrites are inferred by the dominance of very small crystals, the absence of aragonite relicts, and low amounts of Sr (see Lasemi and Sandberg 1984, 1993; Flügel 2010). In this book, we consider well-preserved ancient micritic matrices that lack recrystallized products to have originally been calcitic in composition.

15.1.3 Cretaceous Micritic Deposits in Venezuela

In the Cogollo Group, micritic-dominated microfacies (including dolomitized mudstones) of late Barremian, Aptian and Albian age, corresponding to transgressive deposits separating the black shale intervals (OAEs 1a and 1b), were deposited under mesotrophic-to-eutrophic conditions and were occasionally low in oxygen. Benthic heterotrophic calcifiers (molluscs, echinoids, smaller miliolids, ostracods, bryozoans, sponge spicules, and serpulids) dominate the bioclastic content, followed by photoautotrophic and photosymbiotic calcifiers (calcareous algae and LBF, respectively). Depending on the type of carbonate skeletal components (photozoan, heterozoan, or mixed), the skeletal grain associations identified, the color of the matrix (light or dark), and the TOC content (wt.%, when available), we estimate substantial variations in the contributions of the neritic lime-mud factory (photosynthetically precipitated allomicrite) and the benthic automicrite factory (microbially-mediated micrite precipitated on the seafloor).

On the other hand, automicrite boundstones (i.e. carbonate-dominated black shales) containing PF and low-oxygen SBF (e.g., buliminids) as the dominant skeletal components are a volumetrically significant lithology deposited during the mid-Cretaceous OAEs. These autochthonous micritic microfacies, containing varying amounts of clay minerals, are organic-rich (TOC > 1%) and commonly characterized by varying amounts of authigenic minerals (mainly pyrite and fluorapatite). These lithologic features indicate low-oxygen to anoxic conditions during deposition. In the case of the rock intervals corresponding to the Aptian and Albian black shales, matrices rich in organic carbon vary widely in composition, from micritic- to partially dolomitic-dominated to clay-dominated. In either of those compositional end members, microbial contribution to calcification and organic matter primary production is significant. Regarding Late Cretaceous black shales in the Maracaibo and Eastern Venezuela basins, micritic- and clay-dominated black shales are common in both the La Luna Formation (e.g., Liborius-Parada 2019) and the Querecual Formation (e.g., Truskowski 2006; Caroprese 2009), respectively. However, the visualization of the matrix in those microfacies with the polarizing microscope is, in many cases, partially or totally prevented due to the high content of organic carbon.

15.2 Non-skeletal Grains

Non-skeletal grains (i.e., coated grains, peloids, aggregates and clasts; Folk 1959) are those not obviously derived from the skeletal material of calcareous micro-organisms (e.g., Tucker and Wright 1990). This diverse group of grains is common in ancient tropical carbonate strata and in modern tropical carbonate settings (e.g., Schlager 2003, 2005; Pomar et al. 2004; Bádenas and Aurell 2010; Gischler 2011; Gischler et al. 2013; Campos-Soto et al. 2016, 2019; Michel et al. 2019), and is particularly abundant in facies corresponding to the mid-Cretaceous (e.g., Iryu and Yamada 1999). The strict differentiation of "skeletal" and "non-skeletal" grains, however, has been discarded, because the formation of most coated grains and grain aggregates, as well as some ooids and peloids, is substantially controlled by biotic processes (Flügel 2010). Furthermore, non-skeletal grains contain micritic sediments in varying proportions.

Regarding carbonate factories, as a whoe group, non-skeletal grains are components of the Biochemical T-factory of Michel et al. (2019); this factory typically occurs in shallow environments (including peritidal settings) with high temperatures and salinities, and carbonate-supersaturated waters. Some non-skeletal grains, such as ooids, are common components of the Tropical factory of Schlager (2003, 2005). We consider a significant contribution of the automicrite benthic factory in the formation of microbially-caoted grains, as constructive micrite precipitation (microbially-mediated) is a key process in their formation (see Appendix A for a detailed treatment of the marine carbonate benthic factories considered in this book). Non-skeletal grains tend to be more abundant in transgressive and highstand deposits (especially in ancient epeiric platform/ramps) than in lowstand deposits because they benefit from the amplification of tidal currents and the extensive winnowing by waves on the vast, flooded platform top (Schlager 2005). In modern settings, non-skeletal grains such as ooids, peloids, and aggregates are abundant on platforms from the Atlantic Ocean but scarce or virtually missing in many modern Indo-Pacific reefs and platforms (Gischler 2011).

15.2.1 Pellets and Peloids

Pellets and peloids are non-skeletal grains composed mainly of micrite, but differ in origin. Pellets are lithified organic excrements (fecal products) of carbonate-ingesting organisms. They are rounded to slightly elongate or ovoid/ellipsoidal and are very small (commonly no more than 0.3 mm in length), and characterized by a homogeneous micritic composition lacking internal structure. Pellets are abundant wherever worms, crustaceans, molluscs, and other grazing, burrowing, or swimming invertebrates exist, but are commonly disaggregated/disintegrated into the fine fraction prior to burial. They accumulate mostly in tropical and temperate enclosed lagoons (especially hypersaline and/or lower intertidal ones) and shallow shelves with moderate to low exposure to wind and waves; while pellets originated from fecal products of zooplanktonic and nektonic animals accumulate in deep-water settings associated to coccolithophore test deposits. Rapid hardening and cementation (usually microbially mediated) and early lithification during marine and meteoric diagenesis aid preservation, and so does rapid sedimentation in low-energy settings (see Folk 1959; Tucker and Wright 1990; Scholle and Ulmer-Scholle 2003; Flügel 2010; Kendall and Alsharhan 2011). Fecal pellets are especially prominent in Jurassic and Cretaceous rocks (e.g., Füchtbauer and Tisljar 1975; Longman et al. 1998; Husinec and Sokač 2006; Kietzmann et al. 2009; Campos-Soto et al. 2016), although they are known from middle Paleozoic strata to Holocene sediments (Scholle and Ulmer-Scholle 2003).

Peloid grains, commonly referred to in the literature as grains of uncertain origin, are ubiquitous components in modern and ancient carbonates (e.g., Samankassou et al. 2005). These allochems are composed of cryptocrystalline or microcrystalline calcium carbonate with no restrictions on the size or origin of the grains (McKee and Gutschick 1969). The term "peloid" allows reference to micritic grains without the need to imply any particular mode of origin. It is therefore a useful non-genetic "term of ignorance" covering possible pellets, grains of biotic origin (e.g., algal peloids, bioerosional peloids), reworking of mud and grains (e.g., mud peloids), alteration of skeletal and non-skeletal grains (i.e., micritization and recrystallization), grains of *in situ* formation (e.g., microbial peloids), as well as indistinct intraclasts and aggregate grains. Peloids are varied in shape (irregularly-shaped, rounded, ovoid, angular) and size (average size of 100-500 μm, but often up to >2 mm) depending on their origin. They can be formed and deposited along platforms and ramps under varied hydrodynamic regimes, from low-energy, shallow subtidal and tidal-flat environments and offshore environments (below the fair-weather wave base); to moderate-to-high-energy middle shelves/ramps and open-marine environments (see James et al. 1976; Wanless and Tagett 1989; Tucker and Wright 1990; Scholle and Ulmer-Scholle 2003; Kiessling et al. 2003; Samankassou et al. 2005; Flügel 2010; Bádenas and Aurell 2010).

Peloids are present in all Phanerozoic strata. They are usually more common than ooids, and occurred in high abundance in the Devonian, the Early Carboniferous, the Middle Triassic, the Middle and Late Jurassic, and the Early Cretaceous platforms (Kiessling et al. 2003). A large proportion of peloids are probably derived from microbial activity (e.g., Chafetz 1986; Kazmierczak et al. 1996). In that regard, micritization (i.e., destructive, texture obliterating) is thought to be largely involved in the formation of many peloids in ancient carbonates (see Samankassou et al. 2005; Flügel 2010; Kendall and Alsharhan 2011). Micritization carried out by cyanobacteria and other endolithic bacteria and algae involves early marine (syndepositional) microboring and/or recrystallization of grains and infilling of the boreholes by micritic sediments (in modern settings these micritic sediments are composed of HMC and/or aragonite) (see Bathurst 1966; Kobluk and Risk 1977; Tucker and Wright 1990; Reid et al. 1992; Reid and Macintyre 1998, 2000; Flügel 2010; Kendall and Alsharhan 2011; Garuglieri et al. 2024). Furthermore, due to the organic-rich composition of many micritic peloids, they are well-suited for the precipitation of glauconite, pyrite, and Ca-phosphate. These marine diagenetic minerals may partially or completely replace peloidal grains (including fecal pellets). Microbial activity, combined with transitory low-oxygen conditions, promotes the nucleation of such minerals on the peloid surfaces (phosphatization, pyritization and glauconitization will be dealt on Chaps. 20, 21, and 22, respectively).

Keys to petrographic recognition and differentiation between pellets and peloids are based mainly on the size, as pellets are smaller. Furthermore, because most pellets in any single sediment sample will have been produced by just one or two types of organisms, they are generally very uniform in both shape and size, in contrast to the varied sizes and shapes of peloids (Scholle and Ulmer-Scholle 2003). In this book, we use the term "peloidal" for grains that may be either fecal pellets or peloids of varied/uncertain origins.

15.2.2 Coated Grains (Ooids, Oncoids, and Cortoids)

Coated grains are regularly or irregularly concentric particles consisting of carbonate cortices (layered or not) surrounding a nucleus. The latter can be a peloidal grain, a skeletal or a non-skeletal grain, or an aggregate grain. Ooids and oncoids, and corresponding rocks, i.e., oolites and oncolites, respectively (Peryt 1983a), are the most common coated, non-skeletal grains throughout the Phanerozoic. oic (Kiessling et al. 2003). Accretionary/growth processes associated with the formation of these non-skeletal grains are largely associated with wind-generated wave energy sufficient to subsequently cause rolling and overturning of grains at the water-sediment interface (e.g., Védrine et al. 2007; Duguid et al. 2010).

15.2.2.1 Ooids

Ooids are grains consisting of carbonate microlayers (aragonite or calcite) of relatively uniform thickness and arranged concentrically around a nucleus. This applies to ooids with spherical nuclei, but not to those with angular, elongate and skeletal cores (Richter 1983). Many cortices are composed of carbonate layers and dark interlayers (biofilms). Ooids are commonly between 0.3 to 2 mm in diameter, with many between 0.5 to 1 mm. The most common primary microfabrics of the cortices (arrangement of layers around the nucleus) are the tangential and radial,; micritic (random) microfabrics are mainly diagenetic (see Richter 1983; Tucker and Wright 1990; Scholle and Ulmer-Scholle 2003; Cantrell 2006; Flügel 2010; Lehrmann et al. 2012; Varkouhi and Jaques-Ribeiro 2021).

The biotic or abiotic origin of ooids has been a matter of debate for decades. They were originally considered chemical (inorganic) precipitates (e.g., Peryt 1983a). Schlager (2003, 2005) considers that inorganic precipitation has been the dominant carbonate mode of precipitation for the formation of ooids and oolitic deposits throughout the Phanerozoic; although this author acknowledges the influence of organic matter and life processes in the formation of these non-skeletal grains. Pomar et al. (2022) consider that microbially mediated organomineralization processes are likely involved in the origin of ooids. They can either be biologically induced (e.g., by-products of metabolic activities that increase environmental alkalinity and trigger carbonate precipitation), or biologically influenced (e.g., microbial extracellular polymeric substances (EPS) can serve as templates for carbonate mineralization; see Brehm et al. 2006). Recent studies on modern ooids from different locations on the Bahamas (the Mecca for modern ooid sands; see Harris 1979; Harris et al. 2019) point to different origings: **(a)** Duguid et al. (2010) proposed an abiotic model with an intermediate metastable phase, in which ooid growth alternates between active and stationary phases. Specifically, a veneer of a high-Mg/Ca amorphous abiotic precipitate, purported to be amorphous calcium carbonate (ACC), forms on the outermost lamina while the ooid is at the sediment-water interface (active stage) prior to its crystallization to aragonite needles (stationary stage); microbial-endolithic activity alters the chemistry and textures of the ooids after formation; **(b)** Diaz and Eberli (2019) argue for microbial mediation both as a direct result of metabolic activities inducing changes in the environment conducive to precipitation and as a passive mechanism, in which mucilaginous EPS act as a template for mineral nucleation; ooid accretion occurs through pre-nucleation of biomediated ACC precipitates that act as a metastable phase prior to aragonite crystallization (see also Diaz et al. 2015, 2017; Batchelor et al. 2018). More recently, in their study of ancient ooids, Riaz et al. (2023) propose a biochemical origin that involves an ACC-aragonite precipitation sequence and reinforces the succession of microbial activities and physicochemical processes that govern the active and stationary stages in their formation.

The earliest ooids are known from the Late Archean (2.9–2.7 Ga; Flannery et al. 2018). They are common in the Triassic and Early and Early Cretaceous strata, and abundant in the Late Cambrian-Early Ordovician, the Mississippian, and the Jurassic strata (Kiessling et al. 2003). Most modern marine ooids are composed of aragonite, with tangential microfabrics predominating over radials. In general, ancient aragonite sea systems (e.g., the Triassic) contain both aragonite and HMC ooids, including bimineralic ooids (aragonite and HMC), whereas systems with calcite seas (e.g., Late Jurassic and Cretaceous) contain exclusively calcitic ooids (Wilkinson and Given, 1986; Hardie, 1996; but see Cantrell, 2006).

Ooids have a wide environmental spectrum for formation and final deposition. They are largely associated to modern and ancient tropical skeletal and peloidal carbonate deposits (e.g., normal marine to saline subtidal shoals and lagoons, tidal bars, open-marine and peritidal-marine hypersaline/evaporative settings), where they may form mud-supported, muddy grain-supported, and mud-free lithofacies (see Bartok et al. 1981; Purdy and Gischler 2003; Bádenas and Aurell 2010; Duguid et al. 2010; Gischler 2011; Kendall and Alsharhan 2011; Campos-Soto et al. 2019). The shape of ooids often results from the shape of the nuclei. Ooids size is controlled by the supply of nuclei, growth rate, mobilization and agitation, and abrasion. Their size and sorting provide clues to the hydrodynamic condition (see Flügel, 2010). Because of their rounded shapes, ooids are easily reworked into adjacent environments (Scholle and Ulmer-Scholle, 2003). Some ancient oolite deposits hosting combined primary (interparticle) and secondary porosities are excellent hydrocarbon reservoirs (e.g., Honarmand and Amini 2012; Lehrmann et al. 2012; Qiao et al. 2016; Hollis et al. 2017).

15.2.2.2 Microbially Coated Grains (Oncoids and Cortoids)

A consensus about a microbial origin for the precipitation of the $CaCO_3$ cortices of marine oncoids is fully accepted. We use the term oncoid in reference to coated grains of microbial and microbial/algal origin exhibiting calcimicrobe fabrics (e.g., dense-layered micritic and/or calcified-filamentous laminae cortex). Oncoids are unattached microbialites consisting of a more or less distinct/centralized nucleus (e.g., skeletal or non-skeletal single/aggregate grains and quartz grains) and a thick cortex formed by irregular, concentric or not, partially overlapping micritic/filamentous laminae. In addition to microbially-mediated $CaCO_3$ precipitation, trapping and binding of allochems (carbonate or not-carbonate) in the microbial cortex is a relevant process in the growth of oncoids (see Peryt 1983b; Riding 1983, 2000; Tucker and Wright 1990; Scholle and Ulmer-Scholle 2003; Flügel 2010; Pederson et al. 2015; Amao et al. 2016).

On the other hand, cortoids are carbonate grains (e.g., bioclasts, aggregates) exhibiting thin micrite envelopes. The micritic coatings, derived from constructive automicrite precipitation carried out by endolithic bacteria, are commonly thinner than those of oncoids, which may develop from the overgrowth of cortoids (Flügel 2010). Although far less commonly referred to in the literature (but see Neuweiler et al. 1999; Takayanagi et al. 2012; Abed et al. 2013), cortoids are quite common in shallow facies of the Cogollo Group.

Modern marine oncoids are scarce (Riding 2000). These microbially coated grains are known since the Late Paleoproterozoic (1.9 Ga; Schaefer et al. 2001) and have contributed to the formation of subtidal and intertidal platform carbonates since the Precambrian, but became frequent in the late Paleozoic and peaked in the Jurassic. The youngest marine oncoid-rich deposits are known from the Cretaceous (Riding 2000; Kiessling et al. 2003). The micritic preservation (not neomorphized/recrystallized fabric) of most Paleozoic and Mesozoic oncoids suggests an original calcitic composition. Late Pliocene micrite oncoids from New Caledonia were originally aragonitic and HMC in composition. Rapid mineralogical stabilization of those grains to LMC led to the preservation of their original fabric and texture (Pederson et al. 2015). Modern micritic oncoids from an arid intertidal embayment of the Gulf of Aqaba, Saudi Arabia, are calcitic-dominated (Amao et al. 2016).

Shallow-marine oncoids, associated with both mud-supported and muddy grain-supported lithofacies, are often used as indicators of: high microbial activity during transgressive episodes; and intertidal to shallow subtidal muddy lagoons, shoals and bar environments with enough hydrodynamic energy (sustained, periodic or episodic) to cause frequent overturning and rolling of the grains at the sediment-water interface, with the degree of turbulence being the factor controlling oncoid shapes (see Bartok et al. 1981; Tucker and Wright 1990; Tišljar et al. 2002; Flügel 2010; Bádenas and Aurell 2010; Védrine et al. 2007; Campos-Soto et al. 2016; Sequero et al. 2020; Fossum et al. 2021). Rolling appears to be indicated by subspherical shapes, well-laminated cortices, concentric symmetrical growth patterns, and abrasion of cyanobacterial structures. Reworking of oncoids is indicated by variations in shape and size (Flügel 2010). Alternating conditions during the formation of oncoids, from agitated (growth phase) to calm conditions, are indicated by discontinuous growth/accretion of lamellae (e.g., Campos-Soto et al. 2016); or from high-energy conditions (growth phase) to intermittent periods of wave agitation and partial erosion of the cortex (e.g., Védrine et al. 2007).

15.2.3 Aggregates

Aggregate grains are formed when several carbonate particles become bound and cemented together. They usually range in size from 0.5 to 3 mm and have irregular shapes (e.g., grape-like or rounded lumps). Therefore, these compound or composite grains are generally sand-sized. They may consist of a combination of skeletal and non-skeletal grains as well as quartz grains. Micrite is the main cement that binds individual grains (carbonates and/or siliciclastics) into a single compound grain, hence, many aggregates can be termed as mud-rich or "mudstone" aggregates (see Laycock et al. 2017). Different names characterize different shapes and the variable contribution of organisms and carbonate cements to the formation of these composite grains. *Grapestones* are aggregates of spherical grains (often micritized ooids) whose external shape resembles microscopic clusters of grapes. *Lumps* are aggregates with a smoother outline than grapestones; they commonly have hollow interiors and are often strongly micritized. *Oolitic lumps* are grapestones or lumps with a thin oolitic coating. *Microbial/algal* aggregate grains are sedimentary particles agglutinated by biofilms, cyanobacteria and algae (see Tucker and Wright 1990; Flügel 2010).

15.2.4 Clasts (Intraclasts and Extraclasts)

Limestone clasts (lime-clasts) are defined as grains (single or compound, mainly micritic-dominated aggregates) that have undergone reworking prior to final deposition. By definition, *intraclasts* are fragments of penecontemporaneous, commonly weakly consolidated, carbonate grains that exhibit rounded or angular shapes and have been eroded and redeposited, generally nearby, within the same depositional sequence in which they formed (Folk 1959, 1962). Intraclasts are commonly found in shallow-marine environments, but they are also transported to deepwater. Shallow-marine environments in which intraclasts are formed are characterized by wave-dominated, sand tidal regimes that continuously rework carbonate (see Tucker and Wright 1990; Flügel 2010). An *extraclast* is a detrital grain of lithified carbonate sediment derived from outside the depositional area of the current sedimentation (Folk 1959). The latter scenario may include erosion of marine sediments (e.g., inner and middle platform/ramp deposits), and reworking and redeposition into an adjacent, outer/distal, or downslope environment; or erosion of exposed rocks and redeposition in a newly marine environment. The term *lithoclast* was originally coined for erosional detritus brought from an outside source into the basin (= extraclast), but is now frequently used for both intra- and extraclasts, when the genetic origin is unknown (Flügel 2010). Redeposited clasts and other non-skeletal grains containing authigenic minerals (i.e., glauconite, pyrite, Ca-phosphate) are common on hardground surfaces caused by breaks in the sedimentation and synsedimentary marine lithification, and are typically developed in warm and shallow, agitated tropical waters (e.g., Bromley 1975; Bartok et al. 1981; Flügel 2010; Amorosi 2012; Christ et al. 2015).

Differentiating intraclasts and extraclasts can often be difficult. In general, both types of grains are large in size (from a few millimeters to up to few centimeters). They mostly consist of more than one allochem and micritic sediments binding allochems. Extraclasts may commonly present mineralogical, textural, and compositional differences with respect to the lithology of the final deposit.

15.3 Microbial Carbonates

In this section we deal with microbial carbonates, i.e., microbialites, which are organosedimentary products (µm- to mm-sized precipitates and microstructures) and/or sedimentary deposits (buildups, structures). They result from the activity (or mediation) of benthic microbial communities. Their mode of formation may include four different processes: biologically-induced calcification (dominant), biologically-influenced calcification (secondary), trapping and binding of detrital sediments, and abiotic precipitation of $CaCO_3$ and other minerals (e.g., dolomite, gypsum and halite; see Burne and Moore 1987; Riding 2000; Bosence et al. 2015). The term "microbialite" has been most widely used to describe *in situ* carbonate stromatolites (laminated macrofabric), thrombolites (clotted macrofabric), and similar structures that occur as domes and columns in the shallow waters of seas and lakes. Nevertheless, this term can also be applied to many additional authigenic accumulations in which microbes are locally conspicuous (Riding 2011).

Microbes, most simply defined as all microscopic organisms, are generally considered to comprise the three major domains of life (e.g., Riding 2000; Brocks and Summons 2004; Brocks and Pearson 2005): *Bacteria* (including cyanobacteria, green and purple sulfur bacteria, and sulfate-reducing bacteria), *Eukarya or Eucarya* (such as some benthic microalgae and microscopic fungi), and *Archaea* (such as halobacteria, methanogens, and hyperthermophiles). In modern settings, carbonate deposits rich in microbialites generally occur in meso-eutrophic to hypertrophic (nutrient-rich), and/or hypersaline environments, where organic matter is abundant and the automicrite benthic factory dominates carbonate production.

The principal organisms involved in carbonate organomineralization are bacteria, particularly cyanobacteria (formerly called "blue-green algae") and other endolithic bacteria, microalgae, and fungi (Riding 2000). The term "organomineral" describes mineral products (mostly carbonates and silica) whose precipitation may be induced (active mineralization) or influenced (passive mineralization) by biological (microbial) activity (Perry et al. 2007; Dupraz et al. 2009). In a biologically-induced mechanism, $CaCO_3$ polymorphs form as a result of metabolic activities (e.g., photosynthesis, denitrification, sulfate reduction, anaerobic sulfide oxidation) and metabolic byproducts that react with ions or compounds in the environment, leading to an increase in environmental alkalinity and supersaturation. In a biologically-influenced mechanism, mucilaginous material or EPS plays a crucial role in biofilm development and acts as a template for carbonate nucleation in microbialite systems (see Diaz and Eberli, 2019 and references therein). Details on chemical reactions and microbial pathways involved in microbial carbonate mineralization may be found in: Visscher et al. 1998; Dupraz and Visscher 2005; Baumgartner et al. 2006; Konhauser 2007; Perry et al. 2007; Dupraz et al. 2009; Decho 2010; Stal 2012; Bundeleva et al. 2012; Seifan and Berenjian 2019; Diaz and Eberli, 2019, 2022; Görgen et al. 2021).

15.3.1 Microbial Deposits Through Time

Sedimentary deposits in shallow seas and lakes produced by microbial growth and metabolism have been important components of carbonate sediments since the Archean (Riding 2000). Stromatolites, the fossil evidence of calcified microbial mats, can be viewed as the first process to produce a hard body (Dupraz et al. 2009; Pomar et al. 2022). The oldest known stromatolite dates back to 3700 Ma. Prior to the singular finding of this microbialite structure by Nutman et al. (2016) within a metamorphosed sedimentary rock in Greenland, stromatolites were widely accepted as having first appeared at ~3450 Ma. They were generally diverse and abundant from 2,800 to 1,000 Ma. The beginning of a Proterozoic decline, variously identified at 2,000, 1,000, and 675 Ma, has been attributed to eukaryote competition and/or reduced lithification (Riding 2000). They were also common during the Silurian, the Late Devonian, the Late Carboniferous, and the Middle Permian, and dominant during the Middle and Late Triassic (Kiessling et al. 2003; Riding et al. 2019).

In the case of the mid-Cretaceous OAEs, microbes (mostly bacteria and episodic blooms of archaea) are not only widely recognized as having played a key role in organic matter primary production (see Kuypers et al. 2001, 2002, 2004; Duque-Botero and Maurrasse 2004; Dumitrescu and Brassell 2006; Ohkouchi et al. 2006; Jenkyns et al. 2007; Kashiyama et al. 2008; Jenkyns 2010; Pacton and Gorin 2014; Castro et al. 2019), but the role of microbial communities in calcification (mostly bacteria) is also widely acknowledged (e.g., Duque-Botero and

Maurrasse 2004; Immenhauser et al. 2005; Amodio and Weissert 2017; Núñez-Useche et al. 2020). In general, carbonate factories of the Cenozoic are skeletal-dominated and poor in microbial deposits (see Riding 1991, 1992; Kiessling et al. 2003).

Intervals of marine microbial carbonate abundance and reduced invertebrate diversity correspond to increased carbonate saturation state, elevated temperature, and low concentrations of dissolved O_2. Conversely, increased marine invertebrate diversity and lower microbial carbonate abundance occurred during intervals with lower temperatures and higher concentrations of dissolved O_2. These secular patterns support the inference that, after the Cambrian, microbial carbonate abundance was limited by invertebrate competition and low seawater carbonate saturation state (Riding et al. 2019).

15.3.2 Some Examples of Carbonate Microbialites

Modern and ancient microbial carbonate deposits are highly diverse (see Kennard and James 1986; Riding 1991, 2000, 2011; Konhauser 2007; Desnues et al. 2008; Dupraz et al. 2011; Stal 2012; Frantz et al. 2015). Microbial carbonates do not always preserve direct, or diagenetically robust, evidence of their mode of formation (Bosence et al. 2015). Nevertheless, in the case of the majority of the photographic examples considered in this chapter, the ubiquitous presence of microbes in those tropical carbonate settings (a mid-Cretaceous epeiric platform and a Holocene hypersaline lagoon) suggests a relevant microbial contribution, albeit to varying degrees. This brief review focuses on some of the photographic examples of microbial carbonates shown below: **(a)** Carbonate "microspheres" ($\leq$ 50 μm in diameter) of authigenic origin (early Aptian in age), which are moderately common in the Machiques Member (OAE 1a). These structures are similar to those shown by Kazmierczak and Kempe (1992), Kazmierczak et al. (1996), Tribovillard (1998), Tribovillard et al. (2000), Duque-Botero and Maurrasse (2004), and Robin et al. (2015), and have been interpreted to be of bacterial origin; **(b)** A modern shallow subtidal-intertidal stromatolite structure formed in a perennial hypersaline lagoon of the Los Roques Archipelago and consisting of combined stromatolite (laminated) and thrombolite (clotted) macrofabrics (see Spadafora et al. 2010; Riding 2011; Riding and Virgone 2020; Keim et al. 2020; for similar examples of stromatolites characterized by hybrid macrofabrics and components). Stromatolites are laminated benthic microbial deposits (Riding 1991). The growth/accretion of these microbial bioherms is controlled by the *in situ* precipitation of micritic $CaCO_3$ and by trapping and binding of allochems. It should be noted that trapping and binding of sediments by uncalcified EPS alone does not form a stromatolite. Therefore, the key process in the formation of laminated microbialite bioherms is the precipitation and accumulation of calcified microbes and mat accretion, preservation and lithification in the subtidal zone (see Riding 2000; Reid et al. 2003; Dupraz et al. 2009, 2011). **(c)** Automicrite boundstones (carbonate-dominated black shales; see Sect. 15.1.1), or microfacies dominated by microbial carbonate mud and associated organic matter (plus dolomite and clay minerals); **(d)** microbial cementation via precipitation of micritic envelopes and meniscus bridges (see Sect. 15.1.1; see also Sect. 16.4.2); **(e)** oncoids and cortoids (i.e., microbially coated grains) as representative of unattached microbialites (see Sect. 15.2.2.2).

15.3.3 Modern Microbialites in Los Roques Archipelago

Microbialite structures and deposits are common in modern peritidal, hypersaline environments of the Los Roques Archipelago. The latter consist of perennial and ephemeral lagoons and sabkha flats. Among the variety of organosedimentary structures/deposits, the highlights include: shallow subtidal, gypsum-dominated thrombolites (see Petrash et al. 2012); shallow subtidal-intertidal, carbonate-dominated, laminated stromatolites; and subtidal-intertidal, laminated and non-lithified stromatolites and microbial mats. Photographic examples of some of these modern peritidal settings and microbialites are shown in Chap. 3. However, although only a small number of peritidal environments have been explored in Los Roques, we estimate the occurrence of microbialite structures and deposits in many unexplored modern hypersaline environments of this archipelago.

15.4 Non-skeletal Components: Photographic Set and Descriptive Captions

In all photos, IW = image width (the true length of the horizontal axis of the view)

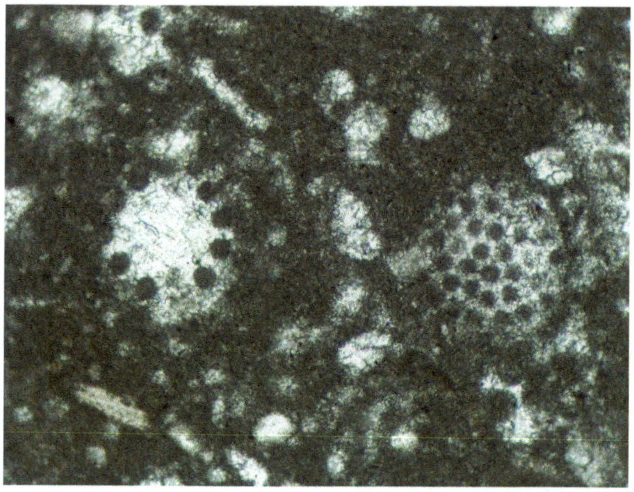

Cretaceous (middle Aptian). Piché Member, Apón Formation. Maracaibo Basin

Organic-lean micritic sediments in this green algae wackestone are the product of $CaCO_3$ precipitation in the water column (allomicrite). The presence of green algae (photoautotrophs) as the dominating skeletal components in this microfacies indicates that lime-mud production occurred in shallow waters of the upper photic zone under well-oxygenated and low-nutrient conditions (probably oligo-to mesotrophic).

IW = 1.3 mm. PPL.

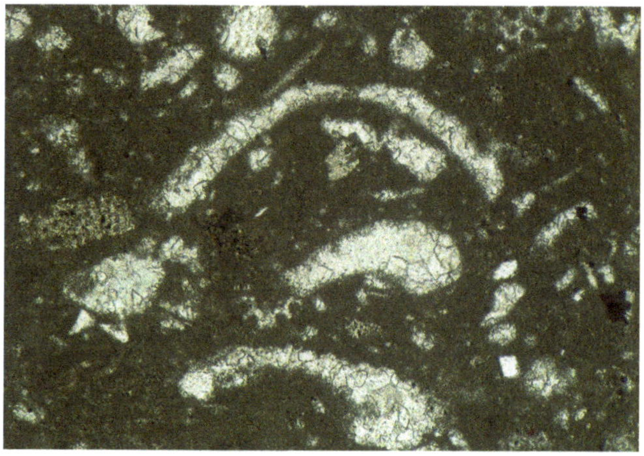

Cretaceous (middle Albian). Lisure Formation. Maracaibo Basin

Allomicrite is a major component of this wackestone, containing bioclasts, the majority of them which were originally aragonitic in composition (mainly molluscs and secondary dasycladalean algae). As in the previous example, the light color of the matrix, combined with the nature of the benthic biota (typical of shallow and well-oxygenated marine environments), serves as the criterion for considering an allomicritic origin for the micritic sediments that make up the matrix of this example.

IW = 2.4 mm. PPL.

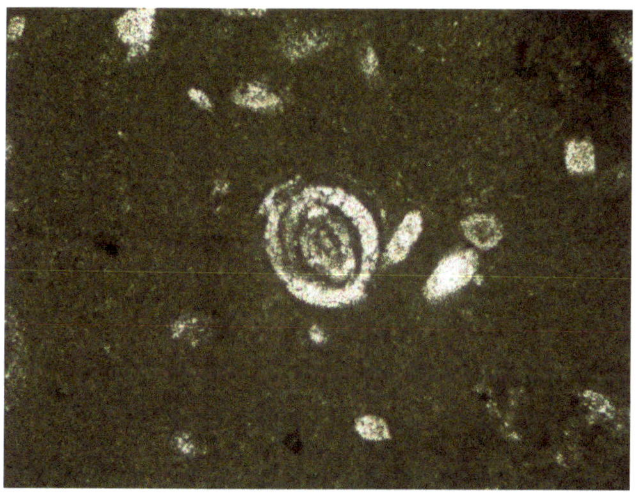

Cretaceous (middle Aptian). Piché Member, Apón Formation. Maracaibo Basin

A mudstone microfacies exhibits a noticeable dark micritic matrix. The smaller miliolid in the center of the image (a symbiont-barren benthic foraminifer common in eutrophic environments within the euphotic zone) is accompanied by small sponge spicules and dolomite crystals. For this sample, we estimate that the micritic sediments that make up the matrix were mainly precipitated on the seafloor (automicrite) due to microbial activity. This example, along with the two previous ones, illustrates the common well-preserved state of micritic matrices of Aptian and Albian ages in the MB, suggesting an original LMC composition.

IW = 1.3 mm. PPL.

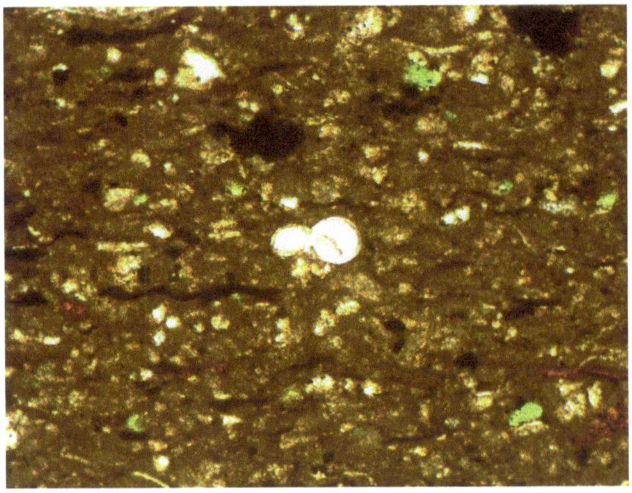

Cretaceous (late Albian). Lisure Formation. Maracaibo Basin

This sample corresponds to a transgressive, muddy deposit in the upper part of the Lisure Formation containing scarce PF (such as the one in the center of the image). The matrix is slightly organic-rich (organic matter lamellae are abundant) and carbonate-dominated. However, clay minerals are present in moderate amounts. Fine sand-sized quartz grains are also moderately common. All of the above, combined with the lack of benthic biota, suggests low-oxygen conditions and a primary contribution of automicrite precipitation at the seafloor for the formation of the micritic sediments comprising the matrix.

IW = 1.3 mm. PPL.

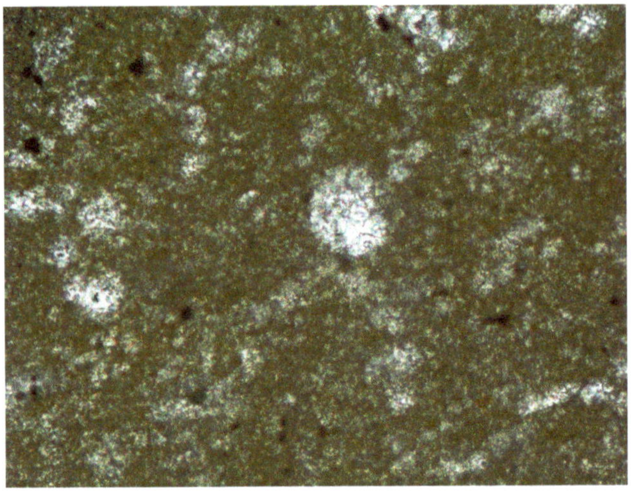

Cretaceous (early Aptian). Machiques Member, Apón Formation. Maracaibo Basin

This sample is an automicrite boundstone with undifferentiated planktonic calcareous microfossils and corresponds to the upper part of the OAE1a in core TOT-3. It has a carbonate content of 92.66% and a TOC content of 0.76%. The high magnification allows visualization of the micropeloidal texture of the matrix. Scattered pyrite framboids (black spots) suggests low-oxygen or anoxic conditions, although the moderate organic enrichment of the sample suggests the former. Microbially-induced micrite dominated the carbonate benthic production during the deposition of the Machiques Member.

IW = 0.5 mm. PPL.

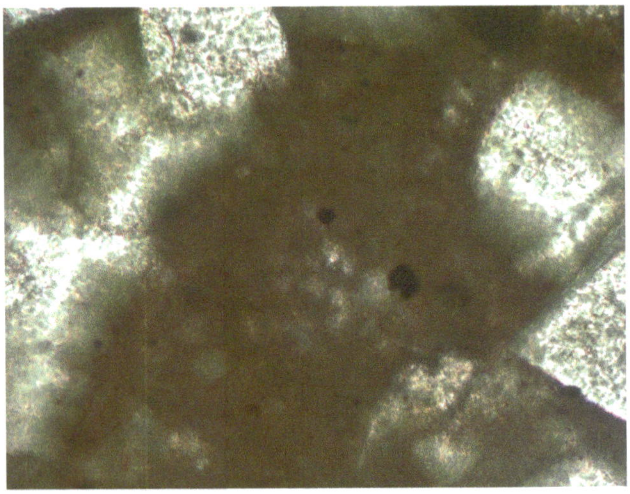

Cretaceous (early Aptian). Machiques Member, Apón Formation. Maracaibo Basin

Another close view of an automicrite boundstone. Note the visually imperceptible size of the crystals that make up the organic-rich matrix (this sample has 1.4% TOC). The presence of pyrite framboids (black spots), combined with the lack of carbonate benthic biota, indicates seafloor anoxia during deposition. The presence of dolomite crystals (white,cloudy areas) suggests hypersaline conditions during early marine diagenesis in a depositional environment within the photic zone. All of the above, added to the timing of deposition (the OAE 1a), suggests a dominant contribution of benthic microbial activity to the production of micritic sediments comprising the matrix.

IW = 0.5 mm. PPL.

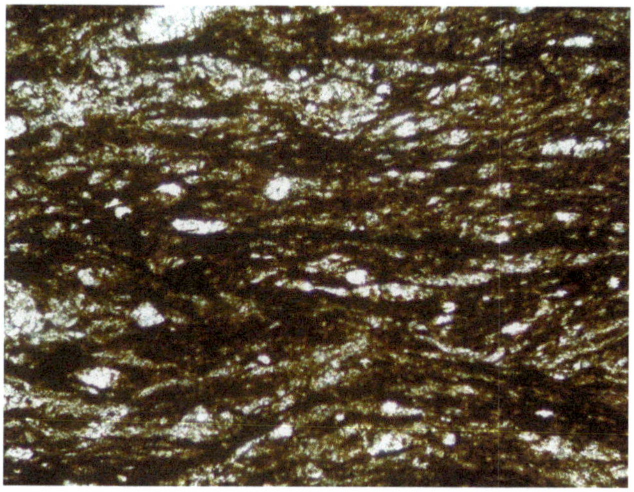

Cretaceous (late Aptian). Piché Member, Apón Formation. Maracaibo Basin

This black shale (13.12% TOC) from the to the OAE 1b first sub-event (upper part of the Piché Member) is an example of strongly microbially-mediated deposition of laminated, organic-rich microfacies, as indicated by organic geochemical proxies (biomarkers). The dark-wavy matrix is composed of organic matter and clay minerals. Light-colored bands are composed of organodolomite. Anoxic-euxinic conditions and a predominance of organic matter of prokaryotic origin (e.g., cyanobacteria) over eukaryotic phytoplankton indicate that benthic microbial processes dominated the deposition of this organic-rich microbialite.

IW = 1.3 mm. PPL.

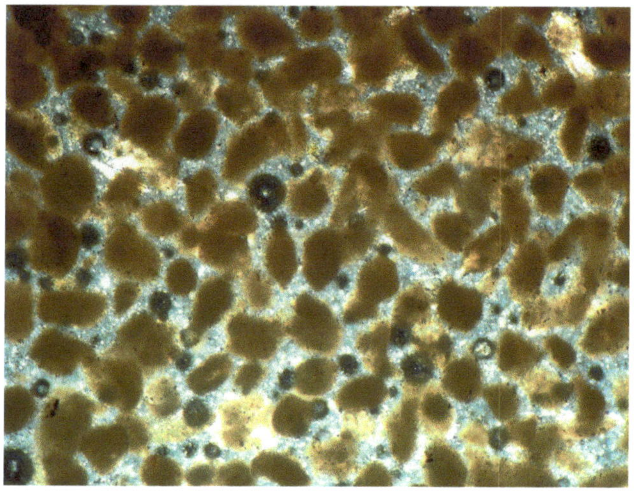

Holocene. Larga Island. Carabobo Sate shoreline

Modern peloidal grains in the image are most likely fecal pellets. The image corresponds to a lithified muddy microfacies in a low-energy, subsaline (i.e., moderately saline) to hypersaline tidal flat environment. The micritic grains are ovoidal to ellipsoidal in shape and more or less equant-sized. They have smooth surfaces and homogeneous texture. Calcium carbonate crystals are visibly imperceptible. The dark-brown color of the grains is typical of carbonate fecal/excrement products. Early surface hardening and lithification of muddy peloidal grains are associated with microbial activity.

IW = 1.3 mm. CPL.

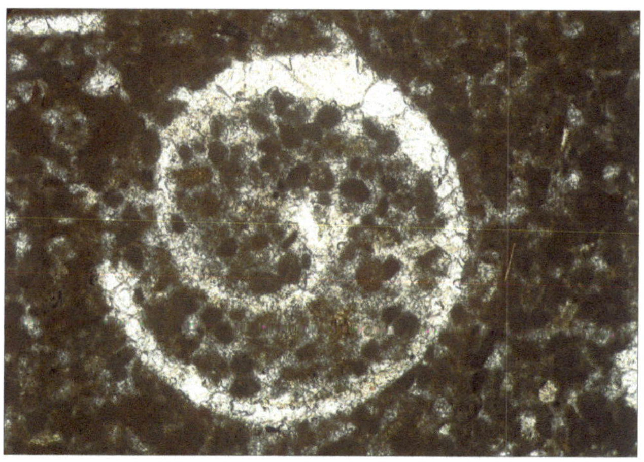

Cretaceous (middle Aptian). Piché Member, Apón Formation. Maracaibo Basin

Peloidal grains in the image, probably fecal pellets, are the main components of this peloidal limestone with a well-preserved mold of a gastropod spiraled shell in the center. The micritic grains are small-sized (< 200 μm) and exhibit more or less uniform shape and size, which in the case of a likely fecal origin, might indicate that they were produced by the same organism. Many peloidal grains formed larger micritic aggregates, thus preventing visual differentiation of individual grains.

IW = 2.4 mm. PPL.

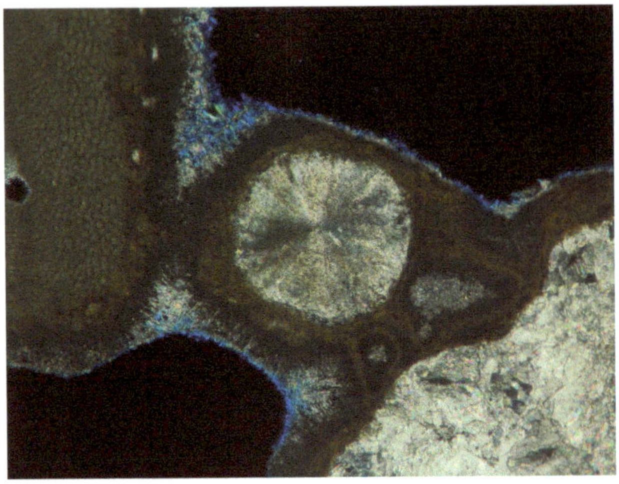

Holocene, Aves Island

Precipitation of automicrite is exemplified in this image by micritic envelopes (dark brown rims) coating grains (allochems). This constructive micrite precipitation is the product of microbial mediation, mainly biologically-induced $CaCO_3$. The envelopes exhibit a filamentous fabric and form meniscus bridges that cement the grains together, thereby transforming loose sediments into lithified rock. In this example, precipitation of micritic cement of HMC composition (as indicated by Energy Dispersive Spectrometry, EDS) occurred in the intertidal (marine–vadose) environment.

IW = 1.1 mm. CPL.

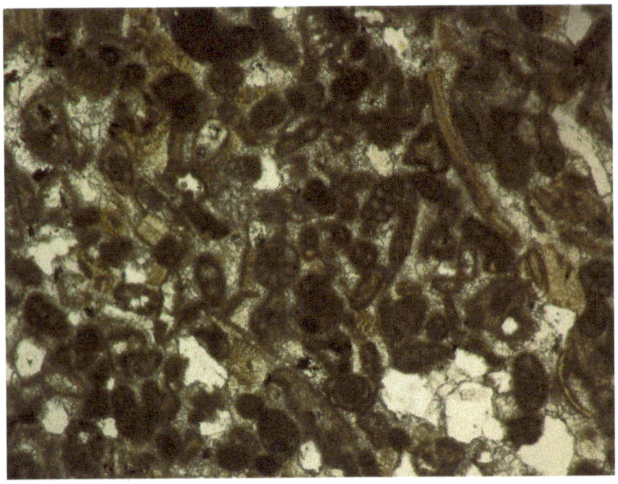

Cretaceous (Barremian). Tibú Member, Apón Formation. Maracaibo Basin

In this sandy grainstone of SBF, micritization during marine diagenesis led to partial or complete obliteration of test morphology. Note that the chamber arrangements remain identifiable in some tests, whereas pervasive micritization imparted a peloidal appearance to many others. Although the LMC cement postdates micritization and fills interparticle porosity during meteoric diagenesis, the microcrystalline crystals composing the micritic grain replacements remained unrecrystallized.

IW = 2.3 mm. PPL.

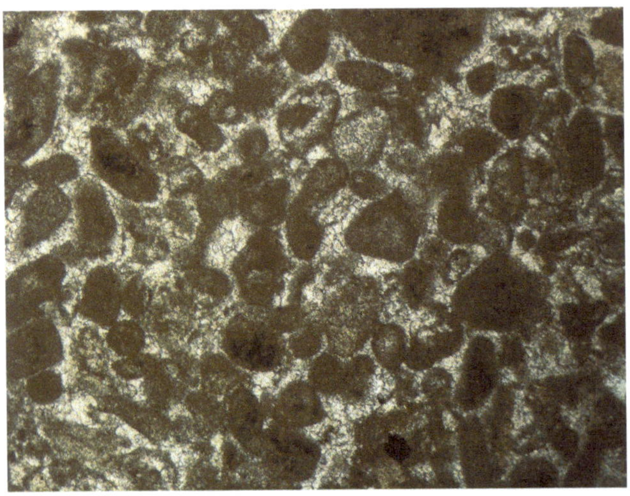

Cretaceous (middle Albian). Lisure Formation. Maracaibo Basin

Micritic peloids in this cemented grainstone exhibit different sizes (poor sorting) and shapes (more or less rounded to elongated). These textureless, micritic grains are the product of micritization of bioclasts. In the Lisure Formation, peloidal grainstones are common in shallow and agitated settings, where grains underwent reworking and redeposition into a adjacent mud-lean environment.

IW = 2.4 mm. PPL.

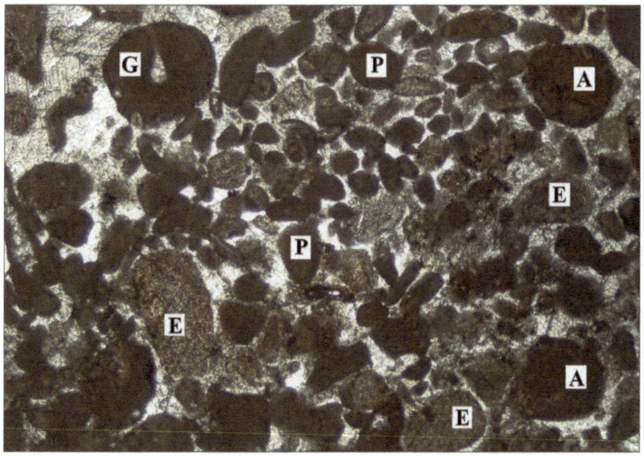

Cretaceous (middle Aptian). Piché Member, Apón Formation. Maracaibo Basin

Peloidal grains in this grainstone have a variety of sizes and origins. The smaller ones (with rounded to elongate shapes) are micritic peloids (P), probably derived from reworking of carbonate mud and/or micritization of green algal grains. Bigger grains consist of several visibly grains (skeletal and/or non-skeletal) cemented together by micritic sediments, thus forming aggregates (A). Some other grains exhibit identifiable features such as the gastropod (G) seen in transverse section, or echinoderms (E) that escaped micritization.

IW = 7.5 mm. PPL.

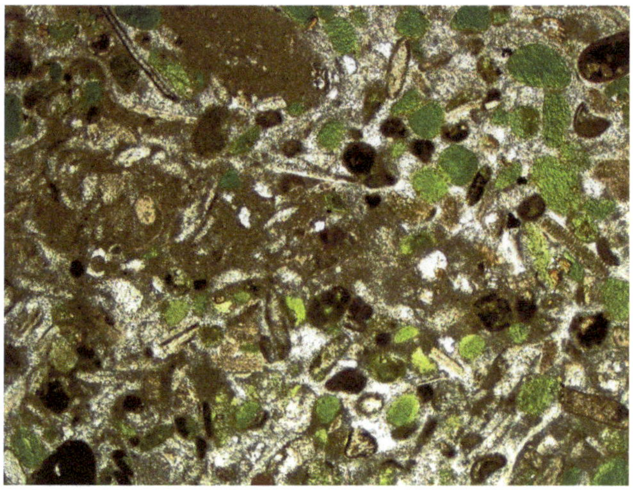

Cretaceous (middle Albian). Lisure Formation. Maracaibo Basin

Peloid grains in this patchy sparry, sparse packstone consist of dark micritized grains (single and aggregates) and glauconite peloids (green grains). The latter were derived from the replacement of the former by glauconite during early marine diagenesis. Transient low-oxygen (dysoxic) conditions and organic-rich micritic substrates are associated with the selective precipitation of the green silicate mineral. Other grains in the image are mainly fragments of echinoderms and molluscs.

IW = 4.2 mm. PPL.

Cretaceous (Campanian). San Antonio Formation. Eastern Interior Range. Outcrop in the State of Anzoátegui

Peloids in this grain-supported fabric are replaced by fluorapatite (Ca-phosphate). Darkening of the grains is due to precipitation of pyrite postdating phosphatization. Prior to these early diagenetic mineralization processes, the ovoidal peloids were most probably ooids and/or micritic peloids, while larger/angular grains were probably skeletal in origin. Phosphatic peloids are common in Coniacian to Campanian facies in Eastern and Western Venezuela. Their formation is associated with low-oxygen conditions and regional upwelling affecting reworked and redeposited shallow carbonate components.

IW = 2.4 mm. PPL.

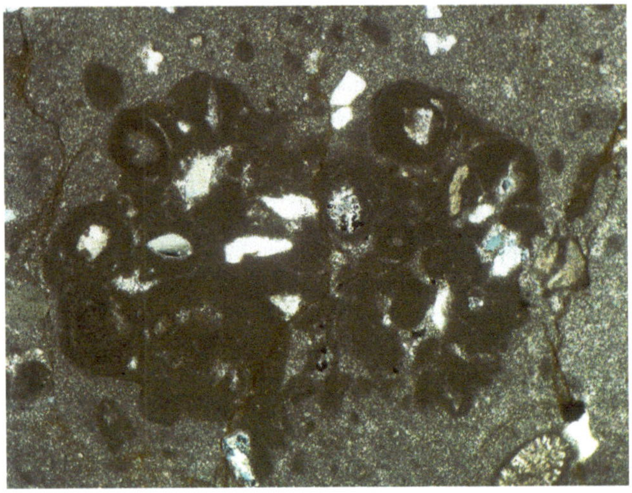

Late Oligocene-Early Miocene. Perla Limestone. Eastern Gulf of Venezuela Sub-Basin

Dark micritic grains in the image are embedded in a muddy matrix that underwent aggrading neomorphism. Note the rounded/spherical shapes of the small grains. Some of these formed around angular quartz grains that served as nuclei, resembling layered ooids cortices. In the Perla Limestone, peloidal muddy microfacies (not volumetrically significant) are interpreted as downflow ramp deposits corresponding to low-energy environments below the base of surface wave action (see Pomar et al. 2015).

IW = 2.4 mm. PPL.

Cretaceous (middle Albian). Lisure Formation. Maracaibo Basin

Close view of a common tangential ooid of the Lisure Formation. The coated grain exhibits a spherical to ovate shape and tangential-concentric arrangement of $CaCO_3$ microlayers (light brown) around a nucleus. The thinner dark interlayers are organic-rich micritic layers reminiscent of microbial activity. The concentric growth of cortices with tangential microfabrics is associated with moderate high-energy settings.

IW = 0.6 mm. PPL.

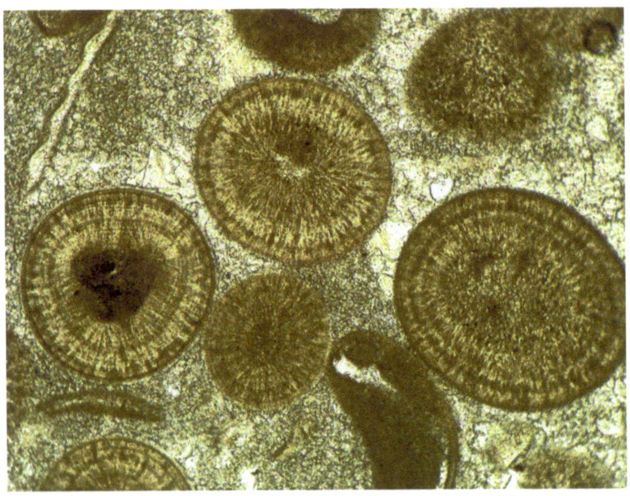

Cretaceous (middle Albian). Lisure Formation. Maracaibo Basin

The four central ooids in the image exhibit a well-preserved radial-concentric microfabric, which is generally associated with lower energy settings compared to the formation of tangential ooids. Note the radially-arranged fibrous crystals and interspersed layers of concentric, very fine crystalline calcium carbonate. The radial-fibrous microfabric of many ancient calcitic ooids is considered to be primary because of its similarities with respect to modern HMC radial ooids from Baffin Bay, Texas (Land et al. 1979), the Amazon Shelf (Milliman and Barretto 1975), and from deep waters of the Great Barrier Reef (Marshall and Davies 1975).

IW = 1.8 mm. PPL.

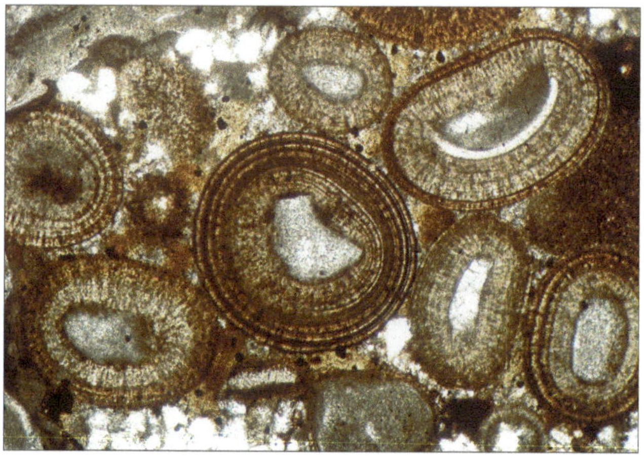

Cretaceous (middle Albian). Lisure Formation. Maracaibo Basin

Ooids in the image exhibit different inner and outer microfabrics. The inner cortices, lighter in color, consist of radial layers; the outer cortices, darker due to prominent micritization, consist of tangential layers. This change in the microfabric indicates a break in ooid growth that might be associated with changes in the hydrodynamic energy of the environment where ooids were formed. These changes ranged from calmer conditions during the formation of the inner-radial layers, to more agitated conditions during the formation of the outer-tangential layers.

IW = 1.8 mm. PPL.

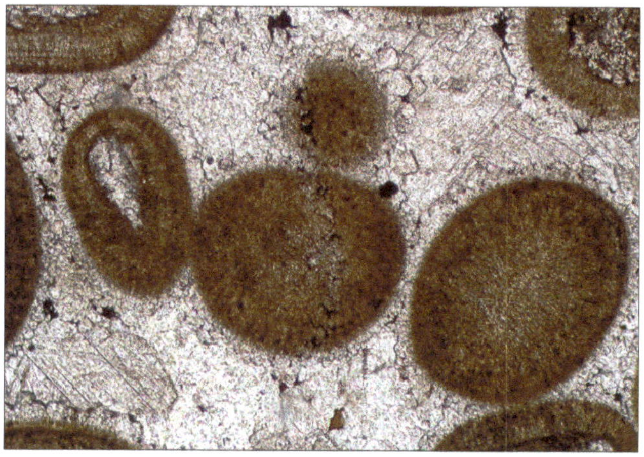

Cretaceous (middle Albian). Lisure Formation. Maracaibo Basin

The three central ooids in this grainstone exhibit a micritic (random) microfaric. Destructive micritization involving microboring, recrystallization, and precipitation of micritic sediments obliterated the original microfabric of the grains. Traces of the concentric arrangement of cortical layers are barely visible. Subsequent meteoric diagenesis, which led to precipitation of calcite cement occluding primary interparticle porosity, did not alter the already mineralogically stabilized ooids.

IW = 1.3 mm. PPL.

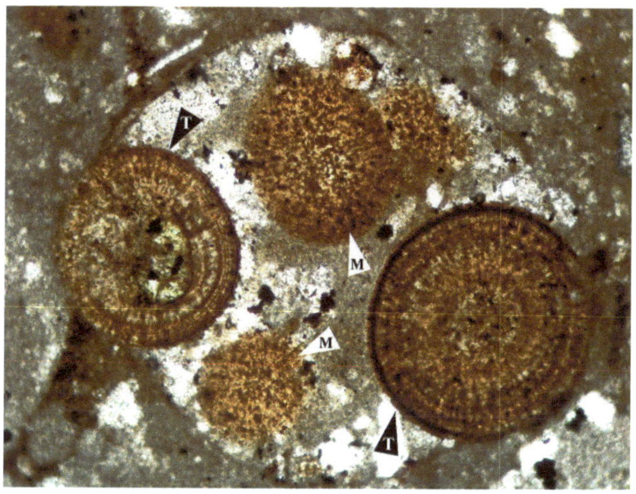

Cretaceous (middle Albian). Lisure Formation. Maracaibo Basin

The degree of micritization in ooids of the Lisure Formation varies, as well exemplified in this sample. Two of the ooids in the image exhibit a micritic (M) random microfabric, or total obliteration of the original microfabric, while partially micritized ooids exhibit traces of the tangential (T) microfabric. Furthermore, note abundant pyrite framboids (black spots) partially replacing these non-skeletal grains and the micritic matrix. The latter indicates both strong bacterial activity postdating micritization and transient low-oxygen conditions during early diagenesis.

IW = 1.3 mm. PPL.

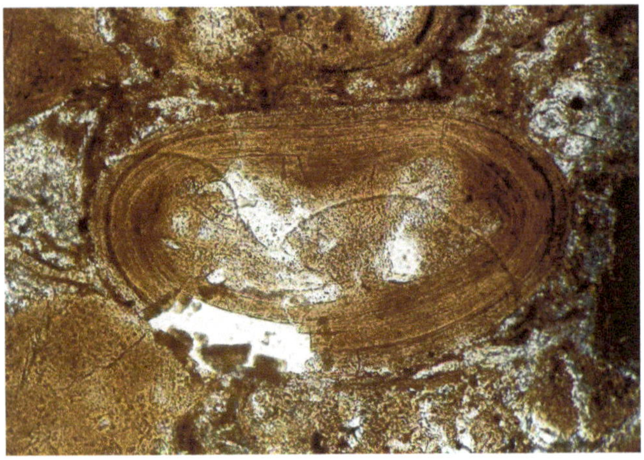

Cretaceous (early Campanian). Tres Esquinas Member, La Luna Formation. Maracaibo Basin

Phosphatized ooid of the Tres Esquinas Member. Note the tangential-concentric microfabric of the ooid. This sample combines a variety of depositional and diagenetic processes. The ooid was originally formed in a shallow setting; the grain was redeposited in a deeper setting in the lower neritic to upper bathyal zone, where phosphatization took place under low-oxygen conditions; the white precipitate that is partially coating the ooid is amorphous silica precipitated during shallow-burial diagenesis.

IW = 0.5 mm. PPL.

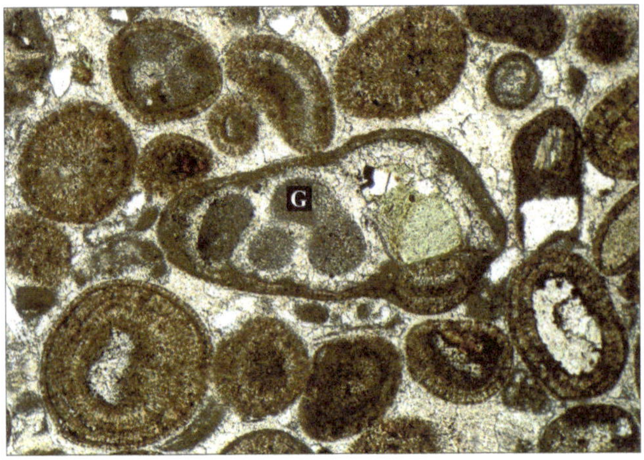

Cretaceous (middle Albian). Lisure Formation. Maracaibo Basin

Ooids in this oolite grainstone consist of a variety of shapes, which at the time of their formation were controlled by the shape of the nuclei around which the cortices were formed (spherical to ovate shapes predominate). Ooid cortices are also commonly formed around nuclei with elongate shapes, such as in the central gastropod (G). In the Lisure Formation, the variety of skeletal grains that serve as nuclei for the precipitation of ooid cortices include bryozoans, echinoderms, molluscs, micritic peloids and quartz grains.

IW = 4.2 mm. PPL.

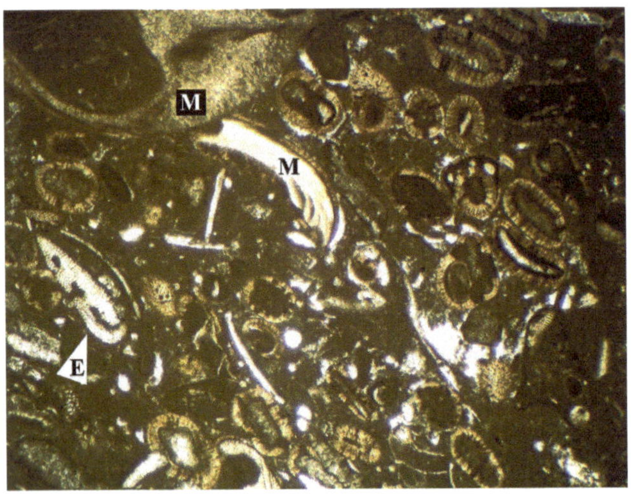

Cretaceous (middle Albian). Lisure Formation. Maracaibo Basin

Muddy packstone in which ooids are major components. These are accompanied by fragments of molluscs (M) and echinoderms (E). In the Lisure Formation, microfacies containing ooids are characterized by the common presence of these heterotrophic calcifiers (plus bryozoans). In addition, oolitic microfacies in this unit, either cemented grainstones (such as the previous example), or muddy fabrics (such as this example), do not host porosity, neither primary (e.g., interparticle pores) nor secondary (e.g., oomoldic pores).

IW = 7.5 mm. PPL.

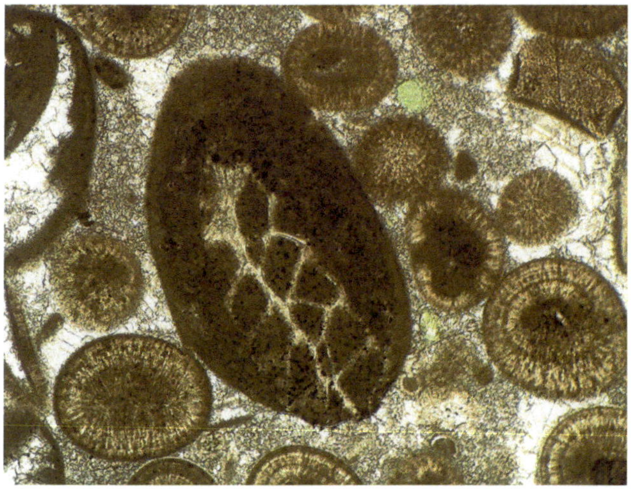

Cretaceous (middle Albian). Lisure Formation. Maracaibo Basin

A cortoid surrounded by several ooids. This microbially coated grain consists of a bryozoan as a nucleus and a thin micritic envelope. Note that the boundary between the grain core and the microbial envelope is indistinct, which is a common feature of cortoids. Microfacies containing abundant non-skeletal grains (mainly peloids, ooids, oncoids and cortoids), such as this example, occurred abundantly in mid-Cretaceous platforms, especially at tropical paleolatitudes (see Iryu and Yamada 1999).

IW = 2.4 mm. PPL.

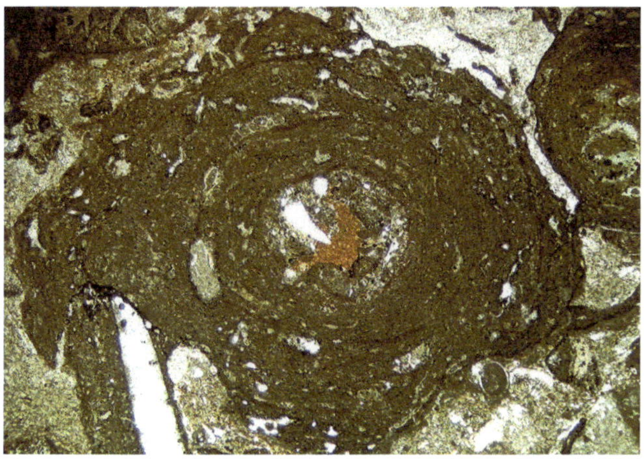

Cretaceous (middle Albian). Lisure Formation. Maracaibo Basin

The oncoid grain in the image consists of a dark-micritic microbial coating around an aggregate grain as a nucleus. The spheroidal microstructure exhibits a more or less concentric arrangement. Microbial carbonate filaments that compose this micritic oncoid overlap each other, forming a dense structure. The irregularly concentric pattern in oncoids (visible in this example) is markedly less apparent compared to that of more symmetrical ooids. Note the light-colored grains of different sizes trapped within the micritic coating, as trapping and binding of small debris is part of the growth/accretion process in these unattached microbialites.

IW = 7.5 mm. PPL.

Cretaceous (middle Albian). Lisure Formation. Maracaibo Basin

This oncoid grain exhibits a more or less concentric arrangement of calcified filaments. As in the previous example, rolling and overturning on the seafloor in an agitated environment led to the formation of this microbialite grain prior to final deposition in a heterogeneous matrix. Individual sheets are composed of microbially-mediated precipitates of a brown-reddish color, which alternate irregularly with lime-mud. The sheets thicken away from the nucleus (a gastropod seen in transverse section), forming a sub-spherical coating. Trapped allochems are abundant within the filamentous cortex.

IW = 4.2 mm. PPL.

Cretaceous (middle Albian). Lisure Formation. Maracaibo Basin

The oncoid grain in the center of the image has an elongate shape inherited from the bryozoan grain that served as a nucleus for the precipitation of the microbial coating. The formation of microbially coated grains in the Lisure Formation is commonly associated with heterozoan deposits dominated by molluscs, echinoderms and bryozoans. Textural differences (color, crystal size and arrangement) among the variety of oncoids and other microbially coated grains in this unit point to origins associated with different benthic microbial communities (bacterial and/or bacterial/algal).

IW = 7.5 mm. PPL.

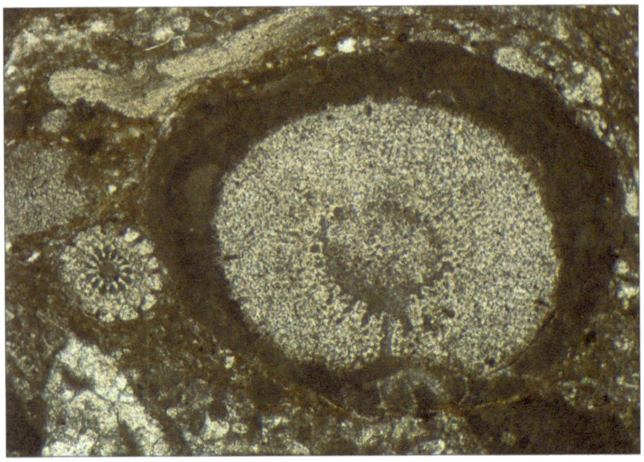

Cretaceous (middle Albian). Lisure Formation. Maracaibo Basin

This microbially coated grain consists of a rounded echinoid grain as a nucleus and a dark microbial coating. The latter is thin relative to the diameter of the nucleus; hence, this grain is classified as a cortoid. The coating is composed of micritic sediments forming a dense structure that does not exhibit any distinctive or apparent layered microstructure or individual sheets. In general, microbial coatings in the Lisure Formation are well-preserved, most of them exhibiting no evidence of having undergone significant neomorphism or recrystallization, which suggests an original LMC composition.

IW = 4.2 mm. PPL.

Cretaceous (middle Albian). Lisure Formation. Maracaibo Basin

This large cortoid grain consists of a thin micritic envelope coating an aggregate grain. The latter is composed of micritic sediments cementing numerous small-sized carbonate grains. Note the abundant pyrite framboids partially covering the gravel-sized, rounded grain, indicating transient low-oxygen conditions during early marine diagenesis, which is very common in microfacies in the Lisure Formation that are rich in microbially coated grains.

IW = 7.5 mm. PPL.

Cretaceous (middle Albian). Lisure Formation. Maracaibo Basin

Packstone of microbially coated grains. Note the varied sizes and shapes of the grains in the image. Microbial precipitates are coating bioclasts (bryozoans) and micritic aggregates. The heterogeneous muddy-bioclastic matrix is characteristic of agitated bottom waters associated with rolling, overturning and formation of these compound grains. As in the previous example, the apparent darkening of the grains due to pyritization is indicative of transient low-oxygen (but not anoxic) and nutrient-rich conditions that promoted microbially-induced micrite precipitation at the seafloor.

IW = 7.5 mm. PPL.

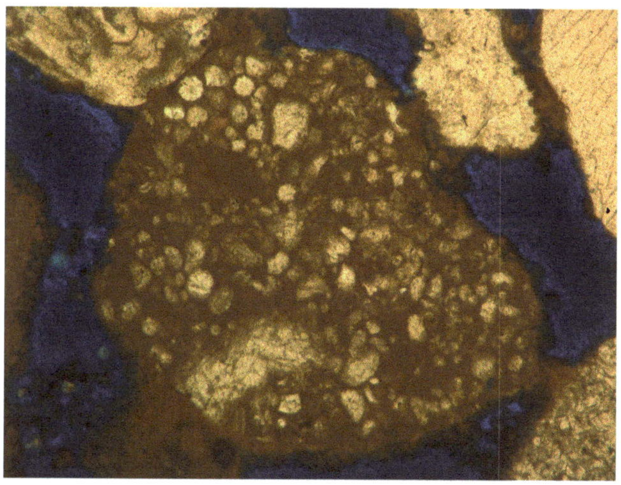

Late Pleistocene. Aves Island

This micritic aggregate, cemented to adjacent grains by micritic meniscus bridges, is found in a porous calcarenite facies. The compound grain, moderately rounded in shape, consists of numerous small carbonate grains bound together by micritic sediments. The aggregate formed in a moderately agitated environment, where it underwent rolling along the seafloor prior to its final deposition in a very shallow, low-energy subtidal environment.

IW = 2.3 mm. PPL.

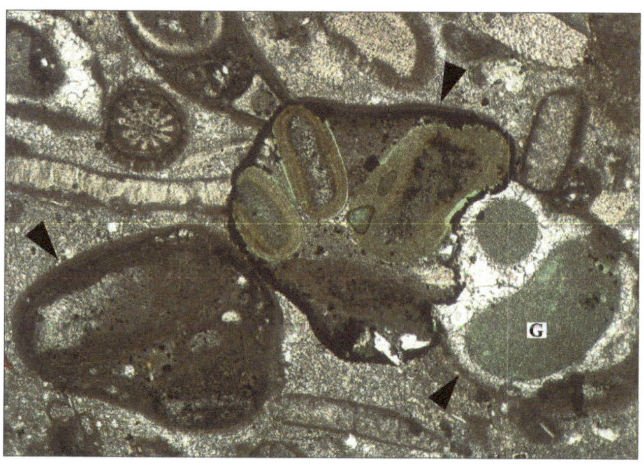

Cretaceous (middle Albian). Lisure Formation. Maracaibo Basin

Carbonate clasts in the image are aggregate grains (intraclasts) exhibiting characteristic grape-like shapes. These compound grains consist of a variety of allochems, including ooids and a gastrotop (G), which were cemented together by micrite and coated by thin micritic layers (black arrows). Note that the larger central aggregate is partly outlined by precipitation of pyrite. The abundance of micritic sediments, both within the aggregate grains and in the matrix, suggests that the clasts were finally deposited near the same depositional sequence in which they formed.

IW = 4.2 mm. PPL.

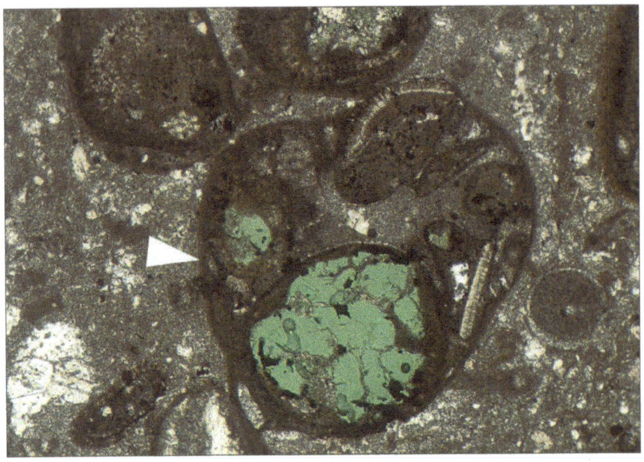

Cretaceous (middle Albian). Lisure Formation. Maracaibo Basin

Another example of a grapestone aggregate (hardground intraclast) embedded in a muddy-bioclastic matrix. The clast consists of several coated grains (micritized ooids and peloidal grains) and glauconitic peloids cemented together by micritic sediments. Note the constructive micritic envelope that is coating this compound grain (white arrow). This coating is the product of microbially-mediated $CaCO_3$. In addition, early marine precipitation of glauconite is common on hardground surfaces in the Lisure Formation.

IW = 4.2 mm. PPL.

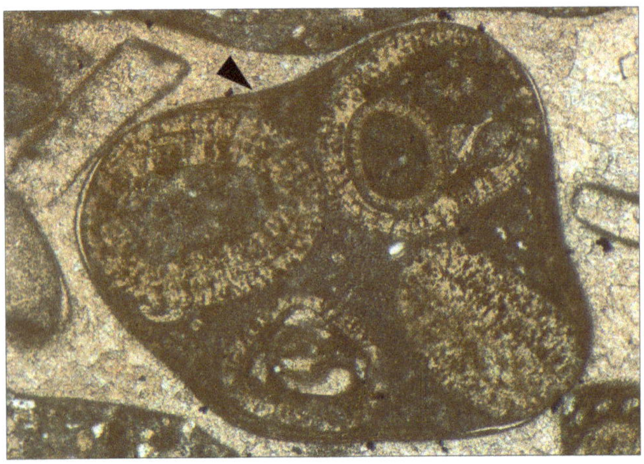

Cretaceous (middle Albian). Lisure Formation. Maracaibo Basin

The four ooids in the image were cemented together by carbonate mud, hence forming an aggregate. The grapestone appearance of this poly-ooid was given by the thin, outer oolitic cortex (black arrow) that coats the entire grain, which can be termed an oolitic lump. Rolling and overturning on a muddy seafloor predated final deposition of this peculiar grain in a mud-lean environment. Therefore, it can also be classified as an extraclast.

IW = 2.4 mm. PPL.

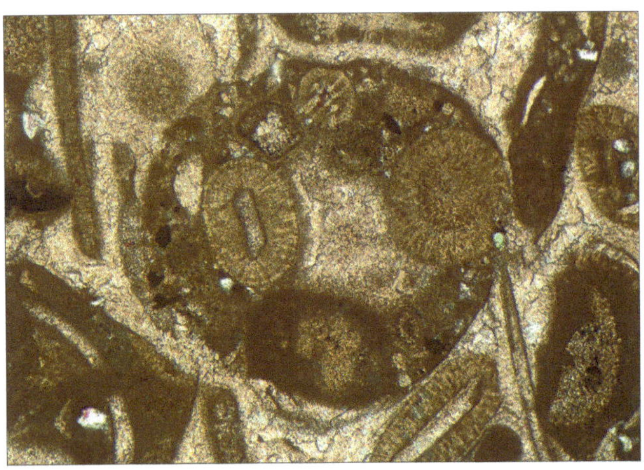

Cretaceous (middle Albian). Lisure Formation. Maracaibo Basin

Grainstone of aggregate clasts. The rounded central aggregate extraclast in the image consists of several ooids and peloidal grains cemented together by micritic sediments. Rounding of the grain involved rolling and reworking prior to final deposition in a mud-lean environment. Later, during meteoric diagenesis, calcite cement precipitated and infilled intraparticle porosity.

IW = 4.2 mm. PPL.

Cretaceous (middle Albian). Lisure Formation. Maracaibo Basin

The rounded hardground extraclast in the center of the image is embedded in a highly packed, mixed sandy matrix. The extraclast consists of several micritized and pyritized grains cemented together, including echinoid spines, ooids, glauconitic peloids, and quartz grains. Glauconitic peloids are present in both the extraclast and in the matrix, which suggests an authigenic origin for these green grains. The heterogeneity of this microfacies evidences environmental differences during the formation of the ooids; the formation of the aggregate grain; the final deposition in a mixed deposit; and the formation of the glauconite peloids.

IW = 4.2 mm. PPL.

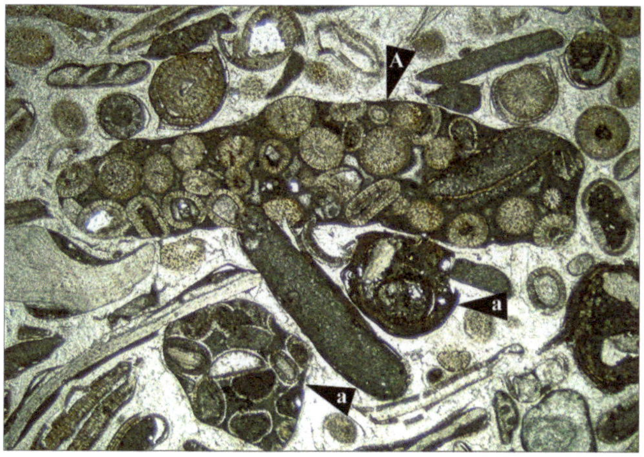

Cretaceous (middle Albian). Lisure Formation. Maracaibo Basin

The large aggregate (A) (an extraclast) in the center of the image is accompanied by smaller, rounded aggregates (a), as well as by numerous ooids and other elongate-coated grains in a grainstone. The abundance of ooids, both cemented by micrite within the aggregates and as individual grains, indicates that the compound grains were derived from an originally oolitic deposit. Moreover, their formation occurred in a muddy environment, and they were later redeposited in a mud-lean environment.

IW = 7.5 mm. PPL.

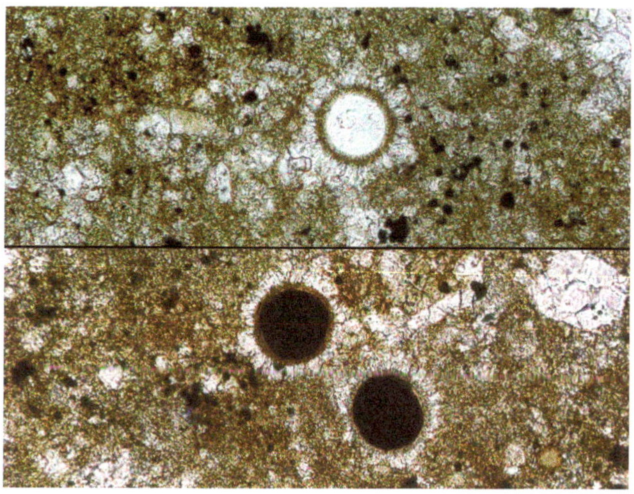

Cretaceous (early Aptian). Machiques Member, Apón Formation. Maracaibo Basin

The spherical precipitates in the images are termed “microspheres”. These carbonate authigenic products are interpreted to be the result of calcification induced by endolithic cyanobacteria in an organic-rich micritic matrix (Florentin Maurrasse, personal communication, 2016). The inner calcitic cores of the two microspheres in the image below were selectively replaced by pyrite. Note that the cores of these minute microbialites are surrounded by a thin rim of microcrystalline calcite reminiscent of a “test”. This brown-colored microstructure is in turn rimmed by an isopachous radial arrangement of fibrous crystals (not always preserved).

IW = 0.5 mm each. PPL.

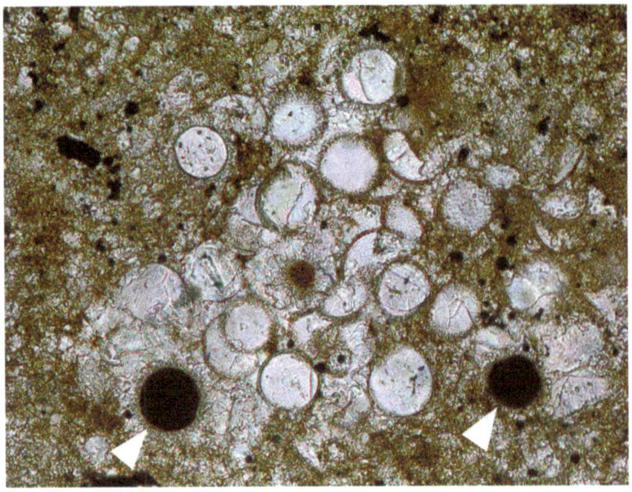

Cretaceous (early Aptian). Machiques Member, Apón Formation. Maracaibo Basin

Microspheres in black shales of the Machiques Member may be either found isolated or in joined clusters, as is seen in the image, with individual microbial products partially overlapping each other. Microbially-induced $CaCO_3$ precipitation occurred as replacement of the micritic matrix. Note abundant pyrite framboids scattered across the matrix, indicating anoxic-euxinic conditions during early marine diagenesis. Bacterial sulfate reduction promoted the selective pyritization of some of the microspheres (white arrows). This sample has 2.6% TOC, suggesting that the cyanobacteria that induced this peculiar Ca-carbonate organomineralization were anoxygenic.

IW = 0.5 mm. PPL.

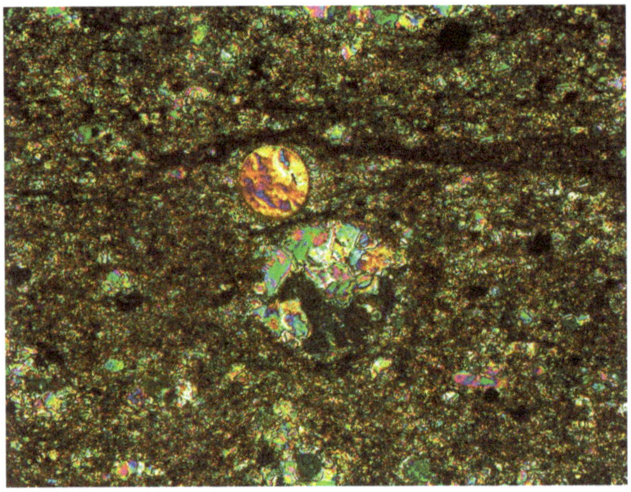

Cretaceous (early Aptian). Machiques Member, Apón Formation. Maracaibo Basin

A microsphere (above) accompanied by a globular planktonic foraminifer (below) in a recrystallized mudstone. The view under CPL allows visualization of the high birefringence and high resistance of these minute microbialites to "wet" diagenesis, which is apparent in both the recrystallized matrix and the coarse-grained blocky calcite filling the globular chamber of the planktonic foraminifer. The latter suggests an original LMC composition for the microsphere.

IW = 0.5 mm. CPL.

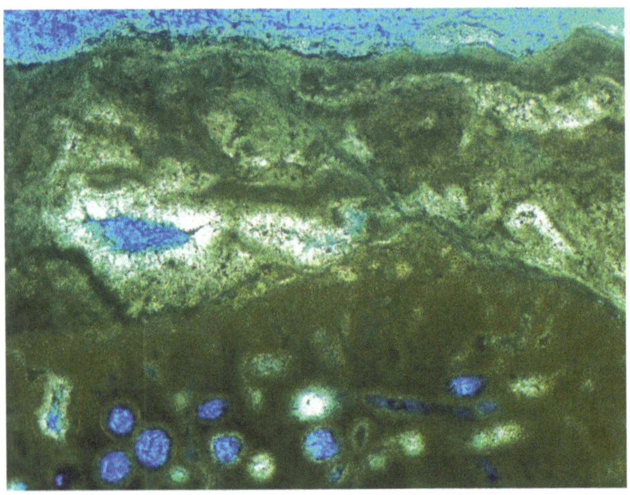

Holocene. Los Roques Archipelago

This and the following two images correspond to a lithified, laminated, intertidal stromatolite formed in the shallower waters of a perennial hypersaline lagoon. This photomicrograph shows the uppermost part of the structure. The fabric in the image consists of wavy calcified filaments at the top (micrite sediments and microbial mats) and a micritized *Halimeda* flake right underneath. Pores are partially filled by early marine cements (white precipitate). This image serves as an example of two key processes involved in the formation of microbial constructions: microbially-mediated calcification and trapping of allochems.

IW = 1.3 mm. PPL.

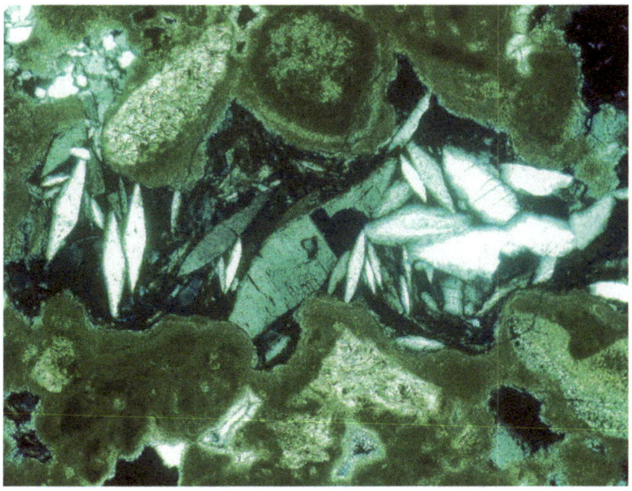

Holocene. Los Roques Archipelago

This image corresponds to the middle-lower part of the laminated stromatolite shown in the previous example. Note the dark-brown automicrite sediments (which are the product of induced calcification by endolithic bacteria) enveloping and cementing grains, and outlining the central subhorizontal pore. The latter is partially filled by lenticular gypsum crystals exhibiting low white/gray birefringence, indicating hypersaline conditions and supersaturation of the shallow marine-evaporated waters with respect to $CaSO_4.2H_2O$.

IW = 2.4 mm. CPL.

Holocene. Los Roques Archipelago

This lower magnification image of the middle-lower part of the intertidal stromatolite allows visualization of microfabric variations, from micritic-laminated in the uppermost part (as shown in the two previous examples) to clotted-thrombolitic in the middle and lower parts (represented in this image). Trapped allochems are mainly *Halimeda* flakes and mollusc fragments cemented together by automicrite. Some of the fenestral and interparticle pores in the image are filled by gypsum cement (white precipitate).

IW = 7.5 mm. PPL.

References

Abed, A. M., Abu-Hamad, A., Abul-Khair, H., Kraishan, G., 2013. Development of calcrete and clinoforms during emergence and flooding of the Late Cenomanian carbonate platform, Jordan. *Facies*, *59*, 829-842.

Amao, A. O., Al-Ramadan, K., Koeshidayatullah, A., 2016. Automated mineralogical methodology to study carbonate grain microstructure: an example from oncoids. *Environmental earth sciences*, *75*(8), 1-11.

Amodio, S., and Weissert, H., 2017. Palaeoenvironment and palaeoecology before and at the onset of Oceanic Anoxic Event (OAE) 1a: Reconstructions from Central Tethyan archives. *Palaeogeography, Palaeoclimatology, Palaeoecology*, *479*, 71-89.

Amorosi, A., 2012. The occurrence of glaucony in the stratigraphic record: distribution patterns and sequence-stratigraphic significance. *International Association of Sedimentologists Special Publications*, *45*, 37-54.

Bádenas, B., and Aurell, M., 2010. Facies models of a shallow-water carbonate ramp based on distribution of non-skeletal grains (Kimmeridgian, Spain). *Facies*, *56*, 89.

Bartok, P., Reijers, T. J. A., Juhasz, I., 1981. Lower Cretaceous Cogollo Group, Maracaibo Basin, Venezuela: sedimentology, diagenesis, and petrophysics. *American Association of Petroleum Geologists, Bulletin 65*, 1110-1134.

Batchelor, M. T., Burne, R. V., Henry, B. I., Li, F., Paul, J., 2018. A biofilm and organomineralisation model for the growth and limiting size of ooids. *Scientific Reports*, *8*, 1-9.

Bathurst, R. G. C., 1966. Boring algae, micrite envelopes and lithification of molluscan biosparites. *Geological Journal*, *5*, 15-32.

Baumgartner, L. K., Reid, R. P., Dupraz, C., Decho, A. W., Buckley, D. H., Spear, J. R., Przekop, K. M., Visscher, P. T., 2006. Sulfate reducing bacteria in microbial mats: changing paradigms, new discoveries. *Sedimentary Geology*, *185*, 131-145.

Brehm, U., Krumbein, W. E., Palinska, K. A., 2006. Biomicrospheres Generate Ooids in the Laboratory. *Geomicrobiology Journal*, *23*(7), 545-550.

Brocks J. J., and Summons R. E., 2004. Sedimentary hydrocarbons, biomarkers for early life. In: Treatise on Geochemistry Vol 8, Biogeochemistry. Schlesinger WH (ed) Elsevier - Pergamon, Oxford, pp. 63-115.

Brocks, J. J., and Pearson, A., 2005. Building the Biomarker Tree of Life. *Reviews in Mineralogy & Geochemistry, 59*, 233-258.

Bromley, R. G., 1975. Trace fossils at omission surfaces. In: The Study of Trace Fossils (R. W. Frey, Ed.). Springer-Verlag, New York, pp. 399–428.

Bundeleva, I. A., Shirokova, L. S., Bénézeth, P., Pokrovsky, O. S., Kompantseva, E. I., Balor, S., 2012. Calcium carbonate precipitation by anoxygenic phototrophic bacteria. *Chemical Geology, 291*, 116-131.

Burne, R. V., and Moore, L. S., 1987. Microbialites: organosedimentary deposits of benthic microbial communities. *Palaios, 2*, 241–254.

Campos-Soto, S., Benito, M. I., Mas, R., Caus, E., Cobos, A., Suarez-Gonzalez, P., Quijada, I. E., 2016. Revisiting the Late Jurassic-Early Cretaceous of the NW South Iberian Basin: new ages and sedimentary environments. *Journal of Iberian Geology, 42*(1), 69-94.

Campos-Soto, S., Benito, M. I., Cobos, A., Caus, E., Quijada, I. E., Suarez-Gonzalez, P., Mas, R., Royo-Torres, R., Alcalá, L., 2019. Revisiting the age and palaeoenvironments of the Upper Jurassic–Lower Cretaceous? Dinosaur-bearing sedimentary record of eastern Spain: implications for Iberian palaeogeography. *Journal of Iberian Geology, 45*, 471-510.

Cantrell, D. L., 2006. Cortical fabrics of Upper Jurassic ooids, Arab Formation, Saudi Arabia: implications for original carbonate mineralogy. *Sedimentary Geology, 186*, 157-170.

Caroprese, C. G., 2009. Bioestratigrafía y paleoecología de la Formación Querecual en el Oriente de Venezuela. Magister Dissertation. Central University of Venezuela. 95 pp.

Castro, J. M., de Gea, G. A., Quijano, M. L., Aguado, R., Froehner, S., Naafs, B. D. A., Pancost, R. D., 2019. Complex and protracted environmental and ecological perturbations during OAE 1a-Evidence from an expanded pelagic section from south Spain (Western Tethys). *Global and Planetary Change, 183*, 103030.

Chafetz, H. S., 1986. Marine peloids; a product of bacterially induced precipitation of calcite. *Journal of Sedimentary Research, 56*, 812-817.

Christ, N., Immenhauser, A., Wood, R. A., Darwich, K., Niedermayr, A., 2015. Petrography and environmental controls on the formation of Phanerozoic marine carbonate hardgrounds. *Earth-Science Reviews, 151*, 176-226.

Cloud, P. E, Jr., 1962. Environment of calcium carbonate deposition west of Andros Island, Bahamas. US Geological Survey Professional Paper 350, 138 pp.

Decho, A. W., 2010. Overview of biopolymer-induced mineralization: what goes on in biofilms? *Ecological Engineering, 36*(2), 137-144.

Desnues C, Rodriguez-Brito, B., Rayhawk, S., Kelley, S., Tran, T., Haynes, M., Liu, H., Furlan, M., Wegley, L., Chau, B., Ruan, Y., Hall, D., Angly, F. E., Edwards, R. A., Li, L., Vega-Thurber, R., Reid, P., Siefert, J., Souza, V., Valentine, D. L., Swan, B. K., Breitbart, M., Rohwer, F., 2008. Biodiversity and biogeography of phages in modern stromatolites and thrombolites. Nature, *452*, 340–43.

Diaz, M. R., and Eberli, G. P., 2019. Decoding the mechanism of formation in marine ooids: a review. *Earth-Science Reviews, 190*, 536–556.

Diaz, M. R., and Eberli, G. P., 2022. Microbial contribution to early marine cementation. *Sedimentology, 69*(2), 798–822.

Diaz, M. R., Eberli, G. P., Blackwelder, P., Phillips, B., Swart, P. K., 2017. Microbially mediated organomineralization in the formation of ooids. *Geology, 45*, 771–774.

Diaz, M. R., Swart, P. K., Eberli, G. P., Oehlert, A. M., Devlin, Q., Saeid, A., Altabet, M. A., 2015. Geochemical evidence of microbial activity within ooids. *Sedimentology, 62*, 2090–2112.

Duguid, S. M., Kyser, T. K., James, N. P., Rankey, E. C., 2010. Microbes and ooids. *Journal of Sedimentary Research, 80*, 236–251.

Dupraz, C., and Visscher, P. T., 2005. Microbial lithification in marine stromatolites and hypersaline mats. *Trends in Microbiology, 13*, 429–438.

Dupraz, C., Reid, R. P., Braissant, O., Decho, A. W., Norman, R. S., Visscher, P. T., 2009. Processes of carbonate precipitation in modern microbial mats. *Earth-Science Reviews, 96*, 141–162.

Dupraz, C., Reid, R.P., Visscher, P.T., 2011. Microbialites, modern, in Reitner, J., and Thiel, V., eds., Encyclopedia of Geobiology, Encyclopedia of Earth Science Series: Berlin, Springer, 653–654.

Duque-Botero, F., and Maurrasse, F., 2004. Cyanobacterial productivity, variations in the organic carbon, and facies of the Indidura Formation (Cenomanian-Turonian), Northeastern Mexico. *Journal of Iberian Geology, 31*, 85–98.

Dumitrescu, M., and Brassell, S. C., 2006. Compositional and isotopic characteristics of organic matter for the early Aptian oceanic anoxic event at Shatsky Rise, ODP Leg 198. *Palaeogeography, Palaeoclimatology, Palaeoecology, 235*, 168–191.

Flannery, D. T., Allwood, A. C., Hodyss, R., Summons, R. E., Tuite, M., Walter, M. R., Williford, K. H., 2018. Microbially influenced formation of Neoarchean ooids. *Geobiology, 17*, 151–160.

Flügel, E, 2010. Microfacies of Carbonate Rocks—Analysis, Interpretation and Application. Springer, Heidelberg-Dordrecht–London–New York, 984 pp.

Folk, R. L., 1959, Practical petrographic classification of limestones. *American Association of Petroleum Geologists Bulletin, 43*, 1–38.

Folk, R. L., 1962. Spectral subdivision of limestone types. In: Classification of Carbonate Rocks (Ed. Ham, W.E.), *American Association of Petroleum Geologists, Memoir* 1, 62–84.

Fossum, K., Dypvik, H., Haid, M. H., Hudson, W. E., van den Brink, M., 2021. Late Jurassic and Early Cretaceous sedimentation in the Mandawa Basin, coastal Tanzania. *Journal of African Earth Sciences, 174*, 104013.

Frantz, C. M., Petryshyn, V. A., Corsetti, F. A., 2015. Grain trapping by filamentous cyanobacterial and algal mats: implications for stromatolite microfabrics through time. *Geobiology, 13*, 409–423.

Füchtbauer, H., and Tisljar, J., 1975. Peritidal cycles in the lower cretaceous of Istria (Yugoslavia). *Sedimentary Geology, 14*, 219–233.

Garuglieri, E., Marasco, R., Odobel, C., Chandra, V., Teillet, T., Areias, C., Sánchez-Román, M., Vahrenkamp, V., Daffonchio, D., 2024. 2024. Searching for microbial contribution to micritization of shallow marine sediments. *Environmental Microbiology, 26*, e16573.

Gischler, E., 2011. Sedimentary facies of Bora Bora, Darwin's type barrier reef (Society Islands, South Pacific): the unexpected occurrence of non–skeletal grains. *Journal of Sedimentary Research, 81*, 1–17.

Gischler, E., Dietrich, S., Harris, D., Webster, J. M., Ginsburg, R. N., 2013. A comparative study of modern carbonate mud in reefs and carbonate platforms: Mostly biogenic, some precipitated. *Sedimentary Geology, 292*, 36–55.

Görgen S, Benzerara K, Skouri-Panet F, Gugger M, Chauvat F, and Cassier-Chauvat C. 2021. The diversity of molecular mechanisms of carbonate biomineralization by bacteria. *Discover Materials, 1*, 1–20.

Hardie, L. A., 1996. Secular variation in seawater chemistry: An explanation for the coupled secular variation in the mineralogies of marine limestones and potash evaporites over the past 600 my. *Geology, 24*, 279–283.

Harris, P. M., 1979. Facies anatomy and diagenesis of a Bahamian Ooid shoal. *Sedimenta, 7*, 163 pp.

Harris, P., Diaz, M. R., Eberli, G. P., 2019. The formation and distribution of modern ooids on Great Bahama Bank. *Annual Review of Marine Science, 11*, 491–516.

Hillgärtner, H., Dupraz, C., Hug, W., 2001. Microbially induced cementation of carbonate sands: are micritic meniscus cements good indicators of vadose diagenesis? *Sedimentology, 48*, 117–131.

Hollis, C., Lawrence, D. A., de Periere, M. D., Al Darmaki, F., 2017. Controls on porosity preservation within a Jurassic oolitic reservoir complex, UAE. *Marine and Petroleum Geology*, *88*, 888–906.

Honarmand, J., and Amini, A., 2012. Diagenetic processes and reservoir properties in the ooid grainstones of the Asmari Formation, Cheshmeh Khush Oil Field, SW Iran. *Journal of Petroleum Science and Engineering*, *81*, 70–79.

Husinec, A., and Sokač, B., 2006. Early Cretaceous benthic associations (foraminifera and calcareous algae) of a shallow tropical-water platform environment (Mljet Island, southern Croatia). Cretaceous Research, 27, 418–441.

Immenhauser, A., Hillgärtner, H., Van Bentum, E., 2005. Microbial-foraminiferal episodes in the Early Aptian of the southern Tethyan margin: ecological significance and possible relation to oceanic anoxic event 1a. *Sedimentology*, *52*, 77–99.

Iryu, Y., and Yamada, T., 1999. Biogeochemical contrasts between mid–Cretaceous carbonate platforms and Cenozoic reefs. *The Island Arc*, *8*, 475–490.

James, N. P., and Jones, B., 2015. Origin of Carbonate Sedimentary Rocks. John Wiley & Sons, Ltd, Chinchester, 464 pp.

James, N. P., Ginsburg, R. N., Marszalek, D. S., Choquette, P. W., 1976. Facies and fabric specificity of early subsea cements in shallow Belize (British Honduras) reefs. *Journal of Sedimentary Research*, *46*, 523–544.

Jenkyns, H. C., 2010. Geochemistry of oceanic anoxic events. *Geochemistry, Geophysics, Geosystems*, *11*(3), Q03004.

Jenkyns, H. C., Matthews, A., Tsikos, H., Erel, Y., 2007. Nitrate reduction, sulfate reduction, and sedimentary iron isotope evolution during the Cenomanian-Turonian oceanic anoxic event. *Paleoceanography*, *22*, PA3208.

Kabanov, P. B., 2009. Benthic carbonate facies of the Phanerozoic: review and example from the Carboniferous of the Russian Platform. *Stratigraphy and Geological Correlation*, *17*, 493–509.

Kashiyama, Y., Ogawa, N. O., Kuroda, J., Shiro, M., Nomoto, S., Tada, R., Kitazato, H., Ohkouchi, N., 2008. Diazotrophic cyanobacteria as the major photoautotrophs during mid–Cretaceous oceanic anoxic events: Nitrogen and carbon isotopic evidence from sedimentary porphyrin, *Organic Geochemistry*, *39*, 532–549.

Kazmierczak, J., and Kempe, S., 1992. Recent cyanobacterial counterparts of Paleozoic *Wetheredella* and related problematic fossils. *Palaios*, *7*, 294–304.

Kazmierczak, J., Coleman, M. L., Gruszczynski, M., Kempe, S., 1996. Cyanobacterial key to the genesis of micritic and peloidal limestones in ancient seas. *Acta Palaeontologica Polonica*, *41*, 319–338.

Kennard, J. M., and James, N. P., 1986. Thrombolites and stromatolites: two distinct types of microbial structures. *Palaios*, 492–503.

Keim, C. N., dos Santos, H. N., Santiago, C. S., Pennafirme, S., Neumann, R., Schnellrath, J., Lima, I., Crapez, M. A. C., Farina, M., 2020. Microstructure and mineral composition of Holocene stromatolites from Lagoa Vermelha, a hypersaline lagoon in Brazil: Insights into laminae genesis. *Journal of Sedimentary Research*, *90*, 887–905.

Kendall, C. G., and Alsharhan, A. S., 2011. Coastal Holocene carbonates of Abu Dhabi, UAE: depositional setting, geomorphology, and role of cyanobacteria in micritization. *Quaternary carbonate and evaporite sedimentary facies and their ancient analogues. International Association of Sedimentologists Special Publication*, *43*, 205–220.

Kiessling, W., Flügel, E., Golonka, J. A. N., 2003. Patterns of Phanerozoic carbonate platform sedimentation. *Lethaia*, *36*, 195–225.

Kietzmann, D. A., Blau, J., Fernández, D. E., Palma, R. M., 2009. Crustacean microcoprolites from the Upper Jurassic—Lower Cretaceous of the Neuquén Basin, Argentina: systematics and biostratigraphic implications. *Acta Palaeontologica Polonica*, *55*, 277–284.

Kobluk, D. R., and Risk, M. J., 1977. Micritization and carbonate-grain binding by endolithic algae. *American Association of Petroleum Geologists Bulletin*, *61*, 1069–1082.

Konhauser, K. O., 2007. Introduction to Geomicrobiology. Blackwell Publishing. 433 pp.

Kuypers, M. M., Blokker, P., Erbacher, J., Kinkel, H., Pancost, R. D., Schouten, S., Sinninghe Damsté, J. S., 2001. Massive expansion of marine archaea during a mid–Cretaceous oceanic anoxic event. *Science*, *293*, 92–95.

Kuypers, M. M., Blokker, P., Hopmans, E. C., Kinkel, H., Pancost, R. D., Schouten, S., Sinninnghe Damsté, J. S., 2002. Archaeal remains dominate marine organic matter from the early Albian oceanic anoxic event 1b. *Palaeogeography, Palaeoclimatology, Palaeoecology*, *185*, 211–234.

Kuypers, M. M. M., van Breugel, Y., Schouten, S., Erba, E., Sinninghe Damsté, J. S., 2004. N_2–fixing cyanobacteria supplied nutrient N for Cretaceous oceanic anoxic events. *Geology*, *32*, 853–856.

Lasemi, Z., and Sandberg, P. A., 1984. Transformation of aragonite-dominated lime muds to microcrystalline limestones. *Geology*, *12*, 420–423.

Lasemi, Z., and Sandberg, P. A., 1993. Microfabric and compositional clues to dominant mud mineralogy of micritic precursors. In: Rezak, R, Lavoie, D.L. (Eds.), Carbonate Microfabrics. Springer Verlag, Berlin, 173–185.

Land, L. S., Behrens, E. W., Frishman, S. A., 1979. The ooids of Baffin Bay, Texas. *Journal of Sedimentary Research*, *49*, 1269–1277.

Laycock, D. P., Pedersen, P. K., Montgomery, B. C., Spencer, R. J., 2017. Identification, characterization, and statistical analysis of mudstone aggregate clasts, Cretaceous Carlile Formation, Central Alberta, Canada. *Marine and Petroleum Geology*, *84*, 49–63.

Lehrmann, D. J., Minzoni, M., Li, X., Yu, M., Payne, J. L., Kelley, B. M., Schaal, E. K., Enos, P., 2012. Lower Triassic oolites of the Nanpanjiang Basin, south China: Facies architecture, giant ooids, and diagenesis—Implications for hydrocarbon reservoirs. *American Association of Petroleum Geologists Bulletin*, *96*, 1389–1414.

Liborius-Parada, A., 2019. Applied petroleum source rock evaluation and high-resolution sequence stratigraphy for unconventional reservoirs in La Luna Formation (Cenomanian-Santonian) Northwest Maracaibo Basin, Venezuela. Doctoral Dissertation. University of Oklahoma, 345 pp.

Longman, M. W., Luneau, B. A., Landon, S. M., 1998. Nature and distribution of Niobrara lithologies in the Cretaceous Western Interior Seaway of the Rocky Mountain region: The Mountain Geologist, 35, pp. 137–170

Lopez-Gamundi, C., Barnes, B. B., Betzler, C., Harris, P. M., Oehlert, A. M., Eberli, G. P., Purkis, S. J., 2025. The sediment budget of Great Bahama Bank—Earth's largest modern carbonate platform. *Geology*, *53*, 748–752.

Marshall, J. F., and Davies, P. J., 1975. High-magnesium calcite ooids from the Great Barrier Reef. *Journal of Sedimentary Research*, *45*, 285–291.

McKee, E. D., and Gutschick, R. E., 1969. History of the Redwall Limestone of Northern Arizona. *Geological Society of America Memoir 114*, 726 pp.

Michel, J., Laugié, M., Pohl, A., Lanteaume, C., Masse, J. P., Donnadieu, Y., Borgomano, J., 2019. Marine carbonate factories: a global model of carbonate platform distribution. *International Journal of Earth Sciences*, *108*, 1773–1792.

Milliman, J. D., and Barretto, H. T., 1975. Relict magnesian calcite oolite and subsidence of the Amazon shelf. *Sedimentology*, *22*, 137–145.

Neuweiler, F., Gautret, P., Thiel, V., Lange, R., Michaelis, W., Reitner, J., 1999. Petrology of Lower Cretaceous carbonate mud mounds (Albian, N. Spain): insights into organomineralic deposits of the geological record. *Sedimentology*, *46*, 837–859.

Núñez-Useche, F., Barragán, R., Torres-Martínez, M. A., López-Zúñiga, P. A., Moreno-Bedmar, J. A., Chávez-Cabello, G., Canet, C., Chacon-Baca, E., 2020. Response of the western proto-North Atlantic margin to the early Aptian Oceanic Anoxic Event (OAE) 1a: an example from the Cupido platform margin-Gulf of Mexico, NE Mexico, Cretaceous Research, 113, 104488.

Nutman, A. P., Bennett, V. C., Friend, C. R., Van Kranendonk, M. J., Chivas, A. R., 2016. Rapid emergence of life shown by discovery of 3,700-million-year-old microbial structures. *Nature*, *537*, 535–538.

Ohkouchi, N., Kashiyama, Y., Kuroda, J., Ogawa, N. O., Kitazato, H., 2006. The importance of diazotrophic cyanobacteria as primary producers during Cretaceous Oceanic Anoxic Event 2. *Biogeosciences*, *3*, 467–478.

Pacton, M., and Gorin, G. E., 2014. Role of microorganisms in oceanic anoxic events (OAEs). *Geology Today*, *30*, 215–221.

Pederson, C. L., McNeill, D. F., Klaus, J. S., Swart, P. K., 2015. Deposition and diagenesis of marine oncoids: implications for development of carbonate porosity. *Journal of Sedimentary Research*, *85*, 1323–1333.

Perry, R. S., Mcloughlin, N., Lynne, B. Y., Sephton, M. A., Oliver, J. D., Perry, C. C., Campbell, K., Engel, M. H., Farmer, J. D., Brasier, M. D., Staley, J. T., 2007. Defining biominerals and organominerals: direct and indirect indicators of life. *Sedimentary Geology*, *201*, 157–179.

Peryt, T. M., 1983a. Classification of coated grains. In: Peryt, T.M. (Ed.), Coated Grains. Springer, Berlin, 3–6.

Peryt, T. M., 1983b. Oncoids: comment to recent developments. In: Peryt, T.M. (Ed.), Coated Grains. Springer, Berlin, 273–275.

Petrash, D. A., Gingras, M. K., Lalonde, S. V., Orange, F., Pecoits, E., Konhauser, K. O., 2012. Dynamic controls on accretion and lithification of modern gypsum-dominated thrombolites, Los Roques, Venezuela. *Sedimentary Geology*, *245*, 29–47.

Pomar, L., and Hallock, P., 2008. Carbonate factories: a conundrum in sedimentary geology. *Earth-Science Reviews*, *87*, 134–169.

Pomar, L., Brandano, M., Westphal, H., 2004. Environmental factors influencing skeletal grain sediment associations: a critical review of Miocene examples from the western Mediterranean. *Sedimentology*, *51*, 627–651.

Pomar, L., Esteban, M., Martinez, W., Espino, D., Castillo de Ott, V., Benkovics, L., Castro Leyva, T., 2015. Oligocene-Miocene carbonates of the Perla field, offshore Venezuela: Depositional model and facies architecture, In: C. Bartolini and P. Mann, eds., Petroleum geology and potential of the Colombian Caribbean Margin: *American Association of Petroleum Geology*, *Memoir 108*, 647–674.

Pomar, L., Hallock, P., Mateu-Vicens, G., Baceta, J. I., 2022. Why do bio-carbonates exist?. *Journal of Marine Science and Engineering*, *10*, 1648.

Purdy, E. G., and Gischler, E., 2003. The Belize margin revisited: 1. Holocene marine facies. *International Journal of Earth Sciences*, *92*, 532–551.

Qiao, Z., Janson, X., Shen, A., Zheng, J., Zeng, H., Wang, X., 2016. Lithofacies, architecture, and reservoir heterogeneity of tidal-dominated platform marginal oolitic shoal: An analogue of oolitic reservoirs of Lower Triassic Feixianguan Formation, Sichuan Basin, SW China. *Marine and Petroleum Geology*, *76*, 290–309.

Reid, R. P., and Macintyre, I. G., 1998. Carbonate recrystallization in shallow marine environments: a widespread diagenetic process forming micritized grains. *Journal of Sedimentary Research*, *68*, 928–946.

Reid, R. P., and Macintyre, I. G., 2000. Microboring versus recrystallization: further insight into the micritization process. *Journal of Sedimentary Research*, *70*, 24–28.

Reid, R. P., Macintyre, I. G., Post, J. E., 1992. Micritized skeletal grains in northern Belize Lagoon; a major source of Mg-calcite mud. *Journal of Sedimentary Research*, *62*, 145–156.

Reid, R. P., Dupraz, C., Visscher, P. T., Decho, A. W., Sumner, D. Y., 2003. Microbial processes forming modern marine stromatolites: microbe–mineral interactions with a three-billion-year rock record. In: Krumbein, W.E., Paterson, D.M., Zavarzin, G.A. (Eds.), Fossil and Recent Biofilms — a Natural History of Life on Earth. Kluwer Academic Publishers, pp. 103–118.

Riaz, M., Banerjee, S., Latif, K., Liu, Z., Ahmed, M. S., Fathy, D., 2023. Understanding the origin of ancient carbonate ooids: recent findings. *International Geology Review*, *66*, 927–946.

Richter, D. K., 1983, Calcareous ooids: a synopsis, In: Peryt, T.M. (Ed.), Coated Grains. Springer, Berlin, 71–99.

Riding, R., 1983. Cyanoliths (Cyanoids). Oncoids formed by calcified cyanophytes. In: Peryt, T.M. (Ed.), Coated Grains. Springer Verlag Berlin, Heidelberg, pp. 276–283.

Riding, R., 1991. Classification of microbial carbonates. In: Riding, R. (Ed.), Calcareous Algae and Stromatolites. Springer, New York, 21–51.

Riding, R., 1992. Temporal variation in calcification in marine cyanobacteria. *Journal of the Geological Society*, *149*, 979–989.

Riding, R., 2000. Microbial carbonates: the geological record of calcified bacterial–algal mats and biofilms. *Sedimentology*, *47*, 179–214.

Riding, R., 2002. Structure and composition of organic reefs and carbonate mud mounds: concepts and categories. *Earth-Science Reviews*, *58*, 163–231.

Riding, R., 2011. Microbialites, stromatolites and thrombolites. In: Reitner J, Thiel V (eds.) Encyclopedia of geology. Encyclopedia of earth science series. Springer, Heidelberg, pp. 635–654.

Riding, R., and Virgone, A., 2020. Hybrid carbonates: in situ abiotic, microbial and skeletal co–precipitates. *Earth-Science Reviews*, *208*, 103300.

Riding, R., Liang, L., Lee, J. H., Virgone, A., 2019. Influence of dissolved oxygen on secular patterns of marine microbial carbonate abundance during the past 490 Myr. *Palaeogeography, Palaeoclimatology, Palaeoecology*, *514*, 135–143.

Robbins, L. L., and Blackwelder, P. L., 1992. Biochemical and ultrastructural evidence for the origin of whitings: A biologically induced calcium carbonate precipitation mechanism. *Geology*, *20*, 464–468.

Robbins, L. L., Tao, Y., Evans, C. A., 1997. Temporal and spatial distribution of whitings on Great Bahama Bank and a new lime mud budget. Geology, 25, 947–950.

Robin, N., Bernard, S., Miot, J., Black-Valleron, M.M., Charbonnier, S., Petit, G., 2015. Calcification and Diagenesis of Bacterial Colonies. Minerals, 5, 488–506.

Samankassou, E., Tresch, J., Strasser, A., 2005. Origin of peloids in early Cretaceous deposits, Dorset, South England. *Facies*, *51*, 264–274.

Schaefer, M. O., Gutzmer, J., Beukes, N. J., 2001. Late Paleoproterozoic Mn–rich oncoids: Earliest evidence for microbially mediated Mn precipitation. *Geology*, *29*, 835–838.

Schlager, W., 2000. Sedimentation rates and growth potential of tropical, cool-water and mud–mound carbonate systems. *Geological Society, London, Special Publications*, *178*, 217–227.

Schlager, W., 2003. Benthic carbonate factories of the Phanerozoic. *International Journal of Earth Sciences*, *92*, 445–464.

Schlager, W., 2005. Carbonate Sedimentology and Sequence Stratigraphy. SEPM Concepts in Sedimentology and Paleontology #8. Society of Economic Paleontologists and Mineralogists, 200 pp.

Scholle, P. A., and Ulmer-Scholle, D., 2003. A Color Guide to the Petrography of Carbonate Rocks: Grains, Textures, Porosity, Diagenesis. *American Association of Petroleum Geologists*, *Memoir 77*, 474 pp.

Seifan, M., and Berenjian, A., 2019. Microbially induced calcium carbonate precipitation: a widespread phenomenon in the biological world. *Applied Microbiology and Biotechnology*, *103*, 4693–4708.

Sequero, C., Aurell, M., Bádenas, B., 2020. Oncoid distribution in the shallow domains of a Kimmeridgian carbonate ramp (Late Jurassic, NE Spain). *Sedimentary Geology*, *398*, 105585.

Spadafora, A., Perri, E., McKenzie, J. A., Vasconcelos, C., 2010. Microbial biomineralization processes forming modern Ca: Mg carbonate stromatolites. *Sedimentology*, *57*, 27–40.

Stal, L. J., 2012. Cyanobacterial Mats and Stromatolites. In Whitton, B. A. (Ed.), The Ecology of Cyanobacteria. Springer, 65–125.

Takayanagi, H., Iryu, Y., Oda, M., Sato, T., Chiyonobu, S., Nishimura, A., Nakazawa, T., Ishikawa, T., Nagaishi, K., 2012. Temporal changes in biotic and abiotic composition of shallow-water carbonates on submerged seamounts in the northwestern Pacific Ocean and their controlling factors. *Geodiversitas*, *34*, 189–217.

Tišljar, J., Vlahović, I., Velić, I., Sokač, B., 2002. Carbonate platform megafacies of the Jurassic and Cretaceous deposits of the Karst Dinarides. *Geologia Croatica*, *55*, 139–170.

Tribovillard, N. P., 1998. Cyanobacterially generated peloids in laminated, organic-matter rich, limestones: an unobtrusive presence. *Terra Nova*, *10*, 126–130.

Tribovillard, N., Trentesaux, A., Trichet, J., Defarge, C., 2000. A Jurassic counterpart for modern kopara of the Pacific atolls: lagoonal, organic matter-rich, laminated carbonate of Orbagnoux (Jura Mountains, France). *Palaeogeography, Palaeoclimatology, Palaeoecology*, *156*, 277–288.

Truskowski, I., 2006. Calibración bioestratigráfica y correlación con datos geoquímicos de la Formación Querecual en su localidad tipo. Specialization Dissertation. Central University of Venezuela. 106 pp.

Tucker, M. E., and Wright, V. P., 1990. Carbonate Sedimentology. Blackwell Publishing. 482 pp.

Varkouhi, S., and Jaques-Ribeiro, L. M., 2021. Bimineralic Middle Triassic ooids from Hydra Island: Diagenetic pathways and implications for ancient seawater geochemistry. *The Depositional Record*, *7*, 344–369.

Védrine, S., Strasser, A., Hug, W., 2007. Oncoid growth and distribution controlled by sea–level fluctuations and climate (Late Oxfordian, Swiss Jura Mountains). *Facies*, *53*, 535–552.

Visscher, P. T., Reid, R. P., Bebout, B. M., Hoeft, S. E., Macintyre, I. G., Thompson, J. A., 1998. Formation of lithified micritic laminae in modern marine stromatolites (Bahamas); the role of sulfur cycling. *American Mineralogist*, *83*, 1482–1493.

Wanless, H. R., and Tagett, M. G., 1989. Origin, growth and evolution of carbonate mudbanks in Florida Bay. *Bulletin of Marine Science*, *44*, 454–489.

Wilkinson, B. H., and Given, R. K., 1986. Secular variation in abiotic marine carbonates: constraints on Phanerozoic atmospheric carbon dioxide contents and oceanic Mg/Ca ratios. *The Journal of Geology*, *94*, 321–333.

Diagenesis

Diagenesis refers to all physical and chemical changes in unconsolidated sediments and lithified sedimentary rocks following deposition and preceding incipient metamorphism at elevated temperatures and pressures (see James and Jones 2015). For all carbonate allochems (grains) considered in this book, processes preceding final deposition (disaggregation/disarticulation, fragmentation, fracturing, and abrasion) were discussed in preceding chapters 5-15. Carbonates are reactive minerals that undergo mineralogical, physical, and/or chemical transformations during different stages of diagenesis. Three major diagenetic environments are distinguished (e.g., Moore 1989; Tucker and Wright 1990; Fig. 16.1): the marine, meteoric, and burial (or subsurface) environments.

The timing of diagenetic processes can be divided into three stages (or zones) (Choquette and Pray 1970). *Eogenetic* (subaerial and near-surface)refers to diagenesis in the time interval between final deposition and burial below the zone of near-surface alteration processes. It involves diagenetic processes in five different zones: the meteoric-vadose, meteoric-phreatic, mixing, marine-vadose, and marine-phreatic (Fig. 16.2), encompassing the marine, meteoric, and shallow-

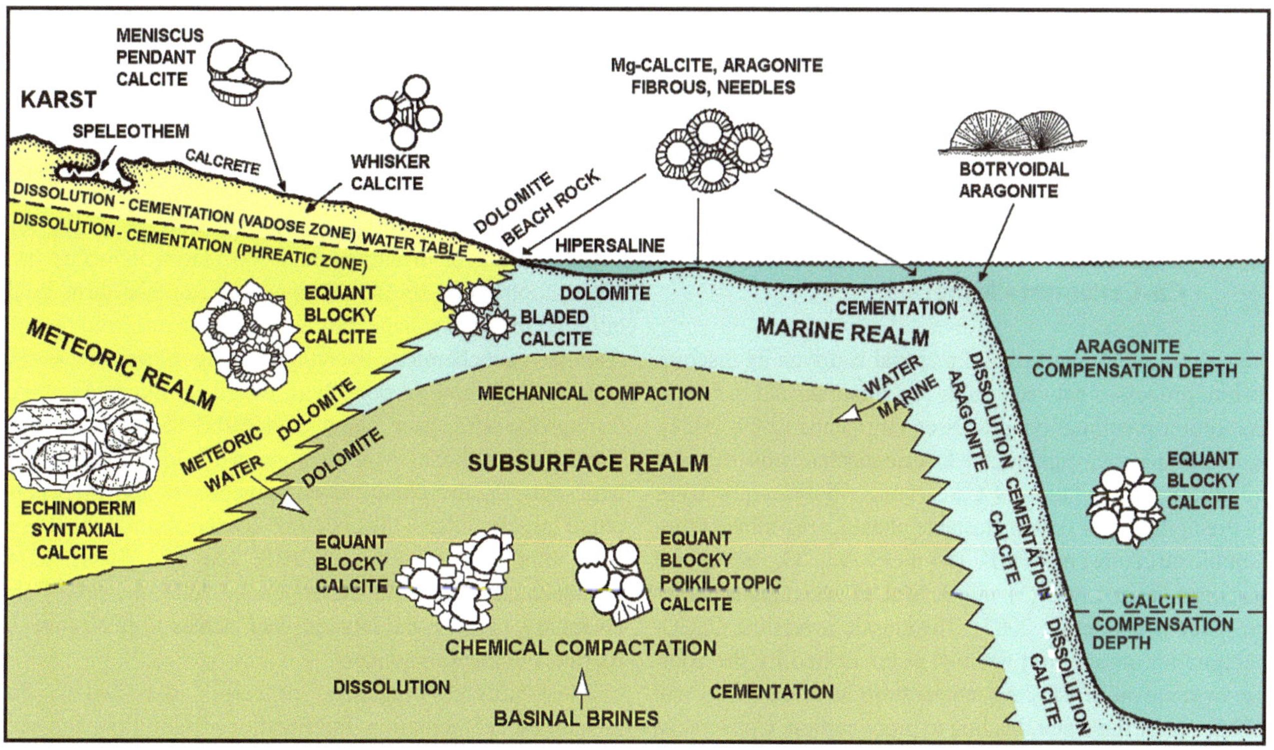

Fig. 16.1 Schematic diagram illustrating the major diagenetic environments. Active diagenetic processes and common cement types are indicated for each of the hydrogeochemically defined realms..Figure modified from Moore (1989)

J. A. Méndez Dot, J. Méndez Baamonde, *Atlas of Tropical Carbonates*, https://doi.org/10.1007/978-3-031-25494-9_16

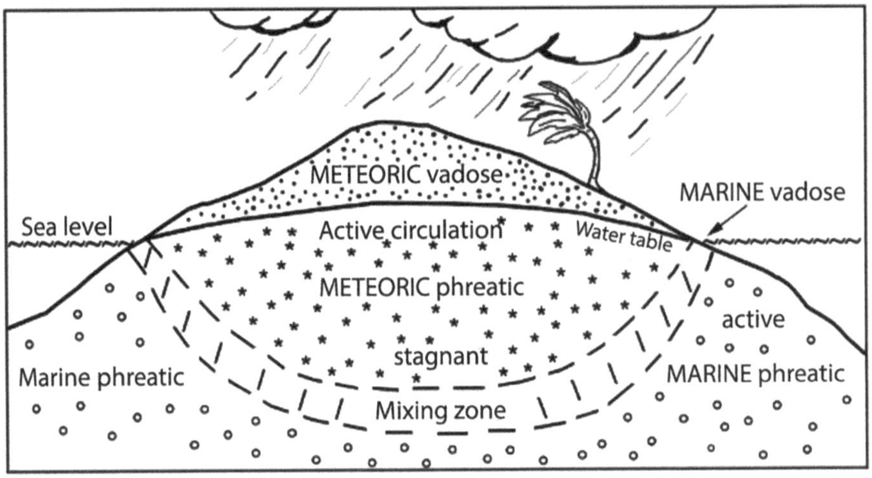

Fig. 16.2 Cross section showing distribution and relation of major diagenetic environments in the shallow subsurface of an ideal permeable carbonate sand island. Original figure modified from Longman (1980). Figure taken from Flügel (2010)

burial environments (the focus of this chapter). *Mesogenetic* (burial) refers to the time period when sediments are buried below the zone where near-surface processes control diagenesis. It involves diagenetic processes in the intermediate- and deep-burial environments. *Telogenetic* refers to the time period when long-buried rocks are brought back to the surface and come under the influence of near-surface processes once again (definitions taken from James and Jones 2015).

Diagenetic studies through thin sections (petrography) focus not only on determining diagenetic environments, processes, and features (or products), but also on establishing paragenetic relations (i.e., the timing of those processes and features). Therefore, determining paragenetic relations (or sequences) is key to understanding the evolution of carbonates, especially in ancient facies, and is one of the main goals of this chapter.

16.1 Dissolution-Precipitation of Marine Ca-Carbonate Minerals

Diagenesis involving fluxes of material is driven by disequilibrium processes between fluids and minerals and results in dissolution-precipitation processes (Immenhauser e 2021). In the marine environment, the kinetic mechanisms that control abiotic precipitation of Ca-carbonate polymorphs favor the precipitation of both metastable phases, aragonite (orthorhombic structure) and HMC (>4 mol% $MgCO_3$; hexagonal structure), over LMC (˂4 mol% $MgCO_3$; hexagonal structure) (see Morse et al. 2007). The classic reversible $CaCO_3$ precipitation-dissolution reaction is controlled by the addition or removal of CO_2. As temperature increases, CO_2 will bubble out of the water, leading to precipitation. Conversely, cooling allows the water to hold more CO_2 and mediates carbonate dissolution by a process known as retrograde solubility. Because of crystal kinetics, dissolution is generally easier than precipitation (see James and Jones 2015).

In solutions supersaturated with respect to both aragonite and calcite (the marine-phreatic environment), the sequence of diagenetic alteration for $CaCO_3$ polymorphs is HMC > aragonite > LMC. In solutions undersaturated with respect to calcite (typical of the meteoric vadose zone), aragonite dissolves more rapidly than HMC. In solutions supersaturated with respect to calcite but still undersaturated with respect to aragonite, mineralogic controls become more significant and HMC containing ≤12 mol% $MgCO_3$ dissolves at rates similar to or greater than those for aragonite (see Flügel 2010). This behavior reflects the fact that the solubility of HMC increases with increasing Mg^{2+} content. Aragonite has about the same solubility in distilled water as HMC with roughly 12 mol % $MgCO_3$. Dolomite is the least soluble of all carbonates (James and Jones, 2015).

16.2 Fluid-Buffered Diagenesis

Diagenesis in the eogenetic zone (marine, meteoric, and shallow-burial environments) is considered to occur in a "fluid controlled" (buffered) open system. The term fluid-buffered refers to a diagenetic scenario characterized by bulk (isotope) geochemical exchange taking place between a significant volume of pore fluid, which continuously passes through the pore space, and a given carbonate host rock in an "open system" behavior. Therefore, in a fluid-buffered diagenetic system, the chemical composition of the diagenetic phase largely reflects the chemical composition of the diagenetic fluid (Higgins et al. 2018; see also Swart 2015; Immenhauser 2021). In the eogenetic zone, the origin and chemistry of pore-fluid types vary across a full range of marine and meteoric waters.

Fluid-buffered diagenetic processes discussed in this chapter are: cementation, micritization, dissolution, and neomorphism (mineralogic inversion, replacement and recrystallization). It should be noted that diagenetic processes in the telogenetic zone are not addressed in this chapter. Diagenesis in the burial (intermediate and deep) environment (a rock-buffered "closed system") will be discussed in Chap. 17.

16.3 Cementation

Abiotic precipitation of carbonate cements may occur in marine, meteoric, and burial environments. From a given fluid (marine, meteoric, or burial water), precipitation of cement, filling primary or secondary pores, as a result of saturation changes in the absence of biological activity, is driven by processes/factors that increase CO_3^{2-} concentrations and induce $CaCO_3$ supersaturation, such as increased temperature and salinity, elevated alkalinity, and water mass mixing (see Fantle et al. 2020). Depending on the diagenetic environment, the composition of pore fluids, and related abiotic-controlling factors (Mg^{2+}/Ca^{2+} ratio, salinity, temperature, primary porosity, size and geometry of pores, water energy, bathymetry, crystals kinetics, mineralogy of the substrates, among others), supersaturation with respect to carbonate minerals may be followed by the precipitation of aragonite, HMC, LMC, or dolomite.

Cement types and cement fabrics are classified according to the morphology and arrangement of crystals. In this context, descriptions are based on the shapes of crystals and subcrystals (e.g., needle-like, microcrystalline, columnar), sizes (from ~5 μm to up to 250 μm; changes in size toward the pore center), width/length ratios, forms (e.g., anhedral, subhedral, syntaxial overgrowth), terminations (e.g., pointed, pyramid-like, acute, flattened), appearance (clear, cloudy), twinning (when present), extinction types (straight, undulose, sweeping), and growth relations with respect to the substrate (e.g., normal, oblique, without preferred orientation). More detailed treatment of carbonate cementation may be found in: Friedman 1964; Bathurst 1975; Longman 1980; Moore 1989, 2001; Tucker and Wright 1990; Sellwood 1994; Morse et al. 2007; Flügel 2010; Fantle et al. 2020; Immenhauser 2021). Biologically-induced cementation (i.e., microbial cementation; Hillgärtner et al. 2001; Diaz and Eberli, 2022) via precipitation of automicrite (precipitated *in situ*) does occur, but to a lesser volumetric extent; furthermore, it is mostly restricted to marine diagenesis in shallow subtidal and intertidal environments (mostly tropical) (see Sect. 16.4.2).

16.4 Marine Diagenesis

Most marine diagenetic processes (at or below the seafloor) occur in the marine-phreatic environment. In this environment, processes occur at ambient temperatures (controlled by bathymetry and latitude), with normal marine water (seawater) or evaporated seawater as the governing pore-fluid type. Normal seawater has an average salinity of 35‰ (ranging from 33‰ to 37‰). Its most abundant ionic components, in decreasing order, are Cl^-, Na^+, SO_4^{2-}, Mg^{2+}, HCO_3^-, Ca^{2+}, and K^+. The Mg:Ca ratio is ~3:1 (e.g., James and Jones 2015). Surface-ocean pH typically ranges between 8.1 and 8.4. Locally, seawater pH can drop down to 7.5 (see Millero et al. 2008; Immenhauser 2021; see also Zeebe 2012). In carbonate-evaporitic settings, with increasing temperature, the pH of evaporated seawater may rise above 9, with inorganic carbon mainly found as CO_3^{2-} (see Petrash et al. 2012; James and Jones 2015). The marine-phreatic environment encompasses settings from shallow subtidal to the deep-sea (Fig. 16.1). Seawater is supersaturated with respect to $CaCO_3$ in shallow seas and undersaturated in cold and deep seas. In general, marine-phreatic diagenesis is commonly less complicated and shorter in duration compared to diagenesis in both meteoric and burial realms (see Flügel 2010; Fantle et al. 2020).

In coastal areas and oceanic carbonate complexes, and associated peritidal environments, marine diagenesis also includes processes occurring in the marine-vadose zone (at the land-sea boundary) and the mixing zone (the interface between seawater and meteoric water; Fig. 16.2). This chemically active zone is characterized by a mixture of different solutions containing $CaCO_3$-saturated seawater and CO_2-rich meteoric water. The mixing zone exhibits marine and meteoric criteria, therefore it encompasses two different settings: (**a**) intertidal and shallow subtidal (land-sea) settings (e.g., upper shoreface and surf zone, mudflats and hypersaline lagoons), where the chemistry of fluids commonly leans toward $CaCO_3$ precipitation (although localized dissolution may occur; e.g., the creation of fenestral pores), and where beachrocks, evaporitic crusts, and shallow subtidal-to-intertidal stromatolites are common carbonate deposits (see Mauz et al. 2015; James and Jones 2015; Warren 2016); (**b**) subsurface near-coastal settings, in which the varied chemistry of fluids may lead to carbonate dissolution, cementation, neomorphism, and dolomitization (this setting and associated processes will be discussed in the section focused on meteoric diagenesis).

The main diagenetic processes involving Ca-carbonate minerals in the marine environment (phreatic and vadose) are the abiotic (inorganic) precipitation of cements (mainly aragonite and HMC in composition) and the precipitation of microbially-induced micrite. See Tucker and Wright (1990), Flügel (2010), Ahm et al. (2018), Higgins et al. (2018), for detailed treatment of marine diagenesis. Dolomitization (or the formation of Ca-Mg carbonate), which is one of the most significant diagenetic process initiated in the marine environment (an in many cases continued during later diagenetic stages), will be discussed in Chap. 18. Deep-sea marine diagenesis, or the processes related to waters undersaturated with respect to $CaCO_3$ and dissolution of carbonate skeletal components below the aragonite and calcite compensation depths are not addressed in this book.

16.4.1 Marine Cementation and Cements

Syndepositional marine cementation is a warm-water phenomenon (not common in cold-water settings; see James and Jones 2015) and occurs only in the eogenetic zone (e.g., Mazzullo, 2004). In shallow subtidal environments where waters are supersaturated with respect to $CaCO_3$ (especially at tropical latitudes), extensive marine cementation is often associated with time intervals of low sedimentation rate resulting in the formation of hardground surfaces (e.g., Swart 2015). The development of such surfaces can be accompanied by authigenic precipitation of glauconite and/or Ca-phosphate. In the case of the intertidal zone and peritidal hypersaline environments (where beachrocks and microbialites are common deposits, respectively), the time needed for cementation is less, as evaporation promotes precipitation (see Tucker and Wright 1990). According to Mauz et al. (2015), lithification of intertidal beachrocks and stromatolites is a function of CO_3^{2-} ion concentration in seawater, degassing of CO_2 from seaward-flowing groundwater, and microbial activity. Particularly, in tropical hypersaline perennial (shallow subtidal) lagoons, carbonate cementation is promoted by elevated temperatures and evaporation rates, and microbial activity, which drive increased alkalinity, pH, and salinities. See the case study of the Pirata lagoon in Los Roques Archipelago by Petrash et al. (2012). See also Chap. 3 for photographic examples of beachrocks and microbialites (stromatolites and thrombolites) in some of the Venezuelan Caribbean carbonate complexes.

Marine cements are mainly composed of aragonite and HMC, which are the dominant carbonate minerals in the marine environment (e.g., Lippmann 1973, 1982; Morse et al. 2007). Common aragonite cements types (primarily abiotic precipitates) are acicular (needles), fibrous, pendant (microstalactitic), and botryoidal. The first two are typical of the marine-phreatic zone but also occur in the marine-vadose zone; pendant cement (composed of aragonitic needles and/or fibrous crystals) is common in the marine-vadose zone; while botryoidal cement is typical of the marine-phreatic zone, particularly reef cavities and steep slopes. Common HMC cements types include microcrystalline crusts or envelopes (e.g., micritic grain envelopes and associated grain-to-grain meniscus bridges; see Hillgärtner et al. 2001; Diaz and Eberli, 2022), as well as fibrous to bladed rims. These isopachous fringes of cements (isopachous refers to crusts or rims of equal thickness) are typical of the marine-vadose zone but also occur in the marine-phreatic zone. Furthermore, radiaxial calcite cement, consisting of three different crystal morphologies (fascicular, radiaxial-fibrous, and radial-fibrous crystals), and mineralogy varying from LMC, to IMC and HMC (see Richter et al. 2011), is typical of the marine-phreatic zone but less common when compared to the abundance of above-mentioned aragonitic and HMC cements. Some fibrous calcite crystals are considered to have formed through neomorphism of HMC to LMC or recrystallization of LMC (see Sect. 16.5.2).

In general, marine cements are abundant in Holocene tropical carbonate microfacies, ranging from normal marine to hypersaline waters. However, they are rarely or poorly preserved in many shallow ancient microfacies, as most marine cements composed of metastable mineral phases (i.e., aragonite and HMC) are prone to being affected and overprinted by posterior meteoric and/or burial alterations. See Bathurst (1974), Tucker and Wright (1990), Scholle and Ulmer-Scholle (2003), and Flügel (2010), for a detailed treatment of marine cementation and marine cement types.

16.4.2 Micritization and Microbial Cementation

Diagenesis involving precipitation of micrite on the seafloor can be roughly divided into two processes: micritization and cementation. Micritization is the alteration of the original texture of grains while on the seafloor or just below it by endolithic algae, fungi and bacteria, thus partially or totally obliterating their original textural features (e.g., Bathurst 1966; Tucker and Wright 1990; Garuglieri et al. 2024). Therefore, it is regarded as a destructive (texture-obliterating) process. Destructive micritization involves the interplay of repeated microboring of carbonate grains and the infilling of the tiny voids by cryptocrystalline HMC or aragonite (e.g., Bathurst 1966; Friedmann et al. 1971). Because primary mineralogy controls the susceptibility of carbonate grains to microborings (Perry 1998), aragonitic bioclasts are often more frequently bored and micritized than calcitic bioclasts. In contrast, accretionary or constructive micrite envelopes, leading to microbial cementation, are mainly the result of both microbially-induced (active mineralization) precipitation of $CaCO_3$ by endolithic microbes colonizing the grain surfaces, and, subordinately, of passive mineralization influenced by mucilaginous EPS (Díaz and Eberli, 2022).2010

In modern shallow tropical settings, both micritization and microbial cementation are common processes. Micritization involving microboring and recrystallization (dissolution and re-precipitation) is carried out mainly by endolithic cyanobacteria (e.g., *Solentia* sp.),. Such micrite precipitation, replacing skeletal and non-skeletal grains (e.g., *Halimeda* flakes and ooids, respectively), occurs especially in lagoonal environments (see Reid et al. 1992; Reid and Macintyre 1998, 2000). Micritization (destructive) and microbial cementation (constructive) are associated, coeval processes in some shallow subtidal to intertidal settings where stromatolites and thrombolites grow, with microbial cementation representing a key process for accretion and lithification of such microbial structures (e.g., Visscher et al. 1998; Macintyre et al. 2000; Kendall and Alsharhan 2011). Microbially-induced cementation is also common in many Holocene beachrocks (marine-vadose) and shallow subtidal (marine-phreatic) grain-supported limestones. Grains in these rocks (mainly sand- and fine gravel-sized) are cemented together by micritic envelopes and micritic bridges (mainly HMC in composition; see Hillgärtner et al. 2001; Vousdoukas et al. 2007; Al-Ramadan, 2014; Shen et al. 2017; Haredy et al. 2019; Díaz and Eberli, 2022).

In ancient facies, micritization (destructive, texture obliterating) is a common marine diagenetic process in Cretaceous rocks (e.g., Burchette and Britton 1985; Strasser 1986; Samankassou et al. 2005). In Aptian and Albian microfacies of the Cogollo Group (MB), micritization involving the formation of micritic peloids and aggregate grains (widely referred to in previous chapters) resulted in micritic replacements highly resistant to neomorphism during meteoric diagenesis. The latter suggests a primary LMC composition for the cryptocrystalline crystals associated to the calcite system of the Cretaceous (Calcite II).

16.5 Meteoric Diagenesis

Meteoric diagenesis refers to the subsurface physical and/or chemical alteration of carbonate sediments and rocks in the presence of water of atmospheric origin (surface and/or groundwater) (e.g., Fantle et al. 2020). The latter, commonly referred to as meteoric water or freshwater, is defined as water (originating from ice, snow, rain, rivers or lakes) that has recently interacted with atmospheric and/or soil-derived CO_2. Therefore, its pH is commonly slightly acidic (c. 5.5 to 7). The high CO_2 and low Ca^{2+} concentrations, respectively, cause meteoric water to be corrosive to all carbonate minerals. Once in the vadose zone, freshwater (now groundwater) dissolves carbonate metastable minerals, eventually leading to supersaturation of the fluids with respect to LMC. The precipitation of this stable carbonate phase (depleted in Sr^{2+}) is kinetically and chemically favored by the low Mg^{2+}/Ca^{2+} ratio of meteoric waters (e.g., Moore 2001; Flügel 2010). Meteoric water inflow into sediments filling marine basins can be substantial and can extend tens of kilometers offshore (see Immenhauser 2021).

The main diagenetic processes involving carbonate minerals in the meteoric environment are neomorphism, dissolution, and cementation. Overall, low-temperature, near-surface and subaerial meteoric diagenesis results in the lithification of the strata. Diagenetic reactions and processes activate rapidly and are relatively short in time in the meteoric environment, however these can extend in time up to a few millions of years. This depends on the length of time spent by a given carbonate strata in the meteoric (near-surface) realm, which in turn depends on the tectonic setting, depositional environment, subsidence rate, carbonate production, sea-level variations, among other factors.

16.5.1 Major Meteoric Diagenetic Environments

The meteoric realm is divided in two major diagenetic environments (Longman 1980): the vadose and phreatic zones. The meteoric-vadose (surface) zone is located above the water table (Fig. 16.2), hence it is commonly referred to as the unsaturated zone, as pores are filled with meteoric water and air. Sediments of the vadose zone, partially or totally subaerially exposed, are directly affected by the input of freshwater by continental fluvial systems (in the case of coastal settings), as well as by rainfall. The most common diagenetic processes in this zone are dissolution (or leaching) of metastable grains and cementation.

The meteoric-phreatic (subsurface) zone is located below the water table (Fig. 16.2), where pores are filled (or saturated) with meteoric waters. The interaction between slightly acidic freshwaters and metastable carbonate minerals makes this zone a very active diagenetic environment. Extensive dissolution and formation of molds and vugs are accompanied by mineralogic inversion of aragonite and HMC to LMC, and by recrystallization (changes in crystal sizes, shapes and orientation). Supersaturation postdating dissolution of metastable mineral phases leads to cementation, occlusion of primary and secondary porosity, and lithification of the strata.

The mixing zone in the subsurface is characterized by the mixing of meteoric water (CO_2-rich) and marine water (calcite-saturated). In ancient and modern shallow platforms and ramps, the occurrence of such lenses of mixed waters was/is common in coastal carbonate settings subjected to high influx of meteoric waters from the continental areas. The mixing of meteoric and marine waters may drive dissolution and recrystallization of metastable minerals (Fantle et al. 2020), or the occurrence of dolomitization (e.g., Badiozamani 1973; Warren 2000; James and Jones 2015; Petrash et al. 2021). The vertical location of the mixing zone between meteoric water and seawater also depends to some extent on sea level (e.g., Swart and Kennedy 2012).

Through meteoric water lenses, the meteoric-phreatic environment can extend to hundreds of meters of burial depth (Flügel 2010). Shallow carbonate successions are prone to undergoing diagenesis in the meteoric environment, as they are reached by meteoric waters, especially during sea-level fall episodes that either reduce the thickness of the water column or expose the uppermost platform/ramp sediments. Grain-supported facies belonging to highstand deposits (deposited in very shallow environments) are more prone to be affected by meteoric diagenesis with respect to facies belonging to transgressive muddy deposits (deposited in deeper environments).

16.5.2 Neomorphism

Neomorphism, a term introduced by Folk (1965), relates to transformations of minerals taking place in the presence of water (i.e., it is a pH-driven "wet" process) through dissolution-reprecipitation and includes the following processes: **(a)** *inversion*, the replacement of a mineral by a more stable polymorph (i.e., aragonite or HMC to LMC); **(b)** *replacement*, the dissolution of one mineral and simultaneous formation of another mineral (e.g., dolomitization and silicification of carbonates; the latter process will be discussed in detail in Chap. 23); **(c)** *recrystallization,* the changes in crystal size, shape, orientation and crystal lattice without a change in mineralogy (i.e., calcite-to-calcite); this term is often used rather loosely, since it is not always possible to demonstrate whether or not a mineralogical change has occurred. Most neomorphism in limestones is of the aggrading type, that is leading to a general increase in crystal size, and this occurs chiefly in fine-grained limestones. Crystal enlargement of micrite may create *microspar* (crystal size between about 5 and 30 μm); while continuing diagenesis, from the meteoric environment to the shallow, intermediate- and deep-burial environments, may produce *pseudospar* (or *pseudosparite*; crystal size >30 μm), with crystals measuring tens to even hundreds of microns in diameter (see also Tucker and Wright 1990; Scholle and Ulmer-Scholle 2003; Flügel 2010).

Crystallization fabrics (or mosaics) of recrystallized carbonates can be described according to equant or non-equant crystal size and crystal fabric depending on crystal shape. We use the definitions of the descriptive terms for crystallization textures and fabrics according to Flügel (2010) (after Friedman 1964, 1965). *Crystallization textures* refer to the shape of mineral crystals and the type of crystal faces at crystal boundaries. Descriptive terms are: *anhedral*, characterized by the absence of crystal faces bounding the mineral grains; *subhedral*, characterized by partly developed crystal faces; and *euhedral*, characterized by crystals that are bound by crystal faces. *Crystallization fabrics* refer to the size and mutual relations of crystals that can be differentiated into: *equigranular fabrics*, which consist of crystals of approximately the same size; and *inequigranular fabrics*, which consist of crystals of various sizes. Fabrics (equigranular and inequigranular) are subdivided into: *xenotopic*, predominantly anhedral crystals; *hypidiotopic*, predominantly subhedral crystals; and *idiotopic,* predominantly euhedral crystals. In addition, inequigranular fabrics are subdivided into: *porphyrotopic,* larger crystals are enclosed in a fine-grained matrix; and *poikilotopic*, crystals are of more than one size, with larger crystals enclosing smaller crystals from another mineral. Terminology for recrystallized carbonates is also commonly used for describing the morphology and fabrics (mosaics) of cement types.

16.5.3 Carbonate Dissolution

In the meteoric environment, undersaturation of pore fluids with respect to carbonate leads to dissolution of metastable carbonate grains (skeletal and non-skeletal) and cements, with aragonite and HMC being metastable minerals and hence more soluble than stable LMC. Dissolution commonly leads to: **(a)** creation of secondary porosity (e.g., molds, vugs); **(b)** cementation, as dissolution of aragonite and/or HMC increases saturation of the water with respect to calcite (or dolomite in some cases) to the point where calcite (or dolomite) ultimately precipitates. The increase in $CaCO_3$ solubility is driven by decreased temperature, salinity and pH; increased pressure; diffusion of pCO_2; and aerobic respiration of organic matter. See Moore (2001), Flügel (2010), Cherns et al. (2011), Moore and Wade (2013), Swart (2015), James and Jones (2015), and Fantle et al. 2020; for a detailed treatment of carbonate dissolution.

Regarding diagenetic alteration aragonitic of grains, according to Granier (2012), many papers state that "aragonite recrystallizes into calcite". However, "recrystallization into calcite" or "preservation as aragonite" are phenomena that are rarely the final state of fossil preservation. This author considers the preservation of aragonitic skeletal material (e.g., green algae, aragonitic bivalves, gastropods, and corals) as neomorphized or recrystallized products to be rare, and leans more towards their dissolution and subsequent filling of the molds by LMC cements. However, we take the conservative approach and consider each case study separately. In many cases, we use the phrase "dissolution and reprecipitation" in a loose manner, i.e., without strictly implying neomorphism (inversion of aragonite to LMC) or dissolution followed by cementation.

16.5.4 Meteoric Cementation

In the meteoric environment, oversaturation with respect to $CaCO_3$, commonly following carbonate dissolution, leads to abiotic (inorganic) precipitation of a variety of LMC cements, which are commonly grouped under the name "orthosparite". To avoid ambiguities in the use of some terms, it is worth mentioning that: the term "sparry calcite" is commonly used in reference to a mosaic of crystals, formed either as cement or by neomorphic/recrystallization processes, that is sufficiently coarsely crystalline to appear fairly transparent in thin section (James and Jones 2015); the term "microspar", commonly used in reference to LMC originating from neomorphism/recrystallization, is also used by some authors in reference to LMC microcrystals filling small pores (e.g., Valencia and Laya 2020).

In addition to the action of meteoric waters, the driving factors that increase CO_3^{2-} concentrations and induce $CaCO_3$

supersaturation in the meteoric environment are elevated temperature and salinity, water mass mixing, fluid flow, and elevated alkalinity (e.g., Given and Wilkinson 1987; Fantle et al. 2020). The common LMC cement types formed in the meteoric environment are (the diagenetic environments of formation are indicated): dogtooth (meteoric-phreatic); bladed, blocky, and granular (meteoric-phreatic and meteoric-vadose); pendant and meniscus (mostly meteoric-vadose, but also meteoric-phreatic); syntaxial (meteoric-phreatic); and drusy and whisker (meteoric-vadose). See Flügel (2010) for a detailed treatment of meteoric cementation and meteoric cement types.

16.6 Shallow-Burial Diagenesis

Shallow-burial diagenesis begins within the first few meters to tens of meters below the water-sediment interface (e.g., Tucker and Wright 1990; Flügel 2010). Shallow-burial diagenetic settings (also called shallow subsurface settings or environments) are similar in many respects, including hydrogeochemistry, to near-surface (meteoric) diagenetic settings. The major differences compared to near-surface settings are that deeper shallow-burial settings generally experience cementation and porosity/permeability loss rather than dissolution and porosity/permeability gain (Machel 2005). It should be noted that shallow carbonates can be affected by meteoric waters not only directly beneath unconformities on land, but also for some distance down-dip into the subsurface (shallow-burial) and for some distance in a seaward direction below sea level, depending on the extent of freshwater lenses (Mazzullo, 2004). On the other hand, Immenhauser (2021) considers a shallow (marine) burial domain down to depths of several hundred meters, in which the pore-fluid type is mainly marine formation waters, although meteoric waters can accumulate in aquifers.20052021

Defining the frontier between meteoric-phreatic diagenesis (e.g., prominent in Early Cretaceous facies of the Cogollo Group) and shallow-burial diagenesis (e.g., prominent in Late Oligocene-Early Miocene facies of the Perla Limestone; see Valencia and Laya 2020) using a polarizing microscope is often a difficult task, as diagenetic products (e.g., LMC cement crystals or recrystallized crystals) are, in many cases, very similar. The consideration of the carbonate depositional system (i.e., platform, ramp), the depositional environment of the facies within such systems (i.e., inner, middle, outer), and estimated paleodepths within the neritic zone (upper, middle, lower) are key for interpretations focused on diagenetic stages and the timing of diagenetic products.

Furthermore, defining the frontier between shallow-burial and intermediate-burial environments is also a subtle matter. Grain reorganization and mechanical compaction, which represent the beginning of burial diagenetic processes in the intermediate-burial environment, begins as soon as there is overlying sediment with simple dewatering, rearrangement of particles, and closer grain packing. This occurs at relatively shallow-burial depths (a few tens of meters), where carbonate dissolution and cementation are already hindered by poor water circulation and the onset of mechanical compaction (Tucker and Wright 1990; Flügel, 2010). In this chapter, we follow this view and consider shallow-burial diagenetic processes (cementation and recrystallization) as those occurring prior to the beginning of apparent mechanical compaction, and hence prior to the destruction of porosity. Within this domain, recrystallization (increase in crystal size) and the formation of pseudospar mosaics occur mainly in micrite sediments making up mud-dominated microfacies with poor clay content. Pseudospar mosaics are particularly common in micritic microfacies deposited during maximum flooding episodes. On the other hand, Machel (2005) and Immenhauser (2021) consider the lower limit of shallow-burial diagenesis to be at burial depths from several hundred meters to 1,000 m.

16.6.1 Shallow-Burial Cements

Common LMC cement types formed in the shallow-burial environment are coarse blocky, bladed, and syntaxial, while poikilotopic, dogtooth, granular, and drusy occur but are less common (see Immenhauser 2021). Crystal sizes are commonly clearer and larger in shallow-burial cements. Recrystallization of pre-existing cements to form new crystalline fabrics consisting of larger crystals (mainly of the blocky and granular types) may also occur in this diagenetic environment. In many cases, shallow-burial cements, either formed by abiotic precipitation or by recrystallization of pre-existing cements, exhibit twin lamellae (or twinning of the calcite; see Sect. 17.3.4 for details on this feature).

In general, cementation in the shallow-burial environment does occur in shallow carbonates deposited in the upper neritic (euphotic) zone, but to a lesser extent, as the availability of pore waters (commonly enriched in Fe^{2+} and Mn^{2+}) is hindered due to predating meteoric cementation. In contrast, carbonate successions deposited in deeper settings (the mesophotic and oligophotic zones) are less affected by meteoric diagenesis, as these facies pass directly from the marine to the shallow-burial environment. In these successions, cementation due to the high rock-to-fluid ratio under shallow-burial conditions tends to be a common process.

16.7 Diagenesis: Photographic Set and Descriptive Captions

In all photos, IW = image width (the true length of the horizontal axis of the view)

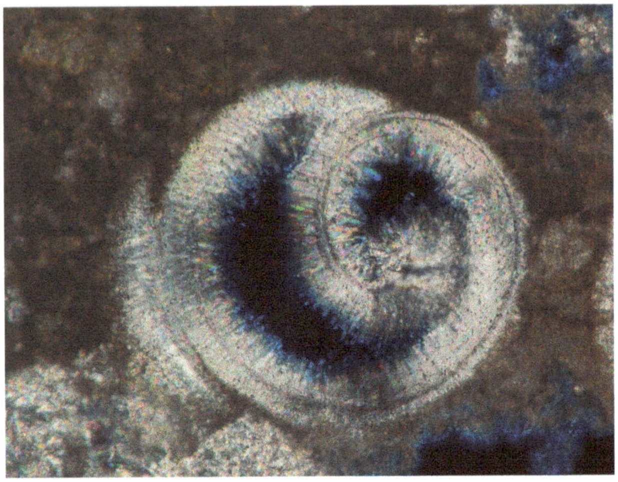

Holocene. Aves Island

Early marine cementation is illustrated in this image by acicular aragonite cement that partially infills a small gastropod. Note the elongate, slender shapes of the needles (i.e., high length/width ratios), which are oriented more or less perpendicularly to the gastropod's inner body wall. In this example, from the Aves Island calcarenite (the unit that forms the carbonate basement beneath the sand sediments of the supratidal island), the aragonite needles are interpreted to have precipitated in a marine-phreatic (shallow subtidal) environment.

IW = 0.6 mm. CPL.

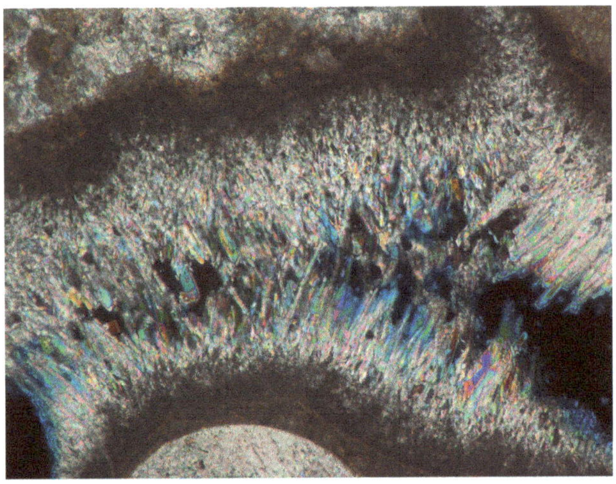

Holocene. Aves Island

A Another example of marine acicular cement. In this case, the needles fill an interparticle pore within a beachrock. As in the previous image, the view under CPL offers a striking visual of the crystals, which represent the most conspicuous type of marine cement in modern tropical carbonate deposits. In this example, the needles are interpreted to have precipitated in the marine-vadose zone (at the land-sea boundary). Notably, they postdate the precipitation of thin micritic rims (dark brown in color) that envelop the grains.

IW = 0.6 mm. CPL.

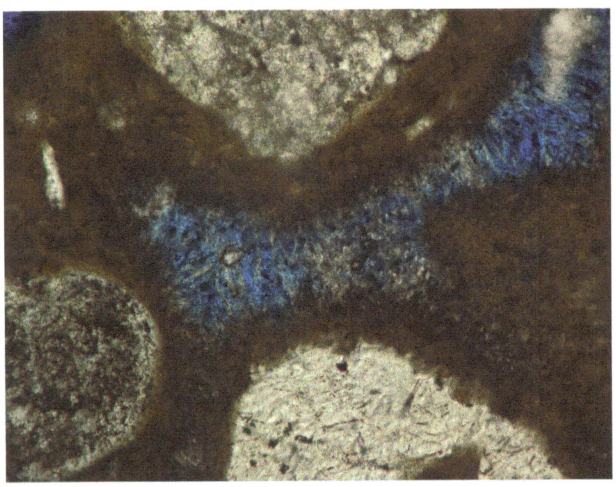

Holocene. Aves Island

This interparticle pore in a beachrock is outlined by micritic envelopes (dark brown rims), which cement the grains together by forming meniscus bridges. These isopachous rims (crusts of uniform thickness) consist of cryptocrystallinee HMC, as indicated by EDS analysis. They result from microbial activity, most likely involving cyanobacteria. This microbially-induced early marine precipitate is responsible for the lithification of Late Pleistocene (basement) and Holocene (beachrocks) carbonate facies on Aves Island.

IW = 0.6 mm. PPL.

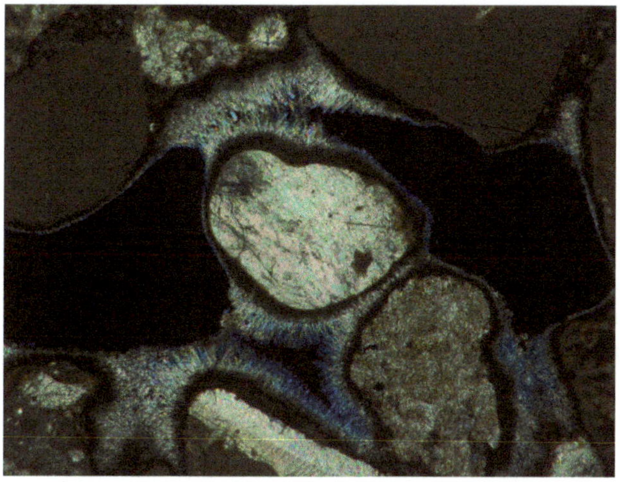

Holocene. Aves Island

This lower-magnification view of a beachrock from Aves Island provides a broader perspective of the grain-supported porous fabric, together with the two types of marine grain-connecting meniscus cements responsible for lithification: the micritic envelopes (dark, microbially-induced rims), which precipitated first and formed meniscus bridges; and acicular aragonite (abiotic), postdating micritic precipitation, which precipitated in pendant, or gravitational, "beard-like" style and formed large bridges through the joining of needle crystals that grew from two opposing grains.

IW = 2.3 mm. CPL

Holocene. Henri Pittier National Park. Aragua State Shoreline

Rounded red algal grains and angular quartz grains are rimmed and cemented together in this porous beachrock by microcrystalline HMC cement. Precipitation of the latter formed isopachous crusts around and between grains, outlining interparticle pores. This cement type is commonly formed in the marine-vadose zone. Although similar to the two previous examples, the microcrystalline texture of the crystals making up the marine cement in this example is clearly differentiated from previous fibrous ones.

IW = 2.4 mm. CPL.

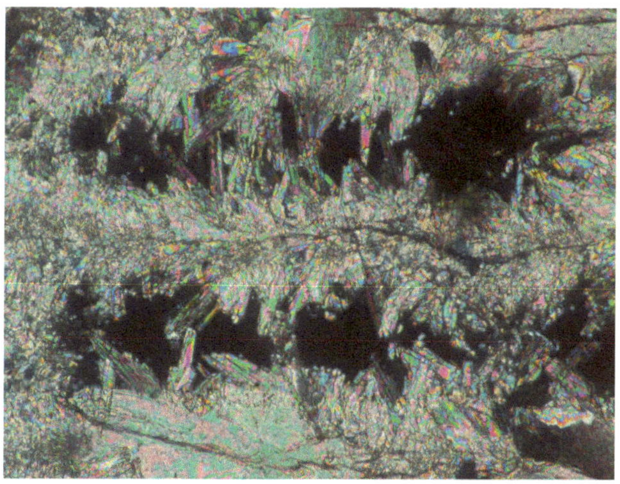

Late Pleistocene. Aves Island

The image depicts fibrous cement partially infilling intraskeletal pores within a scleractinian coral. Note the prismatic shape of the crystals, which are notably thicker than the acicular (needle-like) crystals shown in the previous examples. Their larger sizes allow for the visualization of their polygonal and pointed terminations. Although most likely aragonitic, these marine-phreatic crystals could also be composed of HMC. Scanning Electron Microscopy, coupled with EDS, is necessary to accurately determine their elemental composition, which may provide insights into their mineralogy.

IW = 0.6 mm. CPL

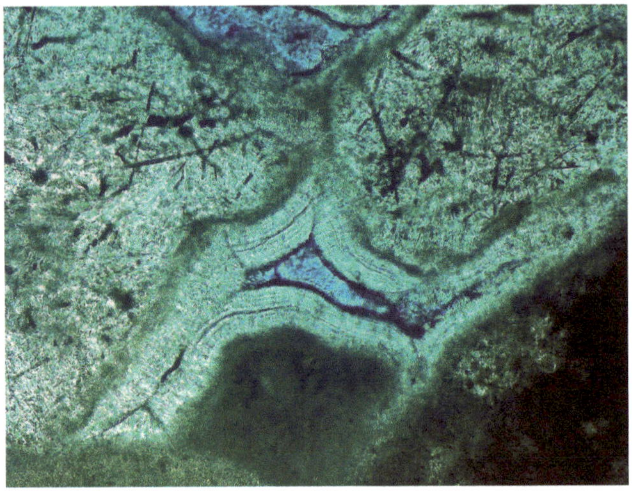

Holocene. Los Roques Archipelago

This and the following image belong to the same thin section, corresponding to a lithified, laminated, intertidal stromatolite. The microbialite structure has formed in the shallower, evaporated waters of a hypersaline lagoon in Pirata cay (located just southeast Gran Roque). Marine microcrystalline HMC crystals in the image are almost completely occluding an interparticle pore. The crystals have grown perpendicular to the substrate, forming an isopachous crust. Note the growth zones and internal differentiations, and the V-shaped pattern, all of which are characteristic features of this type of cement.

IW = 0.5 mm. PPL.

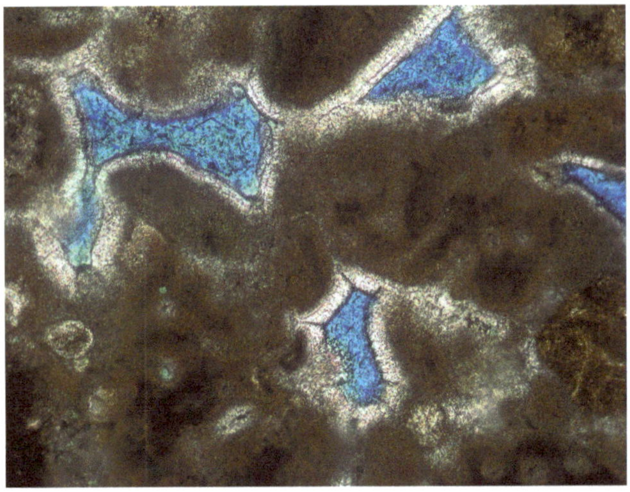

Holocene. Los Roques Archipelago

Isopachous rims of microcrystalline HMC are outlining intraparticle pores in the image. The lower magnification used for this photomicrograph allowed for the capture of several intraparticle pores formed between micritized *Halimeda* grains. These wind-derived allochems (reworked and redeposited grains from an adjacent, shallow subtidal environment into a hypersaline lagoon) served as a substrate for the growth of the HMC rims. These precipitated in very shallow marine evaporated waters with increased salinity, pH, and alkalinity.

IW = 1.3 mm. PPL.

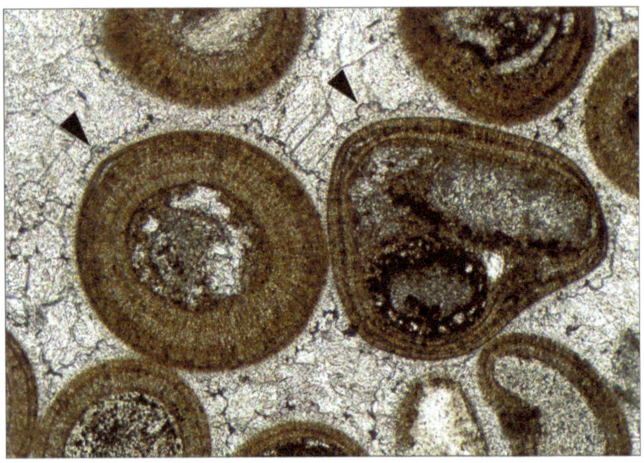

Cretaceous (middle Albian). Lisure Formation. Maracaibo Basin

In ancient shallow carbonates, marine diagenesis is in many cases obliterated by posterior diagenetic processes. Occasioanlly, some marine cements may remain identifiable. This is the case of ooids in the image, which are coated by remnants of rims of early marine crystals (black arrows), most likely originally calcitic in composition. Posterior precipitation of LMC cement filling interparticle pores, combined with successive diagenetic stabilization, as this microfacies corresponds to a deeply-buried core section, obliterated the primary texture of the isopachous rims.

IW = 1.8 mm. PPL.

Cretaceous (early Aptian). Machiques Member, Apón Formation. Maracaibo Basin

Radiaxial calcite crystals in the image exhibit distinctive undulose extinction and radiate away from a calcitic mollusc shell fragment that served as a growth surface. Individual crystals are large and thick and have pyramid-like terminations. For this ancient example (~ 120 Ma), we estimate the crystals are of marine origin and grew in a small depositional cavity developed directly underneath the large shell.

Photomicrograph courtesy of Gilberto Soto.
IW = 1.3 mm. CPL.

Cretaceous (middle Aptian). Piché Member, Apón Formation. Maracaibo Basin

Radiaxial calcite crystals in the image could be either marine or neomorphic in origin. The large crystals (mm-sized with pyramid-like or blunted terminations) radiate away from the growth surface. Note that some of them are converging toward each other, making it difficult to differentiate crystal and sub-crystal boundaries and terminations. They exhibit twin lamellae (parallel lines within the crystals), which are arecommon features of radiaxial calcite crystals. See Richter et al. (2011) for morphological and textural details on types of radiaxial calcite crystals and associated fabrics.

IW = 7.5 mm. CPL

Cretaceous (early Aptian). Machiques Member, Apón Formation. Maracaibo Basin

Although not addressed in this chapter, dolomitization is one of the most significant diagenetic processes initiated in the marine phreatic environment (and in many cases continued during later diagenetic stages) as it forms dolomite, the second most abundant carbonate mineral in the sedimentary record. Euhedral (rhombic) dolomite crystals in the image are the product of the replacement of cryptocrystalline $CaCO_3$ sediments making up the matrix of this dolomitic mudstone. The replacive origin of the crystals is indicated by their cloudy cores.

IW = 0.5 mm. PPL.

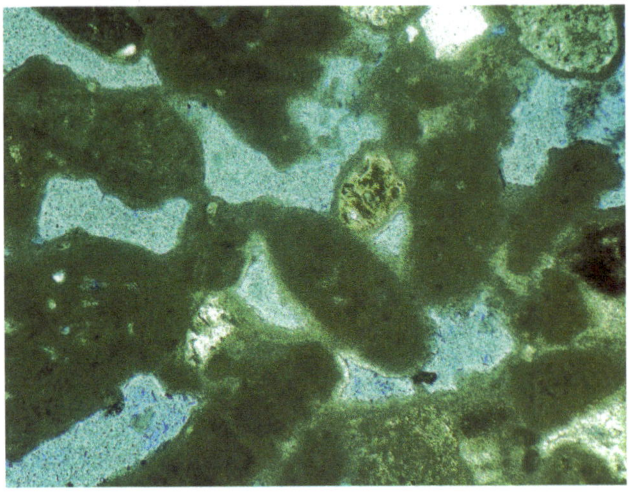

Holocene. Los Roques Archipelago

The photomicrograph shows micritized grains (mainly *Halimeda* flakes) in a highly porous microfacies within a laminated intertidal stromatolite. Note the dark appearance of the grains, whose micrite replacement is composed of cryptocrystalline crystals that are imperceptible under this magnification. In this sample, destructive micritization (texture-obliterating process) is accompanied by precipitation of micritic bridges, a constructive process that cements the grains together, thus promoting stabilization and lithification of the microbialite structure.

IW = 1.3 mm. PPL.

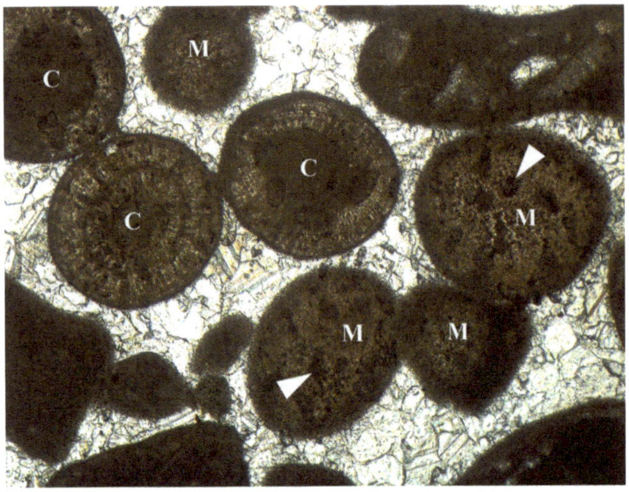

Cretaceous (middle Albian). Lisure Formation. Maracaibo Basin

Micritized ooids in a cemented oolite grainstone. The original concentric (C) microfabric is visible in three of the ooids in the image, while pervasive micritization completely obliterated the original texture of the other ooids, thus forming a micritic (M) fabric. Note the irregularly-shaped microborings filled with microbially-induced micrite (white arrows). Micritization of the ooids during marine diagenesis was followed by precipitation of coarse-grained LMC cement during meteoric diagenesis.

IW = 1.8 mm. PPL

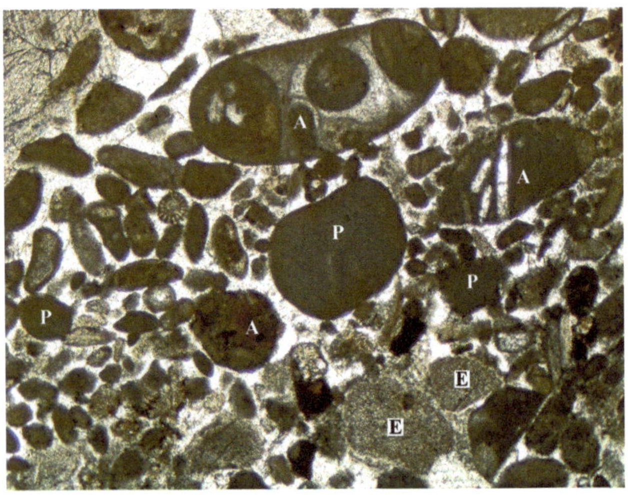

Cretaceous (middle Albian). Lisure Formation. Maracaibo Basin

Micritic peloids (P) and aggregates (A) are accompanied by echinoids (E) in a cemented grainstone. Different-sized micritic peloids are the product of micritization of skeletal grains, mainly green algae. This early marine process also promoted the formation of aggregate grains, consisting of several grains cemented together. On the other hand, many echinoids escaped micritization. A significant feature of micritized grains in the image is their well-preserved state. This suggests an original LMC composition of the micritic crystals, making them highly resistant to postdating meteoric diagenesis.

IW = 7.5 mm. PPL.

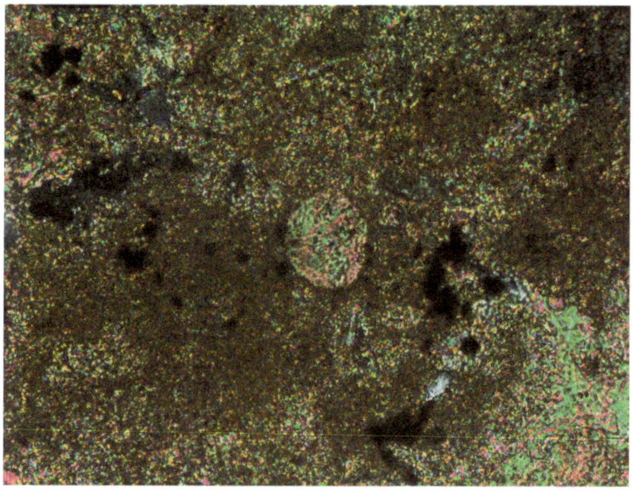

Cretaceous (middle Apbian). Piché Member, Apón Formation. Maracaibo Basin

This and the two following images show examples of recrystallization of fine-grained limestones (a process also commonly termed "aggrading neomorphism"). This image shows a partially recrystallized mudstone belonging to a transgressive deposit. Note the contrast between the dark brown micritic matrix, consisting of LMC cryptocrystals, and visually perceptible microcrystals exhibiting high birefringence, derived from recrystallization. In the center of the image, the small spherical grain is a sponge spicule. Black spots are pyrite framboids cross-cutting recrystallized crystals.

IW = 0.5 mm. PPL

Cretaceous (early Aptian). Machiques Member, Apón Formation. Maracaibo Basin

The original cryptocrystalline micritic matrix of this mudstone underwent recrystallization (calcite-to-calcite), forming a microcrystalline mosaic composed of LMC crystals. We estimate that crystal enlargement probably occurred in the shallow-burial environment, as this sample was deposited during a maximum flooding episode (the OAE 1a), away from direct influence of the meteoric diagenetic environment. The view under CPL allows for better visualization of crystal sizes. A globular planktonic foraminifer in the center of the image is clearly differentiated.

IW = 0.5 mm. CPL

Cretaceous (early Aptian). Machiques Member, Apón Formation. Maracaibo Basin

In this sample, stained with Alizarin Red S, also belonging to the OAE 1a, recrystallization created an inequigranular crystalline fabric consisting of different-sized microspar crystals and patches of pseudosparite (whitish in color). Note that pyrite framboids (black spots) cross-cut the recrystallized fabric, indicating a likely shallow-burial origin for these sulfide precipitates. The image highlights the well-preserved state of a few microspheres (white-limpid) of bacterial origin. These authigenic precipitates might be easily mistaken for globular PF.

IW = 1.3 mm. PPL.

Cretaceous (middle Albian). Lisure Formation. Maracaibo Basin

The original micritic matrix in this sandy-glauconitic, bioclastic limestone is recrystallized to pseudospar crystals. These are forming a packed mosaic consisting of inequigranular subhedral and anhedral calcite crystals. Pseudospar mosaics are the result of recrystallization of micritic-dominated matrices with poor content of clay minerals. This crystal enlargement process may be easily mistaken by dolomitization.

IW = 2.4 mm. PPL.

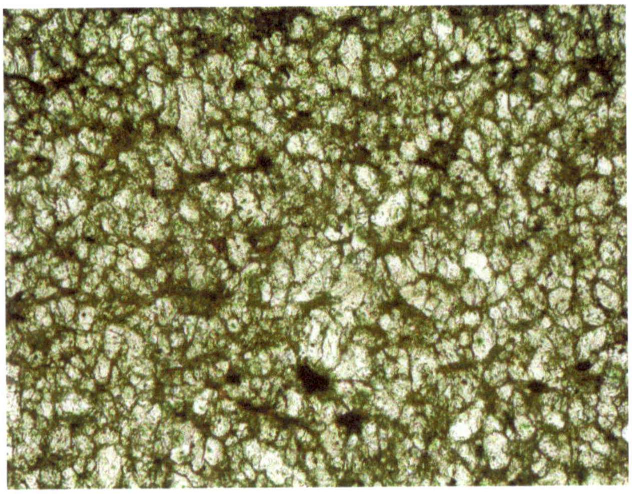

Cretaceous (late Albian). Lisure Formation. Maracaibo Basin

A pseudospar mosaic consisting of inequigranular crystals exhibiting subhedral and anhedral textures and "loaf-shaped" forms. As in the previous example, this diagenetic texture is the product of patchy recrystallization (increase in crystal size) of the micritic matrix (note the remnants of micritic sediments between crystals). We estimate this crystal enlargement process began in a meteoric (near-surface) environment, but probably continued during burial.

IW = 1.3 mm. PPL.

Cretaceous (middle Aptian). Piché Member, Apón Formation. Maracaibo Basin

In this sample, recrystallization (or marked increase in crystal size) of the micritic matrix formed an inequigranular mosaic consisting of anhedral and subhedral crystals, some of which exhibit elongate/prismatic shapes. In the center of the image, two originally aragonitic bioclasts (probably molluscs) suffered dissolution. Subsequent precipitation of coarse-grained LMC filled the molds.

IW = 2.4 mm. PPL.

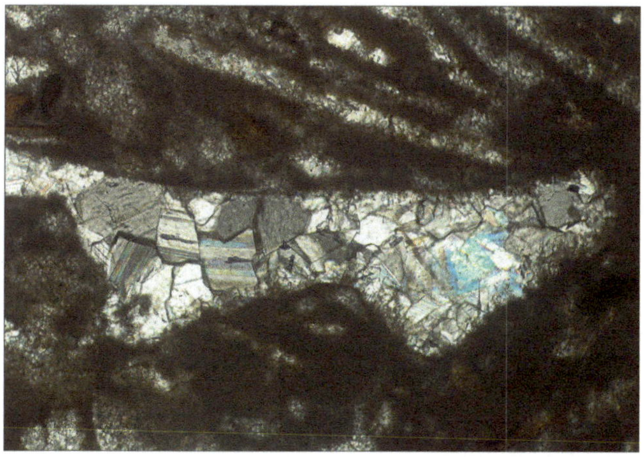

Cenozoic (Paleocene). Guasare Formation. Maracaibo Basin

This and the two following images exemplify processes of diagenetic stabilization of aragonitic skeletal components (neomorphism or dissolution-reprecipitation, or dissolution and filling of the molds by LMC cements). This image shows a mold of a bivalve mollusc valve. Dissolution of the originally aragonitic grain was followed by precipitation of coarse-grained LMC cement. The well-preserved original shape of the thick dorsal margin of the valve, which is embedded in a dark, micritic matrix, indicates that dissolution-reprecipitation occurred as a selective process confined to the metastable aragonitic substrate.

IW = 2.4 mm. CPL.

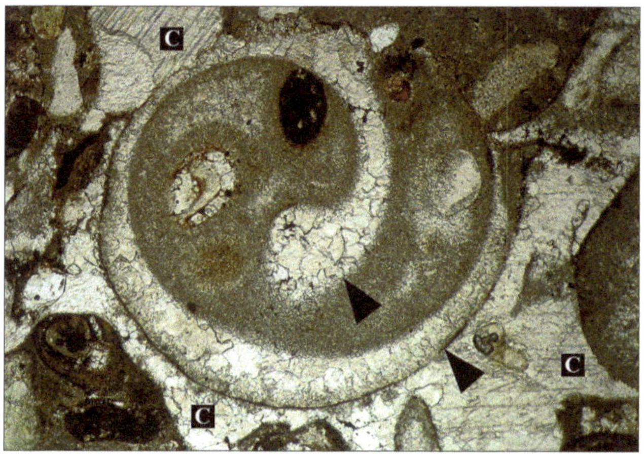

Cretaceous (latest Albian). Maraca Formation. Maracaibo Basin

The original aragonitic gastropod coiled shell (indicated with black arrows) seen in transverse section underwent dissolution-reprecipitation. As in the previous example, this process involved the loss of Sr from the aragonite crystal lattice of the shell. Coarse-grained LMC crystals comprising the mold of the gastropod shell should not be confused with LMC cements (C) in the image, which are filling interparticle pores surrounding the shell.

IW = 2.4 mm. PPL.

Late Oligocene-Early Miocene. Perla Limestone. Eastern Gulf of Venezuela Sub-Basin

The original aragonitic septal structure of the coral grain in this muddy bioclastic packstone underwent dissolution-reprecipitation. Note the whitish color of the diagenetically stabilized LMC walls of the coral, contrasting with the dark micritic infilling of its highly porous internal structure. In this and the two previous examples, the three skeletal components are preserved as pseudospar (or calcite spar) fabrics.

IW = 7.5 mm. PPL.

Cretaceous (middle Aptian). Piché Member, Apón Formation. Maracaibo Basin

Drusy calcite crystals in the image are filling a vuggy pore. This LMC cement type is characterized by coarse-grained, elongate crystals. The packed crystal fabric consists of inequigranular crystals exhibiting anhedral to subhedral textures (diffuse to partly developed boundaries, respectively). Note that the thickness of many crystals increases from the growth surface towards the center of the vug, which is a characteristic feature of this cement type. Drusy mosaics are commonly formed in the vadose (unsaturated) meteoric zone, as well as in burial settings.

IW = 1.3 mm. PPL.

Cretaceous (middle Albian). Lisure Formation. Maracaibo Basin

Precipitation of blocky calcite cement filled primary intergranular porosity in this oolite grainstone. The mosaic is composed of coarse-grained, cloudy LMC crystals. These exhibit straight and curved boundaries, thus forming a xenotopic (anhedral crystals) to hypidiotopic (subhedral crystals) fabric. Primary porosity in oolitic microfacies in the Lisure Formation could have originally reached values up to 40%. In this example, complete occlusion of porosity by blocky calcite occurred in the meteoric phreatic (saturated) zone (although this cement also occurs in the unsaturated zone).

IW = 2.4 mm. PPL

Middle Miocene. Capadare Formation. Falcón Basin

Dogtooth calcite crystals partially filling an interparticle (primary) pore. These crystals are characterized by elongate shapes (high length/width ratio) and bladed (or sharp) terminations, as well as by growing more or less normal to the substrate.tThe precipitation of dogtooth cement is, in many cases, associated to cementation in the meteoric phreatic environment. However, this cement cannot be taken as unequivocal evidence of meteoric environments, because it also occurs in the shallow-burial environment (see Reinhold 1999).

IW = 1.3 mm. CPL.

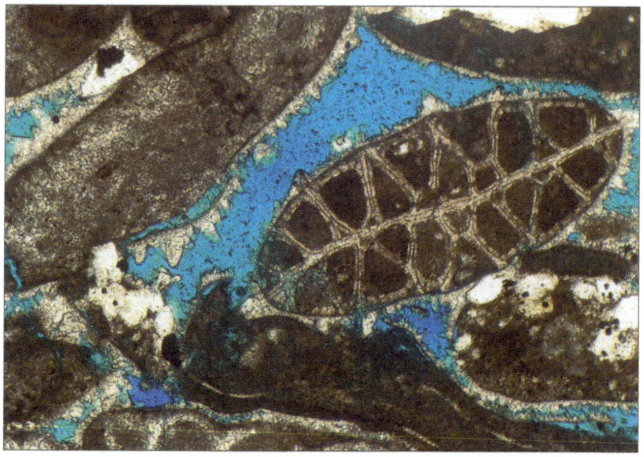

Cretaceous (latest Albian). Maraca Formation. Maracaibo Basin

Grains in this grainstone with interparticle porosity are rimmed by bladed LMC crystals. The latter are texturally characterized by sharp terminations, but crystals are smaller in length and thickness compared to dogtooth crystals. Bladed rims tend to form isopachous crusts and grow normal to the substrate, as it is seen in the image.

IW = 2.4 mm. PPL.

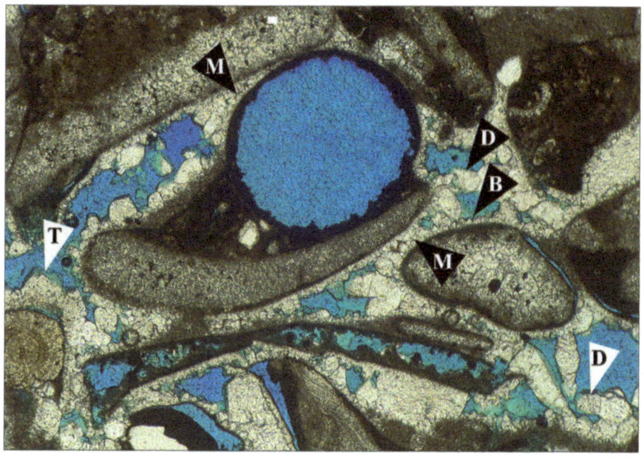

Cretaceous (latest Albian). Maraca Formation. Maracaibo Basin

Different types of meteoric cements are filling interparticle pores in this grainstone. Classification of meteoric cements can occasionally be subjective, as crystals are named according to their texture (shape, length and width) and whether they are substrate-controlled or not. Crusts of bladed (B) and dogtooth (T) crystals (the latter are larger and thicker) are rimming grains. Coarse-grained drusy (D) crystals with no substrate control are also abundant in the image. Meniscus (M) cement, consisting of small crystals at or near grain-to-grain contact, promoted early (near-surface) lithification.

IW = 2.4 mm. PPL.

Cretaceous (middle Albian). Lisure Formation. Maracaibo Basin

Syntaxial calcite overgrowth around echinoids (crinoids and echinoderms) is a common void-filling cement in ancient carbonate microfacies of different ages. The growth of large, cloudy, and highly zoned crystals in the image was controlled by the presence of HMC echinoderms. Alteration of the latter in the meteoric environment led to the formation of calcite overgrowths, which precipitated in optical continuity with the grains. Note the parallel zonation lines crossing both echinoid grains and cement, indicating the substrate control on the precipitation of such overgrowths.

IW = 2.4 mm. PPL.

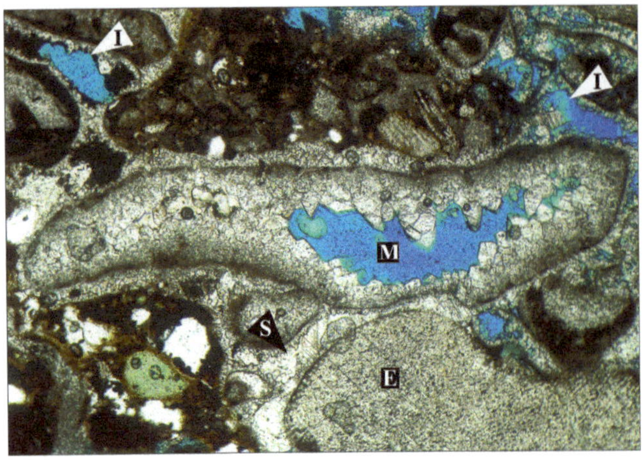

Cretaceous (late Albian). Maraca Formation. Maracaibo Basin

Dissolution and cementation in the meteoric environment are well-represented in this example. Selective dissolution of the aragonitic green algal grain in the center of the image created a moldic (M) pore. Subsequent precipitation of drusy cement partially filled the pore. At the bottom, syntaxial (S) calcite overgrowth precipitated in optical continuity with an echinoid (E) grain and is infilling an interparticle pore (note the zonation lines crossing the echinoid grain). Other interparticle (I) pores in the image are internally rimmed by bladed cement.

IW = 2.4 mm. PPL.

Cretaceous (latest Albian). Maraca Formation. Maracaibo Basin

Selective dissolution (or leaching) of aragonitic grains due to the action of meteoric (undersaturated) waters is prominent in this packstone of oysters and echinoids. Secondary pores in the image are a combination of moldic pores (smaller pores) and vuggy (V) pores. The latter, larger in size and irregularly-shaped, are the product of dissolution-enlargement of preexisting moldic pores. Subaerial exposure and dissolution in the meteoric environment occurred at regional scale during the late Aptian in the MB, thus forming secondary porous lithofacies with great lateral extension.

IW = 7.5 mm. PPL

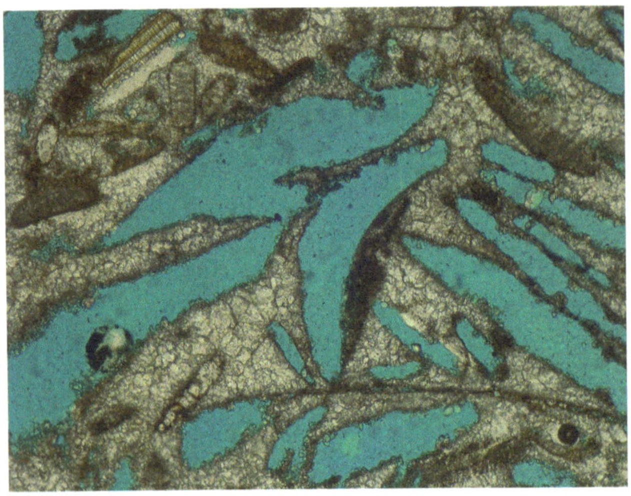

Cretaceous (middle Albian). Piché Member, Apón Formation. Maracaibo Basin

In this wackestone-packstone, two different meteoric diagenetic processes led to the alteration of the original depositional texture. Selective dissolution of aragonitic bivalve molluscs created moldic pores, representing a substantial volume of secondary porosity; while the matrix is recrystallized to pseudospar. Note the inequigranular crystals, which are the product of crystal enlargement due to dissolution-reprecipitation. This sample exemplifies the porous lithofacies that have produced light hydrocarbons in the Piché Member.

IW = 2.4 mm. PPL

Late Oligocene-Early Miocene. Perla Limestone. Eastern Gulf of Venezuela Sub-Basin

This example shows the different response of metastable aragonite and HMC bioclasts to mineralogical stabilization. Selective dissolution of bivalve moluscs formed moldic (M) pores, which are filled with blocky calcite. In contrast, note the well-preserved state of the two central red algal grains. Textural features such as arched-celled filaments and thin, elongate, calcified cells are clearly differentiated. The loss of Mg during dissolution of HMC red algal grains and subsequent precipitation of LMC is a mineralogical stabilization process that is still poorly understood (see Tucker and Wright 1990).

IW = 2.4 mm. PPL.

Late Oligocene-Early Miocene. Perla Limestone. Eastern Gulf of Venezuela Sub-Basin

Precipitation of microspar (white in color) is prominent in this sample. Note the extensive area covered by this LMC precipitate, both replacing the matrix between grains and void-filling small secondary pores. Furthermore, microspar is also found filling small intraparticle pores in the larger red algal grain in the image. Despite this, traces of the original textural features of this alga grain are clearly differentiated.

IW = 2.4 mm. PPL.

Cretaceous (middle-late Cenomanian). La Luna Formation *sensu stricto*. Maracaibo Basin

In this image, the globular chambers of PF tests are filled with coarse blocky cement. Crystal sizes and shapes are better visualized under CPL. These tests had to escape synsedimentary infilling by micritic sediments in order to remain empty for void-filling cementation. Depositional paleodepths estimated for this sample, belonging to the base of the La Luna Formation, are around 80–100 m deep. This suggests that the cementation process occurred in the shallow-burial environment, away from the meteoric environment, and before the crushing of tests in the intermediate-burial realm.

HA = 1.3 mm. CPL.

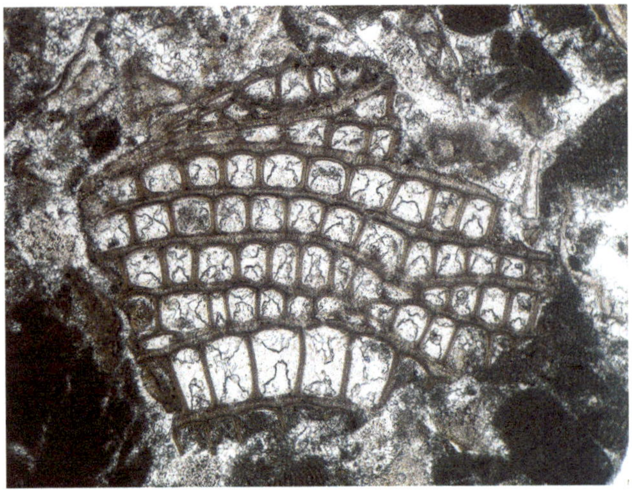

Late Oligocene-Early Miocene. Perla Limestone. Eastern Gulf of Venezuela Sub-Basin

In the Perla Limestone, primary intraparticle (or intraskeletal) pores are well-suited for precipitation of void-filling cements. In this image, the chamberlets within the highly porous skeleton of the larger benthic foraminifer *Heterostegina antillea* are filled by cloudy blocky calcite. As with most cementation processes in this carbonate succession, this void-filling cement has been interpreted to have precipitated in the shallow-burial environment (see Valencia and Laya 2020), prior to mechanical compaction.

IW = 2.4 mm. PPL.

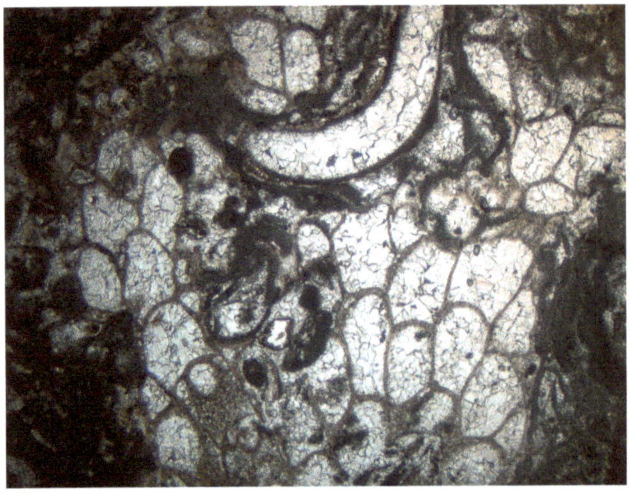

Late Oligocene-Early Miocene. Perla Limestone. Eastern Gulf of Venezuela Sub-Basin

Another example of blocky calcite void-filling intraparticle primary pores, in this case the zooecia cavities of a bryozoan. Note the numerous LMC mosaics, separated by thin zooecia walls. Anhedral crystals with curved boundaries predominate, hence forming a xenotopic fabric. In the Perla Limestone, blocky is the dominant type of calcite cement.

IW = 4.2 mm. PPL.

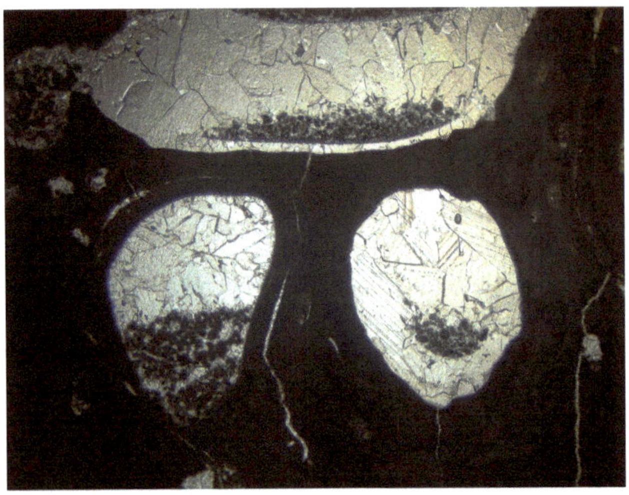

Late Oligocene-Early Miocene. Perla Limestone. Eastern Gulf of Venezuela Sub-Basin

Blocky calcite cement is void-filling large boring pores in a red algal grain. Note that the three secondary pores in the image have geopetal infillings, consisting of micritic peloids at the bottom of the cavities and blocky calcite crystals at the top. These coarse-grained crystals are subhedral to anhedral in texture and some of them exhibit twin lamellae.

IW = 4.2 mm. PPL.

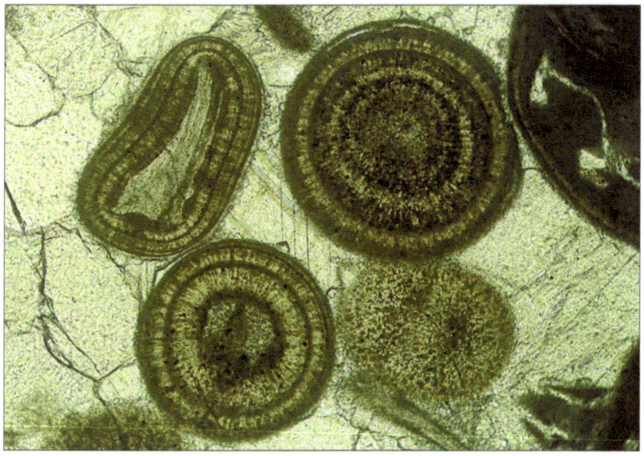

Cretaceous (latest Albian). Maraca Formation. Maracaibo Basin

Poikilotopic cement is filling primary interparticle pores in this oolitic grainstone. The coarse-grained LMC crystals show poorly-defined boundaries, and some of them are zoned, which are characteristic features of this cement type. Note that the grains are in tangential contact, but they are not affected by mechanical compaction, which suggests that cementation occurred in the shallow-burial environment.

IW = 1.8 mm. PPL.

Cretaceous (latest Albian). Maraca Formation. Maracaibo Basin

Poikilotopic cement is void-filling an interparticle pore. The crystals are large (mm-sized) and highly zoned. Note the white-limpid color, which is characteristic of burial, inclusion-free calcite cements. These are also commonly enriched in Fe(II) and Mn(II). In this example, precipitation of LMC occurred in the shallow-burial environment and not in the deep burial realm, where pores between grains (if present)tend to be smaller and narrower.

IW = 1.3 mm. PPL.

Cretaceous (middle Aptian). Piché Member, Apón Formation. Maracaibo Basin

In this photomicrograph taken using CPL, intense twin lamellae (or twinning of the calcite) exhibit a characteristic and striking set of colorful parallel lines. These elements, combined with the large size of the crystals, are used as criteria for differentiating poikilotopic cements. In this case, this shallow burial cement filled interparticle pores between micritized green algal grains and peloidal grains.

IW = 2.4 mm. CPL.

References

Ahm, A. S. C., Bjerrum, C. J., Blättler, C. L., Swart, P. K., Higgins, J. A., 2018. Quantifying early marine diagenesis in shallow-water carbonate sediments. *Geochimica et Cosmochimica Acta, 236*, 140-159.

Al-Ramadan, K., 2014. Diagenesis of Holocene beachrocks: a comparative study between the Arabian Gulf and the Gulf of Aqaba, Saudi Arabia. *Arabian Journal of Geosciences, 7*, 4933-4942.

Badiozamani, K., 1973. The dorag dolomitization model, application to the middle Ordovician of Wisconsin. *Journal of Sedimentary Research, 43*, 965-984.

Bathurst, R. G. C., 1966. Boring Algae, Micrite Envelopes and Lithification of Molluscan Biosparites. Elsevier, Amsterdam, 658 pp.

Bathurst, R. G. C., 1974. Marine diagenesis of shallow water calcium carbonate sediments. *Annual Review of Earth and Planetary Sciences, 2*, 257-274.

Bathurst, R. G. C., 1975. Carbonate Sediments and Their Diagenesis: Amsterdam, Elsevier, 657 pp.

Burchette, T. P., and Britton, S. R., 1985, Carbonate facies analysis in the exploration for hydrocarbons—a case study from the Cretaceous of the Middle East. In: P. J. Benchley and B. P. J. Williams, eds., Sedimentology, recent developments and applied aspects: Oxford, Blackwell, pp. 311–338.

Cherns, L., Wheeley, J. R., Wright, V. P., 2011. Taphonomic bias in shelly faunas through time: early aragonitic dissolution and its implications for the fossil record. In: Allison, P.A., Bottjer, D.J. (Eds.), Taphonomy: Process and Bias Through Time. Springer, Dordrecht, pp. 79–105.

Choquette, P. W., and Pray, L. C., 1970. Geologic nomenclature and classification of porosity in sedimentary carbonates. *American Association of Petroleum Geologists Bulletin, 54*, 207-250.

Diaz, M. R., and Eberli, G. P., 2022. Microbial contribution to early marine cementation. *Sedimentology, 69*(2), 798-822.

Fantle, M. S., Barnes, B. D., Lau, K. V., 2020. The Role of Diagenesis in Shaping the Geochemistry of the Marine Carbonate Record. *Annual Review of Earth and Planetary Sciences, 48*, 549-583.

Flügel, E, 2010. Microfacies of Carbonate Rocks—Analysis, Interpretation and Application. Springer, Heidelberg-Dordrecht-London-New York, 984 pp.

Folk, R. L., 1965. Some aspects of recrystallization in ancient limestones, *in* L. C. Pray, and R. S. Murray, eds., Dolomitization and Limestone Diagenesis: Tulsa, OK, *SEPM Special Publication, 13*, 14-48.

Friedman, G. M., 1964. Early diagenesis and lithification in carbonate sediments. *Journal of Sedimentary Research, 34*, 777-813.

Friedman, G. M., 1965, Terminology of crystallization textures and fabrics in sedimentary rocks. *Journal of Sedimentary Petrology, 35*, 643-655.

Friedmann, G. M., Gebelein, C. D., Sanders, J. E., 1971. Micritic envelopes of carbonate grains are not exclusively of photosynthetic algal origin. *Sedimentology, 16*, 89-96.

Garuglieri, E., Marasco, R., Odobel, C., Chandra, V., Teillet, T., Areias, C., Sánchez-Román, M., Vahrenkamp, V., Daffonchio, D., 2024. 2024. Searching for microbial contribution to micritization of shallow marine sediments. *Environmental Microbiology, 26*, e16573.

Granier, B., 2012. The contribution of calcareous green algae to the production of limestones: a review. *Geodiversitas, 34*, 35-60.

Given, R. K., and Wilkinson, B. H., 1987. Dolomite abundance and stratigraphic age: constraints on rates and mechanisms of Phanerozoic dolostone formation. *Journal of Sedimentary Petrology, 57*, 1068–78.

Haredy, R. A., Ghandour, I. M., Erginal, A. E., Bozcu, M., 2019. Beachrock cementation patterns along the Gulf of Aqaba coast, Saudi Arabia. *Arabian Journal for Science and Engineering, 44*, 479-487.

Higgins, J. A., Blättler, C. L., Lundstrom, E. A., Santiago-Ramos, D. P., Akhtar, A. A., Ahm, A. C., Bialik, O., Holmen, C., Bradbury, H., Murray, S. T., Swart, P. K., 2018. Mineralogy, early marine diagenesis, and the chemistry of shallow-water carbonate sediments. *Geochimica et Cosmochimica Acta, 220*, 512-534.

Hillgärtner, H., Dupraz, C., Hug, W., 2001. Microbially induced cementation of carbonate sands: are micritic meniscus cements good indicators of vadose diagenesis? *Sedimentology, 48*, 117-131.

Immenhauser, A., 2021. On the delimitation of the carbonate burial realm. *The Depositional Record, 00*, 1-51.

James, N. P., and Jones, B., 2015. Origin of Carbonate Sedimentary Rocks. John Wiley & Sons, Ltd, Chinchester, 464 pp. L ARTICLE

Kendall, C. G. S. C., and Alsharhan, A. S., 2011. Coastal Holocene carbonates of Abu Dhabi, UAE: depositional setting, geomorphology, and role of cyanobacteria in micritization. *International Association of Sedimentologists, Special Publication, 43*, 205-220.

Lippmann, F., 1973. Crystal Chemistry of Sedimentary Carbonate Minerals. In: Sedimentary Carbonate Minerals. Minerals, Rocks and Inorganic Materials (Monograph Series of Theoretical and Experimental Studies), vol. 6.; Springer: Berlin, Heidelberg, 1973; pp. 5–96.

Lippmann, F., 1982. Stable and metastable solubility diagrams for the system $CaCO_3$-$MgCO_3$-H_2O at ordinary temperature. *Bulletin de Mineralogie, 105*, 273-279.

Longman, M. W., 1980. Carbonate diagenetic textures from nearsurface diagenetic environments. *American Association of Petroleum Geologists Bulletin, 64*, 461-487.

Machel, H. G., 2005, Investigations of burial diagenesis in carbonate hydrocarbon reservoir rocks: *Geoscience Canada, 32*, 103–128.

Macintyre, I. G., Prufert-Bebout, L., Reid, R. P., 2000. The role of endolithic cyanobacteria in the formation of lithified laminae in Bahamian stromatolites. *Sedimentology, 47*, 915-921.

Mauz, B., Vacchi, M., Green, A., Hoffmann, G., Cooper, A., 2015. Beachrock: a tool for reconstructing relative sea level in the far-field. *Marine Geology, 362*, 1-16.

Mazzullo, S. J., 2004. Overview of porosity evolution in carbonate reservoirs. *Kansas Geological Society Bulletin, 79*, 1-19.

Millero, F. J., Feistel, R., Wright, D.G., Mcdougall, T. J., 2008. The composition of standard seawater and the definition of the reference-composition salinity scale. *Deep Sea Research Part I: Oceanographic Research Papers, 55*, 50–72.

Moore, C. H., 1989. Carbonate Diagenesis and Porosity. Developments in Sedimentology, 46. New York, Elsevier, 338 pp.

Moore, C. H., 2001. Carbonate reservoirs: porosity evolution and diagenesis in a sequence stratigraphic framework, 3. impr. In: Developments in Sedimentology 55. Elsevier, Amsterdam, 444 pp.

Moore, C. H., and Wade, W. J., 2013. Carbonate Reservoirs: Porosity and Diagenesis in a Sequence Stratigraphic Framework: Amsterdam, Elsevier, Developments in Sedimentology 55, 392 pp.

Morse, J. W., Arvidson, R. S., Lüttge, A., 2007. Calcium Carbonate Formation and Dissolution. *Chemical Reviews, 107*, 342-381.

Perry, C. T., 1998. Grain susceptibility to the effects of microboring: implication for the preservation of skeletal carbonates. *Sedimentology, 45*, 39-51.

Petrash, D. A., Gingras, M. K., Lalonde, S. V., Orange, F., Pecoits, E., Konhauser, K. O., 2012. Dynamic controls on accretion and lithification of modern gypsum-dominated thrombolites, Los Roques, Venezuela. *Sedimentary Geology, 245*, 29-47.

Petrash, D. A., Bialik, O. M., Staudigel, P. T., Konhauser, K. O., Budd, D. A., 2021. Biogeochemical reappraisal of the freshwater–seawater mixing-zone diagenetic model. *Sedimentology, 68*, 1797-1830.

Reid, R. P., and Macintyre, I. G., 1998. Carbonate recrystallization in shallow marine environments: a widespread diagenetic process forming micritized grains. *Journal of Sedimentary Research, 68*, 928-946.

Reid, R. P., and Macintyre, I. G., 2000. Microboring versus recrystallization: further insight into the micritization process. *Journal of Sedimentary Research, 70*, 24-28.

Reid, R. P., Macintyre, I. G., Post, J. E., 1992. Micritized skeletal grains in northern Belize Lagoon; a major source of Mg-calcite mud. *Journal of Sedimentary Research, 62*, 145-156.

Reinhold, C., 1999. Dog-tooth cements: indicators of different diagenetic environments: Zentralblatt flir Geologie und Paldontologie, Teil I, v. 1997, p. 1221-1235.

Richter, D. K., Neuser, R. D., Schreuer, J., Gies, H., Immenhauser, A., 2011. Radiaxial-fibrous calcites: A new look at an old problem. *Sedimentary Geology, 239*, 23-36.

Samankassou, E., Tresch, J., Strasser, A., 2005. Origin of peloids in early Cretaceous deposits, Dorset, South England. *Facies, 51*, 264-274.

Shen, J. W., Wang, Y., Zhao, N., Yang, H. Q., Fu, F. X., Jin, Y. B., 2017. Carbonate sedimentary characteristics of the beach rocks around Qilian Islets and Cays, Xisha Islands: Implication for coral reef development and decline. *Palaeogeography, Palaeoclimatology, Palaeoecology, 474*, 264-278.

Scholle, P. A., and Ulmer-Scholle, D., 2003. A Color Guide to the Petrography of Carbonate Rocks: Grains, Textures, Porosity, Diagenesis. *American Association of Petroleum Geologists, Memoir 77*, 474 pp.

Sellwood, B. W., 1994. Principles of carbonate diagenesis. In: Parker, A., Selwood, B. (Eds.), Quantitative Diagenesis: Recent Developments and Applications to Reservoir Geology. NATO ASI Series. Series C, Mathematical and Physical Sciences, vol. 435, pp. 1—32.

Strasser, A., 1986. Ooids in Purbeck limestones (lowermost Cretaceous) of the Swiss and French Jura. *Sedimentology, 33*(5), 711-727.

Swart, P. K., 2015. The geochemistry of carbonate diagenesis: The past, present and future. *Sedimentology, 62*,1233-1304.

Swart. P. K., Kennedy, M. J., 2012. Does the global stratigraphic reproducibility of $\delta^{13}C$ in Neoproterozoic carbonates require a marine origin? A Pliocene–Pleistocene comparison. *Geology* 40(1), 87–90.

Tucker, M. E., and Wright., 1990. Carbonate Sedimentology. Blackwell Publishing. 482 pp.

Valencia, F. L. and Laya, J. C., 2020. Deep-Burial dissolution in an Oligocene-Miocene giant carbonate reservoir (Perla Limestone), Gulf of Venezuela Basin: Implications on microporosity development. *Marine and Petroleum Geology, 113*, 104144.

Visscher, P. T., Reid, R. P., Bebout, B. M., Hoeft, S. E., Macintyre, I. G., Thompson Jr., J., 1998. Formation of lithified micritic laminae in modern marine stromatolites (Bahamas): the role of sulfur cycling. *American Mineralogist, 83*, 1482–149.

Vousdoukas, M. I., Velegrakis, A. F., Plomaritis, T. A., 2007. Beachrock occurrence, characteristics, formation mechanisms and impacts. *Earth-Science Reviews, 85*, 23-46.

Warren, J., 2000. Dolomite: occurrence, evolution and economically important associations. *Earth-Science Reviews, 52*, 1-81.

Warren, J. K., 2016. Evaporites: A Geological Compendium. Second Edition. Springer International Publishing Switzerland. 1813 pp.

Zeebe, R. E., 2012. History of Seawater Carbonate Chemistry, Atmospheric CO2, and Ocean Acidification. *Annual Review of Earth and Planetary Sciences, 40*, 141-165.

17 Burial Diagenesis

Burial diagenesis refers to a realm removed from the near-surface setting (e.g., Choquette and James 1990), being the burial realm the less understood of the major diagenetic environments (e.g., Immenhauser 2021). Generally, diagenetic processes in the burial realm (intermediate- and deep-burial, i.e., the mesogenetic zone) lead to the formation of fitted or condensed fabrics and a reduction of carbonate rock porosity in "semi-closed" systems (Scholle and Halley 1985). With increasing burial depth, the rock-to-fluid ratio shifts in most cases from a fluid-buffered, open system to a carbonate-buffered, closed system (see Pingitore Jr. 1982; Higgins et al. 2018). However, the concept of open (fluid) versus closed (carbonate) system behavior (Higgins et al. 2018) describes the complexity of natural systems only in part (Immenhauser 2021).

Shallow-marine carbonates enter the burial diagenetic realm from one of two standpoints (Tucker and Wright 1990): they may be buried directly from the marine environment, with marine (connate) pore-fluids, or they may be affected by meteoric diagenetic processes first, so that sediment mineralogy is stabilized and pore-fluids are meteoric or mixed marine-meteoric. Generally, shallow carbonates that went through meteoric diagenesis (e.g., cementation and early lithification) are "protected" from burial diagenesis compared to carbonates facies that passed directly from the marine environment to the burial realm.

The time spent by ancient limestones in the deep-burial realm can be up to hundreds of millions of years (10^6–10^8 years). This is the case of Early Cretaceous carbonate units in the subsurface of the MB, which range in age from approximatelly 125 to 100 Ma and were buried after experiencing marine and meteoric diagenesis. Today, the oldest sedimentary rocks in these thick successions (averaging 1300 feet thick = 400 m) are located at depths down to ~19,500 feet (~ 6400 m) in some areas of the subsurface, without the occurrence of incipient low-grade metamorphism. Deep-burial diagenetic settings merge into the metamorphic realm at temperatures around 200 ° C (Machel 2005). Füchtbauer et al. (1988) and Immenhauser (2021) apply a temperature (but no depth) range of between 200 and 250 ° C as the transitional interval between deep-burial diagenesis and very low-grade metamorphism.

As mentioned in the previous chapter, determining paragenetic relations (the timing of diagenetic processes and products) is key to understanding the evolution of carbonate rocks, especially ancient facies that have experienced diagenetic alterations during different stages (eogenetic, mesogenetic, and telogenetic). This is the case for all photographic examples shown in this chapter, which correspond to ancient carbonate facies that were either deeply buried after undergoing meteoric diagenesis or exhumed after undergoing diagenesis in the eogenetic and mesogenetic zones.

17.1 Fluids in the Burial Realm

Before delving into the subject matter of this chapter (burial depth realms and burial diagenetic processes), a brief introduction to the fluids present in buried rocks is necessary. The pore fluids (waters) in buried carbonate or mixed carbonate-siliciclastic rocks include meteoric waters derived from deep recharge, marine connate waters deposited with the sediments (these waters precipitate rapidly as LMC in the shallow-burial environment), mineral-bound water (e.g., evaporite dehydration), fluids derived from compacting shales or evaporites, generated during clay mineral transformations, released during the thermal maturation of organic matter (kerogen), or from processes in the underlying basement. Geochemically, these burial fluids initially have different signatures. However, the composition of a given fluid may be easily modified by its mixing or reaction with any of above-mentioned fluids or with minerals, CO_2, organic acids, hydrocarbons, and hydrogen sulfide (Mazzullo and Harris 1991, 1992; Bjørlykke 1993; James and Jones, 2015; Immenhauser 2021).

J. A. Méndez Dot, J. Méndez Baamonde, *Atlas of Tropical Carbonates*, https://doi.org/10.1007/978-3-031-25494-9_17

Determining the origin of burial fluids is often a difficult task. Some common terms used to describe burial fluids include the following: *Formation water*, or waters resident in pores or fractures of a given rock body, which have no genetic connotation to their origin or age (Bjørlykke 1993). Many cases of burial dissolution are ascribed to formation waters with anomalous high concentrations of CO_2, supplied by thermal decarboxylation of organic matter (kerogen) or hydrous pyrolysis reactions between organic carbon and oxygen in water to yield CO_2 or organic acids (James and Jones, 2015). *Saline brines*, a term used for fluids with salinity significantly higher than that of average seawater, are often depleted in Mg^{2+}, K^+ and SO_4^{2-}, slightly depleted in Na^+, and considerably enriched in Ca^{2+}, Na^+, and Cl^- (see Tucker and Wright 1990; Moore 2001; Immenhauser 2021). *Basement fluids* originated from crystalline and metasedimentary rocks. These are magmatic, crust- and mantle-derived, high-temperature corrosive-aggressive fluids, and are mainly CO_2-rich ($H_2O + CO_2 + CH_4$) or H-rich ($H_2O + CH_4 + H_2$). Some may also yield minor amounts of N_2, C_2H_4, and C_2H_6 (Jin et al. 2004; Immenhauser 2021).

17.2 Burial Depth Realms

As discussed in Chap. 16, burial diagenesis (or mesodiagenesis; Mazzullo and Harris 1992) begins at depths where sediments are no longer affected by processes in the marine and meteoric environments. A separate distinction arises in the subdivision of burial diagenesis into depth realms (or domains): shallow, intermediate, and deep (or deeper) (see Schmoker and Halley 1982; Scholle and Halley 1985; Choquette and James 1987; Machel 2005; Swart 2015; Fantle et al. 2020; Immenhauser 2021). Boundaries between domains of the burial realm are gradual. A wide range of proposed depth windows has been discussed in the literature, but arguments for generally applicable criteria are often lacking. This is because each sedimentary basin has a specific thermal history and rock inventory (see Immenhauser 2021).

17.2.1 Intermediate-Burial

We agree with the suggestion brought forward by Tucker and Wright (1990) and Flügel (2010), and consider that diagenetic processes in the intermediate-burial realm begin at relatively shallow-burial depths (a few tens of meters), where carbonate dissolution and cementation are already hindered by poor water circulation and the onset of mechanical compaction. On the contrary, in the classification of diagenetic settings on the basis of mineralogy, petroleum generation, hydrogeochemistry, and hydrogeology proposed by Machel (2005), the upper limit of this diagenetic environment is placed between 600–1000 m depth. In the intermediate-burial domain, pore-fluids may consist of various formation waters, which increase in salinity with time and ongoing burial. Immenhauser (2021) consider the intermediate-burial domain to encompass depths from hundreds of meters to about 2 km (temperature < 100 °C) and to be characterized by the transition from fluid-buffered to rock-buffered diagenesis.

17.2.2 Deep-Burial

In carbonate burial settings subjected to a normal geothermal gradient (~ 25–30 ° C per Km), the loose boundary delineating the upper limit of the deep-burial diagenetic environment is thought to occur at ~2–3 km depth. This upper boundary is characterized by complete saturation of pore spaces, low fluid flow rates, diagenetic fluids consisting of mixtures of formation waters, and elevated temperature and pressure (Swart 2015; Fantle et al. 2020). Furthermore, this boundary commonly coincides with the onset of the kerogen maturation and hydrocarbon generation, or the so-called "oil window" (Machel 2005). Throughout the deep-burial domain (c. 2–12 km; temperature > 100 °C), diagenetic processes are rock-buffered, unless where fault zones allow for fluid migration. In this domain, marine formation fluids are increasingly modified by rock–fluid interaction and replaced by saline brines. The lower limit of the deep-burial realm varies among basins and tectonic settings. The borehole TS1 (Tarim Basin, northwest China) is one of the deepest petroleum wells and serves as example of deeply-buried carbonates without showing signs of having undergone low-grade metamorphism. It reaches depths near 8500 m and extends through a Cambrian to Middle Ordovician succession of dolomites and limestones up to 2000 m thick (see Wei et al. 2017).

17.3 Burial Diagenetic Processes

Burial (intermediate and deep) diagenetic processes discussed in this chapter are: mechanical compaction (dewatering, reorientation of grains and deformation), chemical compaction (pressure solution), microfracturing, recrystallization, dissolution, and cementation. In the case of recrystallization, dissolution and cementation (previously discussed in Chap. 16), these processes may occur from near-surface (or eogenetic stage) to intermediate- and deep-burial (or mesogenetic stage) environments. Overall, burial diagenesis typically involves net porosity loss, which is often associated with the progressive increase of overburden thickness (e.g., Hamilton 1976; Schmoker and Halley 1982; Shinn and Robbin 1983; Croizé et al. 2013; Worden et al. 2018).

17.3.1 Mechanical Compaction

Mechanical (physical) compaction, strictly related to the increase of lithostatic load and pressure in the upper part of the intermediate-burial environment, begins with dewatering and reorientation of grains. The first, which is the dehydration or the expulsion of fluids (commonly rich in Mg^{2+}), is a process that affects mainly mud-supported fabrics and/or clay-rich facies more than grain-supported fabrics. The second process represents the beginning of physical compaction, starting with reorientation of grains towards parallelism respect to the stratification of the facies (especially noticed in microfacies rich in elongate bioclasts, such as tests of LBF). Carbonate sediments can compact to about one-half of their original thickness under as little as 100 m of overburden by physical compaction alone (Choquette and James 1987, 1990).

During reorientation, grains pass from being in no contact (in the case of mud-supported fabrics), to grains in point contact and subsequent stages of concave-convex, tangential and sutured contacts. Carbonate grains may react to plastic deformation in a ductile way (e.g., flattened red algal grains and peloids) or in a brittle way (e.g., fractured/crushed ooids, mollusc shells and ostracod carapaces). Grain-supported microfacies are more susceptible to be affected by plastic and/or brittle deformation, since grains were already in point contact before burial (see Bathurst 1975; Moore 1989; Choquette and James 1987, 1990; Tucker and Wright 1990; Scholle and Ulmer-Scholle 2003; Vajdova and Baud 2004; Machel 2005; Flügel 2010; for a detailed treatment of mechanical compaction).

17.3.2 Chemical Compaction

Chemical compaction begins during the latter phases of physical compaction and continues into the metamorphic realm (e.g., Choquette and James 1987, 1990). Processes linked to chemical compaction (pressure solution) are known to result in thickness reductions of at least 25–35% in addition to the thickness losses caused by physical compaction (Machel 2005).

17.3.2.1 Pressure Solution

Pressure solution (or pressure dissolution) refers to dissolution of carbonate material (mainly along bedding surfaces and at grain contacts) exposed to high normal load and/or tectonic stress (e.g., Carrio-Schaffhauser et al. 1990; Flügel 2010). In carbonate successions, major controls that influence the inception and course of pressure solution include some of the main lithologic features of carbonate facies and the processes these facies have already undergone. These controlling features/processes are: the fabric of the rock (mud- or grain-supported); the occurrence, or not, of predating meteoric and shallow-burial cementation processes (mud-supported, transgressive facies that escaped from meteoric diagenesis, and uncemented grain-supported facies, are prone to suffer pressure solution with increasing burial depths); the amount of clay minerals (if present) in the carbonate matrix (fine-grained, argillaceous carbonates tend to be more prone to the onset of pressure solution, and at shallower depths, compared to well-washed, grainy carbonates); and the presence or absence of organic matter, including oil and gas (organic matter favors calcite dissolution at high temperatures). According to the above-mentioned controls, pressure solution in carbonates commonly begins perhaps within 200–300 to 500 m of burial (see Machel 2005; Immenhauser 2021).

Pressure solution is a major source of Ca^{2+} and CO_3^{2-} ions for burial cementation (e.g., Kreutzberger and Peacor 1988). At deep-burial depths, cementation and significant porosity reduction are generally observed due to pressure solution (e.g., Immenhauser 2021). In standard petrography, types of pressure solution features leading to the formation of fitted fabrics include intergranular pressure solution, solution seams and stylolites (e.g., Buxton and Sibley 1981).

17.3.2.2 Intergranular Pressure Solution

Intergranular pressure solution (or pressure solution at grain contacts) refers to dissolution involving adjacent grains. The type of interpenetration between grains during pressure solution depends on which minerals are juxtaposed. Where two grains are composed of the same mineral, the pressure solution contact is sutured (i.e., grains are truncated and/or partially to totally interpenetrated), as both grains dissolve to the same extent. However, the contacts are smooth where the mineralogy differs, as one mineral effectively grows at the expense of the other (e.g., dolomite commonly grows into calcite) (see Machel 2005).

17.3.2.3 Solution Seams

A solution seam (also termed microstylolite or horse tail) is defined as a low-relief internal surface produced by subsurface pressure dissolution of soluble carbonate, typically marked by the presence of clays, organic matter or other insoluble material. They are similar to stylolites but marked by lower relief and less obvious insoluble residue and generally occurring as closely-spaced swarms of such dissolution surfaces (definition by Scholle and Ulmer-Scholle 2003).

17.3.2.4 Stylolites

Stylolitization, or the development of stylolites, begins somewhere within the loose boundary between the intermediate- and deep-burial environments. A stylolite is a jagged, columnar surface in carbonate rocks which may be at any orientation relative to bedding. It is produced by pressure-induced dissolution and grain interpenetration and often associated with large amounts of insoluble material (clay minerals, organic matter and localized precipitation of pyrite) accumulated as a result of such dissolution (definition by Scholle and Ulmer-Scholle 2003).

Generally, stylolites cross the rocks (from distances of a few millimeters-to-centimeters to up to hundreds of meters), and in many cases are oriented more or less parallel to the horizontal bedding of the facies (e.g., Petit and Mattauer 1995). They may occur as a single surface or in sets, in which individual stylolites are commonly parallel to each other. Stylolites may exhibit a variety of morphologies (e.g., irregular or smooth, peaks of low- or high-amplitude) and types of sets (e.g., parallel, anastomosing, conjugate), which depend on the original fabric (mud- or grain-supported), among other lithologic features (e.g., organic-matter and/or clay content, type of grains, their mineralogy and sizes, the presence/absence of dolomite crystals). The common denominator of stylolites is a serrated surface, in two-dimensional outcrops visible as an irregular line coated by insoluble residues (stylocumulate) such as clay minerals, organic matter, and pyrite and other non-carbonate minerals (Immenhauser 2021). Stylolite sets, or bedding-parallel stylolites, are good indicators of the principal stress direction, which is usually normal to the orientation of the stylobedding (see Buxton and Sibley 1981; Carrio-Schaffhauser et al. 1990; Petit and Mattauer 1995; Renard et al. 2004; Machel 2005; Koehn et al. 2007; Ebner et al. 2009; Flügel 2010; Heap et al. 2014; Immenhauser 2021; for a detailed treatment of stylolitization).

17.3.3 Microfractures

In geology, fractures are macroscale features (both in length and width) defined as discontinuities in the rocks where cohesion has been lost due to brittle deformation or physical diagenesis (e.g., Nelson 2001; Dashti et al. 2018). On the other hand, microfractures are small, high-aspect-ratio cracks in rocks visible only under magnification, having widths generally less than 0.1 mm (Anders et al. 2014; see also Ezati et al. 2018), but occasionally wider.

Microfractures are features caused by the effects of tectonic stress and overburden and associated application of differential stresses on rocks, developing preferentially in brittle lithologies. These post-compactional, planar discontinuities in the rocks are thought to be formed within a wide depth range in the intermediate- and deep-burial environments. In some cases, the development of microfractures is associated to stylolitization processes (e.g., Alsharhan and Sadd 2000; Morad et al. 2018). Microfracture systems can extend from centimeters to hundreds of meters (and even a few kilometers). They can develop as a single discontinuity, or in parallel or conjugate sets. After their formation, microfractures may be occluded by late-burial calcite cement postdating pressure solution, or may remain preserved as open pores. The latter scenario commonly involves a positive impact on the quality of carbonate reservoirs. See Lucia (2007) and Anders et al. (2014) for a more detailed treatment on the occurrence of microfractures.

17.3.4 Burial Recrystallization

In this chapter, we consider recrystallization (*sensu stricto*) of calcite-to-calcite in the mesogenetic zone as a pH-driven dissolution-reprecipitation process that leads to a significant enlargement of crystals (the Ostwald ripening process, or the growth of larger crystals at the expense of smaller ones; e.g., Folk 1965; Morse and Casey 1988). Continued "wet" diagenesis in the intermediate- and deep-burial environments may operate on both non-recrystallized and previously recrystallized calcite substrates. Generally, recrystallization fabrics formed by in the burial realm are characterized by hypidiotopic and xenotopic fabrics (predominantly subhedral and anhedral crystals, respectively) consisting of large crystals often reaching mm-scale sizes (see definitions of descriptive terms for recrystallization textures and fabrics in Chap. 16, Section 16.5.2). These crystals commonly exhibit twinning lamellae. This feature, also termed twinning of the calcite, is the conspicuous intracrystal "twin" strippy pattern (parallel and closely spaced lines) with striking extinction that forms in response to intracrystalline deformation at low temperatures ($< 200\ ^{\circ}C$) (see Burkhard 1993; Ferrill 1998; Ferrill et al. 2004). In the burial realm, Immenhauser (2021) consider recrystallization a diagenetic process common in the deep-burial environment.

17.3.5 Mesogenetic Dissolution

Dissolution of stable carbonate minerals (LMC and dolomite) in burial settings (away from the influence of acidic waters of meteoric origin) is, in many cases, a localized phenomenon; although each case study is particular. It requires a source or sources for generating fluids sufficiently undersaturated with respect to carbonate. In scenarios where mesogenetic dissolution creates substantial secondary porosity and hence increases the porosity and permeability of a carbonate reservoir, the volume of acidic fluids generated must be sufficiently large to create an effective network for fluid flow.

The importance of burial dissolution creating significant porosity is a matter of a long-standing debate in the literature (e.g., Mazzullo and Harris, 1992; Pöppelreiter et al. 2005; Ehrenberg et al. 2012; Valencia and Laya 2020). Because of the ubiquitous presence and rapid kinetics of dissolution of carbonate minerals, the mesogenetic pore waters in sedimentary basins can be expected to be always saturated and buffered by carbonates (Ehrenberg et al. 2012). Generally, carbonate dissolution in the mesogenetic zone postdates mechanical and chemical compaction, as well as microfracturing. In this context, localized dissolution commonly takes place along stylolites, sutured grain contacts, and microfractures. In many cases, such processes of localized dissolution are followed by cementation. In some study cases, stylolitization is associated to the creation of porosity (i.e., stylolite porosity; see Heap et al. 2014; Hou et al. 2023).

In the mesogenetic zone, the two main models that support the generation of significant volumes of acidic fluids are: the inorganic CO_2-model, in which high abundance of CO_2 are generated by magmatic or metamorphic sources (i.e., basement units) and transported through carbonate strata by hydrothermal CO_2-rich fluids (e.g., Corbella et al. 2004; Beavington-Penney et al. 2008; Valencia and Laya 2020; see also Ehrenberg et al. 2012); and the kerogen maturation model, in which CO_2 and organic acids are formed during hydrocarbon generation (e.g., Mazzullo and Harris 1992; Mazzullo 2004). For both models, acidic fluids may flow through faults, fractures, preexisting permeability, stylolites, unconformities, and bedding planes. In addition, porosity-gain burial dolomitization of large limestone volumes can occur by thermally driven circulation of seawater or evaporatively concentrated seawater to depths of several kilometers (Machel 2004).

The solubility of calcite decreases with increasing temperature, as does the solubility of CO_2 gas in water (Morse and Mackenzie 1990). However, undersaturation respect to $CaCO_3$ and the dissolution of carbonate during burial occurs in the presence of organic matter, as increasing temperature (within a range of ~50–150 ° C) leads to the generation of fluids rich in CO_2, H_2S and organic acids derived from the thermal maturation of the kerogen in organic-rich rocks (see Mazullo and Harris, 1992; Hunt 1996; Mazzullo 2004; Killops and Killops 2005; Machel 2005; Peters et al. 2005; Ahr 2008). In this context, dissolution may occur in both organic-rich limestones and clayey limestones (black shales), as well as in limestone strata directly underlying, overlying, or stratigraphically close to organic-rich intervals.

The localized nature of many mesogenetic dissolution processes is acknowledge by Mazzullo and Harris (1992). These authors state that "dissolution may occur at one depth, and cementation at another depth within a single stratum of carbonate rocks. Likewise, dissolution in one area may be replaced by cementation in another area despite the common burial-depth and temperature histories of strata". Ehrenberg et al. (2012) acknowledge the possibility of relatively small amounts of porosity being formed by late-burial dissolution, although in many cases, dissolution occurs as a localized process restricted to limestones strata in contact with and/or adjacent to acidic fluids. Consequently, the presumption that mesogenetic dissolution produces a net increase in secondary porosity should not be used in the prediction of carbonate reservoir quality. We agree with these authors on the second point but disagree with the first. Mesogenetic dissolution, which includes the enlargement of early pores, can create significant volumes of porosity and have a significant impact in the development of good reservoir zones (e.g., Dravis 1989; Esteban and Taberner 2003; Lambert et al. 2006; Chandra et al. 2014; Jiang et al. 2015; Wei et al. 2017; Valencia and Laya 2020).

In Venezuela, there are two significant cases of mesogenetic dissolution contributing to hydrocarbon production of carbonate reservoirs. In the West Urdaneta oil field, located in western Lake Maracaibo (MB), Pöppelreiter et al. (2005) propose that the creation of dissolution-enlarged (vuggy) pores postdates stylolitization. This secondary porosity creation process occurred in grain-supported fabrics (packstones and grainstones) of the Piché Member (Apón Formation) and the Maraca Formation of late Aptian and late Albian age, respectively. These porous facies have produced light oils generated by Aptian, Albian and Cenomanian source rocks. The Perla Limestone (Late Oligocene-Early Miocene in the EGVSB), the largest gas reservoir in Latin America (Castillo et al. 2017), is a well-documented example of the importance of burial dissolution in some producing carbonate reservoirs. In this red algae and LBF-dominated succession, the highest porosity and permeability values correspond with intervals hosting secondary porosities created in the deep-burial environment due to the action CO_2-rich basement fluids. Among those secondary porosities (e.g., vugs, dissolution-enlarged interparticle and intraparticle pores), the creation of pervasive microporosity stands out due to the intense "blue haze" observed in epoxy-impregnated thin sections. Valencia and Laya (2020) consider that the relative absence of late cements postdating an important burial dissolution process could suggest alternating conditions, from open to closed system, driven by tectonic events that discharged super-saturated fluids out of the reservoir during open-system stages. Secondary porosities in previously mentioned carbonate reservoirs will be discussed in more detail in Chap. 19.

17.3.6 Burial Cementation

The chemistry of the pore-fluids in the burial domain (intermediate and deep), especially the low Mg^{2+}/Ca^{2+} ratio, favors the precipitation of LMC (e.g., Flügel 2010). As previously mentioned, high temperatures in the deep-burial environment promote carbonate precipitation/cementation, as LMC solubility decreases. In simpler terms, the effect of temperature, with increasing burial depth, on carbonate solubility tends to result in carbonate cementation rather than dissolution (Moore 1989). However, in the mesogenetic zone, cementation is, in most cases, localized, as it is a rock-buffered diagenetic processes hindered by the scarcity of pore fluids (see Immenhauser 2021). As discussed in the previous section, mesogenetic dissolution and cementation are closely-related, pH-driven processes that are controlled by the chemistry of pore water. During kerogen maturation and hydrocarbon generation, cementation rather than dissolution instead may occur if the concentrations of acids in pore the fluid do not override the buffering effects (Mazzullo and Harris 1992). Furthermore, pressure solution is a main source of dissolved Ca, Mg and CO_3 (e.g. Immenhauser 2021).

Precipitation of LMC cements in the mesogenetic zone mostly occur as a dissolution-reprecipitation process, post-dating mechanical and chemical compaction, microfracturing, and mesogenetic dissolution. Therefore, LMC cements (of the blocky, poikilotopic, drusy, syntaxial overgrowth, granular, dogtooth types; see Flügel 2010; Immenhauser 2021) are found filling microfractures and grains breakages, partially engulfing grains affected by chemical compaction, and filling partially secondary (dissolution-enlarged) pores. In the intermediate- and deep-burial realms, recrystallization of pre-existing cements, which were originally precipitated in the meteoric or shallow-burial environments, forms new mosaics consisting of larger crystals that commonly exhibit twin lamellae.

17.4 Burial Diagenesis: Photographic Set and Descriptive Captions

In all photos, IW = image width (the true length of the horizontal axis of the view)

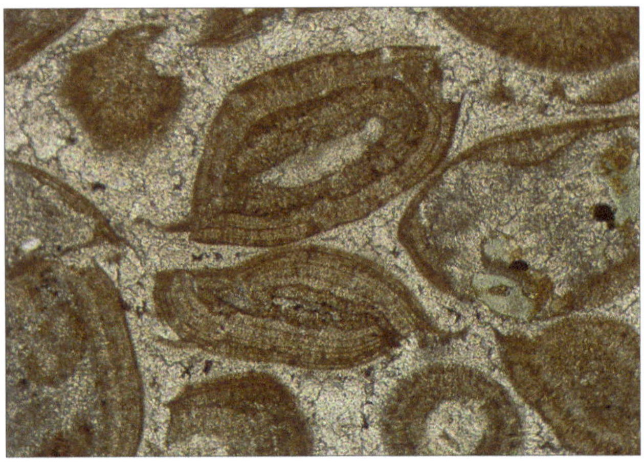

Cretaceous (middle Albian). Lisure Formation. Maracaibo Basin
Mechanical (physical) compaction due to overburden caused brittle fracturing (or crushing) of the ooids in the image. Oolitic cortices are hard and brittle, and prone to collapse at localized points. Although physical compaction of these ooids is consequence of overburden in the intermediate-burial environment, prominent meteoric diagenesis that suffered this oolite microfacies probably weakened and partially leached the concentric structures of these non-skeletal grains.

IW = 1.8 mm. PPL.

Cretaceous (middle Albian). Lisure Formation. Maracaibo Basin
Intergranular pressure solution between ooids and peloidal micritic grains. Note the thin solution seams (white arrows) formed along sutured contacts between deformed grains, evidencing chemical compaction postdating mechanical compaction. Also note the poikilotopic (zoned) cement filling the small, narrow pores. The carbonate source for the precipitation of this post-compactional, LMC precipitate was derived from pressure solution.

IW = 1.8 mm. PPL

Middle Permian (Cisularian to Guadalupian). Palmarito Formation. Venezuelan Andes

Physical compaction due to increasing depth and overburden caused the fracturing of ostracods carapaces in this packed packstone. Some carapaces escaped from crushing due to white-cloudy sparry calcite infilling. Incipient pressure solution is indicated by the accumulation of dark-colored, insoluble material surrounding the valves and along sutured contacts.

IW = 4.2 mm. PPL

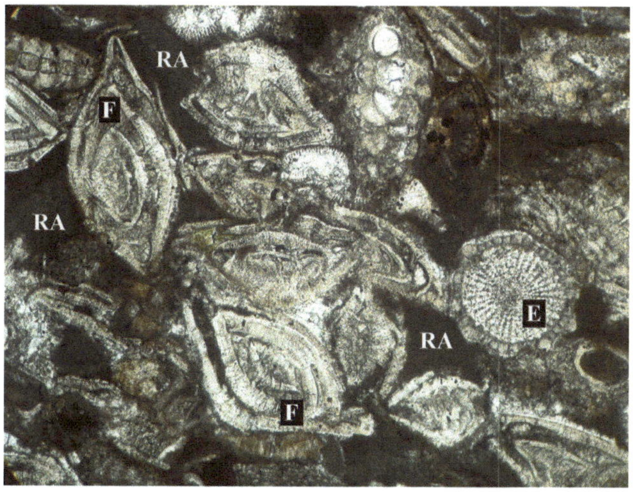

Late Oligocene-Early Miocene. Perla Limestone. Eastern Gulf of Venezuela Sub-Basin

Physical and chemical compaction formed this fitted fabric consisting of tests of LBF (F) and red algal (RA) grains, accompanied by an echinoid (E). Tests are in sutured contact with each other, while the red algae grains are "squeezed" (plastically deformed) between them. Pressure solution is especially apparent in the foraminifera tests in the center of the image, as these are highly deformed and partially interpenetrated.

IW = 2.4 mm. PPL.

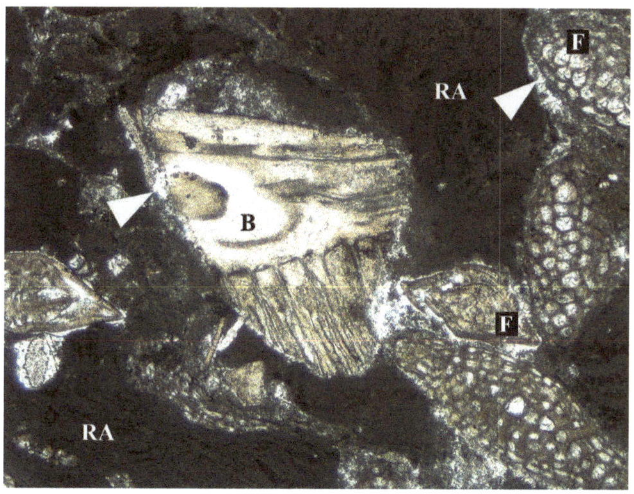

Late Oligocene-Early Miocene. Perla Limestone. Eastern Gulf of Venezuela Sub-Basin

The image shows grains in sutured contact exhibiting different behavior in response to mechanical compaction. Plastic deformation of squeezed red algae (RA) grains contrasts with the brittleness of the central, highly truncated barnacle (B) grain. Less affected LBF (F) tests exhibit characteristic high resistance to physical deformation. Note the calcite cements precipitated along grain contacts (white arrows). This late burial calcite cement postdates mechanical compaction and rearrangement of grains, and indicates localized pressure solution.

IW = 4.2 mm. PPL.

Late Oligocene-Early Miocene. Perla Limestone. Eastern Gulf of Venezuela Sub-Basin

Post-compactional microfractures (grain breakages) in the image are cross-cutting layered red algae. These deformation features are filled by calcite cement (limpid white). The carbonate source for cementation was most likely derived from pressure solution. In the Perla Limestone, most microfractures are associated to grain breakage of red algae, molluscs and barnacles.

IW = 4.2 mm. PPL.

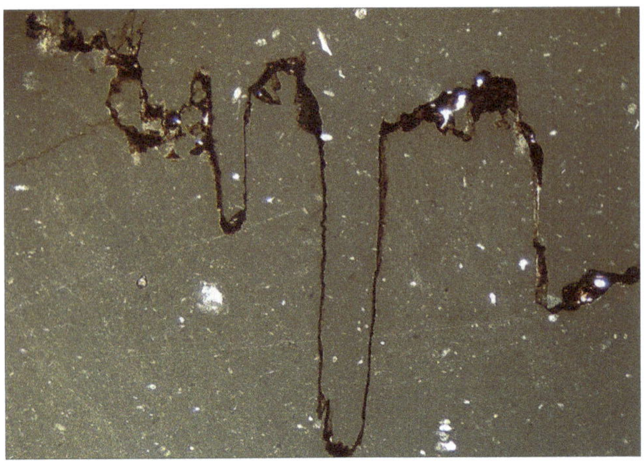

Cretaceous (middle Aptian). Piché Member, Apón Formation. Maracaibo Basin

A stylolite crossing a moderately organic-rich mudstone. The irregular surface was produced by differential vertical movement of insoluble organic matter originally dispersed through the matrix. This single stylolite exhibits high-amplitude peaks, which is a common stylolite type formed in fine-grained mudstones. Localized pressure solution along the rough surface was followed by accumulation of organic matter. This process postdates compaction and dewatering of the homogeneous micritic fabric and expulsion of water films charged with CO_2, which promoted dissolution with increasing temperature and pressure.

IW = 4.2 mm. PPL.

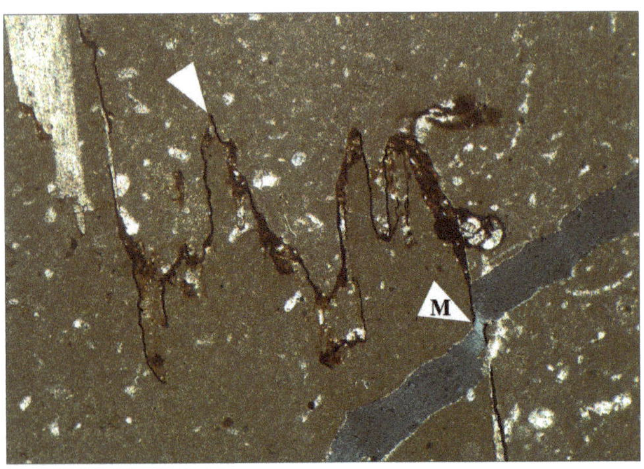

Cretaceous (middle Aptian). Piché Member, Apón Formation. Maracaibo Basin

Another example of a single stylolite crossing a moderately organic-rich mudstone and exhibiting high-amplitude peaks. The formation of the microfracture (M) predates the formation of the stylolite, as the latter is cross-cutting and displacing the former at the intersection point. The stylolite exhibits "teeth-like" terminations (white arrow), which are more or less perpendicularly oriented respect to the direction of propagation of the pressure solution surface. These small sharp peaks are commonly used to decipher the largest principal compressive stress direction.

IW = 2.4 mm. CPL.

Late Oligocene-Early Miocene. Perla Limestone. Eastern Gulf of Venezuela Sub-Basin

In this sample, two deep-burial diagenetic process are clearly visualized. Chemical compaction is evident due to the development of a low-amplitude stylolite (white arrows), which is crossing the entire field of view and delimits the sharp contact between two different microfacies. Above, a micropleoidal matrix exhibits no signs of having been affected by deep-burial dissolution. Below, the same micropeloidal matrix, accompanied by red algal grains, exhibit microporosity development postdating the formation of the stylolite.

IW = 7.5 mm. PPL.

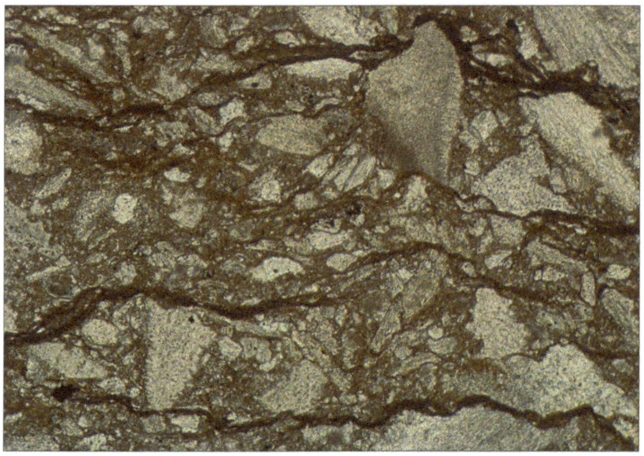

Cretaceous (latest Albian). Maraca Formation. Maracaibo Basin

This packstone exbibits abundant stylolites and microstylolites (or solution seams). Note that the set of low-amplitude pressure solution surfaces are more or less parallel to each other, forming a pseudo-bedding pattern (commonly termed stylobedding). The stylolites were formed parallel to the horizontal bedding and more or less perpendicular to the load applied by overburden. Despite development of such prominent deep-burial diagenetic features, many grains in the image show random ("puzzle-like") orientations. Rapid mechanical compaction might have prevented total reorientation of grains.

IW = 4.2 mm. PPL.

Cretaceous (middle Albian). Lisure Formation. Maracaibo Basin

Facies consisting of abrupt lithologic changes are susceptible to suffer differential compaction and pressure dissolution. In this sample, glauconitic peloids at the top suffered compaction against a dolomitic fabric at the bottom. Note the flattened shapes of the peloids, which undewent ductile ("squeezing") deformation predating the accumulation of organic matter and formation of stylocumulates (mainly consisting of pyrite) along sutured grain contacts. Dolomite fabrics are far more resistant to deformation and dissolution compared to grain-dominated fabrics, although they are more prone to microfracturing.

IW = 2.4 mm. CPL.

Middle Permian (Cisularian to Guadalupian). Palmarito Formation. Venezuelan Andes

This packstone underwent strong mechanical (physical) and chemical compaction, thus forming a condensed (puzzle-like) fabric. Grains in the image, mostly meteorically altered crinoids and aragonitic bioclasts, are more or less horizontally oriented (perpendicular to the load applied by the overburden) and parallel to the stylobedding formed by the irregular and anastomosing set of stylolites. Calcite dissolution along grain contacts is evidenced by abundant truncated and interpenetrated grains.

IW = 4.2 mm. PPL.

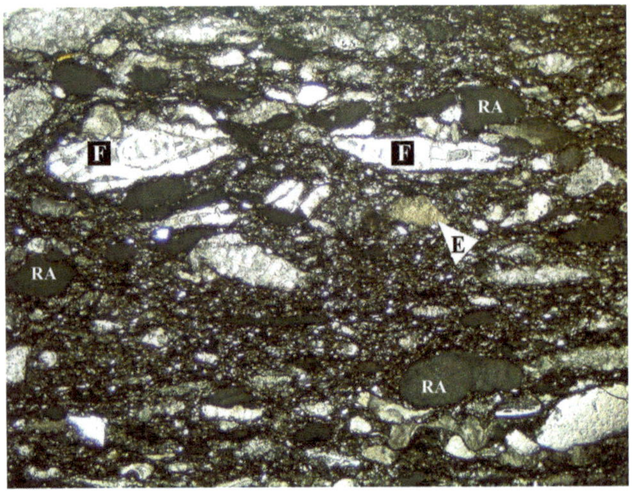

Late Oligocene-Early Miocene. Perla Limestone. Eastern Gulf of Venezuela Sub-Basin

Physical and chemical compaction is prominent in this sample consisting of LBF (F) tests, red algae (RA) grains, subordinate echinoderms (E), and heterogeneous matrix. Mechanical (physical) compaction led to the horizontalization of grains. The development of high-density solution seams (or microstylolites), which severely darken the visual of the image, postdates physical compaction. These features are oriented perpendicular to the direction of main stress, or parallel to the bedding plane. This microfacies corresponds to gravity flow deposits accumulated in the Perla outer ramp (see Pomar et al. 2015).

IW = 7.5 mm. PPL.

Late Oligocene-Early Miocene. Perla Limestone. Eastern Gulf of Venezuela Sub-Basin

This fitted fabric consists mainly of flattened tests of LBF (F) and squeezed (plastically deformated) red algae (RA) grains. Prominent physical and chemical compaction produced an overpacked fabric in which individual grains are difficult to differentiate. The high density of solution seams (white arrows) between grains in sutured contact outlines some of these individual skeletal components, hence facilitating their petrographic recognition.

IW = 7.5 mm. PPL.

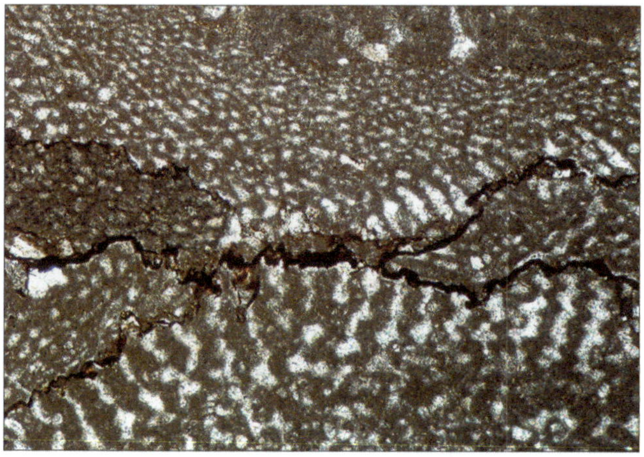

Cretaceous (middle Aptian). Piché Member, Apón Formation. Maracaibo Basin

Mechanical compaction was followed by pressure solution in this fitted fabric consisting of tests of the larger benthic foraminifer *Mesorbitolina*. Deformed tests (originally conical in shape) are recognizable due to numerous, distinctive chamberlets filled with sparry calcite. As in the previous example, pressure solution between tests resulted in accumulation of insoluble organic material along margins of individual tests.

IW = 2.4 mm. PPL.

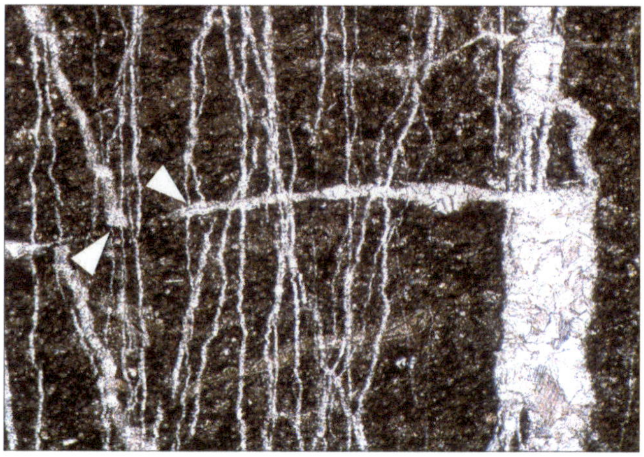

Middle Permian (Cisularian to Guadalupian). Palmarito Formation. Venezuelan Andes

A set of more or less vertical microfractures crosses an organic-rich limestone. These features are filled with burial cements. Therefore, in cases like this, microfractures are commonly referred to as "calcite veins". They were formed due to localized brittle failure and tectonic fracturing. The single horizontal microfracture in the image cross-cuts the vertical set. Note the small displacement at intersection points in some of the vertical microfractures (white arrows), indicating the later formation of the horizontal one.

IW = 4.2 mm. PPL.

Cretaceous (middle-late Cenomanian). La Luna Formation *sensu stricto*. Maracaibo Basin

Microfractures in this muddy limestone containing tests of PF are filled by coarse-blocky calcite cement. A stylolite cross-cuts the microfractures, indicating its occurrence post-dates brittle microfracturing. As in the previous example, the law of "cross-cutting" relationship is key for determining the relative timing of diagenetic features (or paragenetic relations).

IW = 2.4 mm. PPL.

Cretaceous (middle Albian). Lisure Formation. Maracaibo Basin

A close view of coarse-grained blocky calcite cement, filling a post-compactional microfracture. Large, subhedral crystals with well-developed crystal faces form a hypidiotopic fabric. Calcium carbonate dissolution, as a source of Ca^{2+} and CO_3^{2-} ions for burial cementation, was driven by thermal maturation of organic matter contained in the dark micritic matrix surrounding the microfracture.

IW = 1.3 mm. CPL.

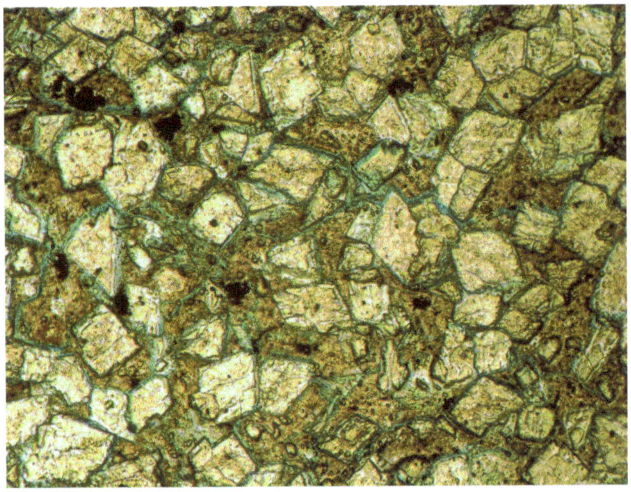

Cretaceous (early Aptian). Apón Formation, Machiques Member. Maracaibo Basin

The development of microfractures is common in dolomitic fabrics, as the dolomite crystals increase the hardness of carbonate rocks. This is the case with this dolomitic shale, in which mechanical compaction and fluid expulsion (dewatering) of the clay-rich matrix were followed by hydraulic fracturing (dilation). Note that the set of microfractures fringes the rhombic crystals. The propagation of such stress-induced cracks occurred along both crystal junctions and crossing the matrix, thus forming a network of straight open discontinuities (see Roehl 1981; for a full review on microfracturing of dolomite facies).

IW = 2.4 mm. PPL.

Cretaceous (early-late Aptian). Piché Member, Apón Formation. Maracaibo Basin

A calcite crystal exhibiting strong development of twin lamellae (or twinning of the calcite). In this particular case, the parallel and closely-spaced intracrystal lines, which exhibit striking extinction, were formed due to burial recrystallization involving dedolomitization (i.e., dolomite replaced by LMC).

IW = 2.4 mm. CPL.

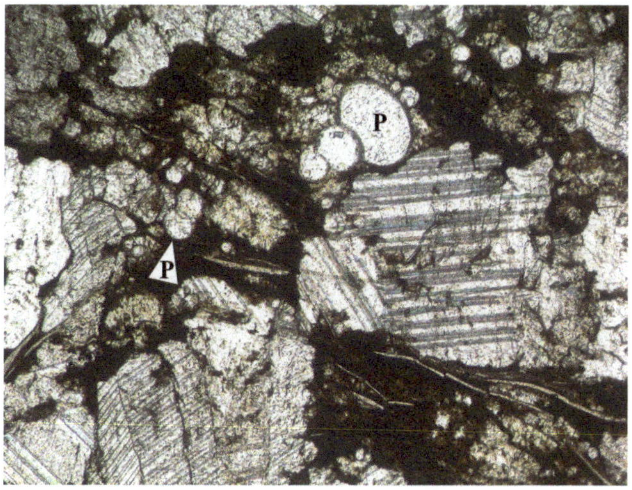

Cretaceous (late Cenomanian). La Luna Formation *sensu stricto*. Maracaibo Basin

This image depicts an organic-rich limestone where carbonate recrystallization is prominent. Note the coarse-grained carbonate crystals exhibiting twinning of the calcite. In this sample, recrystallization postdated kerogen maturation, a process that generated organic acids that dissolved-reprecipitated the LMC tests of PF (P) and their sparry calcite infillings. However, some of these small globular tests escaped dissolution and are therefore identifiable.

IW = 1.3 mm. PPL.

Cretaceous (middle Albian). Lisure Formation. Maracaibo Basin

Recrystallized limestone. Coarse-grained inequigranular crystals (subhedral and anhedral in texture) form a hypidiotopic-xenotopic fabric. Conspicuous, differently oriented twin lamellae, combined with the large size of the crystals, are features associated to recrystallization in the deep-burial realm. In the Lisure Formation, recrystallized fabrics are present as small patches surrounded by micritic-dominated fabrics. Coarse-grained, recrystallized diagenetic textures are commonly the result of more than one recrystallization event.

IW = 2.4 mm. CPL.

Cretaceous (Turonian). O Member, Escandalosa Formation. Barinas Sub-Basin

Another recrystallized limestone consisting of inequigranular crystals forming a hypidiotopic-xenotopic fabric. In the O Member, crystalline mosaics resulting from recrystallization, such as this example, are found as patches enclosed by massive dolomitic fabrics. Prominent twinning of the calcite, combined with abundant hydrocarbon inclusions (dark spots cross-cutting crystals), suggest that the final recrystallization event occurred in the deep-burial diagenetic environment, taking into account that the O Member directly underlies the oil prone La Morita Member of the Navay Formation.

IW = 2.4 mm. PPL.

Cretaceous (Turonian). O Member, Escandalosa Formation. Barinas Sub-Basin

The image depicts a recrystallized poikilotopic fabric consisting of calcite crystals with poorly developed boundaries, exhibiting prominent twinning lamellae, and enclosing a dolomite crystal. The latter exhibits characteristic high birefringence. The formation of this rhombic crystal predates the formation, at higher temperatures, of the enclosing recrystallized fabric. This paragenetic relationship evidences the resistance of dolomite crystals to burial recrystallization compared to calcium carbonate substrates, which are more prone to undergo pH-driven recrystallization.

IW = 2.4 mm. CPL.

Cretaceous (latest Albian). Maraca Formation. Maracaibo Basin

A vuggy pore partially filled with dogtooth crystals. Note the hydrocarbon impregnations in the vug, which might suggest that the vug (dissolution) and LMC crystals (cementation) were formed mesogenetically. For this sample, which belongs to a core section in the Urdaneta West field (western Lake Maracaibo), we consider that localized dissolution and cementation may have occurred in the deep-burial environment. This is based on the results (e.g., $\delta^{18}O$ and $\delta^{13}C$ on LMC cements) obtained by Pöppelreiter et al. (2005) in their study focused on burial diagenesis in Cretaceous reservoirs in that oil field.

IW = 1.3 mm. PPL.

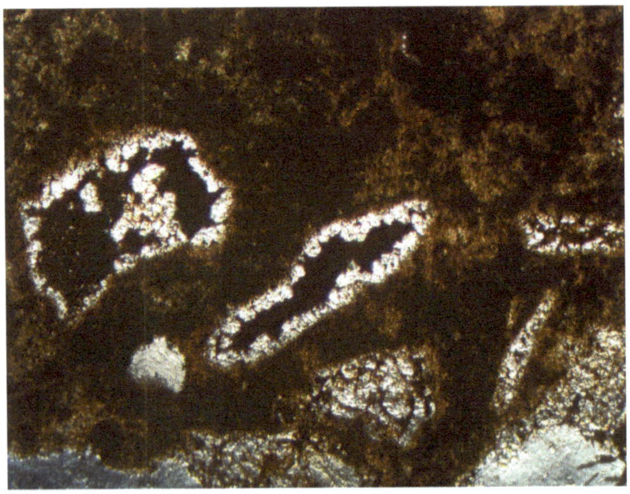

Cretaceous (middle-late Cenomanian). La Luna Formation *sensu stricto*. Maracaibo Basin

Vuggy pores in the image are internally rimmed by LMC crystals. The formation of the vugs postdates kerogen maturation and hydrocarbon generation, as this sample belongs to the base of the thermally mature La Luna Formation. The organic richness of this limestone is indicated by the dark color of the matrix. The rims of cements were precipitated due supersaturation postdating dissolution and formation of the vugs.

IW = 2.4 mm. CPL.

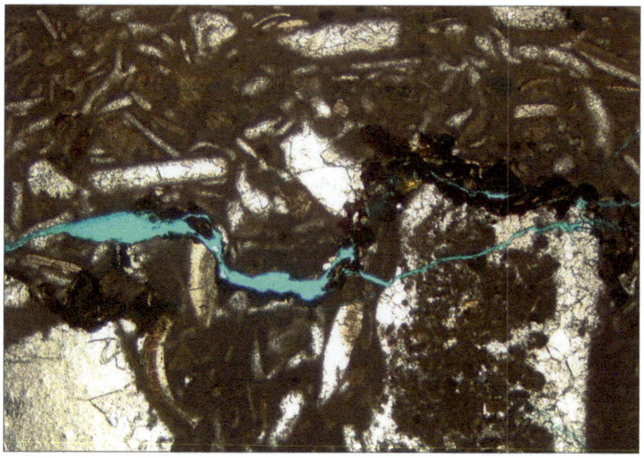

Cretaceous (middle Aptian). Piché Member, Apón Formation. Maracaibo Basin

In this wackestone-packstone of molluscs, localized dissolution occurred along a stylolite surface. Stylolite porosity is not common in shallow carbonates in the MB, as in the majority of cases, dissolution along stylolites (driven by the action of insoluble organic matter) was followed by cementation.

IW = 4.2 mm. PPL.

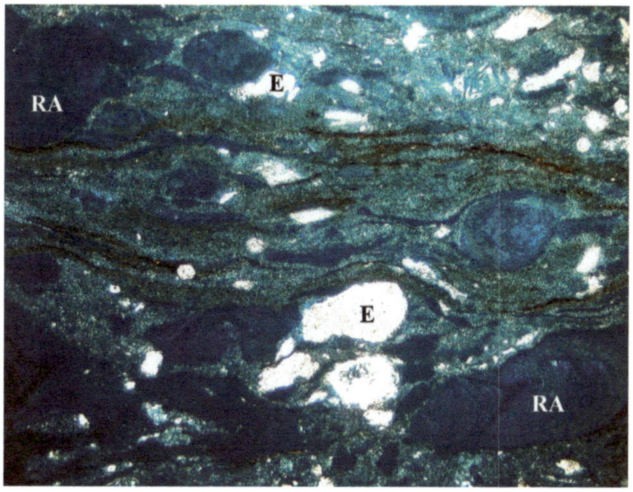

Late Oligocene-Early Miocene. Perla Limestone. Eastern Gulf of Venezuela Sub-Basin

Different stages of burial diagenesis are well exemplified in this dolomitic wackestone-packstone containing red algae (RA) and echinoids (E). Physical compaction (reorientation and plastic deformation of the grains) predates stylolitization and development of stylobedding. Microporosity development, which is thought to postdates stylolitization, is indicated by intense blue stained ("blue haze") of both red algal grains and dolomitic-muddy matrix. This corrosive dissolution process occurred due to the ascent of CO_2-rich fluids coming from the basement of the Perla reservoir (see Valencia and Laya 2020 for more details).

IW = 7.5 mm. PPL.

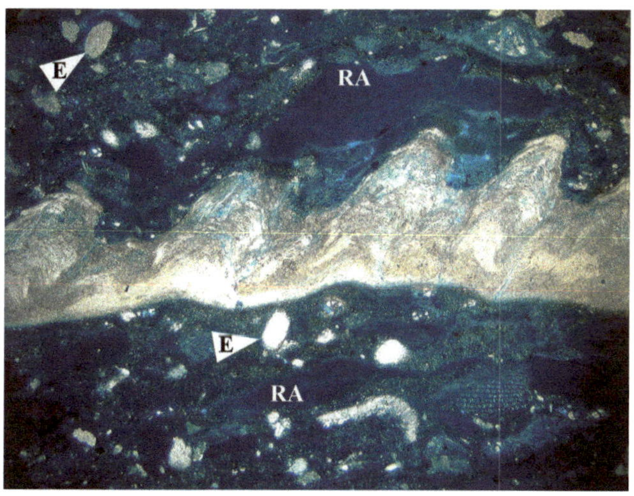

Late Oligocene-Early Miocene. Perla Limestone. Eastern Gulf of Venezuela Sub-Basin

Another example of the development of mechanical compaction predating microporosity development. Note the intense "blue haze" on flattened red algae (RA) grains and dolomitic matrix. This pervasive dissolution contrasts with the noticeably less altered small echinoid (E) grains and the large, plicated central shell. As in the previous example, earlier mineralogical stabilization of HMC echinoids was responsible for their resistance to the posterior action of corrosive burial fluids. The central LMC shell shows traces of corrosion; nevertheless, its original LMC composition prevented total corrosion.

IW = 7.5 mm. PPL.

Late Oligocene-Early Miocene. Perla Limestone. Eastern Gulf of Venezuela Sub-Basin

This skeletal limestone exhibits various types of burial dissolution porosity. In this context, many secondary pores in the image are of the vuggy (V) type. These pores resulted from dissolution-enlarged interparticle pores. In the Perla Limestone, vuggy pores display a wide range of sizes, as seen in the image, reaching the largest diameters of secondary burial dissolution porosities in this unit.

IW = 7.5 mm. PPL.

References

Ahr, W., 2008. Geology of Carbonate Reservoirs. The Identification, Description and Characterization of Hydrocarbon Reservoirs in Carbonate Rocks. Wiley, 277 pp.

Alsharhan A., Sadd J. L., 2000. Stylolites in Lower Cretaceous carbonate reservoirs, U.A.E: Society for Sedimentary Geology Special Publication, 69, 185–207.

Anders, M.H., Laubach, S. E., Scholz, C. H., 2014. Microfractures: a review. *Journal of Structural Geology*, *69*, 377–394.

Bathurst, R. G., 1975. Carbonate Sediments and Their Diagenesis: Amsterdam, Elsevier, 657 pp.

Beavington-Penney, S. J., Nadin, P., Wright, V. P., Clarke, E., McQuilken, J., Bailey, H. W., 2008. Reservoir quality variation on an Eocene carbonate ramp, El Garia Formation, offshore Tunisia: structural control of burial corrosion and dolomitisation. *Sedimentary Geology*, *209*, 42–57.

Bjørlykke, K., 1993. Fluid flow in sedimentary basins. *Sedimentary Geology*, *86*, 137–158.

Burkhard, M., 1993. Calcite twins, their geometry, appearance and significance as stress-strain markers and indicators of tectonic regime: a review. *Journal of Structural Geology*, *15*, 351–368.

Buxton, T. M., and Sibley, D. F., 1981. Pressure solution features in a shallow buried limestone. *Journal of Sedimentary Research*, *51*, 19–26.

Castillo, V., Benkovics, L., Cobos, C., Demuro, D., Franco, A., 2017. Perla field: The largest discovery ever in Latin America, In: R. K. Merrill and C. A. Sternbach, eds., Giant fields of the decade 2000–2010: *American Association of Petroleum Geologists, Memoir 113*, 141–152.

Carrio-Schaffhauser, E., Raynaud, S., Latiere, H. J., Mazerolle, F., 1990. Propagation and localization of stylolites in limestones. *Geological Society, London, Special Publications*, *54*, 193–199.

Chandra, V., Wright, P., Barnett, A., Steele, R., Milroy, P., Corbett, P., Geiger, S., 2014. Evaluating the impact of a late-burial corrosion model on reservoir permeability and performance in a mature carbonate field using near-wellbore upscaling. *Geological Society, London, Special Publications*, *406*, 427–445.

Choquette, P. W., and James, N. P., 1987. Diagenesis in limestones - 3. The deep burial environment: *Geoscience Canada*, *14*, 3–35.

Choquette, P. W., and James, N. P., 1990. Limestones - The burial diagenetic environment, In: McIlreath, I. E., and Morrow, D. A., eds., Diagenesis: Geoscience Canada Reprint Series 4, pp. 75–112.

Corbella, M., Ayora, C., Cardellach, E., 2004. Hydrothermal mixing, carbonate dissolution and sulfide precipitation in Mississippi Valley-type deposits. *Mineral Deposits*, *39*, 344–357.

Croizé, D., Renard, F., Gratier, J. P., 2013. Compaction and porosity reduction in carbonates: A review of observations, theory, and experiments. *Advances in Geophysics*, *54*, 181–238.

Dashti, R., Rahimpour-Bonab, H., Zeinali, M., 2018. Fracture and mechanical stratigraphy in naturally fractured carbonate reservoirs—A case study from Zagros region. *Marine and Petroleum Geology*, 97, 466–479.

Dravis, J. J., 1989. Deep-burial microporosity in Upper Jurassic Haynesville oolitic grainstones, east Texas. *Sedimentary Geology*, *63*, 325–341.

Ebner, M., Koehn, D., Toussaint, R., Renard, F., Schmittbuhl, J., 2009. Stress sensitivity of stylolite morphology. *Earth and Planetary Science Letters*, *277*, 394–398.

Ehrenberg, S. N., Walderhaug, O., Bjorlykke, K., 2012. Carbonate porosity creation by mesogenetic dissolution: reality or illusion? *American Association of Petroleum Geologists, Bulletin 96*, 217–233.

Esteban, M., and Taberner, C., 2003. Secondary porosity development during late burial in carbonate reservoirs as a result of mixing and/or cooling of brines. *Journal of Geochemical Exploration*, *78*, 355–359.

Ezati, M., Azizzadeh, M., Riahi, M. A., Fattahpour, V., Honarmand, J., 2018. Characterization of micro-fractures in carbonate Sarvak reservoir, using petrophysical and geological data, SW Iran. *Journal of Petroleum Science and Engineering*, *170*, 675–695.

Fantle, M. S., Barnes, B. D., Lau, K. V., 2020. The Role of Diagenesis in Shaping the Geochemistry of the Marine Carbonate Record. *Annual Review of Earth and Planetary Sciences*, *48*, 549–583.

Ferrill, D. A., 1998. Critical re-evaluation of differential stress estimates from calcite twins in coarse-grained limestone. *Tectonophysics*, *285*, 77–86.

Ferrill, D. A., Morris, A. P., Evans, M. A., Burkhard, M., Groshong Jr., R. H., Onasch, C. M., 2004. Calcite twin morphology: a low-temperature deformation geothermometer. *Journal of Structural Geology*, *26*, 1521–1529.

Flügel, E, 2010. Microfacies of Carbonate Rocks—Analysis, Interpretation and Application. Springer, Heidelberg-Dordrecht-London-New York, 984 pp.

Folk, R. L., 1965. Some aspects of recrystallization in ancient limestones, *in* L. C. Pray, and R. S. Murray, eds., Dolomitization and Limestone Diagenesis: Tulsa, OK, *SEPM Special Publication*, *13*, 14–48.

Füchtbauer, H., Heling, D., Müller, G., Richter, D. K., Schmincke, H. -U., Schneider, H. -J., Valeton, I., Walther, H. J. Wolf, M., 1988. *Sedimente und sedimentgesteine*. Stuttgart: Schweizerbart, 1141 pp.

Hamilton, E. L., 1976. Variations of density and porosity with depth in deep-sea sediments. *Journal of Sedimentary Research*, *46*, 280–300.

Heap, M. J., Baud, P., Reuschlé, T., Meredith, P. G., 2014. Stylolites in limestones: Barriers to fluid flow? *Geology*, *42*, 51–54.

Higgins, J. A., Blättler, C. L., Lundstrom, E. A., Santiago-Ramos, D. P., Akhtar, A. A., Ahm, A. -S. -C., Bialik, O., Holmden, C., Bradbury, H., Murray, S. T., Swart, P. K., 2018. Mineralogy, early marine diagenesis, and the chemistry of shallow-water carbonate sediments. *Geochimica et Cosmochimica Acta*, *220*, 512–534.

Hou, Z., Fusseis, F., Schöpfer, M., Grasemann, B., 2023. Synkinematic evolution of stylolite porosity. *Journal of Structural Geology*, *173*, 104,916.

Hunt, J. M., 1996. Petroleum Geochemistry and Geology. Second Edition: W. H. Freeman and Company, NY, 743 pp.

Immenhauser, A., 2021. On the delimitation of the carbonate burial realm. *The Depositional Record*, *00*, 1–51.

James, N. P., and Jones, B., 2015. Origin of Carbonate Sedimentary Rocks. John Wiley & Sons, Ltd, Chinchester, 464 pp.

Jiang, L., Pan, W., Cai, C., Jia, L., Pan, L., Wang, T., Li, H., Chen, S., Chen, Y., 2015. Fluid mixing induced by hydrothermal activity in the ordovician carbonates in Tarim Basin, China. *Geofluids*, *15*, 483–498.

Jin, Z., Zhang, L., Yang, L. Hu, W., 2004. A preliminary study of mantle-derived fluids and their effects on oil/gas generation in sedimentary basins. *Journal of Petroleum Science and Engineering*, *41*, 45–55.

Killops, S. D., and Killops, V. J., 2005. An Introduction to Organic Geochemistry. Wiley Blackwell. Second Edition. 393 pp.

Koehn, D., Renard, F., Toussaint, R., Passchier, C. W., 2007. Growth of stylolite teeth patterns depending on normal stress and finite compaction. *Earth and Planetary Science Letters*, *257*, 582–595.

Kreutzberger, M. E., and Peacor, D. R., 1988. Behavior of illite and chlorite during pressure solution of shaly limestone of the Kalkberg Formation, Catskill, New York. *Journal of Structural Geology*, *10*, 803–811.

Lambert, L., Durlet, C., Loreau, J. P., Marnier, G., 2006. Burial dissolution of micrite in Middle East carbonate reservoirs (Jurassic–Cretaceous): keys for recognition and timing. *Marine and Petroleum Geology*, *23*, 79–92.

Lucia, F. J., 2007. Carbonate Reservoir Characterization. second edition Springer-Verlag, Berlin, 342 pp.

Machel, H. G., 2004. Concepts and models of dolomitization: a critical reappraisal. *Geological Society, London, Special Publications*, *235*, 7–63.

Machel, H. G., 2005. Investigations of burial diagenesis in carbonate hydrocarbon reservoir rocks. *Geoscience Canada*, *32*, 103–128.

Mazzullo, S. J. and Harris, P. M., 1991. An overview of dissolution porosity development in the deep-burial environment, with examples from carbonate reservoirs in the Permian Basin. Midland, TX: *West Texas Geological Society*, 91–99.

Mazzullo, S. J., and Harris, P. M., 1992. Mesogenetic dissolution: its role in porosity development in carbonate reservoirs. *American Association of Petroleum Geologists Bulletin 76*, 607–620.

Mazzullo, S. J., 2004. Overview of porosity evolution in carbonate reservoirs. *Kansas Geological Society Bulletin*, *79*, 1–19.

Moore, C. H., 1989. Carbonate Diagenesis and Porosity. Developments in Sedimentology, 46. New York, Elsevier, 338 pp.

Moore, C. H., 2001. Carbonate reservoirs: porosity evolution and diagenesis in a sequence stratigraphic framework, 3. impr. In: Developments in Sedimentology 55. Elsevier, Amsterdam, 444 pp.

Morad, D., Nader, F. H., Morad, S., Al Darmaki, F., Hellevang, H., 2018. Impact of stylolitization on fluid flow and diagenesis in foreland basins: evidence from an Upper Jurassic Carbonate gas reservoir, Abu Dhabi, United Arab Emirates. *Journal of Sedimentary Research*, *88*, 1345–1361.

Morse, J. W., and Casey, W. H., 1988. Ostwald processes and mineral paragenesis in sediments. *American Journal of Science*, *288*, 537–560.

Morse, J. W., and Mackenzie, F. T., 1990. Geochemistry of Sedimentary Carbonates. Elsevier, Amsterdam, 707 pp.

Nelson, R. A., 2001. Geologic Analysis of Naturally Fractured Reservoirs, second edition. Gulf Professional Publishing, Boston. 332 pp.

Peters, K. E., Walters C., Moldowan J. M., 2005. The Biomarker Guide. Biomarkers and isotopes in the environmental and human history. Volume I. Second Edition. Cambridge University Press, 471 pp.

Petit, J. P., and Mattauer, M., 1995. Palaeostress superimposition deduced from mesoscale structures in limestone: the Matelles exposure, Languedoc, France. *Journal of Structural Geology*, *17*, 245–256.

Pingitore, N. E. Jr., 1982. The role of diffusion during carbonate diagenesis. *Journal of Sedimentary Petrology*, *52*, 27–39.

Pomar, L., Esteban, M., Martinez, W., Espino, D., Castillo de Ott, V., Benkovics, L., Castro Leyva, T., 2015. Oligocene-Miocene carbonates of the Perla field, offshore Venezuela: Depositional model and facies architecture, In: C. Bartolini and P. Mann, eds., Petroleum geology and potential of the Colombian Caribbean Margin: *American Association of Petroleum Geology, Memoir 108*, 647–674.

Pöppelreiter, M., Balzarini, M. A., De Sousa, P., Engel, S., Galarraga, M., Hansen, B., Márquez, X., Morell, J., Nelson, R., Rodriguez, F., 2005. Structural control on sweet-spot distribution in a carbonate reservoir: Concepts and 3-D models (Cogollo Group, Lower Cretaceous, Venezuela). *American Association of Petroleum Geologists Bulletin*, *89*, 1651–1676.

Renard, F., Schmittbuhl, J., Gratier, J. P., Meakin, P., Merino, E., 2004. Three-dimensional roughness of stylolites in limestones. *Journal of Geophysical Research: Solid Earth*, *109*, doi:https://doi.org/10.1029/2003JB002555.

Roehl, P. O., 1981. Dilation brecciation—A proposed mechanism of fracturing, petroleum expulsion and dolomitization in the Monterey Formation, California, in R. E. Garrison and R. G. Douglas, eds., The Monterey Formation and related siliceous rocks of California: Pacific Section SEPM, pp. 285–315.

Shinn, E. A., and Robbin, D. M., 1983. Mechanical and chemical compaction in fine-grained shallow-water limestones. *Journal of Sedimentary Research*, *53*, 595–618.

Schmoker J. W., and Halley, R. B., 1982. Carbonate porosity versus depth: a predictable relation for south Florida. *American Association of Petroleum Geology Bulletin*, *66*, 2561–2570.

Scholle, P. A., and Ulmer-Scholle, D., 2003. A Color Guide to the Petrography of Carbonate Rocks: Grains, Textures, Porosity, Diagenesis. *American Association of Petroleum Geologists, Memoir 77*, 474 pp.

Scholle, R. A., and Halley, R. B., 1985. Burial diagenesis: out of sight, out of mind! ln: Schneidermann, N. & Harris, E. M. (eds.) Carbonate Cements. Society of Economic Paleontologists and Mineralogists, Special Publications, 36, 309–334.

Swart, P. K., 2015. The geochemistry of carbonate diagenesis: the past, present and future. *Sedimentology*, *62*. 1233–304.

Tucker, M. E., and Wright, V. P., 1990. Carbonate Sedimentology. Blackwell Publishing. 482 pp.

Vajdova, V., Baud, P., Wong, T.-f, 2004. Compaction, dilatancy and failure in porous carbonate rocks. *Journal of Geophysical Research*, *109*, B05204.

Valencia, F. L. and Laya, J. C., 2020. Deep-Burial dissolution in an Oligocene-Miocene giant carbonate reservoir (Perla Limestone), Gulf of Venezuela Basin: Implications on microporosity development. *Marine and Petroleum Geology*, *113*, 104,144.

Wei, W., Chen, D., Qing, H., Qian, Y., 2017. Hydrothermal dissolution of deeply buried Cambrian dolomite rocks and porosity generation: integrated with geological studies and reactive transport modeling in the Tarim Basin, China. *Geofluids*, *2017*, 9,562,507.

Worden, R. H., Armitage, P. J., Butcher, A. R., Churchill, J. M., Csoma, A. E., Hollis, C., Lander, R. H., Omma, J. E., 2018. Petroleum reservoir quality prediction: overview and contrasting approaches from sandstone and carbonate communities. *Geological Society, London, Special Publications*, *435*, 1–31.

Dolomite and Dolomitization

18

Dolomite [$CaMg(CO_3)_2$] is the second most abundant carbonate mineral in Phanerozoic marine sedimentary rocks (after LMC). The precipitation of dolomite is an authigenic (*in situ*) process that occurs mainly via the replacement of preexisting calcium carbonate ($CaCO_3$) sediments (secondary dolomite), or by direct precipitation from pore fluids (primary dolomite). The formation of secondary dolomite is a diagenetic process responsible for the formation of most dolomite facies in the stratigraphic record. In contrast, the spontaneous nucleation of primary (syngenetic) dolomite (from a solution whose ion activity product of dissolved Ca^{2+}, Mg^{2+} and CO_3^{2-} is greater than the dolomite solubility product constant) is far less common in nature (see Machel and Mountjoy 1986; for details on chemical reactions that exemplify the precipitation of dolomite). Meanwhile, dolomitization is the process by which limestone or its precursor sediment is completely or partly converted to dolomite rock (dolostone) or dolomitic limestone by the replacement of the original $CaCO_3$ by magnesium carbonate, through the action of Mg-bearing water (e.g., Bates and Jackson 1980; Flügel 2010).

18.1 The Dolomite Problem

Dolomite, with a hexagonal crystal structure, is the only abundant double carbonate mineral, containing calcium and magnesium in equal proportion within the crystal lattice of an ideal (stoichiometric) dolomite crystal. This ordered crystal lattice is composed of layers of carbonate ions separated by alternating layers of calcium and magnesium ions (see Lippmann 1973, 1982; Land 1985; Warren 1989, 2000; James and Jones 2015). The crystal kinetics of dolomite are far more complex than those of hexagonal calcite and orthorhombic aragonite, as its tripartite ordered structure requires more thermodynamic energy for precipitation. Such complexity results in the scarcity of dolomite in Holocene sediments, a phenomenon widely referred to in the literature as "The Dolomite Problem" (e.g., Deffeyes et al. 1965; Friedman and Sanders 1967; Hsu and Siegenthaler 1969; Zenger et al. 1980; Baker and Kastner 1981; Land 1985; Machel and Mountjoy 1986; Tucker and Wright 1990; Vasconcelos and McKenzie 1997; Holland and Zimmermann 2000; Warren 2000; Machel 2004; McKenzie and Vasconcelos 2009; James and Jones 2015; Petrash et al. 2017; Pina et al. 2022).

Seawater is the only abundant source of Mg^{2+} capable of forming large and widespread volumes of sedimentary dolomite at or near the Earth's surface, with a Mg/Ca ratio of 3.14 by weight or a molar ratio of 5.2 (Land 1980, 1985; Warren 2000; James and Jones 2015). Although normal marine waters (with salinity and pH values between 33–37‰ and 8.1–8.2, respectively) are supersaturated with respect to dolomite by about two orders of magnitude, the precipitation of large volumes of this carbonate mineral at low temperature is inhibited by the following conditions: mineral kinetics; precipitation rates (the time needed for dolomite nucleation); the hydration barrier, i.e., the strength of electrostatic attraction of Mg^{2+} to the OH^- radical, which is greater than that of Ca^{2+} and much greater than that for the CO_3^{2-} ion; and the low coefficient activity (or effective concentration) of dissolved CO_3^{2-} relative to HCO_3^- (Lippmann 1973; Machel and Mountjoy 1986; Warren 2000; James and Jones 2015). In addition, the presence of sulfate (SO_4^{2-}) is thought to be an inhibitor of the formation of replacement secondary dolomite. This is because dissolved SO_4^{2-} tends to form complexes with Mg^{2+}, thus reducing the effective concentration of this cation (e.g., Baker and Kastner 1981; Kastner 1984). Nonetheless, SO_4^{2-} ions only seem to act as inhibitors at mid-range concentrations, since dolomite precipitation at low temperatures occurs in both sulfate-rich and sulfate-poor solutions, such as in marine evaporitic settings (characterized by high precipitation rates of sulfate-evaporitic minerals) and in the mixing zone, respectively (see Brady et al. 1996; Machel 2004; Sánchez-Román et al. 2009; Petrash et al. 2017).

J. A. Méndez Dot, J. Méndez Baamonde, *Atlas of Tropical Carbonates*, https://doi.org/10.1007/978-3-031-25494-9_18

The above-mentioned inhibiting conditions (or factors) have limited the formation of dolomite facies in modern sedimentary environments. In simple terms, there has not been enough time during the Holocene to form dolomite facies of volumetric relevance. To overcome these inhibiting conditions, the nucleation of low-temperature (˂ 60 ° C) dolomite is chemically favored (thermodynamically and/or kinetically) under the following conditions: high temperature; time (given 10^3–10^6 years); high Mg^{2+}/Ca^{2+} ratios (~ 10:1) driven by the precipitation of $CaSO_4$ or $CaCO_3$, so Ca^{2+} is removed from seawater; breaking Mg-OH bond due to microbial activity or increased temperature; raising pH above 9 so most carbonate in seawater is CO_3^{2-}; sulfate reduction that increases alkalinity by producing HCO_3^-; removing SO_4^{2-} by sulfate mineral precipitation (gypsum and anhydrite), sulfate-reducing bacteria, and freshwater dilution; and substantially higher or lower salinities than that of normal seawater (see Carpenter 1980; Morrow 1982; Machel and Mountjoy 1986; Tucker and Wright 1990; Warren 2000; Machel 2004; James and Jones 2015).

18.2 Types of Dolomites and Dolostones

Regarding exclusively the formation of dolomite (the mineral) and dolostone (the rock) in marine environments (and excluding dolomites of hydrothermal and metamorphic origin), dolomitization can occur on or beneath the shallow seafloor, in zones of brine reflux, and in early to late burial settings. Dolomite may form from seawater, from continental waters, from the mixing of basinal brines, the mixing of hypersaline brine with seawater, or the mixing of seawater with meteoric water. Microbial activity may aid the process of precipitation in settings where sulfate-reducing species flourish and microbial action may influence primary precipitation (Warren 2000).

At the broadest scale, James and Jones (2015) consider three types of dolomites: **(a)** synsedimentary dolomites that form in the depositional environment (muddy tidal flats, organic-rich marine sediments) soon after sediment accumulation; **(b)** early diagenetic dolomites that form in the shallow subsurface (˂ 300 m) by replacing sediment, limestone, or dolostone; and **(c)** late diagenetic dolomites that replace limestone or dolostone in the deep subsurface (> 300 m) or form cement overgrowths on pre-existing crystals.

18.3 Dolomitization Models

We consider the following dolomitization models for dolomite formation in marine environments. These models are based on physical mechanisms pertinent to the hydrological settings. This is a breve review of the dolomitization models that can be applied to dolomite examples shown in the photographic set of this chapter are included.

(a) Sabkha-style dolomite: dolomite formation occurs in the uppermost (1-2 m) tidal flat (sabkha) sediments, as hypersaline (evaporated) seawaters ascend through the porous upper sediments, involving evaporative pumping and seepage reflux, in combination with precipitation of gypsum ($CaSO_4.2H_2O$) and/or anhydrite ($CaSO_4$), and halite (NaCl), thus increasing Mg^{2+}/Ca^{2+} ratio of dolomitizing pore fluids (see Hsu and Schneider 1973; Mckenzie et al. 1980; Mckenzie 1981; Machel and Mountjoy 1986; Hardie 1987; Warren 1991, 2000; Gregg et al. 1992; Wenk et al. 1993; Machel 2004; James and Jones 2015; Sánchez-Román et al. 2023); **(b) Meteoric-seawater mixing zone dolomite (shallow subsurface)**: it involves influx of continental freshwaters, which decreases SO_4^{2-} concentration and increases CO_3^{2-}/HCO_3^- ratio, resulting in undersaturation with respect to aragonite, HMC, and LMC, and driving dolomite precipitation as this process occurs more easily from a dilute solution than from seawater (e.g., Badiozamani 1973; Warren 2000; James and Jones 2015; Petrash et al. 2021); **(c) Reflux dolomite (shallow subsurface)**: this dolomite is formed due to the action of hypersaline brines (formed on tidal fltas and/on in evaporative lagoons) which move basinward through the porous platform carbonates (Adams and Rhodes 1960; Deffeyes et al. 1965; Hsu and Siegenthaler 1969; Warren 1999, 2000; Machel, 2004); **(d) Dolomitization in schizohaline environments**: this dolomitization model is characterized by rapidly changing conditions, from marine-hypersaline to meteoric-hyposaline, due to the episodic and/or seasonal high influx of freshwater (rich in nutrients and carrying oxidized detritus) into shallow lagoons and marine platform settings (Folk and Siedlecka 1974; Petrash et al. 2016); **(e) Burial dolomite**: it is formed in the subsurface (from the intermediate- to deep-burial realms) at high temperatures (ranging from 60 ° C–150 ° C) due to the action of Mg^{2+}-rich fluids derived from compaction and transformation of clay minerals (e.g., montmorillonite-to-illite) (Warren 2000); the action of hydrothermal- and hydrocarbon-associated burial fluids of high ionic strength; and as a by-product of chemical compaction and thermochemical sulfate reduction, a process that involves gypsum dissolution and the reaction of SO_4^{2-} anions with hydrocarbons at elevated temperatures (Machel 1987, 2001, 2005; Scholle and Ulmer-Scholle 2003; Flügel 2010); **(f) Microbial/organogenic dolomite** (it is noteworthy that some authors consider microbially-influenced dolomite and organogenic/methanogenic dolomite as different models of dolomitization): this dolomite is formed from peritidal to shallow-marine to open-marine settings, as active microbial activity (commonly in organic-rich sediments and under eutrophic-to-hypertrophic and low-oxygen to anoxic conditions) sustains high pH and alkalinity and increases Mg^{2+} concentration in pore waters through a series of biochemical pathways (e.g., sulfate reduction, methanogenesis, anoxygenic oxidation of methane) that involve organic matter remineralization (Slaughter and Hill 1991; Vasconcelos and McKenzie 1997; Wright 1999; Mazzullo 2000; Moore et al. 2004; McKenzie and Vasconcelos 2009; Bontognali et al. 2010; James and Jones 2015; Petrash et al. 2017; Sánchez-Román et al. 2023).

Applying dolomitization models on a specific dolomite occurrence often requires the consideration of more than one model and related hydrologic mechanisms for overcoming the kinetic barrier. For example, the combination of seepage reflux and microbial models explain the nucleation of primary and secondary Ca-dolomite (or protodolomite) in Holocene tidal flats and restricted hypersaline lagoons of Los Roques Archipelago. In those peritidal environments, subaerial reflux and evaporative pumping of marine water through the highly porous sabkha sediments (composed of aragonite, HMC, gypsum, halite, and microbial mats) is combined with high rates of gypsum precipitation, which increases Mg^{2+}/Ca^{2+} ratio. The formation of primary (nanometer-scale) Ca-dolomite was determined by Petrash et al. (2015) in the upper 15 cm of the suboxic, non-lithified sediments of an ephemeral hypersaline brine. These authors propose degradation of buried microbial mats and their Ca- and Mg-enriched EPS, combined microbially-mediated Mn(IV) reduction coupled with the diffusion and reoxidation of reduced sulfur species, as potential drivers of dolomite nucleation. The formation of secondary Ca-dolomite occurs within lithified, intertidal-laminated stromatolites formed on the peripheries of a perennial hypersaline lagoon (Méndez-Dot 2011). The high degree of lithification of these microbialite structures suggests that processes for overcoming the kinetic barrier have acted over sufficient time. Precipitation of gypsum (amorphous and lenticular), combined with the fluctuating activity of endolithic-micritizing cyanobacteria (which precipitate micro- to crypto-crystalline $CaCO_3$ serving as substrate for replacement) and sulfate reducing bacteria, acted as divers for replacement process.

18.4 Dolomite Crystal Chemistry

Ancient dolomite crystals have a wide range of composition. Variations in the mol% $CaCO_3$ are known to be from 62 to 48 mol% $CaCO_3$ in sedimentary dolomites (Jones et al. 2001), representing a wide compositional spectrum, from High-Ca dolomites to Low-Ca dolomites, respectively. The rate of crystal solubility increases as the level of excess Ca increases (see Jones 2013 and references therein). Secondary dolomites are derived from the replacement of metastable aragonite and HMC, and stable LMC. Nucleation of very HMC (VHMC) at Earth's surface temperature (also commonly termed protodolomite, Ca-dolomite, or disordered dolomite), composed of 35–40 mol% $MgCO_3$ is rapidly replaced by a crystal composition that leans toward stoichiometry. In the presence of fluids having high Mg^{2+}/Ca^{2+} ratio, crystal compositional evolution after nucleation of VHMC is a function of the crystal growth. The transformation of disordered Ca-dolomite crystals to stable rhombohedral crystals requires successive recrystallization (dissolution-reprecipitation) events, whereby larger crystals grow at the expense of smaller ones. This diagenetic process is known in the literature as "Ostwald Ripening" or "Oswald step rule". Such continuing dolomite crystal growth through ageing until reaching diagenetic stabilization occurs from shallow marine (near-surface) to burial environments (Gregg et al. 1992; Mazzullo 1992; Sibley et al. 1994; Nordeng and Sibley 1994; Warren 2000; James and Jones 2015; Petrash et al. 2017).

Dolomite crystals commonly exhibit compositional zoning, with rhombic crystals that display cores, cortical zones, growth sectors, and growth lines. In general, these crystals evolve from high Ca contents (wt.%) in the cores (Sr commonly being high; 500–1500 ppm) to a progressive growth of outer rims richer in Mg (wt.%) as well as well as in Fe, Mn, and measurable concentrations of other trace elements such as Ni, Co, Cr, Cu (Katz 1971; Reeder and Prosky 1986; Fouke and Reeder 1992; Jones 2013; Petrash et al. 2017). In addition, the occurrence of ankerite, i.e., $Ca(Mg,Fe)(CO_3)_2$, or ferroan dolomite, which can accommodate a maximum of ~70% $CaFe(CO_3)_2$ (Morse et al. 2007), is common of reducing conditions (both at near-surface and in the burial realm) under which dolomitizing fluids become rich in Fe^{2+} and Mn^{2+} (e.g., Tucker and Wright 1990). The manganoan equivalent of dolomite, lying on the $CaCO_3$-$MnCO_3$ join, is termed kutnahorite [$CaMn(CO_3)_2$] (Mucci 1991; Morse et al. 2007).

18.5 Dolomite Crystal Texture

The most widely used dolomite texture classification is that proposed by Sibley and Gregg (1987), based on Gregg and Sibley (1984). We use an expanded classification of those authors (Fig. 18.1), after Wright et al. (2000, 2003) and Wright (2001), who defined an additional crystal texture type, Planar-s to Nonplanar-a (transitional); and after Machel (2004), who uses the term porphyrotopic for describing dolomite fabrics consisting of isolated crystals floating or enclosed in a matrix. This classification is largely descriptive, but it carries genetic implications. Crystal size distributions are classified as "unimodal" or "polymodal", whereas crystal shapes are classified as "planar-e" (euhedral), "planar-s" (subhedral), "nonplanar-a" (anhedral), and "planar-s" to "nonplanar-a" (transitional between subhedral and anhedral). Using this semantic scheme, almost all dolomite texture types can be named, e.g., planar-c (cement), planar-p and nonplanar-p (both porphyrotopic). Saddle dolomite, with its warped crystal faces, is categorized as nonplanar or nonplanar-c (when it is a cement) (Machel 2004).

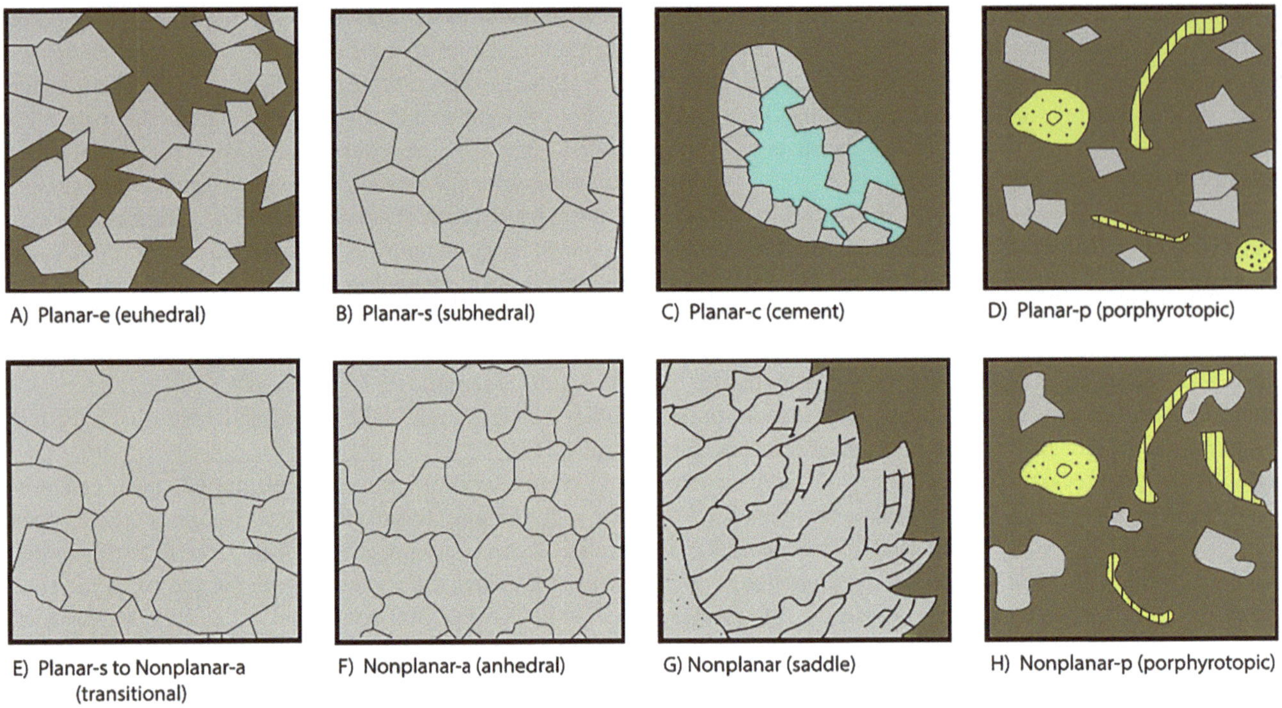

Fig. 18.1 Dolomite textural classification slightly modified from Machel (2004), after Gregg and Sibley (1984), Sibley and Gregg (1987), and Wright (2001). The figure is reproduced with permission from Wright (2001). Drawing for representing nonplanar (saddle) dolomite according to Gregg and Sibley (1984)

Figure 18.1 is a scheme consisting of 8 dolomite types. The primary differentiation between dolomite groups (Planar vs. Nonplanar) is optical extinction: planar-unit, and nonplanar (sweeping or undulose). **(A) Planar-e (euhedral):** most dolomite crystals are euhedral rhombs. Crystals form a supported framework with intercrystalline areas filled by another mineral or porous; **(B) Planar-s (subhedral):** most dolomite crystals are subhedral to anhedral with straight, compromise boundaries and many crystal-face-junctions. Often with low porosity and/or low intercrystalline matrix; **(C) Planar-c (cement):** planar crystals are filling empty pores; **(D) Planar-p (porphyrotopic):** euhedral (rhombic) crystals float in a matrix; **(E) Planar-s to Nonplanar-a (transitional):** crystals are transitional between planar and nonplanar. This dolomite, termed "planar-nonplanar" or "transitional", is characterized by having definable crystal boundaries (planar-s) but exhibiting a modified, often undulatory extinction pattern. Planar and nonplanar crystals occur side by side; **(F) Nonplanar-a (anhedral):** closely-packed anhedral crystals (formed at temperatures >50–60 ° C) with mostly curved, lobate, serrated, or otherwise irregular intercrystalline boundaries. Preserved crystal face junctions are rare and crystals often have undulose extinction under CPL. **(G) Nonplanar (saddle):** crystals (commonly rich in Fe, up to 15 mol% $FeCO_3$; James and Jones 2015) are a subset of nonplanar anhedral (formed at temperatures >50–70 ° C). Crystal faces are lattice warped (or distorted) and highly zoned. Preserved crystal face junctions are rare. Crystals may be clear or turbid and often exhibit sweeping extinction in CPL; **(H) Nonplanar (porphyrotopic):** nonplanar crystals are enclosed in a matrix (Wayne Wright, personal communication 2023; see also Machel 2004).

In addition to the term porphyrotopic, the terms idiotopic, hypidiotopic and xenotopic (proposed by Friedman 1964, 1965; for describing crystallization textures and fabrics) are widely used for describing dolomite fabrics (or mosaics) consisting of euhedral, subhedral, and anhedral crystals, respectively. See Chap. 16 (Sect. 16.5.2) for definitions.

18.6 Some Characteristics of Sedimentary Dolomites

(a) Dolomite selectively replaces fine-grained $CaCO_3$; **(b)** There is a direct relationship between the Mg^{2+}/Ca^{2+} ratio in solution and the Mg/Ca ratio in the dolomite (Warren 2000); **(c)** Dolomite textures are mainly determined by the crystal size of the substrate (Sibley et al. 1994); **(d)** Dolomite crystals with a "dirty/cloudy core and clear rim" in dolostone successions of all ages suggest a cause, inherent to the dolomitization process, that transcends time and geography (Jones 2013); **(e)** Dolomite is a highly birefringent mineral (e.g., Kerr 1977; Jellison Jr. et al. 2018), with birefringence being relatively higher in crystals with clear cores and rims (some of primary origin) compared to those with dark rhombohedral cores (rich in undigested host-rock mineral inclusions).

18.7 Dedolomitization

Dedolomitization (or calcitization of dolomite) is the reverse process of dolomitization. This diagenetic process involves the partial or total replacement of dolomite by calcite due to the action of Ca-rich acidic fluids (or fluids with low Mg^{2+}/Ca^{2+} ratio), often resulting in the creation of secondary porosity. Early (near-surface) or late (burial) diagenetic dissolution of sulfate evaporites leading to the generation of Ca^{2+}-rich fluids capable of dissolving dolomite is thought to be one of the main mechanisms for dedolomitization (e.g., Tucker and Wright 1990; Scholle and Ulmer-Scholle 2003). Near-surface oxidation of pyrite and the reaction of calcium sulfate solutions with dolomite have been proposed as an alternative reaction for dedolomitization (Evamy 1967; Goldberg 1967; Flügel 2010). Many authors consider dedolomitization as a process more common of the burial realm (e.g., Morse and Mackenzie 1990; Immenhauser 2021).

18.8 Dolomite Facies in Venezuela

Regarding the origin of most ancient dolomite facies in Venezuela, we follow the view of Friedman (1980), who states that dolomite is an evaporite mineral, as most dolomites in the rock record were formed under conditions of hypersalinity. Although burial dolomites are present (mainly as patches) in carbonate facies of that country, the thicker dolostone intervals were formed in peritidal (mainly shallow-subtidal) hypersaline environments by evaporative pumping and seepage reflux. We also consider that matrix-selective replacement of shallow water limestones (mostly mudstones and wackestones), which is the most common form of dolomitization (Machel 2004), occurred in most cases under the influence of microbes, as their presence is ubiquitous in shallow-tropical muddy carbonate deposits.

In dolomite microfacies of Venezuela, euhedral crystals are by far the most abundant, forming idiotopic (planar-e) and porphyrotopic (planar-p) fabrics. Subhedral (planar-s) crystals are common, mainly forming closely-packed hypidiotopic mosaics, which are the final product of dolomitization or recrystallization of former euhedral crystals. Planar dolomite crystals filling pores (Planar-c) are present locally. Anhedral (nonplanar-a), nonplanar (saddle) and planar-s to nonplanar-a (transitional) crystals are present as patches.

18.9 Dolomite and Dolomitization: Photographic Set and Descriptive Captions

In all photos, IW = image width (the true length of the horizontal axis of the view)

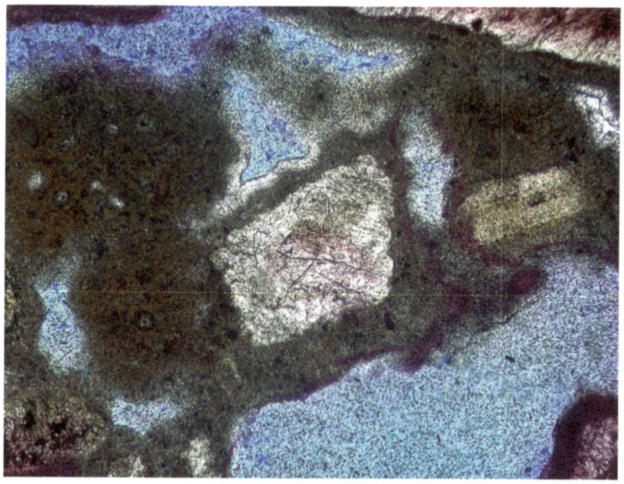

Holocene. Los Roques Archipelago

The white, dusty area at the center of the image is rich in VHMC, as indicated by EDS analysis. The nucleation of this synsedimentary protodolomite phase occurred via replacement of micritic crystals within an intertidal stromatolite. The small crystal size of microbially-induced lime-mud increased the substrate surface area and provided numerous active sites for $CaCO_3$ replacement. The high degree of lithification of the intertidal stromatolite that hosts this protodolomite suggests that diagenetic processes have acted over enough time to overcome the kinetic barrier for Ca-dolomite nucleation.

IW = 0.5 mm. PPL.

Holocene. Los Roques Archipelago

Another example of synsedimentary protodolomite. Note the area with light pink staining, which contrast with the dark staining of the surrounding micritic matrix. The subtle reactive behavior of the VHMC substrate to the Alizarin Red S is due to its Ca-rich composition. VHMC crystalline phases within this microbialite have measured *d-spacing* of the 104 peak ranging from 2.972 to 2.957 Å. These values represent solid solutions between HMC crystals (2.988 Å) within this stromatolite and standard values of stoichiometric dolomite crystals (2.89 Å) (see Zhang et al. 2010; for details on *d104* values and composition of protodolomites).

IW = 0.5 mm. PPL.

Cretaceous (early Aptian). Machiques Member, Apón Formation. Maracaibo Basin

Well-developed euhedral dolomite crystals in the image are replacing the organic-rich micritic matrix of this dolomitic mudstone. Fine-grained micrites are the most common dolomitized substrates in facies of the Cogollo Group. Note the polymodal size of cloudy planar-e and planar-p crystals (forming a supporting framework, and floating in a fine-grained matrix, respectively), which is associated to multiple crystal nucleation events on a unimodal substrate. Hypersaline conditions, photic zone anoxia and microbial mediation are associated to the origin of dolomite microfacies in the Machiques Member.

IW = 2.4 mm. PPL.

Cretaceous (early Aptian). Machiques Member, Apón Formation. Maracaibo Basin

Strongly hydrocarbon impregnated dolostone. Cloudy and zoned dolomite crystals are mainly euhedral (planar-e) in texture, forming an idiotopic fabric (or mosaic). Subhedral (planar-s) crystals are present but less abundant. This dolostone is interpreted to have been deposited in a vast tidal flat (shallow subtidal) environment. Combined hypersaline conditions (as consequence of high evaporation) and strong microbial activity promoted the replacement of $CaCO_3$ by $CaMg(CO_3)_2$.

IW = 2.4 mm. PPL

Cretaceous (early Aptian). Machiques Member, Apón Formation. Maracaibo Basin

Dolomite crystals in the image are mainly subhedral (planar-s) in texture, while euhedral (planar-e) rhombs are present in lesser proportions. At the top of the Machiques Member, subhedral crystals represent the final product of dolomitization, from near-surface to diagenetic stabilization in the shallow-burial environment. Although some crystals show inherent rhombic shapes, many developed irregular shapes. The progressive increase in crystals growth against each other led to the formation of many crystal face junctions, thus forming a hypidiotopic mosaic.

IW = 1.3 mm. PPL.

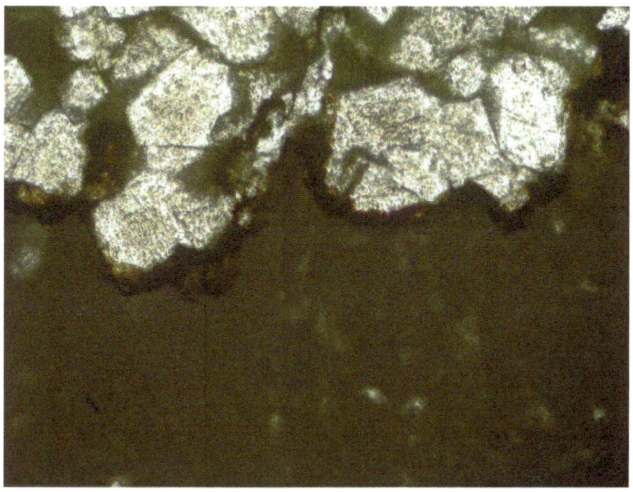

Cretaceous (late Aptian). Piché Member, Apón Formation. Maracaibo Basin

Subhedral (planar-s) dolomite crystals in the image were formed by the replacement of the organic-rich micrite. The latter, cryptocrystalline in texture (imperceptible crystal size), was precipitated under anoxic conditions as an induced product of bacterial activity. The formation of these microbially-mediated (organogenic) dolomite crystals, which are common in all black shale intervals in the Cogollo Group, involved organic matter remineralization within the photic zone.

IW = 0.5 mm. PPL.

Cretaceous (early-late Aptian). Piché Member, Apón Formation. Maracaibo Basin

An example of selective dolomitization. The sutured dolomite mosaic in the image consists predominantly of unimodal subhedral (planar-s) crystals and subordinate euhedral (planar-e) crystals. This closely-packed hypidiotopic fabric was formed due to the replacement of a fine-grained micritic matrix. The large calcitic oyster crossing the field of the photograph remained totally unaltered. Note the well-preserved calcite lamellae and foliated microstructure of the shell, indicating mineralogic stability during dolomitization.

IW = 4.2 mm. PPL.

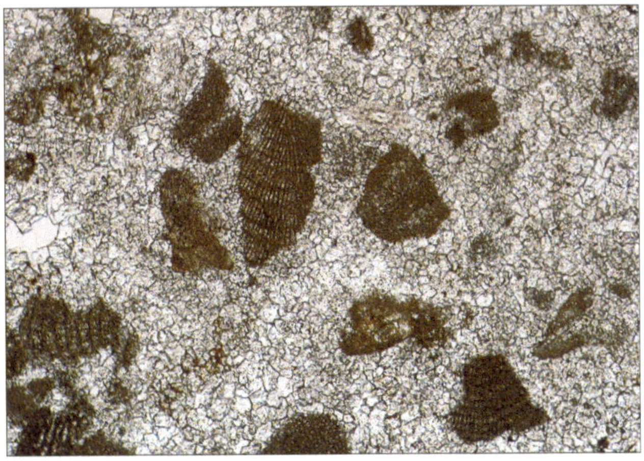

Middle Miocene. Capadare Formation. Falcón Basin

Replacement dolomitization is commonly developed on red algae-rich microfacies. This sutured xenotopic mosaic consists of anhedral (nonplanar-a) and transitional (planar-s to nonplanar-a) crystals. Partial dissolution of metastable red algae increased the Mg^{2+}/Ca^{2+} ratio of the dolomitizing fluids, hence promoting the replacement of HMC skeletons. Non-dolomitized red algae grains are scattered through the sample. The anhedral texture of many dolomite crystals in the image suggests that the final dolomitization event occurred during burial.

IW = 4.2 mm. PPL.

Cretaceous (late Barremian). Tibú Member, Apón Formation. Maracaibo Basin

Dolomitic microfacies in the Tibú Member are common and contain echinoids, molluscs and benthic foraminifera. Replacive dolomite crystals in the image are mainly euhedral (planar-e) in texture. Some echinoids are clearly differentiated, such as the larger central grain. Their HMC composition most likely served as a source of Mg^{2+} for dolomitization. Hypersaline and transient low-oxygen conditions (the latter indicated abundant pyritized grains) are associated to this replacement process.

IW = 7.5 mm. PPL.

Cretaceous (Turonian). O Member, Escandalosa Formation. Barinas sub-Basin

Massive dolostone (high dolomite content relative to any other mineral in the rock) at the top of the Escandalosa Formation are the product of highly evaporative conditions in a tidal flat environment. The sutured mosaic in the image consists of euhedral (planar-e) crystals with cloudy cores forming an idiotopic fabric. Note the unitary extinction that characterizes euhedral crystals when these are seen under CPL. In addition, crystals exhibit low birefringence, as cloudy cores are rich in remnants of precursor micritic substrate.

IW = 1.3 mm. CPL.

Cretaceous (Turonian). O Member, Escandalosa Formation. Barinas sub-Basin

Close view of euhedral (rhombic) dolomite crystals of the O Member. Note the well-defined crystal zoning composed of rhombohedral, Ca-rich cloudy cores (inherited from initial crystal growth as replacement of micrite) and clear (or more limpid) rims or cement overgrowth (more stoichiometric crystal lattice with visibly closely-spaced growth lines). Dark spots between euhedral crystals are hydrocarbon impregnations postdating dolomitization, as this dolostone unit is a hydrocarbon reservoir.

IW = 0.5 mm. CPL.

Cretaceous (Turonian). O Member, Escandalosa Formation. Barinas sub-Basin

Dolomite crystals in the image (euhedral and subhedral in texture) exhibit higher birefringence and more prominent crystal zoning compared to the two previous examples. In addition, note that the cloudy cores are absent. For this sample, these features are interpreted to be derived from successive recrystallization (and associated crystal growth) of precursor replacive crystals towards stoichiometry, thereby removing textural and chemical traces of the original replaced substrate (e.g., isotopes and trace element signatures). Black colored areas are dolomite crystals replaced by amorphous silica.

IW = 1.3 mm. CPL.

Cretaceous (Turonian). O Member, Escandalosa Formation. Barinas sub-Basin

This massive dolostone consists of anhedral (nonplanar-a) crystals with curved faces and secondary transitional (subhedral to anhedral) crystals. Note the strongly turbid appearance of the crystals, a feature inherited from both their replacive origin and the inclusion-rich burial fluids (e.g., metal sulfide and hydrocarbon inclusions) responsible for the anhedral (curved-lobate boundaries) texture. Anhedral-dominated dolomite mosaics are moderately common in this dolomitic unit, appearing as patches enclosed in euhedral dolomite mosaics.

IW = 2.4 mm. PPL.

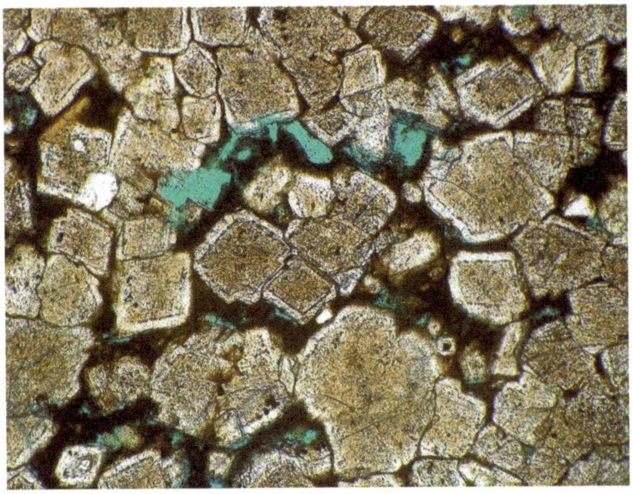

Cretaceous (Turonian). O Member, Escandalosa Formation. Barinas sub-Basin

Porous hydrocarbon-impregnated dolostone from which this reservoir has produced hydrocarbons. Note the cloudy cores of the crystals reminiscent of the original micritic fabric. Massive $CaCO_3$ replacement, combined with selective dissolution of unreplaced micritic sediments remaining between crystals, created intercrystalline porosity and characteristic sucrosic fabric. The term "sucrose" or "sucrosic", widely used in reference to dolomite microfacies with interparticle porosity, was adopted because of the resemblance of the crystals to a lump of sugar (Archie 1952; Murray 1960).

IW = 2.4 mm. PPL.

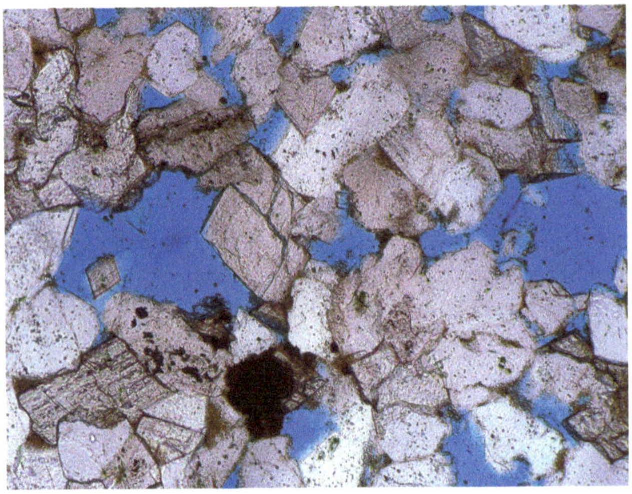

Cretaceous (Turonian). O Member, Escandalosa Formation. Barinas sub-Basin

Porous sandy dolomite. Quartz grains in the image are clearly differentiated from dolomite crystals due to their white limpid look. Lithologic variations from massive dolostones to sandstones characterize the O Member. Gradations between those two end members, such as this example, are common. Pores in the image are, by definition, a combination of intercrystalline (dolomite-related) and intergranular (quartz grains-related).

IW = 2.4 mm. PPL.

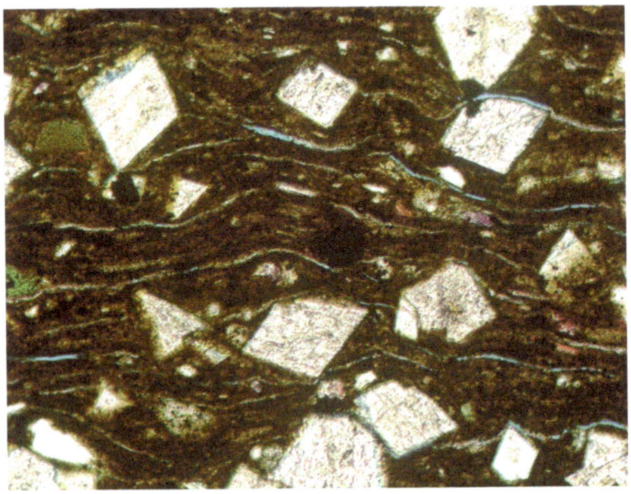

Cretaceous (late Aptian). Piché Member, Apón Formation. Maracaibo Basin

Dolomitic shale exhibiting wavy-laminated matrix. Planar-p (porphyrotopic) dolomite crystals are floating in a clay-dominated matrix. Dolomite nucleation and early crystal growth occurred during short episodes of increased freshwater influx related to the OAE 1b first sub-event hyperthermal spike. Anoxic-euxinic conditions are indicated by abundant pyrite framboids darkening the matrix. Dilution of marine waters and high nutrient influx, coupled with microbial mediation and organic matter remineralization in the photic zone, led to the development of a schizohaline, redox-stratified environment and ideal conditions for dolomite precipitation (see Petrash et al. 2016).

IW = 1.3 mm. PPL

Cretaceous (early Aptian). Machiques Member, Apón Formation. Maracaibo Basin

Another example of a dolomitic shale. This sample belongs to the lower part of the OAE 1a. The close view allows a better visualization of the pyritic-clayey matrix. The highly ordered crystal lattice (nearly stoichiometric) of planar-p dolomite crystals is indicated by their non-reactivity to Red Alizarin S. Note the pink-stained stripes indicating small amounts of carbonate content in the matrix. Such clay dominance suggests a primary origin for the beginning of the nucleation of interstitial, limpid dolomite crystals (lacking of cloudy cores). Dolomitic shales are common in all black shale intervals in the Cogollo Group.

IW = 0.5 mm. PPL.

Cretaceous (Turonian). O Member, Escandalosa Formation. Barinas sub-Basin

Wavy-laminated dolomitic shales are also present in the O Member of the Escandalosa Formation. Note that euhedral (planar-e) and porphyrotopic (planar-p) dolomite crystals in the image do not exhibit cloudy cores, which is optically traduced in a marked high birefringence. The dark, nearly isotropic behavior of the clayey matrix is due to the high contents of fluorapatite and pyrite. These early diagenetic precipitates suggest transient reducing conditions during early marine diagenesis. See a PPL image of this microfacies in Chap. 20.

IW = 1.3 mm. CPL.

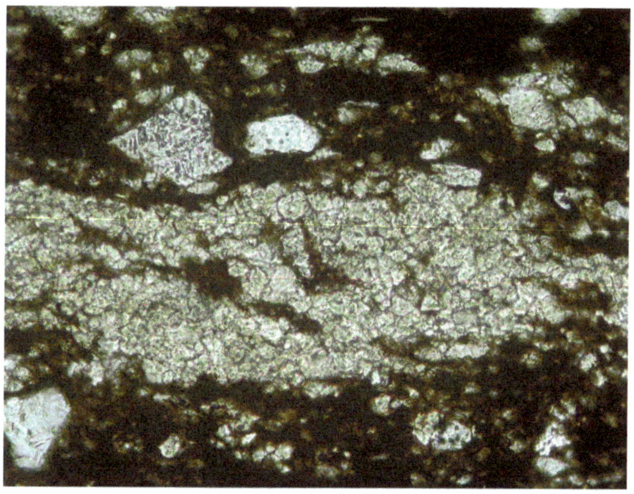

Cretaceous (early Aptian). Machiques Member, Apón Formation. Maracaibo Basin

This sample of the OAE 1a (containing 23% carbonate and 4.2% TOC, and HI of 460) contains transitional (Planar-s to Nonplanar-a) dolomite crystals. Note the patchy "pseudo-spar-like" dolomite mosaic in the center of the image, which consists of small-sized, subhedral to anhedral crystals exhibiting both straight and curved-lobate boundaries. We consider a microbial/organogenic, early diagenetic origin for these crystals. In contrast, anhedral (nonplanar) crystals formed at high temperatures tend to be markedly larger in size.

IW = 0.5 mm. PPL.

Cretaceous (early Aptian). Machiques Member, Apón Formation. Maracaibo Basin

Dual view (PPL and CPL) of a dolomite mosaic. Small-sized dolomite crystals are predominantly subhedral to transitional in texture, with subordinate euhedral crystals. Traces of the matrix, a mixture of clay and micritic sediments, are clearly seen in the image to the left (brown colored spots). Limpid dolomite crystals showing high birefringence suggest a primary origin for this "pseudospar-like" mosaic. These are commonly found as patches in dolomitic shales and argillaceous dolomite microfacies within the black shales of the Cogollo Group.

IW = 0.25 mm each. PPL (left); CPL (right).

Cretaceous (middle Aptian). Piché Member, Apón Formation. Maracaibo Basin

Dolomite crystals in the image are filling a moldic pore derived from dissolution of an aragonitic bioclast. Planar-c (cement) crystals exhibit characteristic rhombic shapes. This dolomite type is not common in shallow carbonate microfacies of Venezuela (both Cretaceous and Cenozoic in age). Primary precipitation from pore fluids with a high Mg^{2+}/Ca^{2+} ratio formed these limpid crystals exhibiting high birefringence.

IW = 2.4 mm. CPL.

Cretaceous (early Aptian). Machiques Member, Apón Formation. Maracaibo Basin

This massive dolostone consists mainly of subhedral and transitional crystals (with straight, compromise boundaries and many face junctions, and straight to curved boundaries, respectively), with euhedral crystals present in minor proportions. Some crystals show undulose extinction, which is better visualized when rotated. Dolomite mosaics dominated by crystals textures different than euhedral appear as small patches enclosed in idiotopic fabrics, both at the top of the Machiques Member and at the lower part of the Piché Member.

IW = 2.4 mm. CPL.

Cretaceous (early-late Aptian). Piché Member, Apón Formation. Maracaibo Basin

This distinctive saddle (nonplanar) fabric, also termed baroque, characterizes some burial-stage dolomites commonly formed at temperatures above 50–70 ° C. This mosaic consists of large-sized, nonplanar crystals exhibiting sweeping extinction. Note the strong internal zonation of crystals due to varying Fe and Mn content. These metals, in a reduced state, were incorporated into the crystal lattice probably in association with hydrocarbons, since this microfacies is overlain in close proximity to the early Aptian, thermally matured black shales.

IW = 7.5 mm. CPL.

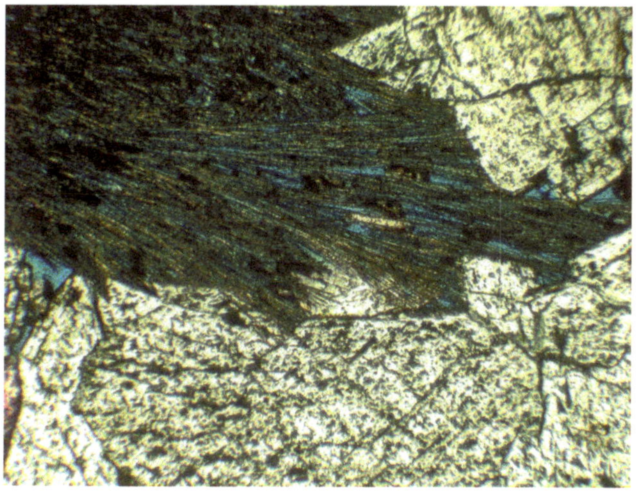

Cretaceous (early Aptian). Machiques Member, Apón Formation. Maracaibo Basin

A saddle (nonplanar) dolomite fabric enclosing fibrous gypsum. Crystals of this evaporite mineral precipitated in a hypersaline, near-surface environment and filled an intercrystalline (secondary) pore. This pore formed due to complete dolomitization of the original fabric. The saddle (warped) dolomite texture was formed during burial. Hydrocarbon associated fluids and/or thermochemical sulfate reduction (the latter process commonly associated to the presence of gypsum crystals) at high temperatures (probably >90 ° C) were probably involved in the origin of lattice-warped and highly zoned crystals.

IW = 2.4 mm. PPL.

Cretaceous (middle Aptian). Piché Member, Apón Formation. Maracaibo Basin

Small-sized, euhedral (planar-e) dolomite crystals in the image are cross-cutting a stylocumulate consisting of clay minerals, pyrite framboids, and organic matter. The crystals most probably postdate the formation of the stylocumulate, as the rest of the sample does not contain dolomite. Contrary to what expected for textures of dolomite crystals formed during burial at high temperatures, these crystals are euhedral. The source of Mg^{2+} for this selective dolomitization was probably derived from the chemical compaction of clay minerals concentrated at the stylocumulate.

IW = 0.5 mm. PPL.

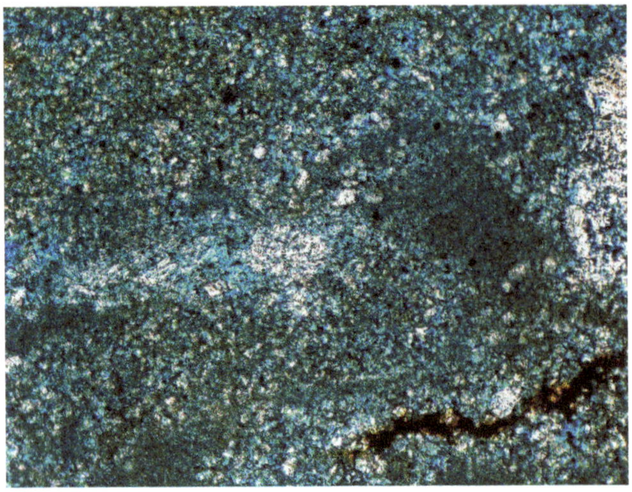

Late Oligocene-Early Miocene. Perla Limestone. Eastern Gulf of Venezuela sub-Basin

High-temperature, microcrystalline euhedral to subhedral dolomite crystals in the image were formed by the replacement of red algae cell-walls. According to Valencia and Laya (2020), the origin of this dolomite and related microporosity (indicated by pervasive "blue haze") was controlled by the occurrence of pressure solution surfaces (such as the stylolite down and to the right), as well as by and by the presence of red algae grains. The later represented the source of Mg^{2+} in the dolomitizing fluids.

IW = 0.5 mm. PPL.

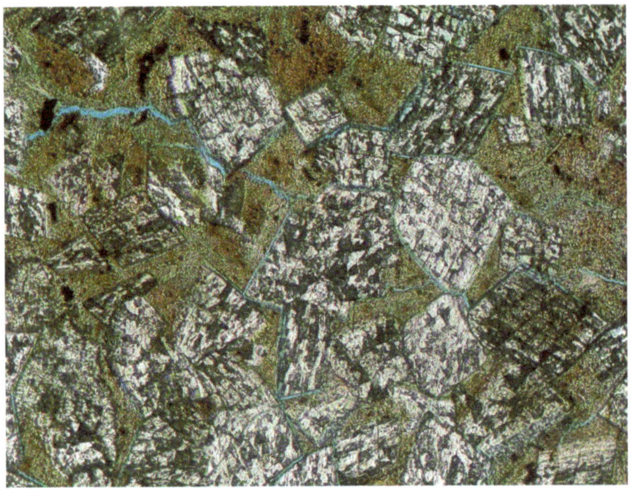

Cretaceous (early Aptian). Machiques Member, Apón Formation. Maracaibo Basin

Well-developed euhedral and subhedral dolomite crystals in the image are embedded in a clay-dominated matrix. Note the dark areas within the highly zoned crystals, which are interpreted to be intercrystalline pores formed during burial by highly corrosive brines. The pores were formed along the rhombohedral cleavage planes of the zoned dolomite crystals. The dark color of these dedolomitized zones is due to fluid inclusions, likely hydrocarbon-rich, filling the intercrystalline pores, contrasting with unfilled, blue-dyed microfractures.

IW = 1.3 mm. PPL.

Cretaceous (early Aptian). Machiques Member, Apón Formation. Maracaibo Basin

The euhedral dolomite crystals in the image suffered localized dedolomitization (or calcitization) due to the action of Ca^{2+}-rich acidic fluids. Note the pink-stained stripes outlining the corroded areas along the rhombohedral cleavage planes within the dolomite crystals. Differentiating between near-surface and burial origins for dedolomitization processes is often difficult only by pretrographic studies. A burial origin seems more likely for this sample, as this unit, which was deposited during a MFS, did not undergo prominent meteoric diagenesis.

IW = 0.5 mm. PPL.

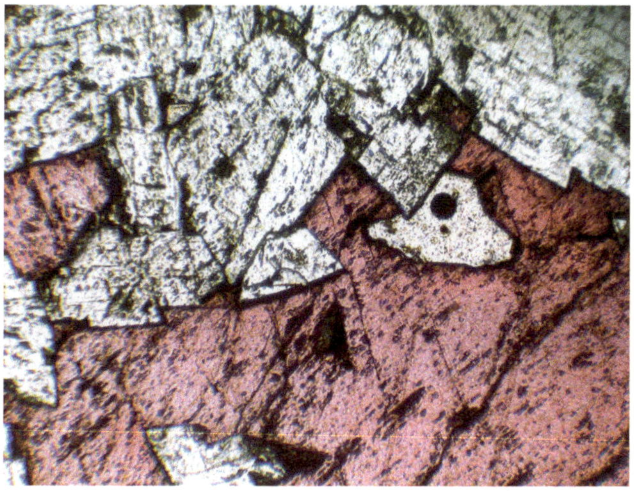

Cretaceous (early Aptian). Machiques Member, Apón Formation. Maracaibo Basin

A burial origin for dedolomitization is considered for this example, as calcitization of pink-stained crystals postdates the formation of saddle (burial) dolomite crystals (above). Calcitization obliterated prominent crystal zonation and lattice warped, which are prominent features of the saddle (unstained) dolomite crystals in the image.

IW = 2.4 mm. PPL.

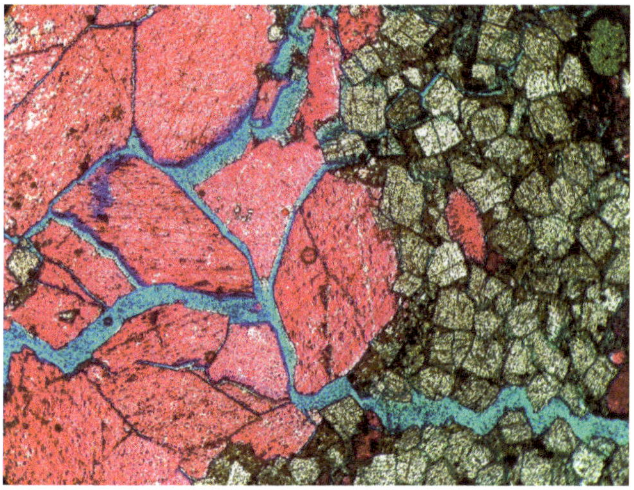

Cretaceous (early Aptian). Machiques Member, Apón Formation. Maracaibo Basin

Sharp contact between dolomite crystals (to the right) and calcitized crystals (pink stained). The latter underwent dedolomitization in the burial realm. The action of acidic Ca-rich fluids not only promoted calcitization and destruction of the dolomite fabric, but also dissolution-enlargement of microfractures. Calcitization of dolomites, in many cases, represents a mechanism for creation of secondary porosity.

IW = 2.4 mm. PPL.

References

Adams, J. F., and Rhodes, M. L., 1960. Dolomitization by seepage refluxion. *American Association of Petroleum Geology Bulletin, 44*, 1912–1920.

Archie, G. E., 1952, Classification of carbonate reservoir rocks and petrophysical considerations: *American Association of Petroleum Geologists Bulletin*, *36*, 278–298.

Badiozamani, K., 1973. The dorag dolomitization model — application to the middle Ordovician of Wisconsin. *Journal of Sedimentary Petrology*, *43*, 965–984.

Baker, P.A., Kastner, M., 1981. Constraints on the formation of sedimentary dolomite. *Science*, *213*, 214–216.

Bates, R. L., and Jackson, J. A., 1980. Glossary of Geology. Second Edition. American Geological Institute. Falls Church, Virginia. 751 pp.

Bontognali, T. R., Vasconcelos, C., Warthmann, R. J., Bernasconi, S. M., Dupraz, C., Strohmenger, C. J., McKenzie, J. A., 2010. Dolomite formation within microbial mats in the coastal sabkha of Abu Dhabi (United Arab Emirates). *Sedimentology*, *57*, 824–844.

Brady, P. V., Krumhansl, J. L., Papenguth, H.W., 1996. Surface complexation clues to dolomite growth. *Geochimica et Cosmochimica Acta*, *60*, 727–731.

Carpenter, A. B., 1980. Dolomite formation I: the stability of dolomite. In: Concepts and Models of Dolomitization (Eds., D. H. Zenger, J. B. Dunham and R. L. Ethington), *SEPM Special Publication*, *28*, 111–121.

Deffeyes, K. S., Lucia, F. J., Weyl, P. K., 1965. Dolomitization of recent and Plio-Pleistocene sediments by marine evaporite waters on Bonaire, Netherlands Antilles. In: Pray, L.C., Mur-ray, R.C. (Eds.), Dolomitization and Limestone Diagenesis. *SEPM Special Publication*, *13*, 71–88.

Evamy, B. D., 1967. Dedolomitization and the development of rhombohedral pores in limestones. *Journal of Sedimentary Research, 37,* 1204–1215.

Flügel, E, 2010. Microfacies of Carbonate Rocks—Analysis, Interpretation and Application. Springer, Heidelberg-Dordrecht-London-New York, 984 pp.

Folk, R. L., and Siedlecka, A., 1974. The "schizohaline" environment: Its sedimentary and diagenetic fabrics as exemplified by Late Paleozoic rocks of Bear Island, Svalbard. *Sedimentary Geology*, 11, 1–15.

Fouke, B. W., and Reeder, R. J., 1992. Surface structural controls on dolomite composition: Evidence from sectoral zoning. *Geochimica et Cosmochimica Acta, 56*, 4015–4024.

Friedman, G. M., 1964. Early diagenesis and lithification in carbonate sediments. *Journal of Sedimentary Research, 34*, 777–813.

Friedman, G. M., 1965, Terminology of crystallization textures and fabrics in sedimentary rocks. *Journal of Sedimentary Petrology, 35,* 643–655.

Friedman, G. M., 1980. Dolomite is an evaporite mineral: evidence from the rock record and from sea-marginal ponds of the Red Sea. In: Zenger, E.H., Dunham, J.B., Ethington, R.L. (Eds.), Concepts and Models of Dolomitization. *SEPM Special Publication, 28,* 69–80.

Friedman, G. M., Sanders, J. E., 1967. Origin and occurrence of dolostones. In: Chilingar, G.V., Bissell, H.J., Fairbridge, R.W.(Eds.), Carbonate Rocks, Origin, Occurrence, and Classification. Elsevier, Amsterdam, pp. 267–348.

Gregg, J. M. and Sibley, D. F., 1984. Epigenetic dolomitization and the origin of xenotopic dolomite texture. Journal of Sedimentary Research, 54, 908–931.

Gregg, J. M., Howard, S. A., Mazzullo, S. J., 1992. Early diagenetic recrystallization of Holocene (< 3000 years old) peritidal dolomites, Ambergris Cay, Belize. *Sedimentology, 39,* 143–160.

Goldberg, M., 1967. Supratidal dolomitization and dedolomitization in Jurassic rocks of Hamakhtesh Haqatan, Israel. *Journal of Sedimentary Research, 37,* 760–773.

Hardie, L. A., 1987. Dolomitization; a critical view of some current views. *Journal of Sedimentary Research, 57,* 166–183.

Holland, H. D., and Zimmermann, H., 2000. The Dolomite Problem Revisited. *International Geology Review, 42,* 481–490.

Hsu, K. J., and Siegenthaler, C., 1969. Preliminary experiments on hydrodynamic movement induced by evaporation and their bearing on the dolomite problem. *Sedimentology, 12,* 11–25.

Hsu, K. J., and Schneider, J., 1973. Progress report on dolomitization hydrology of Abu Dhabi Sabkhas, Arabian Gulf. The Persian Gulf. Springer, New York, pp. 409–422.

Immenhauser, A., 2021. On the delimitation of the carbonate burial realm. *The Depositional Record, 00,* 1–51.

James, N. P., and Jones, B., 2015. Origin of Carbonate Sedimentary Rocks. John Wiley & Sons, Ltd., Chinchester, 464 pp.

Jellison Jr., G. E., Leonard, D. N., Anovitz, L. M., Parish, C. M., Specht, E. D., Rosseel, T. M., 2018. Crystallographic orientation of uniaxial calcite and dolomite determined using reflection generalized ellipsometry. *Journal of Applied Physics, 124,* 223,102.

Jones, B., 2013. Microarchitecture of dolomite crystals as revealed by subtle variations in solubility: implications for dolomitization. *Sedimentary Geology, 288,* 66–80.

Jones, B., Luth, R. W., MacNeil, A. J., 2001. Powder X-ray analysis of homogeneous and heterogeneous dolostones. *Journal of Sedimentary Research, 71,* 791–800.

Kastner, M., 1984. Controls on dolomite formation. *Nature, 311,* 410–411.

Katz, A., 1971. Zoned dolomite crystals. *The Journal of Geology, 79,* 38–51.

Kerr, P. F., 1977. Optical Mineralogy. Fourth Edition. McGraw-Hill, New York. 492 pp.

Land, L. S., 1980. The isotopic and trace element geochemistry of dolomite: the state of the art. In: Zenger, D.H., Dunham, J.B., Ethington, R.L. (Eds.), Concepts and Models of Dolomitization. *SEPM Special Publication, 28,* 87–110.

Land, L. S., 1985. The origin of massive dolomite. *Journal of Geological Education, 33,* 112–125.

Lippmann, F. Crystal Chemistry of Sedimentary Carbonate Minerals., 1973. In: Sedimentary Carbonate Minerals. Minerals, Rocks and Inorganic Materials (Monograph Series of Theoretical and Experimental Studies), vol. 6.; Springer: Berlin, Heidelberg, 1973; pp. 5–96.

Lippmann, F., 1982. Stable and metastable solubility diagrams for the system $CaCO_3$-$MgCO_3$-H_2O at ordinary temperature. *Bulletin de Mineralogie, 105,* 273–279.

Machel, H. G., 1987. Saddle dolomite as a by-product of chemical compaction and thermochemical sulfate reduction. *Geology, 15,* 936–940.

Machel, H. G., 2001. Bacterial and thermochemical sulfate reduction in diagenetic settings—old and new insights. *Sedimentary Geology, 140,* 143–175.

Machel, H. G., 2004. Concepts and models of dolomitization: a critical reappraisal. *Geological Society, London, Special Publications, 235,* 7–63.

Machel, H. G., 2005. Investigations of burial diagenesis in carbonate hydrocarbon reservoir rocks. *Geoscience Canada, 32,* 103–128.

Machel, H. G., and Mountjoy, E. W., 1986. Chemistry and environments of dolomitization—a reappraisal. *Earth-Science Reviews, 23,* 175–222.

Mazzullo, S. J., 1992. Geochemical and neomorphic alteration of dolomite: a review. *Carbonates and Evaporites, 7,* 21–37.

Mazzullo, S. J., 2000. Organogenic dolomitization in peritidal to deep-sea sediments. *Journal of Sedimentary Research, 70,* 10–23.

Mckenzie, J., 1981. Holocene dolomitzation of calcium carbonate sediments from the coastal sabkhas of Abu Dhabi, U.A.E.J. *Geology, 89,* 185–198.

McKenzie, J. A. and Vasconcelos, C., 2009. Dolomite Mountains and the origin of the dolomite rock of which they mainly consist: historical developments and new perspectives. *Sedimentology, 56,* 205–219.

Mckenzie, J., Hsu, K. J., Schneider, J. F., 1980. Movement of subsurface waters under the sabkha, Abu Dhabi, UAE and its relation to evaporative dolomite genesis. *SEPM Special Publications, 28,* 11–30.

Méndez-Dot, J. A., 2011. Modelos del Holoceno en Venezuela para la formación de dolomita singenética en ambientes de sabkha y lagunas hipersalinas del Archipiélago Los Roques y su utilidad para interpretar facies de dolomita del pasado geológico. Magister Dissertation. Central University of Venezuela. 194 pp.

Moore, T. S., Murray, R. W., Kurtz, A. C., Schrag, D. P., 2004. Anaerobic methane oxidation and the formation of dolomite. *Earth and Planetary Science Letters, 229,* 141–154.

Morrow, D. W., 1982. Diagenesis 1. Dolomite - part 1: the chemistry of dolomitization and dolomite precipitation. *Geoscience Canada, 9,* 5–13.

Morse, J. W., and Mackenzie, F. T., 1990. Geochemistry of Sedimentary Carbonates. Elsevier, Amsterdam, 707 pp.

Morse, J. W., Arvidson, R. S., Lüttge, A., 2007. Calcium carbonate formation and dissolution. *Chemical Reviews, 107,* 342–381.

Mucci, A., 1991. The solubility and free energy of formation of natural kutnahorite. *Canadian Mineral, 29,* 113–121.

Murray, R. C., 1960. Origin of porosity in carbonate rocks. *Journal of Sedimentary Petrology, 30,* 59–84.

Nordeng, S. H., and Sibley, D. F., 1994. Dolomite stoichiometry and Ostwald's step rule. *Geochimica et Cosmochimica Acta, 58,* 191–196.

Petrash, D. A., Lalonde, S. V., González-Arismendi, G., Gordon, R. A., Méndez, J. A., Gingras, M. K., Konhauser, K. O., 2015. Can Mn–S redox cycling drive sedimentary dolomite formation? A hypothesis. *Chemical Geology*, *404*, 27–40.

Petrash, D. A., Gueneli, N., Brocks, J. J., Méndez-Dot, J. A., Gonzalez-Arismendi, G., Poulton, S. W., Konhauser, K. O., 2016. Black shale deposition and early diagenetic dolomite cementation during Oceanic Anoxic Event 1: The mid-Cretaceous Maracaibo Platform, northwestern South America. *American Journal of Science*, *316*, 669–711.

Petrash, D. A., Bialik, O. M., Bontognali, T. R., Vasconcelos, C., Roberts, J. A., McKenzie, J. A., Konhauser, K. O., 2017. Microbially catalyzed dolomite formation: From near-surface to burial. *Earth-Science Reviews*, *171*, 558–582.

Petrash, D. A., Bialik, O. M., Staudigel, P. T., Konhauser, K. O., Budd, D. A., 2021. Biogeochemical reappraisal of the freshwater–seawater mixing-zone diagenetic model. *Sedimentology*, *68*, 1797–1830.

Pina, C. M., Pimentel, C., Crespo, Á., 2022. The Dolomite Problem: A Matter of Time. *ACS Earth and Space Chemistry*, *6*, 1468–1471.

Reeder, R. J., and Prosky, J. L., 1986. Compositional sector zoning in dolomite. *Journal of Sedimentary Research*, *56*, 237–247.

Sánchez-Román, M., McKenzie, J. A., Wagener, A. D. L. R., Rivadeneyra, M. A., Vasconcelos, C., 2009. Presence of sulfate does not inhibit low-temperature dolomite precipitation. *Earth and Planetary Science Letters*, *285*, 131–139.

Sánchez-Román, M., Gibert, L., Martín-Martín, J. D., van Zuilen, K., Pineda-González, V., Vroon, P., Bruggmann, S., 2023. Sabkha and salina dolomite preserves the biogeochemical conditions of its depositional paleoenvironment. *Geochimica et Cosmochimica Acta*, *356*, 66–82.

Scholle, P. A., and Ulmer-Scholle, D., 2003. A Color Guide to the Petrography of Carbonate Rocks: Grains, Textures, Porosity, Diagenesis. *American Association of Petroleum Geologists, Memoir 77*, 474 pp.

Sibley, D. F., and Gregg, J. M., 1987. Classification of dolomite rock textures. *Journal of Sedimentary Research*, *57*, 967–975.

Sibley, D. F., Nordeng, S. H., Borkowski, M. L., 1994. Dolomitization kinetics of hydrothermal bombs and natural settings. *Journal of Sedimentary Research*, *64*, 630–637.

Slaughter, M., and Hill, R. J., 1991. The influence of organic matter in organogenic dolomitization. *Journal of Sedimentary Research*, *61*, 296–303.

Tucker, M. E., and Wright, V. P., 1990. Carbonate Sedimentology. Blackwell Publishing. 482 pp.

Valencia, F. L. and Laya, J. C., 2020. Deep-Burial dissolution in an Oligocene-Miocene giant carbonate reservoir (Perla Limestone), Gulf of Venezuela Basin: Implications on microporosity development. *Marine and Petroleum Geology*, *113*, 104,144.

Vasconcelos, C., and McKenzie, J. A., 1997. Microbial mediation of modern dolomite precipitation and diagenesis under anoxic conditions (Lagoa Vermelha, Rio de Janeiro, Brazil). *Journal of Sedimentary Research*, *67*, 378–390.

Warren, J. K., 1989. Evaporite Sedimentology: Importance in Hydrocarbon Accumulation. Prentice-Hall, Englewood Cliffs, NJ, 285 pp.

Warren, J. K., 1991. Sulfate dominated sea-marginal and platform evaporative settings. In: Melvin, J.L. –Ed.., Evaporites, Petroleum and Mineral Resources. Developments in Sedimentology. vol. 50. Elsevier, Amsterdam, pp. 477–533.

Warren, J. K., 1999. Evaporites: Their Evolution and Economics. Blackwell Scientific Publications, Oxford, 438 pp.

Warren, J. K., 2000. Dolomite: occurrence, evolution and economically important associations. *Earth-Science Reviews*, *52*, 1–81.

Wenk, H. R., Hu, M., Frisia, S., 1993. Partially disordered dolomite: microstructural characterization of Abu Dhabi sabkha carbonates. *American Mineralogist*, *78*, 769–774.

Wright, D. T., 1999. The role of sulphate-reducing bacteria and cyanobacteria in dolomite formation in distal ephemeral lakes of the Coorong region, South Australia. *Sedimentary Geology*, *126*, 147–157.

Wright, W. R. 2001. Dolomitization, fluid flow and mineralization of the Lower Carboniferous rocks of the Irish Midlands and Dublin basin. Doctoral Dissertation, University College of Dublin, Belfield, Ireland, 287 pp.

Wright W. R, Somerville I. D, Gregg J. M, Johnson A. W, Shelton K. L., 2000. Dolomitization and Neomorphism of Irish Lower Carboniferous (Early Mississippian): Evidence from Petrographic and Isotopic Data, *Abstract, Permo-Carboniferous Carbonate Platforms & Reefs*, Joint Society of Economic Paleontologists and Mineralogists and International Association of Sedimentologists Research and Field Conference, El Paso, Texas, 149.

Wright, W. R., Somerville, I. D., Gregg, J. M., Johnson, A. W., Shelton, K. L., 2003. Dolomitization and neomorphism of Irish Lower Carboniferous (Early Mississippian) limestones: Evidence from petrographic and isotopic studies. In: Ahr, W., Harris, A. P., Morgan, W. A. & Somerville, I. D. (eds.) Permo-Carboniferous Carbonate Platforms and Reefs. *Society of Economic Paleontologists and Mineralogists, Special Publications*, *78*; *American Association of Petroleum Geologists Memoirs*, *83*, 395–408.

Zenger, D. H., Dunham, J. D. and Ethington, R.L., 1980. Concepts and Models of Dolomitization. Society of Economic Paleontologists and Mineralogists, Tulsa, OK, 320 pp.

Zhang, F., Xu, H., Konishi, H., Roden, E. E., 2010. A relationship between d_{104} value and composition in the calcite-disordered dolomite solid-solution series. *American Mineralogist*, *95*, 1650–1656.

Porosity

19

Porosity is an important rock property because it is a measure of the potential storage volume for hydrocarbons. Carbonate reservoirs store about half of the world"™s oil and gas (e.g., Lucia, 2007; Ahr 2008). These reservoirs are rock bodies that owe their porosity and permeability to processes of deposition, diagenesis, or fracturing, individually or in combination (Ahr 2008). Understanding the origin and history of open primary or secondary pore spaces in carbonate reservoirs, as well as the mechanisms and relative timing of porosity creation, retention, reduction, or destruction, is of prime interest to hydrocarbon explorationists (Scholle and Ulmer-Scholle 2003).

For a better recognition and classification of porosities, the use of thin sections that were pressure-impregnated with color-dyed epoxy is necessary, as blue color reveals open pores. In this chapter we show photographic examples of some of the most representative carbonate porous microfacies in Venezuela. Emphasis is made on the identification of the porosity types, their primary or secondary origin, the rock fabrics and components (matrix, grains, *in situ* organisms, and cements) of such porous microfacies, and paragenetic relations (timing) involved in porosity evolution (creation and modification). The relation between combined porosity types and effective porosity (or permeability) is also discussed.

19.1 Carbonate Porosity Classification

We use the widely accepted classification of porosity proposed by Choquette and Pray (1970) for describing carbonate porosity types in thin sections. This classification is designed to aid in geologic description and interpretations of pore systems and their carbonate host rocks. It is a descriptive and genetic system in which 15 basic porosity types are recognized (Fig. 19.1): seven abundant types (interparticle, intraparticle, intercrystalline, moldic, vug, fracture and fenestral), and eight more specialized types (shelter, growth framework, channel, cavern, breccias, boring, burrow, shrinkage). The classification uses two different approaches to classify the porosity in sedimentary carbonates: **(a)** according to the rock fabric, i.e., fabric-selective and not fabric-selective; **(b)** according to the time of formation, i.e., primary or secondary. When porosity can be specifically related to sedimentary or diagenetic components that constitute the texture or fabric and the pores themselves do not cross those primary boundaries, the porosity is referred to as "fabric-selective". When there is no such dependent relation and porosity cannot be related to these features, the porosity is referred to as "not fabric-selective". These are all porosity patterns that actually or potentially can cross-cut primary grains and depositional fabrics. Other types of porosity are classed as "fabric-selective or not" to avoid ambiguity, as there is no consistent relation for these types. Primary porosity forms during predepositional stage (e.g., intraparticle pores in foraminifera, corals, bryozoans, barnacles) and during depositional stage (e.g., interparticle and growth framework porosities). Secondary porosity is formed during diagenesis, at any diagenetic stage after deposition (eogenetic, mesogenetic and telogenetic). We have added two other porosity types based on their abundance and contribution to hydrocarbon production in some carbonate porous microfacies of Venezuela (Fig. 19.1): intracrystalline porosity (fabric-selective and secondary) and microporosity (not fabric-selective and secondary).

J. A. Méndez Dot, J. Méndez Baamonde, *Atlas of Tropical Carbonates*, https://doi.org/10.1007/978-3-031-25494-9_19

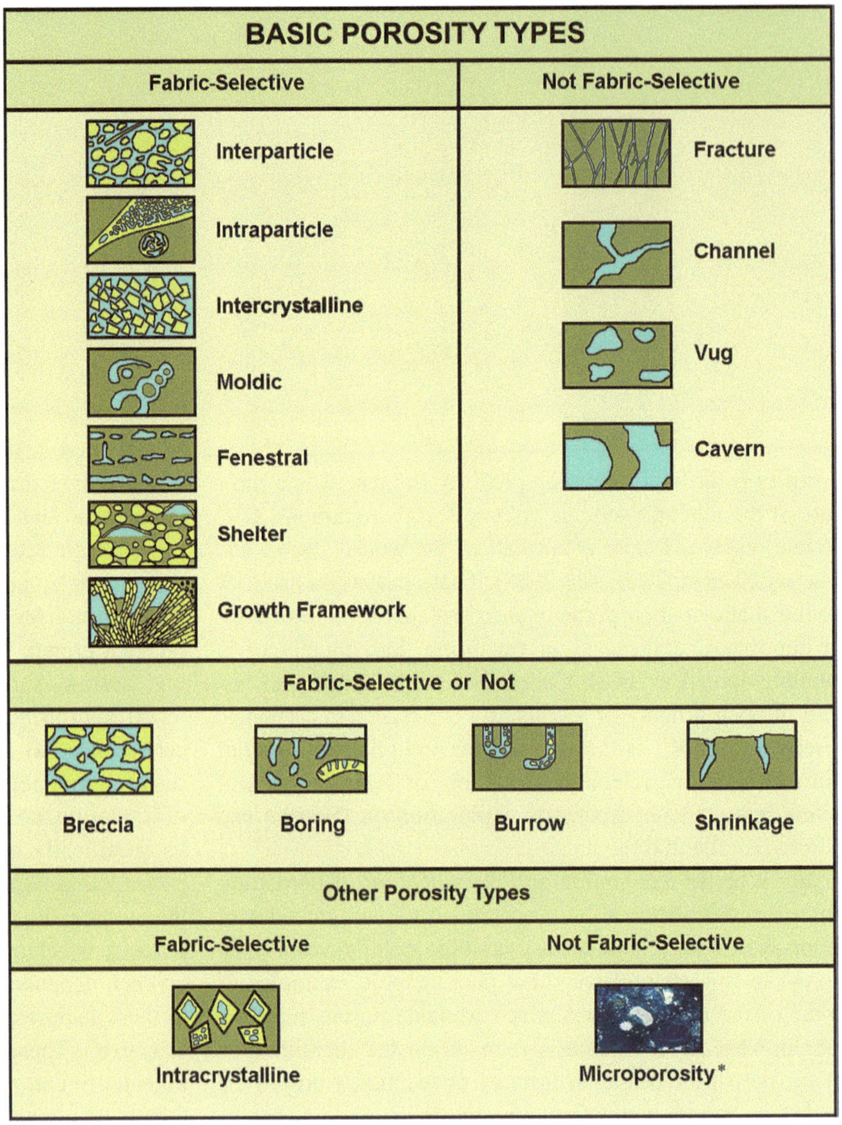

Fig. 19.1 Expanded diagrammatic representation of the basic porosity types used in the Choquette and Pray (1970) carbonate porosity classification. Other porosity types (intercrystalline and microporosity) have been include based on their abundance and contribution to hydrocarbon production in carbonate porous facies of Venezuela (*a real example from the Perla Limestone is used in this diagram for representing microporosity). Figure slightly modified from Choquette and Pray (1970). Reprinted by permission of the AAPG whose permission is required for further use

19.2 Basic Carbonate Porosity Types

In this section we deal with the 15 basic porosity types commonly described in the literature. As previously mentioned, these are classified according to the rock fabric in which the porosity/porosities are preserved or formed, and the timing of their formation.

19.2.1 Fabric-Selective Porosities (Adapted from Choquette and Pray 1970; and Flügel 2010)

Interparticle Porosity: consists of primary pores between individual particles or grains. It corresponds to depositional primary porosity. The preservation of open primary interparticle pores in ancient carbonates is promoted by the absence of water in the pores, which is related to the residence time of a given rock stratum in the active hydrologic eogenetic zone. This porosity type is associated with higher permeability. It is commonly preserved in skeletal (mud-lean) grainstones. Common pore sizes are 0.05 to 1 mm.

Intraparticle Porosity: primary pore spaces corresponding to defined parts of skeletons (e.g., foraminifera, barnacles and corals) and constructional voids created by intra-growth-forms (e.g., nodular coralline red algae). It is commonly preserved in skeletal grainstones. Common pore sizes are <0.01 to 1 mm.

Growth Framework Porosity: primary porosity associated with the growth of autochthonous reef building organisms and stromatolite structures. Growth framework porosity may be high (in modern coral reefs) or low (in reefs dominated by encrusting organisms). Rarely preserved in ancient carbonate facies, as it tends to be filled by micritic sediments or reduced by carbonate cements. This porosity is commonly too large to be recognizable in thin sections.

Fenestral Porosity: primary porosity (also commonly called "birdseye" pores) bound to synsedimentary open-space-structures (e.g., microbialites), and commonly associated to supratidal and intertidal mud-dominated sediments. It is abundant in modern stromatolites and thrombolites due to oxidation (or decayment) of organic matter. Rarely preserved in ancient carbonate facies due to carbonate cementation; although birdseye pores may be occasionally preserved as "bubble-like" vugs due to formation of gas bubbles caused by decaying organic matter. Common pore sizes are from <10 µm to several millimeters.

Shelter Porosity: a type of primary porosity created by the shelter (or "umbrella") effect of relatively large grains (e.g., molluscs, green algae) which prevent the sediment infilling of pore space underneath lying. Pore sizes commonly range from <10 µm to a few millimeters.

Moldic Porosity: results from selective dissolution of grains, mainly of aragonitic composition (e.g., molluscs, green algae). Molds form preferentially in rocks with difference in mineralogies in the meteoric-phreatic environment. It is commonly preserved in grain-supported fabrics (packstones and grainstones). Common pore sizes are <0.2 to several millimeters.

Intercrystalline Porosity: secondary porosity between more or less equant-sized crystals, often related to massive dolomitization processes. It is commonly preserved in massive dolostones. Common pore sizes are from <10 µm to 0.5 mm.

19.2.2 Not Fabric-Selective Porosities (Adapted from Choquette and Pray 1970; and Flügel 2010)

Vug Porosity: secondary porosity caused by irregularly distributed early and late diagenetic dissolution cutting across grains and/or matrix and cement boundaries and creating millimeter-sized pores. Dissolution may start from molds or interparticle pores. It is commonly preserved in grain-supported fabrics with variable amounts of micritic sediments, in packed wackestones, as well as in massive dolostones. Common pore sizes are from <0.5 mm to a few millimeters.

Fracture and Microfracture Porosity: secondary porosity resulted from the presence of openings produced by the post-depositional burial breaking of rocks. Large-scale fracturing is the main porosity type in many naturally fractured carbonate reservoir rocks (see Nelson 1985, 2001), including reservoirs of Cretaceous age in the MB (see Sánchez et al. 1999; Nelson et al. 2000), as open fractures are associated with very permeable pore system. This porosity is too large to be recognizable in thin sections. In this book we refer only to microfractures. These are the result of brittle failure under differential stress, and occur, in most cases, within the brittle domain (at low temperatures and pressures) in the intermediate-burial realm (Ahr 2008). They also occur during later diagenetic stages and are commonly associated with stylolitization. Microfractures increase the connection between separate molds and vugs (if present) and enhance permeability (Lucia 2007). Common in brittle and/or homogeneous carbonates grains and rocks (e.g., mollusc shells, dolostones, recrystallized carbonates). Common pore sizes from <0.5 mm to several centimeters.

Channel Porosity: a system of secondary pores in which the openings are markedly elongate and have developed independently of texture or fabric. Channels are mainly the result of dissolution-enlarged.

microfractures and/or vugs. Pore sizes are of 1 to several millimeters.

Cavern Porosity: not fabric-selective porosity characterized by large caverns. The term cavern applies to man-sized or larger openings of channel or vug shapes formed predominantly by karstic dissolution processes. This porosity is too large to be recognizable in thin sections.

19.2.3 Fabric-Selective or Not (Adapted from Choquette and Pray 1970; and Flügel 2010)

Breccia Porosity: depositional, solution and karst breccias may yield high porosities. Breccia porosity is in most cases the result of dissolution-collapse in the eogenetic and/or telogenetic zones. This secondary porosity is often related to large scale fractures and, therefore, too large to be recognizable in thin sections.

Boring Porosity: micro- and macroborers (e.g., worms, arthropods, algae, sponges, fungi, among others) contribute to the formation of very small to centimeter-sized borings. Commonly preserved in mollusc shells and red algal grains. Common pore sizes are from <0.5 mm to 3 mm.

Burrow Porosity: various organisms create organic burrows in relatively unconsolidated sediment. Porosity representing discrete burrows is relatively uncommon in ancient carbonate rocks.

Shrinkage Porosity: shrinkage of carbonate mud in tidal flats can result in the formation of characteristic pore systems. Although most shrinkage porosity is a specialized type of fracture porosity, it can be formed by shrinkage of individual sedimentary particles. This porosity type is rare in ancient carbonate rocks.

19.3 Other Carbonate Porosity Types

Intracrystalline Porosity: fabric-selective, secondary porosity created as a result of intracrystal dissolution of dolomite crystals. Commonly preserved in massive dolostones (previously reported by Jacka 1984; Purser et al. 1994; Mazzullo 2004; Flügel 2010; Galaup et al. 2012; Ling et al. 2014). Pore sizes (commonly from <10 µm to 1 mm) are dependent on dolomite crystal sizes. This porosity is common in dolostone microfacies of the O Member, Escandalosa Formation in the BSB.

Microporosity: not fabric-selective secondary porosity hosted within a framework of LMC crystals. Limestone microporosity is the intercrystalline pores associated with micro-scale diagenetic crypto- and microcrystals that grow into pore spaces during mineralogical and textural stabilization of metastable carbonate sediments in the meteoric, marine and diagenetic realms. Microporosity is commonly developed on carbonate matrix material (mud) and on the surfaces (rims) and interiors of carbonate allochems (e.g., algae, foraminifera, corals, ooids). The two main microporosity types defined by Kaczmarek et al. (2015) are: *Type I* microporosity, characterized by granular-subhedral crystal fabrics and associated with high porosity and permeability values (> 20%, 1–20 mD); *Type II* microporosity, associated with granular euhedral and clustered fabrics and characterized by intermediate values of porosity and permeability (10–20%, 0.1–1.0 mD). Textural similarities suggest that microporosity in the Perla Limestone (Late Oligocene-Early Miocene in the EGVSB) could be classified as Type I (Valencia and Laya 2020). Microporosity is preserved independently of the rock fabric. Common micropore sizes are from 1 to 10 µm (Moshier 1989; Budd 1989; Kaczmarek et al. 2015; Hashim and Kaczmarek et al. 2019; Valencia and Laya 2020). This porosity is abundant in Oligo-Miocene skeletal limestones of the Perla Limestone.

19.4 How to Name Carbonate Porosities

To name a specific porosity type, the main term ("pore") that corresponds to any of the porosity types is preceded by a genetic modifier or modifiers, direction or stage and the time of formation of porosity (if known or needed) (Fig. 19.2). Multiple terms can be combined, as this nomenclature allows considerable flexibility. Thus, the time of formation of the modified porosity can be replaced by the modified porosity itself (e.g., dissolution-enlarged molds).

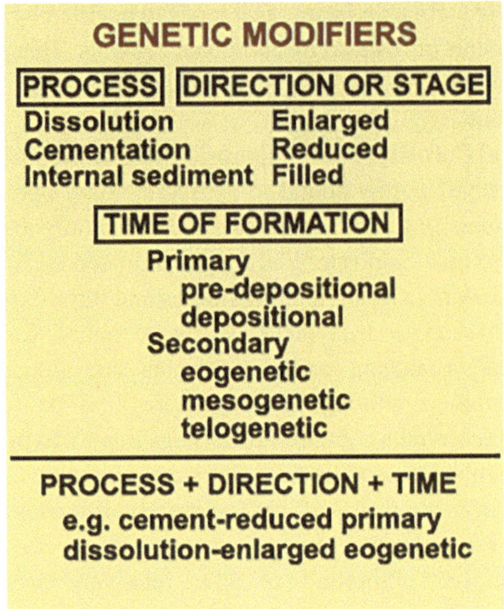

Fig. 19.2 Terminology used to describe the process, direction or stage and time of formation of porosity. Terms originally proposed by Choquette and Pray (1970). Figure slightly modified from Scholle and Ulmer-Scholle (2003). Reprinted by permission of the AAPG whose permission is required for further use

19.5 Pore Systems and Sizes

Gradational pores or pore systems are common and often cannot be clearly differentiated physically and/or genetically. They may be intermediate in characteristics between two types of pores, or they may intergrade in very short distance within a thin section. Compound pore systems are those composed of two or more basic types of pores (Choquette and Pray 1970). Common examples are those pore systems composed of interparticle and intraparticle porosities, of moldic and vug porosities, and of molds/vugs and intercrystalline porosities, or of any of the previous porosity types with microfractures or microporosity. Combined primary porosity types commonly result in higher permeabilities.

We use the pore size classification for carbonates proposed by Luo and Machel (1995), which is applicable for limestones and dolostones (Fig. 19.3). These authors created this classification for hydrocarbon exploration purposes and it is based on textural and petrophysical classification of Archie (1952), Choquette and Pray (1970), and Pittman (1979), as well as on Luo and Machel's (1995) investigations on carbonate reservoir rocks. This classification is practical and easy to apply on thin-section microfacies studies [the

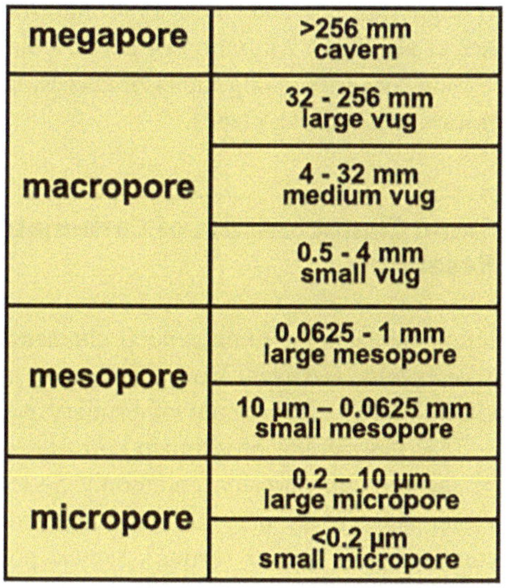

Fig. 19.3 Pore size classification for carbonates. Figure slightly modified from Luo and Machel (1995). Reprinted by permission of the AAPG whose permission is required for further use

reader is also referred to the geological and petrophysical classification of carbonate pore spaces proposed by Lucia 1995]. We have modified the size range of large micropores (originally placed at 1 µm for the upper size limit, and now placed at 10 µm), based on microporosity nature, origin and textural classification proposed by Kaczmarek et al. (2015) and Hashim and Kaczmarek (2019); as well as on the results obtained by Valencia and Laya (2020) on the study of microporosity in the Perla Limestone reservoir.

Luo and Machel's (1995) modified pore size classification has four pore sizes: microporosity (pore diameters between 1 and 10 µm), mesoporosity (pore diameters between 10 µm and 1 mm), macroporosity (pore diameters between 1–256 mm), and megaporosity (pore diameters >256 mm). In many cases, the majority of visibly perceptible carbonate pore types in thin sections fall within the range of mesopores (1 µm to 1 mm) and small macropores (between 1 to 4 mm). Mesoporosity pore types include: interparticle, intraparticle, fenestral, boring, shelter, moldic, intercrystalline, and intracrystalline pores; as well as vuggy pores of varied origins (e.g., dissolution-enlarged intercrystalline, interparticle, and intraparticle pores, and localized matrix dissolution).

The subdivision of macroporosity into three vug categories (small, medium and large) corresponds to terms commonly used in the oil industry for core and sample logging (Luo and Machel 1995), although not all pores larger than 1 mm are genetically of the vuggy type. The lower limit of small vugs is set at 0.5 mm, as dissolution cutting across grains and/or matrix often creates small vugs (even smaller than 0.5 mm). Macroporosity pore types include vugs larger than the average particle size, microfractures, and large primary pores (> 1 mm), such as interparticle and intraparticle (intra-growth forms) pores in rhodoliths, as well as large intraparticle pores in barnacle shells.

Microporosity pore types include small primary pores such as "birdseye" pores in tidal flat deposits, interparticle porosity in grainstones, intraparticle pores within bioclasts (e.g., red algae) and non-skeletal grains (e.g., ooids); secondary pores such as intercrystalline pores between dolomite crystals or calcite cement crystals, and intracrystalline pores within individual dolomite or calcite cement crystals; as well as microporosity in mud-matrix and associated microvugs and microchannels (Luo and Machel 1995; Mazzullo 2004; Hashim and Kaczmarek 2019). Intensively blue-dyed epoxy helps to reveal micropores that could otherwise be overlooked (Scholle and Ulmer-Scholle 2003).

The combined occurrence of different porosity types in a rock interval is the most likely scenario in shallow platform carbonate facies that are deeply buried, which is the case for many of the photographic examples corresponding to the Cogollo Group (Aptian and Albian in the MB), the O Member of the Escandalosa Formation (Turonian in the BSB) and the Perla Limestone. Therefore, differentiating between primary and secondary porosities is key for understanding the timing of porosity evolution.

19.6 Secondary Porosity beneath Unconformities: The Subaerial Meteoric Model

Some of the photographic examples shown in this chapter correspond to shallow-water carbonates (mainly grain-supported fabrics) that lie beneath unconformities in the paleo-vadose and paleo-phreatic meteoric (or freshwater) zones. The subaerial meteoric model (Friedman 1964; Land 1967) is still heavily applied today, especially in sequence stratigraphic-related studies of reservoirs. This model presupposes that if porous carbonate rocks are present beneath unconformities, then that porosity must have been created or enlarged by freshwater dissolution during subaerial exposure of carbonate beds in the eogenetic zone. Of course, meteoric exposure does not always create porosity, and even if it did, that porosity may be occluded during later burial diagenesis. In limestones, common secondary pore types form as a result of post-depositional dissolution, which variously includes exhumed interparticle, intraparticle, fenestral, shelter, and growth-framework pores, all of which are considered to be fabric-selective pores; dissolution enlargement of moldic pores (fabric-selective) and fractures (not fabric-selective); and the creation of not fabric-selective vugs. Meteoric porosity in dolostones (less common) occurs in closely packed dolomite mosaics (e.g., sucrosic textures), if meteoric fluids were highly acidic or acted on the rocks over long periods to create vugs, due to selective dissolution of remaining micritic sediments; it also occurs as dissolution-enlarged intercrystalline porosity due to the selective dissolution of small dolomite crystals (as solubility increases as crystal size decreases); and as intracrystalline porosity due to internal dissolution of cloudy dolomite crystals. In addition, dissolution enlargement of primary or secondary porosities may lead to total obliteration of the pore type that was enlarged. A common example is the dissolution enlargement of molds (commonly involving surrounding micritic sediments) and interparticle pores to create vugs, a process that often makes enlarged porosities indistinguishable (see Mazzullo 2004, for a detailed discussion of meteoric porosity in limestones).

19.7 Alternative Origin of Secondary Porosity: The Mesogenetic Model

In this model, explored in detail by Mazzullo and Harris (1992) and Mazzullo (2004), the dissolution of diagenetically stabilized carbonate minerals and the creation of secondary porosity in the deep-burial diagenetic realm (postdating mechanical compaction and brittle microfracturing) require the action of acidic, corrosive fluids (see Moore 2001; Esteban and Taberner 2003; Ali et al. 2010; Ehrenberg et al. 2012; Wright and Harris 2013; see also Sect. 17.2.5). Dissolution in the mesogenetic environment may occur in both rocks that underwent meteoric diagenesis and in rocks that were buried and never subaerially exposed. We focus on two dissolution models that have contributed to the creation of secondary dissolution porosity and permeability evolution in carbonate hydrocarbon reservoirs in Venezuela. **(a)** The inorganic CO_2 model (e.g., Corbella et al. 2004; Beavington-Penney et al. 2008), responsible for deep-burial dissolution that created pervasive microporosity and coeval vuggy porosity in the Perla Limestone. This corrosive process occurred during the ascent of hydrothermal, CO_2-rich fluids funneled by discontinuity surfaces (faults, fractures and stylolites) that reached the Perla reservoir (Valencia and Laya 2020), thus overprinting earlier diagenesis. **(b)** The kerogen maturation model, responsible for dissolution in carbonate intervals of the Cogollo Group (the Piché Member and Maraca Formation) and the O Member of the Escandalosa Formation, adjacent to or in stratigraphic proximity to thermally mature source rocks (La Luna Formation, La Morita Member of the Navay Formation, and some intervals of the Piché Member of the Apón Formation and the Lisure Formation). Dissolution was driven by the action of CO_2, H_2S and organic acids generated during organic matter maturation and hydrocarbon generation (e.g., Mazzullo and Harris 1992; Mazzullo 2004; Pöppelreiter et al. 2005). Migration of acidic fluids out of the source rocks occurred along faults, fractures, stylolites, unconformities, and bedding planes.

19.8 Some Characteristics of Carbonate Reservoirs

In this section we mention some general characteristics of carbonate reservoirs (adapted from Choquette and Pray 1970; and Ahr 2008): the amount of primary porosity is about 40–70%; the amount of ultimate porosity is a small fraction of the original porosity, commonly 5–15%; pore size (diameter and throat) may not be related to depositional texture (particle size or sorting); typical pore shape varies from strongly related to particle shape to completely unrelated; uniformity of pore size shape and distribution is variable, from fairly uniform to extremely heterogeneous (even within a body made up of a single rock type); fracturing (if present) has major importance in reservoir properties; porosity-permeability relationships are highly varied and may be independent of particle texture.

Characterizing carbonate reservoirs is challenging because of their intrinsic heterogeneity. This fact is a consequence of the wide spectrum of possible depositional environments, diagenetic processes and direct biological imprints that generate a complex multiscale pore systems made of both connected and isolated pores (Borrelli et al. 2019; Maniscalco et al. 2022). The use of log characteristics as indicators of depositional facies (electrofacies mapping) is not reliable because logs cannot generally detect differences in carbonate grain types or textures (Ahr 2008).

Among all rock properties, only rock fabrics, not pore-throat size, permeability-porosity ratio, or flow-zone indicators, have vertical and lateral continuity. Therefore, rock-fabric facies (grainstone, grain-dominated packstone and mud-dominated fabrics) are the basic elements for characterizing a carbonate reservoir (Lucia et al. 2003) (see Lucia 1995, 2007; Moore 2001; Lønøy 2006; Ahr 2008; Moore and Wade 2013; Skalinski and Kenter 2014; Borrelli et al. 2019; Kargarpour 2020; Maniscalco et al. 2022; for a detailed treatment on characterization of carbonate reservoirs).

19.9 Porosity: Photographic Set and Descriptive Captions

In all photos, IW = image width (the true length of the horizontal axis of the view)

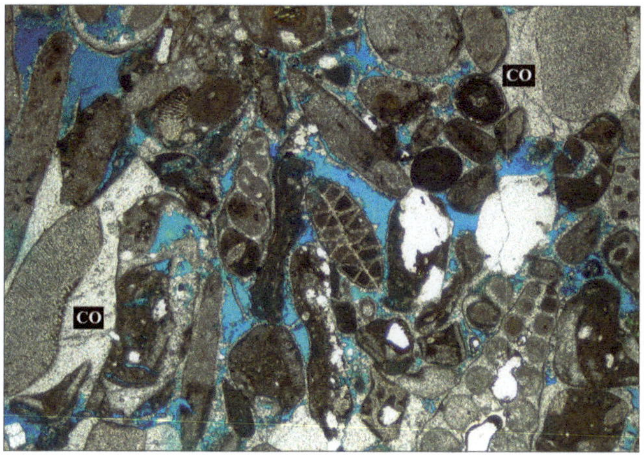

Cretaceous (latest Albian). Maraca Formation. Maracaibo Basin

Interparticle porosity in a grainstone. Grains in the image consist of green algae, echinoderms, gastropods and bryozoans, and were deposited in a mud-lean environment. Interparticle primary pores (mesopores) were formed during deposition of these bioclastic components. The preservation of this fabric-selective porosity was directly linked to the lack of lime-mud during final deposition and the absence of substantial meteoric cementation between grains. Cementation-reduced intraparticle porosity occurred locally, as calcite overgrowth (CO) surrounds some of the echinoid grains.

IW = 7.5 mm. PPL.

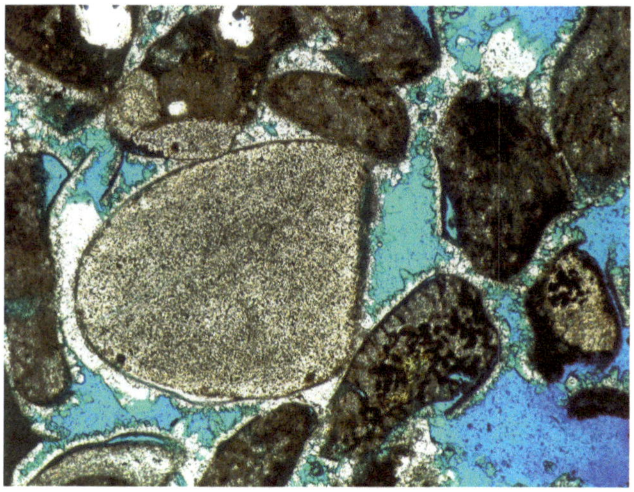

Cretaceous (latest Albian). Maraca Formation. Maracaibo Basin

Green algae and echinoderms grainstone with interparticle (primary) porosity. Meteoric cements (bladed and meniscus) rim the grains and fill the spaces between them, thus occluding a small volume of porosity. As in the previous example, primary porosity was preserved since cementation and porosity reduction most likely occurred in the meteoric-vadose (unsaturated) zone. In the Maraca Formation, secondary pores (molds and vugs) are common, but both higher total and effective porosities are always associated with interparticle primary porosity as the main porosity type.

IW = 2.4 mm. PPL.

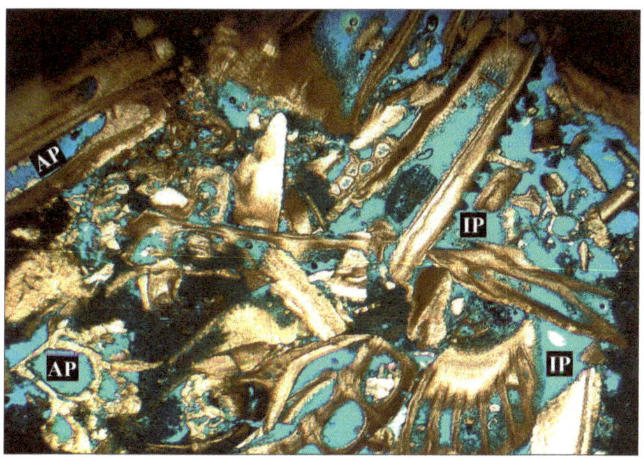

Late Oligocene-Early Miocene. Perla Limestone. Eastern Gulf of Venezuela sub-Basin

Primary interparticle and intraparticle porosities in a barnacle grainstone. Large plate shell fragments were deposited in a mud-free, moderate- to high-energy environment. The combined preservation of interparticle (IP) and intraparticle (AP) pores enhance the effective porosity and hence increase permeability. Pores in the image are mainly large mesopores. A few of them reach the size of a small macropores. Barnacle LMC parietal tubes and large compartments are among the largest primary pores in the Perla reservoir.

IW = 7.5 mm. PPL.

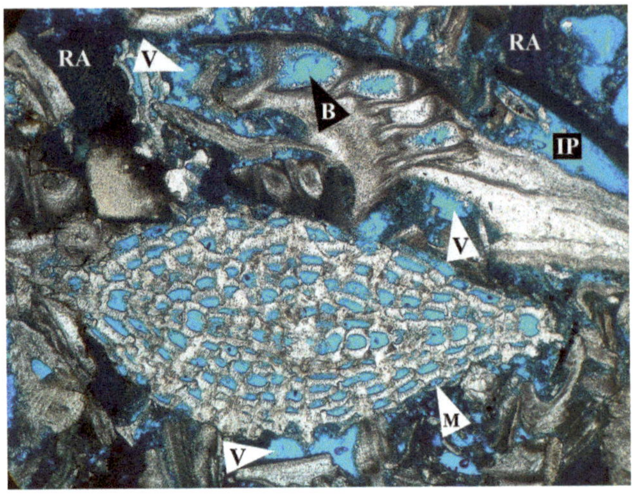

Late Oligocene-Early Miocene. Perla Limestone. Eastern Gulf of Venezuela sub-Basin

This grainstone exhibits both primary and secondary porosities. Most depositional pores in the image (mainly mesopores) are of the intraparticle type. These are well-preserved in a multichambered test of a larger benthic foraminifer *Miogypsina* (M) test, and in a barnacle (B) shell fragment. Interparticle pores (IP) are less abundant, as mechanical compaction substantially reduced voids between grains. Other pores in the image are secondary. These are mainly micropores formed in red algal grains (RA) and associated vugs (V) formed from dissolution enlargement of small interparticle pores.

IW = 2.4 mm. PPL.

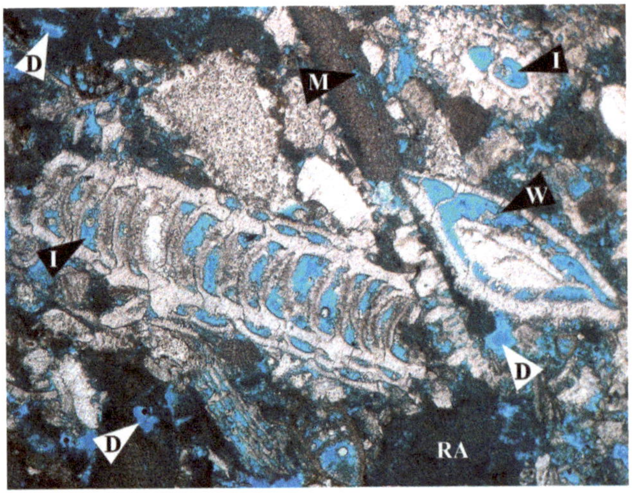

Late Oligocene-Early Miocene. Perla Limestone. Eastern Gulf of Venezuela sub-Basin

A variety of pore types are present in this grainstone. Intraparticle porosity (I) within LBF tests is often preserved in association with localized dissolution of the inner walls (W) of the tests, creating dissolution-enlarged intraparticle pores. Selective dissolution of red algal grains created small moldic (M) pores (or micromolds) and vugs that resulted from dissolution (D) enlargement of these pores. Red algae (RA) grains also show the development of microporosity, which is coeval with the micromolds and vugs.

IW = 2.4 mm. PPL.

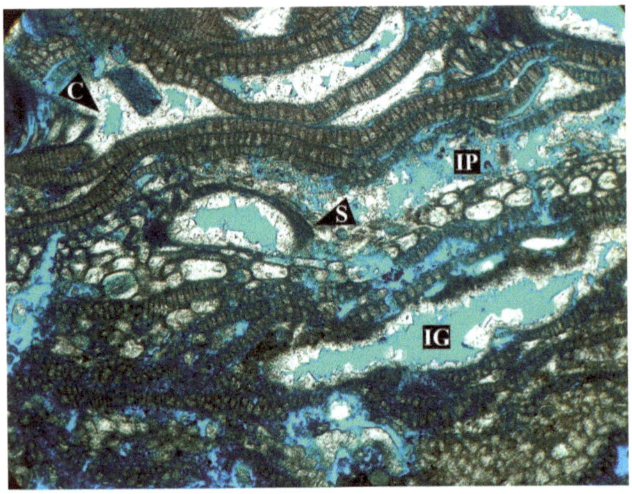

Late Oligocene-Early Miocene. Perla Limestone. Eastern Gulf of Venezuela sub-Basin

Rhodoliths contain some of the largest primary pores in the Perla reservoir, some falling in the range of small macropores. In this image, intraparticle pores are observed in the form of intra-growth-forms (IG) developed within single, cojoined non-geniculate red algae filaments and within serpulid (S) tubes. Other primary pores in the image are of the interparticle (IP) type, formed between filaments of red algae or between these filaments and encrusting benthic foraminifera. All pores shown are partially occluded by calcite cements (C), some of which are of the dogtooth type.

HA = 2.4 mm. PPL.

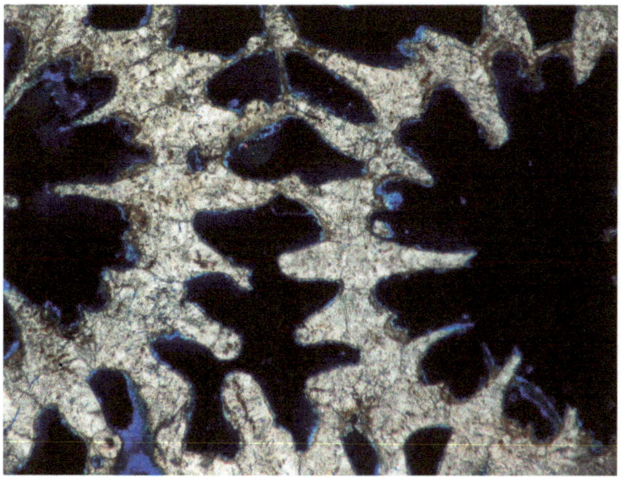

Late Pleistocene. Aves Island Los Roques Archipelago

This image depicts the intraparticle (or intraskeletal) primary porosity of the coral species *Orbicella annularis*. Intraparticle pores occur both within individual corallites and between them (the latter located in the center). This fabric-selective porosity type is abundant in Holocene coral framestones (composed of sedentary carbonate producers). The pore spaces visible the image may reach values up to 35-40% (visual estimation). In ancient reef limestones, however, such high primary porosities are generally substantially reduced, either by micritic sediment infilling in the marine environment or by cementation during meteoric diagenesis.

IW = 4.3 mm. CPL.

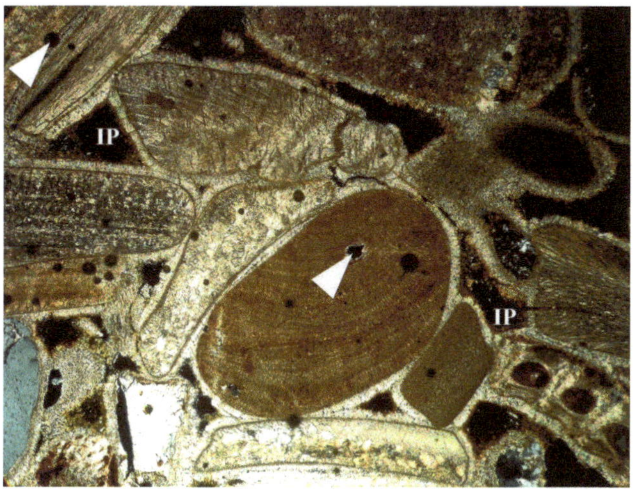

Holocene. Henri Pittier National Park. Aragua State shoreline

This beachrock hosts two types of porosities. Depositional interparticle pores (IP) are the most common type of pores in beachrock deposits. In this image, isopachous crusts of marine cements are rimming the allochems, thus outlining these pores. Primary interparticle porosities in Holocene beachrock fabrics may reach up to 40% (see Strasser et al. 1989; Guerra et al. 2005; Ghandour et al. 2014). The small, spherical perforations within the bioclasts are boring (secondary) pores (white arrows).

IW = 2.4 mm. CPL.

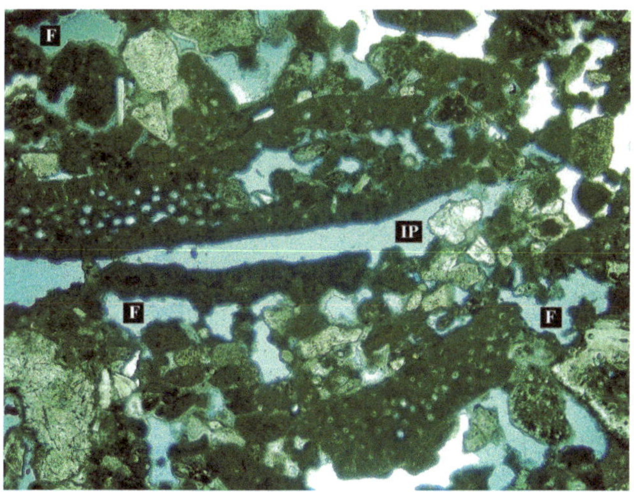

Holocene. Los Roques Archipelago

This stromatolitic boundstone (bindstone) hosts a combination of primary porosities. The large horizontal void in the center of the image is an interparticle pore (IP). Spherical, small intraparticle pores within micritized *Halimeda* grains are abundant. Other pores in the image are of the fenestral (F) type. These exhibit characteristic flat/elongate and/or irregular shapes and varied orientations with respect to the bedding. Fenestral (birdseye) pores were formed due to *in situ* degradation (or respiration) of organic matter, a process that involved localized dissolution of $CaCO_3$.

IW = 2.4 mm. PPL.

Holocene. Los Roques Archipelago

This sample is a variation in the fabric within the same laminated stromatolite presented in the previous example. Depositional interparticle porosity formed due to moderately good sorting of the allochems (bioclasts mainly consisting of dark micritized *Halimeda* grains and light-colored mollusc grains) and the absence of void-filling cements. Grains are cemented together by microbially-induced microcrystalline micrite precipitated at grain contacts.

IW = 2.4 mm. PPL.

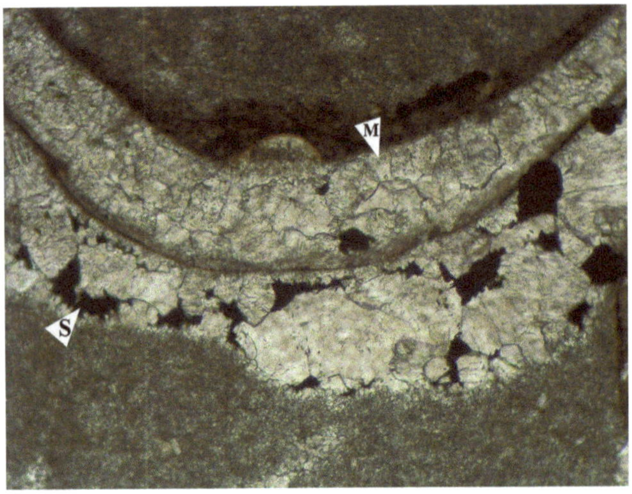

Cretaceous (middle Aptian). Piché Member, Apón Formation. Maracaibo Basin

A large shelter (S) pore (macropore) developed due to the "umbrella" effect created by the leached aragonitic mollusc (M). The shell acted as barrier, preventing the sediment from infilling the large pore underneath. The primary umbrella void was later almost completely occluded by precipitation of coarse-grained calcite cement.

IW = 2.4 mm. CPL.

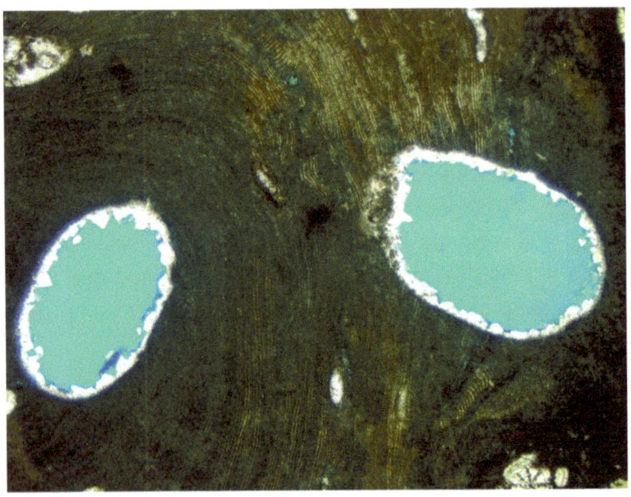

Late Oligocene-Early Miocene. Perla Limestone. Eastern Gulf of Venezuela sub-Basin

Boring pores developed in a red algal grain. These pores are approximately 1 to 1.5 mm in diameter (small macropores). They exhibit characteristic rounded shapes and are internally rimmed by dogtooth cement. In the Perla Limestone, this fabric-selective or not secondary porosity is locally common in red algal grains, with pores with diameters ranging from 0.5 to 2 mm. The bioerosion process was dominated by the activity of sponges (*Entobia*) (see Valencia and Laya 2020).

IW = 4.2 mm. PPL.

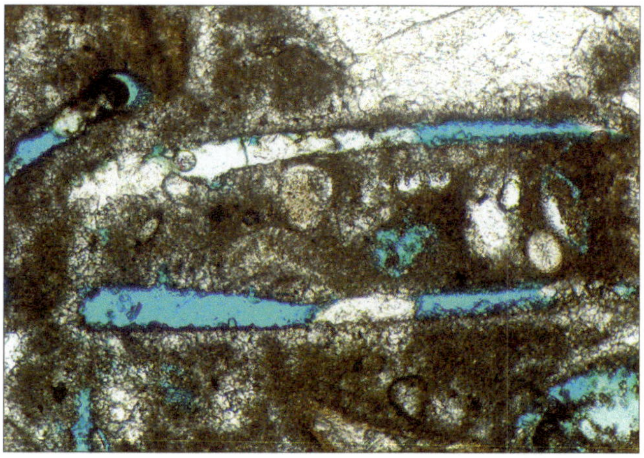

Cretaceous (latest Albian). Maraca Formation. Maracaibo Basin

Close view of biomoldic pores created due to selective dissolution (or leaching) of fragments of aragonitic mollusc valves. The precipitation of LMC cement during subsequent meteoric cementation partially filled these secondary pores. Aragonitic molluscs are among the most commonly leached metastable grains in shallow carbonate microfacies of the MB.

IW = 1.3 mm. PPL.

Middle Miocene. Capadare Formation. Falcón Basin

Selective dissolution of LBF tests (*Amphistegina*) created the moldic pores (M) in the image. Although this dissolution process suggests an original HMC composition of the foraminifera tests, these are rarely completely dissolved, as seen in this example. Other pores in the image are irregular-shaped vugs (V) formed due to dissolution-enlarged molds and localized dissolution of the micritic matrix. Although porous facies in the Capadare Formation are not hydrocarbon reservoirs, dissolution porosities (molds and vugs) are common in some mud- and grain-supported fabrics.

IW = 2.4 mm. PPL.

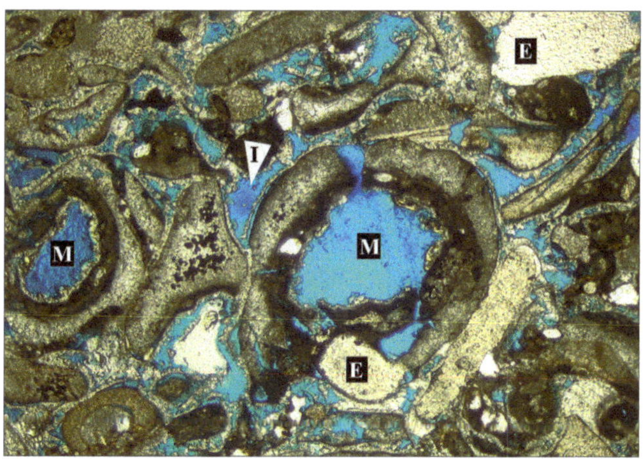

Cretaceous (latest Albian). Maraca Formation. Maracaibo Basin

This grainstone of green algae and echinoids shows a combination of interparticle (I) (primary) and moldic (M) (secondary) pores. The latter are fabric-selective and were formed due to dissolution of the inner medulla regions of metastable green algal grains during meteoric diagenesis. In contrast, HMC echinoid (E) grains do not contribute to secondary porosity creation, as their original highly porous skeletons were occluded by syntaxial calcite cement. In general, the combination of primary and secondary porosities increases permeability.

IW = 4.2 mm. PPL.

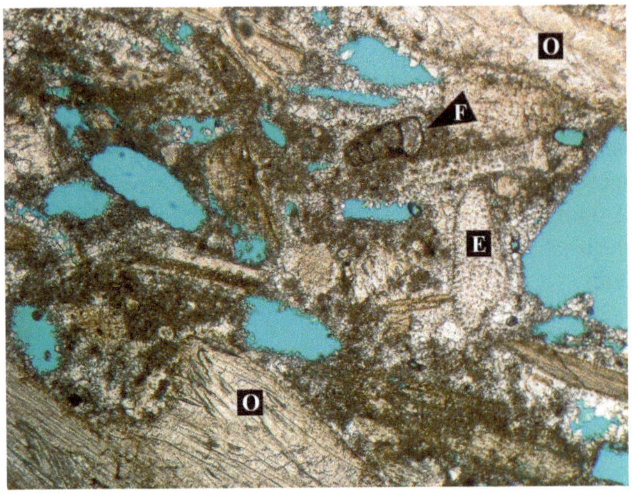

Cretaceous (latest Albian). Maraca Formation. Maracaibo Basin

This packstone exhibits moldic porosity created by the selective dissolution of aragonitic bioclasts. Identifiable grains in the image include oysters (O), echinoderms (E), and a smaller benthic foraminifer (F). Textural variations from mud-lean grainstones (such as the previous example) to packstones are reflected in the absence of primary interparticle pores, as seen in this example. The presence of small amounts of micritic sediments between grains during final deposition led to the absence of primary interparticle porosity.

IW = 4.2 mm. PPL.

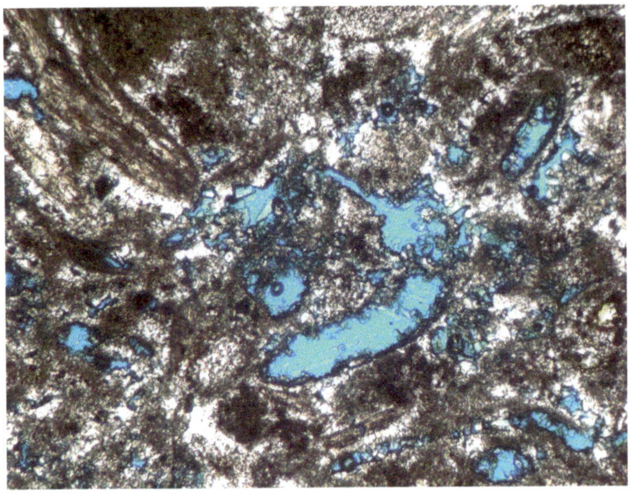

Cretaceous (middle Aptian). Piché Member, Apón Formation. Maracaibo Basin

Another example of moldic porosity in a wackestone-packstone. Meteoric diagenesis resulted in the dissolution (or leaching) of aragonitic grains, thus creating biomoldic pores; patches of microspar (white precipitate) partially replacing the micritic matrix and filling small pores; and precipitation of LMC cements partially rimming the interior of the pores. Note the dark color of micritic sediments, which is due to hydrocarbon impregnation.

IW = 4.2 mm. PPL.

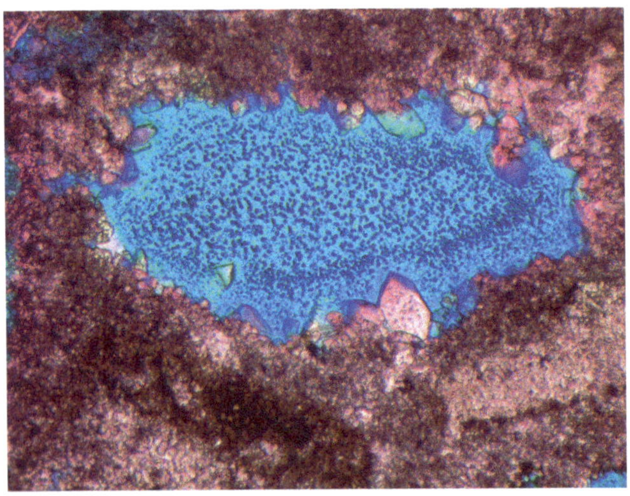

Cretaceous (middle Aptian). Piché Member, Apón Formation. Maracaibo Basin

A large vuggy pore (small macropore; larger than the average particle size) created due to the dissolution enlargement of a moldic pore. Precipitation of coarse-grained LMC crystals postdating the formation of the vug are internally rimming the pore. Dissolution porosities (molds and vugs), occasionally accompanied by microfractures, are the main porosity types in porous intervals of the Piché Member. These intervals are underlain and overlain by source rocks corresponding to OAEs 1a and 1b, respectively.

IW = 2.4 mm. PPL.

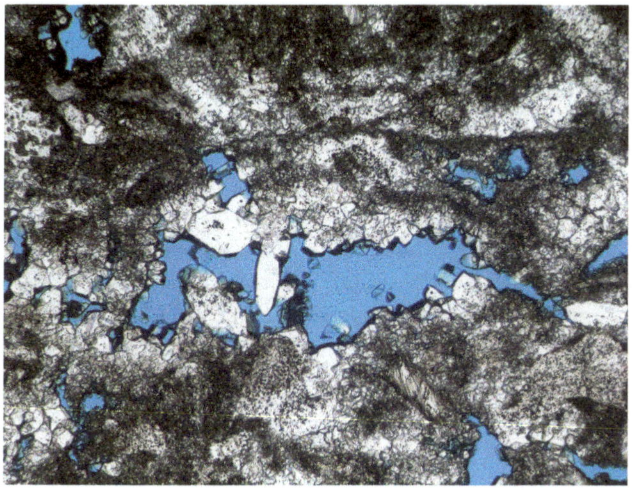

Cretaceous (latest Albian). Maraca Formation. Maracaibo Basin

A dissolution-enlarged vuggy pore in the center of the image, internally rimmed by calcitic cements. Surrounding the vug, the precipitation of microspar (white-dusty areas) contrasts with hydrocarbon-impregnated (dark) areas, which are compositionally micrite-dominated. This example corresponds to the Urdaneta West field, and this packstone with vuggy pores partially filled by void-filling LMC cements matches with those described by Póppelreiter et al. (2005), who proposed late burial dissolution for the formation of vuggy pores in grain-supported fabrics of the Maraca Formation in that oil field.

IW = 2.4 mm. PPL.

Middle Miocene. Capadare Formation. Falcón Basin

In the absence of blue-epoxy impregnation, as in this example, open pores are more easily visualized under CPL, as they appear black. Dissolution-enlarged vuggy (V) pores (some of them fall in the range of small macropores) in this peloidal limestone are internally rimmed by dogtooth calcite cement. Other pores in the image are of the interparticle (I) type.

IW = 2.4 mm. CPL.

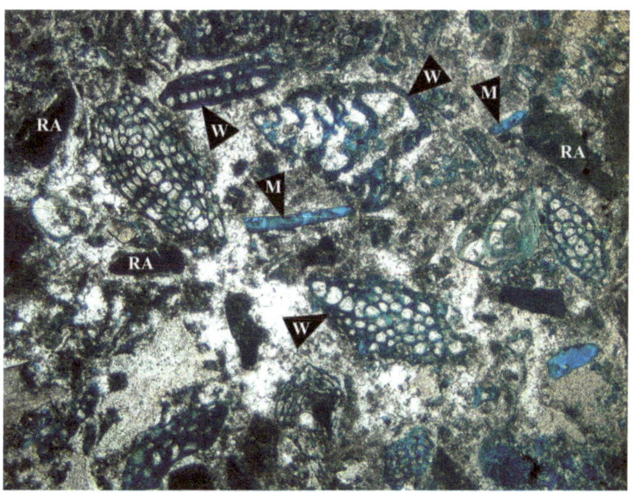

Late Oligocene-Early Miocene. Perla Limestone. Eastern Gulf of Venezuela sub-Basin

This bioclastic packstone exhibits a combination of dissolution porosities. Moldic (M) pores were created due to fabric-selective dissolution of aragonitic fossils. Fabric-selective dissolution of HMC walls (W) of LBF tests created dissolution-enlarged intraparticle pores. Pervasive microporosity was created in red algae (RA) grains.

IW = 7.5 mm. PPL.

Cretaceous (Turonian). O Member, Escandalosa Formation. Barinas sub-Basin

Euhedral to subhedral dolomite crystals in the image are forming a sucrosic fabric hosting intercrystalline porosity. These secondary (fabric-selective) pores were created due to the massive replacement (mole-per-mole) of micritic sediments. This process involves a decrease in the molar size of dolomite compared to the former aragonite or calcite substrate (Murray 1960). Intercrystalline pores consist of characteristic straight shapes derived from porosity gain after total mineralogical replacement. These straight-shaped pores fall in the range of large mesopores, with some of them larger than 0.5 mm.

IW = 1.3 mm. PPL.

Cretaceous (Turonian). O Member, Escandalosa Formation. Barinas sub-Basin

Another example of a porous massive dolostone. In addition to mole-per-mole replacement, the origin of the intercrystalline pore network in this reservoir is also associated with selective dissolution of unreplaced micritic sediments remaining between dolomite crystals. This localized process occurs as the solution becomes undersaturated for calcite after all Mg in excess of dolomite saturation is exhausted (see Machel 2004). Therefore, micrite dissolution postdating massive dolomitization led to the creation of dissolution-enlarged intercrystalline pores in the image. The larger one falls in the range of a small macropore.

IW = 2.4 mm. PPL

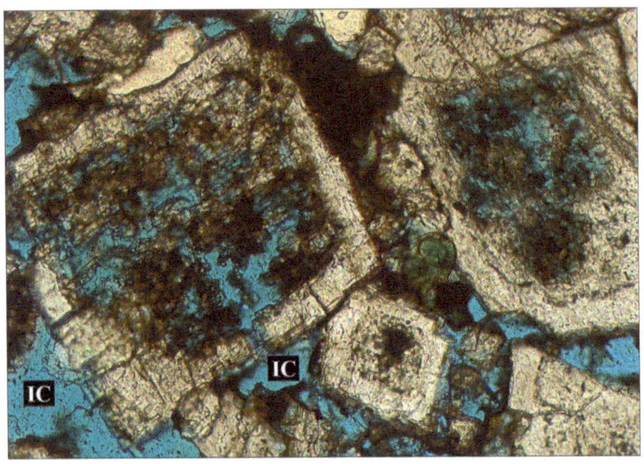

Cretaceous (Turonian). O Member, Escandalosa Formation. Barinas sub-Basin

Intracrystalline (fabric-selective) pores formed within dolomite crystals with cloudy cores. The latter, composed of high-Ca dolomite, are more prone to rapid dissolution compared to more stoichiometric clear rims. In this dolostone microfacies, effective porosity and permeability are enhanced when intracrystal dissolution is combined with intercrystalline (IC) porosity. Note the hydrocarbon impregnations within and between dolomite crystals. This, along with the two previous examples, is representative of dolomitic hydrocarbon reservoirs in the O Member.

IW = 0.5 mm. PPL.

Late Oligocene-Early Miocene. Perla Limestone. Eastern Gulf of Venezuela sub-Basin

This example exhibits a combination of depositional and secondary porosities, the latter created in the deep-burial environment. Intraparticle pores within the central larger benthic foraminifer test are partially occluded by sparry calcite, which infills some of the small rectangular chamberlets. Dolomite-related (intercrystalline) porosity is abundant and easily recognized in this sample. Note that the matrix is highly corroded, consisting mainly of small-sized dolomite crystals formed at high temperatures. Late burial, acid-driven dissolution created microporosity in red algal grains.

IW = 4.2 mm. PPL.

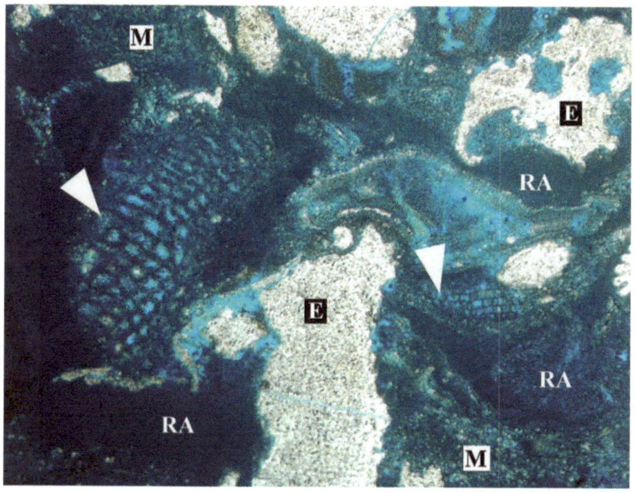

Late Oligocene-Early Miocene. Perla Limestone. Eastern Gulf of Venezuela sub-Basin

Three types of burial dissolution porosities are present in this packed fabric consisting of red algal and echinoids (E). Fabric-selective dissolution of some of the red algae grains in the image formed intraparticle pores, which are easily noticed by the small-scale reticulate cellular structure of these grains (white arrows). Coeval microporosity was developed in squeezed (plastically deformed) red algae (RA) grains and matrix (M). The latter hosts high-temperature dolomite crystals and associated dolomite-related porosity.

IW = 2.4 mm. PPL.

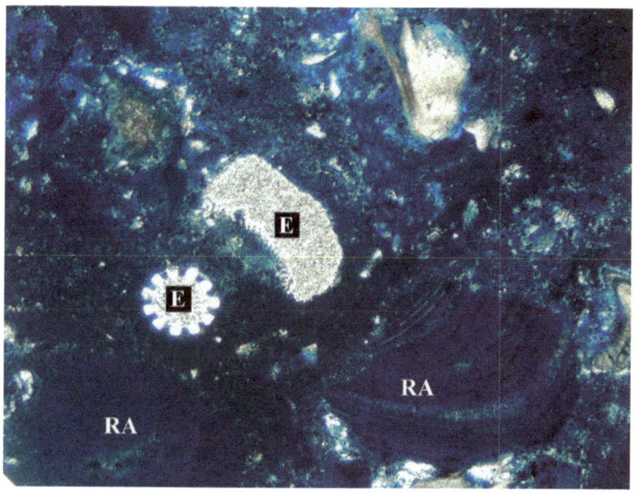

Late Oligocene-Early Miocene. Perla Limestone. Eastern Gulf of Venezuela sub-Basin

Intense "blue haze" staining in this image is due to pervasive, acid-driven dissolution and the development of microporosity in the deep-burial diagenetic environment. This process corroded the matrix and red algal (RA) grains. The latter completely lost their texture, which combined with the intensity of the blue staining, makes their differentiation from the matrix difficult. Intense blue staining contrasts with light-colored echinoid (E) grains, which are generally far less affected by late burial corrosion. Microporosity in the Perla Limestone is associated to porosity values ranging from 15 to 40% (see Valencia and Laya 2020).

IW = 4.2 mm. PPL.

Cretaceous (middle Aptian). Piché Member, Apón Formation. Maracaibo Basin

Boring pores in the image, formed on a calcitic mollusc shell, are connected through open microfractures. In general, calcitic mollusc shells provide a suitable rigid substrate for the occurrence of both early marine boring porosity and burial microfractures. Strong hydrocarbon impregnation of the pores evidences the stratigraphic proximity of the overlying, thermally mature source rock of late Aptian age (the OAE 1b first sub-event).

IW = 2.4 mm. PPL.

Cretaceous (early-late Aptian). Piché Member, Apón Formation. Maracaibo Basin

Porosity in this moderately organic-rich dolomitic mudstone may be classified as a gradation between microfracture and breccia porosities. Brittle microfractures originally formed fringing rhombic dolomite crystals. Subsequent dissolution of the matrix connected the conjugate set. The acidic fluids responsible for the dissolution were derived from the thermal maturation of the organic matter dispersed in the micritic matrix. Dissolution-enlarging is also evident, mainly in the center of the image. This network of interconnected microfractures resembles telogenetic, dissolution-collapse brecciation.

IW = 1.3 mm. PPL.

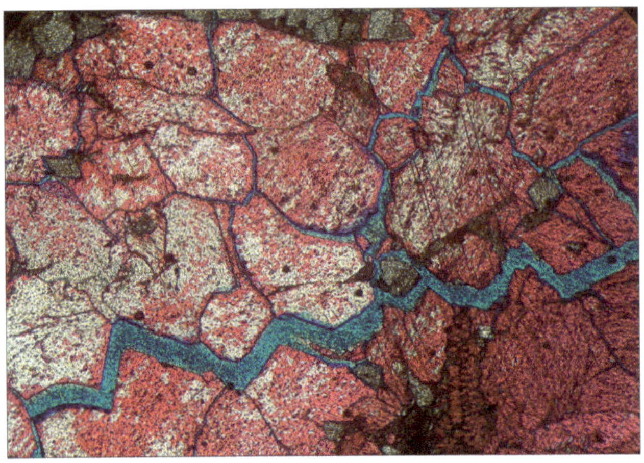

Cretaceous (early Aptian). Machiques Member, Apón Formation. Maracaibo Basin

Combined microfracturing and dedolomitization resulted in the creation of secondary porosity. Recognizing the timing between these two mesogenetic processes is key for understanding the creation of this porosity. We estimate that brittle microfracturing occurred first. Note the "zig-zag" (or jagged) shape of the thicker microfracture, originally formed fringing dolomite crystals. Posterior dedolomitization (or calcitization of dolomite), due to the action of acidic fluids rich in Ca^{2+}, led to localized dissolution along narrower subordinate microfractures, which in turn fringe dedolomitized crystal boundaries.

IW = 4.2 mm. PPL.

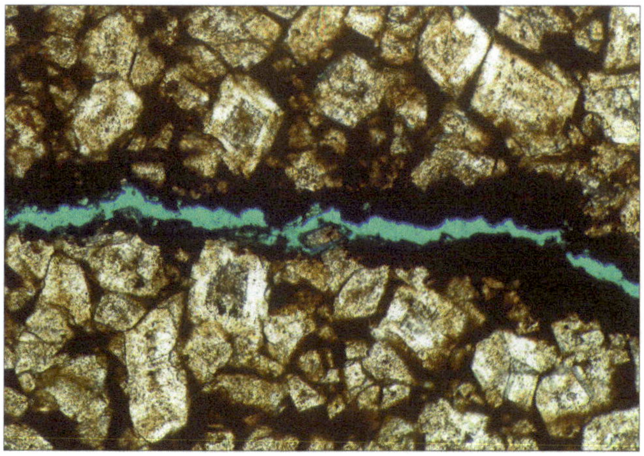

Cretaceous (Turonian). O Member, Escandalosa Formation. Barinas sub-Basin

Brittle microfracturing is also common in massive dolostone microfacies of the O Member. In this case, the microfracture in the image (not fabric-selective) remained open and became impregnated by hydrocarbons generated by the overlying La Morita Member of the Navay Formation. Additionally, impregnations are also visible between dolomite crystals surrounding the microfracture, as hydrocarbons are filling small intercrystalline pores.

IW = 2.4 mm. PPL.

Cretaceous (Turonian). O Member, Escandalosa Formation. Barinas sub-Basin

Dissolution enlargement along microfractures led to the development of channel porosity. Note the marked elongation of pores, indicating that dissolution postdates brittle microfracturing. The action of hydrocarbon-associated acidic fluids in this dissolution-enlargement process is indicated by the dark color of the hydrocarbon-impregnated substrates between dolomite crystals.

IW = 4.2 mm. PPL.

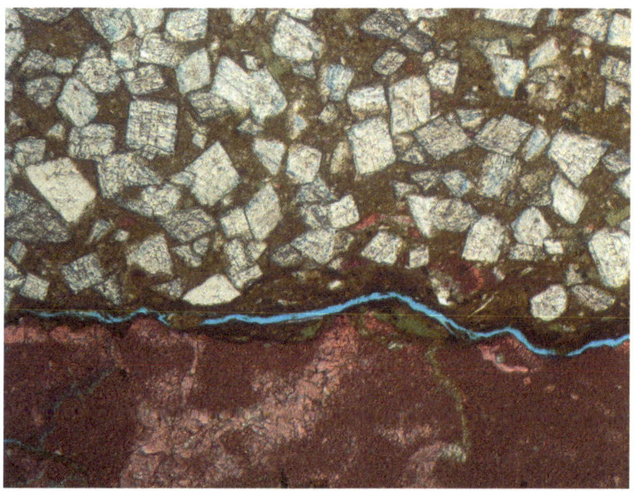

Cretaceous (middle Aptian). Piché Member, Apón Formation. Maracaibo Basin

In this sample, localized dissolution occurred along a stylolite formed at the sharp contact between two different microfacies, a micritic-dominated substrate (pink-stained) and a dolomitic shale (above). Although occasionally present locally in shallow Cretaceous carbonates of the MB (mainly in the Piché Member and the Lisure Formation), stylolite porosity does not contribute to hydrocarbon production in that basin, as this "unconventional" porosity type does not connect open pores.

IW = 2.4 mm. PPL.

References

Ahr, W., 2008. Geology of Carbonate Reservoirs. The Identification, Description and Characterization of Hydrocarbon Reservoirs in Carbonate Rocks. Wiley, 277 pp.

Ali, S.A., Clark, W. J., Moore, W. R., Dribus, J. R., 2010. Diagenesis and reservoir quality. *Oilfield Review*, *22*, 14–27.

Archie, G. E., 1952. Classification of carbonate reservoir rocks and petrophysical considerations. *American Association of Petroleum Geology Bulletin*, *36*, 278–298.

Beavington-Penney, S. J., Nadin, P., Wright, V. P., Clarke, E., McQuilken, J., Bailey, H. W., 2008. Reservoir quality variation on an Eocene carbonate ramp, El Garia Formation, offshore Tunisia: structural control of burial corrosion and dolomitisation. *Sedimentary Geology*, *209*, 42–57.

Borrelli, M., Campilongo, G., Critelli, S., Ida, D. P., Perri, E., 2019. 3D nanopores modeling using TEM-tomography (dolostones-Upper Triassic). *Marine and Petroleum Geology*, *99*, 443–452.

Budd, D. A., 1989. Micro-rhombic calcite and microporosity in limestones: a geochemical study of the Lower Cretaceous Thamama group, U.A.E. *Sedimentary Geology*, *63*, 293–311.

Choquette, P. W., and Pray, L. C., 1970. Geologic nomenclature and classification of porosity in sedimentary carbonates. *American Association of Petroleum Geology Bulletin*, *54*, 207–250.

Corbella, M., Ayora, C., Cardellach, E., 2004. Hydrothermal mixing, carbonate dissolution and sulfide precipitation in Mississippi Valley-type deposits. *Mineral Deposits*, *39*, 344–357.

Ehrenberg, S. N., Walderhaug, O., Bjorlykke, K., 2012. Carbonate porosity creation by mesogenetic dissolution: reality or illusion? *American Association of Petroleum Geologists, Bulletin 96*, 217–233.

Esteban, M., and Taberner, C., 2003. Secondary porosity development during late burial in carbonate reservoirs as a result of mixing and/or cooling of brines. *Journal of Geochemical Exploration*, *78*, 355–359.

Friedman, G. M., 1964. Early diagenesis and lithification in carbonate sediments. *Journal of Sedimentary Research*, *34*, 777–813.

FlÃ¼gel, E, 2010. Microfacies of Carbonate Rocks" Analysis, Interpretation and Application. Springer, Heidelberg-Dordrecht-London-New York, 984 pp.

Galaup, S., Liu, Y., Cerepi, A., 2012. New integrated 2D–3D physical method to evaluate the porosity and microstructure of carbonate and dolomite porous system. *Microporous and Mesoporous Materials*, *154*, 175–186.

Ghandour, I. M., Al-Washmi, H. A., Bantan, R. A., Gadallah, M. M., 2014. Petrographical and petrophysical characteristics of asynchronous beachrocks along Al-Shoaiba Coast, Red Sea, Saudi Arabia. *Arabian Journal of Geosciences*, *7*, 355–365.

Guerra, N. C., Kiang, C. H., Sial, A. N., 2005. Carbonate cements in contemporaneous beachrocks, Jaguaribe beach, Itamaracá island, northeastern Brazil: petrographic, geochemical and isotopic aspects. *Anais da Academia Brasileira de Ciências*, *77*, 343–352.

Hashim, M. S., and Kaczmarek, S. E., 2019. A review of the nature and origin of limestone microporosity. *Marine and Petroleum Geology*, *107*, 527–554.

Jacka, A. D., 1984. Intracrystalline Porosity" A Newly Discovered Pore Type in Dolostone Reservoirs and Implications for Dedolomitization and Pseudospiculites. *American Association of Petroleum Geology Bulletin*, *68*, 491–491.

Kaczmarek, S. E., Fullmer, S. M., Hasiuk, F. J., 2015. A universal classification scheme for the microcrystals that host limestone microporosity. *Journal of Sedimentary Research*, *85*, 1197–1212.

Kargarpour, M. A., 2020. Carbonate reservoir characterization: an integrated approach. *Journal of Petroleum Exploration and Production Technology*, *10*, 2655–2667.

Land, L. S., 1967. Diagenesis of skeletal carbonates. *Journal of Sedimentary Petrology*, *37*, 914–930.

Ling, H. E., Lun, Z., Jianxin, L., Ji, M. A., Ruilin, L. U. I., Shuqin, W. A. N. G., Wenqi, Z. H. A. O., 2014. Complex relationship between porosity and permeability of carbonate reservoirs and its controlling factors: A case study of platform facies in Pre-Caspian Basin. *Petroleum Exploration and Development*, *41*, 225–234.

Lønøy, A., 2006. Making sense of carbonate pore systems. *American Association of Petroleum Geology, Bulletin*, *90*, 1381–1405.

Lucia, F. J., 1995. Rock-fabric/petrophysical classification of carbonate pore space for reservoir characterization. *American Association of Petroleum Geology Bulletin*, *79*, 1275–1300.

Lucia, F. J., Kerans, C., Jennings, J. W., 2003. Carbonate reservoir characterization. *Journal of Petroleum Technology*, *55*, 70–72.

Lucia, F. J., 2007. Carbonate Reservoir Characterization. Second Edition, Springer-Verlag, Berlin, 342 pp.

Luo, P., and Machel, H. G., 1995. Pore size and pore throat types in a heterogeneous dolostone reservoir, Devonian Grosmont Formation, Western Canada sedimentary basin. *American Association of Petroleum Geology, Bulletin 79*, 1698–1720.

Machel, H. G., 2004. Concepts and models of dolomitization: a critical reappraisal. *Geological Society, London, Special Publications*, *235*, 7–63.

Maniscalco, R., Fazio, E., Punturo, R., Cirrincione, R., Di Stefano, A., Di Stefano, S., Forzese, M., Lanzafame, G., Leonardi, G. S., Montalbano, S., et al., 2022. The Porosity in Heterogeneous Carbonate Reservoir Rocks: Tectonic versus Diagenetic Imprint" A Multi-Scale Study from the Hyblean Plateau (SE Sicily, Italy). Geosciences, 12, 149.

Mazzullo, S. J., 2004. Overview of porosity evolution in carbonate reservoirs. *Kansas Geological Society Bulletin*, *79*, 1–19.

Mazzullo, S. J., and Harris, P. M., 1992. Mesogenetic dissolution: its role in porosity development in carbonate reservoirs. *American Association of Petroleum Geologists Bulletin*, *76*, 607–620.

Moore, C. H., 2001. Carbonate reservoirs: porosity evolution and diagenesis in a sequence stratigraphic framework, 3. impr. In: Developments in Sedimentology 55. Elsevier, Amsterdam, 444 pp.

Moore, C. H., and Wade, W. J., 2013. Carbonate Reservoirs: Porosity and Diagenesis in a Sequence Stratigraphic Framework: Amsterdam, Elsevier, Developments in Sedimentology 55, 392 pp.

Moshier, S. O., 1989. Microporosity in micritic limestones: a review. *Sedimentary Geology*, *63*, 191–213.

Murray, R. C., 1960. Origin of porosity in carbonate rocks. *Journal of Sedimentary Petrology*, *30*, 59–84.

Nelson, R. A., 1985. Geologic Analysis of Naturally Fractured Reservoirs. Gulf Publishing, Houston, 320 pp.

Nelson, R. A., 2001. Geologic Analysis of Naturally Fractured Reservoirs, second ed. Gulf Publishing, Houston, 332 pp.

Nelson, R. A., Moldovanyi, E. P., Matcek, C. C., Azpiritxaga, I., Bueno, E., 2000. Production characteristics of the fractured reservoirs of the La Paz field, Maracaibo Basin, Venezuela. *American Association of Petroleum Geology, Bulletin*, *84*, 1791–1809.

Pittman, E. D., 1979. Porosity, diagenesis and productive capability of sandstone reservoir: SEPM Reprint Series 26, 159–173.

Póppelreiter, M., Balzarini, M. A., De Sousa, P., Engel, S., Galarraga, M., Hansen, B., Márquez, X., Morell, J., Nelson, R., Rodriguez, F., 2005. Structural control on sweet-spot distribution in a carbonate reservoir: Concepts and 3-D models (Cogollo Group, Lower Cretaceous, Venezuela). *American Association of Petroleum Geologists Bulletin*, *89*, 1651–1676.

Purser, B. H., Brown, A., Aissaoui, D. M., 1994. Nature, origins and porosity in dolomites, In: B. Purser, M. Tucker, and D. Zenger, eds., Dolomites, a volume in honour of Dolomieu. *International Association of Sedimentologists, Special Publication 21*, 283–308.

Sánchez, M. A., Vásquez, A, R., Van Alstine, D., Butterworth, J., Garcia, J., Carmona, R., Poquioma, W., Ramones, M., 1999. Applications of geomechanics in the development of naturally fractured carbonates of the Mara Oeste field, Venezuela: SPE Paper 54,008, 1–8.

Scholle, P. A., and Ulmer-Scholle, D., 2003. A Color Guide to the Petrography of Carbonate Rocks: Grains, Textures, Porosity, Diagenesis. *American Association of Petroleum Geologists, Memoir 77*, 474 pp.

Skalinski, M., and Kenter, J. A., 2014. Carbonate petrophysical rock typing: integrating geological attributes and petrophysical properties while linking with dynamic behaviour. *Geological Society, London, Special Publications, 406*, 229–259.

Strasser, A., Davaud, E., Jedoui, Y., 1989. Carbonate cements in Holocene beachrock: example from Bahiret et Biban, southeastern Tunisia. *Sedimentary Geology, 62*, 89–100.

Valencia, F. L. and Laya, J. C., 2020. Deep-Burial dissolution in an Oligocene-Miocene giant carbonate reservoir (Perla Limestone), Gulf of Venezuela Basin: Implications on microporosity development. *Marine and Petroleum Geology, 113*, 104,144.

Wright, P. and Harris, P. M., 2013. Carbonate dissolution and porosity development in the burial (mesogenetic) environment. In: AAPG Annual Convention and Exhibition, Search and Discovery. Article #50860, Pittsburgh, PA, 19–22 May.

Ca-Phosphate

20

In marine environments, the two most common Ca-phosphate mineral phases are fluorapatite [$Ca_5(PO_4)_3F$] and hydroxyapatite [$Ca_5(PO_4)_3OH$], with the former being the most abundant. In the literature, this pair of phosphate minerals is commonly grouped under the name of francolite (e.g., Scholle and Ulmer-Scholle 2003; Tribovillard et al. 2006). Nevertheless, many authors use the term "fluorapatite" loosely to refer to marine phosphate mineralization products. Identifying fluorapatite in thin sections is easy due to its striking orange-brownish color and nearly isotropic behavior under CPL. In this chapter, we focus mainly on a variety of Ca-phosphate mineralizations in marine environments (i.e., primary phosphates) that involve the replacement (phosphatization) of carbonate and mixed (carbonate and siliciclastic) sediments and rocks.

We also consider some phosphate deposits formed on tropical islands and coastal localities (mainly in the Caribbean and the Indo-Pacific) where a variety of phosphate minerals are precipitated in partly or totally exposed rocks and sediments in contact with seabird guano. Such phosphatization processes forming secondary phosphates (direct precipitation or weathered/transformed guano phosphates) give rise to the so-called "Insular Phosphates" (see Glenn et al. 1994; Jones, 2020). Monetite, brushite, and leucophosphite are characteristic minerals of guano-derived phosphates (see Stienstra 1985).

20.1 Phosphorus

Ca-phosphate mineralization requires sufficient phosphorus (P) concentration in seawater. P is essential to all forms of life on Earth. It plays a fundamental role as a nutrient and energy carrier for all living organisms. In marine environments, phosphorus in the form of phosphate (PO_4^{3-}) strongly influences primary production (the process by which photosynthetic organisms fix inorganic carbon into cellular biomass) and hence carbon cycling (e.g., Föllmi 1996; Tribovillard et al. 2006; Mackey and Paytan 2009). In addition, phosphate is a major constituent of skeletal material in vertebrates, lingulate brachiopods and several extinct clades (Knoll 2003).

20.2 Phosphate Rocks

Phosphate bearing rocks are named differently with respect to the concentration of the phosphate component. The term "phosphate rocks" defines rock compositions with less than 50% phosphate components of the total rock (< 18% P_2O_5); while the term "phosphorite rocks" defines rocks with more than 50% of phosphate components (> 18% P_2O_5). The 50% boundary roughly divides packed phosphate fabrics from those with fewer phosphate particles. This definition differentiates phosphate-bearing rocks from economically valuable phosphate-rich rocks (Trappe 2001). In this chapter we deal with phosphate rocks whose components (matrix and carbonate allochems) are replaced by Ca-phosphate, but their bulk composition (% by mass) is dominated by carbonate minerals (plus clay minerals and quartz).

20.3 Genesis of Ca-Phosphate

The formation of marine phosphate/phosphorite rocks includes two basic processes (Glenn et al. 1994; Föllmi 1996; Flügel 2010): **(a)** Physicochemical or biochemical precipitation of authigenic (*in situ*) carbonate fluorapatite within the sediment, at the sediment/water interface and/or during early marine diagenesis (i.e., primary phosphogenesis). These non-transported (authigenic) primary phosphates (also termed orthochemical phosphates; Trappe 2001) may occur in any or all of the principal phosphate/phosphorite settings (Glenn et al. 1994) in which fluorapatite commonly

J. A. Méndez Dot, J. Méndez Baamonde, *Atlas of Tropical Carbonates*, https://doi.org/10.1007/978-3-031-25494-9_20

replaces matrix (micritic or clayey) and allochems (skeletal and non-skeletal carbonate or silicilastic grains), as well as fish debris; microbial contribution is important (Krajewski et al. 1994; Soudry 2000; Tribovillard et al. 2006; Konhauser 2007); **(b)** Granular phosphates result from hydraulic and biological reworking of primary phosphorites, by which phosphatic grains are concentrated. Granular phosphorites are accumulations that developed on ancient shelves and epeiric seas and appear to dominate phosphorite occurrence in the Phanerozoic record (Glenn et al. 1994). On the other hand, the formation of insular phosphates from partially to totally emerged limestones is linked to P-rich fluids derived from seabird guano. These fluids, in mixture with other fluids (freshwater or marine), may phosphatize limestones with which they are in direct contact, or may leach or percolate through porous rocks and phosphatize underlying carbonate units (see Jones, 2020 and references therein).

20.4 Environments and Conditions for Ca-Phosphate Formation

Phosphatic sediments are formed not only beneath upwelling systems, but also on oceanic islands, atop seamounts, guyots and plateaus, and on some shelves, slopes and in epeiric environments that may not have experienced upwelling at all (Glenn et al. 1994). In general, the precipitation of fluorapatite (or any other phosphate mineral) replacing calcite or aragonite requires a P concentration high enough to support supersaturation with respect to Ca-phosphate. This enrichment may result from a high organic matter supply in: **(a)** upwelling areas, where phosphate (PO_4^{3-}) is reintroduced into the photic zone and it is recycled for organic matter productivity, but high productivity is not always a prerequisite. In low-productivity portions of the oceans, P enrichment can be achieved through redox cycling of iron, with P sorption onto iron-oxyhydroxide coatings and Fe–P co-precipitation (see Tribovillard et al. 2006 and references therein); **(b)** shelves and ancient epeiric seas with a high influx of phosphate (PO_4^{3-}) derived from continental weathering. In these settings, redox and nutrient conditions associated with the precipitation of fluorapatite may range from oxic and mesotrophic (some benthic photozoan components may be present), through dysoxic-suboxic and eutrophic (benthic heterozoan biota dominates), to anoxic and hypertrophic (planktonic biota dominates); (c) emerged regions (commonly slightly above RSL) in oceanic islands or coastal localities characterized by warm-arid to semi-arid climate and large bird populations, either at present day or in the past. In these regions, seabird colonization produces large quantities of guano (P-rich) that subsequently leaches to produce phosphatizing fluids (Ten Have et al. 1982; Stienstra, 1985; Glenn et al. 1994; Dar and Khan, 2017; Jones, 2020).

Phosphatization processes on carbonate rocks are commonly used as indicators of: marine primary productivity; oxygen deficiency and microbial involvement; upwelling conditions, currents influence and ocean circulation; sea-level changes and sequence stratigraphy, as most phosphatic sediments are associated to transgressive intervals; hiatus surfaces and stratigraphic unconformities; and karst development (see Glenn et al. 1994; Van Cappellen and Ingall 1994; Föllmi 1996; Slomp and Cappellen 2007; Tsandev and Slomp 2009; Jones, 2020).

20.5 Marine Phosphorus Cycling and Regeneration: Implications for the Phosphatization of Carbonates

Continental weathering of phosphate-bearing igneous (basalts) and sedimentary rocks is the most important source of phosphate to aquatic ecosystems (see Föllmi 1996; Gernon et al. 2024; see also Tyrrell 1999; Howarth et al. 2021). Once removed from rocks and released from the soils, phosphate (PO_4^{3-}) is transported by water and air, and eventually transferred into lakes and oceans. Once in the marine environment, almost all reactive and bioavailable phosphate (dissolved or particulate) is used in the surface zone of primary productivity and incorporated into organic matter. Eventually, phosphate is transferred to the deep ocean in the form of partly aggregated organic matter and, upon oxidation, of dissolved inorganic phosphate (Föllmi 1996).

The main source of P to the sediments is the phytoplankton that reaches the sediment–water interface, plus fish scales and bones. The main mechanism for phosphate concentration and precipitation of fluorapatite occurs as P is released (or regenerated) via organic matter remineralization. Usually, P is released as PO_4^{3-} from decaying organic matter during oxic, dysoxic-to-suboxic, and anoxic bacterial degradation below the water-sediment interface (Tribovillard et al. 2006). In anoxic-euxinic conditions, the concentration of reactive soluble phosphorus increases due to its preferential release during bacterial sulfate reduction (e.g., van Cappellen and Ignall, 1994; Poulton et al. 2015).

Generally speaking, the P remineralized to pore waters can either escape from the sediment back to the water column and support phytoplankton productivity in the photic zone, or be precipitated and trapped within the sediments. Therefore, P released can reach relatively high concentrations and authigenic phases can then precipitate, if combined bacterial activity and alkalinity, Ph, Eh conditions are reached (Reimers et al. 1996; Benitez-Nelson 2000).

In shallow (neritic) marine environments, low P burial efficiency under dysoxic-to-anoxic bottom water conditions enhances P regeneration bounded to Fe(III) and Mn(IV) oxyhydroxides, as well as regeneration of organic P (as component of phytoplankton and zooplankton), and mineral-associated-phosphate. Hence, P reductive solubility increases its availability and promotes its recycling as nutrient for primary production, while a fraction of the released P is ultimately incorporated into authigenic fluorapatite (see Van Cappellen and Ingall 1994; Föllmi 1996; Filippelli 2001; Handoh and Lenton 2003; Bjerrum et al. 2006; Slomp and van Cappellen, 2007; Mort et al. 2007; Meyer and Kump 2008; Tsandev y Slomp, 2009; Kraal et al. 2010; Monteiro et al. 2012).

20.6 Cretaceous Oceanic Anoxic Events and Phosphorus as Essential Nutrient

Basaltic rocks are enriched in P (e.g., Horton 2015; Gernon et al. 2016). During the main Cretaceous OAEs (1a and 2), enhanced chemical subaerial and/or submarine weathering of outcropping mafic and LIP-derived lithologies in a Hothouse world (characterized by very high atmospheric CO_2 concentrations due to LIP emplacements) led to an excessive delivery of bioavailable P to the oceans. This caused hypertrophic conditions and stimulated biological primary production, leading to oxygen depletion in the water column and the initiation of anoxia or euxinia (see Gernon et al. 2024).

As mentioned in the previous section, under low-oxygen or anoxic conditions, organic matter remineralization in the photic zone releases P. In this context, recycling of P for sustaining organic matter productivity has been associated with prolonging anoxic conditions during the Cretaceous OAEs (e.g., Handoh and Lenton 2003; Bjerrum et al. 2006; Mort et al. 2007; Meyer and Kump 2008; Kraal et al. 2010; Montero et al. 2012; Beil et al. 2020) and the increase in both TOC content and the quality of the preserved organic matter (high HI values) in the black shales (see Kuypers et al. 2004). The latter organic-rich rocks are the sedimentary expression of the OAEs.

20.7 Ca-Phosphate: Photographic Set and Descriptive Captions

In all photos, IW = image width (the true length of the horizontal axis of the view)

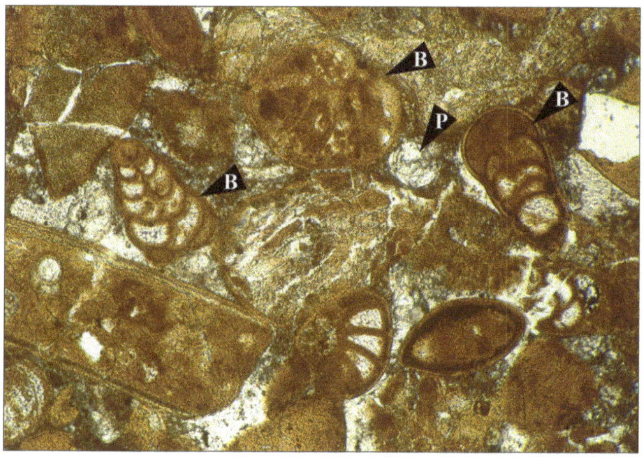

Cretaceous (early Campanian). Tres Esquinas Member, La Luna Formation. Maracaibo Basin

This phosphate limestone (packstone) exhibits characteristic orange/brownish color of fluorapatite. This mineral is replacing tests of SBF (B) and rounded and angular peloidal grains. A non-phosphatized test of a planktonic (P) foraminifer is differentiated among larger tests. The white precipitate that is partially replacing both the matrix and the infilling of foraminifera chambers is silica (chert). This precipitate postdates phosphatization. High concentrations of reactive P for this prominent phosphate mineralization are associated to low-oxygen conditions and intensified regional upwelling.

IW = 1.3 mm. PPL.

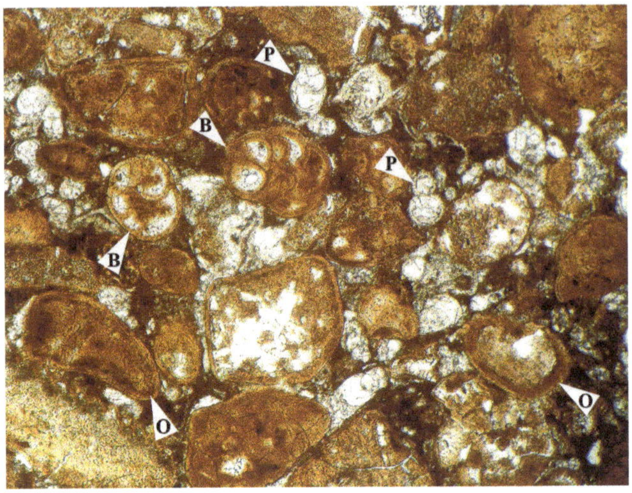

Cretaceous (early Campanian). Tres Esquinas Member, La Luna Formation. Maracaibo Basin

Phosphatized grains in the image are mainly ooids (O) (peloids coated by a thin oolitic cortex), SBF (B) and PF (P). Note that the latter, which are differentiated by their smaller sizes and globular chambers, are not phosphatized, but only affected by postdating authigenic silicification. This observation suggests that phosphatization of SBF and ooids occurred prior to final deposition together with planktonic tests. Therefore, an allochthonous origin for this granular phosphate microfacies is the most plausible hypothesis.

IW = 1.3 mm. PPL.

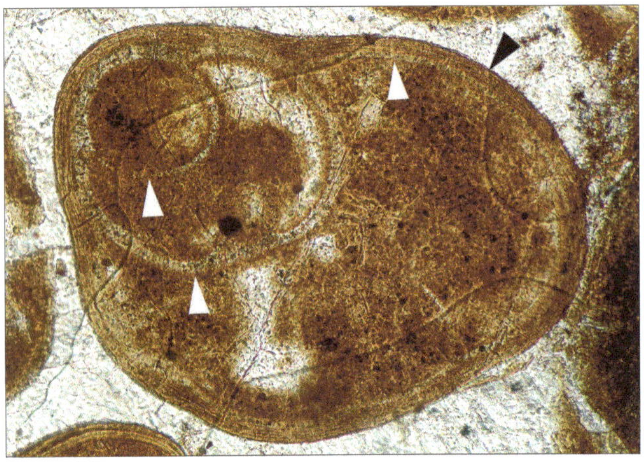

Cretaceous (early Campanian). Tres Esquinas Member, La Luna Formation. Maracaibo Basin

The rounded phosphatized grain in the image is a superficial ooid formed by coating a gastropod. Note the thin oolitic cortex (black arrow) coating the shell. The thin walls of three different whorls of the shell (white arrows) are clearly differentiated. The ooid was formed in an agitated neritic environment. Phosphatization and redeposition of the grain postdate the formation of the coated grain. Ooids are common shallow water components in redeposited microfacies of the Tres Esquinas Member (see Parra et al. 2003).

IW = 0.5 mm. PPL.

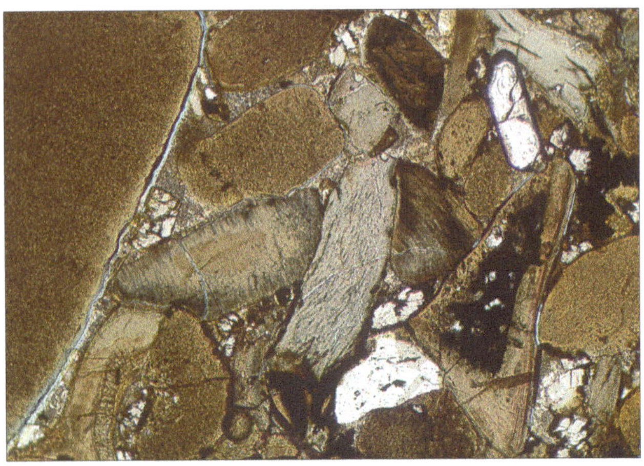

Cretaceous (Campanian). San Antonio Formation. Eastern Interior Range

The original texture of the grains in this packed packstone is obliterated due to combined reworking and intense fluorapatite replacement. Visible foliated texture of the central grain in the image resembles a mollusc shell fragment, while angular nodules and rounded peloids resemble shallow-water carbonate material. Reworking and redeposition of the grains in an outer platform setting influenced by regional upwelling during Campanian times is associated to deposition of this microfacies (see Erlich and Barrett 1992; Erlich et al. 2003).

IW = 2.4 mm. PPL.

Late Oligocene-Early Miocene. Perla Limestone. Eastern Gulf of Venezuela Sub-Basin

This packed grainstone consists of LBF tests, red algae grains (dark components) and phosphatized peloidal grains. The latter were most probably derived from fluorapatite replacement of red algae. Note that precipitation of pyrite postdates phosphatization in some of the peloids, thus causing blackening of the grains. Bioclasts were originally formed in the oligophotic zone and later redeposited downslope by turbulent gravity flows In the Perla Limestone, such phosphate and pyrite precipitation processes are thought to be formed *in situ* and below the sediment water interface at depths below the chemocline (see Pomar et al. 2015).

IW = 4.2 mm. PPL.

Cretaceous (early Aptian). Machiques Member, Apón Formation. Maracaibo Basin

Dual image (PPL and CPL) of phosphatized fish remains surrounding a small planktonic (P) foraminifer test. Note the isotropic (black) behavior of this early marine diagenetic Ca-phosphate mineral (fluorapatite) when it is seen under CPL (below). Low-oxygen or anoxic conditions are associated to this localized fluorapatite mineralization, as this example corresponds to the lower part of the Machiques Member (OAE 1a).

IW = 0.5 mm. PPL (above); CPL (below).

Cretaceous (middle-late Cenomanian). La Luna Formation *sensu stricto*. Maracaibo Basin

Phosphatization of carbonate grains, such as the mollusc shell fragments in the image, is moderately common at the lower part of the La Luna Formation. Contrary to previous examples, redeposition is not estimated for phosphate replacement of these allochems. Low-oxygen conditions (but not anoxic) are associated to deposition of heterotrophic biota in this black shale unit. Organic matter maturation is noticed by abundant stylocumulates, as well as by hydrocarbon impregnation of the shell margins and dolomitic matrix.

IW = 2.4 mm. PPL.

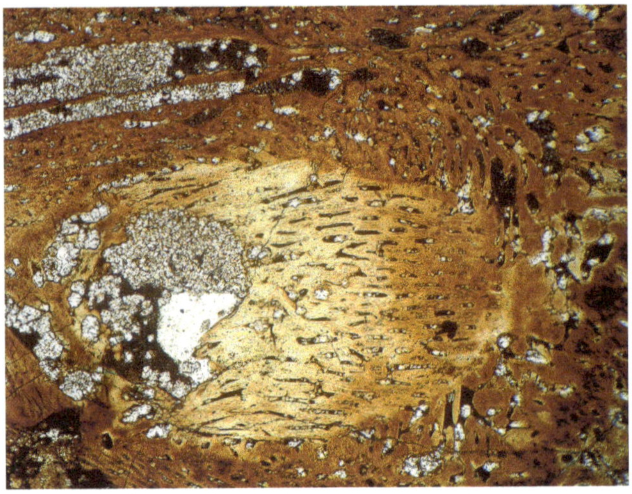

Cretaceous (middle-late Cenomanian). La Luna Formation *sensu stricto*. Maracaibo Basin

Another example of authigenic phosphatization of allochems at the lower part of the La Luna Formation. In this case, fluorapatite is replacing the skeleton of a bryozoan. Note the organic richness and associated organogenic dolomite precipitated within the intraparticle pores. As in the previous example, phosphorus redox solubility and regeneration under low-oxygen conditions as a mechanism for increasing reactive soluble phosphorus concentration during early marine diagenesis is associated to this authigenic precipitation of fluorapatite.

IW = 1.3 mm. PPL

Middle Miocene. Capadare Formation. Falcón Basin

Authigenic precipitation of fluorapatite is moderately common in wackestones of the Capadare Formation, especially replacing mollusc shells, and tests of LBF, such as the ones in the image (miogypsinid to the left and operculinid to the right). Traces of the brown/reddish color of fluorapatite are also visible replacing the slightly organic-rich micritic matrix. Phosphorous regeneration from organic matter within the micritic matrix during transient low-oxygen, bottom-water conditions is the most plausible explanation for phosphorus concentration and fluorapatite precipitation in this example.

IW = 2.4 mm. PPL.

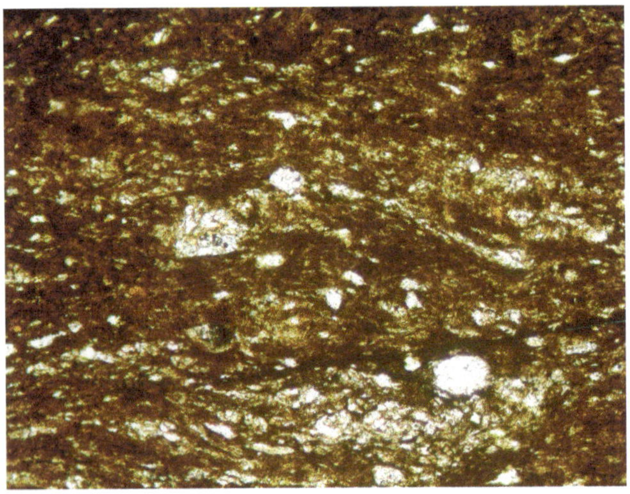

Cretaceous (late Aptian). Apón Formation, Piché Member. Maracaibo Basin

Organic-rich, phosphatized, dolomitic shale of the OAE 1b. Note the prominent precipitation of fluorapatite indicated by the brown/orange color of the wavy matrix. Light-colored dolomitic bands contrast with the dark color of the microcrystalline fluorapatite. For this example, authigenic precipitation of fluorapatite is interpreted to have occurred in association to: anoxic-euxinic conditions, high rates of primary productivity and organic matter preservation and phosphorus regeneration due to its preferential release during organic matter degradation by sulfate reducing bacteria.

IW = 1.3 mm. PPL.

Cretaceous (late Aptian). Apón Formation, Piché Member. Maracaibo Basin

Microcrystalline fluorapatite is coating undulated organic matter filaments in this laminated organic-rich limestone of the OAE 1b. Coeval precipitation of pyrite framboids, which are scattered across the sample, indicates anoxic-euxinic conditions during authigenic mineralization. The nucleation of microcrystalline fluorapatite was locally promoted by benthic microbial activity. Remineralization of organic phosphorus followed by precipitation of Ca-phosphate coating the filamentous microbial mats is a plausible hypothesis for fluorapatite precipitation in this example.

IW = 0.5 mm. PPL.

Cretaceous (Turonian). O Member, Escandalosa Formation. Barinas Sub-Basin

This dolomitic shale exhibits striking phosphatization of the wavy-laminae clayey matrix. Also note abundant pyrite framboids coating fluorapatite thin bands, indicating low-oxygen conditions during early marine authigenic phosphate mineralization. Although dolomite facies of the O Member (Turonian in age) are not organic-rich, the abundance authigenic minerals (fluorapatite, pyrite and glauconite) in some of the microfacies suggests the influence of global anoxia prevailing in Turonian times during deposition of this unit.

IW = 2.4 mm. PPL.

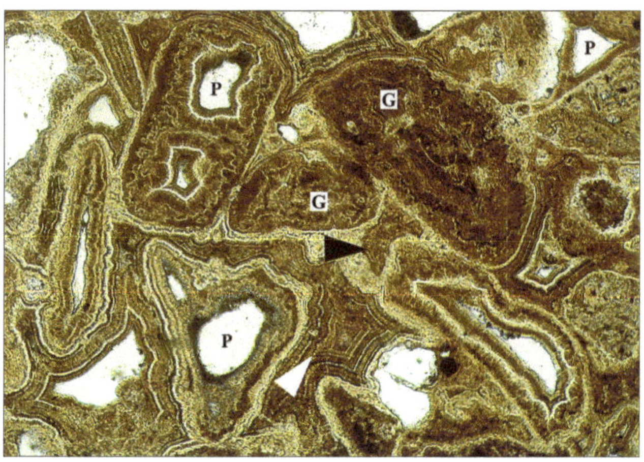

Middle Miocene. Capadare Formation. Falcón Basin

Phosphatization products in this and the following example (Miocene in age) are very similar to guano phosphates reported by Ten Have et al. (1982) and Stienstra (1985) from Curaçao and Aruba, respectively. Note the concentric and irregularly-shaped isopachous phosphatic precipitates. These consist of alternating thin apatite laminae with varying compositions according to their color. Darker laminae are commonly associated to higher Fe content within the apatite type. The concentric precipitates are replacing rounded peloidal (G) grains, outlining pores (P), or completely filling them (white and black arrows).

IW = 2.4 mm. PPL.

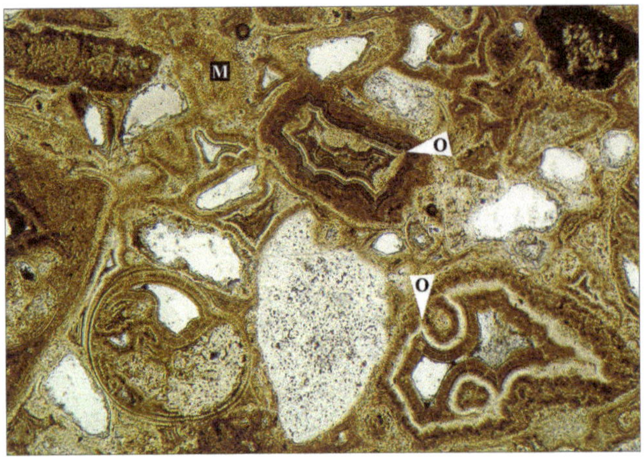

Middle Miocene. Capadare Formation. Falcón Basin
Another example of guano-derived phosphatization. These secondary phosphatic minerals are characterized by sharply outlines (O) with marked contrast in thin, light and dark bands. Such precipitates with more or less concentric arrangements are fringing both pores and grains, contrasting with the light-brown color of the replaced micritic matrix (M). Percolation of mixed phosphate-rich brines and meteoric waters from highly porous sandstones facies confining the phosphatized limestone intervals at the upper part of the Capadare Formation formed such peculiar phosphatic facies (see Albert-Villanueva 2016).

IW = 2.4 mm. PPL.

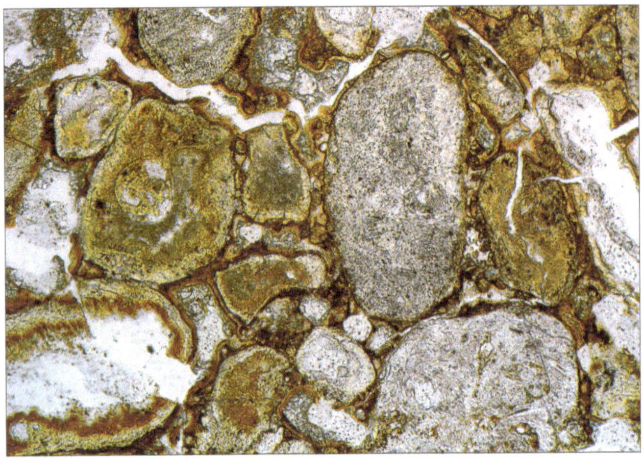

Late Pleistocene-Holocene. Los Roques Archipelago
Guano-derived phosphatization of a calcarenite. Such diagenetic process is currently occurring on lithified facies located slightly above the local sea level in the southwest area of the Gran Roque Island. The original texture of different-sized rounded grains is totally obliterated. The white, amorphous precipitate that is covering some of these grains is the result of their direct contact with guano birds. Apatite mineralization covering grains and matrix is indicated by characteristic brown-orange color. Phosphatizing solutions in this setting are the result of mixing of guano birds with the splashing of marine water.

IW = 2.4 mm. PPL.

References

Albert-Villanueva, E., 2016. Facies y secuencias deposicionales mixtas carbonático-siliciclásticas del Mioceno inferior de la Cuenca de Falcón (noroeste de Venezuela) como modelo exploratorio en el Caribe. Doctoral Dissertation. University of Barcelona. Volumes 1 and 2.

Beil, S., Kuhnt, W., Holbourn, A., Scholz, F., Oxmann, J., Wallmann, K., Lorenzen, J., Aquit, M., Chellai, E. H., 2020. Cretaceous oceanic anoxic events prolonged by phosphorus cycle feedbacks. *Climate of the Past*, *16*, 757–782.

Benitez-Nelson, C. R., 2000. The biogeochemical cycling of phosphorus in marine systems. *Earth-Science Reviews*, *51*, 109–135.

Bjerrum, C. J., Bendtsen, J., Legarth, J. J. F., 2006. Modeling organic carbon burial during sea level rise with reference to the Cretaceous. *Geochemistry, Geophysics, Geosystem*, *7*, 24 pp.

Dar, S. A., Khan, K. F., Birch, W. D., 2017. Sedimentary Phosphates. In W. D. Birch (Ed.), *Reference Module in Earth Systems and Environmental Sciences* (Vol. 20.17, pp. 120–138). Amsterdam: Elsevier.

Erlich, R., and Barrett, S., 1992. Petroleum geology of the Eastern Venezuela Basin, in R. Macqueen and D. Leckie, eds., Foreland basins and fold belts. *American Association of Petroleum Geology*, *Memoir 55*, 341–362.

Erlich, R. N., Villamil, T., Keens-Dumas, J., 2003. Controls on the deposition of Upper Cretaceous organic carbon– rich rocks from Costa Rica to Suriname. In: C. Bartolini, R. T. Buffler, and J. Blickwede, eds., The Circum-Gulf of Mexico and the Caribbean: Hydrocarbon habitats, basin formation, and plate tectonics. *American Association of Petroleum Geology*, *Memoir 79*, 1–45.

Filippelli, G. M., 2001. Carbon and phosphorus cycling in anoxic sediments of the Saanich Inlet, British Columbia. *Marine Geology*, *174*, 307–321.

Flügel, E, 2010. Microfacies of Carbonate Rocks—Analysis, Interpretation and Application. Springer, Heidelberg-Dordrecht–London–New York, 984 pp.

Föllmi, K. B., 1996. The phosphorus cycle, phosphogenesis and marine phosphate-rich deposits. *Earth-Science Reviews*, *40*, 55–124.

Gernon, T. M., Hincks, T. K., Tyrrell, T., Rohling, E. J., Palmer, M. R., 2016. Snowball Earth ocean chemistry driven by extensive ridge volcanism during Rodinia breakup. *Nature Geoscience*, *9*, 242–248.

Gernon, T. M., Mills, B. J. W., Hincks, T. K., Merdith, A. S., Alcott, L. J., Rohling, E. J., Palmer, M. R., 2024. Solid Earth forcing of Mesozoic oceanic anoxic events. *Nature Geoscience*, 1–25.

Glenn, C. R., Föllmi, K. B., Riggs, S. T., Baturin, G. N., Grimm K. A., Trappe, J., Abed, A. M., Calli-Olivier, C., Garrison, R. E., Ilyin, A. V., Jehl, C., Rohrlich, V., Sadaqah, R. M. Y., Schidlowski, M., Sheldon, R. E., Siegmund, H., 1994. Phosphorus and phosphorites: Sedimentology and environments of formation. *Eclogae Geologicae Helvetiae*, *87*, 747–788.

Handoh, I. C., Lenton, T. M., 2003. Periodic mid–Cretaceous oceanic anoxic events linked by oscillations of the phosphorus and oxygen biogeochemical cycles. *Global Biogeochemical Cycles*, *17*, 1092.

Horton, F., 2015. Did phosphorus derived from the weathering of large igneous provinces fertilize the Neoproterozoic ocean?. *Geochemistry, Geophysics, Geosystems*, *16*, 1723–1738.

Howarth, R. W., Chan, F., Swaney, D. P., Marino, R. M., Hayn, M., 2021. Role of external inputs of nutrients to aquatic ecosystems in determining prevalence of nitrogen vs. phosphorus limitation of net primary productivity. *Biogeochemistry*, *154*, 293–306.

Jones, B., 2020. Recycled insular phosphates and coated grains: Case study from Little Cayman, British West Indies. *Sedimentology*, *67*, 1844–1878.

Knoll, A. H., 2003. Biomineralization and evolutionary history. *Reviews in Mineralogy and Geochemistry*, *54*, 329–356.

Konhauser, K.O., 2007. Introduction to Geomicrobiology. Blackwell Publishing. 433 pp.

Kraal, P., Slomp, C. P., Forster, A., Kuypers, M. M. M., 2010. Phosphorus cycling from the margin to abysal depths in the proto-Atlantic during oceanic anoxic event 2. *Palaeogeography, Palaeoclimatology, Palaeoecology, 295*, 42–54.

Krajewski, K. P., Van Cappellen, P., Trichet, J., Kuhn, O., Lucas, J., Martín-Algarra, A., Prévot, L., Tewary, V. C., Gaspar, L., Knight, R. I., Lamboy, M., 1994. Biological processes and apatite formation in sedimentary environments. *Eclogae Geologicae Helvetiae*, *87*, 701–745.

Kuypers, M. M. M., Lourens, L. J., Rijpstra, W. I. C., Pancost, R. D., Nijenhuis, I. A., Sinninghe Damsté, J. S., 2004. Orbital forcing of organic carbon burial in the prono-North Atlantic durind oceanic anoxic event 2. *Earth and Planetary Science Letters*, *228*, 465–482.

Mackey, K. R. M., and Paytan, A., 2009. Phosphorus cycle. *Encyclopedia of Microbiology*, *3*, 322–334.

Meyer, K. M., and Kump, L. R., 2008. Oceanic euxinia in Earth history: Causes and consequences. Annual Review of Earth and Planetary Sciences, 36, 251–288.

Monteiro, F. M., Pancost, R. D., Ridgwell, A., Donnadieu, Y. 2012. Nutrients as the dominant control on the spread of anoxia and euxinia across the Cenomanian-Turonian oceanic anoxic event (OAE2): model-data comparison. *Paleoceanography*, *27*, PA4209.

Mort, H. P., Adatte, T., Föllmi, K. B., Keller, G., Steinmann, P., Matera, V., Berner, Z., Stüben, D., 2007. Phosphorus and the roles of productivity and nutrient recycling during oceanic anoxic event 2. *Geology* 35, 483– 486.

Parra, M., Moscardelli, L., Llorente, M. A., 2003. Late Cretaceous anoxia and lateral microfacies changes in the Tres Esquinas Member, La Luna Formation, Western Venezuela. *Palaios, 18*, 321–333.

Pomar, L., Esteban, M., Martinez, W., Espino, D., Castillo de Ott, V., Benkovics, L., Castro Leyva, T., 2015. Oligocene-Miocene carbonates of the Perla field, offshore Venezuela: Depositional model and facies architecture, *in* C. Bartolini and P. Mann, eds., Petroleum geology and potential of the Colombian Caribbean Margin: *American Association of Petroleum Geology*, *Memoir 108*, 647–674.

Poulton, S. W., Henkel, S., März, C., Urquhart, H., Flögel, S., Kasten, S., Sinninghe Damsté, J. S., Wagner, T., 2015. A continental-weathering control on orbitally driven redox-nutrient cycling during Cretaceous Oceanic Anoxic Event 2. *Geology*, *43*, 963–966.

Reimers, C. E., Ruttenberg, K. C., Canfield, D. E., Christiansen, M. B., Martin, J. B., 1996. Porewater pH and authigenic phases formed in the uppermost sediments of the Santa Barbara Basin. *Geochimica et Cosmochimica*, *Acta 60*, 4037–4057.

Scholle, P. A., and Ulmer-Scholle, D., 2003. A Color Guide to the Petrography of Carbonate Rocks: Grains, Textures, Porosity, Diagenesis. *American Association of Petroleum Geologists*, *Memoir 77*, 474 pp.

Slomp, C. P., and Cappellen, P. V., 2007. The global marine phosphorus cycle: sensitivity to oceanic circulation. *Biogeosciences*, *4*, 155–171.

Stienstra, P., 1985. Geology of a small rock-phosphate deposit, Ceru Colorado, Aruba, Netherlands Antilles. *Journal of Sedimentary Research*, *55*, 96–104.

Soudry, D., 2000. Microbial phosphate sediment. Microbial Sediments. 127–136.

Ten Have, T., Heijnen, W., Nickel, E., 1982. Alterations in guano phosphates and Mio–Pliocene carbonates of table mountain Santa Barbara, Curacao. *Sedimentary Geology*, *31*, 141–165.

Trappe, J., 2001. A nomenclature system for granular phosphate rocks according to depositional texture. *Sedimentary Geology*, *145*, 135–150.

Tribovillard, N., Algeo, T. J., Lyons, T., Riboulleau, A., 2006. Trace metals as paleoredox and paleoproductivity proxies: an update. *Chemical Geology*, *232*, 12–32.

Tsandev, I., and Slomp, C. P., 2009. Modeling phosphorus cycling and carbon burial during Cretaceous Oceanic Anoxic Events. *Earth and Planetary Science Letters*, *286*, 71–79.

Tyrrell, T., 1999. The relative influences of nitrogen and phosphorus on oceanic primary production. *Nature*, *400*, 525–531.

Van Cappellen, P., and Ingall, E. D., 1994. Benthic phosphorus regeneration, net primary production, and ocean anoxia: A model of the coupled marine biogeochemical cycles of carbon and phosphorus. *Paleoceanography*, *9*, 677–692.

21 Pyrite

Pyrite (FeS_2) is a sulfide mineral, opaque (isotropic) in PPL and golden yellow when viewed in reflected light. Authigenic pyrite is common in some carbonate and mixed (carbonate-silicilastic) rocks. It commonly forms under reducing conditions replacing organic-rich sediments or in close proximity to organic material. Its precipitation is indirectly linked to the activity of dissimilatory sulfate (SO_4^{2-}) reduction. Sulfate-reducing bacteria couple the oxidation of organic matter to the reduction of interstitial sulfate, thereby generating dissolved hydrogen sulfide (H_2S) (Konhauser 2007). In continental margins, the H_2S produced by bacterial sulfate reduction (BSR) either migrates to oxygenated portions of the sediments or water column where it is oxidized to sulfate, or reacts with detrital iron to form various sulfides, with the chief end product being FeS_2 (e.g., Berner 1985).

21.1 Pyrite Morphology

Pyrite precipitates as aggregates of minute crystals that form framboids of varied sizes (from less than 5 μm to up to 2 mm in diameter). Crystal habits (cubic euhedral, octahedral, polyhedral, anhedral) tend to be visible with the polarizing microscope as crystals increase in size. Precipitation of pyrite, or pyritization, commonly occurs by replacing fine-grained, organic-rich matrices (micritic and/or clayey), as well as skeletal and non-skeletal grains (e.g., foraminifera tests, peloids) and dolomite crystals.

21.2 Diagenetic Stages for Pyrite Formation

Pyrites can be divided into syngenetic pyrites, early diagenetic pyrites, and late diagenetic pyrites (Liu et al. 2019). Precipitation of near-surface pyrite may occur syngenetically (early marine) if the chemocline (the boundary zone below which a body of water is characterized by the presence of free H_2S, or anoxia-euxinia; Meyer and Kump 2008) lies above the water-sediment interface, or diagenetically if the chemocline lies at or below the water-sediment interface (Tribovillard et al. 2006). In shallow facies, the latter scenario may postdate meteoric diagenesis (near-surface or shallow burial). BSR occurs in low-temperature diagenetic environments (from 0 up to about 60–80 °C) (Machel 2001). Therefore, the formation of pyrite as a by-product of BSR occurs from the upper centimeters of the sediment seafloor to shallow-burial depths. High-temperature pyrite may occur in the intermediate to deeper burial realm (late diagenesis) as by-product of hydrocarbon generation and associated thermochemical sulfate reduction (i.e., the sulfate is inorganically reduced by hydrocarbons, from about 100–180 °C), as well as being associated with chemical compaction (see Machel 2001; Worden et al. 2003; Ghazban and Al-Aasm 2010; Jiang et al. 2018; Liu et al. 2019). Using the polarizing microscope, the "cross-cutting" relationship is key for determining the relative timing of diagenetic features (or paragenetic relations), which in most cases allow one to discern the origin a given pyrite precipitation process.

21.3 Environmental and Biogeochemical Conditions for Pyrite Precipitation

In most cases, the formation of low-temperature (near-surface) pyrite is indicative of anoxic-euxinic conditions (0.0 mL O_2/L H_2O and free H_2S/HS^-). Even so, pyrite can also precipitate, to a lesser extent, under low-oxygen (dysoxic to suboxic) conditions if sufficient reactive iron in the water column is combined with available organic matter for BSR (Berner 1985; Raiswell and Canfield 1998).

The precipitation of pyrite, under sustained or transient anoxia, requires not only the depletion of O_2, but of all oxidizing agents in the water column (the main of which are NO_3^-, NH_4^+, MnO_2, $Mn(OH)_4$, Fe_2O_3 and $Fe(OH)_3$, resulting in the use of SO_4^{2-} as an oxidizing agent for bacterially-mediated

J. A. Méndez Dot, J. Méndez Baamonde, *Atlas of Tropical Carbonates*, https://doi.org/10.1007/978-3-031-25494-9_21

organic matter degradation and reduction of sulfate (SO_4^{2-}) to sulfide (S^{2-}) (Tribovillard et al. 2006; Konhauser 2007; Jenkyns et al. 2007). The redox reactions that led to the complete depletion of such chemical species mainly occur within the redoxcline (i.e., the depth interval in the water column within which redox reactions initiate and the transition from oxic to dysaerobic and anoxic conditions occurs; see Taylor et al. 2006), or below the chemocline (free H_2S/HS^-; Kump et al. 2005; Meyer and Kump 2008).

The two main sources for iron input to the oceans are: lithogenic iron derived from continental weathering and fluvial transport and deposition; and mid-ocean ridge hydrothermally sourced $Fe(II)_{aq}$ (Jenkyns et al. 2007). The source of iron is limited by the influx of terrigenous sediments, mainly Fe-clay minerals (Meyer and Kump 2008). Reductive dissolution of Fe-oxyhydroxides in the redoxcline or below the chemocline increases the concentration of reactive iron. Coupled microbial dissimilatory iron reduction and BSR are thought to be the two main pathways for pyrite precipitation. Primary productivity and organic matter availability also play a key role, as sulfate-reducing bacteria are heterotrophs that use organic compounds as a source of carbon (e.g., Severmann et al. 2006; Tribovillard et al. 2006; Konhauser 2007).

Pyrite precipitation in the shallow-burial environment is associated with BSR. It occurs when reduced iron (Fe^{2+}) and sulfide (S^{2-}) derived from buried seawater, evaporative brines (pore water), and/or from dissolution of gypsum ($CaSO_4.2H_2O$) or anhydrite ($CaSO_4$) are still available. Pyrite precipitation in the intermediate-deeper burial realm (if iron is available, as it is commonly depleted in deeply buried carbonate rocks; Machel 2001) is linked to organic matter maturation, hydrocarbon generation and mechanical-chemical compaction.

21.4 Pyritization Versus Organic Matter Sulfurization

Sulfate (SO_4^{2-}) is the main form of sulfur (S) in the oceans. This ion is, after chloride (Cl^-), the most abundant anion in normal marine waters (Lippmann 1973; James and Jones 2015). Today, riverine delivery, resulting from the dissolution of evaporites and the oxidative weathering of sulfides, represents the main input of sulfate to the ocean (Adams et al. 2010). Once sulfate is reduced, the fate of sulfide (S^{2-}) under anoxic conditions, either to form pyrite or for incorporation into organic matter, will depend on the availability of reactive iron (in the form of Fe^{2+}) and the mineralogical composition of the fine-grained sediments at/or slightly below the water sediments interface. If Fe^{2+} is available, the precipitation of pyrite occurs rapidly (e.g., Berner 1985; Raiswell and Canfield 1998; Tribovillard et al. 2006; Jenkyns et al. 2007), therefore pyritization is promoted by the abundance of reduced iron associated with clay minerals. On the contrary, if the iron supply is limited, sulfide anions incorporate into the kerogen during early diagenesis, a process termed organic matter sulfurization (e.g., Hülse et al. 2019). This is the reason why, in general, sulfur-rich kerogens are more common in organic-rich limestones and cherts compared to organic-rich pyritic shales (see Peters et al. 2005; Killops and Killops 2005; Brumsack 2006; Turgeon and Brumsack 2006; Vandenbroucke and Largeau 2007; Meyer and Kump 2008).

Precipitation of syngenetic pyrite (below the chemocline) was substantially enhanced during the mid-Cretaceous OAEs. Large inputs of volcanic sulfur from submarine LIP eruptions were combined with intensification of the hydrologic cycle (which represented a source of lithogenic Fe, thus promoting prominent pyritization and/or organic matter sulfurization (see Kump et al. 2005; Meyer and Kump 2008; Kidder and Worsley 2010; Bond et al. 2014; Bond and Grasby 2017; Ernst and Youbi 2017). See the case studies of La Luna Formation by Mongenot et al. 1996; of the Querecual Formation by López et al. 2009; of black shales from the Tarfaya Basin, SW Morocco, by Kolonic et al. 2002, 2005; and of black shales from Demerara Rise, offshore Suriname, by Hetzel et al. 2009. Syngenetic pyrite is abundant in all Cretaceous black shales of Venezuela. The high content (wt.%) of organic matter (black-colored and isotropic) in many microfacies of some of these rock units (e.g., La Luna and Querecual formations) commonly prevents the differentiation of this opaque mineral using a polarizing microscope.

21.5 Pyrite: Photographic Set and Descriptive Captions

In all photos, IW = image width (the true length of the horizontal axis of the view)

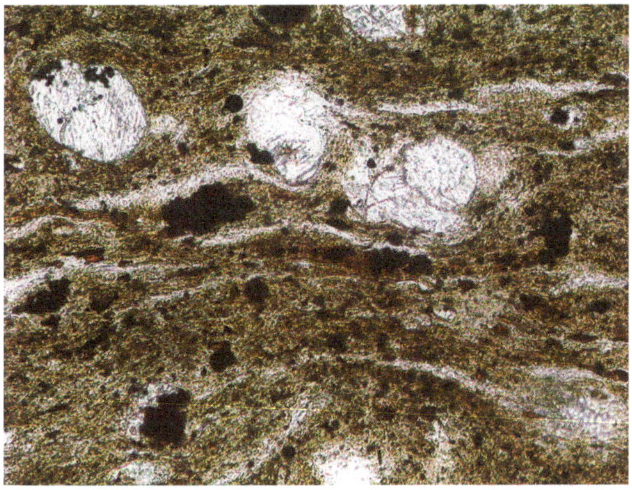

Cretaceous (late Aptian). Piché Member, Apón Formation. Maracaibo Basin

The clayey matrix of this organic-rich (2.1% TOC), calcareous shale containing foraminifera tests, which belongs to the OAE 1b first sub-event, is notoriously rich in pyrite. Note the abundant framboids (irregularly shaped black spots) scattered across the laminated matrix. Syngenetic precipitation of pyrite occurred under anoxic-euxinic conditions linked to sustained BSR and high concentrations of reactive Fe in reduced form (Fe^{2+}) below the chemocline.

IW = 0.5 mm. PPL.

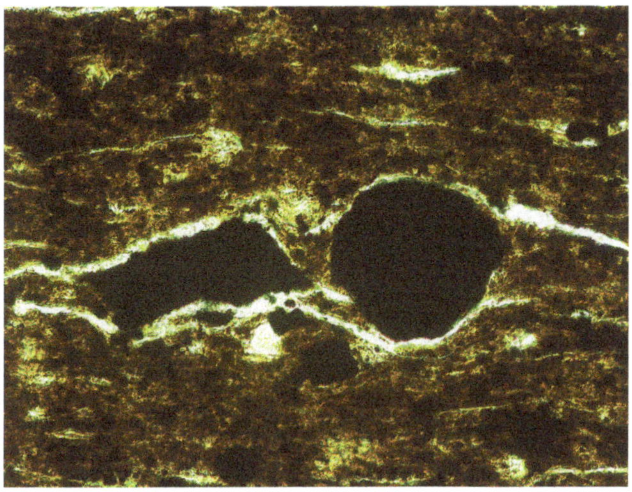

Cretaceous (early Aptian). Machiques Member, Apón Formation. Maracaibo Basin

Pyritized, organic-rich calcareous shale. Combined organic richness (2.6% TOC) and syngenetic pyritization of both the clayey matrix and carbonate components led to severe blackening of this sample belonging to the OAE 1a. Note the pyrite replacement of the central dolomite crystal (to the left) and globular planktonic foraminifer (to the right). Organic matter availability and anoxic-euxinic conditions below a shallow chemocline promoted prominent syngenetic pyritization. As in the previous example, thin and undulated (light-colored) organic matter filaments are replaced by amorphous silica.

IW = 0.5 mm. PPL.

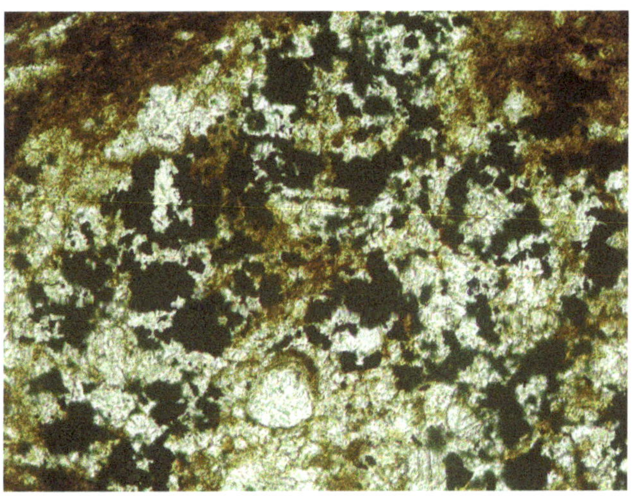

Cretaceous (early Aptian). Machiques Member, Apón Formation. Maracaibo Basin

Early marine syngenetic mineralization is prominent in this black shale. Note that pyrite framboids are partially replacing an organogenic dolomite mosaic. Dolomite crystals are, in turn, replacing a phosphatized matrix (the characteristic brown-orange color of microcrystalline fluorapatite is clearly seen in both upper corners of the image). Framboids are also replacing the phosphatized matrix in these corners.

IW = 0.5 mm. PPL.

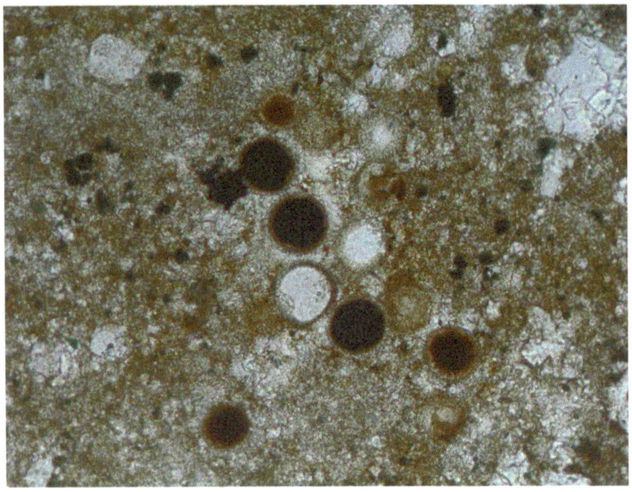

Cretaceous (early Aptian). Machiques Member, Apón Formation. Maracaibo Basin

In this sample (3.34% TOC), syngenetic pyrite precipitation occurred as framboids replacing the clay-dominated matrix and as selective replacement of some calcitic microspheres in the image. These spherical, authigenic products are derived from the activity of endolithic bacteria (probably cyanobacteria). Selective pyritization of these microbially-induced carbonate products is common in Early Cretaceous black shales from Venezuela, as these authigenic precipitates contained sufficient remnant organic matter to fuel bacterial sulfate reduction, making them preferential sites for pyrite precipitation.

IW = 0.5 mm. PPL.

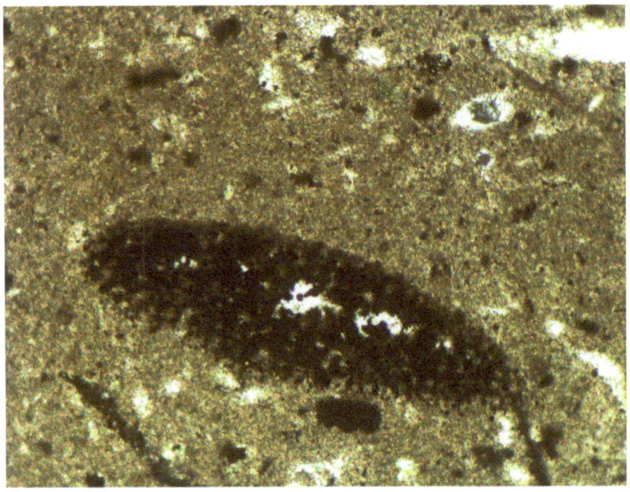

Cretaceous (early Barremian). Tibú Member, Apón Formation. Maracaibo Basin

Diagenetic pyrite framboids are extensively and selectively replacing the test of a larger benthic foraminifer. Note that the framboids, which are also scattered in the micritic matrix, are cross-cutting the sparry-filled central chambers of the test, indicating that pyrite precipitation postdates sparry calcite precipitation. Even though the Tibú Member is not a black shale unit (as it underlies the OAE 1a black shale interval), pyrite is a common accessory mineral in many of its microfacies.

IW = 4.2 mm. PPL.

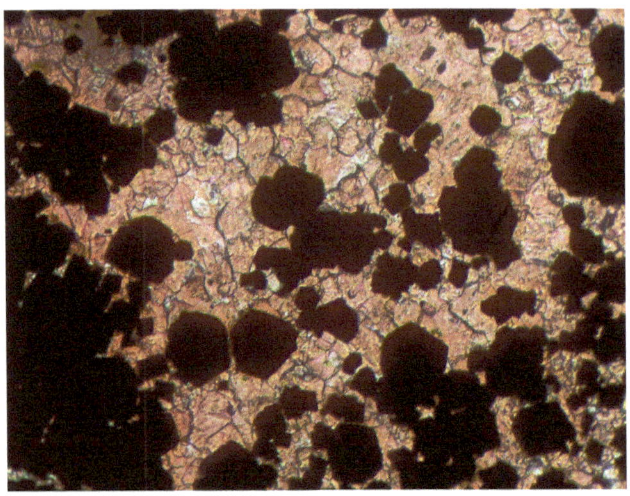

Cretaceous (late Barremian). Tibú Member, Apón Formation. Maracaibo Basin

Diagenetic pyrite framboids in the image are replacing a pseudospar mosaic. Large framboids exhibit cubic and polyhedral habits. Transient anoxia, postdating carbonate recrystallization, probably occurred during shallow burial and was associated with a change in the composition of pore fluids, which eventually became enriched in reduced iron (Fe2+) and sulfide (S2-) ions.

IW = 1.3 mm. PPL.

Cretaceous (Turonian). O Member, Escandalosa Formation. Barinas sub-Basin

Changing conditions during early marine diagenesis are well exemplified in this sample (a dolostone) by three different mineralization processes. Dolomite precipitation occurred during very shallow hypersaline conditions. Subsequent low-oxygen (dysoxic) conditions, most probably linked to an increase in bathymetry, promoted the precipitation of glauconite (green grains), which is replacing dolomite crystals and infilling small intercrystalline pores. A shift towards transient anoxia led to precipitation of pyrite, which is locally replacing iron-rich glauconite.

IW = 2.4 mm. PPL.

Cretaceous (Turonian). O Member, Escandalosa Formation. Barinas sub-Basin

Cubic pyrite crystals in the image are locally replacing dolomite crystals and are also partially infilling secondary intracrystalline pores. This suggests a semi-closed system in the burial realm for pyrite precipitation, likely associated with hydrocarbon generation, as indicated by the abundant hydrocarbon impregnations in the image.

IW = 1.3 mm. PPL.

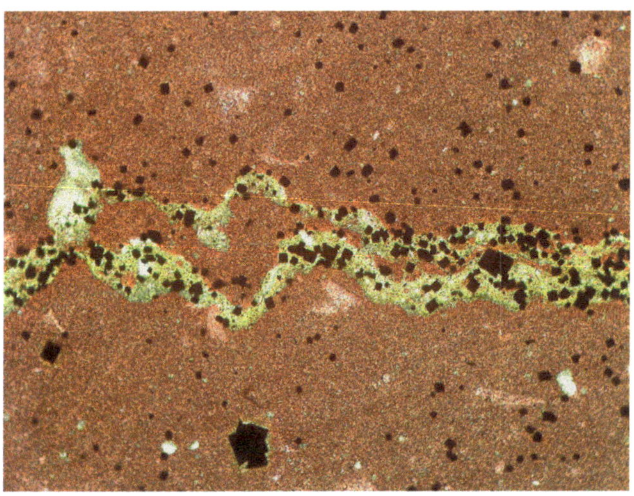

Cretaceous (late Barremian). Tibú Member, Apón Formation. Maracaibo Basin

In this mudstone, stained with Alizarin Red S, pyrite framboids exhibiting characteristic cubic habit are scattered across the image, replacing both the slightly organic-rich micritic matrix and the serrate (jagged) surfaces that cross horizontally across the field view. The latter are stylocumulates, formed by accumulation of insoluble material, in this case mainly clay minerals. The late burial origin of these features, combined with the law of “cross-cutting relationships”, indicates that the precipitation of the framboids postdates stylocumulate development.

IW = 1.3 mm. PPL.

References

Adams, D. D., Hurtgen, M. T., & Sageman, B. B. (2010). Volcanic triggering of a biogeochemical cascade during Oceanic Anoxic Event 2. *Nature geoscience*, *3*(3), 201–204.

Berner, R. A., 1985. Sulphate reduction, organic matter decomposition and pyrite formation. *Philosophical Transactions of the Royal Society of London. Series A, Mathematical and Physical Sciences*, *315*, 25–38.

Bond, D. P., and Grasby, S. E., 2017. On the causes of mass extinctions. *Palaeogeography, Palaeoclimatology, Palaeoecology*, *478*, 3–29.

Bond, D. P., Wignall, P. B., Keller, G., Kerr, A. C., 2014. Large igneous provinces and mass extinctions: an update. *The Geological Society of America, Special Paper 505*, 29–55.

Brumsack, H. J., 2006. The trace metal content of recent organic carbon-rich sediments: implications for Cretaceous black shale formation. *Palaeogeography, Palaeoclimatology, Palaeoecology*, *232*, 344–361.

Ernst, R. E., and Youbi, N., 2017. How Large Igneous Provinces affect global climate, sometimes cause mass extinctions, and represent natural markers in the geological record. *Palaeogeography,Ppalaeo climatology, Palaeoecology*, *478*, 30–52.

Ghazban, F., and Al-Aasm, I. S., 2010. Hydrocarbon-induced diagenetic dolomite and pyrite formation associated with the hormoz Island salt dome, offshore Iran. *Journal of Petroleum Geology*, *33*, 183–196.

Hetzel, A., Böttcher, M. E., Wortmann, U. G., Brumsack, H. J., 2009. Paleo-redox conditions during OAE 2 reflected in Demerara Rise sediment geochemistry (ODP Leg 207). *Palaeogeography, Palaeoclimatology, Palaeoecology*, *273*, 302–328.

Hülse, D., Arndt, S., Ridgwell, A., 2019. Mitigation of extreme ocean anoxic event conditions by organic matter sulfurization. *Paleoceanography and Paleoclimatology*, *34*, 476–489.

Jenkyns, H. C., Matthews, A., Tsikos, H., Erel, Y., 2007. Nitrate reduction, sulfate reduction, and sedimentary iron isotope evolution during the Cenomanian-Turonian oceanic anoxic event. *Paleoceanography*, *22*, PA3208.

Jiang, L., Worden, R. H., Yang, C., 2018. Thermochemical sulphate reduction can improve carbonate petroleum reservoir quality. *Geochimica et Cosmochimica Acta*, *223*, 127–140.

James, N. P., and Jones, B., 2015. Origin of Carbonate Sedimentary Rocks. John Wiley & Sons, Ltd, Chinchester, 464 pp.

Kidder, D. L., and Worsley, T. R., 2010. Phanerozoic large igneous provinces (LIPs), HEATT (Haline Euxinic Acidic Thermal Transgression) episodes and mass extinctions. *Palaeogeography, Palaeoclimatology, Palaeoecology*, *295*, 162–191.

Killops, S., and Killops, V, 2005. Introduction to Organic Geochemistry. Blackwell Publishing. U.S.A, 393 pp.

Kolonic, S., Sinninghe Damsté, J. S., Böttcher, M. E., Kuypers, M. M. M., Kuhnt, W., Beckmann, B., Scheeder, G., Wagner, T., 2002. Geochemical characterization of Cenomanina/Turonian Black Shales from the Tarfaya Basin (SW Morocco). Relationships between paleoenvironmental conditions and early sulphurization of sedimentary organic matter. *Journal of Petroleum Geology*, *25*, 325–350.

Kolonic, S., Wagnet, T., Forster, A., Sinninghe Damsté, J. S., Walsworth-Bell, B., Erba, E., Turgeon, S., Brumsack, H., Chellai, E., Tsikos, H., Kuhnt, W., Kuypers, M. M. M., 2005. Black shale deposition on the northwest African shelf during the Cenomanian/Turonian oceanic anoxic event: Climate coupling and global organic carbon burial, *Paleoceanography*, *20*, PA1006.

Konhauser, K. O., 2007. Introduction to Geomicrobiology. Blackwell Publishing. 433 pp.

Kump, L. R., Pavlov, A., Arthur, M. A., 2005. Massive release of hydrogen sulfide to the surface ocean and atmosphere during intervals of oceanic anoxia. *Geology*, *33*, 397–400.

Lippmann, F. Crystal Chemistry of Sedimentary Carbonate Minerals., 1973. In: Sedimentary Carbonate Minerals. Minerals, Rocks and Inorganic Materials (Monograph Series of Theoretical and Experimental Studies), vol. 6.; Springer: Berlin, Heidelberg, 1973, 5–96.

Liu, Z., Chen, D., Zhang, J., Lü, X., Wang, Z., Liao, W., Shi, X., Tang, X., Xie, G., 2019. Pyrite morphology as an indicator of paleoredox conditions and shale gas content of the Longmaxi and Wufeng shales in the Middle Yangtze area, South China. *Minerals*, *9*, 428.

López, L., Lo Mónaco, S., Escobar, G., Camargo, C., Lugo, P., Rojas, H., González, C., 2009. ESTUDIO DE NÓDULOS DE PIRITA PROVENIENTES DE LA FORMACIÓN QUERECUAL, ESTADO ANZOÁTEGUI (Venezuela) MEDIANTE MICROANÁLISIS CON SONDA DE ELECTRONES. *Acta Microscópica*, *18*(3), 333–343.

Machel, H. G., 2001. Bacterial and thermochemical sulfate reduction in diagenetic settings—old and new insights. *Sedimentary Geology*, *140*, 143–175.

Meyer, K. M., and Kump, L. R., 2008. Oceanic euxinia in Earth history: causes and consequences. *Annual Review of Earth and Planetary Sciences*, *36*, 251–288.

Mongenot, T., Tribovillard, N. P., Desprairies A., Lallier-Vergés, E., Laggoun-Defarge, F., 1996. Trace elements and paleoenvironmental markers in strongly mature hydrocarbon source rock: the Cretaceous La Luna Formation of Venezuela. *Sedimentary Geology*, *103*, 23–37.

Peters, K. E., Walters, C., Moldowan, J. M., 2005. The Biomarker Guide. Biomarkers and isotopes in the environmental and human history. Volume I. Second Edition. Cambridge University Press, 471 pp.

Raiswell, R., and Canfield, D. E., 1998. Sources of iron for pyrite formation in marine sediments. *American Journal of Science*, *298*, 219–245.

Severmann, S., Johnson, C. M., Beard, B. L., MacManus, J., 2006. The effect of early diagenesis on the Fe isotope compositions of pore waters and authigenic minerals in continental margin sediments, *Geochimica et Cosmochimica Acta*, *70*, 2006–2022.

Taylor, G. T., Iabichella-Armas, M., Varela, R., Muller-Karger, F., Lin, X., Scranton, M. I., 2006. Microbial ecology of the Cariaco Basin's oxic/anoxic interface: the U.S.-Venezuelan CARIACO times series program. In: Neretin, L.N. (Ed.), Past and Present Water Column Anoxia. NATO Science Series. Springer, The Netherlands, pp. 473–499.

Tribovillard, N., Algeo, T. J., Lyons, T., Riboulleau, A., 2006. Trace metals as paleoredox and paleoproductivity proxies: An update. *Chemical Geology*, *232*, 12–32.

Turgeon, S., and Brumsack, H. J., 2006. Anoxic vs dysoxic events reflected in sediment geochemistry during the Cenomanian–Turonian Boundary Event (Cretaceous) in the Umbria–Marche Basin of central Italy. *Chemical Geology*, *234*, 321–339.

Vandenbroucke, M., and Largeau, C., 2007. Kerogen origin, evolution and structure. *Organic Geochemistry*, *38*, 719–833.

Worden, R. H., Smalley, P. C., Barclay, S. A., 2003. H_2S and diagenetic pyrite in North Sea sandstones: due to TSR or organic sulphur compound cracking? *Journal of Geochemical Exploration*, *78*, 487–491.

22 Glauconite

Glauconite (often loosely referred to as glaucony) is a silicate mineral of diagenetic origin that occurs exclusively in marine environments and is found in association with both carbonate and siliciclastic sediments. This mineral, of authigenic or detrital origin and easy to distinguish in thin sections due to its conspicuous olive-green color, is commonly found in transgressive deposits and condensed sections, and regarded as good indicator of low-sedimentation rate (e.g., Odin and Matter 1981; Amorosi 1995, 2012; Banerjee et al. 2016a).

22.1 Mineral Glauconite

Glauconite is a potassium- and iron-rich dioctahedral sheet silicate (e.g., Banerjee et al. 2016b). Odin and Létolle (1980) proposed the term "glaucony" to encompass the wide-ranging chemical composition of sand-sized, green grains irrespective of their mineralogical affinity. Subsequently, Odin and Matter (1981) introduced the term 'glauconitic minerals' for high-Fe glaucony (>15% total Fe_2O_3) with glauconitic smectite (3%-5% K_2O) and glauconitic mica (>5% K_2O) as end members (see also Banerjee et al. 2016b). The chemical composition of dioctahedral glauconite can be represented by up to 11 different structural formulae (see Meunier and El Albani, 2007). Chemical variations (specifically in potassium, iron, aluminum, and magnesium content) are inherited from the silicate mineral precursor, the replaced mineral substrate (carbonate or siliciclastic), and the environmental conditions during glauconite formation.

Maturation during glauconitization involves progressive incorporation of potassium and iron and change a of the mineralogical structure toward an end member constituted by a non-expandable, K-rich glauconitic mica (Odin and Matter 1981; Odin and Fullagar 1988), namely glauconite (see Amorosi 2012). According to Huggett (2013), glauconite is an Fe-rich, K-rich dioctahedral mica. Typically, 5–12% of the total iron is ferrous. Changes in the superficial color of glauconite are a function of mineral maturity (%K_2O). K-poor, nascent (K_2O = 2% to 4%) to slightly evolved (K_2O = 2% to 4%) glauconite has a pale to light green color and generally shows obvious traces of the primary substrate of glauconitization. In contrast, evolved (K_2O = 6% to 8%) and highly-evolved (K_2O > 8%) glauconite, in which the substrate has been almost completely replaced, is green to dark green (Amorosi 2012; see also Banerjee et al. 2016a).

22.2 Glauconite Morphology

Morphological varieties of glauconite include (e.g., Triplehorn 1966; Udgata 2007): spheroidal or ovoidal pellets (commonly fine to medium sand-sized); discoidal (flattened) pellets; mammillated pellets (irregular grains separated by sutures); composite pellets (relatively large, up to 3–4 mm); fossil casts and internal molds (e.g., foraminifera, echinoderms, bryozoans); pigmentary glauconite (coating of surfaces of and/or penetrating cracks/cleavage within and/or between other minerals).

Glauconite green grains are featureless and are commonly referred to as glauconitic "pellets" or "peloids". Their textures are most commonly microcrystalline, exhibiting low "greenish" birefringence under CPL, although foliated-micaceous grains and "yellowish" colors under CPL are also moderately common, especially associated with glauconite replacing siliciclastic sediments (see Triplehorn 1966; McRae 1972). Spheroidal/ovoidal grains (of authigenic or detrital origin) may acquire their shape in several ways. They may reflect the reworking and rounding of other grain morphotypes (detrital glauconite) or inherit their shape from precursor grains (Triplehorn 1966; Amorosi 1997). The latter commonly include fecal pellets, micritic peloids and other micritized grains (e.g., aggregates and clasts), and quartz grains. Localized, authigenic glauconitization of organic-rich micritic sediments and dolomite crystals is also common.

J. A. Méndez Dot, J. Méndez Baamonde, *Atlas of Tropical Carbonates*, https://doi.org/10.1007/978-3-031-25494-9_22

22.3 Environment and Conditions for Glauconite Formation

In modern oceans, glauconite forms mainly in the depth range between 200 and 500 m (e.g., Odin and Matter 1981; Amorosi et al. 2007). The same bathymetric interpretation is frequently extrapolated to interpret the depositional settings of ancient glauconite-bearing sections (see Banerjee et al. 2016b; Chattoraj et al. 2016). However, ancient varieties of glauconite are reported from a wide range of depositional conditions, from shallow marine (e.g., Chafetz and Reid 2000; El Albani et al. 2005; Chattoraj et al. 2016; Choudhury et al. 2021) to deep marine (e.g., Berra et al. 2007; Ghabeishavi et al. 2009) and inner to outer shelf (see Banerjee et al. 2016b). In Cretaceous facies of Venezuela, glauconite horizons include shallow-water (Lisure and Escandalosa formations, Albian and Turonian in age, respectively), mid-lower neritic (Chimana Formation of Albian age), and deep-water (e.g., Tres Esquinas Member) varieties.

Despite differences in redox conditions, bathymetry (depth of the seafloor), and depositional setting (carbonate- or siliciclastic-dominated), glauconitization takes place at the sediment/water interface during time intervals of RSL highstand, of very low hydrodynamic energy and of low sedimentation rates. Therefore, glauconite is especially concentrated at hardground surfaces (e.g., Amorosi 1997, 2012; Chafetz and Reid 2000; Scholle and Ulmer-Scholle 2003; Flügel 2010).

Although the formation of glauconite has been reported in environments ranging from moderately anaerobic to strongly oxidizing, many authors hold the view that the formation of authigenic glauconite is favored under slightly reducing to low-oxygen (dysoxic) bottom water conditions charged with abundant cations and facilitated by the presence of decaying organic matter (e.g., McRae 1972; Rudmin et al. 2017, 2019; Tounekti et al. 2021). The formation of glauconite requires simply: a degraded layer of silicate lattice, a plentiful supply of iron and potassium, and a suitable redox potential (see McRae 1972).

In the shallow marine realm, the sources of K and Fe ions are seawater and detrital minerals (mainly in the form of iron oxides and oxyhydroxides), respectively. Fe will be incorporated into either the silicate or the sulfide phases. The silicate phase cannot incorporate iron when sulfide is present; it forms when the sulfide is oxidized. Nonetheless, the creation of reductive microsystems in the vicinity of organic matter promotes glauconite formation (Meunier and El Albani 2007). Incorporation of large amounts of Fe ions into the glauconite structure is favored in a reducing and acidic environment that may be created by the decomposition of organic matter. Consequently, pyrite is a common associate of glauconite (Kelly and Webb 1999; El Albani et al. 2005). Such authigenic mineral association is very common in Cretaceous glauconite varieties from Venezuela.

22.4 Origin of Glauconites: Authigenic or Detrital

Glauconites are formed coevally (authigenically) or non-coevally with sedimentation, and are intrabasinal or extrabasinal. Therefore, the presence of glauconite cannot be regarded *per se* as an indication of marine transgression, since the green grains may experience penecontemporaneous remobilization (parautochthonous glauconite) or reworking from older sedimentary deposits (detrital glauconite), being concentrated in a variety of deposits in the rock record (Amorosi 1997). It is well established that glauconite that forms *in situ* tends to concentrate within the 160–500-μm-size fraction, which corresponds to the ideal size of the semi-confined environment for glauconite formation and evolution (see Amorosi et al. 2011 and references therein). Authigenic glauconite is generally associated with thin stratigraphic horizons, ranging from few cm to up to a few meters in thickness (Odin 1985).

The distinction between authigenic (*in situ*) or allogenic (detrital) origin of glauconitic grains usually involves consideration of the petrology of the rock and the morphology of the grains. Detrital glauconite may have been derived more or less contemporaneously from a nearby zone in the basin of deposition where authigenic glauconite is forming, especially under alkaline pH conditions (McRae 1972). In general, horizons consisting of glauconitic pellets of variable sizes and shapes are associated with authigenic glauconite formation (Odin and Matter 1981; Amorosi and Centineo 1997).

22.5 Age Dating

The K-Ar dating of potassium-rich glauconite grains provides ~40% of the absolute age dates for the geological timescale of the past 250 Myr (e.g., Smith et al. 1998; Clauer et al. 2005; Derkowski et al. 2009; Huggett 2013). However, there are potential problems with this radiogenic, isotopic method (Huggett 2013), such as: argon loss; residual detrital illite or K feldspar will result in incorrect (too old) dates; potassium-rich glauconite, being the most resistant to weathering, is the most likely to survive reworking; hence, ages greater than time of sedimentation may be measured; a single glauconite assemblage may represent a cluster of glauconitization events; hence, the bulk age will be an average of these.

22.6 Glauconite: Photographic Set and Descriptive Captions

In all photos, IW = image width (the true length of the horizontal axis of the view)

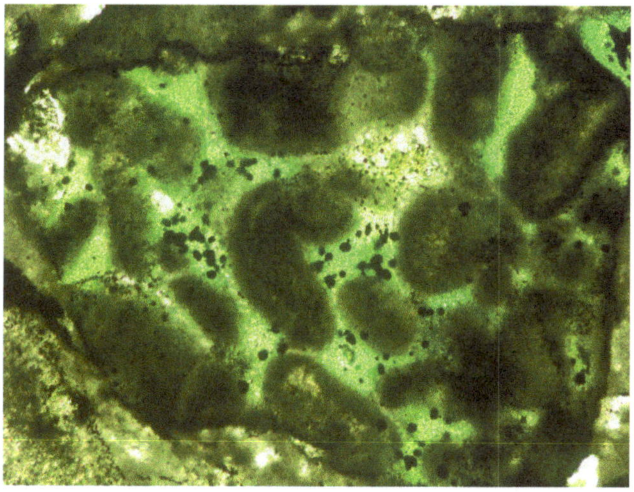

Cretaceous (middle Aptian). Piché Member, Apón Formation. Maracaibo Basin

Peloid grains and matrix in this sample are partially replaced by pigmentary (or surface coating) glauconite. Pyrite framboids postdating glauconitization are also abundant. Organic-rich micritic peloids represent ideal sites for the nucleation of these minerals. Note the soft intensity of the green color of glauconite, indicating that replacive precipitation occurred superficially. This morphologic variety of glauconite is actually not common in Cretaceous facies of Venezuela.

IW = 1.3 mm. PPL

Cretaceous (middle Albian). Lisure Formation. Maracaibo Basin

Glauconitic peloids in the image show characteristic olive-green color. Different-sized grains are visibly flattened and in sutured contact with each other due to mechanical compaction. Dewatering and expulsion of Mg-rich fluids promoted precipitation of small-sized dolomite crystals, which are locally replacing the green grains. We estimate an authigenic origin for the majority of glauconite mineralization in the Lisure Formation, as glauconitic horizons within this unit are present across the subsurface of the MB, as well as in the Perijá Range. The latter points to a stratigraphic control on the origin of such horizons.

IW = 1.3 mm. PPL

Cretaceous (middle Albian). Lisure Formation. Maracaibo Basin

Glauconitic peloids are major components in this muddy dolomitic microfacies. As in the previous example, different-sized green grains were originally most likely rounded micritized peloids whose coatings were enriched in organic matter, hence representing ideal substrates for glauconite precipitation. Note the dark stylocumulates fringing sutured contacts between flattened glauconite peloids. These features, mainly composed of pyrite and organic matter, postdate mechanical compaction and dewatering of iron-rich green grains.

IW = 2.4 mm. PPL

Cretaceous (middle Albian). Lisure Formation. Maracaibo Basin

Glauconitic grains in a hardground. Note that pyrite framboids (black spots) are locally replacing the green grains and micritic sediments cementing reworked clasts in the image, hence indicating that pyritization postdates glauconitization. Low-oxygen conditions (but not anoxic) are associated to ideal localized microenvironments for glauconite precipitation, replacing the organic-rich micritic grains and aggregates. Postdating transient anoxia promoted precipitation of pyrite.

IW = 4.2 mm. PPL

Cretaceous (middle Albian). Lisure Formation. Maracaibo Basin

Another example of a hardground surface in which glauconitic peloids are abundant. Other components in the image are redeposited clasts, aggregate grains and quartz grains (white lean in color). In the Lisure Formation, authigenic mineralization of glauconite and associated pyritization (note the blackening of some grains) is indicative of time intervals of low sedimentation rate. Despite the rounded shapes of glauconitic peloids, which in many cases are inherited from former carbonate peloidal grains, abundant sand-sized quartz grains also represent a suitable substrate for glauconite precipitation.

IW = 4.2 mm. PPL

Cretaceous (early Campanian). Tres Esquinas Member, La Luna Formation

Sandy, glauconitic, foraminiferal packstone. Bioclasts in the image are a combination of SBF and PF, which are embedded in a phosphatized matrix. Deep-water glauconite peloids were formed due to nucleation of this mineral on quartz grains (white-limpid angular grains) and/or foraminifera tests (e.g. *Bolivinoides* and *Orthokarstenia*; see Parra et al. 2003). Dysoxic conditions and upper bathyal paleodepths are associated with deposition of this glauconitic-rich unit.

IW = 2 mm. PPL
Photomicrograph courtesy of Grony Garbán

Cretaceous (middle Albian). Lisure Formation. Maracaibo Basin

Glauconitic grains occasionally preserve some of their original textural features. This is the case of the central echinoid grain in the image, which exhibits characteristic reticulate pattern of the originally highly porous stereom. Glauconite formed as a replacement of organic-rich micritic sediments infilling small intraparticle pores. A very low-energy setting during early marine diagenesis is associated to this authigenic mineralization.

IW = 2.4 mm. PPL

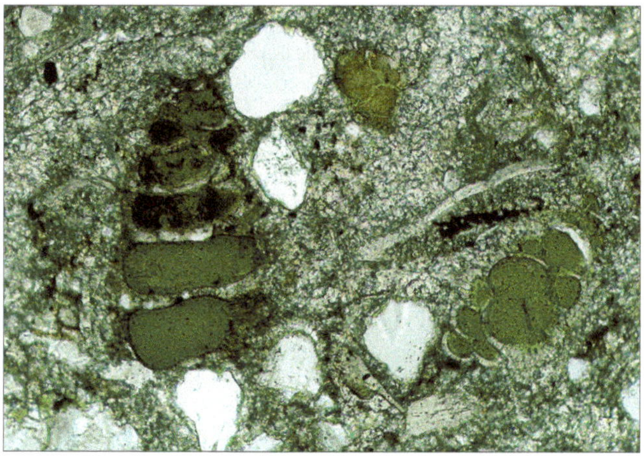

Cretaceous (latest Albian). Maraca Formation. Maracaibo Basin

The chambers of the two benthic foraminifera tests in the image are filled by glauconite. The nucleation of this mineral was promoted due to the development of a microenvironment within the foraminifera chambers. Decaying organic matter and empty foraminiferal tests supplied ideal localized conditions for glauconite precipitation. Although the catalytic activity of marine organisms is not thought to be essential for glauconite genesis (see McRae 1972), its nucleation is particularly common infilling intraparticle cavities within calcareous microfossils (e.g. foraminifera, echinoids and bryozoans).

IW = 1.3 mm. PPL

Cretaceous (Turonian). O Member, Escandalosa Formation. Barinas sub-Basin

Glauconite cement in the image is infilling intercrystalline pores in this inequigranular massive dolostone. Changing conditions during early diagenesis, most probably due to a low-amplitude sea-level rise and deepening of the water column postdating dolomitization, led to ideal conditions for glauconite precipitation. This authigenic porosity-filling glauconite type is moderately common in dolostone microfacies of the O Member, as well as in underlying sandstones of the P Member.

IW = 1.3 mm. PPL

References

Amorosi, A., 1995. Glaucony and sequence stratigraphy: a conceptual framework of distribution in siliciclastic sequences. *Journal of Sedimentary Research*, *65*, 419–425.

Amorosi, A., 1997. Detecting compositional, spatial, and temporal attributes of glaucony: a tool for provenance research. *Sedimentary Geology*, *109*, 135–153.

Amorosi, A., 2012. The occurrence of glaucony in the stratigraphic record: distribution patterns and sequence-stratigraphic significance. *International Association of Sedimentologists Special Publications*, *45*, 37–54.

Amorosi, A., and Centineo, M. C., 1997. Glaucony from the Eocene of the Isle of Wight (southern UK): implications for basin analysis and sequence-stratigraphic interpretation. *Journal of the Geological Society*, *154*, 887–896.

Amorosi, A., Sammartino, I., Tateo, F., 2007. Evolution patterns of glaucony maturity: A mineralogical and geochemical approach. *Deep Sea Research Part II. Topical Studies in Oceanography*, *54*, 1364–1374.

Amorosi, A., Guidi, R., Mas, R., Falanga, E., 2011. Glaucony from the Cretaceous of the Sierra de Guadarrama (Central Spain) and its application in a sequence-stratigraphic context. *International Journal of Earth Sciences*, *101*, 415–427.

Banerjee, S., Bansal, U., Pande, K., Meena, S. S., 2016a. Compositional variability of glauconites within the Upper Cretaceous Karai Shale Formation, Cauvery Basin, India: implications for evaluation of stratigraphic condensation. *Sedimentary Geology*, *331*, 12–29.

Banerjee, S., Bansal, U., Thorat, A. V., 2016b. A review on palaeogeographic implications and temporal variation in glaucony composition. *Journal of Palaeogeography*, *5*, 43–71.

Berra, F., Zanchi, A., Mattei, M., Nawab, A., 2007. Late Cretaceous transgression on a Cimmerian high (Neka Valley, Eastern Alborz, Iran): A geodynamic event recorded by glauconitic sands. *Sedimentary Geology*, *199*, 189–204.

Chafetz, H. S., and Reid, A., 2000. Syndepositional shallow-water precipitation of glauconitic minerals. *Sedimentary Geology*, *136*, 29–42.

Chattoraj, S. L., Banerjee, S., Saraswati, P. K., Bansal, U., 2016. Origin, depositional setting and stratigraphic implications of Palaeogene glauconite of Kutch. *Special Publication of the Geological Society of India*, *6*, 75–88.

Choudhury, T. R., Banerjee, S., Khanolkar, S., Saraswati, P. K., Meena, S. S., 2021. Glauconite authigenesis during the onset of the Paleocene-Eocene Thermal Maximum: A case study from the Khuiala Formation in Jaisalmer Basin, India. *Palaeogeography, Palaeoclimatology, Palaeoecology*, *571*, 110388.

Clauer, N., Huggett, J. M., Hillier, S., 2005. How reliable is the K-Ar glauconite chronometer? A case study of Eocene sediments from the Isle of Wight. *Clay Minerals*, *40*, 167–176.

Derkowski, A., Środoń, J., Franus, W., Uhlík, P., Banaś, M., Zielinski, G., Caplovicova, M., Franus, M. 2009. Partial dissolution of glauconitic samples: implications for the methodology of K-Ar and Rb-Sr dating. *Clays and Clay Minerals*, *57*, 531–554.

El Albani, A., Meunier, A., Fürsich, F., 2005. Unusual occurrence of glauconite in a shallow lagoonal environment (Lower Cretaceous, northern Aquitaine Basin, SW France). *Terra Nova*, *17*, 537–544.

Flügel, E, 2010. Microfacies of Carbonate Rocks—Analysis, Interpretation and Application. Springer, Heidelberg-Dordrecht-London-New York, 984 pp.

Ghabeishavi, A., Vaziri-Moghaddam, H., Taheri, A., 2009. Facies distribution and sequence stratigraphy of the Coniacian–Santonian succession of the Bangestan Palaeo-high in the Bangestan Anticline, SW Iran. *Facies*, *55*, 243–257.

Huggett, J. M., 2013, Minerals: Glauconites and Green Clays, in Reference Module in Earth Systems and Environmental Sciences: Elsevier, https://doi.org/10.1016/B978-0-12-4095489.02893-1.

Kelly, J. C., and Webb, J. A., 1999. The genesis of glaucony in the Oligo-Miocene Torquay Group, southeastern Australia: Petrographic and geochemical evidence. *Sedimentary Geology*, 125, 99–114.

McRae, S. G., 1972. Glauconite. *Earth-Science Reviews*, *8*, 397–440.

Meunier, A., and El Albani, A., 2007. The glauconite–Fe-illite–Fe-smectite problem: a critical review. *Terra Nova*, *19*, 95–104.

Odin, G. S. and Létolle, R., 1980. Glauconitization and phosphatization environments: a tentative comparison. In: Marine Phosphorites (Ed. Y.K. Bentor), SEPM Special Publications, 29, 227–237.

Odin, G. S., and Matter, A., 1981. De glauconiarum origine. *Sedimentology*, *28*, 611–641.

Odin, G. S., 1985. Significance of green particles (glaucony, berthierine, chlorite) in arenites. In: *Provenance of Arenites*, NATO-ASI Series C: Mathematical and Physical Sciences (Ed. by G.G. Zuffa), D. Reidel, Dordrecht, pp. 279–307.

Odin, G. S., and Fullagar, P. D., 1988. Geological significance of glaucony facies, In: Odin, G. S. (Ed.), *Green Marine Clays: Developments in Sedimentology*, *45*, 295–332.

Parra, M., Moscardelli, L., Lorente, M. A., 2003. Late Cretaceous anoxia and lateral microfacies changes in the Tres Esquinas Member, La Luna Formation, Western Venezuela. *Palaios*, *18*, 321–333.

Scholle, P. A., and Ulmer-Scholle, D., 2003. A Color Guide to the Petrography of Carbonate Rocks: Grains, Textures, Porosity, Diagenesis. *American Association of Petroleum Geologists*, *Memoir 77*, 474 pp.

Smith, P. E., Evensen, N. M., York, D., Odin, G. S., 1998. Single-grain 40Ar-39Ar ages of glauconies: Implications for the geologic time scale and global sea level variations. *Science*, *279*, 1517–1519.

Rudmin, M., Banerjee, S., Mazurov, A., 2017. Compositional variation of glauconites in Upper Cretaceous-Paleogene sedimentary iron-ore deposits in South-eastern Western Siberia. *Sedimentary Geology*, *355*, 20–30.

Rudmin, M., Mazurov, A., Banerjee, S., 2019. Origin of ooidal ironstones in relation to warming events: Cretaceous-Eocene Bakchar deposit, south-east Western Siberia. *Marine and Petroleum Geology*, *100*, 309–325.

Tounekti, A., Boukhalfa, K., Choudhury, T. R., Soussi, M., Banerjee, S., 2021. Global and local factors behind the authigenesis of Fe-silicates (Glauconite/Chamosite) in Miocene strata of Northern Tunisia. *Journal of African Earth Sciences*, *184*, 104342.

Triplehorn, D. M., 1966. Morphology, internal structure, and origin of glauconite pellets. *Sedimentology*, *6*, 247–266.

Udgata, D. B. P., 2007. Glauconite as an indicator of sequence stratigraphic packages in a lower Paleocene passive margin shelf succession, central Alabama. Magister Dissertation, Auburn University, Auburn, Alabama. 109 pp.

Silica

23

Silica is a term generally used to refer to morphologic varieties of SiO_2. In this chapter, we consider silicification as the authigenic replacement of original carbonate material accomplished through the concurrent dissolution of calcium carbonate and precipitation of silica (Butts 2014). This diagenetic process involves the replacement, selective or pervasive, of carbonate components (e.g., foraminifera tests, dolomite crystals and matrix) by silica, as well as silica cementation, mainly filling microfractures and intraparticle pores (e.g., chambers of foraminifera). We also consider silicification in mixed in calcareous shales and chert facies.

23.1 Sources of Silica in the Oceans

The major sources of silica in oceans are (Laschet 1984; Lisitzin 1985; Racki and Cordey 2000; Ragueneau et al. 2000; Killops and Killops 2004): (**a**) river input of silica-rich solutions in the form of orthosilicic acid (H_4SiO_4) derived from continental weathering of silicate rocks (lithogenic silica); in present-day oceans, the major silica contribution, about 80%, comes from river runoff, which is very climate dependent, and 75% of the river supply is derived from the humid equatorial zone; (**b**) silica supplied in solution by volcano-hydrothermal systems (endogenous silica).

23.2 Silica Fabrics

The variety of silicification fabrics exhibit white color when viewed under PPL, whereby diagenetic silica may be often confused with diagenetic calcite (microspar, pseudospar, and LMC cements). Therefore, when viewed under CPL, diagenetic silica reveals the crystals texture, which may take the form of (Scholle and Ulmer-Scholle 2003): (**a**) amorphous silica, also known as opal, which is a featureless isotropic precipitate; it occurs mainly as cement, but also as replacement or as small nodules; (**b**) equigranular quartz, characterized by low birefringence (white-gray to straw-yellow); the fabric is termed cryptocrystalline, or so-called "chert", when crystals are <5 μm; microcrystalline for crystals 5–20 μm; and megaquartz (or megagranular quartz) for crystals >20 μm (and up to 10 mm); larger individual crystals are normally euhedral to hexagonal; it occurs mainly as replacement; (**c**) fibrous quartz, termed chalcedony, composed of fibrous-elongate silica crystals forming radial or spherulitic structures; crystals are from a few μm to a few mm; it shows the same birefringence as equigranular quartz, but banded extinction patterns (e.g., zebraic chalcedony) tend to become more prominent as crystal size increases; it is more commonly found as cement than as replacement.

23.3 Environment and Diagenetic Conditions for Silicification of Carbonates

Silicification is a common diagenetic process in shallow inner-to-middle platforms, as well as in outer shelves, slopes and open marine settings. In the former environments, the riverine input is the main mechanism that concentrates silica, thereby promoting silicification of carbonate substrates. In the latter environments, active upwelling of nutrient-rich waters (rich in N, P and Si) concentrates Si for uptake in the form of silicic acid [$Si(OH)_4$] by planktonic siliceous organisms and for precipitation of silica replacements and cements. Diagenetic alteration of amorphous silica skeletal material (sponge spicules, radiolarians and diatoms), which in silica-depleted marine environments is much more unstable than siliciclastic sediments, is the likely source of silica-supersaturated waters (with respect to opal-CT and quartz) for the formation of chert deposits and for diagenetic

J. A. Méndez Dot, J. Méndez Baamonde, *Atlas of Tropical Carbonates*, https://doi.org/10.1007/978-3-031-25494-9_23

replacements of carbonate sediments (e.g., Johnson 1976; Maliva and Siever 1989; Murray et al. 1992; Murray 1994; Moore 2008; Flügel 2010; Garbán et al. 2017).

Precipitation of silica is enhanced by carbonate substrates, while diagenetic silica solubility increases in the presence of siliciclastic sediments (Butts 2014). Silicification of carbonate rocks involves localized dissolution of $CaCO_3$ prior to saturation of pore with waters respect to opal or quartz. In organic-rich sediments, carbonate dissolution and subsequent silica precipitation are linked to the oxidation/decay of organic matter (Maliva and Siever 1989).

In general, silica diagenesis is rarely a very early- or a very late-stage diagenetic event in carbonate rocks. Rather, it is most typically a product of burial diagenesis. This is due to the timing of the conversion of biogenic opal-A, first to opal-CT, and then to stable microquartz or megaquartz, a diagenetic recrystallization that is also dependent on temperature, starting at about ~50 ° C and a depth of <500 m (Scholle and Ulmer-Scholle 2003). Silicification in shallow-marine settings is commonly related to mixing zones (e.g., Maliva and Siever 1989; Flügel 2010). Many ancient silicification processes in carbonate rocks are associated with the shallow-burial diagenetic environment within the fluid-flow domain (e.g., Floquet et al. 2012; Neuweiler et al. 2014).

23.4 Microbial Mediation: An Alternative Origin for Selective Silicification

Microbial-mediated silicification carried out by microbial mat communities may occur under alkaline conditions (pH values between 8 and 9) selectively associated with acidic sites linked to organic matter degradation (see Amores and Warren 2007 and references therein). Microbial activity within the carbonate material lowers the pH, increases CO_2, and promotes dissolution of calcium carbonate, thereby promoting selective silica precipitation (Butts 2014). Acidification due to bacterial organic matter decay (i.e., generation of organic acids, CO_2 and H_2S) and associated silicification is essentially a low-temperature process (<80 °C) that must occur before bacterial activity ceases due to increasing burial (Maliva and Siever 1989).

23.5 Silica: Photographic Set and Descriptive Captions

In all photos, IW = image width (the true length of the horizontal axis of the view)

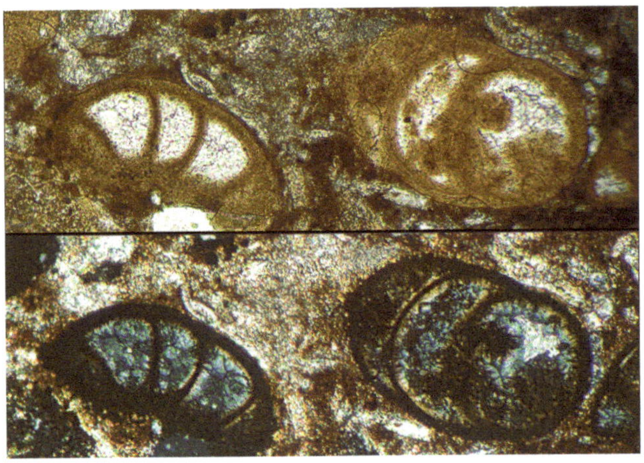

Cretaceous (early Campanian). Tres Esquinas Member, La Luna Formation. Maracaibo Basin

Dual image (PPL and CPL) of two SBF. Calcitic walls of both tests are replaced by fluorapatite, which exhibits a characteristic brown/orange color under PPL (above) and isotropic behavior under CPL (below). Selective chert precipitation (white precipitate infilling chambers) postdates phosphatization. This silica polymorph exhibits a characteristic low birefringence (gray to white) and microcrystalline texture under CPL. Chert is easily confused with sparry calcite under PPL. Thus, CPL is necessary to differentiate these diagenetic products.

IW = 0.5 mm. PPL (above); CPL (below)

Cretaceous (Turonian). O Member, Escandalosa Formation. Barinas sub-Basin

Chert is replacing dolomite crystals in this massive dolostone. Note that the boundaries of chertified crystals are obliterated by diagenetic silica precipitates. The distinctive microcrystalline fabric is highlighted under CPL. As with all silica diagenetic replacements, the formation of this chert fabric is associated with burial depths (most likely in the shallow-burial environment) and minimum temperatures of ~ 50° C.

IW = 1.3 mm. CPL

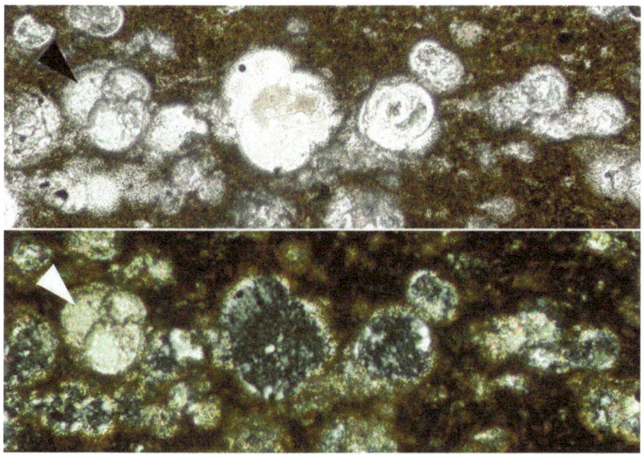

Cretaceous (late Santonian). Táchira Ftanita Member, La Luna Formation. Venezuelan Andes

Dual image of a chert microfacies rich in PF. Many of the globular tests are replaced by microcrystalline chert. Note the low birefringence of chert crystals (dark gray in the image below). This silica precipitate is white limpid under PPL and it partially prevents the visualization of individual chambers of foraminifera. In contrast, tests not replaced by silica (black and white arrows) exhibit well-differentiated chambers. The source of Si for the occurrence of a variety of silica replacements in microfacies of the Táchira Ftanita Member came from diagenetic alteration of radiolarian tests (see Garbán et al. 2017).

IW = 1.3 mm. PPL (above); CPL (bellow)

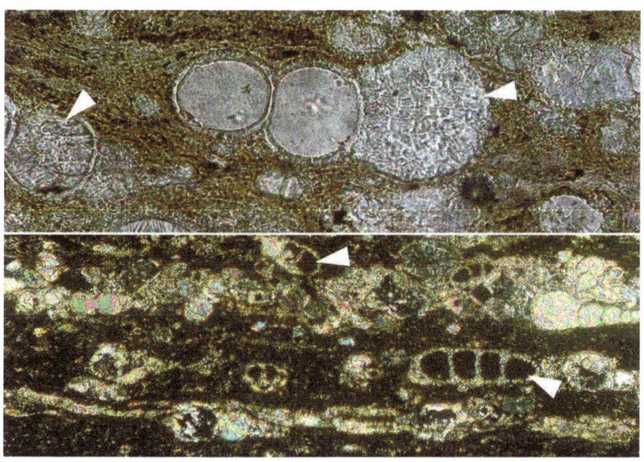

Cretaceous (Coniacian). La Morita Member, Navay Formation. Barinas Sub-Basin

Both photomicrographs correspond to the same thin section (a laminated, pyritic black shale rich in PF). The white arrows point to chambers of planktonic tests whose microspar infillings were replaced by amorphous silica (opal). This diagenetic precipitate is easy to distinguish due to its rusty appearance under PPL (above) and isotropic behavior under CPL (below). The pyritic and organic-rich, clayey-cherty matrix (below) exhibits very similar isotropic black color.

IW = 0.5 mm and PPL (above); 1.3 mm and CPL (below).

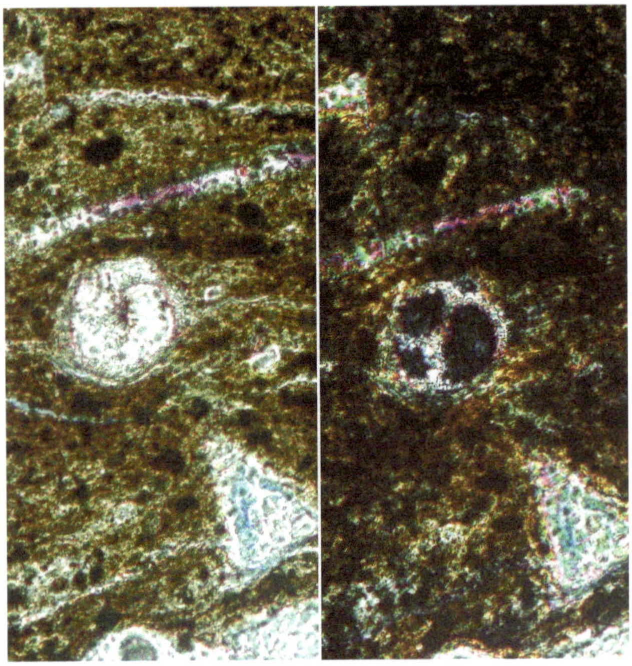

Cretaceous (late Aptian). Piché Member, Apón Formation. Maracaibo Basin

Dual image of a planktonic foraminifer in a black shale of the OAE 1b first sub-event. The chamber infillings of the small globular test were selectively replaced by amorphous silica. Note that this diagenetic precipitate did not replace the calcitic walls of the test. These are pink-colored (stained by Alizarin Red S) under PPL (left) and exhibit characteristic calcite birefringence under CPL (right). Such selective silicification was probably promoted by the organic richness of the synsedimentary infilling of the test.

IW = 0.20 mm each. PPL (left); CPL (right)

Cretaceous (late Aptian). Piché Member, Apón Formation. Maracaibo Basin

Dual image showing pervasive precipitation of amorphous silica replacing dolomite crystals in this organic-rich dolomitic shale (7.2% TOC). Note the white-rusty appearance of barely identifiable rhombic dolomite crystals under PPL (above), while the view under CPL is severely darkened due to the pervasiveness of the silica replacement. In general, silicification, both localized/selective and pervasive, is a moderately common diagenetic process in Cretaceous black shales of Venezuela (deposited under an accelerated hydrological cycle), and rare in shallow, well-oxygenated carbonate microfacies.

IW = 0.5 mm. PPL (above); CPL (below)

Cretaceous (early Aptian). Machiques Member, Apón Formation. Maracaibo Basin

Amorphous silica is occasionally infilling microfractures in dolomite microfacies of the Machiques Member. This dual image allows the visualization of the amorphous (not crystalline) nature and isotropic optical behavior of opal. In this case, silica precipitation postdated microfracturing and occurred as the dolomite strata were still within reach of the Si-rich freshwater lens.

IW = 1.3 mm. PPL (above); CPL (below)

Cretaceous (late Aptian). Piché Member, Apón Formation. Maracaibo Basin

Selective silicification of organic matter filaments is moderately common in laminated black shales of the Cogollo Group (such as this example). Clayey matrix and abundant dolomite crystals, pyrite framboids and PF also characterize the lithology of such microfacies. Bacterial mediation and selective silicification of undulated microbial mats oriented parallel to the bedding plane are a plausible hypothesis for the occurrence of this authigenic mineralization.

IW = 2.4 mm. PPL

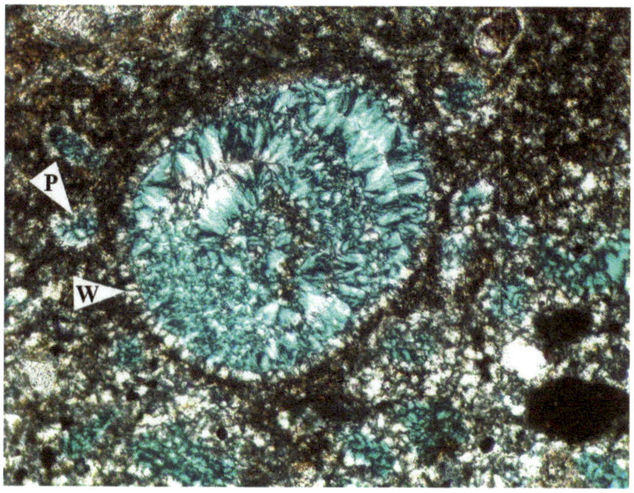

Late Middle Eocene. Jarillal Formation. Falcón Basin

Chalcedony is replacing the synsedimentary infilling of a globular planktonic foraminifer. This SiO_2 polymorph exhibits its characteristic elongate fibers, radial arrangement, low birefringence color (black to blue/gray) and extinction bands (light and dark colored). Smaller tests of PF (P) also exhibit these optical features. Note that the calcitic walls (W) of the central test are replaced by a different SiO_2 polymorph (equigranular quartz), which exhibit "zebraic" extinction.

IW = 0.5 mm. CPL

Cretaceous (early Aptian). Machiques Member, Apón Formation. Maracaibo Basin

The large mollusc shell in the center of the image is replaced by chalcedony. This silica polymorph exhibits characteristic radiating (zebraic) extinction pattern. Chalcedony precipitation postdates recrystallization of aragonitic shells. The latter process is visible in the shell at the top of the image.

IW = 2.4 mm. CPL.

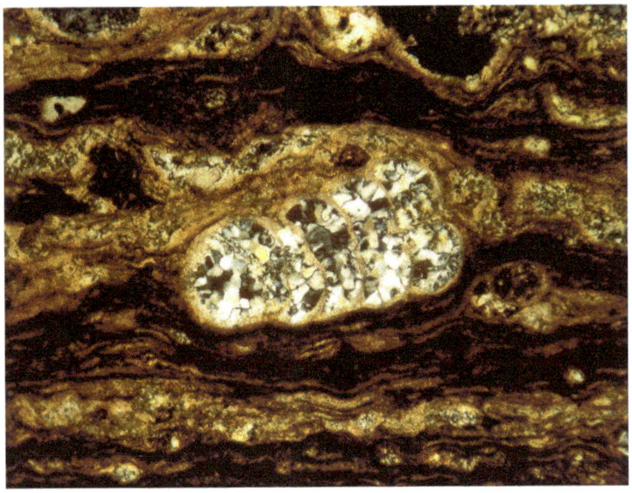

Cretaceous (late Santonian). Táchira Ftanita Member, La Luna Formation. Venezuelan Andes

The synsedimentary infilling of the smaller benthic foraminifer test in the center of the image is replaced by megaquartz crystals (also termed megagranular quartz). Note that individual quartz crystals can be differentiated by their large size and contrasting white-gray-black birefringence. Megaquartz crystals are abundant in this siliceous unit, both infilling tests and comprising the matrix of some microfacies (Garbán 2010).

IW = 0.5 mm. CPL
Photomicrograph courtesy of Grony Garbán

Cretaceous (late Aptian). Piché Member, Apón Formation. Maracaibo Basin

Megaquartz crystals in the image form a thin, light-colored band that is replacing an organic-rich, mixed micritic-clayey matrix. As in the previous example, the size of such crystals suggest that this silica precipitation occurred at a later diagenetic stage compared to previous examples of amorphous silica and chert. The latter assumption is based onthe timing involved in the crystallization of megagranular quartz fabrics, which is dependent on time, temperature and burial depth. In both examples, megaquartz crystals might have formed from recrystallization (crystal enlargement) of chert.

IW = 1.3 mm. CPL

References

Amores, D. R., and Warren, L. A., 2007. Identifying when microbes biosilicify: the interconnected requirements of acidic pH, colloidal SiO_2 and exposed microbial surface. *Chemical Geology*, *240*, 298–312.

Butts, S. H., 2014. Silicification. *The Paleontological Society Papers*, *20*, 15–34.

Floquet, M., Neuweiler, F., Andlé-Onide, P., 2012, The impact of depositional events and burial rate on carbonate–silica diagenesis in a Middle Jurassic stromatactis carbonate mud-mound, Sainte-Baume massif, SE France. *Journal of Sedimentary Research*, *82*, 521–539.

Flügel, E, 2010. Microfacies of Carbonate Rocks—Analysis, Interpretation and Application. Springer, Heidelberg-Dordrecht-London-New York, 984 pp.

Garbán, G. J., 2010. Geoquímica y estratigrafía del Miembro Ftanita de Táchira (Cretácico tardío), Formación La Luna, Venezuela Occidental: Implicaciones Paleooceanográficas y paleoambientales. Doctoral Dissertation. Central University of Venezuela. 402 pp.

Garbán, G., Martínez, M., Márquez, G., Rey, O., Escobar, M., Esquinas, N., 2017. Geochemical signatures of bedded cherts of the upper La Luna Formation in Táchira State, western Venezuela: Assessing material provenance and paleodepositional setting. *Sedimentary Geology*, *347*, 130–147.

Johnson, T. C., 1976. Controls on the preservation of biogenic opal in sediments of the eastern tropical Pacific. *Science*, *192*, 887–890.

Killops, S. D., and Killops, V. J., 2004. An Introduction to Organic Geochemistry. Wiley Blackwell. 408 pp.

Laschet, C., 1984., On the origin of cherts. *Facies*, *10*, 257–289.

Lisitzin, A. P., 1985. The silica cycle during the last ice age. *Palaeogeography, Palaeoclimatology, Palaeoecology*, *50*, 241–270.

Maliva, R. G., and Siever, R., 1989. Nodular chert formation in carbonate rocks. *The Journal of Geology*, *97*, 421–433.

Moore, T., 2008. Chert in the Pacific: biogenic silica and hydrothermal circulation. *Palaeogeography, Palaeoclimatology, Palaeoecology*, *261*, 87–99.

Murray, R., 1994. Chemical criteria to identify the depositional environment of chert: general principles and applications. *Sedimentary Geology*, *90*, 213–232.

Murray, R., Buchholtz ten Brink, M., Gerlach, D., Russ, G., Jones, D., 1992. Interoceanic variation in the rare earth, major, and trace element depositional chemistry of chert: perspectives gained from the DSDP and ODP record. *Geochimica et Cosmochimica Acta, 56*, 1897–1913.

Neuweiler, F., Larmagnat, S., Molson, J., Fortin-Morin, F., 2014. Sponge spicules, silicification, and sequence stratigraphy. *Journal of Sedimentary Research, 84*, 1107–1119.

Racki, G., and Cordey, F., 2000. Radiolarian palaeoecology and radiolarites: is the present the key to the past? *Earth-Science Reviews, 52*, 83–120.

Ragueneau, O., Tréguer, P., Leynaert, A., Anderson, R. F., Brzezinski, M. A., DeMaster, D. J., Dugdale, R. C., Dymond, J., Fischer, G., Francois, R., Heinze, C., Maier-Reimer, E., Martin-Jèzèquel, V., Nelson, D. M., Quéguiner, B., 2000. A review of the Si cycle in the modern ocean: recent progress and missing gaps in the application of biogenic opal as a paleoproductivity proxy. *Global and Planetary Change, 26*, 317–365.

Scholle, P. A., and Ulmer-Scholle, D., 2003. A Color Guide to the Petrography of Carbonate Rocks: Grains, Textures, Porosity, Diagenesis. *American Association of Petroleum Geologists, Memoir 77*, 474 pp.

Appendices

This appendix, divided in two main sections (Appendices A and B), has been written with the intention of introducing some globally-used conceptual models and concepts that we apply throughout the 23 chapters of this book. This document will help the reader understand these principles and how we apply them to the carbonate facies of Venezuela.

Appendix A: Marine Carbonate Factories and Grain Associations

We apply the concepts of marine carbonate factories and grain associations to all the carbonate facies considered in this book. We use a combination of classical and new approaches, some of them originally created from the study of non-tropical (or extratropical) carbonates. In this Appendix, our main objective is to present our vision of the tropical carbonates of Venezuela, based on the implementation of these concepts, and to create a new perspective and a better understanding of tropical carbonates worldwide. The variety of warm-water carbonate facies that were deposited or are currently being deposited in this country in northern most South America includes some features (e.g., skeletal assemblages) that are traditionally considered as typical of the extratropical realm. We aim for the dissemination of this work to contribute to the understanding and differentiation of the broad spectrum between warm- and cool-water carbonates. We also introduce some other definitions that we consider significant in explanatory boxes.

Environmental Controlling Parameters of Shallow-Water Carbonates

Carbonates are produced by organisms that preferentially live under defined environmental conditions and by non-skeletal components that precipitate under specific physicochemical conditions. Thus, carbonate production should occur in defined ranges of environmental parameters or factors (Michel et al. 2019). The main of those factors are: water temperature (either controlled by latitude and/or bathymetry of the seafloor), carbonate saturation, alkalinity and pH, light (as major source of energy for photoautotroph and mixotroph carbonate-producing organisms), salinity, oxygen concentration (as it promotes abundance or absence of benthic fauna; see Box A.1), nutrient concentration (PO_4^{3-} and NO^{3-} control productivity and food supply for light-independent calcifiers and influence the abundance or absence of light-dependent calcifiers; see Box A.2), water transparency, water energy, currents, CO_2 concentration, Mg^{2+}/Ca^{2+} ratio, bathymetry (see Box A.3), siliciclastic input (as limiting carbonate production), substrate requirement, and RSL change (see Mutti and Hallock 2003; Pomar et al. 2004; Schlager 2005; Wilson and Vecsei 2005; Flügel 2010; Westphal et al. 2010; Michel et al. 2019; Reijmer 2021).

Skeletal Grain Associations

Recognizing climatic impact in modulating carbonate production has also emerged as an approach to integrate carbonate mineralogy with skeletal and non-skeletal components (or constituents) and to build conceptual models (Pomar and Hallock 2008). Lees and Buller (1972) and Lees (1975) examined systematically, for the first time, the range of carbonate sediment types from tropical to non-tropical settings. Those authors gave rise to new visions for carbonate models that distinguish between tropical, warm-temperate, cool-temperate, and cool (or cold) carbonates; and to a new compositional terminology, the so-called climate-dependent "skeletal grain associations" (see Box A.4; see also Pomar et al. 2004; Flügel 2010).

Photozoan and Heterozoan Associations

To avoid Cenozoic-biased nomenclatures for skeletal grain associations, James (1997) proposed two climate-independent grain associations, Photozoan and Heterozoan, applicable for benthic carbonate-producing organisms to the entire Phanerozoic. The "Photozoan Association" consists

J. A. Méndez Dot, J. Méndez Baamonde, *Atlas of Tropical Carbonates*, https://doi.org/10.1007/978-3-031-25494-9

of an association of benthic carbonate grains dominated by skeletons of light-dependent organisms: calcareous algae (green and red, photoautotrophs), and scleractinian corals and LBF (both groups of calcifiers are symbiont-bearing mixotrophs with the ability to both feed and photosynthesize; see Hallock 2001). Non-skeletal grains (ooids, peloids, and others) are common components of this association, and plus or minus skeletons from the Heterozoan Association. This photozoan association reflects the existence of shallow (upper photic or euphotic) warm waters, today mostly confined to tropical and subtropical settings. The "Heterozoan Association" consists of an association of benthic carbonate skeletons produced by light-independent, heterotrophic organisms (these feed on organic matter): symbiont-barren benthic foraminifera, molluscs, echinoderms, bryozoans, brachiopods, ostracods, barnacles (balanids), serpulids, trilobites, ammonites, and some calcareous sponges; plus or minus calcareous red algae (photoautotrophs). James (1997) pointed out that while cool-water carbonates are always heterozoan carbonates, the heterozoan association is not indicative of cool-water carbonates. Therefore, this association occurs at any latitude and at any bathymetry (or depth of the seafloor) (see also Westphal et al. 2010).

Box A.1: Common terminology describing the redox conditions in sedimentary environments.
Compiled from Peters et al. (2005), Meyer and Kump (2008), and Flügel (2010). Oxygen content (mL/L) according to Tyson and Pearson (1991).

- **Oxic (aerobic)** = 8–2 mL/L O_2
- **Dysoxic (dysaerobic)** = 2–0.2 mL/L O_2
- **Suboxic (quasi-anaerobic)** = 0.2–0 mL/L O_2
- **Anoxic (anaerobic)** = 0 mL/L O_2
- **Euxinia:** a state of a body of water characterized by the presence of free hydrogen sulfide.
- **Euxinic**: refers to anoxic conditions but with free H_2S in the water column.

Marine Carbonate Factories

"Marine Carbonate Factories" is a conceptual model originally developed by Schlager (2000, 2003, 2005) at global scale that links carbonate precipitation modes (Fig. A.1) to a certain combination of environmental parameters, range of

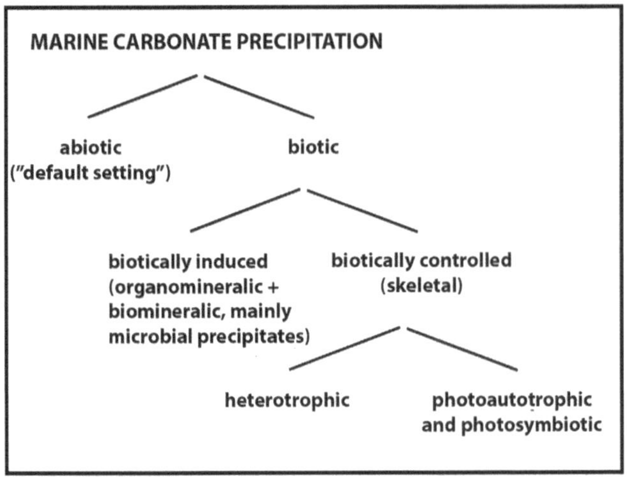

Fig. A.1 Pathways of carbonate precipitation in aquatic environments—a cascade of options governed by the degree of biotic influence. Organisms have the first hand on precipitation in most carbonate settings. Thus, abiotic precipitation is a sort of "default setting" in the carbonate system of modern oceans. Original figure modified from Schlager 2000. Figure slightly modified from Schlager 2003.

sediment composition, depositional architecture, and growth potential.

If one increases the scale of observation from samples and thin sections to mappable formations and beyond, it turns out that all large carbonate sediment accumulations are mixtures of the three main modes of carbonate precipitation (abiotic, biotically-controlled, and biotically-induced; Fig. A.1). These mixtures are not random. They cluster into preferred production systems, or factories, that differ in dominant precipitation mode, mineral composition, depth range of production, as well as growth potential (Schlager 2005).

A carbonate factory is a carbonate precipitation mode that is defined by an ecosystem (i.e., organisms and environment), the sediments which are produced *in situ*, and the early modification of these sediments (Michel et al. 2019; see also James and Jones 2015). Following carbonate mineralization terminology by Dupraz et al. (2009) and Pomar (2020), in this book we use the term "biologically" (instead of "biotically") in reference to controlled mineralization (which forms the skeletons of carbonate-producing organisms) and induced mineralization (microbial active precipitation of $CaCO_3$) processes. It is noteworthy that, there is a fourth mode for $CaCO_3$ precipitation, i.e., influenced mineralization. Biologically-influenced mineralization (passive precipitation of $CaCO_3$) is relevant in the formation of microbial carbonates. Regarding abiotic carbonate precipitation (the least relevant mode in terms of volume of precipitated

Ca-carbonate), this mode is governed by inorganic thermodynamics and reaction kinetics (Schlager, 2003). It is promoted by an increase in temperature, salinity, alkalinity, and/or pH (Fantle, 2020). See Dupraz et al. 2009; Pomar 2020; Görgen et al. 2021; for detailed reviews on carbonate precipitation modes.

Box A.2: Common Terminology Describing Trophic Conditions in Aquatic Environments

(**A**) Terms originally defined for lakes, but widely applied to marine environments (taken from Peters et al. 2005). (**B**) Nutrient gradients in low latitude waters, presented in milligrams of chlorophyll per cubic meter of seawater (taken from Mutti and Hallock 2003).

(A) **Oligotrophic:** describes a lake/sea with few nutrients and abundant dissolved oxygen that supports little primary productivity; **Mesotrophic:** describes a lake/sea at a stage of evolution intermediate between oligotrophic and eutrophic; **Eutrophic** describes a lake/sea containing abundant nutrients that support high primary productivity. Eutrophic lakes commonly have low dissolved oxygen due to high biological oxygen demand.

(B) **Oligotrophic:** chlorophyll concentrations <0.1 mg/m^3 (nutrient deficient); **Mesotrophic:** chlorophyll concentrations between 0.1 and 0.5 mg/m^3 (nutrient intermediate); **Eutrophic:** chlorophyll concentrations between 1–10 mg/m^3 (nutrient abundant); **Hypertrophic:** chlorophyll concentrations between >10 mg/m^3 (nutrient excess). In hypertrophic environments, so much organic carbon is produced in surface waters that oxygen depletion limits the benthos (Hallock 2001). Characteristic sediments in areas of hypertrophication include diatom-rich, organic-rich or phosphatic sediments (Mutti and Hallock 2003).

Schlager's Carbonate Factories

At the scale of the carbonate platform, Schlager (2005) described three carbonate factories based on stratigraphic and diagenetic principles (we comment on each carbonate factory considered in this book by focusing on how we apply different authors' concepts of carbonate factories to the carbonate facies of Venezuela). The Schlager factories are:

The T Factory

T stands for "tropical" and "top-of-the-water-column". Biologically-controlled precipitates dominate. Most abundant among them are light-dependent calcifiers, for instance calcareous algae, scleractinian corals, and LBF, plus certain molluscs. The other characteristic products are abiotic precipitates in the form of marine cements and ooids (although many authors consider ooids to form via biologically-mediated or combined abiotic and biologically-mediated processes). Heterotrophs devoid of photosymbionts are common, but non-diagnostic contributors. The T factory is restricted to the warm, sunlit waters of the ocean that are high in oxygen. In modern oceans, the characteristic settings are the surface waters between 30° N and S of the equator. The northern and southern limit of the T factory closely follows the line where the mean temperature of the coldest month is about 20 °C (Schlager 2005).

Comments on the T Factory

In this book we consider calcareous red algae as common/dominant components of the T factory. This is based on the abundance of these photoautotrophs in Caribbean reef complexes of Venezuela. In these systems, red algae thrive in the euphotic zone in association with scleractinian corals (coralgal association *sensu* Pomar et al. 2012); and with bryopsidalean green algae, benthic foraminifera (larger and smaller), and molluscs (chloralgal lithofacies *sensu* Lees 1975). Coralline (red) algae, commonly associated with heterozoan carbonates (e.g., rhodalgal lithofacies *sensu* Carannante et al. 1988, see Box A.4), and in many cases overlooked in marine tropical settings, are prominent $CaCO_3$-producers and carbonate "engineers" in coral reefs worldwide (see Narayan et al. 2022).

The C Factory

The letter C is derived from "cool-water" and "controlled" precipitation. The products are almost exclusively biologically-controlled precipitates. Heterotrophic organisms dominate. The contribution of photozoan biota in the form of calcareous red algae and LBF is sometimes significant. The C factory extends poleward from the limit of the tropical factory (at about 30°) to polar latitudes. The transition to the T factory is very gradual (Betzler et al. 1997). The C factory also occurs at low latitudes in the thermocline below the warm surface waters and in upwelling areas (Schlager 2005).

Box A.3: Terminology for Referring Bathymetry in Carbonate Platforms and Ramps

(A) **Photic zone bathymetric zonation**

- **Euphotic zone:** this zone encompasses environments where there is sufficient light for high rates of photosynthesis and associated hypercalcification. Its lower limit is defined as the maximum depth of vigorous scleractinian coral growth, which is ~40–50 m in very clear waters, and more commonly about 20–30 m (Pomar 2001; Hallock and Pomar 2008).
- **Mesophotic zone:** it is defined as the zone with sufficient light for coral growth, commonly below wave action (Morsilli et al. 2012). In Pomar (2001), this zone comprises depths between 30 and 80 m where light is sufficient for photosynthesis to support substantial calcification rates, but not true hypercalcification (see also Hallock and Pomar 2008).
- **Oligophotic zone:** this zone is characterized by a decrease in light and temperature. It is defined as the zone with sufficient light for supporting calcification by red algae and very thin flat LBF and corals (Pomar 2001; Morsilli et al. 2012). The lower limit of this zone is typically in the 50–100 + m depth range in clear waters (Pomar 2001), commonly ˂200 m depending on water turbidity (Peters et al. 2005).

The bathymetry of the lower limit of these zones depends on water transparency, and also varies with latitude, being shallower in high latitudes and deeper in low latitudes (see Pomar et al. 2014).

(B) **Lateral classification of carbonate shelves** (Wilson and Jordan 1983) **and carbonate ramps** (Burchette and Wright 1992)

Carbonate shelves (taken from Flügel 2010)

- **Inner shelf**: near-coast tide-dominated zone including peritidal and shallow subtidal environments varying and restricted salinity; sluggish circulation; biota low-diverse.
- **Mid-shelf**: extended shallow subtidal zone between the near-shore area and the shelf break; below fair-weather wave base, but above storm-wave base; water depths between a few tens of meters and 100–200 m.
- **Outer shelf**. Rimmed shelves: a narrow zone near the shelf break, with shoals and reefs. Non-rimmed shelves: a wide zone below normal storm-wave base which may be affected by intruding ocean currents.

Carbonate ramps (taken from Flügel 2010)

- **Inner ramp**: between upper shoreface (beach or lagoonal shoreline) and fair-weather wave base; seafloor more or less constantly affected by wave agitation; includes shoreline deposits, sand-shoals, and back-barrier peritidal sediments.
- **Mid-ramp**: between fair-weather wave base and storm-wave base. The bottom is frequently reworked by storm waves and swells. Sediment composition and textures reflect proximal-distal trends.
- **Outer ramp**: below normal storm-wave base, down to the basin plain. Mud-dominated, but with few storm beds. In deeper zones, restricted bottom conditions may develop.

(C) **Terms used to describe horizontal zonation and deep-water carbonates** (taken from Flügel 2010).

The neritic zone is the water that overlies the continental shelf, today generally with water depth less than 200 m and covering about 8% of the ocean floor.

The oceanic zone is the water column beyond the shelf break, overlying the slope and the deep-sea bottoms, generally with water depths greater than 200 m and down to more than 10,000 m.

Bathyal: sea bottom between about 200 (continental shelf edge, shelf break) and 2000 (4000) m, encompassing the continental slope and continental rise, down to the deep sea.

Abyssal plains: area of deep-sea bottom below 2000 m and between 4000 and 6000 m, including the nearly flat deep-sea plains covered by sediment.

Comments on the C Factory

This factory is not considered in this book, as both ancient and modern carbonate facies of Venezuela do not belong to the marine cool-water realm.

The M Factory

The letter M alludes to mud-mound, micrite, and microbes. The characteristic component is fine-grained carbonate that precipitated *in situ* and was firm or hard upon formation. The precipitation of this fine-grained carbonate was caused by a complex interplay of biotic and abiotic reactions with microbes and decaying organic tissue playing a pivotal role. The term "automicrite" for micrite precipitated *in situ* (Wolf 1965; Reitner et al. 1995) is very useful in instances where the microbial origin is uncertain and the term "microbialite" is not justified.

Box A.4

Climate-dependent (temperature-controlled) skeletal grain associations for benthic calcifiers of warm-water and cool-water shelf carbonates. The list records the name, author and age of the material used, followed by the dominant, plus secondary and minor components, and the observed or inferred setting. H (Heterozoan) and P (Photozoan) at the end of the paragraph indicate the position within the major Grain Associations distinguished by James (1997). Percentage composition corresponds to average values for grains not for bulk samples. Definitions compiled and slightly modified from Carannante et al. (1988), Hayton et al. (1995), Pomar et al. (2004), and Flügel (2010), with the exception of Coralgal association taken from Pomar et al. (2012).

It should be noted that:

- Only associations that are represented in carbonate facies of Venezuela are included.
- The original definitions of Cenozoic- and Modern-biased, heterozoan-dominated skeletal grain associations are ambiguous with respect to benthic foraminifera, as the ecology and environment of these light-dependent/symbiont-bearing or light-independent/symbiont-barren (LBF or SBF, respectively) are not specified. Therefore, we have adapted those definitions according to the assemblages of benthic foraminifera identified by us or by other authors in tropical ancient and modern carbonate facies of Venezuela.

Chlorozoan (Lees and Buller 1972; Modern). Calcareous green algae and scleractinian (zooxanthellate) corals, plus secondary LBF, red algae, molluscs, and non-skeletal grains such as ooids, aggregates and peloids. Tropical. Warm-water. P.

Foramol (Lees and Buller 1972; Modern). LBF, SBF and molluscs, plus secondary echinoids, bryozoans, ostracods, barnacles (balanids), ahermatypic corals and calcareous red algae. Non-tropical. Temperate and cool-water. H.

Chloralgal (Lees 1975; Modern). Calcareous green algae, plus secondary LBF, SBF, calcareous red algae and molluscs. Tropical and warm-temperate zones. Controlled by salinity and water temperature. P.

Molechfor (Carannante et al. 1988; Cenozoic). Molluscs, echinoids, LBF and SBF, plus secondary balanids and serpulids. Non-tropical, warm-temperate environments. H.

Rhodalgal (Carannante et al. 1988; Cenozoic). Calcareous red algae (>80%; encrusting and rhodoliths), plus secondary LBF and bryozoans, and minor SBF, balanids, bivalve molluscs, echinoderms and serpulids. Common in non-tropical warm-temperate settings. High-energetic photic zone. Transitional between chlorozoan-dominated tropical areas and cold-temperate ones. Also in tropical zones where environmental conditions prevent the development of chlorozoan facies. H.

Chloroforam (Beauchamp 1994; Permian). Calcareous green algae, LBF, and SBF, plus secondary non-skeletal grains such as ooids, aggregates and oncoids. Warm-water. Low latitudes. P.

Bryonoderm (Beauchamp 1994; Beauchamp and Desrochers 1997; Permian). Bryozoans and echinoderms (usually crinoids), plus secondary brachiopods, solitary and colonial corals, SBF, LBF, and minor peloids. Regarded as the equivalent of modern bryomol associations. Non-tropical and tropical environments. H.

Bimol (Hayton et al. 1995; Cenozoic). Bivalve molluscs (infaunal and/or epifaunal; >60%, up to 80%), plus secondary bryozoans, LBF, SBF, echinoderms, calcareous red algae and balanids. Non-tropical, but also tropical. H.

Echinofor (Hayton et al. 1995; Cenozoic). Echinoderms (about 40%; mainly echinoids), SBF and LBF (~35%; foraminifera may outspace the echinoderms), plus secondary bryozoans and PF. Non-tropical. H.

Rhodechfor (Hayton et al. 1995; Cenozoic). Calcareous red algae (up to 80%), echinoderms (mostly echinoids), LBF, SBF, bryozoans, plus secondary balanids and bivalves. Non-tropical temperate settings. H.

Coralgal (Pomar et al. 2012; Cenozoic). Scleractinian corals and calcareous red algae, along with other skeletal debris. This term does not correspond to the coralgal defined by Newell et al. (1959) and Purdy (1963), which is equivalent to chlorozoan. Tropical and subtropical. Warm and warm-temperate waters. P.

The typical setting of the M factory in the Phanerozoic is dysphotic (low light) or aphotic (no light), nutrient-rich waters that are low in oxygen (i.e., hypoxic) but not anoxic. In this factory, it is very difficult to distinguish between photically and aphotically formed automicrite unless associated sessile skeletal organisms provide the necessary information. Abiotic marine cement is the second most important component of this factory. Biologically-controlled (skeletal) carbonate may occur but is not characteristic (Schlager 2005).

Comments on the M Factory

We consider "automicrite" (precipitated *in situ*) to be strictly microbially-mediated (induced and/or influenced), and not abiotic. Depending on environmental conditions (e.g., bathymetry, redox and trophic conditions), the contribution of benthic microbial communities to the precipitation of $CaCO_3$ may come from endolithic cyanobacteria, nitrate-reducing bacteria, anoxygenic phototrophic bacteria (e.g., green sulfur and purple sulfur bacteria), or sulfate-reducing bacteria (see Konhauser 2007; Dupraz et al. 2009). We consider dolomite [$CaMg(CO_3)_2$] and some non-carbonate authigenic minerals (pyrite, fluorapatite, glauconite) as common components of this factory. In this book, we do not refer to the M factory, as micrite production is split into two factories described below (the benthic automicrite and the neritic lime-mud factories).

Micritic Carbonate Factories

In their attempt to understand changes in the predominant types of carbonate-producing organisms during the Mesozoic and Cenozoic, Pomar and Hallock (2008) approach integrate estimates of changes in Ca^{2+} concentration in seawater and partial pressure of atmospheric CO_2 (pCO_2), with biological evolution and ecological requirements of characteristic carbonate-producing marine communities. These authors focused mainly on the origin of micritic deposits of the Mesozoic (an era characterized by both high pCO_2 and global temperatures) and their approach is based on the fact that CO_2 is basic to both carbonates and organic matter, and that photosynthesis is a fundamental biological process responsible for both primary production of organic matter and for providing chemical environments that promote calcification.

The Benthic Automicrite Factory

Pomar and Hallock (2008) further consider the Schlanger's mud-mound factory and renamed it the benthic automicrite factory. In this factory, organomineral-boundstone carbonate production (i.e., micrites precipitated on the seafloor by a complex interplay of inorganic and organic chemical reactions between microbes and decaying organic matter) was dominant in marine, low-oxygen (dysoxic) environments of the Triassic and Jurassic. The benthic automicrite factory, although local and sporadic, persisted during Early to mid-Cretaceous times. Neuweiler et al. (1999) concluded that many early Cretaceous mud-mounds were essentially organomineralic in origin, i.e., formed by biologically-induced and/or biologically-influenced mineralization. The substantial depth range of automicrite production indicates that light was not directly required for precipitation of carbonate (Monty 1995; Keim and Schlager 2001). The benthic automicrite factory of Pomar and Hallock (2008) integrates the different loci on which the firm-ground automicrite factories occur, i.e., platform slopes that laterally accrete (slope-forms), as well as mounded structures that vertically accrete above the seafloor (mud-mounds).

Comments on the Benthic Automicrite Factory

In this book we apply the concepts of the benthic automicrite factory primarily for mud-supported fabrics in which the micritic sediments that make up the matrix are dark-colored (qualitative criterion), hence moderately rich to rich in organic carbon. An exception to this criterion is the microbially-mediated precipitation of automicrite as cement (micritic envelopes and micritic bridges) in intertidal beachrocks and some shallow-subtidal grainstones deposited in oligotrophic or mesotrophic environments (i.e., microbial cementation; see Hillgärtner et al. 2001; Diaz and Eberli, 2022). It is widely accepted that precipitation of automicrite is favored in nutrient-rich environments low in oxygen but not anoxic (e.g., Schlager 2005; Flügel 2010). Since nutrient levels exert a primary role controlling precipitation of automicrite, interpretations on nutrient gradients, organic carbon flux and redox conditions commonly involve some geochemical determinations.

The TOC content (wt.%) is among the most useful and instrumentally among the easiest geochemical measures to determine. With a TOC value (when available) one can estimate redox and trophic conditions of a given sample/microfacies. In well-oxygenated environments, sediments rarely preserve more than 0.5% TOC (Demaison and Moore 1981; Arthur and Sageman 1994), and TOC contents above 1% are usually interpreted as signature of dysoxic/anoxic conditions at the seafloor (Schlanger et al. 1987; Lüning et al. 2004; see also Monteiro et al. 2012). Mudstones and wackestones having TOC content of ~0.5% are already visibly dark when these are seen in thin sections. Although there is no TOC cut-off value for assigning micritic sediments an automicritic or an allomicritic origin (precipitated in the benthos or in the water column, respectively), for a given sample/microfacies we consider the contribution of the benthic automicrite factory increases until becomes dominant as TOC content increases. In addition to TOC, the most commonly used geochemical proxies closely associated with nutrient concentrations within the waters are stable isotopes ($\delta^{13}C$ and $\delta^{15}N$) and trace metal concentrations (Ba, Cd) (Mutti and Hallock 2003). Skeletal grains (if present), such as PF and low-oxygen SBF, may also provide the necessary evidence for considering the benthic automicrite factory as dominant in the precipitation of micritic sediments in a given carbonate facies/microfacies.

Regarding some non-skeletal grains, we consider microbially-coated grains (oncoids and cortoids) as components of this factory. The main process involved in the formation of these unattached microbialites is automicrite precipitation (microbially-mediated, constructive) on the seafloor. In addition, micritized grains may also be considered as components of this factory; however, pervasive micritization by endolithic bacteria (living on the seafloor) during early marine diagenesis is a destructive, texture-obliterating process. In addition to low-oxygen (dysoxic to suboxic) conditions prevailing during the precipitation of automicrite (e.g., nitrate-reducing bacteria), we consider automicrite precipitation under anoxic and anoxic-euxinic conditions by anoxygenic phototrophic bacteria, and sulfate-reducing bacteria (see Konhauser 2007; Dupraz et al. 2009; Seifan and Berenjian 2019). Biologically-induced anoxygenic precipitation of micrite was a significant carbonate precipitation mode during deposition of organic-rich rock intervals (the so-called black shales) of the mid-Cretaceous OAEs. Microbially-influenced dolomites formed in hypersaline environments by evaporative pumping and seepage reflux, as well as other non-carbonate authigenic minerals (glauconite formed in low-oxygen conditions, and pyrite and fluorapatite formed under anoxic conditions) are common components of this factory. Furthermore, we include in this factory perennial hypersaline lagoons in Los Roques Archipelago, as well as the variety of microbialite deposits in these shallow-enclosed settings. These lagoons are characterized by highly turbid waters (rich in nutrients and suspended particulate organic matter) and intense microbial activity, both oxygenic and anoxygenic (see Petrash et al. 2012).

The Neritic Lime-Mud Factory

Pomar and Hallock (2008) created the neritic lime-mud factory to account for the production of microcrystalline $CaCO_3$ "rain" in the water column (allomicrite, or "whitings") and within the photic zone. This factory results from biologically-induced and biologically controlled calcification by cyanobacteria and certain eukaryotic phytoplankton (coccolithophores and calcifying dinoflagellates), respectively, via photosynthesis. Such allomicrite carbonate precipitation modes were promoted in the past in oxygenated, low-nutrient marine environments during times of both high pCO_2 and Ca_{2+} concentrations, and are exemplified by the vast amounts of lime mudstones precipitated during the Jurassic and the Early Cretaceous. The coupling between photosynthesis and calcification in shallow, $CaCO_3$ supersaturated warm waters (accelerated by high evaporation rates) resulted in substantial sequestration of CO_2 (the biological pump).

Comments on the Neritic Lime-Mud Factory

In this book we apply the concepts of the neritic lime-mud factory for mud-supported fabrics and muddy packstones in which the micritic sediments, mainly biologically-induced in origin, were precipitated in the photic zone under oxic and oligotrophic to mesotrophic conditions. In modern tropical reef systems such as the Los Roques and Las Aves archipelagos, we consider that a fraction of allomicrite may precipitate abiotically. As previously mentioned, the light or dark color of the micritic sediments is the visual criterion we use for categorizing micrites precipitated by the neritic lime-mud factory or by the benthic automicrite factory, respectively. Therefore, the micritic sediments in samples with low TOC content (light-colored) are most likely precipitated by the former factory. Skeletal grains (if present, such as calcareous algae, LBF, scleractinian corals, and heterozoan biota typical of shallow and well-oxygenated waters) may also provide the necessary evidence for considering the neritic lime-mud factory as the dominant carbonate production system involved in the precipitation of micritic sediments in a given carbonate facies/microfacies.

Regarding ancient allomicritic matrices as significant components of many carbonate microfacies considered in this book, we consider a dominant origin derived from biologically-induced processes, based on: (**a**) the predominance of depositional environments within the photic zone, which favors calcification via photosynthesis; (**b**) the results obtained by Gischler et al. (2013) on a comparative study of modern carbonate mud in several reef and carbonate platforms (Belize, Bahamas, Florida, the Maldives, French Polynesia, and Great Barrier Reef), in which the biologically-derived muds predominate; (**c**) considerations on the biological origin (calcification via photosynthesis) for many Mesozoic micritic deposits by Pomar and Hallock (2008).

The Planktic Factory (P-Factory)

Planktonic carbonate production is taken under consideration by Reijmer (2021), who defined the marine Planktic factory (the P-factory), in which the P relates to planktic carbonate production. This factory comprises carbonate sediment production in the open ocean realm. Its components are PF (many are symbiont-bearing mixotrophs and some are heterotrophs), coccolithophores (photoautotrophs), and pteropods (nektic gastropods), plus radiolarians (silica-secreting planktonic protozoa) (see also the Pelagic marine factory of James and Jones 2015). The term "planktic" is used by Reijmer (2021) instead of "planktonic" following the linguistic argumentation discussed by Emiliani (1952, 1991). The sediment production of the P-factory depends on the oceanic realm, variations in light, temperature, nutrients, water stratification, the chemistry of ambient seawater, and

and sea level, resulting in fluctuating pelagic fall-out. The main sediment production of this factory occurs within the photic zone, and it increases during sea-level highstands.

Comments on the P-Factory

Following definitions of some oceanographic terms (e.g., pelagic, pelagic deposit, hemipelagic deposit and pelagic limestone; see Box A.5), we apply the concepts of the P-factory for mid-Cretaceous black shales and some Middle to Late Eocene microfacies whose most abundant skeletal components are PF (plus SBF and radiolarians). These organic-richmicro facies were deposited under partial to fully developed "pelagic realm-like" conditions that were translated from the open ocean to the shelves during major transgressive episodes. Eutrophic to hypertrophic and low-oxygen to anoxic conditions prevailing in bottom waters during such episodes prevented, in most cases, the flourishing of benthic biota (excluding low-oxygen SBF, and certain molluscs such as inoceramids). It is noteworthy that, in this book we apply the term "planktonic" (and not "planktic") for foraminifera living in the water column, as it is more commonly used in the specialized literature focused on these marine free-living protozoa.

Marine Carbonate Factories: A New Approach

Following the carbonate factory approach of Schlager (2000, 2003, 2005) and the carbonate mineralization terminology of that author and of Dupraz et al. (2009), a new carbonate factory scheme was defined by Michel et al. (2019). This conceptual carbonate scheme relates shallow-water carbonate deposits to oceanographic parameters at a global scale. The model was tested using remote-sensing and *in situ* oceanographic data of modern times, while outputs of paleoceanographic models are utilized for the Lower Aptian (Cretaceous) modeling. These authors split the Photozoan carbonates *sensu* James (1997) into two factories, the Photozoan T-factory and the Biochemical T-factory; and split the Heterozoan carbonates *sensu* James (1997) into two factories, the Photo C-factory and the Heterozoan factory. This conceptual scheme appears to be able to provide a simple, universal model of paleoclimatic zones of shallow-water marine carbonates. Furthermore, Laugié et al. (2019) defined the modern global distribution of the four marine carbonate factories of Michel et al. (2019) based on key oceanographic parameters: SST, sea-surface salinity, marine primary productivity, and water depth.

The Photozoan T-Factory

This factory corresponds to reef platforms, and includes the T factory of Schlager (2005). Depositional settings of the Photozoan T-factory consist of tropical carbonate shelves or detached rimmed platforms ("shelf margin complexes" of Wilson 1975; Tucker and Wright 1990; James 1997). This factory is characterized by low-latitude (warm), oligotrophic waters with high carbonate supersaturation, optimum temperatures and salinities ranging between 25–29 °C and 33–37‰, respectively. Zooxanthellate corals, calcareous algae, symbiont-bearing benthic foraminifera, stromatoporoids, and some sponges, are the main components (plus secondary heterotrophic calcifiers, such as molluscs and echinoids). Light is the main source of energy, while the coupling of photosynthesis and calcification drives high carbonate production rates. Cretaceous rudist-rich platforms are included in this factory as these molluscs were restricted to shallow-tropical, oligotrophic waters. Framestones, bafflestones, and mudstones are characteristic depositional textures. The Caribbean, Great Barrier Reef, and Indo-Pacific islands are the most iconic modern examples (Laugié et al. 2019; Michel *et al.* 2019).

Comments on the Photozoan T-Factory

In this book, and as previously mentioned in comments on the T factory of Schlager (2005), we include calcareous red algae as common to dominant components of the Photozoan T-factory. Furthermore, we consider early diagenetic marine cements, abiotic (e.g., acicular aragonite) and/or microbially-mediated (e.g., micritic envelopes and micritic bridges), components of this factory, when present.

Box A.5: Definition of Some Oceanographic Terms

Taken from Bates and Jackson (1980)

Pelagic: pertaining to the water of the ocean as an environment; said of marine organisms whose environment is the open ocean rather than the bottom or shore areas.

Pelagic limestone: fine-textured limestone formed chiefly by the accumulation of calcareous tests of floating organisms.

Pelagic deposit: marine sediment in which the fraction derived from the continents indicates deposition from a dilute mineral suspension distributed throughout deep-ocean water (Arrhenius 1963).

Hemipelagic deposit: deep-sea sediment in which more than 25% of the fraction coarser than 5 μm is terrigenous, volcanogenic, and/or neritic in origin. Such deposits usually accumulate near the continental margin and its adjacent abyssal plain.

Planktonic: said of that type of pelagic organism which floats.

Nektonic: or nektic, said of that type of pelagic organism that actively swims.

The Biochemical T-Factory

This factory corresponds to flat cratonic areas in clear epeiric seas, areas of moderate subsidence of Wilson (1975), and includes the neritic lime-mud factory of Pomar and Hallock (2008). The Biochemical T-factory is typically a tropical carbonate platform/ramp characterized by a low-angle of repose, comprising extensive evaporitic (sabkha flats), peritidal, and shallow-water microbial deposits. Very high temperatures, salinities, carbonate supersaturation, and evaporation rates, as well as moderate-to-low hydrodynamics, and restricted oceanic circulation, favor the development of this factory. Non-skeletal grains, stromatolites, algal mats, muds, whitings, and bioclasts are typical components. Bindstones, mudstones, and dolomites are characteristic depositional textures. The Persian/Arabian Gulf is the best developed modern example of this factory (Laugié et al. 2019; Michel *et al.* 2019; see also Riegl et al. 2010).

Comments on the Biochemical T-Factory

Since it includes the neritic lime-mud factory, we consider the Biochemical T-factory a low-nutrient carbonate production system. It is noteworthy that Michel et al. (2019) do not consider the benthic automicrite factory (a high-nutrient and low-oxygen carbonate production system) since it is not well constrained in the modern world. Therefore, to avoid ambiguities in the use of different micritic carbonate factories, in this book we apply the concepts of the Biochemical T-factory for organic-lean (light-colored), mud- and/or grain supported fabrics containing allomicrite (lime-mud precipitated in the water column) and/or non-skeletal grains (these fabrics also commonly contain carbonate benthic biota typical of well-oxygenated waters). This factory is well represented by some microfacies in the Palmarito Formation (Middle Permian), the Tibú and Piché members of the Apón Formation (Barremian and Aptian in age, respectively), and the Lisure Formation (Albian). In Los Roques Archipelago, the Biochemical T-factory is represented in many sites of the vast inner lagoon environment where whitings ($CaCO_3$ rain) precipitate, and in ephemeral hypersaline lagoons where waters are clear (not turbid). The uppermost sediments in these lagoons (which become subaerially exposed during the dry season) and in the surrounding sabkhas are included in this factory. These *in situ* sediments consist of evaporitic crusts composed of both abiotic and biologically-induced aragonite and HMC, as well as gypsum and halite.

The Photo C-Factory

This factory corresponds to the modern photozoan-heterozoan transition of Halfar et al. (2004), and the subtropical Early Permian carbonates of Bensing et al. (2008). The Photo C-factory typically forms distally steepened ramps (Pomar 2001) and comprises proximal seagrass-derived bioclastic sediments (Perry and Beavington-Penney 2005; James et al. 2009) and distal red algal accumulations (Pérès and Picard 1964; Bosence and Pedley 1982; Bosence 1983). It typically occurs in the warm-temperate or subtropical province, which shows a marked seasonality. Symbiont-bearing LBF (mixotrophs) and calcareous red algae (photoautotrophs) are the main components of this factory (plus secondary heterotrophic calcifiers, such as echinoids, bryozoans, molluscs, and balanid barnacles). Although light is the steering environmental factor controlling the factory distribution, the Photo C-factory can occur in oligo-mesotrophic conditions (see Michel et al. 2018), as its main components show greater tolerance to turbidity and marine productivity than components of the Photozoan T-factory. Grainstones, rudstones, packstones, and wackestones are characteristic depositional textures. The Mediterranean Sea is the main representative marine realm of the Photo C-factory (Pérès and Picard 1964; Henrich *et al.* 1995; Laugié et al. 2019), which is expected to occur mostly in the Cenozoic Period (Michel et al. 2019).

Comments on the Photo C-Factory

This factory, regarded as typical of warm-temperate (subtropical) carbonates, is common-dominant in Cenozoic tropical carbonates of Venezuela. We apply the concepts of the Photo C-factory to units dominated by the photozoan association between coralline algae and LBF. These units include: (**a**) some Middle Eocene facies of the Santa Rita Formation in the FB deposited in an outer shelf setting at mesophotic depths between ~60 and 80 m (Pitelli and Molina 1989); (**b**) Late Oligocene-Early Miocene carbonates of the Perla Limestone in the EGVSB; deposition in a gentle distally steepened middle-outer ramp occurred in the oligophotic (poor-light) zone (Pomar et al. 2015); (**c**) some Early Miocene facies of the San Luis and Churuguara formations in the FB; carbonate production mainly occurred in the meso-oligophotic zone (40–100 m depth range) of distally-steepened ramps (middle ramp domain) (Albert-Villanueva 2016). Water temperature, ligth availability, and energy level, controlled by bathymetry, were primary local factors controlling deposition of previously mentioned facies in which heterotrophic calcifiers (echinoids, bryozoans, balanids, mollusks, SBF) are secondary components. Although it is well known that coralline algae may thrive in mesotrophic conditions (Halfar et al. 2004; Halfar and Mutti 2005), these photoautotrophs, together with LBF (which are are more common of oligotrophic conditions), are light-dependent calcifiers. Therefore, we consider tropical Cenozoic carbonate lithofacies of Venezuela dominated by the two main components of the Photo C-factory (calcareous red algae and LBF) as Photozoan carbonates, and not Heterozoan carbonates *sensu* James (1997). Prior to the carbonate factories new approach by Michel et al. (2019), many Cenozoic red algae-dominated facies (rhodalgal lithofacies *sensu* Carannante et al. 1988) were described as Heterozoan carbonates *sensu* James (1997), e.g., Brandano et al. (2009), Albert-Villanueva et al. (2017), and Silva-Tamayo et al. (2017).

The Heterozoan Factory

This factory corresponds to the Heterozoan carbonates of James (1997), excluding the Photo C-factory, and the C factory of Schlager (2005). The Heterozoan factory forms ramps, the deposition of which is controlled by the coupling of initial topography and hydrodynamics (Michel et al. 2018). This factory is characterized by heterotrophic calcifiers (molluscs, echinoderms, bryozoans, ostracods, serpulids, balanids, symbiont-barren benthic foraminifera, and some sponges). Components of the Photozoan T-, Photo C-, and Biochemical T-factories can be present. The Heterozoan factory occurs at any latitude, and in mixed carbonate-siliciclastic environments. The marine productivity as food is the source of energy driving the development of this factory. Consequently, it occurs in eutrophic (nutrient-rich) marine environments characterized by high primary productivity, such as upwelling areas and coastal/neritic zones influenced by continental runoff. Grainstones, packstones and wackestones are characteristic depositional textures (Michel et al. 2019). Modern examples where mesotrophic to eutrophic conditions allow the formation of heterozoan carbonates are globally distributed (Laugié et al. 2019). It highlights the Gulf of California (Halfar et al. 2006), the Yucatán Peninsula (Lowery and Rankey 2017), the Pacific coast of Panamá (Reijmer et al. 2012), offshore Mauritania (Klicpera et al. 2015), South Australia (James and Bone 2011), and New Zealand (Nelson 1988).

Comments on the Heterozoan Factory

We apply the concepts of the Heterozoan factory for non-reef heterozoan carbonate facies deposited at tropical paleolatitudes. Although ancient heterozoan-dominated deposits, such as foramol lithofacies *sensu* Lees and Buller (1972), molechfor lithofacies *sensu* Carannante et al. (1988), and bimol lithofacies *sensu* Hayton et al. (1995), are more commonly associated to non-tropical settings, these skeletal grain associations occur at any latitudes and at all depths. During the Cenozoic, including modern settings, meso-oligophotic heterozoan carbonates were deposited in many humid, low-latitude areas characterized by high nutrient input (runoff- or upwelling-sourced; Wilson and Vecsei 2005). Deposition of ancient heterozoan carbonates in Venezuela encompassed the whole bathymetric photic zone spectrum, from euphotic zone to the oligophotic zone. Some examples are: (**a**) inner-mid ramp and outer ramp facies of the Palmarito Formation (ostracod-, bryozoan-, and crinoid-dominated) of Middle Permian age in the Venezuelan Andes (Laya and Tucker 2012); (**b**) inner-mid ramp facies of the Tibú Member (mollusc- and echinoid-dominated) of Barremian age in the MB; (**c**) inner-middle platform facies of the Maraca Formation (oyster-dominated) of latest Albian age in the MB; (**d**) upper ramp facies of the Churuguara Formation (echinoid-dominated) of Early Miocene age in the FB (Albert-Villanueva 2016).

Appendix B: Calcite and Aragonite Seas

This Appendix focuses on understanding the factors that have controlled the mineralogy (aragonite, LMC, or HMC) of carbonate-producing organisms, and their abundance throughout the Phanerozoic. This complex topic is commonly discussed in the specialized literature under the concept of "Calcite and Aragonite Seas".

Inorganic precipitation experiments have shown that the Mg^{2+}/Ca^{2+} ratio of seawater affects the precipitation of $CaCO_3$ polymorphs. A high Mg^{2+}/Ca^{2+} ratio favours the precipitation of aragonite and HMC, while a low Mg^{2+}/Ca^{2+} favours the precipitation of LMC (Morse et al. 1997). Geochemical modelling and proxy data have shown that, during the Phanerozoic, the Mg^{2+}/Ca^{2+} molar ratio of seawater has oscillated on times scales of c. 100 Myr between values of c. 1 and 5 (Stanley and Hardie 1998; Lowenstein et al. 2001). This has resulted in the concept of calcite and aragonite seas, with the skeletal mineralogy of major carbonate producers generally matching these predicted long-term trends (Sandberg 1983; Stanley and Hardie 1998; see also Steuber et al. 2023).

Initially, Spencer and Hardie (1990) showed that small changes in the mid-ocean ridge flux can cause changes in the Mg^{2+}/Ca^{2+}, Na^+/K^+ and Cl^-/SO_4^{2-} ratios of seawater that would result in the observed synchronized shifts in limestone and marine evaporite mineralogy. Hardie (1996) later proposed that such minor changes in mid-ocean ridge flux could be caused by changes in the rate of ocean crust production. Mid-ocean ridges are giant seawater pumping systems. All of the water in the ocean is circulated through the ridges every several million years, but the rate of circulation depends on the rate of seafloor spreading. The ridges act as a huge rock-fluid ion exchange system where basalt is converted to greenstone as it comes in contact hot water, thereby releasing Ca^{2+} and K^+ to the seawater and removing Mg^{2+} and SO_4^{2-} from it. Therefore, mid-ocean ridges act as massive ion exchange systems where high rates of ocean crust production result in relatively high concentrations of Ca^{2+} and low concentrations of Mg^{2+} in seawater (low Mg^{2+}/Ca^{2+} ratios) (see Stanley and Hardie 1998; Ries 2010; James and Jones 2015).

The "calcite seas" intervals correspond to "Calcite I", encompassing nearly all of the Cambrian to the mid-Early Carboniferous; and "Calcite II", from the Middle Jurassic through Eocene. These were time intervals characterized by seawater Mg^{2+}/Ca^{2+} ratio <2, high pCO_2, and favored precipitation of LMC. The "aragonite seas" intervals correspond to "Aragonite I", from the latest Precambrian (late Ediacaran) to the earliest Cambrian (Terreneuvian); "Aragonite II", from the mid-Early Carboniferous to the Middle Jurassic; and "Aragonite III", from the Oligocene to the present. These were time intervals characterized by Mg^{2+}/Ca^{2+} ratio >2, and favored precipitation of aragonite and HMC (see Stanley and Hardie 1998, 1999; Stanley 2006, 2008; Ries 2010; James and Jones 2015). Furthermore, calcite and aragonite seas intervals roughly match intervals defined as Greenhouse (high pCO_2) and Icehouse (low pCO_2) climates, respectively (see Martin 1995; Zhuravlev and Wood 2009; Scotese et al. 2021).

Skeletal Mineralogy of Carbonate-Producing Organisms Throughout the Phanerozoic

Initially, Sandberg (1983) noted Phanerozoic oscillations in primary mineralogies of abiotic carbonates (mainly ooids and early marine cements). This author postulated that tectonically-induced changes in carbon dioxide partial pressure (pCO_2) were responsible for mineralogic changes. He was also the first to recognize the periods when predominantly abiotic calcite or aragonite formed, now known as calcite and aragonite seas (see also Sandberg 1985, 1991). van de Poel and Schlager (1994) and Stanley and Hardie (1998) observed that the oscillations reported by Sandberg (1983, 1985, 1991) were paralleled by oscillations in dominant calcite- and aragonite-secreting, reef-building organisms. Stanley and Hardie (1998) proposed that secular (long-term) variations in seawater Mg^{2+}/Ca^{2+} ratio, or the alternations of calcite and aragonite seas, driven by variations in the global rate oceanic crust production, are believed to have influenced the skeletal mineralogy (i.e., $CaCO_3$ polymorphs) and growth rates of marine calcifiers throughout the Phanerozoic (see also Stanley 2006).

Since the initial recognition of calcite and aragonite seas, there has been considerable speculation about their effects on the skeletal mineralogy of carbonate-producing organisms. Two broad areas of inquiry have been pursued: (**a**) the influence of aragonite and calcite seas on the fates of aragonite- vs. calcite-mineralizing taxa, i.e., whether calcite-shelled organisms suffer disproportionately or not during aragonite seas and vice versa (e.g., Stanley and Hardie 1998, 1999; Steuber 2002; Aguirre and Riding 2005; Hautmann 2006; Kiessling et al. 2008; Key et al. 2022; Cole et al. 2023; Steuber et al. 2023; see also Porter 2010); and (**b**) the role of aragonite and calcite seas in determining the carbonate skeletal mineralogy of clades, i.e., understanding why taxa have the particular skeletal mineralogy they do (e.g., Martin 1995; Erez 2003; Porter 2007, 2010; Zhuravlev and Wood 2008; Porter 2010).

Porter (2010) tested the hypothesis that carbonate skeletal mineralogy is determined by seawater chemistry at the time skeletons first evolve in a clade. According to this study, of the 37 taxa whose mineralogy is known with some confidence (at the time of this study), 25 acquired mineralogies that matched seawater chemistry of the time (although recent studies have changed these results for some taxa), whereas only two taxa acquired non-matching mineralogies. Ten other taxa appeared during times when seawater chemistry is not well constrained (i.e., periods of uncertain calcite-aragonite sea status); these time intervals are: nearly all the Cambrian (encompassing the latest interval of Aragonite I and the earliest interval of Calcite I), the earliest Jurassic (the latest interval of Aragonite II), and the early Paleogene (the latest interval of Calcite II) (see also Sandberg 1983, 1985; Balthasar and Cusack 2015). Few taxa switch mineralogies (from calcite to aragonite or vice versa) despite subsequent changes in seawater chemistry. In addition, Porter (2010) does not distinguish between those taxa that form skeletons from HMC and those that use LMC. The author does not make that distinction for two reasons. First, because Mg content in calcite can vary continuously (e.g., Mackenzie 2003), the distinction between LMC and HMC is arbitrary (in contrast, the distinction between aragonite and calcite is not arbitrary, as they are distinct $CaCO_3$ polymorphs). In fact, the definition of HMC varies in the literature, with some workers making a further distinction between IMC and HMC (see Smith et al. 2006; James and Jones 2015). Second, experimental studies (e.g. Stanley et al. 2002, 2005; Ries 2004, 2006a) have shown that the Mg content of calcitic skeletons can vary with the Mg^{2+}/Ca^{2+} ratio of seawater such that normally HMC taxa (e.g., red algae) can become LMC when the Mg^{2+}/Ca^{2+} ratio is low (see Ries 2010). Evidence from fossil echinoderms (widely considered as original HMC in composition) and rugose corals (extinct Paleozoic corals with original HMC or LMC compositions varying according to seawater chemistry of the time) suggests this happened in the past as well (e.g., Webb and Sorauf 2002; Dickson 2004). In addition, there is no evidence for HMC forms arising during aragonite seas, despite the fact that HMC mineralogy is thought to be compatible with high Mg^{2+}/Ca^{2+} ratios (Stanley 2008; Porter 2010).

Carbonate Skeletons: The *De Novo* Acquisition Hypothesis

In this section we test the "*de novo* acquisition of carbonate skeletons" hypothesis of Porter (2010) using the carbonate-producing taxa (biologically-controlled mineralization) considered in this book. Determining the primary mineralogy of ancient carbonate fossils is not straightforward because aragonite recrystallizes to calcite at Earth's surface pressures. As a result, formerly aragonitic skeletons are often preserved as calcite (Porter 2007). The original skeletal mineralogy of taxa can be inferred from preservational criteria, preserved original biominerals, and chemical data (Kiessling et al. 2008). Here, the primary mineralogy of the taxa considered was drawn from the literature.

It is noteworthy that: (**a**) this hypothesis was tested in 2010; since that year, the taxonomy of some of the carbonate taxa considered in this book has been under revision (e.g., calcareous red algae, barnacles, and serpulids), which has led to the creation of new taxonomic units; (**b**) new findings have changed the age of the first appearance of certain taxa, and thus the age at which those taxa first acquired skeletal mineralogy; (**c**) Porter (2010) considers molluscs as a whole clade, while here we treat the first acquisition of carbonate skeletons by different groups of molluscs as independent appearances; (**d**) we consider some taxa not considered by Porter (2010), i.e., calcareous algae and foraminifera; (**e**) our goal is to test this hypothesis using the updated literature, not to contradict or rebut previous studies; (**f**) for each taxon in this review, we do not discuss the many subjects related to the process of controlled mineralization of $CaCO_3$, such as the energetic costs of calcification, extracellular, intercellular, and intracellular mineralization, or the vital effect (due to physiological processes during biomineralization, some carbonate-producing biota secrete their skeletons in isotopic equilibrium with seawater, and some other do not; this varies among different clades, as well as within a single clade). See Knoll (2003), Weiner and Dove (2003), Pérez-Huerta et al. (2018), and Gilbert et al. (2022), for detailed summaries on the evolution of biologically-controlled mineralization in carbonate-secreting organisms.

Calcareous Algae

Living members of this group of benthic photoautotrophic algae are classified according to their pigments (i.e., green and red algae; aragonitic and calcitic, respectively), among other features. Ancient forms are classified according to their morphology.

Green Algae

Aragonite is the exclusive mineral of marine green algae (Vachard 2021), which are divided in the orders Dasycladales and Bryopsidales (Photos B.1 and B.2, respectively).

Dasycladales

Extant members of this order are aragonitic, and this appears to have been the principal mineralogy of dasycladalean alagae throughout the history of the group (Aguirre and Riding 2005). A discreet number of calcitic species existed contemporaneously in the Early Carboniferous (late in Calcite I), the Late Jurassic, the Early Cretaceous, and the Paleogene (Calcite II) (see Granier 2012). Aragonitic dasycladalean algae appeared during the Middle Ordovician

Photo B.1 Dasycladalean alga (middle Albian). Lisure Formation, Maracaibo Basin
Image width = 2.4 mm. PPL

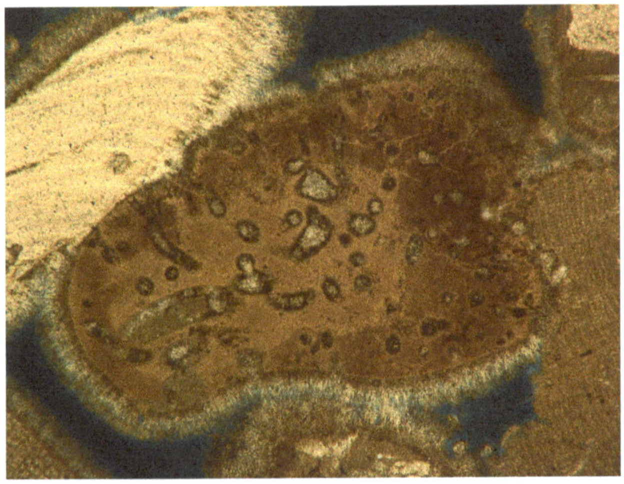

Photo B.2 Bryopsidalean alga (Late Pleistocene). Aves Island
Image width = 1.1 mm. PPL

(Calcite I) (Bucur 1999; Vachard 2021), not matching the seawater chemistry of the time. Recent studies might suggest that the earliest dasycladalean algae are Early Cambrian in age (~520 Ma; Jiang and Zhang 2021); right at the beginning of Calcite I (according to Stanley 2006), during a time when seawater chemistry is not well constrained (according to Porter 2010).

Bryopsidales

Aragonitic bryopsidalean algae, which are dominant in present-day, tropical reef environments, appeared in the Ordovician (Calcite I) (Bucur 1999; Vachard 2021), not matching the seawater chemistry of the time. To our knowledge, no ancient fossil belonging to this order of original calcitic composition has been reported. Experiments using artificial seawater found that, depending on Mg^{2+}/Ca^{2+} ratios, representatives of the genera *Halimeda*, *Penicillus,* and *Udotea* could produce a substantial proportion of skeletal LMC. When lowering the molar Mg^{2+}/Ca^{2+} ratio (from the c. 5.2 in modern "aragonite seawater"), total mineralization decreases, but percentages of LMC produced increases (Ries 2005, 2006b, 2010; Stanley et al. 2010; see also Granier 2012). These results suggest that calcifying bryopsidalean algae inhabiting calcite seas of the geologic past may have secreted a portion or all of their $CaCO_3$ as the calcite polymorph (Ries 2010).

Photo B.3 Coralline alga (Late Pleistocene). Aves Island
Image width = 1.1. CPL

Red Algae

The morphology and taxonomy of calcareous red algae have been under revision for the past two decades, leading to new taxonomic assignments, including the introduction of new families. We consider three groups:

Coralline Algae

The orders Corallinales, Sporolithales, and Hapalidiales are collectively known as coralline algae, or corallines. Modern and ancient members of this group are composed of HMC (Photos B.3 and B.4, respectively). Members of these three orders are widely accepted to have first appeared during the Early Cretaceous (Calcite II) (see Peña et al. 2020). This age is in agreement with Stanley (2008) and Porter (2010), who state that no HMC clade has arisen during aragonite seas. Furthermore, according to Ries (2010), encrusting coralline algae probably produced LMC during the mid-Cretaceous (Calcite II), when seawater Mg^{2+}/Ca^{2+} ratios were near unity.

Photo B.4 Coralline alga (Middle Eocene). Santa Rita Formation, Falcón Basin
Image width = 1.3 mm. PPL

Elianellaceae

This family was introduced by Granier and Dias-Brito (2016a) to embrace *incertae sedis* red algae formerly belonging to the Family Solenoporaceae (a heterogeneous group that includes algae, bryozoans, chaetetid sponges, benthic foraminifera, hydrozoans, and cyanobacteria-like fossils; see Aguirre et al. 2007), such as *Elianella elegans* and *Marinella lugeoni*. The earliest algae so far described as belonging to Elianellaceae is, to our knowledge, the genus *Bija vologdin*, which is Cambrian in age (Vachard et al. 2016); early in Calcite I, during a time when seawater chemistry is not well constrained. This genus is often associated with Cambrian cyanobacteria (see Liu et al. 2016). From their studies of Late Jurassic and mid-Cretaceous specimens, Granier and Dias-Brito (2016a, b) consider that the original mineralogy of members of the Family Elianellaceae is probably HMC (Photo B.5); whereas red algae of the Solenoporaceae are commonly preserved as LMC (Vachard 2021). We find this hypothesis difficult to evaluate for this family because of the cyanobacterial affinity of both its most ancient member (*Bija vologdin*), and one of its early Paleozoic representatives (the genus *Anchisolenopora*, which ranges from the Ordovician to the Late Mississippian; Vachard et al. 2016).

Photo B.5 Red alga of the Family Elianellaceae (middle Albian). Lisure Formation, Maracaibo Basin
Image width = 4.2 mm. PPL

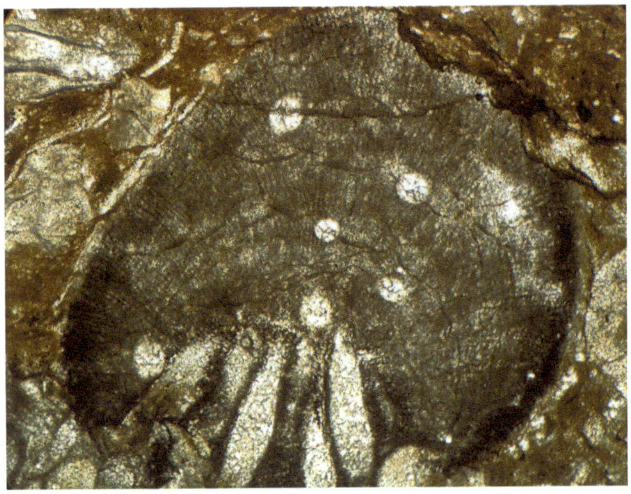

Photo B.6 Red alga of the Order Archaeolithophyllales (Middle Permian). Palmarito Formation, Venezuelan Andes
Image width = 4.2 mm. PPL

Archaeolithophyllales
Members of this order of "ancestral" red algae are known from the Early to Late Carboniferous (the latest interval of Calcite I, or the transition from Calcite I to Aragonite II) and the Permian (Aragonite II) (Photo B.6). The original mineralogical composition of the main representative of this order, the genus *Archaeolithophyllum*, is considered to be HMC (Corrochano et al. 2013), although other members of this group, such as the genus *Neoprincipia* (see Cózar and Vachard 2003), contain aragonite (Vachard 2021). This hypothesis is difficult to evaluate for this small clade for two reasons. First, although the original mineralogy of *Archaeolithophyllum* is considered to be HMC, aragonite also occurs in other members of this order. Second, the earliest occurrence of *Archaeolithophyllum* is recorded (to our knowledge) in the late Viséan (Middle to Late Mississippian; Cózar et al. 2005); in the latest interval of Calcite I, during in the transition from Calcite I to Aragonite II.

According to Stanley et al. (2002), taxa secreting HMC in today's aragonite sea (Aragonite III) will not necessarily have done so in the past (see also Kiessling et al. 2008).

Foraminifera

The vast majority of modern and ancient foraminifera are/were have been calcite producers. Overall, forms composed of LMC dominate those composed of HMC (see Erez 2003; de Nooijer et al. 2009).

Benthic Foraminifera

This is a very varied group consisting of numerous orders. Some of them are discussed in Chap. 6. Benthic foraminifera are divided into two informal groups: LBF (Photo B.7) and SBF (Photos B.8 and B.9) (details on these two groups are addressed in Sects. 6.1, 6.3, and 6.4). The earliest known benthic foraminifera appeared in the Cambrian (Calcite I) and had an organic wall or produced a test by agglutinating

Photo B.7 Larger benthic foraminifer (Early Miocene). Churuguara Formation, Falcón Basin
Image width = 2.4 mm. PPL

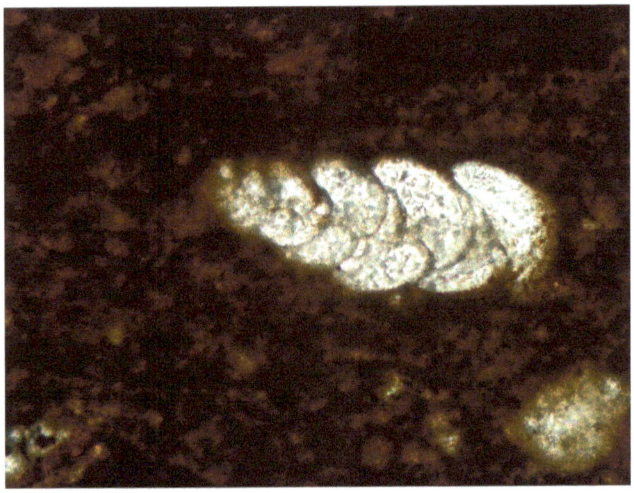

Photo B.8 Smaller benthic foraminifer (late Aptian). Piché Member, Apón Formation, Maracaibo Basin
Image width = 0.5 mm. PPL

Photo B.9 Dry mount of tests of Holocene smaller benthic (B) and planktonic (P) foraminifera. Venezuelan Atlantic Continental Slope
Image width = 8.5 mm

particles within an organic or mineralized (calcareous) matrix. Forms with calcareous tests composed of LMC appeared in the Early Carboniferous (e.g., orders Fusulinida and Palaeotextulariida; see BouDagher-Fadel 2015); late in Calcite I. Although LMC dominates among species of Mesozoic and Cenozoic benthic foraminifera (e.g., most hyaline species), those composed of HMC are also common (e.g., porcelaneous forms of the Order Miliolida and some hyaline species; see de Nooijer et al. 2009). In addition, aragonite occurred in tests of LBF of the Order Involutinida ranging in age from the Early Triassic to the beginning of the Late Cretaceous (see BouDagher-Fadel 2015; Steuber et al. 2023), as well as in SBF of the Order Robertinida ranging in age from the Early Triassic to the Early Cretaceous (see Rigaud et al. 2021).

Planktonic Foraminifera

The earliest planktonic forms (Superfamily Favuselloidea) are known from the Late Triassic (Aragonite II) and were composed of aragonite, matching the seawater chemistry of the time. After the Triassic-Jurassic extinction event, PF recovered and switched to calcite sometime in the Early to Middle Jurassic (see BouDagher-Fadel 2018); during the transition from Aragonite II to Calcite II, or during a time when seawater chemistry is not well constrained. LMC is the dominant $CaCO_3$ polymorph in PF (Photos B.9 and B.10) (e.g., Bender et al. 1975; Scholle and Ulmer-Scholle 2003). Nevertheless, some PF of Early Cretaceous age (Family Favusellidae) were aragonitic (Rigaud et al. 2021).

Corals

We consider scleractinian corals (Subclass Hexacorallia, Order Scleractinia; Pratlong et al. 2017) that harbor endosymbiotic photosynthesizing dinoflagellates (zooxanthellae).

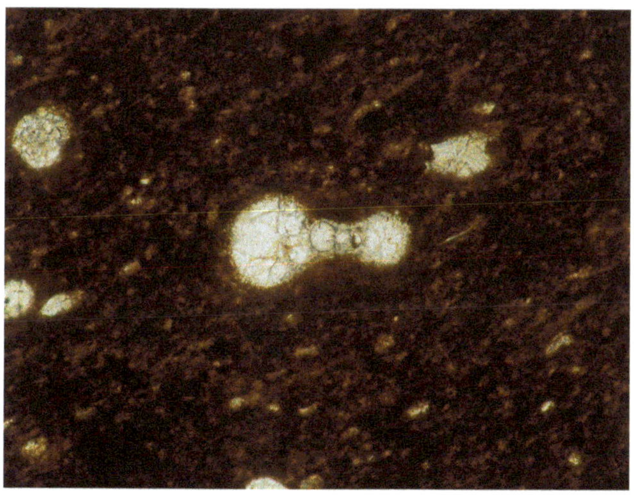

Photo B.10 Planktonic foraminifer (early Aptian). Machiques Member, Apón Formation, Maracaibo Basin
Image width = 0.6 mm. PPL
Photomicrograph courtesy of Gilberto Soto

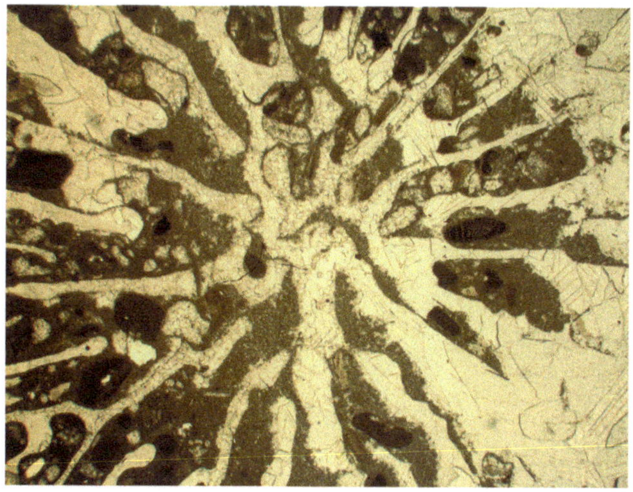

Photo B.11 Corallite of a scleractinian coral (Early Miocene). San Luis Formation, Falcón Basin
Image width = 5 mm. PPL
Photomicrograph courtesy of Eduard Albert-Villanueva

Photo B.12 Modern scleractinian coral (*Diploria labyrinthiformis*). Los Roques Archipelago
Image width = 15 cm

Scleractinian Corals

Zooxanthellate corals appeared during the Middle Triassic (Aragonite II), matching the seawater chemistry of the time (see Stanley 2003; Taylor 2008). Ancient and modern scleractinian corals (Photos B.11 and B.12, respectively) have always been aragonitic, although there is evidence of a calcitic genus (*Coelosmilia* sp.) of Late Cretaceous (~70 Ma) age (Stolarski et al. 2007).

Molluscs

Molluscs compose a very diverse phylum, not only in morphological and taxonomic terms, but also in skeletal mineralogy. We consider the following groups:

Aragonitic Bivalves

Many extinct and extant members of the Class Bivalvia were aragonitic (Photo B.13). The first bivalve mollusc

Photo B.13 Mold fragments of originally aragonitic bivalve molluscs (Early Paleocene). Guasare Formation, Maracaibo Basin
Image width = 4.2 mm. PPL

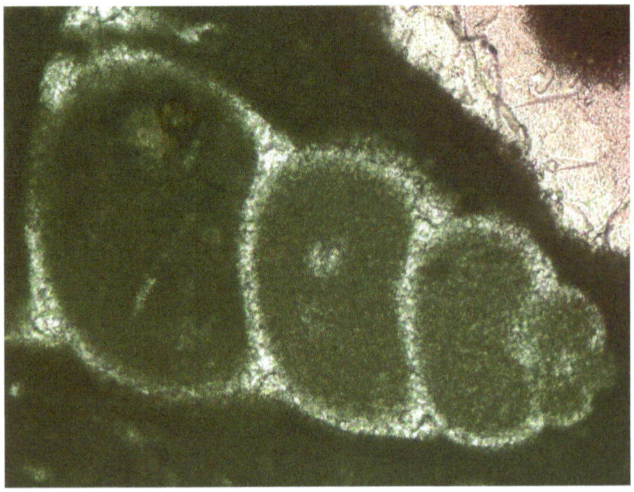

Photo B.14 Mold of a gastropod (middle Albian). Lisure Formation, Maracaibo Basin
Image width = 1.3 mm. PPL

appeared on the eve of carbonate controlled mineralization, sometime between the Precambrian-Cambrian boundary and the earliest Cambrian (Parkhaev 2008, 2017); late in Aragonite I, during a time when seawater chemistry is not well constrained. According to Taylor (2008) and Porter (2010), it is widely held that, ancestrally, molluscan shells (bivalve and univalve) were aragonitic. According to Carter et al. (1998), calcite in bivalve mollusc shells first appeared in the Early Ordovician (families Pterineidae and Ambonychiidae; see also Taylor 2008). We refer to aragonitic molluscs rather than bivalves (or pelecypods) to avoid ambiguities regarding mineralogically "evolved" bivalve molluscs (e.g., oysters, rudist, inoceramids).It is noteworthy that some aragonitic bivalves evolved to have bimineralic shells (including rudists).

Gastropods

Most extinct and extant univalve gastropod shells were predominantly aragonitic (Photo B.14), but like aragonitic bivalves, they evolved bimineralic shells on numerous occasions (Taylor 2008). Again, aragonitic bivales, the first gastropods appeared sometime between the Precambrian-Cambrian boundary and the earliest Cambrian (Parkhaev 2008, 2017), and were aragonitic (see Taylor 2008; Porter 2010). Calcite in univalve molluscs not belonging to the Class Gastropoda first appeared in the Middle Cambrian (monoplacophorans reported by Runnegar 1985; and helcionelloids reported by Vendrasco et al. 2010a, b). The recent finding of shells of helcionellids (a group of univalve molluscs of unclear systematic position) of Terreneuvian age that are entirely composed of LMC proves the existence of aragonite and calcite mineralogy in different earliest Cambrian molluscan taxa (Li et al. 2023).

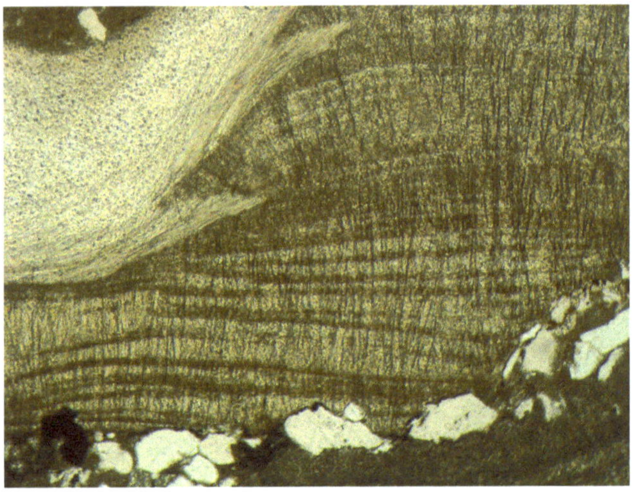

Photo B.15 Fragment of an oyster shell (middle Albian). Lisure Formation, Maracaibo Basin
Image width = 2.4 mm. PPL

Oysters

The first of these bivalves (Class Bivalvia, Order Ostreida, Superfamily Ostreoidea) appeared in the Early Triassic (Aragonite II) and were aragonitic (Hautmann et al. 2017). They switched to calcite in the Early Jurassic (Hautmann 2004; Hautmann et al. 2008); late in Aragonite II, during a time when seawater chemistry is not well constrained, and have remained as calcite producers (mainly LMC; Photo B.15) until the Present.

Inoceramids

Members of the Family Inoceramidae (Class Bivalvia, Order Myalinida; Schneider et al. 2013) acquired a calcitic shell in the Permian (Aragonite II), not matching the seawater chem-

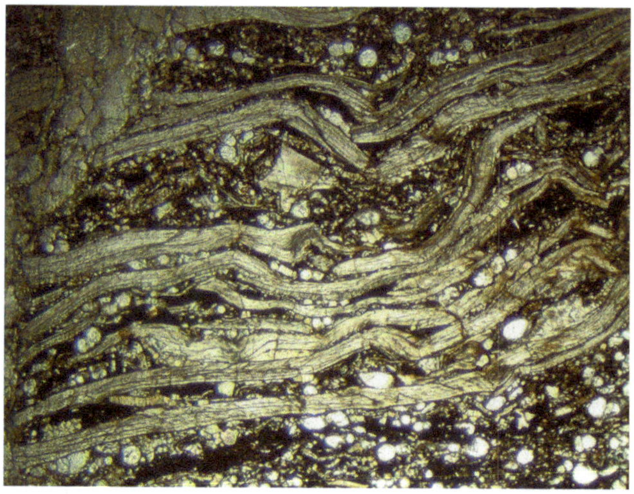

Photo B.16 Shells of inoceramids (middle to late Cenomanian). La Luna Formation, Maracaibo Basin
Image width = 4.2 mm. PPL

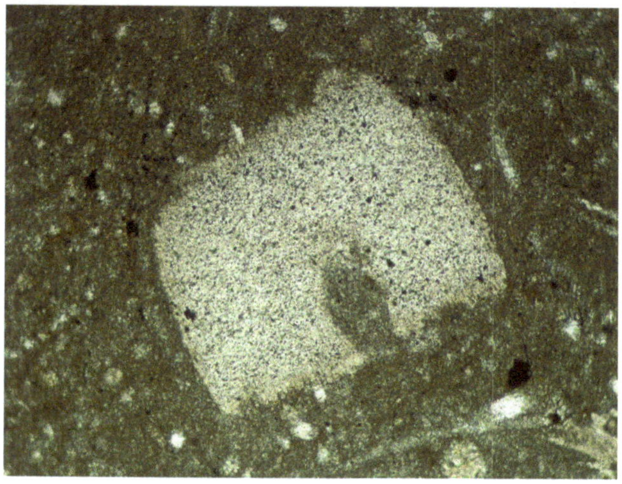

Photo B.17 Crinoid plate fragment (Middle Permian). Palmarito Formation, Venezuelan Andes
Image width = 2.4 mm. PPL

istry of the time. They remained as calcite producers (mainly LMC; Photo B.16) throughout the Calcite II seas until their extinction in the latest Cretaceous (see Hautmann 2006).

Echinoderms

We consider the two most abundant groups of echinoderms in the fossil record: crinoids (Class Crinoidea; Photo B.17) and echinoids (Class Echinoidea; Photo B.18). All echinoderms have always been HMC (e.g., Taylor et al. 2009).

Crinoids

Members of the Crinoidea (HMC) first appeared in the Early Ordovician (Calcite I) (e.g., Guensburg 2012).

Photo B.18 Echinoid spine (middle Albian). Lisure Formation, Maracaibo Basin
Image width = 0.5 mm. PPL

Echinoids

Members of the Echinoidea (HMC) first appeared in the Late Ordovician (Calcite I) (e.g., Thompson et al. 2021).

The age of the first appearance of crinoids and echinoids coincides with the hypothesis that no HMC clade has arisen during aragonite seas, emphasizing that Porter (2010) does not make a distinction between LMC- and HMC-producing taxa. In a study that involved ancient echinoderms from the Cambrian to the Eocene, Dickson (2004) determined that unaltered ossicles (calcitic units comprising the stereom) with high mol% $MgCO_3$ (9–12%) were precipitated during aragonite seas, whereas those with low mol% $MgCO_3$ (3.5–6%) were precipitated during calcite seas.

Bryozoans

We consider several orders of the Class Stenolaemata (Cystoporata, Trepostomata, Fenestrata, and Cyclostomata), as well as the Order Cheilostomata (Class Gymnolaemata).

Stenolaemates

Members of this group, which dominated bryozoan assemblages during the Paleozoic (Photo B.19), first appeared in the Early Ordovician (Calcite I) (e.g., Taylor and Waeschenbach 2015; Ernst 2020), having a primary LMC composition (Zhuravlev and Wood 2008). Stenolaemate bryozoans almost always precipitate LMC (Smith et al. 2006).

Cheilostomes

Members of the most successful post-Paleozoic order of bryozoans (including modern oceans; Photo B.20) first appeared during the Late Jurassic (Calcite II) (e.g., Taylor and Waeschenbach 2015). Their original skeletons were cal-

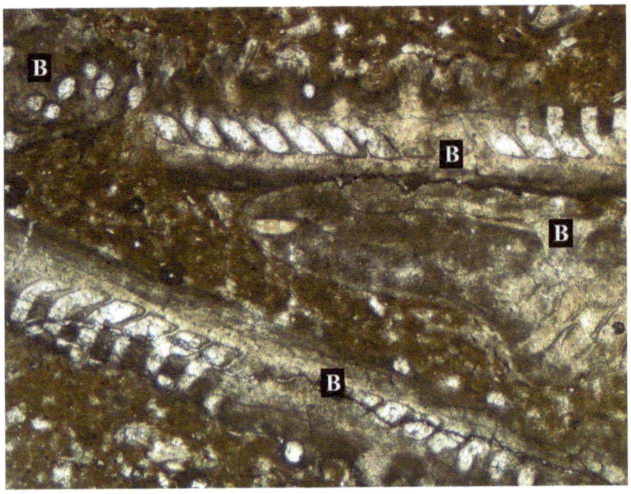

Photo B.19 Fragments of stenolaemate bryozoans (B) (Middle Permian). Palmarito Formation, Venezuelan Andes
Image width = 4.2 mm. PPL

Photo B.21 Plicated shell of a rhynchonelliformean brachiopod (middle Miocene). Capadare Formation, Falcon Basin
Image width = 2.4 mm. PPL

Photo B.20 Holocene cheilostome bryozoan (B). Venezuelan Atlantic Continental Shelf
Image width = 7 mm

Photo B.22 Podocopan ostracod carapace (middle Albian). Lisure Formation, Maracaibo Basin
Image width = 1.3 mm. CPL

citic, but of unknown Mg content. Calcite skeletons dominated throughout the Cretaceous. The first aragonitic skeletons appeared in the Maastrichtian (Taylor et al. 2009). Cenozoic cheilostomes (ancient and modern) have diverse mineralogies (calcitic, aragonitic, and bimineralic). Many are composed of HMC (Smith et al. 2006; Taylor et al. 2009).

Brachiopods

We consider carbonate-producing brachiopods of the Subphylum Rhynchonelliformea (Photo B.21), which is the largest subphylum of the Phylum Brachiopoda.

Rhynchonelliformeans

The first members of this group appeared in the Middle Cambrian (Carlson 2016); early in Calcite I, during a time when seawater chemistry is not well constrained. Ancient rhynchonelliformean brachiopods have a primary LMC composition (Zhuravlev and Wood 2008), and were always LMC (Taylor et al. 2009).

Ostracods

We consider well-calcified ostracods of the Subclass Podocopa (Photo B.22), which includes the orders Palaeocopida, Platycopida, and Podocopida.

Podocopans

Members of the three orders of the Subclass Podocopa first appeared in the Early Ordovician (Calcite I) (see Rodriguez-Lazaro and Ruiz-Muñoz 2012). Their carapaces were originally composed of LMC (Zhuravlev and Wood 2008). Podocopans have always been calcitic, mainly LMC (Gussone and Greifelt 2019).

Barnacles

We consider barnacles of the Order Balanomorpha (Class Thecostraca, Subclass Cirripedia, Infraclass Thoracica, Superorder Thoracicalcarea) (Gale 2015, 2019; Chan et al. 2021).

Balanomorphans

Shells of Paleozoic members of the Thoracica were phosphatic in composition (the Superorder Phosphatothoracica). The change to calcitic mineralization (LMC), and hence the appearance of the Superorder Thoracicalcarea (comprising several orders) occurred in the Middle Jurassic (Chan et al. 2021); early in Calcite II. The Order Balanomorpha appeared in the first half of the Late Cretaceous (Calcite II). Members of this order (balanids), which are the most abundant barnacles in both the fossil record and modern settings, predominantly form LMC shells (Photo B.23). Some cases of HMC have been observed (Ullmann et al. 2018).

Serpulids

We consider serpulid tubes (Class Polychaeta, Order Sabellida) of the Family Serpulidae (Photo B.24), not including Spirorbinae (see Ten Hove and Kupriyanova 2009).

Serpulidae

The first unequivocal serpulid tubes belonging to the Family Serpulidae ($CaCO_3$-secreting polychaete worms) are Permian (Aragonite II) in age (Sanfilippo et al. 2017). Vinn (2020) estimates that Permian serpulid tubes were aragonitic. Their diversification during the Mesozoic Calcite II seas involved the flourishing of calcitic (LMC and HMC) and bimineralic (calcitic and aragonitic) forms. Modern serpulid tubes are composed either of HMC, aragonite, or a combination of both polymorphs (Scholle and Ulmer-Scholle 2003; Smith et al. 2013).

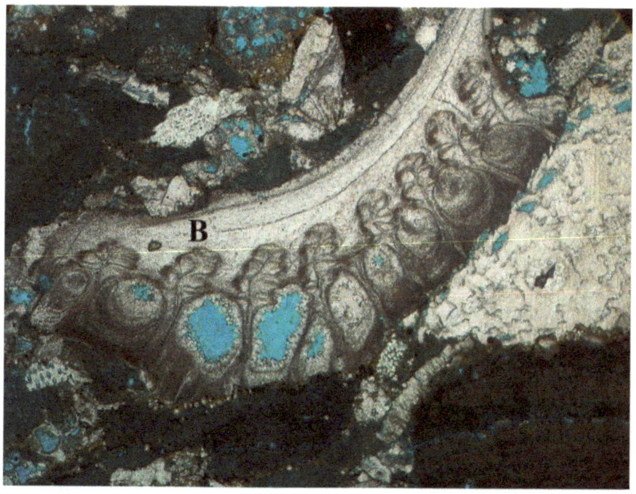

Photo A.23 Balanid barnacle (B) shell fragment (Late Oligocene-Early Miocene). Perla Limestone, Eastern Gulf of Venezuela sub-Basin
Image width = 4.2 mm. PPL

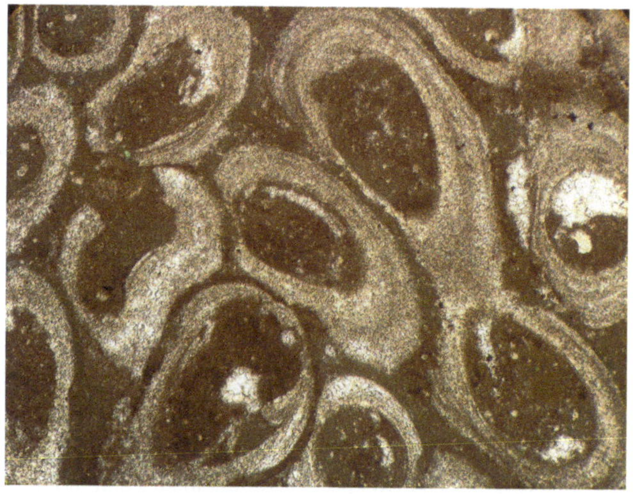

Photo B.24 Serpulid tubes (middle Aptian). Piché Member, Apón Formation, Maracaibo Basin
Image width = 2.4 mm. PPL

Remarks on the *De Novo* Acquisition of Carbonate Skeletons

According to everything discussed in the previous section, of the 20 taxa considered in this book (and whose mineralogy is known with some confidence), 15 acquired skeletal mineralogies that matched seawater chemistry of the time (although aragonitic bivalve molluscs, gastropods, and rhynchonelliformean brachiopods can also be considered as having appeared during a time when seawater chemistry is not well constrained). Three taxa acquired non-matching mineralogies: green algae of the orders Dasycladales and Bryopsidales (although these two orders of green algae can also be considered as a single taxon), and molluscs of the Family Inoceramidae (although this non-matching mineralogy can also be considered as a switch within the mollusc clade). For two taxa, this hypothesis is difficult to evaluate (red algae of the Family Elianellaceae, and red algae of the Order Archaeolithophyllaceae). These results suggest that calcite and aragonite seas do have a strong influence on carbonate skeletal mineralogy at the time mineralized skeletons first evolved, validating the "*De Novo* Acquisition Hypothesis" *Acquisition Hypothesis* that Susannah M. Porter tested in 2010.

Nine taxa switched mineralogies (from aragonite to calcite, or vice versa; the switch also includes changes from monomineralic to bimineralic skeletons). The switch in five of those taxa (some aragonitic bivalve molluscs, gastropods, cheilostome bryozoans, serpulids, and benthic foraminifera) did not occur on the whole-group level, but only in some clades (orders, superfamilies, families, genera, or species). In two taxa, the switch only occurred in a few members (a discrete number of species of dasycladalean algae, and one genus of scleractinian coral). In the case of oysters and PF, both taxa were originally aragonitic in the Aragonite II seas of the Early Triassic and Late Triassic, respectively. The former switched entirely to calcite in the Early Jurassic (although this change in mineralogy can also be considered as a switch within the mollusc clade); the latter also switched entirely to calcite sometime in the Early to Middle Jurassic. Both changes in mineralogy occurred in the transition from Aragonite II to Calcite II, during a time when seawater chemistry is not well constrained.

As discussed by Taylor (2008) and Porter (2010), those taxa that did switch mineralogies do not appear to have done so in response to changing aragonite–calcite seas. There is little evidence to suggest that changes between aragonite and calcite seas were responsible for those switches. The most striking correlation between seawater chemistry and biomineralization occurred during the latest Precambrian-Ordovician, a time interval when metazoans evolving calcareous skeletons for the first time had mineralogies corresponding precisely with seawater chemistry at a time of transition from an aragonite to a calcite sea (Porter 2007, 2010; Zhuravlev and Wood 2008). Some later *de novo* acquisitions of calcareous skeletons also match the seawater chemistry of the time (e.g., thoracicalcarean barnacles and PF), whereas others do not (e.g., inoceramids). Polymorph switches within clades that had already acquired $CaCO_3$ skeletons may either match or contrast with seawater chemistry. There is some indication that the correlation between biomineralization and seawater chemistry declined progressively through the Phanerozoic. Furthermore, although Taylor (2008) and Porter (2010) did not include foraminifera in their reviews, their assertions seem also to be true for these protozoa, which evolved calcareous tests late in the Calcite II seas. For foraminifera, the relation between ocean chemistry and their evolution is not clear and possibly obscured by the existence of different calcification strategies in this group (see de Nooijer et al. 2009).

Calcareous green algae remain as the only whole clade considered here that does not match the seawater chemistry at the time they first acquired skeletons in the Ordovician (Calcite I). In the case of dasycladalean algae, the Phanerozoic distribution of the few calcitic species shows no obvious relationship with the molar Mg^{2+}/Ca^{2+} ratio of seawater (Granier 2012). The *de novo* acquisition of carbonate skeletons in green algae is not addressed by Porter (2007, 2010), Taylor (2008), or Zhuravlev and Wood (2008). Excluding experimental studies focused on bryopsidalean algae (Ries 2005, 2006b, 2010; Stanley et al. 2010), and although it is widely known that calcification in calcareous green algae is largely extracellular *sensu lato* (see Granier 2012; for a review); the topic of why green algae acquired aragonitic skeletons in the Calcite I seas seems to be, to our knowledge, not fully addressed in the available literature. The answer to this question might possibly be linked to a strong physiological control of biomineralization in these photoautotrophs, rather than matching the seawater ambient conditions during Ordovician times.

Remarks on Calcite and Aragonite Seas and the Fates of Carbonate-Mineralizing Taxa Throughout the Phanerozoic

As pioneers in the concept of calcite and aragonite seas and its influence on biologically-controlled carbonate mineralization, Stanley and Hardie (1998) argued that the combination of high Ca^{2+}/Mg^{2+} ratios and higher Ca^{2+} concentrations in seawater, rather than high pCO_2, was the critical factor favoring calcite precipitation over aragonite. Nevertheless, in the following years it was established that aragonite saturation state is heavily dependent upon pCO_2, and high concentrations of this greenhouse gas produce low saturation states of aragonite in shallow tropical seas. In addition, rising pCO_2 may be detrimental to calcification rates across mineralogies, not just to aragonite producers (Hallock 2005; see also Ries 2011).

A comprehensive study by Kiessling et al. (2008) calls into question the traditional calcite-aragonite seas concept controlled by Mg^{2+}/Ca^{2+} ratio in seawater, finding instead a Phanerozoic trend of continuously increasing dominance of skeletal aragonite to the detriment of calcitic taxa, driven more by mass extinctions than by changing seawater chemistry. These authors state that, as previously mentioned: (**a**) the skeletal mineralogy of the early representatives of marine phyla may have depended on ocean chemistry, but their later evolutionary fate was not strongly coupled to gradual changes in Mg^{2+}/Ca^{2+} ratios of seawater; (**b**) the grand timescales over which the Mg^{2+}/Ca^{2+} ratio of sea water fluctuates may just be too large to represent a prominent evolutionary force, or the physiological control of biomineralization may be too strong for changes in Mg^{2+}/Ca^{2+} ratios to significantly affect the fitness of organisms; (**c**) short-term environmental perturbations associated with mass extinctions are evidently more efficient in governing lasting shifts of skeletal mineralogy.

A year later, Zhuravlev and Wood (2009) carried out a quantitative compilation of carbonate skeletal mineralogy through the Phanerozoic. Like Kiessling et al. (2008), their results also show a progressive replacement of LMC by aragonite. This general trend overrides the subsidiary trend of greenhouse intervals favoring biogenic LMC mineralogies (calcite seas), and icehouse intervals facilitating aragonite + HMC mineralogies (aragonite seas). The replacement of LMC by aragonite was, however, achieved episodically at mass extinction intervals. In particular, the end-Permian extinction which occurred during an aragonite sea, both preferentially removed species bearing "unfavorable" LMC (e.g., brachiopods, bryozoans, and tabulate and rugose corals, but also affected HMC crinoids), and allowed the selective radiation of biota with "favorable" aragonite (e.g., dasycladalean algae, bivalve molluscs, and gastropods). These authors suggest that the broad increase of aragonitic biota has been controlled by changes in pCO_2 via carbonate mineral kinetics. Through the Phanerozoic, broadly decreasing pCO_2 led to decreasing total alkalinity and dissolved inorganic carbon, and increasing oceanic pH.

Taking into account the unclear threshold between calcite and aragonite seas intervals, and considering the effect of other parameters that control carbonate saturation, particularly the latitudinal temperature gradient, Balthasar and Cusack (2015) presented an amended view of the calcite-aragonite concept after a quantitative experimental study focused on abiotic precipitation of marine $CaCO_3$. This view includes three new key aspects: (**a**) exclusive aragonite sea and calcite sea conditions are separated by a broad zone (the "gray area") of aragonite and calcite co-precipitation in which the proportions of $CaCO_3$ polymorphs can be quantified as a function of the Mg^{2+}/Ca^{2+} ratio and temperature; (**b**) in warm-water environments (>20 °C), exclusive abiotic calcite precipitation was unlikely during the Phanerozoic; (**c**) abiotic aragonite precipitation was possible in these environments throughout the Phanerozoic. According to these authors, this amended view of aragonite-calcite seas provides a much more realistic environmental framework in which to assess the evolution of calcareous biomineralization.

Throughout the Phanerozoic, the combined effects of the secular variations in seawater Mg^{2+}/Ca^{2+} ratio and atmospheric pCO_2, and mass extinctions appear to have affected mainly the major reef-building (hypercalcifiers) and sediment-producing organisms. On the contrary, "kill mechanisms" triggered by submarine and/or subaerial LIP eruptions (global warming, oceanic anoxia, global cooling, ocean acidification, H_2S and toxic metal poisoning) have been invoked to explain the cause of many of the extinction-related reef crises and mass extinction episodes. LIP eruptions (i.e., igneous pulses of ~1–5 Myr of duration or longer) release vast amounts of volcanic gases (mainly CO_2 and SO_2, but also CH_4, HF, and HCl) and ash. Two of the most severe of these events occurred at the Permo-Triassic boundary (the greatest mass extinction event of all time, triggered by the West Siberian LIP), and the Triassic-Jurassic boundary (triggered by the eruption of the Central Atlantic Magmatic Province). Although certainly catastrophic for both terrestrial and marine biospheres, the end-Cretaceous (or K/T boundary) extinction, caused by a bolide impact, cannot qualify as a reef crisis (see Kiessling 2009; Bond and Wignall 2014; Bond and Grasby 2017; Ernst and Youbi 2017; Scotese et al. 2021; Percival et al. 2024; for detailed reviews on the causes and severities of extinction events throughout the Phanerozoic).

Hypercalcification, carried mainly by symbiont-bearing carbonate producers, is a warm-water phenomenon (e.g., Ries 2010) defined as the development of especially robust calcareous skeletons so that the skeletal-biomass ratio is high (Kiessling 2009). The Phanerozoic distribution of dominant hypercalcifiers and sediment producers is summarized as follows (compiled from Stanley and Hardie 1998, 1999; Stanley 2006, Kiessling 2009; Ries 2010). During Calcite I, calcitic tabulate and rugose corals, and stromatoporoid sponges (possibly calcitic) were the dominant reef builders; while calcitic trepostome and cystoporate bryozoans illustrate the hypercalcification of individual colonial skeletons. During Aragonite II, aragonitic groups of sponges, scleractinian corals, and phylloid algae, and HMC red algae ('solenoporaceans'), became major reef builders; while aragonitic dasycladalean and subordinate aragonitic bivalve molluscs were important sediment producers. Scleractinian corals were hit hard by the end-Triassic Extinction, but they recovered and became important reef-builders during the Late Jurassic and Early Cretaceous (Calcite II), representing probably the major discrepancy between the mineralogy of dominant reef builders and changes in the imputed Mg^{2+}/Ca^{2+} ratio. During the peak of Calcite II (the highest Ca^{2+} concentrations of the Phanerozoic) in mid-Cretaceous times, the bimineralic (but calcitic-dominated) rudist (bivalve molluscs) replaced the aragonitic scleractinian corals as major hypercalcifiers; while calcitic coccolithophores became major sediment producers. During Aragonite III, aragonitic scleractinian corals and HMC red algae once again dominated reef construction; and aragonitic green algae (*Halimeda*, *Penicillus*, and *Udotea*) control sediment production.

Despite everything mentioned above, Phanerozoic distribution (abundance and diversity) and mineralogy of dominant hypercalcifiers and sediment producers must be evaluated for each individual taxon. For dasycladaleans, Aguirre and Riding (2005) concluded that changes in diversity of these photosynthetic algae over the past 350 Myr shows no straightforward relationship with calcite-aragonite sea intervals. Dasycladalean algae were major sediment producers during the Permian and Triassic (Aragonite II). However, at times, they were similarly important during Calcite II (e.g., in the Late Jurassic–Early Cretaceous and Paleocene–Eocene; although they did not attain the role of hypercalcifiers as they did in the Triassic). In fact, the lowest Mg^{2+}/Ca^{2+} ratio during the Calcite II seas (values of c. 1) coincides with the Barremian–Aptian dasycladalean diversification peak, which is the highest for the Mesozoic. These peaks of abundance broadly reflect global temperature and sea-level change. For bimineralic Cretaceous rudist bivalves, patterns of aragonite versus calcite dominance have been reported previously, but the observed pattern is too complex to be explained merely by the seawater Mg^{2+}/Ca^{2+} ratio (Steuber et al. 2023).

Regarding scleractinian corals, time periods of rich and poor fossil coral records coincide with the concept of aragonite-calcite seas. A particular case is the co-occurrence of aragonitic and calcitic corals in the Late Cretaceous (Maastrichtian). The exceptionally preserved aragonitic skeletons of *Micrabacia* from the Late Cretaceous Ripley Formation (~70 Ma, southeastern U.S.A.) exhibit all structural and geochemical properties observed in modern representatives of Micrabaciidae (Janiszewska et al. 2017). In the Cretaceous chalk deposits of eastern Poland (~70 Ma), skeletons of *Micrabacia* are diagenetically altered in a manner consistent with original aragonitic mineralogy. However, in the same chalk deposits, these aragonitic skeletons coexisted with *Coelosmilia* skeletons, the mineralogy of which is inferred to have been originally calcitic, based on well-preserved ultrastructures and consistent geochemical compositions (Stolarski et al. 2007). The co-occurrence in Late Cretaceous chalk deposits of two scleractinian taxa with different skeletal mineralogy indicates taxon-specific biomineralization control (Janiszewska et al. 2017).

Recent studies also show mismatches between the alternation of calcite-aragonite seas and the responses of some heterotrophic calcifiers. Key et al. (2022) consider extinct trepostome bryozoans as active biomineralizers that controlled the mineralogy and robustness of their skeletons regardless of changes in seawater chemistry. The degree of calcitic calcification in trepostome colonies did not decrease across the Calcite I sea to the Aragonite II sea at its transition in the Middle Mississippian.

Taking as a foundation that transitions between calcite-aragonite seas are typically gradual, occurring over tens of millions of years, Cole et al. (2023) explore short-term and long-term responses of echinoderms to shifting calcite-aragonite seas by combining experimental and deep-time biodiversity investigations. While experimental results support a strong relationship between Mg^{2+}/Ca^{2+} ratios and short-term echinoderm regeneration rates, patterns of Phanerozoic echinoderm diversification dynamics show no correspondence with Mg^{2+}/Ca^{2+} ratios or calcite-aragonite sea transitions. This decoupling between short- and long-term responses of echinoderms to seawater Mg^{2+}/Ca^{2+} ratios suggests echinoderms were relatively unaffected by seawater chemistry throughout their evolutionary history, possibly due to their ability to alter skeletal Mg fractionation and/or adapt to gradual shifts in seawater chemistry. According to these authors, these results demonstrate that experimental results should not be extrapolated using a strict uniformitarian approach, since short-term physiological responses of organisms to seawater conditions may not necessarily scale to macroevolutionary outcomes over geological timescales. Instead, the effect of seawater Mg^{2+}/Ca^{2+} ratios should be evaluated for individual clades using both experimental and deep-time biodiversity data in a time series.

Tropical Carbonates of Venezuela: An Example of the Geological Evolution of Shallow Marine Carbonate Systems

In this section, we consider some of the ancient and modern carbonate deposits of Venezuela as examples to evaluate the concepts and hypotheses discussed in the previous sections of this appendix. It is noteworthy that the early Paleozoic (Calcite I) and the Triassic to Jurassic (Aragonite II to Calcite II) intervals are not included in this discussion since marine carbonates of those times are not represented in Venezuela.

From the late Paleozoic to modern Times, carbonate production and deposition in what is today the Venezuelan territory (mainland and offshore) have always occurred in tropical latitudes. Note that, in the headline of this section, we use the phrase "shallow marine carbonate systems" instead of "reef systems" since the use of the term "reef" generally implies the *in situ* growth of organic buildups (calcification dominated by hypercalcifying animals and/or algae) in warm, nutrient-depleted, well-lit waters. Nevertheless, the term "reef" (modern or ancient) can also be used to refer to microbial deposits in which animals and calcareous algae are absent or are only minor components, as well as to cold- and/or deep-water bioconstructions (Kiessling 2009; see also James and Jones 2015). In that context, in the majority of ancient carbonate facies of Venezuela, controlling factors such as trophic conditions and bathymetry (which controls water temperature, energy level, and light availability) led to the deposition of heterozoan-dominated, mixed heterozoan-photozoan, or photozoan-dominated carbonate facies in which the euphotic, photozoan association between calcareous green algae and scleractinian corals is not dominant (i.e., scarce or absent).

Middle Permian (Cisularian to Guadalupian) carbonates of the Palmarito Formation were deposited in a shallow ramp that existed in northwest Gondwana (paleolatitude of about 10° S). Overall, this carbonate succession consists of a combination of heterozoan-dominated, mixed heterozoan-photozoan, and less abundant photozoan-dominated facies, in which the dominant skeletal components are: LMC bryozoans, brachiopods, LBF (fusulinids), and ostracods; HMC crinoids; aragonitic calcareous algae (green and phylloid); and sponge spicules (possibly aragonitic). Although the Palmarito succession was deposited in the Aragonite II seas of the Permian, grain associations suggest that local mesotrophic to eutrophic conditions, combined with paelodepths mostly within the mesophotic zone on a tropical ramp, acted as the main controlling factors in the development of a carbonate benthic factory dominated by those bioclastic components. These factors promoted the dominance of heterozoan associations composed of calcitic biota (both LMC and HMC). Nevertheless, the photozoan association between calcareous algae and fusulinids (these late Paleozoic LBF are considered to have hosted photosymbionts) is also common.

At the global scale, coccolithophores, PF, calcispheres, LBF, corals and rudists bivalves, and species of dasycladalean green algae were the main carbonate producers during the Cretaceous Period (Steuber et al. 2023). In Early Cretaceous (early Barremian to late Albian) shallow carbonates of the MB (the Cogollo Group), PF become moderatelly common only in the organic-rich intervals (black shales = alternation of limestones, marls, and shales; the carbonate content of these rocks varies from 90% to less than 10%) that were deposited during episodes of maximum flooding of the Maracaibo platform associated to the OAEs 1a (early Aptian) and 1b (sub-event 1, late Aptian; and sub-event 2, early Albian). The combination of anoxic-dysoxic and hypertrophic conditions (excess of nutrients) and marine waters charged with excess volcanogenic (LIP-sourced) CO_2 concentrations led to a significant reduction of carbonate benthic biota in these intervals, in which benthic heterotrophic calcifiers are represented only by low-oxygen SBF (mainly buliminids) and molluscs (inoceramids); whereas microbial carbonates (mainly biologically-induced automicrite) are common.

In organic-lean facies of the Cogollo Group (i.e., excluding the OAEs), photosymbiotic LBF are present only in some specific intervals, while photoautotrophic green algae (dasycladales are more common than bryopsidales) can be scarce/absent, moderately common, to locally abundant. In general, mesotrophic to eutrophic conditions prevailing during deposition of these shallow carbonates prevented the full dominance of these light-dependent calcifiers, although both green algae and some LBF are known to be able to thrive under mesotrophic conditions. Shallow carbonates of the Cogollo Group are an example of the demise of scleractinian corals, which are rare or absent. Meso-eutrophic conditions, combined with the low Ma^{2+}/Ca^{2+} ratio and high pCO_2 of that time affected corals more than any other carbonate benthic producers. In the case of the Cogollo Group, the ecological niche of rudists (scarce/absent) was occupied by oysters, which dominated carbonate production in the upper photic zone. Such dominance of these heterotrophic calcifiers represents an example of hypercalcification without symbionts (see Kiessling 2009). The other common calcifiers in this thick, shallow carbonate succession are echinoids (HMC) and aragonitic molluscs (bivalves a more abundant than and gastropods), while bryozoans, ostracods, serpulids, and red algae (HMC) are present but less common. The variety of calcifiers composed of the three $CaCO_3$ polymorphs exemplifies the adaptation of some aragonite and HMC calcifiers to the LMC-favored conditions prevailing during the Calcite II seas.

In Venezuela, the only Cretaceous unit in which corals and rudists are common components is the middle Aptian to early Albian Borracha Formation in the EVB (see Carvajal-Chitty and Rey 2020). This heterozoan-photozoan association is an example of the co-occurrence of these hypercalcifiers during Early Cretaceous times. It is worth noting that a few Cretaceous rudists are considered to have hosted photosymbionts (although this remains unconfirmed (see James and Jones, 2015; Skelton and Gili, 2025), and that rudist-rich deposits are included in the Photozoan T-factory, as these bivalve molluscs were restricted to shallow-tropical, oligotrophic waters (see Michel et al. 2019).

Late Cretaceous carbonate sedimentation in Venezuela was conditioned by the geographic proximity of the Caribbean LIP eruption, which started in the late Albian. Continuous black shale deposition for ~12 Myr in the MB (from the middle-late Cenomanian to late Santonian) and ~17 Myr in the EVB (from the late Albian to middle Santonian) accounted for the deposition of the La Luna and Querecual formations, respectively, encompassing the OAEs 2 and 3, and the OAEs 1d, 2 and 3. Prolonged anoxic-dysoxic and hypertrhophic conditions, which encompassed the Cenomanian-Turonian thermal maximum (the highest temperatures of the Mesozoic, and the second highest of the entire Phanerozoic, only after the Permo-Triassic thermal maximum, see Scotese et al. 2021), led to the deposition of organic-rich expanded sections (black shales) in which carbonate biomineralization is mainly represented by PF. Benthic carbonate-producing biota is mainly represented by subordinate low-oxygen SBF (mainly buliminids), inoceramids, and sponge spicules, plus scarce oysters.

In the Middle Eocene, late in Calcite II, some calcareous facies within the siliciclastic-dominated Santa Rita Formation (FB) consist of the photozoan association between LBF and coralline algae (plus bryozoans and echinoids). At global scale, the former were dominant in tropical and warm-temperate carbonates, and the latter were gradually increasing in abundance. Later, in the Late Oligocene and Early Miocene (Aragonite III), the most significant carbonate units in Venezuela were deposited in ramp settings. Facies of the Perla Limestone (Late Oligocene-Early Miocene in the EGVSB) are largely dominated by the LBF-coralline algae association (up to 80% in terms of volume in some intervals; see Pomar et al. 2015); echinoids, bryozoans, and balanid barnacles are common secondary components. The Early Miocene San Luis and Churuguara formations in north and south FB, respectively, display a more varied grain associations and lateral facies distribution. Corals and green algae, which were gradually becoming dominants in coral reef systems at global scale, are common in the inner ramp facies of the San Luis Formation, although corals did not form reefs, as these grew as mounds in the mesophotic zone. Inner ramp facies of the Churuguara Formation (late Aquitanian in age) are dominated by echinoids, plus molluscs, LBF and red algae. Towards the middle and outer ramp, both units are dominated by LBF and coralline algae (see Albert-Villanueva 2016; for a full review of these units). By the Early Miocene, the latter were dominant hypercalcifiers at global scale.

In general, the four units discussed in the previous paragraph (Santa Rita, Perla, San Luis and Churuguara) have one characteristic in common: they were deposited in ramps where water temperature, energy level, and light availability were controlled by bathymetry. The latter ranged between the mesophotic and oligophotic zones. These controlling factors promoted the dominance of the photozoan association between LBF and red algae (the main components of the Photo C-factory, which is typical of Cenozoic warm-temperate carbonates) to the detriment of corals-calcareous algae photozoan association (typical of Cenozoic coral reefs developed in the euphotic zone). As previously mentioned, the San Luis and Churuguara formations display a more varied biotic associations (shallow-water photozoan and heterozoan biota are common), resulting from varying trophic conditions and deepening of the depositional settings (from littoral to outer ramp settings). Although not commonly considered in the literature as major sediment producers, symbiont-bearing LBF are common to abundant components in the above-mentioned units, which encompass the transition from Calcite II to Aragonite III.

Modern offshore carbonate complexes of Venezuela (e.g., Los Roques and Las Aves archipelagos) are Caribbean examples of globally distributed tropical reefs. At the platform scale, carbonate production is largely dominated by the characteristic modern Aragonite III photozoan biota: hypercalcifiers (symbiont-bearing scleractinian corals and photo-autotrophic coralline algae) and sediment producers (photosynthetic bryopsidalean algae and symbiont-bearing LBF). The contribution of heterozoan biota, mainly molluscs (mostly aragonitic) and subordinate echinoderms (all extant classes are represented, with echinoids and ophiuroids being the most abundant), is significant. Barrier reefs and reef flats at the margins, and fringing reefs, patch reefs, and cays (or supratidal islands) in the inner areas, or isolated reef islands (in the case of Aves Island), are the euphotic expression of biodiverse ecosystems in which biologically-controlled carbonate production by light-independent biota extends to depths beyond those explored.

References

Aguirre, J., and Riding, R., 2005. Dasycladalean algal biodiversity compared with global variations in temperature and sea level over the past 350 Myr. *Palaios*, *20*, 581–588.

Aguirre, J., Baceta, J. I., Braga, J. C., 2007. Recovery of marine primary producers after the Cretaceous–Tertiary mass extinction: Paleocene calcareous red algae from the Iberian Peninsula. *Palaeogeography, Palaeoclimatology, Palaeoecology*, *249*, 393–411.

Albert-Villanueva, E., 2016. Facies y secuencias deposicionales mixtas carbonático-siliciclásticas del Mioceno inferior de la Cuenca de Falcón (noroeste de Venezuela) como modelo exploratorio en el Caribe. Doctoral Dissertation. University of Barcelona. Volumes 1 and 2.

Albert-Villanueva, E., Bover-Arnal, T., Ferràndez-Cañadell, C. Salas, R., 2017. Comment on "Tectonic and environmental factors controlling on the evolution of Oligo-Miocene shallow marine carbonate factories along a tropical SE Circum-Caribbean" by Silva-Tamayo *et al.* (2017). *Journal of South American Earth Sciences*, *91*, 394–396.

Arrhenius, G., 1963. Pelagic sediments. In: The Sea, vol 3. M. N. Hill (Eds.), Interscience, New York, 655–727.

Arthur, M. A., and Sageman, B. B., 1994. Marine black shales: Depositional mechanisms and environments of ancient deposits. *Annual Review of Earth and Planetary Sciences*, *22*, 499–551.

Balthasar, U., and Cusack, M., 2015. Aragonite-calcite seas—Quantifying the gray area. *Geology*, *43*(2), 99–102.

Bates, R. L., and Jackson, J. A., 1980. Glossary of Geology. Second Edition. American Geological Institute. Falls Church, Virginia. 751 pp.

Beauchamp, B., 1994. Permian climatic cooling in the Canadian Arctic. *Geological Society of America Special Paper*, *288*, 229–246.

Beauchamp, B., and Desrochers, A., 1997. Permian warm- to very cold-water carbonates and cherts in Northwest Pangea. In: James, N. P., Clarke J. A. D. (eds.), Cool-water carbonates. *SEPM Special Publication*, *56*, 327–347.

Bender, M. L., Lorens, R. B., Williams, D. F., 1975. Sodium, magnesium and strontium in the tests of planktonic foraminifera. *Micropaleontology*, 448–459.

Bensing, J. P., James, N. P., Beauchamp, B., 2008. Carbonate deposition during a time of mid-latitude ocean cooling: early Permian subtropical sedimentation in the Sverdrup Basin, Arctic Canada. *Journal of Sedimentary Research*, *78*, 2–15.

Betzler, C., Brachert, T. C., Nebelsick, J., 1997. The warm temperate carbonate province: a review of the facies, zonations, and delimitations. *Courier Forschungsinstitut Senckenberg*, *201*, 83–99.

Bond, D. P., and Grasby, S. E., 2017. On the causes of mass extinctions. *Palaeogeography, Palaeoclimatology, Palaeoecology*, *478*, 3–29.

Bond, D. P. G., and Wignall, P. B., 2014. Large igneous provinces and mass extinctions: an update. *Geological Society of America Special Paper 505*, SPE505-02.

Bosence, D. W. J., 1983. Coralline algal reef frameworks. *Journal of the Geological Society of London*, *140*, 365–376.

Bosence, D. W. J., and Pedley, H. M., 1982. Sedimentology and palaeoecology of a Miocene coralline algal biostrome from the Maltese Islands. *Palaeogeography, Palaeoclimatoly, Palaeoecology*, *38*, 9–43.

BouDagher-Fadel, M. K., 2015. Biostratigraphic and Geological Significance of Planktonic Foraminifera (Updated 2nd Edition): UCL Press, London, 298 pp.

BouDagher-Fadel, M. K., 2018. Evolution and geological significance of larger benthic foraminifera. UCL Press, London, 607 pp.

Brandano, M., Frezza, V., Tomassetti, L., Cuffaro, M., 2009. Heterozoan carbonates in oligotrophic tropical waters: the Attard member of the lower coralline limestone formation (Upper Oligocene, Malta). *Palaeogeography, Palaeoclimatology, Palaeoecology*, *274*, 54–63.

Bucur, I. I., 1999. Stratigraphic significance of some skeletalal gae (Dasycladales, Caulerpales) of the Phanerozoic. In: Farinacci, A., Lord, A. R.(Eds.), Depositional Episodesand Bioevents. Università La Sapienza, Rome, pp. 53–104.

Burchette, T. P., and Wright, V. P., 1992, Carbonate ramp depositional systems. *Sedimentary Geology*, 79, 3–57.

Carannante, G., Esteban, M., Milliman, J. D., Simone, L., 1988. Carbonate lithofacies as paleolatitude indicators: problems and limitations. *Sedimentary Geology*, *60*, 333–346.

Carlson, S. J., 2016. The evolution of Brachiopoda. *Annual Review of Earth and Planetary Sciences*, *44*, 409–438.

Carter, J. G., Barrera, E., Tevesz, M. J., 1998. Thermal potentiation and mineralogical evolution in the Bivalvia (Mollusca). *Journal of Paleontology*, *72*(6), 991–1010.

Carvajal-Chitty, H., and Rey, O., 2020. Aportes a la bioestratigrafía del Aptiense–Albiense en la cuenca oriental de Venezuela con base en foraminíferos y nanopláncton calcáreo. *Caribbean Studies*, *48*, 135–163.

Chan, B. K., Dreyer, N., Gale, A. S., Glenner, H., Ewers-Saucedo, C., Pérez-Losada, M., Kolbasov, G. A., Grandall, K. A., Høeg, J. T., 2021. The evolutionary diversity of barnacles, with an updated classification of fossil and living forms. *Zoological Journal of the Linnean Society*, *193*, 789–846.

Cole, S. R., Wright, D. F., Thompson, J. R., 2023. Calcite-aragonite seas as a driver of echinoderm evolution? Experimental insight and deep-time decoupling. *Geology*, *51*, 1091–1095.

Corrochano, D., Vachard, D., Armenteros, I. 2013. New insights on the red alga *Archaeolithophyllum* and its preservation from the

J. A. Méndez Dot, J. Méndez Baamonde, *Atlas of Tropical Carbonates*, https://doi.org/10.1007/978-3-031-25494-9

Pennsylvanian of the Cantabrian Zone (NW Spain). *Facies, 59*, 949–967.

Cózar, P., and Vachard, D., 2003. *Neoprincipia* nov. gen., a new Mississippian red alga, and remarks on the Archaeolithophyllaceae (Rhodophyta). *Geobios, 36*, 505–517.

Cózar, P., Somerville, I. D., Medina-Varea, P., 2005. Note on the earliest occurrences of the calcareous algae *Paraepimastopora* and *Archaeolithophyllum* in Mississippian rocks. *Coloquios de Paleontología, 55*, 7–20.

Demaison, G. J., and Moore, G. T., 1981. Anoxic environments and oil source bed genesis. *Organic Geochemistry, 2*, 9–31.

de Nooijer, L. J., Toyofuku, T., Kitazato, H., 2009. Foraminifera promote calcification by elevating their intracellular pH. *Proceedings of the National Academy of Sciences, 106*(36), 15374–15378.

Diaz, M. R., and Eberli, G. P., 2022. Microbial contribution to early marine cementation. *Sedimentology, 69*(2), 798–822.

Dickson, J. A. D., 2004. Echinoderm skeletal preservation: calcite-aragonite seas and the Mg/Ca ratio of Phanerozoic oceans. *Journal of Sedimentary Research, 74*, 355–365.

Dupraz, C., Reid, R. P., Braissant, O., Decho, A. W., Norman, R. S., Visscher, P. T., 2009. Processes of carbonate precipitation in modern microbial mats. *Earth-Science Reviews, 96*, 141–162.

Emiliani, C., 1952. Nomenclature and grammar. *Journal of the Washington Academy of Sciences, 42*, 137–141.

Emiliani, C., 1991. Planktic/planktonic, nektic/nektonic, benthic/benthonic. *Journal of Paleontology*, 65, 329.

Erez, J., 2003. The source of ions for biomineralization in foraminifera and their implications for paleoceanographic proxies. *Reviews in Mineralogy and Geochemistry, 54*, 115–149.

Ernst, A., 2020. Fossil record and evolution of Bryozoa. In: Schwaha T (ed) Handbook of zoology. Phylum Bryozoa. De Gruyter, Berlin/ Boston, pp. 11–56.

Ernst, R. E., and Youbi, N., 2017. How Large Igneous Provinces affect global climate, sometimes cause mass extinctions, and represent natural markers in the geological record. *Palaeogeography, Ppalaeoclimatology, Palaeoecology, 478*, 30–52.

Fantle, M. S., Barnes, B. D., Lau, K. V., 2020. The Role of Diagenesis in Shaping the Geochemistry of the Marine Carbonate Record. *Annual Review of Earth and Planetary Sciences, 48*, 549–583.

Flügel, E, 2010. Microfacies of Carbonate Rocks—Analysis, Interpretation and Application. Springer, Heidelberg-Dordrecht-London-New York, 984 pp.

Gale A. S., 2015. Origin and phylogeny of the thoracican cirripedes family Stramentidae. *Journal of Systematic Paleontology, 14*, 653–702.

Gale, A. S, 2019. Stalked barnacles (Cirripedia, Thoracica) from the Late Jurassic (Tithonian) Kimmeridge Clay of Dorset, UK; palaeoecology and bearing on the evolution of living forms. *Proceedings of the Geologists' Association, 130*, 355–365.

Gilbert, P. U. P. A., Bergmann, K. D., Boekelheide, N., Tambutté, S., Mass, T., Marin, F., Adkins, J. F., Erez, J., Gilbert, B., Knutson, V., Cantine M., Hernández, J. O., Knoll, A. H., 2022. Biomineralization: integrating mechanism and evolutionary history. *Science Advances, 8*(10), eabl9653.

Gischler, E., Dietrich, S., Harris, D., Webster, J. M., Ginsburg, R. N., 2013. A comparative study of modern carbonate mud in reefs and carbonate platforms: Mostly biogenic, some precipitated. *Sedimentary Geology, 292*, 36–55.

Granier, B., 2012. The contribution of calcareous green algae to the production of limestones: a review. *Geodiversitas, 34*, 35–60.

Granier, B., and Dias-Brito, D., 2016a. On the fossil alga *Elianella elegans* PFENDER & BASSE, 1948, and its so-called lookalikes, with description of *Elianella brasiliana* n. sp Revision of the Juliette PFENDER Collection. Part 1. *Carnets de Géologie*, 213–229.

Granier, B., and Dias-Brito, D., 2016b. On the fosil alga *Marinella lugeoni* Pfender, 1939, nom. cons., and its seven unfortunate avatars. Revision of the Juliette Pfender collection. Part 2. Revision of the Jesse Harlan Johnson collection. Part 2. *Carnets de Géologie, 16*, 231–45.

Görgen, S., Benzerara, K., Skouri-Panet, F., Gugger, M., Chauvat, F., Cassier-Chauvat, C., 2021. The diversity of molecular mechanisms of carbonate biomineralization by bacteria. *Discover Materials, 1*, 1–20.

Gussone, N., and Greifelt, T., 2019. Incorporation of Ca isotopes in carapaxes of marine ostracods. *Chemical Geology, 510*, 130–139.

Guensburg, T. E., 2012. Phylogenetic implications of the oldest crinoids. *Journal of Paleontology, 86*(3), 455–461.

Halfar, J., and Mutti, M., 2005. Global dominance of coralline red-algal facies: a response to Miocene oceanographic events. *Geology, 33*, 481–484.

Halfar, J., Godinez-Orta, L., Mutti, M., Valdez-Holguin, J. E., Borges, J. M. 2004. Nutrient and temperature controls on modern carbonate production: an example from the Gulf of California, Mexico. *Geology, 32*, 213–216.

Halfar, J., Godinez-Orta, L., Mutti, M., Valdez-Holguin, J. E., Borges, J. M., 2006. Carbonates calibrated against oceanographic parameters along a latitudinal transect in the Gulf of California, Mexico. *Sedimentology, 53*, 297–320

Hallock, P., 2001. Coral reefs, carbonate sedimentation, nutrients, and global change. In: Stanley, G.D. (Ed.), The History and Sedimentology of Ancient Reef Ecosystems. Kluwer Academic/ Plenum Publishers, pp. 387–427.

Hallock, P., 2005. Global change and modern coral reefs: new opportunities to understand shallow-water carbonate depositional processes. *Sedimentary Geology, 175*, 19–33.

Hallock, P., and Pomar, L., 2008. Cenozoic evolution of larger benthic foraminifers: paleoceanographic evidence for changing habitats. In *Proceedings of the 11th International Coral Reef Symposium, Ft. Lauderdale, Florida*, 16–20.

Hardie, L. A., 1996. Secular variations in seawater chemistry: An explanation for the coupled secular variation in the mineralogies of marine limestones and potash evaporites over the past 600 m.y: Geology, *24*, 279–283.

Hautmann, M. 2004. Effect of end-Triassic CO_2 maximum on carbonate sedimentation and marine mass extinction. *Facies, 50*, 257–261.

Hautmann, M., 2006. Shell mineralogical trends in epifaunal Mesozoic bivalves and their relationship to seawater chemistry and atmospheric carbon dioxide concentration. *Facies, 52*, 417–433.

Hautmann, M., Benton, M.J., Tomašových, A., 2008. Catastrophic ocean acidification at the Triassic-Jurassic boundary. *Neues Jahrbuch für Geologie und Palaontologie, Abhandlungen, 249*, 119–127.

Hautmann, M., Ware, D., Bucher, H., 2017. Geologically oldest oysters were epizoans on Early Triassic ammonoids. *Journal of Molluscan Studies, 83*, 253–260.

Hayton, S., Nelson C., Hood S., 1995. A skeletal assemblage classification system for non-tropical carbonate deposits based on New Zealand Cenozoic limestones. *Sedimentary Geology, 100*. 123–141.

Henrich, R., Freiwald, A., Betzler, C., Bader, B., Schäfer, P., Samtleben, C., Brachert, T. T. C., Wehrmann, A., Zankl, H., Kühlmann, D. H. H., 1995. Controls on modern carbonate sedimentation on warm-temperate to Arctic coasts, shelves and seamounts in the Northern Hemisphere: implications for fossil counterparts. *Facies, 32*, 71–108.

Hillgärtner, H., Dupraz, C., Hug, W., 2001. Microbially induced cementation of carbonate sands: are micritic meniscus cements good indicators of vadose diagenesis? *Sedimentology, 48*, 117–131.

James, N. P., 1997. The cool-water carbonate depositional realm. In: James N. P., Clarke J (eds.) Cool-water carbonates, vol 56. SEPM Society for Sedimentary Geology, Tulsa, 1–20.

James, N. P., and Bone, Y., 2011. Neritic Carbonate Sediments in a Temperate Realm—South Australia. Springer, Dordrecht Heidelberg London New York, 254 pp.

James, N. P., and Jones, B., 2015. Origin of Carbonate Sedimentary Rocks. John Wiley & Sons, Ltd, Chinchester, 464 pp.

James, N. P., Bone, Y., Brown, K. M., Cheshire, A., 2009. Calcareous epiphyte production in cool-water carbonate seagrass depositional environments; Southern Australia. In: Swart PK, Eberli GP, McKenzie JA (eds.) Perspectives in carbonate geology: a tribute to the career of Robert Nathan Ginsburg, vol 41. Wiley, Hoboken, 123–148.

Janiszewska, K., Mazur, M., Escrig, S., Meibom, A., Stolarski, J., 2017. Aragonitic scleractinian corals in the Cretaceous calcitic sea. *Geology*, *45*, 319–322.

Jiang, H.X., and Zhang, Y. Y, 2021. Earliest calcified green algae from the 520 Ma old Cambrian dolostones in Xinjiang, China. *Biopetrology*, *1*(3), 9–18.

Keim, L., and Schlager, W., 2001. Quantitative compositional analysis of a Triassic carbonate platform (Southern Alps, Italy). Sedimentary Geology 139, 261–283.

Key, M. M., Wyse Jackson, P. N., Reid, C. M., 2022. Trepostome bryozoans buck the trend and ignore calcite-aragonite seas. *Palaeobiodiversity and Palaeoenvironments*, *102*, 253–263.

Kiessling, W., 2009. Geologic and biologic controls on the evolution of reefs. *Annual Review of Ecology, Evolution, and Systematics*, *40*, 173–192.

Kiessling, W., Aberhan, M., Villier, L., 2008. Phanerozoic trends in skeletal mineralogy driven by mass extinctions. *Nature Geoscience*, *1*, 527–530.

Klicpera, A., Michel, J., Westphal, H., 2015. Facies patterns of a tropical heterozoan carbonate platform under eutrophic conditions: the Banc d'Arguin, Mauritania. *Facies*, *61*, 1–24.

Knoll, A. H., 2003. Biomineralization and evolutionary history. *Reviews in Mineralogy and Geochemistry*, *54*, 329–356.

Konhauser, K. O., 2007. Introduction to Geomicrobiology. Blackwell Publishing. 433 pp.

Laugié, M., Michel, J., Pohl, A., Poli, E., Borgomano, J., 2019. Global distribution of modern shallow-water marine carbonate factories: A spatial model based on environmental parameters. *Scientific Reports*, *9*, 1–14.

Laya, J. C., and Tucker, M. E., 2012. Facies analysis and depositional environments of Permian carbonates of the Venezuelan Andes: Palaeogeographic implications for Northern Gondwana. *Palaeogeography, Palaeoclimatology, Palaeoecology*, *331*, 1–26.

Lees, A., 1975. Possible influence of salinity and temperature on modern shelf carbonate sedimentation. *Marine Geology*, *19*, 159–198.

Lees, A., and Buller, A. T., 1972. Modern temperate-water and warm-water shelf carbonate sediments contrasted. *Marine Geology*, *13*, M67–M73.

Li, L., Topper, T. P., Betts, M. J., Dorjnamjaa, D., Altanshagai, G., Enkhbaatar, B., Li, G., Skovsted, C. B., 2023. Calcitic shells in the aragonite sea of the earliest Cambrian. *Geology*, *51*, 8–12.

Liu, L., Wu, Y., Yang, H., Riding, R., 2016. Ordovician calcified cyanobacteria and associated microfossils from the Tarim Basin, Northwest China: systematics and significance. *Journal of Systematic Palaeontology*, *14*(3), 183–210.

Lowery, J. G., and Rankey, E. C., 2017. Nearshore influences of upwelling, waves, and currents on a tropical carbonate ramp: Holocene, northwestern Yucatan shelf, Mexico. *Journal of Sedimentary Research*, *87*, 546–566.

Lowenstein, T. K., Timofeeff, M. N., Brennan, S. T., Hardie, L. A., Demicco, R. V., 2001. Oscillations in Phanerozoic seawater chemistry: evidence from fluid inclusions. *Science*, *294*, 1086–1088.

Lüning, S., Kolonic, S., Belhadj, E. M., Belhadj, Z., Cota, L., Baric, G., Wagner, T., 2004. Integrated depositional model for the Cenomanian-Turonian organic-rich strata in North Africa, *Earth Science Reviews*, *64*, 51–117.

Mackenzie, F. T., 2003. Carbonate mineralogy and geochemistry. In Encyclopedia of Sediments and Sedimentary Rocks (ed. Middleton, G. V). Kluwer Academic Publishers, Boston, pp. 93–99.

Martin, R. E., 1995. Cyclic and secular variation in microfossil biomineralization: clues to the biogeochemical evolution of Phanerozoic oceans. *Global and Planetary Change*, *11*, 1–23.

Meyer, K. M., and Kump, L. R., 2008. Oceanic euxinia in Earth history: Causes and Consequences. *Annual Review of Earth and Planetary Sciences*, *36*, 251–288.

Michel, J., Borgomano, J., Reijmer, J. J., 2018. Heterozoan carbonates: When, where and why? A synthesis on parameters controlling carbonate production and occurrences. *Earth-Science Reviews*, *182*, 50–67.

Michel, J., Laugié, M., Pohl, A., Lanteaume, C., Masse, J. P., Donnadieu, Y., Borgomano, J., 2019. Marine carbonate factories: a global model of carbonate platform distribution. *International Journal of Earth Sciences*, *108*, 1773–1792.

Monteiro, F. M., Pancost, R. D., Ridgwell, A., Donnadieu, Y. 2012. Nutrients as the dominant control on the spread of anoxia and euxinia across the Cenomanian-Turonian oceanic anoxic event (OAE2): model-data comparison. *Paleoceanography*, *27*, PA4209.

Monty, C. L. V., 1995. The rise and nature of carbonate mud-mounds: an introductory actualistic approach. In: Monty, C.L.V., Bosence, D.W.J., Bridges, P.H., Pratt, B.R. (Eds.), Carbonate Mud-Mounds Their Origin and Evolution. *International Association of Sedimentologists Special Publication*, *23*, 11–48.

Morse, J. W., Wang, Q., Tsio, M. Y., 1997. Influences of temperature and Mg:Ca ratio on $CaCO_3$ precipitates from seawater. *Geology*, *25*, 85–87.

Morsilli, M., Bosellini, F. R., Pomar, L., Hallock, P., Aurell, M., Papazzoni, C. A., 2012. Mesophotic coral buildups in a prodelta setting (late Eocene, southern Pyrenees, Spain): a mixed carbonate–siliciclastic system. *Sedimentology*, *59*, 766–794.

Mutti, M., and Hallock, P., 2003. Carbonate systems along nutrient and temperature gradients: some sedimentological and geochemical constraints. *International Journal of Earth Sciences*, *92*, 465–475.

Narayan, G. R., Reymond, C. E., Stuhr, M., Doo, S., Schmidt, C., Mann, T., Westphal, H., 2022. Response of large benthic foraminifera to climate and local changes: Implications for future carbonate production. *Sedimentology*, *69*, 121–161.

Nelson, C. S., 1988. An introductory perspective on non-tropical shelf carbonates. *Sedimentary Geology*, *60*, 3–12.

Neuweiler, F., Gautret, P., Thiel, V., Lange, R., Michaelis, W., Reitner, J., 1999. Petrology of Lower Cretaceous carbonate mud mounds (Albian, N. Spain): insights into organomineralic deposits of the geological record. *Sedimentology*, *46*, 837–859.

Newell, N. D., Imbrie, J., Purdy, E. G. Thurber, D. L., 1959. Organism communities and bottom facies, Great Bahama Bank. *Bulletin of the American Museum of Natural History*, *177*, 717–228.

Parkhaev, P. Y., 2008. The early Cambrian radiation of Mollusca. In Phylogeny and Evolution of the Mollusca (eds W. F. Ponder and D. R. Lindberg), pp. 33–70. University of California Press, Berkeley.

Parkhaev, P. Y., 2017. Origin and the early evolution of the phylum Mollusca. *Paleontological Journal*, *51*(6), 663–686.

Percival, L. M., Matsumoto, H., Callegaro, S., Erba, E., Kerr, A. C., Mutterlose, J., Suzuki, K., 2024. Cretaceous large igneous provinces: from volcanic formation to environmental catastrophes and biological crises. *Geological Society, London, Special Publications*, *544*, SP544-2023-88.

Peña, V., Vieira, C., Braga, J. C., Aguirre, J., Rösler, A., Baele, G., De Clerck, O., Le Gall, L., 2020. Radiation of the coralline red algae (Corallinophycidae, Rhodophyta) crown group as inferred from a multilocus time-calibrated phylogeny. *Molecular phylogenetics and evolution*, *150*, 106845.

Pérès, J. M., and Picard, J., 1964. Nouveau manuel de bionomie benthique de la Mer Méditerranée. *Recueil de la Station Marine d'Endoume*, *31*, 1–137.

Pérez-Huerta, A., Coronado, I., Hegna, T. A., 2018. Understanding biomineralization in the fossil record. *Earth-Science Reviews*, *179*, 95–122.

Perry, C. T., and Beavington-Penney, S. J., 2005. Epiphytic calcium carbonate production and facies development within sub-tropical seagrass beds, Inhaca Island, Mozambique. *Sedimentary Geology*, *174*. 161–176.

Peters, K. E., Walters C., Moldowan J. M., 2005. The Biomarker Guide. Biomarkers and isotopes in the environmental and human history. Volume I. Second Edition. Cambridge University Press, 471 pp.

Petrash, D. A., Gingras, M. K., Lalonde, S. V., Orange, F., Pecoits, E., Konhauser, K. O., 2012. Dynamic controls on accretion and lithification of modern gypsum-dominated thrombolites, Los Roques, Venezuela. *Sedimentary Geology*, *245*, 29–47.

Pitelli, R., and Molina, A., 1989. El Eoceno Medio Tardío y Eoceno Tardío de la parte occidental de la cuenca de Falcón. *Boletín de la Sociedad Venezolana de Geología*, *36*, 5–12.

Pomar, L., 2001. Types of carbonate platforms: a genetic approach. *Basin Research*, *13*, 313–334.

Pomar, L. 2020. Carbonate systems. In: Scarselli, N., Adam, J., Chiarella, D., Roberts, D. G. and Bally, A. W. (eds.) Regional Geology and Tectonics. Volume 1: Principles of Geologic Analysis. 2nd edn. Elsevier, Amsterdam, 235–311.

Pomar, L., and Hallock, P., 2008. Carbonate factories: a conundrum in sedimentary geology. *Earth-Science Reviews*, *87*, 134–169.

Pomar, L., Brandano, M., Westphal, H., 2004. Environmental factors influencing skeletal grain sediment associations: a critical review of Miocene examples from the western Mediterranean. *Sedimentology*, *51*, 627–651.

Pomar, L., Bassant, P., Brandano, M., Ruchonnet, C., Janson, X., 2012. Impact of carbonate producing biota on platform architecture: insights from Miocene examples of the Mediterranean region. *Earth-Science Reviews*, *113*, 186–211.

Pomar, L., Mateu-Vicens, G., Morsilli, M., Brandano, M., 2014. Carbonate ramp evolution during the late Oligocene (Chattian), Salento Peninsula, southern Italy. *Palaeogeography, Palaeoclimatology, Palaeoecology*, *404*, 109–132.

Pomar, L., Esteban, M., Martinez, W., Espino, D., Castillo de Ott, V., Benkovics, L., Castro Leyva, T., 2015. Oligocene-Miocene carbonates of the Perla field, offshore Venezuela: Depositional model and facies architecture, In: C. Bartolini and P. Mann, eds., Petroleum geology and potential of the Colombian Caribbean Margin: *American Association of Petroleum Geology*, *Memoir 108*, 647–674.

Porter, S. M., 2007. Seawater chemistry and early carbonate biomineralization. *Science*, *316*, 1302–1302.

Porter, S. M., 2010. Calcite and aragonite seas and the de novo acquisition of carbonate skeletons. *Geobiology*, *8*, 256–277.

Pratlong, M., Rancurel, C., Pontarotti, P., Aurelle, D., 2017. Monophyly of Anthozoa (Cnidaria): why do nuclear and mitochondrial phylogenies disagree? *Zoologica Scripta*, *46*(3), 363–371.

Purdy, E. G., 1963. Recent calcium carbonate facies of the Great Bahama Bank. 2. Sedimentary facies. *The Journal of Geology*, *71*, 472–497.

Reijmer, J. J., 2021. Marine carbonate factories: Review and update. *Sedimentology*, *68*, 1729–1796.

Reijmer, J. J. G., Bauch, T., Schäfer, P., 2012. Carbonate facies patterns in surface sediments of upwelling and non-upwelling shelf environments (Panama, East Pacific). *Sedimentology*, *59*, 32–56.

Reitner, J., Neuweiler, F., Flajs, G., Keupp, H., Meischner, D., Neuweiler, F., Paul, J., Warnke, K., Weller, H., Dingle, P., Hensen, G., Schaefer, P., Gautret, P., Leinfelder, R. R., Huessner, H., Kaufmann, B., 1995. Mud mounds: a polygenetic spectrum of fine-grained carbonate buildups. *Facies*, *32*, 1–70.

Riegl, B., Poiriez, A., Janson, X., Bergman, K., 2010. The Gulf-Facies belts, physical, chemical and biological parameters of sedimentation on a carbonate ramp. In: Westphal, H., Riegl, B., Eberli, G. (Eds.), Carbonate Depositional Systems: Assessing Dimensions and Controlling Parameters. Springer, Dordrecht, pp. 145–214.

Ries, J. B., 2004. Effect of ambient Mg/Ca ratio on Mg fractionation in calcareous marine invertebrates: a record of the oceanic Mg/Ca ratio over the Phanerozoic. *Geology*, *32*, 981–984.

Ries, J. B., 2005. Aragonite production in calcite seas: effect of seawater Mg/Ca ratio on the calcification and growth of the calcareous alga *Penicillus capitatus*. Paleobiology, *31*(3) 445–458.

Ries, J. B, 2006a. Mg fractionation in crustose coralline algae: geochemical, biological, and sedimentological implications of secular variation in the Mg/Ca ratio of seawater. *Geochimica et Cosmochimica Acta*, *70*, 891–900.

Ries J. B., 2006b. Aragonitic algae in calcite seas: effect of seawater Mg/Ca ratio on algal sediment production. *Journal of Sedimentary Petrology*, *76*, 515–523.

Ries, J. B., 2010. Geological and experimental evidence for secular variation in seawater Mg/Ca (calcite-aragonite seas) and its effects on marine biological calcification. *Biogeosciences*, *7*, 2795–2849.

Ries, J. B, 2011. Skeletal mineralogy in a high-CO_2 world. *Journal of Experimental Marine Biology and Ecology*, *403*, 54–64.

Rigaud, S., Granier, B., Masse, J. P., 2021. Aragonitic foraminifers: an unsuspected wall diversity. *Journal of Systematic Palaeontology*, *19*(6), 461–488.

Rodriguez-Lazaro, J., and Ruiz-Muñoz, F., 2012. A general introduction to ostracods: morphology, distribution, fossil record and applications. *Developments in Quaternary Science*, 17, pp. 1–14. Elsevier.

Runnegar, B., 1985. Shell microstructures of Cambrian molluscs replicated by phosphate. *Alcheringa*, *9*, 245–257.

Sandberg, P. A., 1983. An oscillating trend in Phanerozoic non-skeletal carbonate mineralogy. *Nature*, *305*, 19–22.

Sandberg P. A, 1985. Nonskeletal aragonite and pCO_2 in the Phanerozoic and Proterozoic. In The Carbon Cycle and Atmospheric CO_2: Natural Variations Archean to Present (eds Sundquist ET, Broecker WS). American Geophysical Union, Washington, DC, pp. 585–594.

Sandberg, P. A., 1991. Nonskeletal aragonite and pCO_2 in the Phanerozoic and Proterozoic. In: Sundquist, E. T., Broecker, W. S. (eds.): The carbon cycle and atmospheric CO_2: Natural variations Archean to Present. *Geophysical Monograph*, *32*, 585–594.

Sanfilippo, R., Rosso, A., Reitano, A., Insacco, G., 2017. First record of sabellid and serpulid polychaetes from the Permian of Sicily. *Acta Palaeontologica Polonica*, *62*, 25–38.

Schlager, W., 2000. Sedimentation rates and growth potential of tropical, cool-water and mud-mound carbonate systems. *Geological Society, London, Special Publications*, *178*, 217–227.

Schlager, W., 2003. Benthic carbonate factories of the Phanerozoic. *International Journal of Earth Sciences*, *92*, 445–464.

Schlager, W., 2005. Carbonate Sedimentology and Sequence Stratigraphy. SEPM Concepts in Sedimentology and Paleontology #8. Society of Economic Paleontologists and Mineralogists, 200 pp.

Schlanger, S. O., Arthur, M. A., Jenkyns, H. C., Scholle, P. A., 1987. The Cenomanian-Turonian Oceanic Anoxic Event, I. Stratigraphy and distribution of organic carbon-rich beds and the marine $\delta^{13}C$ excursion. *Geological Society Special Publications*, *26*, 371–399.

Schneider, S., Crampton, J. S., Lukeneder, A., 2013. Propeamussiidae, Inoceramidae, and other Bivalvia from the Lower Cretaceous Puez Formation (Valanginian–Cenomanian; Dolomites, South Tyrol, Italy). *Cretaceous research*, *46*, 216–231.

Scholle, P. A., and Ulmer-Scholle, D., 2003. A Color Guide to the Petrography of Carbonate Rocks: Grains, Textures, Porosity, Diagenesis. *American Association of Petroleum Geologists*, *Memoir 77*, 474 pp.

Scotese, C. R., Song, H., Mills, B. J. W., van der Meer, D. G., 2021. Phanerozoic paleotemperatures: the earth's changing climate during the last 540 million years. *Earth-Science Reviews*, *215*, 103503.

Seifan, M., and Berenjian, A., 2019. Microbially induced calcium carbonate precipitation: a widespread phenomenon in the biological world. *Applied Microbiology and Biotechnology*, *103*, 4693–4708.

Silva-Tamayo, J. C., Lara, M. E., Yobo, L. N., Erdal, Y. D., Sanchez, J., Zapata, P. A., 2017. Tectonic and environmental factors controlling on the evolution of Oligo-Miocene shallow marine carbonate factories along a tropical SE Circum-Caribbean. *Journal of South American Earth Sciences*, 78, 213–237.

Smith, A. M., Key Jr, M. M., Gordon, D. P., 2006. Skeletal mineralogy of bryozoans: taxonomic and temporal patterns. *Earth-Science Reviews*, *78*, 287–306.

Smith, A. M., Riedi, M. A., Winter, D. J., 2013. Temperate reefs in a changing ocean: skeletal carbonate mineralogy of serpulids. *Marine Biology*, *160*, 2281–2294.

Spencer, R. J., Hardie, L. A., 1990. Control of seawater composition by mixing of river waters and mid-ocean ridge hydrothermal brines. In: Spencer, R. J., Chou, I.-M. (Eds.), Fluid–Mineral Interactions: A Tribute to H. P. Eugster. *Geochemical Society Special Publication 19*, 409–419.

Stanley, G. D., 2003. The evolution of modern corals and their early history. *Earth-Science Reviews*, *60*, 195–225.

Stanley, S. M., 2006. Influence of seawater chemistry on biomineralization throughout Phanerozoic time: Paleontological and experimental evidence. *Palaeogeography, Palaeoclimatology, Palaeoecology*, *232*, 214–236.

Stanley, S. M, 2008. Effects of global seawater chemistry on biomineralization: past, present, and future. *Chemical Reviews*, *109*, 4483–4498.

Stanley, S. M., and Hardie, L. A., 1998. Secular oscillations in the carbonate mineralogy of reef-building and sediment-producing organisms driven by tectonically forced shifts in seawater chemistry. *Palaeogeography, Palaeoclimatology, Palaeoecology*, *144*, 3–19.

Stanley, S. M., and Hardie, L. A., 1999. Hypercalcification: paleontology links plate tectonics and geochemistry to sedimentology. *GSA today*, *9*, 1–7.

Stanley, S. M, Ries, J. B, Hardie, L. A., 2002. Low-magnesium calcite produced by coralline algae in seawater of Late Cretaceous composition. *Proceedings of the National Academy of Sciences of the USA*, *99*, 15323–15326.

Stanley, S. M., Ries, J. B., Hardie, L. A., 2005. Seawater chemistry, coccolithophore population growth, and the origin of Cretaceous chalk. *Geology*, *33*, 593–596.

Stanley S. M., Ries J. B., Hardie, L. A., 2010. Increasing production of calcite and slower growth for the major sediment-producing alga *Halimeda* as the Mg/Ca ratio of seawater is lowered to a "calcite sea" level. *Journal of Sedimentary Research*, *80*(1), 6–16.

Steuber, T., 2002. Plate tectonic control on the evolution of Cretaceous platform-carbonate production. *Geology*, *30*, 259–262.

Steuber, T., Löser, H., Mutterlose, J., Parente, M., 2023. Biogeodynamics of Cretaceous marine carbonate production. *Earth-Science Reviews*, 104341.

Stolarski, J., Meibom, A., Przeniosło, R., Mazur, M., 2007. A Cretaceous scleractinian coral with a calcitic skeleton. *Science*, *318*, 92–94.

Skelton, P. W., and Gili, E., 2025. Myths and models of rudist ecology. *Geologica Acta*, *23*, 1–22.

Taylor, P. D., 2008. Seawater chemistry, biomineralization and the fossil record of calcareous organisms. In Origin and Evolution of Natural Diversity (eds Okada K, Mawatari SF, Suzuki N, Gautam P). University of Hokkaido, Sapporo, Japan, pp. 21–29.

Taylor, P. D., and Waeschenbach, A., 2015. Phylogeny and diversification of bryozoans. *Palaeontology*, *58*, 585–599.

Taylor, P. D., James, N. P., Bone, Y., Kuklinski, P., Kyser, T. K., 2009. Evolving mineralogy of cheilostome bryozoans. *Palaios*, *24*, 440–452.

Ten Hove, H. A., and Kupriyanova, E. K., 2009. Taxonomy of Serpulidae (Annelida, Polychaeta): the state of affairs. *Zootaxa*, *2036*, 1–126.

Thompson, J. R., Cotton, L. J., Candela, Y., Kutscher, M., Reich, M., Bottjer, D. J., 2021. The Ordovician diversification of sea urchins: systematics of the Bothriocidaroida (Echinodermata: Echinoidea). *Journal of Systematic Palaeontology*, *19*(20), 1395–1448.

Tucker, M. E., and Wright, V. P., 1990. Carbonate Sedimentology. Blackwell, Oxford. 482 pp.

Tyson, R. V., and Pearson, T. H., 1991. Modern and ancient continental shelf anoxia: an overview. *Geological Society, London, Special Publications*, *58*, 1–24.

Ullmann, C. V., Gale, A. S., Huggett, J., Wray, D., Frei, R., Korte, C., Broom-Fendley, S., Littler, K., Hesselbo, S. P., 2018. The geochemistry of modern calcareous barnacle shells and applications for palaeoenvironmental studies. *Geochimica et Cosmochimica Acta*, *243*, 149–168.

Vachard, D. 2021. Calcareous algae (Rhodophyta and Chlorophyta). In: Alderton, D. and Elias, S.A. (Eds), Encyclopedia of Geology, 389–406. Academic Press.

Vachard, D., Cózar, P., Aretz, M., Izart, A., 2016. Late Viséan-Serpukhovian foraminifers in the Montagne Noire (France): Biostratigraphic revision and correlation with the Russian substages. *Geobios*, *49*, 469–498.

van de Poel, H. M., and Schlager, W., 1994. Variations in Mesozoic–Cenozoic skeletal carbonate mineralogy. *Geologie en Mijnbouw*, 73, 31–51.

Vendrasco, M. J., Porter, S. M., Kouchinsky, A., Li, G., Fernandez, C. Z., 2010a. New data on molluscs and their shell microstructures from the Middle Cambrian Gowers Formation, Australia. *Palaeontology*, *53*(1), 97–135.

Vendrasco M. J., Porter S. M., Kouchinsky A. V., Li, G., Fernandez, C. Z., 2010b. Shell microstructures in early Mollusks. *The Festivus XLII*(*4*), 43–54.

Vinn, O., 2020. Biomineralization of Polychaete Annelids in the Fossil Record. *Minerals*, *10*, 858.

Webb, G. E, and Sorauf, J. E., 2002. Zigzag microstructure in rugose corals: a possible indicator of relative seawater Mg/Ca ratios. *Geology*, *30*, 415–418.

Weiner, S., and Dove, P. M., 2003. An overview of biomineralization processes and the problem of the vital effect. *Reviews in Mineralogy and Geochemistry*, *54*, 1–29.

Westphal, H., Halfar, J., Freiwald, A., 2010. Heterozoan carbonates in subtropical to tropical settings in the present and past. *International Journal of Earth Sciences*, *99*, 153–169.

Wilson, J. L., 1975. Carbonate Facies in Geologic History. Springer Verlag, Heidelberg, 471 pp.

Wilson, J. L., and Jordan, C., 1983. Middle shelf environment. In: Scholle, P. A., Bebout, D. G., Moore, C. H. (Eds.), Carbonate Depositional Environments. *American Association of Petroleum Geologists, Memoir 33*, 297–343.

Wilson, M. E. J., and Vecsei, A., 2005. The apparent paradox of abundant foramol facies in low latitudes: their environmental significance and effect on platform development. *Earth Science Reviews*, *69*, 133–168.

Wolf, K. H., 1965, Gradational sedimentary products of calcareous algae. *Sedimentology*, *5*, 1–37.

Zhuravlev, A. Y., and Wood, R. A., 2008. Eve of biomineralization: controls on skeletal mineralogy. *Geology*, *36*(12), 923–926.

Zhuravlev, A. Y., and Wood, R. A., 2009. Controls on carbonate skeletal mineralogy: Global CO_2 evolution and mass extinctions. *Geology*, *37*, 1123–1126.

Kolibri 360 GmbH
Ferdinand-Jühlke-Straße 7
99095 Erfurt, Deutschland
produktsicherheit@kolibri360.de